HANDBUCH DER EXPERIMENTELLEN PHARMAKOLOGIE

BEGRÜNDET VON A. HEFFTER
FORTGEFÜHRT VON W. HEUBNER

ERGÄNZUNGSWERK

HERAUSGEGEBEN VON

O. EICHLER
PROFESSOR DER PHARMAKOLOGIE
AN DER UNIVERSITÄT HEIDELBERG

UND

A. FARAH
PROFESSOR DER PHARMAKOLOGIE
AN DER STATE UNIVERSITY OF NEW YORK

VIERZEHNTER BAND · TEIL 1

THE ADRENOCORTICAL HORMONES
THEIR ORIGIN, CHEMISTRY,
PHYSIOLOGY, AND PHARMACOLOGY

PART 1

SUB-EDITOR

HELEN WENDLER DEANE

SPRINGER-VERLAG BERLIN HEIDELBERG GMBH

THE ADRENOCORTICAL HORMONES

THEIR ORIGIN · CHEMISTRY PHYSIOLOGY AND PHARMACOLOGY

PART 1

CONTRIBUTORS

M. BLEULER · J. J. CHART · H. W. DEANE · R. I. DORFMAN
R. GAUNT · E. HEFTMANN
E. HOWARD · C. J. MIGEON · J. A. MOORE · W. A. STOLL

SUB-EDITOR

HELEN WENDLER DEANE

WITH 224 FIGURES

SPRINGER-VERLAG BERLIN HEIDELBERG GMBH

ISBN 978-3-642-88387-3 ISBN 978-3-642-88385-9 (eBook)
DOI 10.1007/978-3-642-88385-9

Ursprünglich erschienen bei Springer-Verlag oHG. Berlin · Göttingen · Heidelberg 1962
Softcover reprint of the hardcover 1st edition 1962

Foreword and history

Just over one century ago, THOMAS ADDISON published his classic study revealing the vital nature of the adrenal glands. Until that time, the significance of the adrenals had been a complete mystery. The present volumes aim to review, in as comprehensive a manner as practicable, what has since been learned about the anatomy and physiology of the adrenal cortex, the release of the anterior pituitary hormone (adrenocorticotrophin) that influences much of its secretory activity, the chemistry of the several adrenocortical hormones, their *in vivo* synthesis and catabolism, their normal physiological effects, and the physiological and psychological effects of deficiency or excess of these hormones. Clearly a wealth of information has been accumulated: it is hoped that these volumes will serve both to summarize that knowledge and also to illuminate the areas requiring further research to complete our understanding of the adrenal cortex, the effects of its hormones, and our ability to control their abnormalities.

In the present Handbuch, the adjective "adrenocortical" is used in reference to this tissue not only in mammals, in which it occupies the periphery of the adrenal gland, but also in other vertebrates, in which the homologous tissue (interrenal) is either separate from or irregularly intermixed with the chromaffin (medullary) tissue.

Historical background

As in the study of other endocrine glands, developing knowledge about the adrenal cortex comprised an initial anatomical phase, then a phase in which the characteristics of glandular deficiency were established, followed by a phase in which the principles capable of replacing its effects were extracted. Isolation of the active principles permitted their chemical characterization. In addition, the dependence of the major portion of this gland on the pars distalis (anterior lobe) of the hypophysis was established, and the trophic hormone was identified and characterized. The last decade has seen a flowering of this basic information, with the result that we now have quite detailed knowledge about the physiological control of the gland as well as about the control of the secretion of the adrenocorticotrophic hormone (ACTH), about the chemical nature of hitherto unidentified adrenocortical hormones, about enzymatic processes involved in the formation and breakdown of the various hormones, about some of the intimate processes influenced by the hormones, and about various dysfunctions of the gland.

A few highlights of this evolution of knowledge about the adrenal cortex and its hormones deserve mention here. More detailed surveys can be found in BIEDL (1913), TRENDELENBURG (1929), SWANN (1940), SWINGLE and REMINGTON (1944), GAUNT and EVERSOLE (1949), SAYERS (1950), CELESTINO DA COSTA (1951), BACHMANN (1954), REICHSTEIN (1954) and FARRELL (1958).

Identification. The adrenal glands appear to have gone unnoticed by the ancients; they were first clearly described in man in the 16th century by the gross anatomist EUSTACHIUS. Since then, homologous structures have been identified not only in other mammals but in all classes of the Subphylum Vertebrata (BOURNE 1949, BACHMANN 1954). Thus, by about 1800, adrenal glands had been found in a variety of mammals and birds. Shortly thereafter, adrenocortical bodies were identified in elasmobranch fishes, amphibians, and reptiles. Their identification in teleost and dipnoan fishes, however, was not accomplished until quite recently.

In 1846, ECKER published the first important histological description of the adrenals in a variety of vertebrates, in which he distinguished the cortical and medullary tissues by the different arrangements of their cells and by the presence of lipide droplets in the cortical cells. With the subsequent improvements in histological technique, the morphology and histochemistry of adrenocortical tissue in a large number of different animals has been greatly extended. One early histophysiological observation of interest was that the adrenal cortex becomes hypertrophied in many chronic physiological stresses (BERNARD and BIGART 1902). The anatomical, chemical and physiological information about the adrenal cortex is summarized in Chapter 1.

Vital nature. The function of the adrenal glands remained completely unknown for nearly 3 centuries after their identification. Many observers thought them auxilliary renal, hepatic, or lymphatic tissue. That they were endocrine glands was suspected from their structure and relation to the blood stream by such 19th century physiologists as CLAUDE BERNARD, but this function was not generally accepted until the present century.

THOMAS ADDISON published his justly-famed description of adrenal insufficiency in man in 1855. Starting immediately thereafter, BROWN-SÉQUARD and other physiologists attempted to determine whether or not the adrenal glands are essential for life by experimental extirpation from animals; not until about 1890, however, was their vital nature acceptably established (e.g., LANGLOIS 1893).

Once the essential nature of the adrenal glands was confirmed, debate centered on whether the whole gland or just one of its parts — the cortex or the medulla — was necessary for life. BIEDL reported in 1913 that the cortical tissue constitutes the vital element, since he had found that the removal of the interrenal glands of elasmobranchs (in which the chromaffin tissue lies separately) was fatal. Most workers continued to believe the medullary tissue the indispensible portion, however, until HOUSSAY and LEWIS (1923) convincingly demonstrated that dogs in which one adrenal gland was removed and the medulla of the other was completely destroyed remained healthy, whereas animals from which both adrenals were entirely removed succumbed. In retrospect, it seems that the presence of chromaffin tissue in the paraganglia should have suggested that the medulla was not the vital element.

Adrenocortical hormones. Briefly, *the general signs of adrenocortical insufficiency* are: anorexia, asthenia, a drop in body temperature and in blood pressure, excessive excretion of H_2O and Na^+, retention of K^+ and urea, a retardation of glucose formation, and hypersensitivity to any physiological stress. (These signs are considered in greater detail in Chapters 4 and 8.) By 1930, several groups of physiologists had succeeded in preparing extracts of the adrenal cortex that were capable of repairing these signs in adrenalectomized animals (HARTMAN and BROWNELL 1930, SWINGLE and PFIFFNER 1931a, b).

Successful *isolation and characterization of most of the active principles* — notably those affecting organic metabolism (Chapter 7) — was accomplished in the period of 1935—1940, especially by KENDALL, MASON and co-workers, by WINTERSTEINER and PFIFFNER, and by REICHSTEIN and co-workers (see REICHSTEIN 1954, and Chapters 2 and 3). Also found in adrenocortical extracts were various androgens, estrogens and progestins (Chapters 2, 5). One principle, active in correcting the abnormal electrolyte metabolism of adrenalectomized animals, was not isolated from the previously-termed "amorphous" fraction of adrenocortical extracts until the 1950s (SIMPSON and TAIT 1955, see Chapters 2, 4). And at this writing there is suggestion of a hormone of opposing action, namely one that facilitates the excretion of sodium.

The most recent advance in the chemistry of the adrenocortical hormones concerns the elucidation of the pathways of their synthesis and degradation. This area comprises the subject matter of Chapter 3.

Pituitary control. Concomitantly with the clarification of adrenal function, it was learned that adrenocortical activity depends on the anterior pituitary gland. This conclusion was foreshadowed in the study by ASCOLI and LEGNANI (1912). The definitive experiments by PHILIP SMITH (1920, 1930) demonstrated atrophy of the cortical tissue following hypophysectomy in tadpoles and in rats. A similar syndrome was reported for man in patients suffering from accidental destruction of the pituitary gland (SIMMONDS' disease — see SHEEHAN and SUMMERS 1949, and Chapter 8). Restoration of the adrenal cortex of hypophysectomized animals injected with extracts of the anterior pituitary was demonstrated by P. and I. SMITH in 1923. Quite purified adrenocorticotrophic hormone (ACTH) was available by 1940 (e.g., MILLER and RIDDLE 1939). That hyperadrenocorticism could result not only from an adrenal tumor but also from hyperactivity of the anterior pituitary gland was first emphasized by CUSHING (1932; see Chapter 9).

During the 1940s, SAYERS (1950), in particular, developed the thesis that the level of circulating glucocorticoids governs the secretion of ACTH by the anterior pituitary. Subsequently it became recognized that the nervous system, especially the hypothalamus, plays a significant role in altering that secretory rate. The status of our present knowledge about the control of the output of ACTH is reviewed in Chapter 10.

Functional zonation. HOUSSAY and SAMMARTINO (1933) were first to recognize that the adrenal cortex is only partially inactivated by hypophysectomy and thereafter continues to secrete a factor effective in maintaining electrolyte balance. Subsequent histophysiological studies suggested strongly that the hormone influencing electrolyte balance (mineralocorticoid) is secreted, essentially without pituitary stimulation, by the outermost part of the mammalian cortex, the zona glomerulosa; whereas those hormones influencing organic metabolism (glucocorticoids) are secreted, under the control of ACTH, by the inner portion, the zonae fasciculata and reticularis (GREEP and DEANE 1949, DEANE 1951).

Only recently, since this book was planned and largely completed, FARRELL (1958, 1959) and others have brought forth evidence that the zona glomerulosa may be controlled, at least in part, by a principle secreted by the pineal gland or neighboring tissue. These topics are considered in greater detail in Chapter 1.

Diseases of the adrenal glands. In preceding paragraphs, mention has been made of ADDISON's disease (total adrenal insufficiency), of SIMMONDS' disease (panhypopituitarism, in which the atrophic adrenal glands fail to secrete the hormones affecting organic metabolism) (both considered in Chapter 8), and of CUSHING's disease (in which the adrenal cortex, because of intrinsic hyperplasia or because of the hypersecretion of ACTH, releases an excess of hormones, principally those affecting organic metabolism) (considered in Chapters 7 and 9). In addition, there are two other adrenal abnormalities recognized in human beings. CONN (1956) has described a disease in which aldosterone (mineralocorticoid) is secreted in excessive quantities, either because of adrenocortical tumor or hyperplasia (primary aldosteronism) or in response to certain diseases characterized by edema and ascites (secondary aldosteronism) (Chapters 4, 9). And in human beings, though rarely in other species, hyperplastic or tumorous adrenals may secrete large quantities of sex steroids — either androgens or estrogens. The relatively more frequent of these diseases, in which androgens are secreted in excessive amount, is termed adrenal virilism (adrenogenital syndrome) (Chapters 5, 9). In recent years this has been shown to result from an enzymic deficiency in the adrenal cortex, so that precursor substances are converted almost entirely into C_{19} compounds.

This brief survey indicates in broad outline the topics considered in this Handbuch, albeit in different order. It remains to note that there is growing evidence of an influence of adrenocortical steroids on mental well-being and, contrariwise, of an influence of mental state on adrenal function. These topics are considered in Chapter 6.

May, 1960 HELEN WENDLER DEANE

Acknowledgments

I wish to express my appreciation to the authors of the several chapters in part 1 of this volume for their generous cooperation during preparation of the contents for the press. Professor ALFRED FARAH, one of the Editors, has been a repeated source of valuable counsel. My colleague Dr. BETTY L. RUBIN was of immeasurable aid in the preparation of the subject index. Finally, my appreciation goes to the staff of Springer Verlag, for their constant assistance in problems related to publication and for their expertise in producing a handsome volume. Any errors, of course, remain the responsibility of the authors and me.

October, 1961 H. W. D.

Literature for history

Addison, T.: On the constitutional and local effects of disease of the suprarenal capsules. London: Highley 1855.

Ascoli, G., u. T. Legnani: Die Folgen der Exstirpation der Hypophyse. Münch. med. Wschr. **59**, 518—521 (1912).

Bachmann, R.: Die Nebenniere. Handb. mikr. Anat. Mensch. VI/5, 1—952 (1954).

Bernard, L., et Bigart: Étude anatomo-pathologique des capsules surrénales dans quelques intoxications expérimentales. J. Physiol. Path. gén. **4**, 1014—1029 (1902).

Biedl, A.: The internal secretory organs: Their physiology and pathology (Trans. L. Forster). New York: Wood 1913.

Bourne, G. H.: The mammalian adrenal gland. Oxford: Clarendon Press 1949.

Celestino da Costa, A.: Histophysiologie du cortex surrénal. Ann. Endocr. (Paris) **12**, 361 to 403 (1951).

Conn, J. W.: Aldosterone in clinical medicine—past, present and future. Arch. intern. Med. **97**, 135—144 (1956).

Cushing, H.: The basophil adenomas of the pituitary body and their clinical manifestations (pituitary basophilism). Johns Hopk. Hosp. Bull. **50**, 137—195 (1932).

Deane, H. W.: Physiological regulation of the zona glomerulosa of the rat's adrenal cortex, as revealed by cytochemical observations. In: R. C. Christman, Ed., Pituitary-Adrenal Function, 31—38. Washington, D. C.: Amer. Ass. Advanc. Sci. 1951.

Ecker, A.: Der feinere Bau der Nebennieren beim Menschen und den vier Wirbelthierclassen. Braunschweig: 1846

Farrell, G.: Regulation of aldosterone secretion. Physiol. Rev. **38**, 709—728 (1958).

— Steroidogenic properties of extracts of beef diencephalon. Endocrinology **65**, 29—33 (1959).

Gaunt, R., and W. J. Eversole: Notes on the history of the adrenal cortical problem. Ann. N. Y. Acad. Sci. **50** (6), 511—521 (1949).

Greep, R. O., and H. W. Deane: The cytology and cytochemistry of the adrenal cortex. Ann. N. Y. Acad. Sci. **50** (6), 596—615 (1949).

Hartman, F. A., and K. A. Brownell: The hormone of the adrenal cortex. Proc. Soc. exp. Biol. (N. Y.) **27**, 938—939 (1930).

Houssay, B. A., and J. T. Lewis: The relative importance to life of cortex and medulla of the adrenal glands. Amer. J. Physiol. **64**, 512—521 (1923).

— and R. Sammartino: Histology of the adrenals of hypophysectomized dogs (in Spanish). Rev. Soc. argent. Biol. **9**, 209—219 (1933).

Langlois, P.: Destruction des capsules surrénales chez le chien. Arch. Physiol. norm. path., ser. V, **5**, 488—498 (1893).

Miller, R. A., and O. Riddle: Stimulation of adrenal cortex of pigeons by ant. pituitary hormones and by their secondary products. Proc. Soc. exp. Biol. (N. Y.) **41**, 518—522 (1939).

Reichstein, T.: Die wichtigsten Hormone der Nebennierenrinde. Acta endocr. (Kbh.) **17**, 375—384 (1954).

Sayers, G.: The adrenal cortex and homeostasis. Physiol. Rev. **30**, 241—320 (1950).

Sheehan, H. L., and V. K. Summers: The syndrome of hypopituitarism. Quart. J. Med. **42**, 319—378 (1949).

Simpson, S. A., and J. F. Tait: Recent progress in methods of isolation, chemistry and physiology of aldosterone. Recent Progr. Hormone Res. **11**, 183—210 (1955).

Smith, P. E.: The pigmentary, growth and endocrine disturbances induced in the anuran tadpole by the early ablation of the pars buccalis of the hypophysis. Amer. anat. Mem. **11** (1920).

— Hypophysectomy and a replacement therapy in the rat. Amer. J. Anat. **45**, 205—273 (1930).

— and I. P. Smith: The function of the lobes of the hypophysis as indicated by replacement therapy with different portions of the ox glands. Endocrinology **7**, 579—591 (1923).

SWANN, H. G.: The pituitary-adrenocortical relationship. Physiol. Rev. **20**, 493—521 (1940).
SWINGLE, W. W., and J. J. PFIFFNER: Studies on the adrenal cortex I. The effect of a lipid fraction upon the life-span of adrenalectomized cats. Amer. J. Physiol. **96**, 153—163 (1931a).
— — II. An aqueous extract of the adrenal cortex which maintains the life of bilaterally adrenalectomized cats. Amer. J. Physiol. **96**, 164—179 (1931b).
— and J. W. REMINGTON: The role of the adrenal cortex in physiological processes. Physiol. Rev. **24**, 89—127 (1944).
TRENDELENBURG, P.: Die Hormone, ihre Physiologie und Pharmakologie. Bd. I, 185—209. Berlin: Springer 1929.

Authors and addresses

MANFRED BLEULER, M.D., Psychiatrische Universitätsklinik, Zürich, Switzerland.

JEROME J. CHART, Ph.D., Research Department, CIBA Pharmaceutical Products, Inc., Summit, New Jersey, U.S.A.

HELEN WENDLER DEANE, Ph.D., Departments of Pathology and Anatomy, Albert Einstein College of Medicine, New York 61, New York, U.S.A.

RALPH I. DORFMAN, Ph.D., The Worcester Foundation for Experimental Biology, Shrewsbury, Massachusetts, U.S.A.

ROBERT GAUNT, Ph.D., Research Department, CIBA Pharmaceutical Products, Inc., Summit, New Jersey, U.S.A.

ERIC HEFTMANN, Ph.D., National Institute of Arthritis and Metabolic Diseases, National Institutes of Health, Bethesda 14, Maryland, U.S.A.

EVELYN HOWARD, Ph.D., Department of Physiology, The Johns Hopkins University School of Medicine, Baltimore 5, Maryland, U.S.A.

CLAUDE J. MIGEON, M.D., Department of Pediatrics, The Johns Hopkins University School of Medicine and The Johns Hopkins Hospital, Baltimore 5, Maryland, U.S.A.

JAMES A. MOORE, Ph.D., Department of Chemistry, University of Delaware, Newark, Delaware, U.S.A.

W. A. STOLL, M.D., Psychiatrische Universitätsklinik, Zürich, Switzerland.

Contents

Chapter 1: The anatomy, chemistry, and physiology of adrenocortical tissue

By Helen Wendler Deane

Gross, microscopic, and developmental anatomy of adrenocortical tissue 1

A. Introduction 1

I. Form and location 1
II. Blood and nerve supply 2
III. Histological organization 2
IV. Development 2

B. Survey of glandular structure, by classes 4

I. Class Agnatha 5
a) Order Cyclostomata 5
α) Suborder Mixinoidea 5
β) Suborder Petromyzontia 5

II. Class Chondrichthyes 5
1. Subclass Elasmobranchii 5
2. Subclass Holocephali 6

III. Class Osteichthyes 6
1. Subclass Choanichthyes 6
a) Order Dipnoi 6
b) Order Crossopterygii 7
2. Subclass Actinopterygii 7
a) Superorder Chondrostei 7
b) Superorders Holostei and Teleostei 7

IV. Class Amphibia 8
a) Order Gymnophiona (Apoda) 8
b) Order Urodela 8
c) Order Anura 9

V. Class Reptilia 10
a) Order Chelonia 10
b) Order Crocodilia 10
c) Order Squamata 11
α) Suborder Lacertilia 11
β) Suborder Serpentes (Ophidia) 12

VI. Class Aves 13
1. Gross anatomy 13
2. Blood and nerve supply 13
3. Histological organization 13
4. Cytology 14
5. Development 17

VII. Class Mammalia 17
1. Gross anatomy 17
a) Accessory adrenocortical tissue 21
2. Blood supply 21
3. Lymphatic drainage 24
4. Nerve supply 24
5. Connective-tissue stroma 26
6. Parenchyma: histological organization 27
7. Cytology 30

a) Cell size . . . 30
b) Nucleus . . . 30
c) Lipide droplets . . . 32
d) Subglomerulosal zone . . . 35
e) Cytoplasmic organelles . . . 36
8. Prenatal development . . . 38
a) "Fetal" zone . . . 42
9. Adrenocortical growth and regeneration . . . 45
a) Cell-migration hypothesis . . . 45
b) Zonal hypothesis . . . 46
c) Transformation-field hypothesis . . . 47
10. The X zone of the mouse adrenal . . . 47

Chemistry of adrenocortical tissue . . . 48

A. Introduction . . . 48

B. Adrenocortical lipides . . . 49

I. Total lipides . . . 49
II. Cholesterol . . . 50
III. Steroid hormones . . . 54
IV. Phosphatides . . . 54
V. Plasmalogens and pseudoplasmals . . . 55
VI. Lipofuscins . . . 56

C. Ascorbic acid . . . 58

D. Enzymes . . . 60
I. Steroid-synthesizing enzymes . . . 61
II. Oxidative enzymes . . . 62
III. Hydrolytic enzymes . . . 64
1. Nonspecific alkaline phosphatase . . . 64
2. Nonspecific acid phosphatase . . . 65
3. Nonspecific esterase . . . 65
4. β-Glucuronidase . . . 66
5. Aryl sulfatase . . . 66

E. Miscellaneous compounds . . . 67
I. Glycogen . . . 67
II. Other chemical constituents . . . 68

Physiology of adrenocortical tissue . . . 68

A. Introduction . . . 68

B. Overall physiology of the adrenal cortex . . . 68
I. General dependence on anterior-pituitary secretions . . . 68
II. Nutritional factors influencing adrenocortical tissue . . . 69
1. Proteins . . . 69
2. Carbohydrates . . . 69
3. Lipides . . . 70
4. "Vitamins" . . . 70
III. Other factors influencing adrenocortical tissue . . . 71
1. Species differences . . . 71
2. Age differences . . . 72
3. Sex differences . . . 72
IV. Blood flow through the adrenal gland . . . 72

C. Secretory activity of adrenocortical tissue . . . 73
I. Direct determination of adrenocortical secretions . . . 74
1. Analysis of adrenal-venous blood or peripheral blood . . . 74
2. Analysis of adrenal perfusates . . . 76
3. Analysis of incubation media . . . 77
II. Indirect indices of adrenocortical activity . . . 77
1. Gross indices of altered stimulation . . . 77

a) Adrenal weight 77
α) Atrophy after hypophysectomy 77
β) Atrophy with the exogenous administration of corticoids 82
γ) Hypertrophy of the adrenal cortex with the administration of ACTH . 82
b) Adrenocortical chemistry 83
α) Concentration of cholesterol 83
β) Concentration of ascorbic acid 84
c) Alterations in blood flow 86
2. Cytological and associated chemical indices of adrenocortical activity 86
a) Subglomerulosal zone 87
b) Cellular and zonal volume 87
c) Nuclear volume 88
d) Nucleolar size 90
e) Cytoplasmic ribonucleoprotein 90
f) Mitochondria and oxidative activity 91
g) Golgi apparatus 92
h) Phosphatides 92
i) Triglycerides and steroids 92
j) Pseudoplasmals 94
k) Ascorbic acid 96
l) Enzymes and coenzymes concerned with corticosteroid synthesis 96
m) Hydrolytic enzymes 96

D. Dichotomy of hormone production in the mammalian adrenal cortex 98
I. Factors influencing the inner zones of the cortex and the secretion of glucocorticoids 99
1. Hypophysectomy 99
2. Exogenous administration of cortical hormones 100
3. Adrenal neoplasia with a high output of glucocorticoids 101
4. ACTH administration 101
5. Stimulation of endogenous secretion of ACTH ("stress") 103
a) Species variability 106
b) Estrogens 106
c) So-called ACTH-like action of androgens 107
6. ACTH-producing pituitary tumors 107
7. Locus of secretion of androgens and other sex hormones by the adrenal cortex 108

II. Factors influencing the zona glomerulosa and the secretion of aldosterone . . . 109
1. Hypophysectomy 109
2. Administration of glucocorticoids 110
3. ACTH or "stress" 110
4. Disturbance of electrolyte balance 111
a) Sodium deficiency 111
b) Potassium excess 114
c) Edema and blood volume 114
d) Sodium excess, potassium deficiency, or DOC administration 115
5. Other influences 116
a) Water deprivation; posterior pituitary 116
b) Growth hormone 116
c) Kidney abnormality 116
d) Trauma 116
e) Possible pineal hormone 117
f) Possible specific cytotoxic agents 117
g) Compensatory hypertrophy 118
h) Thiamine deficiency 118

E. Control of adrenocortical regeneration 118
I. Regeneration *in vivo* 118
II. Regeneration *in vitro* 121

F. Prenatal control of the adrenal cortex, mammals 121
I. Secretion of corticoids by fetal adrenals 121
II. Stimulation of fetal adrenals by the fetal pituitary 122
III. Control of the secretion of ACTH by the fetal pituitary 123
IV. Postnatal inactivity of the pituitary 125
V. Control of primate "fetal" zone 125

G. Control of the mouse X zone 127

H. Pathological changes in mammalian glands 128
I. Hyperactivity from increased ACTH 129
1. Cellular hypertrophy and hyperplasia 129
2. Pseudotubular formation 129
3. Hemorrhage and necrosis 131
4. Intracellular infection 132
5. Gross lipide droplets, watery vacuoles, "colloid" droplets, increased pigment 132
II. Adrenocortical hyperplasia and tumors 133
1. Hypersecretion of ACTH 133
2. Hypersecretion of gonadotrophins 134
3. Autonomous adrenocortical tumors 135
III. Inhibitors of adrenocortical function 135
1. Dichloro-diphenyl-dichloroethane (DDD) 135
2. Suramin 136
3. Amphenone B and Su 4885 136
4. Pantothenic-acid deficiency 138
5. Pyridoxine deficiency 139
6. Δ^4-Cholestenone 139
7. Thyroid deficiency 140
8. Inadequacy of the pituitary 140
9. Other causes of adrenal insufficiency 140

I. Physiology of adrenocortical tissue in the lower vertebrates 141
I. Hypophysectomy and adrenalectomy 141
II. ACTH or "stress" 144
III. Other experimental situations 145
Literature 145

Chapter 2: Chemistry of the adrenocortical steroids
By James A. Moore and Eric Heftmann

Chemical nature of adrenocortical steroids 186

A. General characteristics of steroid compounds 186
I. Structure and stereoisomerism of the steroid nucleus 186
1. Nuclear configurations and stereochemical conventions 187
2. Absolute configuration 189
3. Conformational principles 191
II. Nomenclature 193
1. General principles 194
2. Trivial names 196

B. The adrenocortical steroids 197
I. The steroid content of the adrenal cortex 197
1. Isolation of steroids 197
2. Approaches to structure determination 198
3. The properties and structures of the adrenal steroids 201
4. Stereochemistry of the side chains 206
II. Individual structural features 210
1. 17-Ketones and alcohols 210
a) 17-Ketones 210
b) 17-Alcohols 212
2. 20-Ketones and alcohols 213
a) 20-Ketones 213
b) 20-Alcohols 214
3. 17α-Hydroxy-20-ketones and diols 215
a) 17α-Hydroxy-20-ketones 215
b) 17,20-Diols 218
4. 21-Hydroxy-20-ketones and diols 219
a) 21-Hydroxy-20-ketones 219
b) 20,21-Diols 221
5. 17α,21-Dihydroxy-20-ketones and triols 222

a) 17α,21-Dihydroxy-20-ketones . . . 222
b) 17α,20,21-Triols . . . 225
6. 11-Ketones and alcohols . . . 226
a) 11-Ketones . . . 227
b) 11-Alcohols . . . 228
7. Δ^4-3-Ketones, saturated 3-ketones and alcohols . . . 228
a) Δ^4-3-Ketones . . . 229
b) Saturated 3-ketones . . . 230
c) 3-Alcohols . . . 231
d) Δ^5-3β-Hydroxy compounds . . . 231
8. 2-Hydroxyl groups . . . 233
9. 6-Hydroxyl groups . . . 234
10. 16-Hydroxyl groups . . . 236
11. 18-Oxygen functions . . . 236
12. 19-Oxygen functions . . . 238

III. Important reactions and reagents . . . 240
1. Reactions of hydroxyl groups . . . 240
a) Esterification and etherification . . . 240
α) Acetylation . . . 240
β) Esterification with other acids . . . 241
γ) Hydrolysis of acylated steroids . . . 241
δ) Other protective groups . . . 241
b) Oxidation . . . 241
α) Chromic acid . . . 241
β) N-Bromoacetamide (NBA) . . . 242
γ) Oppenauer oxidation . . . 242
δ) Manganese dioxide . . . 242
ε) t-Butylhypochlorite . . . 243
c) Dehydration . . . 243
2. Reactions of carbonyl groups . . . 243
a) Formation of hydrazone derivatives . . . 243
b) Ketals . . . 244
c) Formation of enol derivatives . . . 244
α) Enol esters . . . 244
β) Enol ethers and imines . . . 244
d) Reduction to a hydroxyl group . . . 245
α) Complex hydrides . . . 245
β) Catalytic hydrogenation . . . 246
γ) Reduction with active metals . . . 246
e) Reduction to methylene group . . . 246
α) Wolff-Kishner reduction . . . 246
β) Clemmensen reduction . . . 247
γ) Desulfuration of thioketals . . . 247
3. Reactions of double bonds . . . 247
a) Reduction . . . 247
α) Catalytic hydrogenation . . . 247
β) Partial reduction of unsaturated ketones . . . 248
b) Hydroxylation and epoxidation . . . 248
α) Perbenzoic acid . . . 249
β) Osmium tetroxide . . . 250
γ) Hypobromous acid . . . 250
4. Introduction of isotopic atoms . . . 250

Preparation of adrenocortical steroids and modified hormones . . . 251

A. Synthetic sources of adrenocortical steroids . . . 251
I. General approaches to partial synthesis . . . 251
1. The C-11 oxygen problem . . . 251
2. The side chain problem . . . 253
3. Sequence of steps and completion of synthesis . . . 256
II. Preparative aspects of microbiological transformations . . . 258
1. General scope . . . 258
2. Hydroxylations . . . 259
3. Dehydrogenation and side chain cleavage . . . 260
4. Conditions and procedures . . . 261

III. Total synthesis . . . 261
B. Modified hormones . . . 264
I. Halogen substitution . . . 268
1. 9α-Halo derivatives . . . 268
2. 12α-Halo derivatives . . . 269
3. 21-Fluoro derivatives . . . 270
4. Halogen substitution at other positions . . . 271
a) 4-Halogenated derivatives . . . 271
b) 16-Halogenated derivatives . . . 272
c) 2- and 6-Halogenated derivatives . . . 272
II. Hydroxyl substitution . . . 272
1. 16-Hydroxy derivatives . . . 272
2. Hydroxyl groups at other positions . . . 275
a) 5α-Hydroxy compounds . . . 275
b) 9α- and 12α-Hydroxy compounds . . . 276
c) 14α-Hydroxy derivatives . . . 277
III. Additional double bonds . . . 277
1. Δ^1-Dehydro derivatives . . . 277
2. Δ^6-Dehydro derivatives . . . 278
3. Double bonds at other positions . . . 279
IV. Skeletal modifications . . . 280
1. 19-Nor steroids . . . 280
2. 6α-Methyl derivatives . . . 283
3. 2α-Methyl derivatives . . . 284
4. 16-Methyl derivatives . . . 285
5. 9α-Methyl derivatives . . . 286
6. 17α-Methyl derivatives . . . 286
7. 4-Methyl derivatives . . . 287
8. 1-Methyl-19-nor derivatives . . . 287
9. D-Homo derivatives . . . 288
10. Compounds with abnormal configuration . . . 289
11. Triterpenoid analogues . . . 291
Separation and analysis . . . 291
A. Isolation . . . 292
I. State in biological systems . . . 292
II. Isolation of combined steroids . . . 292
III. Hydrolysis . . . 293
1. Steroidoproteins . . . 293
2. 17-Ketosteroids . . . 293
3. Corticosteroids . . . 295
IV. Extraction . . . 296
1. Solvents . . . 296
2. Extractors . . . 297
3. Isolation from blood and tissues . . . 300
B. Separation . . . 301
I. Preliminary purification of extracts . . . 301
II. Experimental separation methods . . . 303
1. Solvent distribution . . . 304
2. Electrochromatography . . . 304
III. Adsorption chromatography . . . 306
1. Alumina . . . 307
2. Other adsorbents . . . 311
IV. Partition chromatography . . . 313
1. Columns . . . 313
a) Stationary phase: Non-aqueous solvents . . . 313
b) Stationary phase: Aqueous solvents . . . 315
2. Paper . . . 318
a) Stationary phase: Non-aqueous solvents . . . 318
b) Stationary phase: Aqueous solvents . . . 328
C. Identification . . . 334
I. Crystalline properties . . . 335

II. Color reactions . . . 337
1. Reagents . . . 337
a) Strong acids . . . 337
b) Zimmermann reaction and other group tests . . . 337
c) Group tests of side chains . . . 339
2. Applications to paper chromatography . . . 341

D. Instrumental methods . . . 346
1. Spectrophotometry . . . 346
1. Infrared spectrophotometry . . . 347
a) Origin of infrared spectra . . . 348
b) Measurement and presentation of spectra . . . 348
c) State of sample . . . 349
d) Interpretation of spectra . . . 351
α) Carbonyl groups . . . 351
β) Hydroxyl groups . . . 354
γ) Double bonds . . . 354
δ) Carbon-oxygen bonds . . . 355
ε) Carbon-hydrogen bonds . . . 355
ζ) Finger print region and compound comparison . . . 356
η) Quantitative applications . . . 357
2. Ultraviolet and visible spectrophotometry and fluorescence . . . 358
a) Origin of absorption spectra and fluorescence . . . 358
b) Measurement and instrumentation . . . 358
c) Interpretation of ultraviolet spectra . . . 359
α) Isolated carbonyl groups and double bonds . . . 359
β) Conjugated chromophores . . . 360
γ) Spectra in reactive media . . . 362
d) Quantitative applications — colorimetric methods . . . 362
α) 17-Ketosteroids . . . 363
β) Corticosteroids . . . 364
γ) Δ^4-3-Ketosteroids . . . 366
II. Isotope analysis . . . 367
1. Nature and properties of isotopes . . . 367
2. Analysis of isotopically-labeled compounds . . . 368
a) Stable isotopes . . . 368
b) Radioactive isotopes . . . 368
α) Instrumentation . . . 368
β) Preparation of samples and measurement . . . 369
γ) Analytical applications . . . 370
III. Other methods . . . 372
1. Optical rotation . . . 372
a) Measurement . . . 372
b) Applications — molecular rotation differences . . . 373
c) Optical rotatory dispersion . . . 374
2. Nuclear magnetic resonance . . . 376
3. Polarography . . . 380
Literature . . . 381

Chapter 3: Biochemistry of the adrenocortical hormones
By Ralph I. Dorfman

Isolation of steroids and sterols . . . 411

A. Adrenal cortex . . . 411
I. C_{18} steroids . . . 411
II. C_{19} steroids . . . 413
III. C_{21} and C_{27} steroids . . . 416
IV. Nitrogen-containing steroids . . . 418

B. Adrenocortical steroids and derivatives in other tissues . . . 421

C. Adrenocortical steroids and derivatives in bile . . . 422

D. Adrenocortical steroids and derivatives in blood . . . 422
I. Steroid conjugates in blood . . . 424

E. Adrenocortical steroids and related compounds in urine . . . 425
I. C_{18} steroids . . . 425
II. C_{19} steroids . . . 426
III. C_{21} steroids . . . 430
IV. Urinary steroids as conjugates of sulfuric acid and glucuronic acid . . . 434
Biosynthesis of adrenocortical hormones . . . 434
A. Biosynthesis of cholesterol . . . 436
B. Biosynthesis of corticoids . . . 439
I. A non-cholesterol pathway to corticoids . . . 446
C. Biosynthesis of aldosterone . . . 446
D. Biosynthesis of adrenal androgens . . . 449
E. Biosynthesis of adrenal estrogens . . . 451
F. Biosynthesis of steroid conjugates . . . 452
G. Factors influencing biosynthesis of adrenocortical steroids . . . 452
I. Role of ascorbic acid . . . 454
II. Chorionic gonadotrophin . . . 454
III. Δ^4-Cholestenone . . . 455
IV. Estrogens . . . 455
V. Progestational agents . . . 456
VI. Acetylcholine, physostigmine and atropine . . . 456
VII. Amphenone . . . 456
VIII. 2-Methyl-1,2-bis-(3-pyridyl)-1-propanone . . . 457
IX. Phenylethylacetic and diphenylethylacetic acids . . . 457
X. 1,1-Dichloro-2,2-bis(*p*-chlorophenyl)ethane . . . 457
XI. Pantothenic acid . . . 458
XII. Pyrogenic bacterial polysaccharides . . . 458
Catabolism of adrenocortical steroids . . . 458
A. C_{21} steroids . . . 458
I. Pregnenolone and progesterone . . . 463
II. 16α-Hydroxyprogesterone . . . 465
III. 17α-Hydroxyprogesterone . . . 465
IV. 11β-Hydroxyprogesterone and 11-ketoprogesterone . . . 466
V. Deoxycorticosterone . . . 466
VI. 11-Deoxycortisol . . . 467
VII. Corticosterone and 11-deoxycorticosterone . . . 467
VIII. Cortisol and cortisone . . . 471
IX. Daily production of cortisol and corticosterone . . . 472
X. Aldosterone . . . 475
XI. Prednisone and prednisolone . . . 476
B. Adrenal androgens . . . 478
I. $C_{19}O_2$ compounds . . . 478
II. Adrenosterone and 11β-hydroxy-Δ^4-androsten-3,17-dione . . . 480
C. Generalizations on steroid metabolism . . . 481
D. Rate of production of 17-ketosteroid precursors . . . 482
E. Adrenal estrogens . . . 483
Enzymes concerned with steroid metabolism . . . 484
A. Hydroxylating enzymes . . . 484
I. Hydroxylases of C_{18} steroids . . . 486
II. Hydroxylases of C_{19} steroids . . . 487
III. Hydroxylases of C_{21} steroids . . . 488

B. Dehydrogenases and reductases 495
I. 3α-Dehydrogenases and α-orienting 3-keto reductases 495
II. 3β-Dehydrogenases and β-orienting 3-keto reductases 496
III. 11β-Dehydrogenases and β-orienting 11-keto reductases 497
IV. 20-Keto reductases 497
C. Hydrogenases 497
I. Δ^1-Hydrogenases 497
II. Δ^4-5β-Hydrogenases 498
III. Δ^4-5α-Hydrogenase 498
IV. Factors influencing Δ^4-5α- and Δ^4-5β-hydrogenases 498
D. Miscellaneous enzyme systems 499
Literature 499

Chapter 4: Mineralocorticoid action of adrenocortical hormones
By Robert Gaunt and Jerome J. Chart

Introduction 514
History 515
Profiles of major corticoids having mineralocorticoid action 516
A. Extracts of the adrenal cortex 516
B. Deoxycorticosterone (cortexone) 517
C. Cortisol (hydrocortisone) 517
D. Cortisone 518
E. Corticosterone 518
F. 11-Dehydrocorticosterone 518
G. 11-Deoxycortisol (Reichstein's compound S) 518
H. Aldosterone 519
I. Synthetic corticoids 520
J. A sodium-excreting hormone (?) 521
Adrenocortical inhibitors 521
A. Inhibition of the adrenal by corticoids 521
B. Pharmacological inhibition — miscellaneous drugs 522
I. ACTH inhibitors 522
II. Inhibitors of corticoid synthesis 522
1. Amphenone B 522
2. Pinacolones 523
3. DDD 523
4. Hexestrol 523
III. Aldosterone antagonists 524
Relation of mineralocorticoids to life-maintaining functions of the adrenal cortex . . . 524
A. Survival after adrenalectomy — general 525
B. Adrenalectomy in the rat 526
C. Adrenalectomy in other mammalian species 526
D. Effects of different corticoids on survival 527
E. Maintenance with sodium salts 528
The adrenal cortex and mineral metabolism 528
A. Adrenalectomy and electrolyte balance 529
B. Electrolyte replacement therapy 530
C. Steroid replacement therapy and overdosage 530

D. Experimental heart failure 531
E. Eclampsia-like syndrome 532
F. Corticoids and potassium metabolism 532
G. Effects of cortisol, cortisone and derivatives 533
H. Corticoids and brain function 534
I. Diabetes insipidus-like syndrome 534
J. Extra-renal functions of mineralocorticoids 535
I. Balance studies 535
II. Role of sodium in adrenal insufficiency 537
III. Corticoids and fluid compartments 537
IV. Corticoids and tissue electrolytes 538
V. Gastrointestinal functions 540
VI. Sweat 540
VII. Saliva 541
K. Concluding comment 541
Adrenal steroids and water metabolism 542
A. Water diuresis in adrenal insufficiency and replacement therapy 542
B. The adrenal-neurohypophyseal interrelationship 543
C. The adrenals and water intoxication 545
D. The use of corticoids as diuretics 546
Adrenal cortex and hypertension 547
Mineralocorticoids in lower vertebrates 548
A. Cyclostomata 548
B. Elasmobranchii 548
C. Teleostei 549
D. Amphibia 550
I. Effects of exogenous cortical hormones 550
II. Adrenalectomy in amphibians 551
E. Reptilia 552
F. Aves 553
I. Pituitary-adrenal relationship 553
II. Adrenalectomy in birds 554
III. Effects of exogenous cortical hormones 554
G. Chemical nature of corticoids on lower vertebrates 555
Literature 555

Chapter 5: Sex hormone secretion by the adrenal cortex
By Evelyn Howard and Claude J. Migeon

Introduction 570
Androgens 571
A. Nature of adrenal androgens 571
B. Biological properties of the C_{19} steroids of adrenal origin 574
I. Androgen bioassay methods 575
II. Biological studies with adrenal androgens 576
C. Physiological and pathological considerations in man 580
I. Effects of adrenocorticotrophin and adrenalectomy-gonadectomy 580
II. Effects of menstruation and pregnancy 581
III. Effects of age 582

IV. Hyperadrenocorticism 584
V. Regulation of the production of adrenal androgens 585

D. Structural variations in the inner cortex related to hormonal status. 586
I. Primates. 587
1. Fetal reticularis 587
2. Postnatal reticularis 589
3. Metabolic considerations 590
II. Rodents 593
1. The X zone of the mouse adrenal. 593
2. A presumptive X zone in voles. 597
3. A presumptive X zone in the golden hamster 597
4. The rat 597
III. Other mammals 598
IV. Summary 598

E. Experimental studies in animals 598
I. Effects of adrenalectomy in castrated animals 598
1. The rat 599
2. The mouse 602
3. Voles 603
4. The golden hamster 603
II. Effects of adrenals grafted to target organs 604
III. Adrenal transplants in chick embryos 605
IV. Androgen production by adrenocortical tumors in castrated animals 605
V. Effects of ACTH in castrated animals 607

F. Conclusion 609

Progestins 609

A. Metabolic considerations 609

B. Progesterone bioassay methods 611

C. Systemic significance of adrenal progesterone 613

D. Conclusion 615

Estrogens 615

A. Estrogen production in cases of adrenal tumor or hyperplasia in man 615

B. Estrogen bioassay methods 617

C. Estrogen production by experimental tumors in animals 618

D. Does the normal adrenal produce systemically appreciable amounts of estrogens ? 620
I. Estrogen: glucocorticoid antagonism 620
II. Apparent estrogenization following ACTH administration 621
III. Functional relationships between adrenal and ovary 621
IV. Adrenal grafts to the uterus 623

E. Conclusion 623
Epilogue 623
Literature 624

Chapter 6: Psychological manifestations of hyper- and hypoactivity of the adrenal cortex
By M. Bleuler and W. A. Stoll

Psychopathology of hyperadrenocorticism (M. Bleuler) 638

A. Excess of glucocorticoids 638
I. Cushing's syndrome 638
II. Treatment with ACTH or glucocorticoids 640
III. Strong cortical activity within normal limits 642
IV. Discussion 642

V. Therapeutic conclusions 647
B. Excess of adrenocortical androgens 650
I. Virilizing (congenital) adrenocortical hyperplasia in females (with pseudohermaphroditism and pseudopubertas precox) 650
II. Virilizing (congenital) adrenocortical hyperplasia in males (with pseudopubertas precox) 651
III. Adrenogenital syndrome from adrenocortical tumor or acquired adrenocortical hyperplasia 652
IV. Discussion 652
C. Excess of estrogens 653
D. Excess of mineralocorticoids 653
Psychopathology of hypoadrenocorticism (W. A. STOLL) 653
A. Addison's disease 653
B. Mild hypoadrenocorticism 656
Literature 656
Author index 660
Subject index 724

Chapter 1

The anatomy, chemistry, and physiology of adrenocortical tissue*

By

Helen Wendler Deane

With 71 figures

In this chapter, first the normal anatomical and then the normal chemical properties of adrenocortical tissue in the various classes of vertebrates are summarized. In the anatomical section, enough material is now at hand to justify separate descriptions of the gland in the different classes of vertebrates. In the chemical section, no attempt is made to treat the classes separately, since almost all chemical study has been made on mammalian glands. The final section aims to survey what we now know about the function of the gland, both in the so-called "normal" state and when secretion of its several hormones is enhanced or depressed. In this section, functional alterations indicated by morphological and chemical methods will be described in addition to those determined by more traditional physiological techniques. Most of the last section also deals with mammals; a short subsection on non-mammalian classes is appended.

Gross, microscopic, and developmental anatomy of adrenocortical tissue

A. Introduction

I. Form and location

Adrenocortical tissue occurs only in animals of the Subphylum Vertebrata (or Craniata) of the Phylum Chordata. No trace of such tissue has been detected in the more primitive chordates or in any invertebrates. The glandular tissue lies retroperitoneally against or near the kidneys, in close association with the posterior vena cava (or its homologues, the posterior cardinal veins) and the renal veins.

In the fishes and the more primitive amphibians this tissue tends to be diffusely distributed in small islands, but in the anurans and the higher classes of vertebrates it occurs in well-defined, bilateral organs. Moreover, in the lowest vertebrates (Agnatha, Elasmobranchii, Dipnoi) the adrenocortical tissue is quite separate from the chromaffin (medullary) tissue. In teleost fishes and the lower amphibia, some of the chromaffin tissue lies immediately next to or within

* Completion of this manuscript was aided by a grant from the United States Public Health Service, A-3605. The review was finished early in 1960.

The author wishes to thank the following colleagues for reading all or parts of this review and offering valuable suggestions: Dr. Alfred A. Angrist, Dr. Roy O. Greep, Dr. Eric Bloch, Dr. Betty L. Rubin, Dr. Sam Seifter and Dr. Edna H. Sobel.

the interrenal bodies. In anurans, reptiles and birds, the two types of adrenal cells become progressively more intimately intermingled. In mammals, a large mass of chromaffin tissue lies in the center of the adrenal gland surrounded by the cortical tissue. Even where association occurs, however, some chromaffin tissue lies outside the adrenal glands in so-called paraganglia.

In most animals, adrenocortical tissue can be recognized grossly by its white or yellow color, which results from the abundance of lipide droplets in its cells. (In disease or under experimental conditions, of course, the glands may be depleted of lipides and therefore appear brown or red.) In some species, the glands are too small or diffusely distributed to be readily identified, or they may lie embedded in a fatty connective tissue of similar color (e.g. Dipnoi). In a minority of species (Teleostei, ruminant Artiodactyla, and hibernating Rodentia), the glands normally appear red or brown because free lipide droplets are absent.

II. Blood and nerve supply

In the most primitive vertebrates, the adrenocortical tissue lies in the walls of the great trunk veins posterior to the heart. Where the tissue is aggregated into separate organs, it receives a rich arterial supply, with several contributing arteries arising from the aorta, the renal artery, and other nearby vessels. Some reptiles possess a portal venous blood supply as well. One or more large veins empty into the posterior cardiac vein, vena cava, or renal vein.

In those animals with admixed chromaffin tissue, the adrenal glands receive large bundles of preganglionic (myelinated) sympathetic nerve fibers that serve the medullary cells. The few postganglionic (unmyelinated) fibers apparently innervate only the arteries supplying and passing through cortical tissue, and the adrenal vein; there seems to be no significant innervation of adrenocortical tissue proper.

III. Histological organization

The general histological characteristics of adrenocortical tissue are quite similar in all classes of vertebrates, although the detailed organization varies greatly. Where organized into a separate body, the parenchymal tissue is surrounded by a connective tissue capsule, of greater thickness in larger animals. *The parenchyma consists of sheets or cords of epithelial or epithelioid cells supported by a delicate connective tissue and bounded by capillaries.* The parenchymal strands are only one or two cells thick, so that each cell has at least one surface abutting on a blood vessel.

The polygonal cells are generally about $15-20\,\mu$ in diameter. The nucleus is spherical or ovoid, vesicular, and generally centrally-placed. The relatively acidophilic cytoplasm contains a juxtanuclear Golgi apparatus, numerous small spherical or rod-shaped mitochondria and usually (but not always) a variable number of lipide droplets.

IV. Development

Embryologically, adrenocortical tissue typically arises from bilateral masses of cells that proliferate from the dorsal coelomic epithelium at approximately segmental intervals and migrate into the underlying mesenchyme on either side of the root of the dorsal mesentery, just medial to the developing renal and genital tissues (Gruenwald 1946, Vannini and Sabbadin 1954). The mesenchyme in which the cells aggregate has an especially well-developed vascular plexus (Boerner 1952/53). Continued growth of the parenchyma depends both on cell multiplication and on cell enlargement. In most classes of vertebrates, the adrenal glands come to occupy a more caudal position as development proceeds (e.g. BRANIN 1937, GRUENWALD 1943). In the higher vertebrates, the presumptive chromaffin cells invade the embryonic adrenocortical mass after the latter is well-established. Then the enveloping mesenchyme forms a connective-tissue capsule around the gland.

Anatomical methods. Because of the characteristic enlargement and cytological alteration of adrenocortical tissue in animals subjected to physiological or emotional stress, *care must be taken to obtain the organ from unstressed animals*. The animal must be killed quickly and without forewarning. GLICK et al. (1954) recommend shooting the animal unawares! This technique has been used especially to obtain glands from monkeys by GLICK and co-workers and from wild rats (NICHOLS 1950, MOSIER 1957). If the animal is anesthetized, some non-stressful anesthetic like a barbiturate should be used. So far as human glands are concerned, only those obtained from individuals meeting instant accidental death can be considered normal.

Aside from this general caution, the methods used for the study of adrenocortical structure are the same as those for other organs. Some methods must be modified because of the abundance of lipide droplets generally present in the parenchymal cells. This proviso applies particularly to methods for cytoplasmic organelles.

The *gross structure* and relations of adrenocortical tissue may generally be studied with the naked eye or the dissecting microscope, either in the freshly-killed animal or in one preserved in a hardener like formalin. In those species in which the gland cannot be distinguished from surrounding tissues and in small animals and embryos, there may be no alternative to making sections of the perirenal region. Where the gland can be dissected clean it may be weighed, either fresh or after fixation in a fluid that causes little change in weight (e.g., 10% formalin).

For ordinary *histological study*, good fixatives are Bouin's or Heidenhain's susa. Large adrenals should be gently sliced open for fixation, or fixation by gentle perfusion can be used. Paraffin sections may be stained with good nuclear and cytoplasmic stains, such as hematoxylin and eosin or methylene blue and eosin, or by the Feulgen nucleal reaction. The stroma may be demonstrated by an ammoniacal-silver hydroxide method, the periodic acid-Schiff technique, or a triple stain such as Heidenhain's azan mixture.

For distinguishing chromaffin from adrenocortical tissue, a neutral formalin-potassium dichromate fixative like Helly's or Orth's is used; granules in the chromaffin cells are thereby oxidized to insoluble brown complexes (see Fig. 4, p. 8). In general, the chromaffin tissue is far more basophilic than is the adrenocortical.

Serial sections may be used to determine relative *volumes* of cortical and medullary elements and even of the different cortical zones by projecting periodic sections, drawing the outlines and calculating areas planimetrically or by a paper-weight method (SWINYARD 1940, LIEBEGOTT 1944, ERÄNKÖ 1954b, BAHN et al. 1960). It should be self-evident that when such measurements are to have comparative value, all preparations must be fixed and treated alike. This requirement has been demonstrated most rigorously for studies of nuclear size (KRACHT and SPAETHE 1955, MERRIAM 1955).

To preserve and blacken the *lipide droplets*, many workers have used fixatives containing osmium tetroxide (e.g., FLEXNER and GROLLMAN 1939). Though less handsome, preparations that are more reliable from the histochemical point of view are made from blocks fixed in formalin and sectioned on the freezing microtome (p. 50). Sections are then stained with an oil-soluble dye. Methods for demonstrating other properties of the lipide droplets will be listed later (p. 52ff.).

Cell organelles such as mitochondria or the Golgi apparatus require special fixation and stains. Mitochondria are easily demonstrated by one of the special hematoxylin methods (Heidenhain's prolonged iron alum-hematoxylin technique, Mallory's phosphotungstic acid-hematoxylin technique, Baker's acid-hematein method) after formalin-dichromate fixation and postchromation (Fig. 1). Supravital staining with Janus green B must be employed for definite identification. Because of the high lipide content of the cells, the Golgi apparatus is not clearly distinguishable in preparations fixed in osmium tetroxide; a silver technique like the Da Fano or the Aoyama method must be used. The ribonucleoprotein constituent in the cytoplasm can be recognized in preparations stained at approximately p_H 5 in a dilute solution of a basic dye; its identity can be confirmed by prior digestion of control sections with purified ribonuclease.

For *electron microscopic study*, special preparative methods are required (FARQUHAR 1956, SELBY 1959, ZELANDER 1959).

The gross *blood supply* and even the distribution of finer blood vessels can be appreciated to some extent in ordinary gross or histological preparations. For more detailed study, injection via the aorta of an opaque material is required (BENNETT and KILHAM 1940). Needless to say, such techniques require great care, not only that all vessels be filled but also that the walls of the more delicate vessels remain unruptured. Recently HARRISON (1957) has used the radio-opaque compound Thorotrast *in vivo* for observing circulation through the gland under different conditions. Such preparations could then be fixed and sectioned for histological examination.

The gross *innervation* of the gland can be traced, in large animals at least, without special techniques. In smaller animals it might be necessary to employ intravital methylene blue to identify the nerves. The intraglandular nerves should be studied by one of the reliable

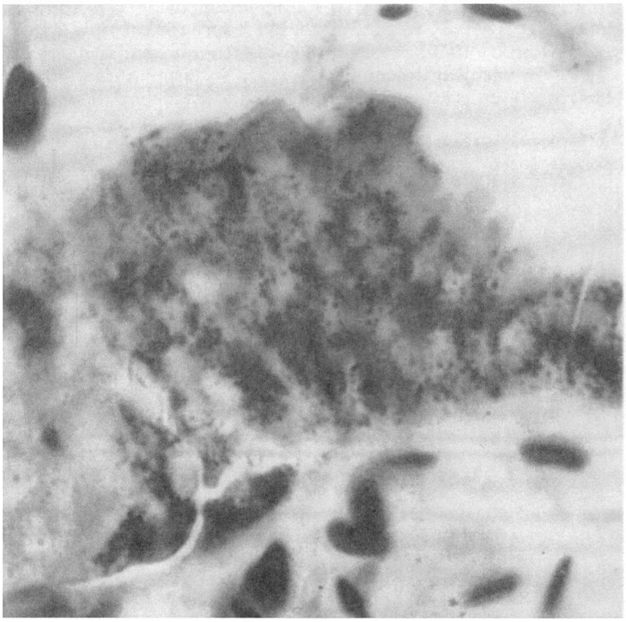

Fig. 1. Mitochondria in the adrenocortical cells of the fresh-water teleost, *Carassius auratus* (goldfish). Frozen section of tissue fixed in formalin-calcium and then postchromated and stained by Baker's acid-hematein technique for phosphatides. ×940. Courtesy Dr. WALTER CHAVIN. In teleost fishes, the adrenocortical tissue lies in the walls of the large veins in the lymphoid "head" kidney (see p. 7)

techniques, like Ranson's pyridine-silver or Bodian's protargol method (MACFARLAND and DAVENPORT 1941). The reticular fibers enveloping the parenchymal cells tend to stain by less specific methods, such as Bielschowsky's.

Except where references have been given above, the methods can be found in texts on anatomical and histological techniques.

B. Survey of glandular structure, by classes

The following survey of the structure of the adrenal cortex will describe its appearance and development in representative animals only. For the anatomical characteristics of the gland peculiar to individual species, see the more extensive reviews of VAN DER SPRENKEL (1934), HARTMAN and BROWNELL (1949), BOURNE (1949), BACHMANN (1954) and CHESTER JONES (1957). Herein are given only particularly instructive references or ones that have appeared since BACHMANN's review (1954).

The classification adopted is that used by ROMER (1945)[1], with emendations for mammals from G. SIMPSON (1945) and for birds from WALLACE (1955).

[1] The staff at the Museum of Comparative Zoology, Cambridge, Massachusetts, kindly checked the names of genera and species found herein.

A valuable source of information on the adrenocortical glands in fishes (Cyclostomata to Teleostei) has been prepared by G. PICKFORD (1957).

I. Class Agnatha

a) Order Cyclostomata

α) Suborder Mixinoidea. No one has yet succeeded in identifying adrenocortical tissue in hagfishes.

β) Suborder Petromyzontia. Only a brook lamprey (*Lampetra planerii*) has been studied in any detail (STERBA 1955), although a marine form (*Petromyzon marinus*) was examined earlier (GIACOMINI 1902). The interrenal (or adrenocortical) tissue occurs as *small islands* in the walls of the posterior cardinal veins and in the segmental arteries and veins along the entire length of the kidney. Nests of chromaffin tissue lie nearby. Some lobules of cortical cells are found immediately underlying the endothelium of the veins; others occur more peripherally in the loose connective tissue, where each strand is enveloped by capillaries. The *cells* are large and polyhedral, with round nuclei. The abundant cytoplasm contains numerous acidophilic granules, small mitochondria and a few lipide droplets.

In *embryonic development*, the cortical cells begin to proliferate from the dorsal coelomic wall in 3.5—4 mm embryos, first at the anterior end of the pronephros and then more caudally. Growth of the tissue is relatively slow in larvae (*Ammocoetes*) but accelerates rapidly during metamorphosis, when the lamprey is 3—4 years old. According to STERBA's (1955) calculations based on the increase in cell number and cell size, the glandular tissue has about 30 times the volume in the adult as in the larva just before metamorphosis.

II. Class Chondrichthyes

1. Subclass Elasmobranchii

The interrenal glands of these cartilaginous fishes have been studied extensively. The gland in sharks (Order Selachii) is generally a single, medial, elongate organ embedded in the caudal half of the fused mesonephroi; occasionally small fragments lie more cranially, separate from the main body. In most skates and rays (Order Batoidea, Families Rajidae, Dasyatidae), the gland is roughly U-shaped and lies between the kidneys, but it may be fragmented into several bodies. In the Family Torpedinidae, however, the glands are more rounded; there may be a pair on either side or a single gland on one side only. HARTMAN et al. (1943, HARTMAN and BROWNELL 1949) dissected out the glands from various elasmobranchs and weighed them. Representative results are shown in Table 1.

Table 1. *Weights of interrenal glands in representative elasmobranchs* (HARTMAN and BROWNELL 1949)

Species	Average body wt.	Interrenal glands, mg/100 g body wt.	
	kg	♂	♀
Mustelus canis (smooth dogfish)	2.5	9.2 (9)[1]	6.6 (6)
Squalus acanthias (spiny dogfish)	3	—	5.7 (20)
Raja erinacea (skate)	0.5	3.6 (17)	3.4 (56)
Dasyatis centrura (sting ray)	50	0.6 (15)	0.7 (22)

[1] Number of specimens in parentheses.

Microscopically, the highly-vacuolated (lipide-rich) cells appear to be arranged in double strands between capillaries (Dittus 1940, Aboim 1944, 1946). *Embryologically*, the prospective interrenal tissue proliferates from the dorsal coelomic epithelium at the cephalic end of the renal blastema in typical fashion in 7—14 mm embryos (*Torpedo, Scyliorhinus* — Chieffi 1952) (Fig. 2).

Chromaffin bodies frequently lie adjacent to the interrenal bodies, but the two types of cells are not intermingled. Young (1933) could find no nerve fibers ending on interrenal cells, although there is a rich nerve supply to the arterioles and the nearby chromaffin bodies.

a

b

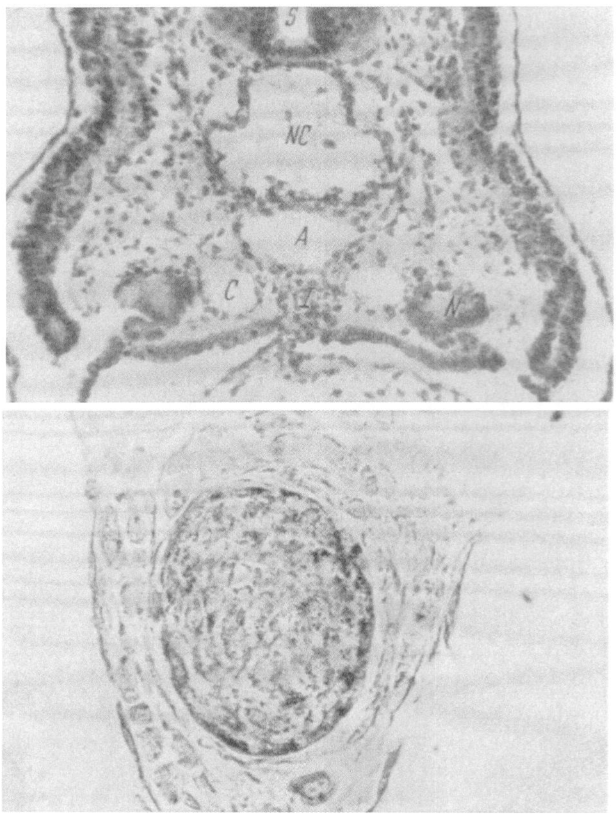

Fig. 2 a & b. Interrenal gland in the embryo shark, *Scyliorhinus canicula*. Courtesy Dr. Giovanni Chieffi. a Cross section of a 14-mm embryo. Paraffin section, hematoxylin and eosin. ×175. The anlage of the medial interrenal gland (*I*.) is arising at the base of the mesentery, between the posterior cardinal veins. *S*., spinal cord; *N.C.*, notochord; *A*., aorta; *C*., cardinal vein; *N*., nephric duct. b Cross section of the root of the mesentery in a 38-mm embryo. Frozen section, Sudan black B. ×350. The cells of the enlarged and encapsulated interrenal gland contain small lipide droplets (which first appear at the 25-mm stage)

2. Subclass Holocephali

There is no reliable description of interrenal tissue in species of the Order Chimaerae.

III. Class Osteichthyes

1. Subclass Choanichthyes

a) Order Dipnoi

What appears to be the interrenal tissue of lungfishes (*Protopterus* spp.) was described only recently (Gérard 1951, W. Holmes and Moorhouse 1956). The adrenocortical cells occur in discontinuous islands embedded in the ventral surface of the kidneys, close to the posterior cardinal veins. Apparently they were not detected at an earlier date because the entire kidney is sheathed in a whitish tissue that masks the adrenocortical cells. This surrounding tissue consists mostly of cells containing abundant acidophilic granules, resembling at least superficially the so-called "summer" cells in anuran interrenal glands (p. 9).

The cytoplasm of the adrenocortical cells proper contains abundant lipide droplets (sudanophilic, osmiophilic). The interdroplet cytoplasm contains small mitochondria.

b) Order Crossopterygii

The interrenals have not been examined in *Latimeria*.

2. Subclass Actinopterygii

a) Superorder Chondrostei

In sturgeons (*Acipenser* spp.), the interrenal tissue forms dorsomedial, semilunar caps in the walls of the posterior cardinal veins (GIACOMINI 1933). The cells are arranged in epithelioid cords and are conspicuously vacuolated (lipides). As in elasmobranchs, there is no intermingling with chromaffin cells.

b) Superorders Holostei and Teleostei

For both the bony ganoid fishes (e.g., *Amia* spp.) and the teleost fishes, *there has been considerable confusion about the identification of the adrenocortical tissue* (GIACOMINI 1922, 1933, VINCENT and CURTIS 1927, G. PICKFORD 1957, FORD 1959). Two different bodies have been implicated, the so-called posterior interrenals or Stannius corpuscles, which lie between the mesonephroi near their caudal ends, and the so-called anterior interrenals, which lie in the vestigial head kidney (a lymphoid organ in adult fishes), near the cranial ends of the posterior cardinal veins and their renal branches. The Stannius corpuscles arise *embryologically* from the pronephric (Wolffian) ducts, whereas the anterior interrenals develop from the coelomic epithelium (GARRETT 1942). Thus *the anterior interrenals are the adrenocortical homologues.* (The cells in the Stannius corpuscles are columnar and organized in ill-defined tubules.)

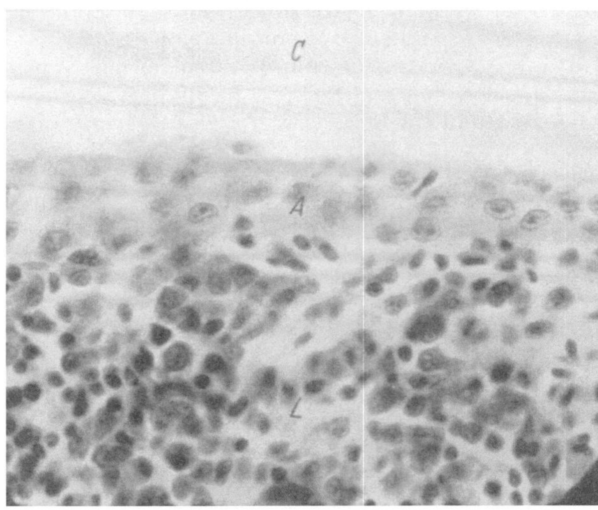

Fig. 3. Adrenocortical tissue of the fresh-water teleost, *Astyanax mexicanus* (characin). Paraffin section. × 800. Courtesy PRISCILLA RASQUIN and the American Museum of Natural History. Two or three layers of large, unvacuolated adrenocortical cells (*A.*) underlie the endothelium of the posterior cardinal vein running through the "head" kidney (lymphatic tissue). *C.*, lumen of posterior cardinal vein; *L.*, lymphatic tissue. In frozen sections stained with oil-soluble dyes, these cells display no free lipide droplets, but the cytoplasm exhibits a diffuse sudanophilia (see RASQUIN 1951)

The *cells* in the anterior interrenals are typically polygonal and arranged in cords lying between the endothelium and tunica media of the veins (Fig. 3). Confusion has arisen in part because the anterior interrenal cells generally *lack free lipide droplets* (BAECKER 1928, RASQUIN 1951, G. PICKFORD 1957), although lipides have been described in the adrenals of a few species (CHESTER JONES 1957). The interrenal cells exhibit conspicuous fuchsinophil granules (mitochondria). In the goldfish, *Carassius auratus*, at least, the cells lie in two layers, a perivenous layer of radially-arranged cells and an outer layer of longitudinally-arranged cells (KRAUTER 1951). In different teleosts, chromaffin tissue is variously associated with interrenal tissue, being merely adjacent in the European flounder,

Pleuronectes flesus, but intimately intermingled in the goldfish, *Carassius auratus* (Krauter 1951).

The interrenal tissue of teleosts is generally too dispersed to be dissected out and weighed, but it is relatively more aggregated in the silver eel, *Anguilla anguilla*. In this species, Fontaine and Hatey (1953) found the glands to comprise about 3.5 mg per 100 g body weight.

IV. Class Amphibia

a) Order Gymnophiona (Apoda)

The interrenal glands of *Ichthyophis glutinosus* (Dittus 1936) consist of scattered *islands*, containing intermingled adrenocortical and chromaffin cells, between the kidneys in the walls of the efferent renal veins and the postcaval vein. The lobules are largest at the cranial end of the kidney and become smaller toward the cloaca. The *cells* are relatively small in the larval stages (7—10 cm specimens) but increase both in number and volume during metamorphosis (10—16 cm specimens). In sexually-mature adults (18—28 cm), the cells remain large, appear granular and contain lipide droplets.

The *anlagen* of the interrenal system develop as paired, segmental proliferations from the coelomic epithelium, beginning at the precardial level and arising subsequently in the more caudal segments to the cloacal region.

b) Order Urodela

The interrenal glands of newts consist of *paired discontinuous islands* situated along the ventromedial surfaces of the mesonephroi, or, in some species, as unpaired medial bodies (Branin 1937). Some patches may lie rostral to the

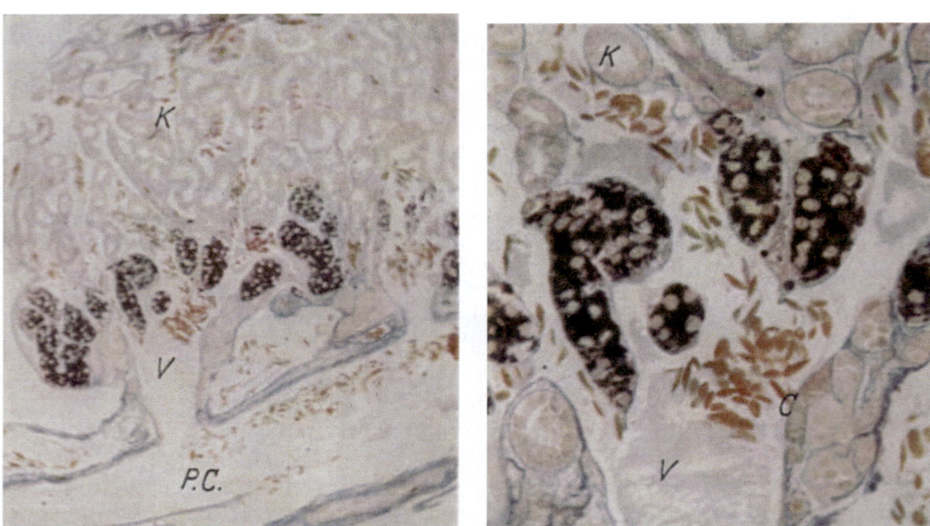

a b

Fig. 4a & b. Adrenocortical tissue of the Western newt, *Taricha torosa*. Courtesy Dr. Malcolm R. Miller. Block fixed in an acetic acid-osmic acid-potassium dichromate solution, stained with Heidenhain's azan mixture. a × ca. 50; b × ca. 150. The cords of adrenocortical cells (A.) lie in the walls and project into the lumens of the efferent renal veins (M. Miller 1953). The abundant lipide droplets in the cells have been blackened and preserved by osmium tetroxide. The catechol amines in the chromaffin tissue (*C.*) lying nearby have been browned by the dichromate. The two kinds of tissue do not intermingle. *K.*, kidney tubules (mesonephros) *V.*, efferent renal vein; *P.C.*, postcaval vein. Note the unusually large size of the cells, which is characteristic of newts kidneys. *The islands occur in the walls of the postcaval vein and the efferent renal veins* (KLOSE 1941) (Fig. 4). In *Taricha* (*Triturus*) many of the cords lie across the lumens of the vessels, covered only by endothelium (M. MILLER 1953) (see Fig. 70, p. 142). Because of the abundance of *cytoplasmic lipide droplets*, the glands characteristically appear yellow or orange. Fuchsinophilic granules (probably mitochondria) occur in the cytoplasm between the lipide droplets when the tissue has been fixed in a bichromate-containing fluid. Chromaffin cells, either single or in clusters, are limited to the peripheries of the cords. The glands are enlarged during the breeding season (M. MILLER and ROBBINS 1955).

Cortical *anlagen* are found in the embryos before hatching, the first occurring cranial to the mesonephroi, the later bodies arising more caudally to these throughout larval development. In the salamander *Hemidactylium scutatum*, the glands exhibit a gradual caudal shift, the first patches arising between the 5th and 8th spinal nerves, the adult glands lying between the 9th and 20th nerves (BRANIN 1937). During metamorphosis, there is considerable coalescence and growth of the islands. Lipide droplets are less abundant in the cells during larval than during adult life.

c) Order Anura

In the anurans (frogs and toads) and all higher vertebrates, the adrenocortical tissue is typically aggregated into a *single pair of glands* rather than being fragmented, as it is in the preceding classes and orders. In the Anura, the two glands lie embedded in the ventral surfaces of the mesonephroi closely adherent to the renal arteries and veins. No connective tissue capsule separates adrenal and renal tissues. This location means that the glands have not often been weighed. SPECTOR (1956, p. 164) gives the relative weight in *Rana catesbiana* as 20 mg-%; RAPELA and GORDON (1956) found the same for *Bufo arenarum.* The glands are relatively enlarged during the breeding season (TUCHMANN-DUPLESSIS et al. 1955).

The glands possess three types of intermingled cells: typical adrenocortical cells, chromaffin cells, and cells possessing abundant acidophilic granules, called "summer" cells.

So-called "summer" cells are present in the glands of most anurans (FUSTINONI and PORTO 1938, VERNE and DELSOL 1956, GABE 1956, CHESTER JONES 1957). These cells possess large, refractile, acidophilic granules in the cytoplasm. They were first named "summer" cells by STILLING (1898) because of their great abundance at this season, but some are present throughout the year. Their nature remains unknown (PATZELT 1955, BURGOS 1959).

The *adrenocortical cells* are characteristically large. Some contain extremely small lipide droplets, others contain much larger ones (PORTO 1940). CATER and LEVER (1954) detected some zonation of cell types in the adrenal of *Rana*: the subcapsular cells appeared smaller and more elongate and contained less lipide than the cells lying deeper in the gland. This tendency to zonation is disputed by CHESTER JONES (1957). The mitochondria appear more concentrated in the cells with small lipide droplets. The Golgi apparatus is a small, juxtanuclear knot in the cells with small droplets, larger and more fragmented in the cells with large droplets. The appearance of these cells in the electron microscope has been described by BURGOS (1959).

The adrenocortical *anlagen* develop as segmental, paired outgrowths from the dorsal coelomic wall, the first arising just caudal to the heart. The masses are associated with the postcaval vein and the efferent renal veins (VANNINI and SABBADIN 1954, WITSCHI 1956) (Fig. 5). "Summer" cells are already present in 18 mm *Rana pipiens* tadpoles, prior to the invasion of chromaffin tissue. By this stage, lipide droplets have appeared in the interrenal cells (STENGER and CHARIPPER 1946, CHIEFFI 1955). Growth of the gland and accumulation of lipide in the cells continues during larval life and even after metamorphosis.

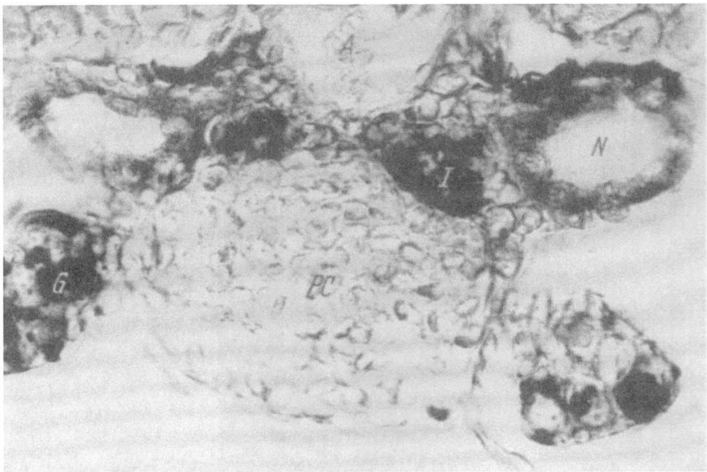

Fig. 5. Anlagen of adrenocortical glands (*I.*) in a 10 mm embryo (tadpole) of the frog, *Rana esculenta*. Frozen section stained with the oil-soluble dye, Sudan black B. ×400. Courtesy Dr. GIOVANNI CHIEFFI. *A.*, aorta; *N.*, nephric duct; *P.C.*, postcaval vein; *G.*, primitive gonad

V. Class Reptilia

In all reptiles, the adrenals are typically unified into a pair of glands, consistently associated with the reproductive organs and the posterior venous system. A valuable survey of the anatomy of the adrenals in all orders has recently been published by HEBARD and CHARIPPER (1955). A peculiarity of the adrenals in the Order Squamata is the presence of a secondary blood supply from the veins of the dorsal musculature or from the afferent renal veins — vessels presumably derived from the embryonic posterior cardinal veins.

a) Order Chelonia

In terrapins and tortoises, the adrenals lie on the ventral surfaces of the kidneys, each usually consisting of a series of partially-fused spherical or irregular masses In sea turtles, the gland is a single medial mass. In the various turtles that have been examined, the relative weight is about 10 mg-%. Each gland is supplied with several small arteries arising from the aorta and the renal and gonadal arteries. The adrenal veins drain into the adjacent efferent renal veins.

The glands are enclosed by a thin connective tissue capsule, from which trabeculae invade and subdivide the parenchyma. *The glandular tissue consists of irregular cords of interrenal cells*, with intermixed clusters of chromaffin tissue. The polygonal adrenocortical cells contain a centrally-placed nucleus and evenly distributed lipide droplets. A description has been published recently by LIU and MANEELY (1959).

b) Order Crocodilia

In *Alligator*, the adrenal glands are discrete, elongated bodies adherent to the midventral surfaces of the kidneys. The paired glands are said to comprise 5—10 mg-% of the body weight. Several pairs of arteries from the aorta enter the dorsomedial surfaces of the glands. The veins drain into the adjacent postcaval vein. The chromaffin tissue occurs for the most part at the surface of the gland, although occasional islets lie intermingled with the strands of adrenocortical cells. The interrenal cords consist to *tall columnar cells*, each with a central nucleus and evenly-distributed, small lipide droplets.. The *anlagen* develop in the 9-mm embryo from the coelomic epithelium and grow rapidly thereafter (FORBES 1940).

c) Order Squamata

α) **Suborder Lacertilia.** Most studies on reptile adrenals have been made on lizards. Particularly valuable are those of RETZLAFF (1949, *Gerrhonotus*), BIMMER (1950, *Lacerta*), M. MILLER (1952, 1955, *Xantusia*), HEBARD and CHARIPPER (1955, *Phrynosoma* and others), and WRIGHT and CHESTER JONES (1957, *Agama*). The glands are typically ovoid and enclosed in the mesorchia or the mesovaria, dorsomedial to the cephalic poles of the gonads, with the right gland lying somewhat cranially to the left. In different species, the glands weigh between 15 and 75 mg per 100 g body weight. The glands are relatively largest during the breeding season (M. MILLER 1955).

There is some variation in *blood supply*. In *Phrynosoma*, each gland receives a single major artery that arises from the gonadal artery. In most other genera, there are several arteries arising from the aorta and the renal and the mesenteric arteries. Some species show *in addition a portal venous supply*. Generally the efferent veins of the left gland drain into the renal veins, while those of the right gland directly enter the postcaval vein.

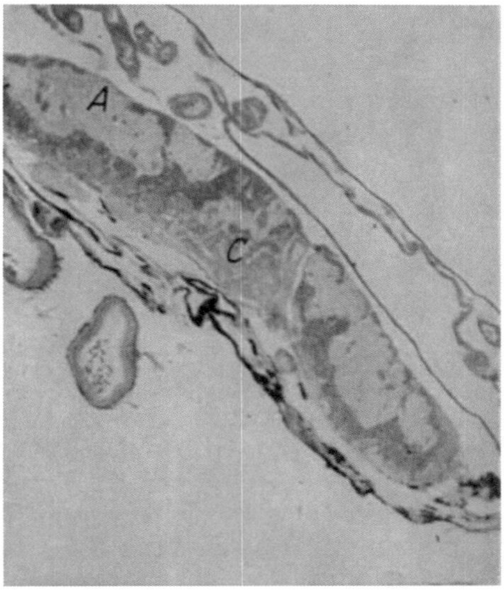

Fig. 6. Longitudinal section of an adrenal gland of the Yucca night lizard, *Xantusia vigilis*. Fixed in Zenker-formalin, stained with Heidenhain's azan mixture. × ca. 75. Courtesy Dr. MALCOLM R. MILLER. In reptiles, the adrenal gland is completely encapsulated and both adrenocortical and chromaffin tissues occur within the capsule. The chromaffin tissue (*C.*), in which the granules have been browned by the fixative, are aggregated along one surface, however. *A.*, adrenocortical cells

Lizards are the only reptiles in which there has been any attempt to trace the *nerve supply*; bundles of fibers enter the glands from the lateral sympathetic ganglia. Many of the fibers are myelinated. Most of the fibers obviously pass to the chromaffin cells, but some small fibers appear to invade the interrenal cords (RETZLAFF 1949, HEBARD and CHARIPPER 1955). Whether these small fibers are nerves or reticulum remains to be determined.

Most of the chromaffin tissue is aggregated along the dorsal surface of the gland, though small clusters also occur among the cortical strands (Fig. 6). According to BIMMER (1950), the cortical tissue comprises about 60% of the gland (*Lacerta*, males). *The interrenal cords generally consist of radially-arranged, tall columnar cells.* In some genera (e.g. *Xantusia*), the nuclei are peripherally placed, against the capillaries, and the lipide droplets lie at the centers of the cords (Fig. 7). In other genera (e.g. *Agama*, *Lacerta*), the nuclei lie centrally in the cords and the lipide droplets are located peripherally

(HEBARD and CHARIPPER 1955). The droplets are said to be largest in the cells lying at the surface of the gland (WRIGHT and CHESTER JONES 1957). Small mitochondria occur in the interstices between the droplets, and there is a juxtanuclear Golgi apparatus.

The *embryological development* is similar to that in the Amphibia (BIMMER 1950).

β) Suborder Serpentes (Ophidia). The general structure and relations of the adrenals in snakes are similar to those in lizards, although, consequent to the elongation of the body, the right gland is more conspicuously anterior to the left

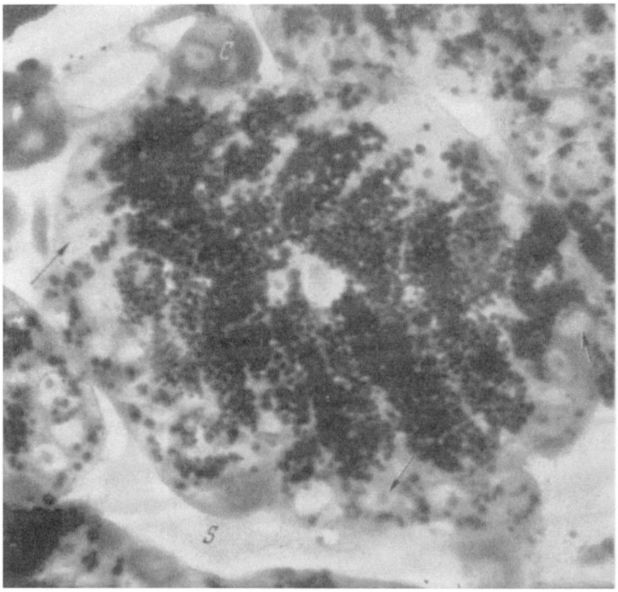

Fig. 7. Cord of adrenocortical cells, *Xantusia vigilis*. Fixed in an acetic acid-osmic acid-potassium dichromate solution, stained with Heidenhain's azan mixture. × ca. 850. Courtesy Dr. MALCOLM R. MILLER. In *Xantusia*, as in some other reptiles, the nuclei (arrows) of the adrenocortical cells lie peripherally in the cells, near the sinusoidal surface; the lipide droplets (here blackened by osmium tetroxide) lie centrally in the cords. In many reptiles however, the nuclei lie near the centers of the cords, as in birds (see Fig. 10). *C.*, chromaffin cells; *S.*, sinusoid (capillary)

(BRAGDON 1953). For small snakes, the relative weight is 25 mg-% (*Natrix* — WRIGHT and CHESTER JONES 1957) to 45 mg-% (*Thamnophis* — HEBARD and CHARIPPER 1955). For larger snakes (*Boa*) it may be as low as 8 mg-% (HARTMAN and BROWNELL 1949, SPECTOR 1956).

Several supplying blood vessels arise from the gonadal arteries or from the aorta. A portal venous supply from the dorsal body wall is conspicuous. Several small efferent veins drain into the renal vein (left gland) or into the posterior vena cava (right gland). Despite this intimacy with vital blood vessels, WRIGHT and CHESTER JONES (1957, CHESTER JONES 1957) report successful adrenalectomy in *Natrix*.

The gland possesses a thin connective tissue capsule, in which lie the major blood vessels, ganglia and nerve fibers. There is a thin layer of chromaffin tissue along the dorsal surface of the gland, and there are also many islets of chromaffin cells among the adrenocortical cords. The *cortical cells* are radially-arranged in short cords that generally appear to run longitudinally in the gland. The nuclei lie centrally in the cords at the base of the cells; the lipide droplets are near the vascular pole.

VI. Class Aves

Among the birds, species of *only a few orders have so far been examined.* Most studies have been made on domestic fowl (Order Galliformes) and pigeons (Order Columbiformes). HARTMAN and BROWNELL (1949) and HARTMAN and ALBERTIN (1951) have reported gross anatomical data for a variety of wild birds in several orders. KNAB (1942) examined the glands in a number of sparrows, finches, etc. (Order Passeriformes), KNOUFF and HARTMAN (1951) have studied the adrenal of the brown pelican (Order Pelecaniformes), and FLICKINGER (1959) described the glands in the California quail (Order Galliformes).

1. Gross anatomy

The glands lie just cranial to the kidneys and dorsal to the gonads. The gland may be a fused, medial structure, or there may be two glands in close contact with each other or distinctly separated (HARTMAN and ALBERTIN 1951). The shape varies considerably, but the organ is much more compact than in the classes considered heretofore. In adult birds, the adrenals vary in *relative weight*, ranging from 7.5 mg-% in the domestic turkey (*Melagris gallopavo*, Order Galliformes), 9 mg-% in the domestic fowl (*Gallus domesticus*) and the California quail (*Lophortyx californica*) and in many passerine birds (e.g., downy woodpecker, *Dryobates pubescens*; redwinged blackbird, *Agelaius phoeniceus*) and 12 mg-% in the domestic duck (*Anas platyrhynchus*, Order Anseriformes) to over 40 mg-% in the brown pelican (*Pelecanus occidentalis*, Order Pelecaniformes) (LATIMER and ROSENBAUM 1926, HARTMAN and BROWNELL 1949, LEROY and BENOIT 1954, FLICKINGER 1959).

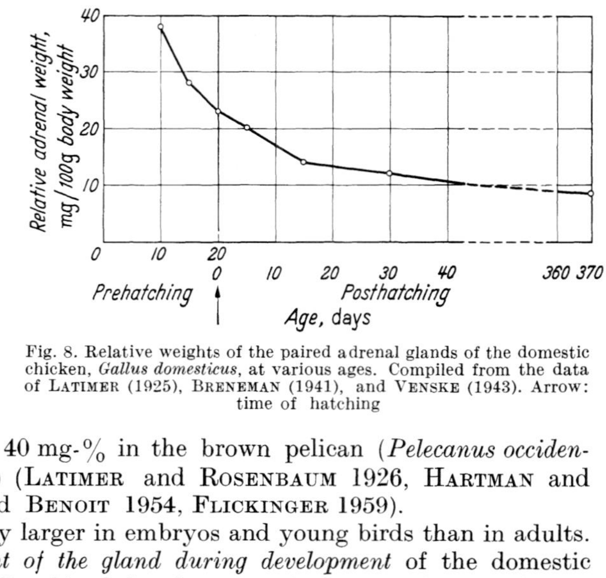

Fig. 8. Relative weights of the paired adrenal glands of the domestic chicken, *Gallus domesticus*, at various ages. Compiled from the data of LATIMER (1925), BRENEMAN (1941), and VENSKE (1943). Arrow: time of hatching

The adrenals are relatively larger in embryos and young birds than in adults. The *decline in relative weight of the gland during development* of the domestic chicken is shown in Fig. 8. In old adults they may be again relatively larger. They are also relatively larger in laying hens than in cocks. According to KORNFELD (1957), the glands are larger in pullets during the summer than during the winter.

2. Blood and nerve supply

The *blood supply* comes from several arteries branching from the aorta and from the intercostal arteries. The one or more veins drain into the postcaval vein Sympathetic nerves and clusters of ganglion cells occur in the capsule and within the gland. Using the Golgi technique, GIACOMINI (1898) could trace no *nerve fibers* to the interrenal cells proper.

3. Histological organization

The gland is surrounded by a loose connective tissue capsule and contains cords of adrenocortical cells with intermingled islets of chromaffin tissue (Fig. 9).

The *proportions of interrenal and chromaffin tissue vary* in different species and in the two sexes. Thus for the pigeon (*Columba*), R. Miller and Riddle (1942) reported that the adrenocortical tissue comprises 65%, the chromaffin tissue, 32%, and the stroma, 3% of the gland. Knouff and Hartman (1951) estimated that in *Pelecanus* the interrenal component occupies as much as 85 to 90% of the gland. In the fowl, Kar (1947) calculated that in males the interrenal tissue composes 40 to 45% of the gland, in females 70%. This sex difference develops

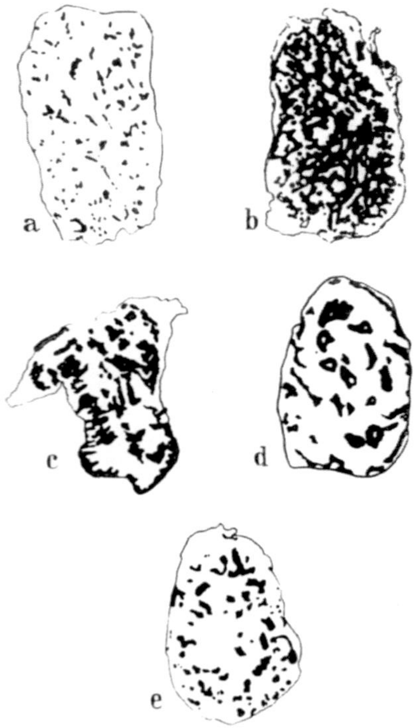

Fig. 9

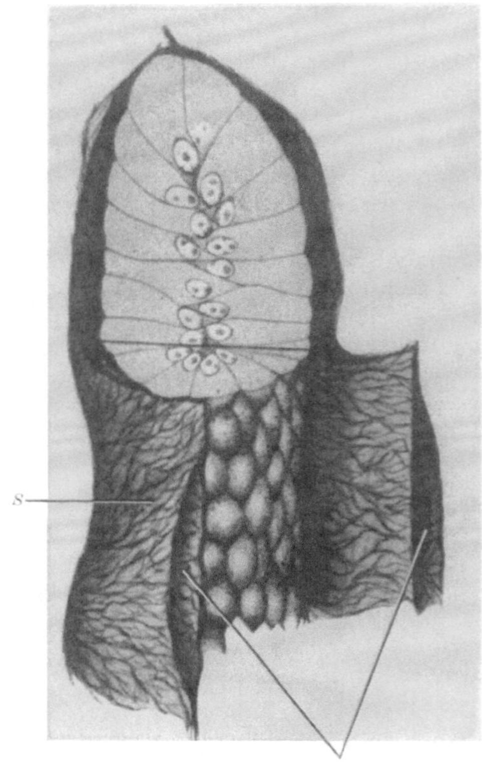

Fig. 10 *V.S.*

Fig. 9. Diagrams of the distribution of chromaffin tissue in the adrenal glands of various birds. Islands of chromaffin tissue (black) lie scattered throughout the gland; in some birds, adrenocortical tissue (white) comprises most of the gland, in others it occupies little. a *Pelecanus occidentalis*, brown pelican (Order Pelecaniformes); b *Accipiter cooperi*, Cooper's hawk (Order Falconiformes); c *Colinus virginianus*, Virginia quail or bob-white (Order Galliformes); d *Parus carolinensisca*, chickadee (Order Passeriformes); e *Melospiza georgiana*, swamp sparrow (same). The outlines are not drawn to scale. Reproduced by permission of the authors and publisher from Plates 1 and 3, Hartman et al.: Anat. Rec. **97**, 217 and 221 (1947)

Fig. 10. A reconstruction of a cord of adrenocortical cells, the pelican adrenal. *V.S.*, vascular sinusoids (capillaries); *S.*, stroma. Reproduced by permission of the authors and publisher from Fig. 14, Knouff and Hartman: Anat. Rec. **109**, 187 (1951)

in 16-day embryo chicks, as soon as there is evidence of androgen secretion by the testis (Maraud and Stoll 1956). The proportion of chromaffin tissue appears much higher in some passeriform birds than in fowl (Hartman and Brownell 1949, Hartman and Albertin 1951).

The *adrenocortical cells* generally lie in cords of radially-arranged cells (appearing as double rows in sections) (Fig. 10). The capillaries surround the cords. In glands in which chromaffin tissue is more-or-less centrally located, as in *Pelecanus* and, to a lesser extent, in *Columba*, the organization seems somewhat analogous to that in mammalian glands, with the peripheral cords arranged in loops (zona glomerulosa), the intermediate cords being straight (zona fasciculata) and the central cords anastomosing (zona reticularis) (Fig. 11). In other orders, zones are not distinguishable (Fig. 12).

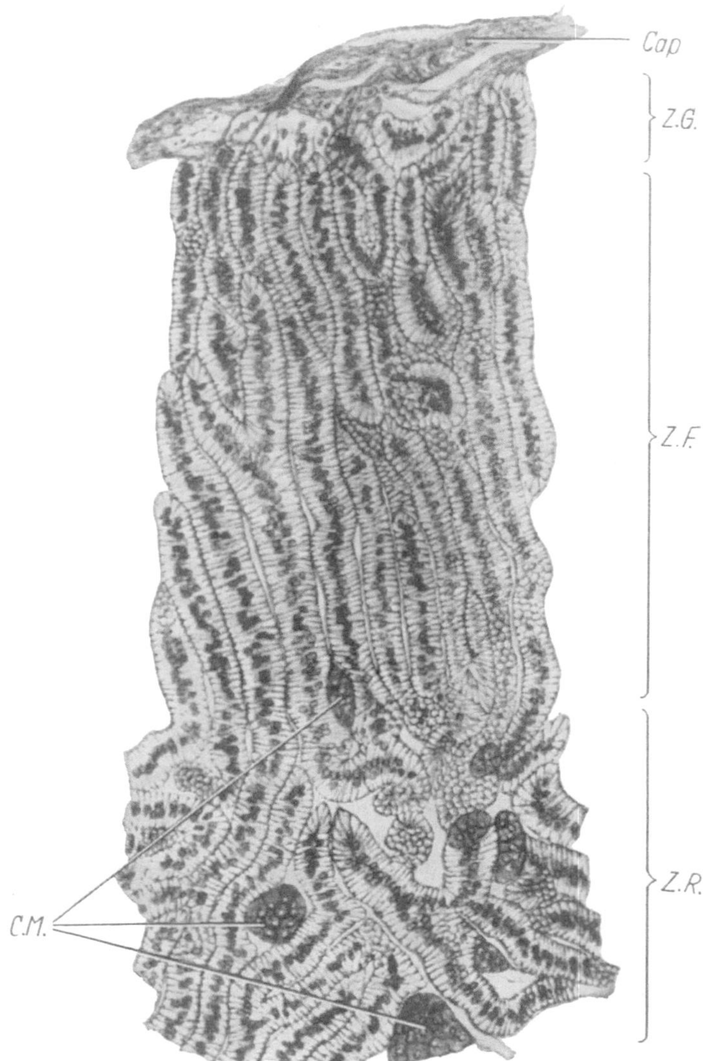

Fig. 11. Drawing of the arrangement of adrenocortical cords in the pelican adrenal, showing the suggestion of a zona glomerulosa (*Z.G.*), zona fasciculata (*Z.F.*) and zona reticularis (*Z.R.*). Cap, connective tissue capsule; *C.M.*, masses of chromaffin tissue. × ca. 170. Reproduced by permission of the authors and publisher from Fig. 1., KNOUFF and HARTMAN: Anat.Rec. **109**, 181 (1951)

4. Cytology

In most orders, the *nuclei* are described as lying centrally in the cords (Fig. 10), but in passerine birds KNAB (1942) found them to be peripheral, at the vascular poles of the cells. *Lipide droplets* occur in the rest of the cytoplasm. These are

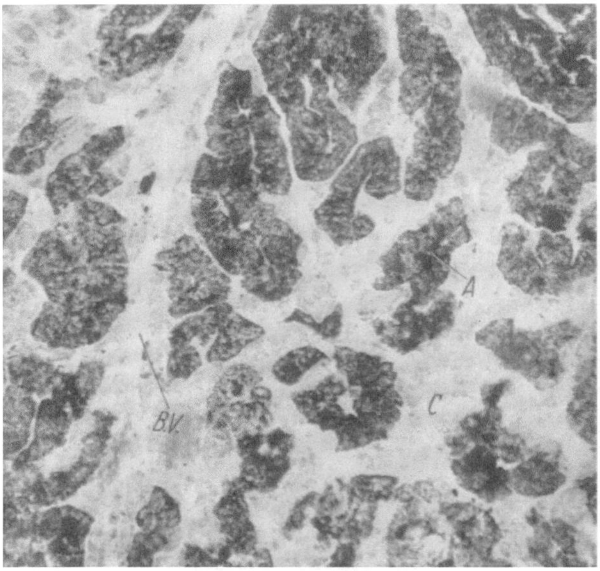

Fig. 12. Section of a chicken adrenal gland showing intermingling of adrenocortical and chromaffin tissues. Formalin-fixed frozen section, stained with Sudan IV for lipides and counterstained lightly with hematoxylin. × 80. Photographed from a slide contributed by Dr. JACQUES PADAWER. Most of the tissue between the adrenocortical cords (*A*) consists of chromaffin cells (*C*). *B.V.*, blood vessel

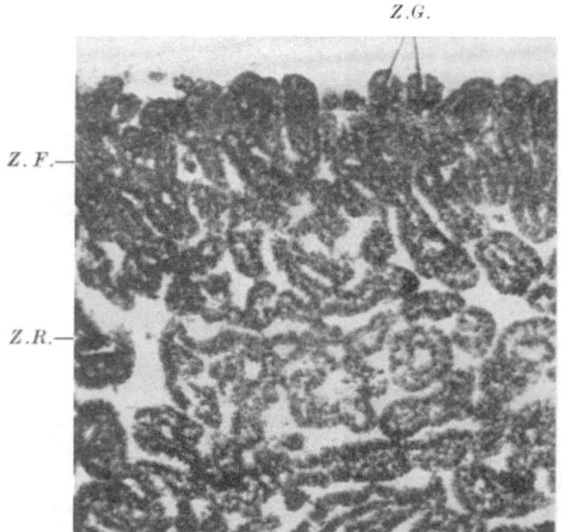

Fig. 13

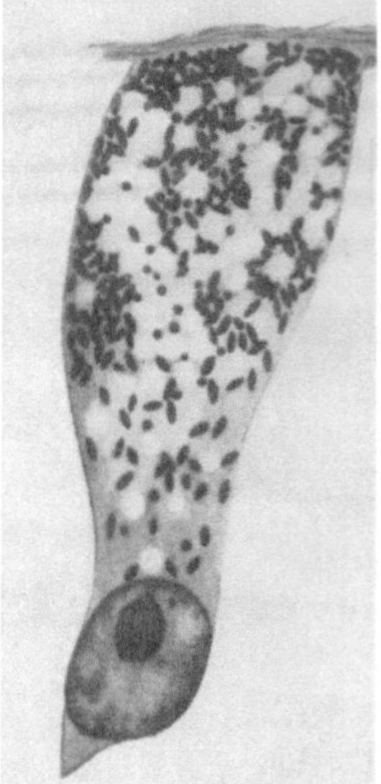

Fig. 14

Fig. 13. A frozen section of a pelican adrenal, stained with oil-soluble Sudan III. × 85. The cells of the zona glomerulosa (*Z.G.*) contain especially abundant lipide droplets. *Z.F.*, zona fasciculata; *Z.R.*, zona reticularis. Reproduced by permission of authors and publisher from Fig. 11, KNOUFF and HARTMAN: Anat. Rec. **109**, 185 (1951)

Fig. 14. Camera-lucida drawing of a pelican adrenocortical cell, showing mitochondria scattered between the lipide vacuoles. This cell contains relatively little lipide, so that the mitochondria appear crowded. Fixed in Zenker-formalin, stained with acid fuchsin and light green. × ca. 3500. Reproduced by permission of the authors and publisher from Fig. 16, KNOUFF and HARTMAN: Anat. Rec. **109**, 187 (1951) small but very abundant in the peripheral zone of the gland, somewhat larger and more abundant in the middle zone and extremely large in the central zone (Fig. 13) (KNOUFF and HARTMAN 1951, PAYNE 1952). The small *mitochondria* occupy the inter-droplet cytoplasm (Fig. 14) and hence are most conspicuous in the cells containing few lipide droplets. After fixation in a solution containing potassium dichromate the latter cells stain intensely with mitochondrial stains, such as acid fuchsin or hematoxylin. These cells have been called the "fuchsinophile" cells (see p. 43), the lipide-rich cells "pale" cells (UOTILA 1939, KAR 1947, R. MILLER and RIDDLE 1942). According to KAR (1947), approximately 60% of the adrenocortical cells in the chicken gland are fuchsinophile (or poor in lipide droplets).

In old birds, particularly chickens and pigeons, the cells accumulate *lipofuscins* (oxidized lipide pigments) with age so that the glands, yellow in young animals, become grossly brown (R. MILLER and RIDDLE 1942, PAYNE 1952).

5. Development

In the domestic chicken (*Gallus*), the gland *anlagen* begin to develop in the 96-hour embryo, as segmental groups of cells on either side of the mesentery along the dorsal peritoneal wall (A. DAWSON 1953, WITSCHI 1956). The groups of cells are immediately vascularized. The isolated groups secondarily fuse into the presumptive gland, which is carried cranially and medially. Chromaffin cells have invaded on the 5th day, and the gland becomes encapsulated by 10 days. Sparse lipide droplets make their appearance during the 6th day, and increase in number thereafter. VERNE et al. (1952) described droplets as still absent at birth in a passerine bird that has a long nesting period (blackbird, *Corvus frugilegus*); only when the birds develop feathers do the cells enlarge, assume their typical organization and become filled with lipide droplets.

VII. Class Mammalia

1. Gross anatomy

In most mammals, the adrenal glands are spherical, ovoid or pyramidal in shape and smooth in contour. In some, however, they are more irregular in form and variably creased on the surface [especially in large primates (*Pan*, *Homo*) and some of the largest ungulates (e.g., *Elephas*, *Rhinoceros*, *Hippopotamus*)] (KOHNO 1925). The glands typically lie capping the cranial poles of the kidneys (Fig. 15), though they may be more distantly located, in closer association with the postcaval vein. They are embedded in the perirenal adipose tissue. The left gland is often slightly larger than the right. The glands of the domestic ox (*Bos*) have been described in detail by SMOLLICH (1958),those of the sheep (*Ovis*) and goat (*Capra*) by FRANZE (1958).

ADOLF (1949) has calculated that the *weight of the paired mammalian adrenal glands* generally bears the following logarithmic relation to body weight:

$$\text{Weight adrenals (mg)} = 1.1\ \text{Body wt. (g)}^{92}.$$

When expressed in direct *relation to body weight*, the ratio varies considerably but generally falls somewhere between 10 and 30 mg per 100 g body weight (Table 2). In domestic cattle (*Bos*), however, the glands weigh only about 5 mg-% (see, however, SMOLLICH 1958), whereas in the guinea pig (*Cavia*) they weigh 50 mg-% or more. The smallest adrenals, relative to body weight, are said to occur in marine mammals, such as whales (Order Cetacea), seals (Order Carnivora)

Table 2. *Relative weights of the two adrenal glands in representative mammals (adults)*

Species	Average body wt.	Adrenal glands, mg/100 g body wt. ♂	♀	Reference
Order Marsupialia				
Didelphis virginiana (Virginia opossum)	—	—	20 (3)[1]	Hartman & Brownell 1949
Order Insectivora				
Blarina brevicauda (shrew)	20 g	20 (29)	20 (39)	Spector 1956, p. 163
Order Chiroptera				
Desmodus rotundus (vampire bat)	28 g	40 (5)		Spector 1956
Myotis lucifugus (brown bat)	—	22 (3)		Hartman & Brownell 1949
Order Primates				
Alouatta palliata (black howler)	ca. 6 kg	10 (28)		Spector 1956
Macaca mulatta (rhesus monkey)	4 kg	15 (6)		Krohn 1955
	7 kg	—	9 (5)	P. Smith 1955
Pan troglodytes (chimpanzee)	ca. 50 kg	30 (5)	28 (8)	Hartman & Brownell 1949
Homo sapiens (man, Caucasian)	60—72 kg	16	19	Spector 1956
		13	—	Uotila & Pekkarinen 1951
		13 (59)	—	Materna 1941
Order Edentata				
Dasypus novemcinctus (armadillo)	ca. 3 kg	30 (12)		Spector 1956
Cholepus hoffmani (2-toed sloth)	2—5 kg	16 (10)	14 (7)	Hartman 1959
Order Lagomorpha				
Oryctolagus cuniculus (rabbit, laboratory)	2—2.5 kg	18 (35)	19 (30)	Latimer & Sawin 1955
Order Rodentia				
Sciurus hudsonicus (red squirrel)	180 g	45 (13)	43 (11)	Hartman & Brownell 1949
Mesocricetus auratus (golden hamster)	100—125 g	18 (65)	13 (72)	Meyers & Charipper 1956
	100 g	20 (25)	11 (17)	Knigge 1954a
Microtus montanus (vole)				
non-breeding	40 g	37 (78)	46 (51)	McKeever 1959
breeding	40 g	19 (74)	96 (62)	McKeever 1959
Rattus norwegicus (Norway rat)				
Wistar strain	210 g	14 (55)	25 (38)	Addis & Gray 1950
wild, gray	200—400 g	28 (6)	47 (12)	Woods 1957a
Mus musculus (mouse)				
Swiss albino strain	25 g	13 (100)	—	Hartman & Brownell 1949
Strong A strain	25 g	11 (45)	17 (17)	Chester Jones 1957
Bagg C albino strain	23 g	14 (55)	26 (28)	Chester Jones 1957
Cavia cobaya (guinea pig)	130—575 g	50 (157)	50 (136)	Mixner et al. 1943
	620—720 g	ca. 65 (34)	—	Mixner et al. 1943
	650 g	51 (100)	—	Latimer 1951
Order Carnivora				
Canis familiaris (dog)	3—25 kg	13 (400)	17 (275)	D. Baker 1937
	10 kg	15 (12)		Osorio & Kraemer 1959
Felis domesticus (cat)	4—5 kg	14 (52)	15 (52)	Latimer 1939b
	2—3 kg	11 (9)	12 (9)	Schwartz & Kling 1960

Table 2: Continuation

Species	Average body wt.	Adrenal glands, mg/100 g body wt. ♂	♀	Reference
Panthera leo (lion)	100—125 kg	10 (4)	10 (3)	SPECTOR 1956
Order Perissodactyla				
Equus caballus (horse)	600—800 kg	ca. 5 (2)		SPECTOR 1956
Order Artiodactyla				
Sus scrofa (pig)	100 kg	—	4 (36)	SPECTOR 1956
Bos taurus (domestic ox)	125 kg	—	5.8 (26)	ROSENFELD 1958a
	515 kg	—	5.5 (19)	GARM 1949
	600 kg	—	6 (198)	SPECTOR 1956
	900 kg	5 (5)	—	SPECTOR 1956

[1] Number of specimens in parentheses

and manatees (Order Sirenia), though this fact is poorly established (LANMAN 1953). STIRLING and KEATING (1958) and STIRLING (1959) claim that the adrenal glands of Negroes living in Jamaica are smaller than those of Caucasians (see p. 69). Wild rats (*Rattus norwegicus*) possess glands about twice as heavy as those of laboratory rats, but large glands do not characterize wild animals in general.

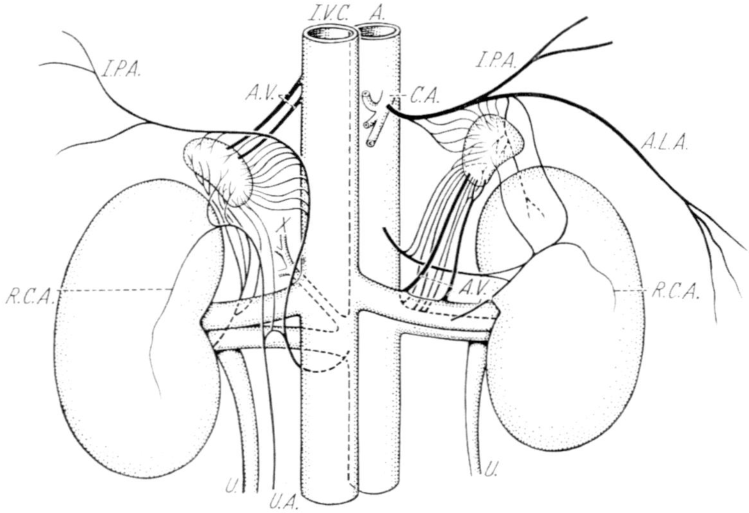

Fig. 15. Diagram of the location and vascular supply of the adrenal glands of the rhesus monkey, *Macaca mulatta*. In this species, the adrenal veins are characteristically paired; in most mammals, there is but one adrenal vein per gland. Note that the veins from the right gland enter the vena cava, those from the left gland the renal vein. *A.*, aorta; *A.L.A.*, anterior lumbar artery; *A.V.*, adrenal veins; *C.A.*, coeliac axis; *I.P.A.*, inferior phrenic artery; *I.V.C.*, posterior (inferior) vena cava; *R.C.A.*, renal capsular artery; *U.*, ureter; *U.A.*, ureteric artery; *X.*, deep artery from the aorta also supplying the adrenal gland. Reproduced by permission of the authors and the Cambridge University Press from HARRISON and ASLING: J. Anatomy **89**, 106 (1955)

Sex differences are common. In the vast majority of species so far examined, *the glands are relatively heavier in females than in males.* One exception to this general rule is the golden hamster (*Mesocricetus auratus*), in which the paired glands weigh nearly 20 mg-% in males, about 12 mg-% in females. This relation appears to hold for other hibernating rodents as well (ZALESKY 1934, DEANE and LYMAN 1954).

Although *the glands* increase in weight with age, they *are usually relatively largest in embryos and young animals*, thereafter gradually declining in relative weight (KORENCHEVSKY 1942). This decline is indicated for man in Table 3. (The acute postnatal involution of the human adrenal cortex will be discussed below under embryology (p. 42).) Fig. 16 illustrates the decline in relative weight of the glands with age in the albino laboratory rat (Wistar strain). Especially noteworthy is the increase in relative weight in females (but not in males) at puberty. Similar data are presented by FOGLIA et al. (1943) for the Buenos Aires strain. ADDIS and GRAY (1950) give formulas for calculating the normal weight of the glands in terms of body weight in the two sexes. In the rat, as in man, there is a slight postnatal involution (Fig. 62, p. 123), though it is nowhere near so conspicuous (DEANE 1956). Relative and absolute weights of adrenals during development have been determined for mice (MOOG et al. 1954, MOOG 1959), for cats (LATIMER 1939a, b), and for dogs (LATIMER 1954).

Table 3. *Absolute and relative weights of the adrenals in human beings from fetal stages to maturity* (SPECTOR 1956, pp. 162, 353)

Mean age	Mean body wt., kg	Mean wt. adrenals, g	Relative wt. adrenals, mg per 100 g body wt.
Fetal life (stillbirths, both sexes)			
5 months	0.5	2.5	500
6 months	1.0	3.3	330
7 months	1.5	4.3	286
8 months	2.0	5.3	264
$8^1/_2$ months	2.5	6.9	276
9 months	3.0	7.6	254
10 months	3.5	9.3	266
Extrauterine life (autopsies)			
0 (newborn), ♂	3.5	9.0	252
♀	3.4	9.0	262
3 months, ♂	5.9	4.7	79
♀	5.6	4.6	82
18 months, ♂	12.2	3.7	32
♀	11.9	3.2	28
$5^1/_2$ years, ♂	20.6	4.9	24
♀	20.0	5.0	25
$10^1/_2$ years, ♂	33.0	6.9	21
♀	34.0	7.4	22
$15^1/_2$ years, ♂	58.0	9.5	16
♀	52.0	11.4	22
35 years, ♂	72.0	11.4	16
♀	60.0	11.6	19
Sudden death (MATERNA 1941, SOKOLOFF et al. 1951)	ca.72.0	9.4	ca.13

The guinea pig appears exceptional in that there is a secondary rise in relative weight of the glands *in both sexes* postpuberally (BIMES et al. 1954, PLANEL and GUILHEM 1957a).

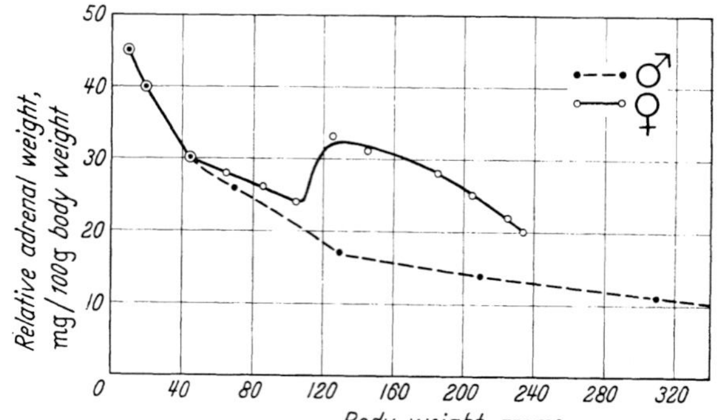

Fig. 16. Relative weight of the paired adrenal glands in *Rattus norwegicus* (laboratory animals, Wistar strain), after birth. Data taken from DONALDSON (1924) and ADDIS and GRAY (1950). The consistent decline in *relative* weight of the glands in males contrasts with the secondary rise in their relative weight in females at the time of puberty

In several species of insectivores and rodents, animals living in the wild show conspicuous enlargement of the adrenals during the *breeding season*, generally in the spring (*Erinaceus* spp. — SKOWRON and ZAJACZEK 1947; *Citellus tridecemlineatus* — ZALESKY 1934; *Cricetus cricetus* — KAYSER and ARON 1950). All of these species are hibernators. In *Microtus montanus*, relative enlargement occurs in the female, while relative shrinkage occurs in the male (McKEEVER 1959).

The relative *proportion of cortex* in the adrenal gland varies somewhat in different mammals. Thus in the golden hamster the cortex (plus capsule) normally comprises about 50% of the volume of the gland (MEYERS and CHARIPPER 1956), in the cow 75% (MACCHI and HECHTER 1954a), in the mouse 75—80% (DEANESLY 1931, CHESTER JONES 1957), in the dog 80—85%, in the cat, rat and man 90 to 95%, and in the guinea pig over 98% (HARTMAN and BROWNELL 1949, RATHER 1951, BACHMANN 1954).

a) Accessory adrenocortical tissue

Extraglandular nodules of adrenocortical tissue (generally without associated chromaffin tissue) occur variably in different species and in different individuals within a species (HARTMAN and BROWNELL 1949). Such nodules become a matter of interest mainly when total adrenalectomy is attempted. Most such accessory bodies are extremely small and lie attached to or in the immediate vicinity of the glands proper (MACFARLAND 1944), so that they are removed with the periadrenal fat in the usual technique of adrenalectomy. But they may occur anywhere along the dorsal body wall from the diaphragm to the pelvic region, sometimes embedded within other organs such as the kidney or gonad.

Such *distantly-located accessory nodules* are reported to be rare in the cat (3—5% of specimens — SWINGLE and PFIFFNER 1931, ROGOFF 1944) and the dog. In most strains of rats they can be found in only about 8% of individuals, although in some strains they apparently occur more frequently, thus permitting an unusually high percentage of survival after bilateral adrenalectomy (GAUNT 1933, GAUNT et al. 1935). Rabbits are reported to exhibit accessory adrenocortical nodules in 70% of individuals, mice in about 50% (SPIERS 1953, HUMMEL 1958), humans in about 30% (DENBER 1949), rhesus monkeys only rarely (SCHMIDT 1956). HUMMEL (1958) has reported that accessory nodules are most frequent in C57 L, BALB/c and C58 mice of 9 inbred strains examined.

GROAT (1944) found that in female thirteen-lined ground squirrels (*Citellus tridecemlineatus*) *new adrenocortical-like tissue developed* rapidly in the ovarian stroma and mesovarium if, after adrenalectomy, the animals were maintained on saline for two weeks; thereafter they required no saline. Correspondingly, MACFARLAND (1945) reported that in rats, if adrenalectomy was performed in two stages, new cortical tissue appeared to arise from the coelomic wall after the first operation, thus permitting survival after the second operation without high salt intake. Secondary development of functional accessories occurs in golden hamsters after adrenalectomy if adequate substitution therapy is given for two weeks (SCHROEDER and BOHLE 1957).

2. Blood supply

Afferent arteries to the adrenal gland are abundant. Several major arteries arise from the larger trunks in the vicinity, especially from the aorta and the phrenic, renal, coeliac, lumbar and superior mesenteric arteries (Fig. 15, p. 19). The precise origin and number vary not only in different species but in different members of the same species. The number of such large supplying arteries is roughly proportional to the size of the animal.

The afferent arteries break up into multiple twigs that penetrate the capsule over its entire surface. In the deeper capsule, the vessels form an extensive plexus. This gives rise to three types of terminal arterioles: the *Arteriae capsulae*, the *A. cortices* and the *A. medullae*. The *A. cortices* receive most of the blood and they directly supply the capillaries that loop around the cell clusters in the zona glomerulosa and then travel radially toward the medulla between the cords or masses of parenchymal cells of the inner zones (Fig. 17). A small number of the terminal *A. cortices* penetrate or loop into the cortex to varying depths before giving rise to capillaries (Fig. 18). The capillaries are of variable diameter; they are

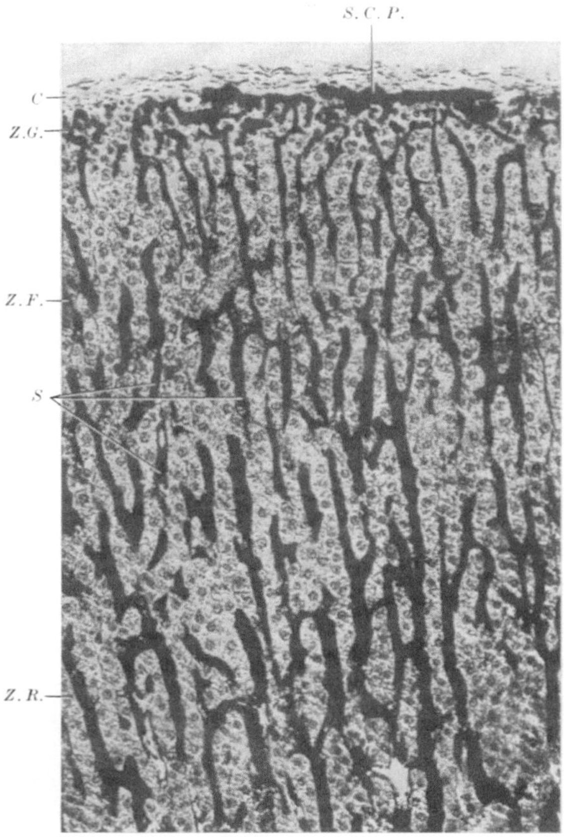

a

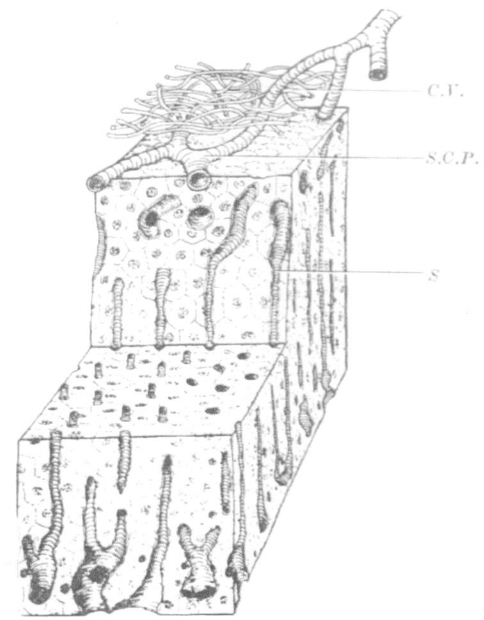

b

Fig. 17. The distribution of blood vessels within the adrenal cortex of the rat, *Rattus norvegicus*. a section of a specimen injected with a fluid containing India ink, counterstained with hematoxylin. × ca. 110. *C.*, capsule; *S.C.P.*, subcapsular arterial plexus; *S.*, sinusoids (capillaries); *Z.G.*, zona glomerulosa; *Z.F.*, zona fasciculata; *Z.R.*, zona reticularis. b stereogram of the cortex, showing that the cortical cells comprise a continuum that is tunneled by the capillanies. *C.V.*, capsular vessels; *S.C.P.*, subcapsular plexus; *S.*, sinusoid (capillary). Reproduced by permission of the author and C. C. Thomas, Publisher, from PAULY: Endocrinology **60**, 247 (1957)

narrow and straight in the zona fasciculata, broader and more tortuous in the zona reticularis. Increase in "bore" of the vessels is of course a necessary consequence of the fusion of the radially-directed vessels near the center of the gland. Anastomoses between adjacent capillaries are infrequent in the outer fasciculata, increasingly abundant inward.

The cortical capillaries drain into the large venous capillaries of the medulla, which in turn enter the central adrenal vein. This vein (sometimes there are two or three) emerges from the hilum and enters the renal or phrenic vein on the left side, the posterior vena cava directly on the right side (Fig. 15). [According to DONÁTH (1957) some small adrenal veins, especially on the left side, may enter the hepatic vein or its tributaries.] In the large species, the adrenal vein possesses a thick layer of longitudially-directed muscle fibers and a sphincter at its entry into the external major vein (KOHNO 1925, BARGMANN 1933, VELICAN 1948a). Within the gland, the adrenal vein lacks valves.

The following papers illustrate the external and internal vessels of the adrenal in different species: rhesus monkey (HARRISON and ASLING 1955); man (ANSON et al. 1947, VELICAN 1948a, BUSCH 1954, ELIAS and PAULY 1954, 1956, GAGNON 1956, JOHNSTONE 1957); rat, mouse and rabbit (GERSH and GROLLMAN 1941, HARRISON 1951a, LEVER 1952, 1954, PAULY 1957); dog (FLINT 1900); cat (BENNETT and KILHAM 1940, HARRISON 1951a).

From light microscopic studies, it was concluded that the small blood vessels in the adrenal cortex are lined by reticuloendothelium and should be termed *sinusoids*. Recent studies with the electron microscope show, however, that the adrenocortical vessels, like those in other endocrine glands, are lined by nonphagocytic cells; *the phagocytes lie in the perivascular spaces.* Of more interest from a physiological point of view is the fact that *the endothelial cells appear to be fenestrated* (BENNETT et al. 1959, ZELANDER 1959). They are separated from the parenchymal cells by an ill-defined periendothelial space (Fig. 19) filled with a homogeneous-appearing material (? basement membrane) and processes (microvilli) of the epithelial cells (also, CARR 1958).

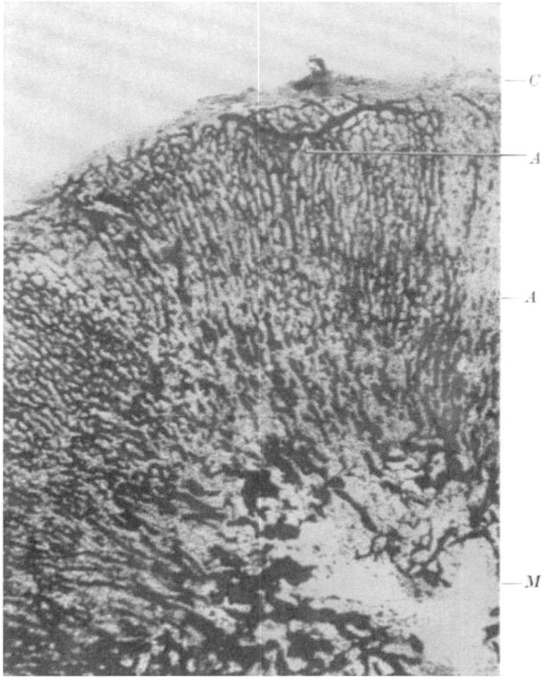

Fig. 18. The vessels within the gland of *Macaca mulatta*, after injection of the arterial supply with a medium containing India ink. × 50. *A.*, cortex; *A.L.*, arterial loop into the cortex; *C.*, capsule; *M.*, medulla. Reproduced by permission of the authors and the Cambridge University Press from HARRISON and ASLING: J. Anatomy 89, 106 (1955)

The phagocytes lying in the perivascular spaces in the adrenal cortex, as in many other sites, *have a hematopoietic potential.* In embryonic life, prior to the differentiation of the bone marrow, the adrenals provide one site for erythro- and myelopoiesis (TOGARI and ITO 1951, BACHMANN 1954). In most species, hematopoiesis does not normally persist here long after birth. Such extramedullary myelopoiesis appears to be quite common in adult cattle, however (A. WEBER et al. 1950, GREENSTEIN and FOLEY 1957), and I have observed it in the Virginia opossum. Furthermore, in any situation in which erythropoiesis or myelopoiesis is markedly stimulated in the adult, the adrenals may again become the site of blood-cell formation (HETT 1945, PEACE 1953).

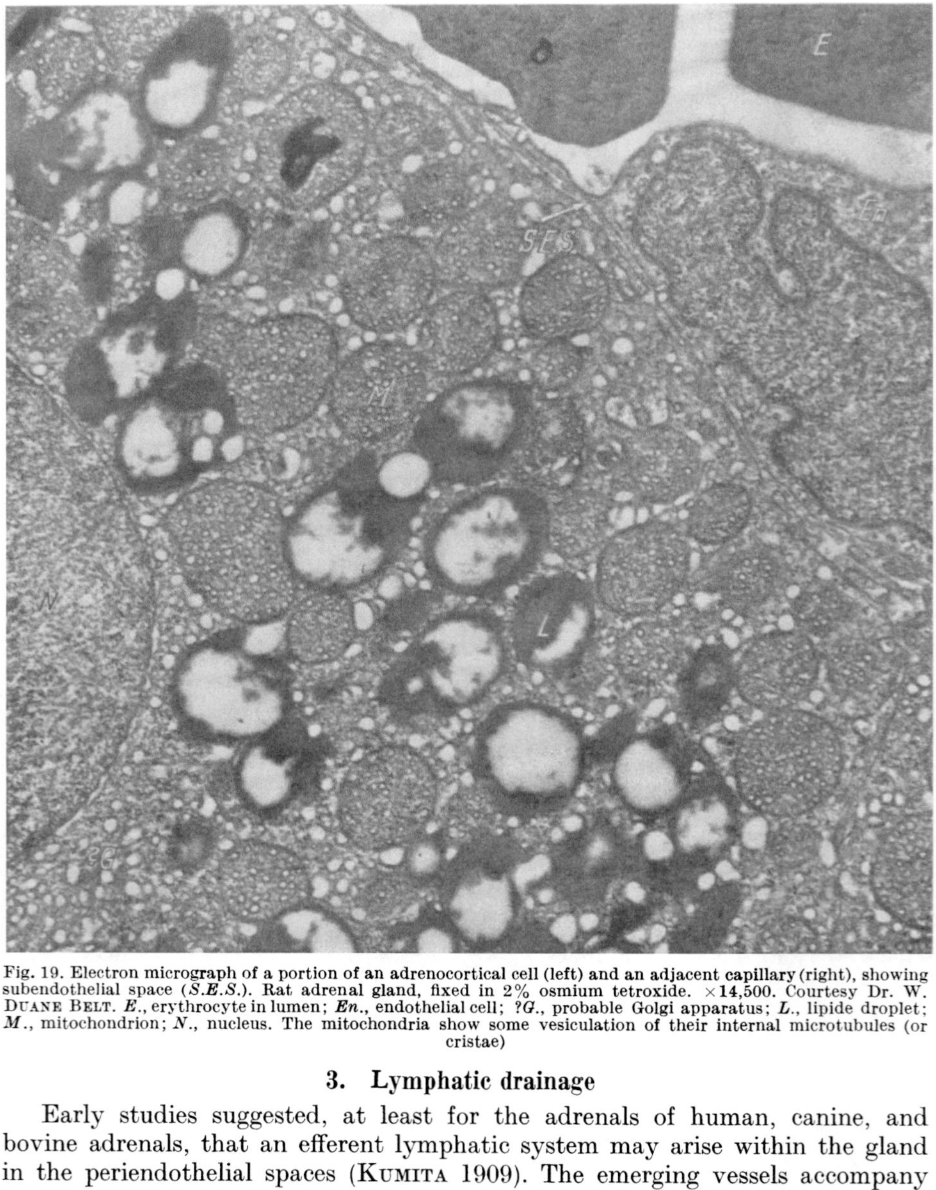

Fig. 19. Electron micrograph of a portion of an adrenocortical cell (left) and an adjacent capillary (right), showing subendothelial space (*S.E.S.*). Rat adrenal gland, fixed in 2% osmium tetroxide. ×14,500. Courtesy Dr. W. DUANE BELT. *E.*, erythrocyte in lumen; *En.*, endothelial cell; *?G.*, probable Golgi apparatus; *L.*, lipide droplet; *M.*, mitochondrion; *N.*, nucleus. The mitochondria show some vesiculation of their internal microtubules (or cristae)

3. Lymphatic drainage

Early studies suggested, at least for the adrenals of human, canine, and bovine adrenals, that an efferent lymphatic system may arise within the gland in the periendothelial spaces (KUMITA 1909). The emerging vessels accompany the afferent arteries or the great veins to the regional lymph nodes. Since KUMITA's time no one appears to have reinvestigated the problem with techniques less likely to produce artifacts. Nevertheless, the subendothelial spaces seen in electron micrographs (Fig. 19) (LEVER 1955c, LUFT and HECHTER 1957, ZELANDER 1959) may exist in life; they bear sufficient similarity to the spaces of Disse in the liver to raise the possibility of significant lymph drainage from the gland.

4. Nerve supply

The nerves to the mammalian adrenal appear to consist solely of sympathetic fibers, originating principally in the greater splanchnic nerve, although a few fibers come from the lesser splanchnic nerve and from the lumbar sympathetics. These nerves form a plexus near the gland, from which fiber bundles enter it, chiefly along its mediocephalic border. The number and size of bundles is correlated with the size of the animal. The majority of fibers are detectably myelinated at the light-microscope level, suggesting that they are preganglionic; a few, however, are not detectably myelinated and are presumably postganglionic. In many species (monkey, man, dog, cat, ox), these postganglionic fibers arise in the small nests of ganglion cells found in the adrenal plexus or in the capsule of the

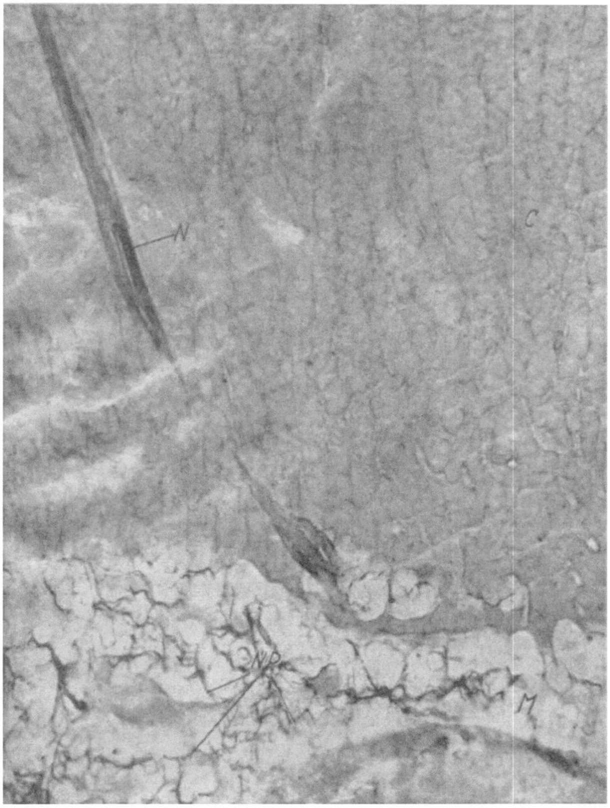

Fig. 20. Nerve fibers running through adrenal cortex to medulla. Rabbit adrenal, fixed by the Nonidez modification of the Cajal silver method, no counterstain. ×125. Photographed from a slide contributed by Dr. PINCKNEY J. HARMAN. *C.*, cortex; *M.*, medulla; *N.*, nerve bundle; *N.P.*, plexus of nerve fibers surrounding clusters of medullary cells. The cortical cells appear uninnervated; the bundle of nerve fibers gives off no branches until it enters the medulla

gland. In several species (especially man), ganglion cells also occur along the nerve bundles in the cortex as well as in the medulla.

There is *no definitive evidence that the cortical cells receive any innervation* (Fig. 20), although of course the subcapsular arterial plexus and any terminal arterioles or arteriolar loops within the cortex are innervated. LEVER (1953, 1955a), however, has reported fine nerve fibers apparently ending on occasional cells in the zona glomerulosa (supravital methylene blue preparations).

Valuable references on the question of adrenal innervation include those of SWINYARD (1937 — cat, human; Ranson's pyridine-silver technique) and MACFARLAND and DAVENPORT (1941 — primates, rodents, carnivores and ruminants;

Bodian's protargol technique). Assertions of rich cortical innervation based on Bielschowsky's ammoniacal-silver nitrate preparations (e.g. STÖHR 1935, KISS 1951, SARTER 1954) are without significance, since this method is incapable of distinguishing between small nerve fibers and reticular fibers. In confirmation of the absence of direct innervation of cortical cells, DEJARDIN (1954) was unable to demonstrate any nerve endings exhibiting specific cholinesterase activity (Koelle acetylthiocholine technique) on cortical cells, although the fibers passing through the cortex as well as the terminals in the medulla were strongly active. With the same method, COUPLAND and HOLMES (1958) described occasional loops of nerve fibers into the glomerulosa (rat, cat, rabbit); they could not determine whether these ended on arterial blood vessels (Fig. 18) or cortical cells.

5. Connective-tissue stroma

The adrenal glands of mammals lie embedded in the perirenal adipose-tissue mass and the renal fascia. The gland proper is surrounded by a tough connective

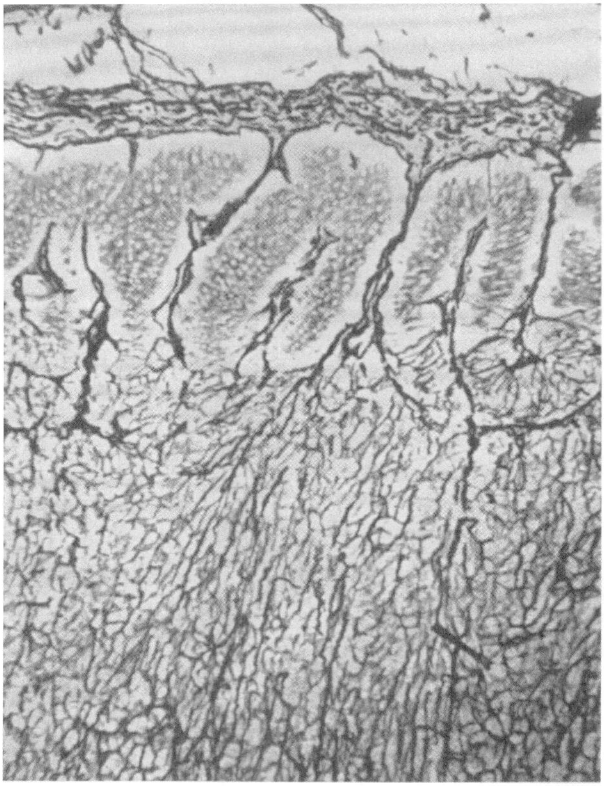

Fig. 21. Reticular fibers in capsule and around the parenchymal cells of adrenal cortex. Dog adrenal gland, fixed in 10% formalin, impregnated by Gomori's silver technique for argyrophilic fibers. × about 200. Courtesy Dr. HANS ELIAS. The reticular network of the capsule is continuous with the strands surrounding the glomerulosal arcades and then passing centrally. *C.*, capsule; *G.*, glomerulosa; *S.G.*, subglomerulosal zone; *F.*, fasciculata

tissue capsule. The outer part of the capsule consists of dense, heavy bundles of collagenous fibers and occasional fibroblasts; the inner part contains looser bundles of reticular (argyrophilic) fibers and abundant fibroblasts or reticular cells. The thickness of the capsule varies directly with the volume of the gland, being far greater around large glands like those of *Homo* and *Bos* than around small glands like those of rodents (KOLMER 1918). The capsule becomes thicker and more fibrous with age (JAYNE 1953, MEYERS and CHARIPPER 1956).

The reticulum of the inner capsule is continuous with what appear to be reticular fibers that pass radially through the cortex between the sheets or cords of parenchymal cells and the adjacent capillaries (Fig. 21). In most species, the reticular fibers appear to form baskets around the arcades or balls of cells in the zona glomerulosa, loose meshworks of fibers supporting the radial cords of cells in the fasciculata, and denser meshworks in the zona reticularis. Some cross fibers seem to pass between adjacent cells in the fasciculata and reticularis but not in the glomerulosa. The fibers tend to aggregate into larger bundles in the reticularis than in the outer zones. This tendency increases with age (e.g. DRIBBEN and WOLFE 1947, ITO 1952, JAYNE 1953).

Valuable descriptions of the connective-tissue stroma in different adrenals have been published by FLINT (1899, 1900 — dog), GRUENWALD and KONIKOV (1944 — many species), DRIBBEN and WOLFE (1947 — rat), BOURNE (1949 — many species), NICANDER (1952 — many species), JAYNE (1953 — rat), BACHMANN (1954 — many species), FRAZÃO (1954 — many species), MEYERS and CHARIPPER (1956 — hamster), ELIAS and PAULY (1954, 1956— human and others), PAULY (1957 — rat).

The newer evidence from electron microscopy does not reveal free reticular fibers in the periendothelial spaces, in the mouse at least — rather, this space is occupied by a homogeneous "basement membrane" (ZELANDER 1959). Probably, thus, it is basement membranes rather than individual reticular fibers that have been revealed by the connective tissue stains employed in the studies listed above. Whether or not these membranes contain minute fibrils, the descriptions listed here must be subjected to review.

In human infants, during the involution of the "fetal" zone after birth (p. 43), there is a transient aggregation of connective tissue to form a "capsule" between the definitive cortex and the medulla (Fig. 37); this tissue is generally resorbed before the end of the second year of postnatal life (BENNER 1940, LANMAN 1953). A similar medullary "capsule" develops in the mouse adrenal following the involution of the X zone (CHESTER JONES 1949a, 1957). In such cases it may be assumed that the stroma that invests the involuting cells collapses to form a dense connective-tissue layer. BOURNE (1949, p. 207) cites instances of such medullary capsules in various other mammals, including a number of marsupials. They are also seen in the glands of monotremes (CHESTER JONES 1957). That such a capsule has formed in all instances as the result of involution of a large cortical zone is not established.

6. Parenchyma: histological organization

In almost all mammals (exceptions noted below), the chromaffin cells are aggregated into a central medulla, and the adrenocortical cells lie in a cortex surrounding this. *The usual arrangement of the adrenocortical parenchyma is in radially-directed cords*, one or two cells thick (Fig. 22), running from the capsule to the medullary border. In rabbits and rodents, however, the cortex appears to be a parenchymal continuum tunneled by the capillaries, much like the mammalian liver (Fig. 17b, p. 22) (GONZALEZ 1955, PAULY 1957). As noted above, the capillaries and the underlying basement membranes form baskets around the outermost clusters of cells (*zona glomerulosa*) and then straighten out (*zona fasciculata*). Generally apparent is a subglomerulosal zone in which there is considerable lateral development of capillaries; here the parenchymal cells are characteristically small (see p. 35). The fasciculate arrangement may continue to the medullary border, or, in the innermost portion of the cortex, the orientation of the cords may become distorted by anastomosis and increase in diameter of the capillaries (*zona reticularis*) (Figs. 17a, 23). There is no clear line of demarcation between the fasciculata and reticularis.

The zona glomerulosa shows great variability in different species in the arrangement of blood vessels and parenchymal masses. In many smaller mammals, exemplified by most rodents, the parenchymal cords form indistinct balls at their outer ends (Figs. 24, 28, 29). In many larger mammals, especially primates (ELIAS and PAULY 1956), carnivores (FLINT 1900, BACHMANN 1954) and ungulates (ELIAS 1948, A. WEBER et al. 1950, FRAZÃO 1954), the outer ends of the cords form distinct arches or folded laminae (Figs. 21, 26, 37). The same situation exists in the large guinea pig adrenal (GRUENWALD and KONIKOV 1944). The capillaries generally surround these tissue masses, but they may send extensions up within the arches.

Unlike the glomerulosa and fasciculata, both of which seem relatively determined in appearance within a species, *the reticularis manifests itself differently from gland to gland within the same species*. The most important factor in the degree of "reticularity" of the zone would seem to be the quantity of blood flow. In an essentially spherical structure, abundant blood flow might be expected to cause distention of the capillaries near the center of the gland, and the presence of many anatosmoses between adjacent capillaries in this region would produce an appearance of broken and distorted fascicles in thin sections. A more moderate blood flow would not produce this distortion. Actually, of course, cutting the blood vessels when the gland is dissected out for preservation allows variable amounts of blood to escape, so that there is no way of knowing how closely the picture seen in a histological preparation mimics the state of the gland *in situ*. And it seems unlikely that the blood flow at the time of death represents a "normal" state. All of these considerations make difficult any evaluation of the size or width of the zona reticularis.

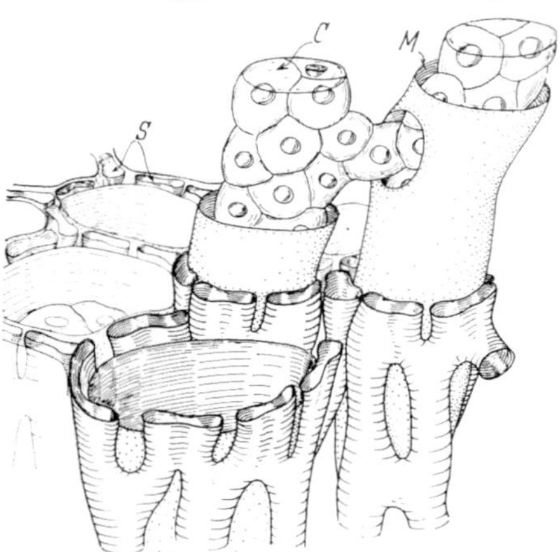

Fig. 22. Stereogram of parenchymal cords (C.), surrounded by "basement membranes" (M.) and sinusoids (S.) (capillaries). Human adrenal, zona fasciculata. Reproduced by permission of the authors and C. C. Thomas, Publisher, from ELIAS and PAULY: Endocrinology 58, 714 (1956)

The relative widths of the three major zones vary considerably in different species. In the dog, the glomerulosa occupies more than 15% of the cortical width, the fasciculata about 70% and the reticularis less than 15% (see Fig. 26). In the guinea pig, the glomerulosa assumes about 7% of the width, the fasciculata 24—41%, the reticularis 52—64% (PLANEL et al. 1957). In rats and other small rodents the glomerulosa is relatively narrower and the reticularis often quite indistinct (Figs. 28, 29) (DEANE and GREEP 1946, DEANE and LYMAN 1954). Relative *volumes* of the zones have been determined in only a few instances. The relative volumes in man (SWINYARD 1940) are: glomerulosa 15%, fasciculata 78%, reticularis 7%. In the dog there are two calculations available: — D.

Baker (1937): glomerulosa 24%, fasciculata 52–67%, reticularis 9–24%; Bahn et al. (1960): glomerulosa 18%, inner zones 82%. In the rat, the glomerulosa occupies 13%, the fasciculata 67% and the reticularis 20% of the cortex (de Groot and Fortier 1959).

In animals with an arcuate zona glomerulosa, nests or laminae of parenchymal cells commonly occur outside the zona glomerulosa proper in the inner portion of the

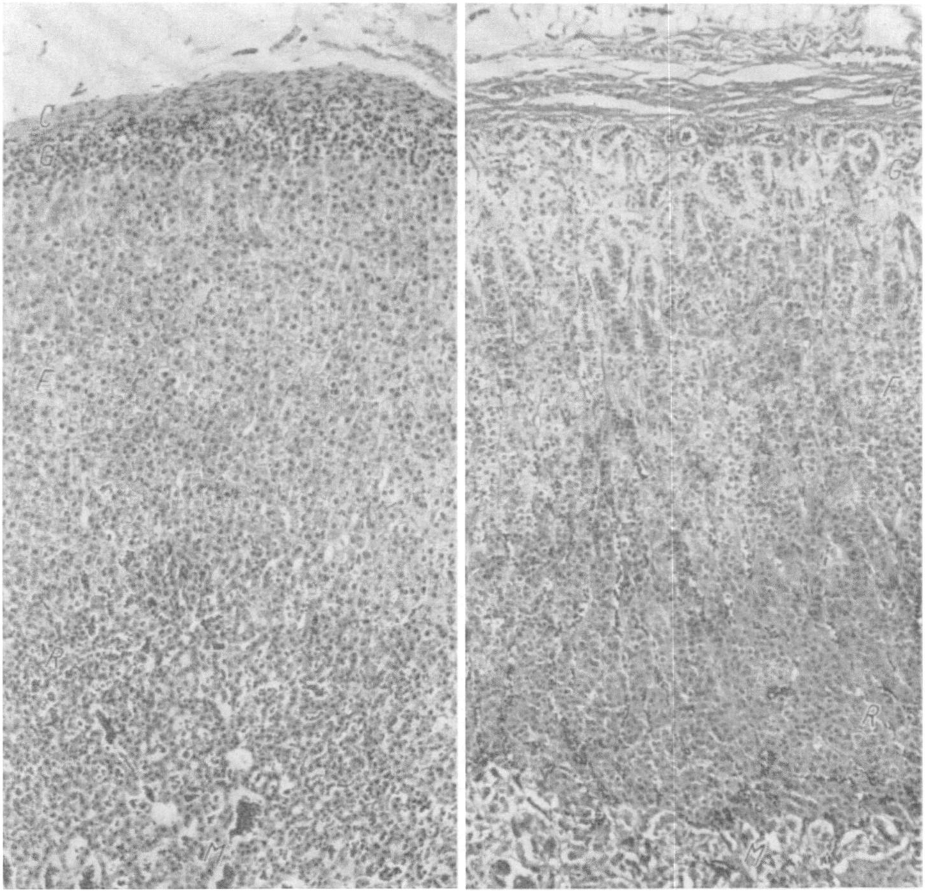

Fig. 23. a The adrenal cortex of an adult male rat. Fixed in Zenker's fluid, stained with hematoxylin and eosin. × 110. Courtesy Dr. Jacob Furth. *C.*, capsule; *G.*, zona glomerulosa; *F.*, zona fasciculata; *R.*, zona reticularis; *M.*, medulla. b The adrenal cortex of an adult human being. Hematoxylin and eosin. × 78. Photographed from a preparation contributed by Dr. David M. Maeir. Zones as in *a*. Note that in both adrenals, but more especially in *b*, the cytoplasm of the cells in the zona reticularis stains more intensely with the acid dye than in the other zones

capsule (see Fig. 38, p. 46); sometimes these lie outside the capsule as extra-adrenal nests. Such capsular nests are only rarely seen in small adrenals like those of most rodents (Dribben and Wolfe 1947, Meyers and Charipper 1956). The significance of these nests will be considered later (p. 45).

Among the primitive orders of mammals, the adrenal glands are not so clearly organized. In the adrenals of species of the Order Monotremata (the platypus, *Ornithorhynchus anatinus*, and the spiny anteater or Australian echidna, *Tachyglossus aculeatus*), the chromaffin tissue lies at the large caudal pole of the gland: it may reach the surface or be covered only by a thin layer of adrenocortical tissue (BOURNE 1949, WRIGHT et al. 1957). In some specimens there occurs distinct intermingling of cortical and medullary tissues, as in reptiles. Similarly, in the Order Marsupialia, several species have adrenals with the chromaffin tissue occupying one pole of the gland rather than the center (e.g., *Dasyurus* sp.). A similar arrangement occurs in fetal insectivores, where the chromaffin tissue matures precociously and retains connections with the extra-adrenal paraganglia (W. ADAMS 1958).

The adrenals of monotremes do not exhibit the typical mammalian organization of the adrenocortical parenchyma into regular cords and zones as described for higher mammals. *In Ornithorhynchus* (CHESTER JONES 1957, WRIGHT et al. 1957), in particular, *the cortical cells fall into several different cytological classes, but the different types occur in separate islands rather than in concentric zones.* Marsupial adrenals, however, even when the chromaffin tissue occupies an eccentric position, seem to be organized into typical zones (CHESTER JONES 1957). The same is true of the adrenals of fetal insectivores (W. ADAMS 1958).

7. Cytology

a) Cell size

In general, mammalian adrenocortical cells are *polygonal*, 10—15μ in diameter, and *mononucleate*. Occasional cells are binucleate. The cytoplasm characteristically contains numerous small *lipide droplets*. There are significant differences between the cells in different zones in a single adrenal, however, and also marked differences between species.

In most species, the cells are largest in the outer portion of the zona fasciculata (spongiocytes); the cells are smaller both in the glomerulosa and reticularis (Figs. 23, 24). Table 4 gives the diameters and calculated volumes of cells in the zones of the adrenal of a young adult male rat. There is an abrupt change in cell size between the glomerulosa and fasciculata, a gradual decline from outer fasciculata to reticularis.

In adrenals possessing an arcuate zona glomerulosa, the cells in this zone are columnar in shape rather than approximately equidimensional like those in the deeper zones.

b) Nucleus

The nucleus is typically spherical and vesicular in appearance, generally exhibiting a single central nucleolus and scattered chromatin bodies. *Nuclear size tends to parallel cell size* and hence is normally somewhat larger in the fasciculata than in the other zones (Fig. 24b). Table 5 lists the average nuclear volumes of adrenocortical cells as determined in mice. JACKSON (1919), R. MILLER (1954) and ANDO (1957) observed comparable differences between the zones in rats, and LAESCHKE (1951) reported the same is true for man. Such differences in nuclear volume seem to be attributable to differing quantities of non-histone proteins (ALFERT et al. 1955, HUTCHISON et al. 1958) rather than to differing amounts of chromatin.

In most species, workers have failed to find different classes of nuclear size that would be indicative of polyploidy. MOSIMANN (1954), however, reported finding 3 classes of nuclei in the glands of goats (*Capra*), with volumes of approximately 75, 150 and 300 μ^3, respectively. In normal females, the *average* size was greatest in the cells of the glomerulosa, since this zone possessed the greatest proportion of the large nuclei.

In a few mammalian species, the nuclei of some or all adrenocortical cells are not spherical but kidney-shaped or frankly polymorphic. WRIGHT et al. (1957, CHESTER JONES 1957) describe cells in *Ornithorhynchus* with kidney-shaped or

Table 4. *Sizes of adrenocortical cells in the rat* [diameters from JACKSON (1919)]

	Average cell diameter, μ	Calculated volume, μ^3
zona glomerulosa	9.0	365
outer zona fasciculata	14.0	1370
zona reticularis	8.2	275

Table 5. *Nuclear volume of adrenocortical cells in mice* (TONUTTI et al. 1954)

	Average nuclear volume, μ^3
zona glomerulosa	81
zona fasciculata	104
zona reticularis	99

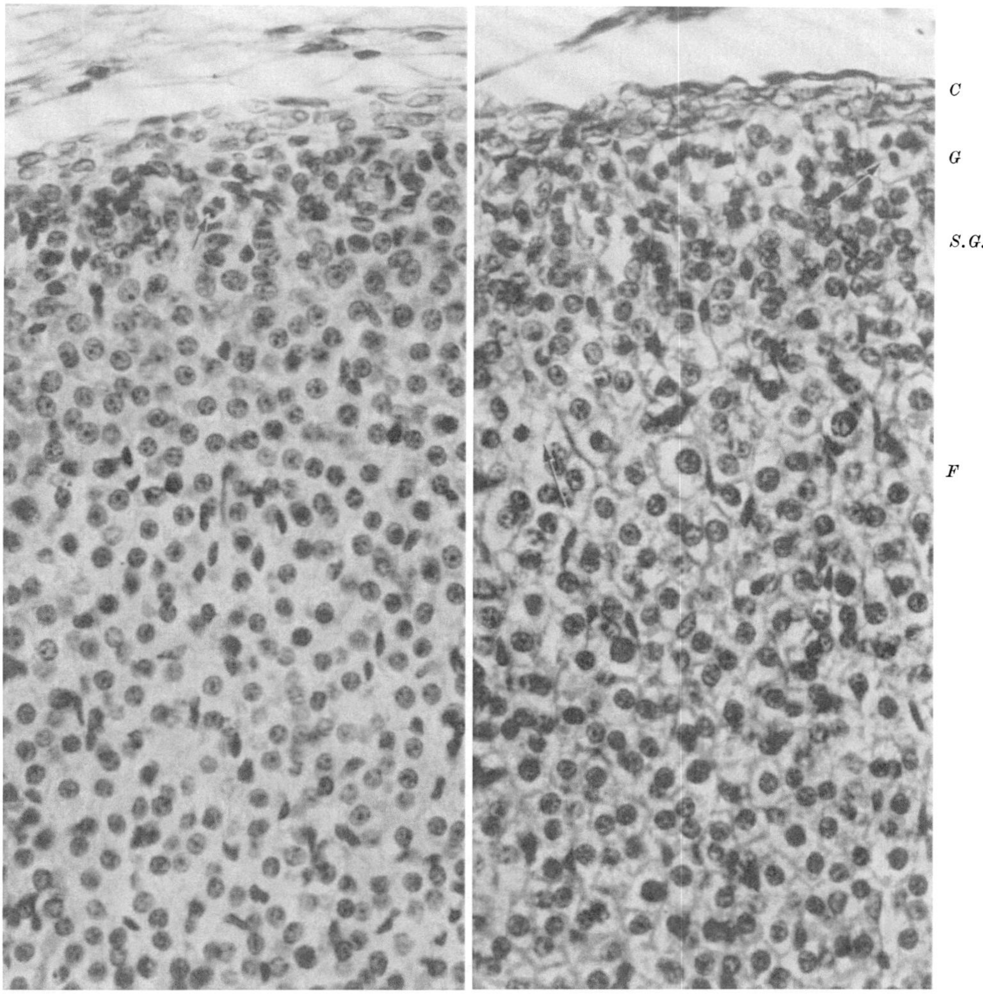

Fig. 24a & b. Adrenals of young rats to show differences in cell and nuclear size in glomerulosa and fasciculata, the occurrence of mitoses, and the presence of a subglomerulosal zone. a 3-day-old female rat, b 29-day-old female rat; hematoxylin and eosin. × 420. From preparations of VAN DORP and DEANE (1950). Arrows, mitotic figures; *C.*, capsule; *G.*, glomerulosa; *S.G.*, subglomerulosal zone; *F.*, fasciculata. Note the increasing size of cells in the fasciculata with maturation of the gland. Hypertrophy accounts for most of the increase in glandular volume and weight (see Fig. 62, p. 123)

elongate nuclei, often containing globules. Even more impressive in their illustrations is the remarkable variability in cell and nuclear size; there is suggestion of polyploidy of some cells. LIEBEGOTT (1944) shows examples of *giant cells*, some with polymorphic nuclei, in human adrenals. ITO (1952) illustrates markedly deformed nuclei in the zona glomerulosa of the guinea pig adrenal. Such deformation was not seen by HAASE (1952), however. From ITO's illustrations, the nuclei in *Cavia* appear larger in the glomerulosa than in the deeper layers. BACHMANN (1954) cites several other instances in which the nuclei of some adrenocortical cells appeared conspicuously enlarged and distorted and contained globules. It seems likely that most of these examples may be considered *pathological*, indicative of prolonged, excessive stimulation (see p. 129).

Mitosis is relatively abundant in the adrenals of rapidly-growing young animals (Fig. 24) and decreasingly frequent thereafter (rat — JACKSON 1919, ENGSTRÖM 1936, MITCHELL 1948, VAN DORP and DEANE 1950, CATER and LEVER 1954; mouse — WHITEHEAD 1933b; guinea pig — BLUMENTHAL 1945, ITO 1952; golden hamster — ITO 1956). A typical result is given in Table 6.

Table 6. *Mitotic rate in the adrenal cortex of rats of increasing age* (VAN DORP and DEANE 1950) (cf. Fig. 62, p. 123)

Age, days	Number of mitoses per section	Area of cortex, mm²	Mitoses per mm² cortex
1	8.2	0.55	14.8
7	14.4	0.54	26.5
14	40.6	1.13	35.0
22	24.7	2.21	11.1
29	18.7	3.05	6.1
42	2.0	3.25	0.6

Mitoses occur most frequently in the cells of the zona glomerulosa and the outer fasciculata (Fig. 24) but are not limited to these regions (above references, plus HOERR 1931, SCHMIDT and SCHMIDT 1938, BAXTER 1946, CATER and STACK-DUNNE 1955a, RACE and GREEN 1955). They tend to be absent from the subglomerulosal zone (MITCHELL 1948, VAN DORP and DEANE 1950, CATER and LEVER 1954). Capsular fibroblasts divide as well. CATER and STACK-DUNNE (1955a) calculated that in an adult rat adrenal having a mitotic ratio of 1.9 per mm², there are approximately 4300 dividing cells in the entire gland.

MÜHLEMANN et al. (1956) and HALBERG et al. (1957, 1959) found, for rats and mice, respectively, that the mitotic rate is twice as great during the night as in the daytime. The first group showed that in 250-g male rats, the average nighttime figure was 0.8 mitoses per mm³, the daytime figure 0.4. Both of these species are nocturnal.

Signs of *cell death* are found in the "normal" adrenal cortex. Cells with dense, pyknotic nuclei are encountered most commonly in the zona reticularis, but they occur in the outer zones as well. BENNETT (1940) found nuclei stainable *intra vitam* with trypan blue (an index of cell death) scattered throughout the cortex (*Felis*), though such cells were most abundant in the reticularis. (See also BAXTER, 1946, *Rattus*.)

c) Lipide droplets

Characteristically, the adrenocortical cells are crowded with *free lipide droplets* that stain readily with Sudan dyes. These are generally small, in the order of 1—2 μ in diameter (range: from the limit of resolution to 5μ). The cytoplasm and its organelles are thereby confined to the interstices between droplets. However, there is variation in distribution of lipides in the gland between species, even ones in the same order or family. BELT (1958) suggests, from an electron microscopic study, that the lipide droplets develop in microbodies possessing a limiting membrane (Fig. 25). These microbodies may arise in the Golgi zone.

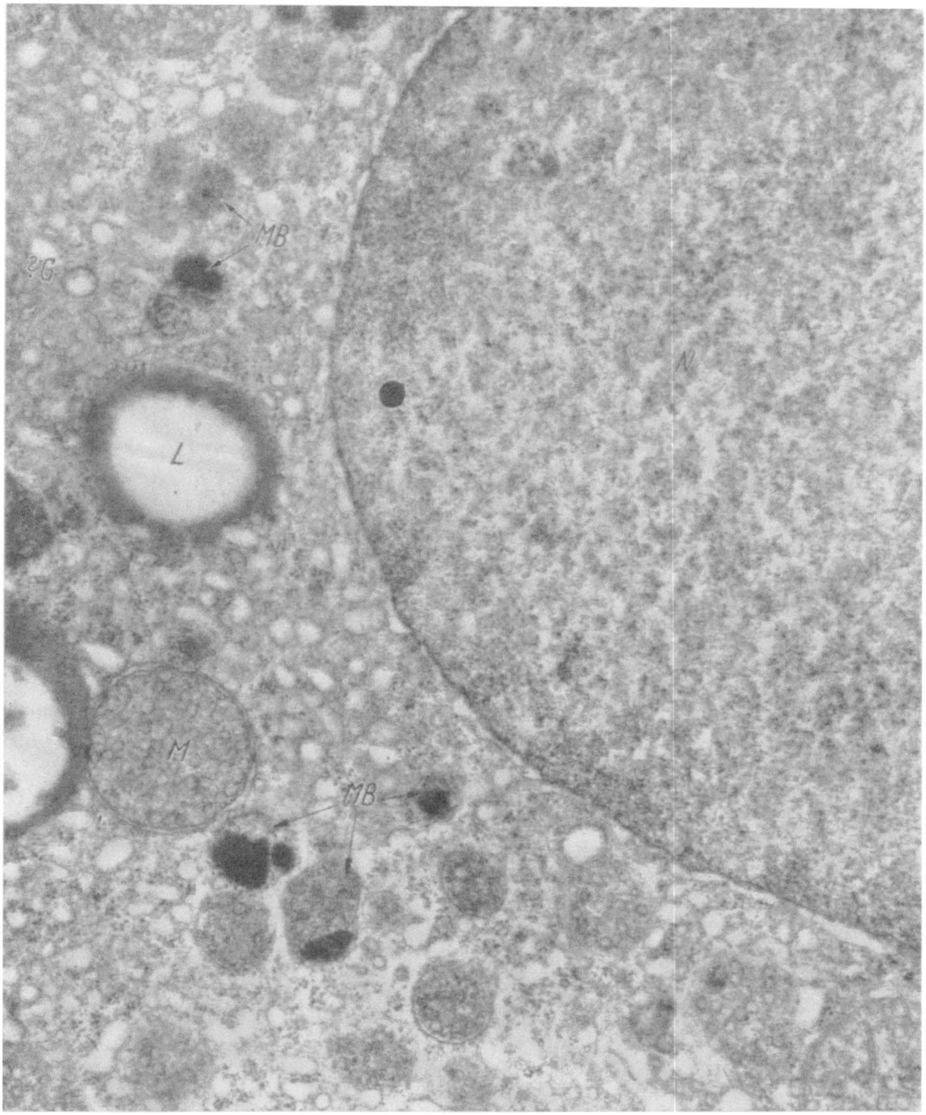

Fig. 25. Apparent development of lipide droplets within "microbodies" (*M.B.*) in an adrenocortical cell. Electron micrograph of a portion of a rat fasciculata cell, prepared as Fig. 19. × 26,000. Courtesy Dr. W. DUANE BELT. *?G.*, possible Golgi membranes; *L.*, lipide droplet; *M.*, mitochondrion; *N.*, nucleus. Two of the microbodies lie in close association with the putative Golgi apparatus

Zonal distribution of lipides. In such genera as *Homo*, *Rattus*, *Canis* and *Equus*, *lipides are abundant in both the glomerulosa and fasciculata* (Fig. 26); generally there is a band of cells between these two layers that contains very little lipide (see Fig. 29). In the inner zones, the concentration of lipide usually tends to decline toward the medulla, although the cells immediately next to the medulla sometimes contain large droplets.

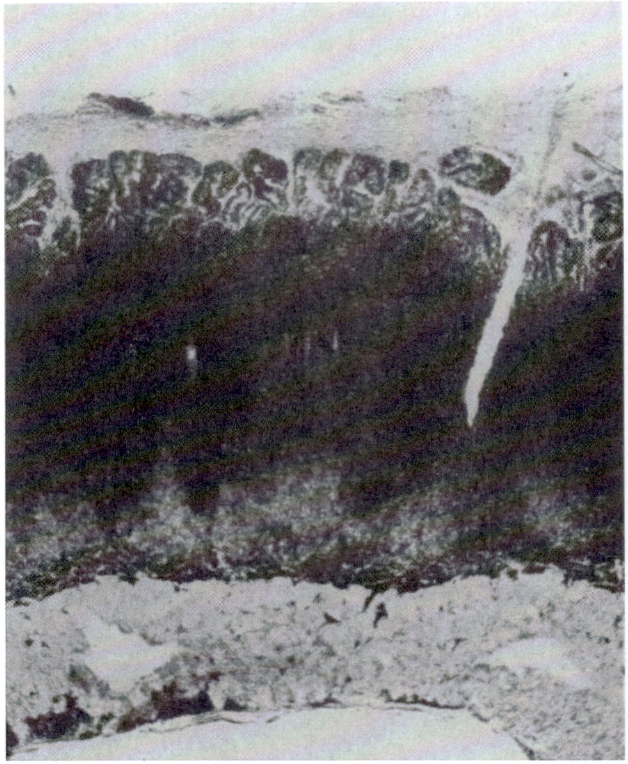

Fig. 26. Lipide droplets occur in virtually all of the adrenocortical cells of the dog, *Canis familiaris*. Frozen section of a formalin-fixed gland, Sudan black B. ×47. From a preparation of DEANE and BARGER (1957). *C.*, capsule; *G.*, glomerulosa; *F.*, fasciculata; *M.*, medulla. Note the arcuate arrangement of the zona glomerulosa

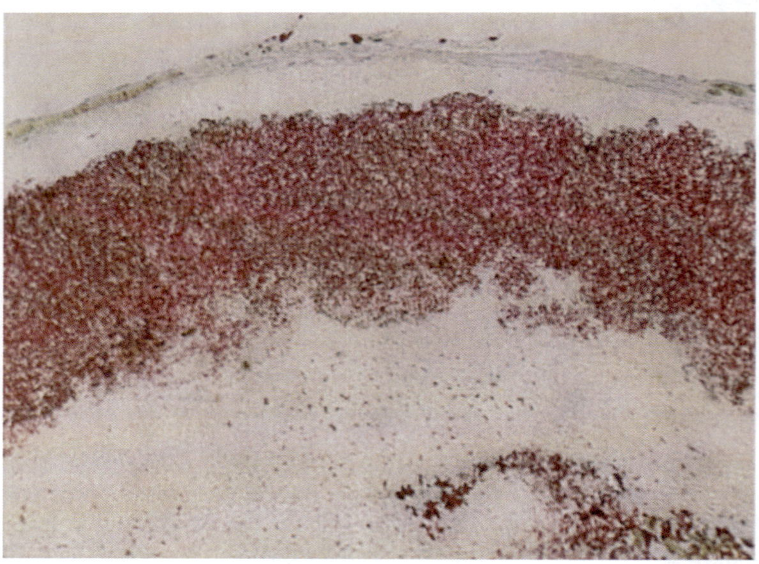

Fig. 27. Lipide droplets are rare in the zona glomerulosa but abundant in the inner zones of the adrenal of the rhesus monkey, *Macaca mulatta*. Formalin-fixed frozen section, Sudan IV. ×45. Micrograph contributed by Dr. ERNEST KNOBIL and ANNA MORSE. *G.*, glomerulosa; *F.*, fasciculata; *M.*, medulla

In another large group of genera, including *Macaca*, *Oryctolagus*, *Microtus*, *Mus*, *Cavia*, *Felis* and *Sus*, *the glomerulosa contains little lipide* (Fig. 27), although *the fasciculata and reticularis present the usual complement*.

A third category of mammals exhibits *free droplets in the glomerulosal cells only*, and even in this zone there may be very little (Fig. 28). Examples include hibernating insectivores and rodents (*Erinaceus*, *Marmota*, *Muscardinus*, *Mesocricetus*, *Cricetus*, *Citellus*), and ruminant Artiodactyla (*Ovis*, *Cervus*, *Bos*, *Capra*). In such adrenals the cytoplasm tends to be more sudanophilic than in glands in which free droplets are present.

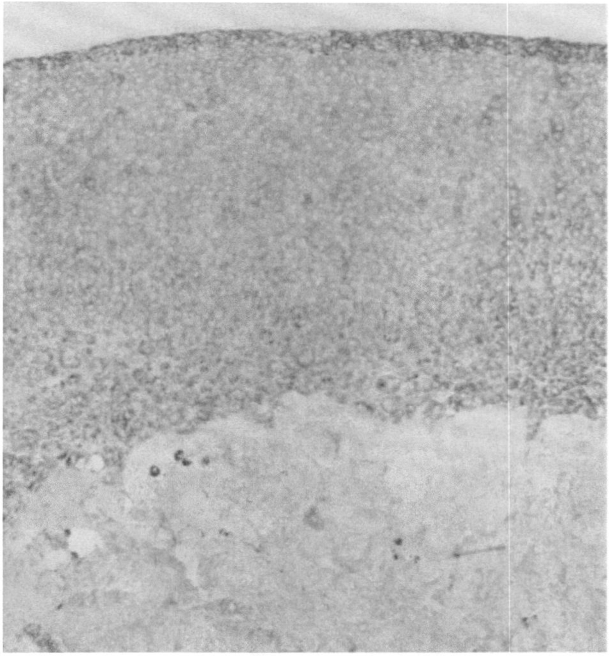

Fig. 28. Detectable (free) lipide droplets are limited to the zona glomerulosa in the adrenal of the golden hamster, *Mesocricetus auratus*. Formalin-fixed frozen section, Sudan black B. × 92. From a preparation of DEANE and LYMAN (1954). *G*., glomerulosa; *F*., fasciculata; *M*., medulla

Further information about lipide distribution in different adrenals may be found in BOURNE (1949), CELESTINO DA COSTA (1951), VERNE and HÉBERT (1951), NICANDER (1952), BACHMANN (1954), DEANE and LYMAN (1954), and FRAZÃO (1954), as well as in the earlier surveys of KOLMER (1918) and KOHNO (1925).

In the one species in which there are well-established inbred strains, *Mus musculus*, there proves to be considerable variation in different strains in both the amount and the distribution of demonstrable lipide (VICARI 1943, ARNESEN 1955, 1956). A sex difference in amount of lipide is especially conspicuous in the guinea pig (WHITEHEAD 1935b, BIMES et al. 1954).

d) Subglomerulosal zone

As implied above, there generally occurs a *narrow band of cells just below the glomerulosa that is relatively deficient in lipides*. The zone is readily evident only in the glands containing lipide droplets in both glomerulosa and fasciculata

(Figs. 24, 29, 33), but it occurs in the others as well. In this region, the capillaries show numerous lateral anastomoses. The cells are typically smaller than those in either the glomerulosa or the outer fasciculata. This zone has received a variety of names, including sudanophobic zone, transitional zone and zona intermedia (VACCAREZZA 1945, GREEP and DEANE 1949a, CATER and LEVER 1954, CHESTER

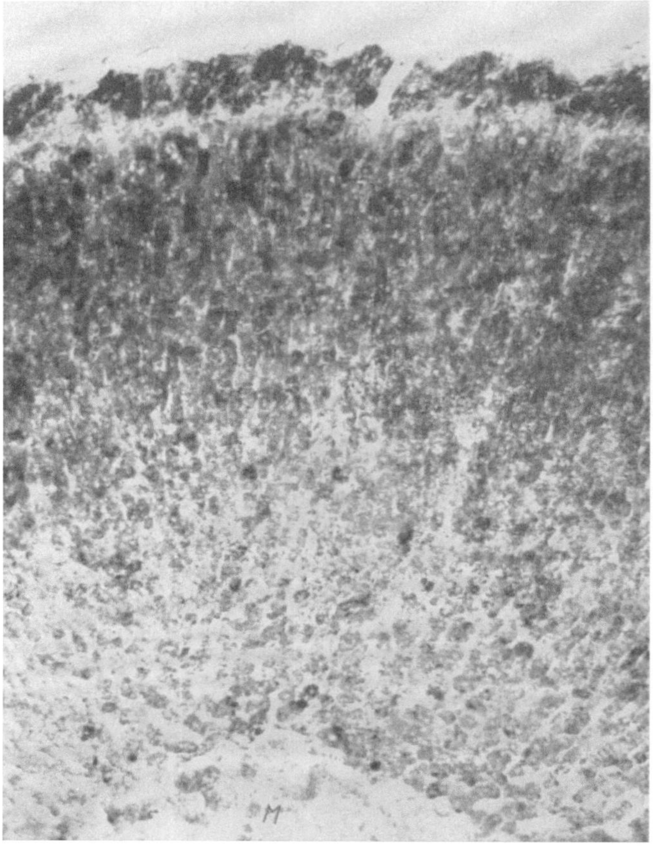

Fig. 29. The subglomerulosal zone (*S.G.*) of a normal male rat adrenal, made conspicuous by staining the lipides. Formalin-fixed frozen section, Sudan IV. × 135. *G.*, glomerulosa; *F.*, fasciculata; *R.*, reticularis; *M.*, medulla. It is typical for the lipide to be abundant in the cells of the outer fasciculata, sparse near the medullary border

JONES 1957), but herein it will be termed the subglomerulosal zone. As will be discussed later (pp. 87, 101, 104), these cells enlarge somewhat and acquire lipide droplets when the animal is subjected to "stress," so that the zone is not always readily detectable.

e) Cytoplasmic organelles

The cytoplasm of adrenocortical cells is typically moderately *acidophilic.* There is very *little cytoplasmic basophilic material* digestible by ribonuclease. When *ribonucleoprotein* is detectable, it *is normally limited to the glomerulosal cells* and is diffusely distributed in the cytoplasm (CELESTINO DA COSTA 1951, CATER and LEVER 1954, FRAZÃO 1954, SYMINGTON and DAVIDSON 1956, SYMINGTON et al. 1956, BURNS and HALE 1959). By electron microscopy, LUFT and

HECHTER (1957) demonstrated, in stimulated adrenocortical cells (*Bos*), a peppering of dense, 150 Å particles (presumably ribonucleoprotein) distributed throughout the cytoplasm. These were generally not associated with the membranes of the sparse endoplasmic reticulum. ZELANDER (1959) described such particles as most abundant in the glomerulosa (mouse).

The *Golgi apparatus*, as revealed by classical silver techniques, appears as a robust, dense juxtanuclear network (Fig. 30) (e.g., REESE and MOON 1938, BENNETT 1940, CAIN and HARRISON 1950, R. MILLER 1950, KNIGGE 1954c). In spongiocytes, processes from this network may extend out between the lipide droplets. Sometimes in ordinary histological preparations, the negative image of the Golgi substance is visible (KNIGGE 1954a, c). In the zona glomerulosa of *Cavia* (HOERR 1936b), *Felis* (BENNETT 1940), and *Equus asinus* (FRAZÃO 1954), where the location of the capillaries surrounding the cell arches is relatively clear-cut, the

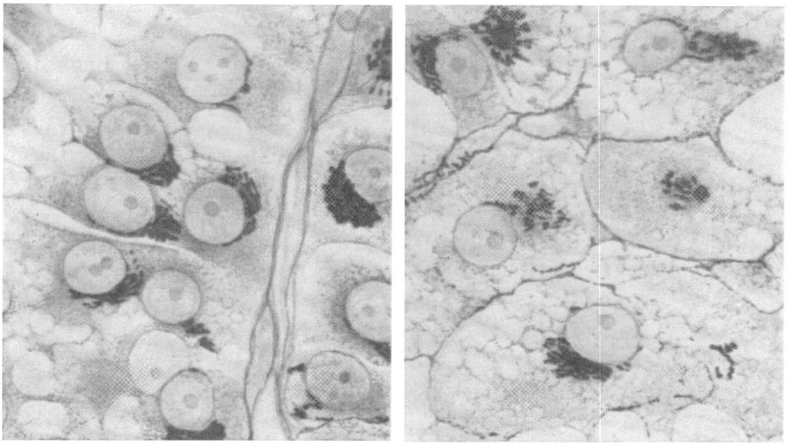

a b

Fig. 30a & b. Golgi apparatus in cat adrenal cortex. a glomerulosa; b inner fasciculata. Golgi material impregnated with silver by the Da Fano technique; drawn by ETTA PIOTTI. × ca. 1250. Reproduced by permission of the author and publisher from Figs. 12 and 13, BENNETT: Amer. J. Anat. **67**, 221 (1940)

Golgi apparatus can be seen to lie on the side of the nucleus nearest the capillary. Such a location is not distinguishable deeper in the cortex, where the relative position of the capillary is less easily appreciated in thin sections. In electron micrographs, the Golgi apparatus consists of the typical array of parallel smooth membranes and associated vacuoles and vesicles (ZELANDER 1959).

The *mitochondria* of adrenocortical cells are spheres or short rods, generally 0.3 to 1 μ in width. A minority within a cell appear even larger, 2—3 μ across. The mitochondria are extremely abundant, crowding the cytoplasm that lies between the lipide droplets (Fig. 31). While they are seemingly more abundant in lipide-free cells such as those in the subglomerulosal zone, counts show them to be actually most numerous in the large spongiocytes of the outer fasciculata (rat — R. MILLER 1953). Satisfactory mitochondrial preparations of various mammalian adrenals are illustrated in the following papers (rat — DEANE and GREEP 1946, CAIN and HARRISON 1950, R. MILLER 1953, YOSHIMURA 1954; mouse — R. MILLER 1950; guinea pig — HOERR 1936b; golden hamster — KNIGGE 1954a, c, ITO 1956, MEYERS and CHARIPPER 1956; cat — BENNETT 1940). Mitochondria prove difficult to fix and stain properly, particularly in fatty tissues; in consequence many other studies, especially those on large adrenals (see ROGOFF 1932), illustrate what are probably artifactitious distortions of of these organelles.

Even greater difficulties in preparation occur in electron microscopy. The first illustrations published showed the adrenocortical mitochondria to possess abundant internal microtubules that were generally swollen and vesiculated (e.g., Figs. 19, 25) (LEVER 1955c, 1956, BELT and PEASE 1956; see also ZELANDER

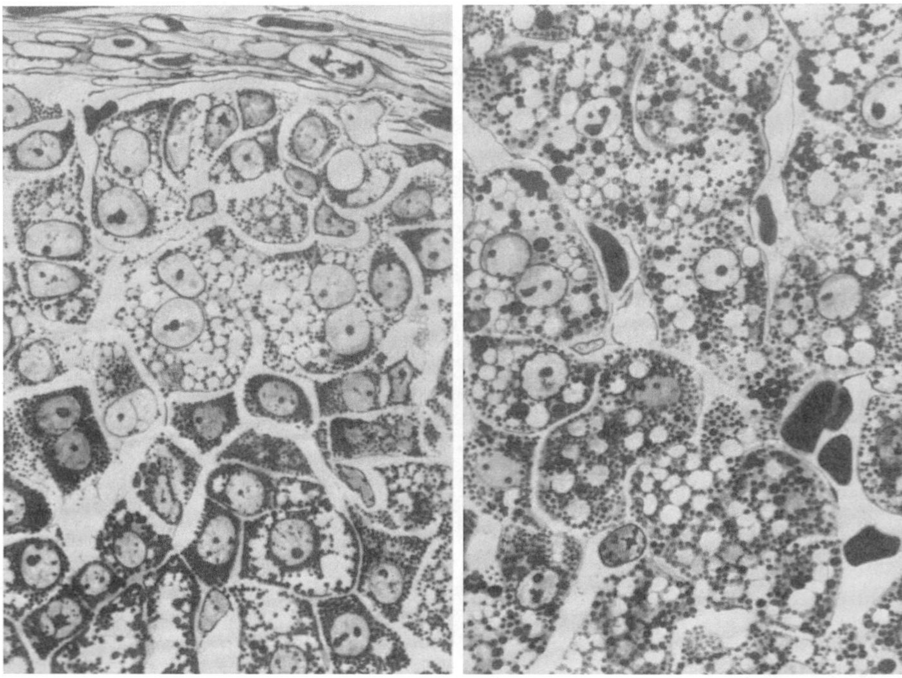

a b

Fig. 31 a & b. Mitochondria of the adrenocortical cells of the rat. a outer cortex, including capsule, zona glomerulosa, subglomerulosal zone (small, dense cells) and outer fasciculata; b mid-fasciculata. Fixed in Zenker-formalin and postchromated; stained by the prolonged iron-alum-hematoxylin method of Heidenhain; drawn by ETTA PIOTTI. × ca. 1200. The mitochondria are small spheres or rods, lying in the cytoplasm between the lipide vacuoles. In occasional cells, large spherules stain. Reproduced by permission of the publisher from Figs. 22 and 23, DEANE and GREEP: Amer. J. Anat. **79**, 145 (1946)

1959). A study by LUFT and HECHTER (1957) raises the possibility that vesiculation of the internal tubules or cristae may be artifactitious; in what was deemed healthy tissue there was no such vesiculation. It must be cautioned, however, that the first studies mentioned were made on rodent adrenals and LUFT and HECHTER's was made on bovine adrenals.

The *surface* of the cell facing the pericapillary space exhibits *microvilli* (CARR 1958, ZELANDER 1959).

8. Prenatal development

The adrenocortical anlagen appear embryonically as bilateral, segmental masses just rostral to the mesonephroi and medial to the degenerating pronephroi. The adrenocortical primordia are contiguous and continuous with the gonadal anlagen (VELICAN 1946/47, VANNINI and CESSI 1949, WITSCHI 1956).

The *prospective parenchymal cells appear to arise by proliferation of the columnar coelomic mesothelial cells* (see, however, WITSCHI 1956). The underlying mesenchyme into which they migrate possesses an extremely rich venous plexus (VELICAN 1946/47, BOERNER 1952/53). In *Homo* the primordia become distinct during the fourth week of gestation (embryos of 6—8 mm crown-rump length) (HEIT 1925, KEENE and HEWER 1927, POLITZER 1937, UOTILA 1940, CLATWORTHY and ANDERSON 1944, VELICAN 1946/47, BOERNER 1952/53). In *Mus* (WARING 1935) and *Rattus* (PANKRATZ 1931, LEVER 1955b) the anlagen appear on about the 13th day of gestation. In *Cavia* (BOERNER 1952/53) they develop at the end of the third week when the embryo is about 8 mm long. The first arteries invade immediately after the aggregates become established (VELICAN 1946/47).

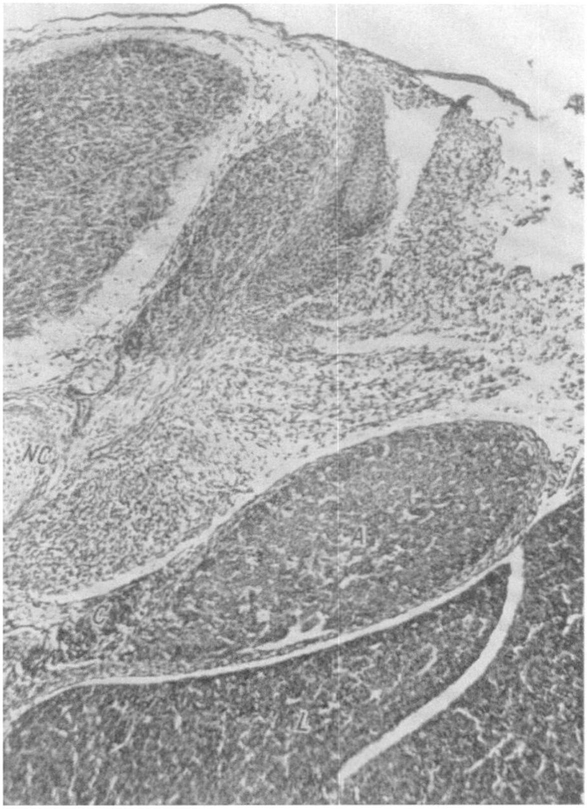

Fig. 32. An adrenal gland (*A.*) of a 17-day rat fetus. The adrenocortical primordium is encapsulated; prospective chromaffin cells (*C.*) lie near the medial border of the gland. Hematoxylin and eosin. × 80. From a preparation of JOSIMOVICH et al. (1954). *L.*, Liver; *N.C.*, notochord; *S.*, spinal cord

In the 9 mm human embryo, the adrenocorticalprimordium is an irregular, elongate aggregation lying between cervical nerve VII and thoracic nerve II. With the subsequent growth of the body, it is displaced, by the time the embryo is 18 mm long, to the level of thoracic nerve X to lumbar nerve I (GRUENWALD 1943). Concomitantly it grows steadily by hyperplasia, it enspherulates (at about 12 mm), it becomes encapsulated (14 mm), and it is then invaded by nerves and clusters of sympathocytoblasts destined to differentiate into chromaffin cells (17 mm) (CLATWORTHY and ANDERSON 1944, BOERNER 1952/53) (see Fig. 32). Owing to its enlargement, the gland soon protrudes into the peritoneal cavity (25 mm). VELICAN (1946/47) provides measurements of the human adrenal during the second month of prenatal development (Table 7).

Table 7. *Growth of the adrenal glands in human embryos* (VELICAN 1946/47)

Gestational age, days	Crown-rump length, mm	Adrenal dimensions: length × width × thickness, mm
30	8	0.5 × 0.3 × 0.1
35	12	0.8 × 0.6 × 0.4
40	15	1.0 × 1.0 × 0.6
45	18	1.2 × 1.0 × 0.8
50	24	1.5 × 1.2 × 1.5
60	30	2.0 × 2.0 × 2.2

The invasion of the prospective chromaffin tissue along the medial surface of the gland continues for some period. Gradually the clusters migrate through the cortical tissue and aggregate in the center of the gland. The process of medullary coalescence is complete in *Cavia* by the 50th day of gestation (PENNACCHIETTI 1938), in insectivores even earlier (W. ADAMS 1958), but in most species the

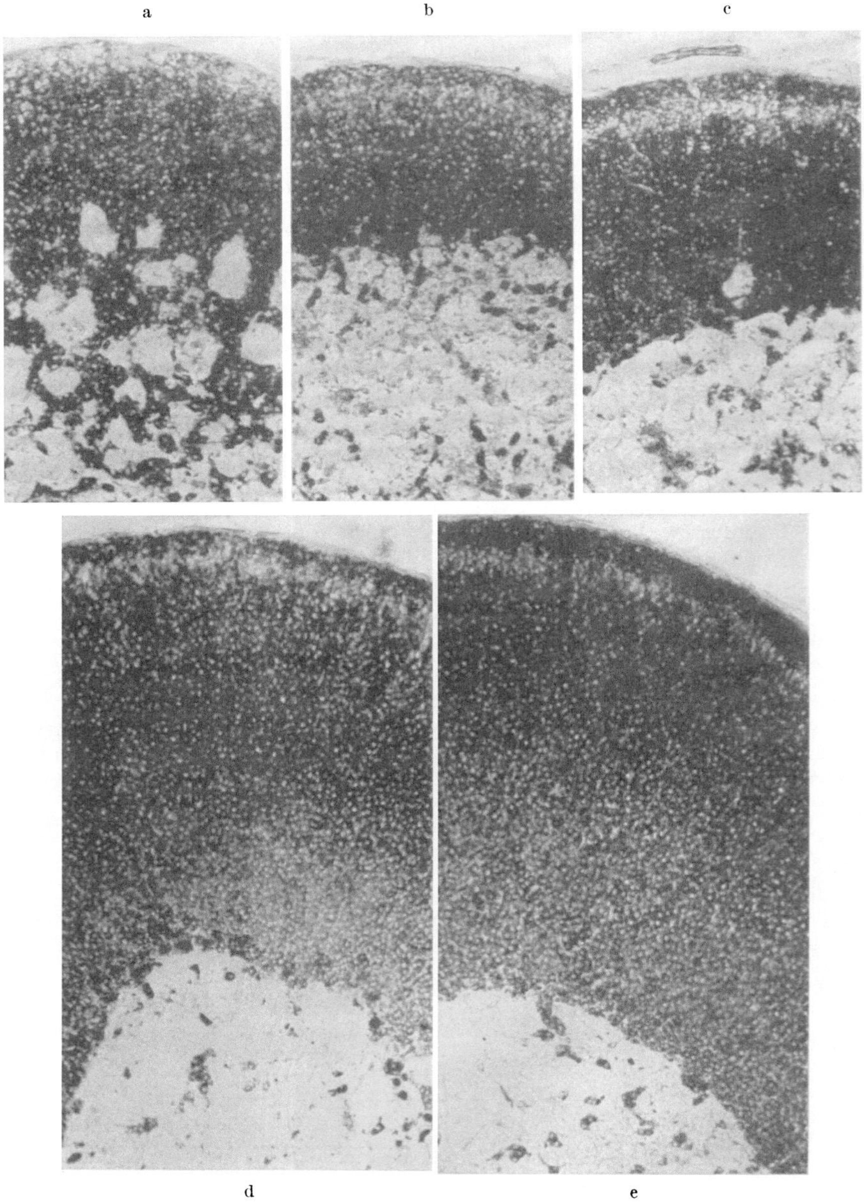

Fig. 33a—e. Maturation and growth of the adrenal cortex in the rat. With coalescence of the medulla postnatally, the central adrenocortical cells ("interlocking zone") disappear. a 1 day old, b 5 days old; c 9 days old, d 16 days old, e 22 days old. Formalin-fixed frozen sections, Sudan black B. × 95. From preparations of VAN DORP and DEANE (1950). Plate reproduced from DEANE (1956), courtesy Dr. LYTT I. GARDNER and Grune & Stratton, Inc.. The cortex in b is narrower than in a; thereafter the cortex grows wider (see Fig. 62, p. 123)

chromaffin and cortical tissues remain intermingled at the center of the gland until after birth. During the final coalescence (Fig. 33), those adrenocortical cells trapped between the medullary clusters appear to undergo destruction and to disappear (JACKSON 1919, VAN DORP and DEANE 1950).

Fig. 34. Definitive and "fetal" zones in the adrenal cortex of a $3^1/_2$-month human fetus. Hematoxylin and eosin. × ca. 100. Micrograph contributed by Dr. KURT BENIRSCHKE. *D.*, definitive zone; *F.*, "fetal" zone. The cells in the "fetal" zone are large and their cytoplasm relatively acidophilic

a) "Fetal" zone

In the higher genera of the Order Primates (Macaca, Pan, Homo), there is a unique constituent of the embryonic adrenal, the "fetal" zone (W. Hill 1933, Lanman 1953, 1957). W. Hill (1930, 1937) has reported that a similar structure occurs in the fetuses of the larger Felidae, especially the leopard (*Panthera pardus*) and lion (*P. leo*), but this observation has never been confirmed.

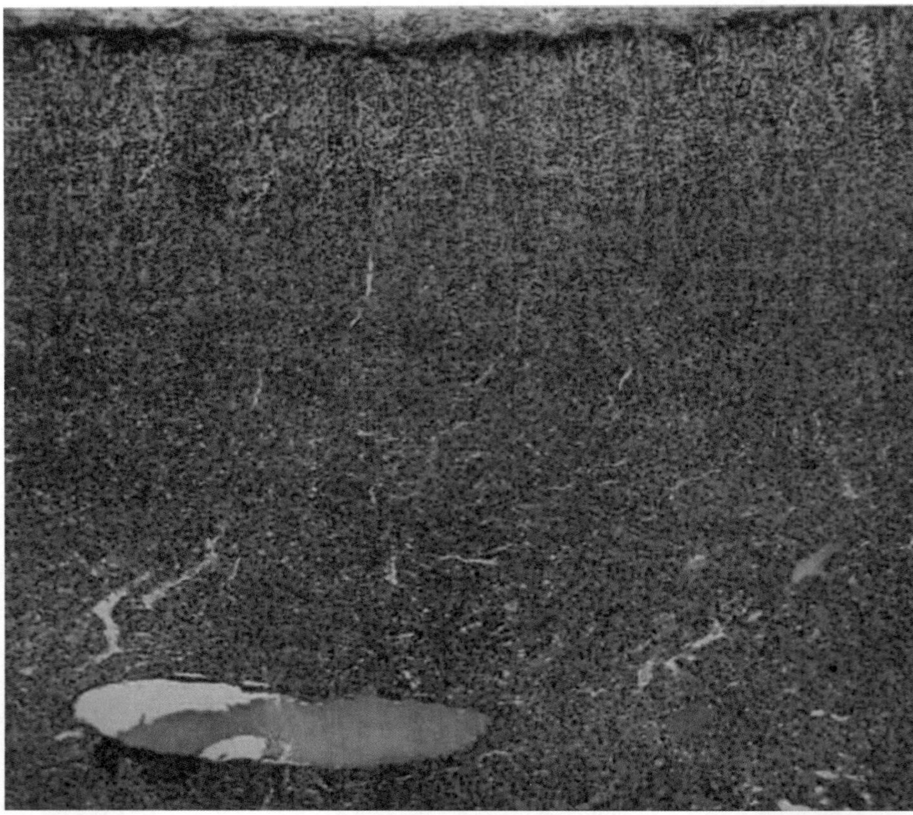

Fig. 35. Adrenal gland of a full-term human fetus, showing the beginning enlargement of the definitive zone (*D.*) and regression of the "fetal" zone (*F.*). Hematoxylin and eosin. × ca. 45. Micrograph contributed by Dr. Kurt Benirschke. Compare with Fig. 36

In the well-developed primate fetus, the "fetal" zone occupies the greater part of the cortex inside a narrow zone destined to be the "definitive" or "permanent" cortex. Its cells are much larger than those of the definitive cortex and more acidophilic (Fig. 34). According to Uotila (1940) and Velican (1946/47, 1948b), the first wave of cells proliferated from the coelomic epithelium gives rise to the "fetal" zone, whereas a second wave gives rise to the surrounding definitive adrenocortical cells. The cross-sectional area occupied by the "fetal" zone is 30—50% in embryos under 2 cm in length and rises to 70% by the time the embryo is 3 cm long (3rd lunar month) (Velican 1948b) and to 85% in older fetuses (Swinyard 1943). *Much of the remarkable size of the human gland in fetal life is therefore attributable to this special tissue.* Signs of involution of the "fetal" zone are already detectable in late fetuses (Figs. 35, 36) (Clatworthy and ANDERSON 1944, ROTTER 1949a). *After birth, there is acute resorption of the cells* in the "fetal" zone (Fig. 37), resulting in a halving of the *absolute weight* of the gland (Table 3, p. 20). According to TÄHKÄ (1951), the involution is rapid for the first 3 postnatal weeks, then slower. Human adrenals reach their lowest absolute weight at about 2.5 months after birth (SWINYARD 1943, TÄHKÄ 1951; see, however, Table 3). All of the measurements cited were of course made on abortuses or infants dying of disease and may not exactly represent the "normal" course of events.

It should be noted here that in the two other species that have been examined in detail there is demonstrable reduction in the absolute as well as the relative

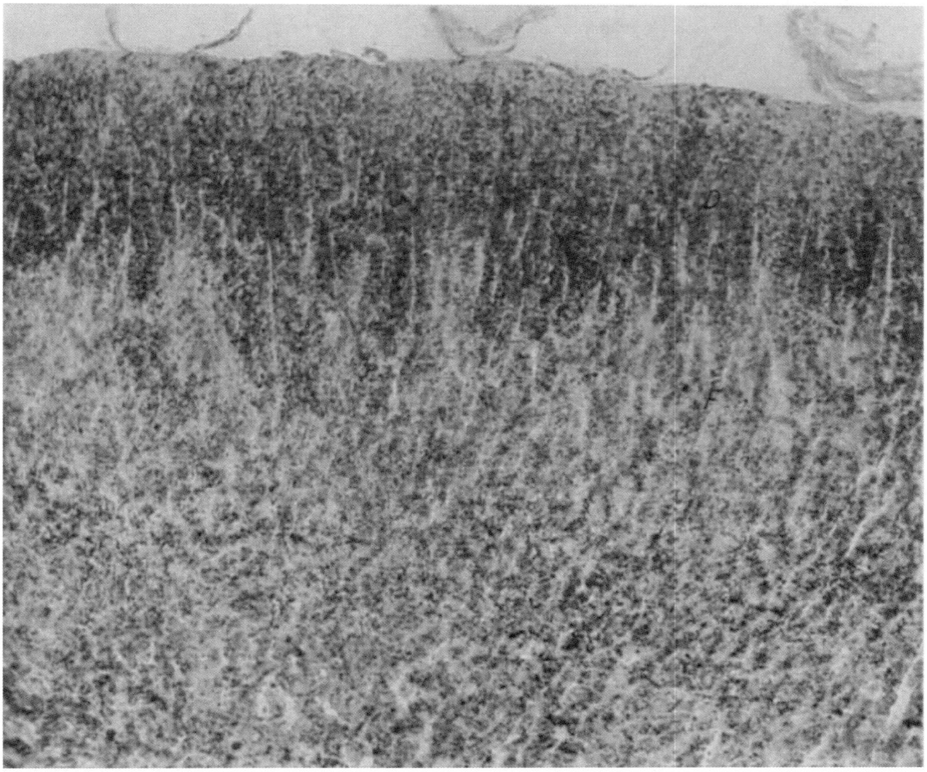

Fig. 36. Adrenal gland of a full-term human fetus, showing abundant lipide in the definitive zone (*D.*), little lipide in the "fetal" zone (*F.*). Formalin-fixed frozen section, Sudan IV and hematoxylin. × ca. 45. Micrograph contributed by Dr. KURT BENIRSCHKE

weight of the gland postnatally (Fig. 62, p. 123) (*Mus* — MOOG et al. 1954, MOOG 1959; *Rattus* — DEANE 1956). The shrinkage lasts for only about a week. A similar decline apparently occurs in some large Felidae (W. HILL 1937), where it may last over a month. As indicated above (p. 41), this involution involves in particular the cortical cells lying intermingled with the medullary cells; these cells do not differ significantly from those in the fasciculata proper, as do the cells of the "fetal" zone of primates.

Fuchsinophilia. Mention should be made of the allegedly peculiar affinity for acid fuchsin of the "fetal" zone and of the adult reticularis, a property attributed to the presence of androgens by VINES (1938) and several subsequent workers (e.g., GOORMAGHTIGH 1940, BLACKMAN 1946, ZINSSER and ZINSSER 1951, LOBBAN 1952, HÉBERT et al. 1954). *This staining property with acid fuchsin is no different from affinity for a variety of acid dyes* (SCHAUMKELL et al. 1957a, b). The cytoplasm of all adrenocortical cells is relatively acidophilic because of the paucity of cytoplasmic nucleoprotein. That the cells of the "fetal" zone appear more acidophilic than those in the "definitive" cortex is demonstrable (Fig. 34). This property seems largely to be associated with the fact that these cells, like those of the adult reticularis, contain few lipide droplets, so that their cytoplasm is compact rather than vacuolated and hence stains more intensely (ROGERS and WILLIAMS 1947, SCHAUMKELL et al. 1957a, b, R. COHEN et al. 1959). There is no basis for a attributing it to an *androgen*. In general, fuchsinophilia parallels affinity for iron hematoxylin (siderophilia) and may be partly caused by

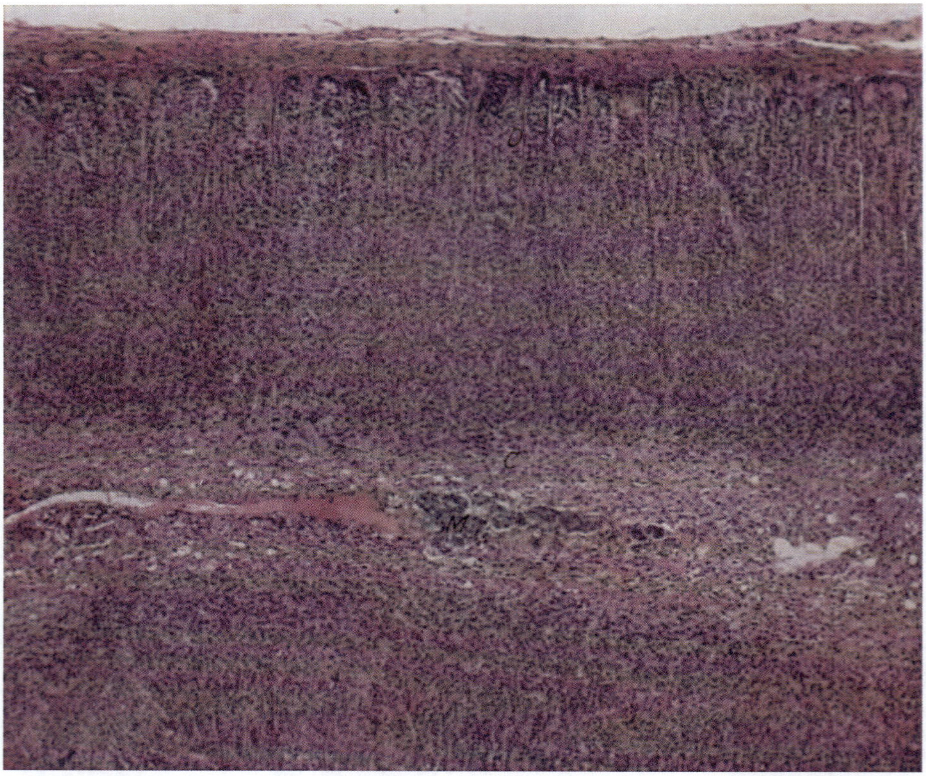

Fig. 37. Adrenal gland of a 30-day old human infant, showing the disappearance of the "fetal" zone and further growth of the definitive zone (*D.*). The disappearance of the cells of the "fetal" zone has left behind a temporary connective-tissue sheet (*C.*), or medullary capsule. *M.*, prospective medullary cells. Hematoxylin and eosin. × ca. 45. Micrograph contributed by Dr. KURT BENIRSCHKE

phosphatides preserved by fixation in potassium dichromate, perhaps those of mitochondria, which are most crowded in lipide-free cells (CELESTINO DA COSTA 1951, 1952, EHRENBRAND 1954, 1959).

In all species, the adrenocortical mass is *initially a cavernous structure*, with cords of parenchymal cells ramifying through irregular, wide blood sinuses. Gradually the parenchyma becomes organized into recognizable zones (glomerulosa and fasciculata) plus an innermost zone that interlocks with the groups of chromaffin cells (Fig. 33) (JOSIMOVICH et al. 1954, A. COHEN 1955, NOUMURA 1959). In the primate adrenal, of course, the definitive cortex is extremely narrow (though of large volume) and consists of a small arcuate glomerulosa and a narrow fasciculata (LANMAN 1953). In the innermost "fetal" zone, the cells are arranged in irregular cords (Fig. 34).

In those species that have been studied, *lipides make their appearance in the adrenocortical cells at about the time the tissue begins to assume its definitive organi-*

zation (Moog 1959). Thus in the rat, free droplets become detectable on the 17th fetal day (Josimovich et al. 1954, Noumura 1959). Rotter (1949a) found lipide in the adrenal of a 6-cm human fetus, the earliest specimen he examined for this feature. Frazão (1954) found only small droplets in the human adrenal up to the fourth fetal month; thereafter there were abundant droplets in the "definitive" zone, but only sparse droplets in the "fetal" zone (Fig. 36). Frazão also found lipide droplets throughout the cortex of fetal cattle, an interesting observation in view of the absence of free droplets from such adrenals (excepting the glomerulosa) in adults.

9. Adrenocortical growth and regeneration

There exist three major hypotheses to account for adrenocortical growth and repair in adult mammalian glands, which may be termed, respectively, the cell-migration hypothesis, the zonal hypothesis, and the transformation-field hypothesis. The absence of direct confirmation of any one of them precludes making final judgment on which one or what combination of them is valid. Brief summaries of each follow.

a) Cell-migration hypothesis

The oldest explanation of glandular growth and maintenance was proposed by Gottschau (1883). As now formulated, it postulates that adrenocortical cells, arising either in the capsule or by division of glomerulosal cells, gradually migrate from the surface of the gland to the medullary border, there to die and disappear. The hypothesis is based on several assumptions. First, it is assumed that continual cell death occurs in the reticularis, because pyknotic nuclei, dying cells vitally stainable by trypan blue, accumulation of lipofuscins and bizarre-shaped mitochondria are more abundant in this zone than elsewhere in the cortex (e.g., Hoerr 1931, Bennett 1940, Baxter 1946). Whether indeed these cells actually die in great numbers or are merely in poor condition by virtue of possessing a relatively poor environment has not been established. By analogy with the liver lobule, the cortical cells might be expected to be healthiest at the pariphery, least healthy at the medullary border because of the nature of the blood supply (Greep and Deane 1949a, Deane 1958).

Second, it is assumed that mitosis is limited to the outer cortex. Indeed, in the adrenals of all mammals so far examined, mitosis is most common in the outer part of the gland, in the zona glomerulosa and in the outer fasciculata (see Fig. 24, p. 31). However, cell division occurs also in the inner fasciculata and reticularis but to a lesser extent (p. 32) (Hoerr 1931, Baxter 1946, van Dorp and Deane 1950). Mitoses are also seen in the capsule, especially at times when the gland is undergoing hypertrophy.

Numerous workers have presented illustrations suggesting that fibroblasts of the inner capsule enlarge, transform into adrenocortical cells and enter the cortex proper (B. Baker 1952, Bachmann 1954, Schaumkell 1955). In some of the larger glands, notably those of *Homo*, *Macaca*, *Cavia*, *Canis*, *Felis* and various ungulates, laminae or cords of cortical cells may be seen in the capsule (Fig. 38) (e.g., I. Dawson 1956, Richter 1958) and it has been suggested that they migrate therefrom into the glomerulosa (Gruenwald and Konikov 1944, Elias 1948, Elias and Pauly 1956). Such capsular nests occur only rarely in the smaller adrenals of *Rattus*, *Mus* or *Oryctolagus*. There is of course no way of testing in a histological section that cells or groups of cells are migrating, let alone in what direction they may be moving. That cords of adrenocortical cells in an enlarging gland might dissect their way into the capsule seems just as likely as migration from capsule to glomerulosa. Migration in any direction is passive, since adrenocortical cells *in vitro* show no signs of motility (WILLIAMS 1947, SCHABERG 1955).

That cells lying in the outer portion of the cortex are capable of multiplication and inward proliferation or migration is demonstrable when the adrenal has been enucleated (Fig. 61, p. 119) (INGLE and HIGGINS 1938a, GEEP and DEANE 1949b, PELLEGRINO 1951) or when the gland, after transplantation, undergoes central

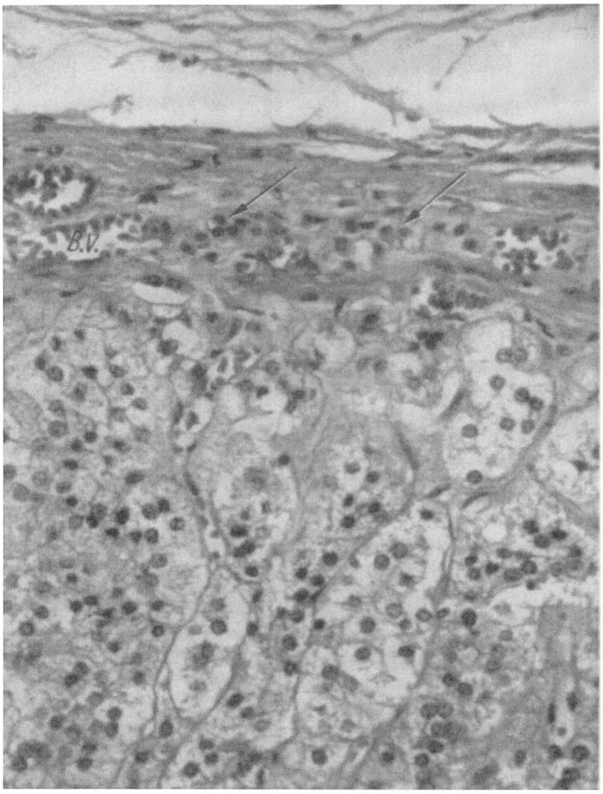

Fig. 38. Clusters of lipide-filled adrenocortical cells (arrows) lying in the capsule of a human adrenal gland. Hematoxylin and eosin. ×280. Photographed from a preparation donated by Dr. DAVID M. MAEIR. *C.*, capsule; *B.V.*, intracapsular blood vessel; *G.*, glomerulosa. Note connective tissue fibers surrounding individual clusters of glomerulosal cells

degeneration and then subsequently regenerates (INGLE and HIGGINS 1938b, WILLIAMS 1947, 1953, MARTINOVITCH 1955, SCHABERG 1955, 1957). How closely such phenomena are related to normal growth and maintenance of the gland remains to be determined. VACCAREZZA (1945), for example, claims that after smaller lesions, cell replacement occurs from nearby cells, wherever the lesion is placed.

Advocacy of the cell-migration hypothesis leaves open the question of whether adrenocortical cells produce different hormones in different zones (p. 98) or whether all cells are capable of producing the same products.

b) Zonal hypothesis

Perhaps first suggested by B. HOUSSAY and SAMMARTINO (1933a, b) and extended by SWANN (1940), SARASON (1943), VACCAREZZA (1945), GREEP and

Deane (1949a) and others, this postulates that cells in the two major zones of the cortex (glomerulosa and inner zones), once formed, normally remain *in situ* and replace themselves as needed. One basis for this hypothesis lies in the demonstration of the morphological dichotomy of the two major zones under different experimental conditions, as described later (p. 98ff.). According to this theory, the cells of the zona glomerulosa produce mineralocorticoids, the cells of the inner zones, glucocorticoids (and androgens). Proponents of this hypothesis emphasize that mitoses occur in both the glomerulosa and fasciculata and that they are absent from the subglomerulosal zone lying between. Obviously, both during embryonic development and during regeneration after major damage, the ability of glomerulosal cells to transform themselves into fasciculata cells must be conceded (Fig. 61, p. 119) (Greep and Deane 1949b, Williams 1953, Schaberg 1957).

c) Transformation-field hypothesis

This hypothesis, put forward by Tonutti (1942, 1951) and recently adopted in essence by Chester Jones (1957), assumes that the fasciculata is the active portion of the cortex, both in terms of secretion and in terms of cell duplication. The glomerulosa and reticularis are considered reserve zones, which transform into fasciculata-like cells during hyperactivity (progressive transformation) and revert to their "normal" appearance when hyperactivity ceases or when the gland becomes frankly hypoactive (regressive transformation).

Like the zonal hypothesis, the transformation-field hypothesis is based on functional theory, in this case that all adrenocortical cells produce or are capable of producing the same hormone(s). Most of Tonutti's observations have been made on mice and guinea pigs, the adrenals of which normally exhibit little lipide in the glomerulosa, so that the distinction between glomerulosa and subglomerulosal zone is not so distinct as in the rat adrenal, for example. Furthermore, as is noted later (p. 87), the cells of the subglomerulosal zone enlarge when the fasciculata is stimulated, thereby masking the morphological separation of the glomerulosa and fasciculata. Certainly Tonutti's claim that inactivity results in a clearly-defined reticularis has not been confirmed by other observers (e.g., Yoffey 1953).

10. The X zone of the mouse adrenal

The adrenal of *Mus musculus* exhibits a special juxtamedullary zone, first clearly described by E. Howard (1927) and Tamura (1927) and generally termed the X zone. This term is as apt today as it was some 30 years ago, although it now seems that the zone is inhibited by androgens and may be stimulated by pituitary LH (luteinizing hormone) (Chester Jones 1957). *It is certainly not stimulated by ACTH.*

In its fully-developed state, the X zone consists of large acidophilic cells with prominent nuclei. Except when degenerating, the cells contain no free lipide droplets. These cells differentiate *postnatally*, at about 2 weeks, from the inner portion of the fasciculata (Whitehead 1933a, Waring 1935). There are some variations in the degree of development in different strains of mice (Westberg et al. 1957, Sakiz 1958, Delost and Chirvan-Nia 1958). In males, the X zone normally begins to involute at puberty, in females during the first pregnancy. It persists in virgin females for several months (Chester Jones 1952, 1957, Sakiz 1958). The involution involves pyknosis of the cell nuclei and profound atrophy of the cells, which sometimes develop cytoplasmic lipide droplets (? fatty degeneration). Ultimately the cells disappear completely, leaving their stroma as a medullary "capsule".

A perhaps analogous zone occurs in the unmated female golden hamster, *Mesocricetus auratus* (LaVelle 1951, W. N. Holmes 1955b); none is detectable in the male. This identification is disputed by Ito (1956). In *Microtus arvalis* and some other voles, a juxtamedullary zone of lipide-free cells occurs in both sexes prepuberally (Delost and Delost 1955, Delost and Chirvan-Nia 1956, Delost 1956). The cells are not so large as they are in the mouse Xzone. This zone atrophies at puberty in males, but hypertrophies in pregnant and lactating females. These and other reasons to be cited later (p. 127) suggest that this zone is not truly homologous with the Xzone in *Mus*. The confusion is only compounded by applying the term wherever a zone of unknown significance is encountered.

Recently Delost (1957) has described a similar zone in the adrenals of two shrews (insectivores). Here the appearance of the cells and their involution with puberty in the male and pregnancy in the female are similar to the mouse.

The frequent equating of the mouse Xzone with the human "fetal" zone (e.g. Bourne 1949) *overlooks, among other things, the great difference in development of the two structures.* Furthermore the cells of the "fetal" zone contain some free lipide droplets, whereas the cells of the Xzone do not develop such droplets until they degenerate (Benua and Howard 1950, Chester Jones 1957). Similarly, an analogy has been drawn between the interlocking zone of adrenocortical and adrenomedullary cells in prenatal and just-postnatal mammalian glands and the "fetal" and Xzones (Roaf 1935, Davies 1937, E. Howard 1938). While these central adrenocortical cells do degenerate and disappear with the coalescence of the chromaffin tissue (Fig. 33, p.40), they clearly resemble the cells lying more externally in the cortex (Whitehead 1936, van Dorp and Deane 1950).

Chemistry of adrenocortical tissue

A. Introduction

The characteristic chemical properties of mammalian adrenocortical cells, distinguishing them from most other cell types, include (1) *a high lipide content*, with a large fraction of the lipide consisting of *cholesterol esters*; (2) a *relatively high concentration of ascorbic acid*; and (3) the presence of *certain enzymes that take part in the synthesis of steroids and the specific cortical hormones*. However, it must be stressed that neither a high lipide content nor a high ascorbic acid content is found in all adrenocortical cells; as noted below, one or both of these constituents is low in many species. The study of specific enzymes has, at this date, been applied only to the glands of a few mammalian species. Indeed, the capacity of the glands in other vertebrates to produce corticoids identical with those produced by mammalian glands has become established only recently with the development of micromethods for steroid characterization (Table 15, p. 75).

This section will survey the available information about the *lipides* of adrenocortical cells, about *ascorbic acid*, and about *enzymes*, some of which are of specific importance to hormone secretion, others of which are of more general nature. In addition to these, data about *miscellaneous compounds*, such as glycogen, salts and nucleic acids, will be given.

The *techniques* that have been employed vary. *Classical chemical methods* may be applied to whole glands, to isolated cortex or to subcellular fractions isolated by centrifugation. Or microtechniques may be applied to single, serial sections, thereby permitting some deductions about the differential distribution of substances in different parts of the gland. This so-called *quantitative histochemical technique*, devised by LINDERSTRØM-LANG, employs thin sections cut in a cryostat (GRUNBAUM et al. 1956). Or *visualizing histochemical techniques* may be applied to appropriate histological sections. The first two kinds of methods permit quantitative assessment, the last permits cellular and sometimes intracellular localization. Where 2 or 3 different approaches have been used, their correlation will be noted.

B. Adrenocortical lipides

I. Total lipides

As indicated in the anatomical survey, *one of the most characteristic properties of adrenocortical cells is the presence in the cytoplasm of free lipide droplets.* Of course, some lipide is present in all protoplasm, even when not demonstrable in droplet form. Such lipide is associated particularly with protoplasmic membranes and is termed "structural" or microscopically "masked" lipide (DEANE 1958). This class probably accounts for the diffuse sudanophilia demonstrable in many types of cells. The free droplet lipide constitutes the "variable" or "metabolic" type. With thorough chemical extraction of an organ or tissue, the chemist obtains a mixture of the two sorts. Separate analyses of the free lipide droplets and of the lipides bound in lipoprotein membranes have not yet been undertaken for the adrenal cortex.

Chemical determinations of the concentration of lipides in adrenals have apparently been made only on mammalian glands, and in general on whole glands rather than separated cortex. This latter fact introduces a variable degree of "dilution" by the adrenal medulla, which usually exhibits little or no free lipide and may possess different structural lipides as well.

Method. To obtain the lipide fraction, the tissue is exhaustively extracted with organic solvents. A valuable solvent system introduced by FOLCH et al. (1951) consists of a mixture of chloroform and methanol employed in the cold. Hydrolysis or other alteration of the compounds is thereby avoided.

Table 8. *Lipide content of mammalian adrenal glands*

Species	Average total lipide, % wet weight of gland	Reference
Homo sapiens		
adult.	20 (recalculated)	E. ADAMS and BAXTER 1949
newborn	4 (recalculated)	STONER et al. 1953
Cavia cobaya	15	OLESON and BLOOR 1941
	20	KNOUFF et al. 1941
	12	CREUTZFELDT et al. 1953
Rattus norwegicus	14	KARNOVSKY and DEANE 1955
Oryctolagus cuniculus	12	MACLACHLAN et al. 1941
Bos taurus	6 (isolated cortex)	J. B. BROWN et al. 1937

In most mammalian species, *lipides comprise approximately 15%* (10—20%) *of the wet weight of glands* taken from healthy, unstressed specimens (Table 8). [Where figures were given in terms of dry weight, the content in terms of wet weight has been recalculated on the assumption that the dry weight is 30% of the wet weight of lipide-rich glands (MCSHAN et al. 1946).] A high lipide content is found for all glands that exhibit free lipide droplets in most of the adrenocortical cells. On the other hand, adrenals that contain few droplets contain only about

5% lipide (human newborn, ox). This value presumably obtains also in the adrenals of hibernating rodents.

Bänder (1950) has described a diurnal cycle in lipide content of mouse adrenals, with most being present in the morning, least at midnight (nocturnal species).

By *the quantitative histochemical technique*, Glick et al. (1955b) showed that most lipide occurs in the outer fasciculata in *Cavia* and *Macaca*, in the zone of the so-called spongiocytes.

Free lipide droplets are made *histologically demonstrable* either by fixing the block in a solution of osmium tetroxide (Figs. 4, 7, 19, 25, 63) or by exposing frozen sections (unfixed or fixed in 10% formalin) to an oil-soluble stain such as Sudan IV, Sudan black B or Oil red 0 (Figs. 12, 13, 26—29, 33, 36, 39, 41, etc.).

OsO_4 blackens and hardens unsaturated lipides. However, it fails to preserve completely-saturated lipides, and it blackens non-lipides having reducing activity (Lison 1953). Proper controls are virtually impossible.

The more reliable technique is to expose frozen sections to a solution of an *oil-soluble stain*. Fixation seems to permit partial leaching-out of lipides, especially of phosphatides (Deane 1958). Triethyl phosphate, the stain solvent introduced by Gomori (1952a), appears to have no tendency to extract even finely-dispersed droplets. The available oil-soluble dyes all possess weakly reactive phenolic hydroxy or secondary amino groups capable of some dye-binding with non-lipides. Thus, even the lipide-soluble dyes possess some limitations. However, if applied with care and if control sections preextracted with acetone are employed, the technique gives valid information about the distribution of lipide droplets.

The distribution of lipide droplets in the adrenocortical tissue of the various classes and orders of vertebrates is to be found in the section on the anatomy of adrenocortical tissue (p. 6ff.). If it is assumed that all adrenals exhibiting free droplets possess a relatively high lipide content (certainly exceeding 5% of the wet weight of the tissue), the adrenals of many lower vertebrates resemble "typical" mammalian adrenals. The adrenocortical cells in representative cyclostomes, elasmobranchs, dipnoans, some ganoid fishes, amphibians, reptiles and birds all contain demonstrable lipide droplets. *Only the bony fishes* (Holostei, Teleostei) *resemble hibernating mammals and the ruminant Artiodactyla in exhibiting only occasional free droplets in the cytoplasm.*

II. Cholesterol

Chemical results. The most conspicuous constituent of adrenal lipides is cholesterol. As a general rule, cholesterol appears to comprise 25—50% of the total lipide extract. The cholesterol is derived in part from the diet, in part from local synthesis (Morris and Chaikoff 1959).

Method. Free cholesterol can be isolated from lipide extracts by precipitation with digitonin; the naturally-esterified cholesterol must first be hydrolyzed before it will precipitate. Usually the total cholesterol is then determined colorimetrically by a modification of the Liebermann-Burchard reaction, wherein cholesterol develops a blue-green color on exposure to a mixture of sulfuric acid and acetic anhydride (e.g., Sperry and Webb 1950). Unless the reaction is applied to a digitonin precipitate, additional chromogenic substances produce falsely high readings (Karnovsky and Deane 1955). There are, however, new color reagents that aparently do not give interfering colors with compounds other than cholesterol and hence do not require digitonin precipitation (Knobil et al. 1954a).

The concentration of cholesterol in the adrenals — even within a given species — *seems to vary considerably.* This variation no doubt reflects a number of factors: differences in chemical technique, in diet, in genetic strain and in age of the animals, and also in the degree of "stress" to the animals during killing.

Thus, in adult rats, the reported results vary from about 2.5 g per 100 g wet weight gland (e.g., Fortier et al. 1950, Simonnet et al. 1951, Poulton and Reese 1957) to about 6% (Tepperman et al. 1947, Nichols 1950, Knobil and

Briggs 1955). There are also age differences: the concentration of cholesterol in the adrenals of term fetal rats is only about one-quarter that found in the maternal glands (J. Jones et al. 1953, Knobil and Briggs 1955).

Results for rat adrenals and the adrenals of some other aximals are compared in Table 9. Most significant is the low cholesterol concentration in adrenals containing little lipide (human fetuses, *Bos*, *Didelphis*). Similarly, Leloup-Hatey (1957) found only 0.3% in the interrenals of various teleost fishes (*Anguilla anguilla*, *Salmo* spp.).

Table 9. *Representative values for total cholesterol concentration in adrenal glands*

Species	Concentration, %	Reference
Homo sapiens		
♂ and ♀	9.5	Sokoloff et al. 1951
♂	6	Rogers and Williams 1948
♀	7	
fetus, last trimester	ca. 1 (recalculated)	Tähkä 1951
newborn	0.5 (recalculated)	
Cavia cobaya	4.5	Sayers et al. 1946
	1.5	Creutzfeldt et al. 1953
Rattus norwegicus		
♂	3.5	Sterling and Longwell 1953a, b
♀, pregnant	4.5	J. Jones et al. 1953
fetus, term	0.8	
♀, pregnant	6.0	Knobil and Briggs 1955
fetus, term	1.8	
laboratory rat	5.4	Nichols 1950
wild rat (shot)	11.0	
Oryctolagus cuniculus	6.0	MacLachlan et al. 1941
	5.6	Elton et al. 1959
normal diet	8—15	Kay and Whitehead 1935
high-cholesterol diet	13—19	
Bos taurus	0.3	J. B. Brown et al. 1937
	0.4	Rosenfeld 1955
Canis familiaris	4.9	Elton et al. 1959
Felis domestica	5.1	
Didelphis virginiana	0.5	
Gallus domesticus	3.6	A. Howard and Constable 1958
Rana sp.	1.1	Elton et al. 1959
Bufo sp.	1.4	

In the adrenal glands of a number of species, *most of the cholesterol is found to be esterified.* In species displaying a large amount of free lipide, the percentage of total cholesterol present in the esterified form seems to be 80% or more (*Homo* — E. Adams and Baxter 1949; *Cavia* — Knouff et al. 1941, Oleson and Bloor 1941; *Rattus* — Tepperman et al. 1947, Fidanza and Bonomolo 1953, Poulton and Reese 1957; *Oryctolagus* — Kay and Whitehead 1935, MacLachlan et al. 1941; *Equus* — Leulier and Revol 1930; *Gallus* — Taylor et al. 1956). *In adrenals containing little free lipide*, however, *less than one-half the cholesterol is esterified* (*Ovis* — Leulier and Revol 1930; *Bos* — Glick and Ochs 1955).

Quantitative histochemistry. A determination of the distribution of cholesterol in sections of different parts of the cortex (Glick and Ochs 1955) showed that, in monkey and rat glands, the highest concentration occurs in the outer fasciculata, a result conforming with that obtained with visualization techniques. In the bovine adrenal, the distribution was almost uniform throughout the gland, including the medulla.

Histological methods. There are two histochemical methods for visualizing cholesterol *in situ.*

The *first* takes advantage of the birefringence of crystallized cholesterol esters. With prolonged fixation at room temperature, the esters tend to precipitate within the lipide

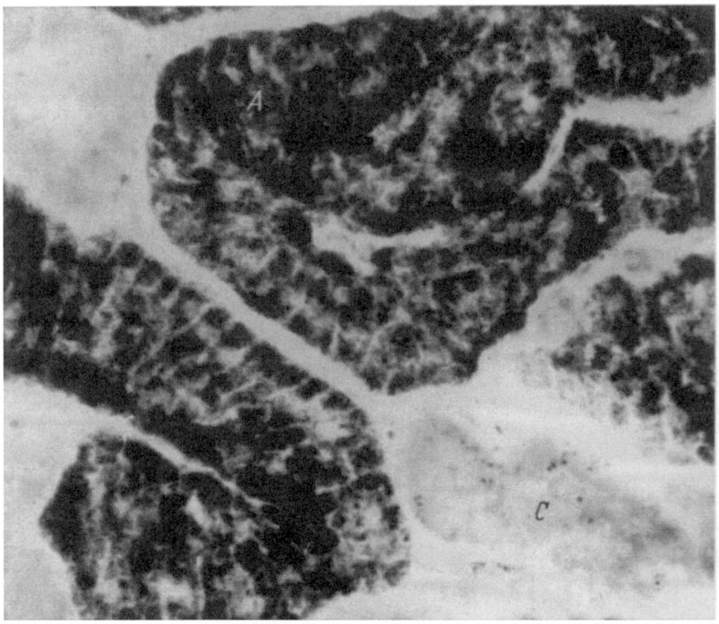

Fig. 39. Frozen section of a chicken adrenal gland stained with Sudan IV and hematoxylin. *A.*, adrenocortical tissue; *C.*, chromaffin tissue. Photographed from a preparation contributed by Dr. JACQUES PADAWER. × 450

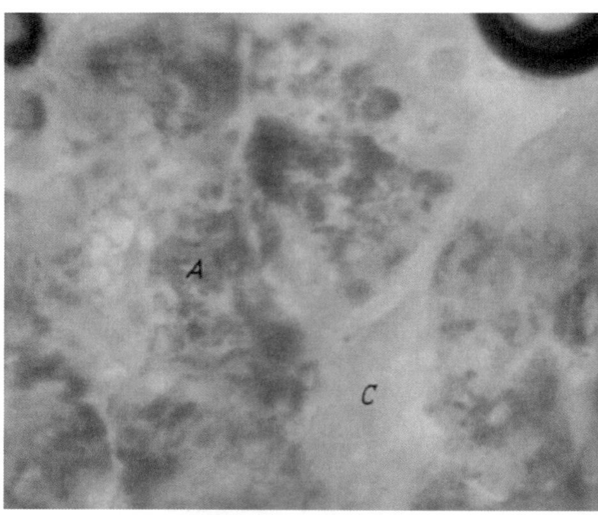

Fig. 40. Frozen section of a chicken adrenal gland, stained for cholesterol by a modification of the Schultz technique. The lipide droplets (Fig. 39) contain conspicuous amounts of cholesterol. *A.*, adrenocortical cells; *C.*, chromaffin cells. Photographed from a preparation made by Dr. JACQUES PADAWER. × 420

droplets. There is no established endpoint for this process, and eventually the crystals tend to aggregate and rupture the cells (YOFFEY and BAXTER 1947). Non-esterified cholesterol can be precipitated by treating sections with a digitonin solution. Since some saturated triglycerides also crystallize at room temperature (CAIN 1950, FAWCETT 1952, LISON 1953), the method is not specific.

The *other visualization method* for cholesterol is the histochemical application of the Liebermann-Burchard color reaction, introduced by SCHULTZ (A. SCHULTZ and LÖHR 1925). This method has proved quite specific (REINER 1953). However, it has only low sensitivity when cholesterol is mixed with other lipides, as it is in adrenals (J. EVERETT 1947, REINER

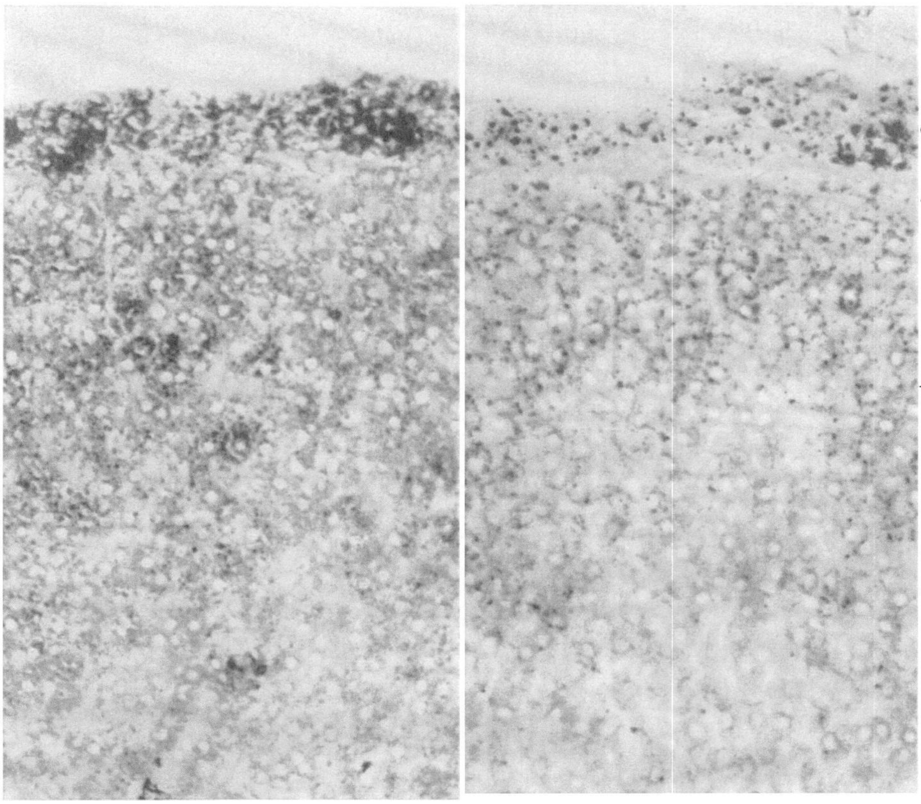

Fig. 41 a & b. Adjacent thin frozen sections of a rat adrenal gland, a stained for lipides with Sudan black B, b stained for cholesterol by the Schultz reaction. × 85. In general, only the larger lipide droplets contain *detectable* amounts of cholesterol (i.e., over approximately 25%). *G.*, glomerulosa; *S.G*, subglomerulosal zone; *F.*, fasciculata

1953, DEANE 1958). Apparently cholesterol must comprise nearly 25% of the lipide to give a detectable color reaction. Thus the birefringence test may be the more sensitive.

In normal specimens, most of the free lipide droplets in adrenocortical cells give these two cholesterol tests. This is true for most mammals (e.g., WHITEHEAD 1934, GREEP and DEANE 1949a, NICANDER 1952, BACHMANN 1954, WRIGHT et al. 1957, R. COHEN et al. 1959) and also for elasmobranchs (ABOIM 1944, GRYNFELTT et al. 1957), dipnoans (GÉRARD 1951, W. HOLMES and MOORHOUSE 1956), anurans (CHIEFFI 1955), reptiles (HEBARD and CHARIPPER 1955, WRIGHT and CHESTER JONES 1957), and birds (R. MILLER and RIDDLE 1942, KAR 1947, KNOUFF and HARTMAN 1951, A. DAWSON 1953). In other words, cholesterol and sudanophilic lipides appear to be roughly coextensive in healthy adrenocortical tissue (Figs. 39, 40, 41). Where lipide droplets are absent (some fetal glands; bovine, hamster and teleost glands), cholesterol is not histologically detectable (DEANE and LYMAN 1954, LELOUP-HATEY 1957).

Several workers have noted, however, that not all lipide droplets within a given section display a positive cholesterol test (e.g., KNOUFF et al. 1941, FELDMAN 1950). As a general rule, *it seems that the larger lipide droplets contain abundant cholesterol* (Fig. 41), whereas the smaller ones contain little detectable cholesterol but contain a higher concentration of unsaturated lipides (pseudoplasmals, see below) (DEANE 1958).

III. Steroid hormones

Only traces of the steroid hormones and their precursors are found in adrenals. With the micromethods now developed, however, their identification is possible in single glands. For example, for term human adrenals, KLOOS and STAEMMLER (1953) report 5.5 mg-% total reducing steroids. In adult rat adrenals, HOLZBAUER (1957a) identified by chromatography about 2 mg-% (23 μg/g) corticosterone. Total reducing steroids are far more concentrated (70 mg-% — DUNCAN 1957). Extensive discussion of the identification of the hormones in adrenocortical tissue is found in Chapter 2.

NEHER (see SIEBENMANN 1959) finds the proportion of the several hormones in human glands to be: cortisol 81%, corticosterone 13%, aldosterone 4%, and cortisone 2%.

WEGMANN (1956) has examined sections of human and rat adrenals with the infrared spectroscope and found conspicuous absorption bands at 1235 cm^{-1}, 1220.5 cm^{-1} and 1017.7 cm^{-1}, which are characteristic for specific groups within the steroid hormones.

IV. Phosphatides

Another conspicuous constituent of adrenal lipide extracts is *phosphatide*, which *typically comprises about* 4% *of the wet weight of the gland* (Table 10).

Table 10. *Phosphatide content of mammalian adrenal glands*

Species	Phosphatide concentration		Reference
	% Total lipide	% Wet wt. of gland	
Homo sapiens	16	3.3	E. ADAMS and BAXTER 1949
Cavia cobaya	35	6	KNOUFF et al. 1941
Rattus norwegicus	30	3.8	KARNOVSKY and DEANE 1955
Oryctolagus cuniculus	27	3.3	MACLACHLAN et al. 1941
Bos taurus	50	3	J. B. BROWN et al. 1937

Undoubtedly most of the phosphatides occur in various cellular membranes — plasma membranes, nuclear membranes, mitochondrial membranes, Golgi membranes, membranes of endoplasmic reticulum. But there is some histological evidence, based on staining sections with hematoxylin after fixation in formalin and potassium dichromate (e.g., the Baker acid-hematein technique), suggesting that some phosphatides may occur in the free droplets as well (HOERR 1936b, DEANE and GREEP 1946, HARRISON and CAIN 1947, CAIN and HARRISON 1950, KNIGGE 1954a, KARNOVSKY and DEANE 1955). BELT (1958) has recently presented evidence that droplets develop within membranes (Fig. 25, p. 33), and these may account for the appearance of the phosphatide reaction in some droplets visible at the light-microscope level. The large granules may, however, be large mitochondria.

In the rat adrenal, the special class of phosphatides termed plasmalogens (see below) occupies only about 0.3 % of the wet adrenal weight; in consequence they comprise only about 10% of the adrenal phosphatides in this species. In all likelihood, most of the adrenal phosphatides consist of lecithin and related compounds.

V. Plasmalogens and pseudoplasmals

This topic requires introduction by reference to the histological literature. In 1924, FEULGEN and VOIT described chemical and histochemical methods for a newly recognized lipide — a phosphatide that on cleavage yields a fatty aldehyde. The released fatty aldehyde is termed *plasmal*, the parent phosphatide, *plasmalogen*. Recent evidence indicates that *plasmalogen contains* not a fatty aldehyde but rather *an α-β unsaturated ether of glycerol*. On treatment with acid this unsaturated ether cleaves to yield an aldehyde (RAPPORT and FRANZL 1957).

Histologically, FEULGEN and VOIT and subsequent workers showed that compounds of this nature can be identified with a carbonyl reagent (e.g., the Schiff reagent) in *unfixed* sections only (VERNE 1929, VOSS 1940, PISCHINGER 1942, HAYES 1949, CAIN 1949a, b, NICANDER 1951, FETZER 1953). The aldehyde is freed by treatment with an $HgCl_2$ or HCl solution; without pretreatment no aldehyde reaction occurs. Control sections are first extracted with an organic solvent. When this method is applied to adrenals, the medulla, in most species, is far more reactive than the cortex. In the cortex, the reaction occurs in the hyaloplasm rather than in the droplets. And *any sort of fixation, especially by formalin, tends to destroy plasmalogen.*

The concentration of plasmalogen in adrenals is not high. MINDER and ABELIN (1954) calculated that the concentration in rat adrenals is about 0.2% of the wet weight. KARNOVSKY and DEANE (1955) found about 0.38% [recalculated on the basis of formulation of RAPPORT and FRANZL (1957)]. From the histological evidence of more intense reactivity in the medulla than in the cortex, it seems probable that most of the plasmalogen derives from the former tissue. In consequence, it is not surprising that no correlation is evident between adrenocortical activity and concentration of plasmalogen (see data of MINDER and ABELIN 1954).

What we will term here the *pseudoplasmal reaction* was described first by VERNE (1929) in the lipide droplets of the adrenal cortex. He noted that sections exposed to any oxidizing agent acquired direct reactivity to the Schiff reagent, i.e., no pretreatment with a hydrolyzing agent was required. He attributed *this reaction to aldehydes produced by the oxidation of unsaturated lipides.*

Apparently unaware of the studies of FEULGEN and of VERNE, BENNETT (1940) applied phenylhydrazine to fixed, frozen sections of cat adrenals and attributed the lipide reactivity to *ketosteroids*. Undoubtedly the consideration most influencing BENNETT's deduction was the then recent recognition of the ketonic nature of adrenocortical hormones. BENNETT's study initiated numerous investigations of the adrenal and other steroid-producing organs with the aim of following the formation and secretion of ketosteroids (review, DEANE and SELIGMAN 1953). In general, frozen sections of *fixed* tissues and any of several carbonyl reagents (the Schiff reagent, phenylhydrazine, or 2-hydroxy-3-naphthoic acid hydrazide) were used. No hydrolyzing agent was required. The droplet reactivity waxed and waned in parallel with the presumed hormone content of the cells.

Abundant evidence was early adduced against the hypothesis that the reactive carbonyls are ketosteroids. The first suggestion was that the reactive compounds are plasmals. When it was recognized that the usual treatment of the sections destroys true plasmalogen, it was suggested that *the droplet reactivity is attributable to aldehydes arising from the autoxidation of unsaturated fatty-acid moieties* (= products of rancidity) (CAIN 1949a, b, LISON 1951, 1953, GOMORI 1952a, b). CAIN termed these *pseudoplasmals*. Chemical evidence for this hypothesis was provided by KARNOVSKY and DEANE (1955), who showed that during fixation ethylene groups disappear while aldehydes increase. Both changes could be prevented by adding a phenolic antioxidant to the fixative (Fig. 42).

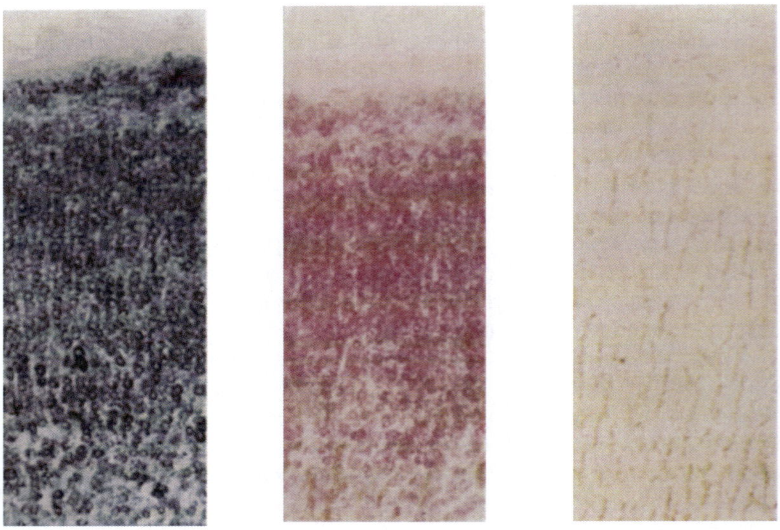

Fig. 42. Segments of rat adrenal cortices to show the pseudoplasmal reaction of the lipide droplets. *a* and *b* fixed 1 week in 10% neutral formalin; a stained with Sudan black B, b stained with the Schiff reagent. c fixed for 1 week in 10% formalin containing 0.1% hydroquinone, stained with the Schiff reagent. All × 200. From the study by KARNOVSKY and DEANE (1955). With ordinary fixation, the lipide droplets develop reactive aldehydic substances. When an antioxidant is included in the fixative, rancidity does not occur

The data of KARNOVSKY and DEANE (1955) reveal that *adrenal lipides contain high concentrations of unsaturated fatty acids*. Thus, in rats, the iodine value of adrenal lipides is about 110, whereas that of lipides extracted from white adipose tissue is about 90. This fact apparently accounts for the greater intensity of the pseudoplasmal reaction in adrenal lipides than in lipide droplets in many other loci (DEANE 1958). Similar conclusions were reached by BELT and HAYES (1956), who exposed tissue to ultraviolet light to accelerate autoxidation. Usually it is the smaller droplets that display the more intense reactivity.

The secondary carbonyl reaction has been described for the droplets in glands of various mammals and also in those of lower vertebrates (e.g., ABOIM 1944).

The conspicuous *osmiophilia* of adrenocortical lipide droplets (e.g., HOERR 1936a, FLEXNER and GROLLMAN 1939) also suggests their unsaturation. Like the pseudoplasmal reaction, osmiophilia of the droplets changes with glandular activity.

VI. Lipofuscins

Brown deposits of insoluble lipides, so-called lipofuscins, are present in adrenocortical cells of some vertebrate species, particularly in old individuals.

Species in which such pigments are conspicuous include the domestic birds *Columba* (R. Miller and Riddle 1942) and *Gallus* (Payne 1952), and the mammals *Homo* (Rogers and Williams 1947, Celestino da Costa 1951, Lüders 1953, Bachmann 1955, 1958, Verne et al. 1956), *Cavia* (Vaccarezza 1945, Ito 1952, Bimes et al. 1955, Planel et al. 1956), and *Mus* (Chester Jones 1948, 1950, Alpert 1953, Bern et al. 1958). In old mice, pigment-laden cells often form juxtamedullary islands, referred to as "brown degeneration." Smaller accumulations of pigment have been described in the adrenals of old rats (Jayne 1957) and old golden hamsters (Ito 1956, Meyers and Charipper 1956).

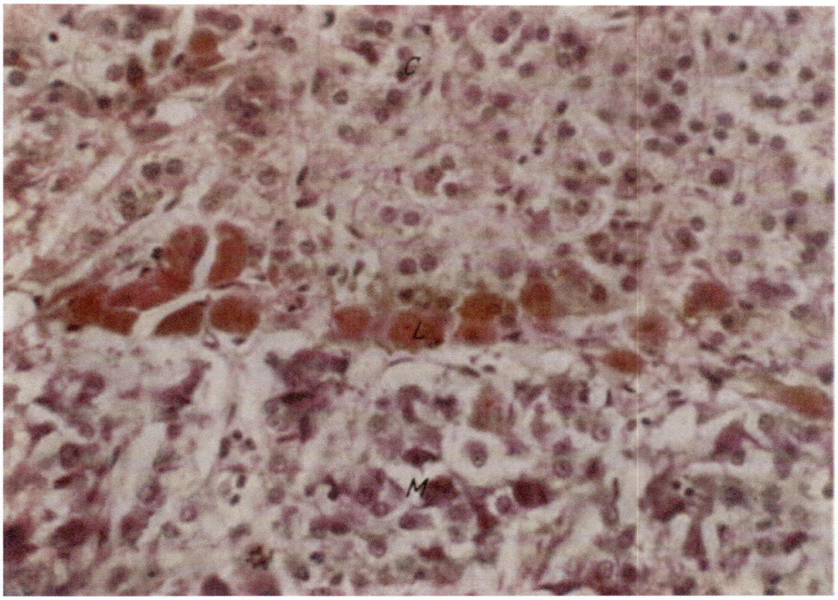

Fig. 43. Lipofuscin in the juxtamedullary cells of the zona reticularis in a human adrenal. *C.*, adrenocortical tissue; *L.*, lipofuscin; *M.*, medullary tissue. Hematoxylin and eosin. × 310. Photographed from a preparation contributed by Dr. Boris Gueft

Lipofuscins have been studied only histologically. Their lipide nature is indicated by their sudanophilia. Verne et al. (1956) report that they are extracted by prolonged treatment with pyridine at 56°. Being insoluble in most organic solvents, however, they persist in paraffin sections and may be stained with oil-soluble dyes in these as well as in frozen sections. *Model studies suggest that lipofuscins are oxidized, polymerized products of unsaturated lipides*, representing the biological equivalent of linoleum (Pearse 1953, Lillie 1954, Deane 1958). It is not surprising that they develop in cells whose lipides are highly unsaturated, like those of the adrenal cortex.

While the staining properties of lipofuscins change as the oxidative process proceeds, they typically display the following reactions: strong basophilia (acid-fastness); reactivity with the Schiff reagent following pretreatment with periodic or performic acid; the capacity to reduce ferric ferricyanide (Schmorl test), and fluorescence attributable to conjugated ethylenic groups. In steroid-secreting cells, cholesterol is frequently trapped in the pigment granules (Tobin and Birnbaum 1947, Deane and Fawcett 1952, Meyers and Charipper 1956, Deane

1958). Proteins and iron pigments are also commonly associated with the lipide granules.

In mammalian adrenals, lipofuscins develop with age. They occur principally in the zona reticularis (Fig. 43), where they are found both in the parenchymal cells and in the phagocytic reticuloendothelial cells. In old birds and guinea pigs, the marked accumulation of lipofuscins gives the gland as a whole a brown coloration. The structure of lipofuscin bodies in electron micrographs has been described by ZELANDER (1959).

C. Ascorbic acid

Mammalian adrenal glands were among the organs found to contain relatively high concentrations of ascorbic acid by SZENT-GYÖRGYI. By histological determination, a high concentration apparently occurs in all mammalian steroid-secreting cells (see A. GIROUD and LEBLOND 1935). However, ascorbate is not so highly concentrated in the adrenocortical tissue of many lower vertebrates.

Chemical method. In mammalian cortices, ascorbic acid exists almost entirely in its reduced form, with little or none of its oxidation products being detectable (GLICK et al. 1953). The technique employed most generally in recent years has been that of ROE and KUETHER

Table 11. *Concentration of ascorbic acid in the adrenal glands of various vertebrates*

Species	Average ascorbic acid concentration, mg-%	Reference
Rattus norwegicus		
adult ♂	540	FORTIER et al. 1950
weanling ♂	465	
nonpregnant ♀	370	KNOBIL and BRIGGS 1955
pregnant ♀	315	
fetus, term	315	
pregnant ♀	340	A. COHEN 1958a
fetus, 17½ days	185	
fetus, 19½ days	248	
fetus, 21½ days	325	
laboratory rat, ♂ and ♀	380	WOODS 1957a
wild rat, ♂ and ♀	360	
Mus musculus	258	ELTON et al. 1959
Mesocricetus auratus	310	W. N. HOLMES 1955a
	143	ELTON et al. 1959
Cavia cobaya	140	SAYERS et al. 1946
Felis domestica	120	ELTON et al. 1959
	87	SCHWARTZ and KLING 1960
Canis familiaris	135	J. H. BROWN et al. 1956
	171	ELTON et al. 1959
Sus scrofa	200	JOHANNES et al. 1956
Bos taurus	150	ROSENFELD 1955
Didelphis virginiana	210	ELTON et al. 1959
Gallus domesticus		
embryo	170	CASE 1952
14-day-old chick	175	JAILER and BOAS 1950
cockerel, 9 weeks	121	A. HOWARD and CONSTABLE 1958
pullets	305	PEREK et al. 1959
laying hens	222	
Colinus virginianus	150	ZARROW and BALDINI 1952
Bufo arenarum	22	PASQUALINI et al. 1949
Rana sp.	40	ELTON et al. 1959
Anguilla anguilla	65	HATEY 1954a, b
Salmo gairdnerii	11	HATEY 1952

(1943). Results with chemical methods employed earlier (see A. GIROUD and RATSIMAMANGA 1942) were generally about one-half those obtained with the Roe and Kuether technique.

The only adrenals that have been chemically analyzed by many different workers for ascorbic acid are those of laboratory rats, where *its concentration has proved a sensitive, though indirect, assay for the degree of glandular stimulation* (SAYERS 1950; see p. 84). In various strains of rats, the adult gland has been found normally to contain between 350 and 500 mg per 100 g fresh tissue (e.g., COVIAN 1949, GERSHBERG et al. 1950, FORTIER et al. 1950, GERSCHMAN and FENN 1954, GUILLEMIN 1955, ZIZINE 1956). In fetal and neonatal rats, the concentration is generally somewhat lower than in adults (JAILER 1950, J. JONES et al. 1953, KNOBIL and BRIGGS 1955, RINFRET and HANE 1955, A. COHEN 1958a).

Some representative results obtained in various vertebrates are cited in Table 11.

Of course, the ascorbate level in the adrenals of any species requiring the compound as a vitamin varies with nutritional state. For example, in young male guinea pigs, DESMARAIS and GAGNON (1955) found the concentration to be about 50 mg-% in animals receiving 5 mg sodium ascorbate *per os* per day, 120 mg-% in those receiving 150 mg. The adrenals of human beings dying accidentally and quickly (Finland) were reported to contain an average of only 29 mg-% (UOTILA and PEKKARINEN 1951). But since these adrenals also contained very little cholesterol, it seems unlikely that they were "normal."

The distribution of ascorbic acid in the adrenal cortex has been studied by the *quantitative technique* of Lindenstrøm-Lang (GLICK and BISKIND 1935a, GLICK et al. 1953, BAHN and GLICK 1954a, b). In the glands of *Macaca, Rattus* and *Bos, the concentration* per section *is relatively low in the zona glomerulosa, very high in the zonae fasciculata and reticularis*, low again in the medulla. A similar pattern was reported by A. GIROUD and RATSIMAMANGA (1942) for the horse adrenal. GLICK and BISKIND (1936) reported a rising titer in the cortices of fetal cattle with increasing age. A. GIROUD and MARTINET (1941) found ascorbate in the "fetal" zone of the human adrenal.

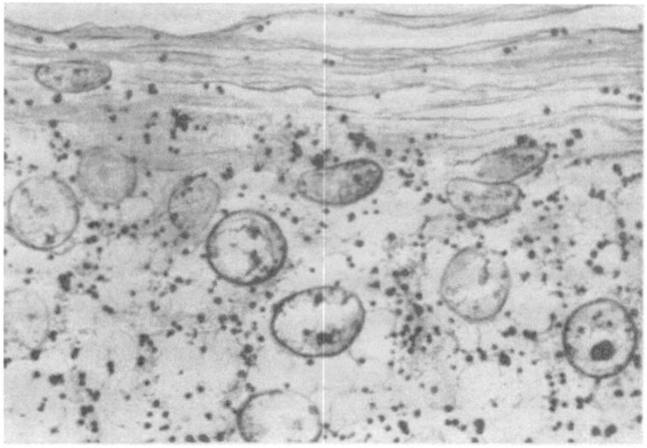

Fig. 44. Camera-lucida drawing of the outermost portion of a rat adrenal gland, showing the distribution of silver precipitated by the action of ascorbic acid. Granules lie scattered in the connective tissue capsule, but are far more abundant in the adrenocortical cells. Gland fixed in an alcoholic acid-$AgNO_3$ solution and treated with sodium thiosulfate; paraffin section counterstained with paracarmine; drawn by ETTA PIOTTI. × 1200. Reproduced by permission of the publisher from Fig. 1, DEANE and MORSE: Anat. Rec. **100**, 139 (1948)

Histologically, the distribution of ascorbic acid in the adrenals of various mammals conforms with that obtained with the quantitative technique.

Method. Ascorbic acid can be visualized because of its capacity to reduce $AgNO_3$ at acid p_H (about 3.5). As far as is known, there are no other substances present in biological specimens capable of reducing activity at such low p_H (REINER 1952). When blocks of tissue or fresh-frozen sections are immersed in an alcoholic acid $AgNO_3$ solution (BARNETT and BOURNE 1940), granules of silver are readily deposited in the adrenal cortex (Fig. 44). Unreduced $AgNO_3$ is removed by treatment with sodium thiosulfate; there appears to be danger of some artifactitious deposition of silver if the tissue is not washed thoroughly before this treatment (HAASE 1952, SOSA 1952, ERÄNKÖ 1954a). REINER (1952) found in model tests that the method was consistently positive only when the concentration reached 70 mg-%, although it was sometimes positive with a concentration of 45 mg-%.

The possibility of accurately localizing the ascorbate intracellularly seems remote. All evidence suggests that most ascorbic acid exists unbound in cytoplasm and that it diffuses readily (Barnett and Fisher 1943, Reiner 1952, Eränkö 1954a, Bourne 1955). The early conclusions that it is associated with the Golgi apparatus or mitochondria (Bourne 1942) must therefore be discounted. The finding of precipitate within the capillaries (Fig. 45) (e.g., Deane and Morse 1948, Bacchus 1951) must also be viewed with skepticism, since the small molecule can probably diffuse through damaged cell membranes. Although ascorbic acid may in fact be secreted into the blood (Slusher and Roberts 1957, Briggs and Toepel 1958), its concentration there appears to be only about 1—2 mg-%, an amount too low to demonstrate by this method.

Fig. 45. Distribution of reduced silver in a rat adrenal cortex. Somewhat less occurs in the glomerulosa (*G.*) than in the fasciculata (*F.*); granules occur in the sinusoids (*S.*) (capillaries) as well as within the cells. Gland prepared as in Fig. 44. × 400. From the study by Deane and Morse (1948)

In most mammals, the amount of silver precipitate is greater in the cells of the fasciculata and reticularis than in those of the glomerulosa (Fig. 45) (*Homo* — Wallraff 1949; *Rattus* — Hoch-Ligetti and Bourne 1948, Deane and Morse 1948, Cater and Lever 1954, Eränkö 1954a, Frazão 1954; *Cavia* — Haase 1952, Roepke 1952, Burns and Hale 1959; *Mus* — Leblond and Gardner 1938, Kar 1953; *Cricetomys* — Quenum and Camain 1959). Often it appears lacking in the glomerulosa, a reflection of the low sensitivity of the method. In *Mus*, precipitate occurs in the Xzone as well as in the cortex proper. There is far less reducing activity in fetal and neonatal glands (*Rattus*) than in the adult gland (Frazão 1954, Lever 1955b).

In other classes of vertebrates, reducing activity resulting in the precipitation of fine silver granules has been reported for the adrenocortical tissue of several birds and reptiles (Knab 1942, Barnett and Bourne 1940, A. Dawson 1953, Kingsbury et al. 1955, Wright and Chester Jones 1957). The amount appears generally lower than that encountered in mammalian adrenal cortices. Chieffi (1955) reported sparse granulation in the interrenal tissue of adult specimens of *Rana esculenta*. Ascorbate is not histologically detectable in teleost interrenal tissue (Chester Jones 1957), but is demonstrable in elasmobranch glands (Grynfeltt et al. 1957).

D. Enzymes

The enzymes found in adrenocortical tissue fall into three classes, (1) *those involved directly in synthesis of steroid hormones*, (2) *those found generally in active cells*, and (3) *those found variably in adrenocortical and other sorts of cells*. The first category includes enzymes involved in steroid synthesis from acetate and the dehydrogenases and hydroxylases involved in specific conversions of precursors to adrenocortical hormones. The second class includes glycolytic enzymes, Krebs-cycle enzymes, hexose-monophosphate shunt enzymes, constituents of the electron transport system and protein-synthesizing enzymes. The activities of some of these contribute, indirectly at least, to hormone synthesis (e.g., ROSENFELD 1956b). The third class includes an assortment of hydrolytic enzymes, such as non-specific alkaline and acid phosphatases, sulfatases, β-glucuronidase, and esterases, the precise importance of which remains obscure.

I. Steroid-synthesizing enzymes

The present consensus about the sequence of steps involved in the synthesis of adrenocortical hormones is presented in Chapter 3. The best-characterized enzymes are the dehydrogenases responsible for oxidizing the 3β-hydroxyl group and the 20β-hydroxyl group the early pregnenediol-like precursor and the three hydroxylases responsible for oxidizing C-11, C-17 and C-21 of progesterone, the probable immediate hormone precursor.

Steroid-3β-ol dehydrogenase does not require aerobic conditions and is a diphosphopyridine nucleotide (DPN)-dependent enzyme (SAMUELS et al. 1951, RUBIN and DORFMAN 1957). According to BEYER and SAMUELS (1956), this enzyme is concentrated in the microsome (ribonucleoprotein) fraction of homogenates. The *20β-ol dehydrogenase* is still illdefined (NABORS and BERLINER 1957).

WATTENBERG (1958) and LEVY et al. (1959a, b) have shown that the distribution of steroid-3β-ol dehydrogenase can be demonstrated *histochemically* by incubating unfixed tissue sections in a medium containing a 3β-hydroxy steroid (dehydroepiandrosterone), DPN and a tetrazolium salt. The reduced indicator, formazan, appears only in the cortical cells of the adrenal gland, indicating that no utilization of endogenous substrates occurs under the conditions used (Fig. 46). The enzyme is highly labile and does not survive fixation in most commonly used fluids, although it remains quite active after treatment of the sections with cold acetone.

Fig. 46. Visualization of the distribution of steroid-3β-ol dehydrogenase in the adrenal cortex. Fresh frozen section of rabbit gland first extracted with acetone and then incubated in a medium containing dehydroepiandrosterone, DPN, nicotinamide and Nitro-Blue Tetrazolium, pH 8.0. Micrograph contributed by Dr. LEE WATTENBERG. × 42. All adrenocortical cells exhibit deposits of formazan; no deposits occur in the capsule or in the medulla

The *hydroxylases* all require molecular, or active, oxygen for activity (HAYANO and DORFMAN 1954, ROSENFELD 1956a). 11β-Hydroxylase apparently resides in mitochondria (HAYANO and DORFMAN 1953, BROWNIE and GRANT 1956). 21-Hydroxylase has been associated both with the microsomal fraction (RYAN and ENGEL 1957) and with the mitochondrial fraction (VILLEE et al. 1959) of homogenates.

Several synthetic enzymes that require *Coenzyme A* (a constituent of which in the B vitamin, pantothenic acid) are believed to be directly involved in the synthesis of lipides, particularly steroids, from acetate (NOVELLI 1953, K. BLOCH 1957, VILLEE et al. 1959). The concentration of Coenzyme A appears to be higher in adrenals (rabbit, rat) than in any organ but liver.

MALMSTRÖM and GLICK (1954) studied the distribution of Coenzyme A in rat adrenals by the Linderstrøm-Lang method. In terms of activity per unit volume of tissue, the enzyme showed peaks of concentration in the outer and inner portions of the fasciculata, with a lower concentration between. Perhaps determinations made in terms of protein would be more meaningful because of the variable lipide content of the cortex. More recently, L. GREENBERG and GLICK (1958) have shown a *diurnal cycle* in Coenzyme A level in the rat cortex, with a higher concentration being present at night than in the daytime.

II. Oxidative enzymes

Chemical methods. General metabolic activity in the adrenal has been assessed an various ways. The Q_{O_2} of various adrenals (beef, guinea pig, rat) without added substrate is between 1 and 3, lower than in kidney, liver or cardiac muscle (SOURKES and HENEAGE 1952, J. GORDON 1957). Similarly, the content of cytochrome oxidase and cytochrome *c* is lower than in muscle, liver, kidney and brain (DANNENBERG 1959).

Krebs-cycle intermediates, such as succinic acid, stimulate O_2 uptake 5 to 10 fold (beef, guinea pig, and rat adrenals) (SOURKES and HENEAGE 1951, 1952, KATSURA 1955, J. GORDON 1957). Homogenates of *isolated cortical* tissue show higher respiratory activity than those of isolated medullary tissue (bovine gland), suggesting that most of the activity observed in slices or homogenates of the whole gland is attributable to cortex (SOURKES and HENEAGE 1952).

Comparable results were obtained by ROSENFELD (1956a), using triphenyltetrazolium chloride as hydrogen acceptor with calf (*Bos*) adrenals perfused anaerobically. Tetrazolium reduction occurs at about the cytochrome *c* level in the oxidative sequence (NOVIKOFF 1959). Formazan deposition appeared distinctly greater in the cortex than in the medulla. Glucose, fructose-6-phosphate, citrate, fumarate, succinate and α-ketoglutarate all enhanced formazan production signficantly over that obtained when no substrate was included in the perfusion medium (calculated in terms of mg formazan per g dry tissue). Amino acids and fatty acids failed to enhance formazan deposition significantly.

KALANT and SELLERS (1954, 1957) have reported that while succinoxidase activity (Warburg technique) is similar for adrenals of young rats of the two sexes, it is greater in adrenals of adult males than in those of adult females. Glucose oxidation was also greater in male than female adult glands, though the difference was not so conspicuous.

ALLEN (1959) has demonstrated a 3—4 fold increase in DPNH oxidation per mg nitrogen by the mouse adrenal in the period from birth to maturity (90 days). More occurs in the female gland than in the male.

The one dehydrogenase that has been shown to be unusually active in adrenals as contrasted with many other organs *is glucose-6-phosphate dehydrogenase*, a triphosphopyridine nucleotide (TPN)-dependent enzyme (GLOCK and McLEAN

1954, KELLY et al. 1955). The latter authors note that TPNH is required for 11β-hydroxylation and suggest thar glucose-6-phosphate may provide the energy. (See also VILLEE et al. 1959.)

KRISCH and STAUDINGER (1958) have identified cytochrome b_5 in adrenal microsomes.

Quantitative histochemical methods. GLICK and co-workers (JARDETZKY and GLICK 1956, GLICK and NAYYAR 1956, GLICK and GREENBERG 1958) have determined the distribution of the succinic dehydrogenase system quantitatively in the gland, using a tetrazolium as hydrogen acceptor. They measured the amount of formazan produced per 16μ standard-bore section in a medium containing succinate. In the steer (*Bos*) gland, maximal activity occurred in sections from the mid-fasciculata. In the rat gland, maximal activity occurred in the reticularis. In the first 2 studies activity was calculated in terms of volume of tissue, in the last in terms of protein N.

Visualization methods. ZWEIFACH and co-workers (ZWEIFACH et al. 1951, CASCARANO and ZWEIFACH 1955) have visualized the location of formazan deposit in thick slices of rat adrenals (not previously frozen) incubated with tetrazolium salts in a medium to which no substrate was added. This procedure presumably reveals the utilization of endogenous metabolites. Much more deposition occurred in the cortex than in the medulla. Formazan appeared most abundant in the glomerulosa and outer half of the fasciculata. The reticularis was pale, and the subglomerulosal zone was virtually inactive. With succinic acid added as substrate, essentially similar distribution was obtained by PADYKULA (1952) and CASCARANO and ZWEIFACH (1955); formazan was of course more abundant. NEUMANN and KOCH (1953) have described the distribution of succinic dehydrogenase in sections of dog and guinea pig adrenals, and SYMINGTON et al. (1956) have demonstrated the dehydrogenase activity of human adrenals.

F. ROSSI et al. (1957) have examined the glands of a few human fetuses (3 mo., 4.5 mo., 5 mo.) for succinic dehydrogenase activity, using Blue Tetrazolium as indicator. Moderate activity was present in the adrenals of all three specimens.

The *tetrazoles* employed in all of the above studies yield formazans readily soluble in lipides (e.g., triphenyltetrazolium chloride, Neotetrazolium, Blue Tetrazolium), so that the formazan becomes dissolved in the lipide droplets rather than being deposited in the cytoplasm (DEANE et al. 1960a). Recently NACHLAS et al. (1957) have developed a *p*-nitrophenyl substituted ditetrazole (Nitro-BT): its formazan binds to protein and hence does not readily diffuse into the lipide droplets. With this, succinic dehydrogenase activity was found to be most conspicuous in the reticularis, less so in the outer zones (rat, dog). The subglomerulosal zone was less active than either the glomerulosa or the outer fasciculata (dog).

Several reports have appeared on the distribution of various DPN-dependent dehydrogenase systems in rat, rabbit, mouse, guinea pig, and hamster glands (WATTENBERG 1958, ALLEN 1959, LEVY et al. 1959a, b). In most species activity is greater in the cortex than in the medulla (lactic dehydrogenase system) (Fig. 55, p. 97). These studies all utilized Nitro-BT.

R. COHEN (1959) has adapted the visualization technique to demonstrate the *glucose-6-phosphate-TPN dehydrogenase system,* which he found most concentrated in the fasciculata and reticularis (rat). Little occurred in the glomerulosa, and none appeared in the subglomerulosal zone. (With TPNH as substrate, activity was greatest in the glomerulosa and fasciculata, with little occurring in either the subglomerulosal zone or the reticularis.) He has correlated the distribution of this enzyme with the presumed association of TPNH with hydroxylation of the parent steroid molecule (see above).

III. Hydrolytic enzymes

The great majority of studies on adrenal enzymic activity have concerned hydrolases, notably nonspecific alkaline and acid phosphatases and nonspecific esterase. *The significance of most of these enzymes in the bodily economy remains obscure.* For the most part these studies have utilized visualization rather than chemical techniques. Visualization techniques for enzymes are reviewed by DEANE et al. (1960a; see also NOVIKOFF 1959).

1. Nonspecific alkaline phosphatase

Because of the early development of methods for demonstrating this enzyme and because of their relative simplicity, studies have been made on adrenals of numerous species. Most of these have been made on ethanol- or acetone-fixed, paraffin-embedded sections — pretreatments that certainly reduce enzymatic activity, perhaps differentially toward different substrates. Only NICANDER (1952), EGER et al. (1953) and ALLEN and SLATER (1956) have used frozen sections

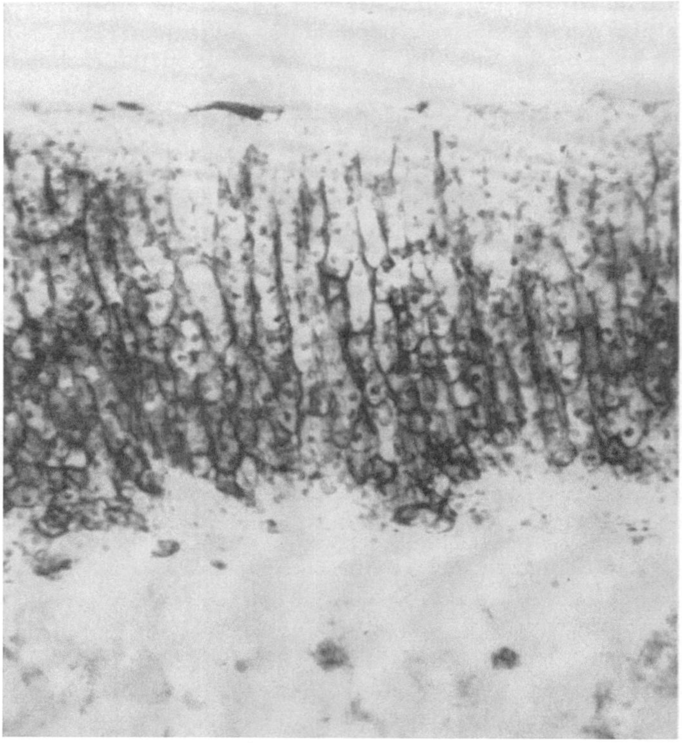

Fig. 47. Distribution of nonspecific alkaline phosphatase in the adrenal of a male mouse. Frozen section of gland fixed overnight in cold neutral formalin incubated for 4 hours in a glycerophosphate-calcium medium. Micrograph contributed by Dr. JOHN M. ALLEN. × 150. The principal activity occurs in the capillary walls of the inner cortical zones. The precipitate in cytoplasm and nuclei in the reticularis is probably artifactitious

of unfixed tissue or of tissue fixed only briefly in cold formalin, wherein activity remains high. Furthermore, in most methods used there is marked tendency for migration of product, so that cellular localization remains questionable.

The most conspicuous fact to emerge from reviewing the large literature on nonspecific alkaline phosphatase in adrenocortical tissue is the tremendous variability in the occurrence and location of activity in different species. In general, apparent concentration is not high. The enzyme is extremely active in the glomerulosa of *Bos*, *Equus*, and *Cavia* (NICANDER 1952, EGER et al. 1953, YOFFEY 1955, PLANEL et al. 1957, BURNS and HALE 1959). This activity appears to reside both in the capillary walls and in the parenchymal cells. *In many* other *species*, however, *nonspecific alkaline phosphatase has much lower activity in the cortex and is located primarily in the capillary walls* (Fig. 47) (*Homo* — YOFFEY 1955, SYMINGTON et al. 1956; *Rattus* — DEMPSEY et al. 1949, NICANDER 1952, EGER et al. 1953; *Mus* — ELFTMAN 1947, ALLEN and SLATER 1956; *Mesocricetus*, *Oryctolagus*, *Sus*, *Ovis*, *Felis* and *Canis* — NICANDER 1952; *Cricetomys* — QUENUM and CAMAIN 1959). Diffusion of product to the parenchymal-cell nuclei is frequently observed with the metal-salt method, however. The distribution of detectable activity in the capillary walls may be essentially uniform throughout the cortex (*Rattus*, *Felis*, *Canis*, *Ovis*), limited to the fasciculata (*Sus*), or maximal in the inner cortex (*Homo*, *Mus*, *Mesocricetus*, *Oryctolagus*). In the goat (*Capra*), the concentration is high in the glomerulosa and reticularis, low in the fasciculata (VERNE and HÉBERT 1951). More activity occurs in the glands of males than females in *Cavia*, *Mus*, *Mesocricetus*, and *Homo*; in other species examined there is no conspicuous sex difference.

In fixed rat glands, BOURNE (1954a, b) has observed localization in the capillary walls with glycerophosphate and assorted sugar phosphates as substrates, but principally in the parenchymal cells with nucleotides and chalcone diphosphates. These differences may merely reflect the slower rate of hydrolysis of the latter group of substrates, permitting diffusion of the product; but they may indicate the occurrence of specific enzymes.

In the adrenals of the pigeon, *Columba*, activity is most conspicuous in the inner portion of the gland and is described in the parenchymal cells (KAR 1950). In *Crocodilus*, alkaline phosphatase is abundant in the adrenocortical cells (ARVY and BONICHON 1958). In *Rana* activity is very slight (BEJDL 1955, TUCHMANN-DUPLESSIS et al. 1955).

2. Nonspecific acid phosphatase

Because of the obviously unsatisfactory nature of histochemical methods for acid phosphatase activity, few studies have been made with them. According to NICANDER (1952), who has performed the most extensive study using unfixed tissues, acid phosphatase is more consistently present in adrenocortical tissue than is alkaline phosphatase. In general, he found that distribution tends to be uniform in the different zones; precipitate occurs in the parenchymal cells rather than in the capillary endothelium. However, in *Equus* the activity is greatest in the glomerulosa; in *Sus*, *Mus* and *Cavia* greatest in the fasciculata; in *Ovis* absent from the glomerulosa. In *Homo*, SELIGMAN and MANHEIMER (1949) and SYMINGTON et al. (1956) have found most activity in the reticularis. With a quite satisfactory naphthyl phosphate substrate, RUTENBURG and SELIGMAN (1955) have found the rat adrenal to show greatest activity in the reticularis and fasciculata; the glomerulosa appears negative. F. ROSSI et al. (1957) report that acid phosphatase occurs in fetal human adrenocortical tissue.

ARVY and BONICHON (1958) observed none in the adrenocortical cells of crocodile glands.

3. Nonspecific esterase

This enzyme splits esters of carboxylic acids (especially acetic acid) and a variety of alcohols. With *homogenates* of whole mouse glands, ALLEN (1958) has shown a nearly 9-fold increase in esterase activity per g N from late fetal stages (17 days of gestation) to maturity (30—90 days). GLICK and BISKIND (1935b) demonstrated its presence throughout the bovine adrenal cortex by the *quantitative histochemical method.*

Several *visualization methods* for demonstrating esterase activity are available — a method utilizing Tween as substrate, methods ultilizing naphthyl acetate, and methods utilizing indoxyl acetate (DEANE et al. 1960a). With such techniques, the esterase activity of adrenocortical tissue has been shown to differ in the various species. Thus, NACHLAS and SELIGMAN (1949), using acetone-fixed, paraffin sections and 2-naphthyl acetate, reported weak activity throughout the rat cortex, activity only in the glomerulosa in rabbit and dog glands, and activity only in the reticularis of guinea pig and human adrenals. Using frozen sections, which retain far more activity, and more satisfactory substrates, subsequent workers have demontrated more conspicuous activity in rat adrenals (e.g., BARRNETT 1952, COUTINHO et al. 1955, PEARSON and DEFENDI 1957). With the Tween technique (relatively insensitive), no activity is demonstrable in the adrenals of adult rats (VERNE 1956), but the activity of glands of neonatal rats is conspicuous (HÉBERT and BARBARIN 1953). ALLEN (1957) has demonstrated moderate activity in all zones of the mouse gland, with somewhat more in the zona glomerulosa (frozen sections, l-naphthyl acetate) (Fig. 56, p. 97). In this species, activity in the cortex inereases with age, and more develops in the female than in the male (ALLEN 1958). The predominance of this enzyme in the reticularis of the human gland has been confirmed repeatedly (MALATY and BOURNE 1954, SYMINGTON et al. 1956). No esterase activity has been demonstrated in the adrenals of human fetuses (paraffin sections, F. ROSSI et al. 1957). A trace occurs in the adrenocortical tissue of the reptile, *Crocodilus* (ARVY and BONICHON 1958).

4. β-Glucuronidase

This enzyme is capable of splitting glucuronides of various phenolic and aliphatic compounds. It varies in concentration in different organs, and its level, especially in target organs of the sex steroids, depends on the degree of stimulation.

Chemically, the activity of this enzyme has been found to be moderately high in adrenal glands of rats (TALALAY et al. 1946, KNOBIL 1953), though lower than in liver, spleen, ovary, uterus, and thyroid.

By a *quantitative histochemical technique*, WATTENBERG and GLICK (1953) found this enzymatic activity of the bovine adrenal to be maximal in the glomerulosa and lower in the inner zones. In several other species (rat, guinea pig, monkeys), NAYYAR and GLICK (1955) observed that activity was maximal in the reticularis. In the rabbit gland, there was little activity, and this showed no zonal variation. In these studies, enzyme activity was calculated in terms of protein nitrogen, so that the differing lipide contents of the zones in the various species did not contribute to the discrepancies.

Enzyme activity has been *visualized* in rat adrenals by two different techniques, one employing 6-bromo-2-naphthyl glucuronide as substrate (SELIGMAN et al. 1954), the other employing 8-hydroxyquinoline glucuronide (FISHMAN and BAKER 1956). All layers of the cortex were moderately stained, while the medulla was only weakly stained. The section illustrated by FISHMAN and BAKER shows maximal activity in the reticularis.

5. Aryl sulfatase

The adrenals of some species possess an enzyme or enzymes capable of hydrolyzing the sulfates of aromatic compounds.

GLICK et al. (1955a) have studied its distribution by the *quantitative technique*, using potassium *p*-nitrophenyl sulfate as substrate. In rat, hog, and dog adrenals, considerable activity occurred in the fasciculata and reticularis, little in the glomerulosa or medulla. In rabbit adrenal, activity was essentially uniform throughout the cortex. In cow and monkey glands, virtually no activity was demonstrable.

Employing potassium 6-benzoyl-2-naphthyl sulfate and homogenates, RUTENBURG et al. (1952) also found considerable activity in rat adrenals. *Histologically*, this substrate was split most actively in the glomerulosa and fasciculata (formalin-fixed frozen sections). The azo dye produced in this method is soluble in lipides, so that the product was taken up by the lipide droplets.

It is perhaps relevant that the rat adrenal appears incapable of conjugating *m*-aminophenol and sulfate (GLICK and STECKLEIN 1955). Therefore it seems unlikely that the aryl sulfatase measured above acts synthetically *in vivo*.

No quantitative or visualization histochemical studies have yet been made on the *phosphorylase* content of the adrenal cortex. This enzyme may be intimately concerned with steroid synthesis (HAYNES 1958).

E. Miscellaneous compounds

I. Glycogen

The concentration of glycogen in adrenocortical tissue is relatively low in all species in which it has been determined (Table 12). NOBLE and PAPAGEORGE (1955a, b) found it unvarying in guinea pigs but to be twice as concentrated in the glands of fasted (8—24 hours) rats as in those of fed rats.

Table 12. *Glycogen content of mammalian adrenal glands*

Species	Average glycogen content, mg-%	Reference
Rattus norwegicus, fed	100	NOBLE and PAPAGEORGE 1955a
fasted	200	
Cavia cobaya	25	
Canis familiaris	80	UTEWSKI and EPSTEIN 1934b
Bos taurus	41	UTEWSKI and EPSTEIN 1934a

Histologically, glycogen is generally sparse in adrenocortical cells. In contrast, this polysaccharide is often abundant in the cytoplasm of adipose-tissue cells, particularly at times of rapid synthesis of triglycerides (FAWCETT 1952). After fixation of the tissue in mixtures of alcohol and picric acid and staining by the sensitive periodic acid-Schiff technique, PLANEL and GUILHEM (1956a, b, 1957b), GRAUMANN (1956), NICANDER (1957), BURNS and HALE (1959), HUNT and HUNT (1959), and QUENUM and CAMAIN (1959) demonstrated traces of glycogen in the glands of man, mouse, hamster, guinea pig, rat, giant rat, rabbit, immature dog, horse and pig. None has been detected in the glands of adult dog, ox, sheep or goat. In general there appears to be more glycogen in the glands of young animals than of older specimens (hamster — GRAUMANN 1956; guinea pig, rabbit, dog — NICANDER 1957; rat — HUNT and HUNT 1959).

True mucopolysaccharides not digestible with saliva apparently do not occur in adrenocortical cells. But it should be noted that *lipofuscins* (p. 57) and some *glycoprotein* colloid droplets (p. 133) may give the periodic acid-Schiff reaction.

II. Other chemical constituents

Water is said to comprise about 80% of the wet weight of adrenal glands in lipide-poor organs (ROSENFELD 1958b) and 70% in lipide-rich glands (MCSHAN et al. 1946). The protein content appears not to have been studied. Relatively *little ribonucleic acid* is present (see pp. 36, 91). JUNKMANN (1957) has reported the following values for *electrolytes*, in mg/100 gm whole fresh human glands: Na^+ 44.1, K^+ 103.1, Ca^{++} 15.6, Mg^{++} 9.9. Using the quantitative histochemical method, GLICK et al. (1955b) have determined the distribution of K^+ in the glands of monkey and guinea pig. And GLICK et al. (1957) found Mg^{++} present at a constant low value throughout the cortex, at a higher level in the medulla (guinea pig, rat).

In addition to ascorbic acid and pantothenic acid, which were mentioned earlier (pp. 58, 62), *vitamin A and riboflavin* (B_2) are relatively concentrated in adrenal glands as compared to most other organs (STARY 1956, JUNKMANN 1957, RUDY 1957). *Glutathione* is also present at a relatively high level (ca. 125 mg-%) (STARY 1956).

Physiology of adrenocortical tissue

A. Introduction

This section undertakes to summarize what is known about the secretory activities of adrenocortical tissue in the various classes of vertebrates and how those activities are governed. As in the preceding section, most available information pertains to mammals, and, among mammals, principally to the rat, cat, dog, and man.

The preliminary portions of this section review the overall physiology of the adrenal cortex and the methods of assessing its activity. Subsequent portions consider the control of the mammalian cortex in adults and in fetuses, the control of regeneration, the control of the mouse X zone, pathological alterations of adrenocortical cells, and the control of adrenocortical tissue in lower vertebrates.

B. Overall physiology of the adrenal cortex

I. General dependence on anterior-pituitary secretions

Subsequent sections discuss in detail the control by the anterior pituitary gland of the anatomical and secretory characteristics of the adrenal cortex. Suffice it to say here that *the volume, or weight, of the gland in all postnatal life is dependent upon the level of circulating adrenocorticotrophic hormone (ACTH).* After hypophysectomy, the adrenal cortex atrophies and the secretory rate remains extremely low in situations that normally enhance it. Of all the species so far studied, only *Bufo arenarum* (a South American toad) shows no gross atrophy of the cortical tissue after hypophysectomy.

The gross effects on the gland after hypophysectomy result not only from a deficiency in ACTH but probably also from deficiencies in *growth hormone* and in *thyroid hormones* (e.g., CATER and STACK-DUNNE 1955c, BOIS and SELYE 1957, LOSTROH and LI 1958, LOSTROH and WOODWARD 1958). Since sex steroids produced by the gonads also influence adrenal weight (p. 106 and p. 107), their absence after hypophysectomy, too, may contribute to the condition of the gland.

It must be emphasized that *even in the absence of the pituitary much of the adrenocortical tissue survives and functions to some degree.* The secretion of the hormone(s) influencing electrolyte distribution (mineralocorticoid(s)) persists only slightly reduced, and even the secretion of the hormones affecting organic metabolism (glucocorticoids) remains detectable.

In the presence of a functional pituitary, many nutritional and environmental factors may increase or decrease the size and possibly the function of the adrenal gland — in most instances by influencing the output of ACTH, but in some instances by direct influence on the adrenal itself, thereby modifying the effects of the pituitary hormone. Later sections will consider in some detail those factors that stimulate ACTH output and hence augment adrenocortical size and function, in what are generally termed "stress" reactions (p. 103). Here note is taken of some factors that may act in part by influencing the cortex directly.

II. Nutritional factors influencing adrenocortical tissue

1. Proteins

The size and responsiveness of the adrenal cortex is affected by numerous nutritional factors. Specifically, *minimum levels of protein in the diet are essential for normal growth and functioning of the gland.* Just what proportion of this influence is attributable to a direct action on the adrenal cortex and what proportion to indirect effects on the anterior pituitary and other organs remain to be determined, although the principal target seems to be the pituitary (MULINOS and POMERANTZ 1941, SLATER 1959). On the one hand, animals on high protein diets (> 20% by weight) have relatively large adrenals (SARASON 1943a, LEATHEM 1947) that are capable of greater than "normal" responsiveness in stressful situations (SKELTON 1951, STERLING and LONGWELL 1953b, DUMM et al. 1955). On the other hand, animals on diets containing little or no protein have smaller than "normal" adrenals (HORN 1955) and show no ability to respond to unilateral adrenalectomy by compensatory hypertrophy of the remaining gland (LEATHEM 1957). Indeed, if the diet contains no protein whatsoever, absolute shrinkage of the contralateral gland may occur. *Dietary proteins of low nutritional value similarly depress the responsiveness of the system* (RINALDI 1950, LEATHEM 1957). Just what constituents of proteins are of crucial importance is at present unknown. And how well weight correlates with hormone production by the gland has not been determined. But the amount and sort of protein must be taken into account in assessing the adrenals of animals or of patients suffering from chronic malnutrition (e.g., rats — JACKSON 1919; guinea pigs — D'ANGELO et al. 1948; humans — SCHABERG and STRAUB 1953). The glands may enlarge or not, depending on the balance of factors. Presumably the shrunken adrenals sometimes observed in anorexia nervosa may result from such generalized inhibition of the pituitary-adrenocortical system. The small and apparently less responsive adrenals found in Jamaican Negroes (STIRLING and KEATING 1958, STIRLING 1959) may be a result of dietary deficiency rather than genetic tendency.

Complete starvation, following either adequate nutrition or chronic malnutrition, *generally results in adrenal hypertrophy* (MULINOS and POMERANTZ 1940, 1941, BLUMENTHAL and LOEB 1942, SARASON 1943, SELYE 1946). Apparently as the result of this severe stress, proteins and other nutrients that are unavailable to the pituitary and/or adrenal in undernutrition are mobilized from other organs and tissues.

2. Carbohydrates

Nutritional factors other than proteins also influence adrenal responsiveness. Glucose administration or the feeding of a high-carbohydrate diet alters the effect of endogenous or exogenous ACTH on the adrenal. Thus, CONSTANTINIDES et al. (1950, SKELTON 1951) found that the oral administration of glucose after an injection of ACTH in hypophysectomized rats accentuated the depletion of cholesterol from the adrenal and the lysis of lymphocytes in the thymus, although it accelerated the restoration of ascorbic acid in the gland. Contrariwise, STERLING and LONGWELL (1953a, b) found that glucose or a high-carbohydrate diet to intact rats exerted a "sparing" effect on cholesterol depletion from the gland when the animals were exposed to an acute stress, as compared to a commercial diet or a high-protein diet.

3. Lipides

The effects of varying the level of triglycerides in the diet have not been analyzed.

A high level of dietary cholesterol appears to stimulate the adrenals. When a diet containing a high concentration of cholesterol was fed to rabbits, there was enlargement of the adrenals and a rise in the percentage of cholesterol in the glands (ca. 30% increase over the controls) (KAY and WHITEHEAD 1935). No studies have been made to determine whether adrenal function is enhanced or depressed. From physiological studies in other situations in which cholesterol is enhanced, VOGT (1957b) has come to the opinion that cholesterol level is directly correlated with secretory potential.

The highly unsaturated aliphatic hydrocarbon, *squalene*, thought to be one precursor of cholesterol (see Chapter 3), may also stimulate the adrenal. When squalene was injected at a dose of 0.2 ml per day into hypophysectomized rats receiving a standard dose of ACTH (1.5 U per day), the adrenals enlarged more than in controls and the thymus gland underwent greater atrophy (KLINE 1957, 1958). Since thymus weight was not influenced by squalene in adrenalectomized rats, the latter observation suggests enhanced secretory activity of the adrenals. No effect was noted in intact animals. Orally-administered squalene failed to stimulate the adrenals of hypophysectomized rats.

Several plant oils induce adrenal hypertrophy. *Rape seed oil* and *turnip seed oil*, when added to the diet of rats at a level of 25%, induced, in 4 weeks, considerable hypertrophy of the adrenal glands and a 3-fold increase in cholesterol concentration (CARROLL 1951). The active constituent of rape seed oil is *erucic acid* (CARROLL 1953), a long-chain fatty acid with one ethylenic group. Such adrenals showed the usual fall in ascorbic acid and cholesterol levels when the animal was given ACTH or subjected to stress (CARROLL and NOBLE 1952). In these studies, there were no peripheral signs of either increased or decreased adrenocortical function.

When erucic acid was given to rats for 7 days by subcutaneous injection (0.4 ml per day), there was similar relative adrenal enlargement, with marked thymic atrophy (KLINE 1957). Whether this atrophy was the result of "stressing" contaminants or indicated that erucic acid directly augments adrenal secretion as well as its cholesterol content remains to be determined.

Nervonic acid also induces adrenal enlargement (CARROLL 1953).

4. "Vitamins"

Ascorbic acid has also been suggested to modify adrenal responsiveness, not only in those species in which it is a vitamin but also in the rat. In the latter species, DUGAL and THERIEN (1949) first reported that the administration of ascorbic acid to stressed animals prevented the usual adrenal enlargement. In a later study (1952), they failed to confirm their earlier results, but found that ascorbic acid given to hypophysectomized rats potentiated the stimulating effect of

ACTH on adrenal weight. However, VOGT (1951) could detect no influence of added ascorbate on the secretory capacity of the isolated canine adrenal gland; KNOBIL and FREGLY (1955) failed to find any influence of administered ascorbic acid on the adrenals (weight or cholesterol concentration) of either hypophysectomized or intact rats; and DURY and DURY (1956) observed no effects of ascorbate on the adrenal or blood chemistry of intact rats given high doses of insulin.

Nevertheless, NICHOLLS (1956) has reported that the injection of ascorbate greatly enhanced the uptake of P^{32} into the inorganic and organic phosphates of the adrenals of intact (unstressed) rats and of the adrenals of hypophysectomized rats treated with a small amount of ACTH. Control injections of succinate had no such influence. Furthermore, DUMM et al. (1955) found that the addition of 1 g ascorbate per 100 g of diet increased the weight gain of the adrenals of stressed rats but inhibited the recovery of normal cholesterol levels following stress.

Glands of scorbutic guinea pigs apparently secrete supranormal amounts of corticoids, as indicated by eosinopenia (EISENSTEIN and SHANK 1951) and an augmented level of urinary corticoids (NADEL and SCHNEIDER 1952, PRUNTY et al. 1955, EISENSTEIN and WENNEKER 1958, NADEL et al. 1958). Moreover, these glands are capable of still further stimulation if ACTH is administered (A. HOWARD and CATER 1959). Nevertheless, these glands are not so "responsive" as the glands of healthy guinea pigs. And there is as yet ill-defined evidence that the secretory products of the scorbutic animal may not be entirely "normal" (e.g., NIGEON-DUREUIL and RATSIMAMANGA 1955, BACCHUS 1957a, b). Thus the question remains open as to whether a high circulating level of ascorbate influences adrenal function.

Other vitamins also influence adrenal reactivity. It now appears probable that pantothenic acid and possibly pyridoxine may act as coenzymes for normal synthetic activities of the adrenal cortex (p. 138 and p. 139). Therefore deficiencies in these vitamins depress secretory activity. Deficiencies of other vitamins appear to act as typical "stresses" (p. 103) and to enhance activity.

III. Other factors influencing adrenocortical tissue

Additional factors influence adrenal responsiveness to ACTH. Thus, in any attempt at assessing the functions of adrenocortical tissue, it must be kept in mind that there are marked species, age, and sex differences. The bases for these differences remain to be ascertained. This broad area has been sketched out only in the barest terms by studies so far undertaken, and this section can only raise the problems. Indeed, in most instances, studies have not clearly indicated whether the influences are on the adrenal cortex per se or on the output of ACTH.

1. Species differences

As will be described in greater detail in subsection C, *there are marked species differences in product* (e.g., BUSH 1953, HOLZBAUER 1957a, CHESTER JONES et al. 1959). The glucocorticoids produced are cortisol, corticosterone, and possibly cortisone. Human, dog, guinea pig, and fish glands produce *cortisol* and possibly cortisone as the major glucocorticoids; corticosterone is produced in relatively low concentration. Rat, mouse, chicken, and snake glands appear to produce *corticosterone* as the principal glucocorticoid. *Aldosterone* may be produced at much different concentrations in different species. And there appear to be considerable differences between species in the output of C_{19} steroids (potential *androgens*) from the adrenal (see Chapter 5). Whether these species differences reflect differences in overall bodily metabolism and hence in requirements for hormones or different patterns of enzymes within the adrenal cells remain open questions.

2. Age differences

In man, at least, shifting patterns of product, both qualitatively and quantitatively, occur with aging (Gallagher 1956/57, Pincus 1956). This difference applies especially to adrenal androgens. Glands with a "fetal" zone form significant quantities of C_{19} steroids (weak androgens) (e.g., E. Bloch et al. 1956, Lanman and Silverman 1957, Solomon et al. 1958). Furthermore there appears to be a postpuberal rise in androgen output (Chapter 5).

The capacity to secrete glucocorticoids appears much lower in the glands of fetuses and new-born infants than in those of children and adults (e.g., Klein et al. 1954). In other species, likewise, there is indication of low hormone production in infancy. In both man and other mammals, however, this low postnatal output seems to reflect a deficiency in ACTH secretion rather than some lack within the adrenal cortex (p. 125).

It is perhaps noteworthy that in rats there is apparently no loss of responsiveness to ACTH, as judged by compensatory hypertrophy, in old age (Verzár and Hügin 1957).

3. Sex differences

In most species (established exceptions, hibernating rodents) *the adrenals of males are smaller than those of females.* In large part, this difference seems to depend on the fact that estrogens stimulate the output of ACTH and thus probably the output of corticoids (p. 106). In the intact animal, androgens generally result in a reduction in adrenal weight (p. 107). However, in one respect at least, androgens stimulate the adrenal: Succinic dehydrogenase activity is enhanced (rat) (Kalant 1958). There is no evidence that this stimulus affects hormone output in any way.

All of the factors listed in subsections II and III probably make their influence felt by altering enzyme balances in and available metabolites for the adrenal. Unraveling these problems will require the development of far more intricate techniques than those so far applied to the study of adrenocortical activity.

IV. Blood flow through the adrenal gland

The mammalian adrenal gland is described as having a higher blood flow than any organ other than a hyperplastic thyroid (Rogoff 1944). In all vertebrates the number of supplying vessels is abundant (see Anatomy). Yet actual measurements of the blood flow have been made rarely, and the results must be viewed with caution, since puncture of the adrenal vein or other procedures for assessing blood flow are inevitably stressful and alter the situation from "normal."

The most reliable measurements are probably those of Hartman et al. (1955) on dogs. They used sodium pentobarbital (Nembutal) as the general anesthetic and Novocaine locally. During the drawing of blood from the adrenal vein, the dog was transfused to compensate for blood loss. The rates determined were mean values for several hours' observation. In a total of 62 dogs, ranging in weight from 5.5 to 21.5 kg, the average blood flow through a single adrenal was 5 ml/g adrenal/minute, 0.3 ml/kg body weight/minute, or 18 ml/gland/kg/hour. The average flow was greater in small dogs than in large ones, as shown in Table 13.

Modified methods for collecting adrenal venous blood have been reported by D. NELSON et al. (1956), MASONI (1957), SCHAPIRO and STJARNE (1958), MCDONALD and REICH (1959).

HARTMAN et al. found no correlation of blood flow with blood pressure. The enhancement in blood flow succeeding the administration of an ACTH preparation appeared to result from contaminating vasopressin. SAPIRSTEIN and GOLDMAN (1959) observed a positive response to ACTH. And it is noteworthy that RICHARDS and PRUITT (1957) observed consistent increases in blood flow through the adrenals of dogs when they were "stressed" by exposure to a high-CO_2 atmosphere. This observation and others to be cited later (p. 86) suggest that *stimulus to the adrenal may be associated with an increase in blood flow through the organ.* The average figures derived by HARTMAN et al. from prolonged observations may represent maximal rather than basal rates.

Table 13. *Blood flow through dog adrenals* (HARTMAN et al. 1955)

Body weight, kg	Number of dogs	Adrenal blood flow	
		ml/g/adrenal/min. ± S.E.	ml/gland/kg body wt./min. ± S.E.
5—10	19	5.3 ± 1.8	0.35 ± .037
10—15	33	5.2 ± 0.8	0.27 ± .003
15—20	10	4.3 ± 1.1	0.21 ± .027

From a calculation of rubidium uptake, SAPIRSTEIN and GOLDMAN (1959) obtained a circulation rate in a single rat adrenal (250 g females) of 1.9 ml/g adrenal/min, or about 0.31 ml/gland/kg/min. This comes to 18.7 ml/gland/kg/hour, the same range reported by SINGER and STACK-DUNNE (1955) (16 ml) and DAS GUPTA and GIROUD (1958) (20—25 ml). For the golden hamster, SCHINDLER and KNIGGE (1959a) observed an average rate of 13.2 ml/gland/kg/hour.

C. Secretory activity of adrenocortical tissue

Until recent years, study of the nature and secretory rate of the adrenocortical hormones has been indirect. No methods were available for detecting the actual products of secretion, and their nature was deduced indirectly by comparing the peripheral changes induced by stimulating the glands with the changes produced in adrenalectomized animals with adrenocortical extracts or, later, with specific hormones. It was such studies that led to the hypothesis that the adrenal cortex produces at least two different types of steroids with differing peripheral action and to the secondary hypothesis that the secretion of these hormones is under separate control (SWANN 1940, SWINGLE and REMINGTON 1944).

The various assays developed for the several adrenocortical hormones may be grouped in three classes. (1) The first, still used in modified form, is life maintenance in the adrenalectomized test animal. By itself of course this is a non-specific test, though the so-called mineralocorticoids appear to be more effective in this regard than glucocorticoids (Chapter 4). (2) The second includes a variety of tests for what we now call *glucocorticoids*, by measurement of increased protein catabolism, increased deposition of glycogen in the liver, the involution of lymphatic tissue and the reduction of circulating lymphocytes, and reduction in the number of circulating eosinophils (see Chapter 7). (3) The third indicates the activity of *mineralocorticoids* and depends on Na^+ retention, as evidenced by an increased Na^+/K^+ ratio in the blood plasma or a decreased Na^+/K^+ ratio in the urine, saliva, or sweat (see Chapter 4). The sensitivity of the latter test was greatly enhanced by the use of radioactive Na^+ and K^+.

The evaluation of the above tests is made in subsequent chapters that discuss the physiological manifestations to the several adrenocortical hormones. Here the methods of assessing adrenocortical activity by direct measurement of the secretory products or by indirect evaluation of other changes in the adrenal cortex itself are summarized. Despite the fact that direct measurementof products has become possible only recently, this will be discussed first.

I. Direct determination of adrenocortical secretions

1. Analysis of adrenal-venous blood or peripheral blood

In 1943, VOGT *introduced the method of analyzing venous blood leaving the adrenal gland* (dogs, cats) to arrive at some measure of the adrenal secretions. She assayed the extracts for life-maintaining ability in adrenalectomized rats. Analysis of the adrenal effluent introduced a powerful tool for direct study of the actual secretions of the adrenal, in contradistinction to study of compounds extractable from glands but perhaps only precursors of released hormones. The subsequent development of chromatographic methods for separating and identifying steroids isolated from blood (or urine) increased the sensitivity and flexibility of the method (BUSH 1953; see Chapter 2).

In recent years, the technique of drawing adrenal venous blood and measuring the output of corticoids per gland per unit time has been applied to a number of different species. Some representative results are listed in Table 14. Two facts are immediately apparent. (1) The rat and rabbit secrete corticosterone as the principal glucocorticoid, whereas the dog secretes cortisol predominantly. (2) The output of corticosterone in the rat and of aldosterone in the dog is relatively low early in the sampling period. These latter observations suggest that most measurements may be too high.

Table 14. *Secretion of cortical hormones in mammals, as measured in the adrenal-venous effluent. Results expressed as μg steroid/gland/kg body weight/hour. Intact animals with no stimulation other than cannulation of the adrenal vein*

Species	Secretion product			Anesthetic	Reference
	Cortisol	Cortico-sterone	Aldosterone		
Dog	20	6	0.25	Nembutal	FARRELL et al. 1956a RAUSCHKOLB et al. 1956
	20	—[1]	—	Ether	GANONG and HUME 1956b
	18	—	—	None[2]	SUZUKI et al. 1957
	—	—	0.1[3]	Nembutal	FARRELL et al. 1956b, c
Rat	0	100	ca. 0.33	Ether and Nembutal	SINGER and STACK-DUNNE 1955
	—	79	—	Urethane	VOGT 1955
	tr.	25[3]	—	Nembutal	BEIGELMAN et al. 1956
Rabbit	—	ca. 30	—	Chloralose	VOGT 1955
	1.2	30	—	Nembutal	REIF and LONGWELL 1958
Golden hamster	21	—	—	Nembutal	SCHINDLER and KNIGGE 1959a

[1] —, not measured
[2] permanently-implanted catheter
[3] results obtained early in the sampling; results obtained later were higher
tr., trace.

(1) *The difference in type of glucocorticoid secreted in various species has been confirmed repeatedly* by analysis of adrenal venous or peripheral blood (D. NELSON 1956). Thus rat, rabbit and mouse glands secrete corticosterone as the principal

glucocorticoid (DONE et al. 1952, BUSH 1953, SINGER and STACK-DUNNE 1955, VOGT 1955, HOLZBAUER 1957a, REIF and LONGWELL 1958, WILSON et al. 1958). Guinea pig, hamster, cat, dog, sheep, rhesus monkey and human glands secrete cortisol in excess of corticosterone (BUSH 1953, ROMANOFF et al. 1953, SWEAT and FARRELL 1954, GOOD et al. 1956, HOLZBAUER 1957a, GRANT et al. 1957, ENDRŐCZI et al. 1958, SCHINDLER and KNIGGE 1959a). (Traces of cortisone, dehydrocorticosterone and deoxycorticosterone are secreted by the glands of some species.)

KASS et al. (1954) presented evidence that in rabbits the cortisol/corticosterone ratio in the effluent changed from about 0.05 in control animals to 4 after prolonged treatment with ACTH (3—4 weeks) (Table 15). GRANT et al. (1957) have found in the human that the ratio of cortisol/corticosterone secreted was higher after administration of ACTH for 3 days prior to operation than if no ACTH had been given.

Aldosterone appears to be the only mineralocorticoid that is released in significant amount. FARRELL (1958) gives additional data on the secretion of aldosterone. In addition, there may be C_{19} steroids, some of which are *weak androgens* ($\triangle^4$-androstene-11β-ol-3,17-dione, dehydroepiandrosterone, androsterone) (ROMANOFF et al. 1953, TOUCHSTONE et al. 1955, BUSH et al. 1955, D. NELSON 1956, GOLD 1957, HARDY et al. 1957, HOLZBAUER 1957a, LOMBARDO et al. 1959). OERTEL and EIK-NES (1959), however, found no C_{19} steroids in dog adrenal blood in excess over levels in peripheral blood. This subject is treated more fully in Chapter 5. *Progesterone* has been detected in adrenal venous blood by BALFOUR et al. (1957), *estrogens* by HARDY et al. (1957). In addition, there are present in the effluent additional, often unidentified compounds of unknown metabolic significance (e.g., FARRELL 1954).

CHESTER JONES et al. (1959) have reported the data listed in Table 15, showing that *in lower vertebrates the products are similar to those in mammals.* In a later paper, PHILLIPS et al. (1959), by extracting about 9 liters of salmon plasma (*Oncorhynchus nerka*), have succeeded in detecting not only cortisol, corticosterone and cortisone, but also aldosterone. The data of LELOUP-HATEY (1958) suggest that many of the measured levels in fish blood are too high, as a result of stress.

Although the method of assaying the adrenal venous blood has the advantage of indicating the products actually released, it is not without its pitfalls. Data presented in Table 14 indicate that the concentration of products tends to rise with time. The operation itself is traumatic (D. NELSON et al. 1956); anesthetics, especially ether, may be "stressful" (BEIGELMAN et al. 1956, ROYCE and SAYERS 1958,

Table 15. *Level of corticoids in adrenal venous blood of lower vertebrates (*CHESTER JONES et al. 1959) *as compared to the rabbit* (KASS et al. 1954). *Results expressed as μg steroid/100 ml blood. Those in parentheses are calculated from levels in peripheral blood*

Species	Corticoid			
	Cortisol	Cortico-sterone	Cortisone	Aldosterone
Dogfish, *Scyliorhinus canicula*	(25)	0	0	0
Carp, *Cyprinus carpio*	(135)	0	0	0
South African toad, *Xenopus laevis*	7	0	0	0
Congo eel, *Amphiuma* sp.	(2.8)	(6)	0	0
Grass snake, *Natrix natrix* (ACTH)	8.5	156	tr.	0
Chicken, *Gallus domesticus* (untreated)	3	312	tr.	0
Chicken (ACTH)	4	66	2.5	1.0
Rabbit (untreated)	7	155	—	—
Rabbit (ACTH)	307	75	—	—

0, not found —, not sought tr., trace.

ENDRŐCZI et al. 1958); reduction in total blood volume increases the output of aldosterone and possibly also of glucocorticoids (FARRELL et al. 1956b, c). *The rates ordinarily measured probably represent maximal secretory capacity* of the adrenal at the time (VOGT 1955, GANONG and HUME 1956b, HOLZBAUER and VOGT 1957a, HILTON et al. 1958). By direct measurement, it seems impossible to determine either the "basal" secretory rate or the proportion of products released. With the gentlest technique yet devised, namely of drawing minute samples of blood from a permanently implanted catheter in the adrenal vein, D. NELSON et al. (1956) calculate that the output of the *two* adrenals in an adult dog (ca. 15 kg) is 6 mg cortisol per day, whereas the secretory rates given in Table 14 would yield about 15 mg. ENDRŐCZI et al. (1958) also obtained low figures in such animals. The "basal" levels of secretion cannot readily be deduced from analysis of peripheral blood, since this concentration is a function of the rate of metabolism and excretion of the products, as well as of their secretion. Analysis of urinary output is even more complicated.

By use of a tracer technique and by measuring excretion levels, AYRES et al. (1958) *have calculated the hormone output in man per day as: cortisol 25 mg, corticosterone 2 mg, aldosterone 0.2 mg, deoxycorticosterone and 17-deoxycortisol* <0.5 *mg.*

The hypophysectomized animal provides a sensitive assay method for ACTH. D. NELSON and HUME (1955) showed that as little as 1 mU ACTH injected intravenously into acutely hypophysectomized dogs (average weight about 15 kg) induced a 10-fold increase in 17-hydroxy corticosteroid (cortisol) output by the adrenal. The recently hypophysectomized rat is a similarly sensitive preparation for assay of ACTH (LIPSCOMB and NELSON 1959).

What information exists on changed rates of secretion of the major types of hormones will be discussed later in the sections on the origin of glucocorticoids and mineralocorticoids (p. 98ff.).

2. Analysis of adrenal perfusates

Another useful technique for studying adrenal secretion is the perfusion of an isolated gland, using blood of the same species. This technique has been applied to the glands of dogs, cats (VOGT 1951, HILTON et al. 1958, 1959), and cattle (e. g., MACCHI and HECHTER 1954a, b, ROSENFELD and BASCOM 1956). VOGT's and HILTON's perfusions were made by way of the aorta, with the glands isolated but left *in situ*. The perfusion of isolated bovine adrenals has been by way of the single adrenal vein; the surface may be lacerated to permit easy drainage.

VOGT has not applied this technique since identification of the products by chromatography has been possible; the level of secretory product (glucocorticoid) was assayed in adrenalectomized rats. Extreme sensitivity of the isolated preparation to ACTH was evident. This has now been confirmed by HILTON et al. (1958), who found that as little as 1 mU ACTH added to the perfusate (blood from a hypophysectomized dog) induces a sharp rise in the "basal" secretory rate (cortisol).

ROSENFELD and BASCOM (1956) identified the major products found in the effluent of stimulated and unstimulated bovine glands. The products were cortisol $>$ corticosterone $>$ (cortisone), the output of all of which was enhanced by ACTH. Such glands have been shown also to produce aldosterone (ROSENFELD et al. 1956). The influence of electrolyte balance of the perfusate on aldosterone output is discussed on p. 113. MACCHI and HECHTER (1954b) found that a detectable rise in corticoid output occurred with the addition of 1 mU ACTH per liter of perfusate. Some glucocorticoid is secreted by both isolated canine and bovine glands without added ACTH, however.

3. Analysis of incubation media

The third technique developed for the study of the secretory products of the adrenal cortex involves incubation of quarters or slices of gland in oxygenated blood or a buffered glucose-Krebs-Ringer medium. The medium is then analyzed for released steroids (SAFFRAN et al. 1952, HAYNES et al. 1952, PÉRON and KORITZ 1958). The earliest studies showed clearly that inclusion of ACTH in the medium enhanced formation of formaldehydogenic steroids (α-ketolic corticosteroids). However, in comparison to glands *in situ* or to whole perfused glands, *incubated adrenals require extremely large concentrations of ACTH for activation.* Thus, even with frequent changing of medium, 50–100 mU ACTH per 100 mg of adrenal (or per 6 ml incubating fluid) is needed to cause significant augmentation of corticoid output (C. GIROUD et al. 1956b, BIRMINGHAM and KURLENTS 1958). Obviously, the problems of penetration and of autolysis of cells deep in the slice make this an insensitive system. Nevertheless the method has the advantage that it can be applied to the adrenals of any species. As an assay method for ACTH it agrees closely with other standard methods (SAFFRAN and SCHALLY 1955, McKERNS and NORDSTRAND 1955, VAN DER VIES 1957, GULLEMIN et al. 1959). And the products produced *in vitro* seem qualitatively similar to those found in the adrenal venous blood of the same species (BUSH 1953, SINGER and STACK-DUNNE 1955, HOFMANN 1956, 1957, SCHINDLER and KNIGGE 1959b, LOMBARDO and HUDSON 1959, PHILLIPS and MULROW 1959, NANDI and BERN 1960). Some differences in product are found depending on the age or on the preconditioning of the donor.

Recently, ROBERTS (1957) reported that a component of the β-globulin fraction of rat plasma stimulates the secretion of corticosterone by rat glands *in vitro*. The active material was present in blood from normal, adrenalectomized and hypophysectomized rats. ACTH appears in the albumin fraction. If substantiated, this interfering substance makes difficult the assay of blood fractions *in vitro*.

II. Indirect indices of adrenocortical activity

Most criteria for judging altered activity of the adrenal cortex are indirect, depending on morphological or chemical alteration of the tissue or cells. Even with the development of methods of direct assay summarized in the preceding section, the indirect techniques must be retained as giving auxilliary information, as well as being more generally applicable. In particular, cytological assessment has the virtue that it makes possible distinguishing the activities of different sorts of cells under conditions in which the secretory rates of the different hormones are varied independently.

1. Gross indices of altered stimulation

a) Adrenal weight

α) Atrophy after hypophysectomy. In general, *growth, maintenance and responsiveness of adrenocortical tissue, following an autonomous period in the embryo, are dependent on the continued secretion of the adrenocorticotrophic hormone (ACTH)* by the adenohypophysis. As will be described later (p. 110), in mammals, at least, the outermost zone of the gland, the glomerulosa, seems to be largely independent of ACTH. Since, however, the inner zones (the fasciculata and reticularis) comprise the bulk of the gland, changes in the level of circulating ACTH produce profound effects on the weight and chemistry of the gland as well as on the structure and function of these inner zones.

In the first part of the 20th century, the dominant role played by the anterior pituitary in the morphological integrity and probably the functioning of several

Table 16. *Effects of hypophysectomy on weight or volume of adrenal glands in various vertebrates*

Species	Size of adrenals in controls	Size of adrenals in experimentals	Period after operation	% Atrophy	Reference
Class Osteichthyes					
Anguilla anguilla	3.3 mg-% (8)*	1.1 mg-% (10)	10 days	67	Hatey 1954a
Class Amphibia					
Hyla sp. tadpoles	383 vol. units (1)	144 vol. units (1)	228 days	62	P. Smith 1920
Class Reptilia					
Agama agama	15.1 mg-% (6)	6.1 mg-% (8)	30 days	60	Wright and Chester Jones 1957
Natrix natrix					
winter	24.1 mg-% (9)	18.7 mg-% (8)	20 days	22	Wright and Chester Jones 1957
summer	27.6 mg-% (13)	19.3 mg-% (10)	30—39 days	30	
Class Aves					
Columba livia	30.2 mg-% (88)	24.2 mg-% (51)	10 days	20	R. Miller and Riddle 1942
Class Mammalia					
Macaca mulatta	718 mg (7)	330 mg (5)	26—126 days	54	Knobil et al. 1954b
	9.1 mg-% (5)	6.9 mg-% (13)	21—617 days	24	P. Smith 1955
Mesocricetus auratus					
♂	19.8 mg-% (25)	10 mg-% (25)	30—60 days	49	Knigge 1954a
♀	11.2 mg-% (17)	7.9 mg-% (22)	30—60 days	29	
Rattus norwegicus					
♂	13.1 mg-% (14)	5.9 mg-% (20)	30 days	55	Cutuly 1936
♀	25.6 mg-% (13)	8.8 mg-% (24)	30 days	66	
♂	21.6 mg-% (12)	9 mg-% (13)	56 days	58	Deane and Greep 1946
♀	27.7 mg-% (10)	8 mg-% (10)	42 days	71	Karnovsky and Deane 1955
♀, infant	41 mg-% (10)	12 mg-% (15)	54 days	71	Walker et al. 1950
Mus musculus					
♂	16.1 mg-% (9)	11.6 mg-% (8)	100 days	28	Chester Jones and Roby 1954
♀	24.8 mg-% (20)	16 mg-% (20)	12—14 days	35	Chester Jones 1949b
Cavia cobaya, ♂	50.5 mg-% (10)	30.1 mg-% (6)	4—6 weeks	40	Schweizer and Long 1950
Canis familiaris	15.7 mg-% (12)	10.2 mg-% (40)	4—343 days	35	B. Houssay and Sammartino 1933a, b

* Number of specimens in parentheses.

endocrine target organs came to be appreciated. In 1912, Ascoli and Legnani reported that the hypophysectomy of dogs resulted in profound atrophy of the inner zones of the adrenal cortex. In the next few years, Simmonds (1919) recorded that the adrenals were shrunken in individuals that had suffered accidental destruction of the pituitary. This dependence has now been confirmed for a wide variety of vertebrates.

The degree of shrinkage of the adrenal glands in various vertebrates after hypophysectomy is indicated in Table 16. Several workers have demonstrated that *removal of the posterior pituitary alone does not produce atrophy* (P. SMITH 1930, B. HOUSSAY and SAMMARTINO 1933a, b, DEANE 1951, CHESTER JONES and WRIGHT 1954b, NOWELL 1959).

In mammals and probably in lower vertebrates, *glandular atrophy results almost entirely from shrinkage of adrenocortical cells*. Changes in volume of chromaffin tissue seem negligible (B. HOUSSAY and SAMMARTINO 1933a, b, SCHAEFFER 1933, CUTULY 1936, H. HOUSSAY 1942, CATER and STACK-DUNNE 1953, KNIGGE 1954a, BURGOS 1959).

In the toad *Bufo arenarum*, total hypophysectomy causes no detectable atrophy of the glands in 15—30 days (PORTO 1940, PENHOS and CARDEZA 1951, RAPELA and GORDON 1956), although it does result in cytological changes indicative of depressed activity (see p. 94). Perhaps the conspicuous accumulation of lipide in the adrenocortical cells of this species following hypophysectomy compensates for a loss of cytoplasm.

The rate of regression varies. In mammals, most of the atrophy has occurred by the end of two weeks (Fig. 48) (P. SMITH 1930, CUTULY 1936, H. HOUSSAY 1942, DEANE and GREEP 1946, SCHWEIZER and LONG 1950). In some poikilothermic species, however, atrophy may be less abrupt, especially if the experiments are conducted in winter when the gland is already relatively atrophic (e.g., *Rana* — SLUITER et al. 1949, BURGOS 1959).

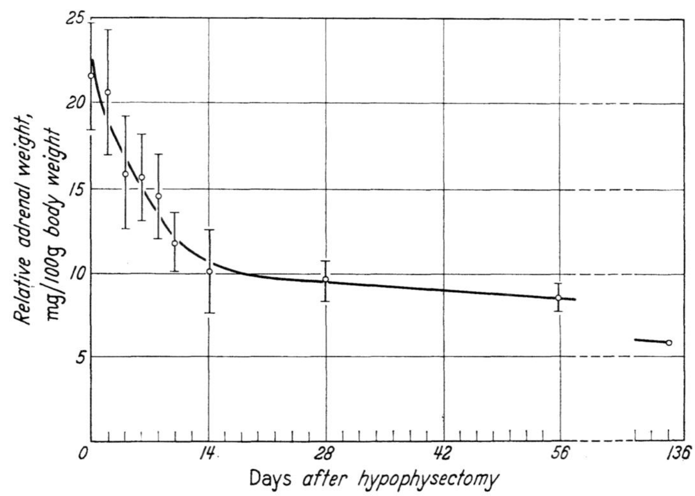

Fig. 48. Time course of adrenal gland atrophy in young male rats following hypophysectomy. The average for each group (about 6 animals per group) and the standard deviation are given; 0 time average represents the gland size in 12 control animals. Reproduced by permission of the publisher from Fig. 1, DEANE and GREEP: Amer. J. Anat. 79, 120 (1946)

It is noteworthy that *destruction of the anterior pituitary must be complete to induce adrenocortical atrophy*. Marked involution of other pituitary target organs (gonads, thyroid) occurs with less than complete removal. For several species, workers have reported that detectable adrenal atrophy fails to occur unless more than 80% of the anterior pituitary is removed (e.g., *Canis* — B. HOUSSAY and SAMMARTINO 1933a, GANONG and HUME 1956a, BAHN et al. 1960; *Rattus* — P. SMITH 1932; *Xantusia* — M. MILLER 1952; *Hyla* tadpole — P. SMITH 1920; *Torpedo* — DITTUS 1940). GANONG and HUME (1956a) and EIK-NES and BRIZZEE

Table 17. *Adrenal changes on the administration of adrenocortical extract or of glucocorticoids*

Species	Treatment	Changes described	Reference
Torpedo spp.	adrenocortical extract, 8 d.	atrophy of adrenocortical cells, loss of lipides	DITTUS 1940
Carassius auratus	adrenocortical extract, 6 d.	atrophy of adrenocortical cells	CHAVIN 1956
Rana temporaria	cortisone, 0.5 mg, 2 × week, 3 weeks	cellular atrophy, loss of lipides, reduction in mitochondria, nuclear pyknosis	TUCHMANN-DUPLESSIS et al. 1955
Xantusia vigilis	cortisone, 25 μg/d., 4d.	cellular atrophy, loss of lipides	M. MILLER 1952
Natrix natrix	cortisone, 5 mg/d., 8 d.	25% atrophy of gland, some cellular atrophy	WRIGHT and CHESTER JONES 1957
Pelecanus occidentalis	cortisone, 5—9 mg/d., 5—11 d.	glandular and cellular atrophy, nuclear pyknosis, depression of mitotic rate, loss of lipides, especially from inner zones	HARTMAN et al. 1954
Macaca mulatta	cortisone, 50 mg/d., 14—68 d.	40% atrophy of gland, cellular atrophy in fasc. and retic., gradual loss of lipides and pseudoplasmal reaction	KNOBIL et al. 1954b
,, ,,	cortisone, 15 mg/d., 14 d.	20% atrophy of gland, atrophy of inner zones, enlargement of lipide droplets and depletion	KROHN 1955
Homo sapiens	cortisone, 4 mo.	distinct atrophy (fasc.), cholesterol level normal	SOKOLOFF et al. 1951
,, ,,	cortisone, prolonged	slight atrophy of fasc., increased lipides in inner zones	LANDING and FERIOZI 1954a
,, ,,	cortisone, 1—3 yrs.	ca. 65% atrophy of cortex, abundant lipide and cholesterol in remaining cells	STONER and WHITELEY 1954
,, ,,	prednisone, 25 mg/d., 7 d.	suppressed rise of circulating 17-OH corticoids on administration of ACTH	CHRISTY et al. 1956b
Rattus norwegicus	cortisone, 3 mg/d., 10 d. and 6 weeks, ♂	40% atrophy at 10 d., 55% at 6 weeks; cholesterol level dropped to 1%, ascorbic acid level depressed	WINTER et al. 1950
,, ,,	cortisone, 1.25 mg/d., 20 d., ♀	21% atrophy, rise in cholesterol level	LEWIS et al. 1950
,, ,,	cortisone, 5 mg/d., 15 d.	61% atrophy in ♂, 35% in ♀	TUCHMANN-DUPLESSIS 1951a
,, ,,	cortisone, 5 mg, 1 inj.	at 5 and 10 d., had suppressed compensatory hypertrophy of remaining gland after unilateral adrenalectomy	GAUNT et al. 1952
,, ,,	cortisone, 0.7 mg/100 g/d., 3—4 weeks	reduction in ascorbic acid concn. in cortex	BAHN and GLICK 1954b
,, ,,	cortisone, 0.7 mg/100 g/d., 3—4 weeks	reduction in cholesterol concn. in cortex	GLICK and OCHS 1955
,, ,,	cortisone, 2.5 mg/d., 7d.	marked atrophy of cells in fasc.	YOSHIMURA 1954
,, ,,	cortisone, 3.5 mg/d., 13 d.	30% atrophy of gland, 25% redn. in nuclear diameter in fasc., loss of lipide from fasc.	HALÁSZ 1955
,, ,,	cortisone, 5 mg/d., 30 d.	adrenals 69% smaller, RNA 66% less than normal	L. MANDEL et al. 1956

Table 17 (cont.)

Species	Treatment	Changes described	Reference
Rattus norwegicus	cortisone, 10 mg/100 g, 1 inj.	prevented decl. in ascorbic acid concn. in remaining gland after unilateral adx. (24 hrs.)	LANGECKER and LURIE 1957
,, ,,	prednisone or prednisolone, 0.5 mg/100 g/d., 15 d.	glandular atrophy, atrophy of fasc., depression of lipides and cholesterol in fasc., reduction of alkaline phosphatase, depression of urinary 17-ketosteroids	BONATI et al. 1957
Mus musculus	cortisone, 0.1 mg/d., 3 weeks, ♀	30% atrophy of glands	FLUX 1954
,, ,,	cortisone or cortisol, 0.5—2.5mg/d., 7—15 d.	involution of fasc., loss of lipides. Cortisol less effective than cortisone	SAKIZ 1955, 1957
,, ,,	cortisone, 0.2 mg/d., 4 weeks	marked involution of fasc., depression of nonspecific esterase	ALLEN 1957
Cavia cobaya	cortisone, 5 mg/d., 4 weeks	adrenals small, fasc. atrophied, lipides abundant, nuclear volume reduced 30%; no adrenal enlargement in response to diphtheria toxin	HOEDE 1956
Oryctolagus cuniculus	cortisone, 5 mg/d., 26—31 d.	glandular atrophy, atrophy of inner zones, increase in lipides and cholesterol	BERGSTRAND 1952
Canis familiaris	cortisone or cortisol, 100 mg/d., 36 d.	ca. 60% atrophy of gland, atrophy of fasc. and retic., reduction in lipides; cortisol output of adrenal 2.5% of normal, corticosterone output 25% of normal	FARRELL et al. 1956a
,, ,,	cortisol infusion before or during stress	depressed secretion of 17-OH corticoids in response to operation or CO_2 exposure	RICHARDS and PRUITT 1957

Key: adx.=adrenalectomy; concn. = concentration; d. = day(s); decl. = decline; fasc. = z. fasciculata; mo. = month(s); redn. = reduction; retic. = z. reticularis.

(1958) have demonstrated that the dog adrenal is capable of hypertrophy and of producing significant, though lower than normal, amounts of corticosteroids if 3—5% of the anterior lobe remains *in situ*. And GOOD et al. (1956) found a persistence of 17-OH corticosteroids in the peripheral blood of hypophysectomized guinea pigs when small amounts of pituitary remained.

β) Atrophy with the exogenous administration of corticosteroids. *Gross shrinkage of the adrenal cortex,* resembling that seen after hypophysectomy, *was reported to follow the continued administration of adrenocortical extracts* to intact rodents in the late 1930s (INGLE et al. 1938, FLEXNER and GROLLMAN 1939). *This response is attributed to a depression of ACTH release* by some sort of feed-back mechanism (INGLE 1950, SAYERS 1950). It has been shown that exogenous administration of cortical hormones has little or no direct effect on the adrenal of the hypophysectomized animal (STEBBINS 1951b, S. GREENBERG and ATERMAN 1954). Thus, with a high level of circulating adrenocortical hormones there is inactivity of the pituitary with respect to ACTH secretion. SAYERS (SAYERS and SAYERS 1949, SAYERS 1950) has summarized the data showing that the inhibitory influence on ACTH liberation is greatest for 17 OH-glucocorticoids, minimal for deoxycorticosterone and deoxycortisol.

Table 17 lists data on compensatory atrophy for a variety of vertebrates. The effects of total adrenocortical extract are included only for fishes, on which pure compounds have not yet been tested. Additional information on cytological, chemical and secretory changes are included here, but will be discussed later.

γ) Hypertrophy of the adrenal cortex with the administration of ACTH. The reverse situation occurs *when ACTH is administered or when its endogenous secretion is enhanced* by "stresses." In these circumstances, *the adrenal cortex gradually hypertrophies.* It also secretes corticoids, particularly glucocorticoids, at higher than normal rates and displays various cytological, chemical, and biochemical changes directly or indirectly associated with this enhanced secretory rate. Some of these facets will be discussed later. First, the adrenal-weight increase accompanying the administration of exogenous ACTH will be summarized as a basis for later discussion of the effects of endogenous ACTH.

Enlargement of the adrenocortical gland on the administration of now highly-purified ACTH is demonstrable in a wide variety of vertebrates. The response occurs both in intact animals and in recently-hypophysectomized animals, the latter providing the more sensitive assay. When animals have been hypophysectomized for a long period, however, the response may be delayed. Some representative results are listed in Table 18. *The weight gain is* almost entirely *the result of hypertrophy of the cortical component,* the medulla being little affected (GOODSELL 1941, SCHNEPPENHEIM and HUHN 1955).

Using a preparation of I^{131}-labeled ACTH (crude), SONENBERG et al. (1951) showed that there was selective uptake of this compound by the adrenal cortex. Significant concentration in the adrenal gland was demonstrable within 30 seconds.

Lower doses are necessary for production of detectable weight gain if the ACTH is injected in a protein vehicle (H. COHEN et al. 1954, EIK-NES et al. 1958), with metallic salts (GEORGES and CAHN 1955) or in beeswax-and-oil (KROHN 1955, LI et al. 1957), rather than in aqueous solution; the carrier permits slow release over many hours.

GANONG (1954) has shown that identical hypertrophy of a gland occurs following ACTH administration in hypophysectomized rats with one or two adrenals. In other words, the response depends on the circulating level of ACTH, not on the amount of adrenocortical tissue present.

Table 18. *Weight changes in adrenals of animals receiving ACTH (purified)*

Species	Condition	Dose level and period	Increase in adrenal weight over control value, %	Reference
Natrix natrix				
75 g	Hypox., 20 d.	0.5 U/d., 10 d.	15	WRIGHT and CHESTER JONES 1957
Gallus domesticus				
chicks	Intact	3 U/d., 5 d.	25	JAILER and BOAS 1950
Macaca mulatta				
3.7 kg	Intact	25—30 U/d,* 14—21 d.	70	KROHN 1955
Canis familiaris				
8.3 kg	Intact	40 U/d., 28 d.	50	RACE et al. 1957
Rattus norwegicus				
120 g	Intact	5 U/d.,* 5 d.	100	H. COHEN et al. 1954
65 g	Intact	2.2 U/d.,* 4 d.	100	LI et al. 1957
120 g	Hypox., 8 d.	1.1 U/d.,* 4 d.	100	
Mus musculus				
15 g	Hypox., 21 d.	3 U/d.,* 7 d.	65	LOSTROH and WOODWARD 1958

* ACTH injected in a protein or beeswax-oil vehicle.

The immediate weight gain following ACTH (or stress) appears to reflect an increase of water in the gland (e.g., CARPENTER et al. 1946, KORPÁSSY et al. 1950, BRODISH and LONG 1960), since the proportionate dry weight declines. Subsequently, however, the weight gain signifies a true increase in protoplasm.

Just what relationship exists between adrenal weight and secretory capacity remains to be determined. Thus there is ample evidence that growth hormone possesses the ability, alone or in combination with ACTH, to stimulate growth of the adrenal gland in hypophysectomized animals (e.g., CATER and STACK-DUNNE 1955c, BOIS and SELYE 1957, LOSTROH 1958, LOSTROH and WOODWARD 1958). SCHINDLER (1959) has presented preliminary evidence that growth hormone alone does not enhance secretory activity (golden hamster). Certainly weight increase of the adrenal is not simply a response to ACTH.

b) Adrenocortical chemistry

α) Concentration of cholesterol. Histologists have reported for over 50 years that changes in apparent cholesterol concentration in the cortex (birefringent crystals, Schultz reaction) accompany hypertrophy or atrophy of the gland (Fig. 49) (ROGOFF 1932, SAYERS et al. 1944). When ACTH became available, this pituitary hormone was shown to cause a rapid decline in the cholesterol concentration in the adrenal gland — a response that preceded detectable change in weight of the gland (SAYERS and SAYERS 1949, SAYERS 1950). Apparently, *with acute stimulation, the cortex utilizes its stores of potential hormone precursor rapidly.* This response has proved to be fairly sensitive and reproducible in assessing the influence of ACTH or stresses in a number of vertebrate species (excepting those that do not store cholesterol). A few examples of the changes observed are listed in Table 19. Depression of the cholesterol concentration with acute or chronic stress has been recorded also for guinea pigs (Oleson and Bloor 1941, Sayers et al. 1946, Creutzfeldt et al. 1953) and for chickens (Taylor et al. 1956, A. Howard and Constable 1958). In many species, however, depletion is not detectable (Elton et al. 1959).

The level of cholesterol found in adrenals more than about 3 hours after the initiation of stress or the administration of ACTH is a composite of the rate of cholesterol synthesis and of its conversion to hormones. Thus there may be further decline in concentration if the stimulus is severe and continuing; the concentration may remain subnormal with moderately severe stimulus; or it may rise to normal or supranormal levels if the stimulus is discontinued or is only moderate (Sayers 1950). Consequently, except in acute situations, the level of cholesterol in the gland does not yield clear-cut information.

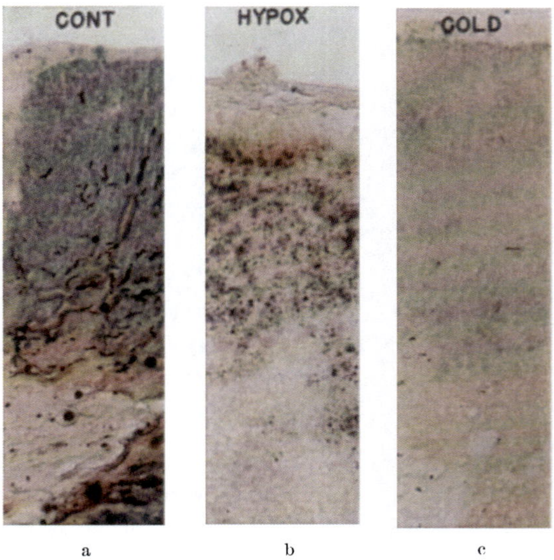

Fig. 49a—c. Changes in amount of cholesterol detectable by the Schultz reaction in rat adrenal glands. a control rat; b rat hypophysectomized 3 weeks earlier; c rat exposed to cold for 9 days. All × 50. The normal gland exhibits cholesterol in most cells. The atrophied gland appears to contain more than normal amounts of steroid, although some of this gives an atypical brownish color. The hypertrophied gland contains almost no detectable cholesterol. [See Deane (1958).]

In inactive glands, such as those of animals subjected to *hypophysectomy* or to repeated injection of adrenocortical hormones, *the concentration of cholesterol in the gland increases for a period* (Fig. 49), but ultimately declines (Rogers and Williams 1948, Winter et al. 1950, Karnovsky and Deane 1955).

β) Concentration of ascorbic acid. *In the laboratory rat, acute stress or the administration of ACTH is associated with a very rapid drop in ascorbic acid concentration.* Thus stimuli that cause a 50% drop in cholesterol concentration over

Table 19. *Acute changes in total adrenal cholesterol concentration with exposure to stress or administration of ACTH*

Species	Stimulus	% Decline in cholesterol concentration	Reference
Rattus norwegicus (laboratory)	Hemorrhage for 1 hr. or *ACTH, 4 mg/100 g, 3 hrs.	50	Sayers and Sayers 1949, Sayers 1950
Rattus norwegicus (wild)	Captivity, 6 hrs.	50	Nichols 1950
Homo sapiens	Peritonitis, 2 days	75	Liebegott 1944, 1952
Macaca irus	*ACTH, 30 mg/kg, 3 hrs.	ca. 50	Glick and Ochs 1955

* Probably about 1 U/mg.

a period of 3 hours cause a 50 or 60% drop in ascorbic acid concentration within 1 hour (SAYERS and SAYERS 1949, SAYERS 1950). This response generally seems more sensitive than other adrenocortical changes with an enhancement of circulating ACTH (e.g., WEXLER et al. 1957). As in the case of cholesterol, the level of ascorbate returns to normal or supranormal levels if the stress is discontinued or only moderate, but it remains low with chronic, extreme stresses. A decline in ascorbic acid in the adrenal of the recently-hypophysectomized rat serves as a standard test for ACTH potency (Sayers test) (see Chapter 10).

KITAY et al. (1959) have compared the standard ascorbic-acid depleting activity of the ACTHs of several species with steroidogenic activity of these same ACTHs on adrenal slices *in vitro*. With rat, bovine and human ACTHs, steroidogenic activity on homologous adrenals was greater than to be expected from the Sayers assay on hypophysectomized rats. The authors suggest that there may be some species specificity for ACTH, much as there is for growth hormone (GESCHWIND 1959).

A recent study by ROYCE and SAYERS (1958a) calls into question the assumption that ACTH is the only agent inducing ascorbic-acid changes in the rat adrenal. They showed that either Pitressin or pure, synthetic arginine-vasopressin causes a depletion of ascorbate from the adrenal glands of rats 24 hours after hypophysectomy. This observation, together with those listed below, suggests that it may not be justifiable to use the ascorbic-acid test as an assay for ACTH activity of unknown substances.

It is insufficiently recognized that *ascorbic-acid depletion fails to occur as a response to stress in most animals other than the rat*. It does seem to occur in those species for which ascorbic acid is a vitamin (e.g., *Cavia* — SAYERS et al. 1946, ROEPKE 1952, R. JONES et al. 1958; *Macaca* — BAHN and GLICK 1954b; *Homo* — UOTILA and PEKKARINEN 1951). But in other animals no consistent decline occurs. For example, no change in adrenal ascorbic acid level occurs as a response to stress in the *wild* Norway rat (WOODS 1957b), in the young chicken (JAILER and BOAS 1950, A. HOWARD and CONSTABLE 1958), or in the Virginia quail (ZARROW and BALDINI 1952). ELTON et al. (1959) have reported finding no significant drop in ascorbic-acid concentration with ACTH or exposure to cold in frog, toad, chicken, rabbit, or cat, although they did obtain significant falls with one or the other stimulus in dog, golden hamster, mouse and opossum. SCHWARTZ and KLING (1960), however, report a moderate but significant effect in the cat. And PEREK et al. (1959) have found a positive effect in adult chicken (see PEREK and EILAT 1960). In incubated slices of hog adrenals, no change in ascorbate level accompanies the increase in corticosteroid output induced by ACTH (JOHANNES et al. 1956). Thus, while the characteristic fall in ascorbic acid in the stimulated rat gland may prove to be a practical assay for ACTH (Chapter 10), it appears to be a rather unusual phenomenon, in no way essential to adrenal responsiveness. Certainly it is unwise to assume that alterations in ascorbic-acid level in a species other than a rat bear any relation to secretory rate (e.g., W. N. HOLMES 1955a).

Contrary to repeated suggestions that ascorbic acid acts in some fashion in hormone synthesis or hormone release (e.g., A. GIROUD 1951), it now appears more likely that *the fall in ascorbic acid accompanying glandular stimulation in the rat results largely*, if not entirely, *from its release into the blood stream* (SLUSHER and ROBERTS 1957, BRIGGS and TOEPEL 1958, LIPSCOMB and NELSON 1960). With marked stimulation, this release precedes that of corticoids. Of course, this fact does not preclude the possibility that some small fraction of the adrenal ascorbate is somehow involved locally in synthetic processes (e.g., KRISCH and STAUDINGER 1959; also, Chapter 3).

Recent studies suggest that the release of ascorbate and the secretion of steroid are separable events in the rat adrenal. GUILLEMIN et al. (1957, 1958) have shown that small doses of ACTH (0.1—0.4 mU) induce as much as a 300% increase in circulating corticosterone without inducing a significant decline in adrenal ascorbic acid. And SLUSHER (1958) found that certain lesions in the hypothalamus (basal tuberal region) inhibited the depletion of adrenal ascorbic acid without depressing corticosterone secretion in response to stress. Other lesions allowed the ascorbic-acid response without a concomitant release of hormone. Such a possible dichotomy makes dangerous the use of ascorbic-acid depletion as an assay for adrenocortical activation, even in the rat.

In the rat, after *hypophysectomy*, there is an initial rise and then a very gradual decline in ascorbic acid concentration in the adrenal. For example, COVIAN (1949) reported the following levels: intact rats, 443 mg-%; 1 day after hypophysectomy, 494 mg-%; 7 days after hypophysectomy, 299 mg-%; 15 days after hypophysectomy, 225 mg-%. Similarly, HESS et al. (1953) found for their control rats 448 mg-%, for animals hypophysectomized 2 weeks 348 mg-%, and for animals hypophysectomized 3 weeks 247 mg-%. WINTER et al. (1950) reported, in rats, with the administration of 3 mg *cortisone* per day, that the concentration fell from a control value of 350 mg-% to 240 mg-% at 6 weeks. HOHLWEG and LASCHET (1959) have similar data. BAHN and GLICK (1954b) have recorded a decline in ascorbate concentration in the adrenal cortex of both monkeys and rats with prolonged cortisone administration.

Other chemical changes occurring in stimulated or inactive glands will be considered together with the cytological indices, below.

c) Alterations in blood flow

Another indirect physiological index of adrenocortical activation has been studied by SJÖSTRAND (1934) and others, in which determinations of the amount of blood in the cortex or the relative area occupied by the sinusoids indicate increased blood flow during stress. SJÖSTRAND made thick sections of mouse adrenals and counted the number of erythrocytes per mm^3 of cortex. In normal animals, the average number was about $450/mm^3$ in males, 600 in females. Two or more times this number were found in the glands of mice subjected to exhausting exercise, severe pain, fever, chilling (1—3°) for 2 hours, morphine, acetylcholine, nicotine, picrotoxin, physostigmine, histamine, strychnine, ether, and ethanol. Injection of physiological saline was without effect.

Similarly, M. DAVIS et al. (1953) measured planimetrically the relative area of cortex occupied by capillaries and observed a drop from normal in the zonae fasciculata and reticularis after the administration of 3 mg/day of cortisone. Contrariwise, during pregnancy there was an enhancement of relative capillary area in the inner zones.

By injecting the radioopaque compound, Thorotrast, and utilizing both gross- and micro-radiography, HARRISON (1957) has recently shown that the injection of adrenaline augments blood flow through the adrenal cortex by causing constriction of the *Arteriae medullae*. After such treatment, all cortical capillaries became filled with blood, whereas under control conditions many were not filled. Since many stresses stimulate the secretion of adrenaline, this local enhancement of blood flow may facilitate adrenocortical response to ACTH.

Although the direct method of determining adrenal blood flow by cannulation of the adrenal vein may itself stimulate blood flow greatly, there is suggestion that ACTH enhances this flow further (p. 73).

MCDONALD and REICH (1959) have reported as reduction in blood flow in sheep fed an Na-deficient diet.

2. Cytological and associated chemical indices of adrenocortical activity

While the histological and cytological changes observable in adrenocortical cells subjected to stimulation or the removal of stimulation are only indirect indices of secretory rate, they have proved extremely valuable. Furthermore *they have the singular advantage of permitting distinction between the activities of different classes of cells (zones) within the gland.* In the Paragraphs below are summarized the major cytological criteria that have been applied successfully to the study of adrenocortical function. This branch of microscopy is termed histo- or cytophysiology.

a) Subglomerulosal zone

In mammals, chronic stimulation of the fasciculata is accompanied by enlargement of the cells in the subglomerulosal zone and accumulation of lipides therein, so that the zone as such disappears. (Acutely, there may be some widening of the zone.) Contrariwise, inactivation of the fasciculata is associated with a broadening of the zone, a decrease in cell size and loss of lipides (M. Simpson et al. 1943, Dalton et al. 1944, Greep and Chester Jones 1950, Cater and Lever 1954, Vasama et al. 1957, Stigliani and Marchi 1957).

b) Cellular and zonal volume

The adrenal cells enlarge when stimulated and shrink when the normal stimulus is reduced. In mammalian

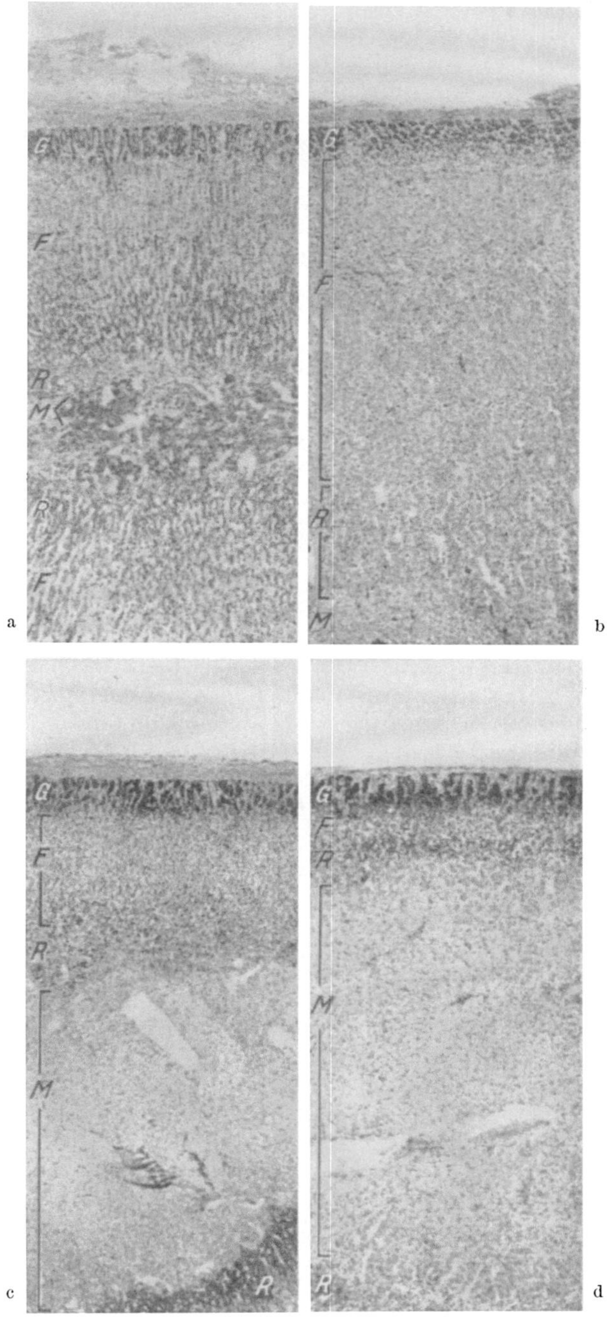

Fig. 50. Changes in the width and hence volume of the inner cortical zones of monkeys subjected to different treatments. a normal control monkey (*Macaca mulatta*); b monkey treated with 5 mg ACTH/day for 34 days; c monkey treated with 50 mg cortisone/day for 31 days; d monkey hypophysectomized 90 days earlier. Fixed in Helly's solution, stained with hematoxylin and eosin. × 36. From the study of Knobil et al. (1954b); courtesy of the authors.
In all, *G.*, glomerulosa; *F.*, fasciculata; *R.*, reticularis; *M.*, medulla. The inner zones hypertrophy when ACTH is given, atrophy when ACTH is absent because of hypophysectomy or cortisone administration. The zona glomerulosa remains about the same width (but not volume) in all

glands, these cellular changes are reflected in an increase or decrease in the width of the zone in which the cells lie (Fig. 50) (e.g., DEANE and GREEP 1946, DEANE et al. 1948, DEANE and MASSON 1951, BACCHUS et al. 1952, EICHNER 1953, DALLWIG 1954, RACE and GREEN 1955, MONTANDON 1957). It seems unfortunate that such calculations have generally not been made in terms of zonal *volume* rather than width, since obviously the width of other zones has an effect on the results. For example, after hypophysectomy the *width* of the zona glomerulosa increases as the inner zones shrink: a recent study indicates that for the dog this increase in width is associated with a decrease in volume of the zone (BAHN et al. 1960).

With cellular hypertrophy there is a decline in the number of cells per unit area, with atrophy there is an increase. Thus quantitative evalution of the degree of change can be obtained from the reciprocal of cell number per unit area (e.g., DEANE and GREEP 1946, OLIVEREAU and FROMENTIN 1954). Actual determination of cell diameter, area, or volume with an ocular micrometer or a planimeter also reveals the direction of change (e.g., DEANE and GREEP 1946, TUCHMANN-DUPLESSIS et al. 1955).

c) Nuclear volume

The most careful measurements yet made have been on nuclear size. *The nuclei appear small and dense when the cells are inactive, large and vesicular when active* (Fig. 70, p. 142) (BOGUTH et al. 1951, HENI et al. 1952, KRACHT and SPAETHE 1953a, CREUTZFELDT et al. 1953, EICHNER 1953, DALLWIG 1954, KNIGGE 1954b, c, KRACHT 1954, TONUTTI et al. 1954, HALÁSZ 1955, KLÄRNER 1955, SCHNEPPENHEIM and HUHN 1955, TUCHMANN-DUPLESSIS et al. 1955, CHAVIN 1956, HOEDE 1956, ANDO 1957, FONTAINE and OLIVEREAU 1957). Some representative results for cells in the zona fasciculata of mammalian glands under condition of enhanced and depressed activity are given in Table 20. All depended on measuring the diameters or areas of 200 nuclei per preparation and calculating the mean volume. Whether the apparent differences between species merely reflect differences in methods of preparation is at present unknown. [Such changes in volume appear to reflect, not changes in deoxyribonucleic acid content, but altered proportions of nuclear histones (BURNS et al. 1956, HUTCHISON et al. 1958).]

A usual concomitant of hypertrophy is hyperplasia, whereas atrophy is assoctated with a decline in mitotic rate and ultimately in cell number (Fig. 51). Thus, DEANESLY (1931) reported increased numbers of mitoses in the adrenals of

Table 20. *Changes in nuclear size in cells of the zona fasciculata with activity and inactivity*

Species	Condition of cortex	Median nuclear diameter, μ	Median nuclear volume, μ³	Reference
Rattus norwegicus	Control	5.52	91.5	KRACHT and SPAETHE 1953a
	Atrophied, hypophysectomy	5.14	71.0	
	Hypertrophied, unilateral adrenalectomy	6.15	125.0	
Mus musculus	Control	—	104.0	TONUTTI et al. 1954
	Atrophied, hypophysectomy	—	72.0	
	Hypertrophied, ACTH	—	141.0	
Cavia cobaya	Control	—	141.0	CREUTZFELDT et al. 1953
	Hypertrophied, starved	—	200.0	

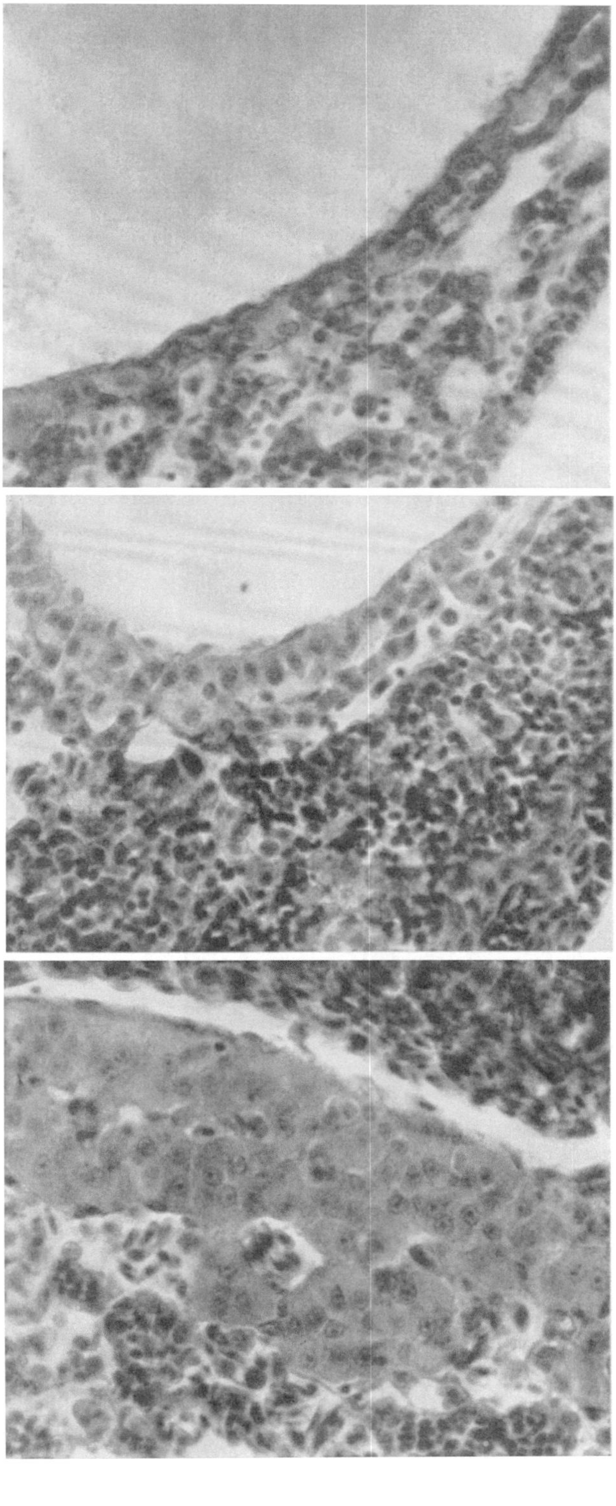

a

b

c

Fig. 51. Changes in adrenocortical cell number with function in the goldfish, *Carassius auratus*. a adrenocortical tissue of a fish hypophysectomized 3 weeks earlier; b normal gland; c gland of a hypophysectomized fish given ACTH, 1U/day for 9 days. Fixed in Bouin's fluid, stained with Masson's trichrome mixture. × 390. From the study by CHAVIN (1956); courtesy of the author. In all, lumen of cardinal vein at top mice stressed by infection or thyroxin, Hoerr (1931) in those of guinea pigs stressed with $CHCl_3$ or diphtheria toxin, Liebegott (1952) in those of human beings given prolonged treatment with ACTH, Klärner (1955) in those of rats given small doses of ACTH for periods of months, Chavin (1956) in those of goldfish given 1 U ACTH per day for 9 days. Cater and Stack-Dunne (1953, 1955a, b, c) have demonstrated a drop in mitotic rate in the rat adrenal following hypophysectomy and a restoration after the administration of relatively pure growth hormone or ACTH. In none of the examples listed was colchicine used. This mitotic arrestor is also a stressing agent, so that its use may yield misleading information (Baxter 1946). Evidence of the increase in cell number associated with hypertrophy is given in Table 22.

d) Nucleolar size

Another frequent concomitant of enhanced cell activity is nucleolar enlargement, of *depressed cell activity, nucleolar shrinkage.* Dittus (1940) described such changes in the adrenocortical cells of *Torpedo* after treatment with ACTH and after hypophysectomy. Other examples are cited by Bachmann (1954). In recent years such changes have been studied quantitatively. For example, Sandritter and Hübotter (1954) found that, in adrenals of stressed rats, cells with large nuclei possessed greatly enlarged nucleoli. When the nuclei were about 60% larger than in controls, by volume, the nucleoli were about 300% larger. Such enlargement was associated with depletion of lipides from the cells (Fig. 53, p. 94). More extensive observations were reported simultaneously by R. Miller (1954) (Table 21). His calculations were made in terms of planimetric readings rather than in terms of actual size. The differences evident between the responses of cells in the glomerulosa and fasciculata will be discussed in later sections. Comparable, though less extensive, observations on nucleolar variation have been made on *Rana* (Tuchmann-Duplessis et al. 1955) and *Salmo* (Fontaine and Olivereau 1957).

Table 21. *Nuclear and nucleolar areas, in arbitrary units², for rat adrenocortical cells under various conditions* (from R. Miller 1954)

Condition	Zona glomerulosa				Zona fasciculata			
	Nucleus	% Control	Nucleolus	% Control	Nucleus	% Control	Nucleolus	% Control
Normal	599		27.6		846		58.9	
Hypophysectomy, 10 days	597	99	27.0	98	513	60.5	21.6	36.7
DOCA*, 56 days	427	71.5	22.6	82	612	72.5	55.4	94
Normal	589		28.2		812		62.8	
ACTH, 1 inj. 10 hrs. previously	604	102	31.6	112	840	102	75.3	120

* DOCA = deoxycorticosterone acetate

e) Cytoplasmic ribonucleoprotein

Associated with nucleolar enlargement is an increase in cytoplasmic ribonucleic acid (RNA), as demonstrated in histological preparations by increased basophilia (Rich and Berthrong 1949, Symington and Davidson 1956, Symington et al. 1956, Burns and Hale 1959). This histological increase has been reported only for human and guinea pig glands. Chemically, a net increase in RNA over deoxyribonucleic acid (DNA), thereby indicating an increase per cell, has been demonstrated in the adrenals of acutely- and chronically-stressed animals as well as of human patients (P. Mandel et al. 1955, Fiala et al. 1956, Symington and Davidson 1956). Some of the data presented by Fiala et al. on the effects of repeated administration of ACTH in the rat are summarized in Table 22.

Table 22. *Effect of repeated injections of ACTH on gland weight, number of nuclei, and amount of RNA and mitochondria of rat adrenals*; 12 *glands per group* (from FIALA et al. 1956)

Condition	Weight of adrenals, mg	Average wt. of glands, mg	Number of nuclei per gland, $\times 10^6$	Microsomal RNA, μg	Mitochondrial N, μg
Normal	116.4	9.7	8.0	245	460
ACTH, 260 U	222.0	18.5	10.8	495	670
increase		+90%	+35%	+102%	+45%
increase per cell				+14%	+8%
ACTH, 425 U	468	39.0	20.3	1450	1393
increase		+300%	+153%	+490%	+200%
increase per cell				+38%	+20%

Such changes are also indicated by the studies of ROSSITER and co-workers (LOGAN et al. 1955, NICHOLLS and ROSSITER 1955a, b, 1956, NICHOLLS et al. 1956a, b, ROSSITER and NICHOLLS 1957). After injections of P^{32}, these workers observed increased incorporation into RNA (as well as into the inorganic phosphates and into phosphatides) in the adrenals of rats exposed to cold for several hours (3°). Uptake into RNA was extremely low in adrenals of hypophysectomized rats and was enhanced in the adrenals of either intact or hypophysectomized rats given ACTH.

On the basis of studies in other cell types (BRACHET 1957), it may be assumed that enlargement of the nucleolus and increase in cytoplasmic RNA are associated with augmented protein synthesis. In a steroid-secreting gland, protein synthesis is not directly related to hormone synthesis but is concerned with the formation of new protoplasm and probably the formation of the enzymes that participate in hormone synthesis. As an indication of the increased protein synthesis in stimulated glands, GANIS et al. (1955) demonstrated that the inclusion of ACTH in the perfusate enhanced the incorporation of lysine-C^{14} into proteins of isolated bovine adrenals. PAVOA and FONTANA (1957) have shown a parallel increase of RNA and protein in stimulated glands.

Contrariwise, L. MANDEL et al. *(1956)* have demonstrated *a significant depression in the concentration of RNA*, as compared with DNA, in the adrenals *of* rats *with adrenal atrophy subsequent to prolonged administration of cortisone*. Thus, the residual cells possessed lower amounts of RNA.

f) Mitochondria and oxidative activity

Typically, the number of mitochondria is augmented over normal in active adrenocortical cells and reduced from normal in inactive cells (PORTO 1940, R. MILLER 1950, 1953, KNIGGE 1954a, YOSHIMURA 1954, LEVER 1956, ZELANDER 1959). In lipide-rich cells, these changes may not be readily evident, since dispersal of the mitochondria makes appreciation of their numbers difficult. R. MILLER (1953) has made the most careful analysis of this aspect of adrenal cytophysiology, by counting the number of mitochondria in mid cross-sections of 16 representative cells in the outer fasciculata of adrenals from animals subjected to different pretreatments. In one typical experiment his results were as follows: mitochondrial number in cells of gland of a control rat, 120; in cells of an untreated hypophysectomized rat, 74; in cells of a hypophysectomized rat treated 13 hours previously with about 1 U ACTH/100 g body weight, 111. A rise in number over control values was seen in the glands of intact rats given either ACTH or epinephrine. The mitochondria appeared to multiply by fission.

Confirmatory evidence of an increase in mitochondrial mass in stimulated adrenals has been presented by FIALA et al. (1956; Table 22). And several workers have found that the activity of the succinic dehydrogenase system (tetrazolium reduction), which apparently resides in the mitochondria, is depressed after hypophysectomy (Glick and Greenberg 1958) and enhanced in stimulated glands (e.g., Perry and Cumming 1952, DesMarais et al. 1955; see, however, Glick and Greenberg 1958). If succinoxidase activity is determined by oxygen consumption, however, no alteration is detectable after cortisone or ACTH treatment (Kalant 1958), indicating that there is no effect on the cytochromes. But there is some increase in overall oxidative activity in glands removed from animals given ACTH (Carpenter et al. 1946).

g) Golgi apparatus

There is also indication that the Golgi apparatus becomes enlarged and more dispersed in active adrenocortical cells, shrunken into a small juxtanuclear knot in inactive ones (e.g., Reese and Moon 1938, R. Miller 1950, Knigge 1954a, c). Such changes have not been studied so extensively in adrenocortical as in other cell types, no doubt because of the difficulties of Golgi demonstration in cells containing numerous lipide droplets.

h) Phosphatides

Like RNA, *phosphatides are more concentrated in active than in inactive glands.* This is demonstrable in sections by virtue of an increased reactivity to the Baker acid-hematein test (e.g., Cain and Harrison 1950). No critical examination of their intracellular localization has been undertaken, but the phosphatides presumably lie in the mitochondria and the membranes of the endoplasmic reticulum and perhaps also in the lipide droplets *per se* (Deane and Greep 1946, Cain and Harrison 1950, Karnovsky and Deane 1955). Droplets larger than typical mitochondria that display a positive phosphatide reaction become more numerous in active adrenals. These are most evident in the adrenals of the golden hamster, where they are not masked by ordinary lipide droplets (e.g., Knigge 1954c). In this species, so-called liposomes, which are not conspicuously sudanophilic but which do give a positive phosphatide test and are osmiophilic, develop in the fasciculata during prolonged stress.

As determined chemically, an increase in relative phosphatide concentration occurs with prolonged stimulation in many species (*Oryctolagus* — MacLachlan et al. 1941; *Cavia* — Oleson and Bloor 1941, Knouff et al. 1941; *Rattus* — Deane 1958; *Homo* — E. Adams and Baxter 1949, Symington and Davidson 1956). A decrease occurs with inactivation (*Rattus* — Karnovsky and Deane 1955). Correlatively, the incorporation of P^{32} into phosphatides is enhanced by the administration of ACTH and depressed by hypophysectomy (Riedel et al. 1954). Hokin et al. (1958), however, were unable to demonstrate any effect of ACTH on P^{32} incorporation into rat adrenal phosphatides *in vitro*.

i) Triglycerides and steroids

With acute *activation* of adrenocortical cells, there is generally a *reduction in the size and number of lipide droplets*, as well as in the amount of detectable cholesterol (Schultz reaction, birefringent crystals) (Figs. 49, 52). If the stimulus remains severe, free lipides may become sparse or disappear completely (Deanesly 1931, Hoerr 1931, Porto 1940, Deane et al. 1948, R. Miller 1950, Eichner 1953, Hagemann 1953, Kar 1953, Sandritter 1954, Eränkö 1955, Laqueur et al. 1955, Symington et al. 1956, Quenum and Comain 1959). If, however, the stimulus is chronic but not overwhelming, sudanophilic lipides and cholesterol may secondarily become unusually abundant (Flexner and Grollman 1939,

DEANE and SHAW 1947, BERGNER and DEANE 1948, SAYERS and SAYERS 1949, SAYERS 1950, CHAN 1950, CREUTZFELDT et al. 1953, KLÄRNER 1955, WEAVER 1956, RITTERSON and MAUER 1957, BURNS and HALE 1959). In this situation of chronic stress, however, the droplets and crystals remain typically smaller than in the "normal" adrenal.

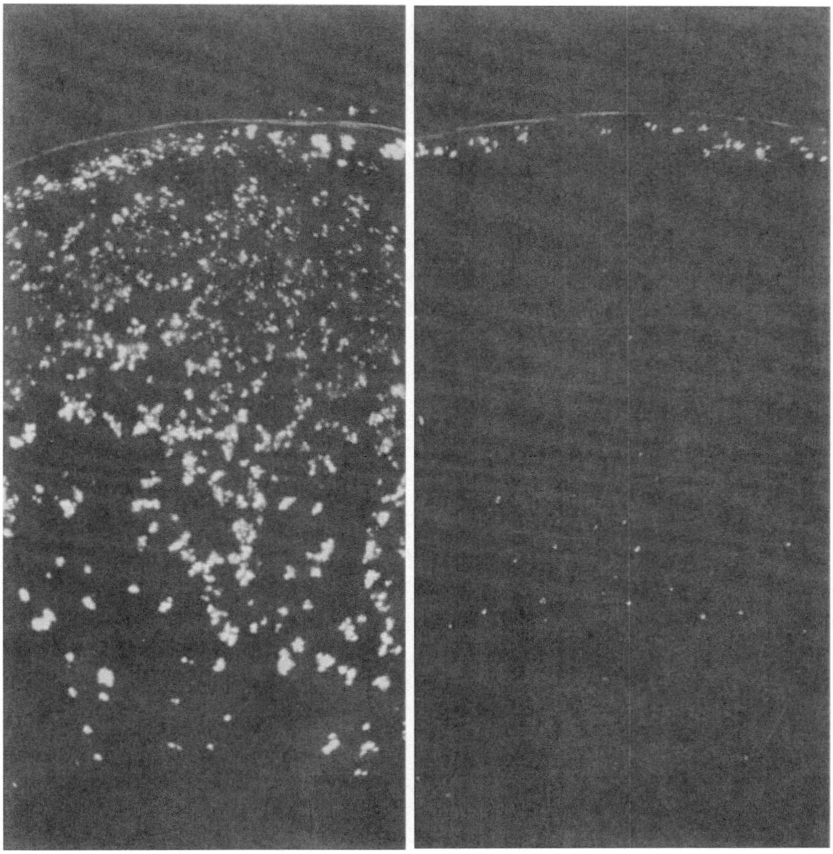

a b

Fig. 52. Loss of cholesterol-ester crystals from the inner cortical zones of the rat adrenal gland, induced by the administration of ACTH. a Normal rat, b rat that received 10 mg of crude ACTH (ca. 10 U) in 4 doses in 18 hours, killed at 24 hours. Formalin-fixed frozen sections photographed under the polarizing microscope. × 100. The line showing the position of the capsule was drawn in. From the study by BERGNER and DEANE (1948). The fasciculata has been severely depleted of its cholesterol stores, the zona glomerulosa is affected much less

BÄNDER (1950) has reported a drop in lipide content of the mouse cortex at midnight, which is a few hours after the peak secretory activity of the adrenals of these nocturnal animals (HALBERG et al. 1959).

SCHÖNBAUM and CASSELMAN (1958) have shown by parallel studies on stimulated rat glands that reduction in size and increase in number of lipide droplets occurs in association with a decline in ascorbate and an augmented production of corticosteroids. Cytochemically, an increase in diffuse sudanophilia of the cytoplasm followed the reduction in size of the lipide droplets.

An accumulation of lipides with chronic stress occurs not only in "typical" lipide-rich adrenals but also in glands that normally lack detectable lipide droplets, such as those of the

European hamster (*Cricetus cricetus*) (KAYSER and ARON 1950), the golden hamster (AGATE 1952, LEATHEM and STAUBER 1952), and the cow (GARM 1949, 1950, BELL and WEBER 1958).

The decline in chemically-determined cholesterol concentration with severe stress and the rebound with continued, moderate stress were noted earlier (p. 83). These chemical shifts depend principally on changes in the inner zones. A few reports in which changes in total lipides as well as cholesterol have been followed during stress include those of NICHOLS (1949) for the laboratory rat, OLESON and BLOOR (1941), KNOUFF et al. (1941) and CREUTZFELDT et al. (1953) for the guinea pig, and LIEBEGOTT (1944, 1952) and E. ADAMS and BAXTER (1949) for the human being. The net situation at any stage is the result of the ratio between synthesis and utilization, so that, except in acute or terminal states, determination of lipides and/or cholesterol may not assist in characterizing the state of the gland.

In the opposite situation, *when the adrenocortical cells become inactive, there are enlargement and confluence of the lipide droplets* (PORTO 1940, M. SIMPSON et al. 1943, DEANE et al. 1948, R. MILLER 1950, EICHNER 1953, M. MILLER 1953, CATER and STACK-DUNNE 1953, SANDRITTER 1954, MOORHOUSE 1956). The birefringent crystals become larger than normal, the Schultz reaction at first more intense (Fig. 49). Secondarily, the lipide droplets may lose detectable amounts of cholesterol. Gradually, if inactivation is complete, the lipide droplets disappear; this process generally takes a month or more. With inactivation, lipide droplets sometimes make their appearance in the adrenocortical cells of the golden hamster which are normally lipide-free (ALPERT 1950, DEANE and LYMAN 1954).

Chemically, a parallel initial increase and secondary decrease in concentrations of lipides and cholesterol in rat adrenals after hypophysectomy was found by KARNOVSKY and DEANE (1955). GLICK and OCHS (1955) observed a fall in cholesterol concentration in the rat cortex, particularly the fasciculata, after prolonged administration of cortisone.

Figure 53 summarizes the typical changes in the size of lipide droplets, nuclei and nucleoli of adrenocortical cells subjected to increased or decreased activity.

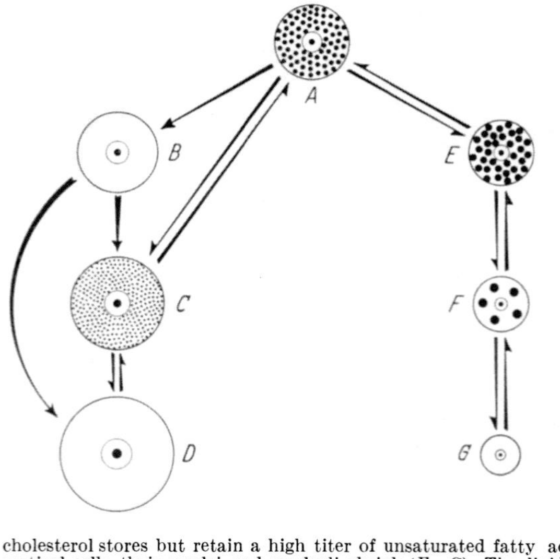

Fig. 53. Diagram of some of the cytological changes in typical lipide-containing adrenocortical cells (*A*) when their activity is stimulated (left) or inhibited (right). Composite of the proposals of DEANE et al. (1948), EICHNER (1953), SANDRITTER and HÜBOTTER (1954), R. MILLER (1954), KARNOVSKY and DEANE (1955). *Left.* With stimulation, adrenocortical cells, their nuclei and nucleoli enlarge. With acute (*B*) or prolonged, severe (*D*) stimulation there may be a loss of lipide stores; with more moderate but chronic (*C*) stimulation the lipide droplets become small, lose detectable cholesterol stores but retain a high titer of unsaturated fatty acids. *Right.* With removal of stimulus, adrenocortical cells, their nuclei and nucleoli shrink (*E*—*G*). The lipide droplets apparently coalesce and gradually disappear. Initially there may be an increase in cholesterol concentration. Gradually both the unsaturated fatty acids and the cholesterol stores decline. Both the severely stimulated and the long-term inactive cell lack lipide stores; they are distinguishable only by size

j) Pseudoplasmals

In the protracted debate over the chemical significance of the carbonyl reactivity of the lipide droplets in fixed adrenals (p. 55), the fact that the reactivity remained high in any residual lipide droplets in stimulated cells but disappeared from the residual lipides of inactive cells (Fig. 54) seemed a strong argument in favor of the reaction's being attributable to ketosteroids (hormones and precursors) (GREEP and DEANE 1949a, DEANE and SELIGMAN 1953). After the

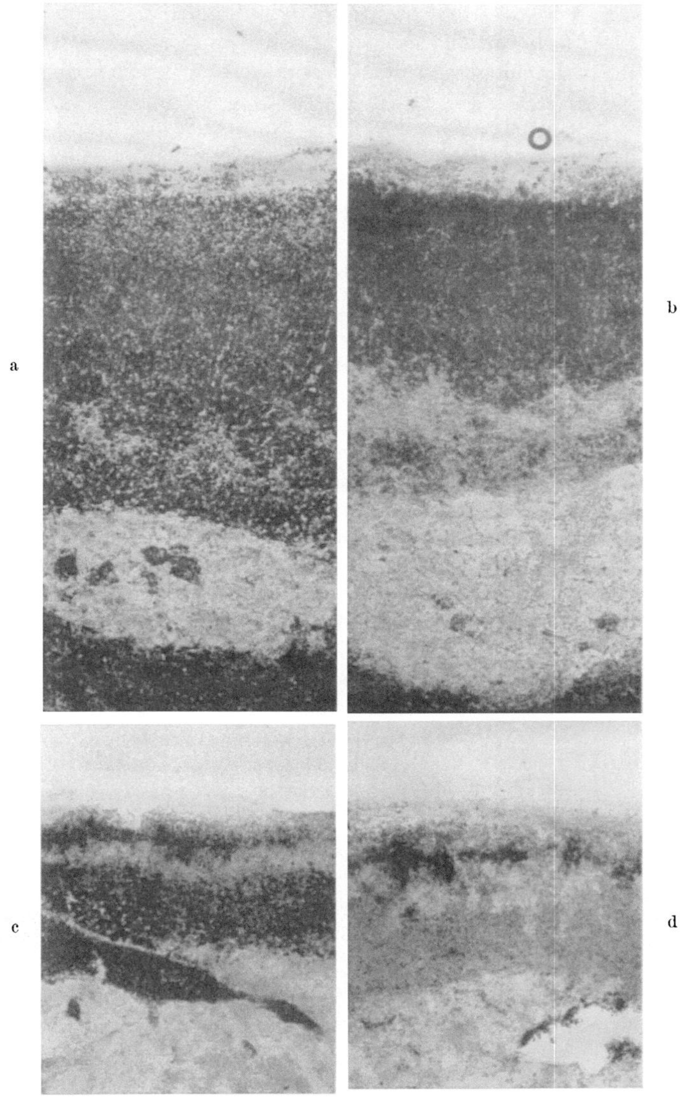

Fig. 54a–d. Disappearance of the pseudoplasmal reaction from the residual lipide droplets in the atrophied fasciculata of the rat after hypophysectomy. Formalin-fixed glands, frozen sections. × 50. a and b control glands; c and d glands from rats hypophysectomized 6 weeks earlier. a and c stained with Sudan black B for lipides; b and d stained with the Schiff reagent for aldehydes (see Fig. 42). The lipide droplets remaining in the atrophied inner zones no longer react at 6 weeks after hypophysectomy, while those in the glomerulosa remain reactive. From the study by KARNOVSKY and DEANE (1955)

chemical demonstration that the reactive compounds are aldehydes formed by the autoxidation, during fixation, of highly-unsaturated fatty-acid moieties, KARNOVSKY and DEANE (1955) found that *the relative concentration of* such *unsaturated compounds declined in the inactive glands of hypophysectomized rats.*

Thus, although the significance of unsaturated fatty acids in adrenocortical lipides remains obscure, a high concentration of unsaturated fatty-acid residues characterizes the lipide droplets of active cells, while a low concentration develops with inactivity.

k) Ascorbic acid

In the laboratory rat, in the mouse, and in the guinea pig, stress or ACTH induces a decline in histologically-demonstrable ascorbic acid in adrenocortical cells (DEANE and MORSE 1948, DUCOMMUN and MACH 1949, BACCHUS 1951, HAASE 1952, ROEPKE 1952, KAR 1953, BURNS and HALE 1959). This parallels the chemical change cited earlier (p. 84). There is a slow decline in detectable ascorbate after hypophysectomy (DEANE and MORSE 1948). Chemically there is a fall after prolonged treatment with cortisone (HOHLWEG and LASCHET 1949).

l) Enzymes and coenzymes concerned with corticosteroid synthesis

Alterations with function in a few of the enzymes and coenzymes reported in the section on adrenocortical chemistry (p.62 ff.) have been sought (see also p. 91ff., and below). Thus, L. GREENBERG and GLICK (1958) found for the rat gland, studied by the quantitative histochemical method, that administration of ACTH or of pantothenate caused increases in the *concentration of coenzyme A* in the adrenal cortex; hypophysectomy caused decreases. Furthermore, in glands removed at various times of day, the level was highest at midnight (cf. BÄNDER 1950, HALBERG et al. 1959). It seems noteworthy that in nocturnal rodents the level of circulating corticosterone is highest in the evening (HALBERG et al. 1959).

When the results were expressed as activity/mg, SAMUELS and HELMREICH (1956) found no significant change in *steroid 3β-ol dehydrogenase* activity after hypophysectomy; the activity per gland dropped significantly as the cortex atrophied. However, visualization of enzyme activity indicated a marked drop in the cells of the inner cortex (LEVY et al. 1959b). Enzyme activity was restored in these cells by ACTH administration (Fig. 55).

RUDOLPH and OLSEN (1956) observed no change in *glucose-6-phosphatase dehydrogenase* activity (per g wet weight) following hypophysectomy. There was a decline, however, in 6-phosphogluconate dehydrogenase activity; the latter enzyme is also TPN-dependent.

m) Hydrolytic enzymes

As noted earlier (p. 64ff.), the distribution and apparently the concentration of various hydrolytic enzymes differ in the various species in which their detection has been attempted. Thus any generalizations about their significance in adrenocortical physiology must be viewed with caution. The following observations about their changes in experimental conditions have been reported.

Alkaline phosphatase. In the rat, a decline in amount of demonstrable nonspecific alkaline phosphatase in the capillary walls of the inner zones of the cortex occurs after hypophysectomy (DEMPSEY et al. 1949, EGER et al. 1953) or the administration of a glucocorticoid (GAUNT et al. 1953, BONATI et al. 1957). The level of phosphatase in the inner zones appears enhanced when ACTH is administered (DEMPSEY et al. 1949, GENESI and MELLONI 1957). VERNE and HÉBERT (1955) have noted its enhancement during regeneration. In the mouse, marked enhancement of activity follows the administration of ACTH or of estradiol (? a stressful agent, p. 106) (ALLEN and SLATER 1956). In the golden hamster, too, alkaline-phosphatase activity rises in the adrenal cortex during a prolonged, chronic stress (infection) involving the accumulation of lipides (LEATHEM and STAUBER 1952). TUCHMANN-DUPLESSIS et al. (1955) report a rise in enzymic activity in the adrenals of frogs given ACTH, a decline in those given cortisone. In the guinea pig, PLANEL et al. (1956) and CLAYTON and HAMMANT (1957) describe an acute fall in alkaline-phosphatase activity in the fasciculata after the

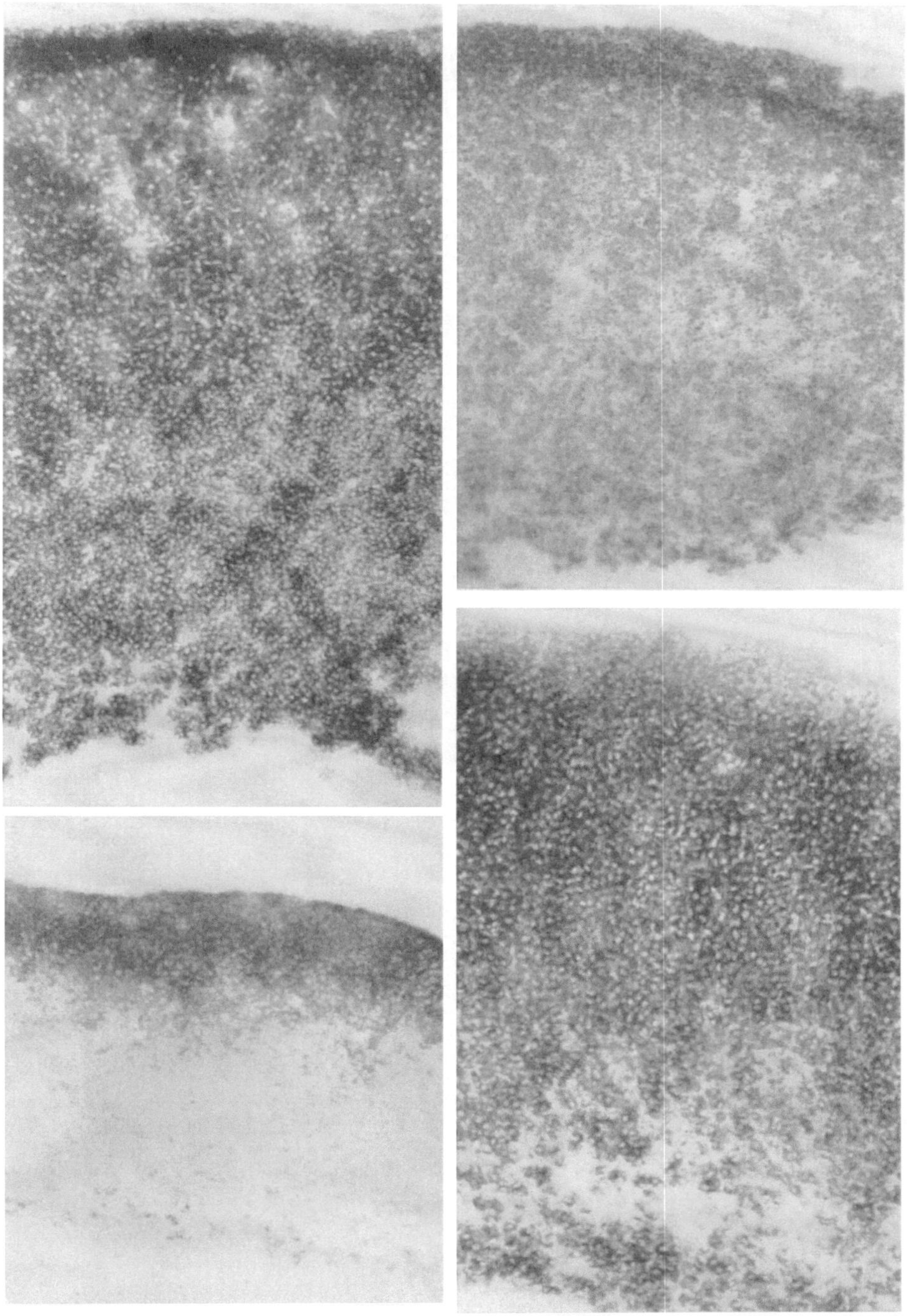

a b c d

Fig. 55a—d. Changes in adrenocortical steroid-3β-ol dehydrogenase activity after hypophysectomy and replacement therapy. Unfixed, cryostat sections incubated in a dehydroepiandrosterone-DPN-Nitro-BT medium; a $\times 80$; b—d $\times 100$. a normal female rat; b 2 weeks after hypophysectomy; c 7 weeks after hypophysectomy; d 11 weeks after hypophysectomy, treated for 1 week with ACTH. Enzymic activity is slowly lost from the inner zones after hypophysectomy, restored by ACTH treatment. From the study by LEVY et al. (1959b) administration of ACTH, whereas BURNS and HALE (1959) describe a rise. Thus, in general there seems to be some parallelism between enzymic activity and presumed secretory activity.

Acid phosphatase. SOULAIRAC et al. (1949) described an increase in the acid phosphatase activity in adrenal cortex of rats injected with insulin — presumably a "stressful" situation.

Nonspecific esterase. In the rat, hypophysectomy induces a decline in esterase activity of the fasciculata (COUTINHO et al. 1955), and in the mouse cortisone administration has a similar effect (ALLEN 1957). Oppositely, X-irradiation (a stress) causes an increase in activity in the fasciculata of the rat, and ACTH-administration has this effect in the mouse (Fig. 56).

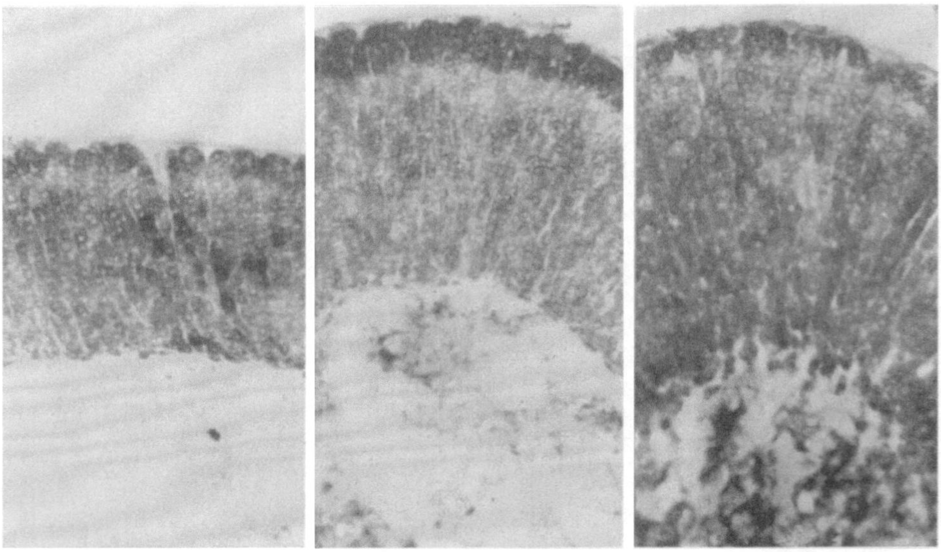

a b c

Fig. 56a—c. Nonspecific esterase activity in the mouse adrenal cortex is enhanced when the cells are stimulated. a normal male mouse; b mouse fed a low-Na diet for 4 weeks; c mouse treated with 8 U ACTH per day for 2 weeks. Formalin-fixed frozen sections, incubated in a 1-naphthyl acetate-Diazo Garnet GBC mixture. × 135. From the study by ALLEN (1957). Courtesy of the author. Esterase is most conspicuous in the cortex of the adrenal gland, where it is more active in the glomerulosa than in the fasciculata. Stimulation of the glomerulosa by low-Na intake enhances activity in this zone. Stimulation of the fasciculata by ACTH stimulates the enzyme in this zone

Diphtheria infection or the administration of a staphylococcus toxin (stresses) induces increased enzymatic activity in the adrenals of guinea pigs until the cortical cells become necrotic, when the reaction disappears (FIORI 1956). On the basis of studies with a relatively insensitive method, using a Tween as substrate, VERNE (1956) reports that no activity is detectable in the cortex of the intact adult *rat* but that activity is demonstrable in the adrenals of young, growing rats and in regenerating adrenals. This result differs from that of ALLEN (1958), using a different method, who observed an increase in activity of the *mouse* cortex with age.

Phenolsulfatase. In quantitative studies of enzyme distribution in the rat cortex, GLICK and STECKLEIN (1956) found some decline in phenolsulfatase activity after hypophysectomy but no significant change after cortisone administration or exposure of the animals to stress.

β-Glucuronidase. No change in concentration of this enzyme occurs with activation or inactivation of the gland (KNOBIL 1953, NAYYAR and GLICK 1955).

D. Dichotomy of hormone production in the mammalian adrenal cortex

During the 1930s and 1940s it became evident from studies on adrenal function undertaken with a variety of techniques that the mammalian cortex forms different steroids affecting organic and inorganic metabolism. For example, after total *adrenalectomy*, most species exhibit both a decline in the plasma sodium/potassium ratio and a conspicuous hypoglycemia, ending in death. When, however, sodium depletion is offset by supplying a high level of dietary sodium, the animal can survive, but hypoglycemia and related signs persist (SWINGLE and REMINGTON 1944). Some corticosteroids, such as deoxycorticosterone and especially aldosterone, repair electrolyte balance and thereby permit survival after adrenalectomy but have little influence on gluconeogenesis and blood sugar levels at doses that may be considered physiological. These hormones have therefore been termed *mineralocorticoids*. Contrariwise, other steroids, notably cortisone and cortisol, stimulate gluconeogenesis and correct hypoglycemia without having significant effect on serum and urinary electrolyte levels (SAYERS and SAYERS 1949, REICHSTEIN 1954, SWINGLE et al. 1954). These have therefore been named *glucocorticoids*. (It should be added that a few corticosteroids, notably corticosterone, influence both types of function. And, in fact, in high doses, typical mineralocorticoids influence organic metabolism, and typical glucocorticoids affect electrolyte distribution. See Chapter 4.)

Further indication of dichotomy is provided by *hypophysectomized* animals, which show nearly as severe disturbances in organic metabolism as do adrenalectomized animals but fail to evince a serious electrolyte imbalance (SWANN 1940, CHESTER JONES and ROBY 1954, EARLE et al. 1951, KNOBIL and GREEP 1958). Correlated with these physiological observations is the histological observation that, after hypophysectomy, the inner zones of the cortex undergo severe atrophy, whereas the zona glomerulosa remains nearly normal (ASCOLI and LEGNANI 1912, B. HOUSSAY and SAMMARTINO 1933a, b, CROOKE and GILMOUR 1938, DEANE and GREEP 1946, KNOBIL et al. 1954b, etc.). In consequence, SWANN (1940) tentatively proposed that *the two major zones of the mammalian cortex produce different types of hormones and are under different control*. He postulated that *the glomerulosa secretes a hormone* or hormones *responsible for maintaining electrolyte balance independently of the anterior pituitary, whereas the inner zones release hormones influencing organic metabolism under the influence of anterior-pituitary ACTH*. Subsequent experimentation has largely borne out Swann's postulates, although it is now clear that there is some overlap in function of the two classes of hormones. Also, the glomerulosa seems partly controlled by the anterior pituitary, while the inner zones may not be entirely inactivated by hypophysectomy. This subject is discussed further on p. 109.

I. Factors influencing the inner zones of the cortex and the secretion of glucocorticoids

1. Hypophysectomy

In mammals, the adrenocortical shrinkage (Fig. 48, p. 79; Table 16, p. 78) and the ultimate decline in the concentration of cholesterol (p. 94) and of ascorbic acid observed after hypophysectomy are almost entirely attributable to changes in the zonae fasciculata and reticularis. For example, DEANE and GREEP (1946) showed by planimetric measurement that the average cross-sectional area of cells in the zona fasciculata declined from about $70\mu^2$ in intact male rats to about $30\ \mu^2$ in long-term hypophysectomized animals, a proportionately greater change than seen in other zones. In addition, there were signs of cell degeneration in the inner zones suggesting that the number of cells here became steadily reduced. Data for the decrease in nuclear and nucleolar size after hypophysectomy are given in Tables 20 (p. 88) and 21 (p. 90). The cells of the inner zones acquire all of the cytological and cytochemical changes cited earlier (p. 86ff.) as signs of inactivity.

A common sign, in most mammals, *of inactivity of the inner zones is broadening of the lipide-free subglomerulosal zone.* After hypophysectomy the number of rows of small, droplet-free cells increases progressively (*Rattus* — M. SIMPSON et al. 1943, SARASON 1943, DEANE and GREEP 1946, CATER and LEVER 1954, LEVER 1954, 1956, YOSHIMURA 1954, LI et al. 1957; *Mus* — CHESTER JONES 1950, R. MILLER 1950, LOSTROH and WOODWARD 1958; *Cavia* — SCHWEIZER and LONG 1950; *Macaca* — P. SMITH 1955).

Associated with these changes, there is a severe *decline in the secretion of glucocorticoids.* This alteration in secretion was suspected earlier because, after hypophysectomy, the thymus fails to show its characteristic relative involution with age, the animals are extremely sensitive to food deprivation and show a tendency to hypoglycemia, and they are unusually sensitive to infection. These physiological deficiencies could be corrected by the administration of moderate doses of glucocorticoids (e.g., corticosterone or cortisone).

More recently, direct evidence of the secretory failure has been provided by assay of the adrenal venous blood. Thus, in dogs the adrenal *output of cortisol is reduced*, within 3 hours after hypophysectomy, to less than one-tenth the "normal" level (DE GURPIDE 1953, SWEAT and FARRELL 1954, FARRELL et al. 1955, RAUSCHKOLB et al. 1956, EIK-NES and BRIZZEE 1958). A comparable reduction in corticosterone output occurs in the rat (SINGER and STACK-DUNNE 1955). Examination of peripheral blood of the guinea pig (GOOD et al. 1956), of man (CHRISTY et al. 1956b, LIPSETT et al. 1957), of the dog (EIK-NES and BRIZZEE 1958), and of the rat (GUILLEMIN et al. 1958) also indicates a reduction in level of circulating glucocorticoids after hypophysectomy.

The inner zones of such glands normally fail to respond to any stimuli other than ACTH (p. 101). In other words, *"stresses" that ordinarily result in hypertrophy of the zones fail to do so in the absence of the pituitary* (e.g., SELYE and COLLIP 1936, SELYE 1946, FELDMAN 1951, GEMZELL 1952, MONEY 1955, STURTEVANT 1955, WEXLER et al. 1957, TANABE and CAFRUNY 1958), indicating that any enhancement of its basal activity is induced by the augmented secretion of ACTH (e.g., M. PICKFORD and VOGT 1951, GANONG and HUME 1956a, GOOD et al. 1956). There may be some hypertrophy in response to growth hormone alone (e.g., CATER and STACK-DUNNE 1953, 1955b, YOSHIMURA 1954, LOSTROH 1958, LOSTROH and WOODWARD 1958), but there is no sign that such adrenals produce enhanced quantities of glucocorticoids (SCHINDLER 1959).

Recently several suggestions of stimulus to the adrenal cortex in the absence of the pituitary have appeared. CHEDID and BOYER (1958a, b, c) have reported that endotoxins isolated from certain Gram-negative bacteria (e.g., *Salmonella typhi*) cause (1) an acute drop in the ascorbic acid content of the adrenals of hypophysectomized rats, (2) stimulation of the adrenals of cortisone-treated rats, and (3) enhanced production of corticosteroids by adrenals incubated *in vitro* without ACTH. A dose of 0.5 mg endotoxin was equivalent to 5 U ACTH.

Furthermore, HILTON et al. (1959) have shown that, in the isolated, perfused adrenal of the hypophysectomized dog, secretion of cortisol is increased 10 fold by addition of 0.4 U of synthetic arginine-vasopressin to the perfusate.

With the *in vitro* method, several substances have been found to enhance corticosteroid synthesis: a β-globulin fraction of blood (ROBERTS 1957), 3:5:3′-triiodothyroacetic acid (ROCHE et al. 1959), and 5-hydroxytryptamine (serotonin) (ROSENKRANTZ 1959).

2. Exogenous administration of cortical hormones

Table 17, p. 80, lists the evidence that, *like hypophysectomy, the administration of glucocorticoids at doses only slightly above the "physiological" level induces atrophy of the cells of the inner cortex* and other signs of inactivity in these zones (see also KRACHT 1958). The subglomerulosal zone becomes conspicuous (e.g., TUCHMANN-DUPLESSIS 1951a, b, GAUNT et al. 1953). This type of chemical "hypophysectomy" has the merit that the only outstanding effect is the depression of ACTH output; other anterior-pituitary hormones continue to be produced. There may, of course, be some additional direct (competitive ?) inhibition by the hormone of the adrenocortical cells (BIRMINGHAM and KURLENTS 1958). It is noteworthy that when glucocorticoid has been administered for a sufficient time to inhibit thoroughly the pituitary-adrenocortical system, a severe stress, like the injection of a bacterial toxin, induces no immediate signs of stimulation in the fasciculata cells (HOEDE 1956, WEXLER et al. 1958). And puncture of the adrenal vein reveals as low an output of glucocorticoids as is found after hypophysectomy (FARRELL et al. 1956a).

When administered at high doses, the mineralocorticoid deoxycorticosterone has similar effects on the fasciculata, albeit in most species atrophy of the glomerulosa is more conspicuous (p. 115) (e.g., SARASON 1943, GREEP and DEANE 1947, KNIGGE 1954b). SAYERS and SAYERS (1949) have shown that this compound serves to depress ACTH output; by weight, its effectiveness is about one-eighth that of cortisone or cortisol, however. (See also WEXLER et al. 1958.) These facts make it unlikely that aldosterone, the natural and far more potent mineralocorticoid, would inhibit the pituitary unless present at pharmacological levels.

3. Adrenal neoplasia with a high output of glucocorticoids

In patients with Cushing's syndrome resulting from adrenal neoplasia rather than hyperactivity of the pituitary, the tumor may be limited to one gland only. In such an instance the other gland exhibits marked atrophy (e.g., CERVIÑO et al. 1947). Recently SIEBENMANN (1959) observed such a case, in which the atrophied adrenal exhibited a normal glomerulosa and only aldosterone in the adrenal extract. Such atrophy of the contralateral adrenal has also been observed in rats developing spontaneous adrenocortical carcinomas (SNELL and STEWART 1959).

A similar phenomenon is demonstrable in rats bearing transplantable adrenocortical tumors that apparently secrete large quantities of glucocorticoids (as indicated by thymic atrophy and by secretory activity *in vitro*) (A. I. COHEN et al. 1957a, SNELL and STEWART 1959). The inner cortical layers of the host's glands virtually disappear after many months (Fig. 57); the glomerulosa persists (MULAY and EYESTONE 1955, A. I. COHEN et al. 1957b). On the other hand, in mice bearing a transplantable tumor that secretes both glucocorticoids and mineralocorticoids (the latter indicated by conspicuous sodium retention), both zones of the host's own glands undergo regression (A. I. COHEN et al. 1957b).

4. ACTH administration

In the opposite situation, *when ACTH is administered, the major effect on the mammalian cortex is stimulation of the cells in the inner zones*. The resultant hypertrophy of the whole gland is indicated in Table 18, p. 83. The subglomerulosal zone becomes inconspicuous or indistinguishable (e.g., BERGNER and DEANE 1948, TUCHMANN-DUPLESSIS 1951b, SCHMIDT and HOFFMANN 1954, KLÄRNER 1955). The fasciculata cells enlarge (p. 87), as to their nuclei and nucleoli (Tables 20 and 21); their lipide droplets and birefringent crystals become small (Fig. 52, p. 93), and they show the other cytological and chemical changes described earlier for hyperactivity. Under these conditions, there is comparatively little change in the zona glomerulosa (e.g., BERGNER and DEANE 1948, DUCOMMUN and MACH 1949, YOFFEY and BAXTER 1949, R. MILLER 1953, 1954, KNOBIL et al. 1954b, KLÄRNER 1955, WEAVER 1956, ALLEN 1957, RACE et al. 1957, LI et al. 1957). The adrenal response to ACTH occurs not only in the intact animal but also after hypophysectomy (M. SIMPSON et al. 1943, LI et al. 1957, LOSTROH and WOODWARD 1958).

In general, any effect of exogenous ACTH on the secretion of glucocorticoids is only indirectly detectable. Thus, there is accelerated involution of the thymus, augmented gluconeogenesis, eosinopenia, a negative nitrogen balance and other

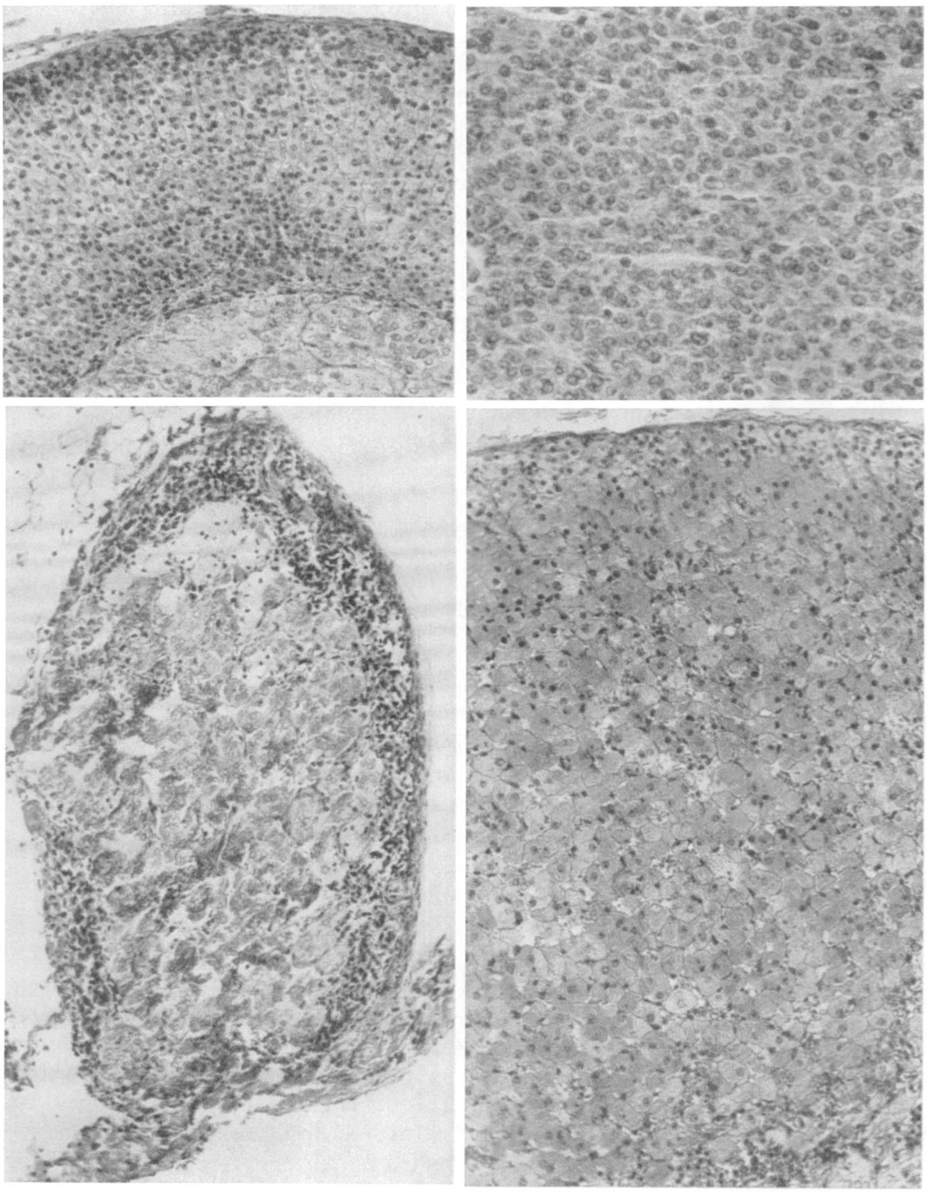

a c

b d

Fig. 57. Changes in mouse adrenal glands induced by the presence of an adrenocortical tumor (b) or an ACTH-producing pituitary tumor (d). a adrenal of normal control male mouse; b adrenal of a mouse bearing a graft of an adrenocortical tumor; c segment of the adrenocortical tumor; d adrenal of a mouse that had developed an ACTH-producing pituitary tumor. Hematoxylin and eosin; a, b and d ×140, *c* ×280. a, b and c from the studies by A. I. Cohen et al. (1957a), *d* from Bahn et al. (1957). All, courtesy Dr. Jacob Furth. The host's own adrenal cortex undergoes atrophy when functional adrenocortical tumor is implanted. The gland hypertrophies conspicuously in the presence of an ACTH-secreting anterior-pituitary tumor. In both instances, the major effect is on the inner zones. Note the remarkable enlargement of fasciculata cells in d.

signs of enhanced glucocorticoid secretion (INGLE et al. 1951, LI et al. 1957). But direct demonstration of enhanced output in the intact animal has not generally been possible, since the operation of puncturing the adrenal vein is so traumatic, even under Nembutal anesthesia, that endogenous secretion of ACTH is already maximal or nearly maximal when the first samples are taken (p. 75). Greatly enhanced output following ACTH administration is demonstrable in animals with permanently-implanted adrenal-vein catheters, however (D. NELSON et al. 1956, ENDRŐCZI et al. 1958, McDONALD and REICH 1959). There is also at least a temporary increase in the glucocorticoid level in the peripheral blood (CHRISTY et al. 1956b, GUILLEMIN et al. 1957, 1958, NABARRO et al. 1958).

The effectiveness of ACTH in raising glucocorticoid output is of course readily demonstrable in hypophysectomized animals, by measurement of the level in either the adrenal venous blood (e.g., D. NELSON and HUME 1955, FARRELL et al. 1955, 1958, SINGER and STACK-DUNNE 1955, LIPSCOMB and NELSON 1959) or in the peripheral blood (EIK-NES and BRIZZEE 1958). Here ACTH can be shown to raise the glucocorticoid output many-fold, while it generally raises aldosterone output much less (see p. 110). Similar effects have been reported for man (CHRISTY et al. 1956b, GOLD 1957).

5. Stimulation of endogenous secretion of ACTH ("stress")

A great variety of physiological and psychological stresses appear to enhance the output of ACTH, as evidenced by enlargement of the adrenal cortex, obliteration of the subglomerulosal zone (Fig. 67, p. 131), cytological signs of stimulation of the cells in the zonae fasciculata and reticularis (Fig. 49, p. 84; Fig. 67), and peripheral signs of increased glucocorticoid output such as acute involution of the thymus, lymphopenia, and eosinopenia. The levels of glucocorticoids in blood and urine rise. In addition, the gross chemical changes cited earlier (p. 83ff.) as characteristic of stimulated glands (especially in acute or terminal stages), with respect to total lipide and cholesterol concentrations, ascorbic-acid concentration, phosphorous turnover, etc., occur in stress situations as after ACTH administration. Since the inner zones comprise most of the gland, it is their chemical alteration that is dominant. In Table 23, some examples of the effects of stresses are cited to illustrate the various parameters that have been examined. None of these has any effect in hypophysectomized animals.

The variety of stimuli having these effects is surveyed in comprehensive reviews by SELYE (1946), SAYERS (1950), and BACHMANN (1954). Only a few of the major categories will be indicated here. In a wide variety of mammalian species, some or all of the following insults appear to act as stimuli to the pituitary-adrenocortical system, thus enhancing the output of glucocorticoids: (1) pyrogenic infections or bacterial toxins, (2) severe physical trauma or hemorrhage, (3) toxic substances such as formalin, tannic acid, histamine, ether, or chloroform, (4) starvation, (5) severe exercise, (6) anoxia, (7) severe cold, (8) X-irradiation (ROSENFELD 1958a, b), (9) scurvy and several other vitamin deficiencies, (10) hyperthyroidism, (11) the removal of one of the two glands (this does not increase the general body level of corticoids), (12) hyperestrogenism, (13) diabetes mellitus, (14) sympathomimetic drugs, (15) transplanted tumors, (16) hypervitaminosis D (BRUNE and EGER 1954), (17) psychic stress, (18) severe crowding, (19) morphine administration (TANABE and CAFRUNY 1958), and still other severe stimuli. References have been given in the above list only where an example is not cited in Table 23.

Table 23. *Representative signs of adrenocortical activation, in the presence of the anterior pituitary, by "stresses"*

Sign	Species	Stimulus	Reference
Adrenocortical hypertrophy, enlargement of cells in inner zones, and increased mitotic rate	*Mus musculus*	bacterial toxin or thyroxine, daily s.c. injections, 4 days	DEANESLY 1931
	Mus musculus	crowding	CHRISTIAN 1959
	Cavia cobaya	thyroxine, or thyroid stimulating hormone	SCHMIDT and SCHMIDT 1938, EHRENBRAND 1954
	Cavia cobaya	$1^1/_2$ hrs. swimming per day or starvation, 4-6 days	CREUTZFELDT et al. 1953
	Cavia cobaya	scorbutic diet	A. HOWARD and CATER 1959
	Canis familiaris	iodinated protein in diet, 5 months	E. SMITH et al. 1953
	Rattus norwegicus	transplanted Walker tumor 256, 3 weeks	HUHN and SCHNEPPENHEIM 1954
	Macaca mulatta	scorbutic diet, 4 months	MUKERJEE and BANERJEE 1954
Reduction of subglomerulosal zone	*Rattus norwegicus*	1% thyroid powder in food	DEANE and GREEP 1947
	Rattus norwegicus	estradiol benzoate	GREEP and CHESTER JONES 1950
Hypertrophy of nuclei in zona fasciculata	*Cavia cobaya*	swimming or starvation	CREUTZFELDT et al. 1953
	Rattus norwegicus	adrenaline, 80 μg/100 g, 4 hrs. previously	R. MILLER 1954
	Rattus norwegicus	KCN, 8—22 days	SCHNEPPENHEIM and HUHN 1955
	Mesocricetus auratus	starvation, 8—10 days	KNIGGE 1954 c
Increased RNA in fasciculata cells	*Homo sapiens*	fatal peritonitis	SYMINGTON et al. 1956
Increased number of mitochondria in fasciculata cells	*Rattus norwegicus*	adrenaline, 80 μg/100 g, 10 hrs. previously	R. MILLER 1953
Depletion of lipide from fasciculata (acute or terminal)	*Homo sapiens*	fatal infections	LIEBEGOTT 1944, 1952, STONER et al. 1953
	Cavia cobaya	starvation, 6 days	CREUTZFELDT et al. 1953
	Cavia cobaya	insulin, 33 U/kg, 6—9 hrs.	HAGEMANN 1953
	Felis domesticus	adrenaline or laparotomy	LAQUEUR et al. 1955
	Cricetulus griseus	*Trichinella* infection, 3 days	RITTERSON and MAUER 1957
Depletion of cholesterol from fasciculata (acute or terminal)	*Homo sapiens*	fatal infections or severe traumas	LIEBEGOTT 1944, 1952, STONER et al. 1953
	Rattus norwegicus	complete water deprivation, 8—10 days	NICHOLS 1949
	Rattus norwegicus	0°, 48 hrs., or adrenaline, 10 mins.	F. ROBINSON and YOFFEY 1950
	Cavia cobaya	electric shock	BARUCCI and MORI 1951
Depletion of ascorbic acid from fasciculata (acute or terminal)	*Rattus norwegicus*	2°, 1 hr.	BACCHUS 1951
	Cavia cobaya	1°, 28 days	HAASE 1952

Table 23 (cont.)

Sign	Species	Stimulus	Reference
Enhanced level of glucocorticoids			
(1) in gland	*Rattus norwegicus*	ether anesthesia	HOLZBAUER 1957a
	Rattus norwegicus	audiogenic seizure, 1—4 hrs. earlier	DUNCAN 1957
(2) in venous effluent	*Canis familiaris*	unilateral adrenalectomy, 1 week earlier	GANONG and HUME 1956b
	Canis familiaris	severe hypothermia	EGDAHL and RICHARDS 1956, SUZUKI et al. 1957
(3) in peripheral blood	*Homo sapiens*	typhoid vaccine	CHRISTY et al. 1956a
	Macaca mulatta	"anxiety"	MASON et al. 1957
(4) in urine	*Cavia cobaya*	scorbutic diet, more than 10 days	NADEL and SCHNEIDER 1952, PRUNTY et al. 1955, EISENSTEIN and WENNEKER 1958
	Cavia cobaya	L-thyroxine	LEVIN and DAUGHADAY 1955
Severe thymic involution	*Rattus norwegicus*	early pantothenic acid deficiency	DEANE and McKIBBIN 1946
	Rattus norwegicus	1% thyroid powder in food, 4—8 weeks	DEANE and GREEP 1947
	Microtus agrestis	fighting (and inanition)	CLARKE 1953
	Mesocricetus auratus	starvation, 7—8 days	KNIGGE 1955
	Mus musculus	ethanol (3 g/kg)	SANTISTEBAN and SWINYARD 1956
Enhanced glycogen deposition in liver	*Rattus norwegicus*	estradiol benzoate	GEMZELL 1948
	Rattus norwegicus	L-thyroxine or triiodothyronine	TIMIRAS and WOODBURY 1955
	Mus musculus	early pantothenic acid deficiency	MELAMPY et al. 1951
Lymphopenia	*Rattus norwegicus*	early pantothenic acid deficiency	WINTERS et al. 1952
	Cavia cobaya	scorbutic diet	EISENSTEIN and SHANK 1951
Eosinopenia	*Rattus norwegicus*	early pantothenic acid deficiency	WINTERS et al. 1952
	Felis domesticus	adrenaline	LAQUEUR et al. 1955

a) Species variability

Situations that are stressful for one animal species may not be for another. To take just one example, exposure to cold or other stimuli that induce signs of adrenal activation in the *laboratory* rat fail to do so in the *wild* Norway rat (WOODS 1957a, b). [Injection of a very high dose of ACTH (40 U) is required to reduce adrenal ascorbate in the wild variety.] Similarly, secretory activity in the dog is apparently unaffected by reducing the ambient temperature to −10° (D. NELSON et al. 1956). To evoke increased secretory activity, the air temperature must be dropped to about −75° (EGDAHL and RICHARDS 1956). And temperatures near 0°, which enhance adrenal activity in most small rodents, apparently fail to do so in hibernating ones (DEANE and LYMAN 1954, KNIGGE 1957). (See also ELTON et al. 1959, FLICKINGER 1959.)

b) Estrogens

The stimulation of the pituitary-adrenocortical system by estrogens requires amplification. With respect to the natural estrogens, there seems ample evidence that these serve to activate the system (COURRIER et al. 1953). First of all, as noted above (Table 2, p. 18), the adrenal cortex of the female is relatively larger than that of the male in most mammalian species so far examined (with the exception of the hibernating rodents). Ovariectomy reduces relative adrenal weight, while the administration of estrogens such as estrone or estradiol to gonadectomized or intact animals of either sex enhances adrenal weight (e.g., DEL CASTILLO and SAMMARTINO 1937, PINTO 1941, 1945a, b, GEMZELL 1948, 1952, GREEP and CHESTER JONES 1950, RENNELS et al. 1953, UUSPÄÄ and JÄRVINEN 1955, CAMPBELL et al. 1956, GOMPERTZ 1958). Such changes require the presence of the pituitary (SELYE and COLLIP 1936, GEMZELL 1952). [It is perhaps noteworthy that not all strains of mice respond in this fashion (WESTBERG et al. 1957), and the guinea pig appears to be relatively unresponsive to estrogen (CLAYTON and HAMMANT 1959).] The adrenals are also enlarged during pregnancy, although the degree varies in different species (e.g., D. BAKER 1938, DUREY 1954, DELOST and DELOST 1955, P. SMITH 1955, WHITELEY and STONER 1957, POULTON and REESE 1957). Animals in estrus, during pregnancy, or after administration of natural estrogens show signs of increased glucocorticoid output, such as thymic involution (at doses at which the estrogen itself has no effect) (GREEP and CHESTER JONES 1950), augmented levels of liver glycogen (GEMZELL 1948), high levels of circulating glucocorticoids (H. ROBINSON et al. 1955, JAILER et al. 1959), and increased responsiveness *in vitro* to ACTH (MCKERNS 1957). The latter effect is not one of direct activation of the adrenal, since responsiveness is not enhanced *in vitro* if the estrogen is added to the incubation medium.

One explanation offered for the stimulatory effect of estrogen on corticosteroid output is that in its presence the corticoids are bound to plasma protein more than usual and therefore their biological activity is reduced (DAUGHADAY 1959, JAILER et al. 1959, SANDBERG and SLAUNWHITE 1959). In addition there may be depressed clearance of corticoids (CHRISTY et al. 1959).

For some wholly unexplained reason, when the adrenal of a unilaterally-adrenalectomized female rat is transplanted to the portal venous bed (spleen), it undergoes more marked enlargement than does a single adrenal left *in situ* (BERNSTEIN 1958, 1959). This effect is offset by ovariectomy. Male rats show no such differential. There is no evidence of difference in responsiveness to ACTH by adrenals in the portal circulation. Estrogen administered to castrated rats simulates the effect of the ovaries. Why an adrenal located in the portal circulation is so sensitive to the influence of estrogen remains unexplained.

A somewhat more complicated situation appears to exist with some synthetic estrogens and substituted natural estrogens. Like estrone or estradiol, diethylstilbestrol, *p*-oxypropiophenone, hexestrol, and some substituted estrogens induce adrenal hypertrophy (e.g., SKELTON et al. 1949, KRACHT and SPAETHE 1953b, GILLISSEN and GILLESSEN 1955, STURTEVANT 1955, VOGT 1955, 1957a, b, McKERNS 1957). Yet for diethylstilbestrol, hexestrol, and ethinyl estradiol, at least, there is evidence that the hypertrophied adrenal (rat) has reduced ability to secrete glucocorticoids. Thus VOGT (1955, 1957a, b) found that the enlarged adrenals of the hexestrol-treated *rat* released subnormal amounts of glucocorticoids, as measured in the adrenal venous effluent. [This effect was not found in the *rabbit*. But hexestrol does not cause adrenal hypertrophy in the rabbit (BERGSTRAND 1952).] HOLZBAUER (1957a, b) observed signs of reduced responsiveness in such rat adrenals. Secretory ability was restored if the animals were infused with blood from normal animals (HOLZBAUER and VOGT 1957b). McKERNS (1957) observed that the addition of diethylstilbestrol or of ethinyl estradiol to incubation media depressed formation of corticoids by rat glands *in vitro*. It thus appears that several unnatural estrogens may somehow inhibit synthetic processes in the adrenal cortex. This is reminiscent of the inhibition by amphenone (p. 136).

The opposite responses of the adrenals of hibernating rodents to gonadectomy and to estrogen and androgen administration (e.g., HARTMAN and BROWNELL 1949, ALPERT 1950, DEANE and LYMAN 1954) remain unanalyzed.

c) So-called ACTH-like action of androgens

In intact or gonadectomized animals, a high circulating titer of testosterone or other androgens generally causes a depression in relative adrenal weight (e.g., GREEP and CHESTER JONES 1950, BERGSTRAND 1952, RENNELS et al. 1953, COURRIER et al. 1953, ZIZINE 1956; see, however, ROY et al. 1958). The adrenals appear somewhat inactive, in that the subglomerulosal zone is broad and the cells in the fasciculata are smaller than normal. However, in contrast to the corticoids, androgens *reduce* the size of the lipide droplets and depress the cholesterol level in the inner zones (GREEP and CHESTER JONES 1950, RENNELS 1952; see also SIMONNET et al. 1951). Thus the shrinkage is not associated with all the characteristic signs of cellular inactivity. And biochemical studies indicate that androgen treatment enhances the phosphorus turnover (BERGSTRAND 1952) and the succinoxidase activity of adrenals (KALANT and SELLERS 1954, 1957, KALANT 1958).

In rats, moreover, many C_{19} steroids of varying androgenic potency appear to offset the atrophy of the adrenal subsequent to either hypophysectomy (LEATHEM 1944, ZIZINE et al. 1950, C. ROSSI and MONTANARI 1956) or cortisone administration (ZIZINE 1953a, b, WINTER et al. 1953, GAUNT et al. 1953, KAR et al. 1957, JOST and GRANDOU 1958, LLAURADO et al. 1959). The latter effect has not been reproduced in the monkey (KNOBIL et al. 1953). There is no evidence to date that such adrenals are at all stimulated with respect to secretion. (See also CARTER et al. 1958.) *In all probability the effect of androgen is a local anabolic one, evident with respect to adrenal weight (rats) only in the absence of ACTH.*

6. ACTH-producing pituitary tumors

Pituitary tumors that produce excessive quantities of ACTH occur spontaneously in man and are apparently responsible for Cushing's syndrome in some instances (CUSHING 1932, V. MARKS 1959). In other cases Cushing's syndrome arises as the result of hyperreactivity of the adrenal cortex to ACTH or even autonomy of the cortex from ACTH (neoplasia) (e.g., JAILER et al. 1954, NABARRO et al. 1958). There is hypertrophy of the inner zones of the cortex (R. COHEN et al.

1959), associated with a massive increase in output of glucocorticoids. (There may also be an increase in output of adrenal androgens — e.g., NABARRO et al. 1958.) Frequently, the glomerulosa appears almost normal, and there is little sign of abnormal output of electrolyte-regulating hormones.

Of interest for the experimental analysis of Cushing's syndrome caused by high ACTH-output is the recent induction, by radiation, of ACTH-producing pituitary tumors in mice (UPTON and FURTH 1955). In the original mice and in host mice to which portions of the pituitary tumor were transplanted, the inner zones of the adrenals hypertrophied greatly (Fig. 57, p. 102), the thymuses atrophied, and there was fatal susceptibility to infection (BAHN et al. 1957). There was little histological change in the zona glomerulosa. Analysis of peripheral blood and urine of recipients of such pituitary tumors indicated high output of corticosterone and of a non-androgenic C_{19} steroid (WILSON et al. 1958). Aldosterone was not detectable in the amounts of fluid analyzed.

7. Locus of secretion of androgens and other sex hormones by the adrenal cortex

For many years it has been customary to assign production of androgens by the adrenal cortex to the zona reticularis. This concept was advocated by VINES (1938), who argued that the conspicuousness of this zone in patients with virilism of adrenal origin indicated the production of androgens there. It was VINES who asserted that the distinct acidophilia (fuchsinophilia) of the zone constituted a histochemical test for androgenicity (p. 43). Numerous subsequent workers have adopted this hypothesis, of whom some of the more recent are BOTELLA LLUSIA (1949), HÉBERT et al. (1954), YOFFEY (1955), and RACE et al. (1957). The case for this hypothesis is presented in Chapter 5.

How much androgen is normally secreted by the adrenal is still a matter of question. But presumably most of the 17-ketosteroids excreted by women and a large portion of those excreted by men are of adrenal origin (Chapters 3, 5), although much may be produced by peripheral metabolism of C_{21} steroids (OERTEL and EIK-NES 1959). Certainly the large quantities excreted by patients with virilizing adrenal hyperplasia arise therein. The evidence for the *secretion* of 17-ketosteroids by adrenals of species other than man is less definite, although the *excretion* of such cempunds by dogs (e.g., J. H. BROWN et al. 1956, RACE et al. 1957), rats (BONATI et al. 1957) and mice (WILSON et al. 1958) is established. A. I. COHEN et al. (1957a) have demonstrated the formation of 11-hydroxyandrostenedione by normal and tumorous mouse adrenals *in vitro*. PRICE and INGLE (1957) have provided evidence for the production of small amounts of androgen by the rat adrenal.

The physiological evidence that the secretion of androgens and other C_{19} steroids is stimulated by ACTH (GARDNER 1956) and the biochemical evidence that adrenocortical cells may produce C_{19} steroids on an alternative pathway to that for C_{21} steroids (see Chapters 3,5) raise an alternative possibility for the site of production of adrenal adrogens that requires investigation.

It seems most likely to this reviewer that adrenocortical cells of the inner zones release C_{19} steroids as well as glucocorticoids, under the influence of ACTH. This hypothesis would account for the usual finding that the excretion of C_{19} as well as C_{21} steroids increases with ACTH administration or stress. Discrepancies in relative secretion of the two hormone types may reflect differences in enzymic activity in different patients or at different ages. The frequent association of a conspicuous reticularis in adrenals from patients with adrenal virilism may mean simply that the adrenal cortex is excessively stimulated, depleted of lipide (hence acidophilic) and engorged. This hypothesis with reference to the *postnatal* adrenal of course does not preclude the possibility that the "fetal" zone of the primate adrenal may secrete 17-ketosteroids *prenatally* (p. 125).

There is no evidence whatsoever for the *locus* of production of estrogens or progesterone in the adrenal. (See Chapter 5.)

II. Factors influencing the zona glomerulosa and the secretion of aldosterone

The accumulated evidence suggests strongly that *in mammals the normal glomerulosa secretes a sodium-retaining corticoid (aldosterone) and that this secretion is not dominantly governed by ACTH.* The glomerulosa remains essentially unaffected by either depression or enhancement of the level of circulating ACTH (Figs. 49, 50, 52, 54, 55), and most evidence indicates that mineralocorticoid output is not markedly altered in these conditions. On the other hand, the glomerulosa and the secretion of aldosterone are dramatically affected by alteration in the ratio of electrolytes in the diet, by diseases producing edema, and by the administration of a potent mineralocorticoid (Figs. 56, 58). The important consideration is that these may have their influence even in the absence of the pituitary (Fig. 59). Thus, the appearance of the glomerulosa and the secretion of aldosterone can be altered independently of alteration of the fasciculata and the secretion of glucocorticoids.

The most direct evidence that the glomerulosa is the source of aldosterone comes from incubation studies. C. Giroud et al. (1956a) found that whole rat glands incubated *in vitro* formed aldosterone at a rate of 0.66 μg/100 mg tissue per hour; whereas isolated capsule with attached glomerulosa formed it at a rate of 1.8 μg/100 mg tissue per hour, while the remainder of the gland (inner zones plus medulla) produced negligible quantities. The inner zone produced corticosterone. Comparable separability of function was observed in slices of bovine gland (Ayres et al. 1956, C. Giroud et al. 1958, Stachenko and Giroud 1958a, b).

1. Hypophysectomy

Unlike the inner zones of the cortex (zonae fasciculata and reticularis), the outermost zone of the mammalian cortex, the zona glomerulosa, appears to be largely independent of the pituitary. Thus, after hypophysectomy, the volume of the glomerulosa appears to change little: its width in cross-section, of course, may increase as a result of the shrinkage of the inner zones. There is no gross shrinkage of the cells, their nuclei or nucleoli (Table 21, p. 90). Lack of any gross effect on the glomerulosa after hypophysectomy has been reported repeatedly for the rat (Crooke and Gilmour 1938, Sarason 1943, Deane and Greep 1946, Cater and Stack-Dunne 1953, Hartroft and Hartroft 1955, Ando 1957) and also for the dog (B. Houssay and Sammartino 1933a, b, Rauschkolb et al. 1956, Bahn et al. 1960), for the mouse (Chester Jones 1950, R. Miller 1950), for the guinea pig (Schweizer and Long 1950), for the golden hamster (Knigge 1954a) and for the monkey (Knobil et al. 1954b). In general, the cytological characteristics and histochemical properties of the cells remain essentially normal (above references, and Dempsey et al. 1949, Coutinho et al. 1955, Karnovsky and Deane 1955, Allen 1957, Levy et al. 1959b).

Several workers, however, have reported signs of moderate inactivity of glomerulosal cells after hypophysectomy. For example, in rats, Deane and

Greep (1946) and more especially Lever (1954) noted shrinkage of the glomerulosal cells with long-term hypophysectomy. Chester Jones (1949b, 1950) reported the loss of lipide from the glomerulosal cells in mice, and Schweizer and Long (1950) noted loss of lipide in the guinea pig. Cater and Stack-Dunne (1955a) recorded a depression of mitosis in this zone in rats. Lever (1956) observed in rats, 9 weeks after hypophysectomy, signs of disruption of the mitochondria (electron microscopy). In dogs hypophysectomized for several months, Lane and de Bodo (1952) reported quite profound atrophy of the zone (see, however, Bahn et al. 1960). Of course some of these changes observed following hypophysectomy might depend on factors other than the loss of ACTH, possibly the lack of growth hormone or the depression of thyroid secretion.

In both the rat (Singer and Stack-Dunne 1955) and the dog (Farrell et al. 1955, Rauschkolb et al. 1956, Ganong et al. 1959), hypophysectomy reduces the secretion of aldosterone significantly as measured in adrenal venous effluent. In rats, 2—44 days after removal of the pituitary, the secretory rate was only about 20% of that obtained in intact animals. In dogs, 6 hours or one week after hypophysectomy, the rate was about 65% of normal, later it was reduced further. In both species, of course, hypophysectomy depressed the output of glucocorticoids much more severely. Luetscher and Curtis (1955), Venning et al. (1956), MacLean et al. (1957) and Llaurado (1957) report that the excretion of aldosterone is within the normal range in human beings lacking pituitary function.

2. Administration of glucocorticoids

Chemical hypophysectomy with respect to ACTH secretion can be produced by chronic administration of a glucocorticoid at doses above the physiological level (Sayers and Sayers 1949). Such treatment, like hypophysectomy, produces profound atrophy of the inner cortex, but similarly spares the zona glomerulosa (rat — Gaunt et al. 1953, Bonati et al. 1957; human — O'Donnell et al. 1951; monkey — Knobil et al. 1954b; dog — Farrell et al. 1956a). And, unlike hypophysectomy, glucocorticoid administration produces no reduction in aldosterone output in the dog (Farrell et al. 1956a) or in the human being (Venning et al. 1956). This result suggests strongly that the effects of surgical hypophysectomy on the glomerulosa may not be attributable to the loss of ACTH, but to the loss of other secretions (see Llaurado 1957).

3. ACTH or "stress"

Contrariwise, exposure of a mammal to most sorts of physiological stress or treatment with commercially-available ACTH generally induces no conspicuous change in the glomerulosa, either in the intact or hypophysectomized individual (ACTH: rat — Bergner and Deane 1948, Ducommun and Mach 1949; mouse — Allen 1957; monkey — Knobil et al. 1954b; stress: rat — Greep and Deane 1949a; human being — Stoner et al. 1953).

However, *the available ACTHs do cause some signs of alteration of the glomerulosa of hypophysectomized animals.* Thus Chester Jones (1949b) reported a reappearance of lipide droplets in the glomerulosa of hypophysectomized mice given ACTH. Wexler et al. (1955) and Wexler and Rinfret (1955) found that a few samples of commercial ACTH, when administered in low doses (< 1 U) were capable of depleting lipides for a few days from the glomerulosa of hypophysectomized rat adrenals; higher doses had no such effect. Other samples of ACTH had no effect at any dose level. Cater and Stack-Dunne (1953, 1955b, c) found some increase in mitotic rate in the glomerulosa of rats hypophysectomized for 14 days and treated with ACTH for 3 days (although an increase was far more striking with growth hormone). In electron micrographs, LEVER (1956) reported "repair" of glomerulosal cells by ACTH in rats hypophysectomized many weeks earlier. With the most highly purified ACTH yet available, LOSTROH and WOODWARD (1958) and LOSTROH (1958) have demonstrated changes in the lipide content of the glomerulosa in both mouse and rat.

Similarly, moderate alterations of the appearance of glomerulosal cells of intact animals or human beings occur when ACTH is administered. Thus R. MILLER (1954; Table 21, p. 90) found enlargement of the nuclei in this zone in rats given a single injection of ACTH a few hours previously. LANDING and FERIOZI (1954b) noted some enlargement of cells and depletion of lipide from the glomerulosa in patients treated with ACTH. KLÄRNER (1955) reported some depletion of lipides from the zone in rats treated for prolonged periods with low daily doses of ACTH (0.2—1 U). RACE and GREEN (1955) observed some increase in glomerulosal width in intact rats treated for 2 days with ACTH, as did RACE et al. (1957) in intact dogs treated for 4 weeks. In all instances, the stimulation of the fasciculata was of course even more conspicuous.

By direct analysis of adrenal venous blood, SINGER and STACK-DUNNE (1955) observed a doubling of aldosterone output with administration of ACTH to rats hypophysectomized 2—44 days. Similarly, FARRELL et al. (1955) observed enhancement of aldosterone secretion in acutely hypophysectomized dogs given ACTH. In both experiments, the impetus to glucocorticoid secretion was much more striking. In a more recent study, however, FARRELL and co-workers (1958) have found a significant stimulation of aldosterone output in acutely decerebrate dogs given highly purified ACTHs. In fact, the relative potencies of these ACTHs were nearly parallel in aldosterone-stimulating power and in glucocorticoid-stimulating power. Whether the influence was attributable to a contaminant, however, could not be determined. (See also VENNING et al. 1958.)

In contrast, the excretion of aldosterone in human urine shows only a transitory rise with chronic ACTH administration (LIDDLE et al. 1956). It still remains a question whether ACTH has more than a temporary effect on aldosterone secretion.

In vitro, the effect of ACTH on aldosterone secretion is generally negligible. In perfused bovine adrenals, ACTH fails to cause any unequivocal rise in aldosterone output (ROSENFELD et al. 1956, C. GIROUD et al. 1958). STACHENKO and GIROUD (1959b) could demonstrate no stimulus to aldosterone output by bovine glomerulosa. Similarly, with incubated rat adrenals, the increase in aldosterone formation is relatively much less than in glucocorticoid formation (C. GIROUD et al. 1956b, MIALHE-VOLOSS and BAULIEU 1958). However, some ACTHs stimulate aldosterone production *in vitro* by both rat and human adrenal tissue (VENNING et al. 1958).

All of these evidences suggest strongly that the anterior pituitary influences this function to some degree. The major influence on the glomerulosa and aldosterone secretion, however, appears to be exerted in some fashion by alteration in electrolyte balance or blood volume, or both, as indicated in the sections below.

4. Disturbance of electrolyte balance

a) Sodium deficiency

Dramatic changes in the appearance of the zona glomerulosa and also in the output of aldosterone can be produced by altering the Na^+/K^+ ratio in the diet (Figs. 56, 58). *Prolonged intake of a diet containing little or no Na^+ causes marked* *hypertrophy of the zone*, reflecting a marked enlargement of the cells. This phenomenon was reported simultaneously for rats by BADINEZ and CROXATTO (1948), DEANE et al. (1948), and NICHOLS (1948). Numerous subsequent workers have confirmed the observation, not only for the rat (BACCHUS 1950, EICHNER 1953, PESCHEL and RACE 1954, RACE and GREEN 1955, HARTROFT and HARTROFT

a b c

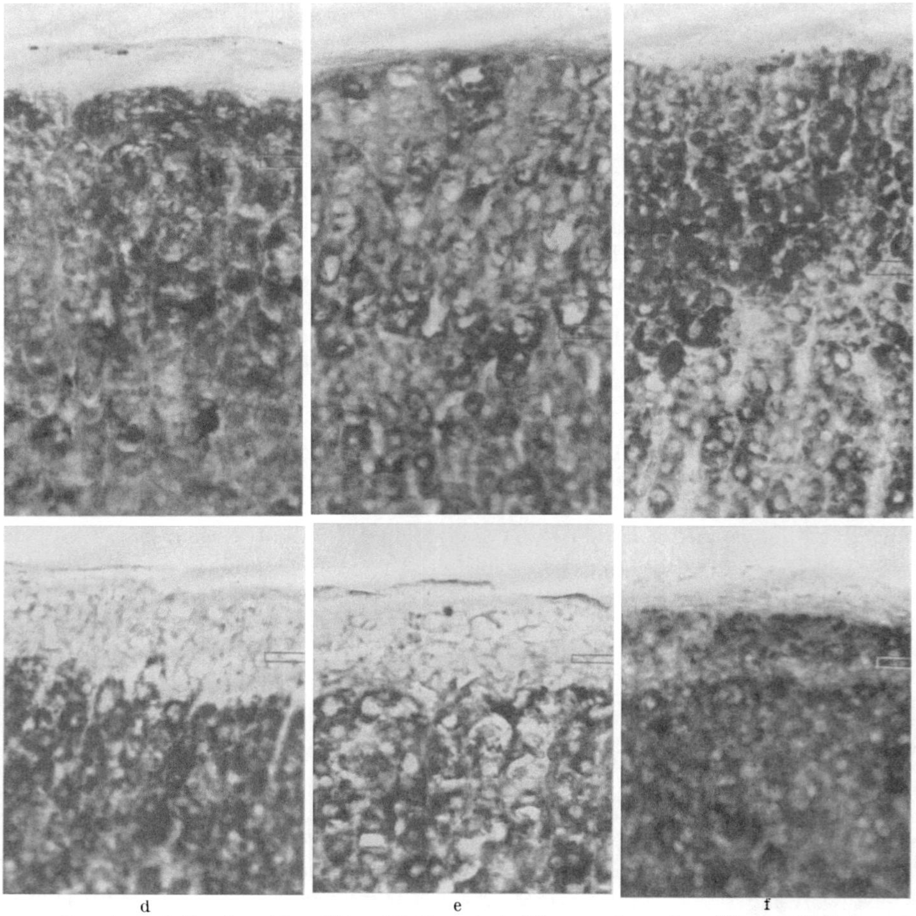

d e f

Fig. 58. Experimental alteration of the width and lipide content of the zona glomerulosa of the rat. All photographs are of formalin-fixed frozen sections stained with Sudan black B. ×240. a Control adult rat; b rat fed a sodium-deficient diet for 4 weeks; c rat with both kidneys wrapped with silk for 120 days (silk perinephritis); d rat given 2.5 mg deoxycorticosterone acetate daily for 4 weeks; e rat given 1% saline as drinking water for $4^1/_2$ weeks; f rat fed a potassium-deficient diet for $5^1/_2$ weeks. In all, the inner border of the zona glomerulosa is marked by a bar. The zone broadens conspicuously with sodium deficiency (b) or kidney disease (e); when more prolonged, sodium deficiency may result in lipide depletion as well (see Fig 53, p. 94). The zone shrinks and gradually loses its lipide inclusions when a mineralocorticoid (d), excess Na^+ (e) or deficient K^+ is provided. a, c and f from the study by DEANE et al. (1948); b, d and e from that by DEANE and MASSON (1951)

1955, GOLDMAN et al. 1956, HARTROFT and EISENSTEIN 1957), but also for the dog (RACE et al. 1957) and for human beings dying in state of hyponatremia (NICHOLS 1956b, PITCOCK and HARTROFT 1958). EICHNER (1953) and KOVÁCS and DÁVID (1959) have recorded nuclear enlargement in such preparations. Usually, with prolonged deficiency, there is reduction in droplet size and eventual loss of all lipides from the glomerulosa (DEANE et al. 1948, EICHNER 1953, HARTROFT and HARTROFT 1955, HUNGERFORD 1959). Sometimes the cells show "watery" vacuoles, a sign of excessive stimulation (p. 132). Such indices of glomerulosal stimulation occur in hypophysectomized rats fed a low-sodium diet as well as in intact animals (Fig. 59) (DEANE et al. 1948). GOLDMAN et al. (1956) and KOVÁCS and DÁVID (1959) showed that the hypertrophied zone shrank detectably and regained lipide within 48 hours after deficient rats were resupplied with NaCl.

That *a dietary regimen deficient in* Na^+ *enhances urinary aldosterone levels* in man is now well established (LUETSCHER and CURTIS 1955, LIDDLE et al. 1956). Such response occurs even after hypophysectomy (MACLEAN et al. 1957). Similarly, the level recorded in adrenal venous blood of experimental animals is above that

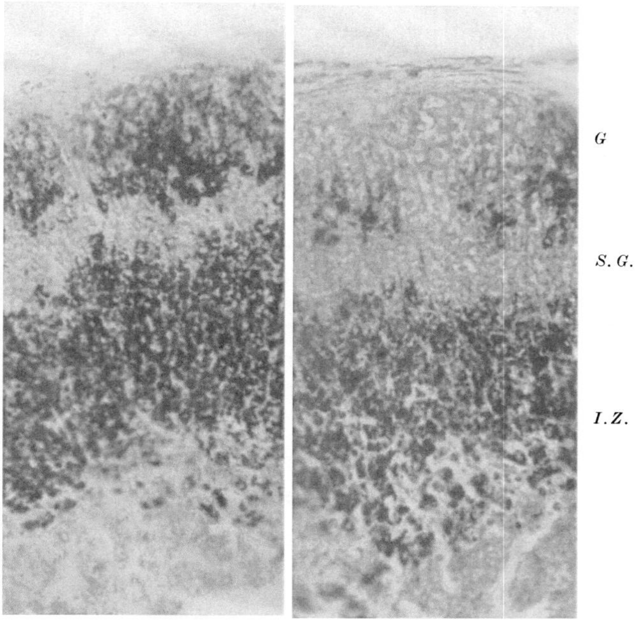

Fig. 59 a & b. Stimulation of the zona glomerulosa of the hypophysectomized rat by feeding a low-sodium diet. a Adrenal of a female rat hypophysectomized for 6 weeks; b gland of a female rat hypophysectomized for 6 weeks and fed a sodium-deficient diet for last 4 weeks. Formalin-fixed frozen sections, Sudan black B. × 150. From the study by DEANE et al. (1948). In both, the inner zones (*I.Z.*) are markedly atrophied, the subglomerulosal zone (*S.G.*) is broad, as is the glomerulosa (*G.*). However, in b the glomerulosa is broader and also depleted of lipides relative to a

in controls (ROSNAGLE and FARRELL 1956, MCDONALD and REICH 1959, SINGER 1959). Moreover, EISENSTEIN and HARTROFT (1957) found that the adrenals from rats chronically fed a severely Na-deficient diet produced a greatly enhanced amount of aldosterone when incubated *in vitro*.

Acute *in vitro* experiments have provided little evidence for a direct influence of electrolyte balance on aldosterone production. ROSENFELD et al. (1956) were able to increase the output of aldosterone (measured by bioassay) of perfused bovine adrenals only by lowering the Na^+ level in the perfusing medium to 125 meq/l and raising the K^+ level to 25 meq/l (ratio 5). C. GIROUD et al. (1956b) similarly were able to enhance aldosterone formation in incubated rats' adrenals only by a gross shift in the Na^+/K^+ ratio of the incubation medium (Na^+, 96.5 meq/l; K^+, 5.4 meq/l; ratio 18). Of course, such gross shifts do not occur *in vivo*. For example, SINGER and STACK-DUNNE (1955) analyzed the blood plasma of rats fed the diets employed by DEANE et al. (1948) and found that in the low-Na diet the Na^+ level was 144 meq/l, the K^+ level 5.8 meq/l (ratio 25), as contrasted with a normal ratio of 30—35 (A. I. COHEN et al. 1957b). These comparisons suggest that the influence of altered electrolyte ratio may occur at some receptor site other than the glomerulosal cell.

With an extremely Na-deficient diet that induced marked hyponatremia (Na^+ approximately 130 meq/l, K^+ 6.5 meq/l, ratio 20) and also atrophy of the fasciculata in young rats, HARTROFT and EISENSTEIN (1957) found that the glomerulosal cells, although greatly enlarged, retained abundant lipide in extremely large droplets. This alteration suggests that the cells were abnormal (p. 132). It may be pertinent that SCHÖNBAUM et al. (1956) found that a low Na^+ level in the environment inhibits the formation of glucocorticoids by adrenals incubated *in vitro*. Severe, prolonged hyponatremia may therefore not only inhibit the fasciculata, as evidenced by its atrophy and reduced ability to produce glucocorticoids (EISENSTEIN and HARTROFT 1957), but also affect the glomerulosa adversely. HUNGERFORD (1959) has demonstrated that this response to a low-Na^+ diet is a function of age.

b) Potassium excess

A similar enlargement of the glomerulosa occurs with *prolonged* intake of a high-K diet in rats (BACCHUS 1950, BACCHUS et al. 1952, ANDO 1957). ANDO demonstrated that the cellular hypertrophy is accompanied by nuclear enlargement as well. (He obtained nearly comparable enlargement of the zone and cells when water containing high levels of Ca or Mg salts was supplied.) LARAGH and STOERK (1957) demonstrated a high secretory rate of aldosterone in dogs with a high plasma K^+ level.

Aside from the chronic experiments cited above, several workers have reported an acute hypertrophy of the glomerulosa after the administration of large doses of hypertonic salt solutions — not only of KCl (DEANE et al. 1948) but also of NaCl (e.g., SELYE 1953, ERÄNKÖ 1955). As a result of the latter observation, ERÄNKÖ (1955) has seriously questioned the hypothesis that the glomerulosa secretes the sodium-retaining hormone.

c) Edema and blood volume

Another situation in which the excretion and presumably the secretion of aldosterone is stimulated in human patients is that assortment of diseases characterized by edema and ascites — nephrosis, cirrhosis, and cardiac insufficiency (LUETSCHER and CURTIS 1955, LIDDLE et al. 1956, FARRELL 1958, WOLFF et al. 1958). In rats with experimental nephrosis, SINGER (1957) and DAS GUPTA and GIROUD (1958, 1959) demonstrated enhanced aldosterone output. And in dogs with chronic cardiac insufficiency, DEANE and BARGER (1957) found that there was hypertrophy of the glomerulosa. DEANE et al. (1960b) have shown the same for nephrotic rats. J. DAVIS et al. (1956, 1957) reported enhanced excretion and secretion of aldosterone in such experimental animals. DRISCOL et al. (1957) concluded that ascites may develop prior to any rise in aldosterone secretion, but this is denied by the acute results of J. DAVIS et al. (1958).

One suggestion to explain the secondary aldosteronism associated with edema has been that changes in intravascular volume may somehow affect aldosterone secretion. There need be no alteration in plasma electrolytes accompanying changes in aldosterone secretion. BARTTER et al. (1956, 1958) have shown with human subjects that *an increase of intravascular volume* (albumin injection) *depresses aldosterone excretion*, while *a reduction of intravascular volume* (phlebotomy) *enhances it*. With dogs, FARRELL (1956b, c) found that severe hemorrhage causes a 2—3 fold increase in the secretion of aldosterone and that replacement of volume with a dextran solution prevents this enhancement. ANDERSON et al. (1959) and BAÏSSET et al. (1959) found that distention of the right atrium reduces secretion. Moreover, J. DAVIS and BALL (1958) have reported that putting a dog with ascites in a cast (which reduces extravascular fluid and raises blood pressure) reduces excretion of aldosterone and restores urinary Na^+ level to normal.

The whole question of the nature and locus of volume receptors remains open (MILLS et al. 1958, FARRELL 1958). GANN et al. (1959) believe them to lie in the common carotid veins. Whether or not stimuli to such receptors affect the glomerulosa by altering the output of the putative pineal glomerulotrophin (p. 117) also remains to be demonstrated.

d) Sodium excess, potassium deficiency, or DOC administration

Histological suppression of the zona glomerulosa occurs when an animal (rat, mouse) *is chronically-fed a high-Na^+ diet*, as, for instance, by providing 1—2.5 % NaCl as drinking water (BADINEZ and CROXATTO 1948, KNOWLTON et al. 1949, BACCHUS 1950, DEANE and MASSON 1951, EICHNER 1953, HARTROFT and HARTROFT 1955, ALLEN 1957, ANDO 1957, MOSIER and RICHTER 1958). The glomerulosa atrophies, the cells and their nuclei shrink, the lipide inclusions at first enlarge and then gradually disappear (Fig. 58), the esterase activity of the cells declines. Calves (*Bos*) repeatedly injected with 5% NaCl or sprayed with sodium acetate also show, within 2—4 days, accumulation of large lipide droplets in the glomerulosa (A. WEBER et al. 1956).

Atrophy, without complete loss of lipide, of the glomerulosa occurs when the diet is deficient in potassium (DEANE et al. 1948, NICHOLS 1948). In such animals, the blood plasma contained 148 meq/l Na^+, 3.3 meq/l K^+ (ratio 45) (SINGER and STACK-DUNNE 1955).

Atrophy also follows when a potent mineralocorticoid such as deoxycorticosterone, DOC, is administered chronically (Fig. 58) (*Rattus* — SARASON 1943, KNOWLTON et al. 1947, GREEP and DEANE 1947, 1949b, BACCHUS 1950, DEANE and MASSON 1951, R. MILLER 1954, HARTROFT and HARTROFT 1955, ANDO 1957; *Mus* — R. MILLER 1950, ALLEN 1957). (See Table 21.) If the dose of mineralocorticoid is high, there may be atrophy of the fasciculata as well (p. 101), but the more sensitive zone is the glomerulosa. And the effect of DOC on the glomerulosa is evident in the absence of the pituitary (GREEP and DEANE 1947), so that the suppressive influence does not involve ACTH. WEXLER (1959) has demonstrated similar inhibition of the rat glomerulosa by treatment with a synthetic corticoid possessing strong mineralocorticoid as well as glucocorticoid activity.

No effect on the glomerulosa has been detected with DOC administration in *Macaca mulatta* (KNOBIL et al. 1954b) or in *Mesocricetus auratus* (KNIGGE 1954b). No other species appears yet to have been examined.

As noted earlier disuse atrophy of the glomerulosa (as well as of the fasciculata) occurs in mice bearing transplantable adrenal tumors apparently secreting significant amounts of mineralocorticoid (Fig. 57, p. 102) [plasma Na^+/K^+ ratio, 39 in experimental animals, 30 in controls (A. I. COHEN et al. 1957b)].

In parallel with the histophysiological signs, *depression of aldosterone excretion occurs* in man *when the NaCl intake is very high* or when K^+ intake is drastically lowered (GARROD et al. 1956, LIDDLE et al. 1956). SINGER and STACK-DUNNE (1955) have shown that either DOC administration or K^+ deficiency suppresses the output of aldosterone by rat adrenals. EISENSTEIN (1959) observed a return to normal secretory rate within 48 hours when salt-deprived rats were given NaCl.

5. Other influences

a) Water deprivation; posterior pituitary

Total deprivation of water has been found to cause shrinkage of the glomerulosa in rats; this was associated with reduction in nuclear area and reduction in cell division in this zone (DALLWIG 1954). Correlatively, GARROD et al. (1956) cite evidence that doses of pitressin sufficient to produce severe water retention depress the excretion of aldosterone. (See discussion of intravascular volume, p. 114.)

Reports conflict on what happens in the absence of the posterior pituitary. Thus GREEP and DEANE (reported by DEANE 1951) observed hypertrophy of the glomerulosa in rats with diabetes insipidus produced by removal of the posterior pituitary, and BAÏSSET et al. (1954) reported the same for dogs with diabetes insipidus following section of the supraoptico-posterior pituitary tract. Contrariwise, CHESTER JONES and WRIGHT (1954b) described atrophy or disappearance of the glomerulosa in rats with long-term diabetes insipidus.

b) Growth hormone

Growth hormone has frequently been suggested as the anterior pituitary hormone capable of stimulating the glomerulosa and the output of a salt-retaining corticoid. Solution of this question has been plagued by the fact that growth hormone, in contrast to most other anterior pituitary hormones, appears to possess considerable species or at least order specificity (BECK et al. 1957, KNOBIL et al. 1958, GESCHWIND 1959). With bovine growth hormone, SINGER and STACK-DUNNE (1955) observed no effect on the secretion of aldosterone by the rat adrenal, and C. GIROUD et al. (1956b) could detect no influence on rat glands incubated *in vitro*. However, with growth hormone of primate origin, BECK et al. (1957) found some increase in the *excretion* of aldosterone by human subjects. That this involved increased secretion remains to be determined.

c) Kidney abnormality

Numerous observations point to an influence of a product of the kidney on the glomerulosa. Thus, OLSON and DEANE (1949) observed marked enlargement of the zone, accompanied by lipide depletion, in weanling rats exhibiting renal lesions induced by choline deficiency. The glomerulosa promptly returned to normal when the kidneys recovered. DEANE and MASSON (1951) recorded zonal hypertrophy in rats, both after renal encapsulation producing hypertension (Fig. 58) and after injection of partially purified "renin".

BACCHUS et al. (1952) questioned the increase in size of the glomerulosa with renal encapsulation, but their results also show such an increase, although the ratio of *width* of glomerulosa to width of fasciculata remained normal because of concomitant stimulation of the fasciculata. Indeed, in such enlarged adrenals, the increase in *volume* of the glomerulosa must have been highly significant.

d) Trauma

HENI et al. (1952) described glomerulosal enlargement and lipide depletion in hypophysectomized rats of which a foot was burned (80° H_2O, 7 seconds) on two successive weeks. No stimulus to the atrophied inner layers was detectable.

LLAURADO and WOODRUFF (1957) have found that the Na^+ retention and K^+ excretion that occur postoperatively (humans) are associated with a transient rise in urinary aldosterone. This presumably reflects the rise in ACTH, however, since no such rise occurs in human beings after the operation of hypophysectomy (LLAURADO 1957).

e) Possible pineal hormone

Suggestive evidence that a hormone derived from the diencephalic region of the brain stimulates the output of aldosterone came from the work of RAUSCHKOLB and FARRELL (1956). Other evidence for diencephalic control has been reviewed by KLOTZ and KAHN (1957). YANKOPOULOS et al. (1959) demonstrated that blood from dogs with ascites (experimental cardiac failure) induces increased secretion of aldosterone by isolated perfused adrenal glands of normal dogs. Since little augmentation of glucocorticoid output occurred, the stimulating factor was presumed not to be ACTH.

According to FARRELL (1959a, b, FARRELL et al. 1959) *extracts of the pineal region (ox) stimulate the output of aldosterone by the adrenal of the decerebrate dog, whereas pinealectomy depresses its secretion.* Neither treatment appeared to affect cortisol secretion. FARRELL has dubbed the putative hormone "*glomerulotropic hormone*" (GTH).

f) Possible specific cytotoxic agents

Atrophy of the glomerulosa has been described in man for a variety of diseases, some of which fail to affect the inner zones. WALTHARD (1950) described such selective atrophy as

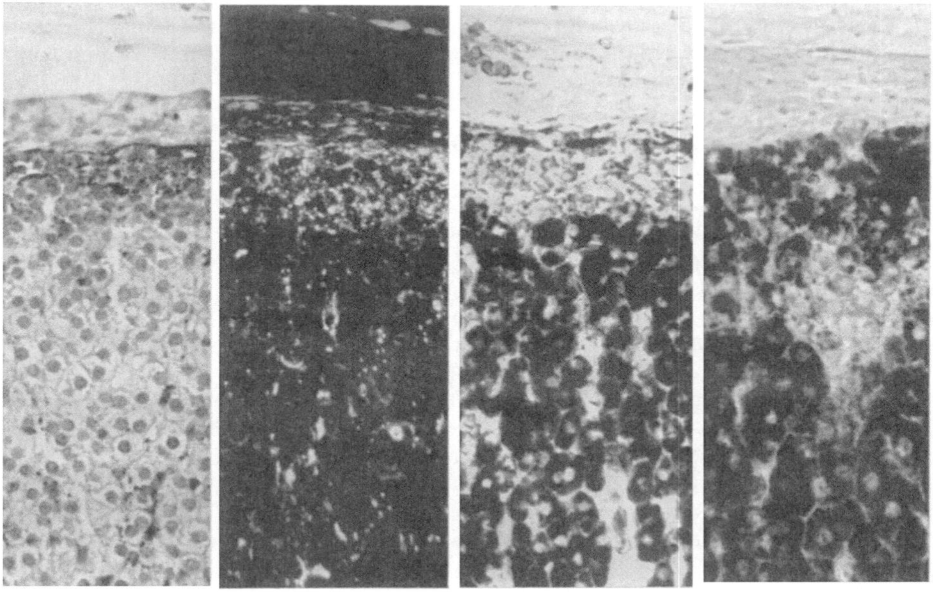

a b c d

Fig. 60a—d. Atrophy of glomerulosal cells following deposition of intracellular silver in a rat given 0.15% $AgNO_3$ in its drinking water for 18 months. a, b and c rat treated with $AgNO_3$; d control rat. a formalin-fixed paraffin section counterstained with paracarmine; b unstained paraffin section photographed under the polarizing microscope; c and d formalin-fixed frozen sections stained with Sudan black B. All ×385. From the study by DEANE et al. (1955). Silver granules are detectable in the atrophied glomerulosa as well as in reticuloendothelial cells in the fasciculata. These granules show up better with polarized light (b) than with ordinary illumination (a). The atrophied glomerulosa cells lack the lipide stores present in the normal gland

occurring in hemochromatosis, lues, metastatic cancer, and endemic cretinism. It was WALTHARD's thesis that the destruction of the glomerulosa in these diseases is responsible for the increased skin pigmentation. The recent evidence that ACTH contains a series of the same amino acids as intermedin and possesses some ability to enhance pigmentation (e.g., LI et al. 1957, GESCHWIND 1959) obviously calls WALTHARD's hypothesis into question.

Be that as it may, it has been observed repeatedly that *certain diseases* do *involve the deposition of foreign particles in the glomerulosal cells,* the ultimate atrophy of these cells, *and concomitant disturbances in electrolyte balance.* In the paper cited above, WALTHARD (1950) described deposition of iron pigments in the glomerulosal cells in patients with hemochromatosis and Addisonian symptoms. I have been shown several such adrenals by pathologists. KAPLAN et al. (1950) and A. HILL et al. (1950) observed that polysaccharide antigens from pneumonia-producing bacteria are deposited in this zone in mice. Patients with pneumonia commonly exhibit disturbances in electrolyte balance suggestive of transient adrenal insufficiency (HEFFRON 1939, p. 500). Finally, DEANE et al. (1955) found that rats given 0.15% $AgNO_3$ in their drinking water displayed, after many months, a deposition of reduced silver within the glomerulosal cells, atrophy and lipide depletion of the zone (Fig. 60), and a moderate decline from control values in the plasma Na^+/K^+ ratio (from 37.8 to 31.6).

g) Compensatory hypertrophy

R. MILLER (1950) described glomerulosal as well as fasciculata hypertrophy in mice subjected to unilateral adrenalectomy.

h) Thiamine deficiency

DELOST et al. (1959) have reported a selective atrophy of the glomerulosa in rats made deficient of thiamine. This involution could be prevented by concomitant administration of ascorbic acid.

E. Control of adrenocortical regeneration

I. Regeneration in vivo

In those species in which the operation is feasible, enucleation of the adrenal glands is followed by extensive regeneration of cortical tissue, apparently from the residual cortical cells attached to the capsule (Fig. 61). [There remains a school of thought believing that regeneration occurs from transformed capsular cells (e.g., B. BAKER 1952, HADHÁZY and KARMAZSIN 1953; see p. 45).] Moreover, if an adrenal gland or a portion thereof is transplanted to another site in the body, there is initial central necrosis followed by restoration of glandular parenchyma after vascularization is accomplished. The species in which such regeneration, after one or both operative techniques, has been studied include the rat (VACCAREZZA 1945, GREEP and DEANE 1949b, INGLE 1951, BRENNER et al. 1953, PELLEGRINO and MONTELLA 1954, SCHABERG 1955, MARTINOVITCH 1955, DE GROOT and FORTIER 1959), the mouse (MEDAWAR and RUSSELL 1958), the rabbit (WILLIAMS 1953, COUPLAND 1957), the dog (CINTI and MIGNANI 1949, DEMPSTER 1955, GROLLMAN 1958), and man (EISEMAN et al. 1955).

Like other examples of adrenocortical hyperplasia and hypertrophy, such *growth does not occur in the absence of the pituitary* (INGLE 1951). For an enucleated gland or autotransplant to exhibit significant hyperplasia, it usually seems necessary to induce relative hypoadrenocorticism and hence increased ACTH output by removal of all or most of the remaining gland(s) (N. EVERETT 1949, INGLE 1951, WILLIAMS 1953, COUPLAND 1957). An enucleated or transplanted gland does not grow if the animal is injected with adrenocortical extract or glucocorticoids (INGLE 1951, WOODRUFF and BOSWELL 1953). Repeated injections of ACTH have little detectable effect on enucleated or transplanted glands in intact animals (INGLE 1951, COUPLAND 1957), but if ACTH is infused by continuous drip, massive hyperplasia and hypertrophy occur (INGLE 1951). In INGLE's experiments, it was apparent from the severe weight loss and glycosuria that the regenerated glands were secreting pharmacological amounts of gluco-corticoids.

In a rat with an intact pituitary but no exogenous ACTH, an enucleated adrenal requires 3—4 weeks to regain approximately normal appearance (Fig. 61) (Greep and Deane 1949b, Pellegrino 1951, Brenner et al. 1953, Pellegrino

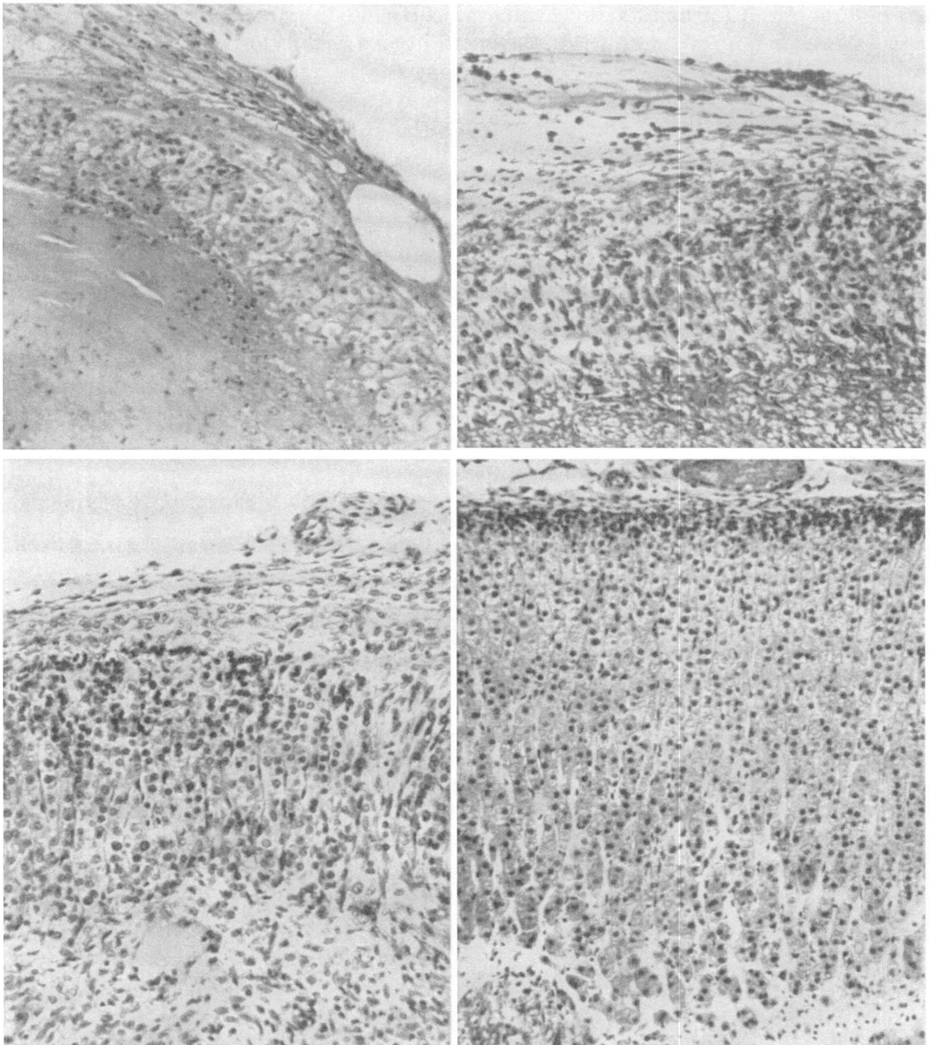

a b c d

Fig. 61a—d. Regeneration of adrenal cortex from glomerulosal cells following enucleation of the gland. a gland 1 day after operation; b 3 days; c 8 days; d 32 days. Hematoxylin and eosin; a—c ×160; d ×130. From preparations of Greep and Deane (1949b). The adrenocortical cells remaining attached to the capsule after the operation undergo hyperplasia and restore the normal glandular structure in about 1 month (rats)

and Montella 1954). *The final weight of cortical tissue appears always to remain subnormal* (M. Gordon 1950, Pellegrino and Montella 1954, Pellegrino and Torcigliani 1957, de Groot and Fortier 1959). Detectable lipides are at first absent, suggesting rapid release of all formed products; later the usual amount of lipide reappears. According to PELLEGRINO and co-workers (1954, 1957), the cholesterol and ascorbic acid levels remain subnormal for as long as 3 months. Initially an augmentation of histologically demonstrable alkaline phosphatase activity occurs (VERNE and HÉBERT 1955).

Physiologically, regenerating glands appear to be adequate with respect to secretion of mineralocorticoid since, after the initial post-operative period, the animals do not require additional salt or mineralocorticoid therapy (previous references, plus KOMRAD and WYMAN 1951, CHESTER JONES and WRIGHT 1955a). By 1—2 weeks after enucleation or transplantation, the secretion of glucocorticoids is adequate to maintain normal homeostasis and to permit successful resistance to various stresses (GREEP and DEANE 1949b, PELLEGRINO 1950, KOVÁCS and JAKOBOVITS 1953, BROGI 1954, PELLEGRINO and CARAMASSA 1955). Toward histamine, resistance is somewhat reduced, but this appears to result from the deficiency of medullary hormones (INGLE 1937, R. HOUSSAY 1955, BROGI 1956).

Recently several workers have studied the secretory activity of regenerating glands in rats by analysis of the adrenal effluent (MASSON et al. 1958, BROGI and PELLEGRINO 1959, FORTIER and DE GROOT 1959). The first two studies indicated that such glands secrete both aldosterone and corticosterone after an initial period, but that the secretory rate remains somewhat subnormal for many months. FORTIER and DE GROOT (1959), however, reported supranormal secretion from 16 days onward. The basis for these discrepancies is not apparent.

Regeneration hypertension. SKELTON (1955, 1959) was the first to show that if the adrenals of young female rats are demedullated and the animals are then given 1% saline to drink, hypertension develops. Early interpretations suggested that regenerated glands secrete disproportately more mineralocorticoid than glucocorticoids. Recent analysis has shown this interpretation to be incorrect (MASSON et al. 1958, BROGI and PELLEGRINO 1959; see, however, FORTIER and DE GROOT 1959). GROLLMAN (1958) showed that the defect apparently arises because the period of temporary insufficiency renders the tissues hyperresponsive to corticoids. Supportive therapy with glucocorticoids for the first few post-operative days prevented the lesion. This subject is discussed further in Chapter 4.

The *immunological factor* assumes major importance in the regeneration of transplants. In general, autologous grafts, whether to the eye, brain, muscle, gonads, kidney, or portal system, survive and grow if the animal is rendered somewhat adrenal insufficient. Transplants to other animals (homologous grafts) usually fail, except within a highly inbred strain (e.g., N. EVERETT 1949, WOODRUFF and BOSWELL 1953, DEMPSTER 1955, PARKES 1956). The only exception is the anterior chamber of the eye, where the transplant is protected by the blood-brain barrier. Using inbred strains of mice, MEDAWAR and RUSSELL (1958) have shown that the failure of homografts to survive depends on an immune response. Homografts survived successfully if the host was made "tolerant" to the donor strain by injection of donor lymphocytes during infancy. "Resistance" could be reinstated by injection of autologous lymphocytes from a non-tolerant donor.

Two techniques have been employed that apparently overcome the immunity to homotransplants. One, alluded to above, is the use of the anterior chamber of the eye as the site of implantation (e.g., COUPLAND 1957). Using this site, MARTINOVITCH (1955) obtained successful regeneration of glands from infant rats which had been cultivated *in vitro* for as long as 7 weeks. (The possibility that this external cultivation weakened the antigenic properties of the grafts could be tested by transplantation to another site.)

A second procedure that reduces the antigenicity of adrenal grafts appears to be deep freezing after partial dehydration. PARKES (1956) showed that, in many cases, rat glands soaked in glycerol-saline and then frozen at —79° "took" successfully when thawed and implanted (kidney), even in animals other than the donor. Recently JOSHI and BLUMENTHAL (1958) found that glycerol treatment followed by deep freezing lowers the foreign-body response of the recipient to a homotransplant.

II. Regeneration in vitro

Explantation of adrenals has proved especially successful with glands from infant rats (SCHABERG 1955, MARTINOVITCH 1955). Without special treatment, such glands survive and exhibit considerable cell division, but absolute growth is limited. When cultured in association with anterior pituitaries or in a medium containing added ACTH, however, significant growth is demonstrable (SCHABERG 1957, SCHABERG and DE GROOT 1958). When no ACTH is added, the juxtacapsular adrenocortical cells appear healthy, but little fasciculata-like tissue is formed. With added ACTH, the glomerulosal cells form a conspicuous fasciculata (as in a gland regenerating *in situ*).

F. Prenatal control of the adrenal cortex, mammals

So far as has been ascertained, the adrenocortical anlagen form and aggregate autonomously (e.g., WITSCHI 1956, RAYNAUD and FRILLEY 1950). Some time during fetal development, however, the tissue comes under the influence of ACTH and perhaps other hormones. In mammals, analysis of the control of the adrenal cortex is complicated by the facts that not only the fetal pituitary but possibly also placental (and maternal pituitary) hormones may influence the fetal adrenals.

To this reviewer, a survey of available information on the adrenal cortices of mammalian fetuses suggests the following. (1) *The adrenal cortex in fetal life secretes corticoids*, although probably at a lower rate than postnatally. (2) *The major control of the adrenal cortex in fetal life is exercised by the fetal pituitary* (ACTH), although there is some possibility that placental ACTH is also effective on the fetal adrenals. (3) *The secretory activity of the fetal pituitary is governed by the level of corticoids* emanating from the fetal adrenals and those of maternal origin crossing the placenta. (4) *Perhaps because of excessive depletion during birth, the pituitary requires a period of several days, postnatally, before it begins to secrete at its adult rate.* Because of this, and possibly because of the loss of the supplementary supply of ACTH derived from the placenta, the adrenal cortex in the first few postnatal days undergoes some involution. The data suggesting these conclusions are reviewed below. (See also JOST 1954, MOOG 1959, WELLS 1959.)

A subsequent section will consider the special problems of control of that inner cortical zone apparently peculiar to primates, the "fetal" zone.

I. Secretion of corticoids by fetal adrenals

There is increasing evidence that the adrenals secrete corticoids while the fetus is still *in utero*. The evidence is clearest for rodents, less well established for primates.

(1) Histologically, *adrenocortical tissue acquires the characteristics of a secretory tissue before birth* (MOOG 1959). The cells acquire typical steroid-containing lipide droplets. For example, in the rat, cholesterol is detectable histochemically in the cells of the fasciculata by the 19th day of gestation (JOSIMOVICH et al. 1954, NOUMURA 1959), in keeping with the significant amount measurable chemically (KNOBIL and BRIGGS 1955). The droplets are small, suggesting that the cells may be secretory (JOSIMOVICH et al. 1954, A. COHEN 1954, 1955). In the human gland, birefringent crystals appear in the outer, definitive cortex by the time the fetus is 2 cm long (ca. 7th week of gestation) (ROTTER 1949a, FRAZÃO 1954, BENIRSCHKE et al. 1956). Nevertheless, in both rat and man, the cells are distinctly smaller in fetal glands than they become postnatally. Possibly this signifies lower activity.

(2) In rabbits and rats, JOST and JACQUOT (1955) and JACQUOT (1955) have shown that decapitation of fetuses, which results in atrophy of the adrenal cortex

(see below), prevents accumulation of glycogen in the liver. Liver glycogen is restored by injecting ACTH into the fetus.

(3) In human glands, 17-hydroxy corticosteroids are detectable in extracts of the gland by the 20th week of gestation (Kloos and Staemmler 1953, E. Bloch et al. 1956). Fetal glands are also capable of forming corticosteroids *in vitro* (e.g., Villee et al. 1959). This constitutes presumptive evidence that such glands secrete glucocorticoids.

Other indications of secretion by the fetal adrenal cortex are included below.

II. Stimulation of fetal adrenals by the fetal pituitary

(1) The evidence that *the fetal pituitary stimulates the fetal adrenal glands* comes principally from the fact that removal of the fetal pituitary, either by X-irradiation (mouse — Raynaud and Frilley 1950) or by decapitation (rat —

Table 24. *Endocrine control of output of ACTH by the fetal pituitary, as indicated by the size of the fetal adrenal at term (see text for explanations)*

Species and Treatment	No. fetuses	Fetal adrenal size	Relative fetal adrenal size	Reference
Rattus norwegicus				
None	62	2.1 mg (2)	49 mg-%	
Mother adx. on day 16	114	2.4 mg (2)	67 mg-%	
Mother hypox. on day 16	76	1.8 mg (2)	49 mg-%	
Mother hypox. on day 12, adx. on day 16	138	2.1 mg (2)	66 mg-%	Christianson & Chester Jones 1957
Mother adx. on day 16, ACTH days 19—22	131	2.1 mg (2)	63 mg-%	
Fetus decap. on day 19	9	1.1 mg (2)	31 mg-%	
Mother adx., fetus decap.	9	1.0 mg (2)	30 mg-%	
Rattus norwegicus				
None	14	.39 mm³ (1)	8.5, vol./ body wt.	Kitchell & Wells 1952a
Fetus $^1/_2$ adx. on day 19	14	.52 mm³ (1)	11.7, vol./ body wt.	
Macaca mulatta				
None	10	461 mg	125 mg-%	Schmidt & Hoffmann 1954
Mother, ACTH for 21—67 days	4	247 mg	64 mg-%	
Rattus norwegicus				
None	108	2.0 mg (2)	35 mg-%	
Mother stressed days 18—21	98	1.4 mg (2)	30 mg-%	Holland 1958
Mother adx., day 12	105	2.5 mg (2)	52 mg-%	
Mother adx., stressed	98	2.3 mg (2)	72 mg-%	
Rattus norwegicus				
None	53	3.1 mg (2)	71 mg-%	
Fetus decap., day 16—20	30	1.3 mg (2)	36 mg-%	
Mother, cortisone, 3 mg/day, 18—20 days	27	1.9 mg (2)	50 mg-%	
Mother, ACTH, 1 U/day, 18—21 days	53	2.4 mg (2)	61 mg-%	Noumura 1959
Mother adx., day 16	75	3.7 mg (2)	93 mg-%	
Mother adx. and ACTH.	34	3.1 mg (2)	85 mg-%	
Mother adx. and cortisone	29	2.0 mg (2)	58 mg-%	
Mother hypox., day 14	22	2.7 mg (2)	73 mg-%	
Mother hypox. and adx..	29	2.7 mg (2)	84 mg-%	

Key: adx. = adrenalectomized; $^1/_2$ adx. = unilaterally adrenalectomized; hypox. = hypophysectomized; decap. = decapitated (hypophysectomized).

WELLS 1948, KITCHELL and WELLS 1952b, A. COHEN 1955, DOMM and LEROY 1955, CHRISTIANSON and CHESTER JONES 1957, NOUMURA 1959; rabbit — JOST 1948, 1953), halts or slows growth of the adrenal cortex (Table 24). Growth of such adrenals can be stimulated by injection of ACTH into the fetus (KITCHELL and WELLS 1952b, CHRISTIANSON and CHESTER JONES 1957, NOUMURA 1959). A. COHEN (1958a, b) has shown that ascorbic acid accumulates in the adrenals of fetal rats, that this accumulation can be prevented by decapitation, and that the latter effect is overcome if ACTH is given to the decapitated fetus. Similarly, epinephrine injection into 19-day rat fetuses causes depletion of adrenal ascorbic acid, as compared to levels in uninjected latter-mate controls (MILKOVIĆ and MILKOVIĆ 1958).

In human beings, anencephaly with absence or dysfunction of the pituitary appears to result in reduction in size of the definitive cortex as well as in absence of the "fetal" zone (BENIRSCHKE 1956, TUCHMANN-DUPLESSIS and LARROCHE 1958). This conclusion has not been established by volumetric studies. However, KLOOS and STAEMMLER (1953) also found a lower than normal concentration of reducing steroids in glands of anencephalic monsters.

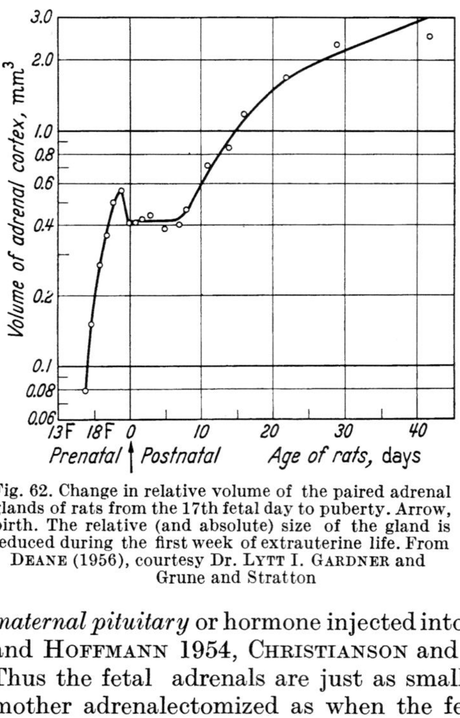

Fig. 62. Change in relative volume of the paired adrenal glands of rats from the 17th fetal day to puberty. Arrow, birth. The relative (and absolute) size of the gland is reduced during the first week of extrauterine life. From DEANE (1956), courtesy DR. LYTT I. GARDNER and Grune and Stratton

(2) Whether or not there is any additional stimulation of the *fetal* glands by the putative ACTH of placental origin (LUNDLIN and HOLMDAHL 1957) is difficult to determine. However, the postnatal involution of the adrenals (Fig. 62) suggests this as one explanation (DEANE 1956). Alternatively, the regression may reflect pituitary exhaustion at birth (MILKOVIĆ and MILKOVIĆ 1958).

There is certainly *no conclusive evidence that ACTH derived from the maternal pituitary* or hormone injected into the mother *crosses the placenta* (SCHMIDT and HOFFMANN 1954, CHRISTIANSON and CHESTER JONES 1957, NOUMURA 1959). Thus the fetal adrenals are just as small when the fetus is decapitated and the mother adrenalectomized as when the fetus is decapitated but the mother left intact (Table 24). Presumably in the first instance the level of circulating maternal ACTH is greatly enhanced. It is noteworthy, however, that HOLLAND (1958) and NOUMURA (1959) produced greater *relative*, though not absolute, enlargement of the adrenals in fetuses of adrenalectomized mothers subjected to stress or given ACTH than in those whose mothers were not so treated.

III. Control of the secretion of ACTH by the fetal pituitary

The secretory activity of the fetal pituitary appears to be regulated by the level of circulating corticoids (Table 24). *These apparently derive both from the fetus and from the mother.* Thus, it has been shown repeatedly that adrenalectomy of the mother results in great enlargement of the fetal adrenals, presumably as the result of the loss of suppressing activity of maternal corticoids on the fetal pituitary

(Knobil and Briggs 1955, Christianson and Chester Jones 1957, Holland 1958, Noumura 1959, plus earlier workers). However, the fetal adrenals do *not* enlarge if the mother is hypophysectomized (Yakaitis and Wells 1956, Christianson and Chester Jones 1957, Noumura 1959). Apparently the decline in maternal corticoids must be complete to affect the fetal pituitary detectably.

In contrast, injection of glucocorticoids or ACTH into an intact mother reduces the size of the fetal adrenals significantly, probably because more maternal corticoids reach the fetal pituitary (Courrier et al. 1951, M. Davis and Plotz

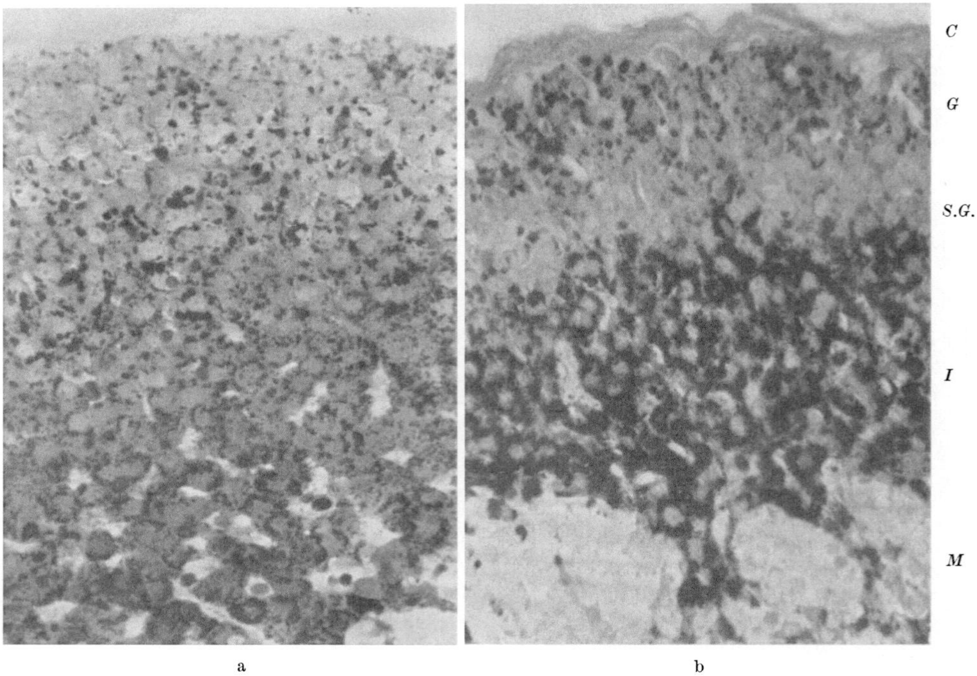

Fig. 63 a & b. Appearance of normal full-term (22-day gestation) fetal rat adrenal (a) and gland from a litter-mate that received a pellet of 1 mg cortisol *in utero* on day 19 (b). Fixed in Champy's solution. ×420. From the study by Yakaitis and Wells (1956), courtesy of the authors. The lipide droplets have been blackened by osmium tetroxide. In b the inner zones are narrower, the cells smaller, and the lipide droplets are larger than in a. A subglomerulosal lipide-free zone is conspicuous. *C.*, capsule; *G.*, glomerulosa; *S.G.*, subglomerulosal zone; *I.*, inner zones: *M.*, medulla

1954, Schmidt and Hoffmann 1954, Noumura 1959). Similarly, subjection of the mother to a stress that causes enlargement of her adrenals causes relative atrophy of the fetal adrenals (Picon 1957, Holland 1958).

In addition, there is suggestion that the fetal gland itself may secrete enough hormone to influence the fetal pituitary. Kitchell (1950, Kitchell and Wells 1952a) showed that unilateral adrenalectomy of the fetus on day 19 resulted in increased volume of the remaining gland 2 days later, in comparison to the size of a single gland in litter-mate controls (Table 24). Noumura (1959) has repeated this result.

As might be expected, injection of corticoids into a fetus likewise induces adrenal atrophy, in comparison to the glands of litter-mate controls (Fig. 63) (Jost et al. 1955, Yakaitis and Wells 1956). Kitchell and Wells (1952a)

and NOUMURA (1959) found that the administration of cortisone to or hypophysectomy of a unilaterally-adrenalectomized fetus prevented the hypertrophy seen otherwise in the remaining gland.

IV. Postnatal inactivity of the pituitary

Abundant evidence for a variety of species suggests that the output of ACTH by the pituitary in the immediate postnatal period is low. This possibility was indicated by JAILER's (1949, 1950) studies showing that application of stress to neonatal rats failed to induce signs of adrenal activation (ascorbic acid depletion). Responsiveness developed only at 8 days. (See also BACQ and FISCHER 1956, ENDRŐCZI et al. 1957.) That this failure is not attributable to insensitivity of the adrenal is indicated by the responsiveness of the adrenals to exogenous ACTH (MOON 1940, JAILER 1950, LADMAN and DEANE 1955). Similarly, in neonatal humans there is little corticoid in the circulating blood or in urine during the first week of extrauterine life (VENNING 1951, KLEIN et al. 1954, GARDNER 1956). However, with severe stress, the pituitary is capable of activation, even at this early period (VENNING 1951, RINFRET and HANE 1955, KLEIN and ROVNANEK 1956).

Grossly the adrenals of newborn mice (MOOG et al. 1954, MOOG 1959) *and rats* (VAN DORP and DEANE 1950, JOSIMOVICH et al. 1954, DEANE 1956) *shrink* absolutely and relatively after birth and only begin to grow again about a week later (Fig. 62; Fig. 33, p. 40). Cytologically, as well, these adrenals contain large lipide droplets and appear inactive. During the second postnatal week there is a decline in lipide content, suggesting glandular stimulation. MOOG (1959) has adduced considerable evidence from the responses of several target organs to indicate that the adrenal secretes actively by the end of the second week.

Whether the activity of the pituitary of the newborn is lower than it was during fetal life, or whether the adrenal regresses because of the removal of another source of ACTH (? the placenta) remains to be finally determined. It seems possible that one factor in depression of the fetal pituitary may be the rise in maternal glucocorticoids just prior to and during parturition, which may effectively inhibit the susceptible fetal pituitary for some days. Alternatively, as suggested above, the fetal pituitary or hypothalamus may be over-taxed and depleted at birth (MILKOVIĆ and MILKOVIĆ 1958).

V. Control of the primate "fetal" zone

The "fetal" zone of the adrenals of higher primates has a distinctive origin (p. 42) and *appears also to have different control from the definitive cortex.* Thus the postnatal administration of ACTH does not prevent the regression of the "fetal" zone, although is accelerates the growth and activation of the definitive zone (e.g., STONER et al. 1953, LANMAN 1953). Furthermore, the "fetal" zone regresses in typical fashion in infants with a high titer of endogenous ACTH associated with infectious disease (STONER et al. 1953, 1955) or adrenal virilism (BENIRSCHKE et al. 1956).

The cells in the "fetal" zone develop birefringent lipides in young fetuses, although generally by the end of gestation lipides have virtually disappeared (KEENE and HEWER 1927, BENIRSCHKE et al. 1956). During the postnatal involution, however, lipides exhibiting birefringence generally reappear (STONER et al. 1953). These characteristics suggest, although they cannot prove, that the "fetal" zone secretes a steroid hormone. And the high titer of 17-ketosteroids in human fetal blood (see GARDNER 1956; Chapter 5) and in extracts of fetal glands (E. BLOCH et al. 1956) certainly makes plausible the belief that this portion of the gland secretes weak androgens. The zone also forms androgens readily *in vitro* (SOLOMON et al. 1958).

The "fetal" zone develops and grows for about 5 months even in anencephalic monsters in which the pituitary is absent or nonfunctional (MEYER 1912, KEENE and HEWER 1927, BENIRSCHKE 1956, NICHOLS 1956a, TUCHMANN-DUPLESSIS and LARROCHE 1958). Only thereafter does it regress. By the time of birth, it it completely absent in such specimens (Figs. 64 and 65). These facts suggest thas

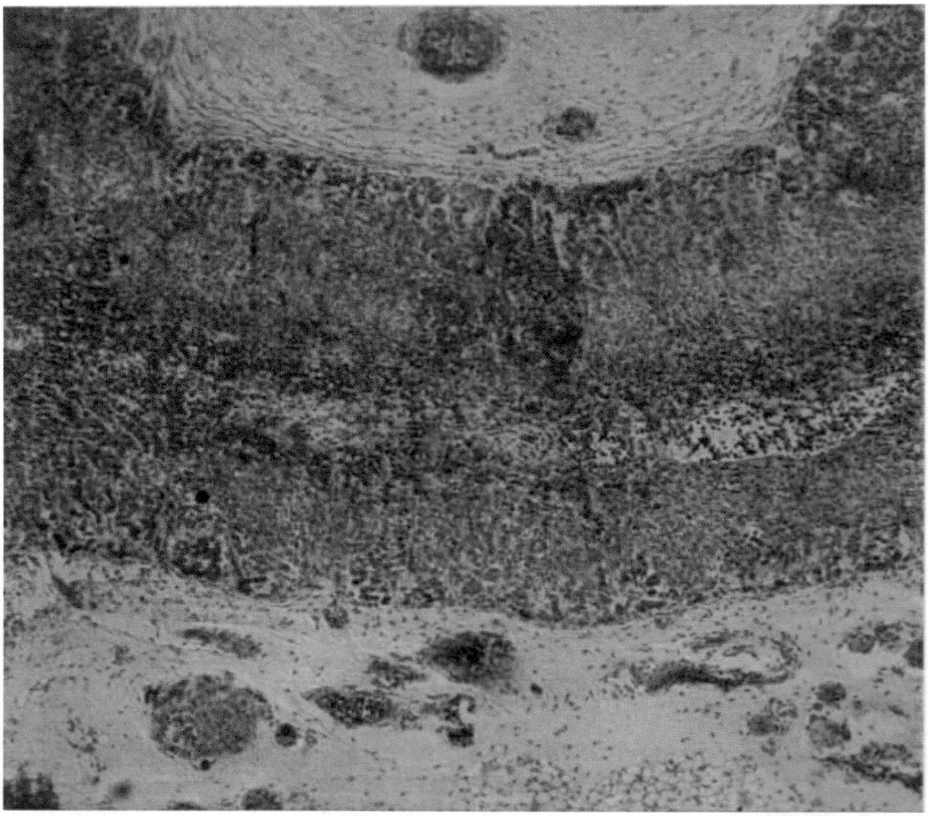

Fig. 64. Absence of a "fetal" zone in the adrenal gland of a full-term hydrocephalic stillborn human infant. Hematoxylin and eosin. × ca. 47. Micrograph contributed by Dr. KURT BENIRSCHKE. Compare Fig. 35, p. 42

(1) *the initial development of the "fetal" zone may be autonomous or dependent on some factor derived from the placenta*, but (2) *its stimulation after the 5th month of gestation requires the activity of the fetal pituitary*. Many students of the problem have suggested that *chorionic gonadotrophin*, secreted abundantly during the early part of gestation in primates, *may stimulate the zone during early fetal life* (e.g., ROTTER 1949a, b, BENIRSCHKE 1956, GARDNER 1956, NICHOLS 1956a). *During the second half of fetal development, the proposed trophic hormone is LH* secreted by the fetal pituitary, the output of which is presumably stimulated by estrogen of placental origin (GARDNER and WALTON 1954, GARDNER 1956). If this latter hypothesis is valid, it would account for the rapid postnatal involution and also for the prenatal involution in anencephalics. In this connection it is

of interest that 17-ketosteroids are virtually absent from the plasma of anencephalic monsters, in contrast to normal newborn infants (NICHOLS et al. 1958).

It thus appears likely that the "fetal" zone secretes weak androgens, not under the stimulus of ACTH but under the stimulus of gonadotrophins. This situation differs radically from the output of androgens under the influence of ACTH in some cases of adrenal hyperplasia (p. 133).

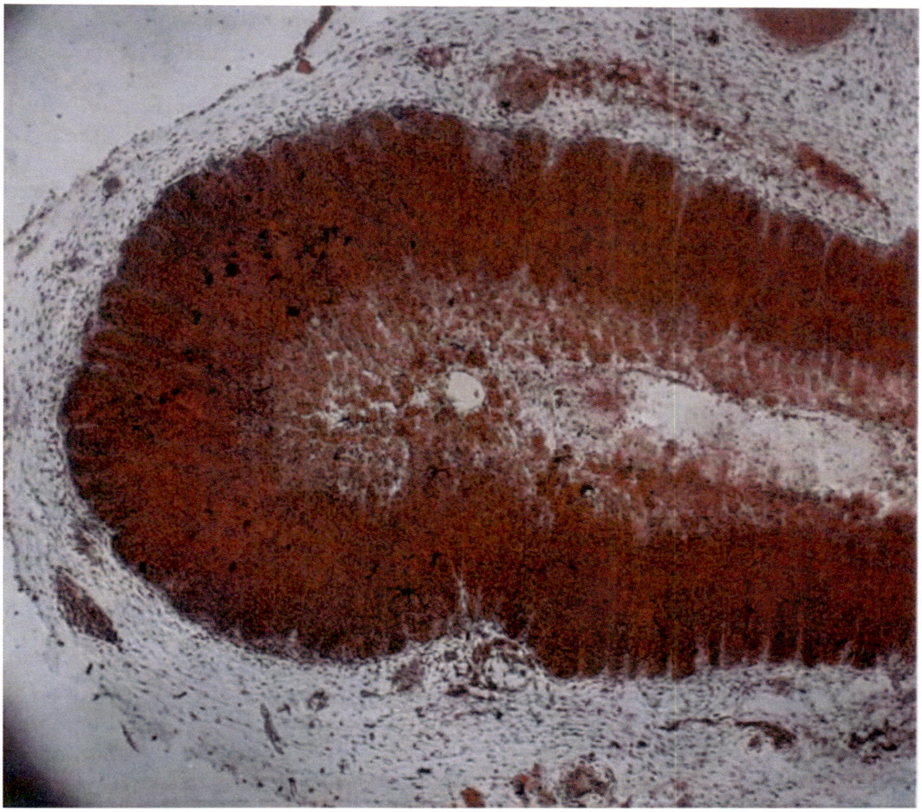

Fig. 65. Approximately normal quantities of lipide occur in the definitive adrenocortical tissue of a full-term anencephalic human infant. Sudan IV and hematoxylin. × ca. 47. Micrograph contributed by Dr. KURT BENIRSCHKE. Compare Fig. 36, p. 43. The "fetal" zone is entirely absent

G. Control of the mouse X zone

Like the "fetal" zone of the primate adrenal, the X zone of the mouse adrenal seems not to be under the control of ACTH. Thus, if the mouse is hypophysectomized, the X zone, like the fasciculata, involutes. But the administration of ACTH fails to cause regeneration of the X zone (CHESTER JONES 1949b, 1957). If only the output of pituitary ACTH is inhibited (by the administration of cortisone) the fasciculata involutes but the X zone remains healthy and may even hypertrophy (SAKIZ 1955, DELOST and CHIRVAN-NIA 1956).

On the other hand, if a hypophysectomized male mouse was gonadectomized as well, the injection of a gonadotrophin preparation with strong LH activity maintained the X zone, while having no detectable action on the fasciculata (CHESTER JONES 1949b). (The administration of gonadotrophin to a hypophysectomized mouse with intact testes did not restore the zone, presumably because of stimulation of androgen output.) It thus appears likely that the appearance the X zone in early postnatal life (p. 47) may depend on the rising titer of LH at this time. A secondary X zone develops in a postpuberally castrated male mouse (CHESTER JONES 1949b, 1957, ITO 1954) or in a parous female after ovariectomy (McPHAIL and READ 1942).

The X zone appears to be what might be called *a negative target organ of androgens.* The zone involutes at puberty in the male mouse and will regress earlier if androgen production is stimulated by the administration of LH or if androgen is administered to a mouse of either sex (BISHOP and LEATHEM 1946, LEATHEM 1949, CHESTER JONES 1949b). In female mice, regression occurs abruptly during the first pregnancy or gradually in older nulliparous animals. Both LEATHEM (1949) and CHESTER JONES (1952, 1957) have suggested that this regression may reflect the output of andromimetic steroids by the mouse ovary (probably from the corpus luteum). CHESTER JONES (1952) showed that eggs implanted in the anterior chamber of the eye or under the kidney capsule of mature virgin mice formed trophoblasts, stimulated the corpora lutea, caused androgenic stimulation of the ventral prostate and induced X-zone involution. No involution occurred in ovariectomized mice with implanted eggs. The only evidence in conflict with this hypothesis comes from WARING (1942), who found that the injection of high doses of chorionic gonadotrophin (human) into gonadectomized mice properly primed with estrone was followed by X-zone degeneration. And recently DEANESLY (1958) demonstrated X-zone regression in ovariectomized mice given ACTH; it is her hypothesis that adrenocortical rather than ovarian androgens cause involution. This interpretation has been challenged by SAKIZ (1959).

At present there is no definite evidence that the mouse X zone is secretory. Its behavior in relation to androgens has been interpreted to mean that it is androgen-secreting, undergoing disuse atrophy when the titer of circulating androgens rises and reappearing when the level of circulating androgens declines (Chapter 5). However, many tissues atrophy under the influence of hormones without themselves being a source of such hormones (e.g., the Müllerian duct in males). The cells of the X zone do not show the usual characteristics of steroid-secreting cells, since at no time do the cells develop lipide droplets containing detectable amounts of cholesterol. The only positive evidence comes from ALLEN (1960), who observed a trace of steroid-3β-ol-dehydrogenase activity in the X zones of female, though not of male, mice.

So-called X zone of other rodents. A zone somewhat resembling the mouse X zone has been described in several meadow mice (DELOST 1955), in young female golden hamsters (W. N. HOLMES (1955b), in the European hamster (KAYSER and ARON 1950), and in some shrews (DELOST 1957). That these zones are in all cases equivalent to the mouse X zone is far from established.

In the vole, *Microtus arvalis*, at least, the response of this zone to experimental intervention differs from that of the mouse X zone (DELOST 1956, DELOST and CHIRVAN-NIA 1956). Thus, while the zone disappears in males at puberty, it enlarges in females during pregnancy. Moreover, this X zone involutes when cortisone is administered. It seems probable that what is here being called an X zone is really just part of the true cortex, responding to changing titers of androgen, estrogen and corticoids like the inner cortical zones of other mammalian species (pp. 100, 106, 107). In two shrews (*Sorex araneus* and *Crocidura russula*), however, the so-called X zone responds to maturation as it does in the ordinary mouse (DELOST 1957).

H. Pathological changes in mammalian adrenal glands

The succeeding sections on pathology of the adrenal cortex do not aim to present a complete survey of all abnormalities encountered. Such accounts may be found in textbooks of pathology, and FASSBENDER (1956) has recently surveyed the subject. Here an attempt is made to examine the underlying causes of certain commonly-encountered abnormalities. SELYE and STONE (1950) reported other speculations about adrenocortical pathology some years ago.

Suffice it to say that many pathological findings in adrenals at autopsy failed to produce any signs or symptoms during life. The reserve of the adrenals is so great that a large portion of the cortical tissue must be affected before gross signs develop. Small adenomas, even though functional, may be "silent." Contrariwise, a scar within the gland may bespeak an earlier episode of hemorrhage or necrosis that occupied so small a region that fatal cortical insufficiency did not ensue.

I. Hyperactivity from increased ACTH

Destructive changes occur frequently in overstimulated adrenals. From present evidence, it is impossible to determine what proportion of the change is attributable to continued bombardment by ACTH and what to other influences, such as toxins. *In general it may be assumed that without a high level of circulating ACTH, toxins would not have such effects on the adrenals.* Direct cytotoxic action seems to be rare (see DDD, p. 135; Suramin, p. 136). This conclusion is supported by the study of HOEDE (1956), who found that if cortisone was administered to guinea pigs for a month prior to the injection of diphtheria toxin, no hemorrhages or other signs of necrosis developed in the adrenal cortex. TONUTTI (1953) has made similar observations.

1. Cellular hypertrophy and hyperplasia

One pathological change seen in chronically-overstimulated adrenals is *bizarre hypertrophy of the cells.* This has been observed most often in human glands. For example, giant adrenocortical cells (100 μ in diameter) have been described in the glands of human patients that received prolonged treatment with ACTH (LIEBEGOTT 1952), suffered from severe cachexia (LÜDERS 1953), or exhibited adrenal hyperplasia with virilism (LOWENTHAL et al. 1958). Such irregular cellular hypertrophy also occurs in adrenal tumors and carcinomas (e.g., SOBEL et al. 1953, WOOD et al. 1957, SHERMAN et al. 1958, R. COHEN et al. 1959) and in the "fetal" zone (KAMPMEIER 1927, CRAIG and LANDING 1951). Similar giant cells have been recorded in adenomas in adrenal glands of goats (RICHTER 1958) and in a transplantable carcinoma of rat cortex (SNELL and STEWART 1959).

Hyperplasia with consistent cellular hypertrophy is the more frequent occurrence in highly-stimulated glands. This is the picture seen in mice bearing ACTH-producing pituitary tumors (BAHN et al. 1957), in which the cells of the inner zones are greatly enlarged but uniformly so (Fig. 57, p. 102). In human beings following a prolonged course of ACTH administration, hyperplastic adenomas are frequently reported. Similarly, adenomas are found in the adrenals of gonadectomized rodents (see p. 134).

2. Pseudotubular formation

This pathological change is observed in acute situations. Pseudotubular formation occurs principally in primate adrenals, though related changes have been described for rodents. The relative species specificity may depend in part on structural patterns. In 1918/19, GOORMAGHTIGH reported that the adrenals of soldiers dying of gangrene displayed *lumina within the cell columns of the fasciculata.* Since then, the observation has been made repeatedly that severe pyrogenic infections, especially from meningococcus, pneumococcus, streptococcus or the diphtheria bacillus, resultin this destructive change if the disease is fatal within a period of 24—72 hours (RICH 1944, ZAMCHECK 1948). The centers of the cords contain a lumen that is filled with flocculent debris (Fig. 66). Often most of the cell cytoplasm appears to have been shed into this lumen, and the cells are reduced to thin squamous sheets. Many of the nuclei are pycnotic. Leucocytes may invade the lumina, or small hemorrhages may occur. If the patient survives for a longer period before death, signs of regeneration, like mitosis (Fig. 66), are evident in the more peripheral part of the cortex (SCHABERG 1956). Presumably such drastic destruction of the adrenocortical parenchyma may lead to structural collapse and the Waterhouse-Friderichsen syndrome. BACHMANN (1958) has

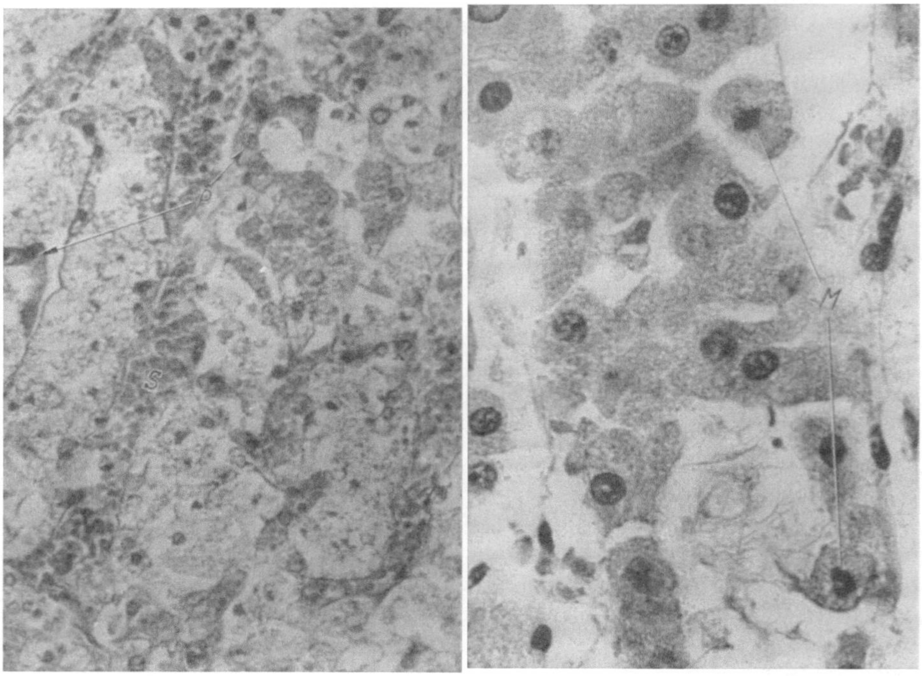

a b

Fig. 66 a & b. Pseudotubular formation in the human adrenal gland. a Patient dying of diphtheria after 7 days. ×425. Most of the parenchyma (*P.*) has disappeared and the residual spaces are filled with cellular debris. *S.*, sinusoidal capillary, filled with erythrocytes. b Patient dying 12 days after onset of meningococcemia. ×1050. Active cell division (*M.*) is occurring in the adrenocortical cells.
Courtesy Dr. NORMAN ZAMCHECK

described such "tubule" formation in the adrenals of patients with adrenal virilism or Cushing's disease.

Similar changes have been described in the adrenals of monkeys with malaria (GOLDEN and OVERMAN 1949). And somewhat similar changes have occasionally been produced in the adrenals of severely-stressed rats, or of rats given massive doses of ACTH (SELYE and STONE 1950, RABINOVICI 1951, WILBUR and RICH 1953, WILBUR 1955, HUHN and SCHNEPPENHEIM 1954).

The parenchymal tubules described frequently in human adrenals obtained at autopsy (e.g., MOERI 1951, ELIAS and PAULY 1954, 1956) are probably of this damaged type, not normal structures. It is noteworthy that they may occur in the glomerulosa as well as in the fasciculata; they generally do not occur in both zones of a single gland.

3. Hemorrhage and necrosis

An abnormality akin to pseudotubular formation, in which the initial site of damage appears to be the capillary endothelium rather than the parenchymal cells, results in hemorrhage and hence areas of necrosis in the cortex. These have

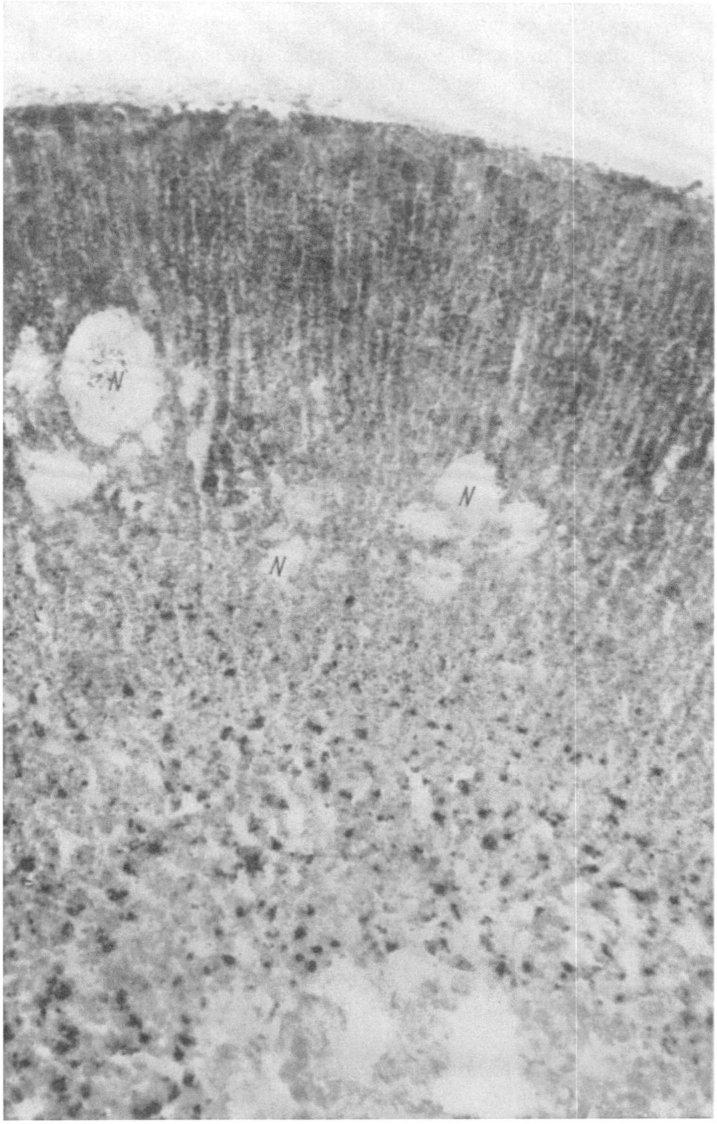

Fig. 67. Adrenal of a rat that had been severely "stressed." Areas of necrosis (*N.*) are scattered throughout the inner zones. Note that no subglomerulosal zone is evident. Adult female rat exposed to cold, 5 hours daily, for 64 days and treated with 6.5 U ACTH daily for the last 58 days. Formalin-fixed frozen section stained with oil red 0 and hematoxylin. × 110. From an experiment by Drs. MASIMICHI OKA and ALFRED A. ANGRIST

been observed repeatedly in the adrenals of severely stressed rats (Fig. 67) (DEANE and MCKIBBIN 1946, DEANE and MASSON 1951, STIGLIANI and MARCHI 1957) and guinea pigs (GOLDEN and OVERMAN 1949, HOEDE 1956), as well as in human glands. When restricted in area, this phenomenon does not result in detectable adrenal insufficiency, but if extensive it may lead to acute insufficiency.

4. Intracellular infection

FRENKEL (1956a, b) has described the intracellular parasitization of the adrenals of *Mesocricetus auratus* infected with the protozoan, *Besnoitia jellisoni*. This led to necrosis of the cells and adrenal insufficiency. Adrenal necrosis was preventable by inactivating the adrenal either by hypophysectomy or treatment with glucocorticoids. These observations suggested that the adrenals must be

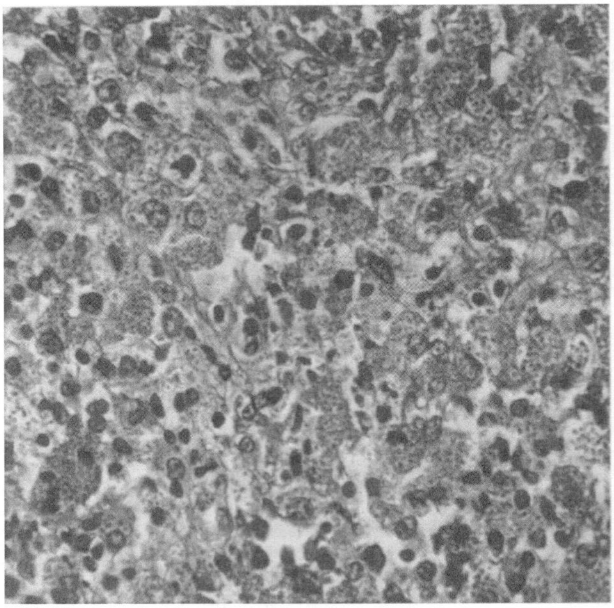

Fig. 68. Intracellular infection of human adrenocortical cells by the fungus, *Histoplasma capsulatum*. Photographed from a preparation contributed by Dr. BORIS GUEFT; hematoxylin and eosin. × 400

active for invasion to occur. Thus *the local anti-inflammatory action of glucocorticoids apparently favors the proliferation of the parasites*. FRIEDMAN (1948) and FRENKEL (1956a, b) cite several pathological reports that have suggested that intracellular infection by a variety of agents may lead to destruction of the adrenal gland and to Addison's disease (see p. 140). Intracellular infection of human adrenocortical cells by *Histoplasma* is illustrated in Fig. 68.

5. Gross lipide droplets, watery vacuoles, "colloid" droplets, increased pigment

Assorted cytological changes of unknown significance have been reported for adrenals subjected to severe, chronic stimulation. In some instances, similar changes occur in inactive glands.

(1) Despite the usual occurrence of small lipide droplets (if any) in stimulated adrenocortical cells, *large confluent droplets sometimes develop*. For example, DEANE and GREEP (1947) illustrated cells in the inner zones of the cortex of rats

made thyrotoxic by feeding thyroid powder for 1—2 months, in which some droplets were extremely large. And HARTROFT and EISENSTEIN (1957) have reported extremely large lipide droplets in the glomerulosal cells of rats fed a severely Na-deficient diet (p. 112). WHITEHEAD (1935a) noted what he termed fatty degeneration of cells in the reticularis of very old rabbits. Whether such changes bespeak a pathological response to prolonged stimulation or altered responsiveness of the cells because of an unfavorable environment remains to be determined.

(2) Highly stimulated cells that are depleted of lipide often display, in frozen sections, non-fatty, clear vacuoles (presumably "watery"). These have been described repeatedly for glands of humans dying of severe disease (e.g., LIEBEGOTT 1952). Similar vacuoles have been observed in the adrenals of other species subjected to infection of prolonged treatment with ACTH (SADOWNIKOW 1949, HUHN and SCHNEPPENHEIM 1954, KLÄRNER 1955, SCHNEPPENHEIM and HUHN 1955).

(3) An associated phenomenon described for glands as a result of stress is the development of intracellular "colloid" droplets — presumably proteinaceous, because of their fixation properties, but possibly also containing carbohydrates or insoluble lipides (SELYE and STONE 1950, RATHER 1951, LIEBEGOTT 1952, WILBUR and RICH 1953, SCHNEPPENHEIM and HUHN 1955, KLÄRNER 1955, PLANEL and GUILHEM 1956b, LOUSTALOT 1957). But such "colloid" droplets also develop in what are presumably inactive cells (SELYE and SALGADO 1955, STUTINSKY 1955, LOUSTALOT 1957).

(4) In several species, development of lipofuscin aggregates in the adrenocortical (and reticuloendothelial) cells also occurs not only in older individuals (WHITEHEAD 1935b, BIMES et al. 1954, VERNE et al. 1956, MEYERS and CHARIPPER 1956, JAYNE 1957) but also with prolonged stimulation in younger individuals (LIEBEGOTT 1952, BACHMANN 1955, 1958, PLANEL et al. 1956, 1958, BERN et al. 1958), or after surgical or chemical hypophysectomy (CUTULY and CUTULY 1937, BERN et al. 1958, ZELANDER 1959). The typical aggregation of lipofuscin in the human reticularis is illustrated in Fig. 43, p. 57. Presumably, because of some adverse change, the cells are no longer able to prevent oxidation of unsaturated lipides.

II. Adrenocortical hyperplasia and tumors

1. Hypersecretion of ACTH

Cushing's syndrome. In human beings and in animals, adrenocortical hyperplasia and hyperactivity may result from a pituitary tumor that secretes excessive amounts of ACTH (CUSHING 1932, UPTON and FURTH 1955, BAHN et al. 1957, V. MARKS 1959). The histological characteristics of the adrenals in such cases resemble those seen with prolonged administration of exogenous ACTH (R. COHEN et al. 1959). Usually such glands produce excessive quantities of glucocorticoids (e.g., NABARRO et al. 1958).

Adrenal virilism. In humans, ACTH output rises sharply when there is an enzymatic deficiency in the adrenal cortex resulting in inadequate production of glucocorticoids (Chapter 3). This syndrome results in marked hyperplasia of the adrenal and usually in augmented secretion of adrenal androgens (adrenal virilism, virilizing congenital adrenal hyperplasia). These androgens lack any significant capacity to inhibit ACTH output (GARDNER 1956). Histologically, such adrenals resemble hyperplastic glands capable of producing glucocorticoids, and presumably the cells of the inner zones (fasciculata and reticularis) are able to produce both types of hormone (p. 108).

2. Hypersecretion of gonadotrophins

Some adrenocortical tumors developing spontaneously in humans produce predominantly sex steroids (KENYON 1944, WALLACH et al. 1957, SNAITH 1958). Since techniques for establishing whether or not adrenal tumors are autonomous, or, if dependent, for determining what pituitary hormone stimulates them are only now in the process of refinement, it is unknown whether some of these are dependent on gonadotrophins rather than ACTH. The fact that gonadotrophin-dependent tumors have been shown to develop in some species of rodents raises this possibility for primates and other orders of mammals.

In several species of rodents, gonadectomy results, after many months, in the development of adrenocortical tumors that secrete sex steroids. In mice, where this phenomenon has been most studied, the incidence of such tumors varies with strain (e.g., FRANTZ and KIRSCHBAUM 1949, WOOLLEY 1949, WOOLLEY et al. 1953). A list of representative observations is given in Table 25. B. HOUSSAY et al. (1953, 1955) and A. I. COHEN et al. (1957b) give additional references. In some strains of mice, spontaneous estrogen-producing tumors arise in the adrenals of non-castrate animals (KIRSCHBAUM et al. 1946). The tumors are pituitary dependent (ATKINSON and DICKIE 1958, BROWNING 1958).

Table 25. *Apparent production of sex steroids by adrenal glands with high gonadotrophin stimulation after gonadectomy*

Species	Treatment and Results	Reference
Cavia cobaya	♂♂ castrated when young. At $2^1/_2$—4 years, 3 exhibited stimulation of the secondary sex organs and adrenocortical adenomas.	SPIEGEL 1939
Rattus norwegicus (Buenos Aires strain)	♀♀ and ♂♂ castrated when young. About 33% developed adrenocortical hyperplasia or adenomas, at 1—2 years. ♀♀ exhibited estrus.	B. HOUSSAY et al. 1953, 1955, CARDEZA 1954
,,	Estrogens or androgens inhibited the development of adenomas, cortisone did not; endogenous or exogenous ACTH inhibited.	A. HOUSSAY et al. 1954
Mus musculus strain DE × strain DBA, F_1 hybrids	Gonadectomized at 1—3 days of age. All developed adrenocortical tumors by 6—12 months. ♀♀ showed signs of hyperestrogenism, ♂♂ signs of androgenic or estrogenic stimulation.	DICKIE and LANE 1956
strain A	♂♂ gonadectomized 1—3 days of age. By 20 months, 42% had cortical nodules and most showed stimulation of mammary glands.	SILBERBERG et al. 1953
strain C3H	♀♀ ovariectomized at 21—23 days. Untreated mice all developed adenomas, those treated with diethylstilbestrol failed to.	KING et al. 1949

The induced tumors arising after gonadectomy appear to arise from hyperplastic foci lying beneath the capsule (WOOLLEY 1949, B. HOUSSAY et al. 1955, DICKIE and LANE 1956). As these nodules grow, both by hyperplasia and cellular hypertrophy, they dissect their way through the cortex and also through the capsule. In some strains of mice (e.g., DBA — WOOLLEY 1949; Strong A — FLAKS 1949), the nodules may become metastatic.

3. Autonomous adrenocortical tumors

In human beings, Cushing's syndrome may arise from adrenal neoplasia that is independent of hypersecretion of ACTH. Little depression of glandular secretion occurs when exogenous cortisone is administered (JAILER et al. 1954, NABARRO et al. 1958), and no conspicuous rise in steroid output follows ACTH administration (SOBEL et al. 1958). When removed, at operation or at autopsy, such glands are extremely large, sometimes metastatic. The cells are generally both extremely large and hyperplastic. The whole gland may be involved, or there may be localized hyperplastic adenomas, with the rest of the cortex appearing either essentially normal or atrophied (CERVIÑO et al. 1947, E. WEBER and MENTEN 1948, SOBEL et al. 1953, 1958). The secretory products may consist of any one of the 3 major types of hormones, glucocorticoids (SIEBENMANN 1959), a sex steroid (KENYON 1944, GRIBOFF 1952), or aldosterone (CONN 1956, VON BUCHEM et al. 1956, THERIEN et al. 1959), or any combination thereof (see Chapter 9).

Recently, comparable tumors have been produced in rats and mice, apparently as the result of total-body radiation (A. I. COHEN et al. 1957a, b). These tumors have been transplanted through several generations of animals. A rat tumor appears to secrete principally glucocorticoids, whereas a mouse tumor appears to produce both glucocorticoids and mineralocorticoid (Fig. 57, p. 102). And a report has appeared on the spontaneous appearance of adrenocortical tumors in rats of the Osborne-Mendel strain; such tumors also produce glucocorticoids, and possibly aldosterone and sex steroids (SNELL and STEWART 1959). They do not require the presence of the pituitary to grow, but when grown in hypophysectomized hosts their cytological and secretory nature has altered.

IGLESIAS and MARDONES (1958) have described another transplantable adrenocortical tumor that arises spontaneously in a certain hydrid strain of rats. It is not clear whether or not this tumor is independent of pituitary stimulation.

III. Inhibitors of adrenocortical function

There now seem to be several fairly well-established mechanisms whereby adrenocortical function can be depressed. In some of these, insufficiency develops despite normal or even enhanced pituitary function. The best-established of these are summarized below. Further discussions will be found in Chapters 3, 4 and 8.

1. Dichloro-diphenyl-dichloroethane (DDD)

When *fed to dogs, this insecticide* (marketed as Rhothane) *produces gross atrophy of the adrenals and degeneration of cells in the inner cortex* (zonae fasciculata and reticularis) (A. NELSON and WOODWARD 1949, NICHOLS and GARDNER 1951, NICHOLS and SHEEHAN 1952, KÜCHMEISTER et al. 1955, LARSON et al. 1955, OSOROIO and KRAEMER 1959). The dogs die with signs of adrenal insufficiency. The dose of technical grade DDD required is 50—200 mg/kg body weight per day. The drug results in almost no detectable damage to the adrenals of rats, mice, rabbits, cattle, or humans (D'ARCY 1954, LARSON et al. 1955, ZIMMERMANN et al. 1956, NICHOLS and HENNIGAR 1957a, b, BLEIBERG and LARSON 1957, A. WEBER et al. 1958). No satisfactory explanation for the species specificity is at hand. In dogs DDD is stored extensively in lipides of various organs including the adrenal (CUETO and BROWN 1958b).

Recently CUETO and BROWN (1958a, b) have separated a highly active fraction from technical-grade DDD that produces marked atrophy of the dog adrenal at a dose of 4 mg/kg per day. This component was identified as *o, p'* DDD (2-*o*-chlorophenyl-2-*p*-chlorophenyl-1,1-dichloroethane) (CRISTOL and HALLER 1948), rather than *p, p'* DDD [2,2-bis(*p*-chlorophenyl)-1,1- dichloroethane], thought to be the active component by most workers. Nichols and Hennigar (1957a) have also produced evidence that the active agent is a contaminant of commercial DDD. Vilar and Tullner (1959) have shown that feeding *o, p'* DDD for as short a period as 2 days depresses cortisol output by the dog adrenal. The effects of this more active compound should be tested in other species, since equivocal evidence for adrenocortical hypofunction in several species besides the dog has been produced with crude DDD (cf. Nichols and Hennigar 1957a, b).

In dogs, DDD destroys the cells of the inner cortex; the morphological change thus resembles that seen after hypophysectomy. That *the atrophy is not the result of deficiency of ACTH*, however, is indicated by the fact that exogenous ACTH does not prevent the atrophy or signs of adrenal insufficiency (Nichols and Green 1954, J. H. Brown et al. 1956, Cobey et al. 1956, Little et al. 1956). As in the hypophysectomized dog, there is evidence of severe deficiency of glucocorticoids. The animals are more sensitive than normal to insulin (Nichols and Gardner 1951, J. H. Brown et al. 1956), fail to develop hyperglycemia when given alloxan (Nichols and Sheehan 1952), exhibit a drop in urinary corticoids and 17-ketosteroids (Küchmeister et al. 1955, J. H. Brown et al. 1956), and fail to show a rise in plasma 17-hydroxy corticosteroids or a drop in the number of circulating eosinophils on the administration of ACTH (Cobey et al. 1956, Cueto and Brown 1958b). (See also Vilar and Tullner 1959, Tullner and Hertz 1960.)

With the usual dose of DDD (50 mg/kg per day), there is no visible disturbance of the structure of the glomerulosa. In line with this, no signs of Na^+ deficiency (Nichols and Gardner 1951) or hypersensitivity to K^+ (Nichols and Sheehan 1952) are demonstrable. However, with higher doses, signs of glomerulosal damage and of disturbance of electrolyte balance do develop (J. H. Brown et al. 1956, Nichols and Hennigar 1957b).

A close relative of DDD, 2,2-bis(*p*-ethyl phenyl)-1,1-dichloroethane (DDE, marketed as Perthane), was shown by Larson et al. (1955) to produce similar atrophy of the inner zones of the cortex in dogs and by Cobey et al. (1956) similarly to prevent a rise of plasma 17-hydroxy corticosteroids in response to ACTH Taliaferro and Leone (1957) have presented evidence that this compound inhibits the output of glucocorticoids by the human adrenal to a small degree, and Bleiberg and Larson (1957) noted some effect on rats. Cobey et al. (1958) have found that several other compounds having the 2,2-diphenyl-1,1-dichloroethane nucleus inhibit the dog adrenal gland.

2. Suramin

The only other cytotoxic agent that appears to have been identified is Suramin (Germanin, Bayer 205), which has been used in treatment of some infectious diseases. A patient treated with this compound died in adrenal insufficiency, and the adrenals exhibited cytotoxic atrophy (Humphreys and Donaldson 1941). When the latter workers administered the compound to guinea pigs, rats, and rabbits at a dose of 20 mg/kg every other day for 20—60 days, progressive degeneration of the fasciculata occurred. Dogs showed no damage at this dosage level. Recently, however, Frisch and Gardner (1958) reported that they were unable to repeat this observation with rats.

3. Amphenone B and SU 4885

Amphenone B, now thought to be 1,1-bis(*p*-aminophenyl)-1 methyl propanone (Bencze and Allen 1957), causes enlargement of the uterus, the thyroid and the adrenal glands of intact rats (Hertz et al. 1950, 1951). *Most of the available evidence suggests that the adrenal hypertrophy is a consequence of a lowered secretion of glucocorticoids which induces an increased output of ACTH* (HERTZ et al. 1955 (see, however, VOGT 1957a).

In most experiments on rats, the drug has been administered by stomach tube at a level of 20 mg/100 g body weight daily, or mixed with the diet (0.5 %). By 10—14 days, the adrenals are approximately double the normal absolute (and relative) weight (HERTZ et al. 1955, NICHOLS and HENNIGAR 1957b). Similar enlargement occurs in the adrenals of dogs (immature) fed amphenone (NICHOLS and HENNIGAR 1957b). In rats, there is no decline in thymus weight, suggesting that secretion of glucocorticoids is not enhanced. The concomitant administration of cortisone (2 mg daily) offsets any adrenal enlargement in rats, and hypophysectomy prevents it likewise. Histologically, the cortical enlargement results from a conspicuous accumulation of lipide in the cells of the fasciculata and reticularis, producing a great increase in cell volume. The glomerulosa remains essentially normal in appearance, and the animals do not need Na^+ supplements.

There is a marked increase in the total lipide and cholesterol concentration, so that the lipide level rises to about 25% and the cholesterol concentration to 15% or more of the wet weight of the gland after a month's treatment (HERTZ et al. 1955, GREER and ERWIN 1956). Exposure of treated animals to cold fails to cause any significant drop in the cholesterol concentration, although there is a drop in ascorbic acid concentration in the gland, as in untreated controls (SCHRICKER et al. 1952). Such observations have led HERTZ and co-workers (1955) to postulate that amphenone depresses some enzyme(s) acting in the conversion of cholesterol to cortical hormones. That the depression is not complete is indicated by the fact that amphenone-treated rats survive exposure to cold under conditions in which adrenalectomized animals succumb.

Some supporting evidence for a depression in 17 hydroxycorticosteroid secretion has been provided by acute experiments with dogs (HERTZ et al. 1955, TULLNER et al. 1956, JENKINS et al. 1959). In either intact or hypophysectomized animals, an intravenous infusion of amphenone severely depressed but did not obliterate the effect of injected ACTH in raising the level of 17 hydroxycorticosteroids in the adrenal venous blood. THORN et al. (1956), HERTZ et al. (1957), RENOLD et al. (1957) and SOBEL et al. (1958) have reported a drop in urinary 17-hydroxycorticoids, urinary 17-ketosteroids and aldosterone in human patients treated with amphenone for adrenal carcinoma. ROSENFELD (1956a) and ROSENFELD and BASCOM (1956) found about a 75% depression in corticoid output of isolated, perfused bovine adrenals when amphenone (7.5×10^{-4} *M*) was added to the perfusion medium. There was little depression of general dehydrogenase activity, but there was reduced ability to oxidize the C-3 hydroxy group or to hydroxylate the 11, 17 or 21 carbons of added substrates. These effects were readily reversible if fresh perfusate was supplied. GALLAGHER (1958) has presented some evidence that the primary effect of amphenone is a depression of C-11-oxidation, but this has been denied by JENKINS et al. (1959).

These results on secretory activity of dog, human, and bovine glands stand in contrast with those on the rat's. VOGT (1957) found that the adrenal effluent of rats chronically treated (10—17 days) with amphenone (20 mg/100 g per day) contained about twice as much corticosterone as did the effluent of untreated rats. Intravenous administration of amphenone, however, proved extremely toxic but did cause a drop in corticosterone secretion. Thus the precise effects of amphenone in the rat remain unclear.

With golden hamsters, B. MARKS et al. (1958) have shown that there is neither adrenal enlargement nor inhibition of hormone production. They suggest that the synthetic pathway differs in hamsters from that in more "typical" mammals.

SU 4885, 2-methyl-1,2-bis(3-pyridyl)-1-propanone, was recently shown to depress the output of cortisol by the adrenals of the dog (JENKINS et al. 1958, 1959) and man (LIDDLE et al. 1958). Clearly *here the principal effect is to inhibit C-11 oxygenation*, since large quantities of 11-deoxycortisol and deoxycorticosterone are secreted (AUGUST et al. 1958, JENKINS et al. 1959).

Also reported recently as depressants of corticosteroid secretion are two derivatives of triazine (COSTA et al. 1959).

It is important to relate the effects of these drugs to *natural diseases*. In human congenital hyperplasia with virilism, deficiency or absence of one or more of the specific enzymes involved in production of normal glucocorticoids results in the secretion of excessive quantities of abnormal steroids. With such deficiencies the adrenals appear depleted of their lipide stores. Contrariwise, PRADER and SIEBENMANN (1957) and SIEBENMANN (1957) have recently described cases in which infants and children dying of adrenocortical insufficiency were found to have hyperplastic adrenals in which there were great accumulations of cholesterol. This may reflect an enzyme deficiency earlier in the system, wherein conversion of cholesterol to more immediate hormone precursors is inhibited.

4. Pantothenic-acid deficiency

Early studies of pantothenic-acid deficiency in rats revealed hypertrophy, hemorrhage and necrosis of the adrenals. Histologically, the changes observed were those characteristic of any severe stress — hypertrophy and hyperplasia of the cells in the fasciculata and reticularis, disappearance of free lipide droplets from these zones, and ultimate necrosis (DEANE

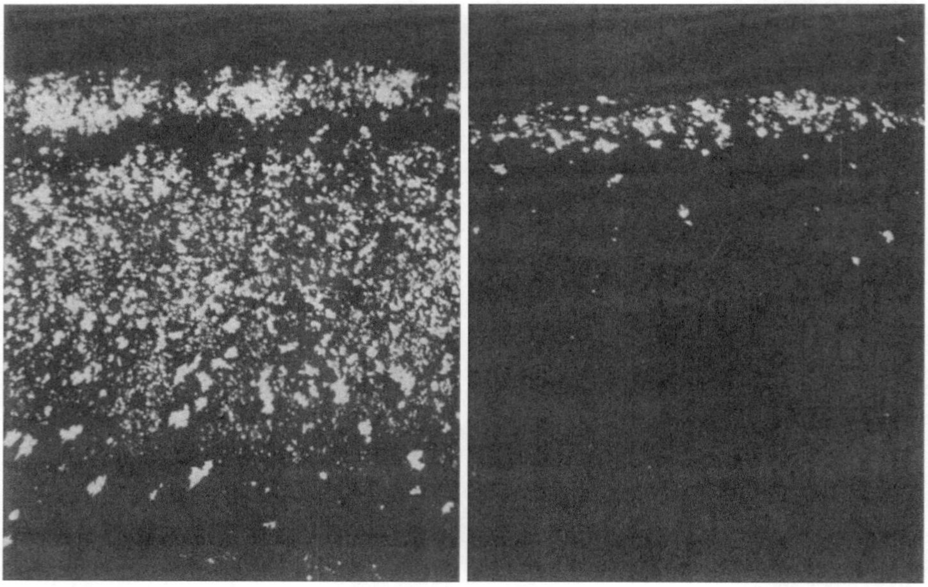

a b

Fig. 69 a & b. Disappearance of cholesterol esters from fasciculata of rat fed a pantothenic-acid deficient diet for 5 weeks (b), as compared with its control (a). Unstained formalin-fixed frozen sections, photographed under the polarizing microscope. ×95. From the study by DEANE and MCKIBBIN (1946). In the gland from the deficient rat, the glomerulosa may contain less birefringent material than normal, but the fasciculata is far more depleted

and MCKIBBIN 1946, MCQUEENEY et al. 1947, R. SCHULTZ et al. 1952). The glomerulosa appeared unaffected in rats fed a deficient diet for 6 weeks (Fig. 69). When puppies became moribund, however, they showed depletion of lipides from the glomerulosa as well as from the inner zones (McKibbin and Deane, see Greep and Deane 1949a). The adrenal enlargement suggested an increased secretion of ACTH, and a higher than normal level of ACTH in the pituitaries of deficient rats has been demonstrated (Butler and Morgan 1955). Even greater enlargement of the adrenals could be induced in deficient animals by treatment with ACTH (R. Schultz et al. 1952).

Subsequent studies suggested that *this stimulated adrenal cortex is*, however, *subnormal in its secretory capacity* (Ershoff 1952). Thus, animals placed on a severely-deficient diet at a very early age or kept on the deficient diet for prolonged periods reach a stage at which they no longer exhibit the lymphopenia characteristic of early deficiency (Melampy et al. 1951). The injection of ACTH or adrenaline fails to induce a depletion of circulating eosinophils or lymphocytes (Dumm et al. 1949, Winters et al. 1952, Dumm and Ralli 1954); stress fails to induce the usual deposition of glycogen in the liver (Hurley and Morgan 1952); there is a decreased ability to excrete a water load (Gaunt et al. 1946, Guggenheim 1954); and sensitivity to insulin is enhanced (Hazelwood et al. 1956). In humans, Bean et al. (1955) have shown a positive rise in the level of circulating eosinophils after 14 days on a deficient diet, increased insulin sensitivity and reduced 17-ketosteroid excretion. (Unfortunately, these subjects developed multiple vitamin deficiencies.) Recently Eisenstein and Wenneker (1958) have shown decreased secretion of 17-hydroxy corticosteroids in deficient guinea pigs given ACTH, and Longwell et al. (1958) and Goodman (1960) found depressed levels of corticoids in the adrenal venous effluent of rats (made deficient by the vitamin antagonist, ω-methylpantothenate). Several of the above workers demonstrated that the signs noted could be repaired by glucocorticoid administration. The hypothesis was formulated that *pantothenic-acid deficiency results in a lack of coenzyme A in the cortex and thereby reduces steroid production* (see p. 62).

Two studies with adrenals from deficient rats indicate a restricted ability to form corticosteroids *in vitro* when stimulated by ACTH (Angelico and Quintilani 1955, Eisenstein 1957). In both studies, the adrenals of severely-deficient rats produced less than 50% as much Δ^4 steroid as did those of control rats.

In the opposite direction, Dumm et al. (1955) have adduced evidence that a diet containing a high level of pantothenic acid (40 mg/100 g diet) accelerates hypertrophy and cholesterol restoration in the adrenals of rats subjected to stress, as compared to rats receiving the "normal" complement (1 mg/100 g diet). L. Greenberg and Glick (1958) have demonstrated that feeding pantothenate enhances the concentration of coenzyme A in the adrenal cortex.

5. Pyridoxine deficiency

Severe deficiency of pyridoxine in rats, produced by feeding a deficient diet containing the vitamin-antagonist deoxypyridoxine (2 mg/100 g diet), induces conspicuous absolute and relative hypertrophy of the adrenals (Stebbins 1951a, Olsen and Martindale 1954). Butler and Morgan (1954) found that the adrenals of such animals fail to show the customary decline in cholesterol when exposed to stress. Stebbins (1951c) and Guggenheim (1954) found that such animals, like those deficient in pantothenic acid, show lowered ability to excrete a water load. Since there was no sign in these animals of any inability to inactivate Pitressin, the lesion was suggested to be a lower-than-normal secretion of adrenocortical hormones.

Since the functions of pyridoxine in metabolism remain obscure (Sebrell and Harris 1954), no specific interpretation can be placed on these observations at present. The principal role of pyridoxine seems to be in the realm of protein synthesis. Perhaps the deficiency lies in the failure to produce enzymes in general, including those required for hormone synthesis.

6. Δ^4-Cholestenone

Cholestenone is thought to act as a cholesterol-analogue and hence *a competitive inhibitor of cholesterol synthesis.* Rats fed a diet containing this compound at a evel of 1% exhibit a 7-fold increase in adrenal weight in 6 weeks (FREDRICKSON et al. 1958). Such glands contain little cholesterol and ascorbic acid, although there is abundant lipide (STEINBERG et al. 1958). Thymus weight is relatively increased. The output of corticosterone in the adrenal venous blood is reduced greatly, both in terms of body weight (20% of normal) and in terms of gland weight (3% of normal) (FREDRICKSON et al. 1958). Aldosterone secretion appears slightly subnormal. Moderate deficiency was already observed in 12 days. STEINBERG et al. (1958) reported that the substance was too toxic to be used to depress adrenocortical activity in human patients.

7. Thyroid deficiency

In a variety of mammals, there is some involution of the adrenal cortex in thyroid insufficiency, whether this is produced by thyroidectomy or by the administration of one of many anti-thyroid drugs (DEANE and GREEP 1947, LI VOTI 1949, MAQSOOD 1950, 1954, ZARROW and ZARROW 1951, KRACHT and SPAETHE 1953a, MONEY 1955, DE VISSCHER and BECKERS 1957, ZARROW et al. 1957). (BROWN-GRANT (1959) could demonstrate no such effect in the laboratory rabbit, however.) EIK-NES and BRIZZEE (1956) showed that the level of circulating 17-OH corticosteroids in thyroidectomized dogs was 1—2 μg/100 ml plasma (normal, 5 μg-%). [With some anti-thyroid drugs, such as thiouracil, the cortical atrophy is sometimes masked by medullary enlargement (DEANE and GREEP 1947).] The atrophy is attributable to shrinkage of cells in the inner zones of the cortex. The cells give every evidence of reduced stimulation. *The depression induced by thyroid deficiency apparently resides principally in the pituitary,* since the adrenals respond typically to ACTH (ZARROW and ZARROW 1951, MONEY 1955, EIK-NES and BRIZZEE 1956, KOWALEWSKI 1958). LAZO-WASEM (1959) has found a reduced content of ACTH in the pituitary. This depression in ACTH-output can be overcome, however, if the animal is exposed to severe or chronic stress (e.g., DEANE and GREEP 1947). Thus the condition differs from the preceding ones, in which the deficiency lies at the adrenal rather than at the pituitary level.

The effects of the goitrogen amphenone are quite different, in that the adrenal is greatly hypertrophied, apparently as the result of an increased output of ACTH (p. 136). This reaction demonstrates clearly that the pituitary is capable of producing large quantities of ACTH in thyroid insufficiency.

8. Inadequacy of the pituitary

In Chapter 10, a number of conditions that result in depressed activity of the anterior pituitary with respect to ACTH release are considered. All of these will of course lower adrenocortical activity.

9. Other causes of adrenal insufficiency

Local ischemia and necrosis may result from vascular occlusion (FRIEDMAN 1948). If sufficiently extensive, this may lead to Addison's disease. In an attempt to simulate the adrenal insufficiency following gross vascular occlusion, ROGOFF (1944) studied the effects of ligation of the adrenal vein in dogs and cats. The majority of animals showed chronic, subacute insufficiency. Their cortices showed variable degrees of atrophy, with patches of regenerating cells. Focal lesions in the adrenal also result from inadvertent damage to the arteries during nephrectomy (HARRISON 1951b). This has been reported for human glands as well (FRIEDMAN 1948).

Circulatory failure. In irreversible hemorrhagic shock, circulation through the adrenal fails. In consequence, such an adrenal does not show the expected signs of stimulation and retains a normal complement of lipides (RABINOVICI 1951).

Gross invasion and consequent destruction by *metastasizing carcinomas* occur occasionally (RIBAUDO 1958).

The basis for the so-called *adrenocortical contraction* leading to Addison's disease is poorly understood (FRIEDMAN 1948). No doubt there are multiple causes. Sometimes there may be pseudotubular formation (p. 129) resulting from hyperstimulation of the adrenal, associated with some toxic condition. Sometimes there may be direct infection of the adrenocortical cells (Fig. 68, p. 132) (FRENKEL 1956a, b). Indeed, FRENKEL has suggested that the local concentration of glucocorticoids may facilitate the growth of organisms within adrenocortical cells.

Whether there exist, in addition, still unknown cytotoxic agents for adrenocortical cells that result in glandular collapse is still undetermined. The *origin of Addison's disease is obscure in many instances* (FRIEDMAN 1948). Suggestions of genetic tendency exist (MOSIER 1956, SHEPARD et al. 1959). Isoimmunity is a possible cause (COLOVER and GLYNN 1958). HADLOW (1953) described 3 cases of Addisonianism in dogs, in which there was generalized destruction of adrenocortical cells. It seemed unlikely that the destruction in these cases was caused by the chance ingestion of DDD (p. 135), since the glomerulosal cells were destroyed as well as the cells in the inner zones of the cortex.

I. Physiology of adrenocortical tissue in the lower vertebrates

Earlier sections have cited evidence that in lower vertebrates (1) adrenocortica tissue atrophies grossly after hypophysectomy (Table 16, p. 78; Fig. 51a, p. 89; Fig. 70) or the administration of glucocorticoids (Table 17, p. 80), (2) hypertrophies during the administration of ACTH (Table 18, p. 83; Fig. 51c), and (3) secretes cortisone, corticosterone and possibly cortisone and aldosterone (Table 15, p. 75) (FONTAINE and HATEY 1954, LELOUP-HATEY 1958, CHESTER JONES et al. 1959). The cells undergo typical changes when subjected to stimulation or loss of stimulation (p. 77ff.). *In such overall terms,* then, *the adrenocortical tissue of various lower vertebrates resembles that of mammals.* The present section summarizes any additional information on the control and secretory activity of the gland in these animals. Other information is given in Chapter 4.

I. Hypophysectomy and adrenalectomy

In those lower vertebrates in which both hypophysectomy and adrenalectomy have proved practicable, hypophysectomy generally appears the less severe operation in terms of survival time (CHESTER JONES 1957). Thus, in an elasmobranch, *Torpedo* sp., adrenalectomy is fatal within 4 days, whereas hypophysectomy permits survival up to 24 days (DITTUS 1940). In adrenalectomized skates (*Raja*), there is the characteristic fall in the plasma Na^+/K^+ ratio as a result of a marked rise in plasma K^+ (CHESTER JONES 1957).

Adrenalectomy is not feasible in teleost fishes because of the diffuse distribution of the glandular tissue in the walls of major blood vessels, but hypophysectomy permits survival for at least a month (e.g., OLIVEREAU and FROMENTIN 1954, CHAVIN 1956, G. PICKFORD 1957). The same situation obtains in lungfish (MOORHOUSE 1956).

Among the amphibians, hypophysectomy generally permits prolonged survival (*Taricha torosa* — M. MILLER 1953; anuran tadpoles — P. SMITH 1920, CHANG and WITSCHI 1955; *Bufo arenarum* — PORTO 1940, B. HOUSSAY 1949,

a

b

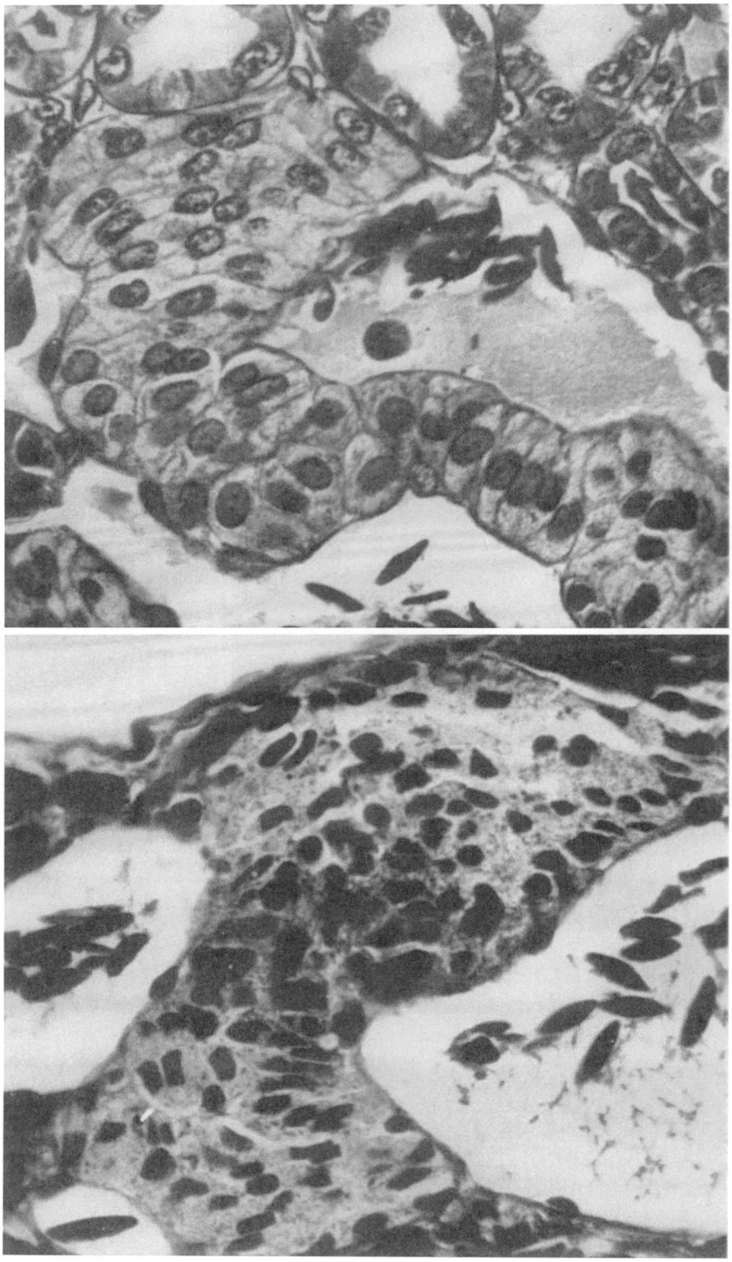

Fig. 70a & b. Atrophy of adrenocortical cells in the newt, *Taricha torosa*, after hypophysectomy. a normal animal; b animal hypophysectomized 115 days earlier; Zenker-formalin, Heidenhain's azan. Both ca. ×450. Courtesy Dr. MALCOLM R. MILLER; see MILLER (1953)

Penhos and Cardeza 1951; *Rana esculenta* (winter) — Sluiter et al. 1949; *Rana pipiens* — Levinsky and Sawyer 1952, Burgos 1959). Hypophysectomy in *Rana temporaria* is survived satisfactorily in the winter but is rapidly fatal during the summer (Chester Jones 1957). Adrenalectomy is extremely difficult to perform in amphibians because the interrenal is imbedded in the kidney and removal generally damages critical blood vessels as well as the nephric duct. There are, however, a few reports of successful operations. Maes (1937) reported that frogs adrenalectomized in the winter (*Rana temporaria*) survived adrenalectomy for prolonged periods if kept in the cold. In contrast, "summer" frogs did not survive adrenalectomy for more than a day, unless kept in physiological saline (Chester Jones 1957). Thus in these species the situation after adrenalectomy does not differ significantly from that after hypophysectomy. Chester Jones (1957) speculates that the susceptibility of *R. temporaria* and *R. esculenta* to hypophysectomy may result from the requirement for some adenohypophyseal product for proper water and salt transport through the skin. Clark et al. (1944) successfully adrenalectomized specimens of *Bufo marinus*; unless given supportive therapy the toads died in 3—6 days. Adrenalectomized toads (*B. arenarum*) die much earlier than do hypophysectomized ones (Fustinoni 1938). In both *R. temporaria* and *B. arenarum* there is a more conspicuous decline in muscle and blood Na^+ after adrenalectomy than after hypophysectomy.

Among reptiles, only *Thamnophis* sp. (Schaeffer 1933), *Xantusia vigilis* (M. Miller 1952), *Agama agama* and *Natrix natrix* (Wright and Chester Jones 1957) seem to have been hypophysectomized. Such animals survive long periods. Adrenalectomy of *Natrix* is fatal within a week to 10 days (Chester Jones 1957) and is associated with a 50% decline in muscle Na^+ concentration. Adrenalectomy of a terrapin is fatal in 2—3 days unless the animal is given cortisone injections (Combescot 1957).

In domestic birds, hypophysectomy has been accomplished repeatedly (e.g., R. Miller and Riddle 1939, 1941, 1942, Nalbandov and Card 1943, Case 1952). Apparently the animals survive for considerable periods. Complete adrenalectomy has proved rapidly fatal in most species (e.g., Herrick and Torstveit 1938, Bülbring 1940, R. Miller and Riddle 1943), although Leroy and Benoit (1954) report prolonged survival (over 1 month) of several ducks in which no gross remnants could be found. (No microscopic examination was made for remnants or accessory nodules.) Herrick and Torstveit (1938) and R. Miller and Riddle (1943) found that life could be prolonged in adrenalectomized chickens and pigeons, respectively, by including large quantities of NaCl in the diet. In pigeons, 1 mg deoxycorticosterone per day also maintained life.

Several of these observers have reported that the cellular atrophy after hypophysectomy is not uniform throughout the adrenocortical tissue. In amphibians, Sluiter et al. (1949) and M. Miller (1953) noted that after hypophysectomy a few of the cells retained lipide droplets long after the majority had been depleted. (Such a difference is denied by Tuchmann-Duplessis et al. 1955.) Similarly, in the reptiles *Xantusia* (M. Miller 1952), *Agama* and *Natrix* (Wright and Chester Jones 1957), groups of cells are said to retain their nearly normal appearance many months after hypophysectomy. And in the bird *Columba*, R. Miller and Riddle (1942) reported survival of the outermost zone of cells. None of the students of fish adrenals has reported such differences.

Thus *there is some evidence for different types of cells in the adrenocortical tissue certain lower vertebrates, with some cells not dependent on the anterior pituitary,* reminiscent of the difference between the zona glomerulosa and the inner zones in most mammalian adrenals (p. 98ff.). Only in birds, however, is there any tendency for zonation of the different classes of cells (see below).

The combined data suggest that most but not all lower vertebrates survive longer after hypophysectomy than after adrenalectomy. Whether this difference in survival time depends especially on the persistent secretion of a mineralocorticoid or merely on the secretion of a basal level of cortical hormones in general remains to be elucidated.

II. ACTH or "stress"

Enlargement of the adrenal glands or cytological evidence of stimulation occurs in lower vertebrates in situations of stress or following the administration of ACTH (fishes — Dittus 1940, Chavin 1956, Moorhouse 1956, G. Pickford 1957, G. Pickford and Kosto 1957, Robertson and Wexler 1957, 1959; amphibians — P. Smith and Smith 1923, Porto 1940, M. Miller 1953, Chang and Witschi 1955, Tuchmann-Duplessis et al. 1955, Burgos 1959; reptiles — M. Miller 1952, Wright and Chester Jones 1957; birds — R. Miller and

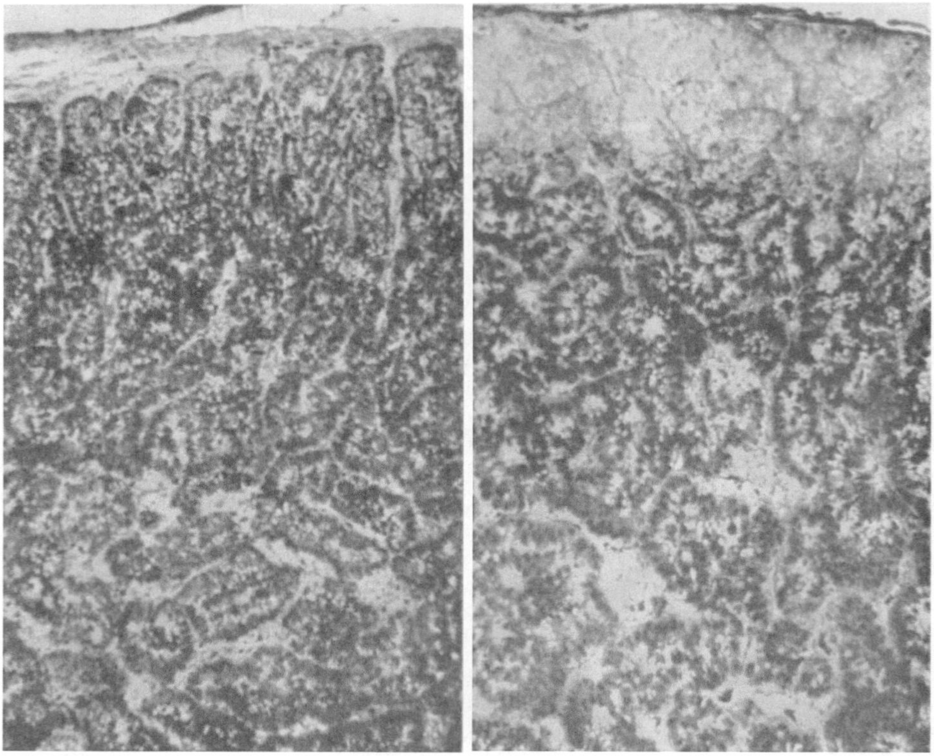

a b

Fig. 71 a & b. Depletion of osmiophilic lipide droplets from "zona glomerulosa" of the adrenal of the pelican when treated with ACTH. a control bird; b bird injected on 3 successive days with ACTH; glands fixed in Helly's fluid and osmium tetroxide. × ca. 130. The only zone that appears to be detectably affected is the glomerulosa. Reproduced by permission of the authors and publisher from Figs. 4 and 5, Hartman et al.: Anat. Rec. **120**, 491 (1954).

Riddle 1942, Jailer and Boas 1950, Zarrow and Baldini 1952, Dulin 1953, Hartman et al. 1954, Taylor et al. 1956, A. Howard and Constable 1958, Flickinger 1959). With excessive doses of ACTH, cell fragmentation not unlike that observed during pseudotubular formation in mammalian glands (p. 129) has been described (e.g., DITTUS 1940, M. MILLER 1952, 1953, HARTMAN et al. 1954, ROBERTSON and WEXLER 1959).

Of especial importance with relation to the possibility of equating the zones in bird adrenals with those in mammalian adrenals is the observation of HARTMAN et al. (1954) that, in *Pelecanus*, either ACTH or stress induces lipide depletion more rapidly from the so-called glomerulosa than from the inner zones (Fig. 71). Thus *there is no similarity to the response observed in rodent and carnivore glands.*

III. Other experimental situations

An unusual observation was made by R. MILLER and RIDDLE (1941, 1942). They found that, in hypophysectomized as well as in intact pigeons, the administration of insulin or formalin induced signs of hyperactivity in the adrenocortical cells: there was glandular hypertrophy plus an increase in the number of mitochondria, hypertrophy of the Golgi apparatus and a depletion of lipide from the cells. Recently RAPELA and GORDON (1957) have reported that insulin (20 U/100 g body weight) causes hypertrophy in the interrenal cells of hypophysectomized toads, as of intact specimens. *The possibility of some direct responsiveness of the adrenal tissue in submammalian vertebrates requires further investigation.*

As in mammals, the glands of lower vertebrates shrink and the cells show signs of hypoactivity after the injection of deoxycorticosterone or glucocorticoids (Table 17, p. 80) (M. MILLER 1952, 1953, R. MILLER and RIDDLE 1942, HARTMAN et al. 1954). Also, in hypothyroid chick embryos (thiourea), the adrenocortical cells shrink and appear inactive (KINGSBURY et al. 1955). VAN OVERBEEKE (1958) has recently shown that adrenocortical cells of the Siamese fighting fish, *Betta splendens*, hypertrophy conspicuously when the animals are kept in water containing amphenone (ca. 5 mg-%).

Literature

ABOIM, A. N.: L'organe interrénal des sélaciens. Arch. port. Sci. biol. 7, 89—134 (1944).
— L'organe interrénal des cyclostomes et des poissons. Port. Acta biol. A 1, 353—383 (1946).
ADAMS, E., and M. BAXTER: Lipid fractions of human adrenal glands. A.M.A. Arch. Path. **48**, 13—26 (1949).
ADAMS, W. E.: The development of the adrenals and sympathetic paraganglia in certain Madagascar insectivores. J. Embryol. exp. Morph. **6**, 575—588 (1958).
ADDIS, T., and H. GRAY: Body size and suprarenal weight. Growth **14**, 81—92 (1950).
ADOLF, E. F.: Quantitative relations in the physiological constitution of mammals. Science **109**, 579—585 (1949).
AGATE, F. J., JR.: The possible role of the adrenal cortex in fluid shifts in the hamster following renal damage. Ann. N. Y. Acad. Sci. **55** (Art. 3), 404—411 (1952).
ALFERT, M., H. A. BERN and E. ALFERT: Hormonal influence on nuclear synthesis. III. On the effect of fixation on nuclear volume. Z. wiss. Mikr. **62**, 323—326 (1955).
ALLEN, J. M.: A chemical and histochemical study of aliesterase in mouse adrenal cortex under hormonal influences and altered sodium intake. Endocrinology **61**, 368—374 (1957).
— A chemical and histochemical study of aliesterase in the adrenal gland of the developing mouse. Anat. Rec. **132**, 195—207 (1958).
— A chemical and histochemical study of DPN-diaphorase activity in the adrenal cortex of the developing mouse. Anat. Rec. **134**, 385—395 (1959).
— Some enzymatic properties of the X-zone of the mouse adrenal gland studied by cytochemical methods. Anat. Rec. **136**, 153 (1960).
— and J. J. SLATER: Influence of hormones upon the distribution of alkaline phosphatase in mouse adrenal cortex. J. Histochem. Cytochem. **4**, 112—117 (1956).
ALPERT, M.: Observations on the histopathology of the adrenal gland of the golden hamster. Endocrinology **46**, 166—176 (1950).
— Hormonal induction of deposition of ceroid pigment in the mouse. Anat. Rec. **116**, 469 to 494 (1953).

ANDERSON, C. H., M. MCCALLAY and G. L. FARRELL: The effects of atrial stretch on aldosterone secretion. Endocrinology **64**, 202—207 (1959).

ANDO, T.: Adrenal structure and function with special reference to the relationship between zona glomerulosa and zona fasciculata. Endocr. jap. **4**, 35—56 (1957).

ANGELICO, R., and M. QUINTILANI : Effect of deficiency of pantothenic acid on the production *in vitro* of corticosteroids by the adrenal gland of the rat (in Italian). Boll. Soc. ital. Biol. sper. **31**, 1308—1310 (1955).

ANSON, B. J., E. W. CAULDWELL, J. W. PICK and L. E. BEATON: The blood supply of the kidney, suprarenal gland, and associated structures. Surg. Gynec. Obstet. **84**, 313—320 (1947).

ARNESEN, K.: Constitutional difference in lipid content of adrenals in two strains of mice and their hybrids. Acta endocr. (Kbh.) **18**, 396—401 (1955).

— The adrenothymic constitution and susceptibility to leukemia in mice. Acta path. microbiol. scand., Suppl. **109** (1956).

ARVY, L., et A. BONICHON: Contribution à l'histoenzymologie de la gland surrénale chez *Crocodylus niloticus* Laurenti. C. R. Acad. Sci. (Paris) **246**, 1759—1761 (1958).

ASCOLI, G., u. T. LEGNANI: Die Folgen der Exstirpation der Hypophyse. Münch. med. Wschr. **59**, 518—521 (1912).

ATKINSON, W. B., and M. M. DICKIE: Essential role of hypophysis in hypercorticism and hypeovarianism in DBA × CE and reciprocal mice. Proc. Soc. exp. Biol. (N. Y.) **99**, 267—269 (1958).

AUGUST, J. T., D. H. NELSON and G. W. THORN: Aldosterone. New Engl. J. Med. **259**, 917 to 923, 967—971 (1958).

AYRES, P. J., R. P. GOULD, S. A. SIMPSON and J. F. TAIT: The *in vitro* demonstration of differential corticosteroid production within the ox adrenal gland. Biochem. J. **63**, 19P (1956).

— J. BARLOW, O. GARROD, S. A. S. TAIT, J. F. TAIT and G. WALKER: Correlation of adrenal steroid concentrations and changes in electrolyte metabolism. Scand. J. clin. lab. Invest. Suppl. **31**, 29—49 (1958).

BACCHUS, H.: Cytochemical study of the adrenal cortex of the rat under salt stresses. Amer. J. Physiol. **163**, 326—331 (1950).

— Cytological distribution of cholesterol and ascorbic acid in the adrenal cortex of the rat exposed to cold. Anat. Rec. **110**, 495—504 (1951).

— *In vitro* conversion of desoxycortisol to cortisol by adrenal tissues in ascorbic acid deficiency. Amer. J. Physiol. **188**, 297—302 (1957a).

— *In vitro* formation of α-β-unsaturated ketones by adrenal tissues in ascorbic acid deficiency. Amer. J. Physiol. **188**, 302—307 (1957b).

— M. H. HEIFFER and N. ALTSZULER: Cytochemical study of the adrenal cortex of the rat with experimental cardiac hypertrophy. Endocrinology **50**, 264—268 (1952).

BACHMANN, R.: Die Nebenniere. Handb. mikr. Anat. Mensch. **VI/5**, 1—952 (1954).

— Perjodsäure-Leukofuchsinreaktion (PAS-Reaktion) an der Nebenniere im Falle von Virilisms. Endokrinologie **32**, 257—264 (1955).

— Struktur und Funktion in der Grenzsituation (am Beispiel Nebennierenrinde). Verh. anat. Ges., Jena **55** (Anat. Anz. **105** Suppl.), 194—205 (1958).

BACQ, Z. M., et P. FISCHER: Action du rayonnement X sur les surrénales du rat nouveau-né. Rev. belg. Path. **25**, 384—387 (1956).

BADINEZ, O., and H. CROXATTO: Histological modifications of the adrenal cortex in the rat treated with potassium chloride and sodium chloride (in Spanish). Bol. Soc. biol. Santiago **5**, 40—43 (1948).

BAECKER, R.: Über die Nebenniere der Teleostier. Z. mikr.-anat. Forsch. **15**, 204—273 (1928).

BAHN, R. C., and D. GLICK: The quantitative histological distribution of ascorbic acid in the adrenal gland of the rat and monkey. J. Histochem. Cytochem. **2**, 103—109 (1954a).

— — Effects of stress conditions, ACTH, cortisone and desoxycorticosterone on the quantitative histological distribution of ascorbic acid in the adrenal glands of the rat and monkey. Endocrinology **54**, 672—684 (1954b).

— J. FURTH, E. ANDERSON and E. GADSDEN: Morphologic and functional changes associated with transplantable ACTH-producing pituitary tumors of mice. Amer. J. Path. **33**, 1075—1097 (1957).

— H. E. STORINO and R. W. SCHMIT: The mass of the zona glomerulosa following complete anterior and posterior hypophysectomy and subtotal removal of the pars tuberalis in the dog. Endocrinology **66**, 403—408 (1960).

BAÏSSET, A., L. DOUSTE-BLAZY, H. PLANEL et P. MONTASTRUC: Cortex surrénal et diabète insipide expérimental. C. R. Soc. Biol. (Paris) **148**, 1867—1870 (1954).

— — et P. MONTASTRUC: Réduction de la sécrétion d'aldostérone sous l'effet d'une distention auriculaire. J. Physiol. (Paris) **51**, 393—394 (1959).

Baker, B. L.: A comparison of the histological changes induced by experimental hyperadrenalcorticalism and inanition. Recent Progr. Hormone Res. **7**, 331—373 (1952).

Baker, D. D.: Studies of the suprarenal glands of dogs. 1. Comparison of the weights of suprarenal glands of mature and immature male and female dogs. Amer. J. Anat. **60**, 231—252 (1937).

— Comparison of the weights of suprarenals of dogs in oestrus, pregnancy and lactation. J. Morph. **62**, 3—15 (1938).

Balfour, W. W., R. S. Comline and R. V. Short: Secretion of progesterone by the adrenal gland. Nature (Lond.) **180**, 1480—1481 (1957).

Bänder, A.: Die Beziehungen des 24-Stunden-Rhythmus vegetativer Funktionen zum histologischen Funktionsbild endokriner Drüsen. Z. ges. exp. Med. **115**, 229—250 (1950).

Bargmann, W.: Über den Bau der Nebennierenvenen des Menschen und der Säugetiere. Z. Zellforsch. **17**, 118—138 (1933).

Barnett, S. A., and G. Bourne: The distribution of ascorbic acid (vitamin C) in the early stages of the developing chick embryo. J. Anat. (Lond.) **75**, 251—264 (1940).

— and R. B. Fisher: Experiments on the silver nitrate method for the histological demonstration of ascorbic acid (vitamin C). J. exp. Biol. **20**, 14—15 (1943).

Barrnett, R. J.: The distribution of esterolytic activity in the tissues of the albino rat as demonstrated with indoxyl acetate. Anat. Rec. **114**, 577—600 (1952).

Bartter, F. C., G. W. Liddle, L. E. Duncan, jr., J. K. Barber and C. Delea: The regulation of aldosterone secretion in man: The role of fluid volume. J. clin. Invest. **35**, 1306 to 1315 (1956).

— E. G. Biglieri, P. Pronove and C. S. Delea: Effect of changes in intravascular volume on aldosterone secretion in man. In : A. F. Muller and C. M. O'Connor, Eds., An International Symposium on Aldosterone, 100—110. Boston: Little, Brown 1958.

Barucci, M., and F. Mori: Modifications in cholesterol content of the adrenal of the guinea pig after electroshock (in Italian). Boll. Soc. ital. Biol. sper. **27**, 1031—1032 (1951).

Baxter, J. S.: The growth cycle of the cells of the adrenal cortex in the adult rat. J. Anat. (Lond.) **80**, 139—146 (1946).

Bean, W. B., R. E. Hodges and K. Daum: Pantothenic acid deficiency induced in human subjects. J. clin. Invest. **34**, 1073—1084 (1955).

Beck, J. C., E. E. McGarry, I. Dyrenfurth and E. H. Venning: Metabolic effects of human and monkey growth hormone in man. Science **125**, 884—885 (1957).

Beigelman, P. M., M. A. Slusher, G. G. Slater and S. Roberts: Effect of anesthetics and collection time on corticosteroid secretion by rat adrenal. Proc. Soc. exp. Biol. (N. Y.) **93**, 608—611 (1956).

Bejdl, W.: Vergleichende Untersuchungen über die Verteilung der alkalischen Phosphatase in der Ur- und Nachniere sowie in der Nebenniere. Z. mikr.-anat. Forsch. **61**, 533—540 (1955).

Bell, J. T., jr., and A. F. Weber: Lipid accumulation in the bovine adrenal cortex. Anat. Rec. **130**, 270 (1958).

Belt, W. D.: The origin of adrenal cortical mitochondria and liposomes: A preliminary report. J. biophys. biochem. Cytol. **4**, 337—340 (1958).

— and E. R. Hayes: An ultraviolet-Schiff reaction for unsaturated lipids. Stain Tech. **31**, 117—122 (1956).

— and D. C. Pease: Mitochondrial structure in sites of steroid production. J. biophys. biochem. Cytol. **2**, Suppl., 369—374 (1956).

Bencze, W. L., and M. J. Allen: Structure of amphenone B and related amphenones. J. org. Chem. **22**, 352—353 (1957).

Benirschke, K.: Adrenals in anencephaly and hydrocephaly. Obstet. and Gynec. 8, 412 to 425 (1956).

— E. Bloch and A. T. Hertig: Concerning the function of the fetal zone of the human adrenal gland. Endocrinology **58**, 598—625 (1956).

Benner, M. C.: Studies on the involution of the fetal cortex of the adrenal glands. Amer. J. Path **16**, 787—797 (1940).

Bennett, H. S.: The life history and secretion of the cells of the adrenal cortex of the cat. Amer. J. Anat. **67**, 151—227 (1940).

— and L. Kilham: The blood vessels of the adrenal gland of the adult cat. Anat. Rec. **77**, 447—472 (1940).

— J. H. Luft and J. C. Hampton: Morphological classification of vertebrate blood capillaries. Amer. J. Physiol. **196**, 381—390 (1959).

Benua, R. S., and E. Howard: A carbonyl reaction differentiating the fetal zona reticularis of the human adrenal cortex from the mouse X zone. Johns Hopk. Hosp. Bull. **86**, 200—213 (1950).

BERGNER, G. E., and H. W. DEANE: Effects of pituitary adrenocorticotropic hormone on the intact rat, with special reference to cytochemical changes in the adrenal cortex. Endocrinology **43**, 240—260 (1948).
BERGSTRAND, C. G.: The influence of testosterone propionate on the adrenal cortex of the infantile rabbit. Acta endocr. (Kbh.) Suppl. **8** (1952).
BERN, H. A., S. NANDI, R. A. CAMPBELL and L. E. PISSOTI: The effects of hormones and other agents on weight change and on ceroid deposition induced by oestrogen administration and by hypophysectomy in the adrenal glands of BALB/c mice. Acta endocr. (Kbh.) **31**, 349—383 (1958).
BERNSTEIN, D. E.: Autotransplantation of the adrenal of the rat to the portal circulation: Effect of administration of sex hormones. Endocrinology **63**, 765—767 (1958).
— ... Effect of administration of ACTH. Endocrinology **65**, 343—346 (1959).
BEYER, K. F., and L. T. SAMUELS: Distribution of steroid-3β-ol-dehydrogenase in cellular structures of the adrenal gland. J. biol. Chem. **219**, 69—76 (1956).
BIMES, C., H. PLANEL et A. GUILHEM: Variations pondérales et structurales de la surrénale du cobaye suivant l'age. C. R. Ass. Anat. **41**, 738—752 (1954).
— — — et P. GIRARD: Etude histologique et histophysiologique des lipofuscines de la glande surrénale du cobaye. C. R. Ass. Anat. **42**, 325—338 (1955).
BIMMER, E.: Metrische Untersuchungen über die Entwicklung der Nebenniere und der ihr benachbarten Organe bei Eidechsen. Anat. Anz. **97**, 276—311 (1950).
BIRMINGHAM, M. K., and E. KURLENTS: Inactivation of ACTH by isolated rat adrenals and inhibition of corticoid formation by adrenocortical hormones. Endocrinology **62**, 47—60 (1958).
BISHOP, D. M., and J. H. LEATHEM: Response of prepuberal male mice to equine gonadotropin. Anat. Rec. **95**, 313—320 (1946).
BLACKMAN, S. S.: Concerning the function and origin of the reticular zone of the adrenal cortex. Johns Hopk. Hosp. Bull. **78**, 180—208 (1946).
BLEIBERG, M. J., and P. S. LARSON: Studies on derivatives of 2,2-bis(p-chlorophenyl)-1,1-dichloroethane (DDD, TDE) with special reference to their effects on the adrenal cortex. J. Pharmacol. **121**, 421—431 (1957).
BLOCH, E., K. BENIRSCHKE and E. ROSEMBERG: C_{19} steroids, 17α-hydroxycorticosterone and a sodium retaining factor in human fetal adrenal glands. Endocrinology **58**, 626—633 (1956).
BLOCH, K.: The biological synthesis of cholesterol. Vitam. and Horm. **15**, 119—151 (1957).
BLUMENTHAL, H. T.: Aging processes in the endocrine glands of the guinea pig. I. The influence of age, sex and pregnancy on the mitotic activity and the histologic structure of the thyroid, parathyroid and adrenal glands. A.M.A. Arch. Path. **40**, 264—269 (1945).
— and L. LOEB: Two antagonistic effects of underfeeding on the adrenal cortex of the guinea pig. Amer. J. Path. **18**, 615—632 (1942).
BOERNER, D.: Die Beziehungen der ersten Anlage der Nebenniere und ihre Gefäße zueinander. Z. mikr.-anat. Forsch. **59**, 137—160 (1952/1953).
BOGUTH, W., H. LANGENDORF u. E. TONUTTI: Zellkerngröße als Indikator der Funktionsbeziehung Hypophyse-Nebennierenrinde. Med. Welt **20**, 408—414 (1951).
BOIS, P., and H. SELYE: Enhancement by thyroxine of the adrenal-stimulating effect of somatotrophic hormone. J. Endocrin. **15**, 171—179 (1957).
BONATI, B., G. C. MANFREDI and G. B. PANCATI: Metabolic and histochemical aspects of the adrenal after experimental inhibition by prednisone or prednisolone (in Italian). Minerva Med. (Torino) **48**, 1908—1914 (1957).
BOTELLA LLUSIA, J.: Histophysiologie de la zone sexuelle de l'écorce surrénale. C. R. Soc. franç. Gynéc. **19**, 17—24 (1949).
BOURNE, G. (H.): Mitochondria and Golgi apparatus. In: G. BOURNE, Ed., Cytology and Cell Physiology. London: Oxford Univ. Press 1942.
— The Mammalian Adrenal Gland. Oxford: Clarendon Press 1949.
— The histochemical dephosphorylation of riboflavin phosphate (flavin mono-nucleotide) and pyridoxal phosphate (co-decarboxylase). Quart. J. micr. Sci. **95**, 359—369 (1954a).
— The histochemical dephosphorylation of certain sugar phosphates, α-naphthohydroquinone diphosphate and two chalcone diphosphates. Acta anat. **22**, 289—300 (1954b).
— Aspects of the histochemistry of the adrenal cortex. Ciba Found. Coll. Endocrin. **8**, 1—14 (1955).
BRACHET, J.: Biochemical Cytology, Chap. 6. New York: Academic Press 1957.
BRAGDON, D. E.: A contribution to the surgical anatomy of the water snake, Natrix sipedon; the location of the visceral endocrine organs with reference to ventral scutellation. Anat. Rec. **117**, 145—161 (1953).
BRANIN, M. L.: The development of the cortical adrenal in the four-toed salamander Hemidactylum (Schlegel). J. Morph. **60**, 521—561 (1937).

BRENEMAN, W. R.: Growth of the endocrine glands and viscera in the chick. Endocrinology **28**, 946—954 (1941).

BRENNER, R. M., D. I. PATT and L. C. WYMAN: Cellular changes during adrenocortical regeneration in the rat. Anat. Rec. **117**, 759—771 (1953).

BRIGGS, F. N., and W. TOEPEL: The effect of ACTH on the ascorbic acid concentration of adrenal venous plasma of the rat. Endocrinology **62**, 24—29 (1958).

BRODISH, A., and C. N. H. LONG: Characteristics of the adrenal ascorbic acid response to adrenocorticotrophic hormone (ACTH) in the rat. Endocrinology **66**, 149—159 (1960).

BROGI, M. P.: Resistance of rats with enucleated adrenals to intraperitoneal injection of a hypertonic glucose solution. J. Endocrin. **11**, 344—349 (1954).

— Role of the adrenal cortex regenerated after enucleation in the resistance against histamine shock. J. Endocrin. **13**, 274—278 (1956).

— and C. PELLEGRINO: The secretion of corticosterone and aldosterone by the rat adrenal cortex regenerating after enucleation. J. Physiol. (Lond.) **146**, 165—178 (1959).

BROWN, J. B., R. A. KNOUFF, M. M. CONLIN and B. M. SCHNEIDER: Comparative lipid analyses of cortex and medulla of beef adrenals. Proc. Soc. exp. Biol. (N. Y.) **37**, 203—205 (1937).

BROWN, J. H. U., J. B. GRIFFIN, R. B. SMITH, III, and A. ANASON: Physiologic activity of an adrenocorticolytic drug in the adult dog. Metabolism (Baltimore) **5**, 594—600 (1956).

BROWN-GRANT, K.: Thyroid function and adrenal weight in the rabbit. J. Endocrin. **17**, 197—200 (1958).

BROWNIE, A. C., and J. K. GRANT: The *in vitro* enzymic hydroxylation of steroid hormones. 3. An investigation of fumarate and pyridine nucleotides in the enzymic 11β-hydroxylation of 11-deoxycorticosterone. Biochem. J. **62**, 29—33 (1956).

BROWNING, H. C.: Hormonal dependence of adrenal cortical tumors of CE and BALB/c mice in serial intraocular transfer. Cancer Res. **18**, 781—789 (1958).

BRUNE, H., u. W. EGER: Der Einfluß von Adsorbentien (Bolus alba) auf wachsende Ratten bei der chronischen Vitamin D_3-Vergiftung. Arch. Tierenähr. **5**, 244—269 (1954).

BUCHEM, F. S. P. VAN, H. DOORENBOS and H. S. ELINGS: Primary aldosteronism due to adrenocortical hyperplasia. Lancet *ii*, 335—337 (1956).

BÜLBRING, E.: The relation between cortical hormone and the size of the testis in the drake, with observations on the effect of different oils as solvents, and of desoxycorticosterone acetate. J. Pharmacol. **69**, 52—63 (1940).

BURGOS, M. H.: Histochemistry and electron microscopy of the three cell types in the adrenal gland of the frog. Anat. Rec. **133**, 163—185 (1959).

BURNS, J. K., and A. J. HALE: The distribution of ribonucleic acid and some other substances in the adrenal cortex after administration of adrenocorticotrophin. J. Physiol. (Lond.) **146**, 465—471 (1959).

— — and W. C. HUTCHISON: The effect of adrenocorticotrophin on adrenal nuclei. J. Physiol. (Lond.) **135**, 12—13 (1956).

BUSCH, W.: Die arterielle Gefäßversorgung der Nebenniere. I. Virchows Arch. Path. Anat. **324**, 688—699 (1954).

BUSH, I. E.: Species differences in adrenocortical secretion. J. Endocrin. **9**, 95—100 (1953).

— J. SWALE and J. PATTERSON: Analysis of steroids in adrenal venous blood and in urine in a case of virilism. Biochem. J. **62**, 16 P. (1956).

BUTLER, L. C., and A. F. MORGAN: Weight, ascorbic acid and cholesterol changes in adrenals of pyridoxine-deficient adult male rat. Proc. Soc. exp. Biol. (N. Y.) **86**, 264—266 (1954).

— — The content of adrenocorticotropic hormone in the pantothenic acid deficient female rat. Endocrinology **56**, 322—326 (1955).

CAIN, A. J.: On the significance of the plasmal reaction. Quart. J. micr. Sci. **90**, 75—86 (1949a).

— A critique of the plasmal reaction, with remarks on recently proposed techniques. Quart. J. micr. Sci. **90**, 411—426 (1949b).

— The histochemistry of lipoids in animals. Biol. Rev. **25**, 73—112 (1950).

— and R. G. HARRISON: Cytological and histochemical variations in the adrenal cortex of the albino rat. J. Anat. (Lond.) **84**, 196—226 (1950).

CAMPBELL, R. A., H. A. BERN and K. B. DEOME: The adverse effect of oestrogens on the resistance of mice to stress. Acta endocr. (Kbh.) **23**, 49—59 (1956).

CARDEZA, A. F.: Histology of adrenal tumors in castrate rats (in Spanish). Rev. Soc. argent. Biol. **30**, 200—209 (1954).

CARPENTER, R. K., L. D. MACLEOD and M. REISS: The effect of adreno-corticotrophic hormone on the respiratory metabolism of the adrenal cortex of the rat. J. Physiol. (Lond.) **105**, 231—235 (1946).

CARR, I. A.: Microvilli of the cells of the human adrenal cortex. Nature (Lond.) **182**, 607—608 (1958).

CARROLL, K. K.: Effect of dietary fats and oils on adrenal cholesterol. Endocrinology **48**, 101—110 (1951).
— Erucic acid as the factor in rape oil affecting adrenal cholesterol in the rat. J. biol. Chem. **200**, 287—292 (1953).
— and R. L. NOBLE: Effects of feeding rape oil on some endocrine functions of the rat. Endocrinology **52**, 476—486 (1952).
CARTER, A. C., S. WEISENFELD and M. G. GOLDNER: Failure of 17 α-ethyl-19-nortestosterone to affect plasma 17-hydroxycorticosteroids and ACTH responsiveness in man. Proc. Soc. exp. Biol. (N. Y.) **98**, 593—594 (1958).
CASCARANO, J., and B. W. ZWEIFACH: Comparative histochemical and quantitative study of adrenal and kidney tissue by tetrazolium technique. J. Histochem. Cytochem. **3**, 369—381 (1955).
CASE, J. F.: Adrenocortical-anterior-pituitary relationships during embryonic life. Ann. N. Y. Acad. Sci. **55** (Art. 2), 147—158 (1952).
CATER, D. B., and J. D. LEVER: The zona intermedia of the adrenal cortex. A correlation of possible functional significance with development, morphology and histochemistry. J. Anat. (Lond.) **88**, 437—454 (1954).
— and M. P. STACK-DUNNE: The histological changes in the adrenal of the hypophysectomized rat after treatment with pituitary preparations. J. Path. Bact. **66**, 119—133 (1953).
— — The rate of decay of mitotic activity in the adrenal cortex of the rat following hypophysectomy. J. Physiol. (Lond.) **127**, 265—272 (1955a).
— — The time of onset of mitotic activity in the adrenal cortex of the hypophysectomized rat following injections of growth hormone. J. Physiol. (Lond.) **127**, 273—279 (1955b).
— — The effects of growth hormone and corticotrophin upon the adrenal weight and adrenocortical mitotic activity in the hypophysectomized rat. J. Endocrinol. **12**, 174—184 (1955c).
CELESTINO DA COSTA, A.: Histophysiologie du cortex surrénal. Ann. Endocr. (Paris) **12**, 361—403 (1951).
— La sidérophilie des cellules cortico-surrénales; sa signification cytochimique. Verh. anat. Ges. Jena **50**, 137—139 (1952).
CERVIÑO, J. M., J. M. MANARO, E. POLLAK and A. PROTO: Cushing's syndrome from an adrenal tumor (in Spanish). Medicina (B. Aires) **7**, 468—484 (1947).
CHAN, T. K.: Experimental adrenal hypertrophy. Acta physiol. pharm. neerl. **1**, 413—487 (1950).
CHANG, G.-Y., and E. WITSCHI: Independence of adrenal hyperplasia and gonadal masculinization in the experimental adrenogenital syndrome of frogs. Endocrinology **56**, 597—605 (1955).
CHAVIN, W.: Pituitary-adrenal control of melanization in xanthic goldfish, *Carassius auratus* L. J. exp. Zool. **133**, 1—45 (1956).
CHEDID, L., et F. BOYER: Corticotrophines d'origine bactérienne. Mise en évidence de leur action, *in vivo*, chez le rat hypophysectomisé et *in vitro* par le test de Saffran. C. R. Acad. Sci. (Paris) **246**, 2664—2667 (1958a).
— — Pouvoir corticotrope spécifique des endotoxines. C. R. Acad. Sci. (Paris) **246**, 2801 to 2804 (1958b).
— — Pouvoir corticotrope spécifique d'un produit non toxique extrait d'une souche rugueuse de *Salmonella typhi*. C. R. Acad. Sci. (Paris) **246**, 2937—2940 (1958c).
CHESTER JONES, I.: Variation in the mouse adrenal cortex with special reference to the zona reticularis and to brown degeneration, together with a discussion of the 'cell migration' theory. Quart. J. micr. Sci. **89**, 53—74 (1948).
— The action of testosterone on the adrenal cortex of the hypophysectomized prepuberally castrated mouse. Endocrinology **44**, 427—438 (1949a).
— The relationship of the mouse adrenal cortex to the pituitary. Endocrinology **45**, 514—536 (1949b).
— The effect of hypophysectomy on the adrenal cortex of the immature mouse. Amer. J. Anat. **86**, 371—404 (1950).
— The disappearance of the X zone of the mouse adrenal cortex during first pregnancy. Proc. roy. Soc. B **139**, 398—410 (1952).
— The Adrenal Cortex. Cambridge: University Press 1957.
— and C. C. ROBY: Some aspects of zonation and function of the adrenal cortex. I. The effects of hypophysectomy on the adrenal cortex of the adult male mouse. J. Endocrin. **10**, 245—250 (1954).
— and A. WRIGHT: III. Self selection after adrenal enucleation in rats. J. Endocrin. **10**, 262—265 (1954a).
— — IV. The histology of the adrenal in rats with diabetes insipidus. J. Endocrin. **10**, 266—272 (1954b).

Chester Jones, I., J. G. Phillips and W. N. Holmes: Comparative physiology of the adrenal cortex. In : A. Gorbman, Ed., Comparative Endocrinology, 582—612. New York: Wiley 1959.
Chieffi, G.: The organogenesis of the interrenal and the medulla of the gonad in *Torpedo ocellata* and *Scylliorhinus canicula* (in Italian). Pubbl. Staz. zool. Napoli **23**, 185—199 (1952).
— Contribution to the study of embryonic sex development. I. Histochemical study of the interrenal and the genital ridge in the tadpole of *Rana esculenta* (in Italian). Riv. Biol. **47**, 439—452 (1955).
Christian, J. J.: Roles of endocrine and behavioral factors in the growth of mammalian populations. In: A. Gorbman, Ed., Comparative Endocrinology, 71—97. New York: Wiley 1959.
Christianson, M., and I. Chester Jones: The interrelationships of the adrenal glands of mother and foetus in the rat. J. Endocrin. **15**, 17—42 (1957).
Christy, N. P., A. Donn and J. W. Jailer: Inhibition by aminopyrine of adrenocortical activation caused by pyrogenic reaction. Proc. Soc. exp. Biol. (N. Y.) **91**, 453—456 (1956a).
— E. Z. Wallace and J. W. Jailer: Comparative effects of prednisone and of cortisone in suppressing the response of the adrenal cortex to exogenous adrenocorticotropin. J. clin. Endocrin. **16**, 1059—1074 (1956b).
— — W. E. L. Gordon and J. W. Jailer: On the rate of hydrocortisone clearance from plasma in pregnant woman and in patients with Laennec's cirrhosis. J. clin. Invest. **38**, 299—305 (1959).
Cinti, G., and E. Mignani: Phenomena of regeneration and behavior of the "subcapsular blastema" following aseptic wounds of the adrenal cortices (in Italian). Arch. De Vecchi Anat. pat. **12**, 949—966 (1949).
Clark, W. G., E. L. Brackney and R. A. Miliner: Adrenalectomy in frogs and toads. Proc. Soc. exp. Biol. (N. Y.) **57**, 222—224 (1944).
Clarke, J. R.: The effect of fighting on the adrenals, thymus and spleen of the vole *(Microtus agrestis)*. J. Endocrin. **9**, 114—126 (1953).
Clatworthy, H. W., jr., and R. G. Anderson: Development and growth of the human embryo and fetus. Amer. J. Dis. Child. **67**, 167—175 (1944).
Clayton, B. E., and J. E. Hammant: Histochemical responses in the adrenal gland of the guinea pig to ACTH and gonadotrophins. J. Endocrin. **15**, i (1957).
— — Effects of oestrogens and ACTH on the adrenals of the guinea-pig compared with those in the rat. J. Endocrin. **18**, 90—97 (1959).
Cobey, F. A., I. Taliaferro and H. H. Haag: Effect of DDD and some of its derivatives on plasma 17-OH-corticosteroids in the dog. Science **123**, 140—141 (1956).
— — — Further observations on effect on plasma 17-OH-corticosteroids in the dog of derivatives of 2,2-bis-(*p*-chlorophenyl)-1,1-dichloroethane (DDD, TDE). Proc. Soc. exp. Biol. (N. Y.) **97**, 491—494 (1958).
Cohen, A.: Variations de la teneur en lipides de la corticosurrénale foetale du rat, au cours du développement. C. R. Soc. Biol. (Paris) **148**, 321—324 (1954).
— Répartition des lipides dans le cortex surrénal d'embryons de rats décapités. Arch. Anat. micr. Morph. exp. **44**, 265—276 (1955).
— Évolution de la concentration de l'acide ascorbique dans les surrénales d'embryons de rat, au cours des derniers jours de la gestation. C. R. Acad. Sci. (Paris) **246**, 166—167 (1958a).
— Influence de la décapitation du foetus de rat sur la teneur en acide ascorbique de ses surrénales. Action de la corticostimuline. C. R. Acad. Sci. (Paris) **246**, 1917—1920 (1958b).
Cohen, A. I., E. Bloch and E. Celozzi: *In vitro* response of functional experimental adrenal tumors to corticotropin (ACTH). Proc. Soc. exp. Biol. (N. Y.) **95**, 304—309 (1957a).
— J. Furth and R. F. Buffett: Histologic and physiologic characteristics of hormone-secreting transplantable adrenal tumors in mice and rats. Amer. J. Path. **33**, 631—651 (1957b).
Cohen, H., H. H. Freedman, W. Kleinberg, N. R. Schechter and G. J. Martin: Enhancemeent of corticotrophic activity: Effect on adrenal and thymus weights. Endocrinology **54**, 81—83 (1954).
Cohen, R. B.: Histochemical localization and metabolic significance of glucose-6-phosphate dehydrogenase system in adrenal cortex. Proc. Soc. exp. Biol. (N. Y.) **101**, 405—407 (1959).
— W. B. Chapman and B. Castleman: Hyperadrenocorticism (Cushing's disease): A study of surgically resected adrenal glands. Amer. J. Path. **35**, 537—561 (1959).
Colover, J., and L. E. Glynn: Experimental iso-immune adrenalitis. Immunology **1**, 172—178 (1958).
Combescot, C.: Surrénales et survie de la tortue d'eau algérienne. C. R. Soc. Biol. (Paris) **151**, 292—295 (1957).

CONN, J. W.: Aldosterone in clinical medicine — past, present and future. Arch. intern. Med. **97**, 135—144 (1956).
CONSTANTINIDES, P., C. FORTIER and F. R. SKELTON: The effect of glucose on the adrenal response to ACTH in hypophysectomized rats. Endocrinology **47**, 351—363 (1950).
COSTA, P. J., D. D. BONNYCASTLE and I. J. SELIGMAN: Adrenocortical suppressive activity of several triazine derivatives in the guinea pig. Fed. Proc. **18**, 379 (1959).
COUPLAND, R. E.: Factors affecting the survival of the adrenal medulla and associated cortical cells in the anterior chamber of the eye. J. Endocrin. **15**, 162—170 (1957).
— and R. L. HOLMES: The distribution of cholinesterase in the adrenal glands of the rat, cat and rabbit. J. Physiol. (Lond.) **141**, 97—106 (1958).
COURRIER, R., A. COLOGNE et M. BACLESSE: Action de la cortisone administrée à la mère sur la surrénale du foetus de rat. C. R. Acad. Sci. (Paris) **233**, 333—336 (1951).
— M. BACLESSE et M. MAROIS: Rapports de la corticosurrénale et de la sexualité. J. Physiol. (Paris) **45**, 327—374 (1953).
COUTINHO, H. B., B. L. BAKER, J. F. KENT, E. C. PLISKE and J. G. VAN DYKE: Esterases of the adrenal gland after hypophysectomy and total body roentgen irradiation. Univ. Michigan med. Bull. **21**, 391—399 (1955).
COVIAN, M. R.: Variations of adrenal ascorbic acid in the hypophysectomized white rat (in Spanish). Rev. Ass. argent. Biol. **25**, 26—30 (1949).
CRAIG, J. M., and B. H. LANDING: Anaplastic cells of fetal adrenal cortex. Amer. J. clin. Path. **21**, 940—949 (1951).
CREUTZFELDT, W., M. HUSTEN u. K. HAAGER: Zur histologischen Funktionsdiagnostik der Nebennieren. Beitr. path. Anat. **113**, 428—429 (1953).
CRISTOL, S. J., and H. J. HALLER: The isolation of 1,1-dichloro-2-*o*-chlorophenyl-2-*p*-chlorophenylethane from technical TDE. J. Amer. chem. Soc. **70**, 1323—1324 (1948).
CROOKE, A. C., and J. R. GILMOUR: A description of the effect of hypophysectomy on the growing rat, with the resulting histological changes in the adrenal and thyroid glands and the testicles. J. Path. Bact. **47**, 525—544 (1938).
CUETO, C., and J. H. U. BROWN: The chemical fractionation of an adrenocorticolytic drug. Endocrinology **62**, 326—333 (1958a).
— — Biological studies on an adrenocorticolytic agent and the isolation of the active components. Endocrinology **62**, 334—339 (1958b).
CUSHING, H.: The basophil adenomas of the pituitary body and their clinical manifestations (pituitary basophilism). Johns Hopk. Hosp. Bull. **50**, 137—195 (1932).
CUTULY, E.: Quantitative study on the adrenals of hypophysectomized rats. Anat. Rec. **66**, 119—122 (1936).
— and E. C. CUTULY: Pigmented cells in adrenals and testes of hypophysectomized rats. Proc. Soc. exp. Biol. (N. Y.) **36**, 335—337 (1937).
DALLWIG, R.: Mitoseaktivität und Kerngrößen in der Nebennierenrinde der Ratte nach Dursteinwirkung. Z. mikr.-anat. Forsch. **61**, 138—154 (1954).
DALTON, A. J., E. R. MITCHELL, B. F. JONES and V. B. PETERS: Changes in the adrenal glands of rats following exposure to lowered oxygen tension. J. nat. Cancer Inst. **4**, 527 to 536 (1944).
D'ANGELO, S. A., A. S. GORDON and H. CHARIPPER: The effect of inanition on the anterior pituitary-adrenocortical interrelationship in the guinea pig. Endocrinology **42**, 399—411 (1948).
DANNENBERG, H.: Biochemie der Tumoren. In: B. FLASCHENTRÄGER und E. LEHNARTZ, Herausgeb., Physiologische Chemie, Bd. II/2c, 342—460. Berlin: Springer 1959.
D'ARCY, P. F.: An investigation into the effects of 2 : 2-bis-(*p*-chlorophenyl)-1 : 1-dichloroethane (D. D. D.) on the mouse adrenal cortex. J. Pharm. (Lond.) **6**, 625—628 (1954).
DAS GUPTA, D., and C. J. P. GIROUD: Experimental aminonucleoside nephrosis. I. Action of cortisone on aldosterone and corticosterone production. Proc. Soc. exp. Biol. (N. Y.) **98**, 334—339 (1958).
— — Effect of an aldosterone antagonist on the fluid retention of aminonucleoside nephrosis. Endocrinology **65**, 500—503 (1959).
DAUGHADAY, W. H.: Steroid protein interactions. Physiol. Rev. **39**, 885—902 (1959).
DAVIES, S.: The development of the adrenal gland of the cat. Quart. J. micr. Sci. **80**, 81—98 (1937).
DAVIS, J. O., and W. C. BALL, JR.: Effects of a body cast on aldosterone and sodium excretion in dogs with experimental ascites. Amer. J. Physiol. **192**, 538—542 (1958).
— M. J. GOODKIND, M. M. PECHET and W. C. BALL, JR.: Increased excretion of aldosterone in urine from dogs with right-sided congestive heart failure and from dogs with thoracic vena cava constriction. Amer. J. Physiol. **187**, 45—50 (1956).

DAVIS, J. O., M. M. PECHET, W. C. BALL, JR., and M. J. GOODKIND: Increased aldosterone secretion in dogs with right-sided congestive heart failure and in dogs with thoracic inferior vena cava constriction. J. clin. Invest. **36**, 689—694 (1957).
— B. KLIMAN, N. A. YANKOPOULOS and R. E. PETERSON: Increased aldosterone secretion following acute constriction of the inferior vena cava. J. clin. Invest. **37**, 1783—1790 (1958).
DAVIS, M. E., and E. J. PLOTZ: The effects of cortisone acetate on intact and adrenalectomized rats during pregnancy. Endocrinology **54**, 384—395 (1954).
— — and E. PLOTZ: A quantitative study of the vascularity of the adrenal gland of the rat during pregnancy and after cortisone treatment. Endocrinology **52**, 164—172 (1953).
DAWSON, A. B.: Histochemical evidence of early differentiation of the suprarenal gland of the chick. J. Morph. **92**, 579—596 (1953).
DAWSON, I. M. P.: Changes in the adrenal cortex in essential and renal hypertension. J. Path. Bact. **72**, 393—409 (1956).
DEANE, H. W.: Physiological regulation of the zona glomerulosa of the rat's adrenal cortex, as revealed by cytochemical observations. In: R. C. CHRISTMAN, Ed., Pituitary-Adrenal Function, 31—38. Washington, D. C.: Amer. Ass. Adv. Sci. 1951.
— Histophysiology of the adrenal cortex in fetal and prepuberal rats, with comments on the carbonyl reactivity of the cortical lipide droplets. In: L. I. GARDNER, Ed., Adrenal Function in Infants and Children, 4—20. New York: Grune and Stratton 1956.
— Intracellular lipides: Their detection and significance. In: S. L. PALAY, Ed., Frontiers in Cytology, 227—263. New Haven: Yale Univ. Press 1958.
— and A. C. BARGER: Histophysiology of the adrenal cortex in dogs with mild and severe cardiac damage. Endocrinology **61**, 758—764 (1957).
— and D. W. FAWCETT: Pigmented interstitial tissue showing "brown degeneration" in the ovaries of old mice. Anat. Rec. **113**, 239—246 (1952).
— and R. O. GREEP: A morphological and histochemical study of the rat's adrenal cortex after hypophysectomy, with comments on the liver. Amer. J. Anat. **79**, 117—146 (1946).
— — A cytochemical study of the adrenal cortex in hypo- and hyperthyroidism. Endocrinology **41**, 243—247 (1947).
— and C. P. LYMAN: Body temperature, thyroid and adrenal cortex of hamsters during cold exposure and hibernation, with comparisons to rats. Endocrinology **55**, 300—315 (1954).
— and G. M. C. MASSON: Adrenal cortical changes in rats with various types of experimental hypertension. J. clin. Endocrin. **11**, 193—208 (1951).
— and J. M. MCKIBBIN: The chemical cytology of the rat's adrenal cortex in pantothenic acid deficiency. Endocrinology **38**, 385—400 (1946).
— and A. MORSE: The cytological distribution of ascorbic acid in the adrenal cortex of the rat under normal and experimental conditions. Anat. Rec. **100**, 127—142 (1948).
— and A. M. SELIGMAN: Evaluation of procedures for the cytological localization of ketosteroids. Vitam. and Horm. **11**, 173—204 (1953).
— and J. H. SHAW: A cytochemical study of the responses of the adrenal cortex of the rat to thiamine, riboflavin, or pyridoxine deficiencies. J. Nutr. **34**, 1—20 (1947).
— — and R. O. GREEP: The effect of altered sodium or potassium intake on the width and cytochemistry of the zona glomerulosa of the rat's adrenal cortex. Endocrinology **43**, 133—153 (1948).
— F. G. HOFMANN, A. K. SOLOMON and G. B. WISLOCKI: Selective uptake of silver by the adrenal zona glomerulosa. Anat. Rec. **121**, 283 (1955).
— R. J. BARRNETT and A. M. SELIGMAN: Histochemical methods for the demonstration of enzymatic activity. In: W. GRAUMANN and K. NEUMANN, Herausgeb., Handbuch der Histochemie, Vol. VII/1. Stuttgart: Fischer 1960 (a).
— R. H. SCHNEIWEISS and L. I. GIDEZ: Response of rat zona glomerulosa in experimental nephrosis. Proc. Soc. exp. Biol. (N. Y.) **104**, 417—419 (1960b).
DEANESLY, R.: The histology of adrenal enlargement under experimental conditions. Amer. J. Anat. **47**, 475—509 (1931).
— Secretion of androgen by the adrenal cortex of the mouse. Nature (Lond.) **182**, 262—263 (1958).
DEJARDIN, M.: Localisation histochimique des cholinestérases dans la surrénale. C. R. Soc. Biol. (Paris) **148**, 1898—1900 (1954).
DEL CASTILLO, E. B., and R. SAMMARTINO: Action of large prolonged doses of estrone on the endocrines and sexual organs of the rat (in Spanish). Rev. Soc. argent. Biol. **13**, 455—475 (1937).
DELOST, P.: Nouvelles données sur la zone X surrénalienne. J. Physiol. (Paris) **47**, 164—167 (1955).
— Les corrélations genito-surrénaliennes chez le campagnol des champs (*Microtus arvalis* P.). Ann. Sci. nat., Zool. 8, 387—567 (1956).

Delost, P.: Recherches sur le cortex surrénal et la zone X des muscaraignes (*Sorex araneus* L. et *Crocidura russula* H.). C. R. Soc. Biol. (Paris) **151**, 484—487 (1957).
— et P. Chirvan-Nia: Données expérimentales comparées sur la zone X surrénalienne. Action de la cortisone et de la testostérone. C. R. Soc. Biol. (Paris) **150**, 1330—1332 (1956).
— — Différences raciales dans l'involution de la zone X surrénalienne chez la souris adulte vierge. C. R. Soc. Biol. (Paris) **152**, 453—455 (1958).
— et H. Delost: Modifications pondérales et histologiques des glands surrénales du campagnol agreste (*Microtus agrestis* L.) en fonction de l'état sexuel. C. R. Soc. Biol. (Paris) **149**, 910—914 (1955).
— — et T. Terroine: Involution de la zone glomerulée surrénalienne dans la carence en B_1. J. Physiol. (Paris) **51**, 445—446 (1959).
Dempsey, E. W., R. O. Greep and H. W. Deane: Changes in the distribution and concentration of alkaline phosphatases in tissues of the rat after hypophysectomy or gonadectomy, and replacement therapy. Endocrinology **44**, 88—103 (1949).
Dempster, W. J.: The transplanted adrenal gland. Brit. J. Surg. **42**, 540—552 (1955).
Denber, H. C. B.: Non-encapsulated adrenal cortical tissue in the peri-renal fat. Amer. J. Path. **25**, 681—688 (1949).
Des Marais, A., et P. M. Gagnon: Influence de l'exposition au froid sur l'activité thyroïdienne et sur le poids et le teneur en acide ascorbique des surrénales chez le cobaye carencé ou non en acide ascorbique. C. R. Soc. Biol. (Paris) **149**, 1714—1718 (1955).
— L.-P. Dugal et P.-M. Gagnon: Effect de l'hypophysectomie, de l'administration d'ACTH et de l'ACTH endogène sur le poids et l'activité métabolique de la surrénale du rat blanc. Rev. canad. Biol. **14**, 108—112 (1955).
DeVisscher, M., et C. Beckers: La gland thyroïde et l'équibre endocrinien. J. Physiol. (Paris) **49**, 439—520 (1957).
Dickie, M. M., and P. W. Lane: Adrenal tumors, pituitary tumors, and other pathological changes in F_1 hybrids of strain DE × strain DBA. Cancer Res. **16**, 48—52 (1956).
Dittus, P.: Interrenalssystem und chromaffine Zellen im Lebensablauf von Ichthyophis glutinosus L. Z. wiss. Zool. **147**, 459—512 (1936).
— Histologie und Cytologie des Interrenalorgans der Selachier unter normalen und experimentellen Bedingungen. Z. wiss. Zool. **154**, 40—124 (1940).
Domm, L. V., and P. Leroy: A method for fetal hypophysectomy. Anat. Rec. **123**, 183—199 (1955).
Donaldson, H. H.: The Rat: Data and Reference Tables, Ed. 2. Amer. anat. Mem. **6** (1924).
Donáth, T.: Experimentelle Beiträge zum venösen Kreislauf der Nebenniere. Anat. Anz. **104**, 412—418 (1957).
Done, A. K., R. S. Ely, R. B. Raile and V. C. Kelley: Species differences in 17-hydroxycorticosteroid concentrations. Proc. Soc. exp. Biol. (N. Y.) **81**, 667—669 (1952).
Dorp, A. W. V. van, and H. W. Deane: A morphological and cytochemical study of the postnatal development of the rat's adrenal cortex. Anat. Rec. **107**, 265—281 (1950).
Dribben, I. S., and J. M. Wolfe: Structural changes in the connective tissue of the adrenal glands of female rats associated with advancing age. Anat. Rec. **98**, 557—585 (1947).
Driscol, T. E., M. M. Maultsby, G. L. Farrell and R. M. Berne: Aldosterone secretion in experimental congestive heart failure. Amer. J. Physiol. **191**, 140—144 (1957).
Ducommun, P., et R. S. Mach: Effet de l'hormone adrénocorticotrope sur la morphologie du cortex surrénalien, son contenu en acide ascorbique et en esters de cholésterol chez le rat normal. Acta endocr. (Kbh.) **3**, 17—26 (1949).
Dugal, L.-P., and M. Therien: The influence of ascorbic acid on adrenal weight during exposure to cold. Endocrinology **44**, 420—426 (1949).
— — L'acide ascorbique et l'activité de la surrénale chez le rat normale et chez le rat hypophysectomisé. Ann. Endocr. (Paris) **13**, 578—588 (1952).
Dulin, W. E.: The effects of adrenocorticotropin on the white leghorn cockerel and capon. Endocrinology **53**, 233—235 (1953).
Dumm, M. E., and E. P. Ralli: A critical analysis of the eosinophile response in rats to ACTH and cortisone. Endocrinology **54**, 71—80 (1954).
— P. Orvando, P. Roth and E. P. Ralli: Relation of pantothenic acid to white blood cell response of rats following stress. Proc. Soc. exp. Biol. (N. Y.) **71**, 368—371 (1949).
— B. Laken and E. P. Ralli: Factors influencing adrenal weight and adrenal cholesterol in rats following stress. J. Nutr. **56**, 517—531 (1955).
Duncan, I. W.: The effect of audiogenic seizures in rats on the adrenal weight, ascorbic acid, cholesterol, and corticosteroids. J. biol. Chem. **229**, 563—568 (1957).
Durey, J.-M.: Variations de l'acid ascorbique dans les surrénales chez la rate pendant la gestation. Ann. Endocr. (Paris) **15**, 284—286 (1954).
Dury, M., and A. Dury: Effects of ascorbic acid pretreatment on the response of blood glucose and adrenal cholesterol in the intact rat to insulin. Endocrinology **58**, 671—674 (1956).

EARLE, D. P., JR., R. C. DE BODO, E. L. SCHWARTZ, S. J. FARBER, M. KURTS and J. GREENBERG: Effect of hypophysectomy on electrolyte and water balance in the dog. Proc. Soc. Exp. Biol. (N. Y.) **76**, 608—612 (1951).

EGDAHL, R. H., and J. B. RICHARDS: Effect of extreme cold exposure on adrenocortical function in the unanesthetized dog. Amer. J. Physiol. **185**, 239—242 (1956).

EGER, W., W. GOTTESLEBEN u. M. TIEFJEN: Über das Verhalten der alkalischen und sauren Phosphatase bei funktioneller Belastung der Nebenniere. Virchows Arch. path. Anat. **324**, 173—192 (1953).

EHRENBRAND, F.: Beiträge zur Morphokinese der Nebennierenrinde bei experimenteller Hyperthyreose. Anat. Anz. **101**, 41—63 (1954).

— Sind d e sog. ,,fuchsinophilen" Zellen der Nebennierenrinde spezifische Androgenbildner? Acta histochem. Jena **7**, 1—73 (1959).

EICHNER, D.: Über den morphologischen Ausdruck funktioneller Beziehungen zwischen Nebennierenrinde und neurosekretorischem Zwischenhirnsystem der Ratte. Z. Zellforsch. **38**, 488—508 (1953).

EIK-NES, K., and K. R. BRIZZEE: Adrenocortical activity and metabolism of 17-hydroxycorticosteroids in thyroidectomized dogs. Amer. J. Physiol. **184**, 371—375 (1956).

— — Adrenocortical activity in hypophysectomized dogs. Amer. J. Physiol. **193**, 403—407 (1958).

— A. A. SANDBERG, F. H. TYLER and L. T. SAMUELS: Changes in plasma level of 17-hydroxycorticosteroids induced by ACTH. III. Effects of various vehicles and modes of administration. J. clin. Endocrin. **18**, 882—885 (1958).

EISEMAN, B., R. HUGHES, W. SUMMERS and F. TRAYLOR: Human adrenal cortical autograft into the portal circulation. An experimental and clinical study. Ann. Surg. **142**, 961—966 (1955).

EISENSTEIN, A. B.: Pantothenic acid and adrenocortical hormone secretion. Endocrinology **60**, 298—302 (1957).

— Effect of sodium chloride feeding on adrenocortical hormone secretion of salt deprived rats. Proc. Soc. exp. Biol. (N. Y.) **101**, 850—852 (1959).

— and P. M. HARTROFT: Alterations in the rat adrenal cortex induced by sodium deficiency: Steroid hormone secretion. Endocrinology **60**, 634—640 (1957).

— and R. E. SHANK: Relationship of ascorbic acid to secretion of adrenocortical hormones in guinea pigs. Proc. Soc. exp. Biol. (N. Y.) **78**, 619—624 (1951).

— and A. S. WENNEKER: Excretion of 17-hydroxycorticosteroid hormones by pantothenic acid deficient guinea pigs. Proc. Soc. exp. Biol. (N. Y.) **97**, 844—847 (1958).

ELFTMAN, H.: Response of the alkaline phosphatase of the adrenal cortex of the mouse to androgens. Endocrinology **41**, 85—91 (1947).

ELIAS, H.: Growth of the adrenal cortex in domesticated Ungulata. Amer. J. vet. Res. **9**, 173—189 (1948).

— u. J. E. PAULY: Strukturelemente in menschlichen Nebennierenrinden. Anat. Anz. **100**, 134—147 (1954).

— — The structure of the human adrenal cortex. Endocrinology **58**, 714—738 (1956).

ELTON, R. L., I. G. ZARROW and M. X. ZARROW: Depletion of adrenal ascorbic acid and cholesterol: A comparative study. Endocrinology **65**, 152—157 (1959).

ENDRŐCZI, E., J. SZALAY u. K. LISSÁK: Untersuchungen über die Entwicklung der Funktion des Hypothalamus-Hypophysen-Nebennierenrinden-Systems bei neugeborenen Ratten mit chronischen Tiefenelektroden. Endokrinologie **34**, 331—336 (1957).

— G. BATA u. J. MARTIN: Untersuchungen über die Sekretion von Nebennierenrindenhormonen. Endokrinologie **35**, 280—290 (1958).

ENGSTRÖM, H.: Studien über die postnatale Entwicklung der Nebennierenrinde in der weißen Ratte. Anat. Anz. **83**, 1—19 (1936).

ERÄNKÖ, O.: Histochemical demonstration of ascorbic acid in the adrenal gland with a critical study of the method. J. Histochem. Cytochem. **2**, 167—177 (1954a).

— Estimation of the volumes of the cortex and medulla from adrenal sections. Ann. Med. exp. Fenn. **32**, 377—391 (1954b).

— Short-term effects of potassium, sodium and ACTH on the histochemistry of the adrenal glands of rats. Acta endocr. (Kbh.) **18**, 189—200 (1955).

ERSHOFF, B. H.: Nutrition and the anterior pituitary, with special reference to the general adaptation syndrome. Vitam. and Horm. **10**, 79—140 (1952).

EVERETT, J. W.: Hormonal factors responsible for deposition of cholesterol in the corpus luteum of the rat. Endocrinology **41**, 364—377 (1947).

EVERETT, N. B.: Autoplastic and homoplastic transplants of the rat adrenal cortex and medulla to the kidney. Anat. Rec. **103**, 335—348 (1949).

FARQUHAR, M. G.: Preparation of ultrathin tissue sections for electron microscopy. Lab. Invest. **5**, 317—337 (1956).

FARRELL, G. L.: Steroids in adrenal venous blood of the dog: Venous-arterial differences across the adrenal. Proc. Soc. exp. Biol. (N. Y.) **86**, 338—341 (1954).
— Regulation of aldosterone secretion. Physiol. Rev. **38**, 709—728 (1958).
— Steroidogenic properties of extracts of beef diencephalon. Endocrinology **65**, 29—33 (1959a).
— Glomerulotropic activity of an acetone extract of pineal tissue. Endocrinology **65**, 239—241 (1959b).
— E. W. RAUSCHKOLB and P. C. ROYCE: Secretion of aldosterone by the adrenal of the dog. Effects of hypophysectomy and ACTH. Amer. J. Physiol. **182**, 269—272 (1955).
— R. C. BANKS and S. KOLETSKY: The effect of corticosteroid injection on aldosterone secretion. Endocrinology **58**, 104—108 (1956a).
— R. S. ROSNAGLE and E. W. RAUSCHKOLB: Aldosterone secretion during bleeding. Fed. Proc. **15**, 60—61 (1956b).
— — — Increased aldosterone secretion in response to blood loss. Circulation Res. **4**, 606—611 (1956c).
— R. B. FLEMING, E. W. RAUSCHKOLB, F. M. YATSU, M. MCCALLY and C. H. ANDERSON: Steroidogenic properties of purified corticotropins. Endocrinology **62**, 506—512 (1958).
— S. KOLETSKY and L. W. LAPHAM: Decreased aldosterone secretion following pinealectomy. Fed. Proc. **18**, 44 (1959).
FASSBENDER, H. G.: Pathologische Anatomie der endokrinen Drüsen. B. Die Nebenniere. In: E. KAUFMANN, Herausgeb., Lehrbuch der speziellen pathologischen Anatomie, Bd. I/2, 1475—1510. Berlin: de Gruyter 1956.
FAWCETT, D. W.: A comparison of the histological organization and cytochemical reactions of brown and white adipose tissue. J. Morph. **90**, 363—406 (1952).
FELDMAN, J. D.: Histochemical reactions of adrenocortical cells. Anat. Rec. **107**, 347—358 (1950).
— Endocrine control of the adrenal gland. Anat. Rec. **109**, 41—69 (1951).
FETZER, S.: Zur Verwendung formolfixierten Materials zur Ausführung der Plasmal-Reaktion. Mikroskopie 8, 189—191 (1953).
FEULGEN, R., u. K. VOIT: Über einen weitverbreiteten festen Aldehyd. Seine Entstehung aus einer Vorstufe, sein mikrochemischer und mikroskopisch-chemischer Nachweis und die Wege zu seiner präparativen Darstellung. Pflügers Arch. ges. Physiol. **206**, 389—410 (1924).
FIALA, S., E. E. SPROUL and A. FIALA: The action of corticotropin (ACTH) on nucleic acids and subcellular elements of the adrenal cortex. J. biophys. biochem. Cytol. **2**, 115—126 (1956).
FIDANZA, A., and A. BONOMOLO: The behavior of cholesterol in the adrenals of rats fed a diet deficient in pantothenic acid plus antibiotics (in Italian). Boll. Soc. ital. Biol. sper. **29**, 1732—1733 (1953).
FIORI, C.: Behavior of lipase activity during parenchymal degenerative changes from bacterial toxin (in Italian). Boll. Soc. ital. Biol. sper. **32**, 1583—1584 (1956).
FISHMAN, W. H., and J. R. BAKER: Cellular localization of β-glucuronidase in rat tissues. J. Histochem. Cytochem. **4**, 570—587 (1956).
FLAKS, J.: Adrenal cortical carcinoma with metastases in an ovariectomized Strong A mouse. J. Path. Bact. **61**, 266—269 (1949).
FLEXNER, L. B., and A. GROLLMAN: The reduction of osmic acid as an indicator of adrenal cortical activity in the rat. Anat. Rec. **75**, 207—222 (1939).
FLICKINGER, D. D.: Adrenal responses of California quail subjected to various physiologic stimuli. Proc. Soc. exp. Biol. (N. Y.) **100**, 23—25 (1959).
FLINT, J. M.: Reticulum of the adrenal. Anat. Anz. **16**, 1—13 (1899).
— The blood-vessels, angiogenesis, organogenesis, reticulum, and histology, of the adrenal. Johns Hopk. Hosp. Bull. **9**, 153—230 (1900).
FLUX, D. S.: The effect of adrenal steroids on the growth of the mammary glands, uteri, thymus and adrenal glands of intact, ovariectomized and oestrone-treated ovariectomized mice. J. Endocrin. **11**, 238—254 (1954).
FOGLIA, V. G., B. GÜNTHER and R. M. PINTO: Weight of the body and of organs in the white rat of the Instituto de Fisiología de Buenos Aires (in Spanish). Rev. Soc. argent. Biol. **19**, 384—396 (1943).
FOLCH, J., I. ASCOLI, M. LEES, J. A. MEATH and F. N. LEBARON: Preparation of lipide extracts from brain tissue. J. biol. Chem. **191**, 833—841 (1951).
FONTAINE, M., et J. HATEY: Recherches sur le controle hypophysaire de l'interrénal antérieur d'un poisson téléostéen, l'anguille (*Anguilla anguilla* L.). Variations pondérales de l'interrénal antérieur. C. R. Soc. Biol. (Paris) **147**, 217—220 (1953).
— — Sur la teneur en 17-hydroxycorticostéroides du plasma de saumon (*Salmo salar* L.). C. R. Acad. Sci. (Paris) **239**, 319—321 (1954).

FONTAINE, M., J. HATEY, et M. OLIVEREAU: Interrénal antérieur et smoltification chez *Salmo salar* (L.). J. Physiol. (Paris) **49**, 174—176 (1957).
FORBES, T. R.: Studies on the reproductive system of the alligator. IV. Observations on the development of the gonad, the adrenal cortex, and the Müllerian duct. Contr. Embryol. Carneg. Inst. **28**, 129—155 (1940).
FORD, P.: Some observations on the corpuscles of Stannius. In: A. GORBMAN, Ed., Comparative Endocrinology, 728—734. New York: Wiley 1959.
FORTIER, C., and J. DE GROOT: Adenohypophysial corticotrophin and plasma free corticosteroids during regeneration of the enucleated rat adrenal gland. Amer. J. Physiol. **196**, 589—592 (1959).
— F. R. SKELTON, P. CONSTANTINIDES, P. O. TIMIRAS, M. HERLANT and H. SELYE: A comparative study of some of the chemical and morphological changes elicited in the adrenals by stress and purified ACTH. Endocrinology **46**, 21—29 (1950).
FRANTZ, M. J., and A. KIRSCHBAUM: Sex hormone secretion by tumors of the adrenal cortex of mice. Cancer Res. **9**, 257—266 (1949).
FRANZE, F.: Formen, Masse, Gewicht und Lage der Nebennieren von Schaf und Ziege. Vet.-med. Diss. Berlin: Humboldt-Univ. 1958.
FRAZÃO, J. V.: The Adrenocortical Gland (in Portuguese). Lisbon: Livraria Portugália 1954.
FREDRICKSON, D. S., R. E. PETERSON and D. STEINBERG: Inhibition of adrenocortical steroid secretion by Δ^4-cholestenone. Science **127**, 704—705 (1958).
FRENKEL, J. K.: Effects of hormones on the adrenal necrosis produced by Besnoitia jellisoni in golden hamsters. J. exp. Med. **103**, 375—398 (1956a).
— Potentiating effects of endogenous corticoids in human adrenocortical infection and necrosis. Amer. J. Path. **32**, 626—627 (1956b).
FRIEDMAN, N. B.: The pathology of the adrenal gland in Addison's disease with special reference to adrenocortical contraction. Endocrinology **42**, 181—200 (1948).
FRISCH, E., and L. I. GARDNER: Failure of suramin (Bayer 205; germanin) to cause changes in the adrenal cortex of the rat. Endocrinology **63**, 500—501 (1958).
FUSTINONI, O.: L'asthénie des insuffisances hypophysaire et surrénal du crapaud. C. R. Soc. Biol. (Paris) **129**, 1252—1253 (1938).
— and J. PORTO: Morphology of the adrenal glands of the toad *Bufo arenarum* (Hensel) (in Spanish). Rev. Soc. argent. Biol. **14**, 315—320 (1938).
GABE, M.: Caractères histochimiques des cellules d'été de la glande surrénale chez la grenouille (*Rana temporaria* L.). Ann. Histochim. **1**, 256—263 (1956).
GAGNON, R.: The venous drainage of the human adrenal gland. Rev. canad. Biol. **14**, 350—359 (1956).
GALLAGHER, T. F.: Steroid hormone metabolism and the control of adrenal secretion. Harvey Lect. **52**, 1—22 (1956/57).
— Adrenocortical carcinoma in man. The effect of amphenone on individual ketosteroids. J. clin. Endocrin. **18**, 937—949 (1958).
GANIS, F. M., L. L. MILLER and L. R. AXELROD: Effect of ACTH on protein synthesis in isolated perfused beef adrenal gland. Proc. Soc. exp. Biol. (N. Y.) **89**, 634—639 (1955).
GANN, D. S., I. H. MILLS and F. C. BARTTER: The hemodynamic parameter mediating increase in aldosterone secretion. Progr. 41st Meet. Endocr. Soc. 52—53 (1959).
GANONG, W. F.: The effect of ACTH on adrenal size in hypophysectomized rats after removal of one adrenal. Endocrinology **55**, 117—120 (1954).
— and D. M. HUME: The effect of graded hypophysectomy on thyroid, gonadal, and adrenocortical function in the dog. Endocrinology **59**, 293—301 (1956a).
— — The effect of unilateral adrenalectomy on adrenal venous 17-hydroxycorticosteroid output in the dog. Endocrinology **59**, 302—305 (1956b).
— A. H. LIEBERMAN, W. J. R. DAILY, V. S. YUEN, P. J. MULROW, J. A. LUETSCHER, JR., and R. E. BAILEY: Aldosterone secretion in dogs with hypothalamic lesions. Endocrinology **65**, 18—28 (1959).
GARDNER, L. I.: Adrenocortical metabolism of the fetus, infant and child. Pediatrics **17**, 897—924 (1956).
— and R. L. WALTON: Plasma 17-ketosteroids of the human fetus: demonstration of concentration gradient between cord and maternal circulations. Helv. paediat. Acta **9**, 311—316 (1954).
GARM, O.: A study of bovine nymphomania. Acta endocr. (Kbh.) Suppl. **3** (1949).
— Changes in the blood, adrenals, and hypophysis in bovine parturient paresis and eclampsia. A preliminary report. Acta endocr. (Kbh.) **5**, 413—424 (1950).
GARRETT, F. D.: The development and physiology of the corpuscle of Stannius in ganoid and teleostean fishes. J. Morph. **70**, 41—68 (1942).
GARROD, O., S. A. SIMPSON and J. F. TAIT: Some factors affecting aldosterone secretion. Proc. roy. Soc. Med. **49**, 888—889 (1956).

GAUNT, R.: Adrenalectomy in the rat. Amer. J. Physiol. **103**, 494—510 (1933).
— J. H. GAUNT and C. E. TOBIN: Colony differences in survival of adrenalectomized rats. Proc. Soc. exp. Biol. (N. Y.) **32**, 888—892 (1935).
— M. LILING and C. W. MUSHETT: Disturbances of water metabolism in vitamin deficiencies and effects of adrenal cortical hormones. Endocrinology **38**, 127—132 (1946).
— J. H. LEATHEM, C. HOWELL and N. ANTONCHAK: Some biological properties of different esters of desoxycorticosterone. Endocrinology **50**, 521—530 (1952).
— C. H. TUTHILL, N. ANTONCHAK and J. H. LEATHEM: Antagonists to cortisone: An ACTH-like action of steroids. Endocrinology **52**, 407—423 (1953).
GEMZELL, C.-A.: The influence of corticotrophic hormone and oestrogen on liver glycogen content and phosphate metabolism in the adrenal cortex. Acta endocr. (Kbh.) Supp. **1** (1948).
— Increase in the formation and the secretion of ACTH in rats following administration of oestradiol-monobenzoate. Acta endocr. (Kbh.) **11**, 221—228 (1952).
GENESI, M., and G. F. MELLONI: Histochemical modifications in endocrines by tropic pituitary hormones. II. Action of corticotropin on thyroid, ovary, uterus and adrenal of the rat (in Italian). Biol. latina **10**, 71—91 (1957).
GEORGES, G., et J. CAHN: Étude de différentes corticotrophines à action prolongée au moyen des variations du taux de l'acide ascorbique et du poids des surrénales chez le rat hypophysectomisé. C. R. Soc. Biol. (Paris) **149**, 1245—1247 (1955).
GÉRARD, P.: Sur la cortico-surrénale du protoptère (*Protopterus Dolloi* Blgr.). Arch. Biol. (Paris) **62**, 371—378 (1951).
GERSCHMAN, R., and W. O. FENN: Ascorbic acid content of adrenal glands of rat in oxygen poisoning. Amer. J. Physiol. **176**, 6—8 (1954).
GERSH, I., and A. GROLLMAN: The vascular pattern of the adrenal gland of the mouse and rat and its physiological response to changes in glandular activity. Contr. Embryol. Carneg. Inst. **29**, 111—125 (1941).
GERSHBERG, H., E. G. FRY, J. R. BROBECK and C. N. H. LONG: The role of epinephrine in the secretion of the adrenal cortex. Yale J. Biol. Med. **23**, 32—51 (1950).
GESCHWIND, I.: Species variation in protein and polypeptide hormones. In: A. GORMAN, Ed., Comparative Endocrinology, 421—443. New York: Wiley 1959.
GIACOMINI, E.: Sur les terminations nerveuses dans les capsules surrénales des oiseaux. Arch. ital. Biol. **29**, 482—483 (1898).
— Contribution to the knowledge of the adrenal cortex in cyclostomes. On the adrenal cortex of petromyzonts (in Italian). Monit. zool. ital. **13**, 143—162 (1902).
— On the microscopic anatomy and development of the adrenal cortex of lophobranchiate fishes (in Italian). Arch. ital. Anat. Embriol. **18**, Suppl., 548—565 (1922).
— The interrenal system and Stannius bodies in ganoids (in Italian). Boll. Soc. ital. Biol. sper. **8**, 636—640 (1933).
GILLISSEN, G., u. I. GILLISSEN: Die morphokinetische Beeinflussung der Nebennierenrinde durch Sexualhormone. Endokrinologie **33**, 45—53 (1955).
GIROUD, A.: Acide ascorbique et fonction cortico-surrénale. Ann. Endocr. (Paris) **12**, 733—741 (1951).
— et C.-P. LEBLOND: Localisations électives de l'acide ascorbique ou vitamine C. Arch. Anat. micr. Morph. exp. **31**, 111—142 (1935).
— et M. MARTINET: Valeur fontionnelle de la zone foetale de la cortico-surrénale. C. R. Soc. Biol. (Paris) **135**, 1028—1030 (1941).
— et A. R. RATSIMAMANGA: Acide ascorbique — vitamine C. Actualités sci. industr. **921**, III, 1—212 (1942).
GIROUD, C. J. P., J. STACHENKO and E. H. VENNING: Secretion of aldosterone by the zona glomerulosa of rat adrenal glands incubated *in vitro*. Proc. Soc. exp. Biol. (N. Y.) **92**, 154—158 (1956a).
— M. SAFFRAN, A. V. SCHALLY, J. STACHENKO and E. H. VENNING: Production of aldosterone by rat adrenal glands *in vitro*. Proc. Soc. exp. Biol. (N. Y.) **92**, 855—859 (1956b).
— J. STACHENKO and P. PILETTA: *In vitro* studies of the functional zonation of the adrenal cortex and of the production of aldosterone. In: A. F. MULLER and C. M. O CONNOR, Eds., An International Symposium on Aldosterone, 56—72. Boston: Little, Brown 1958.
GLICK, D., and G. R. BISKIND: The histochemistry of the adrenal gland. I. The quantitative distribution of vitamin C. J. biol. Chem. **110**, 1—7 (1935a).
— — II. The quantitative distribution of lipolytic enzymes. J. biol. Chem. **110**, 575—582 (1935b).
— — The quantitative distribution of vitamin C in the adrenal gland at various stages of development. J. biol. Chem. **115**, 551—555 (1936).
— and L. J. GREENBERG: The quantitative histological distribution of the succinic dehydrogenase system in the rat adrenal and the influence of hypophysectomy, ACTH, and cortisone. Endocrinology **63**, 778—784 (1958).

GLICK, D., and S. N. NAYYAR: Further studies on the determination of succinic dehydrogenase in microgram quantities of tissue and distribution of the activity in the bovine adrenal. J. Histochem. Cytochem. **4**, 389—396 (1956).
— and M. J. OCHS: Quantitative histological distribution of cholesterol in adrenal glands of the cow, rat and monkey, and effects of stress conditions, ACTH, cortisone and desoxycorticosterone. Endocrinology **56**, 285—298 (1955).
— and H. R. STECKLEIN: Micromeasurement of enzymatic conjugation of *m*-aminophenol and lack of activity in rat adrenal tissue. Arch. Biochem. **54**, 520—522 (1955).
— — Effects of stress conditions, ACTH, cortisone, desoxycorticosterone and hypophysectomy on the quantitative histological distribution of phenol sulfatase and acetylcholinesterase in the rat adrenal. Endocrinology **58**, 573—581 (1956).
— M. ALPERT and H. R. STECKLEIN: The determination of L-ascorbic acid, and dehydro-L-ascorbic acid plus diketo-L-gulonic acid in microgram quantities of tissue. J. Histochem. Cytochem. **1**, 326—335 (1953).
— R. H. SWIGART, R. C. BAHN and D. W. HASTINGS: A technique for killing animals to obtain tissues for studies on stress effects. Endocrinology **54**, 706—708 (1954).
— H. R. STECKLEIN and B. G. MALMSTROM: Distribution of phenolsulfatase in the adrenal of various species. A correction. Arch. Biochem. **54**, 513—519 (1955a).
— R. H. SWIGART, S. N. NAYYAR and H. R. STECKLEIN: Flame photometric determination of potassium in microgram quantities of tissue and the distribution of potassium and lipid in the adrenal of the monkey and guinea pig. J. Histochem. Cytochem. **3**, 6—15 (1955b).
— E. F. FREIER and M. J. OCHS: Microdetermination of magnesium and its histological distribution in the adrenal in various functional states. J. biol. Chem. **226**, 77—82 (1957).
GLOCK, G. E., and P. MCLEAN: Levels of enzymes of the direct oxidative pathway of carbohydrate metabolism in mammalian tissues and tumours. Biochem. J. **56**, 171—175 (1954).
GOLD, J. R.: Blood corticoids: Their measurement and significance. J. clin. Endocrin. **17**, 296—316 (1957).
GOLDEN, A., and R. R. OVERMAN: Experimental cortical necrosis of adrenal gland in the monkey in diphtheria and in malaria. Amer. J. clin. Path. **19**, 899—906 (1949).
GOLDMAN, M. L., E. RONZONI and H. A. SCHROEDER: The response of the adrenal cortex of the rat to dietary salt restriction and replacement. Endocrinology **58**, 57—61 (1956).
GOMORI, G.: Microscopic Histochemistry. Chicago: Univ. Chicago Press 1952(a).
— The histochemistry of lipid carbonyl compounds. J. Lab. clin. Med. **39**, 649—659 (1952b).
GOMPERTZ, D.: The effect of sex hormones on the adrenal gland of the male rat. J. Endocrin. **17**, 107—113 (1958).
GONZALEZ, I. E.: Comparative histological study of the adrenal gland in native rabbits. Anat. Rec. **121**, 397—398 (1955).
GOOD, T. A., R. S. ELY, L. R. HEISELT, A. K. DONE and V. C. KELLEY: Studies of 17-hydroxycorticosteroids: Adrenalectomy and hypophysectomy in guinea pigs and their effects on 17-OHCS concentrations. Endocrinology **58**, 651—658 (1956).
GOODMAN, A. D.: Studies on the effect of omega-methylpantothenic acid on corticosterone secretion in the rat. Endocrinology **66**, 420—427 (1960).
GOODSELL, J. E.: Weight changes in the cortex and the medulla of the adrenal gland of the dog in acute vitamin-B_1 deficiency. Amer. J. Physiol. **134**, 119—124 (1941).
GOORMAGHTIGH, N.: Contribution à l'étude du fonctionnement de la capsule surrénale humaine à l'état normal et dans les états infectieux, en particulier dans les gangrènes gazeuses. Arch. Méd. exp. **28**, 277—321 (1918/19).
— The cytology of functioning adrenal cortex tumors. Amer. J. Cancer **38**, 32—40 (1940).
GORDON, J. J.: Observations on the respiratory metabolism of the rat-adrenal cortex. Biochem. J. **67**, 701—704 (1957).
GORDON, M. L.: An immediate response of the demedullated adrenal gland to stress. Endocrinology **47**, 13—18 (1950).
GOTTSCHAU, M.: Struktur und embryonale Entwicklung der Nebennieren bei Säugetieren. Arch. Anat. Physiol. Lpz. **1883**, 412—458.
GRANT, J. K., A. P. M. FORREST and T. SYMINGTON: The secretion of cortisol and corticosterone by human adrenal cortex. Acta endocr. (Kbh.) **26**, 195—203 (1957).
GRAUMANN, W.: Topochemische Untersuchungen über die perjodatreaktiven Stoffe in der Nebenniere des Goldhamsters. Acta histochem. (Jena) **2**, 237—258 (1956).
GREENBERG, L. J., and D. GLICK: Quantitative histochemical distribution of coenzyme A in the rat adrenal in various functional states. Endocrinology **63**, 909—915 (1958).
GREENBERG, S. M., and K. ATERMAN: Effects of prolonged cortisone administration to hypophysectomized rats. Endocrinology **55**, 700—704 (1954).
GREENSTEIN, J. S., and R. C. FOLEY: The incidence of extramedullary myelopoiesis in bovine adrenal glands. J. Anim. Sci. **16**, 341—348 (1957).

GREEP, R. O., and I. CHESTER JONES: Steroid control of pituitary function. Recent Progr. Hormone Res. **5**, 197—261 (1950).
— and H. W. DEANE: Cytochemical evidence of the cessation of hormone production in the zona glomerulosa of the rat's adrenal cortex after prolonged treatment with desoxycorticosterone acetate. Endocrinology **40**, 417—425 (1947).
— — The cytology and cytochemistry of the adrenal cortex. Ann. N. Y. Acad. Sci. **50** (Art. 6), 596—615 (1949a).
— — Histological, cytochemical and physiological observations on the regeneration of the rat's adrenal gland following enucleation. Endocrinology **45**, 42—56 (1949b).
GREER, M. A., and H. L. ERWIN: Evidence of separate hypothalamic centers controlling corticotropin and thyrotropin secretion by the pituitary. Endocrinology **58**, 665—670 (1956).
GRIBOFF, S. I.: The rationale and clinical use of steroid hormones in cancer. Arch. intern. Med. **89**, 812—852 (1952).
GROAT, R. A.: Formation and growth of adrenocortical-like tissue in the ovaries of the adrenalectomized ground squirrel. Anat. Rec. **89**, 33—42 (1944).
GROLLMAN, A.: The pathogenesis of "adrenal regeneration" hypertension. Endocrinology **63**, 460—463 (1958).
DE GROOT, J., and C. FORTIER: Quantitative and histological aspects of adrenal cortical regeneration in the male albino rat. Anat. Rec. **133**, 565—577 (1959).
GRUENWALD, P.: The normal changes in the position of the embryonic kidney. Anat. Rec. **85**, 163—176 (1943).
— Embryonic and postnatal development of the adrenal cortex, particularly the zona glomerulosa and accessory nodules. Anat. Rec. **93**, 391—422 (1946).
— and W. M. KONIKOV: Cell replacement and its relation to the zona glomerulosa in the adrenal cortex of mammals. Anat. Rec. **89**, 1—22 (1944).
GRUNBAUM, B. W., J. R. GEARY, JR., and D. GLICK: The design and use of improved apparatus for the preparation and freezing-drying of fresh-frozen sections of tissue. J. Histochem. Cytochem. **4**, 555—560 (1956).
GRYNFELTT, E., J. TURCHINI et S. TURCHINI: Recherches histochimiques sur les corps interrénaux des sélaciens. C. R. Ass. Anat. **43**, 382-383 (1957).
GUGGENHEIM, K.: Studies on water metabolism of pyridoxine and pantothenic acid deficient rats. Endocrinology **55**, 156—162 (1954).
GUILLEMIN, R.: A re-evaluation of acetylcholine, adrenaline, nor-adrenaline and histamine as possible mediators of the pituitary adrenocorticotrophic activation by stress. Endocrinology **56**, 248—255 (1955).
— G. W. CLAYTON, J. D. SMITH et H. S. LIPSCOMB: Étude de la fonction corticosurrénalienne par la mesure simultanée de la corticostérone dans le sang périphérique et de l'acide ascorbique surrénalien. C. R. Acad. Sci. (Paris) **245**, 1834—1836 (1957).
— — — — Measurements of free corticosteroids in rat plasma: Physiological validation of a method. Endocrinology **63**, 349—358 (1958).
— C. FORTIER and H. S. LIPSCOMB: Comparison of *in vitro* and *in vivo* assaying procedures for rat adenohypophysial corticotrophin. Endocrinology **64**, 310—312 (1959).
DE GURPIDE, E. H. G.: Influence of adrenocorticotrophin and hypophysectomy on the adrenal secretion of 17-hydroxycorticosterone. Rev. Soc. argent. Biol. **29**, 15—20 (1953).
HAASE, J.: Das Verhalten der histochemisch nachweisbaren Ascorbinsäure in der Nebennierenrinde von Meerschweinchen nach einseitiger Adrenalektomie, Kälteeinwirkung, Wasserentzug und Hunger. Endokrinologie **29**, 1—22 (1952).
HADHÁZY, C., and L. KARMAZSIN: Regeneration of the adrenal cortex. Acta morph. Acad. Sci. hung. **3**, 427—435 (1953).
HADLOW, W. J.: Adrenal cortical atrophy in the dog. Amer. J. Path. **29**, 353—361 (1953).
HAGEMANN, U.: Das Verhalten der Hypophyse und Nebennieren bei experimentell induzierter kontrainsulinärer Adaptation. Beitr. path. Anat. **113**, 121—132 (1953).
HALÁSZ, B.: Die Ruckwirkung von Cortisonzufuhr auf den Bau der Nebennierenrinde nach Hypothalamusläsion. Acta morph. Acad. Sci. hung. **6**, 119—127 (1955).
HALBERG, F., M. J. FRANTZ and J. J. BITTNER: Phase difference between 24-hour rhythms in cortical adrenal mitoses and blood eosinophils in the mouse. Anat. Rec. **129**, 349—356 (1957).
— R. E. PETERSON and R. H. SILBER: Phase relations of 24-hour periodicities in blood corticosterone, mitoses in cortical adrenal parenchyma, and total body activity. Endocrinology **64**, 222—230 (1959).
HARDY, J. D., V. B. WARD, M. D. TURNER and L. P. SAMPSON: Further studies of human adrenal vein blood: Secretion of estrogens and 17-ketosteroids. Surgical Forum **8**, 109—111 (1957).

HARRISON, R. G.: A comparative study of the vascularization of the adrenal gland in the rabbit, rat and cat. J. Anat. (Lond.) **85**, 12—23 (1951a).
— The influence of unilateral nephrectomy on the production of experimental arthritis in the albino rat. Brit. med. J. **1951 b II**, 1299—1304.
— The adrenal circulation in the rabbit. J. Endocrin. **15**, 64—71 (1957).
— and C. W. ASLING: The anatomy and functional significance of the vascularization of the adrenal gland in the rhesus monkey *(Macaca mulatta)*. J. Anat. (Lond.) **89**, 106—113 (1955).
— and A. J. CAIN: Variations in the distribution of lipoids in the adrenal cortex of the albino rat. J. Anat. (Lond.) **81**, 286—299 (1947).
HARTMAN, F. A.: Notes on the adrenal of the sloth. Anat. Rec. **133**, 105—113 (1959).
— and R. H. ALBERTIN: A preliminary study of the avian adrenal. Auk (Ithaca, N.Y.) **68**, 202—209 (1951).
— and K. A. BROWNELL: The Adrenal Gland. Philadelphia: Lea and Febiger 1949.
— F. F. SHELDEN and E. L. GREEN: Weights of interrenal glands of elasmobranchs. Anat. Rec. **87**, 371—378 (1943).
— R. A. KNOUFF, A. W. MCNUTT and J. E. CARVER: Chromaffin patterns in bird adrenals. Anat. Rec. **97**, 211—221 (1947).
— — and G. A. HOWARD: Response of the pelican adrenal to various stimuli. Anat. Rec. **120**, 469—493 (1954).
— K. A. BROWNELL and T. Y. LIU: Blood flow through the dog adrenal. Amer. J. Physiol. **180**, 375—377 (1955).
HARTROFT, P. M., and A. B. EISENSTEIN: Alterations in the adrenal cortex of the rat induced by sodium deficiency: Correlation of histologic changes with steroid hormone secretion. Endocrinology **60**, 641—651 (1957).
— and W. S. HARTROFT: Studies on renal juxtaglomerular cells. II. Correlation of the degree of granulation of juxtaglomerular cells with width of the zona glomerulosa of the adrenal cortex. J. exp. Med. **102**, 205—212 (1955).
HATEY, J.: Interrénal et acide ascorbique des poissons téléostéens. C. R. Soc. Biol. (Paris) **146**, 566—568 (1952).
— Recherches sur la contrôle hypophysaire de l'interrénal antérieur d'un poisson téléostéen: l'anguille *(Anguilla anguilla* L.). Variation de l'acide ascorbique de l'interrénal antérieure consécutive à l'hypophysectomie. C. R. Soc. Biol. (Paris) **148**, 231—234 (1954a).
— Variation de l'acide ascorbique de l'interrénal antérieur après injection d'hormone corticotrope. C. R. Soc. Biol. (Paris) **148**, 324—327 (1954b).
HAYANO, M., and R. I. DORFMAN: The enzymatic C-11 β-hydroxylation of steroids. J. biol. Chem. **201**, 175—188 (1953).
— — On the mechanism of the C-11 β-hydroxylation of steroids. J. biol. Chem. **211**, 227—235 (1954).
HAYES, E. R.: A rigorous re-definition of the plasmal reaction. Stain Tech. **24**, 19—23 (1949).
HAYNES, R. C., JR.: The activation of adrenal phosphorylase by the adrenocorticotropic hormone. J. biol. Chem. **233**, 1220—1222 (1958).
— K. SAVARD and R. I. DORFMAN: An action of ACTH on adrenal slices. Science **116**, 690 to 691 (1952).
HAZELWOOD, R. L., L. L. BENNETT and M. M. NELSON: Effect of pantothenic acid deficiency on insulin sensitivity and response to ACTH of intact and diabetic rats. Endocrinology **58**, 427—434 (1956).
HEBARD, W. B., and H. A. CHARIPPER: A comparative study of the morphology and histochemistry of the reptilian adrenal gland. Zoologica, N. Y. **40**, 101—124 (1955).
HÉBERT, S., et Y. BARBARIN: Étude histochimique de la cortico-surrénale du rat nouveau-né. C. R. Soc. Biol. (Paris) **147**, 1886—1888 (1953).
— TRAN-VY et J. VERNE: L'activité androgène du cortex surrénal chez le rat nouveau-né. Ann. Endocr. (Paris) **15**, 493—498 (1954).
HEFFRON, R.: Pneumonia. New York: Commonwealth Fund 1939.
HENI, F., P. GRUNER u. H. MAST: Untersuchungen über die Eigenleistung der Nebennierenrinde hypophysektomierter Ratten. Acta endocr. (Kbh.) **11**, 229—244 (1952).
HERRICK, E. H., and O. TORSTVEIT: Some effects of adrenalectomy in fowls. Endocrinology **22**, 469—473 (1938).
HERTZ, R., M. J. ALLEN and W. W. TULLNER: Effects of Amphenone "B" on thyroid, adrenals, and genital tract of the female rat. Proc. Soc. exp. Biol. (N. Y.) **75**, 627—630 (1950).
— W. W. TULLNER and M. J. ALLEN: Selective inhibition of adrenal and thyroid stimulating effect of Amphenone B by cortisone and thyroxine. Proc. Soc. exp. Biol. (N. Y.) **77**, 480 to 481 (1951).
— — J. A. SCHRICKER, F. G. DHYSE and L. F. HALLMAN: Studies on amphenone and related compounds. Recent Progr. Hormone Res. **11**, 119—147 (1955).

HERTZ, R., D. M. BERGENSTAL, H. A. LUBS and S. J. JACKSON: Observations on adrenal carcinoma with special reference to the effects of amphenone. Cancer (N..Y) **10**, 765—776 (1957).
HESS, M., E. G. RENNELS and J. C. FINERTY: Response of preputial and adrenal glands of hypophysectomized rats to ACTH. Endocrinology **52**, 223—227 (1953).
HETT, J.: Ein Beitrag zur Histogenese der menschlichen Nebenniere. Z. mikr.-anat. Forsch. **3**, 179—282 (1925).
— Blutbildung in den Nebennieren. Z. Zellforsch. **33**, 389—404 (1945).
HILL, A. G. S., H. W. DEANE and A. H. COONS: Localization of antigen in tissue cells. V. Capsular polysaccharide of Friedländer bacillus, type B, in the mouse. J. exp. Med. **92**, 35—44 (1950).
HILL, W. C. O.: Observations on the growth of the suprarenal cortex. J. Anat. (Lond.) **64**, 479—502 (1930).
— The suprarenal cortex in monkeys of the genus Pithecus. J. Anat. (Lond.) **68**, 19—38 (1933).
— The suprarenals of the larger Felidae. J. Anat. (Lond.) **72**, 71—82 (1937).
HILTON, J. G., D. C. WEAVER, C. MUELHEIMS, V. V. GLAVIANO and R. WÉGRIA: Perfusion of the isolated adrenals *in situ*. Amer. J. Physiol. **192**, 523—530 (1958).
— L. F. SCIAN, C. D. WESTERMANN and O. R. KRUESI: Direct stimulation of adrenocortical secretion by synthetic vasopressin in the dog. Proc. Soc. exp. Biol. (N. Y.) **100**, 523—524 (1959).
HOCH-LIGETI, C., and G. H. BOURNE: Changes in the concentration and histological distribution of the ascorbic acid in the ovaries, adrenals and livers of rats during the oestrous cycle. Brit. J. exp. Path. **29**, 400—407 (1948).
HOEDE, K.: Über den Einfluß langfristigen Cortisonvorbehandlung auf die hämorrhagische Nekrose der Nebennierenrinde nach Diphtherietoxinvergiftung. Endokrinologie **33**, 279—286 (1956).
HOERR, N. L.: The cells of the suprarenal cortex in the guinea pig, their reaction to injury and their replacement. Amer. J. Anat. **48**, 139—197 (1931).
— On osmic acid as a microchemical reagent with special reference to lipins. Anat. Rec. **66**, 149—171 (1936a).
— A cytological analysis of the liposomes in the adrenal cortex of the guinea pig. Anat. Rec. **66**, 317—342 (1936b).
HOFMANN, F. G.: Observations on *in vitro* adrenal steroid synthesis in the albino mouse. Endocrinology **59**, 712—715 (1956).
— The *in vitro* hydroxylation of 21-carbon steroids by rat adrenal glands. Endocrinology **60**, 382—389 (1957).
HOHLWEG, W., u. U. LASCHET: Zur Frage der Cortison- und Corticotrophin-Wirkung auf das Hypophysen-Nebennierenrinden-System. Acta endocr. (Kbh.) **32**, 437—447 (1959).
HOKIN, M. R., L. E. HOKIN, M. SAFFRAN, A. V. SCHALLY and B. U. ZIMMERMANN: Phospholipides and the secretion of adrenocorticotropin and of corticosteroid. J. biol. Chem. **233**, 811—813 (1958).
HOLLAND, R. C.: The effect of hypoxia on the fetal rat adrenal. Anat. Rec. **130**, 177—195 (1958).
HOLMES, W., and D. E. MOORHOUSE: The peri-renal tissue of *Protopterus*. Quart. J. micr. Sci. **97**, 123—154 (1956).
HOLMES, W. N.: Variations in adrenal ascorbic acid concentration of the hamster during the pre-hibernatory period. Endocrinology **57**, 409—413 (1955a).
— Histological variations in the adrenal cortex of the golden hamster with special reference to the X zone. Anat. Rec. **122**, 271—293 (1955b).
HOLZBAUER, M.: The corticosterone content of rat adrenals under different experimental conditions. J. Physiol. (Lond.) **139**, 294—305 (1957a).
— The effects of oestrogens on the secretory capacity of the rat adrenal cortex *in vivo*. J. Physiol. (Lond.) **139**, 306—315 (1957b).
— and M. VOGT: Functional changes produced in the adrenal cortex of the rat by the administration or by release of corticotrophin. J. Physiol. (Lond.) **138**, 449—459 (1957a).
— — Restoration of normal secretion in the hexoestrol-inhibited adrenal cortex of the rat. J. Physiol. (Lond.) **138**, 460—472 (1957b).
HORN, E. H.: Nutritional and hormonal influences upon reproductive maturation, organ weights and histochemistry of the immature male rat. Endocrinology **57**, 399—408 (1955).
HOUSSAY, A. B., A. F. CARDEZA, B. A. HOUSSAY et R. M. PINTO: Facteurs susceptibles de modifier l'apparition et la fonction des tumeurs surrénales chez les rats castrés. C. R. Soc. Biol. (Paris) **148**, 2123—2125 (1954).
HOUSSAY, B. A.: Hypophyseal functions in the toad *Bufo arenarum* Hensel. Quart. Rev. Biol. **24**, 1—27 (1949).

HOUSSAY, B. A., and R. SAMMARTINO: Histology of the adrenals of hypophysectomized dogs (in Spanish). Rev. Soc. argent. Biol. **9**, 209—219 (1933a).
— — Modifications histologiques de la surrénale chez les chiens hypophysoprivés ou a tuber lisé. C. R. Soc. Biol. (Paris) **114**, 717—721 (1933b).
— A. B. HOUSSAY, A. F. CARDEZA, V. G. FOGLIA and R. M. PINTO: Adrenal tumors in gonadectomized rats. Acta physiol. lat.-amer. **3**, 125—130 (1953).
— — — et R. M. PINTO: Tumeurs surrénales oestrogéniques et tumeurs hypophysaires chez les animaux castrés. J. suisse Méd. **85**, 291—308 (1955).
HOUSSAY, H. E. J.: Change in the thyroid and adrenal after hypophysectomy of the rat (in Spanish). Rev. Soc. argent. Biol. **18**, 410—415 (1942).
HOUSSAY, R. H.: Sensitivity to histamine in rats with accessory or demedullated adrenals (in Spanish). Rev. Soc. argent. Biol. **31**, 131—146 (1955).
HOWARD, A. N., and D. B. CATER: The adrenal cortex and the effect of ACTH and cortisone in scorbutic guinea pigs. J. Endocrin. **18**, 175—185 (1959).
— and B. J. CONSTABLE: The metabolism of adrenocorticotrophic hormone and ascorbic acid in the chick. Biochem. J. **69**, 501—505 (1958).
HOWARD, E. (HOWARD-MILLER): A transitory zone in the adrenal cortex which shows age and sex relationships. Amer. J. Anat. **40**, 251—293 (1927).
— The representations of the adrenal X zone in rats, in light of observations on the X zone variability in mice. Amer. J. Anat. **62**, 351—374 (1938).
HUHN, A., u. P. SCHNEPPENHEIM: Über die Morphologie der Nebenniere von Ratten bei Transplantationstumoren. Virchows Arch. path. Anat. **326**, 46—72 (1954).
HUMMEL, K. P.: Accessory adrenal cortical nodules in the mouse. Anat. Rec. **132**, 281—295 (1958).
HUMPHREYS, E. M., and L. DONALDSON: Degeneration of the adrenal cortex produced by Germanin. Amer. J. Path. **17**, 767—775 (1941).
HUNGERFORD, G. F.: The effect of age in determining the adrenal response to a sodium low diet. Anat. Rec. **133**, 394—395 (1959).
HUNT, T. E., and E. A. HUNT: Glycogen in the adrenal gland of rats at different ages. Anat. Rec. **133**, 537—551 (1959).
HURLEY, L. S., and A. F. MORGAN: Carbohydrate metabolism and adrenocortical function in the panthothenic acid-deficient rat. J. biol. Chem. **195**, 583—590 (1952).
HUTCHISON, W. C., J. K. BURNS and A. J. HALE: The deoxyribonucleic acid content of guinea-pig adrenal nuclei after administration of corticotrophin. Exp. Cell Res. **14**, 193 to 200 (1958).
IGLESIAS, R., and E. MARDONES: Influence of hormones on the growth of a transplantable suprarenal tumour. Brit. J. Cancer **12**, 28—31 (1958).
INGLE, D. J.: Resistance of the rat to histamine shock after destruction of the adrenal medulla. Amer. J. Physiol. **118**, 57—63 (1937).
— The biologic properties of cortisone: A review. J. clin. Endocrin. **10**, 1312—1354 (1950).
— Control of regeneration of the adrenal cortex in the rat. In: R. C. CHRISTMAN, Ed., Pituitary-Adrenal Function, 49—55. Washington: Amer. Ass. Advance Science 1951.
— and G. M. HIGGINS: Regeneration of the adrenal gland following enucleation. Amer. J. med. Sci. **196**, 232—239 (1938a).
— — Autotransplantation and regeneration of the adrenal gland. Endocrinology **22**, 458—464 (1938b).
— — and E. C. KENDALL: Atrophy of the adrenal cortex in the rat produced by administration of large amounts of cortin. Anat. Rec. **71**, 363—372 (1938).
— M. C. PRESTRUD and C. H. LI: Effects of administering adrenocorticotrophic hormone by continuous injection to normal rats. Amer. J. Physiol. **166**, 165—170 (1951).
ITO, T.: Histology and histogenesis of the adrenal cortex in the guinea pig. Okajimas Folia anat. jap. **24**, 269—289 (1952).
— The histology of the adrenal cortex of the postpuberally castrated male mouse. Okajimas Folia anat. jap. **26**, 271—283 (1954).
— Postnatal histogenesis of the adrenal cortex of the golden hamster. Okajimas Folia anat. jap. **29**, 67—92 (1956).
JACKSON, C. M.: The postnatal development of the suprarenal gland — and the effects of inanition upon its growth and structure in the albino rat. Amer. J. Anat. **25**, 221—289 (1919).
JACQUOT, R.: Recherches sur le controle hormonal du glycogène hépatique chez le foetus de rat. J. Physiol. (Paris) **47**, 857—866 (1955).
JAILER, J. W.: The pituitary-adrenal relationship in the infant rat. Proc. Soc. exp. Biol. (N. Y.) **72**, 638—639 (1949).
— The maturation of the pituitary-adrenal axis in the newborn rat. Endocrinology **46**, 420—425 (1950).

JAILER, J. W., and N. F. BOAS: The inability of epinephrine or adrenocorticotropic hormone to deplete the ascorbic acid content of the chick adrenal. Endocrinology **46**, 314—318 (1950).
— J. J. GOLD and E. Z. WALLACE: Evaluation of the "cortisone test" as diagnostic acid in differentiating adrenal hyperplasia from adrenal neoplasia. Amer. J. Med. **16**, 340—345 (1954).
— N. P. CHRISTY, D. LONGSON, E. Z. WALLACE and W. E. L. GORDON: Further observations on adrenal cortical function during pregnancy. Amer. J. Obstet. Gynec. **78**, 1—10 (1959).
JARDETZKY, C. D., and D. GLICK: Determination of succinic dehydrogenase in microgram amounts of tissue and its distribution in rat adrenal. J. biol. Chem. **218**, 283—292 (1956).
JAYNE, E. P.: Cytology of the adrenal gland of the rat at different ages. Anat. Rec. **115**, 459—483 (1953).
— Histochemical and degenerative changes in the adrenal cortex of the rat with age. J. Geront. **12**, 2—8 (1957).
JENKINS, J. J., J. W. MEAKIN, D. H. NELSON and G. W. THORN: Inhibition of adrenal steroid 11-oxygenation in the dog. Science **128**, 478—479 (1958).
— — — A comparison of the inhibitory effects of 2-methyl-1,2-bis(3-pyridyl)-1-propanone and amphenone B on adrenocortical secretion in the rat. Endocrinology **64**, 572—578 (1959).
JOHANNES, A., H. STAUDINGER u. H. WILZ: Stoffwechsel der Nebennierenrinde und Biosynthese der Corticosteroide. X. Wirkung von überlebenden Nebennierengewebe. Biochem. Z. **327**, 484—490 (1956).
JOHNSTONE, F. R. C.: The suprarenal veins. Amer. J. Surg. **94**, 615—620 (1957).
JONES, J. M., C. W. LLOYD and T. C. WYATT: A study of the interrelationships of maternal and fetal adrenal glands of rats. Endocrinology **53**, 182—191 (1953).
JONES, R. S., Y. C. MAYNE and K. EIK-NES: C^{14} labeled bacterial polysaccharide and adrenal cortical function in the guinea pig. Endocrinology **62**, 843—852 (1958).
JOSHI, R. A., and H. T. BLUMENTHAL: Alteration of homotransplantability of rat adrenal and aorta by dehydration freezing. Lab. Invest. **7**, 19—29 (1958).
JOSIMOVICH, J. B., A. J. LADMAN and H. W. DEANE: A histophysiological study of the developing adrenal cortex of the rat during fetal and early postnatal stages. Endocrinology **54**, 627—639 (1954).
JOST, A.: Influence de la décapitation sur le développement du tractus genital et des surrénales de l'embryon de lapin. C. R. Soc. Biol. (Paris) **142**, 273—275 (1948).
— Problems of fetal endocrinology: The gonadal and hypophyseal hormones. Recent Progr. Hormone Res. **8**, 379—418 (1953).
— Hormonal factors in the development of the fetus. Cold Spr. Harb. Symp. quant. Biol. **19**, 167—181 (1954).
— et P. GRANDOU: Action opposée sur la surrénale du rat, de l'hydrocortisone et de la 9-α-fluoro-11-β-hydroxy-17-α-methyltestostérone. C. R. Soc. Biol. (Paris) **152**, 896—898 (1958).
— et R. JACQUOT: Recherches sur les facteurs endocriniens de la charge en glycogène du foie foetal chez le lapin (avec des indications sur le glycogène placentaire). Ann. Endocr. (Paris) **16**, 849—872 (1955).
— — et A. COHEN: Sur les interrelations entre les hormones corticosurrénaliennes maternelles et la surré ale du foetus de rat. C. R. Soc. Biol. (Paris) **149**, 1319—1322 (1955).
JUNKMANN, K.: Die physiologische Chemie der inneren Sekretion. In: B. FLASCHENTRÄGER u. E. LEHNARTZ, Herausgeb., Physiologische Chemie, Bd. II/2b, 425—627. Berlin: Springer 1957.
KALANT, O. J.: The effects of testosterone propionate, ACTH and cortisone acetate on the succinoxidase activity of the adrenal gland of the rat. Endocrinology **62**, 237—244 (1958).
— and E. A. SELLERS: The influence of age and sex on the succinoxidase activity of the adrenal gland of the rat. Endocrinology **55**, 777—781 (1954).
— — The effects of sex hormones on the succinoxidase activity of the adrenal gland of the rat. Endocrinology **61**, 264—270 (1957).
KAMPMEIER, O. F.: Giant epithelial cells of the human fetal adrenal. Anat. Rec. **37**, 95—102 (1927).
KAPLAN, M. H., A. H. COONS and H. W. DEANE: Cellular deposition of pneumococcal polysaccharides types II and III in the mouse. J. exp. Med. **91**, 15—30 (1950).
KAR, A. B.: The adrenal cortex testicular relations in the fowl: The effect of castration and replacement therapy on the adrenal cortex. Anat. Rec. **99**, 177—197 (1947).
— Studies in the distribution and concentration of alkaline phosphatase in the adrenal cortex of normal and of sex hormone treated pigeons. Endocrinology **46**, 363—366 (1950).

KAR, A. B.: Changes in the distribution of ascorbic acid and sudanophilia in the adrenal cortex of mice after prolonged stimulation with adrenocorticotrophic hormone (ACTH). Anat. Anz. **100**, 39—46 (1953).
— J. N. KARHUN and N. N. DE: The effect of 19-nortestosterone on the adrenal cortex of the rat. Acta endocr. (Kbh.) **25**, 238—248 (1957).
KARNOVSKY, M. L., and H. W. DEANE: Aldehyde formation in the lipide droplets of the adrenal cortex during fixation, as demonstrated chemically and histochemically. J. Histochem. Cytochem. **3**, 85—102 (1955).
KASS, E. H., O. HECHTER, I. A. MACCHI and T. W. MOU: Changes in patterns of secretion of corticosteroids in rabbits after prolonged treatment with ACTH. Proc. Soc. exp. Biol. (N. Y.) **85**, 583—587 (1954).
KATSURA, H.: Comparative studies on the respiratory enzyme activities of thyroid, adrenal cortex and adrenal medulla. Endocr. jap. **2**, 155—162 (1955).
KAY, W. W., and R. WHITEHEAD: The suprarenal in cholesterol-fed rabbits. J. Path. Bact. **41**, 293—301 (1935).
KAYSER, C., et M. ARON: Le cycle saisonnier des glandes endocrines chez les hibernants. Arch. Anat. (Strasbourg) **33**, 21—42 (1950).
KEENE, M. F. L., and E. E. HEWER: Observations on the development of the human adrenal gland. J. Anat. (Lond.) **61**, 302—324 (1927).
KELLY, T. L., E. D. NIELSON, R. B. JOHNSON and C. S. VESTLING: Glucose-6-phosphate dehydrogenase of adrenal tissue. J. biol. Chem. **212**, 545—554 (1955).
KENYON, A. T.: Adrenal cortical tumors — physiologic considerations. Surgery **16**, 194—232 (1944).
KING, J. T., C. B. CASAS and M. B. VISSCHER: The influence of estrogen on cancer incidence and adrenal changes in ovariectomized mice on caloric restriction. Cancer Res. **9**, 436—437 (1949).
KINGSBURY, J. W., S. L. EMERY and A. E. ADAMS: Effects of thiourea on the adrenal glands of chick embryos. Endocrinology **56**, 299—304 (1955).
KIRSCHBAUM, A., M. FRANTZ and W. L. WILLIAMS: Neoplasms of the adrenal cortex in noncastrate mice. Cancer Res. **6**, 707—711 (1946).
KISS, T.: Experimentell-morphologische Analyse der Nebenniereninnervation. Acta anat. **13**, 81—89 (1951).
KITAY, J. I., D. A. HOLUB and J. W. JAILER: Evidence for species specificity of human, beef and rat ACTH demonstrated by assay in vitro. Progr. 41st. Meet. Endocr. Soc., 34—35 (1959).
KITCHELL, R. L.: Compensatory hypertrophy of the intact adrenal of fetal rats subjected to unilateral adrenalectomy. Proc. Soc. exp. Biol. (N. Y.) **75**, 824—827 (1950).
— and L. J. WELLS: Reciprocal relation between the hypophysis and adrenals in fetal rats: Effects of unilateral adrenalectomy and of implanted cortisone, DOCA and sex hormones. Endocrinology **50**, 83—93 (1952a).
— — Functioning of the hypophysis and adrenals in fetal rats: Effects of hypophysectomy, adrenalectomy, castration, injected ACTH and implanted sex hormones. Anat. Rec. **112**, 561—591 (1952b).
KLÄRNER, P.: Veränderungen an Nebenniere und Hypophyse der Ratte nach vielmonatiger Zufuhr von ACTH. Beitr. path. Anat. **115**, 488—513 (1955).
KLEIN, R., and A. ROVNANEK: ACTH response in newborn infants. In: L. I. GARDNER, Ed., Adrenal Function in Infants and Children, 46—61. New York: Grune and Stratton 1956.
— J. FORTUNATO and C. PAPADATOS: Free blood corticoids in the newborn infant. J. clin. Invest. **33**, 35—37 (1954).
KLINE, I. T.: Potentiation by squalene of adrenal and thymic responses to corticotropin. Endocrinology **61**, 85—92 (1957).
— Studies on squalene potentiation of thymic response to corticotropin. Endocrinology **63**, 335—344 (1958).
KLOOS, K., u. H.-J. STAEMMLER: Zur Morphologie und Pathophysiologie der Nebennierenrinde von Feten und Neugeborenen. Ein Beitrag zum Problem der diaplacentaren Regulation des Interrenalsystems. Virchows Arch. path. Anat. **324**, 285—306 (1953).
KLOSE, H.-G.: Über den Einfluß der Kastration auf Schilddrüse, Hypophyse und Interrenalsystem der Urodelen. Z. wiss. Zool. **155**, 46—108 (1941).
KLOTZ, H. P., et F. KAHN: A propos du controle central de la sécrétion d'aldostérone. Ann. Endocr. (Paris) **18**, 587—591 (1957).
KNAB, J.: Untersuchungen über den histochemischen Nachweis von Vitamin C in der Niere und Nebenniere von Sauropsiden. Z. mikr.-anat. Forsch. **52**, 418—439 (1942).
KNIGGE, K. M.: The effect of hypophysectomy on the adrenal gland of the hamster (Mesocricetus auratus). Amer. J. Anat. **94**, 225—272 (1954a).

KNIGGE, K. M.: Response of the hamster's adrenal cortex to desoxycorticosterone acetate, with observations on the kidney, thyroid and hypophysis. Endocrinology **55**, 731—744 (1954b).
— The effect of starvation on the adrenal cortex of the hamster. Anat. Rec. **120**, 555—582 (1954c).
— Influence of DCA and cortisone on adrenal glands of fed and acutely starved hamsters. Proc. Soc. exp. Biol. (N. Y.) **88**, 348—351 (1955).
— Influence of cold exposure upon the endocrine glands of the hamster, with an apparent dichotomy between morphological and functional response of the thyroid. Anat. Rec. **127**, 75—96 (1957).
KNOBIL, E.: Effects of hypophysectomy, ACTH, cortisone and estradiol on the β-glucuronidase activity of adrenal gland in the rat. Proc. Soc. exp. Biol. (N. Y.) **83**, 769—770 (1953).
— and F. N. BRIGGS: Fetal-maternal interrelations: The hypophyseal-adrenal system. Endocrinology **57**, 147—152 (1955).
— and M. J. FREGLY: Effect of ascorbic acid on the adrenal gland after hypophysectomy and after exposure to cold. Endocrinology **56**, 614—616 (1955).
— and R. O. GREEP: Serum electrolytes in the hypophysectomized rhesus monkey. Endocrinology **62**, 61—63 (1958).
— F. G. HOFMANN and R. O. GREEP: Failure of methylandrostenediol to prevent compensatory atrophy of the adrenal in the rhesus monkey. Endocrinology **53**, 242—244 (1953).
— M. G. HAGNEY, E. J. WILDER and F. N. BRIGGS: Simplified method for determination of total adrenal cholesterol. Proc. Soc. exp. Biol. (N. Y.) **87**, 48—50 (1954a).
— A. MORSE, F. G. HOFMANN and R. O. GREEP: A histologic and histochemical study of hypophyseal-adrenal cortical relationships in the rhesus monkey. Acta endocr. (Kbh.) **17**, 229—238 (1954b).
— — R. C. WOLF and R. O. GREEP: The action of bovine, porcine and simian growth hormone preparations on the costochondral junction in the hypophysectomized rhesus monkey. Endocrinology **62**, 348—354 (1958).
KNOUFF, R. A., and F. A. HARTMAN: A microscopic study of the adrenal of the Brown Pelican. Anat. Rec. **109**, 161—188 (1951).
— J. B. BROWN and B. M. SCHNEIDER: Correlated chemical and histological studies of the adrenal lipids. I. The effect of extreme muscular activity on the adrenal lipids of the guinea pig. Anat. Rec. **79**, 17—38 (1941).
KNOWLTON, A. I., E. N. LOEB, H. C. STOERK and B. C. SEEGAL: Desoxycorticosterone acetate. The potentiation of its activity by sodium chloride. J. exp. Med. **85**, 187—198 (1947).
— — B. C. SEEGAL and H. C. STOERK: Desoxycorticosterone acetate: Studies on the reversibility of its effect on blood pressure and renal damage in rats. Endocrinology **45**, 435—445 (1949).
KOHNO, S.: Zur vergleichenden Histologie und Embryologie der Nebenniere der Säuger und des Menschen. Z. Anat. Entwickl.-Gesch. **77**, 419—480 (1925).
KOLMER, W.: Zur vergleichenden Histologie, Zytologie und Entwicklungsgeschichte der Säugernebenniere. Arch. mikr. Anat. **91** (I), 1—139 (1918).
KOMRAD, E. L., and L. C. WYMAN: Histamine tolerance in rats with intramuscular adrenocortical transplants. Endocrinology **49**, 481—483 (1951).
KORENCHEVSKY, V.: Natural relative hypoplasia of organs and the process of ageing. J. Path. Bact. **54**, 13—24 (1942).
KORNFELD, W.: Gonadal effects upon the growth of pituitary, adrenals, thyroids, oviduct and comb of immature fowl. Anat. Rec. **128**, 419—431 (1957).
KORPÁSSY, B., J. TÖRÖK u. K. KOVÁCS: Endokrine Veränderungen bei experimenteller akuter Gerbsäurevergiftung, mit besonderer Rücksicht auf die Nebennierenrinde. Acta physiol. Acad. Sci. hung. **1**, 113—124 (1950).
KOVÁCS, K., and M. A. DÁVID: Effect of cortisone on the morphological reactions of the adrenal cortex due to changes in the Na/K intake. Acta anat. **36**, 169—181 (1959).
— u. A. JAKOBOVITS: Über die Funktion der in die Milz überpflanzten Nebenniere. Endokrinologie **30**, 303—310 (1953).
KOWALEWSKI, K.: Effect of corticotrophin on the plasma 17-hydroxycorticosteroids in normal and thiouracil treated guinea pigs. Acta endocr. (Kbh.) **27**, 257—261 (1958).
KRACHT, J.: Zur Dauer der Nebennierenrinden-Aktivierung nach ACTH und ACTH-Depot. Naunyn-Schmiedebergs Arch. exp. Path. Pharmak. **223**, 355—361 (1954).
— Sogenannte Rebound-Phänomen an der Nebennierenrinde nach Zufuhr von Glukocorticoiden. Endokrinologie **35**, 290—296 (1958).
— u. M. SPAETHE: Über Wechselbeziehungen zwischen Schilddrüse und Nebennierenrinde. I. Der Thyreo-corticotrope Phasenwechsel in der Sekretionsbiologie des Hypophysenvorderlappens. Virchows Arch. path. Anat. **323**, 174—193 (1953a).

KRACHT, J., u. M. SPAETHE: II. Untersuchungen über den „Hypophysenhemmstoff" *p*-Oxy-propiophenon. Virchow Arch. path. Anat. **323**, 629—644 (1953b).
— — Die Karyometrie der Nebennierenrinde und ihre Fehlerquellen. Z. wiss. Mikr. **62**, 227—233 (1955).
KRAUTER, D.: Zur Histologie der Nebennieren der Knochenfische. Mikrokosmos **41**, 10—12 (1951).
KRISCH, K., u. H. STAUDINGER: Isolierung von Cytochrom b_5 aus Nebennieren. Biochem. Z. **331**, 37—45 (1958).
— — Zur Wirkung von Ascorbinsäure und Cytochrom b_5 auf die mikrosomal DPNH-Oxydation. Biochem. Z. **331**, 195—208 (1959).
KROHN, P. L.: The effect of ACTH and cortisone on the survival of skin homografts and on the adrenal glands of monkeys *(Macaca mulatta)*. J. Endocrin. **12**, 220—226 (1955).
KÜCHMEISTER, H., G. PLIESS u. W. WILHELM: Untersuchungen über die fraktionierte Schädigung der Nebennierenrinde des Hundes mit 2,2-Bis(parachlorophenyl)-1,1-dichloroethane (D.D.D.). Acta endocr. (Kbh.) **20**, 39—49 (1955).
KUMITA: Über die parenchymatösen Lymphobahnen der Nebenniere. Arch. Anat. Physiol. Lpz. **1909**, 321—326.
LADMAN, A. J., and H. W. DEANE: The effect of the administration of ACTH or chorionic gonadotrophin on the adrenal cortex of the neonatal rat. Anat. Rec. **122**, 481 (1955).
LAESCHKE, R.: Die physiologischen und vom Verhalten der Keimdrüsen abhängenden Veränderung der Nebennierenrinde des erwachsenen Menschen. Z. mikr.-anat. Forsch. **57**, 1—84 (1951).
LANDING, B. H., and D. FERIOZI: Effects of ACTH and cortisone on the adrenals and pituitaries of children with acute leukemia. J. clin. Endocr. **14**, 901—921 (1954a).
— — Effect of ACTH on the adrenals in the nephrotic syndrome and rheumatic fever. J. clin. Endocr. **14**, 1023—1028 (1954b).
LANE, N., and R. C. DE BODO: Generalized adrenocortical atrophy in hypophysectomized dogs and correlated functional studies. Amer. J. Physiol. **168**, 1—19 (1952).
LANGECKER, H., u. R. LURIE: Die Hemmung der Corticotropin-Sekretion durch Steroide. Acta endocr. (Kbh.) **25**, 54—58 (1957).
LANMAN, J. T.: The fetal zone of the adrenal gland. Its developmental course, comparative anatomy, and possible physiologic functions. Med cine (Baltimore) **32**, 389—430 (1953).
— The adrenal fetal zone: Its occurrence in primates and a possible relationship to chorionic gonadotropin. Endocrinology **61**, 684—691 (1957).
— and L. M. SILVERMAN: *In vitro* steroidogenesis in the human neonatal adrenal gland, including observations on human adult and monkey adrenal glands. Endocrinology **60**, 433—445 (1957).
LAQUEUR, G. L., S. M. MCCANN, L. H. SCHREINER, E. ROSEMBERG, D. M. RIOCH and E. ANDERSON: Alterations of adrenal cortical and ovarian activity following hypothalamic lesions. Endocrinology **57**, 44—54 (1955).
LARAGH, J. H., and H. C. STOERK: A study of the mechanism of secretion of the sodium-retaining hormone (aldosterone). J. clin. Invest. **36**, 383—392 (1957).
LARSON, P. S., G. R. HENNIGAR, J. K. FINNEGAN, R. B. SMITH, JR., and H. B. HAAG: Observations on the relation of chemical structure to the production of adrenal cortical atrophy or hypertrophy in the dog by derivatives of 2,2-bis-(p-chlorophenyl)-1,1-dichloroethane (DDD, TDE). J. Pharmacol. **115**, 408—412 (1955).
LATIMER, H. B.: The relative postnatal growth of the systems and organs of the chicken. Anat. Rec. **31**, 233—253 (1925).
— The prenatal growth of the cat. IX. The ponderal growth of the hypophysis, thyroid, thymus and suprarenal glands. Growth **3**, 337—347 (1939a).
— The weights of the hypophysis, thyroid and suprarenals in the adult cat. Growth **3**, 435—445 (1939b).
— Weights, percentage weights and correlations of the endocrine glands of the adult male guinea pig. Anat. Rec. **111**, 299—316 (1951).
— Growth of the hypophysis, thyroid and suprarenals in the fetal dog. Anat. Rec. **120**, 495—514 (1954).
— and J. A. ROSENBAUM: A quantitative study of the anatomy of the turkey hen. Anat. Rec. **34**, 15—23 (1926).
— and P. B. SAWIN: Morphogenetic studies of the rabbit. XIII. The influence of the dwarf gene upon organ size and varability in race X. Anat. Rec. **123**, 447—466 (1955).
LA VELLE, F. W.: A study of hormonal factors on the early sex development of the golden hamster. Contr. Embryol. Carneg. Inst. **34**, 19—54 (1951).
LAZO-WASEM, E. A.: Pituitary ACTH concentration during adrenal involution following thiouracil. Fed. Proc. **11**, 270 (1959).

LEATHEM, J. H.: Influence of testosterone propionate on the adrenals and testes of hypophysectomized rats. Anat. Rec. **89**, 155—161 (1944).

— Plasma protein concentrations and organ weights of rats as related to a high protein diet. Proc. Soc. exp. Biol. (N. Y.) **64**, 90—92 (1947).

— The response of the mouse adrenal gland to steroid hormones. Trans. N. Y. Acad. Sci. **2**, 239—243 (1949).

— The adrenal and dietary protein. Arch. int. Pharmacodyn. **110**, 103—113 (1957).

— and L. A. STAUBER: Adrenal lipid and alkaline phosphatase in hamsters infected with *Leishmania donovani*. Endocrinology **50**, 311—317 (1952).

LEBLOND, C. P., and W. U. GARDNER: Distribution of vitamin C in the adrenal gland of the mouse with reference to the nature of the X zone. Anat. Rec. **72**, 119—129 (1938).

LELOUP-HATEY, J.: Le cholestérol de l'interrénal antérieur de quelques poissons téléostéens. Bull. Soc. Chim. biol. (Paris) **39**, 1311—1320 (1957).

— Influence de l'agitation motrice sur la teneur du plasma en 17-hydroxy-corticostéroïds d'un téléostéen: la carpe *(Cyprinus carpio* L.). C. R. Acad. Sci. (Paris) **246**, 1088—1091 (1958).

LEROY, P., et J. BENOIT: Surrénalectomie bilaterale chez la canard mâle adulte: Action sur le testicule. J. Physiol. (Paris) **46**, 422—428 (1954).

LEULIER, A., et L. REVOL: Sur la répartition du cholestérol et de ses éthers dans les capsules surrénales. C. R. Acad. Sci. (Paris) **190**, 657—659 (1930).

LEVER, J. D.: Observations on the adrenal blood vessels in the rat. J. Anat. (Lond.) **86**, 459—467 (1952).

— Nerve fibres in the adrenal cortex of the rat. Nature (Lond.) **171**, 882 (1953).

— Vascular zoning in the adrenal cortex of the normal and hypophysectomized rat with observations on the distribution of lipids. J. Endocrin. **10**, 133—146 (1954).

— Cellular-vascular relationships in the adrenal cortex, as studied in the rat. Ciba Foundn. Coll. Endocrinol. 8, 42—48 (1955a).

— Adreno-cortical histogenesis in the rat: with observations on lipid and ascorbic acid distribution. J. Anat. (Lond.) **89**, 293—300 (1955b).

— Electron miscroscopic observations on the adrenal cortex. Amer. J. Anat. **97**, 409—430 (1955c).

— Cytological studies on the hypophysectomized rat adrenal cortex: the alterations of its fine structure following ACTH administration and on lowering the Na/K ratio. Endocrinology **58**, 163—180 (1956).

LEVIN, M. E., and W. H. DAUGHADAY: The influence of the thyroid on adrenocortical function. J. clin. Endocrin. **15**, 1499—1511 (1955).

LEVINSKY, N. G., and W. H. SAWYER: Influence of the adenohypophysis on the frog waterbalance response. Endocrinology **51**, 110—116 (1952).

LEVY, H. H. W. DEANE and B. L. RUBIN: Observations on steroid-3β-ol-dehydrogenase activity in steroid-producing glands. J. Histochem. Cytochem. **7**, 320 (1959a).

— — — Visualization of steroid-3 β-ol-dehydrogenase activity in tissues of intact and hypophysectomized rats. Endocrinology **65**, 932—943 (1959b).

LEWIS, R. A., E. ROSEMBERG and L. WILKINS: The effects of 17-hydroxy-11-dehydrocorticosterone upon the adrenals of normal and of hypophysectomized rats maintained with adrenocorticotropin. Endocrinology **47**, 414—417 (1950).

LI, C. H., P. FØNSS-BECH, I. I. GESCHWIND, T. HAYASHIDA, G. F. HUNGERFORD, A. J. LOSTROH, W. R. LYONS, H. D. MOON, W. O. REINHARDT and M. SIDEMAN: Corticotropins (ACTH). X. Biological investigations of α-corticotropin. J. exp. Med. **105**, 335—360 (1957).

LIDDLE, G. W., L. E. DUNCAN, JR., and F. C. BARTTER: Dual mechanism regulating adrenocortical function in man. Amer. J. Med. **21**, 380—386 (1956).

— D. ISLAND, E. M. LANCE and A. P. HARRIS: Alterations of adrenal steroid patterns in man resulting from treatment with a chemical inhibitor of 11 β-hydroxylation. J. clin. Endocrin. **18**, 906—912 (1958).

LIEBEGOTT, G.: Studien zur Orthologie und Pathologie der Nebennieren. Beitr. path. Anat. **109**, 93—178 (1944).

— Die Pathologie der Nebennieren. Verh. dtsch. path. Ges. **36**, 21—68 (1952).

LILLIE, R. D.: Histopathologic Technic and Practical Histochemistry, Ed. 2. New York: Blakiston 1954.

LIPSCOMB, H. S., and D. H. NELSON: Measurement of corticosterone in rat adrenal venous plasma as a bioassay for ACTH. Fed. Proc. **18**, 95 (1959).

— — Dynamic changes in ascorbic acid and corticosteroids in adrenal vein blood after ACTH. Endocrinology **66**, 144—146 (1960).

LIPSETT, M. B., C. D. WEST, J. P. MACLEAN and O. H. PEARSON: Adrenal function after hypophysectomy in man. J. clin. Endocrin. **17**, 356—363 (1957).

LISON, L.: La valeur des réactions histochimiques des cétostéroïdes. Acta clin. belg. **6**, 633 to 638 (1951).
— Histochimie et Cytochimie Animales, Ed. 2. Paris: Gauthier-Villars 1953.
LITTLE, J. M., W. M. KELSEY and E. H. YOUNT, JR.: Influence of the adrenal cortex on renal hemodynamics in the dog. Effects of ACTH and adrenal atrophy induced by Rhothane. Amer. J. Physiol. **185**, 159—166 (1956).
LIU, H.-C., and R. B. MANEELY: Some aspects of adrenal histology and histochemistry in the soft shelled turtle of South China (Chelonia amyda). J. Endocrin. **19** 1—9 (1959).
LI VOTI, P.: Influence of methylthiouracil on the suprarenal glands (in Italian). Arch. Sci. med. **87**, 76—84 (1949).
LLAURADO, J. G.: Aldosterone excretion following hypophysectomy in man: Relation to urinary Na/K ratio. Metabolism **6**, 556—563 (1957).
— and M. F. A. WOODRUFF: Postoperative transient aldosteronism. Surgery **42**, 313—324 (1957).
— J. B. TRUNNELL and J. L. CLAUS: Some effects of simultaneous administration of norethandrolone and cortisone in the rat. Acta endocr. (Kbh.) **32**, 536—544 (1959).
LOBBAN, M. C.: Structural variations in the adrenal cortex of the adult cat. J. Physiol. (Lond.) **118**, 565—574 (1952).
LOGAN, J. E., F. C. HEAGY and R. J. ROSSITER: Phosphorus metabolism of the adrenal gland: Effect of hypophysectomy and administration of ACTH on the incorporation of radioactive phosphate into RNA nucleotides. Endocrinology **56**, 455—460 (1955).
LOMBARDO, M. E., and P. B. HUDSON: The biosynthesis of adrenocortical hormones by the human adrenal gland *in vitro*. Endocrinology **65**, 417—425 (1959).
— C. MCMORRIS and P. B. HUDSON: The isolation of steroidal substances from human adrenal vein blood. Endocrinology **65**, 426—432 (1959).
LONGWELL, B. B., A. E. REIF and E. HANSBURY: The effects of pantothenic acid deficiency on the secretion of corticosteroids by the albino rat. Endocrinology **62**, 565—672 (1958).
LOSTROH, A. J.: Synergistic effect of somatotropin with α-corticotropin on the adrenal of hypophysectomized rats. Proc. Soc. exp. Biol. (N. Y.) **97**, 828—832 (1958).
— and C. H. LI: Effect of growth hormone and thyroxine on body weight of hypophysectomized C_3H mice. Endocrinology **62**, 484—492 (1958).
— and P. WOODWARD: Changes in the adrenal of the hypophysectomized C_3H mouse with α-corticotropin and growth hormone. Endocrinology **62**, 498—505 (1958).
LOUSTALOT, P.: Über das Vorkommen von PAS-positiven Tröpchen in der Nebennierenrinde der Ratte. Schweiz. Z. Path. Bact. **20**, 716—723 (1957).
LOWENTHAL, M., H. E. LESZYNSKY, M. MARCUS and H. ZONDEK: Congenital adrenocortical hyperplasia with giant cells in a twelve-year old child. J. Endocrin. **16**, 429—434 (1958).
LÜDERS, C. J.: Zur Frage der pigmentierten Nebennierenrindenadenome. Virchows Arch. path. Anat. **324**, 123—135 (1953).
LUETSCHER, J. A., JR., and R. H. CURTIS: Observations on aldosterone in human urine. Fed. Proc. **14**, 746—751 (1955).
LUFT, J., and O. HECHTER: An electron microscopic correlation of structure with function in the isolated, perfused cow adrenal, preliminary observations. J. biophys. biochem. Cytol. **3**, 615—620 (1957).
LUNDLIN, P. M., and S. HOLMDAHL: Corticotrophin activity of human placenta. Acta endocr. (Kbh.) **26**, 388—394 (1957).
MACCHI, I. A., and O. HECHTER: Studies of ACTH action upon perfused bovine adrenals: Corticosteroid biosynthesis in isolated glands maximally stimulated with ACTH. Endocrinology **55**, 387—402 (1954a).
— — Minimal ACTH concentration requisite for maximal glandular response. Endocrinology **55**, 426—433 (1954b).
MACFARLAND, W. E.: Adrenal cortical function independent of direct nervous action. A neurological study of normal, denervated, and transplanted adrenal glands of albino rats. J. exp. Zool. **95**, 345—360 (1944).
— The vital necessity of adrenal cortical tissue in a mammal and the effects of proliferation of cortical cells from dormant coelomic mesothelium. Anat. Rec. **93**, 233—249 (1945).
— and H. A. DAVENPORT: Adrenal innervation. J. comp. Neurol. **75**, 219—234 (1941).
MACLACHLAN, P. L., H. C. HODGE and R. WHITEHEAD: The lipids of the adrenal gland in normal and fasting rabbits. J. biol. Chem. **139**, 185—191 (1941).
MACLEAN, J. P., M. B. LIPSETT, M. C. LI, C. D. WEST and O. H. PEARSON: Salt metabolism after hypophysectomy in man. J. clin. Endocrin. **17**, 346—355 (1957).
MAES, J.: Étude du syndrome surrénoprivé de la grenouille. Arch. int. Physiol. **45**, 135—188 (1937).
MALATY, H. A., and G. H. BOURNE: The distribution of "simple" esterase in human tissues. Acta anat. **21**, 249—258 (1954).

MALMSTRÖM, B. G., and D. GLICK: Determination of millimicrogram quantities of coenzyme A and its distribution in the rat adrenal. J. biol. Chem. **211**, 677—686 (1954).
MANDEL, L., A. BABER et P. MANDEL: Action de la cortisone sur les acides nucléiques des surrénales du rat blanc. C. R. Soc. Biol. (Paris) **150**, 424—427 (1956).
MANDEL, P., M. JACOB et L. MANDEL: Évolution des acides nucléiques des surrénales au cours du jeûne total et "le phénomène d'agression" (stress). J. Physiol. (Paris) **47**, 236—238 (1955).
MAQSOOD, M.: The effect of varying levels of thyroidal stimulation on adrenals under different environmental temperatures. J. Endocrin. **7**, 82—85 (1950).
— Effects of the thyroid state, season and castration on the adrenal glands in male rabbits. J. Endocrin. **11**, 103—109 (1954).
MARAUD, R., et R. STOLL: Sur l'apparition du dimorphisme sexual dans la surrénale de l'embryon de poulet. C. R. Soc. Biol. (Paris) **150**, 687—689 (1956).
MARKS, B. H., M. ALPERT and F. A. KRUGER: Effect of amphenone upon steroidogenesis in the adrenal cortex of the golden hamster. Endocrinology **63**, 75—81 (1958).
MARKS, V.: Cushing's syndrome occurring with pituitary chromophobe tumours. Acta endocr. (Kbh.) **32**, 527—537 (1959).
MARTINOVITCH, P. N.: Infantile rat adrenal transplanted into the anterior eye chamber of adrenalectomized hosts after cultivation in vitro. J. exp. Zool. **129**, 99—127 (1955).
MASON, J. W., J. V. BRADY and M. SIDMAN: Plasma 17-hydroxycorticosteroid levels and conditioned behavior in the rhesus monkey. Endocrinology **60**, 741—752 (1957).
MASONI, A.: Catheterisation of the right adrenal vein in man. Acta med. scand. **159**, 225—234 (1957).
MASSON, G. M. C., S. B. KORITZ and F. G. PERON: Corticosteroid formation in regenerating rat adrenals. Endocrinology **62**, 229—233 (1958).
MATERNA, A.: Neue Untersuchungen über das Gewicht der Nebennieren. Beitr. path. Anat. **106**, 158—177 (1941).
McDONALD, I. R., and M. REICH: Corticosteroid secretion by the autotransplanted adrenal gland of the conscious sheep. J. Physiol. (Lond.) **147**, 33—50 (1959).
McKEEVER, S.: Effects of reproductive activity on the weight of the adrenal glands in Microtus montanus. Anat. Rec. **135**, 1—6 (1959).
McKERNS, K. W.: The effect of some natural and synthetic sex hormones on adrenal cortical secretion *in vitro*. Endocrinology **60**, 130—135 (1957).
— and E. NORDSTRAND: The *in vitro* adrenal response to corticotrophin subtypes. Canad. J. Biochem. **33**, 681—686 (1955).
McPHAIL, M. K., and H. C. READ: The mouse adrenal. I. Development, degeneration and regeneration of the X-zone. Anat. Rec. **84**, 51—73 (1942).
McQUEENEY, A. J., L. L. ASHBURN, F. S. DAFT and R. FAULKNER: Tissue changes and sodium balance in pantothenic acid-deficient rats. Endocrinology **41**, 441—450 (1947).
McSHAN, W. H., W. F. ERWAY and R. K. MEYER: Succinic dehydrogenase activity of certain endocrine tissues of the rat. Arch. Biochem. **9**, 69—74 (1946).
MEDAWAR, P. B., and P. S. RUSSELL: Adrenal homografts in mice, with special reference to "immunological" adrenalectomy. Immunology **1**, 1—12 (1958).
MELAMPY, R. M., D. W. CHANG and L. C. NORTHRUP: Effect of pantothenic acid deficiency upon adrenal cortex, thymus, spleen, and circulating lymphocytes in mice. Proc. Soc. exp. Biol. (N. Y.) **76**, 24—27 (1951).
MERRIAM, R. W.: A critique of karyometric methods. Z. wiss. Mikr. **62**, 401—411 (1955).
MEYER, R.: Nebennieren bei Anenzephalie. Virchows Arch. path. Anat. **210**, 158—164 (1912).
MEYERS, M. W., and H. A. CHARIPPER: A histological and cytological study of the adrenal gland of the golden hamster *(Cricetus auratus)* in relation to age. Anat. Rec. **124**, 1—25 (1956).
MIALHE-VOLOSS, C., et E.-E. BAULIEU: Étude qualitative des corticostéroïdes liberés par la surrénale du rat in vitro sous l'influence des hormones corticotropes antehypophysaires. C. R. Acad. Sci. (Paris) **246**, 639—642 (1958).
MILKOVIĆ, K., and S. MILKOVIĆ: The reactivity of the rat fetal pituitary-adrenal system during the last days of pregnancy. Arch. int. Physiol. **66**, 534—539 (1958).
MILLER, M. R.: The normal histology and experimental alteration of the adrenal of the viviparous lizard, Xantusia vigilis. Anat. Rec. **113**, 309—324 (1952).
— Experimental alteration of the adrenal histology of the urodele amphibian, Triturus torosus. Anat. Rec. **116**, 205—226 (1953).
— Cyclic changes in the thyroid and interrenal glands of the viviparous lizard, Xantusia vigilis. Anat. Rec. **123**, 19—32 (1955).
— and M. E. ROBBINS: Cyclic changes in the thyroid and interrenal glands of the urodele amphibian, Taricha torosa (Triturus torosus). Anat. Rec. **122**, 79—104 (1955).

MILLER, R. A.: Cytological phenomena associated with experimental alteration of secretory activity in the adrenal cortex of mice. Amer. J. Anat. **86**, 405—437 (1950).
— The relation of mitochondria to secretory activity in the fascicular zone of the rat's adrenal. Amer. J. Anat. **92**, 329—360 (1953).
— Quantitative changes in the nucleolus and nucleus as indices of adrenal cortical secretory activity. Amer. J. Anat. **95**, 497—522 (1954).
— and O. RIDDLE: Stimulation of adrenal cortex of pigeons by ant. pituitary hormones and by their secondary products. Proc. Soc. exp. Biol. (N. Y.) **41**, 518—522 (1939).
— — Cellular response to insulin in suprarenals of pigeons. Proc. Soc. exp. Biol. (N. Y.) **47**, 449—453 (1941).
— — The cytology of the adrenal cortex of normal pigeons and in experimentally induced atrophy and hypertrophy. Amer. J. Anat. **71**, 311—341 (1942).
— — Effects of prolactin and cortical hormones on body weight and food intake of adrenalectomized pigeons. Proc. Soc. exp. Biol. (N. Y.) **52**, 231—233 (1943).
MILLS, I. H., A. CASPER and F. C. BARTTER: On the role of the vagus in the control of aldosterone secretion. Science **128**, 1140—1141 (1958).
MINDER, W. H., u. I. ABELIN: Über die hormonale und nicht hormonale Beeinflussung der Acetalphosphatide (Plasmalogene) der Rattenorgane. Hoppe-Seylers Z. physiol. Chem. **298**, 121—131 (1954).
MITCHELL, R. M.: Histological changes and mitotic activity in the rat adrenal during postnatal development. Anat. Rec. **101**, 161—186 (1948).
MIXNER, J. P., A. J. BERGMAN and C. W. TURNER: Relation of certain endocrine glands to body weight in growing and mature guinea pigs. Endocrinology **32**, 298—304 (1943).
MOERI, E.: Les surrénales chez le foetus, le nouveau-né, le nourrisson et l'enfant. Acta endocr. (Kbh.) **8**, 259—311 (1951).
MONEY, W. L.: The interrelation of the thyroid and the adrenals. Brookhaven Symp. Biol. **7**, 137—168 (1955).
MONTANDON, A.: Quantitative und qualitative Zellveränderungen im Hypophysenvorderlappen bei therapeutischen Hypercorticismus. Virchows Arch. path. Anat. **330**, 629—650 (1957).
MOOG, F.: The development of function in the adrenal cortex. In: A. GORBMAN, Ed., Comparative Endocrinology, 624—638. New York: Wiley 1959.
— C. J. BENNETT and C. M. DEAN, JR.: Growth and cytochemistry of the adrenal gland of the mouse from birth to maturity. Anat. Rec. **120**, 873—891 (1954).
MOON, H. D.: Effect of adrenocorticotropic hormone in 4-day-old rats. Proc. Soc. exp. Biol. (N. Y.) **43**, 42—44 (1940).
MOORHOUSE, D. E.: Experimental alteration of the peri-renal tissue of *Protopterus*. Quart. J. micr. Sci. **97**, 519—534 (1956).
MORRIS, M. D., and I. L. CHAIKOFF: The origin of cholesterol in liver, small intestine, adrenal gland, and testis of the rat: Dietary *versus* endogenous contributions. J. biol. Chem. **234**, 1095—1097 (1959).
MOSIER, H. D.: Hypoplasia of the pituitary and adrenal cortex. J. Pediat. **48**, 633—639 (1956).
— Comparative histological study of the adrenal cortex of the wild and domesticated wild rat. Endocrinology **60**, 460—469 (1957).
— and C. P. RICHTER: Response of the glomerulosa layer of the adrenal gland of wild and domesticated Norway rats to low and high salt diets. Endocrinology **62**, 268—277 (1958).
MOSIMANN, W.: Vergrößerung der Kernvolumina in der Nebennierenrinde von Ziegen nach intensiver Behandlung mit Stilben-Derivaten. Z. mikr.-anat. Forsch. **61**, 93—98 (1954).
MÜHLEMANN, H. R., T. M. MARTHALER u. K. H. RATEISCHAK: Mitoseperiodik in der Nebennierenrinde, Schilddrüse, im Duodenal- und Mundhöhlenepithel der Ratte. Acta anat. **28**, 331—341 (1956).
MUKERJEE, A. K., and S. BANERJEE: Studies on histological changes in experimental scurvy. Anat. Rec. **120**, 907—915 (1954).
MULAY, A. S., and W. H. EYESTONE: Transplantable adrenocortical adenocarcinomas in Osborn-Mendel rats fed a carcinogenic diet. J. nat. Cancer Inst. **16**, 723—739 (1955).
MULINOS, M. G., and L. POMERANTZ: Pseudo-hypophysectomy. A condition resembling hypophysectomy produced by malnutrition. J. Nutr. **19**, 493—504 (1940).
— — Hormonal influences on the weight of the adrenal in inanition. Amer. J. Physiol. **132**, 368—374 (1941).
NABARRO, J. D. N., A. MOXHAM and G. WALKER: Stimulation and suppression of the adrenal cortex in Cushing's syndrome. J. clin. Endocrin. **18**, 586—599 (1958).
NABORS, C. J., JR., and D. L. BERLINER: A C-20 β-reductase in the pancreas and a C-20-dehydrogenase in the adrenal. Arch. Biochem. **70**, 298—300 (1957).
NACHLAS, M. M., and A. M. SELIGMAN: The comparative distribution of esterase in the tissues of five mammals by a histochemical technique. Anat. Rec. **105**, 677—695 (1949).

NACHLAS, M. M., K.-C. TSOU, E. DE SOUZA, C.-S. CHENG and A. M. SELIGMAN: Cytochemical demonstration of succinic dehydrogenase by the use of a new p-nitrophenyl substituted ditetrazole. J. Histochem. Cytochem. **5**, 420—436 (1957).
NADEL, E. M., and J. J. SCHNEIDER: Excretion of formaldehydogenic (FG) substances by normal and scorbutic guinea pigs. Endocrinology **51**, 5—11 (1952).
— B. G. YOUNG, A. C. HILGAR and S. BURSTEIN: Urinary corticosteroids in normal, scorbutic and stilbestrol-treated guinea pigs before and after administration of corticotrophin. Acta endocr. (Kbh.) **28**, 283—292 (1958).
NALBANDOV, A. V., and L. E. CARD: Effect of hypophysectomy of growing chicks. J. exp. Zool. **94**, 387—413 (1943).
NANDI, J., and H. A. BERN: Corticosteroid production by interrenal tissue in teleost fishes. Endocrinology **66**, 295—303 (1960).
NAYYAR, S. N., and D. GLICK: The quantitative histochemical distribution of β-glucuronidase in the adrenal of various species and the influence of stress conditions, hormonal treatments and hypophysectomy. Endocrinology **56**, 67—79 (1955).
NELSON, A. A., and G. WOODWARD: Severe adrenal cortical atrophy (cytotoxic) and hepatic damage produced in dogs by feeding 2,2 bis(parachlorophenyl)-1,1-dichloroethane (DDD or TDE). A.M.A. Arch. Path. **48**, 387—394 (1949).
NELSON, D. H.: The significance of plasma corticosteroids as a measure of adrenal cortical function. In: L. I. GARDNER, Ed., Adrenal Function in Infants and Children, 171—183. New York: Grune and Stratton 1956.
— and D. M. HUME: Corticosteroid secretion in the adrenal venous blood of the hypophysectomized dog as an assay for ACTH. Endocrinology **57**, 184—192 (1955).
— R. H. EGDAHL and D. M. HUME: Corticosteroid secretion in the adrenal vein of the non-stressed dog exposed to cold. Endocrinology **58**, 309—314 (1956).
NEUMANN, K., u. G. KOCH: Übersicht über die feinere Verteilung der Succino-Dehydrogenase in Organen und Geweben verschiedener Säugetiere, besonders des Hundes. Hoppe-Seylers Z. physiol. Chem. **295**, 35—61 (1953).
NEWMAN, A. E., E. S. REDGATE and G. FARRELL: The effects of diencephalic-mesencephalic lesions on aldosterone and hydrocortisone secretion. Endocrinology **63**, 723—736 (1958).
NICANDER, L.: The plasmal reaction of Feulgen and Voit with special reference to the adrenal body. Acta anat. **12**, 174—197 (1951).
— Histological and histochemical studies on the adrenal cortex of domestic and laboratory animals. Acta anat. **14**, Suppl. **16** (1952).
— A histochemical study of adrenal glycogen. Acta anat. **31**, 388—397 (1957).
NICHOLLS, D.: Effect of ascorbic acid on the phosphorus metabolism of the adrenal gland. Canad. J. Biochem. **34**, 919—926 (1956).
— and R. J. ROSSITER: Effect of cold stress on the phosphorus metabolism of the adrenal gland. Canad. J. Biochem. **33**, 233—247 (1955a).
— — The role of the pituitary and thyroid glands in the phosphorus metabolism of the adrenal gland during cold stress. Endocrinology **56**, 547—559 (1955b).
— — Phosphorus metabolism of the adrenal gland of the rat acclimatized to cold. Amer. J. Physiol. **187**, 11—14 (1956).
— F. C. HEAGY and R. J. ROSSITER: Phosphorus metabolism of the adrenal gland of the rat. Effect of exposure to a cold environment for eight days on the amounts and P^{32} labelling of phospholipid and ribonucleic acid. Canad. J. Biochem. **34**, 543—553 (1956a).
— — — Effect of cold stress on the labelling of phospholipid and RNA nucleotides with radioactive inorganic phosphate. Endocrinology **58**, 587—597 (1956b).
NICHOLS, J.: Effects of electrolyte imbalance on the adrenal glands. A.M.A. Arch. Path. **45**, 717—721 (1948).
— The effects of deprivation of water on the adrenal glands of rats. Amer. J. Physiol. **25**, 301—307 (1949).
— Effects of captivity on adrenal gland of wild Norway rat. Amer. J. Physiol. **162**, 5—9 (1950).
— Observations on the adrenal of the premature anencephalic fetus. A.M.A. Arch. Path. **62**, 312—317 (1956a).
— Studies of the adrenal glands of patients with low plasma sodium. A.M.A. Arch. Path. **62**, 419—424 (1956b)
— and L. I. GARDNER: Production of insulin sensitivity with the adrenocorticolytic drug DDD (2,2-bis(parachlorophenyl)-1,1-dichloroethane). J. Lab. clin. Med. **37**, 229—238 (1951).
— and H. D. GREEN: Effect of DDD treatment on metabolic response to ACTH injection. Amer. J. Physiol. **176**, 374—376 (1954).
— and G. HENNIGAR: Studies on DDD, 2,2-bis(parachlorophenyl)-1,1-dichloroethane. Exp. Med. Surg. **15**, 310—316 (1957a).
— — Studies on the adrenal during simultaneous administration of DDD and Amphenone. Endocrinology **61**, 226—234 (1957b).

— and H. L. Sheehan: Effect of partial adrenal cortical atrophy on the course of alloxan diabetes. Endocrinology **51**, 362—377 (1952).
— O. L. Lescure and C. J. Migeon: Levels of 17-OH-CS in maternal and cord plasma in anencephaly. J. clin. Endocrin. **18**, 444—452 (1958).
Nigeon-Dureuil, M., et A. R. Ratsimamanga: Preuves biologiques de la déviation du métabolisme du système enzymatique de las urrénale scorbutique. C. R. Soc. Biol. (Paris) **149**, 21—24 (1955).
Noble, N. L., and E. Papageorge: Adrenal glycogen in the guinea pig and the white rat. J. Nutr. **56**, 15—24 (1955a).
— — Loss of adrenal glycogen in the rat following stress or treatment with various hormones. Endocrinology **57**, 492—497 (1955b).
Noumura, T.: Development of the hypophyseal-adrenocortical system in the rat embryo in relation to the maternal system. Jap. J. Zool. **12**, 279—299 (1959).
Novelli, G.: Metabolic functions of pantothenic acid. Physiol. Rev. **33**, 525—543 (1953).
Novikoff, A. B.: The intracellular localization of chemical constituents. In: R. C. Mellors, Ed., Analytical Cytology, Ed. 2, 69—168. New York: McGraw-Hill 1959.
Nowell, N. W.: Studies in the activation and inhibition of adrenocorticotrophin secretion. Endocrinology **64**, 191—201 (1959).
O'Donnell, W. M., S. S. Fajans and J. G. Weinbaum: Human adrenal cortex after administration of ACTH and cortisone. Arch. intern. Med. **88**, 28—35 (1951).
Oertel, G. W., and K. B. Eik-Nes: Plasma levels of dehydroepiandrosterone in the dog. Endocrinology **65**, 766—776 (1959).
Oleson, M. C., and W. R. Bloor: The adrenal lipids of fasted guinea pigs. J. biol. Chem. **141**, 349—354 (1941).
Olivereau, M., et H. Fromentin: Influence de l'hypophysectomie sur l'histologie de l'interrénal antérieur de l'anguille (*Anguilla anguilla* L.). Ann. Endocrin. (Paris) **15**, 805—826 (1954).
Olsen, N. S., and W. E. Martindale: Studies on chronic vitamin B_6 deficiency in the rat. II. J. Nutr. **53**, 329—340 (1954).
Olson, R. E., and H. W. Deane: A physiological and cytochemical study of the kidney and adrenal cortex during choline deficiency in weanling rats. J. Nutr. **39**, 31—56 (1949).
Osorio, J. A., and A. Kraemer: Influence of atrophy of the adrenal cortex produced by Rhothane (DDD) on the submaxillary gland of the dog (in Portuguese). Acta physiol. lat.-amer. **9**, 128—132 (1959).
Overbeeke, A. P. van: A preliminary note on the effect of Amphenone "B" in *Betta splendens* Regan. Proc. Acad. Sci. (Amst.) C **61**, 78—88 (1958).
Padykula, H. A.: The localization of succinic dehydrogenase in tissue sections of the rat. Amer. J. Anat. **91**, 107—146 (1952).
Pankratz, D. S.: The development of the suprarenal gland in the albino rat with a consideration of its possible relation to the origin of foetal movements. Anat. Rec. **49**, 31—49 (1931).
Parkes, A. S.: Viability of adrenocortical tissue transplanted after freezing and thawing. Proc. roy. Soc. (Lond.) **144** B, 314—328 (1956).
Pasqualini, R. Q., C. D. de Pasqualini and J. C. Garberi: Determination of ascorbic acid in the suprarenal gland of the toad (in Spanish). Rev. Soc. argent. Biol. **25**, 255—262 (1949).
Patzelt, V.: Über die chromotropen Zellen der Nebenniere vom Wasserfrosch. Z. Zellforsch. **41**, 460—473 (1955).
Pauly, J. E.: Morphological observations on the adrenal cortex of the laboratory rat. Endocrinology **60**, 247—264 (1957).
Pavoa, H., jr., and P. Fontana, jr.: Influence of ACTH on ribonucleic acid and protein nitrogen content of rat adrenal gland. Acta physiol. lat.-amer. **7**, 86—89 (1957).
Payne, F.: Cytological changes in the cells of the pituitary, thyroids, adrenals and sex glands of ageing fowl. In: A. I. Lansing, Ed., Cowdry's Problems of Ageing, Ed. 3, 381—402. Baltimore: Williams and Wilkins 1952.
Peace, R. F.: Myelonecrosis, extramedullary myelopoiesis, and leuko-erythroblastosis. A mesenchymal reaction to injury. Amer. J. Path. **29**, 1029—1057 (1953).
Pearse, A. G. E.: Histochemistry, Theoretical and Applied. Boston: Little, Brown 1953.
Pearson, B., and V. Defendi: A comparison between the histochemical demonstration of non-specific esterase activity by 5-bromoindoxyl acetate, α-naphthyl acetate and naphthol AS acetate. J. Histochem. Cytochem. **5**, 72—83 (1957).
Pellegrino, C.: The regulation of lymphatic tissue after enucleation of the adrenals (in Italian). Folia endocr. (Pisa) **3**, 689—694 (1950).
— Some histological and histochemical aspects of the adrenal regenerating after enucleation (in Italian). Atti (P. V.) Soc. tosc. Sci. nat. (Pisa) **58** B, 5—14 (1951).
— and F. Caramazza: Influence of fasting on blood sugar in the rat at various periods after enucleation (in Italian). Riv. Biol. (Perugia) **47**, 267—273 (1955).

PELLEGRINO, C., and A. MONTELLA: Cholesterol and ascorbic acid in the regenerating adrenal (in Italian). Arch. Sci. biol. (Napoli) **38**, 338—348 (1954).
— and A. TORCIGLIANI: A cytochemical study of the adrenal cortex regenerating after enucleation. J. Physiol. (Lond.) **135**, 536—549 (1957).
PENHOS, J. C., and A. F. CARDEZA: Endocrine glands of the fed, hypophysectomized toad (in Spanish). Rev. Soc. argent. Biol. **27**, 173—177 (1951).
PENNACCHIETTI, M.: The particular time of organogenesis of the adrenal gland of mammals (in Italian). Monit. zool. ital. **49**, 39—49 (1938).
PEREK, M., and A. EILAT: Effect of removal of bursa Fabricii on depletion of adrenal ascorbic acid in ACTH-treated chicks. Endocrinology **66**, 304—305 (1960).
— B. ECKSTEIN and Z. ESHKOL: The effect of ACTH on adrenal ascorbic acid in laying hens. Endocrinology **64**, 831—832 (1959).
PERON, F. G., and S. B. KORIZ: On the exogenous requirements for the action of ACTH in vitro on rat adrenal glands. J. biol. Chem. **233**, 256—259 (1958).
PERRY, W. F., and G. R. CUMMING: Adrenal succinic dehydrogenase activity determined by the reduction of tetrazolium salt by adrenal homogenate. Endocrinology **50**, 385—387 (1952).
PESCHEL, E., and G. J. RACE: Studies on the adrenal zona glomerulosa of hypertensive patients and rats. Amer. J. Med. **17**, 355—364 (1954).
PHILLIPS, J. G., and P. J. MULROW: Corticosteroid production in vitro by the interrenal tissue of killifish, *Fundulus heteroclitus* (Linn.). Proc. Soc. exp. Biol. (N. Y.) **101**, 262—264 (1959).
— W. N. HOLMES and P. K. BONDY: Adrenocorticosteroids in salmon plasma *(Oncorhynchus nerka)*. Endocrinology **65**, 811—818 (1959).
PICKFORD, G. E.: The adrenal and corticotrophin. In: G. E. PICKFORD and J. W. ATZ, Eds., The Physiology of the Pituitary Gland of Fishes, 60—83. New York: N. Y. Zool. Soc. 1957.
— and B. KOSTO: Hormonal induction of melanogenesis in hypophysectomized killifish *(Fundulus heteroclitus)*. Endocrinology **61**, 176—196 (1957).
PICKFORD, M., and M. VOGT: The effect of adrenaline on secretion of cortical hormone in the hypophysectomized dog. J. Physiol. (Lond.) **112**, 133—141 (1951).
PICON, L.-O.: Influence de l'hyperfonctionnement des surrénales maternelles sur le développement des surrénales foetales chez le rat. C. R. Soc. Biol. (Paris) **151**, 1104—1106 (1957).
PINCUS, G.: Aging and urinary steroid secretion. In: E. T. ENGLE and G. PINCUS, Eds., Hormones and the Aging Process, 1—20. New York: Academic Press 1956.
PINTO, R. M.: Action of the ovary on the adrenal, experimental study (in Spanish). Buenos Aires: Ferrari 1941.
— Interrelations of adrenal and sex glands in parabiotic rats. Amer. J. Physiol. **144**, 652—657 (1945a).
— Direct and indirect action of estrogens on the adrenal glands (in Spanish). Rev. Soc. argent. Biol. **21**, 136—143 (1945b).
PISCHINGER, A.: Über den Einfluß der histologischen Technik auf die Azetalphosphatide in den Geweben. Z. mikr.-anat. Forsch. **52**, 530—551 (1942).
PITCOCK, J. A., and P. M. HARTROFT: The juxtaglomerular cells in man and their relationship to the level of plasma sodium and to the zona glomerulosa of the adrenal cortex. Amer. J. Path. **34**, 863—883 (1958).
PLANEL, H., et A. GUILHEM: Étude histochimique du glycogène du cortex surrénal. C. R. Soc. Biol. (Paris) **150**, 1274—1276 (1956a).
— — Le glycogène du cortex surrénal: Étude histologique, variations expérimentales et rapport avec la colloid surrénalienne. Ann. Histochim. **1**, 237—248 (1956b).
— — Recherches expérimentales sur le dimorphisme sexuel de la glande surrénale du cobaye. C. R. Soc. Biol. (Paris) **151**, 1256—1260 (1957a).
— — Action de l'ACTH sur le glycogène du cortex surrénal. C. R. Soc. Biol. (Paris) **152**, 165—168 (1957b).
— — et P. BEC: Variations expérimentales des pigments du cortex surrénal. Action de l'A.C.T.H. Ann. Histochim. **1**, 68—77 (1956).
— — — Phosphatase alcaline du cortex surrénal du cobaye. Action de la castration et des hormones génitales. C. R. Soc. Biol. (Paris) **151**, 784—787 (1957).
— — et P. BANIDE: Action de l'inanition sur le cortex surrénal du cobaye. Variations de la pigmentation. C. R. Soc. Biol. (Paris) **152**, 853—855 (1958).
POLITZER, G.: Über die Frühentwicklung der Nebennierenrinde beim Menschen. Z. Anat. Entwickl.-Gesch. **106**, 40—48 (1937).
PORTO, J.: Relations between the hypophysis and the adrenal in the toad (in Spanish). Rev. Soc. argent. Biol. **16**, 489—494 (1940).
POULTON, B. R., and R. P. REESE: The activity of the pituitary-adrenal cortex axis during pregnancy and lactation. Endocrinology **61**, 217—225 (1957).

PRADER, A., u. R. E. SIEBENMANN: Nebenniereninsuffizienz bei kongenitaler Lipoidhyperplasie der Nebennieren. Helv. paediat. Acta **12**, 569—595 (1957).
PRICE, D., and D. J. INGLE: Androgenic effects of autotransplants of adrenals in the accessory reproductive glands of adult castrated rats. Rev. suisse Zool. **64**, 743—755 (1957).
PRUNTY, F. T. G., B. E. CLAYTON, R. R. MCSWINEY and I. H. MILLS: Studies of the interrelationship between the adrenal cortex and ascorbic acid metabolism. Ciba Foundn. Coll. Endocrin. 8, 324—339 (1955).
QUENUM, A., et R. CAMAIN: Étude histochimique du cortex surrénal de *Cricetomys gambianus*. C. R. Soc. Biol. (Paris) **153**, 697—701 (1959).
RABINOVICI, N.: The effect of stress upon the ketosteroid and non-carbonyl lipoid content of the adrenal cortex of the rat. Endocrinology **49**, 579—588 (1951).
RACE, G. J., and R. F. GREEN: Studies on zonation and regeneration of the adrenal cortex of the rat. A.M.A. Arch. Path. **59**, 578—586 (1955).
— W. M. NICKEY, P. S. WOLF and E. J. JORDAN: Studies on functional zonation of the adrenal cortex. A.M.A. Arch. Path. **64**, 312—323 (1957).
RAPELA, C. E., et M. F. GORDON: Action de la testostérone sur la surrénale du crapaud. C. R. Soc. Biol. (Paris) **150**, 1617—1618 (1956).
— — Effect of insulin on the adrenal gland of the toad. Acta physiol. lat.-amer. **7**, 58—67 (1957).
RAPPORT, M. M., and R. E. FRANZL: The structure of plasmalogen. III. J. Neurochem. **1**, 303—310 (1957).
RASQUIN, P.: Effects of carp pituitary and mammalian ACTH on the endocrine and lymphoid systems of the teleost Astyanax mexicanus. J. exp. Zool. **117**, 317—357 (1951).
RATHER, L. J.: Differences of cortical/medullary ratio in the adrenals of normal rats, normal rats treated with ACTH, and renal hypertensive rats. Proc. Soc. exp. Biol. (N. Y.) **77**, 651—653 (1951).
RAUSCHKOLB, E. W., and G. L. FARRELL: Evidence for diencephalic regulation of aldosterone secretion. Endocrinology **59**, 526—531 (1956).
— — and S. KOLETSKY: Aldosterone secretion after hypophysectomy. Amer. J. Physiol. **184**, 55—58 (1956).
RAYNAUD, A., et M. FRILLEY: Développement intra-utérin des embryons de souris dont les ébauches de l'hypophyse on été détruites, au moyen des rayons X, au 13e jour de la gestation. II. Développement des capsules surrénales. C. R. Acad. Sci. (Paris) **230**, 331—333 (1950).
REESE, J. D., and H. D. MOON: The Golgi apparatus of the cells of the adrenal cortex after hypophysectomy and on the administration of the adrenocorticotropic hormone. Anat. Rec. **70**, 543—556 (1938).
REICHSTEIN, T.: Die wichtigsten Hormone der Nebennierenrinde. Acta endocr. (Kbh.) **17**, 375—384 (1954).
REIF, A. E., and B. B. LONGWELL: Extraction of corticosteroids from adrenal vein blood of rats and rabbits. Endocrinology **62**, 573—586 (1958).
REINER, C. B.: A critical examination of the histochemical staining technique for ascorbic acid. Proc. Soc. exp. Biol. (N. Y.) **80**, 455—457 (1952).
— The Schultz histochemical reaction for cholesterol. Lab. Invest. **2**, 140—151 (1953).
RENNELS, E. G.: An experimental study of cytoplasmic inclusions in adrenal cortical cells of the immature rat. Anat. Rec. **112**, 509—527 (1952).
— M. HESS and J. C. FINERTY: Response of preputial and adrenal glands of the rat to sex hormones. Proc. Soc. exp. Biol. (N. Y.) **82**, 304—309 (1953).
RENOLD, A. E., J. CRABBÉ, L. HERMANDO-AVENDANO, D. H. NELSON, E. J. ROSS, K. EMERSON, JR., and G. W. THORN: Inhibition of aldosterone secretion by amphenone in man. New Engl. J. Med. **256**, 16—21 (1957).
RETZLAFF, E. W.: The histology of the adrenal gland in the alligator lizard, Gerrhontus multicarinatus. Anat. Rec. **105**, 19—33 (1949).
RIBAUDO, C. A.: Addison's disease and diabetes insipidus. A. M. A. Arch. int. Med. **102**, 478—483 (1958).
RICH, A. R.: A peculiar type of adrenal cortical damage associated with acute infections, and its possible relation to circulatory collapse. Johns Hopk. Hosp. Bull. **74**, 1—15 (1944).
— and M. BERTHRONG: Evidence for the presence of ribonucleic acid in the cytoplasmic bodies that appear in the hepatic and adrenal epithelial cells of man in acute infection. Johns Hopk. Hosp. Bull. **85**, 327—343 (1949).
RICHARDS, J. B., and R. L. PRUITT: Hydrocortisone suppression of stress-induced adrenal 17-hydroxycorticosteroid secretion in dogs. Endocrinology **60**, 99—104 (1957).
RICHTER, W. R.: Adrenal cortical adenomata in the goat. Amer. J. vet. Res. **19**, 895—901 (1958).

RIEDEL, B. E., J. E. LOGAN and R. J. ROSSITER: Phosphorus metabolism of the adrenal gland: Effect of hypophysectomy and administration of ACTH on the incorporation of radioactive phosphate into phospholipid. Endocrinology **55**, 219—227 (1954).
RINALDI, L. M.: Effect of malnutrition as compared with hypophysectomy on organ weight of the albino rat. J. Anat. (Lond.) **84**, 262—271 (1950).
RINFRET, A. P., and S. HANE: Depletion of adrenal ascorbic acid following stress in the infant rat. Endocrinology **57**, 497—499 (1955).
RITTERSON, A. L., and S. I. MAUER: Adrenal lipid response in Chinese hamsters infected with Trichinella spiralis. Science **126**, 1293—1294 (1957).
ROAF, R.: A study of the adrenal cortex of the rabbit. J. Anat. (Lond.) **70**, 126—134 (1935).
ROBERTS, S.: Corticoid-releasing activity in blood. Ciba Foundn. Coll. Endocr. **11**, 167—190 (1957).
ROBERTSON, O. H., and B. C. WEXLER: Pituitary degeneration and adrenal tissue hyperplasia in spawning Pacific salmon. Science **125**, 1295—1296 (1957).
— — Hyperplasia of the adrenal cortical tissue in Pacific salmon (genus *Oncorhynchus*) and rainbow trout (*Salmo gairdnerii*) accompanying sexual maturation and spawning. Endocrinology **65**, 225—238 (1959).
ROBINSON, F. B., and J. M. YOFFEY: Histochemical changes produced by cold and adrenaline in the suprarenal cortex of the adult male rat. J. Anat. (Lond.) **84**, 32—37 (1950).
ROBINSON, H. J., W. G. BERNHARD, H. BRUGIN, H. WANNER, G. W. SEWEKOW and R. H. SILBER: 17,21-Dihydroxy-20-ketosteroids in plasma during and after pregnancy. J. clin. Endocrin. **15**, 317—323 (1955).
ROCHE, J., R. MICHEL et P. JOUAN: Action de l'acide 3:5:3'-triiodothyroacétique et de l'ACTH sur la sécrétion des corticostéroïdes par les capsules surrénales du rat *in vitro*. C. R. Soc. Biol. (Paris) **153**, 255—258 (1959).
ROE, J. H., and C. A. KUETHER: The determination of ascorbic acid in whole blood and urine through the 2,4-dinitrophenylhydrazine derivative of dehydroascorbic acid. J. biol. Chem. **147**, 399—407 (1943).
ROEPKE, M.-L.: Das Verhalten der histochemisch nachweisbaren Askorbinsäure in der Nebennierenrinde von Meerschweinchen nach Formalin-Stress. Z. mikr.-anat. Forsch. **58**, 404—428 (1952).
ROGERS, W. F., JR., and R. H. WILLIAMS: Correlations of biochemical and histologic changes in the adrenal cortex. A.M.A. Arch. Path. **44**, 126—137 (1947).
— — Cholesterol of the human adrenal gland. A.M.A. Arch. Path. **46**, 451—466 (1948).
ROGOFF, J. M.: The suprarenal bodies. In: E. V. COWDRY, Ed., Special Cytology, Ed. 2, Vol. II, 871—931. New York: Hoeber 1932.
— Experimental pathology and physiology of the adrenal cortex. A.M.A. Arch. Path. **38**, 392—409 (1944).
ROMANOFF, E. B., P. HUDSON and G. PINCUS: Isolation of hydrocortisone and corticosterone from human adrenal venous blood. J. clin. Endocrin. **13**, 1546—1548 (1953).
ROMER, A. S.: Vertebrate Paleontology, Ed. 2. Chicago: U. Chicago Press 1945.
ROSENFELD, G.: Steroidogenesis by intact calf adrenals perfused *in vitro*. Endocrinology **56**, 649—666 (1955).
— Metabolic studies of intact perfused calf adrenals using tetrazolium. Arch. Biochem. **62**, 125—140 (1956a).
— Anaerobic studies of steroidogenesis using perfused calf adrenals. Proc. Soc. exp. Biol. (N. Y.) **92**, 66—70 (1956b).
— Functional studies of the calf adrenal cortex following total-body exposure to a single lethal dose of γ-radiation. Amer. J. Physiol. **192**, 232—240 (1958a).
— The functional capacity of the adrenal cortex immediately before radiation sickness death. J. Lab. clin. Med. **51**, 198—210 (1958b).
— and W. D. BASCOM: The inhibition of steroidogenesis by Amphenone B: Studies *in vitro* with the perfused calf adrenal. J. biol. Chem. **222**, 565—580 (1956).
— E. ROSEMBERG, F. UNGAR and R. I. DORFMAN: Regulation of the secretion of aldosterone-like material. Endocrinology **58**, 255—261 (1956).
ROSNAGLE, R. S., and G. L. FARRELL: Alterations in electrolyte intake and adrenal steroid secretion. Amer. J. Physiol. **187**, 7—10 (1956).
ROSSI, C. A., and L. MONTANARI: Influence of esterification on the ACTH-like action of methylandrostenediol (in Italian). Arch. Sci. biol. (Bologna) **40**, 98—103 (1956).
ROSSI, F., G. PESCETTO and E. REALE: Enzymatic activities in human ontogenesis: Fir synoptic tables of histochemical research. J. Histochem. Cytochem. **5**, 221—235 (1957).
ROSSITER, R. J., and D. NICHOLLS: Phosphorus metabolism of the adrenal gland of rats exposed to a cold environment. Rev. canad. Biol. **16**, 249—268 (1957).
ROTTER, W.: Die Entwicklung der fetalen und kindlichen Nebennierenrinde. Virchows Arch. path. Anat. **316**, 590—618 (1949a).

ROTTER, W.: Das Wachstum der fetalen und kindlichen Nebennierenrinde. Z. Zellforsch. **34**, 547—561 (1949b).

ROY, S. N., J. N. KARKUN and B. MUKERJI: Effect of prolonged administration of high dose of testosterone propionate on the adrenal cortex of young male rats. Arch. int. Pharmacodyn. **116**, 402—409 (1958).

ROYCE, P. C., and G. SAYERS: Extrapituitary interaction between Pitressin and ACTH. Proc. Soc. exp. Biol. (N. Y.) **98**, 70—74 (1958a).

— — Blood ACTH: Effects of ether, pentobarbital, epinephrine and pain. Endocrinology **63**, 794—800 (1958b).

RUBIN, B. L., and R. I. DORFMAN: Quantitative assay of the 3 β-ol-dehydrogenase system in whole adrenal homogenate. Endocrinology **61**, 601—610 (1957).

RUDOLPH, G. G., and N. S. OLSEN: Effect of hypophysectomy on dehydrogenase activity of rat tissues. Biochim. biophys. Acta **19**, 382—383 (1956).

RUDY, H.: Vitamine. In: B. FLASCHENTRÄGER u. E. LEHNARTZ, Herausgeb., Physiologische Chemie, Bd. II/2b, 627—865. Berlin: Springer 1957.

RUTENBURG, A. M., and A. M. SELIGMAN: The histochemical demonstration of acid phosphatase by a post-incubation coupling technique. J. Histochem. Cytochem. **3**, 455—470 (1955).

— R. B. COHEN and A. M. SELIGMAN: Histochemical demonstration of aryl sulfatase. Science **116**, 539—543 (1952).

RYAN, K. J., and L. L. ENGEL: Hydroxylation of steroids at carbon 21. J. biol. Chem. **225**, 103—114 (1957).

SADOWNIKOW, W.: Über die Veränderungen der Nebennieren bei den akuten toxischen Infektionen (Diphtherie, Botulismus und Tetanus). Virchows Arch. path. Anat. **317**, 315—341 (1949).

SAFFRAN, M., and A. V. SCHALLY: *In vitro* bioassay of corticotropin: Modification and statistical treatment. Endocrinology **56**, 523—532 (1955).

— B. GRAD and M. J. BAYLISS: Production of corticoids by rat adrenals *in vitro*. Endocrinology **50**, 638—643 (1952).

SAKIZ, E.: Nouvelles remarques à propos de l'action de la cortisone sur le cortex surrénal de la souris. C. R. Soc. Biol. (Paris) **149**, 513—515 (1955).

— Action comparative de la cortisone et de l'hydrocortisone sur la corticosurrénale. C. R. Soc. Biol. (Paris) **151**, 1857—1860 (1957).

— La zone X de la cortico-surrénale chez les souris de souches pures. C. R. Acad. Sci. (Paris) **246**, 1461—1463 (1958).

— Action sur la zone X surrénalienne de la souris, de la corticostimuline ou d'une agression. C. R. Soc. Biol. (Paris) **153**, 405—407 (1959).

SAMUELS, L. T., and M. L. HELMREICH: The influence of chorionic gonadotropin on the 3 β-ol-dehydrogenase activity of testes and adrenals. Endocrinology **58**, 435—442 (1956).

— — M. B. LASATER and H. REICH: An enzyme in endocrine tissues which oxidizes Δ^5-3 hydroxy steroids to α—β unsaturated ketones. Science **113**, 490—491 (1951).

SANDBERG, A. A., and W. R. SLAUNWHITE: Transcortin: Characteristics and levels in various conditions. Progr. 41st Meet. Endocr. Soc. 23—24 (1959).

SANDRITTER, W.: Über die Bedeutung des Nukleinsäure-Eiweißstoffwechsels in der Nebennierenrinde. Verh. anat. Ges. Jena **52**, 55—60 (1954).

— u. F. HÜBOTTER: Über die Bedeutung des Nucleolus in der Nebennierenrinde. Frankfurt. Z. Path. **65**, 219—229 (1954).

SANTISTEBAN, G. A., and C. A. SWINYARD: The effect of ethyl alcohol on adrenocortical activity in mice. Endocrinology **59**, 391—397 (1956).

SAPIRSTEIN, L. A., and H. GOLDMAN: Adrenal blood flow in the albino rat. Amer. J. Physiol. **196**, 159—162 (1959).

SARASON, E. L.: Morphologic changes in the rat's adrenal cortex under various experimental conditions. A.M.A. Arch. Path. **35**, 373—390 (1943).

SARTER, J.: Histologische Studie über die Innervation der Nebennierenrinde. Z. Zellforsch. **40**, 207—221 (1954).

SAYERS, G.: The adrenal cortex and homeostasis. Physiol. Rev. **30**, 241—320 (1950).

— and M. A. SAYERS: The pituitary-adrenal system. Ann. N. Y. Acad. Sci. **50** (Art. 6), 522—539 (1949).

— — E. G. FRY, A. WHITE and C. N. H. LONG: The effect of the adrenotrophic hormone of the anterior pituitary on the cholesterol content of the adrenals. Yale J. Biol. Med. **16**, 361—392 (1944).

— — T.-Y. LIANG and C. N. H. LONG: The effect of pituitary adrenotrophic hormone on the cholesterol and ascorbic acid content of the adrenal of the rat and the guinea pig. Endocrinology **38**, 1—9 (1946).

SCHABERG, A.: Regeneration of the adrenal cortex in vitro. Anat. Rec. **122**, 205—221 (1955).

SCHABERG, A.: The morphology of the adrenal cortex in acute fulminating diseases. Proc. Acad. Sci. (Amst.) C **59**, 229—239 (1956).

— The influence of ACTH and the anterior lobe of the hypophysis on the adrenal cortex *in vitro*. Proc. Acad. Sci. (Amst.) C **60**, 463—470 (1957).

— and C. A. DE GROOT: The influence of the anterior hypophysis on the morphology and function of the adrenal cortex *in vitro*. Exp. Cell Res. **15**, 475—483 (1958).

— and M. STRAUB: The adrenals in hunger cachexia. Docum. Med. geogr. trop. (Amst.) **5**, 49—55 (1953).

SCHAEFFER, W. H.: Hypophysectomy and thyroidectomy of snakes. Proc. Soc. exp. Biol. (N. Y.) **30**, 1363—1365 (1933).

SCHAPIRO, S., and L. STJARNE: A method for collection of intermittent samples of adrenal vein blood. Proc. Soc. exp. Biol. (N. Y.) **99**, 414—415 (1958).

SCHAUMKELL, K. W.: Das subkapsuläre Blastem der Rattennebenniere unter experimentellen Stressbedingungen. Anat. Anz. **102**, 292—300 (1955).

— H. H. STANGE u. P. DÖRFFLER: Methodische Untersuchungen über die p_H-abhängige Aufnahme saurer und basischer Hellfeld-Diachrome in der Nebennierenrinde. Mikroskopie **12**, 227—232 (1957a).

— — — Zum Problem der Säurefuchsinophilie „Dunkler Zellen" in der Nebennierenrinde. Z. Zellforsch. **46**, 610—618 (1957b).

SCHINDLER, W. J.: The influence of chronic treatment with corticotropin and/or growth hormone on *in vitro* and *in vivo* steroidogenesis by hamster adrenals. Anat. Rec. **133**, 334—335 (1959).

— and K. M. KNIGGE: Adrenal cortical secretion by the golden hamster. Endocrinology **65**, 739—747 (1959a).

— — *In vitro* studies and adrenal steroidogenesis by the golden hamster. Endocrinology **65**, 748—765 (1959b).

SCHMIDT, I. G.: An accessory adrenal cortex in the kidney of a rhesus monkey and the effects of pyrimethamine on this tissue. Endocrinology **59**, 454—457 (1956).

— and R. A. HOFFMANN: Effects of ACTH on pregnant monkeys and their offspring. Endocrinology **55**, 125—141 (1954).

— and L. H. SCHMIDT: Variations in the structure of adrenals and thyroids produced by thyroxine and high environmental temperatures. Endocrinology **23**, 559—565 (1938).

SCHNEPPENHEIM, P., u. A. HUHN: Die Nebenniere im Sauerstoffmangel. Beitr. path. Anat. **115**, 119—156 (1955).

SCHÖNBAUM, E., and W. G. B. CASSELMAN: Correlated studies of steroid formation in vitro, ascorbic acid concentrations, and cytological changes in adrenal glands of rats given histamine. Canad. J. Biochem. **36**, 571—574 (1958).

— M. K. BIRMINGHAM and M. SAFFRAN: Metabolism of glucose and steroid formation by rat adrenals in vitro. Canad. J. Biochem. **34**, 527—533 (1956).

SCHRICKER, J. A., G. R. FISHER, W. W. TULLNER and R. HERTZ: The effect of Amphenone-B on the cholesterol and ascorbic acid content of the rat adrenal. Endocrinology **51**, 394—400 (1952).

SCHROEDER, W., u. K. BOHLE: Der Goldhamstertest und seine Anwendung zur Klärung biologischer Probleme. Acta endocr. (Kbh.) **24**, 8—19 (1957).

SCHULTZ, A., u. G. LÖHR: Zur Frage der Spezifität der mikrochemischen Cholesterinreaktion mit Eisessig-Schwefelsäure. Zbl. allg. Path. path. Anat. **37**, 529—533 (1925).

SCHULTZ, R. B., R. W. WINTERS and W. A. KRAHL: The adrenal cortex of the pantothenic acid deficient rat: Modification of the lesion by ACTH and cortisone treatment. Endocrinology **51**, 336—343 (1952).

SCHWARTZ, N. B., and A. KLING: Stress-induced adrenal ascorbic acid depletion in the cat. Endocrinology **66**, 308—310 (1960).

SCHWEIZER, M., and M. E. LONG: Partial maintenance of the adrenal cortex by anterior pituitary grafts in fed and starved guinea pigs. Endocrinology **46**, 191—206 (1950).

SEBRELL, W. H., JR., and R. S. HARRIS, Eds.: The Vitamins, Vol. III, Chap. 14, Pyridoxine and related compounds. New York: Academic Press 1954.

SELBY, C. C.: Electron microscopy: Techniques and applications in cytology. In: R. C. MELLORS, Ed., Analytical Cytology, Ed. 2, 273—342. New York: McGraw-Hill 1959.

SELIGMAN, A. M., and L. M. MANHEIMER: A new method for the histochemical demonstration of acid phosphatase. J. nat. Cancer Inst. **9**, 427—434 (1949).

— K.-C. TSOU, S. H. RUTENBURG and R. B. COHEN: Histochemical demonstration of β-D-glucuronidase with a synthetic substrate. J. Histochem. Cytochem. **2**, 209—229 (1954).

SELYE, H.: The general adaptation syndrome and diseases of adaptation. J. clin. Endocrin. **6**, 117—230 (1946).

— Adrenal changes produced by the parenteral administration of highly hypertonic solutions. Acta neuroveg. **6**, 212—219 (1953).

SELYE, H., and J. B. COLLIP: Fundamental factors in the interpretation of stimuli influencing endocrine glands. Endocrinology **20**, 667—672 (1936).
— and E. SALGADO: PAS-tingible bodies in the adrenocortical cells of rats treated with various steroid hormones. Acta anat. **24**, 197—207 (1955).
— and H. STONE: On the Experimental Morphology of the Adrenal Cortex. Springfield, Ill.: C. C. Thomas 1950.
SHEPARD, T. H., B. H. LANDING and D. G. MASON: Familial Addison's disease. A. M. A. J. Dis. Child. **97**, 154—162 (1959).
SHERMAN, F. E., L. W. BASS and G. H. FETTERMAN: Congenital metastasizing adrenal cortical carcinoma associated with cytomegaly of the fetal adrenal cortex. Amer. J. clin. Path. **30**, 439—446 (1958).
SIEBENMANN, R. E.: Die kongenitale Lipoidhyperplasie der Nebennierenrinde mit Nebennierenrinden-Insuffizienz. Schweiz. Z. allg. Path. **20**, 77—84 (1957).
— Zur Lokalisation der Aldosteronbildung in der menschlichen Nebennierenrinde. Schweiz. med. Wschr. **89**, 837—841 (1959).
SILBERBERG, R., M. SILBERBERG and M. OPDYKE: Effects of anterior hypophysis on mammary glands and adrenals. A.M.A. Arch. Path. **55**, 506—515 (1953).
SIMMONDS, M.: Zwergwuchs bei Atrophie des Hypophysisvorderlappens. Dtsch. med. Wschr. **45**, 487—488 (1919).
SIMONNET, H., L. THIEBLOT et V. SEGAL: Action du propionate de testostérone sur la teneur en cholestérol et en acide ascorbique des surrénales du rat blanc. Ann. Endocr. (Paris) **12**, 701—703 (1951).
SIMPSON, G. G.: The principles of classification and a classification of mammals. Bull. Amer. Mus. nat. Hist. **85**, 1—350 (1945).
SIMPSON, M. E., H. M. EVANS and C. H. LI: Bioassay of adrenocorticotropic hormone. Endocrinology **33**, 261—268 (1943).
SINGER, B.: Aldosterone in the adrenal vein blood of nephrotic rats. Endocrinology **60**, 420—422 (1957).
— Effect of aldosterone-antagonist, SC 8109, on the secretion of aldosterone in normal rats. Endocrinology **65**, 512—514 (1959).
— and M. P. STACK-DUNNE: The secretion of aldosterone and corticosterone by the rat adrenal. J. Endocrin. **12**, 130—145 (1955).
SJÖSTRAND, T.: On the capillary circulation of the blood in the suprarenal body of mice under physiological conditions and the influence of drugs. Skand. Arch. Physiol. **71**, 85—122 (1934).
SKELTON, F. R.: On certain factors conditioning the action of the pituitary-adrenal system. In: R. C. CHRISTMAN, Ed., Pituitary-Adrenal Function, 39—48. Washington, D. C: Amer. Ass. Advance Sci. 1951.
— Development of hypertension and cardiovascular-renal lesions during adrenal regeneration in the rat. Proc. Soc. exp. Biol. (N. Y.) **90**, 342—346 (1955).
— Adrenal regeneration and adrenal-regeneration hypertension. Physiol. Rev. **39**, 162—182 (1959).
— C. FORTIER and H. SELYE: Some chemical and morphological changes elicited in the adrenal by stilbestrol and LAP. Proc. Soc. exp. Biol. (N. Y.) **71**, 120—122 (1949).
SKOWRON, S., et S. ZAJACZEK: Modifications histologiques des glandes endocrines durant le cycle annuel chez le hérisson. C. R. Soc. Biol. (Paris) **141**, 1105—1107 (1947).
SLATER, G. G.: Influence of galactoflavin and inanition on the adrenal ascorbic acid response to stress in rats. Endocrinology **65**, 731—738 (1959).
SLUITER, J. W., J. C. A. MIGHORST and G. J. VAN OORDT: The changes in the cytology of the adrenals of Rana esculenta following hypophysectomy. Proc. Acad. Sci. (Amst.) C **52**, 1214—1219 (1949).
SLUSHER, M. A.: Dissociation of adrenal ascorbic acid and corticosterone responses to stress, in rats with hypothalamic lesions. Endocrinology **63**, 412—419 (1958).
— and S. ROBERTS: Fate of adrenal ascorbic acid: Relationship to corticosteroid secretion. Endocrinology **61**, 98—105 (1957).
SMITH, E. M., M. L. CALHOUN and E. P. REINEKE: The histology of the anterior pituitary, thyroid and adrenal of thyroid-stimulated purebred English bulldogs. Anat. Rec. **117**, 221—239 (1953).
SMITH, P. E.: The pigmentary, growth and endocrine disturbances induced in the anuran tadpole by the early ablation of the pars buccalis of the hypophysis. Amer. anat. Mem. **11** (1920).
— Hypophysectomy and a replacement therapy in the rat. Amer. J. Anat. **45**, 205—273 (1930).
— The secretory capacity of the anterior hypophysis as evidenced by the effect of partial hypophysectomies in rats. Anat. Rec. **52**, 191—207 (1932).

SMITH, P. E.: The endocrine glands in hypophysectomized pregnant rhesus monkeys (Macaca mulatta) with special reference to the adrenal glands. Endocrinology **56**, 271—284 (1955).
— and I. P. SMITH: The function of the lobes of the hypophysis as indicated by replacement therapy with different portions of the ox glands. Endocrinology **7**, 579—591 (1923).
SMOLLICH, A.: Gestalt, Topographie, Masse und Gewichtsverhältnisse der Nebennieren des Rindes. Anat. Anz. **105**, 205—221 (1958).
SNAITH, A. H: A case of feminizing adrenal tumor in a girl. J. clin. Endocrin. **18**, 319—322 (1958).
SNELL, K. C., and H. L. STEWART: Variations in histologic pattern and functional effects of a transplantable adrenal cortical carcinoma in intact, hypophysectomized, and newborn rats. J. nat. Cancer Inst. **22**, 1119—1155 (1959).
SOBEL, E. H., C. M. LEE, JR., V. M. ESSELBORN and L. C. CLARK, JR.: Functioning adrenal tumors in childhood. A. M. A. J. Dis. Child. **86**, 733—751 (1953).
— A. E. RENOLD, J. E. BETHUNE, J. J. HOET, W. J. REDDY and G. W. THORN: Effects of hypophysectomy and of amphenone administration in a child with functioning metastatic adrenal carcinoma. Amer. J. Med. **24**, 482—492 (1958).
SOKOLOFF, L., J. T. SHARP and E. H. KAUFMAN: The adrenal cortex in rheumatic disease. Arch intern. Med. **88**, 627—639 (1951).
SOLOMON, S., J. T. LANMAN, J. LIND and S. LIEBERMAN: The biosynthesis of Δ^4-androstenedione and 17 α-hydroprogesterone from progesterone by surviving fetal adrenals. J. biol. Chem. **233**, 1084—1088 (1958).
SONENBERG, M., A. S. KESTON and W. L. MONEY: Studies with labelled anterior pituitary preparations: Adrenocorticotropin. Endocrinology **48**, 148—161 (1951).
SOSA, J. M.: Vitamin C microscopic demonstration and Golgi apparatus. Exp. Cell Res. **3**, 184—191 (1952).
SOULAIRAC, A., P. DESCLAUX et J. TEYSSEYRE: Variations de l'activité phosphatasique de la corticosurrénale au fonction de modifications endocrines. Ann. Endocr. (Paris) **10**, 285—287 (1949).
SOURKES, T. L., and P. HENEAGE: The effects of cortisone in vivo on the oxidative metabolism of the rat adrenal. Endocrinology **49**, 601—605 (1951).
— — Some studies on the oxidative metabolism of the adrenal gland. Endocrinology **50**, 73—82 (1952).
SPECTOR, W. S., Ed.: Handbook of Biological Data. Philadelphia: Saunders 1956.
SPERRY, W. M., and M. WEBB: A revision of the Schoenheimer-Sperry method for cholesterol determination. J. biol. Chem. **187**, 97—106 (1950).
SPIEGEL, A.: Über das Auftreten von Geschwülsten der Nebennierenrinde mit vermännlichender Wirkung der frühkastrierten Meerschweinmännchen. Virchows Arch. path. Anat. **305**, 367—393 (1939).
SPIERS, R. S.: Eosinopenic activity of epinephrine in adrenalectomized mice. Amer. J. Physiol. **172**, 520—526 (1953).
SPRENKEL, H. B. VAN DER: Nebenniere und Paraganglien. In: L. BOLK, E. GÖPPERT, E. KALLIUS u. W. LUBOSCH, Herausgeb., Handbuch der vergleichenden Anatomie der Wirbeltiere, Bd. II/1, 777—816. Berlin: Urban & Schwarzenberg 1934.
STACHENKO, J., and C. J. P. GIROUD: Functional zonation of the adrenal cortex: Pathways of corticosteroid biogenesis. Endocrinology **64**, 730—742 (1959a).
— — ... Site of ACTH action. Endocrinology **64**, 743—752 (1959b).
STARY, Z.: Leber und Galle. In: B. FLASCHENTRÄGER u. E. LEHNARTZ, Herausgeb., Physiologische Chemie, Bd. II/2a, 1—569. Berlin: Springer 1956.
STEBBINS, R. B.: Cytochemical changes in the adrenal cortex of the rat in pyridoxine deficiency. Endocrinology **49**, 25—35 (1951a).
— Failure of cortisone to reduce adrenal weight in hypophysectomized rats. Endocrinology **49**, 671—672 (1951b).
— Impaired water metabolism in pyridoxine deficiency and effects of pyridoxine and adrenal cortical hormone. Amer. J. Physiol. **166**, 538—540 (1951c).
STEINBERG, D., D. S. FREDRICKSON and J. AVIGAN: Effects of Δ^4-cholestenone on animals and in man. Proc. Soc. exp. Biol. (N. Y.) **97**, 784—790 (1958).
STENGER, A. H., and H. A. CHARIPPER: A study of adrenal cortical tissue in Rana pipiens with special reference to metamorphosis. J. Morph. **78**, 27—42 (1946).
STERBA, G.: Das Adrenal- und Interrenalsystem im Lebensablauf von *Petromyzon planeri* Bloch. I. Morphologie und Histologie einschließlich Histogenese. Zool. Anz. **155**, 151—168 (1955).
STERLING, R. E., and B. B. LONGWELL: The effect of oral and intravenous glucose administration on the adrenal response to cold exposure. Endocrinology **53**, 98—105 (1953a).
— — The effect of the type and quantity of food intake on the response of the adrenal gland to cold exposure. Endocrinology **53**, 106—113 (1953b).

STIGLIANI, R., u. P. MARCHI: Wirkung der Implantation von Hypophysenvorderlappen trächtiger und nichtträchtiger Tiere in der Nähe der Nebennieren. Virchows Arch. path. Anat. **330**, 525—537 (1957).
STILLING, H.: Zur Anatomie der Nebennieren. Arch. mikr. Anat. **52**, 176—195 (1898).
STIRLING, G. A. The adrenal glands in Jamaican infants. J. Path. Bact. **77**, 555—564 (1959).
— and V. J. KEATING: Size of adrenals in Jamaicans. Brit. med. J. **1958** ii, 1016—1018.
STÖHR, P., JR.: Zur Innervation der menschlichen Nebenniere. Z. Anat. Entwickl.-Gesch. **104**, 475—490 (1935).
STONER, H. B., and H. J. WHITELEY: The effects of cortisone and corticotrophin on the human adrenal cortex. Lancet **1954** ii, 992—994.
— — and J. L. EMERY: The effect of systemic disease on the adrenal cortex of the child. J. Path. Bact. **66**, 171—183 (1953).
— J. N. P. DAVIES, H. J. WHITELEY and J. L. EMERY: The adrenal X-zone and Kwashiorkor. Brit. med. J. **1955** i, 1260.
STURTEVANT, F. M.: The effect of two steroid analogs on organ weights of intact and hypophysectomized rats. Endocrinology **56**, 256—260 (1955).
STUTINSKY, F.: Production expérimentale de colloïde dans le cortex surrénal du rat. Ann. Endocr. (Paris) **16**, 900—904 (1955).
SUZUKI, T., K. YAMASHITA and T. MITAMURA: Effect of hypothermia on the secretion of 17-hydroxycorticosteroids of the adrenal glands in non-anesthetized dogs. Tohoku J. exp. Med. **66**, 144 (1957).
SWANN, H. G.: The pituitary-adrenocortical relationship. Physiol. Rev. **20**, 493—521 (1940).
SWEAT, M. A., and G. L. FARRELL: Decline of corticosteroid secretion following hypophysectomy. Proc. Soc. exp. Biol. (N. Y.) **87**, 615—618 (1954).
SWINGLE, W. W., and J. J. PFIFFNER: Studies on the adrenal cortex. I. The effect of a lipid fraction upon the life-span of adrenalectomized cats. Amer. J. Physiol. **96**, 153—163 (1931).
— and J. W. REMINGTON: The role of the adrenal cortex in physiological processes. Physiol. Rev. **24**, 89—127 (1944).
— R. MAXWELL, M. BEN, C. BAKER, S. J. LE BRIE and M. EISLER: A comparative study of aldosterone and other adrenal steroids in adrenalectomized dogs. Endocrinology **55**, 813—821 (1954).
SWINYARD, C. A.: The innervation of the suprarenal glands. Anat. Rec. **68**, 417—430 (1937).
— Volume and cortico-medullary ratio of the adult human suprarenal gland. Anat. Rec. **76**, 69—79 (1940).
— Growth of the human suprarenal glands. Anat. Rec. **87**, 141—150 (1943).
SYMINGTON, T., and J. N. DAVIDSON: The effect of exogenous ACTH and conditions of stress on the chemical composition of the human adrenal gland. Scot. med. J. **1**, 15—31 (1956).
— W. P. DUGUID and J. N. DAVIDSON: Effect of exogenous corticotropin on the histochemical pattern of the human adrenal cortex and a comparison with the changes during stress. J. clin. Endocrin. **16**, 580—598 (1956).
TÄHKÄ, H.: On the weight and structure of the adrenal glands and the factors affecting them, in children of 0—2 years. Acta paediat. (Stockh.) Suppl. **81** (1951).
TALALAY, P., W. H. FISHMAN and C. HUGGINS: Phenolphthalein glucuronic acid as a substrate for the assay of glucuronidase activity. J. biol. Chem. **166**, 757—772 (1946).
TALIAFERRO, I., and L. LEONE: Inhibitory effect of perthane [2,2-bis(paraethylphenyl)-1,1-dichloroethane] on adrenocortical function in human subjects. New Engl. J. Med. **257**, 855—860 (1957).
TAMURA, Y.: Structural changes in the suprarenal gland of the mouse during pregnancy. J. exp. Biol. **4**, 81—92 (1927).
TANABE, T., and E. J. CAFRUNY: Adrenal hypertrophy in rats treated chronically with morphine. J. Pharmacol. **122**, 148—155 (1958).
TAYLOR, D. J., J. GREENBERG, E. S. JOSEPHSON and E. M. NADEL: Histochemical sudanophilia and cholesterol concentration in the adrenal glands of chicks parasitized with *Plasmodium gallinaceum*. Acta endocr. (Kbh.) **22**, 173—178 (1956).
TEPPERMAN, J., H. M. TEPPERMAN, B. W. PATTON and L. F. NIMS: Effects of low barometric pressure on the chemical composition of the adrenal glands and blood of rats. Endocrinology **41**, 356—363 (1947).
THERIEN, B., R. C. MELLINGER, J. R. CALDWELL and P. J. HOWARD: Primary aldosteronism due to adrenal hyperplasia. A. M. A. J. Dis. Child. **98**, 90—99 (1959).
THORN, G. W., A. E. RENOLD, A. GOLDFEIN, D. H. NELSON, W. J. REDDY and R. HERTZ: Inhibition of corticosteroid secretion by amphenone in a patient with adrenocortical carcinoma. New Engl. J. Med. **254**, 547—551 (1956).
TIMIRAS, P. S., and D. M. WOODBURY: Adrenocortical function after administration of thyroxine and triiodothyronine. J. Pharmacol. **115**, 144—153 (1955).

TOBIN, C. E., and J. P. BIRNBAUM: Some factors influencing brown degeneration of the adrenal gland in the Swiss albino mouse. A.M.A. Arch. Path. **44**, 269—281 (1947).
TOGARI, C., and T. ITO: Hematopoiesis in the adrenal gland of the embryonic guinea-pig. Nagoya J. med. Sci. **14**, 54—57 (1951).
TONUTTI, E.: Die Umbauvorgänge in den Transformationsfeldern der Nebennierenrinde als Grundlage der Beurteilung der Nebennierenrindenarbeit. Z. mikr.-anat. Forsch. **52**, 32—86 (1942).
— Über die strukturelle Funktionsanpassung der Nebennierenrinde. Endokrinologie **28**, 1—15 (1951).
— Experimentelle Untersuchungen zur Pathophysiologie der Nebennierenrinde. Verh. dtsch. path. Ges. **36**, 123—158 (1953).
— F. BAHNER u. E. MUSCHKE: Die Veränderungen der Nebennierenrinde der Maus nach Hypophysectomie und nach ACTH-Behandlung, quantitativ betrachtet am Verhalten der Zellkernvolumina. Endokrinologie **31**, 266—284 (1954).
TOUCHSTONE, J. C., L. GLAZER, D. Y. COOPER and J. M. ROBERTS: The isolation of Δ^4-androstene-11 β-ol-3,17-dione from human adrenal incubates. J. clin. Endocrin. **15**, 382—384 (1955).
TUCHMANN-DUPLESSIS, H.: Action de la cortisone sur le système hypophyso-surrénalien du rat. C. R. Soc. Biol. (Paris) **145**, 559—562 (1951a).
— Rôle de la surrénale dans la régulation de la sécrétion antéhypophysaire corticotrope. Presse méd. **59**, 1749—1752 (1951b).
— et J. LARROCHE: Anencéphalie et atrophie corticosurrénale. C. R. Soc. Biol. (Paris) **152**, 300—302 (1958).
— C. DELGATTE et F. DELGATTE: L'action de l'ACTH et de la cortisone sur l'aspect cytologique de la gland interrénale de *Rana temporaria*. C. R. Soc. Biol. (Paris) **149**, 553—556 (1955).
TULLNER, W. W., and R. HERTZ: Effect of intravenous o,p' DDD on plasma 17-hydroxycorticoid secretion in the dog. Endocrinology **66**, 494—496 (1960).
— M. M. GRAFF and R. HERTZ: Amphenone inhibition of adrenal corticosteroid output in the hypophysectomized dog. Endocrinology **58**, 802—807 (1956).
UOTILA, U. U.: On the fuchsinophile and pale cells in the adrenal cortex tissue of the fowl. Anat. Rec. **75**, 439—449 (1939).
— The early embryological development of the fetal and permament adrenal cortex in man. Anat. Rec. **76**, 183—203 (1940).
— and A. PEKKARINEN: The relation of human adrenal glands to the pathological changes produced by intense continuous stress ending in death. Acta endocr. (Kbh.) **6**, 23—50 (1951).
UPTON, A. C., and J. FURTH: Spontaneous and radiation-induced pituitary adenomas of mice. J. nat. Cancer Inst. **15**, 1005—1021 (1955).
UTEWSKI, A., and S. EPSTEIN: Der intermediäre Stoffwechsel in den endokrinen Drüsen. II. Glykolyse in Gl. suprarenalis und Corpus luteum. Biochem. Z. **273**, 359—361 (1934a).
— — III. Gly olyse und Proteolyse in Gl. suprarenalis bei erhöhter Hormonproduktion. Biochem. Z. **273**, 362—364 (1934b).
UUSPÄÄ, V. J., and P. A. JÄRVINEN: Effects of progesterone, desoxycorticosterone and estradiol on the pituitary and adrenal glands in oophorectomized guinea pigs. Ann. Med. exp. Fenn. **33**, 262—270 (1955).
VACCAREZZA, A. J.: Histophysiology of the adrenal cortex (in Spanish). Medicina (B. Aires) **5**, 425—445; **6**, 46—73 (1945).
VANNINI, E., and T. CESSI: Common origin of the adrenocortical blastema and the gonadal medullary tissue in the embryo of the guinea pig (in Italian). Atti (R. C.) Accad. naz. Lincei, Ser. 8, **6**, 650—656 (1949).
— and A. SABBADIN: The relation of the interrenal blastema to the origin of the somatic tissues of the gonad in frog tadpoles. J. Embryol. exp. Morph. **2**, 275—289 (1954).
VASAMA, R., L. KALLIOMAKI, R. VASAMA and E. NÄÄTANEN: The adrenal cortex and administration of ACTH, cortisone and a combination of ACTH and cortisone. Ann. Med. exp. Fenn. **35**, Suppl. 1, 1—18 (1957).
VELICAN, C.: Embryogenèse de la surrénale humaine. Arch. Anat. micr. **36**, 316—333 (1946/47).
— Le dispositif sphinctéro-propulseur de la surrénale. Arch. Anat. micr. **37**, 28—40 (1948a).
— La zone transitoire de la cortico-surrénale humaine. Arch. Anat. micr. **37**, 73—81 (1948b).
VENNING, E. H.: Adrenal function in the newborn. In: R. C. CHRISTMAN, Ed., Pituitary-Adrenal Function, 149—157. Washington, D. C.: Amer. Ass. Advance. Sci. 1951.
— I. DRYENFURTH and J. C. BECK: Effect of corticotropin and prednisone on the excretion of aldosterone in man. J. clin. Endocrin. **16**, 1541—1553 (1956).
— O. J. LUCIS, I. DRYENFURTH and J. C. BECK: Effect of various pituitary and hypothalamic extracts on aldosterone secretion. Progr. 40th Meet. Endocr. Soc. 58—59 (1958).

VENSKE, W. G.: Endocrine gland weights of chick embryos. Growth **7**, 265—271 (1943).
VERNE, J.: Étude histochimique des substances aldéhydiques formées au cours du métabolisme des corps gras. Ann. Physiol. Physicochim. biol. **5**, 245—267 (1929).
— Les étapes de nos conaissances sur l'histochimie de la surrénale. Ann. Histochim. **1**, 53—63 (1956).
— et M. DELSOL: Les cellules positives à la réaction de MacManus (sic) dans la surrénale de *Rana esculenta* L. Ann. Histochim. **1**, 91—97 (1956).
— et S. HÉBERT: Les types lipidiques de la cortico-surrénale chez les mammifères. Ann. Endocr. (Paris) **12**, 192—198 (1951).
— — Étude histochimique du cortex de la surrénale au cours de la régénération suivant son énucléation chez le rat blanc. Ann. Endocr. (Paris) **16**, 279—282 (1955).
— F. BOURLIERE et M. DERAMOND: Sur le développement post-embryonnaire de la cortico-surrénale chez les oisseaux. C. R. Soc. Biol. (Paris) **146**, 1859 (1952).
— M. GABE et B. SCHRAMM: Étude histochimique d'une lipofuscine du cortex surrénal humain. Ann. Histochim. **1**, 98—101 (1956).
VERZÁR, F., u. F. HÜGIN: Einfluß des Alters auf die Entwicklung der Arbeitshypertrophie von Organen. Acta anat. **30**, 918—927 (1957).
VICARI, E. M.: The adrenal lipids of mice with high and low mammary gland tumor incidences. Anat. Rec. **86**, 523—543 (1943).
VIES, J. VAN DER: Experience with an assay of adrenocorticotropic hormone based on the steroid output of rat adrenals *in vitro*. Acta physiol. pharm. neerl. **5**, 361—384 (1957).
VILAR, O., and W. W. TULLNER: Effects of o,p′ DDD on histology and 17-hydroxycorticosteroid output of the dog adrenal cortex. Endocrinology **65**, 80—86 (1959).
VILLEE, D. B., L. L. ENGEL and C. A. VILLEE: Steroid hydroxylation in human and fetal adrenals. Endocrinology **65**, 465—474 (1959).
VINCENT, S., and F. R. CURTIS: A note on the teleostean adrenal bodies. J. Anat. (Lond.) **62**, 110—114 (1927).
VINES, H. W. C.: The adreno-genital syndrome: some histological observations. In: L. R. BROSTER, C. ALLEN, H. W. C. VINES, J. PATTERSON, A. W. GREENWOOD, G. F. MARRIAN and G. C. BUTLER, The Adrenal Cortex and Intersexuality, 137—197. London: Chapman and Hall 1938.
VOGT, M.: The output of cortical hormone by the mammalian suprarenal. J. Physiol. (Lond.) **102**, 341—356 (1943).
— Cortical secretion of the isolated perfused adrenal. J. Physiol. (Lond.) **113**, 129—156 (1951).
— Inhibition by hexoestrol of adrenocortical secretion in the rat. J. Physiol. (Lond.) **130**, 601—614 (1955).
— The effects of hexoestrol and of "amphenone B" on morphology and function of the rat adrenal cortex. Yale J. Biol. Med. **29**, 469—479 (1957a).
— Morphological changes in the adrenal cortex in relation to concentration of steroids in adrenal vein blood. Ciba Foundn. Coll. Endocr. **11**, 193—199 (1957b).
VOSS, H.: Vergleichende histotopochemische Untersuchungen über das Verhalten der Nebenniere zur Plasmalreaktion. Z. Zellforsch. **31**, 43—53 (1940).
WALKER, D. G., M. E. SIMPSON, C. W. ASLING and H. M. EVANS: Growth and differentiation in the rat following hypophysectomy at 6 days of age. Anat. Rec. **106**, 539—554 (1950).
WALLACH, S., H. BROWN, E. ENGLERT, JR., and K. EIK-NES: Adrenocortical carcinoma with gynecomastia. J. clin. Endocrin. **17**, 945—958 (1957).
WALLACE, G. J.: An Introduction to Ornithology. New York: Macmillan 1955.
WALLRAFF, J.: Histochemische Untersuchungen an den Nebennieren des erwachsenen Menschen. Z. Zellforsch. **34**, 362—427 (1949).
WALTHARD, B.: Funktionen der Nebennierenrinde, inbesondere der Zona Glomerulosa. Acta endocr. (Kbh.) **5**, 61—85 (1950).
WARING, H.: The development of the adrenal gland of the mouse. Quart. J. micr. Sci. **78**, 329—366 (1935).
— Effects of hormones on degeneration of the X-zone in the mouse adrenal. J. Endocrin. **3**, 123—131 (1942).
WATTENBERG, L. W.: Microscopic histochemical demonstration of steroid-3 β-ol-dehydrogenase in tissue sections. J. Histochem. Cytochem. **6**, 225—232 (1958).
— and D. GLICK: Determination of β-glucuronidase in microgram quantities of tissue and its distribution in the cow adrenal. J. Histochem. Cytochem. **1**, 1—7 (1953).
WEAVER, J. A.: A comparison of the NAHD reaction with lipoid stains in the adrenal cortex of the rat. J. Endocrin. **13**, 189—194 (1956).
WEBER, A. F., S. H. MCNUTT and B. B. MORGAN: Structure and arrangement of zona glomerulosa cells in the bovine adrenal. J. Morph. **87**, 393—415 (1950).

WEBER, A. F., W. R. PRITCHARD and A. F. SELLERS: The production of lipid accumulation and other histological and cytological changes in the zonae glomerulosae of calves. Amer. J. vet. Res. **17**, 402—409 (1956).
— J. T. BELL, JR., and A. F. SELLERS: The histological and cytochemical effects of the administration of 1,1-dichloro-2,2-bis-(p-chlorophenyl)ethane on the adrenal cortices of dairy calves. Amer. J. vet. Res. **19**, 51—57 (1958).
WEBER, E. J., and M. L. MENTEN: Histologic studies on a virilizing tumor of the adrenal cortex. Amer. J. Path. **24**, 293—303 (1948).
WEGMANN, R.: Spectrographie dans l'infra-rouge. III. Histospectrographie de la surrénale. Ann. Histochim. **1**, 30—38 (1956).
WELLS, L. J.: Some experimental evidence of production of adrenotrophin by the fetal hypophysis. Proc. Soc. exp. Biol. (N. Y.) **68**, 487—488 (1948).
— Functioning of the anterior hypophysis in the fetal rat. In: A. GORBMAN, Ed., Comparative Endocrinology, 444—451. New York: Wiley 1959.
WESTBERG, J. A., H. A. BERN and E. B. BARNAWELL: Strain difference in the response of the mouse adrenal to oestrogen. Acta endocr. (Kbh.) **25**, 70—82 (1957).
WEXLER, B. C.: Lipid depletion of the zona glomerulosa of the rat adrenal cortex after administration of 9-α-fluorohydrocortisone. Endocrinology **64**, 300—304 (1959).
— and A. P. RINFRET: Histological observations on the zona glomerulosa of the rat adrenal cortex following administration of corticotropin preparations. Endocrinology **57**, 608—620 (1955).
— — A. C. GRIFFIN and H. L. RICHARDSON: Evidence for pituitary control of the lipid content of the zona glomerulosa of the rat adrenal cortex. Endocrinology **56**, 120—123 (1955).
— A. E. DOGLIN and E. W. TRYCZYNSKI: Effects of a bacterial polysaccharide (Piromen) on the pituitary-adrenal axis: Adrenal ascorbic acid, cholesterol and histologic alteration. Endocrinology **61**, 300—308 (1957).
— — J. F. ZAROSLINSKI and E. S. TRYCZYNSKI: Effects of a bacterial polysaccharide (Piromen®) on the pituitary-adrenal axis: Cortisone blockade of Piromen-induced release of ACTH. Endocrinology **63**, 201—204 (1958).
WHITEHEAD, R.: The involution of the transitory cortex of the mouse suprarenal. J. Anat. (Lond.) **67**, 387—392 (1933a).
— Growth and mitosis in the mouse suprarenal. J. Anat. (Lond.) **67**, 399—408 (1933b).
— The Schultz cholesterol reaction in the suprarenal cortex. J. Path. Bact. **39**, 443—447 (1934).
— Involution of the permanent cortex of the rabbit suprarenal by fatty change. J. Path. Bact. **41**, 305—309 (1935a).
— The sex difference in the proportion of the suprarenal cortex occupied by lipoid in guinea-pigs over one year old. J. Anat. (Lond.) **70**, 123—125 (1935b).
— Variations in the cortical lipoid of the rabbit suprarenal with sex and age. J. Anat. (Lond.) **70**, 380—385 (1936).
WHITELEY, H. J., and H. B. STONER: The effect of pregnancy on the human adrenal cortex. J. Endocr. **14**, 325—334 (1957).
WILBUR, O. M., JR.: A study of the effect of thyroxin on the development of tubular degeneration of the adrenal cortex produced by ACTH. Johns Hopk. Hosp. Bull. **97**, 47—52 (1955).
— and A. R. RICH: A study of the role of adrenocorticotropic hormone (ACTH) in the pathogenesis of tubular degeneration of the adrenals. Johns Hopk. Hosp. Bull. **93**, 321 to 347 (1953).
WILLIAMS, R. G.: Studies of the adrenal cortex: Regeneration of the transplanted gland and the vital quality of autogenous grafts. Amer. J. Anat. **81**, 199—232 (1947).
— Studies of autoplastic and homoplastic grafts in rabbits. Amer. J. Anat. **93**, 1—24 (1953).
WILSON, H., J. J. BORRIS and R. C. BAHN: Steroids in the blood and urine of female mice bearing an ACTH-producing pituitary tumor. Endocrinology **62**, 135—149 (1958).
WINTER, C. A., R. H. SILBER and H. C. STOERK: Production of reversible hyperadrenocorticism in rats by prolonged administration of cortisone. Endocrinology **47**, 60—72 (1950).
— H. L. HOLLINGS and R. B. STEBBINS: The effect of androgenic hormones upon the adrenal atrophy produced by cortisone injections and upon the anti-inflammatory action of cortisone. Endocrinology **52**, 123—134 (1953).
WINTERS, R. W., R. B. SCHULTZ and W. A. KREHL: The adrenal cortex of the pantothenic acid-deficient rat: Eosinophile and lymphocyte responses. Endocrinology **50**, 379—384 (1952).
WITSCHI, E.: Development of Vertebrates. Philadelphia: Saunders 1956.
WOLFF, H. P., K. R. KOCZOREK and E. BUCHBORN: Aldosteronuria in oedema. In: A. F. MULLER and C. M. O'CONNOR, Eds., An International Symposium on Aldosterone, 193—206. Boston: Little, Brown 1958.

WOOD, K. F., F. LEES and F. D. ROSENTHAL: Carcinoma of the adrenal cortex without endocrine effects. Brit. J. Surg. **45**, 41—48 (1957).
WOODRUFF, M. F. A., and T. BOSWELL: The effect of cortisone and ACTH on adrenal transplants in the rat. J. Endocr. **10**, 86—93 (1953).
WOODS, J. W.: The effects of long-term exposure to cold upon adrenal weight and ascorbic acid content in wild and domesticated Norway rats. J. Physiol. (Lond.) **135**, 384—389 (1957a).
— The effects of acute stress and of ACTH upon ascorbic acid and lipid content of the adrenal glands of wild rats. J. Physiol. (Lond.) **135**, 390—399 (1957b).
WOOLLEY, G. W.: The adrenal cortex and its tumors. Ann. N. Y. Acad. Sci. **50** (Art.6), 616—626 (1949).
— M. M. DICKIE and C. C. LITTLE: Adrenal tumors and other pathological changes in reciprocal crosses in mice. II. An introduction to results of four reciprocal crosses. Cancer Res. **13**, 231—245 (1953).
WRIGHT, A., and I. CHESTER JONES: The adrenal gland in lizards and snakes. J. Endocr. **15**, 83—99 (1957).
— — and J. G. PHILLIPS: The histology of the adrenal glands of the Prototheria. J. Endocr. **15**, 100—107 (1957).
YAKAITIS, A. A., and L. J. WELLS: Hypophysis-adrenal system in the fetal rat. Adrenals in fetuses subjected to cortisone, DCA, hypophyseoprivia and growth hormone. Amer. J. Anat. **98**, 205—229 (1956).
YANKOPOULOS, N. A., J. O. DAVIS, B. KLIMER and R. E. PETERSON: Evidence that a humoral stimulates the adrenal cortex to secrete aldosterone in experimental secondary hyperaldosteronism. Fed. Proc. **18**, 174 (1959).
YOFFEY, J. M.: The suprarenal cortex: The structural background. In: J. M. YOFFEY, Ed., The Suprarenal Cortex, 31—38. London: Butterworths 1953.
— Some observations in the problem of cortical zoning. Ciba Foundn. Coll. Endocrin. 8, 18—26 (1955).
— and J. S. BAXTER: The formation of birefringent crystals in the suprarenal cortex. J. Anat. (Lond.) **81**, 335—347 (1947).
— — Histochemical changes in the suprarenal gland of the adult male rat. J. Anat. (Lond.) **83**, 89—98 (1949).
YOSHIMURA, F.: Étude cytologique sur la corticosurrénale à l'hyper- et à l'hypofonction chez le rat. Endocrin. jap. **1**, 177—194 (1954).
YOUNG, J. Z.: The autonomic nervous system of selachians. Quart. J. micr. Sci. **75**, 571—624 (1933).
ZALESKY, M.: A study of the seasonal changes in the adrenal gland of the thirteen-lined ground squirrel (Citellus tridecemlineatus), with particular reference to its sexual cycle. Anat. Rec. **60**, 291—321 (1934).
ZARROW, M. X., and J. T. BALDINI: Failure of adrenocorticotropin and various stimuli to deplete the ascorbic acid content of the adrenal gland of the quail. Endocrinology **50**, 555—561 (1952).
— and I. G. ZARROW: Mechanism of adrenal involution in the rat after treatment with thiouracil. Proc. Soc. exp. Biol. (N. Y.) **76**, 620—623 (1951).
— L. M. HORGER and J. L. MCCARTHY: Atrophy of adrenal gland following thiouracil and vitamin B_{12}. Proc. Soc. exp. Biol. (N. Y.) **97**, 348—349 (1957).
ZELANDER, T.: Ultrastructure of the mouse adrenal cortex. J. Ultrastruct. Res. Suppl. **2** (1959).
ZIMMERMANN, B., H. S. BLOCH, W. L. WILLIAMS, C. R. HITCHCOCK and B. HOELSCHER: The effects of DDD [1,1-dichloro-2,2-bis (p-chlorophenyl)ethane] on the human adrenal. Cancer (N. Y.) **9**, 940—948 (1956).
ZINSSER, A. D., and H. H. ZINSSER: Fuchsinophilia in the adrenal cortex. A.M.A. Arch. Path. **51**, 393—398 (1951).
ZIZINE, L. (A.): Action comparative de doses variées de testostérone sur l'atrophie surrénalienne produite par la cortisone. C. R. Soc. Biol. (Paris) **147**, 26—28 (1953a).
— Contribution a l'étude de l'action de la testostérone sur l'atrophie cortico-surrénalienne du rat. Ann. Endocr. (Paris) **14**, 758—763 (1953b).
— Inhibition par la testostérone de l'hypertrophie surrénalienne compensatrice. C. R. Soc. Biol. (Paris) **150**, 1333—1334 (1956).
— M. E. SIMPSON and H. M. EVANS: Direct action of male sex hormones on the adrenal cortex. Endocrinology **47**, 87—101 (1950).
ZWEIFACH, B. W., M. M. BLACK and E. SHORR: Evaluation of tetrazolium as a histochemical index of adrenal cortical activity. Proc. Soc. exp. Biol. (N. Y.) **76**, 446—454 (1951).

Chapter 2

Chemistry of the adrenocortical steroids*

By

James A. Moore and Erich Heftmann

With 103 Figures

Chemical nature of adrenocortical steroids

A. General characteristics of steroid compounds

The adrenocortical hormones are one of about eight more or less clearly-defined groups of steroid compounds, and historically are one of the most recent groups to be included in the steroid complex. In the past decade, however, they have become the focus of one of the most concerted and vigorous research efforts in modern organic chemistry, and a major part of the activity in steroid chemistry has been devoted to the adrenocortical hormones and their relatives. Before discussing the chemical aspects of adrenal steroids, which are of particular interest in this volume, it is necessary to summarize the basic features characteristic of steroid compounds in general. The outstanding treatise of Fieser and Fieser (1949, 1959), and the summaries of Lettré et al. (1954), Nazarow and Bergel'son (1955), Reuber et al. (1955), and Klyne (1957), are recommended for fuller discussions and references to the original literature. A more recent survey, with references to current reviews, has been prepared by Heftmann and Mosettig (1960).

I. Structure and stereoisomerism of the steroid nucleus

The generic term *steroid* embraces a family of some 400 chemical compounds of ubiquitous occurrence in the plant and animal kingdoms, plus several thousands of synthetic substances, all of which are related by a common structural unit, the steroid nucleus. This nucleus is a skeleton of four saturated carbocyclic rings, substituted at two ring junctions by methyl groups, in a unique steric arrangement.

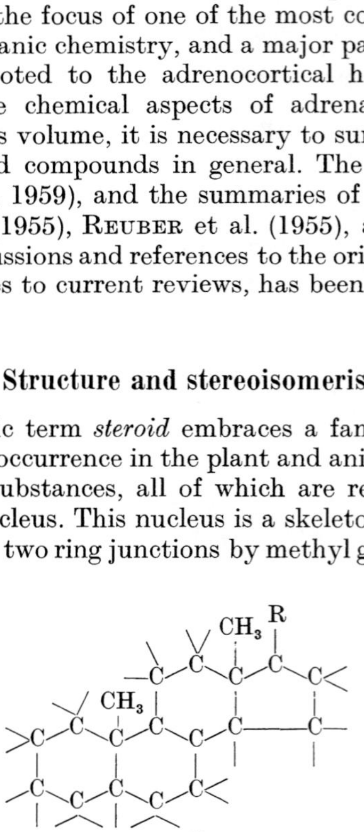

I

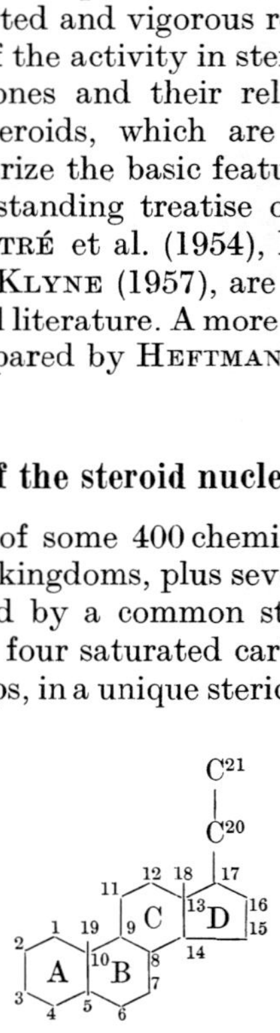

II

* The coverage in this chapter extends, with a few later additions, only to August, 1958, when the manuscript was completed.

The steroid family is conveniently subdivided into several series in which the group R in (I) represents a distinctive side chain. In the plant sterols and cholesterol, R is a branched chain of eight to ten carbon atoms; in the bile acids, a five-carbon carboxylic acid; and in the steroidal sapogenins, cardiotonic glycosides, and alkaloids, one of several heterocyclic structures. In the sex hormones, R is an oxygen atom; in progesterone, an acetyl group, and in the adrenocortical hormones proper, an oxygenated two-carbon chain.

The nucleus is usually represented by the abbreviated formula (II), with the carbon atoms omitted and the methyl groups indicated by dashes. The rings are designated by capital letters, and the carbon atoms by the numbering shown in (II). The only departure from this skeleton among naturally-occurring steroids is encountered in the estrogens, in which C-19 is absent and ring A is aromatic. The multiplicity of compounds based on this single nuclear structure arises from substitution by double bonds or oxygen atoms, which have been found in natural steroids with varying frequency at every possible position, i. e., excepting C-10 and C-13.

Evidence establishing the size and points of attachment of the rings and of the side chain in the steroid skeleton was adduced by systematic and painstaking degradative investigations in the period 1900—1930, notable among them being the studies on cholesterol by WINDAUS and on the bile acids by WIELAND. The characterization of oxidation products of these steroids in which one or more rings were opened, and of aromatic hydrocarbons derived by selenium dehydrogenation and finally X-ray crystallographic measurements led to the formulations (III) and (IV) for cholesterol and deoxycholic acid, respectively, presented in 1932 by WIELAND and

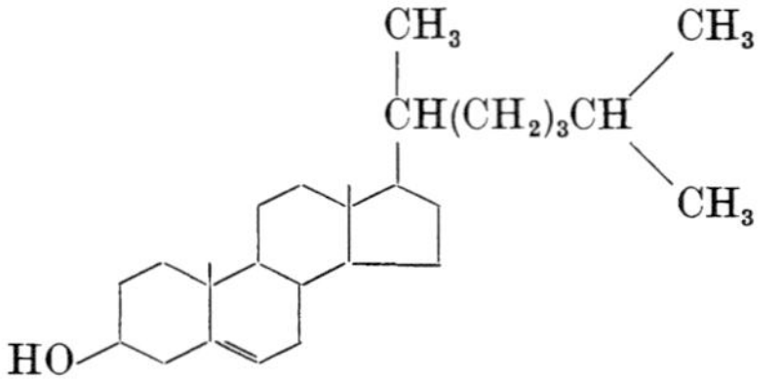

III

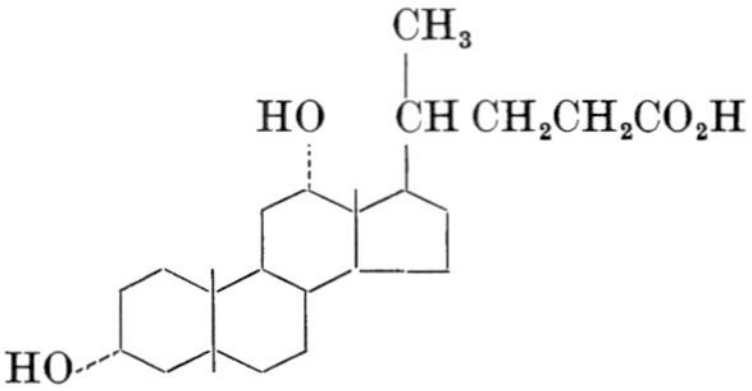

IV

by ROSENHEIM and KING. In the meantime, several diverse groups of natural products were found to be structurally related to the two key substances, (III) or (IV), or their degradation products. Thus a variety of plant sterols were shown to differ only slightly from cholesterol, and several types of alkaloids and glycosides were shown to be steroids. As an expanding catalogue of steroid compounds became available, it was possible to establish within a few months of their isolation that the estrogens, androgens and adrenocortical hormones all possessed this same tetracyclic ring system.

1. Nuclear configurations and stereochemical conventions

As a consequence of the tetrahedral structure of the carbon atom, each of the ring junctions at carbon atoms 5, 10, 9, 8, 14 and 13 in the saturated steroid nucleus is a center of asymmetry, in which the substituent at these positions, a hydrogen atom or a methyl group, may be in one of two non-equivalent *configurations*. For example, the projection of the molecule could be written with the angular methyl group at C-10 in front of the plane of the ring system, or in the opposite configuration, extending from the rear face of the molecule. The presence of six asymmetric centers, each with two possible orientations, means that 2^6 or 64 stereoisomers of this basic nucleus are possible. Each additional substituent on a ring carbon atom, such as the side chain at C-17, gives rise to an additional asymmetric center. A double bond extending to a ring junction, as in ring B in cholesterol, renders the ring junction planar and no longer asymmetric.

To define the complete structure of a steroid, the configuration of each substituent on the ring system, whether at a ring junction or not, must be specified with respect to the position relative to the plane of projection of the molecule. A substituent in front of the plane is written with a solid valence bond and designated as β; one behind the plane is written with a broken valence bond and designated α. Thus the configuration of the hydroxyl group at C-3 in cholesterol (III, above) is β, and that in deoxycholic acid (IV) is α.

Since the same key degradation products, such as the 17-ketone (I) or the 20-carboxylic acid (II) were obtained by simplification of the side chain structures of different groups of naturally-occurring steroids, it was early recognized that

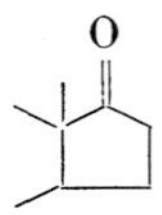
I

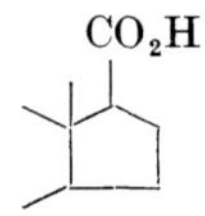
II

the configurations of all of the nuclear asymmetric centers with the exception of that at C-5, and, in the cardiotonic glycosides, at C-14, must be the same. X-ray measurements of the molecular dimensions of a number of steroids showed the skeleton to be rather flat, with the four rings extended in a lath shape. This information permits certain conclusions to be drawn regarding the relative configurations of the ring junctions.

Each pair of contiguous rings in the nucleus, e. g. rings B and C, can be fused in two ways, with the substituents at the coannular atoms C-8 and C-9 in the same *(cis)* (III) or opposite *(trans)* (IV) configurations. Similarly, the configurations of the two adjacent ring residues attached to a central ring, as at C-8 and C-14 in ring C, may be the same (V) or opposite (VI).

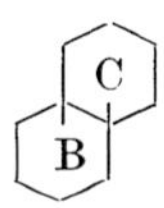

III

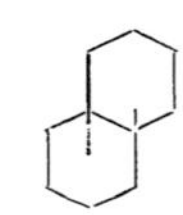
IV

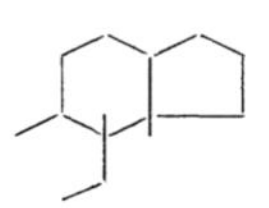
V

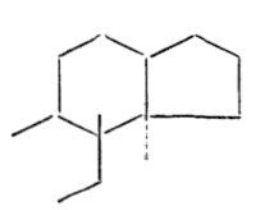
VI

The relative configurations at two asymmetric centers common to only one ring, as at C-8 and C-14, are designated *syn* (V) and *anti* (VI) to distinguish the relationship from that involving two atoms common to two rings (e. g., C-8 and C-9). These different stereochemical arrangements about the central "backbone" impart considerably different shapes to the molecule. Only with an *anti-trans-anti* relationship of carbon atoms 10, 9, 8, and 14 (VII) is the backbone completely extended as required by the physical measurements.

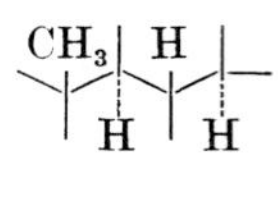
VII

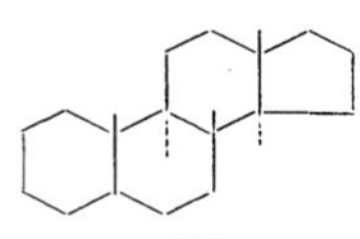
VIII

To establish a standard structural expression for the steroid nucleus, the methyl group (C-19) at C-10 was arbitrarily written in the β-configuration; the other ring junctions were then assigned the relative configurations 9α, 8β, and 14α (VIII) (14β in the cardiotonic glycosides).

The C/D ring fusion was eventually established as *trans* (*cis* in the 14β-glycosides) and the C-13 methyl group is thus in the β-configuration in all natural steroids. The A/B ring fusion, unlike all others, is variable, and compounds in both the *cis* (5β-hydrogen) and *trans* (5α-hydrogen) series, as well as compounds with a double bond at 4—5 or 5—6, are found among natural steroids. In structural formulas, the configurations at C-9, C-8 and C-14 are usually not specifically designated, but *the central chain always comprises the 10β, 9α, 8β, 14α, 13β configurations, unless explicitly indicated otherwise.*

These nuclear configurations have all been confirmed by chemical evidence of several types. One of the most reliable criteria for the configuration at a nuclear center depends upon the fact that a *trans* fusion of two six membered rings is a more stable system than a *cis* fusion (this is not necessarily the case with a six-membered ring fused to a five-membered one). A *cis* decalin will isomerize irreversibly to the *trans* compound if the barrier between them is overcome, as in the alkaline treatment of a *cis*-decalone (IX), in which the anion (X) reverts to the more stable *trans* ketone (XI). Thus in a steroid substituted at C-7 or C-11 by a carbonyl group (XII), the hydrogen atom at C-8 or C-9, respectively, is removable by base, and isomerization would occur in either case if the ring fusion were *cis*. A second source of information concerning configurational relationships is the formation, or non-formation, of a ring by interaction of two suitably disposed groups. For example, the facile closure of an oxide ring between C-3 and C-9 (XIII) which has been observed in certain derivatives of

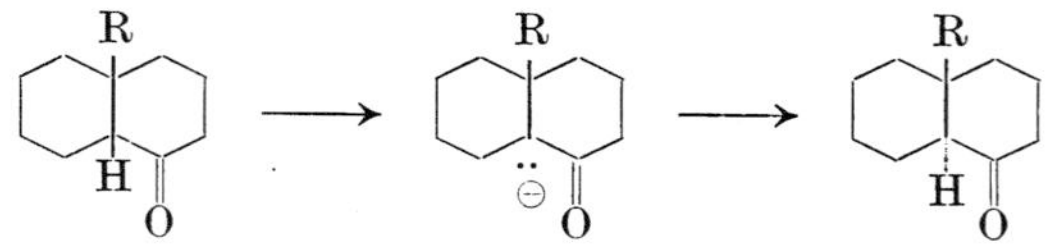

IX X XI

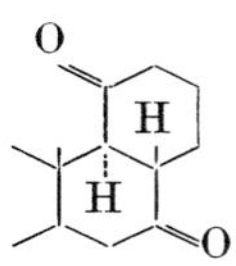

XII

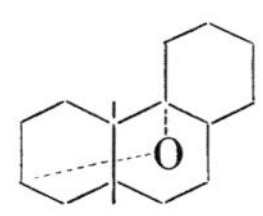

XIII

deoxycholic acid (IV, p. 187) requires not only that the configurations at both points of attachment be the same, and moreover both α (otherwise, interaction with the β-oriented C-10 methyl group would be prohibitive), but also that the A/B ring fusion be *cis* to bring the two centers within bonding distance. Still another approach to the diagnosis of configuration is the application of conformational principles (p. 191), which have led to an increasing understanding of the stereochemical requirements of elimination and addition reactions. A summary of the basis for the configurational assignment at various asymmetric centers has been compiled by TURNER (1949).

2. Absolute configuration

It was noted that the stereochemical convention expressed in the projection formula (I) is based on the arbitrary assignment of the β-configuration at the reference point, C-10, with the nucleus written with ring A to the left. The choice of the α-configuration at this point, with C-19 in back of the projection and the other centers in the corresponding *anti-trans-anti-trans* relationship, would have expressed equally well the relationships which were established by the evidence available at the time, but would have resulted in the mirror image (II). One of these two enantiomeric formulations, which differ in absolute configuration, must

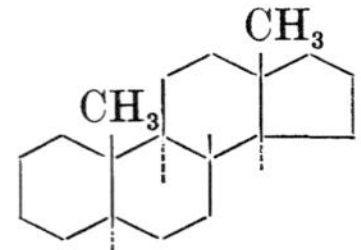

I

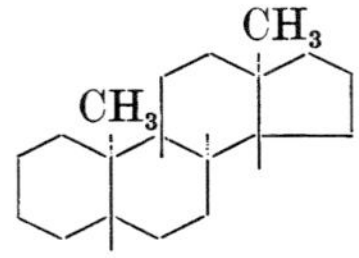

II

represent the actual spatial arrangement; a decision as to which is correct was until recently impossible. It is evident from the structural interrelationships discussed in the preceding section that *all natural steroids belong to a single configurational series.* Both enantiomeric forms are produced in a total synthesis (p. 261), however, and these can be separated as pure individuals that differ only in the sign of optical rotation and in biological activity. *In the case of enantiomeric hormone structures, only the form having the absolute configuration or "hand" of the natural series* (I) *is biologically active.*

A closely related and equally important question bearing on absolute configuration concerns the correspondence of the accepted formulation (I) with the primary reference standard for configuration, namely, glyceraldehyde. This second question is of especial significance in studies of enzymatic reactions since a comparison of the action of a stereospecific dehydrogenase, for example, in the steroid and carbohydrate series would be meaningless if the stereochemical conventions in the two series were different.

This problem of the correlation of configurational conventions in the steroid and carbohydrate series, both of which were, independently, based upon an arbitrary choice, is perhaps best visualized by considering the first type of solution attempted. In this approach (Fig. 1), a single asymmetric carbon in the steroid, whose relative configuration with respect to the reference point C-10 is known, is separated from the molecule by reactions which do not affect the asymmetric center. The configuration of the asymmetric center in this fragment is then correlated with that of a reference substance.

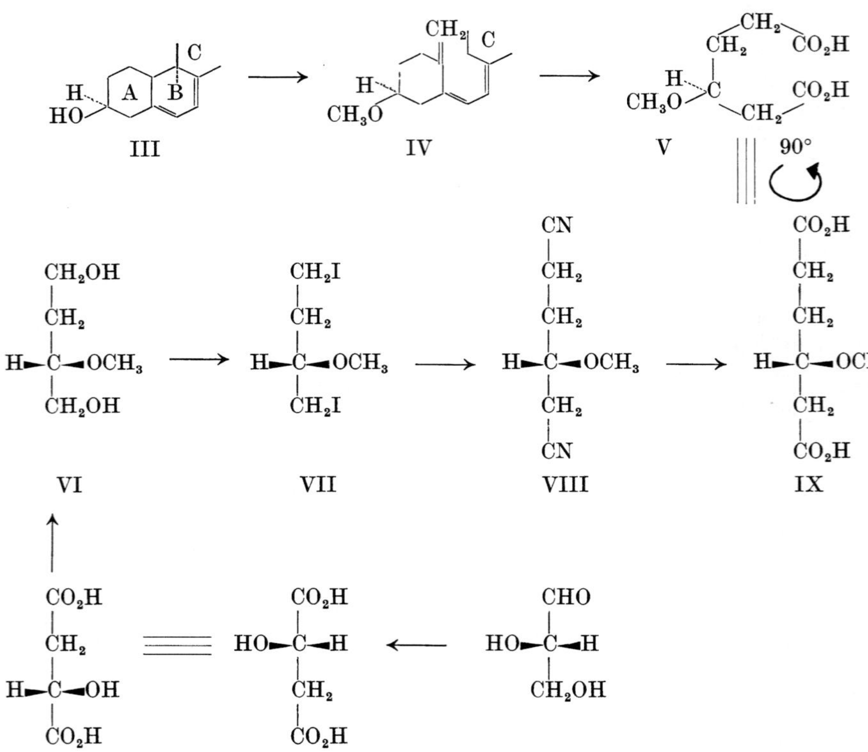

Fig. 1. Correlation of steroid configuration with D-glyceraldehyde

The compound selected for study was calciferol (Fig. 1, IV), obtained by irradiation of ergosterol (III), in which the β-configuration of the C-3 hydroxyl group was established, and which was readily cleaved, by oxidation of the methyl ether, to β-methoxyadipic[1] acid (V).

[1] The index β refers to the position in the acid and has no significance with respect to the configuration in the steroid.

If the two conventions agree, the β-methoxyadipic acid obtained in this degradation, shown in Fischer projection in (IX), should have the same configuration, and therefore sign of rotation, as the L-β-methoxyadipic acid prepared from L-malic acid (XI). The configurational identity of the asymmetric center in L-malic acid and L-glyceraldehyde (XII) has been rigorously established. The first results obtained by this series of transformations (LARDON and REICHSTEIN 1949) suggested that derivatives of the acids from the two sources were of opposite rotation, and that the steroid convention was therefore inconsistent, but the observed rotations were very small, and the conclusion was strictly tentative. Recent investigations in two laboratories (VISCONTINI and MIGLIORETTO 1955, BRENNEISEN et al. 1956) have led to a very clear demonstration that the acids (IX) from the two sources are indeed identical, and that the conventions are therefore consistent.

This same type of chemical correlation of configuration has been successfully applied to fragments containing other asymmetric centers of the steroid molecule (RINIKER et al. 1954, CORNFORTH et al. 1954). Indirect methods, based upon the analysis of molecular rotations (p. 373) and upon asymmetric synthesis have also been used to establish correspondence of the steroid and aliphatic series. The first solid evidence that the usual steroid projection is correctly written was obtained by the latter approach (W. DAUBEN et al. 1953) (Fig. 2). The glyoxylate esters (XIII or XVI) of several hydroxy steroids were treated with methyl magnesium bromide and the products (XIV or XVII) hydrolyzed, giving the hydroxy acids (XV or XVIII). As indicated by the arrows, attack occurs in the two epimeric esters (XIII and XVI) essentially at opposite sides of the carbonyl chain so that hindrance to approach of the reagent by the bulky group L is minimized. The new asymmetric center which is created by addition of the Grignard reagent is therefore of opposite configuration in the two cases (the reaction is not completely stereospecific, but a preponderance of one configuration is formed if the difference in size of the two groups L and M is sufficiently large). The sign of rotation of the acid (XV or XVIII) provides a configurational link between the original steroidal alcohol and simpler alicyclic and aliphatic alcohols of known absolute configuration.

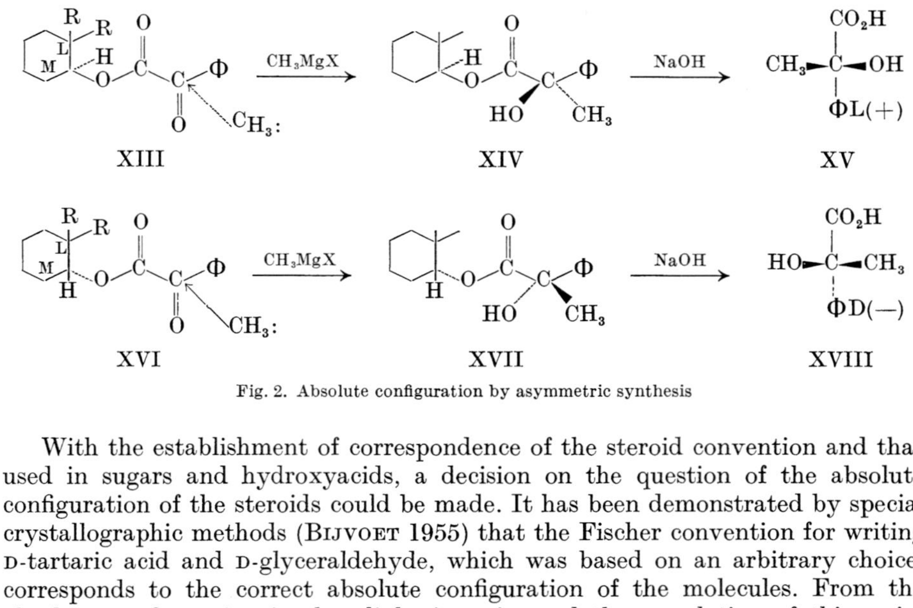

Fig. 2. Absolute configuration by asymmetric synthesis

With the establishment of correspondence of the steroid convention and that used in sugars and hydroxyacids, a decision on the question of the absolute configuration of the steroids could be made. It has been demonstrated by special crystallographic methods (BIJVOET 1955) that the Fischer convention for writing D-tartaric acid and D-glyceraldehyde, which was based on an arbitrary choice, corresponds to the correct absolute configuration of the molecules. From the absolute configuration in the aliphatic series and the correlation of this series with the steroids, it follows that *the usual steroid notation* (I) and not the enantiomorph (II) *represents the actual "hand" of the molecule.*

3. Conformational principles

The term *conformation* refers to the spatial arrangement of the atoms in a molecule of given structure and configuration. Although a multiplicity of conformations might be conceived in a molecule that can assume different shapes, it is necessary to consider only a few that are thermodynamically most favorable, i. e., those of lowest energy content. The lowest energy level, and thus the preferred conformation, is achieved when the steric strain, due to various factors including non-bonded interactions of the several atoms and groups in the molecule, is minimal.

One of the most fruitful applications of conformational principles has been in the chemistry of six-membered rings, which are not planar but may assume one

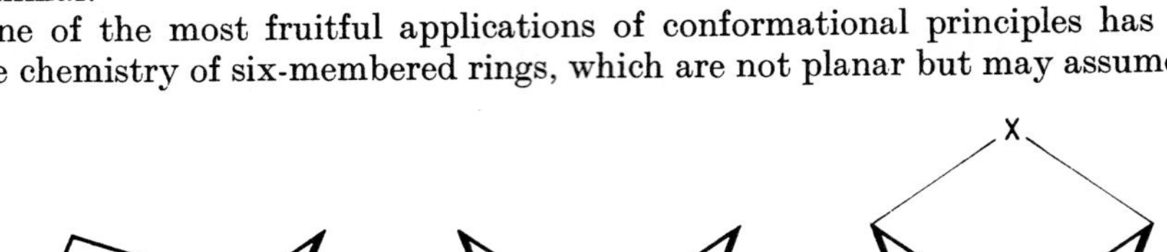

I II III

of two conformations, corresponding to the general shape of a chair (I) or a boat (II). There is a considerable body of evidence that the chair form is the more stable in the absence of special features, as in a bridged ring such as (III), which is constrained by the requirements of the bond angles into a boat form. This greater stability of the chair form is a consequence of lesser steric interference of the substituent atoms or groups on the ring. It must be understood that the

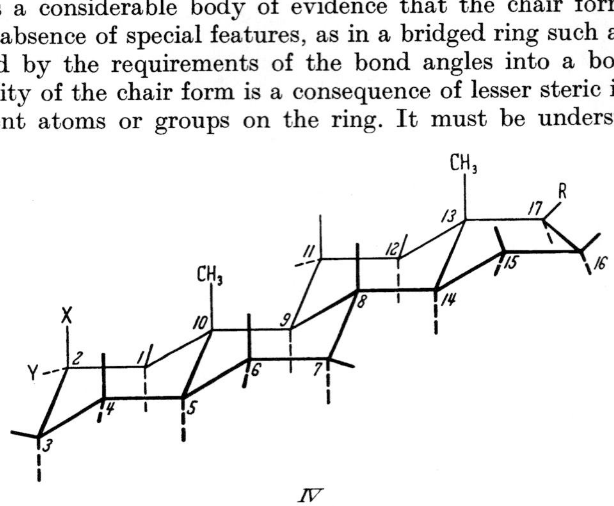

IV

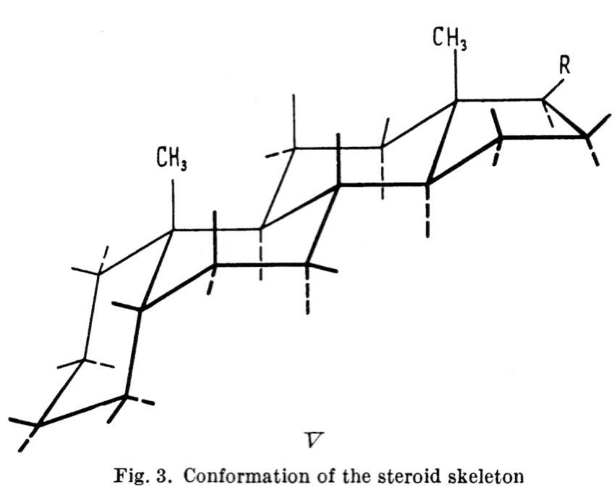

V

Fig. 3. Conformation of the steroid skeleton

interconversion of these two forms, which does not involve the breaking of bonds, is energetically a very facile process, and "conformational isomers" are unknown. *In a relatively rigid fused-ring system such as the steroid nucleus, the rings are locked into an "all chair" conformation,* which can be represented for the case of 5α compounds (A/B trans) as (IV, Fig. 3), and for 5β (A/B cis) as (V). In the case of a double bond at 4—5 or 5—6, the conformation of ring A approximates that in (IV). In a steroid in which one of the "backbone" atoms has the unnatural configuration, e. g., 8α (8-iso) or 9β (9-iso), either ring B or ring C is forced to assume a boat form; a corollary of this is the conclusion that *the natural steroid skeleton is in the most stable conformation,* a fact abundantly confirmed by chemical evidence.

In the formulations (IV) and (V) (in which the solid and broken valence bonds have the usual *configurational* significance), it will be seen that one of the two substituents at each ring atom is directed perpendicularly away from the projection plane of the ring, e. g., group "X" at C-2 in (IV); this orientation, in this case in the β-configuration, is termed *axial* (formerly polar). The other substituent extends *in* the projection plane, e. g., group "Y", the α substituent, and is in the *equatorial* conformation. *Proceeding around a ring, substituents in the same configuration will be alternately axial and equatorial.* It will also be noted that *the change of the A/B ring junction from trans to cis reverses the conformations of substituents at C-1, 2, 3, and 4.*

The significance of the conformation of steroid molecules, and especially the concept of a difference of geometry of ring substituents and the consequent differences in their behavior have been discussed by BARTON (1953) and subsequently by KLYNE (1954) and BARTON and COOKSON (1956). These principles have led to a much better understanding of the chemical properties of steroids, and *the prediction of the stereochemical outcome of reactions and interpretation of the considerable differences in reactivity of functional groups at various centers have become possible.*

Due to the spatial crowding between a group in an axial conformation at a given position and axial groups at other centers that are in a 1,3-relationship, a substituent is appreciably less stable in the axial than the corresponding equatorial conformation. Examples of such axial orientations are those at C-11β (1,3-interactions with axial methyl groups at C-10β and C-13β, and hydrogen at C-8β) and at C-6β (1,3-axial methyl at C-10β, axial hydrogen at C-4β and C-8β). As a result of these "non-bonded interactions" *an epimerization* (reversal of configuration) *will always occur in the sense axial → equatorial.* By the same token, an equatorial group, being less hindered, is more reactive in any process in which the ease of reaction depends primarily on the accessibility of the substituent itself to attack by a reagent, as in the esterification of a hydroxyl group. The most important requirement for an ionic elimination reaction, on the other hand, is an antiparallel (*trans* and coplanar) arrangement of the groups involved. This condition is more nearly fulfilled when the two adjacent groups are both axially disposed, as with a hydroxyl or tosyloxy group at C-7α and the hydrogen atom at C-8β. *An axial group will therefore be more susceptible, in general, to elimination than an equatorial one.* A further very useful generalization relates to *ring-opening reactions of epoxides,* in which *the newly formed substituents always assume axial conformations.* Further reference to these and other consequences of conformation will be made in subsequent sections, especially as they apply to the problem of the assignment of configuration.

II. Nomenclature

As the number and variety of steroidal compounds multiplied during the past twenty years, a formal system of nomenclature developed which, with numerous variations in detail, has been universally adopted. In 1951 a set of recommendations for a standardized steroid nomenclature was published (CIBA Foundation 1951), embracing existing usuage as far as possible. (A slightly different set of proposals is presented in an unpublished report of the National Research Council

Subcommittee on Steroid Nomenclature, available from Chemical Abstracts Service.) One of the principal aims of this standardized system was to provide a unique and unambiguous name for every steroid to meet increasingly formidable indexing problems. Such systematic names are, however, inevitably cumbersome for even moderately complex compounds, and abbreviated or trivial names will certainly retain a continuing importance in the literature and are widely used in this volume. Before considering the problem of trivial names, the main features of the recommended system (and important departures from it) will be briefly set forth.

1. General principles

The systematic name of any steroid is based upon the modification, by prefixes and suffixes, of one of a small number of names of parent hydrocarbons. The parent compounds of importance in the adrenocortical series are indicated in Fig. 4. The convention for designating the variable configuration at C-5 is undergo-

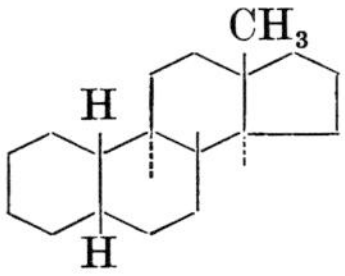

I

Estrane

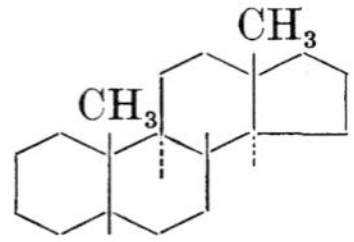

II

Etiocholane
(Testane)
(5 β-androstane)

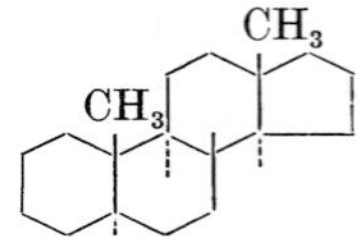

III

Androstane
(5 α-androstane)

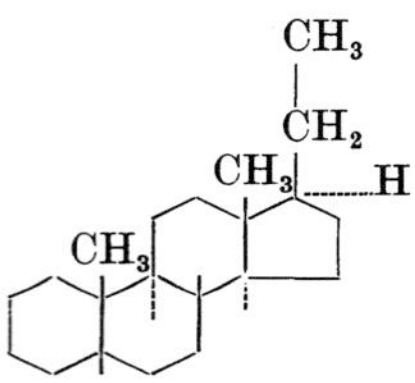

IV

Pregnane
(5 β-pregnane)

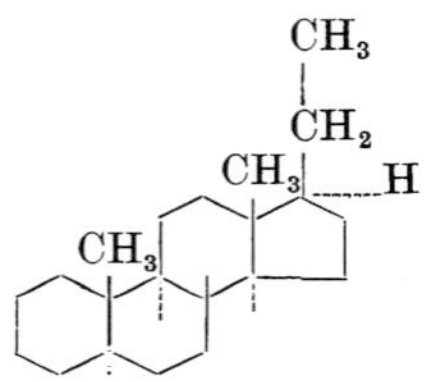

V

Allopregnane
(5 α-pregnane)

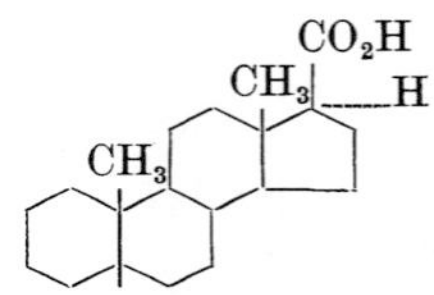

VI

Etianic acid
(5 α configuration,
alloetianic acid)

Fig. 4. Parent hydrocarbons

ing a transition. In the bulk of the American and German literature up to 1959 the terms *pregnane* and *allopregnane* were used for compounds IV and V, respectively, the prefixing syllable *allo* denoting the 5 α-configuration. A recent trend has been the use of *5 β-pregnane* and *5 α-pregnane* as parent names for IV and V, and of *5 β-androstane* and *5 α-androstane* for compounds II and III, respectively. The prefixes 5 β- for pregnane and 5 α- for androstane were specifically implied in the traditional usage. The newer convention has been adopted by Fieser and Fieser (1959), but the names *allopregnane* and *etiocholane* have been retained in this volume. (The name "testane" for II, recommended in the CIBA report, has not gained acceptance, and is not used in this chapter.) In either convention, the names in cursive in Fig. 4 (III and IV) are the ones used as stem names in case asymmetry at C-5 is absent due to the presence of a double bond.

Regardless of the system used for naming C-5 isomers, the configuration at all other nuclear centers is specified as that of the natural steroid series in all of these names. If an asymmetric center in the carbon skeleton is inverted, i. e., of opposite configuration to that in the natural series, the configuration at this center is specified, as in 17α-pregnane (VII) or 9β-androstane (VIII) (Fig. 5). A common alternative convention is the use of "17-iso-" or "9-iso-." Modification of the nucleus by the removal or addition of a carbon atom is designated by the prefixes "nor" and "homo," respectively, with the appropriate position or ring being designated, as in 19-norpregnane (IX) or D-homoandrostane (X).

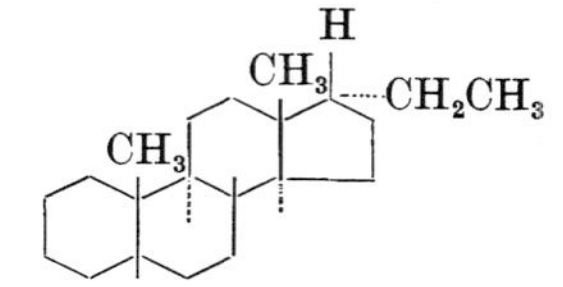

VII

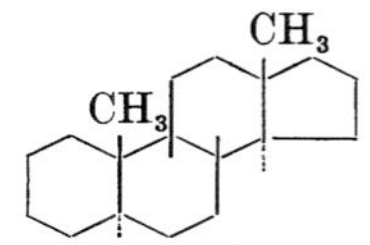

VIII

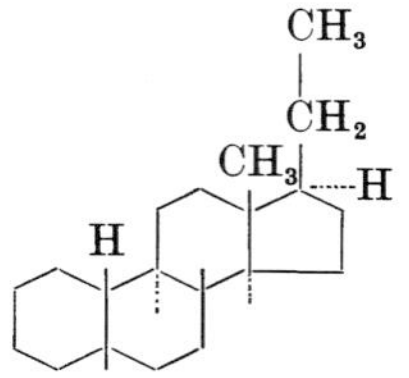

IX

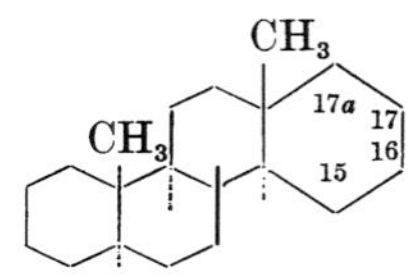

X

Fig. 5. Modified skeletons

The terminal ending "-ane" is changed to "-ene", "-diene," etc., if one or more double bonds are present. Other functional groups, such as hydroxy, keto, or aldehydo, are designated by these names used as prefixes, or "-ol, -diol," "-one, -dione" or "-al," respectively, as suffixes.

The prefix or suffix is always *preceded* by the number of the carbon atom which is substituted and by the configurational index, α or β, when pertinent. The significance of these configurational indices is different from that defined on p. 188 when applied to a substituent at C-20; the stereochemical convention at C-20 is indicated in Fig. 13 (p. 208). In the case of a double bond, only the lower numbered of the two carbon atoms is designated, unless the second is not the next higher numbered, in which case the number of the second atom is given in parentheses. In American and continental European usage, the number designating the position of a double bond precedes the parent name; in British usage the number is inserted in the middle.

The recommended system requires that no more than one functional group, excluding double bonds, be named in a suffix, and this practice is followed in the systematic names used in this chapter. This rule is a departure from a widely used, and still prevalent, practice of using two or more suffixes (e. g., pregnane-3β-ol-20-one). The latter usage gives rise to shorter names and will probably continue to be widely employed.

A problem that has only recently arisen is a means of specifying unambiguously the configurational relationships in steroids of the unnatural stereochemical series which arise from total synthesis (p. 261). In a total synthesis of 3α-hydroxy-androstane-17-one, for example, natural androsterone and its mirror image will be formed.

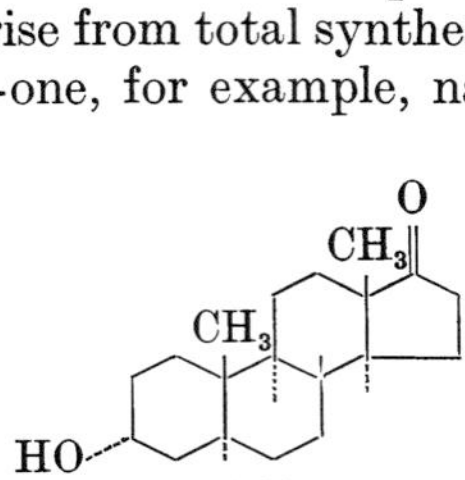

XI

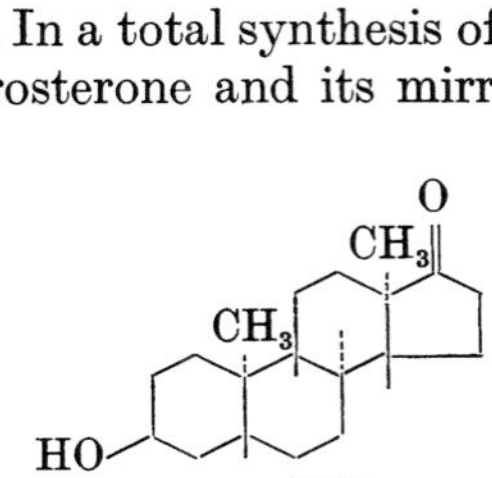

XII

The two enantiomorphs have been distinguished by the prefixes *d*- and *l*- (v. Euw et al. 1955), or by "naturalo-" and "enantio-," respectively (Horeau et al. 1955). The name *l*-3α-hydroxyandrostane-17-one for (XII), however, is inadmissable if the indices α and β are used in the traditional sense, since every asymmetric center in the mirror image (XII) is inverted. To circumvent this difficulty, Lardon et al. (1957) have suggested that the indices α and β be given a relative significance, as in the standard practice of carbohydrate chemistry. An α substituent would then be defined as one in the configuration opposite *(anti)* to the methyl group at the C-10 reference point (or the C-13 methyl group in 19-nor compounds), and a β substituent as one in the same configuration as the angular methyl groups. This suggestion seems to have much merit, and it has received unofficial approval.

2. Trivial names

In the course of their isolation and characterization, nearly all of the steroid hormones were christened with descriptive names, and most of these have become an integral, if officially unsanctioned, part of steroid nomenclature. In Table 1 is given a list of the trivial names of the hormones that are widely used and recognized, together with the systematic names.

All of these names are structurally and stereochemically specific and are universally recognized. It is unfortunate that several inconsistencies have arisen. Thus the generic ending "-erone" would preferably be reserved as an indication of the 4-ene-3-keto function, but it has been applied in two cases to compounds with the ketonic function at C-17. Particularly glaring is the paradox that has arisen with the name "cortisol." After the name cortisone had gained wide currency, the equally-important 11-hydroxy counterpart was termed "hydrocortisone." This led to considerable confusion with the 4,5-dihydro derivatives of cortisone, and the name "hydrocortisone" has been widely replaced by "cortisol." This is completely at variance with the traditional ending for steroid hormones, but the name has gained widespread usage and is employed in this volume.

Table 1. *Common names and synonyms*

Trivial name	Systematic name
Estrone	3-Hydroxy-1,3,5(10)-estratrien-17-one
Estradiol	1,3,5(10)-Estratriene-3,17β-diol
Estriol	1,3,5(10)-Estratriene-3,16α,17β-triol
Androsterone	3α-Hydroxyandrostan-17-one
Testosterone	17β-Hydroxy-4-androsten-3-one
Dehydroepiandrosterone (Androstenolone)	3β-Hydroxy-5-androsten-17-one
Adrenosterone	4-Androstene-3,11,17-trione
Progesterone	4-Pregnene-3,20-dione
11-Deoxycorticosterone (DOC, Cortexone). . . .	21-Hydroxy-4-pregnene-3,20-dione
Corticosterone	11β,21-Dihydroxy-4-pregnene-3,20-dione
11-Dehydrocorticosterone .	21-Hydroxy-4-pregnene-3,11,20-trione
Cortisone	17α,21-Dihydroxy-4-pregnene,3,11,20-trione
Cortisol	11β,17α,21-Trihydroxy-4-pregnene-3,20-dione
Aldosterone	11β,21-Dihydroxy-3-keto-4-pregnen-18-al

As these trivial names gained currency, they were often used, with modifications, as parent names, sometimes with unfortunate results. The indication of a substituent or other modification in one of these hormones by the use of the trivial name with a sufficiently precise prefix including designation of position and configuration is accepted practice, and leads to convenient and unambiguous names, as 6β-hydroxycortisol.

However, the practice of designating a specific structure, as of a urinary metabolite or other naturally occurring substance, by using only a trivial or systematic parent name together with the number and nature of substituents can lead to confusion and is to be discouraged. The most frequently encountered examples are "etiocholanolone" (3α-hydroxyetiocholan-17-one), "isoandrosterone" (3β-hydroxyandrostan-17-one), "pregnanediol" (usually implies 3α,20α), "tetrahydrocortisone," and analogous names for derivatives of cortisol and corticosterone. These last names, and the corresponding "allotetrahydro-" series, are used for urinary reduction products of the hormones, and all imply the 3α-hydroxypregnane or 3α-hydroxyallopregnane structures. The names "urocortisone" and "urocortisol" have been suggested for the respective 3α-hydroxypregnane compounds.

The prefix "epi-" invariably denotes inversion of configuration of a substituent (usually a hydroxyl group), as in "11-epicortisol" (11α, 17α, 21-trihydroxy-4-pregnene-3,20-dione). The prefix "iso-" should be, but has not always been, reserved to denote inversion at a nuclear center. The prefix "desoxy" or "deoxy" refers to the absence of a hydroxyl group; if the position is not indicated, C-11 is implied.

A final complication in the nomenclature of the adrenal steroids is the continued use of alphabetical designations. Several of the early letter designations of the corticosteroids have gained wide currency as convenient abbreviations; these are indicated in Table 2, p. 202. This practice can lead to confusion, since several series of alphabetical designations have been used, and the name of the author is not always specified. Reichstein's compound S (17α,21-dihydroxy-4-pregnene-3,20-dione) and Kendall's compound E (cortisone) and F (cortisol) are most frequently encountered and have been used as parent names, as in "tetrahydro-E" (urocortisone) or "epi-F" (11-epicortisol).

B. The adrenocortical steroids

I. The steroid content of the adrenal cortex

1. Isolation of steroids

Pioneering investigations in the isolation and characterization of the principle responsible for the vital function of the adrenal cortex were begun in the period 1932—1935 in the United States by Kendall and by Pfiffner and Wintersteiner and in Switzerland by Reichstein. Outstanding progress was made in the span of a very few years; this earlier work has been discussed in several reviews, among them by Reichstein (1938), Pfiffner (1942), Reichstein and Shoppee (1943), and Heard (1948), and only a brief summary is included here. A detailed description of isolation procedures is presented in a later section (p. 292ff.).

The earliest isolation studies were evaluated by assays based upon the life-maintenance of adrenalectomized dogs and by a muscle work test (Everse de Fremery); the two methods furnished roughly comparable values of the "cortin" content of extracts. Most of the procedures developed for the preparation of crude extracts were based upon whole beef adrenals, since dissection of the cortex is impractical on a large scale. The glands were extracted with cold alcohol or acetone, and after a sequence of distributions between immiscible solvent phases to remove fats and other lipides, the "cortin" acticity was obtained in an aqueous solution containing approximately 60 g of solids per metric ton of fresh glands. Such an extract had to be treated with permutit or sulfite to remove epinephrine before administration. Further purification of the active material was effected by exhaustive extraction with ether or ethyl acetate and reextraction into water.

Chemical fractionation became possible with the observation that the total activity could be precipitated by phenylhydrazine. A very efficient separation of ketonic and non-ketonic material was accomplished with the use of Girard's reagent (p. 302); regeneration of the biologically-active ketonic material from the Girard derivative required very cautious treatment with acid, since the activity was rapidly destroyed by strong acid or alkali.

A number of pure substances were then isolated in rapid succession by crystallization of these highly purified fractions. The first of these was biologically inactive, but several compounds were soon obtained which possessed in varying degress the characteristic "cortin" activity. These products were very similar, with compositions in the range $C_{21}H_{28-30}O_{4-5}$. A number of the compounds showed strong reducing properties; it was further recognized that all of the biologically-active compounds were α,β-unsaturated ketones. The first evidence of a relationship to the known steroid hormones arose from a $C_{19}H_{24}O_3$ compound that was isolated from extracts and also as an oxidation product of several of the C_{21} compounds. This substance was found to be a highly active androgen in the capon test, and a close structural similarity to androsterone was apparent, from the empirical formula and androgenic activity. The structures of the C_{21} compounds were then rapidly established by the approaches outlined in the following section.

A very important advance in the isolation work was made with the introduction of chromatography of the acetylated corticosteroids on alumina. Previously uncrystallizable mixtures were resolved by this technique, and a number of less highly oxygenated compounds, most of them inactive, were obtained. By 1940, 27 steroids of established structure, including 6 substances that were active in some or all of the assays of adrenocortical function, had been isolated from adrenal extracts. In spite of this remarkable accomplishment, a major part of the life-maintenance activity remained in amorphous fractions of the extracts, and the nature of the constituents of this material remained obscure for more than a decade.

With the development of several revolutionary techniques, notably partition chromatography and an elegant urinary sodium-potassium ratio assay based upon radioactive tracers, a renewed effort in the isolation problem became possible. This was climaxed by the isolation by SIMPSON et al. (1953, 1954a), MATTOX et al. (1953) and KNAUFF et al. (1953) of the highly potent aldosterone, which, although present in extremely small concentration, is responsible for the major fraction of the activity of adrenal extracts measured by replacement assays in adrenalectomized animals. Continuation of this work, with application of the most refined chromatographic techniques to the fractionation of large amounts of adrenal extract, has resulted in the isolation of 16 further steroids whose structures have been established by NEHER and WETTSTEIN (1956b), v. EUW et al. (1958) and NEHER et al. (1959).

2. Approaches to structure determination

Before discussing the individual structures of the adrenal steroids and the chemistry of the several functional groups, it is desirable to summarize the evidence on which these structures are based. In the following paragraphs is presented a very brief account of the main lines of evidence and correlation that led to the structural formulation of the large group of "prewar" corticosteroids. This work is due principally to REICHSTEIN, and has been reviewed by REICHSTEIN and SHOPPEE (1943).

Following the suggestion of a relationship of one of the adrenal constituents to the androgenic steroids, rapid progress was made in the structure elucidation of the compounds of the adrenal series. The compounds were conveniently classified according to the number of oxygen atoms present, and the members of the $C_{21}O_5$ group, and separately the $C_{21}O_4$ group, were first interrelated.

Each of the $C_{21}O_5$ compounds could be oxidized with chromic acid to one of two $C_{19}O_3$ derivatives, eventually shown to be (I) and (II) (Fig. 6), thus establishing the presence of three oxygen atoms associated with a two-carbon side chain and, further, the common location of the side chain and the two other oxygen atoms. The 2 C_{19} compounds differed only in the presence of a conjugated double bond, and an interrelationship was established by catalytic hydrogenation. Drastic Clemmensen reduction of the saturated C_{19} compound to a mixture of androstane (IV) and androstan-17β-ol (III) established the *steroidal skeleton* and the *position of the side chain.*

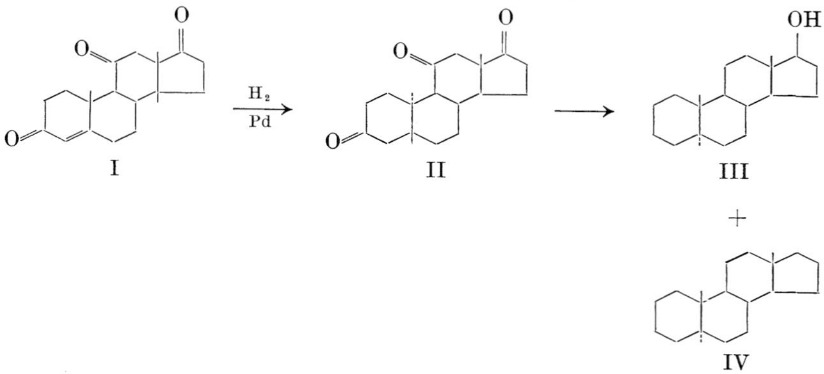

Fig. 6. Conversion of cortical derivatives to known steroid structures

The nature of the functional groups in which the 5 oxygen atoms were involved was inferred from analytical data obtained with the hydroxy compounds and the acetates and the results of oxidation studies. A number of the compounds had reducing properties typical of aldehydes or α-ketols; these were degraded by periodic acid to acidic products (VI) with the loss of one carbon atom and were thus ketols. All of the non-reducing steroids in the group were oxidized by periodate to C_{19} compounds. This information, taken together with the results of chromic acid oxidation, established *the two side chain structures as the dihydroxy ketone* (V) *and triol* (VII), respectively.

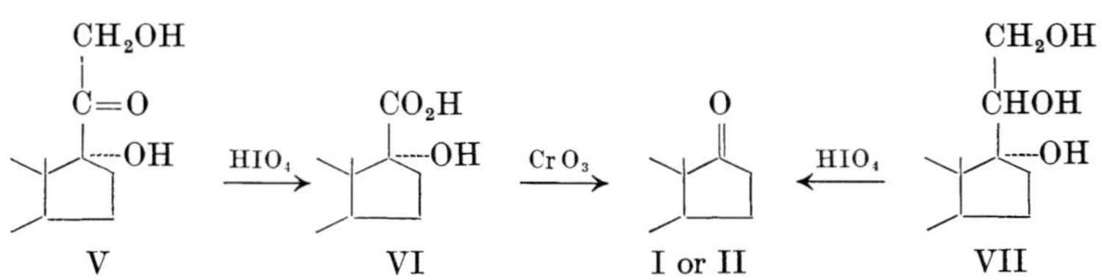

Fig. 7. Side chain degradation

The location of one of the nuclear oxygen atoms at C-3 was inferred by the close relationship to the androgens, but the nature and position of the fifth oxygen atom presented special difficulties. It was recognized that this was present as a secondary hydroxyl group in some of the compounds and the corresponding ketone in others. The hydroxyl group could not be esterified, but it was readily oxidized to the very unreactive keto group, or removed by dehydration with mineral acid. Catalytic hydrogenation of the derived anhydro product (IX) (Fig. 8) of one of the C_{19} compounds (VIII) led to the known androstane-3β,

17 β-diol (X), *establishing* firmly the position of *the 3-hydroxyl group*. The location of the unreactive oxygen function was tentatively assigned as C-11 by a process of elimination. Since the triketone (II) did not exhibit the properties expected of an α-diketone or a β-diketone, only positions 6,7 and 11 remained; the first two were considered unlikely since carbonyl groups at these positions are not unusually inert. Confirmation of the C-11 position was not obtained until the synthesis of 11-dehydrocorticosterone was accomplished by LARDON and REICHSTEIN (1943).

The foregoing data established a *pattern of substitution* that comprised *either* the *Δ^4-3-ketone or 5α-3β-hydroxyl* grouping in ring A, a *keto or hydroxyl group at C-11* and a *triol or dihydroxyketone side chain* at C-17. With this information the structures of the eight individual members of the $C_{21}O_5$ series were resolved

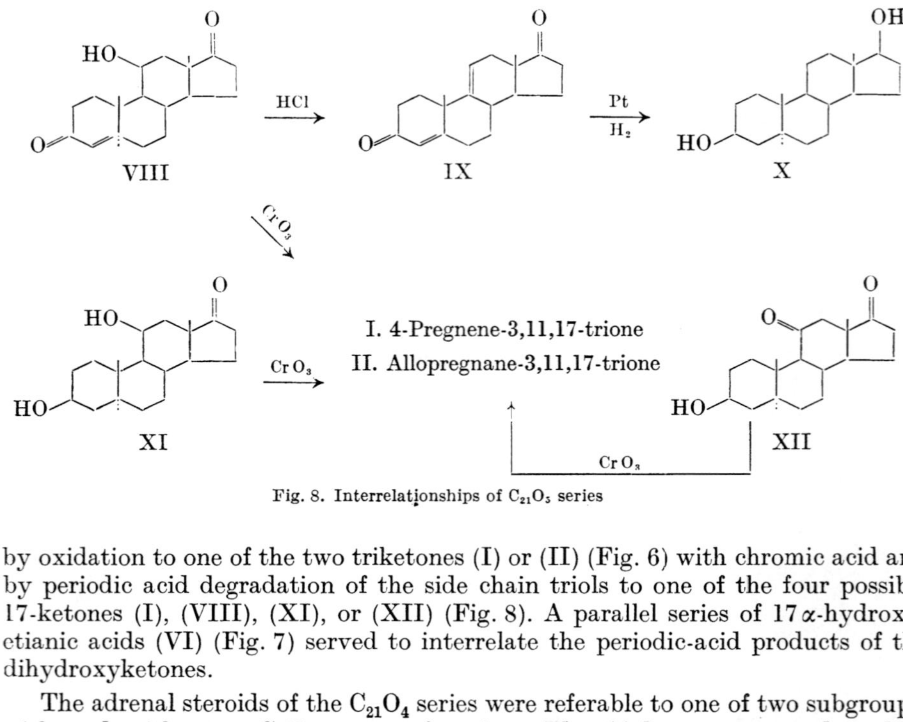

Fig. 8. Interrelationships of $C_{21}O_5$ series

by oxidation to one of the two triketones (I) or (II) (Fig. 6) with chromic acid and by periodic acid degradation of the side chain triols to one of the four possible 17-ketones (I), (VIII), (XI), or (XII) (Fig. 8). A parallel series of 17α-hydroxyetianic acids (VI) (Fig. 7) served to interrelate the periodic-acid products of the dihydroxyketones.

The adrenal steroids of the $C_{21}O_4$ series were referable to one of two subgroups, with and without a C-11 oxygen function. *The 11-deoxy compounds, which contained oxygen atoms at C-3, 17, 20 and 21,* could be immediately related by side-chain degradation to known 17-ketones or 17-hydroxyetianic acids. *The 11-oxygenated compounds* of the $C_{21}O_4$ series were degraded by periodic acid to etianic acids (XIV) (Fig. 9), which resisted further oxidation, and *were thus characterized as 17-deoxy counterparts* (XIII) *of the $C_{21}O_5$ group*. A relationship between these two series was established by means of the Serini reaction (p. 218). A 3,17,20,21-tetrol-11-one triacetate (XV) from the $C_{21}O_5$ series was converted to the 17-iso-ketol (XVI) and thence to the stable isomer (XIII), which was identified with a member of the $C_{21}O_4$ series.

These degradative interconversions were reinforced by partial syntheses, which began for the most part with 17-ketones, and later, as synthetic techniques improved, with 20-ketones. By 1940, the structures of nearly all of the 11-deoxy

$C_{21}O_4$ compounds and the members of the $C_{21}O_3$ and $C_{21}O_2$ series had been confirmed by synthesis. With the advent of methods for the preparation of 11-oxygenated compounds, synthesis of most of the $C_{21}O_5$ compounds became possible. At least one, and in a few cases literally dozens, of synthetic sequences from a standard precursor, e. g., progesterone or 11-ketoprogesterone, can now be traced in the literature for all except one (3α, 11β, 17α, 21-tetrahydroxyallopregnan-20-one, Substance C) of the "prewar" corticosteroids. The reactions employed in these partial syntheses are discussed in subsequent sections; an account documenting each individual synthesis has been eschewed. With this fabric of reciprocal interconversions and partial syntheses, all of the structural features of the corticosteroids, with the exception of certain stereochemical uncertainties, were firmly established by 1943, and the stage was set for the great upsurge of activity in metabolic studies that began in 1946.

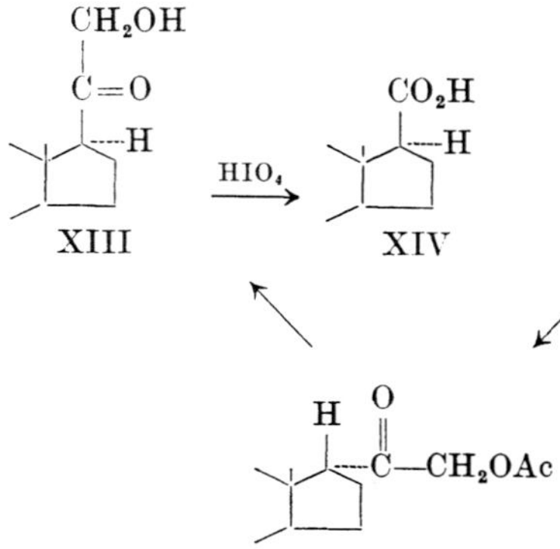

Fig. 9. Side-chain interconversions of $C_{21}O_4$ group

Considerably streamlined approaches have evolved in the structural elucidation of the more recently isolated adrenal steroids and the numerous incubation and perfusion products and urinary metabolites. With increasingly refined methods of separation, many new substances have been isolated in extremely minute amounts, and classical degradation studies are often precluded. The structural parameters of natural steroids have become increasingly well understood, however, and with extensive collections of reference steroids and catalogues of infrared spectra available, identification can be made rapidly on a micro scale. It is sometimes possible to deduce the structure of a new steroid from the data obtained during paper chromatography and confirm it by comparison with an authentic sample prepared for the purpose. Partial synthetic methods are now available for the modification or introduction of functional groups at all but a few positions in the steroid molecule. When a novel and synthetically-inaccessible structural feature is encountered, however, as in the case of aldosterone, a systematic degradation is, of course, still required.

3. The properties and structures of the adrenal steroids

A list of all of the steroids that have been isolated from adrenocortical tissue extracts is given in Table 2, with the more important physical properties. In addition to these 45 compounds, a large number of steroids of adrenal origin have been isolated or detected in urine and blood, especially in perfusion and incubation

Table 2. *Steroids of the adrenal cortex*

Systematic and common name	Letter designations: Reichstein	Letter designations: Kendall	Ref. to isolation	Free steroid M. P.°	Free steroid $[\alpha]_D$[1]	Acetate position	Acetate M. P.	Acetate $[\alpha]_D$[1]	IR[3]
C_{18} and C_{19} Compounds									
1. 3-Hydroxy-1,3,5(10)-estratrien-17-one, *estrone*			a	260	+160 (E)[2]	3—	126		5 6 Ac
2. 4-Androstene-3,17-dione			a	174	+190 (E)[2]				113
3. 3 β-Hydroxy-5-androsten-17-one, *dehydro-epiandrosterone*			b	153	+11 (C)	3—	171—172[g]		139 140 Ac
4. 17 β-Hydroxy-4-androsten-3-one, *testosterone*			c	155	+109 (E)[2]	17—	139—141		130 131 Ac
5. 4-Androstene-3,11,17-trione, *adrenosterone*	G		a	223—224	+262 (E)				123
6. 11 β-Hydroxy-4-androstene-3,17-dione			d	190—195	[2]				192
7. 3 β-Hydroxyandrostane-11,17-dione			n						
8. 3 β,11 β-Dihydroxyandrostan-17-one			a	234—235	+84 (E)	3—	228—229	+70 (D)	166 Ac
$C_{21}O_2$ Compounds									
9. 4-Pregnene-3,20-dione, *progesterone*			a	121; 129	+192 (E)[2]				119
10. 3 β-Hydroxyallopregnan-20-one			a	194	+91 (E)	3—	144	+80 (E)	150 151 Ac
11. 16 α-Hydroxy-4-pregnene-3,20-dione, *16 α-hydroxy-progesterone*			q	225—227	+158 (C)	16—	134—135	+107 (C)	
$C_{21}O_3$ Compounds									
12. 17 α-Hydroxy-4-pregnene-3,20-dione, *17-hydroxy-progesterone*			a	222—223	+104 (C)	17—	243—244	+56 (D)	204
13. 21-Hydroxy-4-pregnene-3,20-dione, *11-deoxycortico-sterone* (*DOC; Cortexone*)	Q		a	141—142	+178 (E)[2]	21—	157—159	+177 (E)	206
14. 3 α,16 α-Dihydroxyallopregnan-20-one			p	209—211; 227—229	+60(M-C)				
15. 3 β,16 α-Dihydroxyallopregnan-20-one			o	260—262	+64(M-C)	3,16-Di	175—177	+20(M-C)	
16. 3 β,17 α-Dihydroxyallopregnan-20-one	L		a	264	+38 (E)	3 — 3,17-Di[h]	188—190 198	+16 (An) —9 (D)	170 Ac

Table 2 *cont.*

Systematic and common name	Letter designations		Ref. to isolation	Free steroid M. P.°	Free steroid $[\alpha]_D$[1]	Acetate position	Acetate M. P.	Acetate $[\alpha]_D$[1]	IR[3]
17. Allopregnane-3 β,17 α,20 α-triol	O		a	222—223	—13 (M)	3,20-Di	250	—30 (An)	84 Ac
18. Allopregnane-3 β,17 α,20 β-triol	J		a	216—217	—8 (E)	3,20-Di	159—160	+25 (An)	85 Ac
$C_{21}O_4$ Compound									
19. 21-Hydroxy-4-pregnene-3,11,20-trione, 11-*dehydrocorticosterone*		*A*	a	178—180	+235 (D)[2]	21—	179—181	+234 (D)	225
									226 Ac
20. 3 β,21-Dihydroxyallopregnane-11,20-dione	N	H	a	189—191	+94 (E)	3,21-Di	144—145	+86 (D)	211 Ac
21. 11 β,21-Dihydroxy-4-pregnene-3,20-dione, *corticosterone*	H	*B*	a	180—182	+222 (E)[2]	21—	145; 153	+200 (D)	213
									214 Ac
22. 17 α,21-Dihydroxy-4-pregnen-3,20-dione, *11-deoxycortisol (17-hydroxy DOC)*	*S*		a	213	+132 (C)[2]	21—	239—241	+116 (An)	215 Ac
						17,21-Di[h]	220—221	+50 (D)	
23. 19,21-Dihydroxy-4-pregnene-3,20-dione, *19-hydroxy DOC*			d	165—168[i]	+180 (C)	19—[i]	189—190	+215 (C)	
			e			21—	197—199	+178 (C)	
						19,21-Di	127	+210 (C)	
24. 20 β,21-Dihydroxy-4-pregnene-3,11-dione	T		a	224[j]	+176 (E)	20,21-Di[j]	207—208	+170 (E)	212 Ac
25. 3 β,11 β,21-Trihydroxyallopregnan-20-one	R		a	202—204		3,21-Di	173—174	+84 (D)	183 Ac
26. 3 β,17 α,21-Trihydroxyallopregnan-20-one	P		a	230—239	+48 (E)	3,21-Di	208—209	+45 (D)	184 Ac
27. Allopregnane-3 β,17 α,20 β,21-tetrol	K		a	198—200	—1 (E)	3,20,21 Tri-	178—179	+53 (An)	90 Ac
$C_{21}O_5$ Compounds									
28. 6 β,21-Dihydroxy-4-pregnene-3,11,20-trione			d	206—208		6,21-Di	178—183		
29. 11 β,21-Dihydroxy-3,20-diketo-4-pregnen-18-al, *aldosterone*			f	110—114;	+152 (An)	21—	198—199	+122 (C)	Ref. f
				153—158		18,21-Di	Ca. 70		
30. 17 α,21-Dihydroxy-4-pregnene-3,11,20-trione, *cortisone*	Fa	*E*	a	212—215	+209 (E)[2]	21—	239—241	+218 (C)	231 Ac
						17,21-Di[h]	234	+113 (D)	
31. 3 α,17 α,21-Trihydroxyallopregnane-11,20-dione			n	212—214		3,21-Di	221—223	+94 (D)	
32. 3 β,17 α,21-Trihydroxyallopregnane-11,20-dione	D	G	a	238—242	+66 (E)	3,21-Di	223—224	+72 (D)	218 Ac
33. 11 β,17 α,21-Trihydroxyallopregnane-3,20-dione			d	219—224		21—[k]	210—212	+69 (C)	
34. 6 β,11 β,21-Trihydroxy-4-pregnene-3,20-dione, *6 β-hydroxycorticosterone*			d	225—227	+118 (D)	6,21-Di	182—188		
35. 6 β,17 α,21-Trihydroxy-4-pregnene-3,20-dione			d	229—233[l]	+44 (C)	6,21-Di	192—195	+62 (C)	
36. 11 β,17 α,21-Trihydroxy-4-pregnene-3,20-dione, *cortisol*	M	*F*	a	207—210	+167 (E)[2]	21—	223—225	+163 (C)	221 Ac
						11[m]—	113—118	+167 (C)	
						11,21-Di[m]	192		

Table 2 *cont.*

Systematic and common name	Letter designations		Ref. to isolation	Free steroid M. P.°	Free steroid $[\alpha]_D$[1]	Acetate position	Acetate M. P.	Acetate $[\alpha]_D$[1]	IR[3]
37. 11 β,19,21-Trihydroxy-4-pregnene-3,20-dione, *19-hydroxycorticosterone*			d	187—192		19,21 Dibenzoate	177—182		
38. 17 α,19,21-Trihydroxy-4-pregnene-3,20-dione			d	235—241	+119 (D)	19,21-Di	229—232		
39. 17 α,20 α,21-Trihydroxy-4-pregnene-3,11-dione			d	240—243	+141 (D)	20,21-Di	277—280	+99 (D)	
40. 17 α,20 β,21-Trihydroxy-4-pregnene-3,11-dione	U		a	208		20,21-Di	252—253	+178 (An)	219 Ac
41. 3 α,11 β,17 α,21-Tetrahydroxyallopregnan-20-one	C	C	a	273—276	+59 (D)	3,21-Di	204—205	+74 (D)	187 Ac
42. 3 β,11 β,17 α,21-Tetrahydroxyallopregnan-20-one	V		a	220—225	+51 (D)	3,21-Di	225—227	+63 (D)	188 Ac
43. 11 β,17 α,20 β,21-Tetrahydroxy-4-pregnen-3-one	E		a	125	+87 (E)	20,21-Di	229—230	+163 (An)	189 Ac
44. Allopregnane-3 β,11 β,17 α,20 β,21-pentol	A	D	a	221—222	+16 (E)	3, 20, 21-Tri	219—220	+74 (An)	91 Ac
45. 20,20,21-Trihydroxy-3-keto-4-pregnen-18-oic acid 18 → 20 lactone, *deoxycorticosterone 18-acid lactone*			d	215—219	+88 (D)				
C_{27} Compounds									
46. Cholesterol			a	149	—39 (C)	3	115	—47 (C)	

[1] Rotation in solvent indicated (C, chloroform; E, ethanol; D, dioxane; M, methanol; An, acetone). [2] Rotatory dispersion recorded. [3] Number of infrared curve in DOBRINER et al. (1953); Ac indicates spectrum of acetate.

References a) REICHSTEIN and SHOPPEE (1943). b) BLOCH et al. (1956). c) (Isolated from adrenal tumor) ANLIKER et al. (1956). d) NEHER and WETTSTEIN (1956b). e) MATTOX (1955). f) SIMPSON et al. (1954a). g) RUZICKA and WETTSTEIN (1935). h) TURNER (1953a). i) BARBER and EHRENSTEIN (1954). j) SARETT (1946). k) PATAKI et al. (1952). l) FLOREY and EHRENSTEIN (1954). m) OLIVETO et al. (1953). n) v. EUW et al. (1958). o) NEHER et al. (1958). p) NEHER et al. (1959). q) WETTSTEIN et al. (1959).

experiments. A summary of these substances and quantitative data on the secretion products of the adrenal cortex in various species are presented in Chapter 4.

The compounds in Table 2 are arranged according to the number of carbon and oxygen atoms, and then with increasing number of hydroxyl groups; unsaturated compounds precede saturated ones. Strictly systematic and also important trivial names are given, as well as the original alphabetical designations of REICHSTEIN and of KENDALL; the letters which have retained currency are indicated in italics. A complete bibliography of the original literature on the earlier isolation studies is given by REICHSTEIN and SHOPPEE (1943), who have also summarized all of the values for the physical constants of the corticosteroids up to that date; the data on the older compounds in the Table are taken from this review. Unless otherwise indicated, other constants are taken from the reference in which the isolation was described. The infrared spectrum of a steroid is perhaps the ultimate physical property, and the comparison of infrared spectra plays a commanding role in the identification of a steroid. The spectra of a large number of steroids have been compiled in an atlas by DOBRINER et al. (1953), and the numbers in the column headed "IR" in the Table refer to the spectrum number in this atlas. The optical rotations of a few cortical steroids have been determined at 5461 Å (mercury green line); these values are not of great utility and have been omitted. Recently, complete optical rotatory dispersion spectra have been recorded for some of the hormones by FOLTZ et al. (1955); this is indicated in the Table.

The term "corticosteroid" has been used to refer to any of the steroids present in the adrenal cortex and also to urinary steroids of adrenal origin. In chemical parlance, the term is usually restricted to the thirty compounds in Table 2 in the $C_{21}O_3$, $C_{21}O_4$ and $C_{21}O_5$ series, the distinguishing feature being a side chain bearing two or three oxygen atoms. The other unique structural characteristic of the adrenocortical steroids is the frequent occurrence of an oxygen substituent at C-11.

The corticosteroids that possess adrenocortical-hormone activity fall into two side-chain categories, 21-hydroxy-20-ketones and 17α,21-dihydroxy-20-ketones. In addition, all of the hormones contain, in common with testosterone and progesterone, the unsaturated ketone grouping in ring A. These features are present in fourteen of the corticosteroids, among them the seven compounds shown in Fig. 10, which includes the 6 "classical" adrenal hormones and aldosterone. Six of the other compounds possessing these structural units contain additional hydroxyl groups at C-6β or C-19; any of these substances that have been evaluated have a low order of hormonal activity.

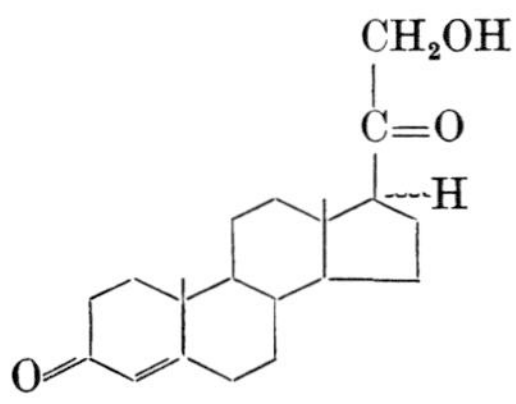

I

11-Deoxycorticosterone

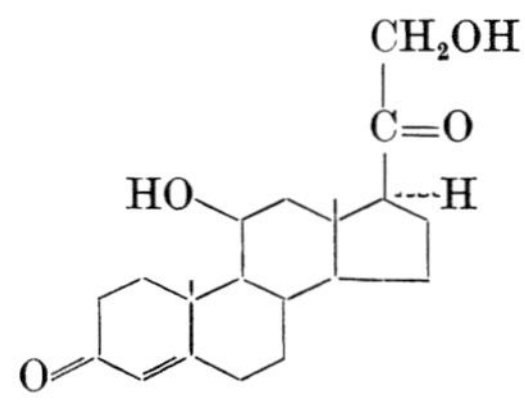

II

Corticosterone

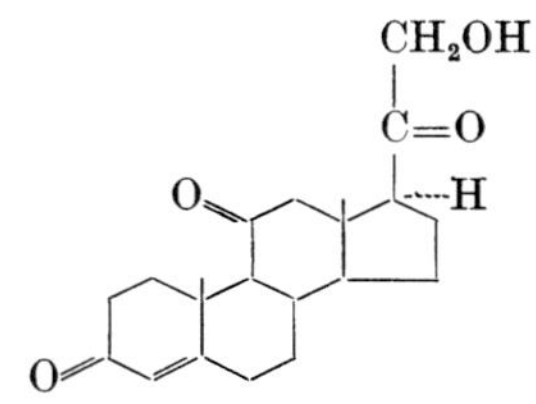

III

11-Dehydrocorticosterone

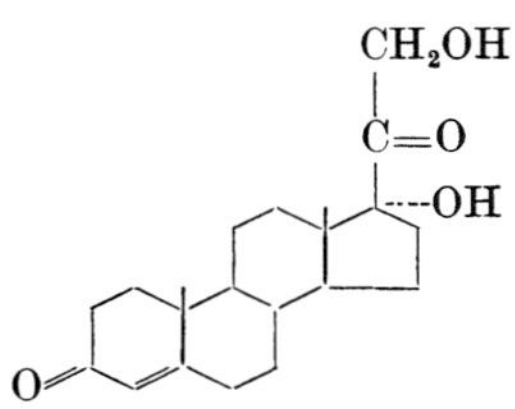

IV

11-Deoxycortisol

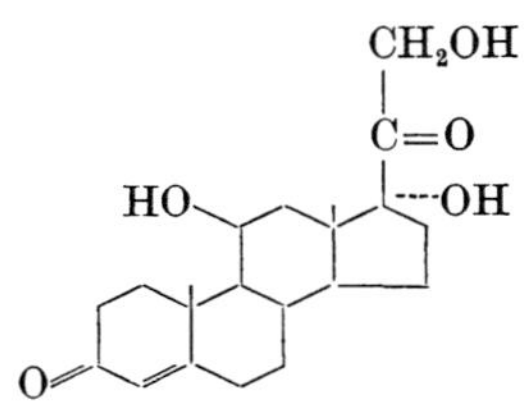

V

Cortisol

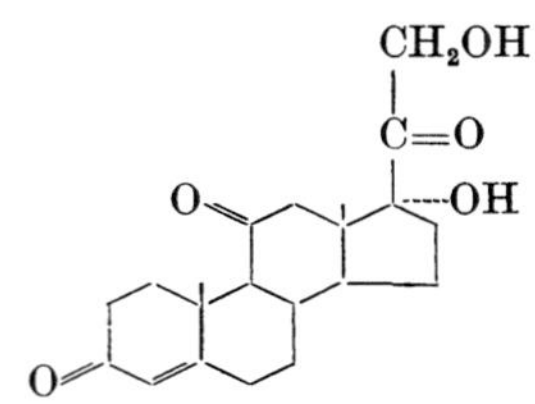

VI

Cortisone

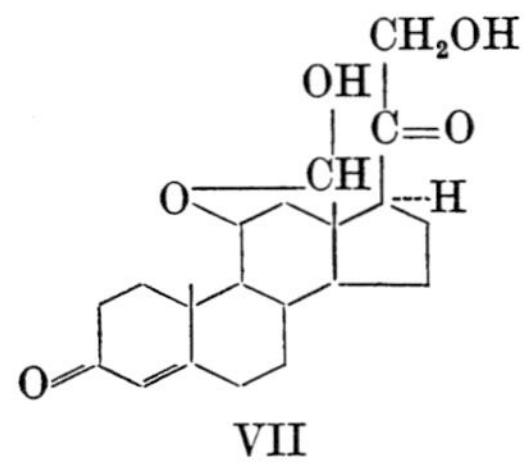

VII

Aldosterone

Fig. 10. Structures of the active corticosteroids

Precise evaluations of the relative glucocorticoid and mineralocorticoid activities of the several adrenal steroids as measured by different pharmacological methods are provided in subsequent chapters; it will suffice here to mention a few of the chemical features that are associated with these two general types of activity. A keto or β-hydroxyl group at C-11 is a prerequisite for glucocorticoid activity, and a 17α-hydroxyl group reinforces this property; cortisol and cortisone are the most active glucocorticoids among the natural hormones, and corticosterone and the corresponding 11-ketone the next most active. These generalizations are valid also for the modified corticosteroids with greatly enhanced activity.

The structural features that impart mineralocorticoid-type activity are less readily summarized. The sodium-retaining activity of aldosterone is some 100 fold greater than that of any the other corticosteroids, and the 18-aldehyde group is thus of paramount importance in the intensification of this property. On the other hand, the lactone (VIII, Fig. 44, p. 237) derived from aldosterone by oxidation of the 18-aldehyde function to a carboxyl group is reported to have a specific sodium *diuretic* action (WETTSTEIN et al. 1957). The 3β,16α-dihydroxyallopregnan-20-one isolated from the adrenal by NEHER et al. (1958), as well as a number of synthetic 16α-hydroxy compounds, are also reported to cause sodium excretion under certain conditions. Among the 6 classical hormones (I)—(VI), the oxygen functions at C-11 and C-17 reduce mineralocorticoid activity, and the simplest compound, 11-deoxycorticosterone, is the most active. Any conclusions on the effect of substituents on the activity of corticosteroids must also take into account the extremely high mineralocorticoid and glucocorticoid activities of some of the synthetic derivatives that are discussed in a later section (pp. 264 to 291).

The other 18 "companion steroids" with three or more oxygen atoms are essentially devoid of hormonal activity. They are formally related to the hormones as reduction products, 8 are C-20 alcohols, and most are reduced in ring A. Although both the C-5 hydrogen atom and the C-3 hydroxyl group are found in both α and β configurations in the urinary steroids, all of the saturated steroids of the adrenal cortex are allopregnane derivatives and with the exception of one saturated ketone and three 3α-hydroxy compounds, they are 3β-hydroxy derivatives.

4. Stereochemistry of the side chains

The stereochemistry of the substituents at C-17 and C-20 in the cortica. steroids presents some special features that will be considered in this section before discussion of the individual side chains. As noted on p. 188, the configuration of the side chain at C-17 in the several classes of steroids was shown to be the same by numerous degradations and interconversions with 20-keto pregnanes and etianic acid derivatives. This correlation was very readily established in the adrenal steroids having no hydroxyl group at C-17 by the interconversions indicated in Fig. 11; the methods used are indicated in subsequent sections. In the earlier literature the configuration of the C-17 side chain in naturally-occurring steroids was considered to be α, i. e., *trans* with respect to the angular methyl group at C-13. Inconsistencies in some of the older data and new observations, principally on lactone formation of isomeric 12-hydroxy etianic acids (SORKIN and REICHSTEIN 1946), led to revision to the 17β-configuration shown in Fig. 11.

A correlation of the C-17 configuration of the side chain in the 17-deoxycorticosteroids and the 17-hydroxy compounds was based on several lines of evidence relating to the configuration of the carbon chain and that of the hydroxy group. It seemed probable, a priori, that the carbon chain was in the same configuration in both series (accordingly, the 17-hydroxyl group in the corticosteroids is described as 17 β in the earlier literature; see HEARD 1948). This assumption was supported by a comparison of the reactivities of the 17-hydroxyl groups in the androstane and pregnane series, which indicated that the configuration of the

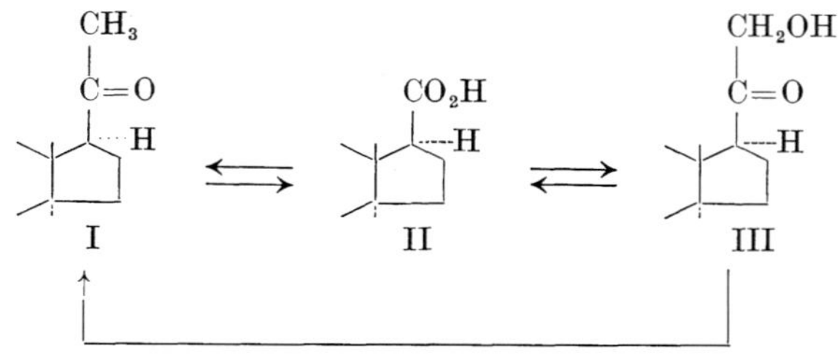

Fig. 11. Interconversion of 17 β-side chains

hydroxyl group in the 17-hydrocorticosteroids was opposite to that of the 17 β-hydroxyl group in testosterone. Rigorous proof of the correspondence of the side chain configurations in the 17-deoxy and 17-hydroxy series was obtained when glycols with the opposite configuration at C-17 (VIII) became available by synthesis from 17-ketones (Fig. 12). Interconversion of the two series, with inversion of configuration at C-17, was accomplished by the Serini reaction (cf. Fig. 27, p. 219), in which either of the triols (V) or (VIII) is converted to the 20,21-ketol of opposite configuration at C-17.

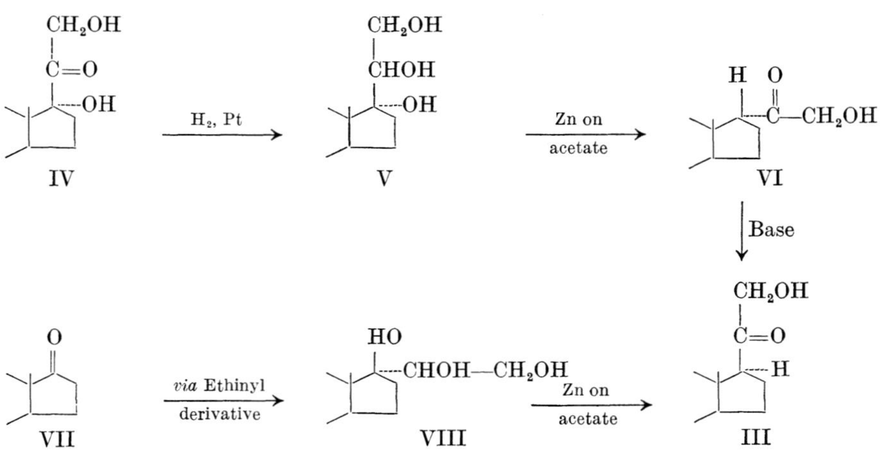

Fig. 12. Interrelationship of side chains by Serini reaction

In order to clearly distinguish the "unnatural" 17 α-series in structural formulas, the 17 α side chain is often written horizontally as in (VI) and (VIII) in Fig. 12. There are important differences in the reactivity and behavior of the side chain substituents in the two C-17 configurations. The 17 β-hydroxyl group (VIII) in the unnatural series is more readily esterified, and the 20-keto group in (VI) is much less reative than in the natural series. A mixture of isomers is obtained by alkali equilibration of either 17-isomer of simple 20-ketones having an enolizable hydrogen atom at C-17 (i. e., easily removed by base); the mixture usually contains predominately the 17 β compound (III) (see p. 213). This greater stability of the natural isomer, and the differences in reactivity noted above, are evidently reflections of the steric hindrance exerted by C-18 and by C-12 respectively (FIESER and FIESER 1949).

When C-20 is an asymmetric center, as in 20-hydroxy compounds, the stereochemical relationship of the two isomers is of a different nature than that at a nuclear center, in which the configuration is defined with respect to the plane of the ring. The configuration of a C-20 substituent must be described in terms of Fischer projection, as used in a purely aliphatic compound. In the convention proposed by FIESER and FIESER (1948) and now universally accepted, the side chain (in either configuration at C-17) is rotated to place C-21 in the rearmost position (IX, Fig. 13). *The configuration of a substituent at C-20 is then expressed in terms of its position in a projection of this conformation; a group on the right is designated α* (XI), *and one on the left β* (X).

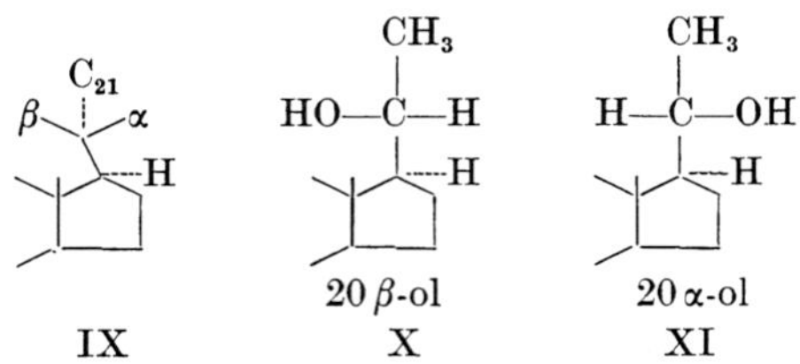

Fig. 13. Stereochemical convention at C-20

The configurations defined by this convention were originally assigned on the basis of the stereochemical specificity of the hydroxylation of a Δ^{17}-double bond (Fig. 14). In the formation of Δ^{17}-olefins by most methods of preparation (see Fig. 34, p. 266), one of the two geometrical isomers is formed exclusively or in preponderant amounts, and this (more stable) isomer can with confidence be assigned the *trans* structure (XII), with the bulky C-21 group farthest removed from the angular methyl group and ring C. Attack on (XII) by osmium tetroxide occurs almost exclusively at the rear of the ring system (p. 211), and since the reaction proceeds via the cyclic *cis* osmate ester (XIII), the glycol formed must have the configuration (XIV). The major product of this sequence is seen by rotation about the 17—20 bond to be the 17α, 20β-diol (XV). These arguments are reinforced by the demonstration by HOGG et al. (1955a) that the less stable *cis* olefin (XVI) leads by the same process to the 17α, 20α-diol (XIX).

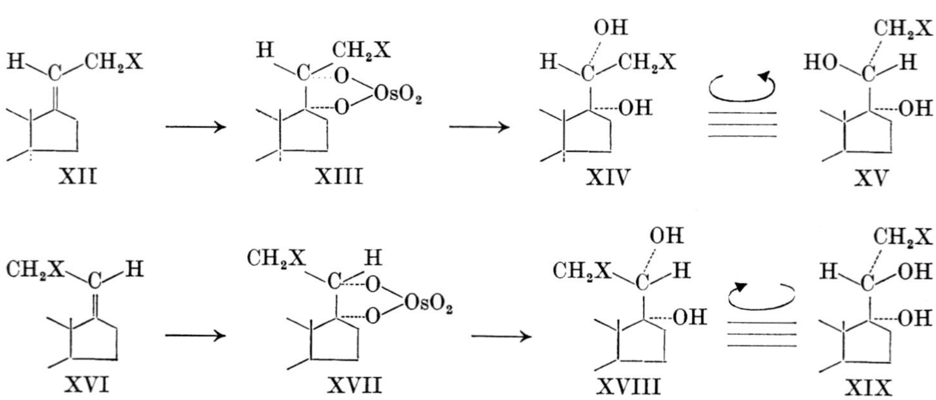

Fig. 14. Stereospecific hydroxylation of Δ^{17}-pregnenes

Proof that the Fieser convention represents the correct absolute configuration at C-20 in simple C-20 alcohols has been obtained by the application of the method of asymmetric synthesis (p. 191) to the keto ester (XXIII, Fig. 15) derived from a 20β-ol (PRELOG and TSATSAS 1953). A correlation of the C-20 configuration in 20-alcohols and in the sterols and bile acids was established by configurational linkage of the 20α-hydroxyl substituent with a 22-ketone (XXI) derived from cholesterol by side chain degradation (P. WIELAND and MIESCHER 1949). The ketone (XXI) and the isomeric ketone (XXIV) obtained by isomerization of (XXI) with alkali were converted to the 20α (XXII) and 20β (XXV) acetoxy derivatives, respectively, by perbenzoic acid oxidation, a reaction that proceeds with retention of configuration (p. 212).

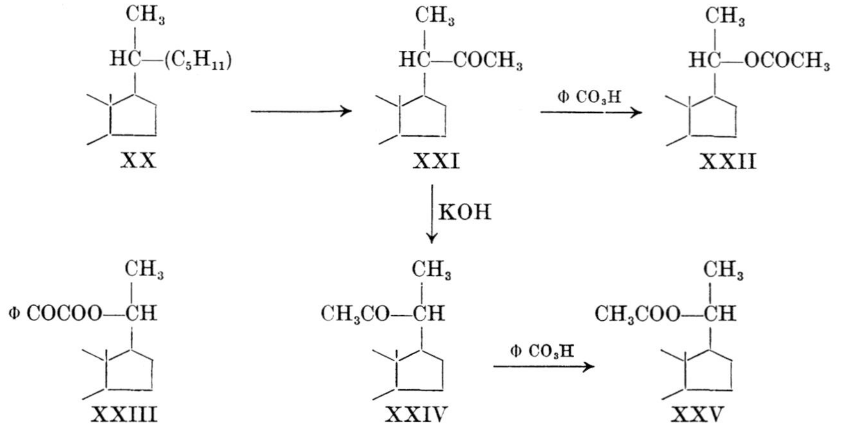

Fig. 15. Correlation of configuration at C-20

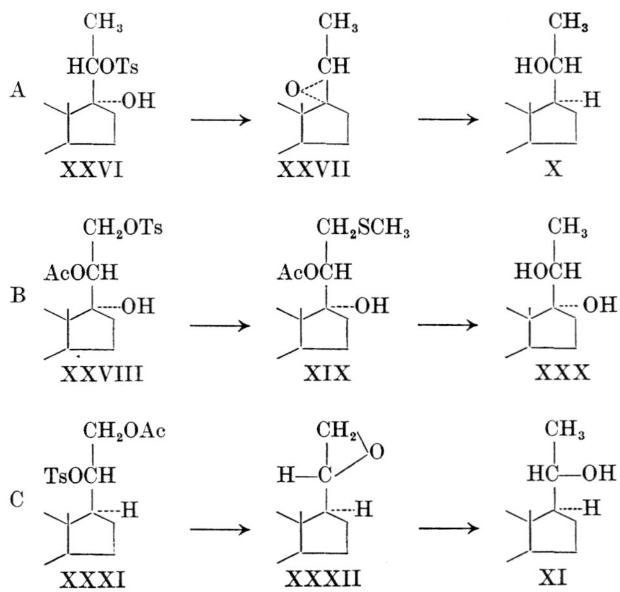

Fig. 16. Configurational interconversions of diol and triol side chains

Several trivial designations, without configurational significance, were used in the earlier literature to describe naturally-occurring 20-hydroxy compounds. MARKER et al. (1937) used the index α arbitrarily to indicate the configuration of the 20-hydroxyl group of the pregnandiol from human pregnancy urine, and β for the epimeric 20-hydroxy compounds that are the predominant products of reduction of 20-ketones. PRINS and REICHSTEIN (1940) designated the C-20 configuration of the 20-epimeric corticosteroids "O" and "J" by α and β, respectively; other 20-hydroxy corticosteroids were related to these. A chain of configurational links was forged by SARETT (1949a, b) which showed that, fortuitously, these arbitrary designations were mutually consistent and also consistent with the Fieser convention.

This conclusion was based upon a series of interconversions of 20-hydroxy compounds, 17,20-diols, 20,21-diols, and 17,20,21-triols by the reactions summarized in Fig. 16. The crux of the interconversions involved displacement reactions of the three toluenesulfonate esters (Ts) (XXVI), (XXVIII), and (XXXI), which were prepared from the corresponding diols or monoacetates. In the two sequences A and C, the displacement was attended by inversion of configuration at C-20, interrelating the two diol types with monohydroxy compounds of opposite configuration at C-20. Inversion is precluded in sequence B, permitting direct correlation of the 17α,20β-diols (XXX) with the 17α,20β,21-triols (XXVIII).

II. Individual structural features

In the following sections, the most important functional groups present in the adrenal steroids and also in the numerous urinary metabolites are treated individually, with emphasis on typical properties and general reactions. There is a definite hazard in such a presentation, since a change in environment can cause significant and sometimes drastic departures from the "normal" course of reaction of a given functional group. The reduction of a double bond in ring A, for example, or a rearrangement in the side chain may be profoundly affected by the nature of the substituent at C-11. The same difficulty obtains with generalizations concerning synthetic procedures, since the presence of a sensitive group completely remote from the center in question may preclude the use of a certain reaction or reagent. It would be quite impossible, however, to deal adequately with individual compounds, and the pitfalls of "typical behavior" and "general methods" must be accepted. The more important instances of complications due to other groups or differing stereochemistry are noted when possible.

An attempt has been made to indicate the most useful methods for the recognition and characterization of the groups that may be encountered in a newly-isolated compound. Details on methods and identification procedures are given in the last section of the chapter (p. 334ff.). When possible, information on the relative reactivities of functional groups at various positions is presented. Characteristic rearrangement reactions which may be encountered in chemical manipulations of sensitive compounds are discussed. The most useful synthetic routes available for the introduction of a particular functional group are presented briefly; the major lines of synthesis of the more important hormones are discussed separately.

For more extensive coverage of the material in these sections, reference is again made to the steroid treatises and the excellent review by H. HIRSCHMANN (1955). Emphasis has been placed primarily on the most recent work; no attempt has been made to document all of the earlier references.

1. 17-Ketones and alcohols

a) 17-Ketones

All of the C_{18} and C_{19} steroids of the adrenal cortex and most of the urinary C_{19} steroids contain a carbonyl group at C-17. As indicated earlier, 17-keto compounds are of great importance as key degredation products in structural studies, providing an interrelationship of a number of diverse pregnane derivatives which have a common nuclear structure. Drastic oxidation of sterols and bile acids leads, in very low yields, to 17-ketones; in early investigations, these degradations served both preparative and correlative ends. For synthetic purposes, 17-ketones are most conveniently produced by several methods from the abundant 20-keto-pregnanes or their derivatives. SONDHEIMER et al. (1956) have summarized these approaches and have described a very practical method (Fig. 17), consisting of the Beckmann rearrangement of oximes (II) of Δ^{16}-20-ketopregnenes (I), which are directly available from sapogenins (Fig. 51, p. 254).

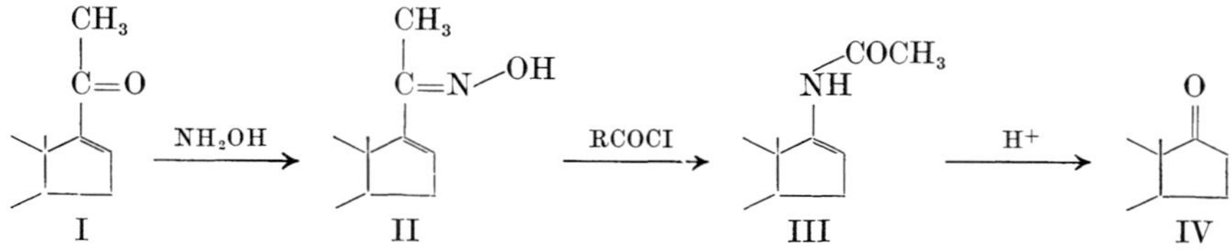

Fig. 17. 17-ketones from Δ^{16}-20-ketones

The presence of a 17-keto group is recognized by the formation of a color in the Zimmermann test (p. 337), coupled with the frequency of the carbonyl band in the infrared spectrum at 1745—1742 cm^{-1}, characteristic for a five-membered ring ketone. The carbonyl group at position 17, flanked by a quaternary carbon atom at C-13, suffers steric hindrance to carbonyl addition reactions, and *a high degree of selectivity can be achieved in a reaction at C-3*, for example, *in a 3,17-dione*. It has been observed (HERR and HEYL 1953) that the presence of an 11-keto group increases the reactivity of a 17-ketone toward formation of an enol-imine. This hindrance does not affect reactions dependent upon the availability of the adjacent methylene group; thus 17-ketosteroids are among the most effective Zimmermann chromogens.

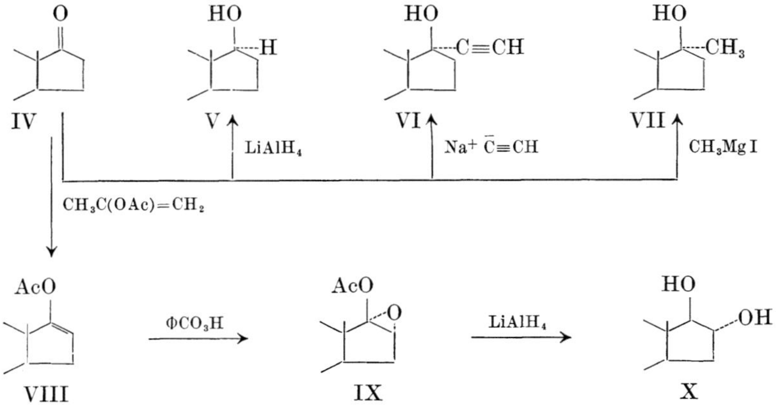

Fig. 18. Reactions of 17-ketones

Addition reactions of the 17-keto group lead to a number of useful transformation products (Fig. 18). In all of these reactions one isomer is formed exclusively or in great preponderance. GALLAGHER and KRITCHEVSKY (1950) have summarized the course of such additions in the "rule of the rear," which states that *the entering group always assumes the α-configuration.* This steric selectivity, arising from the shielding of the β-face of rings C and D by the β-oriented angular methyl groups, is observed with many reaction of Δ^{16}, $\Delta^{17\,(20)}$ and Δ^{11} double bonds, as well as carbonyl additions at C-17 and C-12.

Reaction of 17-ketones with metallic hydrides, acetylides and Grignard reagents furnishes the 17β-ol (V), 17α-ethinyl-17β-ol (VI), and 17α-alkyl-17β-ol (VII), respectively, all in high yields. The ethinyl compounds are versatile intermediates in synthesis of side chain ketols of the unnatural 17 β-hydroxy-17-isopregnane series (Fig. 12, p. 207), and both the ethinyl and alkyl derivatives are of value as clinically-useful relatives of testosterone and progesterone. Treatment of the 17-keto group with isopropenyl acetate and sulfuric acid gives the enol acetate (VIII), which with perbenzoic acid yields the 16 α,17 α-oxide (IX), attack of reagent being from the rear (LEEDS et al. 1954). Reduction of the oxide gives the 16 α,17 β-diol (X), which is the glycol system present in estriol and in several metabolites found in normal and adrenal carcinoma urines.

b) 17-Alcohols

In the naturally-occurring 17-hydroxy steroids having no side chain at C-17 (which include the important hormones testosterone and estradiol), with the exception of a few 17 α-hydroxy metabolites, the hydroxyl group is in the β orientation. This configuration, which, as noted on p. 206, is the reverse of that assigned in the earlier literature, is borne out by the mechanisms of two methods of formation of 17-hydroxy compounds. The reduction of a 17-ketone by a variety of reagents, including sodium and alcohol, catalytic hydrogenation, fermenting yeast and, most important, lithium aluminum hydride, leads to the 17 β-ol in good yield. As discussed above, the attack of hydride ion with the last reagent occurs from the rear, generating the β-hydroxyl group. Another reaction, which has been of value in correlating configurations both at C-17 and at C-20, is the oxidation of a ketone with persulfuric or perbenzoic acid. This oxidation converts a methyl ketone to an acetate ester, and is known (TURNER 1950 b) to proceed with retention of configuration; thus, the 20-ketone (XI) is converted to the 17 β-acetoxy compound (XII) (GALLAGHER and KRITCHEVSKY 1950).

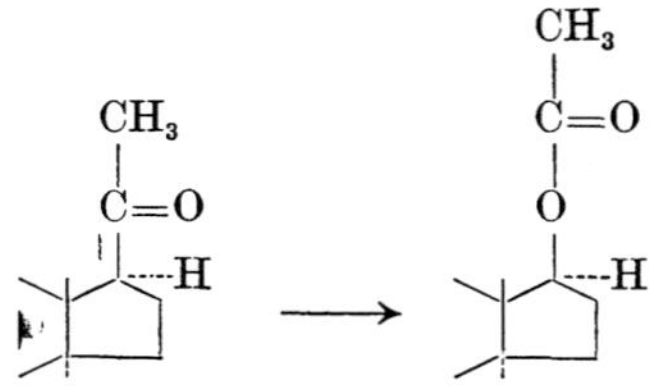

XI XII

The 17 β-hydroxyl group undergoes the usual reactions of oxidation and esterification; an indication of the steric influence of the C-13 angular methyl group and the methylene group at C-12 on reactions at C-17 is seen in the relative ease of hydrolysis of acetates: C-3 > C-17 β > > C-17 α. Attempts to introduce a Δ^{16} double bond by removal of the 17 β-hydroxyl group or to perform a displacement reaction [e. g., solvolysis of the 17 β-tosyloxy group (XIII) with formation of a 17 α-substituent] lead mainly to characteristic rearrangement products, in which the angular methyl group migrates to C-17, with the formation of the ψ-androstene derivative (XIV) (MIESCHER and KÄGI 1949, MADAEVA and LUR'I 1952).

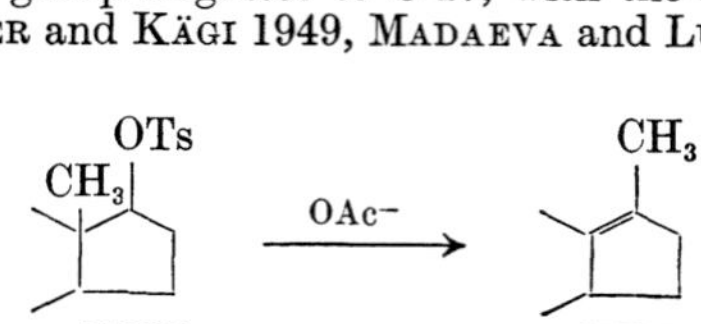

XIII XIV

The epimeric 17 α-hydroxy compounds are less well known, and only recently has a practical source become available (Fig. 19). The 17 α-ol (XVII) is formed in good yield by reduction of the 16 α, 17 α-oxide (XVIII) (HEUSSER et al. 1950 b), which in turn is readily obtained from the Δ^{16} olefin (XVI). The major difficulty lay in the formation of the Δ^{16} double bond, which is a structural feature of a few naturally-occurring steroids, and which, curiously, can impart a strong smell of urine to the compound. A convenient method for preparing Δ^{16} olefins is now available by the decarboxylation of Δ^{16} etienic acids (XV) (SONDHEIMER et al. 1955).

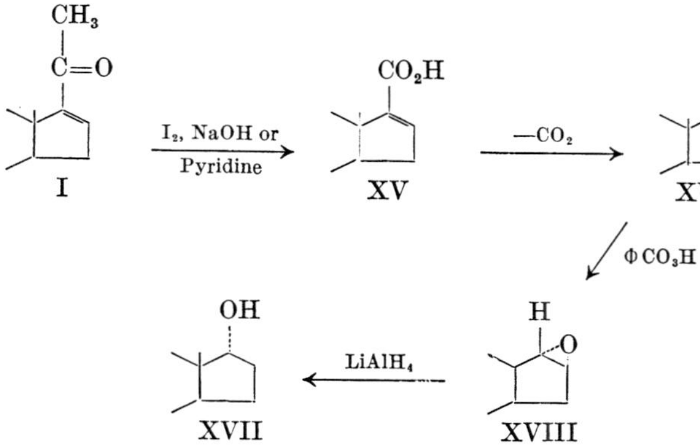

I XV XVI XVII XVIII

Fig. 19. Formation of 17 α-hydroxy group

2. 20-Ketones and alcohols

a) 20-Ketones

Two adrenal cortex constituents and several urinary steroids contain a simple C-20 keto group; the main interest in this side chain is the fact that it is the starting point for most of the practical synthesis of the cortical diols, triols, and ketols. As discussed in the section on synthesis (p. 253), 20-ketones are intermediates in the conversion of most of the important steroid raw materials to hormones and are formed in the degradation of a number of side chain structures. The acetyl side chain has been built up from a 17-ketone by cyanohydrin formation, dehydration and Grignard condensation (Fig. 20). Reduction of a Δ^{16}-20-ketone (VI) gives exclusively the 17β-acetyl isomer (V).

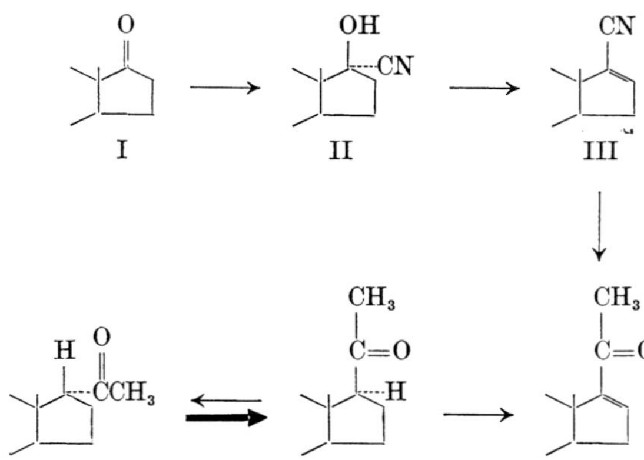

I II III IV V VI

Fig. 20. Formation of 20-ketone side chain

As discussed on p. 207, isomerization of 20-ketones with base furnishes an equilibrium mixture of the 17α- and 17β-acetyl side chains. In steroids with the usual *trans* C/D ring fusion, *the 17 β side chain predominates* at equilibrium except in the case of 18-oxygenated derivatives. The reverse is true for 14-*iso* compounds, in which the 17α-acetyl side chain is the more stable. Isomerization of the side chain can occur under very mild conditions, as in the case of 11α-hydroxyprogesterone (FLOREY and EHRENSTEIN 1954) or compounds with a carbonyl substituent at C-18 (p. 238). Although this equilibration affords the most convenient method of prepa ation of 17-*iso* derivatives, it is more often an undesired complication in hydrolytic reactions.

The reactivity of a 20-ketone toward carbonyl reagents is comparable to that of 17-ketones, and *numerous examples of selective reactions at C-3 in 3,20-diones are known.* In the bromination of allopregnane-3,20-dione, the first bromine atom is introduced at C-2, and the second at C-17 (M. RUBIN et al. 1951); C-21 is substituted at elevated temperatures. The C-20 carbonyl is converted to the Δ^{17}-enol acetate (IX, Fig. 21) with acetic anhydride and *p*-toluenesulfonic acid; the isomeric Δ^{20}-enol acetate (X) is formed by treatment with *iso*propenyl acetate. Condensation of a 20-ketone with ethyl oxalate or benzaldehyde in alkaline media furnishes the 21-ethoxalyl and 21-benzylidene derivatives, respectively (Fig. 28, p. 220). The application of these diverse transformation products in the synthetic elaboration of ketol and glycol side chains is illustrated in subsequent sections.

Reduction of simple C-20 ketones by catalytic hydrogenation or with complex hydrides nearly always yields the 20β-hydroxy compound (XI) as the main product; sodium and alcohol reduction furnishes chiefly the 20α-epimer (VIII). KLYNE (1953) has summarized these findings and suggested that the predominant formation of 20β-hydroxy compounds can be accounted for by a preferred conformation of the acetyl side chain. The interesting observation was made by SHAPIRO et al. (1955) that a Δ^{16}-20-ketone gave mainly the unsaturated 20α-hydroxy compound.

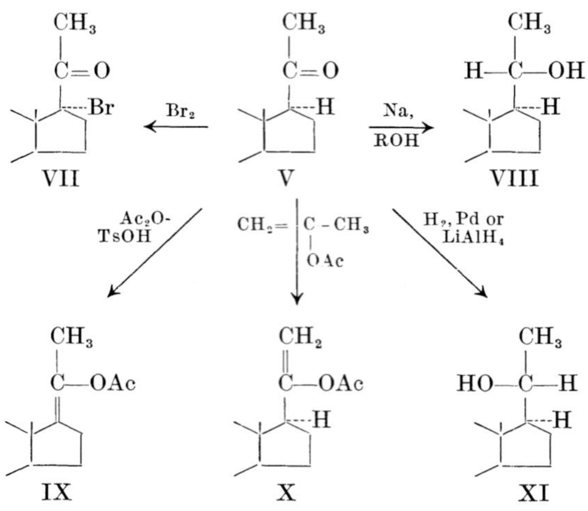

Fig. 21. Reactions of 20-ketones

b) 20-Alcohols

A number of compounds with the simple 20-hydroxy side chain are found among urinary steroids, with the 20α-configuration more prevalent. C-20 hydroxyl groups exhibit no exceptional properties; the reactivity is markedly less than that of a hydroxyl group at C-3, and monoacetylation at C-3 in a 3,20-diol can be accomplished in good yield (KIRK et al. 1957).

The preparation of 20β-ols by catalytic hydrogenation or lithium aluminum hydride reduction is usually a satisfactory synthetic method, but difficultly-separable mixtures of epimers may result. R. BROOKS et al. (1953). have found that the chromatographic separation of the benzoate esters is an effective means of resolving such mixtures. Sodium-alcohol reduction of the ketone or hydride reduction of the unsaturated ketone followed by hydrogenation afford synthetic approaches to the 20α-ols.

The differentiation of 20α- and β-epimers, in simple alcohols and in diols and triols, can be accomplished with a high degree of confidence by application of molecular rotation differences (p. 374). SARETT (1949c) has analyzed a large number of cases, and found that the *change in molecular rotation on acetylation of a 20-hydroxyl group is positive for the 20β-epimer and negative for the 20α.*

3. 17α-Hydroxy-20-ketones and diols

Only two 17α-hydroxy-20-ketones are present in the adrenal steroid complex, but a number of metabolic products with this side chain have been isolated. There has been increasing interest in the possible pathological significance of these 21-deoxy metabolites, and a number of recent studies of reactions of 17,20-ketols and glycols have been reported.

a) 17α-Hydroxy-20-ketones

In contrast to the isomeric 20,21-ketols, the 17α-hydroxy-20-ketone system does not reduce the reagents, such as tetrazolium salts, periodic acid or sodium bismuthate, that are commonly used to characterize corticosteroids. The 17α-hydroxy-20-ketone group is, in fact, the only one of the six main types of cortical side chain which is not attacked by sodium bismuthate (C. Brooks and Norymberski 1953), and this behavior has been used as a basis for the determination of metabolites with this side chain in urinary mixtures (p. 339). A unique band at 1693 cm^{-1} in the carbonyl region of the infrared spectrum, and peculiarities in the hydroxyl region, attributed to hydrogen bonding in 17α-hydroxy-20-ketones, are useful in the recognition of this side chain system (R. N. Jones et al. 1952b).

The 17α-hydroxy-20-ketone side chain is readily accessible by synthesis, and a number of compounds of this type have been prepared. A very general method (Fig. 22) comprises epoxidation of the Δ^{17}-20 enol acetate (I) of a 20-ketone (cf. Fig. 21), followed by hydrolysis of the epoxy ester (Kritchevsky and Gallagher 1951). Other approaches to the formation of the 17α-hydroxy-20-ketone system are indicated in Figures 52 (p. 255) and 53 (p. 255). Methods have also been devised for removal of the 21-hydroxyl group of the 17α,21-diol-20-one system (Fig. 33, p. 224) and this is sometimes a convenient source of the 17α-hydroxy-20-ketone.

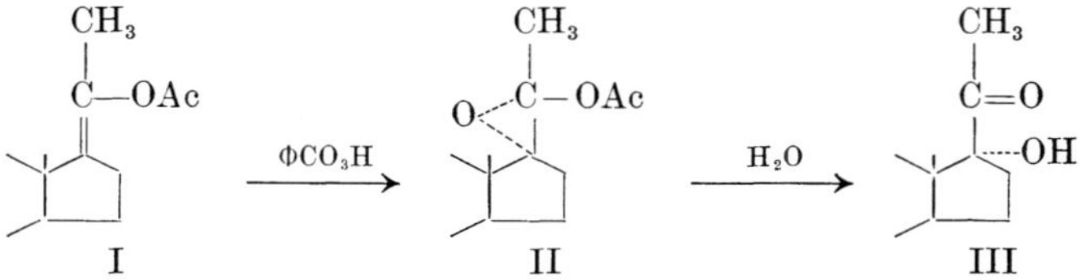

Fig. 22. Formation of 17α-hydroxy-20-ketones

As indicated above, the direct degradation of a 17α-hydroxy-20-ketone to the 17-ketone, a transformation that is very important in structure determination, is not accomplished with the usual ketol cleavage reagents. The resistance of this ketol system to cleavage most probably arises from the inability of the tertiary hydroxyl group to undergo esterification with the oxidizing agent. Degradation to a 17-ketone has often been accomplished by reduction of the ketol to the glycol and cleavage of the latter (Fig. 23). Facile removal of the side chain has also been effected by Beckmann rearrangement of the ketol oxime (IV) (Schmidt-Thomé 1957).

The reactivity of the ketol in typical carbonyl reactions is depressed by the presence of the adjacent hydroxyl group, but derivatives such as ketals or hydrazones can be formed. Bromination of the 17α-hydroxy-20-ketone at C-21 proceeds smoothly, but less rapidly than in a saturated 3-ketone. Reduction of the ketol by catalytic hydrogenation gives mainly the 17α,20β-diol (VI). Several cases have been cited (Klyne 1953, Turner 1953b) in which the major product of reduction of a 17α-hydroxy-20-ketone (unsubstituted at C-11) with lithium aluminum hydride is the 17α,20α-diol (VII). On the other hand, two examples have been reported (OLIVETO and HERSHBERG 1953, POOS 1956) of the formation of a somewhat larger amount of the 17α,20β-diol by hydride reduction of 17α,20-ketols (both containing an 11β-hydroxyl group). It may be that the nature of the ring C substituent is a controlling factor in these reductions, but no generalization can be made. In contrast to the 17α,20-ketols, *hydride reduction of the 17α,21-dihydroxy-20-ketone system* (Fig. 33, p. 224), *regardless of substitution in ring C, leads to the 17α,20β,21-triol in good yield* (JULIAN et al. 1951).

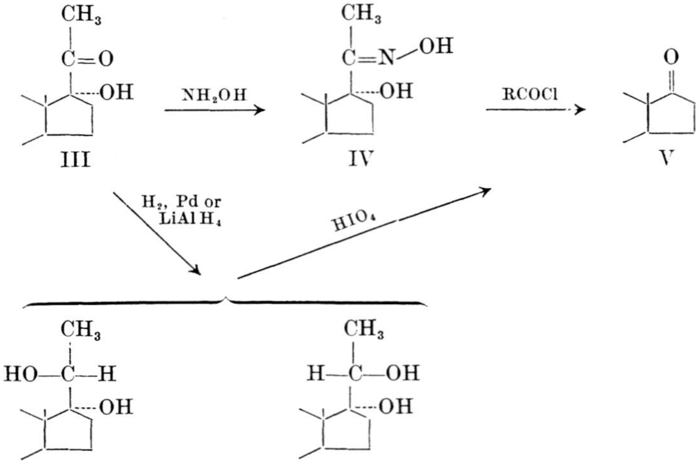

Fig. 23. Transformation products of 17α-hydroxy-20-ketones

Reductive removal of the 17α-hydroxyl group to give the 20-ketone (Fig. 24) has been accomplished in high yield by ROSENFELD (1957) by treatment of the 17-acetate with zinc in acetic acid. The free ketol is largely unchanged by the same treatment (NORYMBERSKI and STUBBS 1956), whereas the 17-hydroxyl group is eliminated readily in a 17α-hydroxy-21-acetoxy-20-ketone (Fig. 33).

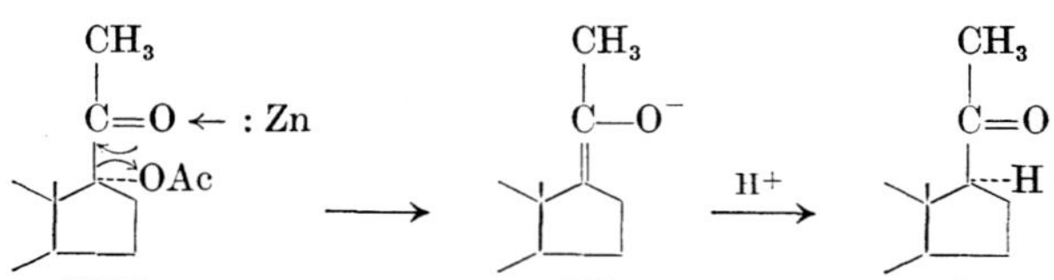

Fig. 24. Deacetoxylation of 17α-acetoxy-20-ketone

The hydroxyl group in these tertiary ketols (and in the diolones) is not acetylated under the usual pyridine-acetic anhydride conditions, a property shared with only one other hydroxyl group present in the corticosteroids, namely that at C-11β. The failure to react can be ascribed primarily to the shielding effect of the methylene group at C-12; the tertiary hydroxyl group at C-17β in the unnatural 17-*iso*-20-ketopregnanes is less resistant toward esterification. Acetylation of the 17α-hydroxyl group can be accomplished in good yield by treatment with acetic anhydride and a strong acid catalyst (TURNER 1953b) or with refluxing acetic anhydride (HUANG-MINLON et al. 1952). The latter conditions, however, have also been found (OLIVETO et al. 1957b) to give principally D-homo rearrangement products.

A characteristic reaction of 17α-hydroxy-20-ketones (and also the epimeric 17β-hydroxy-17-*iso*-20-ketones) is a carbon skeleton *rearrangement to form a D-homo steroid*. This ring-expansion, or D-homoannulation, reaction occurs in a variety of circumstances, including the routine isolation of urinary metabolites and has been the subject of several recent detailed investigations. Compounds of the D-homo series were first encountered in some of the earlier work on cortical side chain synthesis, and the skeletal structures were established by classical methods (FIESER and FIESER 1959). The formation of the D-homo compound involves migration of either the 13—17 or 16—17 bond to C-20, which becomes C-17a of the D-homo product (Fig. 25). In addition to the rearrangement of a 17α-hydroxy-20-ketone, several other reactions involving C-17 lead to D-homo derivatives (p. 288).

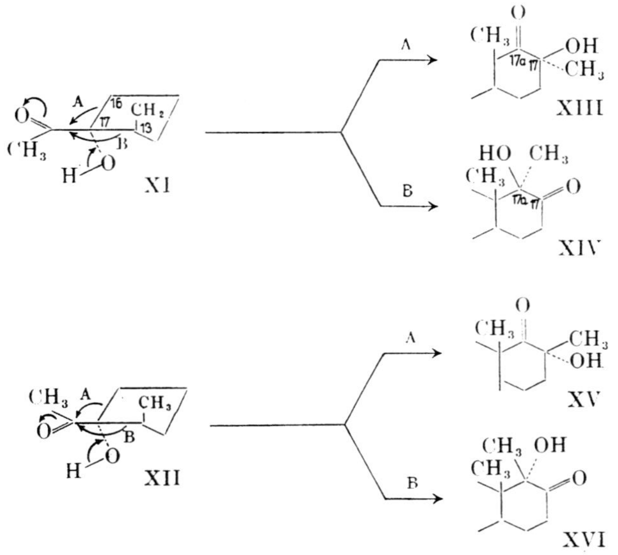

Fig. 25. D-homo rearrangement of 17α-hydroxy-20-ketones

In the rearrangement of a 17α, 20-ketol, four isomeric D-homo compounds could result. Thus the ketol can react by two paths in either of two conformations, depending on whether the reagent, as with strong base, tends to separate the two oxygen atoms (XI), or, as with electron-deficient reagents, tends to coordinate them (XII). This interpretation of the reaction was suggested by TURNER (1953a) to account for the products formed under basic and acidic conditions and has been amplified by FUKUSHIMA et al. (1955), who have carefully examined the products formed from several 17α-hydroxy-20-ketones. With strong alkali, the two products (XIV) and (XVI), formed by path B, are formed, with (XIV) predominating. Treatment with aluminum alkoxides (Oppenauer oxidation conditions) or boron trifluoride in acetic acid-acetic anhydride or simply refluxing acetic anhydride, on the other hand, furnishes mainly (XV), the product of 16—17 bond migration. Thermal rearrangement, which is responsible for the indefinite melting points of 17α-hydroxy-20-ketones, leads to (XV) and (XVI). The 17a-ketones (XIII) and (XV) can be distinguished from the 17-ketones (XIV) and (XVI) by the Zimmermann test, which is positive only for the latter pair, which contain a methylene group adjacent to the carbonyl. The structures of all four isomeric ketols in one series have been conclusively established by synthesis (WENDLER et al. 1956).

The D-homoannulation of 17α,21-dihydroxy-20-ketones has not been so thoroughly studied, since strong acid and base cause other changes with this system (Figs. 31 and 32). Rearrangement does occur, however, with aluminum alkoxides, and the main product obtained has been shown by WENDLER and TAUB (1957a) to possess the 17α-hydroxy-17β-hydroxymethyl-17a-ketone structure (type XV, Fig. 25). The course of the reaction thus parallels that found in the 21-deoxy series.

The occurrence of D-homo compounds among urinary steroids has been noted on several occasions. FUKUSHIMA and GALLAGHER (1956) have identified three metabolites isolated by LIEBERMAN et al. (1948) as D-homo ketols corresponding to types (XIV), (XV), and (XVI) and have observed that the conditions required for the artificial formation of rearrangement products from 17α,20-ketols are not severe. It seems quite certain that these urinary D-homo steroids, as well as those reported by H. Hirschmann and Hirschmann (1947) and by A. Miller and Dorfman (1950), were formed during the isolation procedure; the corresponding 17α-hydroxy-20-ketone precursor was also present in every case. Mineral-acid hydrolysis, chromatography on alumina and even prolonged exposure to silica gel can produce detectable or isolable quantities of D-homo compounds from 17α,20-ketols. No reports of D-homo artifacts arising from 17α,21-dihydroxy-20-ketones have appeared. In contrast to these urinary D-homoketols, the 17α-methyl-D-homoandrostane-3β,17α-diol of mare's urine (Klyne 1950) appears to be a genuine urine constituent; Klyne (1953) has suggested that the compound might arise from a 17,20-diol by rearrangement *in vivo*.

b) *17,20-Diols*

Several 17,20-glycols (Fig. 26), both 17α, 20α (VI) and 17α, 20β (VII), have been isolated from adrenal tissue, adrenal perfusates and urine. These diols are oxidized by periodic acid to the 17-ketone and acetaldehyde; the formation of the latter fragment occurs only with the 17,20-diol side chain (Table 17, p. 334), and has been used for both qualitative and quantitative determination (Cox 1952). Both of the 20-epimeric diols are converted to the 20-monoacetate in the usual acetylation conditions, and the molecular rotation difference accompanying this change furnishes strong presumptive evidence for the configuration at C-20. The 17α-hydroxyl group in these diols exhibits the expected resistance to acetylation, and it is interesting that a 17α,20-diol (probably 20β) has been found to form an acetonide (XVII) with acetone and acid catalyst under mild conditions (Adams et al. 1954).

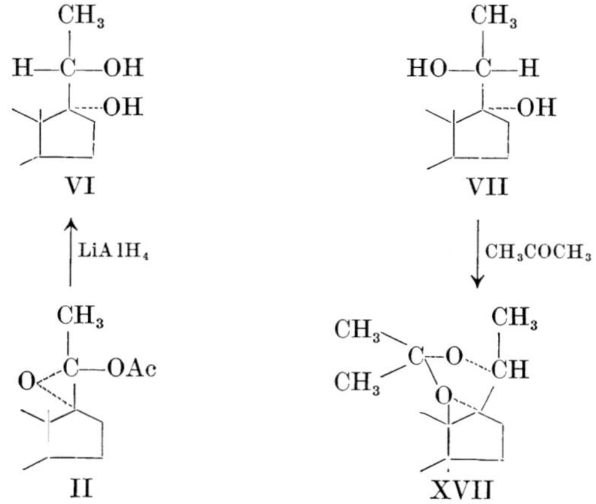

Fig. 26. Formation and reactions of 17,20-diols

The formation of 17,20-diols by reduction of the ketols has been discussed (Fig. 23). In general, a satisfactory yield of the 17α, 20β-diol is realized by hydrogenation of the ketol, and in certain cases (p. 216) hydride reduction furnishes a suitable preparation of the 17α, 20α-diol. A stereoselective route to the 20α-epimer which may be of general applicability lies in the lithium aluminum hydride reduction of the epoxy acetate (II, Fig. 26) (Soloway et al. 1954); this same intermediate on hydrolysis furnishes the ketol (Fig. 22).

The 17,20-diol systems undergo several rearrangement reactions of a basically similar nature (Fig. 27), all of which are characteristic of glycol systems with tertiary hydroxyl groups. The best known of these was first described by Serini (see Shoppee 1949). This reaction comprises a pinacol-type rearrangement of the diol 20-monoacetate, under the influence of zinc dust in an inert medium, to furnish a ketone with inversion of the side chain. A typical example is the transformation of the 17α,20β-diol 20-acetate (XX) to the 17-*iso*-20-ketone (XVIII). The same 20-ketone is formed from either 20-epimer; the reaction is successful also with 17α-hydroxy-20,21-diacetoxy compounds, leading to the 17-*iso*-20,21-ketol 21-acetate. This rearrangement was of importance in establishing correlations between different side chain structures in the early phases of adrenal steroid chemistry, and was a very useful synthetic tool in preparing ketols of the natural series from intermediates in the 17-*iso* series obtained via 17-ketone addition reactions (cf. Fig. 12).

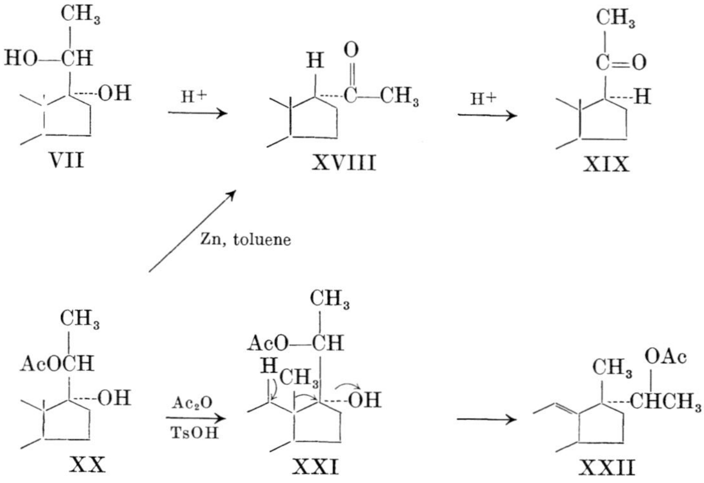

Fig. 27. Rearrangements of 17,20-glycol system

Another form of the pinacol rearrangement of 17,20-diols is encountered on treatment with mineral acid, which leads in low yield to the 20-ketone (XIX), with overall retention of configuration of the side chain. This transformation was first observed by Steiger and Reichstein (1938); Klyne (1953) has discussed a reinvestigation of the problem that confirmed the formation of the ketone and which suggested further products that may have arisen from migration of the angular methyl group. Fukushima and Gallagher (1957) have found that the yield of the ketone from the 20β-epimer is much higher than from the 20α, and that no D-homo products are formed. It is probable that the rearrangement actually proceeds with inversion at C-17; the 17-*iso* product (XVIII) would be expected to suffer isomerization to the more stable 17β-acetyl side chain under the reaction conditions. Renewed interest in this rearrangement was supplied by the suggestion of Rosselet et al. (1957) that *the reaction can lead to artifacts in urinary extracts.* This was proven to be the case by Fukushima and Gallagher (1957), who demonstrated that the *20-ketone "metabolite" formed on administration of a labeled 17α-hydroxy-20-ketone was found only in that portion of the extract that had been treated with strong acid.*

A different rearrangement has been found to occur when attempts are made to acetylate the tertiary hydroxyl group in a 17α-hydroxy-20-acetoxy pregnane derivative (XX) with acetic anhydride and toluene-sulfonic acid (Herzog et al. 1957b). The major product obtained is a Δ^{12}-17β-methyl-17-*iso* compound (XXII), resulting from migration of the angular methyl group. This transformation has a number of counterparts in related systems, including the formation of ψ-androstene derivatives (p. 212) and the rearrangements attending the cleavage of a 16α,17α-epoxide (Fig. 64, p. 274).

4. 21-Hydroxy-20-ketones and diols

a) 21-Hydroxy-20-ketones

All of the adrenal hormones proper contain the strongly reducing α-ketol grouping at C-20,21, either with or without an additional hydroxyl group at C-17. Several methods for the detection of this grouping have been developed (p. 339), and the strongly reducing properties of the primary ketol grouping are frequently utilized in the quantitative determination of corticosteroids.

A number of approaches are available for the formation of the 20,21-ketol grouping (Fig. 28). Removal of the cholanic acid side chain can be carried out to furnish the ketol acetate directly (Fig. 50, p. 253) and the most effective procedure for degradation of the butenolide side chain system of the cardiac glycosides (I) is ozonization and hydrolysis of the glyoxalyl ester (II) to the ketol (III) (LARDON and REICHSTEIN 1952). The ketol side chain is readily reconstituted from the etianic acid (IV) by formation and acetolysis of the diazoketone (VI), the method which was used by STEIGER and REICHSTEIN (1937) in the first partial synthesis of an adrenal hormone. Several routes are available starting from the 20-ketone. The Δ^{20}-enol acetate (VIII) may be epoxidized and the oxide rearranged to the ketol acetate; alternatively (VIII) has been treated with N-iodosuccinimide to furnish the 21-iodoketone (XI) (DJERASSI and LENK 1953a). The latter intermediate is also formed from the 21-ethoxalyl compound (X) (RUSCHIG 1955) and is smoothly converted to the ketol acetate (XII) with potassium acetate. Finally, in certain cases the direct acetoxylation of the 20-ketone with lead tetraacetate leads to the ketol acetate in useful yield.

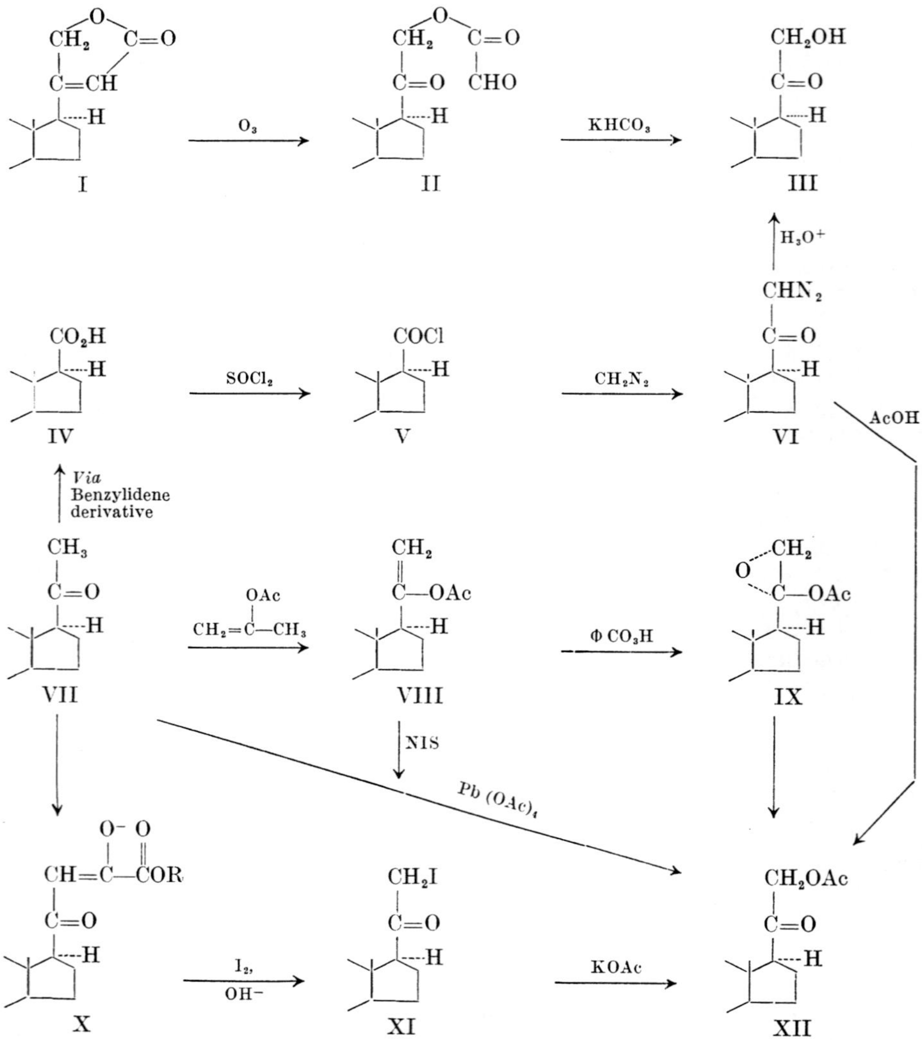

Fig. 28. Formation of 20,21-ketol side chain

The 21-hydroxyl group in the α-ketol side chain is very easily esterified, and in many of the synthetic routes leading to these side chains the ketol acetate is first formed. Very often the 21-acetate is used in pharmacological and clinical studies, and a distinction between the effects of the acetate and of the free ketol is not always clearly drawn. A number of 21-acyl derivatives of the important hormones have been prepared with the aim of obtaining derivatives with longer duration of activity following administration (see JUNKMANN 1954).

The hydrolysis of 21-acetates is complicated by the sensitivity of the α-ketol and especially the 17α,21-dihydroxy-20-ketone groups to alkaline and acid treatment. This is an important transformation, since earlier isolation procedures frequently involved acetylation, and as noted above, most synthetic routes furnish the 21-acetate. The hydrolysis procedure recommended by REICHSTEIN and v. EUW (1938) employed aqueous methanolic potassium bicarbonate. Hydrolysis of the 20,21-ketol acetate with potassium hydroxide in the absence of air is satisfactory (HERZIG and EHRENSTEIN 1951), and base-catalyzed ester exchange in methanol solution has been used for both mono- and dihydroxy ketol acetates (RINGOLD et al. 1956b). MATTOX and KENDALL (1951) introduced an acid-catalyzed hydrolysis procedure that can, however, lead to rearrangement products if not carefully controlled. An improvement in this method is the use of perchloric acid in methanol (FRIED and SABO 1957).

One of the important characterization reactions of the α-ketol system is the degradation to an etianic acid. This is accomplished quantitatively with periodic acid or sodium bismuthate; the same oxidation also occurs upon alkali treatment in the presence of air (VELLUZ et al. 1947), giving the etianic acid and formic acid. The α-ketol or the 21-acetate furnish typical carbonyl derivatives (XIII, Fig. 29) with reagents such as semicarbazide or 2,4-dinitrophenylhydrazine (REICH and SAMUELS 1954); under vigorous conditions, the ketol is converted to the osazone (XIV). The cyanohydrin (XV) of a 20,21-ketol acetate is readily obtained, and can be dehydrated to the 21-acetoxy-Δ^{17}-20-cyano compound (XVI), which is an intermediate in the elaboration of the 17α,21-dihydroxy-20-ketone system (Fig. 54, p. 256). The action of various reducing agents on the α-ketol side chain, as in the case of the simple 20-ketone, gives mainly the 20β,21-diol (XXIa, Fig. 30).

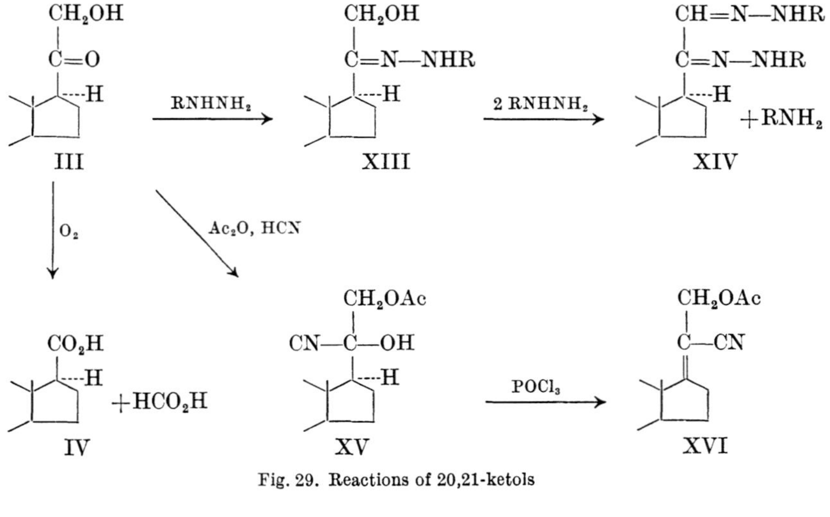

Fig. 29. Reactions of 20,21-ketols

b) 20,21-Diols

Only one 20,21-diol has been isolated from adrenal tissue; this compound has been assigned the 20β-configuration on the basis of the molecular rotation of a synthetic preparation. The 20-epimeric diols are obtained in approximately equal amounts by the hydroxylation of a Δ^{20}-pregnene derivative (XXII, Fig. 30), obtained in turn by deamination of a 20-amino compound (XVIII), and a pure 20α,21-diol (XX) has been prepared in this way (SARETT 1946a). A good yield of the 20β,21-diol (XXIa) can be realized by reduction of the ketol with complex hydrides (C. BROOKS and NORYMBERSKI 1953). It has been reported by TAUB et al. (1959) that borohydride reduction of a 20,21-ketol 21-acetate is accompanied by acetyl migration, leading to the 20-monoacetate of the 20β,21-diol (XXIc). The 21-monoacetate (XXIb) has been prepared by selective acetylation of the diol.

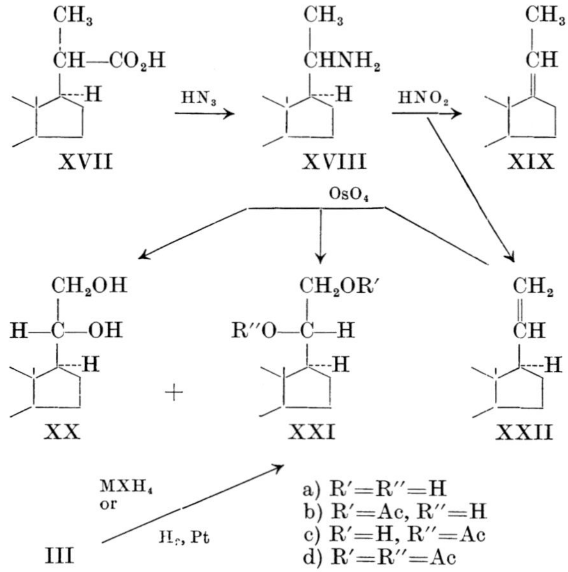

Fig. 30. Formation of 20,21-diol side chains

5. 17α,21-Dihydroxy-20-ketones and triols

a) 17α,21-Dihydroxy-20-ketones

Many of the properties of the dihydroxyketone side chain are referable to either the 17α,20- or 20,21-ketol systems, and have been discussed in the preceding sections. Approaches to the synthesis of the dihydroxyketone side chain are based for the most part on combinations of the various methods used for the individual ketol groups. The emphasis in devising synthetic methods for the elaboration of this side chain system has been placed upon procedures that can be adapted for large scale operation. The selection of the route depends upon the functional groups furnished during the side-chain degradation of the starting material, and on the sequence of the main stages of the synthesis. Some illustrative procedures are outlined in Figures 53 and 54 (pp. 255 and 256); again, most of the procedures lead to the 21-acetate.

One of the most significant properties of compounds containing the dihydroxy ketone side chain is the sensitivity to acid, base, and oxidation. The 21-ketol grouping is responsible for the reduction of tetrazolium salts; the 17α-hydroxyl group does not interfere with this reaction, although it has been stated that a 16α, 17α-oxido function prevents reduction (BEYLER and HOFFMAN 1956). Periodic acid furnishes the 17α-hydroxyetianic acid (II, Fig. 31), but this intermediate is further oxidized by sodium bismuthate to the 17-ketone (III). Advantage has been taken of this property of the latter reagent in the degradation of biosynthetically labeled cortisol, C-20 and C-21 being cleaved simultaneously to carbon dioxide and formaldehyde respectively (CASPI et al. 1956).

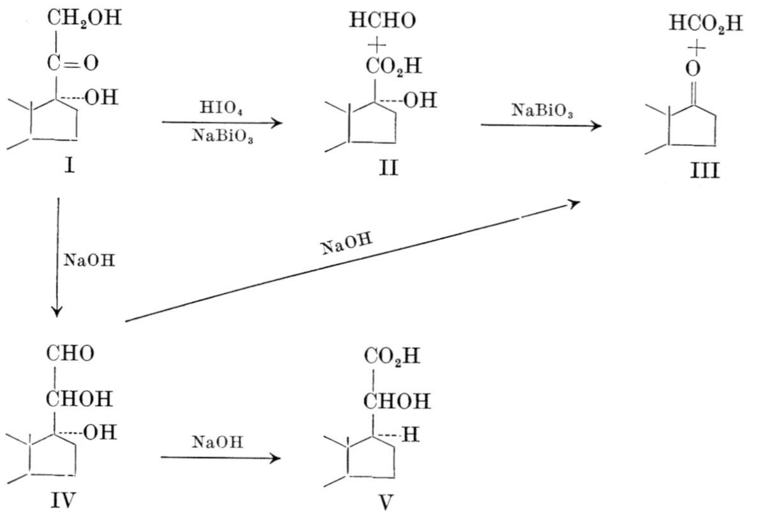

Fig. 31. Degradation of 17α,21-dihydroxy-20-ketone side chain

The susceptibility to oxidation is greatly diminished in the 21-acetate; periodic acid, sodium bismuthate, and even such powerful oxidants as chromium trioxide and ozone can be employed under suitable conditions without appreciable effect on the 17α-hydroxy-21-acetoxyketone. The sensitivity to acid and alkali of the free diolone and the 21-acetate are essentially equivalent, of course, since the 21-acetyl group is removed under very mild conditions, as discussed in the preceding section.

The nature of the products resulting from exposure of the side chain system to alkali in the absence of oxygen was suggested by MASON and KENDALL (1938), and has recently been examined more closely (WENDLER and GRABER 1956) (Fig. 31). The first reaction is the establishment of the aldose-ketose equilibrium; the dihydroxyaldehyde (IV) then leads to two products, the 17-ketone by a reverse aldol condensation and the 20-hydroxy-21-acid (V) by a saccharinic acid rearrangement.

The instability of crude adrenocortical extracts to mineral acid noted in early work was probably due at least in part to loss of the 11β-hydroxyl group in the hormones containing this functional group, but *the diolone side chain is altered by relatively mild acid treatment* (Fig. 32). The reaction has been studied by MATTOX (1952), who found that cortisone is converted in good yield by 0.2 *N* methanolic hydrogen chloride to the dimethyl acetal (VIII) of the 20-keto-21-aldehyde. The same product is also formed from the Δ^{16}-20,21-ketol; the reaction can be formulated as an enolization (VI) followed by loss of the elements of water and acetal formation. This rearrangement has been used to effect the conversion of the diolone side chain to a 20,21-ketol (TAUB et al. 1954). The glyoxal derivative (VIII) is reduced at C-20 and the hydroxy acetal (XI) is hydrolyzed and treated with base to form the ketol.

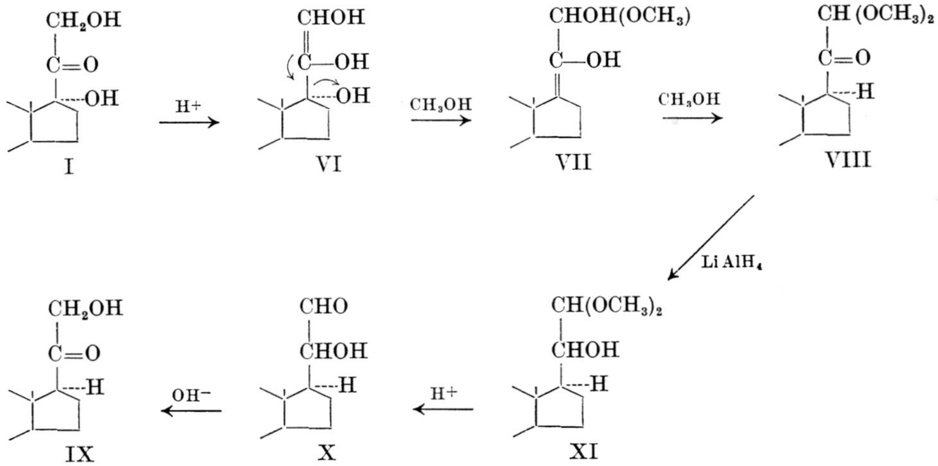

Fig. 32. Acid-catalyzed rearrangement of 17α,21-dihydroxy-20-ketone

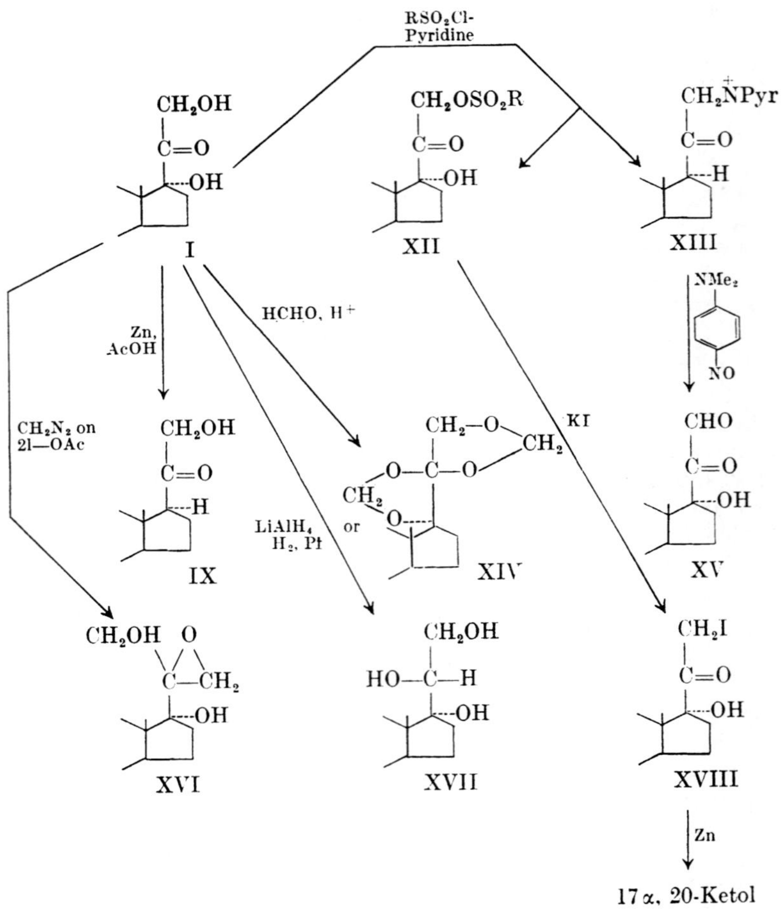

Fig. 33. Transformation products of 17α,21-dihydroxy-20-ketones

The carbonyl reactivity of the 17α,21-dihydroxy-20-ketone system is comparable to that of the 20,21-ketols. Thus 2,4-dinitrophenylhydrazones are formed normally, although a 21-acetoxy compound such as cortisone acetate can be converted to the 3-monohydrazone or 3-monosemicarbazone in good yield (REICH and SAMUELS 1956). The carbonyl group in the 21-acetate reacts with semicarbazide under somewhat more vigorous conditions (R. E. JONES and ROBINSON 1956). The lower reactivity of the 21-acetate is also observed in ketal formation with ethylene glycol; the ketal is obtained from the dihydroxyketone, but not from the acetate (ANTONUCCI et al. 1953). Another interesting carbonyl addition reaction of the 21-acetoxy-17α-hydroxyketone is the formation of the epoxide (XVI, Fig. 33) on prolonged treatment with diazomethane (NUSSBAUM and CARLON 1957). BEYLER et al. (1958) have reported the formation of spiroketal derivatives (XIV) of 17,21-dihydroxy-20-ketones by treatment with formaldehyde and strong acid; this bismethylenedioxy grouping, which is removed by hydrolysis with aqueous acetic acid, is recommended as a protective group for the side chain.

Treatment of cortisone or related diolones with *p*-toluenesulfonyl chloride or methanesulfonyl chloride in pyridine furnishes either the ester (XII) or the pyridinium salt (XIII), depending on the conditions used. The salt is used in the preparation of the 21-aldehyde (XV) (LEANZA et al. 1954); the methanesulfonate can be converted to the iodide (XVIII) and thence by reduction to the 17α,20-ketol (POOS 1956) (Fig. 33). Transformation to the 20,21-ketol (IX) is readily accomplished by mild treatment of the free diolone (the acetate does not react) with zinc in dilute acetic acid (NORYMBERSKI 1956). Reduction of the dihydroxyketone side chain by either catalytic hydrogenation or hydrides gives the 17α,20β,21-triol system (XVII) as the main or exclusive product in nearly every case.

b) 17α,20,21-Triols

Several 17α,20β,21-triols are present among the adrenal cortex constituents; only one triol with the 20α configuration has been isolated. The 20β triol system is more readily accessible by synthesis, since the reduction of a 17α,21-dihydroxy-20-ketone with sodium borohydride or lithium aluminum hydride affords a general route to pure 17α,20β,21-triols (JULIAN et al. 1951, NORYMBERSKI 1956).

The other main approach to triol side chains is the oxidation of a 21-acetoxy-Δ^{17}-pregnene derivative, which has been used in several modifications to effect stereoselective syntheses of both 20-epimeric triol systems. The mechanism of the *cis* hydroxylation of a 17—20 double bond by osmium tetroxide has been discussed previously (Fig. 14, p. 208); a 21-acetoxy olefin with the *trans* geometry (XXVIII, Fig. 34) furnishes the 17α,20β,21-triol and the *cis* olefin (XXVI) leads to the 17α,20α,21-triol. The more stable *trans* olefins are obtained either by allylic rearrangement of a 17β-hydroxy-17α-vinyl compound (XXIV) (FIESER and FIESER 1949), or by alkali-induced rearrangement of a 17,21-dibromo-20-ketone (XXIII) followed by hydride reduction of the resulting acid (XXVII) (WAGNER and MOORE 1950). Application of the latter sequence of reactions to the 21,21-dibromo-20-ketone (XXI) leads to the unstable *cis* olefin and thence by hydroxylation to the 17α,20α,21-triol (HOGG et al. 1955).

An ingenious and practical route to the 20α-triol has been described by FUKUSHIMA et al. (1956). This comprises an inversion of configuration at C-20 by a "neighboring group" displacement reaction (XXXIII) in a 17α,21-diacetoxy-20β-toluenesulfonoxy system (XXXII), obtainable from the 17α,20β,21-triol 21-acetate (XXXI) by successive tosylation and acetylation. Another avenue is furnished by acetolysis, again with inversion, of the 17α,20β-oxide (XXIX) (SOLOWAY et al. 1954).

The 17,20,21-triol side chain is converted to the 20,21-diacetate under the usual acetylation conditions. The 21-monoacetate, which is formed in several of the synthetic sequences, can be obtained in good yield by partial acetylation (SARETT 1946b). The 20,21-acetonide is readily formed by treatment with acetone and copper sulfate. The triol 20,21-diacetate undergoes the Serini reaction, but the triol system has not been so thoroughly investigated as the 17α,20-diols, and the other acid-catalyzed rearrangements characteristic of the latter have not been reported in the triol series.

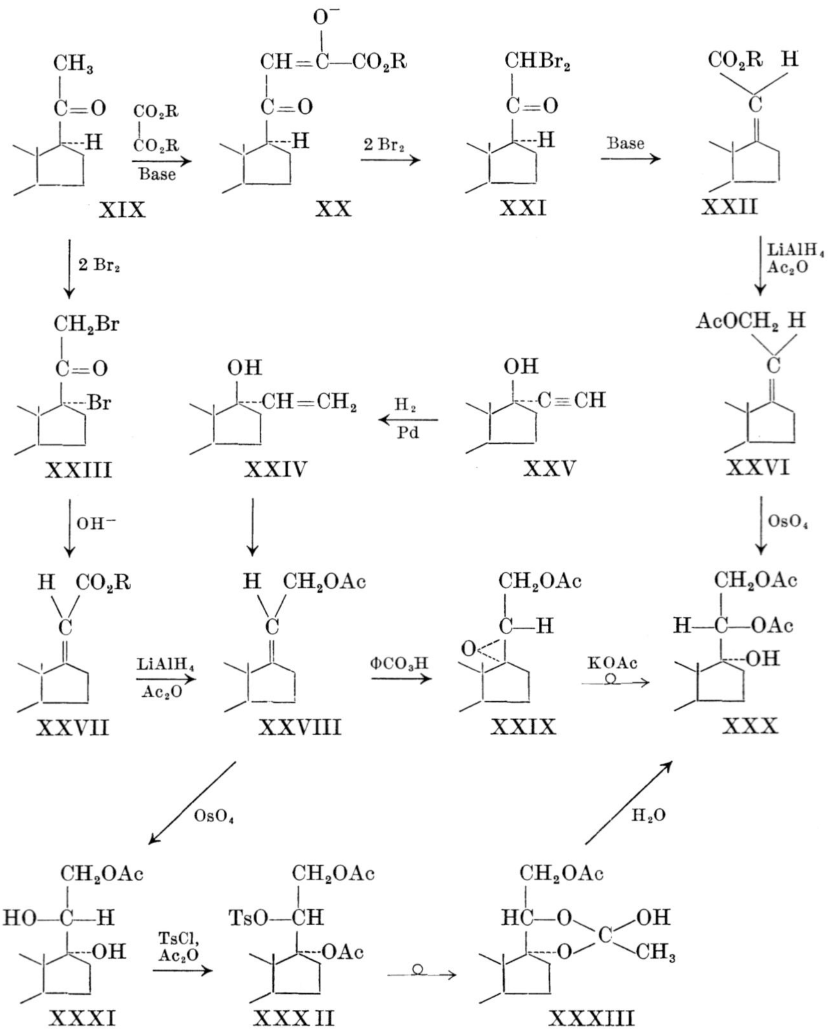

Fig. 34. Formation of 17α,20,21-triol side chains

6. 11-Ketones and alcohols

One of the most important and unique structural features of the adrenocortical steroids is the oxygen atom at C-11 which is present in more than half of the compounds. It is probable (Chapter 3) that all of the 11-oxygenated urinary steroids are of adrenal origin. The oxygen atom may be either an 11-keto or an 11 β-hydroxyl group, and frequently both an 11-ketone and the corresponding alcohol occur together, both in adrenal extracts and in urine. The only other naturally occurring 11-oxygenated steroids are a few relatively rare cardiac glycosides, and experience in the chemistry of functional groups at C-11 has been gained almost wholly in investigations centering about the adrenal steroids. In most synthetic sequences leading to 11-oxygenated adrenal steroids, the 11-oxygen function is established at an early stage; several approaches to the introduction of an oxygen atom at C-11 are discussed on p. 252ff.

a) 11-Ketones

Substituents at C-11 in the normal steroid skeleton are subject to a high degree of steric hindrance due to the axial substituents in a 1,3-relationship at C-10, C-8, and C-14 (I). *A keto group at C-11 is less reactive than at any other*

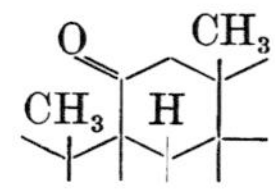

I

position in the molecule and can be preserved intact throughout numerous transformations at other positions; this diminished reactivity is frequently advantageous in the development of partial synthetic sequences. An 11-ketone is completely inert to carbonyl reagents such as 2,4-dinitrophenylhydrazine or hydroxylamine. The formation, in low yield, of a cyclic ketal at C-11 has been reported (MAGERLEIN and LEVIN 1955), but keto groups at C-3 and C-20 can be converted to ketals without reaction at C-11. The Clemmensen reduction of an 11-ketone has been accomplished by REICHSTEIN (1936) under drastic conditions; removal of the 11-keto group by Wolff-Kishner reduction has also been realized (MOFFET and HUNTER 1951), but selective reduction of other carbonyl groups in the presence of an 11-ketone function is obtained without difficulty.

Reduction of the 11-keto group by hydrogenation or complex hydride reduction occurs rather more readily, and although selective reduction of carbonyl groups at C-3 and C-20 can be effected with sodium borohydride, some reduction at C-11 occurs (OLIVETO and HERSHBERG 1953). The product of hydrogenation or hydride reduction is almost exclusively the 11 β-hydroxy compound (III, Fig. 35), in accordance with the generalization that highly hindered ketones give rise to the axial alcohol. In the earlier literature it was erroneously considered that catalytic hydrogenation of an 11-ketone led to a hydroxyl group of configuration opposite to that of the natural 11-hydroxy corticosteroids, and these reduction products were assigned the 11α-hydroxy configuration. This discrepancy was clarified by v. EUW and REICHSTEIN (1947). Small amounts of the 11α-hydroxy product (II, Fig. 35) are obtained with lithium aluminum hydride (LEVIN et al. 1954). Reduction of an 11-ketone with metal-ammonia or metal-alcohol combinations leads to the equatorial 11α-hydroxy compound (HERZOG et al. 1953, SONDHEIMER et al. 1953c).

The 11-keto group can be converted to a Δ^9-11-enol acetate (VII); again, selective enol esterification at C-20 can be achieved (H. ANDERSON et al. 1954). The configuration at C-5

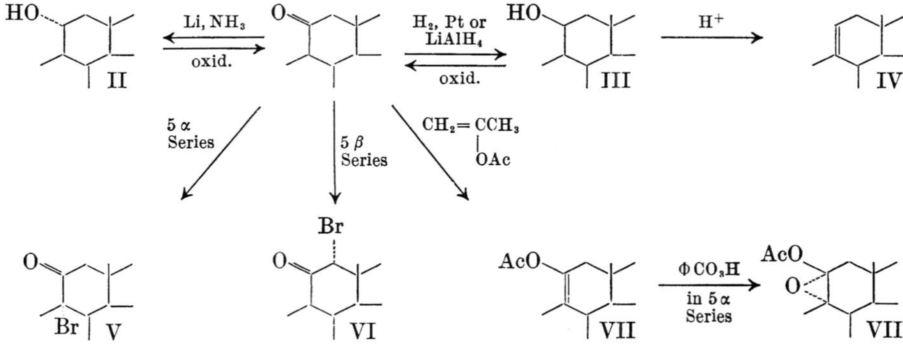

Fig. 35. Reactions of 11-ketones

exerts an important influence on the course and ease of reactions involving the C-11 system, as illustrated by the oxidation of the Δ^9-11 enol acetate system. In the 5β-series, the 9α, 11α-oxide (VIII) is formed in good yield, whereas in the 5β-series, the $\Delta^{9(11)}$-enol acetate system resists oxidation (KRITCHEVSKY et al. 1952), evidently because the *cis* A—B ring fusion causes ring A to be folded back, shielding the unsaturated system in ring C (BARTON et al. 1954). The bromination of an 11-ketone is much slower than that of a 3- or 20-ketone, and again the influence of the nature of the A—B ring junction is observed. In the 5α-series, bromination furnishes the 9α-bromoketone (V) (HENBEST et al. 1955); the 12α-bromo derivative (VI) is obtained in the 5β-series (TURNER et al. 1946).

b) 11-Alcohols

The β-configuration of the 11-hydroxyl group in the adrenal steroids was originally inferred from the resistance to esterification, since a β-oriented hydroxyl group is subject to much greater steric interference than one in the α-configuration. The configurations of substituents at C-11 and C-12 have been confirmed by an abundance of evidence relating the several 11,12-oxides and bromohydrins (TURNER 1949). *Under the customary acetylation conditions* using acetic anhydride and pyridine, *an 11β-acetate is not formed,* although acetylation has been observed during heating or prolonged reaction time (KEMP et al. 1954). Acid-catalyzed esterification of an 11β-hydroxyl group proceeds smoothly under the conditions used for acetylation of a 17α-hydroxyl group (OLIVETO et al. 1953a). Retention of an 11β-hydroxyl group during an oxidation step by protection as the acetate is impractical since rather vigorous conditions are required for subsequent hydrolysis (CALLOW and JAMES 1956). This purpose can be met by the use of the formate or trifluoroacetate esters, which are more readily formed and much more readily hydrolyzed than the acetate (LARDON and REICHSTEIN 1954, OLIVETO et al. 1955). The acetylation of an 11α-hydroxyl group and hydrolysis of the acetate proceed under conditions and at rates comparable to those of a hydroxyl group at C-3.

The two other important reactions of secondary alcohols, oxidation and dehydration, are not impeded in the case of the 11β-hydroxyl group but are actually enhanced, since in both cases the non-bonded interactions of the axial hydroxyl group are relieved by formation of a trigonal carbon atom at C-11. *An 11β-hydroxyl group is oxidized to the 11-ketone with chromic acid more rapidly than any other secondary hydroxyl group in the molecule* (SCHREIBER and ESCHENMOSER 1955), although steric factors retard oxidation by the Oppenauer procedure (p. 242).

Dehydration of an 11β-hydroxy compound proceeds readily on warming with mineral acid or Lewis acids, with the formation of a $\Delta^{9(11)}$-double bond (IV). It is probable but not proved that the $\Delta^{9(11)}$-unsaturated metabolites that have been isolated from urine are artifacts formed during acid hydrolysis (BITMAN et al. 1957). The dehydration is favored by the axial orientation of both the 11β-hydroxyl group and the 9α-hydrogen atom; for preparative purposes, the reaction is conveniently brought about with phosphorus oxychloride in pyridine (LITTEL and BERNSTEIN 1956). The 11β-hydroxyl function has been retained without protection during the conversion of an 11β-hydroxyetianic acid to the chloride (OLIVETO et al. 1956b). Removal of the elements of water from either an 11β- or 11α-hydroxyl group by any process, including elimination reactions of 11α-*p*-toluenesulfonates, invariably furnishes the $\Delta^{9(11)}$ product, indicating a significant barrier to the formation of $\Delta^{11(12)}$ double bond (CRAWSHAW et al. 1954).

7. Δ^4-3-Ketones, saturated 3-ketones and alcohols

Four systems of functional groups involving C-3 and C-5 are encountered among the non-aromatic adrenal steroids and their metabolites — Δ^4-3-ketone,

Δ^5-3β-hydroxyl, saturated 3-ketone and saturated 3-hydroxyl. In compounds obtained from adrenal tissue, the saturated ketones and alcohols are all members of the 5α (allo) series, and, with two exceptions, the hydroxyl group at C-3 is in the β-configuration. Ketones and alcohols with all combinations of configurations at C-3 and C-5 are found among the urinary steroids. The reactions and steric relationships of functional groups at C-3 have been exhaustively studied, and only a few general remarks can be made here. Since oxygen substitution at this position is present in every class of steroids, synthetic problems are not concerned with the introduction of an oxygen atom, but rather with devising the most direct approach to the required system from the most readily available starting material.

a) Δ^4-3-Ketones

The Δ^4-3-keto grouping, characteristic of all of the adrenocortical hormones, is immediately recognized experimentally by the ultraviolet absorption maximum at 238—240 mμ; no other α,β-unsaturated carbonyl chromophore is present among the adrenal steroids or their metabolites. The unsaturated ketone system forms the corresponding Δ^4-hydrazone derivatives with carbonyl reagents such as 2,4-dinitrophenylhydrazine and the Girard reagents. The 2,4-dinitrophenylhydrazones or semicarbazones are intermediates in many syntheses of the adrenocortical hormones (Fig. 55, p. 257), and regeneration of the Δ^4-3-ketone system is an important reaction. It is accomplished in nearly quantitative yield by treatment of the hydrazone derivative with excess pyruvic acid in acetic acid (McGuckin and Kendall 1952).

It is frequently necessary to protect temporarily the Δ^4-3-keto system against reaction, especially during hydride reduction at another carbonyl group, e. g., at C-11. This can be accomplished by preparing $\Delta^{3,5}$-dienol derivatives (Fig. 36), which can usually be formed selectively at C-3, leaving carbonyl groups at C-11, C-17, and C-20 (in the case of the less reactive 17- or 21-hydroxy-20-ketones) free

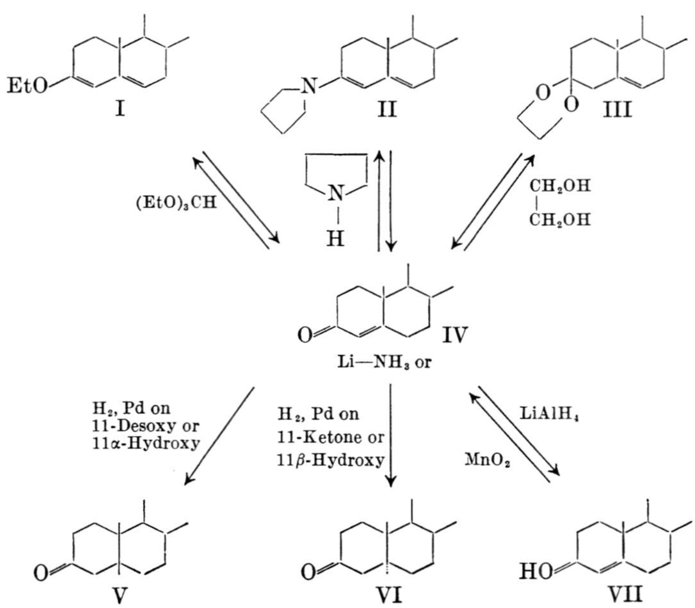

Fig. 36. Transformation products of Δ^4-3-ketone system

and available for reduction. Among the derivatives that have been used for this purpose are the enol ether (I)(JULIAN et al. 1951) and the imine (II), which is especially convenient (J. JOHNSON et al. 1956). Another widely used protective group is the cyclic ketal (ANTONUCCI et al. 1953), in which the double bond is shifted to the Δ^5-position. All of these derivatives are reconverted to the Δ^4-3-ketone by mild hydrolytic treatment.

The reduction of a Δ^4-3-ketone to the saturated ketone has been studied extensively in connection with both structural determination and synthetic efforts in the adrenal steroid field. Hydrogenation with a platinum catalyst causes at least partial reduction of the carbonyl group, but use of a palladium-carbon catalyst permits selective reduction of the double bond. In the earlier literature (FIESER and FIESER 1949) this reduction was reported to furnish varying amounts of the 5α and 5β isomers. It has been established with adrenal steroids that a C-11 substituent exerts a marked effect on the steric course of hydrogenation of a Δ^4 double bond, presumably by controlling the availability of the α and β faces for adsorption on the catalyst surface. Predominantly 5α (allo) derivatives (VI) are obtained when an 11-keto or 11β-hydroxyl group is present (PATAKI et al. 1952); with no substituent or with an 11α-hydroxyl group, the 5β isomer (V) is the major or exclusive product (MANCERA et al. 1953). The presence of alkali or other strong base in the hydrogenation medium has been shown by WILDS et al. (1950) to favor strongly formation of the *cis* product in the catalytic reduction of an α,β-unsaturated ketone, and yields of 99% of the 5β-3-ketone have been obtained by this modification (SLOMP et al. 1955). Chemical reduction of the Δ^4-double bond with lithium in liquid ammonia furnishes high yields of the more stable 5α-3-ketone (BOWERS et al. 1957).

Preferential reduction of the carbonyl group is effected with lithium aluminum hydride or sodium borohydride, although some reduction of the double bond has been observed with the latter reagent (SONDHEIMER et al. 1954b). The mixture of Δ^4-3β-ol and Δ^4-3α-ol, in which the former epimer predominates, can be separated with digitonin (GUT 1956). These allylic alcohols are readily dehydrated to the $\Delta^{3,5}$-diene with acid, and can be selectively reoxidized to the Δ^4-3-ketone with manganese dioxide (SONDHEIMER et al. 1953a).

In general, the Δ^4-3-ketone system is not highly susceptible to oxidation, and does not require protection, for example, during a mild chromium oxide oxidation at another site. It is possible to attack selectively a Δ^{16}-20-ketone system with a number of reagents such as osmium tetroxide (ALLEN and BERNSTEIN 1956a) or hypobromous acid (LÖKEN et al. 1956), without affecting the Δ^4-3-keto grouping. Reaction of the Δ^4-3-keto system can be brought about with more vigorous conditions, and several of these oxidative transformations have been investigated recently in the preparation of ring A-substituted analogues of the adrenal hormones (Fig. 62, p. 271).

b) Saturated 3-ketones

The 3-ketones in both the 5α and 5β series can be obtained in high yield from either of the corresponding epimeric alcohols. In addition to the standard methods of oxidation, several procedures have been developed for selective oxidation at C-3 (pp. 241—243). Among these may be mentioned N-bromoacetamide, t-butyl hypochlorite, and catalytic dehydrogenation with platinum and oxygen (SNEEDEN and TURNER 1955b).

The carbonyl group in saturated 3-ketones of both the 5α and 5β series is among the least hindered and most reactive in the steroid molecule, and it is possible to effect most reactions at C-3 selectively in the presence of carbonyl groups at C-11, C-17, or C-20. The most important difference in the behavior of the C-5 isomeric 3-ketones is the direction of enolization, which occurs from the 2-position in the 5α (allo) series, and the 4-position in the 5β series. Thus the 5α-(VI, Fig. 37) and 5β-3-ketones (V) give respectively the 2α-bromo (X) and 4α-bromo (XV) derivatives on bromination, the Δ^2- (IX) and Δ^4-enol esters (XIV) and acetates (DAUBEN et al. 1952), and the 2,3- and 3,4-dicarboxylic acids on vigorous oxidation. The preference for reaction at C-4 in the *cis* 5β series is presumably a consequence of the folding of ring A, with shielding of C-2 by ring C.

The isomeric ketones also exhibit contrasting behavior on reduction, since the conformations of the 3-hydroxy compounds in a given configuration are opposite in the two C-5 series. With complex hydrides, which give rise to the equatorial alcohol (in an unhindered ketone), the 5α-3-ketone (VI) gives mainly the 3β-hydroxy compound (VIII), and the 5β ketone (V) is reduced to the 3α-ol (XIII) (SHOPPEE and SUMMERS 1950). On the other hand, catalytic reduction in an acidic medium, which gives mainly the axial alcohol, leads to the 3α epimer (XI) in the 5α series, and the 3β-ol (XVI) in the 5β series.

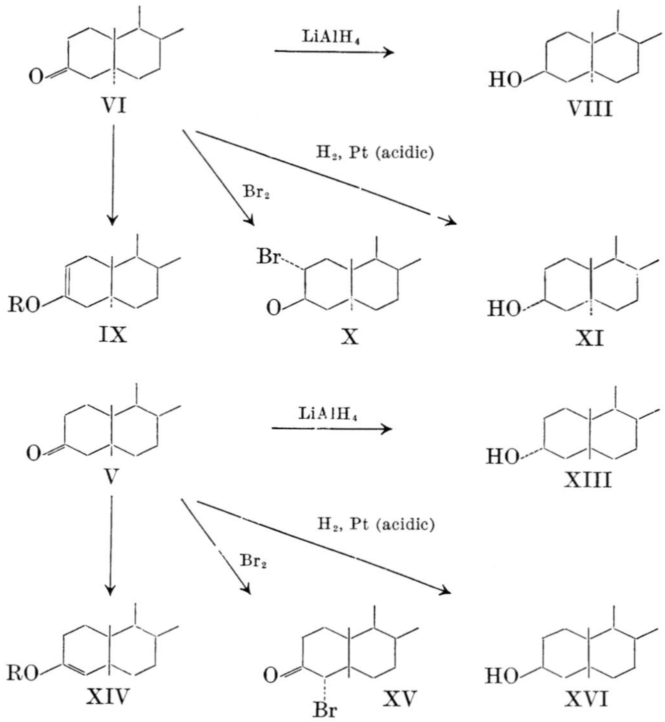

Fig. 37. Transformation products of 3-ketones

c) 3-Alcohols

Compounds representative of all four isomeric 3-alcohols are found among adrenal and urinary steroids, and mixtures of epimeric alcohols are frequently encountered. A separation of the 3α- and 3β-ols in either C-5 series can be achieved by precipitation of the 3β epimers with digitonin (p. 303). This steroidal saponin forms a very sparingly soluble molecular complex with all 3β-hydroxy steroids and with a few other substances (FIESER and FIESER 1949); 3α-hydroxy steroids are not precipitated. Digitonin precipitation has been studied recently by HASLAM and KLYNE (1953), who found that 5α-, 3β-ols (VIII, Fig. 31) are precipitated more completely and with lower concentrations of digitonin than the 3β-hydroxy compounds (XVI) of the 5β series.

As noted above, the hydroxyl groups in the $5\alpha,3\alpha$-ol and $5\beta,3\beta$-ol are axial, while those in the $5\alpha,3\beta$-ol and $5\beta,3\alpha$-ol are equatorial; the characteristic differences in the two pairs have been summarized by BARTON (1953). A typical example is the solvolysis of the *p*-toluenesulfonate esters, which in the case of the axial compounds gives a mixture of the Δ^2 and Δ^3 compounds in good yield, but with the equatorial epimers results largely in displacement with inversion of configuration (NACE 1952, EVANS and SHOPPEE 1953). The axial and equatorial alcohols and especially their respective acetates have distinctive bands in the infrared spectrum which are useful in characterization (R. N. JONES et al. 1951).

d) Δ^5-3β-Hydroxy compounds

The Δ^5-3β-hydroxy system (Fig. 38) characteristic of cholesterol and dehydroepiandrosterone is of interest in the chemistry of the adrenal steroids, inasmuch as it represents an important metabolic way station and is also present in several of the important raw materials used in corticosteroid synthesis. The equatorial 3 β-hydroxyl group and the double bond both undergo the expected reactions, and in addition there are several transformations unique to this homoallylic alcohol system. Among these may be mentioned the "i-steroid rearrangement", in which solvolysis of the *p*-toluenesulfonate under certain conditions leads to 3α,5α-cyclo-6β-substituted products (XVII) (for a review, see SHOPPEE and SUMMERS 1952), and a number of oxidation reactions, which furnish an impressive array of products involving C-3, 4, 5, 6 and 7 (FIESER and FIESER 1949, FIESER 1953a).

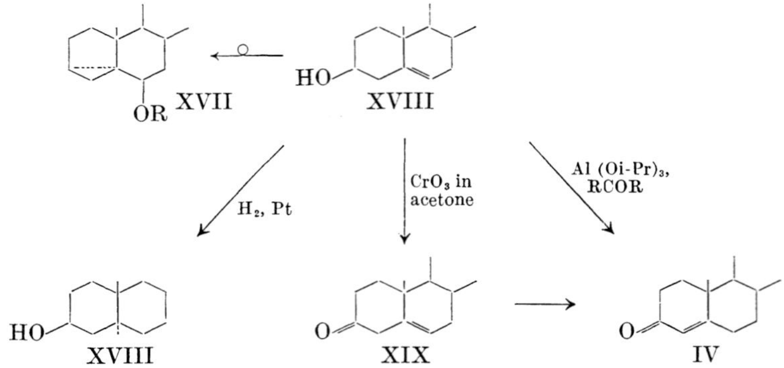

Fig. 38. Reactions of Δ^5-3β-hydroxy system

Reduction of the Δ^5-double bond by catalytic hydrogenation gives almost exclusively the 5α-compound (XVIII, Fig. 38). The system can be protected from oxidation by formation of the acetate 5,6-dibromide, from which the double bond is regenerated by treatment with zinc or sodium iodide. One of the most important transformations of the Δ^5-3β-ol is the dehydrogenation, with concurrent migration of the double bond, by Oppenauer oxidation to furnish the Δ^4-3-ketone. RINGOLD et al. (1956b) have found that this oxidation can be performed directly with the Δ^5-3β-formate ester, which can be used as a blocking group during construction of the cortical side chains. The Δ^5-3-ketone system (XIX) (DJERASSI et al. 1956a), which can be obtained by oxidation of the Δ^5-3β-ol with chromium oxide in acetone, is very labile, and rearranges to the conjugated ketone (IV) with great ease.

Although Δ^5-3β-hydroxy compounds are abundant and serve as starting materials for many purposes, there has been occasion to develop synthetic methods leading to this system, primarily for the incorporation of an isotopic carbon atom into ring A. A labeled atom may be introduced at C-4 by a sequence of reactions devised by TURNER (1950a) and FUJIMOTO (1950) (Fig. 39). Ozonization of the Δ^4-3-ketone to the keto acid (XX) and dehydration with acetic anhydride furnish the enol lactone (XXII), which on treatment with labeled methyl

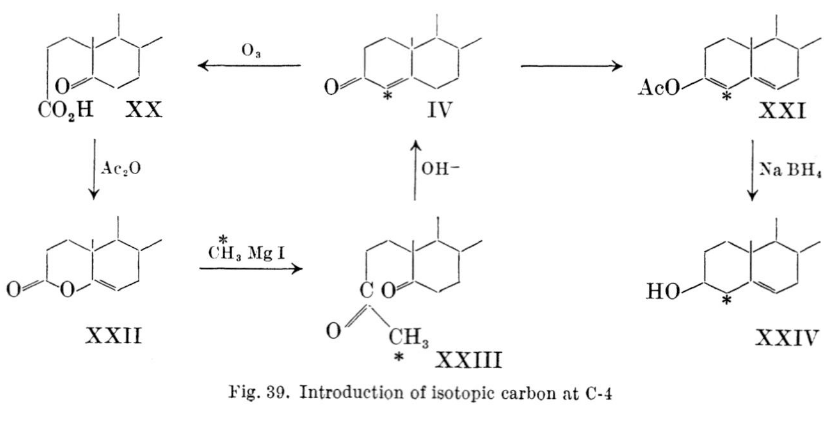

Fig. 39. Introduction of isotopic carbon at C-4

magnesium iodide and then base is reconverted to the unsaturated ketone. The most convenient route to the Δ^5-3β-ol involves treatment of the dienol acetate (XXI) with sodium borohydride (DAUBEN and EASTHAM 1951b, SCHWENK et al. 1951).

8. 2-Hydroxyl groups

Until the establishment by BURSTEIN (1956b) of the structure of a urinary metabolite as 2α-hydroxycortisol, the only naturally occurring steroids known to have a hydroxyl group at C-2 were a few sapogenins of plant origin. A few 2β-hydroxylated compounds have subsequently been isolated as metabolic or microbiological oxidation products. The following discussion will be restricted to the 2-hydroxy-Δ^4-3-ketone system encountered in hormone metabolites.

The assignment of configurations of the hydroxyl groups in the epimeric 2-hydroxy-Δ^4-3-ketones rests largely upon the facile conversion of one to the other under mild alkaline conditions (SONDHEIMER et al. 1953b). This change must comprise the epimerization of an axial hydroxyl group to an equatorial one, and the unstable and stable epimers may be assigned the 2β (II) and 2α (I) configurations, respectively (Fig. 40). In conformity with this assignment is the finding that the 2β-acetoxyl but not the 2α-acetoxyl group is smoothly removed by reductive elimination with zinc and acetic acid (SONDHEIMER et al. 1953b), a reaction known (ROSENFELD and GALLAGHER 1955) to require axial disposition of the acetoxy group.

The presence of the 2-hydroxy-Δ^4-3-ketone system can be inferred from the very characteristic ultraviolet spectrum in alkaline solution (A. MEYER 1955b); due to the rapid inversion of the 2β-hydroxyl group, the same spectrum is obtained from either 2-epimer, and this holds also for spectra in sulfuric acid. The 2β-ketol is reported (DODSON et al. 1957) to give a positive tetrazolium reaction. Introduction of a 2β-hydroxyl group is accompanied by a very large negative shift in rotation (ΔM_D—600°).

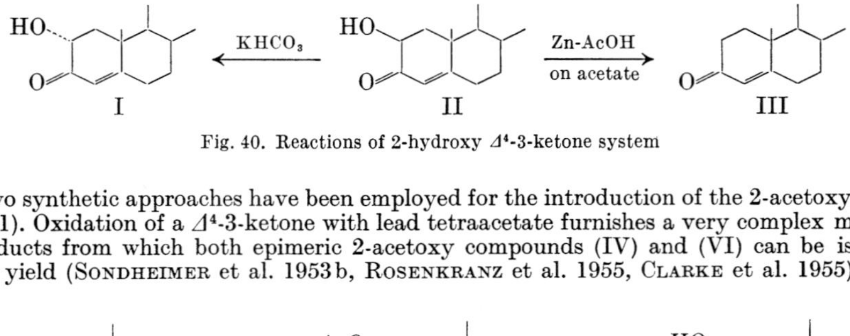

Fig. 40. Reactions of 2-hydroxy Δ^4-3-ketone system

Two synthetic approaches have been employed for the introduction of the 2-acetoxy group (Fig. 41). Oxidation of a Δ^4-3-ketone with lead tetraacetate furnishes a very complex mixture of products from which both epimeric 2-acetoxy compounds (IV) and (VI) can be isolated in low yield (SONDHEIMER et al. 1953b, ROSENKRANZ et al. 1955, CLARKE et al. 1955). This

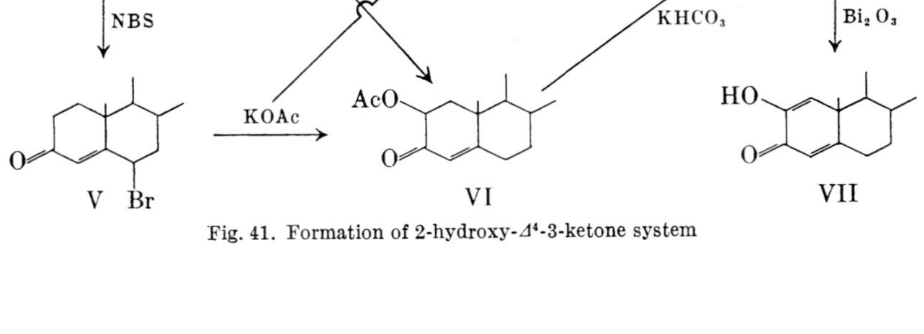

Fig. 41. Formation of 2-hydroxy-Δ^4-3-ketone system

reagent normally cleaves the dihydroxy ketone side chain, which can be protected as the 17,21-diacetate; the direct 2α-acetoxylation of cortisol 21-acetate in unspecified yield has been reported by BURSTEIN (1956). The other method comprises acetolysis with potassium acetate of a 6-bromo-Δ^4-3-ketone (V), obtained from the ketone by treatment with N-bromo succinimide. This reaction gives none of the 6-acetoxy ketone, and the latter is not an intermediate (FIESER and ROMERO 1953). Both 2β- (CLARKE et al. 1955) and 2α- (ROSENKRANZ et al. 1955) epimers have been obtained by this method; the yields are very poor. Since hydrolysis of either acetate furnishes the 2α-hydroxy derivative, the 2β-hydroxy-Δ^4-3-ketone system is not yet available by chemical synthesis. The oxidation of 2-hydroxy-Δ^4-3-ketones to the enolic 2-hydroxy-$\Delta^{1,4}$-3-ketone system (VII) has been described by BARAN (1958).

9. 6-Hydroxyl groups

In addition to the three 6β-hydroxy-Δ^4-3-keto adrenal steroids, several metabolites and microbiological oxidation products contain 6α- or 6β-hydroxyl groups. A few other naturally occurring 6-hydroxy compounds are found among the bile acids and steroidal sapogenins; the chemical behavior of 6-oxygen functions has been extensively investigated.

The 6β-hydroxyl group imparts a characteristic hypsochromic shift to the ultraviolet absorption maximum of the Δ^4-3-ketone, and *a maximum at ca. 237 mμ is diagnostic of the presence of this system.* Since a 6β-substituent is axially disposed, regardless of unsaturation or the configuration at C-5, the reactivity of a 6β-acetoxy or 6β-hydroxy group is somewhat depressed, and it is possible to effect the partial hydrolysis of a 3β,5α,6β-triol 3,6-diacetate to the 6-monoacetate (EHRENSTEIN and STEVENS 1941) or the partial acetylation of a 6β,21-diol to the 21-monoacetate (SONDHEIMER et al. 1954a). Under the usual conditions, a hydroxyl group in either the 6α- or 6β-configuration is acetylated. A characteristic reaction of 6-hydroxy-Δ^4-3-ketones (I, Fig. 42) is isomerization to the 5α-3,6-dione (II), and this rearrangement is a complicating factor in the hydrolysis of a 6β-acetoxy group. Hydrolysis with aqueous carbonate leads to extensive isomerization (HERZIG and EHRENSTEIN 1951); hydrolysis of the 6β-acetate with a limited amount of cold alcoholic potassium hydroxide is usually quite satisfactory, although several products have been detected under these conditions (BURSTEIN and DORFMAN 1956). Isomerization of the 6-hydroxy unsaturated ketones to the 3,6-diones also occurs in acid media; as expected, the reaction proceeds more readily with the axial 6β-ol (FLOREY and EHRENSTEIN 1954). Advantage has been taken of this facile isomerization to remove 6β-hydroxydeoxycorticosterone from a mixture with the 18-hydroxy isomer. The mixture, which could not be separated by paper chromatography, was refluxed with bicarbonate; the resulting mixture of the 3,6-dione and unchanged 18-hydroxy compound was readily resolved (KAHNT et al. 1955).

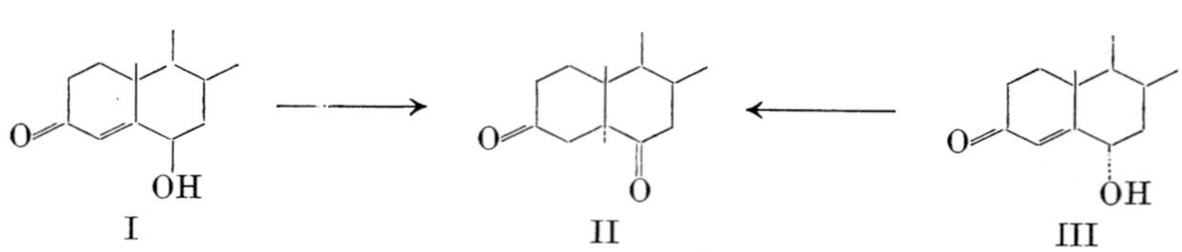

Fig. 42. Isomerization of 6-hydroxy-Δ^4-3-ketone system

The introduction of an oxygen atom at C-6 is a relatively simple matter since several types of starting material with a Δ^5 double bond are readily available, and a number of methods have been devised to meet the requirements of individual cases (Fig. 43). The most direct approach is oxidation of a Δ^5-3β-ol (VI) with chromium oxide, which furnishes the Δ^4-3,6-diketone (IV) in useful yields (FIESER 1953a). The latter system is readily reduced with zinc in acetic acid to the 5α-3,6-dione (II). Reduction of the saturated diketone with complex hydrides furnishes mainly the 3 β,6 β-diol (V), the unhindered carbonyl group at C-3 giving rise to an equatorial hydroxyl group and the relatively hindered 6-carbonyl the axial hydroxyl (ROMO et al. 1954).

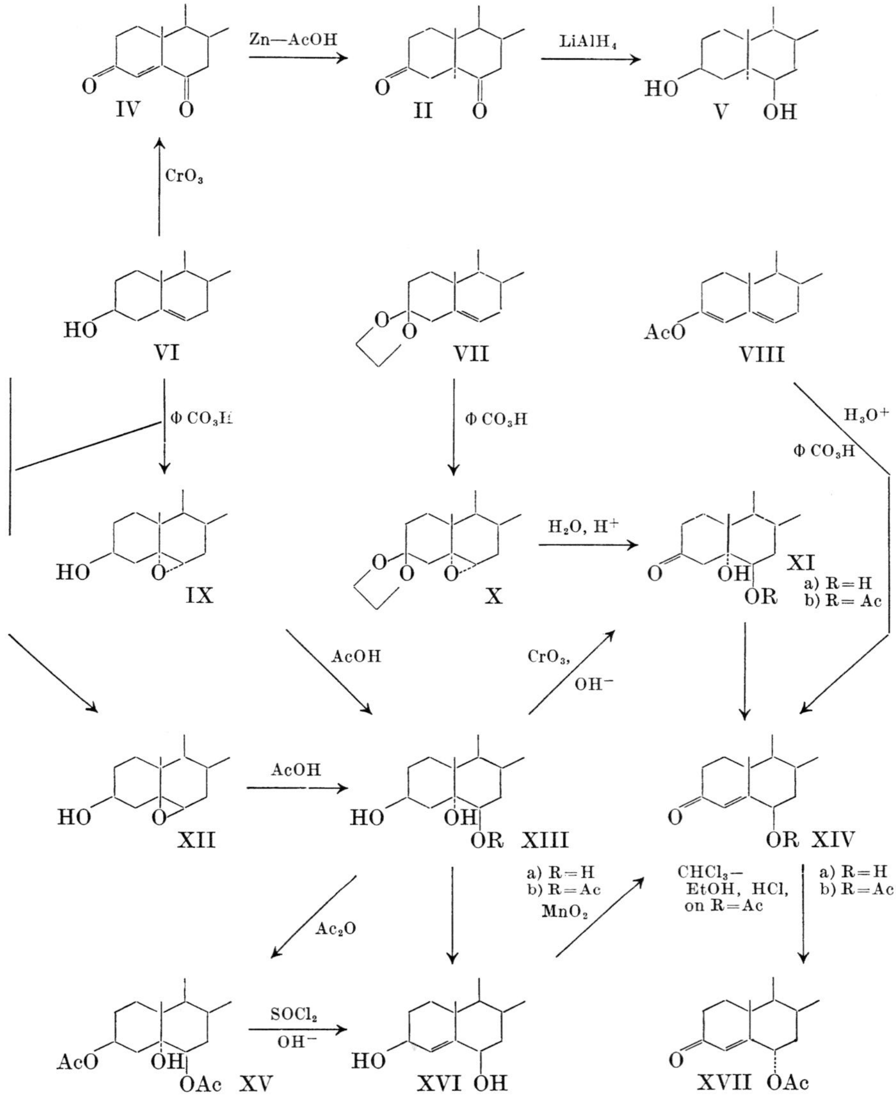

Fig. 43. Formation of 6-hydroxy compounds

Nearly all of the synthetic routes to the important 6 β-hydroxy-Δ^4-3-ketone system involve initial epoxidation of a Δ^5-double bond; the starting material may be the Δ^5-3 β-ol, the $\Delta^{3,5}$-dienol acetate (VIII) or, in the most efficient approach, the Δ^5-3-ethylenedioxy ketal (VII). The oxide obtained is largely the 5 α,6 α-isomer (IX, X), but it is unnecessary to purify the mixture, since either oxide furnishes the same diaxial 5 α,6 β-diol on hydrolysis, or 5 α,6 β-diol-6-monoacetate on acetolysis. The 3 β,5 α,6 β-triol has been used in several sequences leading to the 6 β-hydroxy-Δ^4-3-ketone. Oxidation of the 6-monoacetate (XIII b) furnishes the 3-ketone (XI b), which can be dehydrated with thionyl chloride in pyridine, or better with hydrogen chloride in ethanol-free chloroform, to the 6 β-acetoxy-Δ^4-3-ketone (XIV b) (HERZIG and EHRENSTEIN 1951). The use of ordinary chloroform (containing 0.7% ethanol) causes epimerization of the axial 6 β-acetoxy group, and leads to the 6 α-acetoxy-Δ^4-3-ketone (XVII) in high yield. A different approach comprises dehydration of the triol (XIII a) (or the diacetate) to the Δ^4-3 β,6 β-diol (XVI) and selective oxidation of the latter with manganese dioxide to (XIV a) (AMENDOLLA et al. 1954). The ketal (VII) is a particularly suitable starting point since hydrolysis of the oxide gives directly the 5 α,6 β-dihydroxyketone (XI a), which is acetylated and then dehydrated to the 6 β-acetoxy Δ^4-3-ketone (SONDHEIMER et al. 1954 a). Alternatively, the ketal can be hydroxylated with osmium tetroxide to give the 5 β,6 β-diol, which is simultaneously dehydrated and hydrolyzed to the Δ^4-6 β-ol (BERNSTEIN et al. 1955).

10. 16-Hydroxyl groups

Three C-16 oxygenated steroids have been obtained from the adrenal cortex, and a number of C_{19} and C_{21} compounds with a 16 α-hydroxyl group have been isolated from urine. Steroids with 16 β-oxygen substituents are widespread in the plant kingdom. The stereochemical relationships and properties of the urinary 16-hydroxylated pregnanes have been summarized by H. HIRSCHMANN (1955). A great deal of attention has recently been focused on synthetic 16 α-hydroxycorticosteroids with modified hormonal activity; the chemistry of these compounds is discussed on p. 272.

11. 18-Oxygen functions

Two compounds containing oxygen substituents at C-18, aldosterone (I), the 18-aldehyde of corticosterone, and the 18-carboxylic acid (III) of 11-deoxycorticosterone have been obtained from adrenal tissue; in addition 18-hydroxy-11-deoxycorticosterone (II) has been isolated from the incubation of 11-deoxycorticosterone with adrenal homogenates (KAHNT et al. 1955). A metabolite of aldosterone has been reported by ULICK and LIEBERMAN (1957) which is apparently an 18-hydroxy derivative; 18-aldehyde structures have been suggested by NOWACZYNSKI et al. (1957 b) for two other steroids of urinary origin. The C-18 oxygen groups in these compounds are very uncommon structural features in steroids, and prior to the structural elucidation of aldosterone, the chemical properties of such compounds were virtually unknown. Only one other C-18 oxygenated steroid, the 18-hydroxy alkaloid isorubijervine, has been found in nature; a few other alkaloids contain a basic nitrogen substituent at C-18.

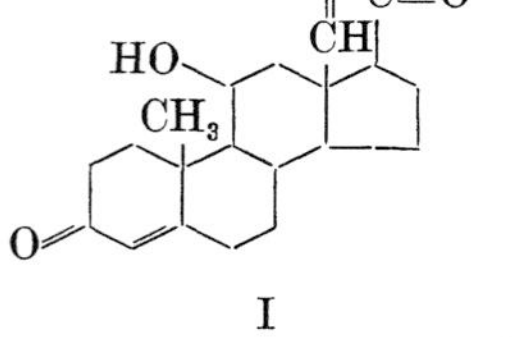

I

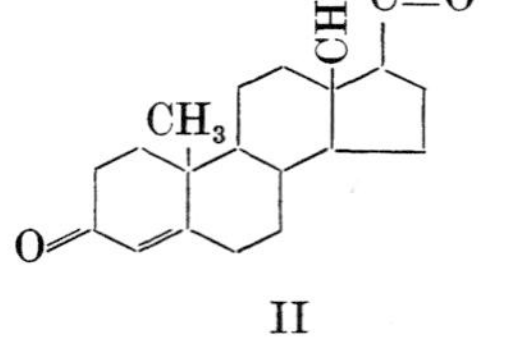

II

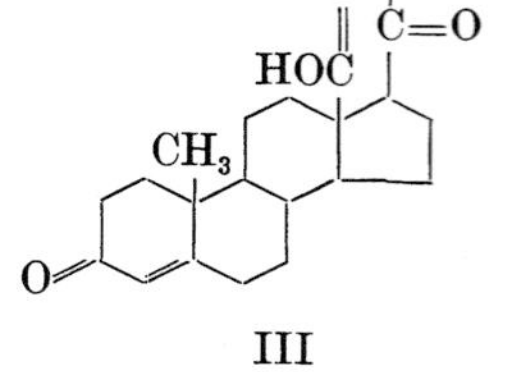

III

The partial synthesis of C-18 oxygen-substituted steroids from C-18 unsubstituted compounds requires the introduction of a functional group at the unactivated and inert angular methyl group. Several ways of accomplishing this difficult step have been developed. The direct substitution of an oxygen atom has been achieved by vigorous treatment of a 20 β-hydroxy compound with lead tetraacetate, which leads to the 18,20-oxide (IV) (CAINELLI et al. 1959); further oxidation furnishes the lactone (V). Decomposition of a 21-diazo-20-ketone with copper oxide leads to the cyclic ketone (VI), which can be degraded to 18-oxygenated derivatives (GREUTER et al. 1958). One of the most promising routes is the conversion of cyclic 18,20-amines such as (VII) to 18-hydroxy compounds (BUZZETTI et al. 1959, PAPPO 1959). The alkaloid conessine contains this cyclic amino grouping; similar amines have also been obtained synthetically by ring closure of N-chloro-20-methyl-amino derivatives (COREY and HERTLER 1958, BUCHSCHACHER et al. 1958).

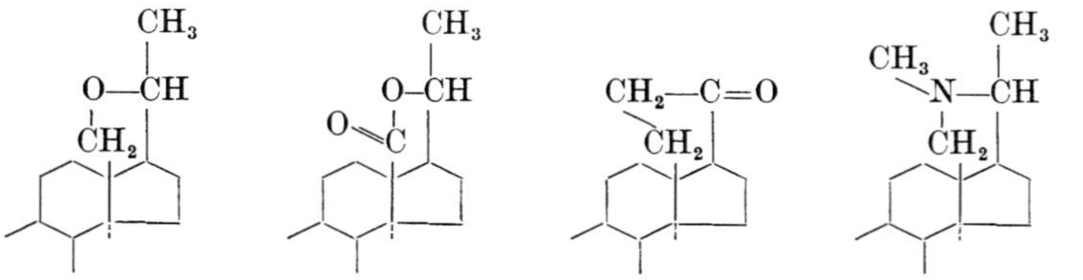

IV V VI VII

Hydroxyl and carbonyl substituents at C-18 undergo the usual reactions of primary alcohols and aldehydes, respectively, but data on the relative reactivities of groups at C-18 are very fragmentary. Acetylation of aldosterone with one equivalent of acetic anhydride furnishes the 21-monoacetate; the 18,21-diacetate of the hemi-acetal tautomer (VIII, Fig. 44) is obtained with excess reagent. Both the primary alcohol and the aldehyde at C-18 can be oxidized to the carboxylic acid; partial oxidation of the alcohol to the aldehyde with chromium trioxide has been accomplished in high yield in the case of *iso*rubijervine (WEISENBORN and BURN 1953). The ester of an 18-carboxylic acid is exceptionally resistant to

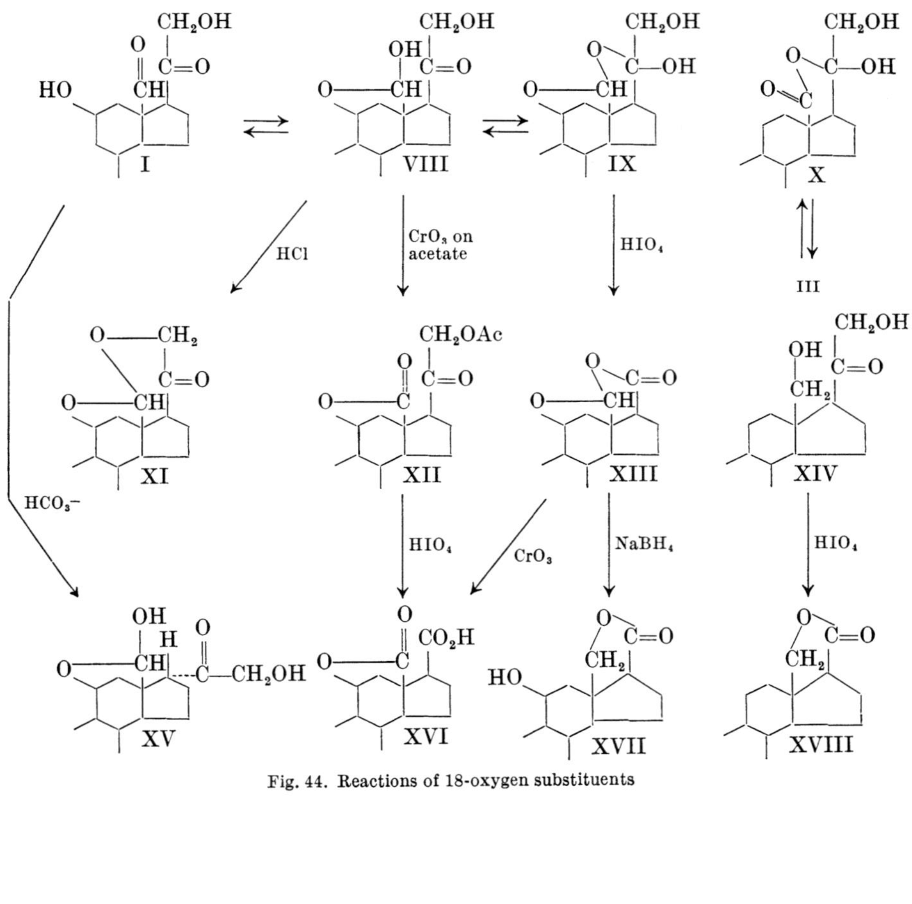

Fig. 44. Reactions of 18-oxygen substituents

hydrolysis (SATO and JACOBS 1951). An extremely important factor in the chemistry of the 18-oxygenated corticosteroids is the fact that substituents on the axial C-18 methyl group are within bonding distance of C-11 β, C-20, and C-21. The formulation of the structures of aldosterone by SIMPSON et al. (1954b) and of the deoxycorticosterone derivatives (II) and (III) was based in large part on the formation of lactones and lactols with compounds in various oxidation states at C-18 and C-20.

The infrared spectra and reactions of aldosterone show clearly that the compound exists, at least in solution, entirely in the tautomeric lactol (hemiacetal) (VIII) and dilactol (IX) forms. The participation of the C-20 carbonyl group in lactol formation in aldosterone (IX) and in the 18-carboxylic acid (X) is indicated by the very weak Tollen's reaction (ammoniacal silver ion) (NEHER and WETTSTEIN 1956); the tetrazolium reaction of the masked ketol group is not depressed. The presence of the 18→20 lactone group in (X) is also confirmed by the typical γ-lactone infrared carbonyl band. Periodate oxidation of aldosterone furnishes the very stable lactol-lactone (XIII), and similarly, cleavage of 18-hydroxy-11-deoxycorticosterone (XIV) gives the lactone (XVIII) (KAHNT et al. 1955). Chromic acid oxidation of aldosterone monoacetate leads to the 18→11 lactone (XII). In the aldosterone degradation reported by HAM et al. (1955), the lactol ring of the lactol-lactone (XIII) was opened by reduction with sodium borohydride, whereas oxidation of (XII) gave the lactone acid (XVI).

Aldosterone is quite sensitive to both acid and base. Exposure to very mild acid treatment leads to the formation of several products; the major one is the acetal (XI) formed by participation of the 21-hydroxyl group. The mono- and diacetates can be hydrolyzed with potassium bicarbonate, but even under carefully controlled conditions a certain amount of the 17-*iso* derivative (XV) is formed. It has been found by SCHMIDLIN et al. (1957) that in both aldosterone and the lactone (XII), in contrast to C-18 unsubstituted 20-ketones, the 17 α-side chain isomer is more stable than the natural 17 β isomer.

The direct correlation of a C-18-oxygenated steroid with a compound of established structure required the elimination of the C-18 oxygen function. This was achieved in the case of aldosterone by Wolff-Kishner reduction of the ring A reduction product of the lactone (XII). The reaction was complicated by the formation of side products, but the 3,11-dihydroxy etianic acid was obtained in amount sufficient for identification. Because of the very limited amounts of material, correlation with a known compound, and thus rigorous proof of the carbon skeleton was not accomplished with the 11-deoxy-18-oxygenated derivatives (II) and (III). The structures of these compounds are based on the assumption of a structural relationship with the other corticosteroids and the exclusion of alternative formulations.

12. 19-Oxygen functions

Attention was drawn to 19-hydroxycorticosteroids in 1955 with the demonstration that 19-hydroxylation of progesterone (LEVY and KUSHINSKY 1955), 4-androstene-3,17-dione (A. MEYER 1955a) or 11-deoxycorticosterone (ZAFFARONI et al. 1955) could be effected with adrenal enzyme systems in perfusion or incubation experiments. Two 19-hydroxy compounds have since been isolated from adrenal extracts (Table 2, p. 202); 19-aldehydes analogous to aldosterone have not been encountered. The only other 19-hydroxylated steroids of natural occurrence are a few butenolide glycosides.

The C-19 primary hydroxyl group in a 19-hydroxy-Δ^4-3-ketone (I, Fig. 45) is normally reactive, furnishing the acetate under the usual conditions. Acetylation produces a shift of the Δ^4-3-ketone ultraviolet absorption maximum from 242 mμ to 239 mμ (BARBER and EHRENSTEIN 1954), providing one indication of the presence of this functional group. Oxidation of 19-hydroxy-4-androstene-3,17-dione to the aldehyde has been accomplished with chromium trioxide in pyridine (A. MEYER 1955a), although this reagent has been reported to fail with other 19-hydroxy compounds (BARBER and EHRENSTEIN 1955). Chromic acid has been used to obtain the aldehyde, or, with excess reagent, the 19-carboxylic acid. An interesting reaction which has been used to characterize 19-hydroxy-Δ^4-3-ketones, and establish structural correlations, is the retroaldol reaction which is brought about with strong alkali (A. MEYER 1955a) (Fig. 45). Formaldehyde is eliminated and the 19-nor-Δ^4-3-ketone (III) is obtained in low yield.

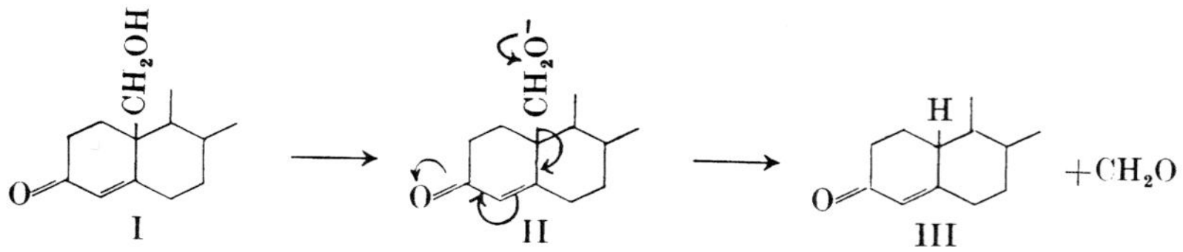

Fig. 45. Conversion of 19-hydroxy-Δ^4-3-ketone to 19-norsteroid

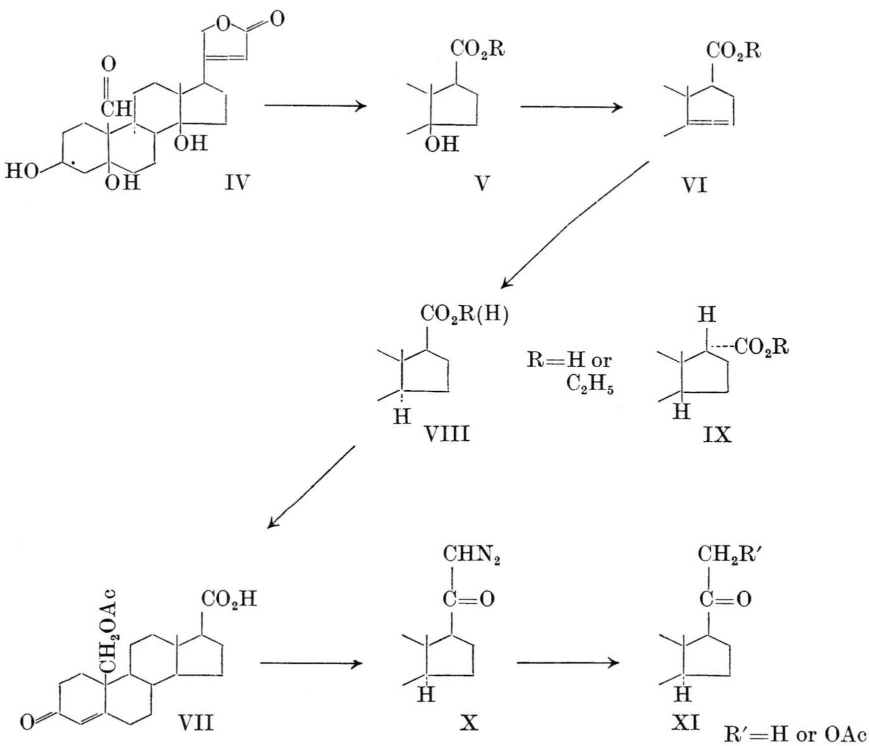

Fig. 46. Formation of 19-hydroxypregnane derivatives

The reactions of functional groups at C-19, like those at C-18, may be strongly influenced by interactions with substituents at other centers that are within bonding distance. HUNGER and REICHSTEIN (1952) have observed markedly reduced reactivity of C-19 functions in a saturated 5α (A/B *trans*) compound. Hemi-acetal formation of C-19 aldehydes can occur with hydroxyl groups at C-3β or C-9α in either C-5 series. An 11β-hydroxy-19-aldehyde would be expected to exhibit lactol formation analogous to that encountered in aldosterone.

Microbiological hydroxylation at C-19 has not yet been realized, and all partial synthetic routes to 19-hydroxy compounds are based on the 19-oxygenated butenolides, of which strophanthidin (IV, Fig. 46) is the most abundant and useful. The transformation of this aglycone to various 19-oxygenated hormone derivatives has been reported in a series of papers by EHRENSTEIN and co-workers (see DÜNNENBERGER and EHRENSTEIN 1956). Although the butenolide side chain can be degraded smoothly to a 20,21-ketol (Fig. 28), the preferred path begins with permanganate oxidation to the etianic acid (V, Fig. 46). The general synthetic approach comprises dehydration or the C-5 β and C-14 β hydroxyl groups and establishment of the *trans* C/D ring junction (VIII) by catalytic hydrogenation of (VI). Caution must be exercised at this stage to avoid formation of the unnatural 14 β,17 α-isomer (IX; cf. Fig. 80, p. 290). Necessary operations to form the Δ^4-3-ketone system are performed and the pregnane side chain is finally elaborated via the diazoketone (X). The C_{19} series is entered by perbenzoic acid oxidation of the 20-ketone.

III. Important reactions and reagents

In the course of interconversions and partial syntheses of the multifunctional adrenal steroids, certain simple reactions are performed repeatedly. Optimum conditions, special reagents and convenient techniques have been developed for most of these transformations to achieve maximum yields and frequently a high degree of selectivity. In this section, a few of the more common reactions that may be encountered in the characterization of corticosteroids are briefly discussed; when possible, an indication of the specificity is given. No attempt has been made to include all of the transformations used in partial synthetic studies. An excellent survey of many important reagents and experimental conditions is presented in the manual by FIESER (1955).

1. Reactions of hydroxyl groups

a) Esterification and etherification

α) Acetylation. Probably *the most common reaction in steroid chemistry is the acetylation of a hydroxyl group.* This reaction not only renders the hydroxyl group resistant to oxidation, a necessary first step in many operations, but furnishes direct evidence of the number of primary and unhindered secondary hydroxyl groups. The number of acetyl groups introduced can usually be determined unambiguously by combustion analysis, but it is sometimes possible to infer this from the change in polarity, evaluated by paper chromatography, or on ultra-micro scale, as in the structure proof of aldosterone (SIMPSON et al. 1954a), by the introduction of radioactive acetyl groups and measurement of specific activity.

The standard procedure for acetylation comprises solution of the steroid in acetic anhydride and anhydrous pyridine at room temperature. The solution is allowed to stand for 4—16 hours and then evaporated *in vacuo*, or alternatively treated with ice; the steroid is extracted with ether or chloroform and the solution washed with dilute acid and base. Comparable results are usually obtained by heating the reaction mixture for 30 minutes. The relative amounts of acetic anhydride and pyridine are not critical; the two reagents react slowly, and a mixture cannot be stored.

Under these conditions, any primary or secondary hydroxyl group except that at C-11 β is acetylated. Tertiary hydroxyl groups, except that at C-17 β (17-*iso* series), usually do not react.

Other conditions that have been used for acetylation are refluxing acetic anhydride or acetic acid. Refluxing acetic anhydride (140°) is equivalent to acetic anhydride-pyridine with respect to primary and secondary hydroxyl groups, but this treatment can also effect acetylation of tertiary hydroxyl groups, e. g., 17 α-hydroxycorticosteroids. Refluxing acetic acid is less often used, but most primary and secondary hydroxyl groups (including C-11 α) are smoothly acetylated; this method is useful for the acetylation of hydroxy carboxylic acids, which form mixed anhydrides with acetic anhydride.

β) Esterification with other acids. It is occasionally necessary to protect a hydroxyl group with a very readily hydrolyzed acyl grouping (p. 228); the formyl and trifluoroacetyl esters meet this requirement. Formylation can be accomplished with formic acid-acetic anhydride-pyridine mixtures (REBER et al. 1954) or by warming with 85% formic acid (RINGOLD et al. 1956b). The trifluoroacetate is obtained by treatment with trifluoroacetic anhydride at low temperature. *p*-Toluenesulfonate esters and esters of higher molecular weight acids, which are frequently prepared to obtain greater lipide solubility or longer duration of steroid hormones, are usually formed by treatment with the corresponding acid chloride in pyridine (GOULD et al. 1957a).

γ) Hydrolysis of acylated steroids. The saponification of steroid acetates is usually carried out with alcoholic potassium hydroxide at reflux temperature; all except hindered acetoxy groups such as that at C-11β are hydrolyzed under these conditions. Removal of an acyl group can be accomplished by base-catalyzed ester exchange, using less than one equivalent of alkali (RINGOLD et al. 1956b). Excellent yields can often be achieved in partial saponifications. Despite the frequently cited instability of ketols to alkali, ketol acetates can be hydrolyzed very efficiently with alkali when proper precautions are observed.

In the hydrolysis of 6β-hydroxycortisone 6,21-diacetate (SONDHEIMER et al. 1954a), the diacetate was treated with 3 moles of potassium hydroxide in methanol solution in the absence of air; after one hour at room temperature the solution was acidified with acetic acid and evaporated and diluted with water. An 88% yield of 6β-hydroxycortisone was obtained.

Other methods for the hydrolysis of ketol acetates have been mentioned on p. 221. Formate or trifluoroacetate esters are much more readily hydrolyzed, benzoates rather less readily, than the acetates.

δ) Other protective groups. It is sometimes advantageous to convert a hydroxyl group to a derivative that is stable to alkaline hydrolysis or hydride reduction. This can be accomplished by formation of an easily cleaved ether, such as the 2′-tetrahydropyranyl ether. This derivative, which is actually an acetal, is readily formed by reaction of the hydroxy steroid with dihydropyran in the presence of acid catalyst (OTT et al. 1952). The derivative is susceptible only to acid-catalyzed

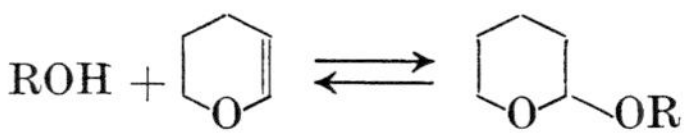

hydrolysis, which smoothly regenerates the hydroxyl group. Another grouping which has been used for this purpose is the triphenylmethyl ether (HIRSCHMANN and HIRSCHMANN 1950). This group is very sensitive to acid; it may also be removed by catalytic hydrogenation.

b) Oxidation

Another very common reaction in steroid chemistry is the interconversion of a secondary alcohol group and the corresponding ketone; in a number of cases, oxidation of a new alcohol to the known ketone is the main step in the establishment of structure.

α) Chromic acid. The most general oxidant for the dehydrogenation of a secondary alcohol is chromic acid, which can be used in one of several forms.

The most frequently used reagent is chromium trioxide (chromic anhydride); this reagent is hygroscopic and readily soluble in acetic acid containing 10% water. One mole of reagent oxidizes 1.5 moles of alcohol:

$$3\ RCHOHR + 2\ CrO_3 + 6\ AcOH \rightarrow 3\ RCOR + 2\ Cr(OAc)_3 + 6\ H_2O$$

The oxidation is usually carried out in aqueous acetic acid at room temperature with a slight stoichiometric excess of the oxidant. Another procedure for chromic acid oxidations suggested by FIESER (1953a) uses the non-hygroscopic sodium dichromate, $Na_2Cr_2O_7 \cdot 2\,H_2O$, which is soluble in glacial acetic acid.

A highly satisfactory method for the oxidation of acid-sensitive compounds employs the chromic anhydride-pyridine complex. The reagent is prepared by the cautious addition of chromic anhydride to pyridine (not the reverse!) (POOS et al. 1953); the compound to be oxidized is then added to the pyridine suspension.

By limiting the amount of reagent, highly selective oxidations are possible, as exemplified by the conversion of allopregnane-3β,11β,20β-triol 3-monoacetate to the 20-hydroxy-11-ketone in over 90% yield (v. EUW et al. 1944). Quantitative data on the rate of oxidation of hydroxyl groups at a number of positions have been determined (SCHREIBER and ESCHENMOSER 1955).

β) **N-Bromoacetamide (NBA).** This reagent, $CH_3CONHBr$, is used both as a selective oxidizing agent for secondary alcohols and as a source of hypobromous acid for the conversion of an olefin to the bromohydrin.

The oxidation of an alcohol with NBA is usually carried out in a t-butyl alcohol or dioxane solution containing a little water. The reaction is allowed to proceed in the absence of light at room temperature or below; the extent of reaction can be determined by removing samples and titrating excess hypobromite. One equivalent of NBA is consumed and one equivalent of acid is liberated:

$$RCHOHR + CH_3CONHBr \rightarrow RCOR + CH_3CONH_2 + HBr.$$

In the oxidation of an acid sensitive compound such as a ketal, pyridine can be added to prevent hydrolysis (SARETT 1949b).

The generality of this reagent for the oxidation of secondary alcohols was demonstrated by REICH and REICHSTEIN (1943), who obtained excellent yields of mono- or bis-oxidation products. A remarkable feature of NBA is the specificity for secondary hydroxyl groups even in the presence of the unprotected ketol side chain; KRITCHEVSKY et al. (1952) oxidized 3α,17α,21-trihydroxypregnane-11,20-dione to the 3,11,20-trione in high yield. Under the usual reaction conditions, and using a limited amount of reagent, a conjugated double bond (e.g. Δ^4-3-keto grouping) is not affected; an isolated double bond may interfere. Simultaneous oxidation of a 3-hydroxyl group and bromination, giving the 4α-bromo-3-ketone, has been accomplished with NBA (HERSHBERG et al. 1952). Other haloamides, particularly N-bromosuccinimide, can be used in a similar manner (FIESER and RAJAGOPOLAN 1950).

γ) **Oppenauer oxidation.** This oxidation procedure comprises the aluminum alkoxide-catalyzed exchange of hydrogen between an alcohol and a ketone such as acetone or cyclohexanone:

$$RCHOHR + CH_3COCH_3 \xrightarrow{Al(O\text{-}iPr)_3} RCOR + CH_3CHOHCH_3.$$

Removal of the product alcohol by distillation drives the reaction to completion. The scope of the reaction and experimental procedures have been reviewed (DJERASSI 1951). An example of the utility of this method is the selective oxidation of allopregnane-3α,11β,20β-triol 3-monoacetate to the 20-monoketone (v. EUW et al. 1944); this is the reverse of the above-mentioned selective chromic acid oxidation of the same compound to the 11-ketone.

δ) **Manganese dioxide.** Allylic alcohols, $R{-}\overset{\overset{OH}{|}}{C}H{-}\overset{|}{C}{=}\overset{|}{C}{-}$, are smoothly and selectively oxidized to the α,β-unsaturated ketones by stirring in chloroform solution at room temperature with freshly precipitated manganese dioxide (ATTENBURROW et al. 1952).

ε) t-Butylhypochlorite. This reagent, $(CH_3)_3COCl$, has been used by FONKEN et al. (1955) to oxidize 3-hydroxy steroids to the ketones in very high yields. The selectivity of this oxidation method has not been thoroughly studied; simultaneous chlorination of the ketone can be effected in one step if desired.

c) Dehydration

Removal of a hydroxyl group with formation of a double bond is occasionally an important step in the establishment of structure (Fig. 8) or in partial synthesis (Fig. 43). The reaction is usually effected with strong acids or inorganic acid chlorides such as hydrogen chloride in acetic acid, phosphorus oxychloride, or thionyl chloride. These reagents function by ionic mechanisms involving attack on the hydroxyl group. Base-catalyzed dehydration is possible in the case of a β-hydroxyketone in which a proton can be abstracted.

One of the most generally applicable procedures uses phosphorus oxychloride in pyridine. A solution of the hydroxysteroid in pyridine is treated at 0° or below with phosphorus oxychloride and allowed to stand at room temperature overnight or longer. The reaction mixture is then quenched with ice and the product filtered or extracted. W. S. ALLEN and BERNSTEIN (1955) have found that thionyl chloride in pyridine is a more powerful dehydrating agent; an 11β,17α-dihydroxy-20-ethylene ketal was converted to the Δ^9-derivative with $POCl_3$ in pyridine and to the $\Delta^{9,16}$-*bis*anhydro compound by the use of $SOCl_2$.

Another procedure for the removal of a secondary hydroxyl group that is often superior to direct dehydration comprises formation and elimination of the toluenesulfonate or methanesulfonate ester. The elimination is usually accomplished by heating the sulfonate with a base such as collidine. An alternative method for removing the elements of water from a secondary hydroxyl group is pyrolysis of an ester such as a benzoate at temperatures in excess of 200°; the yield of olefin is usually low.

2. Reactions of carbonyl groups

a) Formation of hydrazone derivatives

Derivatives of steroidal ketones with substituted amines have been prepared for a variety of purposes. The hydrazones formed with the Girard reagents and with 2,4-dinitrophenylhydrazine are useful as water-soluble and highly-colored derivatives, respectively, and are discussed in a later section on isolation and identification (p. 338). Semicarbazones and oximes have found some use as protective groups, although ketal formation is in general more suitable for this purpose. Steroidal semicarbazones are usually high melting and very sparingly soluble, and it is sometimes advantageous to isolate and purify ketonic steroids via the semicarbazones.

The reaction of hydrazone formation is dependent upon pH; the optimum conditions vary with the reagent. Semicarbazones and oximes are best prepared in neutral or slightly acidic media, whereas strong acid is required for reaction with 2,4-dinitrophenylhydrazine, which is a very weak base. OLIVETO et al. (1956a) have studied the conditions for the reaction of cortisone acetate with several reagents. They obtained quantitative yields of the 3,20-bis-semicarbazone by refluxing the steroid with excess semicarbazide hydrochloride and an equivalent amount of pyridine in 80% aqueous methanol for 15 hours.

The order of reactivity of carbonyl groups in corticosteroids usually observed in hydrazone formation and most other carbonyl reactions is — saturated 3-ketone > Δ^4-3-ketone > 17- or 20-ketone > 21-hydroxy-20-ketone > 21-acetoxy-20-ketone ≅ 17α,21-dihydroxy-20-ketone > 17α-hydroxy-21-acetoxy-20-ketone. BARTON (1953) has deduced a scale of relative hindrance of carbonyl groups at various nuclear centers on the basis of steric factors; in the 5α (allo) series, the sequence of increasing hindrance, and presumably decreasing reactivity, is: C-3 < C-2 < C-7 < C-4 = C-6 < C-1 = C-12 ≪ C-11. Hydrazone formation at C-11 has not been observed.

b) Ketals

Ketal derivatives of carbonyl compounds play an important role as protective groups in corticosteroid transformations as shown by the numerous examples in the preceding and subsequent sections. Although dimethyl and diethyl ketals have occasionally been used for this purpose, the cyclic ketals obtained with ethylene glycol (dioxolanes) are more suitable and have been much more widely employed.

In the standard procedure for ketal formation, the ketone is refluxed in benzene solution with ethylene glycol and a catalytic amount of *p*-toluenesulfonic acid. The water formed is removed by distillation. Saturated carbonyl groups at C-3, C-17, and C-20 and Δ^4-3-ketones are readily ketalized under these conditions; a 17,21-dihydroxy-20-ketone forms the 20-ketal, but the reaction is prevented by acetylation at C-21. In a superior method for ketalization of 17,20-dihydroxy-20-ketones the benzene is omitted; excess ethylene glycol is used as solvent and is distilled from the reaction under reduced pressure (W. S. Allen et al. 1954). The acid-catalyzed exchange reaction of a steroidal ketone and the ketal of butanone has been recommended for selective dioxolane formation (H. Dauben et al. 1954). Regeneration of the steroid ketone from these derivatives has been accomplished by warming with 50—90% aqueous acetic acid or by exchange reaction of acetone with the steroid dioxolane.

c) Formation of enol derivatives

Derivatives of the enol form of ketones, both saturated and unsaturated, are frequently useful both as protective groups and as intermediates in partial syntheses. Free enols exist only in special cases such as the diosphenol tautomers (II) of α-diketones (I), but the esters (III), ethers (IV), and imino derivatives (V) are readily formed and hydrolyzed.

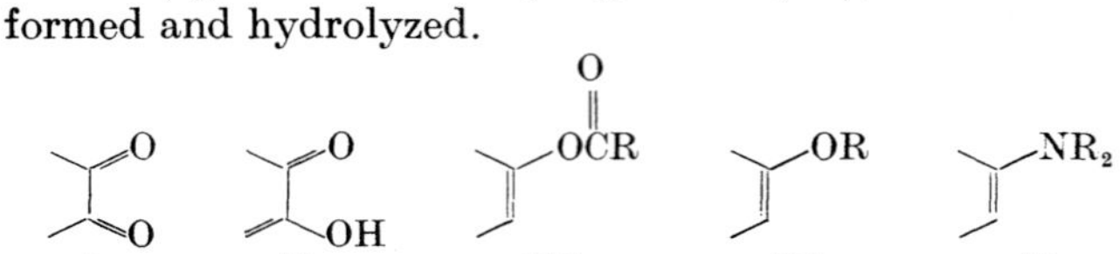

I II III IV V

α) Enol esters. Two reagents, acetic anhydride and *iso*propenyl acetate ($CH_2{=}C(CH_3)OCOCH_3$), the enol acetate of acetone, are most often used for the conversion of steroidal ketones to the enol acetate esters. In both cases a strong acid catalyst is used to promote tautomerism; the equilibrium is then displaced by esterification.

In the conversion of cholestenone to the 3,5-diene-3-ol acetate (W. Dauben and Eastham 1951 a), a solution of the steroid in *iso*propenyl acetate is treated with a trace of concentrated sulfuric acid and heated to reflux; some of the liquid is allowed to distill to remove acetone and drive the reaction to completion. When acetic anhydride is used, the catalyst is usually *p*-toluenesulfonic acid; a solution of the ketone is refluxed and then processed as in any acetylation.

When two enol derivatives can be formed, the two reagents may furnish different esters; this has been noted in the case of saturated 20-ketones (Fig. 21). Enol acetates have been used as intermediates in both oxidation (Fig. 22) and reduction (Fig. 39) sequences. Very mild alkaline hydrolysis suffices to regenerate the ketone.

β) Enol ethers and imines. One of the most important functions of these derivatives is the temporary protection of a carbonyl group toward hydride reduction (Fig. 36). The 3,5-diene-3-ol ethyl ether has also been utilized to stabilize the Δ^4-3-keto system against strongly alkaline conditions (C. Engel and Just 1954).

The ether (IV) is obtained by warming a solution of the unsaturated ketone and ethyl orthoformate ($HC(OC_2H_5)_3$) in dioxane or alcohol with a trace of concentrated sulfuric acid

(JULIAN et al. 1951). The ether is unaffected by basic treatment but is readily hydrolyzed by mild acid hydrolysis. The imine (V) is obtained from a Δ^4-3-ketone simply by dissolution in anhydrous pyrrolidine. The structure and properties of the 3,5-diene-3-amine system have been studied by J. JOHNSON et al. (1956). Regeneration of the ketone is best accomplished by mild alkaline treatment.

d) Reduction to a hydroxyl group

Practically every synthetic sequence leading to a hydroxy steroid involves at some stage or other the reduction of a carbonyl group to the alcohol. By far the most important method for this transformation is reduction with complex hydrides. These reagents have largely displaced older methods such as the Meerwein-Pondorff reduction and other chemical reduction procedures.

α) **Complex hydrides.** In the preceding paragraph the term "complex hydrides" has been used generically to include several compounds of the type MXH_4, in which M is lithium or sodium and X is aluminum or boron. These reagents bring about reduction by effecting the addition of a hydride ion, $H:^-$, to the carbonyl group. One of the most important virtues of the complex hydrides is the fact that carbon-carbon double bonds are not reduced except in occasional cases of conjugated unsaturated ketones. The two compounds which are by far the most important in the steroid field are lithium aluminum hydride, $LiAlH_4$, and sodium borohydride, $NaBH_4$; $LiBH_4$ has found limited use. There have been numerous reviews of the literature of hydride reductions; notable among them is the comprehensive treatise of GAYLORD (1953).

$LiAlH_4$ is the most powerful reducing agent of the group, and is capable of reducing any carbonyl group in the steroid molecule, including that in a carboxylic acid. The reagent reacts violently with water, alcohols and other "active hydrogen" compounds, with the liberation of hydrogen. Reductions are usually carried out by adding an ether solution of the steroid to a solution of the hydride in ether or tetrahydrofuran; rigorous precautions against air, moisture, and peroxide-containing solvents should be taken. The reaction is usually mildly exothermic and quite rapid. The reduction furnishes the lithium aluminum alkoxide, which may precipitate from the reaction mixture. Excess hydride is cautiously destroyed by the addition of ethyl acetate or moist ether, and the reaction mixture is then acidified and the product isolated from the ether solution.

$$4\,R\overset{\overset{\displaystyle O}{\|}}{C}R' + LiAlH_4 \rightarrow (RR'HCO)_4^-\ LiAl^{+4} \xrightarrow{H^+} RR'CHOH$$

$$2\,R\overset{\overset{\displaystyle O}{\|}}{C}OR' + LiAlH_4 \rightarrow (RCH_2O)_2^-\ (R'O)_2^-\ LiAl^{+4}$$

In contrast to $LiAlH_4$, sodium borohydride can be used in methanol, water or pyridine solution. When the reduction is complete, excess reagent is destroyed by the addition of acid, and the solution is concentrated and the product extracted with ether. Reductions are sometimes carried out in refluxing methanol in the case of a hindered carbonyl group.

Sodium borohydride is a less powerful and more selective reducing agent than is $LiAlH_4$, and in general only the carbonyl groups of aldehydes or ketones are reduced with the former. Since an ester carbonyl group is reduced to the primary alcohol with $LiAlH_4$, deacetylation will always occur, i. e., an acetoxy ketone furnishes the diol. With $NaBH_4$, on the other hand, an acetoxy ketone is reduced to the acetoxy alcohol, permitting a further selective reaction of the newly formed hydroxyl group.

Although the two hydrides in most cases give much the same results in the reduction of a ketone, differences in the ratio of the epimeric alcohols have been observed. As a rule, ketones that are highly hindered, as at C-11, are reduced with either reagent to the axial hydroxy compound, while with unhindered ketones, as at C-3, the main product is the equatorial alcohol. W. DAUBEN et al. (1956a, b) have discussed the theoretical basis for this behavior, and have noted that $NaBH_4$, which has somewhat greater steric requirements than $LiAlH_4$, in some cases leads to a larger amount of the axial epimer.

The selectivity that can be realized in hydride reductions of polycarbonyl compounds is not as high as in reactions such as ketal formation. The partial reduction of a 20-keto group in the presence of Δ^4-3-keto and 11-keto groups has been accomplished in moderate yield with a limited amount of $NaBH_4$ at 0° (NORYMBERSKI and WOODS 1955). These authors have reviewed the literature and indicate a sequence of decreasing ease of reduction as: saturated 3-ketone > 17 or 20-ketone > Δ^4-3-ketone > 11-ketone. Non-reduction of 11-keto groups with $LiAlH_4$ has been reported, but this is probably due to precipitation of the incompletely reduced compound as a metal alkoxide.

β) Catalytic hydrogenation. Hydrogenation in the presence of platinum catalysts is an effective method for reducing carbonyl groups; concomitant reduction of double bonds occurs except in highly resistant ditertiary olefins such as a $\Delta^{8(14)}$ compound. An important feature of the catalytic reduction of ketones is the steric course of the reaction, which has been discussed by BARTON (1953). Rapid hydrogenation, which is achieved in an acidic medium such as glacial acetic acid, gives predominately the axial alcohol from all ketones, and this furnishes a method for obtaining, in the case of unhindered ketones such as that at C-3, the alcohol of opposite configuration to that formed by hydride reduction (cf. Fig. 37). The steric course of hydrogenations in neutral or alkaline media, which proceeds at a slower rate, parallels that observed in hydride reductions.

γ) Reduction with active metals. The reduction of ketones or esters with excess sodium in boiling alcohols is a classical method that has been widely used in the steroid series to obtain equatorial alcohols from ketones. This procedure gives a larger amount of the 20α-epimer from simple 20-ketones than any other reduction technique. An important variation of this method is the use of lithium in liquid ammonia-alcohol, which is described in a later section (p.248). These metal reduction procedures provide the only method for the conversion of an 11-ketone to the 11α-hydroxy compound (HERZOG et al. 1953, SONDHEIMER et al. 1953c).

e) Reduction to methylene group

The reductive elimination of a carbonyl group is occasionally a crucial step in the establishment of a steroid structure for a novel substance, as in the case of aldosterone. It is usually more practical to reduce a carbonyl group than the corresponding hydroxyl group, and in a number of cases the removal of a hydroxyl group has been accomplished via the ketone.

α) Wolff-Kishner reduction. This reaction comprises the decomposition of the hydrazone of a ketone or aldehyde with very strong alkali:

$$R{-}\overset{\overset{\displaystyle O}{\|}}{C}{-}R + NH_2NH_2 \rightarrow R{-}\overset{\overset{\displaystyle NNH_2}{\|}}{C}{-}R \rightarrow R{-}CH_2{-}R + N_2$$

The process is most conveniently carried out by a procedure described by HUANG MINLON (1949). The carbonyl compound is treated with excess aqueous hydrazine and concentrated alkali hydroxide in a high boiling solvent such as triethyleneglycol [$HO(CH_2)_2O(CH_2)_2O(CH_2)_2OH$]. The reaction mixture is heated to 200° by distilling out water and maintained at this temperature until gas evolution is complete. It is sometimes advantageous to prepare the hydrazone or semicarbazone first and then effect the decomposition. A modification of the procedure using anhydrous hydrazine has been found to be superior for the reduction of highly hindered carbonyl compounds resistant to the Huang-Minlon conditons (BARTON et al. 1955).

The reduction proceeds in high yields with most simple ketones; migration of the double bond frequently complicates the reduction of conjugated unsaturated ketones. With aldehydes, Cannizarro disproportionation:

$$RCHO \rightarrow RCO_2H + RCH_2OH$$

is a competing reaction. Selectivity in the Wolff-Kishner reduction of polycarbonyl compounds is limited; preferential reduction of other carbonyl groups without attack at C-11 is readily obtained.

β) **Clemmensen reduction.** This method, in which the carbonyl group is reduced to methylene with zinc in strong acid, has not found extensive use in steroid work. The ketone is refluxed with zinc foil in alcohol and a mixture of alcohol and concentrated hydrochloric acid is slowly added. The zinc foil can be amalgamated to obtain more vigorous reducing conditions. With the use of unamalgamated zinc, the C-3 carbonyl group in 3,12- or 3,20-diones can be selectively reduced in high yields.

γ) **Desulfuration of thioketals.** A very general reduction process, which has been used in several variations in the steroid series, is the hydrogenolysis of carbon-sulfur bonds with Raney nickel (Mozingo reaction). The reaction is applicable to practically any sulfur-containing functional groups, including thioethers, thioketals and hemithioketals and thioacid derivatives.

I $R{-}S{-}R' \rightarrow RH$

II 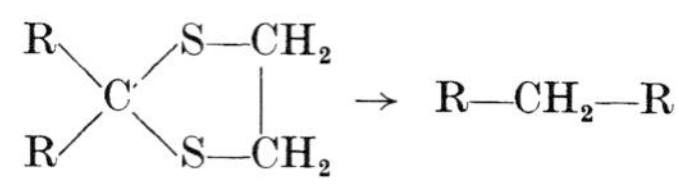

III 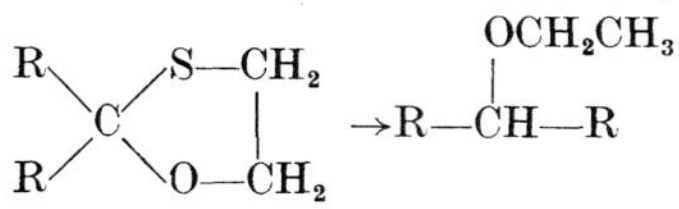

IV $R{-}\overset{O}{\overset{\|}{C}}{-}SR \rightarrow R{-}\overset{O}{\overset{\|}{C}}H$

Desulfuration of the thioketal (II) provides an efficient and relatively gentle means for the removal of a carbonyl group, in contrast to the drastic alkaline and acid conditions of the Wolff-Kishner and Clemmensen reductions, respectively.

In a typical example, the thioketal derived from a 15-keto compound was refluxed in ethanol solution for 12 hours with 20 times the weight of Raney nickel (Klass et al. 1955). The usual nickel hydrogenation catalyst can be employed; sufficient hydrogen is absorbed in the metal to effect the hydrogenolysis. In the desulfuration of thioesters (IV), the catalyst is deactivated by refluxing with acetone to avoid further reduction of the aldehyde (Spero et al. 1948).

3. Reactions of double bonds

a) Reduction

α) **Catalytic hydrogenation.** The most general method for saturation of a double bond is catalytic hydrogenation with hydrogen and a platinum or nickel catalyst.

The reaction is commonly carried out in acetic acid solution at room temperature and 1—3 atm. hydrogen pressure with shaking or stirring. Commercially available platinum oxide (Adams' catalyst) is the most frequently used catalyst; the oxide consumes two moles of hydrogen and furnishes a very active platinum surface. The amount of catalyst is usually 1—10% of the weight of olefin, but with small amounts of steroids it is often convenient to use a much larger ratio of catalyst.

All except highly hindered double bonds can be reduced under these conditions. Double bonds vary considerably in ease of hydrogenation, and it is possible to effect very clean-cut partial reductions by interrupting the reaction after the uptake of the desired volume of hydrogen. As a general rule, the more highly substituted a double bond, the slower the hydrogenation. The reduction of sluggish double bonds can be accelerated by addition of hydrochloric acid. In some cases, this functions by isomerizing the double bond to a more susceptible position. Aromatic rings, as in the estrogens, are reduced under these conditions, and halogen atoms are replaced by hydrogen. Hydrogenolysis of an allylic hydroxyl group, as in a Δ^5-7-hydroxy compound, may occur.

Catalytic hydrogenation of a double bond leads to the *cis* addition of hydrogen; thus a $\Delta^{8,9}$-olefin is reduced to the $8\beta,9\beta$-dihydro (9-*iso*) compound. The stereochemical outcome of a hydrogenation depends on the relative ease of approach of the two sides of the molecule to the catalyst surface and may be profoundly influenced by "remote" substituents (see Fig. 36). Some of the conformational factors governing the steric course of catalytic reductions have been discussed by HADLER (1955).

***β*) Partial reduction of unsaturated ketones.** A double bond conjugated with a carbonyl group is much more readily reduced than an isolated one, and several procedures are available for the selective reduction of an α,β-unsaturated ketone to the saturated carbonyl compound.

Selective hydrogenation is readily effected with the use of supported palladium catalysts such as 10% Pd on barium sulfate or charcoal. The reduction is carried out in a neutral medium under the conditions outlined above, and furnishes high yields of the saturated ketone. The reduction of the Δ^4-3-keto system has been discussed in some detail on pp. 229 to 230 (Fig. 36).

Metal-ammonia combinations provide another approach to the partial reduction of unsaturated ketones. This method was originally developed for the related reduction of aromatic ethers to dihydro derivatives (Birch reduction, p. 280). The same general conditions with certain important modifications are used for both purposes.

In the reduction of an unsaturated ketone, a solution of the steroid in ether or dioxane is added to a solution of lithium in liquid ammonia at the normal boiling point (—35°). The presence of excess lithium is evidenced by an intense blue color. After a short time, ammonium chloride is added to neutralize the solution and the saturated ketone is isolated. If reduction to the equatorial hydroxy compound is desired, the lithium is added to a solution of the steroid in ether and liquid ammonia containing a small amount of methanol. In the reduction of a phenol derivative, the steroid is added to the lithium-ammonia solution and alcohol is then added until the excess metal is consumed.

Another form of chemical reduction is applicable to unsaturated diketones such as Δ^4-3,6-diones, which are reduced to the saturated diones in high yield with zinc in warm acetic acid.

Both of these chemical reduction procedures invariably lead to the more stable isomer of the saturated ketone. The Δ^4-3-keto system is reduced exclusively to the 5α (allo) pregnane, and a $\Delta^{8(9)}$-11-ketone to the natural 8β, 9α configuration.

b) Hydroxylation and epoxidation

The conversion of a double bond to the oxide or glycol is a common synthetic operation, and stereospecific methods are available for the preparation of the several isomeric oxides and glycols. Since these transformations are closely interrelated they have been treated collectively. The reactions are illustrated in Fig. 47 for the case of a Δ^2-unsaturated steroid in the 5α (allo) series. Two considerations govern the stereochemical outcome of these oxidations; the direction of initial attack of the reagent and the mechanism of the reaction. In the Δ^2-5α-olefin, attack occurs chiefly at the rear face of the ring, as with practically all unsaturated steroids (an exception is the Δ^2-double bond in the 5β-series, which gives mainly products of β-attack).

With perbenzoic acid or osmium tetroxide, attack of the double bond in (VI) leads directly to the $2\alpha,3\alpha$-oxide (X) and the $2\alpha,3\alpha$-diol (V), respectively, the latter arising via the cyclic osmate ester (cf. Fig. 14, p. 208). In the reaction of (VI) with bromine or hypobromous acid, an intermediate bromonium ion (II) is formed which then leads to the diaxial bromohydrin (III) or dibromide (IV) by attack of water or bromide ion respectively. A similar process is operative in the Prevost hydroxylation with iodine-silver acetate-moist acetic acid; the overall reaction involves two inversions with formation of the *cis* 2β,3β-diol (I). The diaxial *trans* diol (XI) is obtained from either oxide by hydrolysis or acetolysis. The correctness of the configurational assignments can be confirmed by reduction of the two oxides with lithium aluminum hydride to the respective axial alcohols (VIII) and (IX).

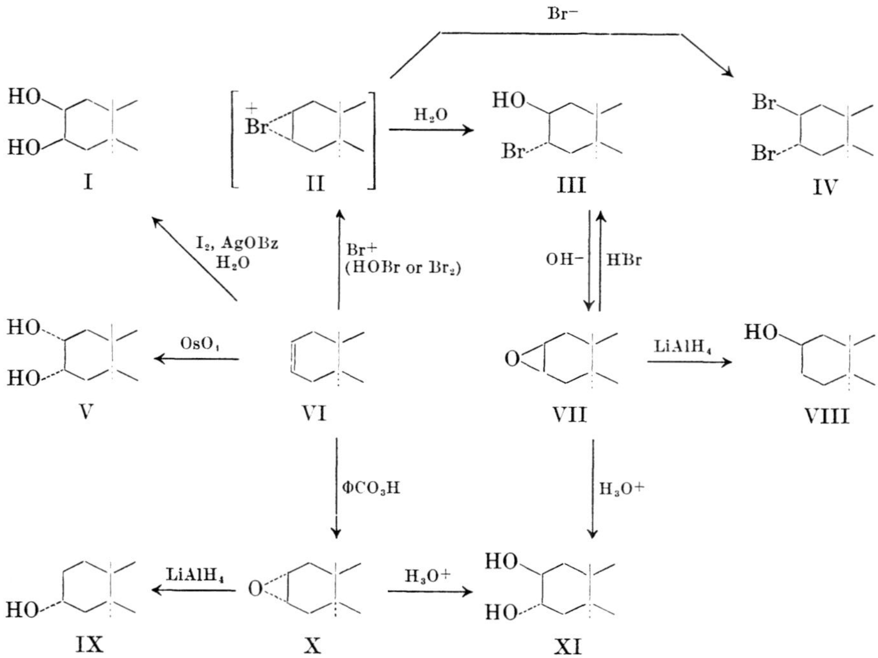

Fig. 47. Conversion of double bond to oxides and glycols

The stereochemical course of these oxidation reactions indicated in Fig. 47 has been observed with double bonds at a number of different positions. The yields are quite good if the environment of the double bond is such that attack is predominately from one side. The reagents used in these double bond oxidations are in general quite specific, and the reactions can be applied to fairly complex molecules, as illustrated in the section on modified steroids. Few data are available on the relative ease of reaction of double bonds at different locations in these oxidations, but it can be stated that the order of reactivity does *not* parallel that of hydrogenation. An example is the ditertiary $\Delta^{8(9)}$-double bond, which is highly resistant to hydrogenation but is readily attacked by osmium tetroxide (CASTELLS and MEAKINS 1956). It is also generally true that the double bond in conjugated olefinic ketones, such as the Δ^4-3-keto system, is less readily attacked by the reagents under discussion than is an isolated double bond. The epoxidation of an unsaturated ketone is readily effected with alkaline hydrogen peroxide.

α) Perbenzoic acid. This reagent (C_6H_5COOOH) is prepared and used in chloroform solution; chloroform solutions are stable for a limited time if stored at 0°. The preparation and use of the reagent have been reviewed by SWERN (1953). The epoxidation reaction is carried out at room temperature or below with a slight excess of the reagent. The course of the reaction can be followed by removing aliquots and titration of the excess perbenzoic acid by reduction with sodium iodide. Excess reagent can be destroyed in this way before isolating the product. Free hydroxyl groups need not be protected, but cleavage of a ketone with formation of an ester or lactone (cf. Fig. 15) can occur with prolonged reaction times at room temperature.

B. RUBIN et al. (1953a) and BITMAN et al. (1957) have utilized epoxidation with perbenzoic acid to convert Δ^9-dehydro compounds to the oxides in the analysis of chromatographically inseparable mixtures of urinary Δ^9-steroids with their saturated and 11-oxygenated counterparts, respectively.

β) Osmium tetroxide. Hydroxylations with osmium tetroxide (OsO_4) can be performed simply by dissolving equimolar quantities of the reagent (vapors poisonous) and olefin in ether or benzene and storing the solution for a week at room temperature. The reaction proceeds more rapidly and often more completely when pyridine is added as a catalyst; the osmate ester usually separates as a brown or black pyridine complex (CRIEGEE et al. 1942). The osmate ester is hydrolyzed with sodium sulfite or an alkaline solution of mannitol and the diol extracted. Osmium tetroxide is essentially without effect on other functional groups. The *cis* hydroxylation of olefins can also be effected with other oxidants such as *t*-butyl hydroperoxide and a catalytic amount of OsO_4.

γ) Hypobromous acid. The most convenient method for the conversion of a double bond to the bromohydrin utilizes N-bromoacetamide in the presence of a strong acid as a source of bromonium ion:

$$CH_3CONHBr + H^+ \rightarrow CH_3CONH_2 + Br^+$$

FRIED and SABO (1957) have found dilute perchloric acid to be superior; dioxane is a suitable solvent.

4. Introduction of isotopic atoms

The labeling of a steroid with isotopes of carbon or hydrogen is of considerable importance in studies of biochemical transformations and metabolic pathways. The incorporation of isotopes can be accomplished by chemical methods or by enzymatic synthesis. In the latter method a randomly labeled steroid is obtained, as in the preparation of cholesterol-^{14}C by incubation of labeled acetic acid with liver tissue (SCHWENK and WERTHESSEN 1952). The insertion of a carbon isotope at a specific position involves of necessity a partial synthesis in which a single carbon atom is introduced, preferably at as late a stage of the synthesis as possible to conserve isotope. An example of this approach is the labeling of steroids at C-4 (Fig. 39, 232); the diazoketone method for the 20,21-ketol side chain (Fig. 28, p. 220) is applicable to the introduction of isotope at C-21 by the use of labeled diazomethane.

The chemical introduction of deuterium or tritium can be effected by reduction procedures or by direct exchange of hydrogen atoms with isotope. The catalytic hydrogenation or hydrogenolysis reactions already discussed have frequently been used for the insertion of deuterium or tritium at the site of a double bond or halogen atom. A solvent such as acetic acid should be enriched in the isotope in such reductions (EIDINOFF et al. 1952).

The platinum-catalyzed exchange of hydrogen isotopes with steroids has been studied extensively by FUKUSHIMA and GALLAGHER (1952). The reaction is carried out by agitating a solution of the steroid in an isotope-enriched medium at 130—150° with a relatively large amount of platinum prepared by pre-reduction of Adams' catalyst with the isotope. Maximum exchange is obtained in 70% acetic acid, 30% water. For deuterium exchange, pure CD_3COOD and D_2O are used; tritium-enriched water and HT-reduced catalyst are used in the exchange reaction to obtain radioactive steroids (FUKUSHIMA et al. 1952).

Isotopic atoms that are introduced by exchange at positions activated by a carbonyl group are labile and can be removed by treatment with base. After deuterium exchange with

4-androstene-3,17-dione, nearly one-half of the isotope was lost, from C-2, C-4, C-6 and C-16, by prolonged base-catalyzed enolization. The extent and distribution of isotope substitution in stable positions depends on the structure. With cholesterol, nearly all of the isotope was found at C-6 and the four terminal carbon atoms of the side chain; with 3α-acetoxypregnane-11,20-dione, the non-enolizable isotope was predominantly at C-4 and C-16.

Preparation of adrenocortical steroids and modified hormones

A. Synthetic sources of adrenocortical steroids

Activity in the synthesis of the adrenocortical hormones and related steroids has been more extensive by far than in all other branches of steroid chemistry taken together, and the accomplishments in this field comprise an outstanding chapter in the annals of contemporary organic chemistry. The major impetus of this program was, of course, the demand for quantities of the hormones for therapeutic use, which has been a matter of great importance since the beginnings of steroid-hormone therapy. The available quantities of naturally occurring hormones, either pure or as glandular extracts, is always extremely limited. In the case of estrogens and the adrenal hormones, efficient methods for the preparation of extracts were developed and are still being operated, but the increasing clinical use of adrenocortical hormones has created requirements greatly exceeding the supply from natural sources.

I. General approaches to partial synthesis

A comprehensive survey of the voluminous literature of adrenocortical hormone synthesis is beyond the scope of this chapter; moreover, a number of excellent reviews on this subject are available (Djerassi 1953, Rosenkranz and Sondheimer 1953, and Shoppee 1953). In the following sections are sketched the principle problems encountered in the preparation of the corticosteroids and some of the approaches that have been developed. Emphasis is placed on the most recent work, and apologies are tendered to the many investigators whose contributions have not been noted.

1. The C-11 oxygen problem

When the need for synthetic sources of substantial amounts of cortisone and cortisol arose, attention was centered on sufficiently abundant steroids that could be transformed in practical synthetic sequences to the desired hormones. One of the most pressing requirements was the necessity of having a functional group in the starting material that could be used for the establishment of an 11-keto or 11β-hydroxyl group. Since only two rare glycosides, obtained from an African plant and a toad secretion, were known, besides the adrenal steroids, to possess an oxygen function at C-11 the first candidates for starting materials were the 12-hydroxylated bile acids.

Several effective procedures for the translocation of a keto group from C-12 to C-11 have been developed, originally for application to ketones derived from bile acids, and later to hecogenin (X, Fig. 51), a 12-keto sapogenin found in quantity in certain tropical *Agave* species. Illustrative of these is the method of Gallagher (1946) (Borgstrom and Gallagher 1949), in which bromination of a 12-ketone is followed by alkali treatment to furnish the 12β-hydroxy-11-ketone (III,

Fig. 48). The ketol can be converted to the bromoketone (IV) and thence to the 11-ketone by zinc reduction; a very effective method for removing the superfluous C-12 hydroxyl group is the reductive deacetoxylation of the acetate (VI) with calcium in liquid ammonia (CHAPMAN et al. 1956).

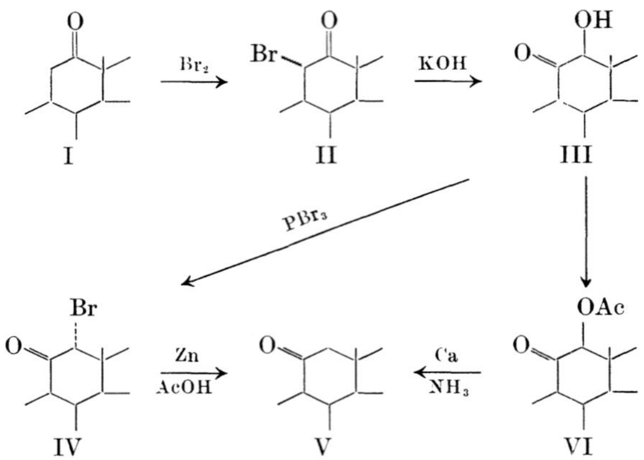

Fig. 48. Translocation of oxygen from C-12 to C-11

As requirements for 11-oxygenated corticosteroids multiplied, an urgent need arose for synthetic approaches utilizing the much more plentiful steroids having no functional group in ring C. The principal compounds available were diosgenin (VII), stigmasterol (VIII) and ergosterol (IX), obtained, respectively, from various abundant sub-tropical *Dioscorea* species, from soy beans and from yeast. All of these compounds satisfied the requirement of facile degradation of the side chain to pregnane derivatives and contained either a Δ^5 double bond or, in the case of ergosterol, a 5,7-diene system.

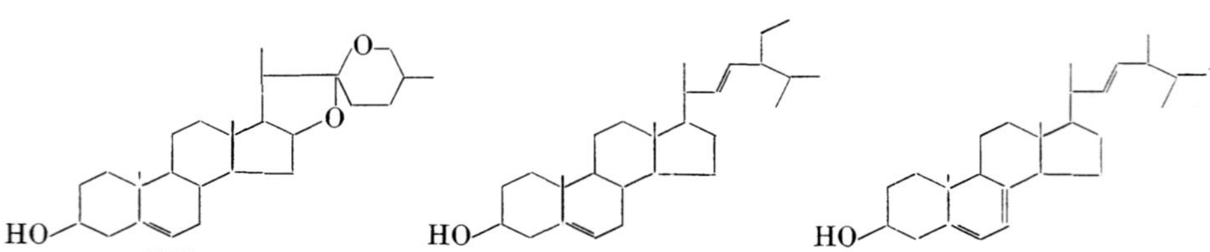

VII VIII IX

A number of research groups mounted vigorous programs to develop methods for the conversion of these compounds, or the pregnane intermediates, into C-11 oxygenated derivatives. The approach in most of these efforts was the same (Fig. 49). A "ladder" was constructed from ring B to reach C-11, and when the necessary group was in place, this scaffolding was removed. A standard procedure was formation of the 5,7-diene (XI) system (already present in ergosterol), removal of the Δ^5 double bond, dehydrogenation to the 7,11-diene (XV) and oxidation to the 9α,11α-oxide (XIV). Several alternatives were then possible; in a typical sequence the oxide was hydrolyzed with accompanying prototropic rearrangement to the Δ^8-11-ketone (XIII), and this was reduced with lithium and ammonia to the saturated ketone.

Ironically, these ingenious methods, evolved at tremendous effort, became of largely historical interest very shortly following their development. The key problem of introduction of the C-11 oxygen function was resolved by a single-step microbiological oxidation, and the problem of corticosteroid production was placed in an entirely new perspective.

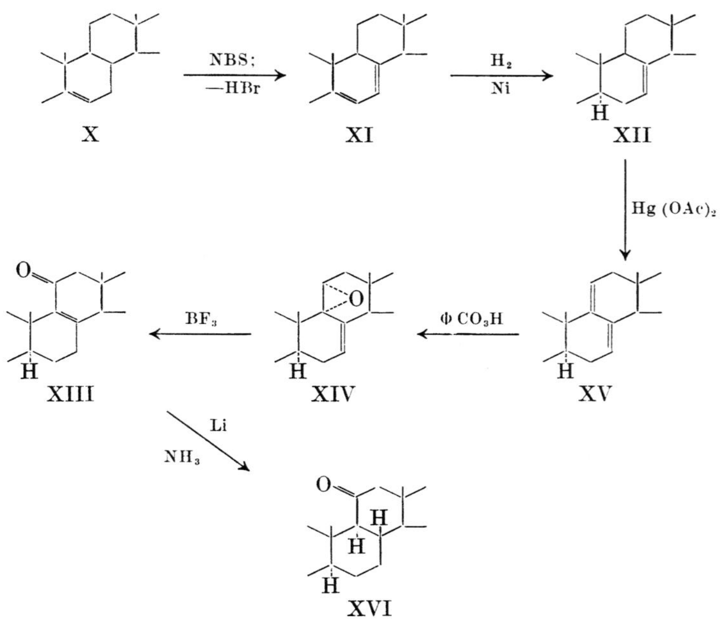

Fig. 49. Introduction of C-11 oxygen

2. The side chain problem

A second major stage in corticosteroid synthesis is the conversion of the starting material to a C_{21} compound followed by construction of the cortical side chain. A prerequisite for a suitable raw material is the presence of a functional group in the side chain to furnish a point of attack for the degradation. This restriction has, up to the present, prevented the utilization of the very cheap and widely distributed plant sitosterols, which have saturated alkyl side chains. Procedures for the side chain alteration of all of the steroid raw materials presently in use were available when the concentrated corticosteroid synthesis programs

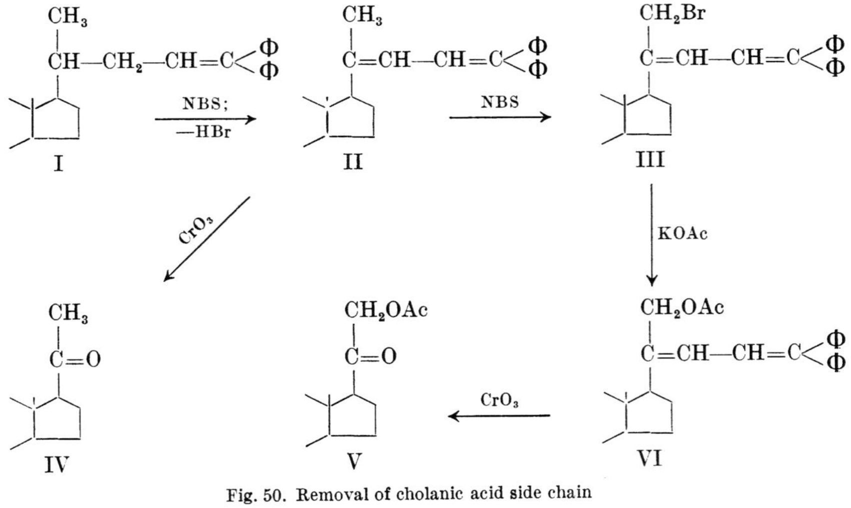

Fig. 50. Removal of cholanic acid side chain

were launched, but a number of refinements have been introduced to tailor these methods more closely to fit the special problems of the cortical side chain.

Degradation of the cholanic acid side chain (Fig. 50) is accomplished very efficiently by conversion of the methyl ester to the diphenyl cholene (I) with phenylmagnesium bromide, followed by treatment with N-bromosuccinimide and dehydrobromination to give the diene (II) (MEYSTRE et al. 1944). The 21-bromo diene (III) can be prepared in a single operation from (II) by bromination in allyl bromide solution (HEER and WETTSTEIN 1953), and (III) is then converted to the acetoxydiene (VI) and thence to the ketol acetate (V).

The spiroketal side chain (VII, Fig. 51) of the sapogenins diosgenin and hecogenin (X) is exceptionally suitable for degradation to pregnane intermediates by a simple sequence developed by MARKER (1940) for the preparation of progesterone. Isomerization of the spiroketal to the pseudosapogenin (VIII) is accomplished with hot acetic anhydride, and mild oxidation of (VIII) followed by mild hydrolysis furnishes the Δ^{16}-20-ketone in very good yield. An important feature of this degradation is the fact that the nuclear double bond does not require protection and is retained unchanged in the product.

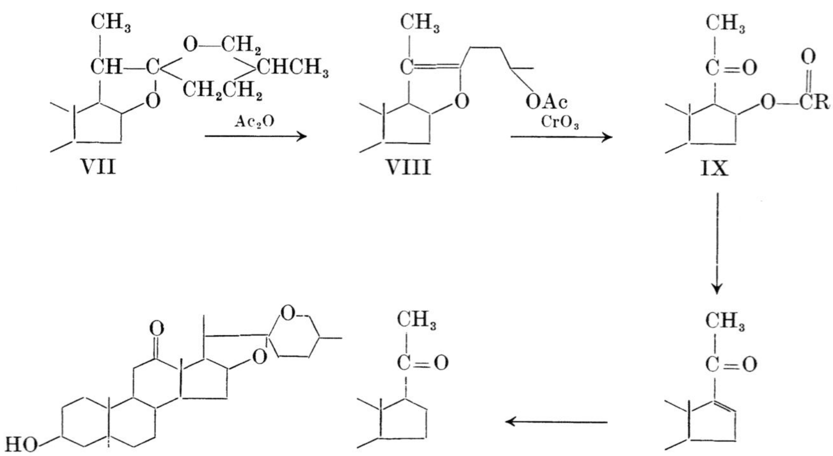

Fig. 51. Conversion of sapogenins to pregnane series

The side-chain degradations (Fig. 52) of stigmasterol (XIII) and ergosterol (XVI) proceed by oxidation, usually with ozone, of the Δ^{22} double bond; the nuclear unsaturation must be reduced or protected in this operation. The aldehyde (XIV) is then converted to an enol derivative (XV) (HERR and HEYL 1952) followed by further oxidation to the 20-ketone (XVIII). Alternately, the 22-aldehyde (XIV) can be converted directly to the 17α-hydroxy-20-ketone (XXI) by bromination and dehydrobromination, oxidation of the unsaturated aldehyde (XIX) with monoperphthalic acid to the epoxyformate (XX) and hydrolysis (CHAMBERLIN et al. 1957).

A number of routes to the 17α,21-dihydroxy-20-ketone side chain from the several degradation products available have been devised (Fig. 53). In syntheses based upon sapogenin starting materials, advantage may be taken of the Δ^{16}-20-keto function for introduction of the 17α-hydroxyl group. Epoxidation of the double bond proceeds smoothly and selectively with alkaline hydrogen peroxide and the oxide (XXIII) is then cleaved with hydrogen bromide to furnish the bromohydrin (XXIV), which is reduced with nickel to the ketol (XXVII) (JULIAN et al. 1949, 1950). Several other approaches also lead to the 17α-hydroxy-20-ketone (p. 215). The 21-hydroxyl function is obtained simply by bromination at C-21 (XXVI) followed by hydrolysis (KRITCHEVSKY and GALLAGHER 1951) or displacement with potassium acetate (ROSENKRANZ et al. 1950).

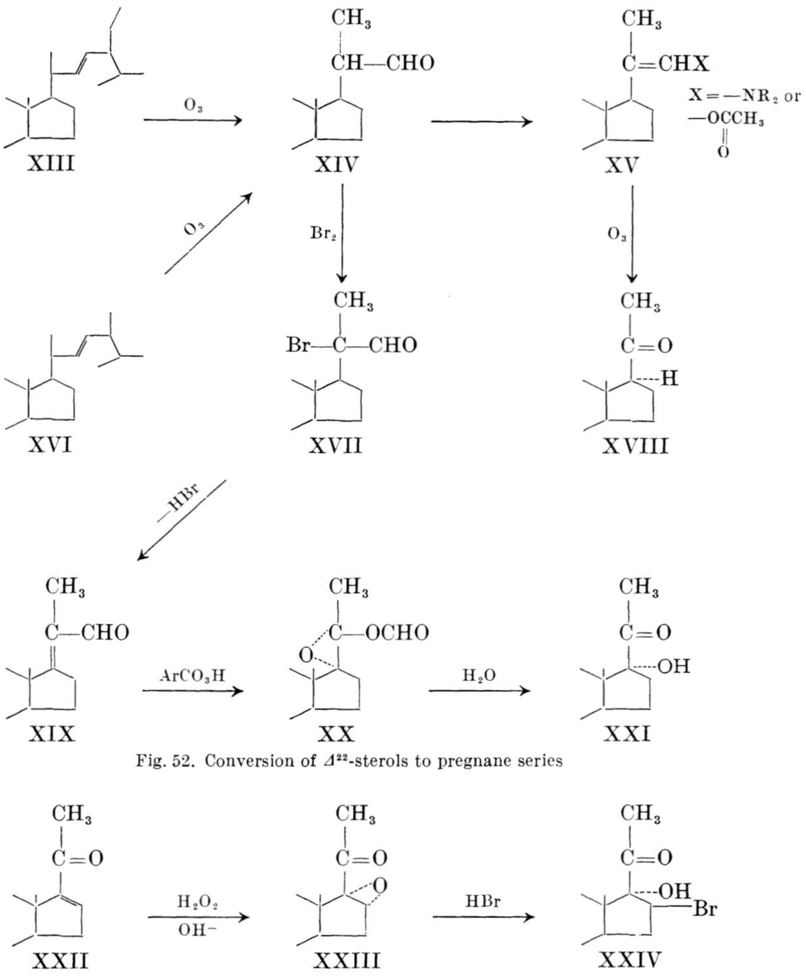

Fig. 52. Conversion of Δ^{22}-sterols to pregnane series

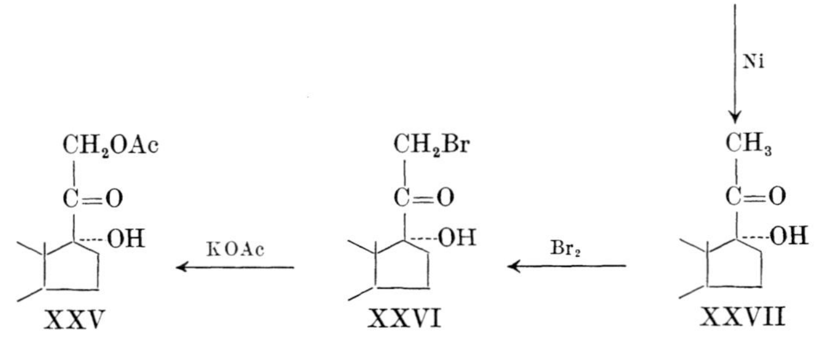

Fig. 53. Formation of dihydroxyketone side chain from 20-ketones

An entirely different approach to the dihydroxyketone side chain involves the simultaneous introduction of the substituents at C-17 and C-20 by oxidation of Δ^{17}-21-acetoxypregnene derivatives (Fig. 54). Hydroxylation with osmium tetroxide of a Δ^{17}-21-acetoxypregnene containing a cyano group or, less satisfactory, a bromine atom at C-20 leads directly to the 17α,21-diol-20-one 21-acetate. The 20-cyano pregnene (XXXII) is obtained by dehydration of the 20,21-ketol cyanohydrin (XXXI) (SARETT 1948), and was an intermediate in the first practical cortisone synthesis. The 20-bromo derivative (XXXIII) is obtained via the rearrangement of the 17,21,21-tribromo-20-ketone (XXIX) and reduction of the resultant 20-bromo-17-pregnenoic acid (XXX) (WAGNER and MOORE 1950). An important variation of this approach is the *oxidative* hydroxylation of the more readily accessible unsubstituted Δ^{17}-pregnene (XXXIV) (cf. Fig. 34) with reagents such as hydrogen peroxide (MIESCHER and SCHMIDLIN 1950) or phenyl iodoso acetate (HOGG et al. 1955) in the presence of catalytic rather than equimolar amounts of osmium tetroxide.

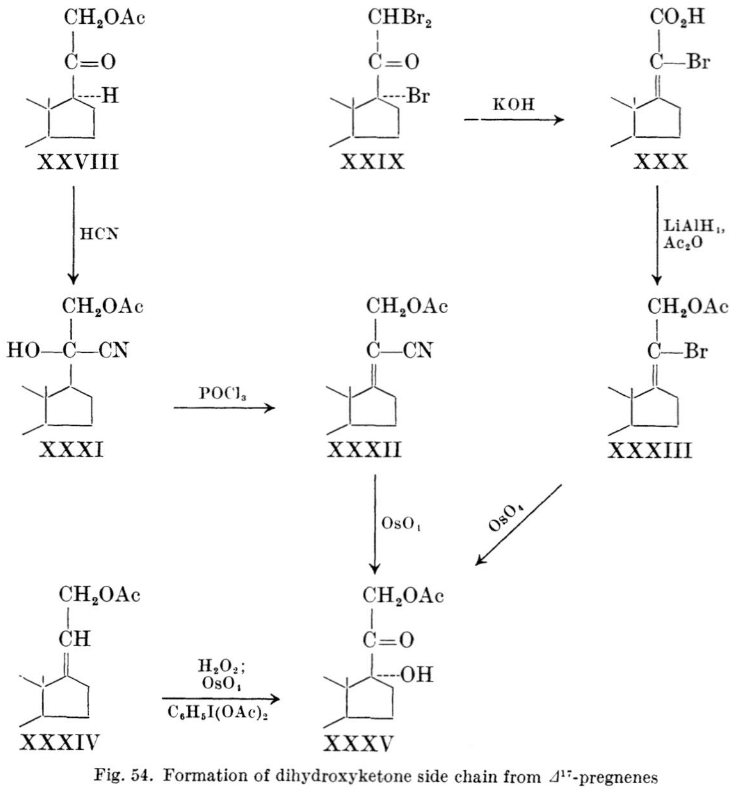

Fig. 54. Formation of dihydroxyketone side chain from Δ^{17}-pregnenes

3. Sequence of steps and completion of synthesis

Most of the earlier routes to cortisone were designed with the three major structural units, 11-ketone, dihydroxyketone side chain and Δ^4-3-keto grouping being introduced in that order. This sequence minimized the interference of groups already established with subsequent operations, since methods were developed for the formation of the ring A system that did not seriously affect the side-chain or C-11 functions. Intermediates obtained from bile acid starting materials, which are 3α-hydroxypregnane derivatives, can be selectively oxidized at C-3 and the resulting ketone brominated at C-4 and dehydrobrominated.

Direct dehydrobromination can be accomplished by treatment of the bromoketone with lithium chloride (Holysz 1953). Another process which has been extensively used involves the conversion of the bromoketone (II, Fig. 55) to the 2,4-dinitrophenylhydrazone (III, R=$C_6H_3(NO_2)_2$—) or semicarbazone (III, R=NH_2CO-) with concomitant loss of hydrogen bromide; the unsaturated ketone is regenerated by treatment of the derivative with pyruvic acid (McGuckin and Kendall 1952). Conversion of saturated ketones in the allopregnane series to the Δ^4-3-keto grouping is somewhat more cumbersome, since bromination of (IV) occurs at C-2. This transformation was an important one, since the chemical introduction of the C-11 oxygen function by the paths shown in Fig. 49 led in most cases to allopregnane derivatives. A method has been developed (Rosenkranz et al. 1950a) which proceeds via the 2,2-dibromoketone (VII) to the 2,4-dibromide (VIII) and thence to the 2-iodo-Δ^4-3-ketone (V) and the unsaturated ketone.

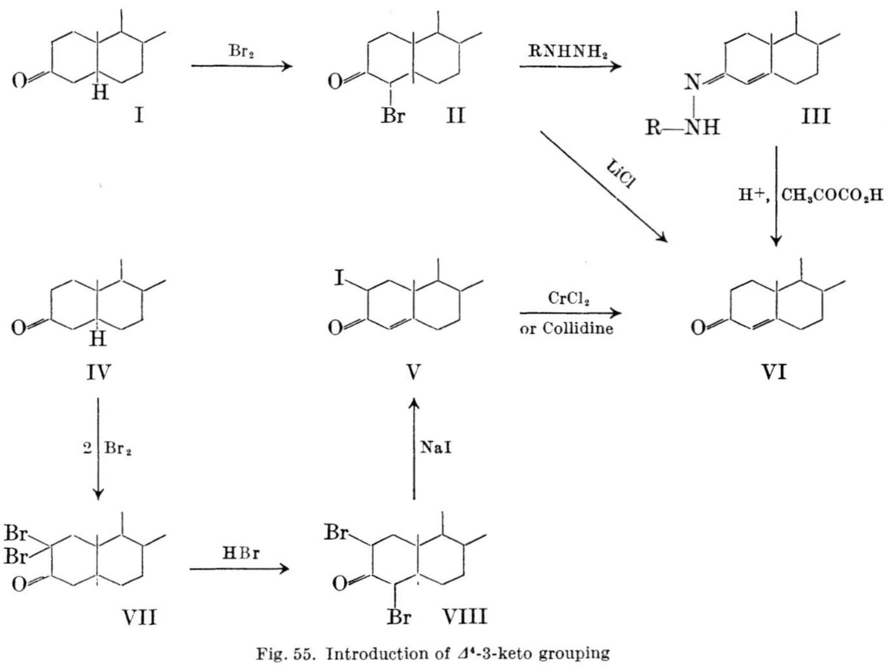

Fig. 55. Introduction of Δ^4-3-keto grouping

More direct and efficient synthetic routes (Fig. 56), in which the Δ^4-3-ketone system is introduced at an early stage, have largely displaced the various methods that proceed via saturated intermediates. Two important factors led to this revised approach. Two of the cheapest raw materials, diosgenin and stigmasterol, both contain the Δ^5-3β-ol system, and it is, of course, highly advantageous to preserve and utilize this grouping rather than to destroy and then regenerate the unsaturation; secondly, the most satisfactory substrates for microbiological oxidation are Δ^4-3-ketones. Either diosgenin (IX) or stigmasterol (X) can be converted in a few steps to progesterone (XV) or 17α-hydroxyprogesterone (XIII). One of the most efficient microbiological processes is the 11α-hydroxylation of progesterone, and a direct synthesis of either cortisol or the modified corticosteroids from 11α-hydroxyprogesterone (XVI), using the enamine protecting group in ring A, has been described (Hogg et al. 1955a). A very important intermediate is 11-deoxycortisol (XI), which can be prepared from 17α-hydroxyprogesterone (XIII) in one step without protection of the Δ^4-3-ketone system (RINGOLD and STORK 1958); microbiological oxidation to cortisol or 11-epicortisol is accomplished in good yield. The success achieved in the development of practical preparative methods for cortisone and its congeners can be realized by noting the cost of cortisone acetate in the United States, which dropped from $ 234 per gram to $ 3.50 in the period 1949 to 1957.

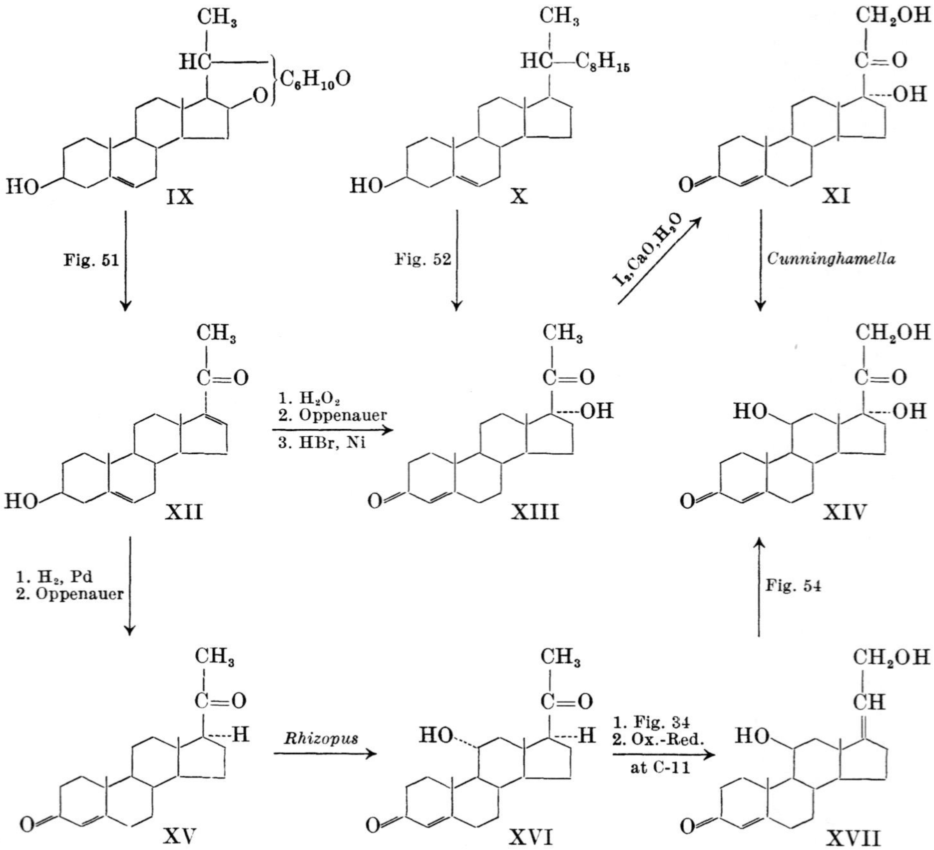

Fig. 56. Synthetic routes to cortisol

II. Preparative aspects of microbiological transformations

1. General scope

No single development in the last decade has had a greater impact on the methodology of adrenocortical steroid synthesis than microbiological transformations of steroids. In addition to their tremendous importance in the commercial production of 11-oxygenated compounds, microbiological conversions have made available for study a number of hitherto inaccessible substituted steroids. A number of reviews on microbiological transformations are available; noteworthy among them are those of FRIED et al. (1955c), WETTSTEIN (1955), SHULL (1956), and the very comprehensive essay by EPPSTEIN et al. (1956).

The use of the enzymatic apparatus of a microorganism in a specific synthetic problem in the steroid series was first reported by MAMOLI and VERCELLONE (1938), who developed a highly practical method for the conversion of dehydroepiandrosterone (I) to testosterone (III). Both oxidation of the 3β-hydroxyl group and selective reduction of the 17-ketone function were performed with yeast, under aerobic and anaerobic conditions, respectively.

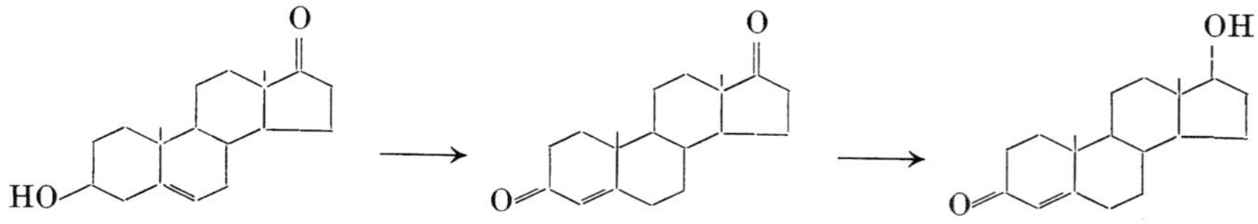

I II III

The full significance of microbiological process in steroid synthesis was revealed by the report of PETERSON and MURRAY (1952), describing the introduction of an 11α-hydroxyl group into progesterone by the mold *Rhizopus arrhizus*. A large number of publications from several groups closely followed this report, describing a multiplicity of reactions with different steroidal substrates. A summary of the most important types of transformations is given in Table 3. More than one reaction may occur on incubation of a steroid with a microorganism, and examples of these multiple conversions are included. With the exception of the references indicated, all of the data are taken from the review of EPPSTEIN et al. (1956), where a comprehensive bibliography may be found.

2. Hydroxylations

The substance most thoroughly studied in hydroxylation reactions has been progesterone, which is one of the cheapest steroid starting materials and is dissimilated by a wide spectrum of microorganisms. Although most conversions have been carried out with C-19 and C-21 compounds containing the Δ^4-3-ketone grouping, this is in most cases not a prerequisite. Since the hydroxylating organisms possess efficient esterase systems, 21-acetoxy steroids are in general as suitable for incubation as the free ketols, and hydrolysis of the 21-acetate group is accomplished as an added feature.

From a preparative standpoint, hydroxylations at positions 11α, 11β, 16α, 17α, and 21 are of the greatest importance. All of these reactions have been performed with a variety of substrates, and several in remarkably high yields. The conversion of 11-deoxycortisol to cortisol has been accomplished in 77% yield (O'CONNELL et al. 1955), and the 11α-hydroxylation of progesterone is practically quantitative (PETERSON et al. 1952). It is impossible in this short discussion to indicate the numerous synthetic pathways to corticosteroids that can and have been assembled by combined chemical and microbiological methods. Some operations, such as the introduction of the 17α-hydroxyl group, can be performed very efficiently and conveniently by chemical steps, but intricate partial synthetic sequences requiring the introduction and removal of protective groups are largely passé. A very interesting aspect of microbial hydroxylation is the ability of organisms to dissimilate such bizarre compounds as 1-dehydro-9α-fluorocortisol (p. 277), as in the 16α-hydroxylation of this compound reported by THOMA et al. (1957).

Table 3. *Microbiological transformations of preparative importance*

Type of reaction	Site of reaction
1. Hydroxylation: C—H → C—OH	
a) Monohydroxylation	Introduction of —OH at position 1ξ[a,c], 2β[a,b,c], 6β, 7α, 7β, 8(?), 9(?), 10ξ[d], 11α, 11β, 12[e], 14α, 15α, 15β, 16α, 17α, 21.
b) Dihydroxylation	Introduction of —OH at positions 6β and 11α, 6β and 17α, 7β and 11α, 11β and 14α, 11α and 17α.
2. Dehydrogenation at a carbon-carbon bond —CH—CH→ —C=C—	Δ^4-3-one to $\Delta^{1,4}$-dione-3-one 5α-H-3-one to $\Delta^{1,4}$-diene-3-one
3. Dehydrogenation of alcohol H —C—OH → —C=O	
a) Simple dehydrogenation	Hydroxyl groups at 3α, 3β, 7α, 12α, 17α to corresponding ketones.
b) Dehydrogenation accompanied by hydroxylation	17β-ol to 17-one with introduction of —OH at 6β and 15α.
4. Fission of carbon-carbon bond —C—C— → —C=O	
a) Simple fission	Cleavage at 17—20 or 20—22 to form 17- or 20- one, cleavage at 13—17 to form lactone.
b) Accompanied by Δ^1-dehydrogenation	Cleavage at 13—17 or 17—20 with Δ^4-3-one to $\Delta^{1,4}$-diene-3-one.
5. Epoxidation of double bond	Δ^9 double bond to 9β,11β-oxide, Δ^{14} double bond to 14α,15α-oxide.
6. Reductions	
a) Of double bond	Δ^4-3-one to 5α-H-3-one and 5β-H-3-one.
b) Of ketones	17-one to 17β-ol, 20-one to 20β-ol, 3-one to 3α-ol.
c) Reduction of double bond accompanied by hydroxylation	Δ^4-3-one to 5α-H-3-one or 5α-H-3-one with introduction of 11α-hydroxyl.
d) Reduction of ketone accompanied by Δ^1-dehydrogenation	20-one to 20β-ol with Δ^4-3-one to $\Delta^{1,5}$-dien-3-one.

Key. a) DODSON et al. 1957. — b) HERZOG et al. 1957a. — c) GREENSPAN et al. 1957. — d) In 19-nortestosterone. — e) SCHUBERT et al. 1957.

3. Dehydrogenation and side chain cleavage

The other very important synthetic operation accomplished by microbiological dissimilation is selective dehydrogenation at the 1,2 position with formation of a $\Delta^{1,4}$-dien-3-one (II). In the first cases in which this reaction was observed, the most conspicuous reaction was degradation of the side chain, which occurred with the 20-ketone, 20,21-ketol and dihydroxyketone, furnishing the 17-ketone (III), and also the 13 → 17 lactone (IV) (FRIED et al. 1953, PETERSON et al. 1953, VISCHER and WETTSTEIN 1953). 1-Dehydro derivatives of these C-19 products were also obtained, and the conversion of progesterone to 1,4-androstadien-3-one was achieved in very high yield. This transformation represents two of the three stages in the partial synthesis of estrone, and is of prime preparative importance. Subsequently, several organisms were found which brought about dehydrogenation without destruction of the side chain, furnishing a very facile route to the 1-dehydrocorticoids (p. 277).

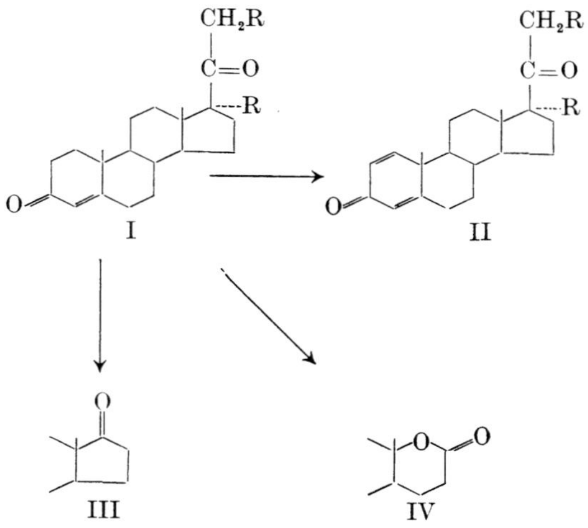

4. Conditions and procedures

The microbial conversions discussed in the preceding paragraphs are oxidative reactions and are carried out in submerged cultures with agitation and aeration. Equipment and procedures developed for the preparation of antibiotics require little modification. In the usual practice in steroid conversions, a sterile solution of the substrate in a water-miscible solvent such as alcohol, dioxane or propylene glycol is added to the medium after inoculation with the organism. Traditional fermentation media, both synthetic and natural, have been employed. The concentration of steroid is usually in the range or 0.1 to 1 g per l. The selectivity and yield in a given transformation can be strikingly improved by additions to the culture medium, as in the 11 β-hydroxylation of 11-deoxycortisol (MANN et al. 1955). It was found that the concomitant formation of cortisone by dehydrogenation, which was a serious practical disadvantage, was suppressed by the addition of 10% of ethanol. The course of the conversion can be monitored by paper chromatography, which has also been invaluable in the qualitative evaluation of the capacities of a new organism. The steroids are recovered by extraction, frequently with methylene chloride, and separated by chromatography if necessary.

III. Total synthesis

The total synthesis of a steroid comprises the construction of the entire carbon skeleton from the simplest fragments by consecutive condensation reactions of various types. The first success in this problem was recorded by BACHMAN et al. (1939), with the total synthesis of equilenin, a weakly estrogenic urinary steroid, which, by virtue of the naphthalenic A—B ring system, possesses only two asymmetric centers. Some years later, a total synthesis of estrone was completed (ANNER and MIESCHER 1948). The total synthesis of a hydroaromatic steroid, which presents a much more complex stereochemical problem, was announced simultaneously by WOODWARD et al. (1951, 1952) and ROBINSON and co-workers (CARDWELL et al. 1951). A considerable number of alternative syntheses of both aromatic and hydroaromatic steroids have subsequently been devised. In this section, only a very small fraction of the brilliant work on total synthesis related directly to the cortical hormones can be described; reviews have been presented by CORNFORTH (1955) and in the several monographs on steroid chemistry.

The execution of a total synthesis involves the successive fusion of the carbocyclic rings in the correct stereochemical relationship, with provision being made for the desired functional groups. As each asymmetric center is created, two diastereoisomers may be formed, and the isomer with the correct configuration must be separated and carried through the subsequent steps. When possible, reactions and conditions are chosen that furnish the desired isomer in preponderant amounts, but this cannot always be done. Eventually a racemic mixture (a pair of mirror images) is obtained, the *d*-enantiomer (see p. 190) of which is identical with a steroid of natural origin. As far as is known at present, the *l*-enantiomers of the steroid hormones are devoid of biological activity.

The WOODWARD synthesis led to the methyl ester of 3-keto-4,9,16-etiatrienic acid (I) by a series of reactions in which the rings were generated in the sequence D ← C → B → A. The ROBINSON synthesis employed a relay; a tricyclic intermediate (II) comprising rings A, B and C was prepared, resolved into the optical mantiomorphs, and one of these identified with an abundantly available degradation product of cholesterol. Material of natural origin could then be used to complete the synthesis, which led to *epi*androsterone (III). Partial synthetic paths can be traced in the literature leading directly or indirectly from these two total synthetic compounds to virtually the entire spectrum of naturally occurring steroids. These total syntheses are thus tantamount, in a formal sense, to the total synthesis of all of these compounds.

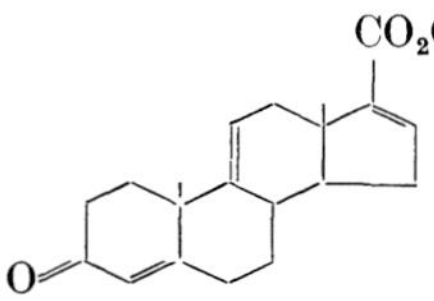

I

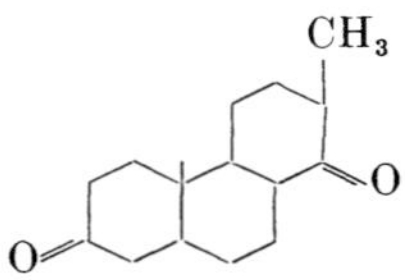

II

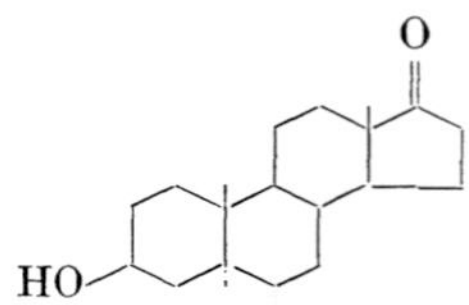

III

These and the numerous improved subsequent syntheses are outstanding achievements from both theoretical and practical standpoints; several are quite efficient, considering the large number of steps. No total synthesis, however, provides a route for the production of classical steroid hormones that is competitive at present with partial synthetic transformations starting from abundant plant steroids, especially with the availability of microbiological oxidations. The cost of the simple raw materials for these total syntheses is not negligible, since the overall yield in these very lengthy sequences, based on the first reaction product, or first monocyclic intermediate, is of the order of a few per cent at best. The likely potential candidates for production by total synthesis would be compounds such as estrone, with an aromatic ring A. This feature simplifies a total synthesis, since the number of asymmetric centers is reduced but requires several additional steps in any partial synthesis from a plant steroid.

In the case of aldosterone or any related 18-oxygenated compounds, however, total synthesis represented until recently the only means for supplementing the extremely limited supply obtainable from adrenal glands. The total synthesis of aldosterone was reported by SCHMIDLIN et al. (1955), scarcely a year after the announcement of the structure of the compound.

Such rapid progress in this very difficult field was possible because of the adaptability to the aldosterone problem of a total synthesis of cortisone described by SARETT et al. (1952). This elegant total synthetic route was noteworthy in several respects. It was designed to lead directly to cortisone, and was a highly stereospecific synthesis; in each reaction leading to a fixed asymmetric center, the ratio of the desired isomer, having the configuration of the end product, to other isomers was greater than 8 : 1. The starting point of the synthesis, the major stages of which are outlined in Fig. 57, was the *cis*-decalin derivative (IV), formed by a Diels-Alder condensation. Catalytic reduction of the enedione, followed by hydride reduction and acid hydrolysis gave the dihydroxy decalone (V). Ring A was attached in one step by the Robinson method using methyl vinyl ketone in a combined Michael-aldol condensation. The unsaturated ketone grouping was then protected as the ketal, and the hydroxyl group at C-1 (phenanthrene numbering) in (VI) was selectively oxidized to a carbonyl. Base treatment then effected isomerization at C-10a to the desired *anti-trans* geometry (IX).

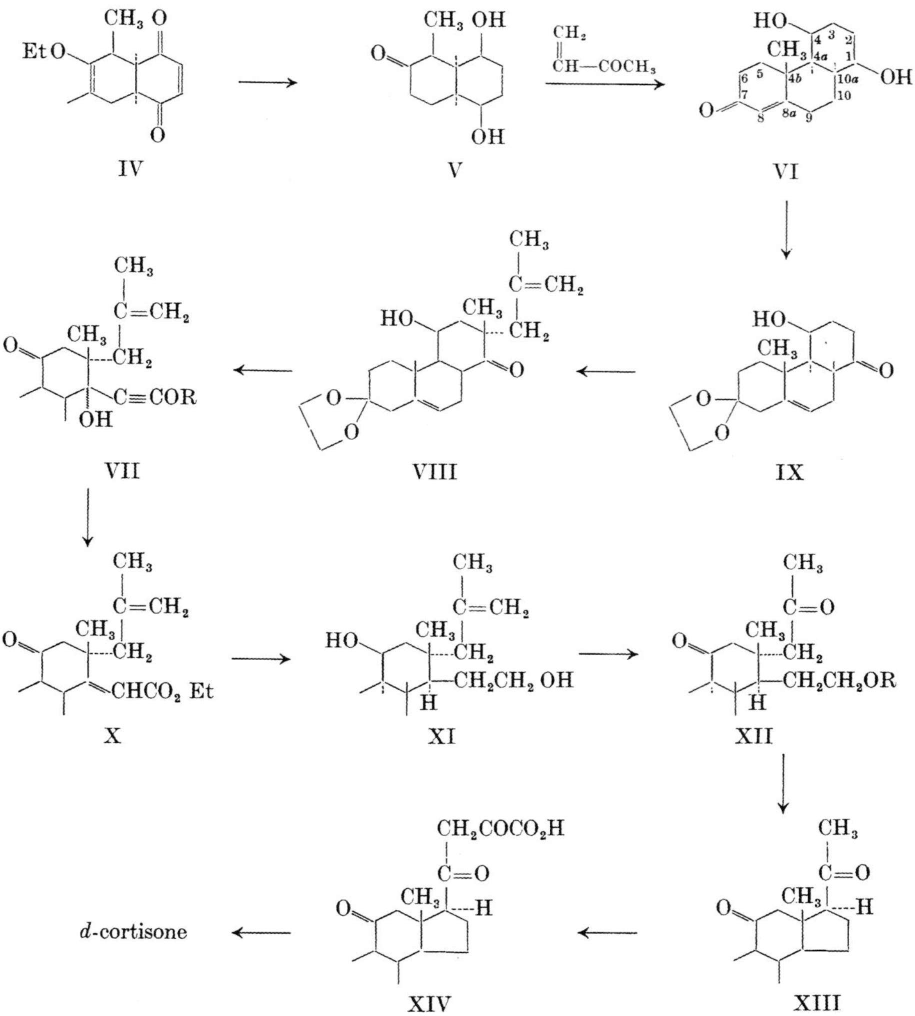

Fig. 57. Total synthesis of cortisone

The resulting tricyclic ketone was further converted to cortisone by SARETT, and also served as the point of departure for the aldosterone synthesis.

The next phase of the cortisone synthesis comprised successive alkylation of the ketone with methyl iodide and methallyl iodide; alkylation occurred at the rear of the anion, so that the first group introduced assumed the β-configuration and the second, the α-position (VIII). The remaining carbon atoms were introduced by condensation of the ketone with ethoxyacetylene. The ring D residue with the proper configuration was obtained by acid-catalyzed rearrangement of the ethinyl carbinol (VII) to the unsaturated ester (X), followed by selective reduction of the exocyclic double bond to the more stable (equatorial) ester (XI). The functional groups required for ring closure were then provided by hydride reduction of the carbethoxyl group and formation of the tosylate, and hydroxylation of the terminal double bond followed by periodate cleavage. Internal alkylation of (XII) followed by alkali equilibration of the initially formed 17-isopregnane led to the ketal of *dl*-11-ketoprogesterone (XIII). The 21-hydroxyl group was introduced via the oxalyl acid (XIV) which also supplied a suitable acidic function for resolution of the racemic mixture as the strychnine salt. The synthesis was completed by hydroxylation of the ketol cyanohydrin.

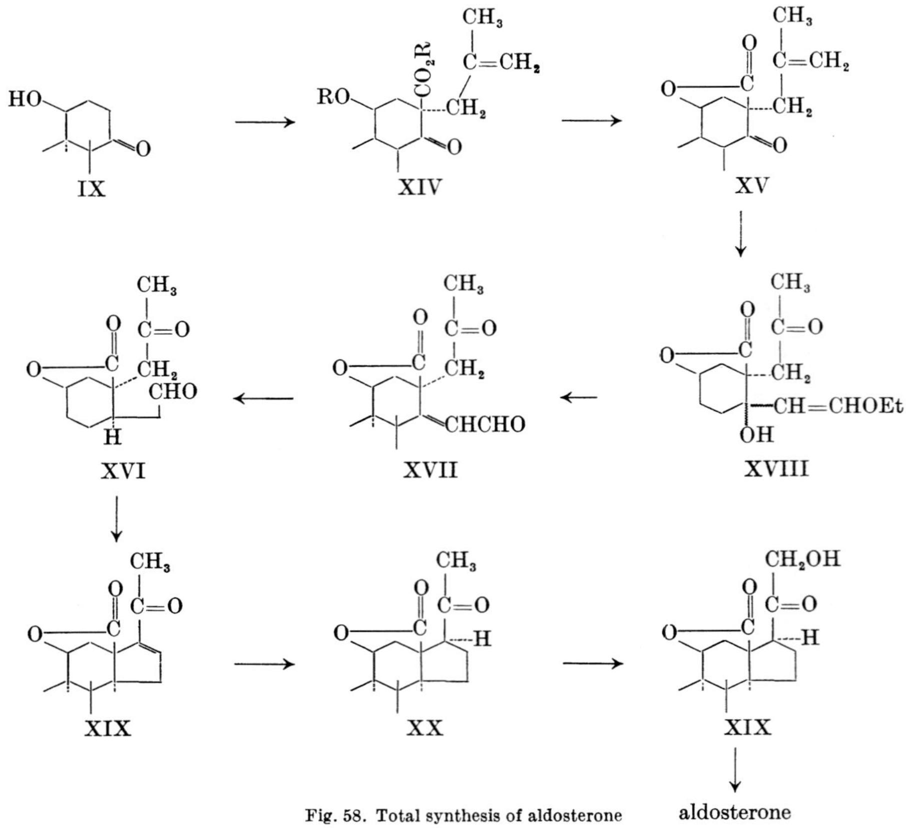

Fig. 58. Total synthesis of aldosterone

The aldosterone synthesis proceeded by much the same pattern as that described for cortisone (Fig. 58). The tricyclic ketone (IX) was condensed with ethyl formate and then alkylated with methallyl iodide to furnish the potential aldehyde group and the ring D residue. The keto lactone (XV), easily obtained by hydrolysis of the ester (XIV), was then condensed with ethoxyacetylene. To eliminate the necessity for selective reduction, the methallyl group was hydroxylated prior to the reduction of the ethinyl group to vinyl. Rearrangement of the ethoxyvinyl ketone (XVIII) led to the unsaturated aldehyde (XVII), which was reduced to furnish the required β-oriented two-carbon chain (XVI). Ring D was closed by aldol condensation and the unsaturated ketone (XIX) was reduced. At this stage, three further operations were required: introduction of the 21-hydroxyl group, reduction of the lactone to a lactol, and resolution of the racemate.

An extraordinarily simple and effective method for carrying out the two most laborious of these steps was discovered by VISCHER et al. (1956). Microbiological oxidation at C-21 of the methyl ketone (XX) by *Ophiobolus herpotrichus* occurred exclusively with the natural *d*-isomer, leading to a readily separated mixture of the *l*-isomer of (XX) and the desired *d*-isomer of the 21-hydroxy lactone (XIX). Reduction of (XIX) as the 20-ketal with lithium aluminum hydride (v. EUW et al. 1955) completed the synthesis. A detailed description of the synthesis, and chemical resolution, of the lactone (XIX) has been given (SCHMIDLIN et al. 1957).

B. Modified hormones

The effect of structural changes in the pharmacological properties of drugs and hormones has long attracted the interest of pharmacologists and chemists alike.

Systematic investigations in a number of diverse types of pharmacologically important substances have furnished more effective therapeutic agents, and must inevitably lead to a clearer understanding of the interaction of a drug or hormone and an enzyme site. Some of the notable successes in this line of endeavor have been made in the steroid hormones series.

It was early discovered that estrogenic activity is remarkably non-specific; a variety of compounds, some bearing only the most superficial structural resemblance to the steroid estrogens, were found to possess estrogenic activity much greater than that of the natural hormones. In the case of the androgenic and progestational hormones, the permissible structural variations do not appear to be nearly as wide, but numerous pharmacologically active steroids, some with rather deep-seated structural changes, have been described. Removal of C-19, the insertion of a 17α-methyl group or the enlargement of ring D have all been found to potentiate one or more of the actions of the parent hormone.

There have been a number of publications describing rather drastic modifications of the adrenocortical steroid structure, such as mono-, bi-, and tricyclic substances with appropriate side chains, but in no case has there been a convincing report of the duplication of the characteristic biological properties of the adrenocortical hormones in these simple structural analogues. TISHLER (1952), LOGEMAN and GIRALDI (1953) and POOS and SARETT (1956) have summarized some of the literature in this area.

Modification of the basic adrenocortical hormone structure by the introduction of certain additional substituents, however, has led to a series of compounds of the greatest importance. The report by FRIED and SABO (1953) of the enhanced liver glycogen deposition by 9α-halo derivatives of cortisol marked the opening of a new era in the chemistry of the adrenal steroids. A great deal of activity has been devoted to the systematic introduction of groups, chiefly halogen atoms and additional double bonds, hydroxyl groups, and methyl substituents, all of which, singly and more strikingly in combination, have been found to potentiate, one or more of the physiological properties of the parent hormones. At the time of this writing, much of the literature has been reported only in preliminary notes; it is to be expected that much important progress will be recorded in the next few years.

The structural elaborations that have been found so far to cause the *greatest increases in the glycogenic activities* of corticosteroids are *9α-fluoro* and *2α- and 6α-methyl* substituents and *Δ^1-dehydrogenation*. The 9α-fluoro and 2α-methyl substituents also effect substantial increases in sodium-retaining activity. These modifications have been most thoroughly studied in the "complete" glucocorticosteroids, i. e., compounds with a full complement of oxygen atoms at C-11, 17 and 21. Many lacunae exist in the pharmacological evaluation of compounds whose synthesis has been reported. Not all of the variations of the corticosteroid molecule that have been examined have proved fruitful; thus halogen substitution at several other positions has been shown to abolish activity, and certain modifications that have been successful in the androgen series decrease corticosteroid activity. An attempt has been made in the following sections to discuss all of the modifications shown to be compatible with some degree of adrenal hormone function. Several structural modifications, while not causing dramatic changes in activity, are of interest for comparison of the effects on the activities of androgens, progesterone and corticosteroids, and have been included.

A critical discussion of the pharmacological properties of these modified hormones in various assay procedures is presented elsewhere in this work (Chapter 9), and only a crude semi-quantitative orientation has been essayed in

Table 4. *Modified corticosteroids*

Compound[1]	M. P.	$[\alpha]_D$[2]	Biological activity: Glucocorticoid	Biological activity: Mineralocorticoid	Ref.
9α-Halo derivatives					
9α-Chloro-11β-hydroxyprogesterone	204	205 (C)	(+)	—	a
9α-Fluoro-11β-hydroxyprogesterone	216	197 (C)	+	+	a
9α-Fluoro-11-ketoprogesterone	174	209 (C)	+	++	a
9α-Fluorocorticosterone Ac	212	192 (C)	++	+++	a
9α-Fluoro-11-dehydrocorticosterone Ac	208	189 (C)	++	+++	a
9α-Chlorocortisol Ac	200	139 (C)	++		b
9α-Chlorocortisone Ac	257	252 (C)	++		b
9α-Fluorocortisol Ac	233	123 (C)	+++		c
9α-Fluorocortisone Ac	254	155 (C)	+++		c
12α-Halo derivatives					
12α-Fluoro-11β-hydroxyprogesterone	182	193 (C)	+	+	d
12α-Chlorocorticosterone Ac	230	179 (C)	(+)		e
12α-Fluorocorticosterone	190		++	+	e
21-Fluoro-21-deoxy derivatives					
21-Fluoro-21-deoxycortisol	243	163 (E)	(+)		f
21-Fluoro-21-deoxy-9α-fluorocortisol	269	147 (Di)	++		f
16α-Hydroxy derivatives					
16α-Hydroxycortisol	236	119 (M)	+	—	g
16α-Hydroxy-9α-fluorocortisol-16,21-DiAc	238	70 (C)	++	—	h
Δ¹-Dehydro derivatives					
Δ¹-Dehydrocortisol (Prednisolone)	240	102 (Di)	++	+	i
Δ¹-Dehydrocortisone (Prednisone)	234	172 (Di)	++	+	i
Δ¹-Dehydro-4,5α-dihydro-9α-fluoro cortisol Ac	237	35 (An)	(+)		j
Δ¹-Dehydro-9α-fluoro-21-deoxycortisol	313	47 (Py)	++	—	k
Δ¹-Dehydro-9α-chlorocortisol Ac	242	145 (E)	+++	++	k
Δ¹-Dehydro-9α-fluorocortisol Ac	243	99 (An)	+++	++	j, k
Δ¹-Dehydro-9α-chlorocortisone Ac	262	244 (C)	+++		k
Δ¹-Dehydro-9α-fluorocortisone Ac	275	158 (E)	+++		k
Δ¹-Dehydro-21-fluoro-21-deoxycortisol	260	86 (Di)	++		f
Δ¹-Dehydro-9α,21-difluoro-21-deoxycortisol	267	111 (E)	+++		f
Δ¹-Dehydro-16α-hydroxy-9α-fluorocortisol 16,21-DiAc	188	22 (C)	+++	—	h
Δ⁶-Dehydro derivatives					
Δ⁶-Dehydro-9α-fluoro-21-deoxycortisol	295	112 (Di)	(+)	—	k
Δ⁶-Dehydro-9α-fluorocortisol Ac	216	123 (E)	++	+++	k
Δ¹,⁶-Bisdehydrocortisol	239	100 (Di)	++	+	l
	210	131 (Di)	++		m
Δ¹,⁶-Bisdehydrocortisone	235	246 (Di)	++	+	l
6α-Methyl derivatives					
6α-Methylcortisol	205	114 (An)	++	—	n
6α-Methyl-9α-fluorocortisol Ac	219	113 (An)	++	—	o
6α-Methyl-21-fluoro-21-deoxycortisol	220		++	—	o
6α-Methyl-9α,21-difluoro-21-deoxycortisol	210	189 (An)	++	—	o
6α-Methyl-Δ¹-dehydrocortisol	230	83 (Di)	+++	—	n
6α-Methyl-Δ¹-dehydro-9α-fluorocortisol Ac	237	87 (Di)	++++	—	o
6α-Methyl-Δ¹-dehydro-21-fluoro-21-deoxycortisol Ac	218		++	—	o
6α-Methyl-Δ¹-dehydro-9α,21-difluoro-21-deoxycortisol	265	71 (An)	++	—	o

Table 4. (cont.)

Compound[1]	M.P.	$[\alpha]_D$[2]	Biological activity		Ref.
			Glucocorticoid	Mineralocorticoid	
2α-Methyl derivatives					
2α-Methyl-21-deoxycortisol			+		p
2α-Methyl-9α-fluoro-21-deoxycortisol			+		p
2α-Methylcorticosterone Ac			(+)		p
2α-Methylcortisol Ac	133, 171, 197	164 (C)	++	++	p, q
2α-Methylcortisone Ac	205	170 (An)	—		p, q
2α-Methyl-9α-fluorocortisol Ac	236	167 (Di)	+++	+++	p, q
2α-Methyl-9α-fluorocortisone Ac	228	167 (Di)	(+)		p, q
2α-Methyl-Δ^6-dehydrocortisol Ac			+		p
2α-Methyl-16α-hydroxycortisol	202	145 (C)	(+)	—	r
2α-Methyl-16α-hydroxy-9α-fluorocortisol	232	115 (Py)	++	(—)	r

[1] Ac = 21-Acetate.

[2] Rotation in solvent indicated (C, chloroform; E, ethanol; Di, dioxane; M, methanol; An, acetone; Py, pyridine). All rotations positive.

References to Table 4. a) FRIED et al. (1955b). — b) FRIED and SABO (1953). — c) FRIED and SABO (1954). — d) HERZ et al. (1956b). — e) TAUB et al. (1957a). — f) HERZ et al. (1956a). — g) ALLEN and BERNSTEIN (1956a). — h) BERNSTEIN et al. (1956). — i) HERZOG et al. (1955b). — j) R. HIRSCHMANN et al. (1955). — k) FRIED et al. (1955a). — l) GOULD et al. (1957b). — m) AGNELLO and LAUBACH (1957). — n) SPERO et al. (1956). — o) SPERO et al. (1957). — p) DULIN et al. (1957). — q) HOGG et al. (1955b). — r) BERNSTEIN et al. (1959a).

this chapter. It should be noted that several of the structural changes described in the sequel have been shown to impart anti-hormonal or antagonistic properties of one type or another to certain compounds; a few were investigated primarily for this purpose. No discussion of these properties has been included; the compounds of interest in this area are of very diverse chemical nature.

A summary of the most active modified corticosteroids reported as of Dec., 1957, plus a few more recent compounds, is presented in Table 4. Only compounds for which some biological data have been recorded are included, and a number of substances of a low order of activity have been omitted. The order of presentation is arbitrary and corresponds to the order in which the functional groups are discussed in the following sections. In each successive grouping in the table, the compounds are arranged by increasing complexity, with all of the preceding modifications included in the same sequence. Thus a number of 9α-fluoro derivatives are included in later groups, each of which treats an additional substituent. This arrangement has been adopted to facilitate the comparison of members of the same series, and to emphasize the effects of multiple substitution. The names have been chosen to emphasize the relationship to the parent compound; this has led to inelegant, but it is hoped descriptive, nomenclature. The ratings of activity are based on the following scale, with cortisol (or acetate) *or* cortisone (or acetate) as the standard (unity) for glucocorticoid, and deoxycorticosterone (or acetate) as the standard for mineralocorticoid activity.

Rating scale in Table 4

—	Negative or less than 0.2 fold standard
(+)	0.2–1.0 fold; "positive but less than standard"
+	1–2 fold; "approximately equal to," "somewhat greater than standard"
++	2–10 fold; "considerably greater than standard"
+++	10–100 fold standard
++++	Over 100 fold standard.

The ratings in Table 4 correspond to the ranges of activity indicated, or to various qualitative statements which were arbitrarily assigned on the basis of adverbial strength. In adopting this scale, an attempt was made to furnish a guide to the relative activities as reported, for the most part in the *chemical* literature, with proper regard to the variation in the same bioassay technique under differing laboratory conditions. All of the glucocorticoid activities indicated are based upon rat liver glycogen assays, and all of the mineralocorticoid values on sodium retention assays. In view of the necessity of assigning values to the verbal ratings noted above, distinctions of detail in assay technique, mode of administration, and the varying use of cortisol and cortisone as a standard seem unjustified and quite beyond the intended scope of the data.

I. Halogen substitution

1. 9α-Halo derivatives

The first success in the preparation of modifications of the adrenal steroids possessing equal or superior biological activity was recorded by FRIED and SABO (1953), who described the 9α-bromo and -chloro derivatives of cortisol and cortisone, and the 3–4 fold increase in liver glycogen deposition accompanying the latter substitution. Since the activity increased in the order I < Br < Cl, *the fluoro derivatives were* prepared (FRIED and SABO 1954) and found to be *9–10 fold more potent.* A similar enhancement of activity has been observed upon 9α-fluoro substitution in nearly all of the subsequent analogues studied, and this structural feature has become ubiquitous among the "superglycogenic" steroids. Particularly striking is the glycogenic activity, approximately equal to that of cortisol, of 9α-fluoro-11β-hydroxyprogesterone. Sodium retention activity is similarly potentiated; 9α-fluorocorticosterone acetate has about 25 times the activity of deoxycorticosterone acetate (FRIED et al. 1955b).

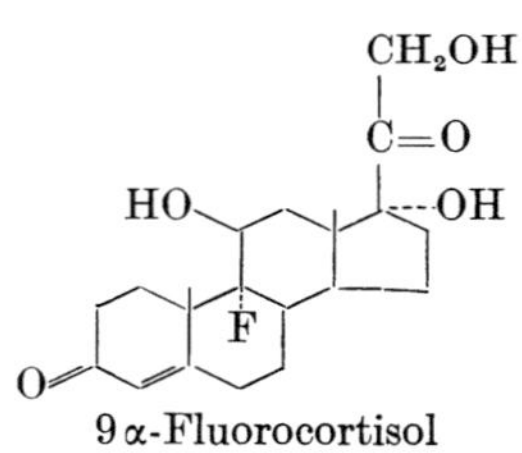

9α-Fluorocortisol

The efficient synthetic approach (Fig. 59) first used for the introduction of the 9α-halogen atom has undergone little change in ensuing applications; it has been described in detail (FRIED and SABO 1957). The procedure is noteworthy in that it is particularly well adapted to the conversion of an 11α-hydroxy steroid, obtained as the main product from many microbiological oxidations, to the 9α-halo-11β-hydroxy system. The sequence of operations proceeds without complications in the presence of the diolone acetate side chain, Δ^4-3-ketone or $\Delta^{1,4}$-diene-3-one systems.

The first step is the introduction of the $\Delta^{9,11}$ double bond, proceeding by solvolytic elimination of an 11α-toluenesulfonate or by dehydration (thionyl chloride) of an 11β-hydroxyl group. The olefin (II) is then oxidized with N-bromoacetamide in perchloric acid giving the 9α-bromo-11β-ol (III) directly in high yield. In order to obtain the 9α-chloro and -fluoro derivatives (V), which cannot be prepared directly, the bromohydrin is converted by mild base to the 9β,11β-oxide (VI). The re-opening of this ring by the appropriate hydrogen halide then furnishes the chloro- or fluorohydrin (V). The reaction of the oxide with hydrogen fluoride requires carefully specified conditions (R. HIRSCHMANN et al. 1956b), particularly the presence of the proper amount of a Lewis base such as tetrahydrofuran to furnish the necessary concentration of fluroide ion, otherwise a by-product (IV) is formed in important amounts.

The 9α-halo-11-ketones (VII) are obtained by oxidation of the 11β-hydroxyl in the usual way. The 9α-bromo and -chloro, but not the -fluoro, 11β-hydroxy compounds are highly sensitive to base, the oxide (VI) being promptly regenerated. The 9α-fluoro-11β-hydroxy system is unaffected by saponification of the 21-acetate and even the introduction of the $\Delta^{1,4}$-diene system by collidine dehydrobromination.

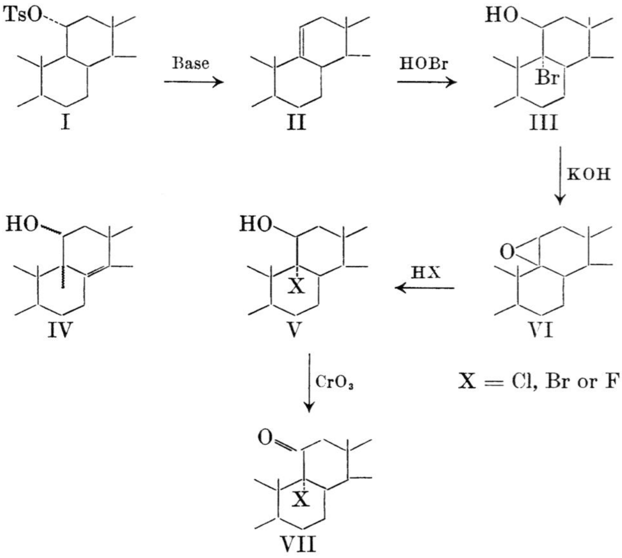

Fig. 59. Introduction of 9α-halogen

The basis for the assignment of the structures and stereochemistry of these halohydrins lies primarily in the reductive debromination of III to cortisol (11β-hydroxyl) and the facile ring opening and closure of the oxide ring, which requires *trans* (9α-halogen) geometry. It should be noted that $\Delta^{9,11}$-olefins have figured prominently in a number of earlier investigations and may be formed in a variety of ways (cf. Fig. 49, p. 253). Most applications to the synthesis of 11-oxygenated steroids have proceeded by way of the 9α,11α-oxides, which are formed by peracid oxidation of the olefin (Fig. 49), but a 9β,11β-oxide has been obtained by permanganate oxidation (CONSTANTIN and SARETT 1952). It is of interest that with the advent of microbiological oxidations 11-hydroxy compounds have become readily available starting materials, and the direction of this portion of previous synthetic routes has actually been reversed.

2. 12α-Halo derivatives

The important effect of 9α-halogen substitution on the activity of 11-oxygenated adrenocortical hormones led promptly to the investigation of the 12β-halo derivatives, in which the C-11 substituent is subject to the similar influence of an electron-withdrawing adjacent axial halogen atom. The data available so far indicate that *the introduction of a chlorine, bromine, or fluorine atom in the 12α-position has nearly the same effect on liver glycogen and mineral activity of the parent compound as does the corresponding 9α-halogen substitution.* The 12α-halo series has been less extensively investigated, and the activities of the 12α-fluoro derivatives of cortisol or cortisone have not been recorded.

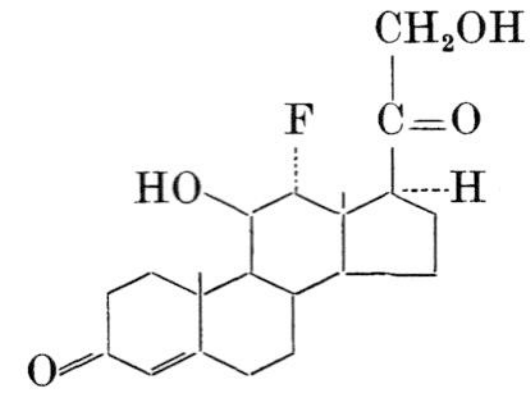

12α-Fluorocorticosterone

The preparation of 12α-halogenated derivatives closely parallels that of the 9α-substituted series (Fig. 60). The intermediate 11β,12β-oxides (III) are much less readily available, however, and have not been reported in the 17α,21-dihydroxy-20-ketone series. Addition of hypobromous acid (N-bromoacetamide in aqueous acid) to Δ^{11}-dehydroprogesterone (HEGNER and REICHSTEIN 1943) furnishes the diaxial 11β,12α-bromohydrin (II), which readily reverts to the 11β,12β-oxide with base (HERZ et al. 1956b). In the corticosterone series (TAUB et al. 1957a), the 12α-bromo-11-ketone was reduced (after protection of carbonyl groups at C-3 and C-20) to the bromohydrin. The 12α-chloro- and 12α-fluoro-11β-hydroxy compounds are obtained by fission of the oxide with the appropriate hydrogen halide. As in the 9α-fluoro series, the 12α-fluoro derivatives are not sensitive to base-catalyzed oxide formation.

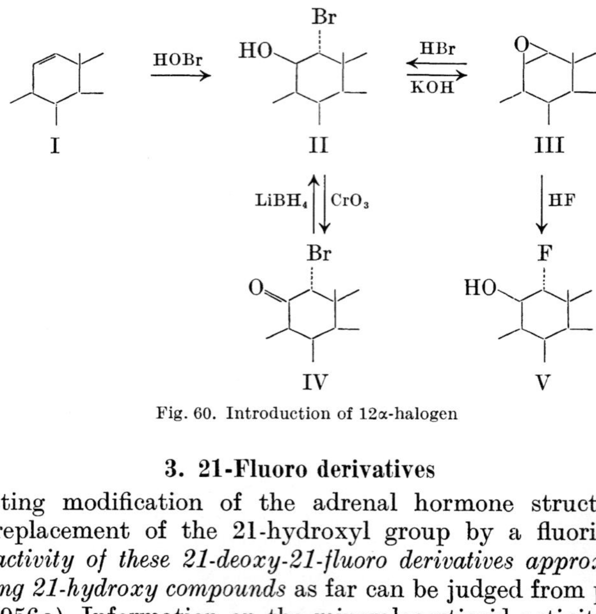

Fig. 60. Introduction of 12α-halogen

3. 21-Fluoro derivatives

An interesting modification of the adrenal hormone structure has been obtained by replacement of the 21-hydroxyl group by a fluorine atom. The *glucocorticoid activity of these 21-deoxy-21-fluoro derivatives approximates that of the corresponding 21-hydroxy compounds* as far can be judged from published data (HERZ et al. 1956a). Information on the mineralocorticoid activity is not available; it is noteworthy that 21-chloro-21-deoxy-9α-fluorocortisol is inactive in the liver glycogen assay.

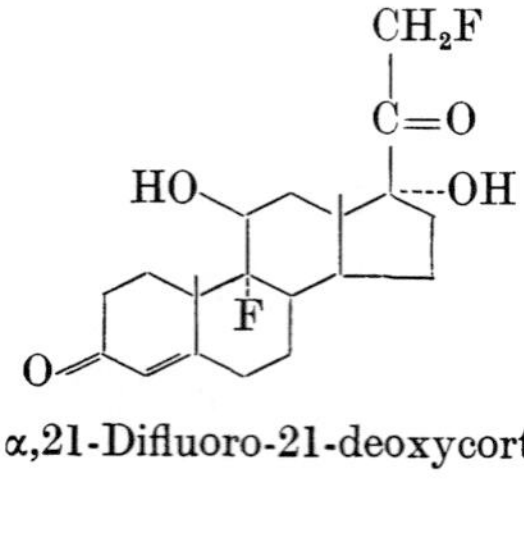

9α,21-Difluoro-21-deoxycortisol

The first 21-fluoro-20-ketones were reported by TANNHAUSER et al. (1956), who prepared several 21-deoxy-21-fluorocorticosteroids from the 21-iodo derivatives by treatment with silver fluoride in acetonitrile solution (Fig. 61). The iodo compounds (III) are most readily prepared by displacement of the mesyl (methanesulfonyl) group with iodide ion. HERZ et al. (1956a) obtained the 21-fluoro-21-deoxy derivatives directly from the 21-mesylate (II) by reaction with potassium fluoride in polar solvents such as dimethylformamide or dimethylsulfoxide. The replacement of the 21-hydroxyl group by a fluorine atom can be accomplished in the presence of the other functional groups characteristic of the highly active glucocorticoids, so that a large variety of 21-fluoro derivatives are potentially available. Several other 21-deoxy-21-fluorocortisol derivatives with additional functional groups have been described in preliminary communications (SPERO et al. 1957).

A by-product obtained in these displacement reactions of 21-mesylates is the 17,21-oxide (V), formed by internal displacement with the 17α-hydroxyl group (HERZ et al. 1956a; R. HIRSCHMANN et al. 1956a). The strained oxide structure rearranges in acid medium to the 18-nor-17β-methyl-13α,21-oxide structure (IV). It is rather interesting that the 17α,21-oxide (IV) derived from 9α-fluorocortisol has a liver glycogen activity comparable to that of cortisol (R. HIRSCHMANN et al. 1956a).

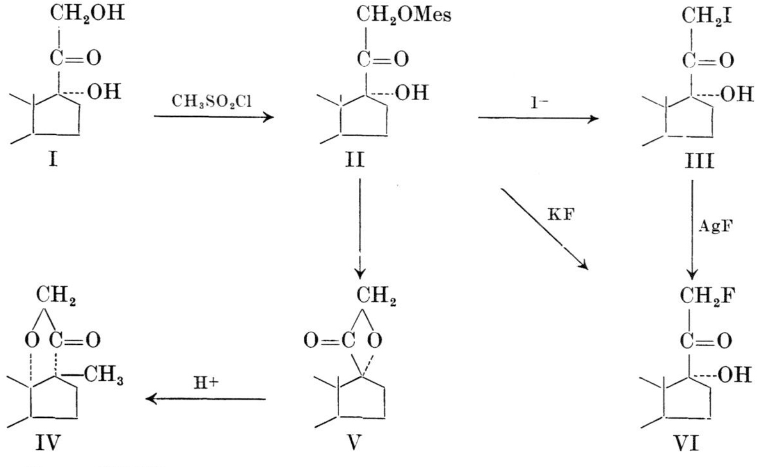

Fig. 61. 21-Deoxy-21-fluoro derivatives

4. Halogen substitution at other positions

a) 4-Halogenated derivatives

Interest in 4-halo-Δ^4-3-ketosteroids stems from the increased anabolic/androgenic ratio which has been reported for a number of 4-halogenated androgen

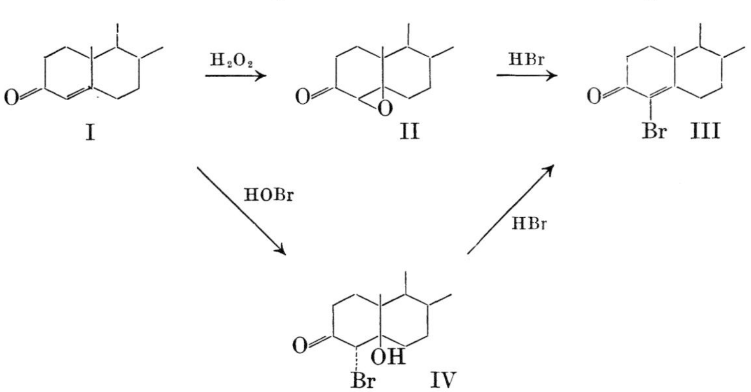

Fig. 62. Introduction of 4-halogen

derivatives (CAMERINO et al. 1956). Several groups have subsequently reported the preparation of 4-halogenated derivatives of various corticosteroids; biological data are lacking.

The 4-halo-Δ^4-3-ketone system (III, Fig. 62) is obtained by hydrogen halide cleavage of the 4β,5β-oxide (II); the intermediate bromohydrin (IV) is not isolated. A mixture of α- and β-oxides is obtained by treatment of the Δ^4-3-ketone with strong alkaline peroxide (CAMERINO et al. 1956, RINGOLD et al. 1956a); the ketol side chain is protected as the 21-tetrahydropyranyl ether. The 4-bromo- and 4-chlorocortisone derivatives have been obtained by treatment of cortisone acetate with hypobromous acid followed by hydrogen halide (OLIVETO et al. 1957). The yields in all of these methods are low.

b) *16-Halogenated derivatives*

One 16β-halo derivative, 16β-chlorocortisone, has been prepared (BEYLER and HOFFMAN 1956) and found to be inactive in the liver glycogen assay. The compound was obtained by hydrogen chloride cleavage of the 16α,17α-oxide (see Fig. 64) followed by introduction of the Δ^4-3-keto grouping.

c) *2- and 6-Halogenated derivatives*

A few glucocorticoids with the 2- or 6-bromo-Δ^4-3-ketone system have been described incidentally in connection with partial synthetic transformations; a 2-bromo derivative of 9α-fluorocortisol was reported by FRIED et al. (1955) to have one-third the liver glycogen activity of cortisone. Heightened interest in substitution at C-6 was generated by the very high glucocorticoid activity of 6α-methyl homologues (p. 283), and methods for the preparation of 6α- and β-fluoro-Δ^4-3-ketones via the 5,6-oxides have been reported by two groups (BOWERS and RINGOLD 1958, HOGG et al. 1958). Remarkably high glucocorticoid activities have been recorded in this series, the 6α-fluoro derivatives of cortisol, Δ^1-cortisol and Δ^1-9α-fluorocortisol having respectively 11, 81 and 427 times the glucocorticoid activity of cortisol (HOGG et al. 1958). KISSMAN et al. (1959) have prepared 2α-fluorocortisol, and found the glycogenic activity to be less than that of cortisol.

II. Hydroxyl substitution

1. 16-Hydroxy derivatives

The "physiological significance" of 16α-hydroxy substituents, which was inferred from the presence of a 16α-hydroxyl group in several urinary steroids and briefly from an erroneous suggestion of a 16-hydroxyl substituent in aldosterone (SIMPSON and TAIT 1953), stimulated the investigation of 16-hydroxylated corticosteroids. These studies culminated in the important finding (BERNSTEIN et al. 1956) that *the presence of a 16α-hydroxyl group leads to some reduction of glycogenic activity but abolishes the electrolyte activity of corticosteroids.* Sodium retention is a serious drawback in the clinical utilization of many of the "super glycogenic" corticosteroids. The 16α-hydroxy derivatives of several modified corticosteroids have been prepared; the more important members are listed in Table 4. In contrast, introduction of a 16β-acetoxy group into 9α-fluorocortisol or the 1-dehydro compound results in loss of all glycogenic properties (BERNSTEIN et al. 1959b).

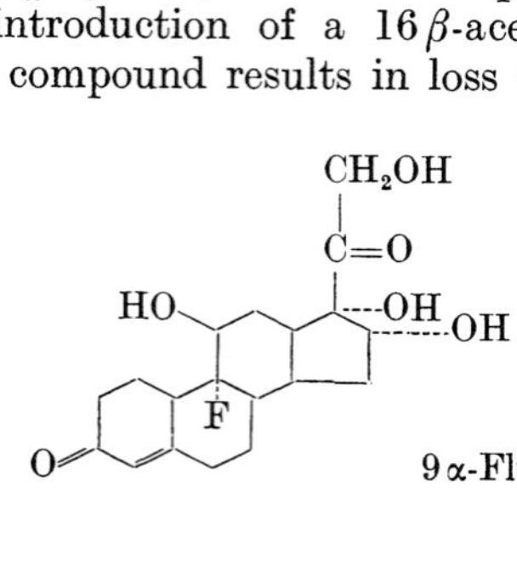

9α-Fluoro-16α-hydroxycortisol

The chemistry of 16-hydroxylated pregnane derivatives, especially the preparation of 16-hydroxycorticosteroids, has been the subject of much recent work. Simple 16β,20α- and 16β-20β-diols (V) can be prepared directly from sapogenins (I, Fig. 63) (H. HIRSCHMANN et al. 1949), but most synthetic routes to 16α-hydroxy- or 16α or β,17α-dihydroxy-20-ketones proceed from Δ^{16}-20-ketones (II) or the corresponding 16α,17α-oxide (VIII). The Δ^{16}-20-ketone system, readily obtained by two-step degradation of the sapogenin side chain (Fig. 51, p. 254), is very reactive, and furnishes the oxide with alkaline peroxide, or the 16α-benzyloxy derivative (III) by the addition of benzyl alcohol (H. HIRSCHMANN et al. 1952). Hydrogenolysis of (III) gives the 16α-hydroxy-20-ketone (IV), also obtained by reduction of the oxide (VIII) with chromous acetate (W. COLE and JULIAN 1954). Addition of hypobromous acid to the Δ^{16}-20-ketone leads to the 16β-hydroxy-17α-bromo-20-ketone (VI), which is reductively debrominated to give the 16β-hydroxy-20-ketone (VII) (LÖKEN et al. 1956). The 16β-hydroxy-20-ketone and particularly the acetate are very readily reconverted to the unsaturated ketone; the acetate cannot be hydrolyzed to the free alcohol under any conditions.

Hydroxylation of the Δ^{16}-20-ketone with potassium permanganate (Fig. 64) has been studied by PETROW and co-workers, who obtained the 16α,17α-dihydroxy-20-ketone (IX) in moderate yields (COOLEY et al. 1955); the Δ^{14}-16α,17α-dihydroxy-20-ketone was also obtained (ELLIS et al. 1955). The 16α,17α-diol system in (IX) readily forms an acetonide (X); hydride reduction of the 20-carbonyl group followed by periodate oxidation gives the 16α-hydroxy-17-ketone (XI). Hydroxylation with osmium tetroxide-pyridine furnishes the osmate ester (XVIII) in high yield, but reductive cleavage of the latter is accompanied by extensive D-homo rearrangement. Another route to the 16α,17α-dihydroxy-20-ketone system has been described by ROMO and DE VIVAR (1956). The 16β-bromo-17α-acetoxy-20-ketone (XII), obtained from the oxide (VIII) via the bromohydrin, is treated with sodium acetate, and the 17α-acetoxy group provides anchimeric assistance (XIII) in a displacement reaction leading to the 16α-acetoxy-17α-hydroxy-20-ketone (XIV).

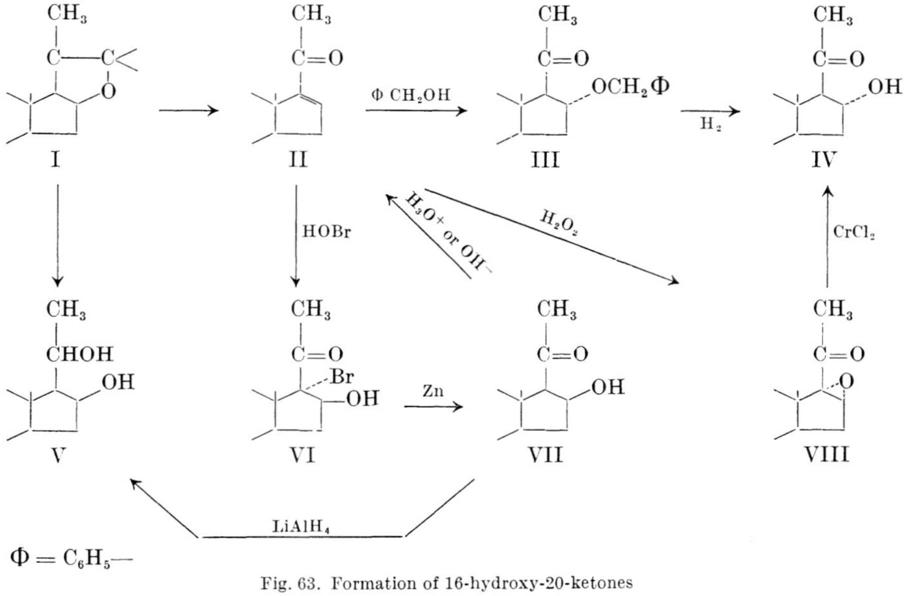

Fig. 63. Formation of 16-hydroxy-20-ketones

The 16β-acetoxy-17α-hydroxy-20-ketone is formed by acetolysis of the oxide (VIII) (HEUSLER and WETTSTEIN 1954). Both the 16α, 17α- and 16β,17α-dihydroxy-20-ketone systems are exceptionally prone to rearrangement. Attempted hydrolysis of either of the 16-acetoxy derivatives (XIV) or (XVI) leads to a D-homo compound, and similar rearrangements to isomeric D-homo derivatives occur on alumina chromatography of the diols or hydrolysis of the osmate esters. The exact structures of some of these compounds are still in doubt; it has been shown by WENDLER and TAUB (1957b) that the alumina product is the 17β-methyl-16α, 17α-dihydroxy-17a-one (XVII). Another type of rearrangement has been observed in the cleavage of the oxide (VIII). With acetic anhydride-toluenesulfonic acid, an 18-nor-17β-methyl-17α-acetyl derivative (XV, R = O) is obtained (HEUSLER and WETTSTEIN

1954); reduction of the oxide with fermenting yeast furnishes the corresponding 20-alcohol (XV, R = H,OH) (Camerino and Vercellone 1956).

Much the same preparative methods have been applied to 16,17α,21-trihydroxy-20-ketones (Fig. 65); practically all of the work in this series has been carried out with 11β-hydroxy or 11-keto compounds. A general method for the preparation of Δ^{16}-20,21-ketols (XX) is afforded by dehydration of the 17α,21-dihydroxy-20-ketal (XIX) (W. S. Allen and

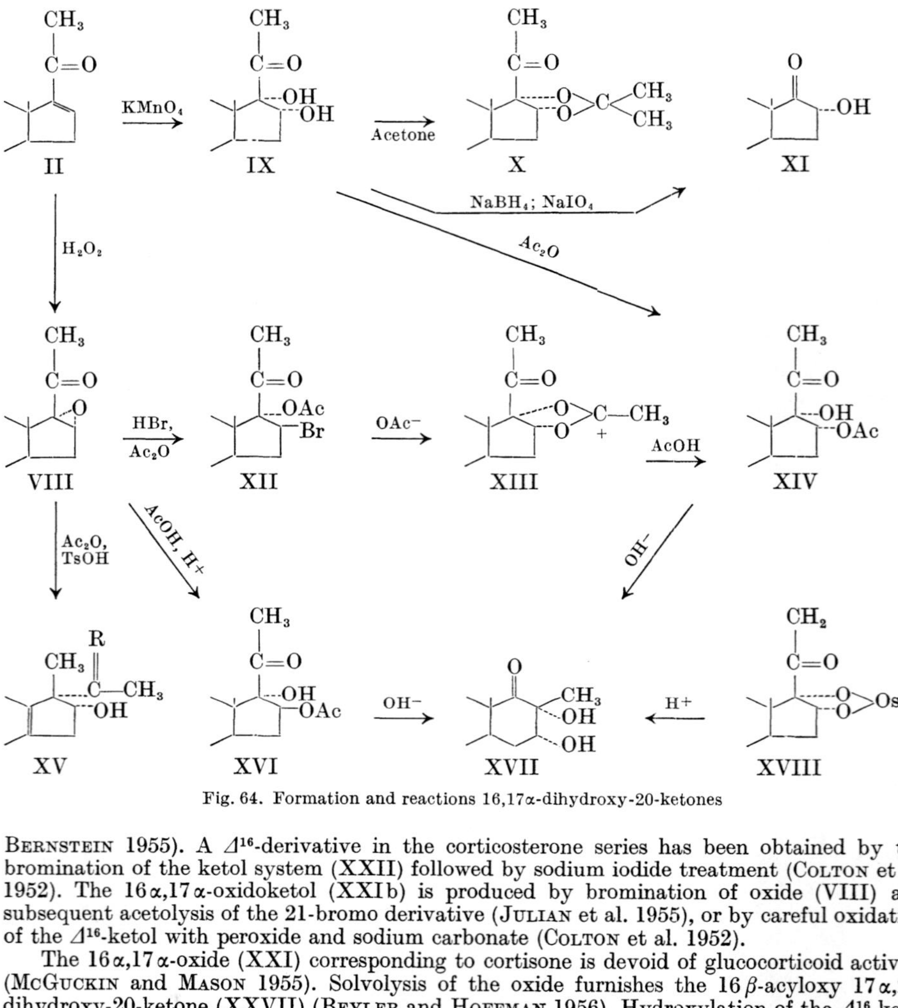

Fig. 64. Formation and reactions 16,17α-dihydroxy-20-ketones

Bernstein 1955). A Δ^{16}-derivative in the corticosterone series has been obtained by tribromination of the ketol system (XXII) followed by sodium iodide treatment (Colton et al. 1952). The 16α,17α-oxidoketol (XXIb) is produced by bromination of oxide (VIII) and subsequent acetolysis of the 21-bromo derivative (Julian et al. 1955), or by careful oxidation of the Δ^{16}-ketol with peroxide and sodium carbonate (Colton et al. 1952).

The 16α,17α-oxide (XXI) corresponding to cortisone is devoid of glucocorticoid activity (McGuckin and Mason 1955). Solvolysis of the oxide furnishes the 16β-acyloxy 17α,21-dihydroxy-20-ketone (XXVII) (Beyler and Hoffman 1956). Hydroxylation of the Δ^{16}-ketol with permanganate (Ellis et al. 1955) or with osmium tetroxide (W. S. Allen and Bernstein 1956a) leads to the 16α,17α,21-trihydroxyketone (XXIV). In addition to chemical methods for the introduction of a 16α-hydroxyl group, very effective microbiological 16α-hydroxylations have been reported (Table 3). Illustrative of the utility of this approach is the conversion of 9α-fluorocortisol to the 16α-hydroxy derivative in 50% yield by the organism *Streptomyces roseochromogenus* (Thoma et al. 1957).

The 11-oxygenated 16α or β,17α,21-trihydroxy-20-ketones appear to be rather less susceptible to rearrangement reactions than the simple 16,17α-dihydroxy-20-ketones. Thus, in the former series the 16-acetates can be hydrolyzed to the free 16,17α,21-triol-20-ones, and the osmate complexes can be cleaved in satisfactory yields.

Allen and Bernstein (1956b) have observed an interesting cyclization on hydrolysis of the 16-mesylate-21-acetate of the 16α,17α,21-trihydroxy-20-ketone, when the 16β,21-oxide (XXV) is formed in excellent yield. Fried et al. (1958) have reported that several of *the 16α,17α-acetals and ketals* (XXVI) obtained by treating 16α-hydroxy-9α-fluorocortisol or the corresponding Δ^1-dehydro compound with lower-molecular-weight aldehydes and ketones *surpass the parent alcohols in glucocorticoid activity.*

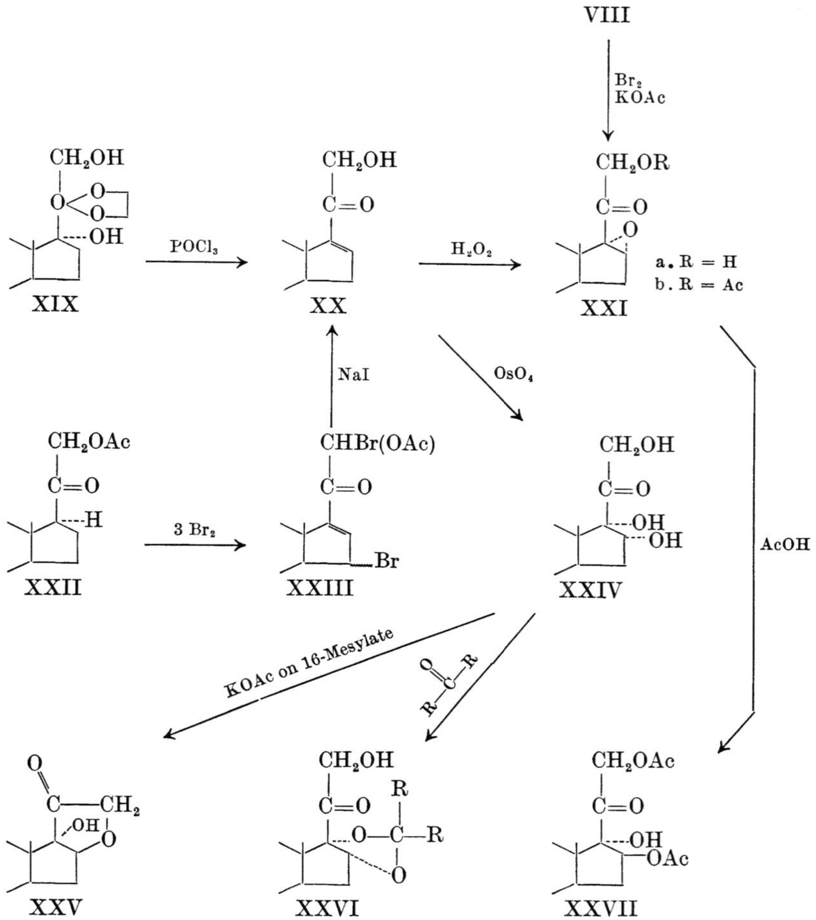

Fig. 65. Formation of 16,17α,21-Trihydroxy-20 ketones

2. Hydroxyl groups at other positions

a) 5α-Hydroxy compounds

5α-Hydroxy-3-ketones (V, Fig. 66) have occasionally figured as intermediates for generation of the Δ^4-3-keto system in corticosteroid syntheses (Bladon et al. 1953), and particularly in synthetic routes to 6-oxygenated derivatives (Fig. 43). The interesting observation has been reported by Bernstein and Lenhard (1955) that 5α,11β,17α,21-tetrahydroxypregnane-3,20-dione has appreciable activity in a thymus-involution assay. Compounds with the 5α-hydroxy-3-ketone system (V) are available by lithium aluminum hydride reduction of the 5α,6α-oxides (II) derived from Δ^5-3-ketals (I) followed by hydrolysis of the ketal grouping.

Since 5α-hydroxy-3-ketones possess a tertiary hydroxyl group in an axial orientation with respect to two rings, and in a 1,3-relationship to a carbonyl group, they are readily dehydrated by mild acid or base (Julia et al. 1952). It seems possible that this facile elimination reaction occurs upon administration of the compound, and that the biological activity observed in due to the derived Δ^4-3-ketone.

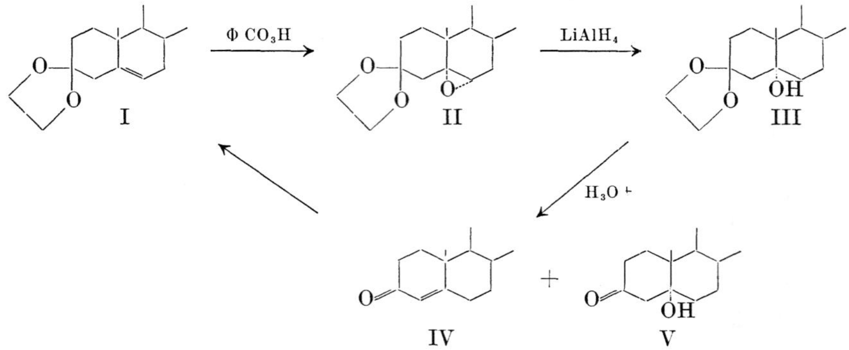

Fig. 66. 5-Hydroxy-3-ketones

b) 9α- and 12α-Hydroxy compounds

The 9β,11β- and 11β,12β-oxides, which are intermediates in the synthesis of the important 9α- and 12α-halo-11β-hydroxy derivatives (Figs. 59 and 60), also lend themselves to the preparation of the corresponding 9α,11β- and 11β,12α-dihydroxy compounds (Fig. 67).

The oxides (VI) and (VIII) are cleaved with aqueous perchloric acid to furnish the 9α,11β-dihydroxy (VII, R=H) and 11β,12α-dihydroxy (IX) derivatives, respectively (Fried and Sabo 1957, Taub et al. 1957a). Solvolysis of the 9β,11β-oxide (VI) with alcoholic acid gives

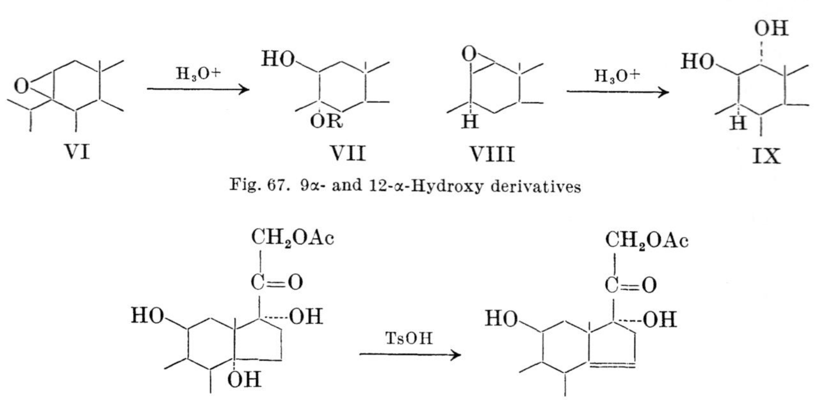

Fig. 67. 9α- and 12-α-Hydroxy derivatives

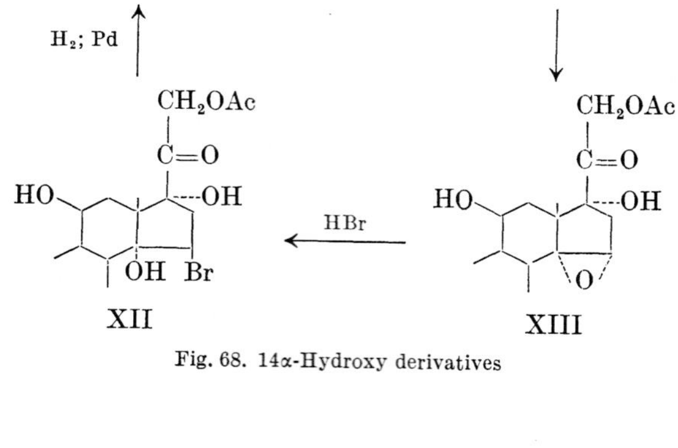

Fig. 68. 14α-Hydroxy derivatives

the 9α-alkoxy compounds (VII, R=CH_3 or C_2H_5). A more circuitous route to the 9α-hydroxy-11-ketone and 9α,11β-diol, involving osmium tetroxide hydroxylation of the Δ^9-olefin and subsequent oxidation of the 11α-hydroxyl group, has been reported by LITTEL and BERNSTEIN (1956).

The glycogenic activity of 9α-hydroxycortisol is of the order of one-half that of cortisol; the electrolyte activity is comparable to that of cortisol. The 9α-alkoxy substitution causes almost complete loss of glucocorticoid activity.

c) 14α-Hydroxy derivatives

A product obtained by the microbiological dihydroxylation of 11-deoxycortisol by the organism *Curvularia lunata* has been assigned the 14α-hydroxy cortisol structure (X, Fig. 68) by AGNELLO et al. (1955). The compound is more active than cortisol in a thymus-involution assay and is an active anti-inflammatory agent. The compound was selectively dehydrated to furnish the Δ^{14}-derivative (XI), from which the 14α-hydroxy group was regenerated via the oxide (XIII) and bromohydrin (XII).

III. Additional double bonds

1. Δ^1-Dehydro derivatives

Another modification of the basic corticosteroid structure of major importance in enhancing glycogenic activity has been found by HERZOG et al. (1955a) to be a Δ^1-double bond. *Conversion of cortisone and cortisol to the Δ^1-dehydro derivatives* (which have been named prednisone and prednisolone, respectively) *causes a 4–8 fold increased in the glycogenic activity and anti-inflammatory effect*; the increase in electrolyte activity is slight. It has been found by several groups (R. HIRSCHMANN et al. 1955, HOGG et al. 1955c, FRIED et al. 1955a) that the effect of Δ^1-dehydrogenation and 9α-fluorine substitution are multiplicative, and several highly glycogenic steroids combining these two structural features have been reported (cf. Table 4).

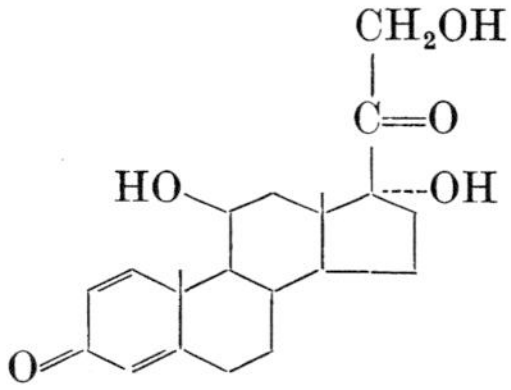

Prednisolone

Simpler steroids with the $\Delta^{1,4}$-diene-3-one system (V) had long been known, and had been studied particularly in the androstane series in connection with the aromatization to furnish estrogens (Fig. 71). The classical approach to the dienone system involved dibromination of a 3-ketone, usually in the 5α-series, followed by collidine dehydrobromination of the 2,4-dibromoketone (II). This method was described by HERZOG et al (1955b) for the preparation of Δ^1-dehydrocortisone in low yield.

The very desirable biological properties of the Δ^1-dehydrocorticoids stimulated considerable interest in preparative routes, and several superior methods for the introduction of a Δ^1-double bond have been reported. Several strains of microorganisms have been found to bring about selective dehydrogenation of a Δ^4-3-ketone. VISCHER et al. (1955) used molds of the *Fusarium* genus, and NOBILE et al. (1955) described the use of *Corynebacterium simplex*. Chemical introduction of a Δ^1-double bond in a Δ^4-3-ketone, or formation of the $\Delta^{1,4}$-dienone from a saturated 3-ketone, can be smoothly accomplished with selenium dioxide in *t*-butanol with a trace of acid (MEYSTRE et al. 1956).

Several types of rearrangements of steroidal dienones have been observed. With 1,4-androstadiene-3,17-dione, for example, strong mineral acids bring about the formation of the phenols (III) and (VI) (Fig. 69). Deep-seated rearrangements of the $\Delta^{1,4}$-dienone system, which must be reckoned with in considering the stability of solutions of Δ^{1}-dehydrocorticosteroids, are brought about by exposure to ultraviolet radiation. In ethanol solution, Δ^{1}-dehydrocortisone (prednisone) (V) is converted to an isomer which has been named lumiprednisone and assigned the structure (IX) (BARTON and TAYLOR 1958). This compound, which arises by rupture of the 9—10 bond and shift of an electron pair from the 4—5 double bond, is converted to the cyclopentenone (X) by traces of acid. Irradiation of (V) in acidic solution leads to (VIII), with migration of the 5—10 bond to C-1 (BARTON and TAYLOR 1957). A total of eight products, including phenols and ketones, have been obtained by the irradiation of Δ^{1}-dehydrotestosterone acetate (DUTLER et al. 1957).

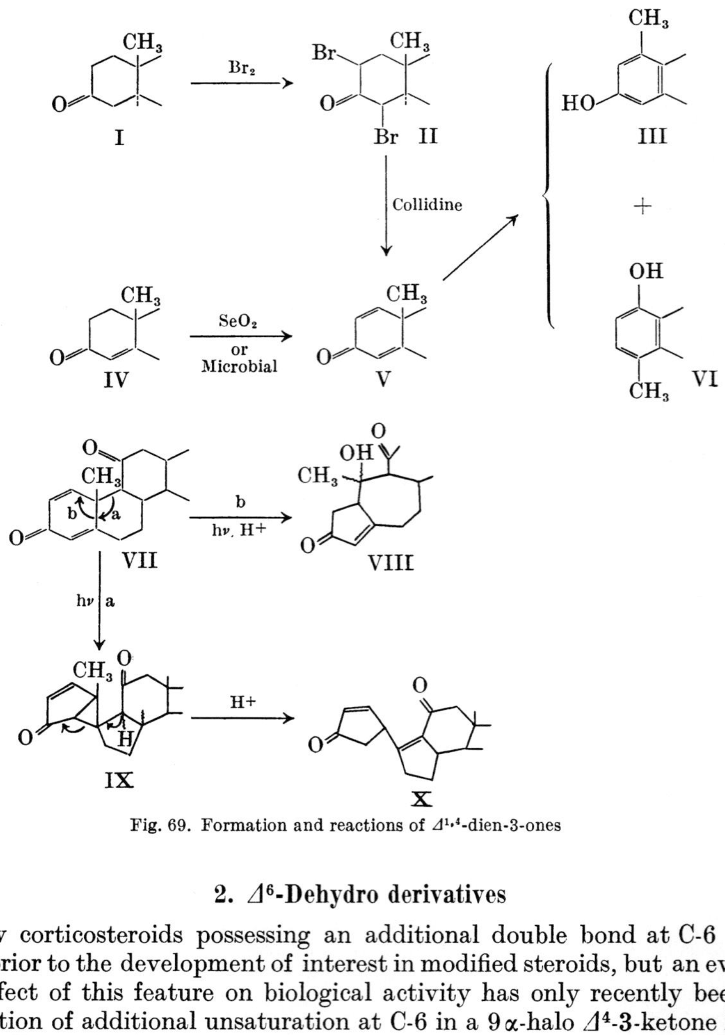

Fig. 69. Formation and reactions of $\Delta^{1,4}$-dien-3-ones

2. Δ^{6}-Dehydro derivatives

A few corticosteroids possessing an additional double bond at C-6 were described prior to the development of interest in modified steroids, but an evaluation of the effect of this feature on biological activity has only recently been made. Introduction of additional unsaturation at C-6 in a 9α-halo Δ^{4}-3-ketone has been found by FRIED et al. (1955a) to decrease slightly the glycogenic potency and to increase sharply the electrolyte activity. Dehydrogenation of a $\Delta^{1,4}$-dien-3-one to the $\Delta^{1,4,6}$-trienone appears to have only a slight effect on the activity of the former (GOULD et al. 1957b).

Δ^{6}-Dehydrocortisone was encountered as a by-product in the preparation of cortisone (MATTOX et al. 1952); the compound arose in the bromination-dehydrobromination sequence used for generation of the Δ^{4}-3-keto system. The $\Delta^{4,6}$-dien-3-one system (III, Fig. 70) can be formed directly in the Oppenauer oxidation of a Δ⁵-3-hydroxy compound if quinone is used as the hydrogen acceptor (WETTSTEIN 1940). Introduction of a 6,7-double bond in a Δ⁴-3-ketone or a Δ¹,⁴-dienone can be effected by bromination at C-6 with N-bromosuccinimide followed by treatment with refluxing collidine (KAUFMANN et al. 1950; GOULD et al. 1957b).

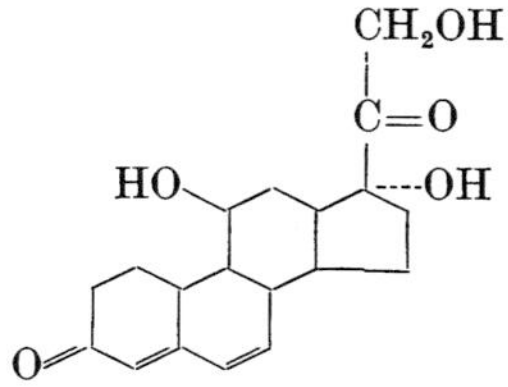

Δ⁶-Dehydrocortisol

Dehydrogenation of a Δ⁴-3-ketone has been accomplished by vigorous treatment with manganese dioxide (SONDHEIMER et al. 1953a). A very convenient and versatile dehydrogenation reagent is chloranil (tetrachlorobenzoquinone), which converts a Δ⁴-3-ketone initially to the Δ⁴,⁶-dienone, and at higher temperatures directly to the Δ¹,⁴,⁶-trienone (V) (AGNELLO and LAUBACH 1957). The customary functional groups of the glucocorticoids are not affected by this treatment, and a variety of trienones thus is potentially available.

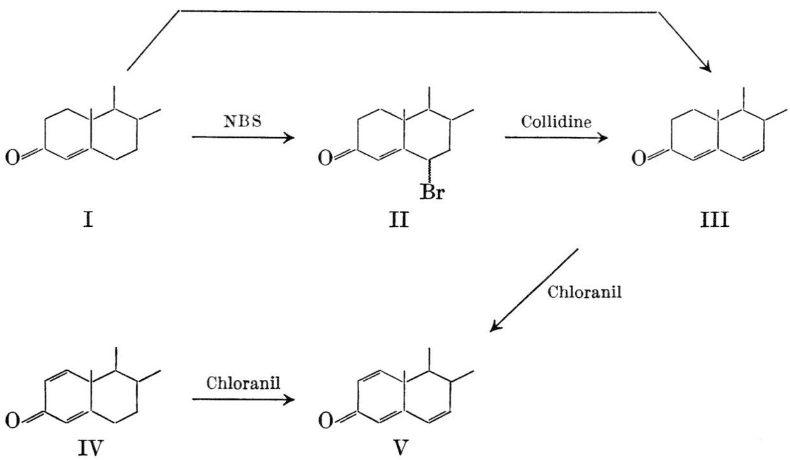

Fig. 70. Formation of Δ¹,⁴,⁶-trienones

3. Double bonds at other positions

The remaining positions available for additional unsaturation in the "complete" corticosteroid molecule, i. e., oxygenated at C-11 and C-17, are 7, 8, 14 and 15. Of these, Δ⁸⁽⁹⁾-(I) and Δ⁸⁽¹⁴⁾-dehydro (II) derivatives of cortisol have been discussed by WENDLER et al. (1957), and Δ¹⁴-dehydrocortisol (III) by AGNELLO et al. (1955); biological data have not been reported. It should be noted that the nuclear skeleton in these derivatives is somewhat flattened and distorted since a nuclear asymmetric center is disrupted.

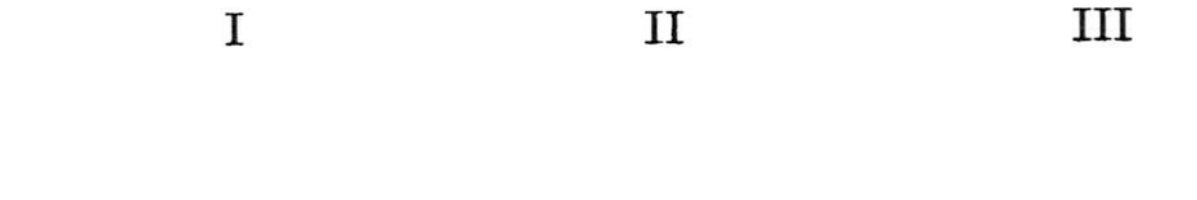

Both $\Delta^{9(11)}$ (IV) and Δ^{11}-dehydro (V) derivatives of 11-deoxycorticosterone and $\Delta^{9(11)}$-dehydro-17α-hydroxycorticosterone have been prepared as degradation products or intermediates in syntheses described in other sections. The activity of these compounds should be compared with that of the corresponding 11-deoxy-hormone, since it is probably more meaningful to consider them as *dehydro* derivatives rather than *anhydro* derivatives of their 11β-hydroxy counterparts. The mineralocorticoid activity of (IV) (CASANOVA 1953) and (V) (MEYSTRE and WETTSTEIN 1948) is of the same order as that of 11-deoxycortisterone. It is noteworthy that introduction of a Δ^{11}-double bond markedly enhances the activity of progesterone (MEYSTRE 1948).

It has been reported by McGUCKIN and MASON (1955) that the Δ^{16}-dehydro derivative of 11-dehydrocorticosterone (VI) is devoid of glycogenic activity. The mineralocorticoid activity was not recorded.

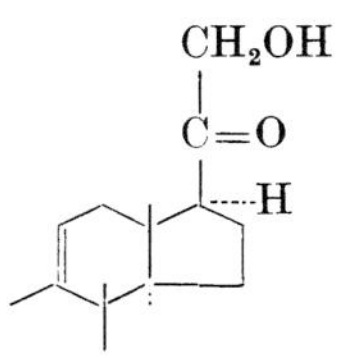

IV

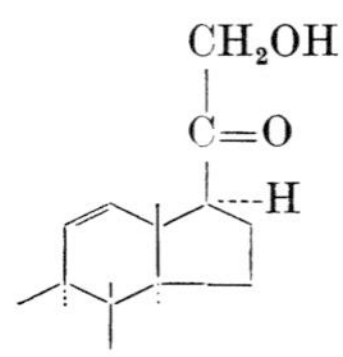

V

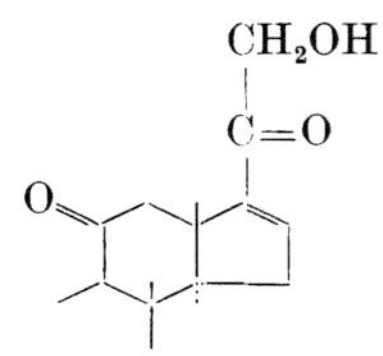

VI

IV. Skeletal modifications

In this section are discussed compounds in which one or more changes in the carbon skeleton of the natural corticosteroids have been made and in which adrenocortical activity is retained or enhanced. Three important types of alteration have been described: removal of a carbon atom (C-19), addition of a carbon atom as a methyl substituent, and inversion of an asymmetric center.

1. 19-Nor steroids

The possibility that at least a certain degree of physiological activity might reside in saturated steroid hormones lacking C-19, the angular methyl group at the A/B ring junction, was adumbrated by the investigations of DIRSCHERL et al. (1936) on reduction products of estrone. With the development of a superior metal-ammonia reduction procedure for the reduction of aromatic rings (BIRCH 1950), the investigation of these 19-nor derivatives was rapidly extended. Although 19-nortestosterone, the first compound to be investigated (WILDS and NELSON 1953b), shows less than half the androgenic potency of testosterone, the progestational activity of 19-norprogesterone (DJERASSI et al. 1953) and of 19-nor-17α-ethinyltestosterone (DJERASSI et al. 1954) is several times that of the parent hormones. In the corticosteroids, removal of C-19 from 11-deoxycorticosterone

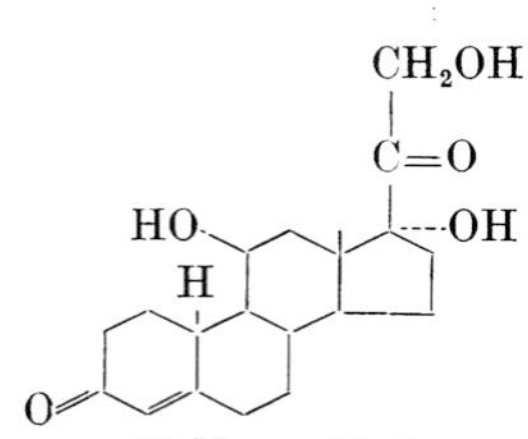

19-Norcortisol

doubles the mineralocorticoid activity (SANDOVAL et al. 1955). The glucocorticoid activity of the more difficultly accessible 11-oxygenated 19-nor derivatives appears to be of a low order; MAGERLEIN and HOGG (1958) report the activity of 19-norcortisol as one-tenth that of cortisol. It should be noted that the potentiation of activity in the 19-nor series is not the only effect observed; significant changes in the androgenic: anabolic ratio and anti-estrogenic properties have been observed in various modified 19-nor androgens. It seems likely that interesting qualitative differences in activity may be encountered in the 19-norcorticosteroids.

The configuration of the hydrogen atom at C-10 in all of the 19-nor steroids thus far examined is the same as that of the normal carbon skeleton, with the substituent in the more stable 10 β (C-10, C-9 *anti*) configuration. This follows from the fact that the Birch reduction of a ring A phenol and the retroaldol cleavage of a 19-hydroxy-Δ^4-3-ketone (Fig. 45, p. 239), which lead to the same C-10 series, both proceed via a carbanion (I) which will lead to the more stable 10 β isomer (II) on protonation. This assignment is fully supported by molecular rotatory dispersion data (DJERASSI et al. 1956c).

I II

The only important preparative procedure for 19-norhormone derivatives is the reduction of an intermediate ring A aromatic compound. The aromatization of ring A in a hydroaromatic steroid is a key step in the large scale synthesis of estrogens from plant steroids and has been intensively studied. The pioneering work of INHOFFEN demonstrated that pyrolysis of a steroidal 1,4-diene-3-one (Fig. 71) proceeded in favorable cases with the extrusion of C-19 as a methylene radical and aromatization of ring A. With improved conditions (mineral oil solvent, temperatures of 500—600°) (HERSHBERG et al. 1950) this pyrolysis leads to very respectable yields; an example is the conversion of 1,4,16-pregnatriene-3,20-dione to 3-hydroxy-19-nor-1,3,5(10)-pregnatriene-20-one in 30% yield (DJERASSI et al. 1951).

This thermal aromatization procedure has been applied in one case to an 11 β-hydroxy-$\Delta^{1,4}$-diene-3-one; a 15% yield of the corresponding phenol was obtained (MAGERLEIN and HOGG 1958). This is a remarkably high yield considering that elimination of the labile 11 β-hydroxyl group would be facilitated by conjugation of a $\Delta^{9(11)}$ double bond with the aromatic ring. The phenol obtained in this reaction was further transformed into 3,11 β,17 α,21-tetrahydroxy-19-nor-1,3,5(10)-pregnatriene-20-one (III), the aromatic analogue of cortisol, possessing about one-tenth its glucocorticoid activity.

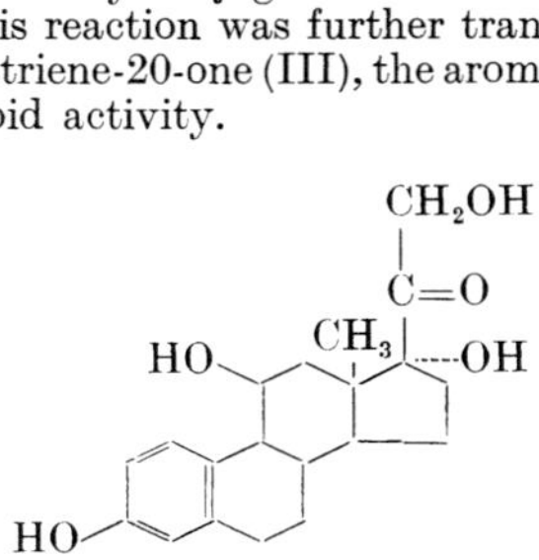

III

Reduction of the methyl ethers of these phenolic steroids by means of lithium in ammonia followed by alcohol (WILDS and NELSON 1953a) furnishes the 3-methoxy-$\Delta^{2,5(10)}$-diene system (VI, Fig. 71). Cautious acid hydrolysis leads to the $\Delta^{5(10)}$-3-ketone (VII), which is readily isomerized to the 19-nor-Δ^4-3-ketone (II). Carbonyl groups at other centers must be blocked during the reduction or regenerated by subsequent oxidation. Standard procedures have been applied for the elaboration of the cortical side chains in the 19-nor intermediates. The introduction of an 11 α-hydroxyl group into 19-nortestosterone has been accomplished by microbiological oxidation (PEDERSON et al. 1956). The very interesting observation has been reported by ZAFFARONI et al. (1954) that adrenal enzymes can effect 11 β-hydroxylation of 19-norsteroids; 19-norcorticosterone and 19-norcortisol were prepared by incubation of the respective 11-deoxy compounds with adrenal homogenates.

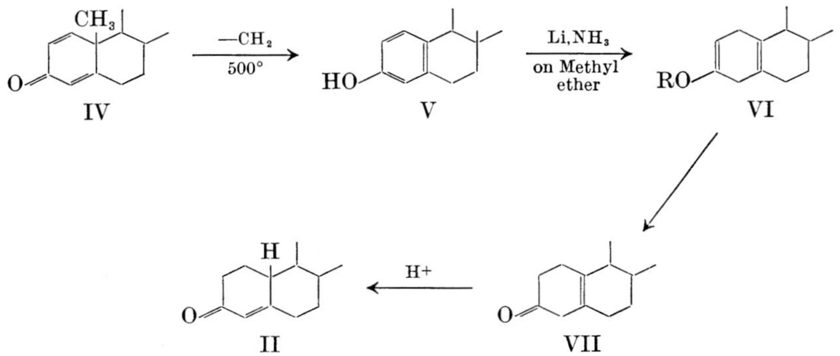

Fig. 71. Formation of 19-nor derivatives

Another, less general, source of 19-nor derivatives is provided by transformations of a few butenolide glycosides, which contain a 19-oxygen substituent (Fig. 72). EHRENSTEIN and co-workers have devoted much attention to the conversion of the aglycone strophanthidin (VIII) to modified hormones. Removal of C-19 can be effected by pyrolysis of the derived 5β-acetoxy-19-carboxylic acid (IX) to furnish the $\Delta^{5(10)}$ derivative (X) which is rearranged to the 19-nor-Δ^4-3-ketone (BARBER and EHRENSTEIN 1957). A considerable number of additional steps are required to reach the 19-norpregnane skeleton. Another reaction sequence has been used for the removal of the C-19 function in the highly hydroxylated aglycone ouabagenin (XI) (SNEEDEN and TURNER 1955a). Alkali treatment of the 3-ketone (XII)

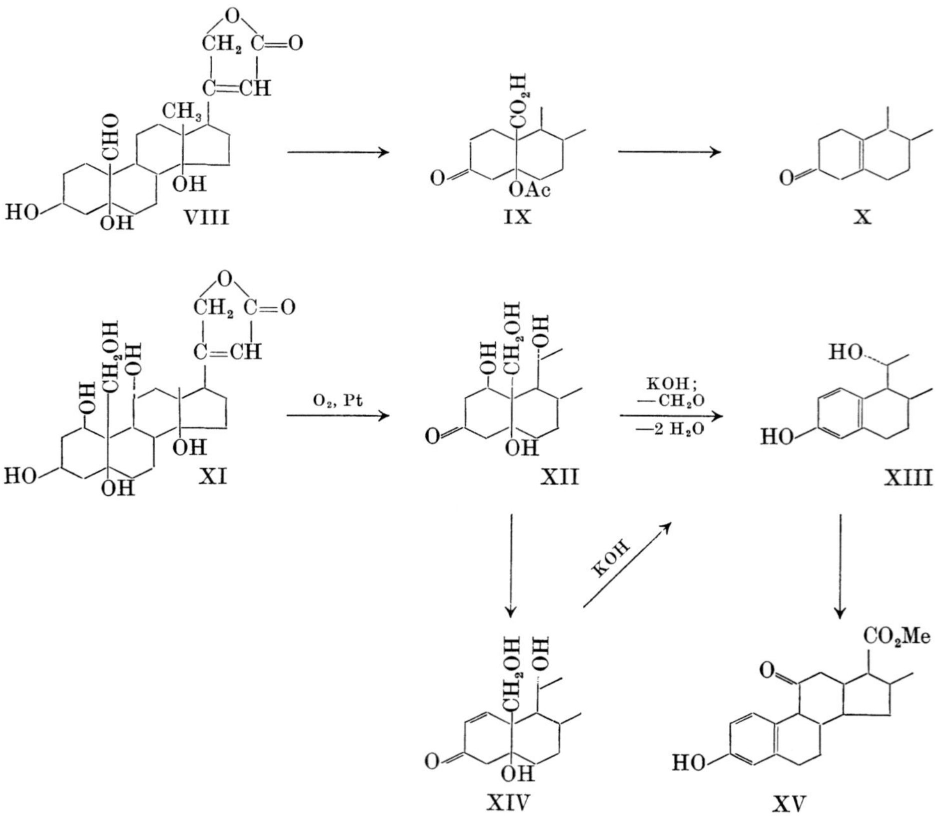

Fig. 72. 19-Nor derivatives from butenolide aglycones

causes successive dehydration and reverse aldol cleavage with loss of two molecules of water and formaldehyde with formation of the phenol (XIII). TAMM et al. (1957) performed the same reaction with the 5β,19-dihydroxy-Δ^1-3,11-dione (XIV) and eventually obtained the 11-keto-estratriencarboxylic ester (XV). By a suitable sequence of operations, it should be possible to develop a reasonably direct synthesis of 19-nor-11-oxygenated corticosteroids from ouabagenin.

Finally, it may be mentioned that any of the existing total synthetic paths to 11-oxygenated steroids (p. 262) could be readily modified to furnish 19-nor derivatives, should sufficient impetus arise.

An important difference in behavior of 19-nor steroids is the lessened steric hindrance of 11-oxygen functions. It has been found by MAGERLEIN and HOGG (1958) that 19-norcortisol forms the 11β,21-diacetate under very mild conditions, and the 11-keto group in the ester (XV) readily forms carbonyl derivatives (C. TAMM et al. 1957). Another consequence of the absence of the methyl group at C-10 is a tendency, with suitable substitution, for aromatization of ring A. The interesting 10-hydroxy-19-nortestosterone (XVI) obtained by microbiological oxidation of 19-nortestosterone (PEDERSON et al. 1956), reverts to estradiol with mild acid treatment (Fig. 73). Similarly, a 19-nor-$\Delta^{1,4}$-diene-3-one system (XVII) would be expected to isomerize spontaneously to the phenol.

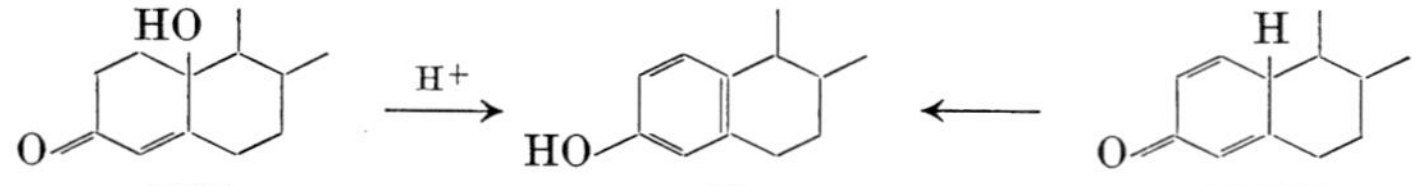

Fig. 73. Aromatization of 19-nor derivatives

2. 6α-Methyl derivatives

One of the most interesting structural modifications of the adrenal steroids that has been studied, from the standpoint of glucocorticoid activity, is the insertion of a methyl group. In two preliminary notes, SPERO et al. (1956, 1957) have reported the preparation of 6α-methylcortisol, 6α-methylcortisone and the corresponding Δ^1-dehydro-, 9α-fluoro- and 21-fluoro-21-deoxy-derivatives. *The 6α-methyl group imparts a 4—16 fold increase in the glycogenic activity of cortisol*; introduction of the Δ^1- and 9α-fluoro-features furnishes the remarkably active 6α-methyl-Δ^1-dehydro-9α-fluorocortisol, which is reported to have 100—200 times the gluco-corticoid activity of the parent hormone. All of these 6α-methyl derivatives are reported to have negligible sodium-retaining properties.

Introduction of the 6α-methyl group was accomplished by conversion of the Δ^4-3-ketone function in 11α-acetoxyprogesterone to the ketal and thence to the 5α,6α-oxide (II), which was treated with methyl magnesium iodide (Fig. 74). Addition of the Grignard reagent led

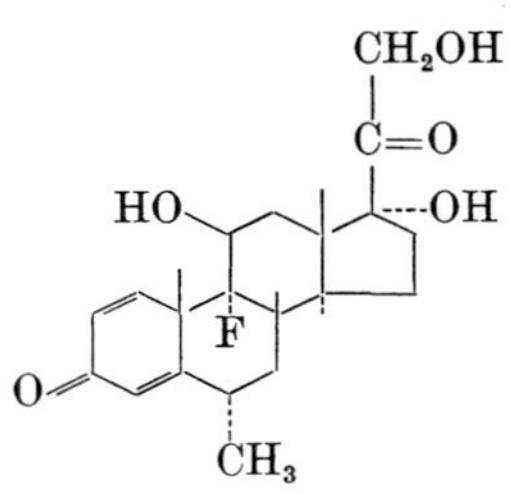

6α-Methyl-9α-fluoro-
Δ^1-dehydrocortisol

to the 6β-methyl-5α-hydroxyketal, from which the ketone (III) was regenerated by mild acid hydrolysis. Base-catalyzed dehydration of (III) followed by oxidation at C-11 led to 6β-methyl 11-ketoprogesterone (IV), isomerized by alkali to the 6α-methyl derivative. Epimerization of the axial 6β-methyl group occurred concurrently with dehydration in the case of the 5-hydroxy 3,11-dione (VI), giving directly 6α-methyl-11-ketoprogesterone (V). The configurational assignment follows that made by TURNER (1952) in the 6-methylcholestenone series. Subsequent operations were then carried out for the side chain elaboration via the 21-oxaloester and Δ^{17}-pregnen-21-oic acid (Fig. 34, p. 226). The Δ^{1}-double bond was introduced by microbiological dehydrogenation.

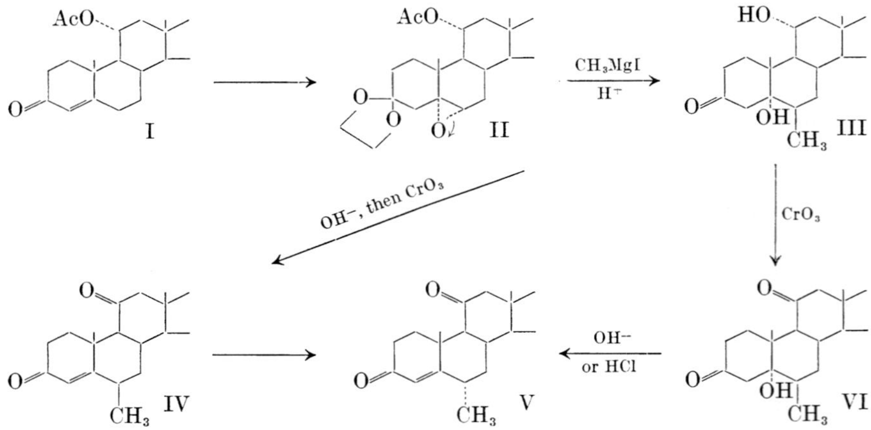

Fig. 74. Formation of 6-α-methyl derivatives

3. 2α-Methyl derivatives

The introduction of a 2-methyl group into the corticosteroid skeleton was first reported by HOGG et al. (1955b), who described the 2α-methyl derivatives of cortisol, cortisone, and the respective 9α-fluoro compounds. The glucocorticoid activity of the 11β-hydroxy derivatives is enhanced approximately ten fold, but even greater potentiation of the mineralocorticoid activity (as much as 90 times that of 11-deoxycorticosterone in the case of 2α-methyl-9α-fluorocortisol) is brought about by this homologation. Unexpectedly, a marked *decrease* in glucocorticoid activity is observed in the 2α-methyl-11-keto derivatives (DULIN et al. 1957). The very high mineralocorticoid activity of 2α-methyl-9α-fluorocortisol is practically abolished by introduction of a 16α-hydroxyl group (BERNSTEIN et al. 1959a).

The introduction of the 2-methyl substituent is effected by alkylation, with methyl iodide and potassium carbonate, of the 2-ethoxyoxalyl derivative (II) of the Δ^{4}-3-ketone (Fig. 75). This operation has been carried out both with the Δ^{17}-21-ol side chain intermediate or the

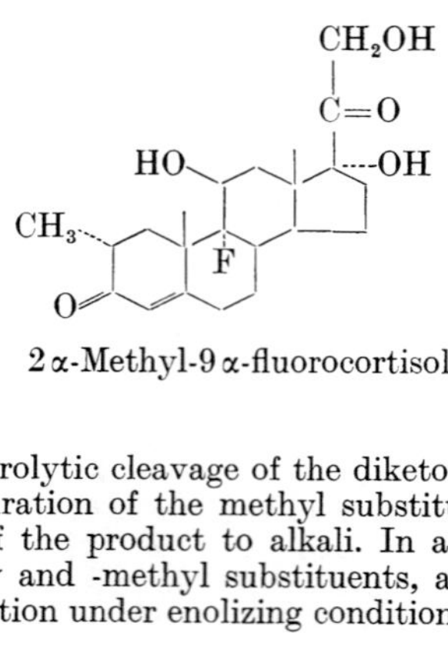

2α-Methyl-9α-fluorocortisol

17,21-dihydroxy-20-ketal. Hydrolytic cleavage of the diketone (IV) furnishes the 2α-methyl-Δ^{4}-3-ketone (III). The configuration of the methyl substituent is based upon the mode of formation and the stability of the product to alkali. In analogy to the axial 2β-hydroxy (p. 233) and axial 6β-hydroxy and -methyl substituents, an axial 2β-methyl group would be expected to suffer epimerization under enolizing conditions.

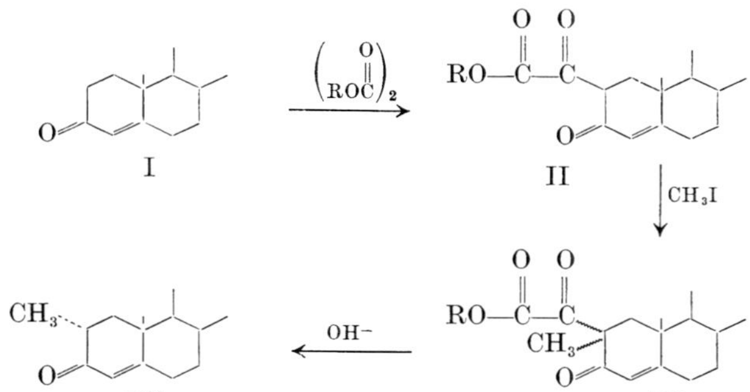

Fig. 75. Formation of 2α-methyl derivatives

4. 16-Methyl derivatives

Another position in which a methyl substituent has been found to have a strongly enhancing effect on glucocorticoid activity is C-16. The 16α-methyl derivatives of cortisol, Δ^1-cortisol, and the respective 9α-fluoro compounds have been prepared by ARTH et al. (1958a, b); the glycogenic and anti-inflammatory activities of all of the compounds surpass those of cortisol. The mineralocorticoid activity of the 16α-methyl derivatives is negligible; the fully substituted compound, 16α-methyl-Δ^1-9α-fluorocortisol, is reported to be 28—40 times more active than cortisol as an anti-inflammatory agent in man, and to cause no sodium retention (ARTH et al. 1958a). It is interesting that the 16α-methyl group has the same mineralocorticoid-abolishing effect as the 16α-hydroxyl group.

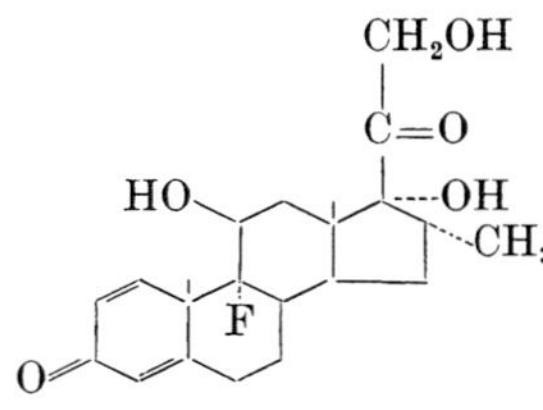

16α-Methyl-Δ^1-9α-fluorocortisol

Introduction of the 16α-methyl group was accomplished by 1,4-addition of methyl magnesium iodide to the Δ^{16}-20-keto system to give the 16α-methyl-20-ketone. The cortical side chain was then elaborated by standard methods involving epoxidation of the 20-enol acetate and hydrolysis to the 16α-methyl-17α-hydroxy-20-ketone (see Fig. 22, p. 215) followed by bromination at C-21 and treatment with potassium acetate (Fig. 53, p. 255). The 9α-fluoro and Δ^1-groups were introduced by standard chemical procedures. The 16α-methyl group reduces the reactivity of the 20-keto group in the dihydroxyketone; it has been suggested that the increased glucocorticoid activity of the 16α-methyl derivatives is due to a reduction in the rate of *in vivo* destruction of the substituted corticosteroids.

The 16β-methyl derivatives of cortisol, prednisolone and 9α-fluoroprednisolone have also been prepared (OLIVETO et al. 1958a, 1958b, TAUB et al. 1958). These compounds exhibit enhanced glycogenic activity of the same order as the 16α-methyl homologues, and are similarly devoid of mineralocorticoid activity. The 16β-methyl group, which is the first β-oriented substituent at any position that has been found to potentiate glycogenic activity, is introduced by hydrogenation of a 16-methyl-Δ^{16}-steroid. This precursor is obtained by pyrolysis of the 16,17-pyrazoline; the latter is prepared by addition of diazomethane to a Δ^{16}-20-ketone.

5. 9α-Methyl derivatives

The 9α-methyl derivatives of cortisol and prednisolone have been prepared by HOFFMAN et al. (1958). Introduction of the methyl group was accomplished by reaction of a suitably protected 9α-bromo-11-ketone with methyl magnesium iodide. The glycogenic activity of these compounds was substantially less than that of the parent steroids. This finding is consistent with an empirical generalization that can be drawn from the effects of other 9α-substituents, namely that an electron-withdrawing group such as fluorine causes enhancement of activity, whereas electron-releasing groups such as alkoxyl or methyl bring about a decrease in biological activity.

6. 17α-Methyl derivatives

Interest in 17α-methylcorticosteroids was stimulated by the high hormonal activity of 17α-methyl derivatives of the androgens and especially the high progestational activity of 17α-methylprogesterone (HEUSSER et al. 1950a). The 17α-methyl homologues of 11-deoxycorticosterone and of corticosterone have been described by C. ENGEL and JUST (1954) and C. ENGEL (1956, 1957). Replacement of the 17α-hydrogen atom in these hormones by a methyl group results in compounds of a relatively low order of activity. The mineralocorticoid activity of the 11-deoxycorticosterone derivative (I) is one-fifth that of the parent hormone; the glucocorticoid activity of 17α-methylcorticosterone is about half that of cortisone. It is interesting that the progestational activity of (I) is much higher than that of the parent corticoid, equaling that of progesterone.

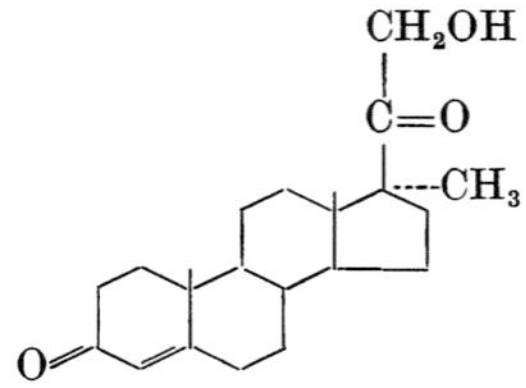

I

The preparation of this type of compound involves the Favorskii rearrangement of either the 17α-bromo- (II) or 21-halo-20-ketone (V) with base (Fig. 76). The 17α-methyl-17-etianic acid ester (III) is obtained in good yield from (II); the 21-halo ketone gives a mixture of the isomeric 17-methyletianic esters. Drastic saponification of the tertiary ester leads to the acid (IV), which is transformed to the diazoketone and, thence, under forcing conditions, to the ketol (VI).

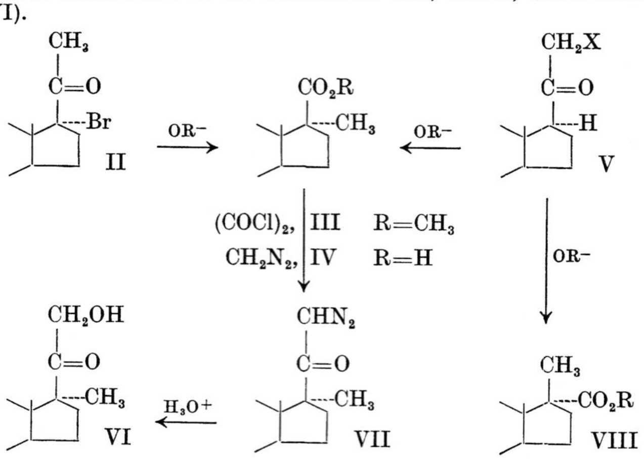

Fig. 76. Formation of 17β-methyl derivatives

7. 4-Methyl derivatives

The introduction of a methyl group at C-4 has been described in the cases of testosterone and progesterone by SONDHEIMER and MAZUR (1957) and by ATWATER (1957). The 4-methyl derivatives exhibit about one-half the androgenic and progestational activities, respectively, of the parent hormones. The 4-methylated homologues of cortisol and cortisone have been reported by N. STEINBERG et al. (1958) to be devoid of glucocorticoid activity.

The direct alkylation of a Δ^4-3-ketone proceeds rapidly to the 4,4-dimethyl-Δ^5-3-ketone (II), and only moderate yields of the monoalkylation product (V) are obtained (Fig. 77). An indirect procedure that has been used for the preparation of 4-methyl-Δ^4-3-ketones (V) is a variant of the enol-lactone used for the incorporation of C^{14} into ring A. The lactone (III) obtained by ozonolysis of the Δ^4-3-ketone is reacted with ethyl magnesium bromide; base treatment furnishes the 4-methyl-Δ^4-3-ketone together with by-products resulting from reaction of the lactone with two moles of Grignard reagent.

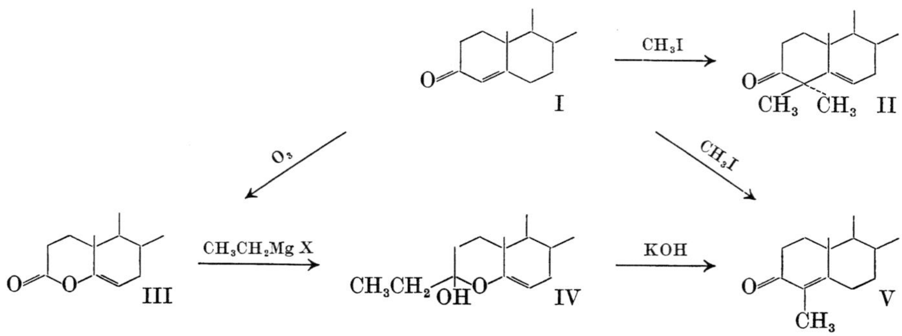

Fig. 77. Formation of 4-methyl derivatives

8. 1-Methyl-19-nor derivatives

The preparation of 1-methyl steroid hormone analogues has not been described, but the 1-methyl-19-nor counterparts of testosterone and 17α-ethinyltestosterone (RINGOLD et al. 1956c) and of progesterone (DJERASSI et al. 1956b) have been prepared. Stereochemical uncertainties preclude any generalizations on the effect of this structural change on the biological properties of the parent hormone; a compound that is very probably 1α-methyl-19-norprogesterone is an active progestational agent.

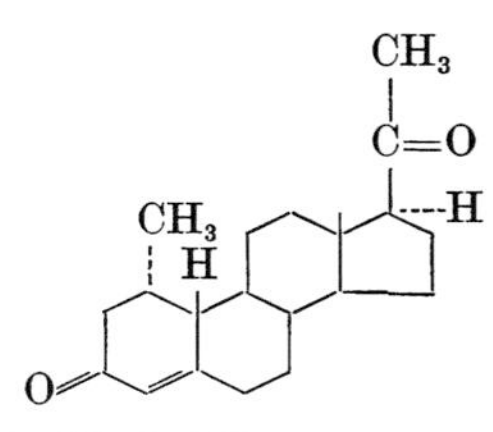

1α-Methyl-19-norprogesterone

The formation of the 1-methyl-19-nor structure is accomplished by Birch reduction of the 1-methyl-3-phenols, which are obtained by acid-catalyzed rearrangement of $\Delta^{1,4}$-dien-3-ones or $\Delta^{1,4,6}$-trien-one (Fig. 78). Rearrangement of a dienone can occur in two ways. In anhydrous media such as acetic anhydride containing *p*-toluene sulfonic acid or zinc chloride, the so-called heterophenol (VI) is the main product (WOODWARD et al. 1953), arising from rupture of the 9—10 bond (III), with rotation about the 5—6 bond prior to the reestablishment of ring B. With aqueous mineral acid, a separable mixture of the heterophenol (VI) and the 1-methyl-3-phenol (V), formed by methyl migration (II), is obtained (DREIDING et al. 1953). The presence of an additional double bond at C-6 suppresses formation of the heterophenol, and rearrangement of the $\Delta^{1,4,6}$-trienone (IV) gives high yields of the Δ^{6}-1-methyl-3-phenol (VII), which is readily converted to (V) by hydrogenation (DJERASSI et al. 1950). It has been stated by KIRK et al. (1957) that the dienone-phenol rearrangement does not take place when an 11-keto group is present in the $\Delta^{1,4}$-dien-3-one or $\Delta^{1,4,6}$-trienone; 11β-hydroxy derivatives have not been studied.

Reduction of the phenol (V) with lithium in ammonia furnishes a mixture of isomeric 1-methyl-19-nor-Δ^{4}-3-ketones. Evidence has been presented by DJERASSI et al. (1956) that both isomers possess the 10β-configuration, and differ in the configuration of C-1. It has been suggested that interaction of a 1β-methyl group with the C-11 methylene group results in deformation of ring B with resultant abnormalities in the rotatory dispersion spectrum. On this basis, the more dextrorotatory and biologically-active isomer of 1-methyl-19-nor-progesterone is assigned the 1α-methyl configuration.

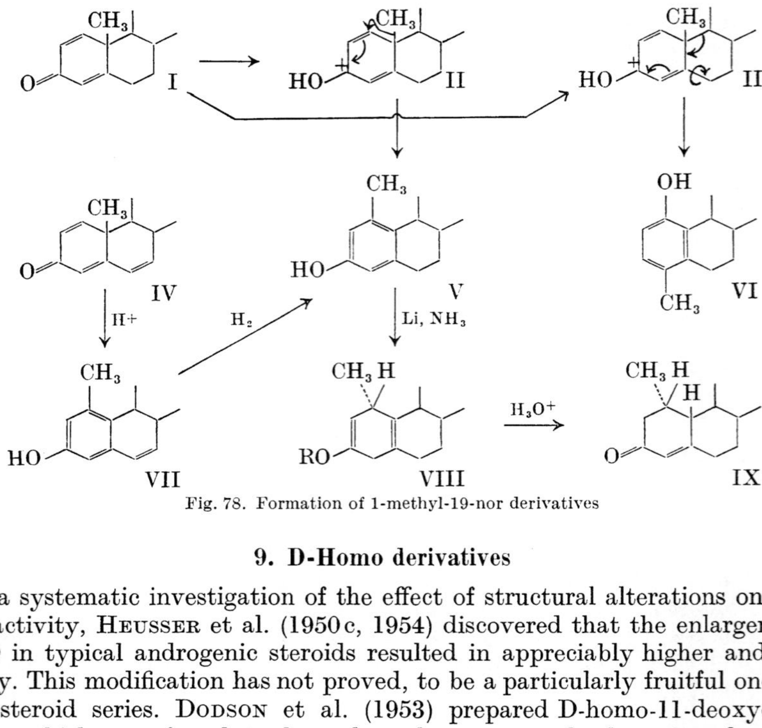

Fig. 78. Formation of 1-methyl-19-nor derivatives

9. D-Homo derivatives

In a systematic investigation of the effect of structural alterations on androgenic activity, HEUSSER et al. (1950c, 1954) discovered that the enlargement of ring D in typical androgenic steroids resulted in appreciably higher androgenic activity. This modification has not proved, to be a particularly fruitful one in the corticosteroid series. DODSON et al. (1953) prepared D-homo-11-deoxycorticosterone, which was found to have less than one-tenth the mineralocorticoid activity of the parent hormone. Very recently, in a more extensive investigation of D-homocorticosteroids, CLINTON et al. (1958b) have reported the synthesis of D-homocortisone, which possesses about one-half the activity of cortisone.

The most useful preparative approach to D-homo-17a-ketones, which represent the starting materials for all of these D-homologated hormones, is the deamination of a 17-amino-methyl-17-hydroxy steroid, obtained by reduction of the cyanohydrin (II and III, Fig. 79). Contrary to predictions based upon the sterochemical requirements of the Demjanov rearrangement, each of the epimeric amines (V) and (VI) furnishes the same mixture of 17a-ketone (VIII) and 17-ketone (IX), in which the former predominates, and the sequence of reactions is best carried out without isolation or separation of intermediates (WENDLER et al. 1955).

The course of addition reactions at the C-17a carbonyl group is less predictable than in the case of the 17-carbonyl in normal steroids (p. 211), due to the different steric environments in the two series. Thus methyl Grignard reagent gives rise to the 17aβ-methyl-17aα-hydroxy compound (IV) (RUZICKA et al. 1952), indicating frontal attack of the reagent, whereas sodium acetylide furnishes a mixture containing mainly the 17aα-ethinyl-17aβ-hydroxy isomer (CLINTON et al. 1958a).

In the elaboration of the D-homo-17aβ,21-dihydroxy-20-ketone side chain, the 17aβ-ethinyl compound (XI) was hydrated to furnish the 17aα-hydroxy-20-ketone (XII). This reaction sequence is precluded in the pregnane series by the unavailability of the 17β-ethinyl isomer. Bromination and replacement by acetate completed the side chain steps. Alternately, the 17aα-ethinyl compound was converted to the vinyl derivative (X) and thence to the Δ^{17}a-D-homopregnene (XIII), which was oxidatively hydroxylated to the dihydroxyketone.

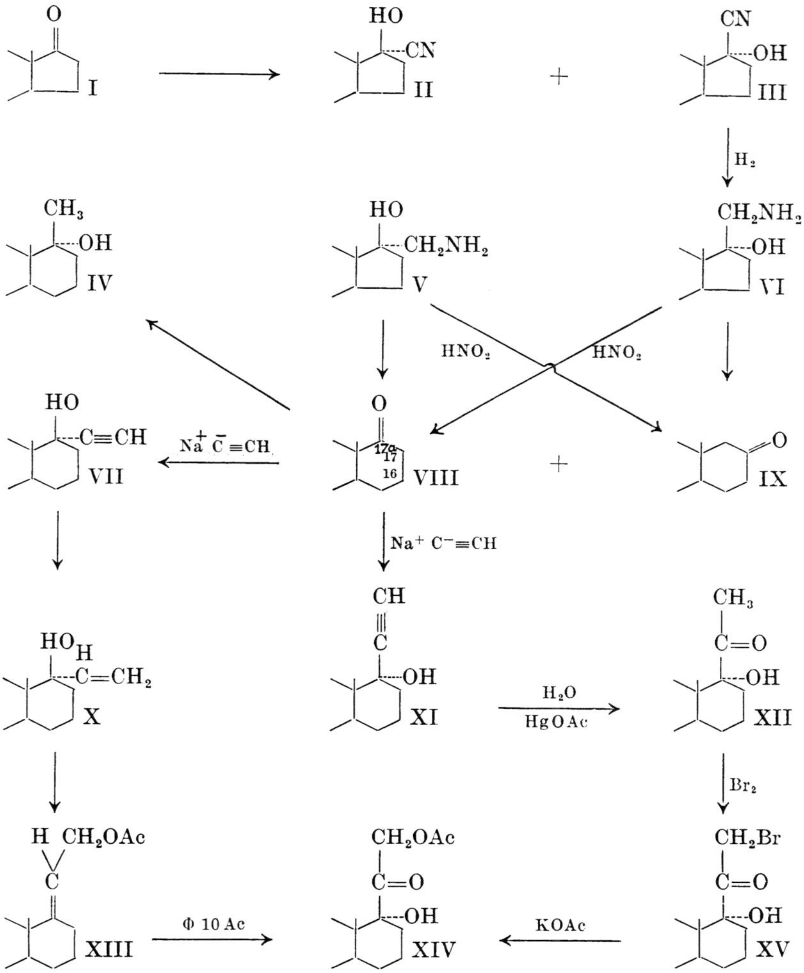

Fig. 79. Formation of D-homo derivatives

10. Compounds with abnormal configuration

Although it has been known for a number of years that certain structural changes, such as alkyl substitution or additional unsaturation, were compatible with biological activity in the steroid hormones, it was generally thought that inversion of an asymmetric center, such as C-17 in progesterone or the adrenocortical hormones, resulted in the complete loss of hormonal properties. More recently, however, several compounds with unnatural configuration at a nuclear center have been prepared which possess an important degree of activity. DJERASSI et al. (1957) have reported the synthesis of 8α-progesterone (I) and 8α-testosterone, both of which have about one-half the potency of the corresponding hormone with the natural 8β-configuration. BARBER and EHRENSTEIN (1957) have described 19-nor-14β,17α-11-deoxycorticosterone and the corresponding 19-nor-14β, 17α-progesterone (II). Although the deoxycorticosterone analogue had negligible mineralocorticoid activity, the norprogesterone isomer, with the unnatural configuration at both C-14 and C-17, was 8 times more active than progesterone, and is one of the most active progestational agents known.

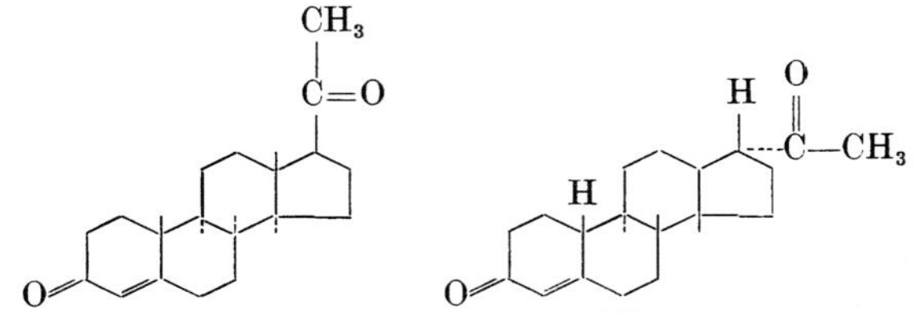

I II

It must be emphasized that the inversion of configuration at a nuclear center is not a minor structural change, since quite a different geometry of the ring system results. Inversion at C-8 requires that ring B assume a boat conformation, with considerable buckling of the nucleus. This alteration is reflected in prominent changes in the optical rotatory dispersion spectrum (DJERASSI et al. 1956c). The distortion of the molecular geometry accompanying inversion at C-14 is somewhat less marked.

In the synthesis of the 8α-hormone isomers, the unnatural configuration at C-8 was generated by catalytic hydrogenation of a $\Delta^{8(9)}$-11-ketone (III, Fig. 80) in the sapogenin series, which furnished the *cis* 8α, 9α B/C ring junction (IV). The 11-keto group was then removed by Wolff-Kishner reduction and the subsequent steps followed the established pattern. Variations of this approach should afford an opportunity for the preparation of several types of isomeric hormone structures, including 9-iso and 11-oxygenated compounds. The preparation of the 14β,17α-19-nor derivatives proceeded from strophanthidin (VI) by a path similar to that outlined in Fig. 72 (p. 282). The 14β,17α-configuration was established

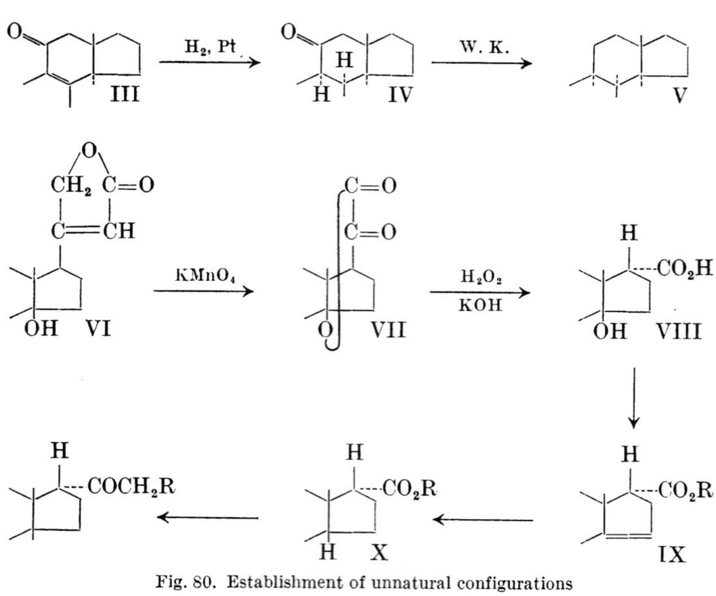

Fig. 80. Establishment of unnatural configurations

by isomerization at the 14β-ol-21-acid lactone stage (VII); the 17α-configuration of the side chain is the more stable when the C/D ring fusion is *cis*, and the α-oriented side chain directs hydrogenation of the Δ^{14} double bond to the 14β-configuration.

11. Triterpenoid analogues

A number of compounds bearing more or less structural resemblance to the adrenocortical steroids have been prepared; a few have been reported to possess in varying degrees some of the biological properties of the adrenal hormones. One of these analogues (I) deserves mention for the interesting structural relationship as well as the biological activity, which is reported equal to that of cortisone in an anti-inflammatory assay (LOGEMAN 1957). The compound is derived from glycerrhetic acid (II), a naturally occurring triterpene. Triterpenoids arise biologically by a path similar to that of the steroids, and possess the same absolute configuration as well as relative configurations at most of the nuclear centers as the steroids. The ketol (I) can thus be looked upon as a genuine corticosteroid homologue, albeit a rather distorted one. The activity of the compound suggests that much more extensive manipulation of the corticosteroid structure than has been explored to date might be compatible with interesting pharmacological activity. Triterpenes with a variety of structural features in ring E are readily available as starting materials for such investigations.

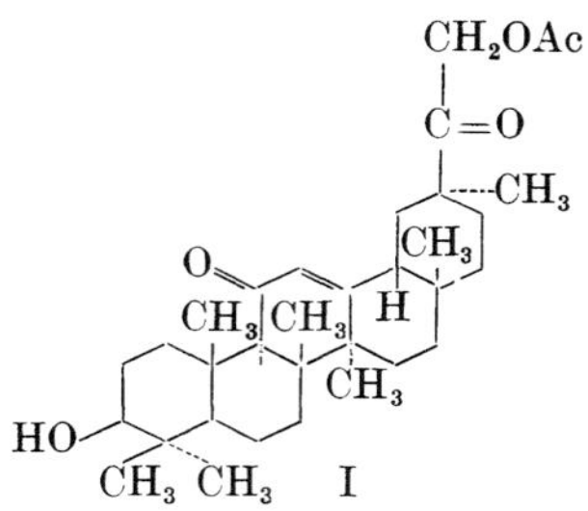
I

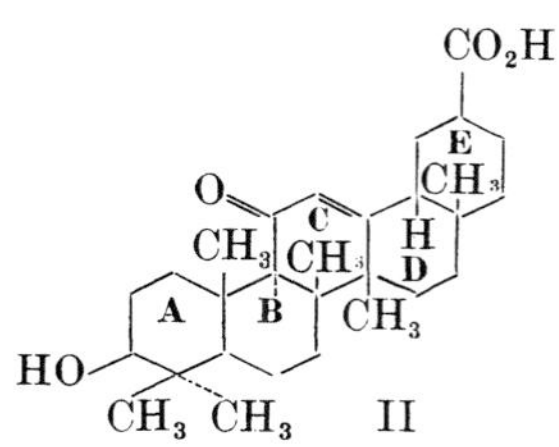
II

Separation and analysis

The structural similarity of members of the same class of steroids makes their separation and specific assay particularly difficult. In addition, the concentrations of steroids in biological extracts are rather low and, to make matters worse, the ratios of the amounts present may be in different orders of magnitude. The reliability of chemical and physical assays depends on their specificity and on the purity of the steroid samples. While it may not be practicable to isolate individual steroids quantitatively, methods are evolving that permit separation of the individual compounds from other compounds that may interfere with their analysis. These have greatly contributed to research on the metabolism of individual adrenal steroids.

In this section an attempt has been made to review the more recent and significant work in this field. Methods representative of current laboratory practice have been described in detail. Review articles of special interest are quoted in the appropriate subsections. Other recent reviews of steroid analysis include a chapter on the isolation, identification, and determination of steroids by OERTEL et al. (1955) and a monograph by ZIMMERMANN (1955). Several review articles on special topics have appeared in *Recent Progress of Hormone Research*, particularly in Vol. IX (1954).

A. Isolation

I. State in biological systems

Steroids of some kind apparently occur in all animal tissues and body fluids, but usually in very small amounts and invariably in mixture with other steroids. Relatively little is known about the chemical form in which they exist originally, but the fact that they can be at least partially extracted by solvents indicates that they are, to some extent, present in the uncombined, free state, i. e., either in solution or in loose combination.

The binding between steroids and proteins in the animal organism has been reviewed by Devis (1955) and by Sandberg et al. (1957). While there is little information concerning the chemical nature of the combination, several methods have been proposed for the isolation of steroids from the steroidoprotein complex. In body fluids, especially in urine, steroids also occur in the form of conjugates. Conjugation involves a hydroxyl group of steroids (usually the C-3 hydroxyl) and another compound such as sulfuric acid or glucuronic acid to form the more water-soluble steroid sulfates or glucosiduronates. Two examples of conjugated steroids are shown in (I) and (II). There is evidence that other, unknown forms of conjugates exist.

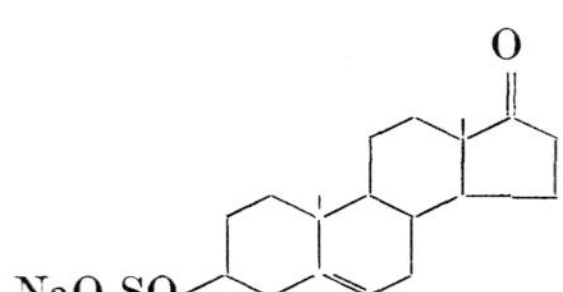

I

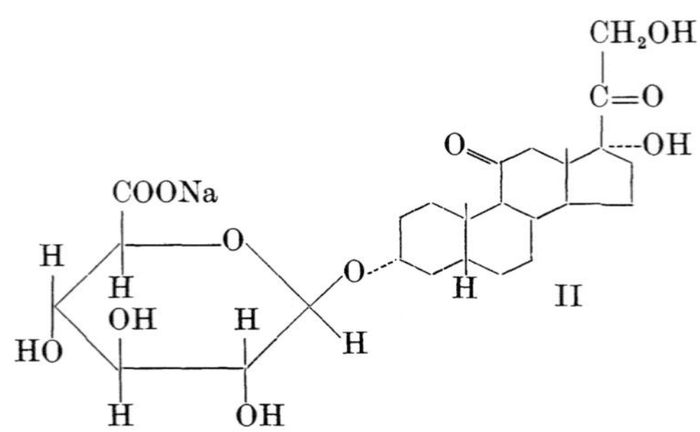

II

Sodium dehydroepiandrosterone sulfate

Sodium salt of tetrahydrocortisone glucosiduronate

II. Isolation of combined steroids

In the isolation of steroids from tissues and body fluids, advantage is taken of their solubility in organic solvents. Combined steroids may be separated as such, or the steroids may be liberated by hydrolysis prior to extraction with organic solvents. The various methods of lipoprotein fractionation are also applicable to the isolation of steroidoproteins, e. g., ammonium sulfate fractionation (Macheboeuf 1929), electrophoresis (Blix et al. 1941), alcohol precipitation at low temperatutes (Oncley et al. 1950), ultracentrifugation (Gofman et al. 1949), and electrochromatography (Boyd 1954, Westphal et al. 1956). Steroids in the form of water-soluble conjugates and uncombined steroids may be isolated from urine by means of butanol extraction. The extraction by butanol is incomplete unless the urine is acidic. Thus Bitman and Cohen (1951) found that only 60% of the 17-ketosteroids present in urine were extracted by butanol at p_H 7.0, while extraction was nearly quantitative at p_H values below 3.

As an example of this technique the extraction of conjugated urinary corticoids (Reddy et al. 1952) as modified by Reddy (1954) may be quoted:

Ten ml of urine in a glass-stoppered centrifuge tube is adjusted to p_H 1.0, saturated with sodium sulfate, shaken with 10 ml of butanol for 5 minutes, and then centrifuged. Five ml of the butanol layer is pipetted into another test tube, shaken with 0.5 g of anhydrous sodium carbonate, and centrifuged. One ml of this solution is used for the Porter-Silber test (p. 341).

JAYLE and CRÉPY (1952, 1954) obtained some fractionation of steroid conjugates by extracting the urine at various p_H values with butanol. A method of extracting conjugated steroids, which does not involve adjustment of p_H and employs low-boiling solvents, was proposed by EDWARDS et al. (1953):

Without adjustment of the p_H, 50 g of ammonium sulfate is added to 100 ml of urine. The urine is extracted three times with 100 ml each of ether-ethanol mixture (3 : 1). The combined extracts are evaporated to dryness *in vacuo*. The residue is dissolved in ethanol and filtered to remove salts.

To separate 17-ketosteroid glucosiduronates from 17-ketosteroid sulfates these extracts may be acetylated (KELLIE and WADE 1956). This treatment renders the glucosiduronates soluble in organic solvents and separable from the water-soluble sulfates.

The isolation of conjugated steroids by ion exchange (A. ANDERSON and WARREN 1951) is an interesting possibility, which has not yet been sufficiently explored. B. LEVIN and DAVIES (1956) have recently reported the separation of steroid sulfates, glucusiduronates, and free steroids on a small scale by electrochromatography on paper.

III. Hydrolysis

1. Steroidoproteins

By far the most widely used approach to the analysis of combined steroids is hydrolysis followed by extraction. If a separate analysis of free steroids is desired, they may be extracted prior to hydrolysis. The method of hydrolysis depends on the nature of the combination and on the nature of the steroids.

Thus, cholesterol and its esters may be liberated from the lipoproteins of blood, milk, etc., by simply mixing and shaking serum, etc., with 3 volumes of a mixture of 12.5 volumes of 70% w/w sulfuric acid and 1 volume of *iso*amyl alcohol and centrifuging in a special tube (HERRMANN et al. 1934). A less drastic procedure involves coagulation of the proteins by brief boiling or homogenizing in at least 20 volumes of solvent mixtures such as ethanol-ether (3 : 1) or chloroform-methanol (1 : 1 or 2 : 1) (SPERRY 1955). For the isolation of neutral steroids bound to serum proteins, J. TAMM et al. (1957) precipitate by heating to 40° with 5 volumes of ethanol. REDDY et al. (1956) recommend the precipitation of serum proteins by zinc hydroxide prior to the butanol extraction of conjugated and unconjugated corticosteroids from the protein-free filtrate. Tissues may be homogenized and dehydrated with acetone prior to extraction (SPERRY 1955). Progesterone was extracted from placental tissue after hydrolysis with 100 ml of 5% sodium hydroxide per 100 g of tissue for 5 days (PEARLMAN and CERCEO 1952). Of course, alkaline hydrolysis is not suitable for adrenocortical hormones. A very gentle method of liberating neutral steroids from plasma proteins is to digest the protein moiety with papain (DEVIS 1955).

2. 17-Ketosteroids

The hydrolysis of urinary 17-ketosteroid conjugates has received a good deal of attention. Much of this work has been reviewed by LIEBERMAN et al. (1954) and by L. ENGEL (1954). Since 17-ketosteroids are relatively stable to hot acid treatment and since such alterations in their structure as do occur have no effect on the colorimetric assay the hydrolysis is rather simple.

Sulfates of 17-ketosteroids are commonly hydrolyzed by acid treatment, either at elevated temperatures or at room temperature while simultaneously extracting the urine with an organic solvent, preferably ether (LIEBERMAN et al. 1954). BUEHLER et al. (1951) have found that 92% of dehydroepiandrosterone sulfate is hydrolyzed in 24 hours at room temperature when an aqueous solution of the steroid sulfate at p_H 0.7 is continuously extracted with ether, and KELLIE and WADE (1957) have reported that 94.6% of dehydroepiandrosterone sulfate is hydrolyzed in 48 hours, if the aqueous solution is adjusted to pH 1.0 and continuously extracted with ether. On the other hand, FRIEDGOOD et al. (1943) compared the yield of 17-ketosteroids from urine hydrolyzed with and without simultaneous extraction and concluded that independent hydrolysis and carbon tetrachloride extraction was superior to simultaneous hydrolysis and carbon tetrachloride extraction.

JENSEN and TÖTTERMAN have compared various hydrolytic procedures (1952a) and proposed a double hydrolysis method (1952b):

A two-hundredth aliquot of a 24-hour urine sample is heated to 100° overnight and then extracted with ether. The extracted aliquot is now heated to 100° for 30 minutes with 0.1 volume of 70% (v/v) sulfuric acid and again extracted with ether. Double hydrolysis gives considerably higher 17-ketosteroid values than single hydrolysis.

POND (1953) has compared this method with the hydrolysis of urine by boiling with one-tenth volume of hydrochloric acid for 10 minutes, followed by extraction with ether. Her results indicate that, although the yield of 3β-hydroxy-17-ketosteroids is greater in the JENSEN-TÖTTERMAN method, there is a decrease in the yield of 3α-hydroxy-17-ketosteroids.

Acid hydrolysis does not involve Walden inversion (LIEBERMAN et al. 1948a), but at elevated temperatures saturated sulfates are converted to Δ^2 (or Δ^3) derivatives. β, γ-Unsaturated 3β-hydroxysteroid sulfates, such as that of dehydroepiandrosterone, heated in neutral medium, form i-steroids, in this case 6β-hydroxy-3,5-cycloandrostan-17-one. In weakly acid medium (acetate buffer pH 5.8 or 4.7) β,γ-unsaturated 3β-hydroxy-steroids are selectively hydrolyzed in 4 hours at 100° (BITMAN and COHEN 1951). An artifact formed during this hydrolysis is the 3β-chloro-5-androsten-17-one. Some of the artifacts produced by acid hydrolysis of 17-ketosteroids at elevated temperatures are shown in Fig. 81.

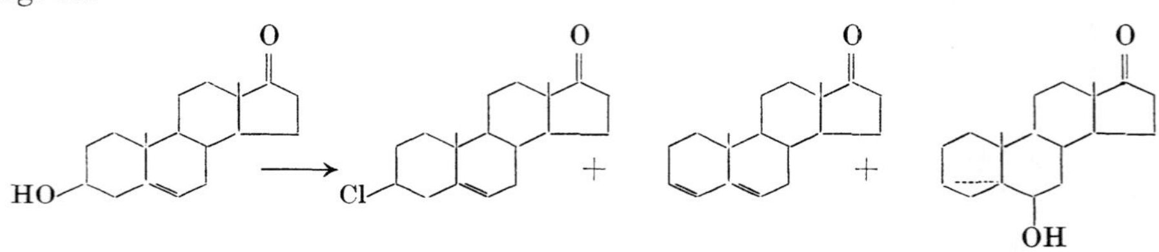

Dehydroepiandrosterone — 3β-Chloro-5-androsten-17-one — 3,5-Androstadien-17-one — 6β-Hydroxy-3,5-cycloandrostan-17-one

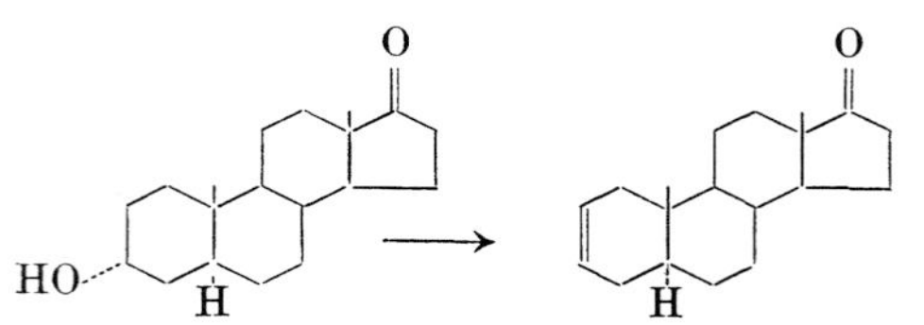

Androsterone — 2-Androsten-17-one

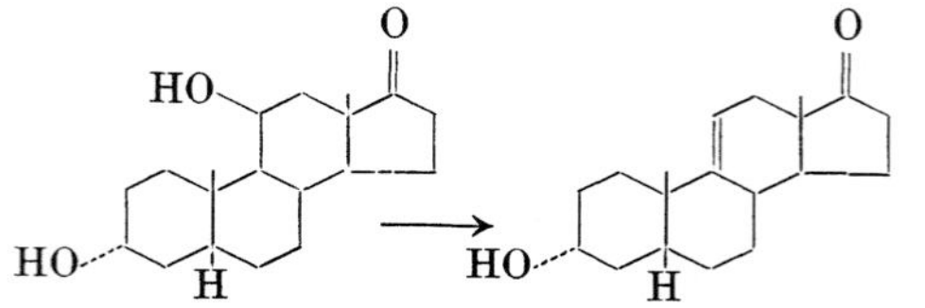

3α,11β-Dihydroxyetiocholan-17-one — $9^{(11)}$-Etiocholen-3α-ol-17-one

Fig. 81. Hydrolysis artifacts of neutral 17-ketosteroids

Buehler et al. (1951) have reported that both sulfates and glucosiduronates of 17-ketosteroids are hydrolyzed with a minimum of destruction by making the urine 7.2 N in acid and extracting continuously with ether. On the other hand, Cohen and Oneson (1953) have obtained better yields, at least of β,γ-unsaturated 3β-hydroxysteroids, by the following procedure:

The urine is adjusted to p_H 2.5—3.0 and continuously extracted overnight with 0.5 volume of butanol. The butanol extract is washed with 4×0.1 volume of water and evaporated to dryness. Concentrates equivalent to 100—400 ml of urine are hydrolyzed either overnight at room temperature or by boiling for 10 minutes with 10 ml of dioxane containing 10% w/v of trichloroacetic acid. The dioxane solution is then neutralized with sodium hydroxide and evaporated to dryness *in vacuo*.

For the hydrolysis of 17-ketosteroids in plasma Gardner (1953) boiled a 6 to 10-ml sample of plasma with 25 ml of 10% hydrochloric acid for 12 minutes.

In view of the large array of hydrolytic procedures for urinary 17-ketosteroids in the literature (L. Engel 1954) and the dependence of results on the hydrolysis method used, the Medical Research Council of Great Britain (1951) proposed the method of Robbie and Gibson (1943), which is rapid and convenient, as a standard method for clinical use:

A 100-ml sample of urine is brought to boiling under reflux over a Bunsen flame. Concentrated hydrochloric acid (10 ml) is added down the condenser and boiling is maintained for 10 minutes. The urine is allowed to cool somewhat and 30 ml of carbon tetrachloride is added down the condenser. The contents of the flask are again kept boiling for 10 minutes. The flask is cooled, the carbon tetrachloride layer is removed and replaced by another 30 ml of carbon tetrachloride, and the mixture is again refluxed for 10 minutes. The two carbon tetrachloride layers are combined.

Hydrolysis of ketosteroid glucosiduronates by acid requires either elevated temperature or high acidity and continuous extraction. The concomitant alterations in the steroid structure may be avoided by use of β-glucuronidase. Several procedures have been described for the hydrolysis of urinary ketosteroid glucosiduronates (Buehler et al. 1951, Cohen 1951, Glenn and Nelson 1953, Henry et al. 1952). Kappas and Gallagher (1955) have combined the glucuronidase and acid hydrolysis with continuous extraction in a two-stage procedure:

Stage 1. The urine is adjusted to p_H 5 with sulfuric acid, and 10% by volume of sodium acetate buffer at p_H 5 together with 300,000 units of beef liver β-glucuronidase per liter are added. After incubation at 37° for 120 hours the mixture is adjusted to p_H 1 with sulfuric acid and continuously extracted with ether for 48 hours. The ether extract is washed with 2N sodium hydroxide, followed by 5% sodium chloride until the washes are neutral.

Stage 2. The spent urine is combined with the washes from Stage 1, adjusted to an acidity equivalent to 1 N sulfuric acid, and continuously extracted with ether for 72 hours.

Jayle and Baulieu (1954) have used the digestive juice of *Helix pomatia*, which contains both β-glucuronidase and sulfatase. In the enzymatic hydrolysis of urinary neutral 17-ketosteroids by the method of Stitch et al. (1956), Stitch and Halkerston (1956) the urine was incubated with an enzyme preparation from *Patella vulgata*, containing sulfatase and β-glucuronidase.

3. Corticosteroids

The hydrolysis of urinary corticosteroids has been reviewed by Mason (1954). The sensitivity of corticosteroids to strong acids and alkalis poses a special problem in the hydrolysis of their conjugates. Although boiling with one-tenth volume (Vanek and Devis 1954) or one-fifth volume (Tompsett 1953) of concentrated hydrochloric acid and even more drastic conditions (Rivoire and Paszkowski 1952, Tompsett and Smith 1955a) or hydrochloric acid-zinc dust treatment (Stimmel 1951b) have been suggested, boiling acid produces artifacts (Fukushima and Gallagher 1957), and in laboratories not using enzymatic hydrolysis the urine is now extracted at room temperature with or without acidification (MARRIAN and ATHERDEN 1953, MARRIAN 1954). Since the chemical assay methods for corticosteroids vary in their specificity, the magnitude of the increase in reactive material produced by a given hydrolytic procedure may vary with the method of determination.

Extraction of urine at p_H 1.0 with chloroform yields not only free adrenal steroids, but apparently also some conjugated material that is not hydrolyzable with glucuronidase (MARRIAN et al. 1953). While extraction of untreated urine (TALBOT et al. 1945) presumably yields only unconjugated steroids, it was found (HEARD et al. 1946) that the yield of corticosteroids increases progressively as the acidity of the urine is increased to p_H 1. It has been common practice to adjust the urine to p_H 1.0 and extract it either immediately (DAUGHADAY et al. 1948) or after allowing it to stand at room temperature (ROBINSON and NORTON 1951), which increases the yield. While some of the material is slowly released, there is apparently some destruction of material extractable immediately after acidification (BAYLISS 1952). DYRENFURTH and VENNING (1957) found that continuous extraction of urine at p_H 1.5 and room temperature with chloroform for 48 hours was more effective than β-glucuronidase hydrolysis for the isolation of aldosterone.

KLEIN et al. (1955) have observed that additional corticosteroids may be extracted from laked blood, which has been extracted according to the method of NELSON and SAMUELS (1952) (see below), after shaking it with concentrated hydrochloric acid for 15 minutes.

The discovery that the enzyme β-glucuronidase splits steroid glucosiduronates has virtually revolutionized the analysis of corticosteroids. This method is now a standard procedure in most laboratories (COHEN 1951, KINSELLA and GLICK 1953, GLENN and NELSON 1953, HORWITT 1954, BONGIOVANNI et al. 1954). As an example, the GLENN-NELSON procedure is given, which is applicable to the glucosiduronates of 17-ketosteroids and corticosteroids:

The urine is adjusted to p_H 4.5 (if incubated at 37.5° for 28 hours) or to p_H 5.0 (if incubated at 47° for 15 hours) by addition of acetate buffer. For a 30-ml aliquot 10,000 Fishman units of commercial beef liver β-glucuronidase and 25,000 units of penicillin are added.

For the determination of glucuronic acid-bound corticosteroids in urine STAUDINGER and BAUER (1954) combined glucuronidase hydrolysis with dialysis:

A dialysis bag, 5 cm in diameter, is filled with 20 ml of urine, buffered at p_H 4.5, to which 1500 units of spleen or liver glucuronidase has been added. The bag is placed in a beaker containing 80 ml of acetate buffer at p_H 4.5 and 30 ml of chloroform. The beaker is incubated at 37° for 72 hours, fresh buffer and chloroform being substituted after 24 and 48 hours. The combined dialyzates are extracted 5 times with 30-ml portions of chloroform.

IV. Extraction

1. Solvents

Free steroids or steroids liberated by hydrolysis are usually isolated from tissues and body fluids by solvent extraction. The general principles and methods of extraction have been discussed by CRAIG et al. (1956). For technical information on solvents and their purification the reader is referred to WEISSBERGER et al. (1955). Only some aspects germane to steroid analysis will be dealt with here.

Far too little work has been done on partition coefficients of steroids, and the choice of solvents is usually based on personal preference rather than on their efficiency. A comparison of the efficiency of chloroform and ethyl acetate in the extraction of hydrophilic steroids from aqueous solutions (BURSTEIN 1956a) sheds doubt on analytical results based on chloroform extraction and on the practice of washing chloroform extracts (see also TOMPSETT and SMITH 1955b). Ideally, a solvent used in the isolation of steroids from aqueous media should not only be efficient but it should also: 1) have a boiling point that is conveniently low for evaporation without destruction of the extractives, 2) have a density sufficiently different from water to allow separation with a minimum of emulsification,

3) be nontoxic, nonexplosive, and preferably nonflammable and 4) be free of interfering impurities. Thus ether should be free of peroxides, chloroform free of phosgene, alcohol free of aldehydes, etc.

The extraction is usually carried out by shaking the aqueous solutions with organic solvents in separatory funnels. The number of extractions and the volume ratio of aqueous to organic phase depend on the partition ratios of the steroids. Emulsions, which are frequently encountered, are broken by addition of more solvent, centrifugation, additions of salts, alcohol, etc., or by drying the extracts with sodium sulfate, etc. Stopcock lubricants may interfere with the analyses. Tube-separatory funnels eliminate shaking and venting and minimize the formation of emulsions (NADEL and HIGHHOUSE 1954) (Fig. 82).

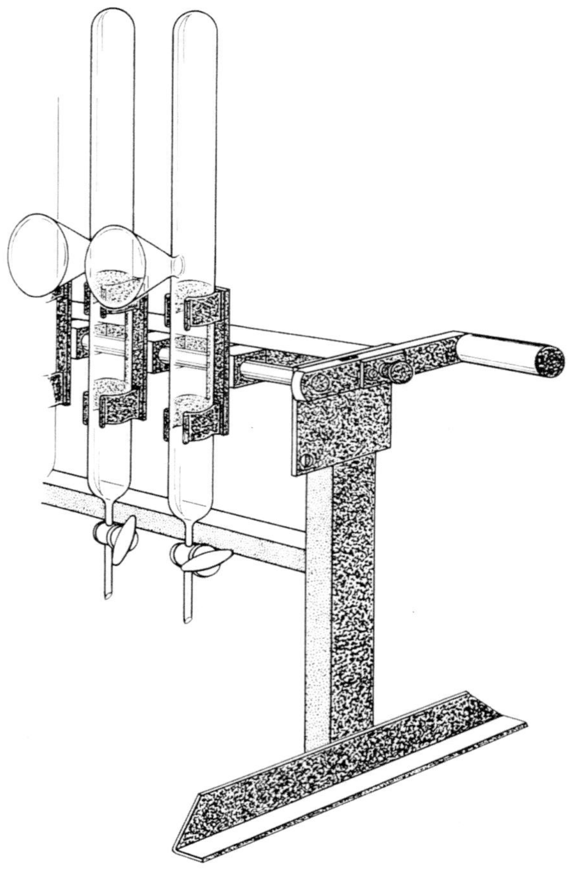

Fig. 82. Tube-separatory funnels (NADEL and HIGHHOUSE 1954). A series of tubes is mounted on a rack and can be rocked simultaneously either by hand or by a motor. The funnel serves for filling as well as venting the tubes. Layers are withdrawn through the stopcock. Redrawn by permission of the authors and J. T. Baker Chemical Co. from the Chemist-Analyst **43**, 25 (1954)

2. Extractors

A number of extractors have been described for the isolation of steroids, which offer advantages of automatic or continuous operation, absence of emulsions, low operating temperature, etc. HILL (1952) used a glass tube 3×120 cm, provided with a stopcock at the bottom and containing 300 ml of ether to extract 17-ketosteroids from 300 ml of hydrolyzed urine, which was passed through the ether in a fine stream 3 times. ROBINSON and NORTON (1951) (Fig. 83) and COOK and ROOKS (1952a) (Fig. 84) have designed tubes that can be used with solvents heavier than urine. The tubes are filled with chloroform and the urine is allowed to pass through the columns several times. COHEN (1950) (Fig. 85) described a continuous extraction apparatus suitable for extraction of urine with either heavier or lighter solvents. The lighter liquid rises in dispersed form in a column containing the heavier one and collects on top. It is drawn into a dispensing chamber above the column by vacuum through a tube so arranged that the vacuum is broken when the transfer is completed. Similar apparatus has been described by NEWHOUSE (1953) and by HADD and PERLOFF (1957). While these extractors have the advantage of operating at room temperature, the extraction proceeds only to an equilibrium between urine and solvent which depends on the partition coefficient.

Continuous extraction with fresh solvents gives more nearly complete extraction, since it is not an equilibrium process. Simultaneous hydrolysis and extraction (LIEBERMAN et al. 1948a and 1954, BUEHLER et al. 1951) is also advantageous. Many continuous extractors for organic solvents lighter (WYLIE and SCHMULOVITZ 1936, TALBOT and LANGSTROTH 1939, NETERVAL 1940, WILSON et al. 1940) or heavier [HERSHBERG and WOLFE 1940, 1941 (Fig. 86), REPETTO and DREWES 1947, PIERCE and PETERSON 1956] than urine have been described.

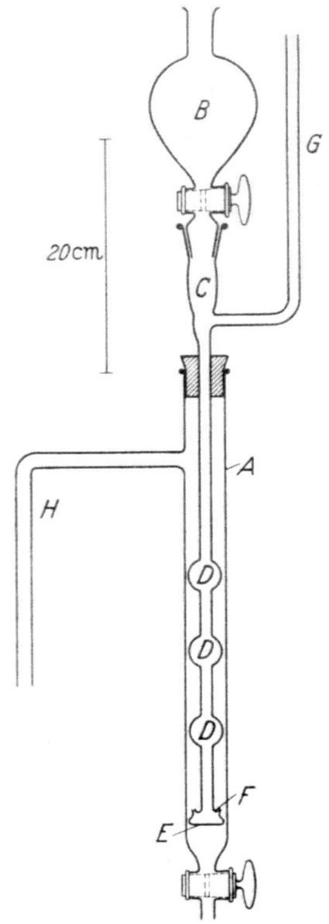

Fig. 83

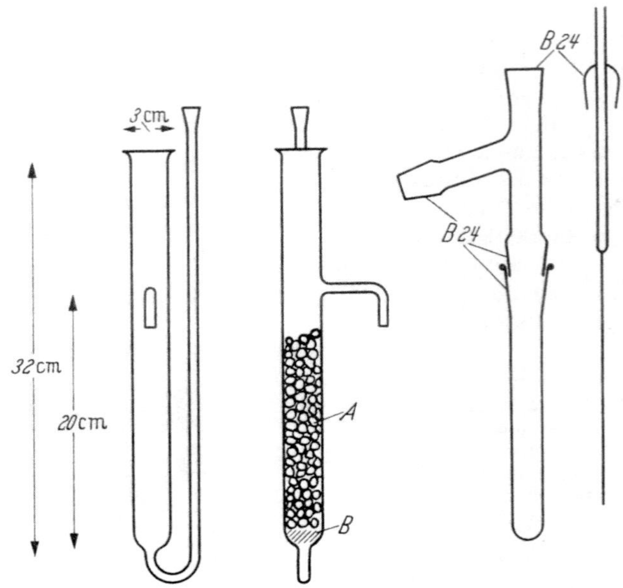

Fig. 84

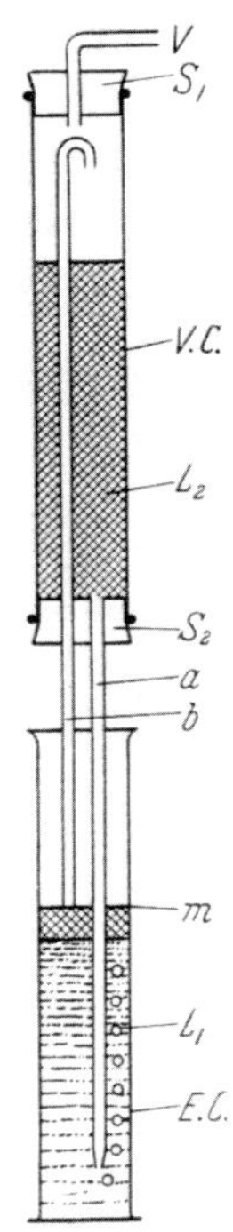

Fig. 85

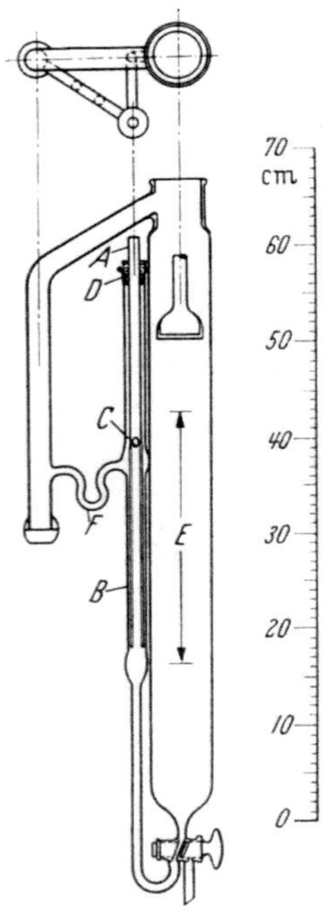

Fig. 86

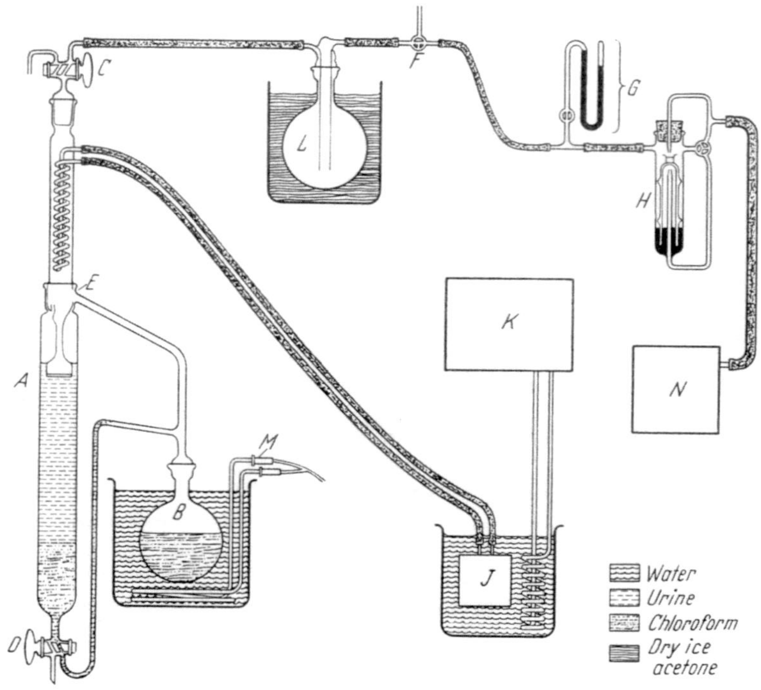

Fig. 87

Fig. 83. Extraction apparatus (ROBINSON and NORTON 1951). Tube A is filled with chloroform, the inner tube C is fitted, and urine is allowed to flow slowly from funnel B until C is full. The stopcock is then opened fully and air locks are removed through G. The urine issues from 4 or 5 jets, F, in the terminal bulb, E, rises past the baffles, D, and overflows through H. Urine is passed through the apparatus at least 10 times. Redrawn by permission of the authors and of Cambridge University Press from the Journal of Endocrinology **7**, 323 (1951)

Fig. 84. Extractor and distillation head (COOK and ROOKS 1952a). Two side views of the extractor (left) and two details of the distillation head (right) are shown. The inlet is covered with a pad of glass wool, B, and the tube is partially filled with glass beads, A. The beads are covered with 40 ml of chloroform, and 100 ml of urine is passed in through the bottom and out through the side arm 20 times. The distillation head is fitted with a glass capillary (extreme right), which acts as an air leak when the chloroform extract is evaporated in the tube under vacuum. Redrawn by permission of the authors and the Society for Analytical Chemistry from The Analyst **77**, 274 (1952)

Fig. 85. Extraction apparatus (COHEN 1950). The heavier liquid L_1 is placed in the extraction column E. C. and the lighter liquid L_2 poured on top of L_1. Gentle vacuum applied at V draws L_2 through tube b into the vacuum chamber V. C. until its level falls below level m. Then L_2 passes in a thin stream from the orifice in tube a through L_1 and again collects on top. This closes tube b, again creates a vacuum in V. C., and the cycle is repeated. Redrawn by permission of the author and the C. V. Mosby Company from The Journal of Laboratory and Clinical Medicine **36**, 770 (1950)

Fig. 86. Continuous extractor (HERSHBERG and WOLFE 1941). The urine in E, which may be heated, is continuously extracted by droplets of chloroform issuing from a sintered glass disc below its surface. The extract passes through B and F into a boiling flask (cf. Fig. 87), and the chloroform vapor passes through the side arm into a condenser (cf. Fig. 87). Liquid chloroform again drips through the sintered glass disc, completing the cycle. This apparatus is improved by an overflow device, consisting of a barrel B, loosely fitting cork D, and plunger A with an overflow hole C, which permits compensation for differences in specific gravity of solvents. Redrawn by permission of the authors and of the American Society of Biological Chemists, Inc., from the Journal of Biological Chemistry **141**, 216 (1941)

Fig. 87. Continuous low-temperature extractor (HEFTMANN and JOHNSON 1953). The Hershberg-Wolfe extractor (cf. Fig. 86) is modified by operation at reduced pressure. The assembly consists of a Hershberg-Wolfe extractor (A—E), a vacuum system (vacuum pump N, manostat H, manometer G, vent F, and trap L), a bath heated by the immerson heater M, and a cooling system (refrigerator K and circulating pump J). Redrawn by permission of the American Chemical Society from Analytical Chemistry **25**, 1578 (1953)

Attempts have made to overcome the disadvantage of these extractors, viz., prolonged heating of the extract, either by placing the extractor under vacuum so that the solvent can be boiled at low temperature (HEFTMANN and JOHNSON 1953) (Fig. 87), or by flash evaporation of the extract (DOBSON and RANDALL 1951) (Fig. 88).

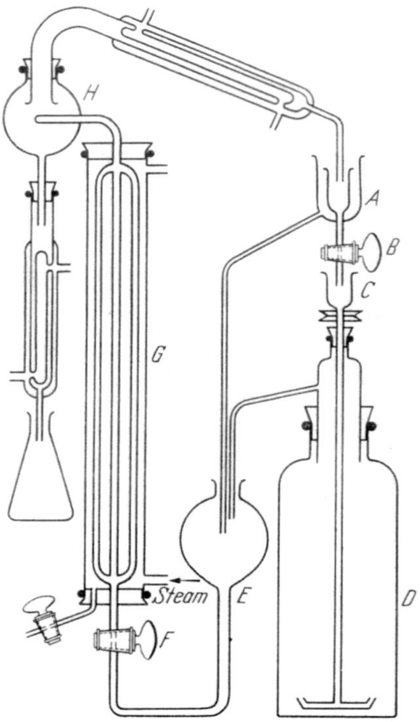

Fig. 88. Continuous extractor and evaporator (DOBSON and RANDALL 1951). A solvent lighter than urine passes from *A* through *B* into the extractor *D*, where it is dispersed by the hollow turbine stirrer *C*. The extract collects at the top and overflows into the reservoir *E*. Extract and excess solvent from *A* are admitted to the climbing-film evaporator *G* through stopcock *F*. Steam heat evaporates the solvent and the spray passes into a cyclone *H*. The centrifugal effect separates the concentrate from the vapor. The former is cooled and collected in a flask, while the latter rises in the cyclone and then condenses to complete the cycle. Redrawn by permission of the authors and Cambridge University Press from the Biochemical Journal **49**, 399 (1951)

3. Isolation from blood and tissues

The isolation of corticosteroids from blood has been reviewed by LEVY and KUSHINSKY (1954). Their work on the adsorption of corticosteroids in adrenal perfusates on activated charcoal (Darco G-60) and subsequent elution with dichloromethane and benzene showed that up to 25% of the steroids are irreversibly adsorbed or destroyed by the charcoal treatment. LOMBARDO et al. (1955b) found that various C_{19} and C_{21} steroids could be quantitatively recovered from activated charcoal (Mallinckrodt) by elution with 40% benzene in ethanol. This treatment does not produce artifacts and effectively removes urinary pigments.

It is also possible to extract the steroids in blood directly with solvents. NELSON and SAMUELS (1952) give the following directions:

Blood is laked by diluting with an equal volume of water and then extracted 4 times with an ether-chloroform (4 : 1) mixture by stirring with a glass rod in centrifuge bottles. The extracts are separated by centrifugation, combined, and brought to dryness. The residue is taken up in 20 ml of 70% aqueous ethanol and washed 3 times with 15 ml of hexane. For the extraction of plasma 3 portions of 1.5 volumes of chloroform are used. The recovery of added 17-hydroxycorticosterone in this method ranges from 83 to 112%.

MEYER (1953) recovered about 90% of cortisone added to whole blood-adrenal perfusates by ethyl acetate extraction. This solvent gives less trouble with emulsions than the ether-chloroform mixture.

AXELROD and ZAFFARONI (1954) (see also ZAFFARONI 1953, NOWACZYNSKI and GENEST 1955, DE VENUTO et al. 1957) have made a study of the isolation of corticosteroids from blood tissues by dialysis.

Separatory tubes, 5.5×26 cm (Fig. 89) are filled with 200 ml of 40% v/v aqueous methanol and 50 ml of chloroform. Dialysis bags, containing 150—165 ml of a blood-methanol-water mixture (1 : 1 : 1), or tissue homogenate (1 part adrenal gland per 3 parts of buffer) instead of blood, and a glass bead are agitated in the tubes by inversion on a wheel rotating around a horizontal axis for 48 hours at 20°. The chloroform extract is drawn off, and the dialyzate further extracted with 3 portions of 50 ml chloroform. Recoveries of added corticosteroids range from 60 to 91%.

LOMBARDO et al. (1955a) have combined dialysis with continuous extraction:

A Hershberg-Wolfe extractor is charged with 1—4 dialysis bags of 100—125-ml capacity and 60% v/v aqueous methanol. With blood-water-methanol mixtures (1 : 1 : 1) inside the dialysis bags continuous extraction by dichloromethane yielded 84.4—103.9% of added corticosteroids.

Generally, tissues may be extracted by methods similar to those already described for blood after they have been homogenized. For the isolation of adreno-cortical hormones from tissues the procedure given by HAINES (1952) may serve as an example. The purpose of the process is the removal of lipides, which are relatively more soluble in petroleum ether, and of acidic and basic impurities.

Frozen hog adrenals were ground into acetone, extracted for 6 days and filtered. The acetone filtrate was diluted with water first to 50, then to 40, and finally to 20% and extracted each time with petroleum ether (boiling range 60—70°). The corticosteroids in the aqueous acetone fraction were extracted with 1,2-dichloroethane, which was then evaporated. The residue was taken up in ethanol and partitioned in succession between petroleum ether and 70, 50, and finally 30% aqueous ethanol. From the aqueous ethanol the corticosteroids were then extracted by ethyl acetate, which was washed, in succession, with dilute alkali, dilute acid, and water.

BONGIOVANNI et al. (1954, 1955b) have demonstrated an increase in extractable corticosteroids after incubation of plasma with β-glucuronidase:

The free corticosteroids were isolated by precipitation of plasma with 2.5 volumes of 95% ethanol, filtering, evaporating the alcohol almost to dryness at 45°, adding to the residue a volume of water equal to that of the plasma used and extracting the aqueous solution 3 times with two-thirds volume of dichloromethane. The conjugated corticoids were liberated by incubating the aqueous solution at p_H 4.5 and 37° for 48 hours with 300—500 units of β-glucuronidase per ml and then extracting with dichloromethane as above.

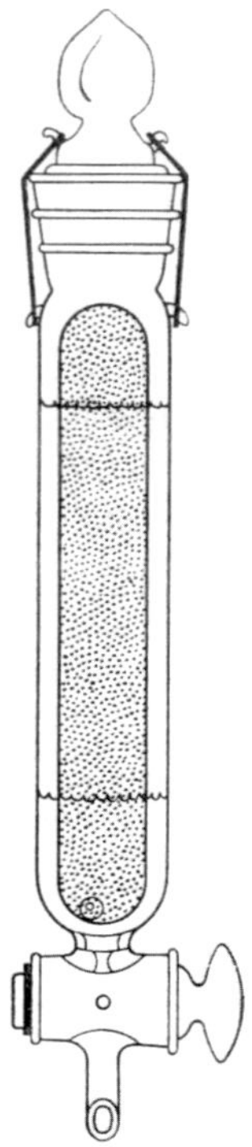

Fig. 89. Separatory tube (AXELROD and ZAFFARONI 1954). Tubes of about 450 ml capacity are mouted on a wheel with horizontal axis and can be inverted by rotation of the wheel. The dialysis bags contain the blood or tissue homogenate and a glass bead. The stopper is held in place by metal springs. A stopcock permits withdrawal of the dialyzate. Redrawn by permission of the authors and Academic Press, Inc., from Archives of Biochemistry and Biophysics **50**, 348 (1954)

B. Separation

I. Preliminary purification of extracts

The purpose of the preliminary fractionation of steroid extracts is to remove as much as possible substances that will interfere with the subsequent analysis. The degree of purification required depends on the assay procedure. The resulting fractions are still rather crude, but should contain a group of substances having some common chemical characteristic in quantitative yield.

A fractionating scheme of general applicability was developed by DOBRINER et al. (1948) for the isolation of urinary steroids:

The ether extract of hydrolyzed urine is extracted with 10% sodium bicarbonate solution and then with 2*N* sodium hydroxide solution to remove, respectively, carboxylic acids and phenols. The neutral ether extract is washed, brought to dryness, and separated into a ketonic and non-ketonic fraction with Girard's reagent T. The neutral non-ketonic fraction is further subdivided into an alcoholic and a non-alcoholic fraction by treatment with phthalic anhydride. Both the ketonic and the non-ketonic alcoholic fraction are then divided into 3α- and 3β-hydroxy steroid fractions with digitonin.

Modifications and improvements of this basic scheme will now be described in detail. In the determination of neutral 17-ketosteroids in urine the object of the extraction with alkali is the removal of carboxylic acids and phenols, e. g., estrone, which may interfere with the colorimetric determination. Usually 1 *N* sodium hydroxide solution is used. Washing with 10% sodium hydrosulfite solution in 1 *N* sodium hydroxide is effective in removing much of the urinary pigments. DREKTER et al. (1952) have recommended shaking a 1,2-dichloroethane extract of urine with sodium hydroxide pellets for this purpose. While 17-ketosteroids are resistant to alkali treatment, the washing of extracts for corticosteroid assay must be carried out with more dilute (0.1 *N*), preferably ice-cold, sodium hydroxide solution (TALBOT et al. 1945).

The separation into a ketonic and a non-ketonic fraction by Girard's reagent T depends on the conversion of fat-soluble ketonic compounds to water soluble hydrazones

$$Me_3\overset{+}{N}\overset{-}{Cl}CH_2CONHNH_2 + O{=}C< \longrightarrow Me_3\overset{+}{N}\overset{-}{Cl}CH_2CONHN{=}C<$$

The non-ketonic steroids remain in the organic solvent, while the hydrazones go into the aqueous phase. The separation should be carried out rapidly in the cold, since the hydrazones are readily hydrolyzed. The ketones can be recovered by solvent extraction after acid hydrolysis. PINCUS and PEARLMAN (1941) proceed as follows:

The extract from 100 ml of urine is thoroughly dried in a vacuum desiccator over calcium chloride. Then 0.5 ml of glacial acid and approximately 100 mg of recrystallized Girard's reagent T are added and the mixture is heated in an oil bath at 90—100° for 20 minutes. After it has cooled, 15 ml of ice water is added and the reaction mixture is immediately transferred to a cold separatory funnel. Sufficient 10% sodium hydroxide solution is added to neutralize nine-tenths of the acetic acid and 3 extractions with 20-ml portions of ether are made. The ether extracts are combined and washed once with 10 ml of ice water. The wash water is combined with the rest of the aqueous ketonic fraction. The ether extract, containing the non-ketonic fraction, is washed with 10 ml of 2.5% sodium carbonate and three further 10-ml portions of water. To hydrolyze the hydrazones the ketonic fraction is acidified with 3 ml of concentrated hydrochloric acid, allowed to stand at room temperature for 2 hours, and then extracted 3 times with 20-ml portions of ether. The ether extract, containing the ketonic fraction, is washed with 10 ml of 2.5% sodium carbonate and 3 10-ml portions of water.

For the separation of steroids into an alcoholic and a non-alcoholic fraction various half-esters have been used, e. g., hemiphthalates and hemisuccinates. These can be separated from the mixture by extraction with alkali. If necessary, the esters may be saponified to regenerate the original alcoholic steroids.

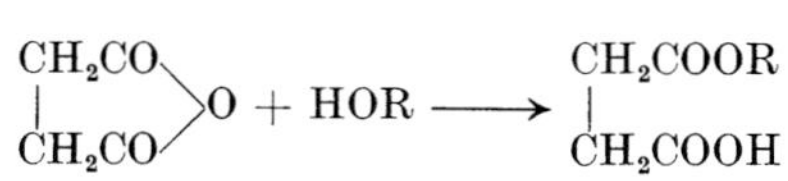

PINCUS and PEARLMAN (1941) give the following directions for preparing hemisuccinates of urinary alcoholic steroids:

The residue from the extraction of 100 ml of urine is thoroughly dried over calcium chloride in a vacuum desiccator. Approximately 0.5 g of succinic anhydride and 2 ml of dry pyridine are added. The mixture is heated at 90—100° for 2 hours in an oil-bath with occasional stirring. About 10 ml of water is then added with stirring and the reaction mixture is allowed to stand in the oil-bath for another half-hour. After it has cooled, the mixture is extracted with 20 ml and then twice with 10-ml portions of ether. The combined ether layers are washed 4 times with 5-ml portions of 2 *N* hydrochloric acid, then twice with 5-ml portions of water. The hemisuccinates are extracted from the ether solution with one 20-ml and then with three 5-ml portions of 0.1 *N* potassium carbonate, containing 5% sodium chloride. The ether, containing the non-alcoholic steroids, is washed 4 times with 5 ml of water each. The potassium carbonate extract is made acid to Congo red with 2 *N* hydrochloric acid and extracted once with 20 ml, then twice with 10 ml each of ether. The ether extract of hemisuccinates is washed with four 5-ml portions of water.

Hemiphthalates of urinary alcoholic steroids were prepared by DOBRINER et al. (1948) as follows:

The dried urine extract is dissolved in 1 ml of anhydrous pyridine per 100 mg of residue and 100 mg of freshly-fused phthalic anhydride is added. The mixture is heated on an oil bath at 130° for 3 hours under a reflux condenser protected with a calcium chloride tube. Next, the reaction mixture is poured into ice-water and extracted 3 times with ethyl acetate. The combined extracts are washed with 10% sulfuric acid. To separate the hemiphthalates from non-alcoholic steroids the ethyl acetate is extracted 6 times with $2N$ sodium hydroxide and washed several times with water. The combined sodium hydroxide extracts, containing the hemiphthalates, are made acid to Congo red with 10% sulfuric acid and extracted with 3 portions of ethyl acetate. The ethyl acetate solution is washed 5 times with water and evaporated under reduced pressure.

The hemiphthalates in the residue are hydrolyzed as follows:

Per 100 mg of residue, 5 ml of benzene and 1 ml of absolute ethanol are added and the solution is heated to boiling. Then, 1.5 ml of a hot 10% sodium methoxide solution is added and the heating is continued under reflux for 10 minutes with frequent agitation. After it has cooled, an excess of water is added to the mixture and 3 extractions with ethyl acetate are made. The combined extracts are washed several times with $2N$ sodium hydroxide and with water. The ethyl acetate is finally dried over sodium sulfate and then evaporated *in vacuo*.

L. ENGEL et al. (1950) have used a similar method in the determination of alcoholic steroids from urine in the form of their 3,5-dinitrohemiphthalates:

To the thoroughly dried urine extract (equivalent to one-tenth of the 24-hour collection from a person) 24 mg of 3,5-dinitrophthalic anhydride and 0.2 ml of pyridine are added. The mixture is heated on the boiling water-bath for 15 seconds, then cooled, diluted with 10 ml of $0.1\,N$ hydrochloric acid, and extracted once with 10 ml and 4 times with 5 ml of ether. The combined ether extracts are washed with 3 5-ml portions of $0.1N$ hydrochloric acid and 5 5-ml portions of water and used for subsequent analysis.

The separation of alcoholic steroids into α-hydroxy and β-hydroxy compounds by digitonin depends on the formation of an insoluble digitonide by steroids containing a free 3β-hydroxyl group. The method works well for urinary dehydroepiandrosterone, but some 3β-hydroxy compounds are not precipitated. Moreover, some 3α-hydroxy compounds are also known to form insoluble digitonides. FRAME (1944) gave the following directions for the digitonin separation of urinary 17-ketosteroids:

The urine extract, containing 1 mg of neutral 17-ketosteroids, but less than 0.5 mg of the 3β-hydroxy fraction, is dried in a centrifuge tube and 1 ml of warm 1% digitonin solution in 90% ethanol is added. After warming to effect solution, the tube is allowed to stand overnight. The digitonides are precipitated by addition of 10 ml of anhydrous ether, added in small portions with stirring. The precipitate is centrifuged off and washed with 3 10-ml portions of ether, stirring and centrifuging after each addition. The combined supernatant liquids, containing the 3α-hydroxy steroids, are washed 3 times with 25-ml portions of water, then evaporated and dried. The precipitated digitonides are decomposed by dissolving in 0.5 ml of dry pyridine and heating in a water-bath to 60—70° for 3 minutes. After cooling, 10 ml of anhydrous ether is added to precipitate the digitonin. This is centrifuged and the pyridine-ether treatment of the precipitate is repeated. The remaining precipitate is washed with 2 portions of 10 ml of ether each, stirring and centrifuging as before. The combined supernatant liquids containing the 3α-hydroxy steroids are washed 3 times with 20 ml of $2N$ sulfuric acid and 3 times with 25 ml of water, then dried and evaporated.

A similar procedure was also described by BUTT et al. (1948).

II. Experimental separation methods

In this section are considered two methods of separation that have not yet found general application in the analysis of neutral steroids except for some individual experimental work: countercurrent distribution and electrochromatography. The possibilities of these two methods have been discussed by SCHROEDER and VOIGT (1955), and VOIGT et al. (1955) have reviewed electrochromatography of steroids in comparison with other chromatographic methods.

1. Solvent distribution

Separation of steroids from other lipide-soluble material or resolution of steroid mixtures by partition between partially miscible solvents is less destructive than other fractionation methods, and solvent distribution is a valuable step in isolation work. However, unless there is a large difference in the partition coefficient of the substances that are to be separated, a small number of distributions is inadequate. In view of the dearth of information on the partition coefficients of nonphenolic steroids in various solvent systems, one may well doubt the validity of quantitative procedures in which solvent partion is used for either the purification or fractionation of extracts. It has been reported (e. g., VENNING et al. 1953) that a portion of the corticosteroids is removed by the customary sodium hydroxide wash of urine extracts, and MASON (1950) has pointed out that some cortisone is lost when urine extracts are partitioned between water and benzene, as is recommended for the separation of more polar from less polar steroids (TALBOT et al. 1945, LLOYD and LOBOTSKY 1950, EIK-NES 1957). Similarly, the defatting of steroid extracts by use of petroleum ether, and the separation of 11-hydroxy from 11-deoxycorticosteroids (WEISSBECKER and STAUDINGER 1951) by partition between petroleum ether and water are not likely to give clear-cut separations.

On the other hand, it is possible to exploit even slight differences in partition coefficients for efficient separation, provided the number of distributions is sufficiently great. This can be accomplished either manually with separatory funnels or mechanically in countercurrent distribution machines of various design. Countercurrent distribution (CRAIG et al. 1956, HECKER 1955) has not been used very extensively in the separation of nonphenolic steroids, although it is a very useful tool for the preliminary fractionation of relatively large quantities of complex mixtures (PEARLMAN 1954, L. ENGEL et al. 1955).

Table 5 summarizes partition coefficients of adrenal hormones from the literature. Table 6 contains partition coefficients at $25 \pm 3°$ of some 17-ketosteroids from the work of L. ENGEL et al. (1954).

2. Electrochromatography

Ionized solutes can be separated in an electric field on the basis of their electrical charge and mobility. When a mixture containing various ionic species is applied to a narrow zone of a stabilized electrolyte (e. g., a filter paper strip impregnated with buffer solution), the passage of an electric current through the electrolyte will cause the components of the mixture to migrate. If their migration rates are sufficiently different, the components will form separate zones (an electrochromatogram, electropherogram, etc.), which may be analyzed by suitable methods. For the general techniques of electrochromatography, especially paper electrophoresis, the reader is referred to the various textbooks in that field (e. g., BLOCK et al. 1955, McDONALD 1955, LEDERER 1955).

While the neutral steroids are not separated by electrochromatography in nonaqueous solvents (PAUL and DURRUM 1952), the mobility of water-soluble derivatives in the electric field has been investigated. In anionic detergent solution cortisone apparently moves in the electric field in the form of a steroid-detergent complex (SCHMID and VOIGT 1954). Hemiphthalates of deoxycorticosterone and dehydroepiandrosterone are separated by paper electrophoresis in a phosphate buffer at p_H 7.5 containing 20% ethanol (VOIGT and BECKMANN 1953, 1954). McKINLEY (1955) found the following electrophoretic mobilities for the hydrazones of steroids with Girard's reagent T in 0.05 M borate buffer, relative to that of

Table 5. *Partition coefficients of adrenal hormones in various solvent systems*

Solvent system	Temp.	21-Hydroxy-4-pregnene-3,20-dione	21-Hydroxy-4-pregnene-3,11,20-trione	11 β,21-Dihydroxy-4-pregnene-3,20-dione	17 α,21-Dihydroxy-4-pregnene-3,20-dione	17 α,21-Dihydroxy-4-pregnene-3,11,20-trione	11 β,17 α,21-Trihydroxy-4-pregnene-3,20-dione	11 β,21-Dihydroxy-3,20-diketo-4-pregnen-18-al	Ref.
Petroleum ether/water	room	9.0		0.25	3.0	0.11			a
Chloroform/water	room						6.4		b
Ethyl acetate/water	room						12.2		b
30% Ethyl acetate in cyclohexane/ 30% ethanol in water	25±3°	4.4					0.31		c
30% Ethyl acetate in cyclohexane/ 50% ethanol in water	25±3°	0.67					0.08		c
30% Ethyl acetate in cyclohexane/ 70% ethanol in water	25±3°	0.21					0.04		c
50% Ethyl acetate in cyclohexane/ 30% ethanol in water	25±3°	8.6					1.1		c
50% Ethyl acetate in cyclohexane/ 50% ethanol in water	25±3°	1.8					0.39		c
50% Ethyl acetate in cyclohexane/ 70% ethanol in water	25±3°	0.47					0.16		c
60% Ethyl acetate in cyclohexane/ 30% ethanol in water	25±3°	10					1.8		c
60% Ethyl acetate in cyclohexane/ 50% ethanol in water	25±3°	2.4					0.59		c
60% Ethyl acetate in cyclohexane/ 70% ethanol in water	25±3°	0.82					0.31		c
70% Ethyl acetate in cyclohexane/ 30% ethanol in water	25±3°	17					3.1		c
70% Ethyl acetate in cyclohexane/ 50% ethanol in water	25±3°	3.2					1.2		c
Benzene/water	22°			10.1	9.0	0.92	0.32		d
Benzene/water	27°				10.2	1.01	0.39		d
Benzene/50% methanol in water	22°	16	2.6	1.57	1.22	0.26	0.14		d
Benzene/50% ethanol in water	22°	7	2.23	1.27	1.63	0.64	0.30		d
Hexane/water	22°	2.45	0.03						d
20% Ethanol in 0.05% aqueous sodium chloride/ chloroform-hexane (1 : 1)	21°					0.43	2.45	0.59	e
Hexane-benzene (3 : 2)/ 5% ethanol in water	22°		1.33	1.08	1.13				f
20% Ethanol in 0.05% aqueous sodium chloride/ carbon tetrachloride-hexane (7 : 3)	22°			4.26	1.94				d
20% Ethanol in 0.05% aqueous sodium chloride/ carbon tetrachloride	23°		0.75	1.86	0.79				d
2-Butanol-hexane (1 : 1)/water	21°					2.93	2.83	1.44	e
2-Butanol-hexane (65 : 35)/water	21°		8.0			6.7		3.5	e
Iso-octane-2-butanol-water (3 : 2 : 4)	room	17.8	1.91		3.73	1.55	1.40		g
Carbon tetrachloride-chloroform-water-methanol (1 : 1 : 1 : 1)	room	5.0				1.32	0.5		h
2-Butanol-hexane (1 : 3)/water	20 to 21°		0.51	0.96	2.85	0.32	0.24		f
Ethyl acetate-hexane (1 : 1)/water	20 to 21°			4.3	6.1	1.38	1.00		f
Hexane-benzene (1 : 3)/water	20 to 21°			3.3		0.28	0.19		f
Hexane-benzene (1 : 1)/water	20 to 21°		3.1	1.06	2.20				f
20% ethanol in 0.05% aqueous sodium chloride/ carbon tetrachloride-chloroform (9 : 1)	20 to 21°			0.31	0.82	3.5			f

References. a) PFEFFER et al. (1952). — b) BURSTEIN (1956a). — c) L. ENGEL et al. (1954). — d) CARSTENSEN (1955). — e) CARSTENSEN (1956a). — f) CARSTENSEN (1956b). — g) TALBOT et al. (1955). — h) SCHROEDER et al. (1955).

Table 6. *Partition coefficients of some 17-ketosteroids in a quaternary solvent system* (L. ENGEL et al. 1954)

Solvent system: % Ethyl acetate in cyclohexane	Solvent system: % Ethanol in water	3α-Hydroxy-androstan-17-one	3α-Hydroxy-etiocholan-17-one	3β-Hydroxy-androstan-17-one	3β-Hydroxy-5-androsten-17-one	Androstane-3,17-dione	1-Androstene-3,17-dione
20	40	7.3	2.8	2.9	2.7	5.8	1.9
20	60	0.89	0.52	0.56	0.45	1.2	0.54
20	70	—	0.28	0.31	—	0.60	0.28
20	80	0.29	—	—	0.21	—	0.25
40	40	12.2	5.2	4.1	6.1	8.4	3.4
40	60	1.6	1.0	0.97	0.92	1.8	0.92
40	70	—	0.51	0.52	—	0.88	0.51
40	80	0.54	—	—	0.41	—	0.46
60	40	24	12.3	10	13	17	8.3
60	60	2.4	2.0	1.9	1.7	2.5	1.3
60	70	1.1	1.2	0.95	0.84	1.2	0.89

progesterone = 1.00: androsterone 0.85, deoxycorticosterone 0.83, testosterone 0.75, estrone 0.53. Ketosteroid hydrazones with phenylhydrazine-*p*-carboxylic acid have also been separated by electrochromatography in borate buffer of p_H 9.0 (MARIANI and VICARI 1956). CAVINA (1955a) (see also CAVINA and TENTORI 1956) prepared the sulfates of a number of steroids and determined their electrophoretic mobility in 0.05 *M* borate buffer. The values, expressed in 10^{-4} cm/sec/volt/cm, for the monosulfates (unless otherwise stated) were: dehydroepiandrosterone 4.78, androsterone 4.70, testosterone 4.91, estrone 4.57, estradiol 4.08 (disulfate 10.30), 5-androstene-3β,17β-diol 4.21 (disulfate 9.68), pregnanediol (glucosiduronate) 3.38. Conjugated steroids have also been separated by ANDREČ and STRAUCH (1956), who used one- and two-way paper electrophoresis.

III. Adsorption chromatography

Chromatography is a process that permits resolution of mixtures by differential migration. The migration is a result of a continuous exchange of solute between a mobile phase (eluent, developing solvent, etc.) and a stationary phase. The most widely-used system of classifying chromatography is based on the nature of the interaction with the stationary phase: Adsorption on a solid (adsorption chromatography), solution in a stationary liquid (partition chromatography, including paper chromatography), and chemical interaction with a solid (chemisorption, including ion exchange). When a mixture is applied to the stationary phase in a narrow zone, the movement of a suitable mobile phase over this zone effects a separation of its components, each of which migrates to an extent determined by the solubility in the mobile phase and the affinity for the stationary phase. In the resulting chromatogram the components are concentrated in zones at various distances from the original place of application. The general principles and techniques of adsorption chromatography are described in various textbooks (for references, see E. LEDERER and LEDERER 1957).

Adsorption chromatography has been used extensively for the preliminary purification of steroid mixtures as well as for the separation of individual steroids. It is especially well suited for preparative work with rather impure steroids of relatively low polarity.

1. Alumina

Alumina is a very useful adsorbent for steroid chromatography, although it may induce a variety of changes in the steroids, especially if it contains alkali (E. LEDERER and LEDERER 1957, p. 61). Thus, 21-hydroxypregnan-20-ones or 17α,21-dihydroxypregnan-20-ones are often altered to a considerable extent, even when neutralized alumina is used. In this case it is best to acetylate the steroids prior to adsorption chromatography (see. p. 240).

Alumina chromatography of steroids by fractional elution is described in detail by REICHSTEIN and SHOPPEE (1949). The preparation, activation, and standardization of neutralized alumina can be found in that article, but this adsorbent is now commercially available.

Chromatography tubes with stopcocks that require no lubrication and that are preferably provided with a sintered glass disc at the bottom are also commercially available, but buret-type glass tubes will also serve, if the constriction is filled with cotton or glass-wool and a layer of purified sand is placed on it to level the bottom of the adsorbent. The following sizes are recommended (Table 7):

Table 7. *Dimensions of columns for alumina chromatography* (REICHSTEIN and SHOPPEE 1949)

Amount of alumina, g	Internal diameter, mm	Usable length, mm	Bore of tap, mm
1	8	110	3
2	10	130	3
4	13	160	4
8	16	200	4
15	20	250	4
30	25	300	6
60	32	400	6
125	40	500	6
250	50	600	8
500	65	750	8

Reprinted by permission of the authors and the Faraday Society from Discussions of the Faraday Society 7, 307 (1949).

The tube should be less than half-full when packed with alumina. The weight of alumina used in a separation should be at least 30 times that of the mixture to be chromatographed. The first solvent in the solvent sequence (Table 8) used should be one in which the mixture barely dissolves. This solvent is used to fill the chromatographic tube before introducing the alumina. Even packing is accomplished by pouring the alumina slowly in a fine stream into the tube while it is gently tapped. The stopcock is opened and the solvent is allowed to run out dropwise, taking care to keep the alumina covered with solvent at all times. By refilling the tube repeatedly with the filtrate or by applying gentle pressure to the top a firmly packed column is finally produced. The top of the column is leveled by tapping or tamping. A layer of cotton or glass-wool or sand on top will prevent disturbance of the adsorbent by the addition of solvents. Finally, the solvent is allowed to run out until the alumina just remains covered and the mixture, dissolved in the first solvent in the series (Table 8) is added in as concentrated a solution as possible. The following series of solvents is suitable for steroid chromatography, but the choice of solvents may be modified according to the nature of the mixture. However, it is not advisable to change from one pure solvent to another without using a mixture of these first, since bands may be distorted and channeling may occur in the column.

Chloroform may be used in place of ether. Two portions of each eluent, each 100 ml per g of mixture, are used in succession, except in case the weight of the eluted material is increasing. The eluate fractions are collected separately, evaporated to dryness and weighed. Tabulation or a graph of the residue weights will indicate the location of the chromatographic bands. Bands weighing more than 10% of the original mixture should be rechromatographed. It is desirable to complete a chromatographic separation without interruption, since diffusion of bands on the column or channeling may vitiate the separation.

The separation of urinary ketosteroids by adsorption chromatography has been studied extensively by LIEBERMAN et al. (1948b, 1950). On the basis of their experience with alumina chromatography these authors have established the order of elution of various steroids:

"1. C_{21} pregnane derivatives are eluted prior to their corresponding C_{19} androstane derivatives.

2. Compounds with Ring A—B trans juncture (androstane and allopregnane) are eluted before their corresponding isomers with Rings A—B cis (etiocholane and pregnane). One

Table 8. *Solvent sequence for alumina chromatography* (REICHSTEIN and SHOPPEE 1949)

Petroleum ether (boiling range 30—60°)
1% Benzene in petroleum ether
2.5% Benzene in petroleum ether
5% Benzene in petroleum ether
10% Benzene in petroleum ether
25% Benzene in petroleum ether
50% Benzene in petroleum ether
Benzene
10% Ether in benzene
25% Ether in benzene
50% Ether in benzene
Ether
0.5% Methanol in ether
2.5% Methanol in ether
5% Methanol in ether
10% Methanol in ether
20% Methanol in ether
Ethyl acetate-chloroform-methanol (1 : 1 : 1) +1% acetic acid
Ethyl acetate-chloroform-methanol (1 : 1 : 1) +2.5% acetic acid
Ethyl acetate-chloroform-methanol (1 : 1 : 1) +5% acetic acid

Reprinted by permission of the authors and the Faraday Society from Discussions of the Faraday Society **7**, 307 (1949).

exception to this rule has been noted: 3β-hydroxyetiocholan-17-one is eluted from alumina before 3α-hydroxyandrostan-17-one. This inversion in the expected sequence of elution may be a special case because these substances also differ in their configuration at C-3. The configuration of the 3-hydroxyl group seems to be more influential in determining the chromatographic sequence than does the steric juncture of Rings A and B.

3. Saturated compounds are eluted before their unsaturated analogues, except in the case of 3α-hydroxyetiocholan-17-one, which is eluted after 3α-hydroxy-9(11)-etiocholen-17-one. 1-Androstene-3,17-dione, as expected, is eluted after its saturated analogue, androstane-3,17-dione. 1-Androstene-3,17-dione precedes its isomer 4-androstene-3,17-dione in the sequence of elution.

4. The diketomonohydroxy compounds are eluted before their corresponding dihydroxymonoketo compounds with the important exception that the dihydroxymonoketones having one hydroxyl group on C-11 are eluted before those diketomonohydroxy compounds with one ketonic group on C-11. Thus, 3α,11β-dihydroxyandrostan-17-one and 3α,11β-dihydroxyetiocholan-17-one are eluted before their corresponding C-11 carbonyl derivatives, 3α-hydroxyandrostane-11,17-dione and 3α-hydroxyetiocholane-11,17-dione."[1]

The over-all sequence was monoketones, monoketo monoacetoxy compounds, diketones, and monohydroxy monoketones. More polar compounds were chromatographed on a mixture of equal parts of magnesium silicate and celite. On the latter adsorbent the order of elution was dihydroxy monoketones of C_{19} steroids, monohydroxy diketones of C_{19} steroids, and dihydroxy monoketones of C_{21} steroids. In some cases the order of elution on the magnesium silicate columns was the reverse of that expected from alumina chromatography, e. g., 3α-hydroxyetiocholane-11,17-dione was eluted before 3α-hydroxyandrostane-11,17-dione, and dehydroepiandrosterone was eluted before 3β-hydroxyandrostan-17-one.

Generally, in both adsorption and partition chromatography axial alcohols are more readily eluted than their equatorial counterparts (BARTON 1953), but exceptions to this rule have been observed (R. BROOKS et al. 1953).

Systematic alumina chromatography of urinary 17-ketosteroids, developed by DINGEMANSE et al. (1946, 1952) (see also ROBINSON 1954) and modified by various authors (ZYGMUNTOWICZ et al. 1951, MARTI 1951, BAUER and KARL 1952, POND 1954), has been recently reviewed by HUIS IN'T VELD (1956). DINGEMANSE et al. (1952) proceeded as follows:

[1] Reprinted by permission of the authors and of the American Society of Biological Chemists, Inc., from the Journal of Biological Chemistry **182**, 302—303 (1950).

Fifteen g of alumina, which has been kept equilibrated at the vapor pressure of 58% sulfuric acid is slurried in benzene and poured into a glass tube, 15×500 mm. A solution of urine extract, containing less than 6 mg of ketosteroids in 50 ml of benzene is poured through the column. This is followed by 50-ml portions of the solvents listed in Table 9. The 17-ketosteroids in the 50-ml eluate fractions from urine of a normal man are also given.

Table 9. *Alumina chromatogram of urinary 17-ketosteroids* (DINGEMANSE et al. 1952)

Fraction No.	Label in Fig. 90	Solvent	17-Ketosteroids in the eluate
1—3	I	Benzene	2-Androsten-17-one, 3 β-chloro-5-androsten-17-one, 3,5-androstadien-17-one
4—5		Benzene	none
6—8	II	Benzene	6 β-Hydroxy-3,5-cycloandrostan-17-one
9—10		Benzene +0.1 vol. % EtOH	none
11—13	III	Benzene +0.1 vol. % EtOH	3 β-Hydroxy-5-androsten-17-one, 3 β-hydroxy-androstan-17-one
14		Benzene +0.1 vol. % EtOH	none
15—19	IV	Benzene +0.1 vol. % EtOH	3 α-Hydroxyandrostan-17-one, 3 α-hydroxy-9-androsten-17-one
20—21		Benzene +0.1 vol. % EtOH	none
22—30	V	Benzene +0.1 vol. % EtOH	3 α-Hydroxyetiocholan-17-one, 3 α-hydroxy-9-etiocholen-17-one
31—32		Benzene +0.1 vol. % EtOH	none
33—38	VI	Benzene +0.5 vol. % EtOH	unidentified compound, 3 α,11 β-Dihydroxy-androstan-17-one
39—40		Benzene +0.5 vol. % EtOH	none
41—43	VII	Benzene +0.5 vol. % EtOH	3 α,11 β-Dihydroxyetiocholan-17-one
44		Benzene +2.0 vol. % EtOH	none
45—48	VIII	Benzene +2.0 vol. % EtOH	unidentified 17-ketosteroids

Reprinted by permission of the authors and Charles C. Thomas, Publisher, from The Journal of Clinical Endocrinology and Metabolism **12**, 67 (1952).

A plot of 17-ketosteroids in the eluate fractions from the urine of a normal man is shown in Fig. 90.

A difficulty which chromatography on alumina shares with that on other adsorbents is the so-called tailing. This phenomenon, which is due to the non-linearity of adsorption isotherms, causes the front of chromatographic bands to

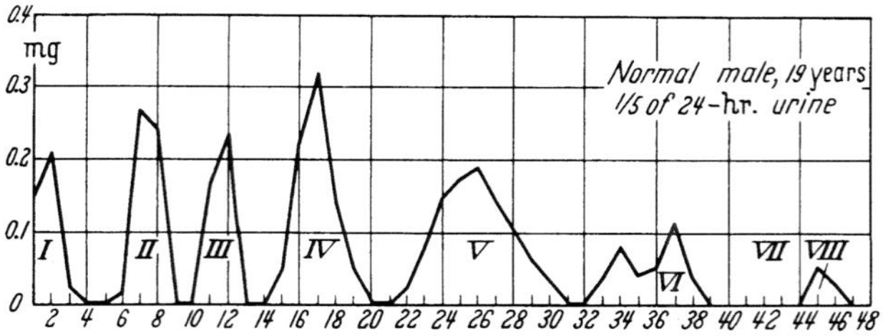

Fig. 90. Alumina chromatogram of urinary 17-ketosteroids (DINGEMANSE et al. 1952). The elution pattern was obtained by plotting fraction numbers (Table 9) against weight. For the meaning of the Roman numerals, see Table 9. Reproduced by permission of the authors and Charles C.Thomas, Publisher, from The Journal of Clinical Endocrinology and Metabolism **12**, 68 (1952)

be sharper than the tail-end. This increases the chance that the tail of one band will overlap the front of the next one and consequently make the separation incomplete. If the polarity of the eluent is gradually and continually increased, tailing is diminished because the tail-end of the band comes under the influence of a more strongly eluting solvent than the front. This process, called gradient elution, has been applied to the separation of urinary ketosteroids by LAKSHMANAN and LIEBERMAN (1954) and resulted in a substantial increase in resolution. They employed partially deactivated alumina as an adsorbent and benzene with an increasing concentration of ethanol as eluent. KELLIE and WADE (1957) have recently published further details of this method. In their case the elution gradient was produced by adding a solution of 2% ethanol in benzene to a solution of 0.2% ethanol in bezene. The theory (DRAKE 1955) and practice (BOCK and LING 1954) of gradient elution chromatography have recently been discussed.

Various authors have proposed the preparation of derivatives as a means of obtaining clear-cut separation of steroids by adsorption chromatography. Thus COFFMAN (1941) has separated various hydroxysteroids in the form of their colored azoates, prepared by esterification with *p*-phenylazobenzoyl chloride, by chromatography on alumina. Steroid benzoates have been separated on alumina by R. BROOKS et al. (1953) and *p*-I^{131}-benzoates on silicic acid-Celite columns by STOKES et al. (1955). Epimeric steroids are well separated by the former technique, especially C-20 epimers. KELLIE and WADE (1953) prepared 3,5-dinitrobenzoates of urinary hydroxysteroids and separated them on columns of alumina or silica. JOHNSTON (1947), HILMER and HESS (1949), REICH et al. (1952), and ZWINGELSTEIN et al. (1953) have described the separation of 2,4-dinitrophenylhydrazones of various steroids by chromatography on alumina. The method of REICH et al. (1952) will be quoted as an example of these techniques.

A dried sample containing 5 μg to 1 mg of progesterone and 3β-hydroxy-5-pregnan-20-one is dissolved in 0.1—0.2 ml of absolute ethanol. To this is added 0.3 ml of a solution containing 125 mg of 2,4-dinitrophenylhydrazine of highest purity in a mixture of 15 ml of absolute ethanol and 0.4 ml of concentrated hydrochloric acid (dissolved by gentle heating). The mixture is kept at room temperature overnight. In order to destroy the excess dinitrophenylhydrazine 0.4 ml of Benedict's qualitative reagent and 0.4 ml of water are added, and the mixture is heated for 10 minutes in a bath at 95°. The suspension is transferred to a small separatory funnel with a small amount of water and extracted with about 5 ml of chloroform. The chloroform layer is separated, washed in a second funnel with 0.5 volume of water, and dried with anhydrous sodium sulfate. A second portion of 5 ml of chloroform is passed through the two aqueous solutions, dried with the same sodium sulfate, and combined with the first portion. The sodium sulfate is washed with small amounts of chloroform until colorless and the combined chloroform solutions are evaporated to dryness.

The residue is dissolved in 5—20 ml of hexane-benzene (1 : 1), depending on the amounts of 2,4-dinitrophenylhydrazones present, and the solution is chromatographed on alumina. If the amount expected is 5—200 μg, 300—400 mg of the adsorbent is used in a column of 4 mm inside diameter. For 0.2—1.0 mg, a column of 10 mm inside diameter, containing 3 g of alumina, is used. The small column is washed with a minimum of four 5-ml portions of benzene, while 20-ml portions are used for the large column. If the last fraction is still yellow, the elution with benzene is continued until 2 colorless fractions are obtained. Now at least 4 fractions of 5 or 20 ml of benzene-chloroform (9 : 1) are passed through the column and, again, elution is continued if the last fraction is colored. If the band moving with this eluant is intense, a benzene-chloroform (4 : 1) mixture gives more rapid elution. Progesterone bis-dinitrophenylhydrazone is eluted by benzene and the pregnenolone dinitrophenylhydrazone by benzene-chloroform mixtures. Recoveries from the column are nearly quantitative.

Conjugates of urinary 17-ketosteroids have also been separated by adsorption chromatography. Thus, EDWARDS and KELLIE (1956) have separated the glucosiduronates of 3α-hydroxyandrostan-17-one and 3α-hydroxyetiocholan-17-one on silicic acid columns containing 0.6 ml of 0.1 N sodium acetate per g by elution with chloroform containing 2% ethanol and 0.4% acetic acid. CRÉPY et al. (1957) separated sulfates from glucosiduronates by alumina chromatography. Butanol containing increasing amounts of water eluted first the former and then the latter.

On a micro scale, adsorption chromatography may also be carried out on filter paper impregnated with alumina (BUSH 1950, 1952, SHULL et al. 1952), stearato chromic chloride (KRITCHEVSKY and KIRK 1952a), or glass-fiber paper, impregnated with silicic acid (DIECKERT and REISER 1954).

2. Other adsorbents

Silica, especially in the form of silica gel, is very useful for the preliminary purification of steroid extracts and in the separation of individual steroids. Being ordinarily less active and less destructive than alumina, it has been found especially valuable for analytical work with corticosteroids. Thus, HECHTER et al. (1951) have used silica gel for the preliminary purification of adrenal perfusates, and FILLERUP and MEAD (1953) have concentrated plasma sterols by silica

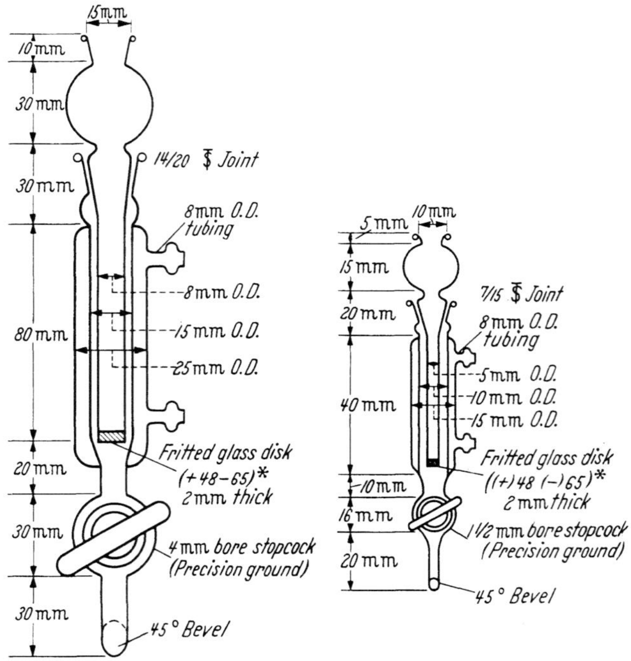

Fig. 91. Silica gel microcolumns (SWEAT 1954). The column on the left is used for 0.5 g, the one on the right for 0.1 g of silica gel. Both are water-jacketed to cool the silica gel. Redrawn by permission of the author and of the American Chemical Society from Analytical Chemistry **26**, 1965 (1954)

chromatography. Silica gel is particularly useful for the rapid separation of steroids from fat and cholesterol in extracts of blood and plasma (BUSH and SANDBERG 1953).

ROMANOFF and WOLF (1954) have been able to separate corticosteroid metabolites in urine into three major fractions by eluting silica gel columns with 1) benzene-ether mixtures, 2) ether-ethyl acetate mixtures, and 3) acetone and methanol. SWEAT (1954b) has proposed the following method for the resolution of as little as 0.25 μg each of corticosterone and cortisol (see also ELY et al. 1958):

The procedure given here applies to the smaller column (Fig. 91). The fritted glass disc is first covered with a 5-mm layer of powdered glass. Next, 0.1 g of purified silica gel, activated by heating at 500° for 3 hours, is suspended in chloroform and poured into the column. It is first washed by passing through, in succession, first 4 ml, then 2 ml of 10% ethanol in chloroform, then 2 portions of chloroform, 2 ml each. Next, the sample, in a small amount of chloroform, is applied to the column, and then the chromatogram is developed as indicated in Table 10.

For a preliminary fractionation of urinary neutral 17-ketosteroids, SAVARD et al. (1953) have used a silica gel column, which was eluted in succession with benzene, benzene-ethyl acetate mixtures, ethyl acetate, and methanol. Similarly,

B. RUBIN et al. (1953b) have fractionated the neutral ketonic α-hydroxy steroid fraction of urine by silica gel chromatography. A 1000 : 1 ratio of silica gel : 17-ketosteroids in the sample was used and the polarity of the eluent was increased step-wise from benzene-petroleum ether (1 : 3) to ether-benzene (1 : 1). Gradient elution has been used by TALALAY et al. (1955): 4-Androstene-3,17-dione was eluted before testosterone on a silicic acid column developed with hexane-chloroform (1 : 1) which was gradually changed to either pure chloroform or 1% methanol in chloroform.

Table 10. *Silica gel chromatogram of corticosteroids* (SWEAT 1954b)

Fraction No.	Amount of eluent, ml	Composition of eluent: %ethanol in chloroform	Compound eluted
1	4	0	
2	2	0	
3	2	1	Corticosterone
4	6	1	Corticosterone
5	2	1	Corticosterone
6	2	4	Cortisol
7	2	4	Cortisol
8	2	4	Cortisol

Reprinted by permission of the author and of the American Chemical Society from Analytical Chemistry **26**, 1966 (1954).

Silica gel columns, containing absolute ethanol, have been used by KATZENELLENBOGEN et al. (1954) for the separation of acetates of corticosteroids. The chromatographic technique, which has also been applied to the isolation of 17-ketosteroids (MILLER et al. 1956), was as follows:

The chromatographic tube has a diameter of 1 cm and is provided with a water jacket for cooling it during preparation and operation. The mobile phase consists of dichloromethane, containing 1—5% of ethanol. Silica gel is mixed with 4 ml of absolute ethanol per 10 g by shaking. After it has cooled, the powder is poured into the chromatographic tube, previously filled with the mobile phase, while tapping the column with a rubber hammer. The gel is supported by a layer of sand and the top of the column is protected by another layer of sand. The amount of solvent-equilibrated silica gel recommended is 400 times the weight of mixture to be separated. The sample, dissolved in a minimum amount of mobile phase, is added to the column and fractions are collected in an automatic fraction collector.

Whether adsorption or partition is the preponderant mechanism of separation depends on the amount of water in the system (LINFORD 1956). Adsorption chromatography on filter paper has been suggested as a method of separating adrenal hormones (SAKAL and MERRILL 1953) and CHEN et al. (1955) have used chromatography on a paper strip for the preliminary purification of corticosteroids in blood:

The sample is applied near the bottom end of a strip of filter paper and hexane is allowed to rise to the top of the paper strip. The hexane carries away interfering lipides. The corticosteroids may be concentrated by development with methanol, stopping the ascent near a line in the middle of the strip.

Displacement analysis of steroids on Darco G-60-Hyflo (2 : 1), with 0.5% cholesterol as a displacer, has been described by HAMILTON and HOLMAN (1952).

Florisil, a synthetic magnesium silicate, is used in the NELSON and SAMUELS (1952) method of corticosteroid analysis in blood. The specificity of the colorimetric assay in this case depends largely on the removal of interfering chromogens.

The adsorbent must be thoroughly washed with 95% ethanol, rinsed with absolute ethanol, dried at 120° for 24 hours, and then activated at 600° for 4 hours. Prolonged storage may cause it to absorb moisture and perform poorly (EIK-NES et al. 1953). The column is prepared by packing a glass column, 11 mm in inside diameter, with dry adsorbent to a height of 70 mm. After wetting the column with chloroform, a chloroform solution of the sample is added, and the chromatogram is developed, using 25 ml of chloroform, 25 ml of 2% methanol in chloroform, and 45 ml of 25% methanol in chloroform. The last fraction contains the 17-hydroxycorticosteroids estimated by this method.

IV. Partition chromatography

Partition chromatography differs from adsorption chromatography in the nature of the sorptive process. Whereas the stationary phase in the latter is a solid adsorbent, the stationary phase in the former is a liquid solvent. During partition chromatography distribution of solutes between the stationary and mobile phases takes place continuously. The process is akin to countercurrent distribution and the theory may be treated on the basis of a cascade process of discrete distribution steps (HEFTMANN 1957), leading to a relationship between the distribution coefficient and the rate of flow of a given solute:

$$\alpha = \frac{A_P}{A_W}\left(\frac{1}{R_F} - 1\right)$$

where the distribution coefficient α is the ratio of the concentration of the solute, in moles per l, in the stationary phase to the concentration in the mobile phase, A_P and A_W are related to the thickness of the layers of the mobile phase and stationary phase, respectively, and the rate of flow, R_F, is the ratio of the distance traveled by the solute to the distance traveled by the solvent. Compared to countercurrent distribution, partition chromatography is not only more convenient but also more efficient, because of the great number of distributions that take place during the chromatographic development. However, larger amounts can generally be fractionated by countercurrent distribution than by partition chromatography.

Several textbooks are available in which technical details of partition chromatography can be found (E. LEDERER and LEDERER 1957, BLOCK et al. 1955). Technical details and apparatus applicable to steroids are also discussed in various papers, e. g., ZAFFARONI (1953), HAINES and KARNEMAAT (1954), TENNENT et al. (1951), SCHWARZ (1953b), OERTEL (1956a and 1956b), ARX and NEHER (1956), AYRES et al. (1957a) and EBERLEIN (1957). The application of partition chromatography to steroids has been reviewed by BUSH (1954b) and by HEFTMANN (1955). One of the problems that must be overcome, especially in working with the less polar steroids, is the choice of a stationary phase in which they are reasonably soluble. Best results have been obtained with a stationary phase of either a relatively polar organic solvent that is partially miscible with the relatively non-polar mobile phase, or with the incorporation of organic solvents in the stationary water phase.

1. Columns

Larger quantities of less pure materials can be separated by column chromatography than by paper chromatography. Paper chromatography, on the other hand, is more convenient and more informative for qualitative analysis. For quantitative work column chromatography is not only more accurate, but also offers the advantages 1) that small amounts of material can be measured in the presence of large amounts of other substances, 2) that several assays can be performed on the same fraction, and 3) that interfering extractives present in eluates from filter paper may be avoided.

a) Stationary phase: Non-aqueous solvents

BAKER et al. (1951) have used a column of powdered cellulose, impregnated with 25% v/v propylene glycol in chloroform for the isolation of cortisol and cortisone from hog adrenal extract. After allowing the chloroform to evaporate, the column was packed with the propylene glycol-impregnated cellulose powder and irrigated with toluene. Similarly, DIRSCHERL et al. (1956) have impregnated cellulose powder with propylene glycol in methanol. After evaporation of the methanol, the impregnated cellulose was homogenized with toluene to make a slurry suitable for packing the chromatographic column. MATTOX et al. (1953) have used the toluene/propylene glycol solvent system for the isolation of aldosterone from beef adrenal extract on silica gel columns. A formamide-impregnated Celite column with benzene or benzene-isoöctane as the mobile phase has been proposed for the assay of adrenal extracts by BANES (1953):

About 450 ml of benzene is mixed with 10 ml of formamide and 10 ml of water. After equilibration, 8 g of Celite 545 is mixed with the benzene layer and 4 ml of the aqueous layer is added drop-wise with vigorous stirring. The solid is transferred to a 200×25 mm chromatographic tube and packed down in portions. After the supernatant liquid is drained off, the adrenocortical hormones are transferred to the column in benzene solution and development with the benzene phase is started. Dehydrocorticosterone and corticosterone emerge in the first 70 ml of eluate, cortisone in the next 100-ml fraction, and cortisol in the following 230-ml fraction.

The separation of dehydrocorticosterone and corticosterone is carried out on a second column, prepared as above, but using a benzene-isoöctane (3 : 2)/formamide-water (4 : 1) solvent system. After a 70-ml forerun, dehydrocorticosterone is obtained in the next 80-ml fraction and corticosterone in the following 160-ml fraction.

KATZENELLENBOGEN et al. (1954) have used a formamide-impregnated silica gel column and cyclohexane as the eluent for separating androsterone from 3α-hydroxyetiocholan-17-one. In 1953, J. JONES and STITCH proposed the following method for the separation of neutral 17-ketosteroids on partition columns (see also COOK et al. 1954):

Ten g of powdered silicic acid is mixed with 5 ml of nitromethane and slurried in the mobile phase, which consists of a 3% (v/v) solution of chloroform in petroleum ether (boiling range 60—80°), saturated with nitromethane. The slurry is poured into a chromatographic tube 20×300 mm and the solid is compressed with nitrogen until a flow rate of about 1—1.5 ml/min is obtained. The steroid mixture, dissolved in a minimum quantity of eluent, is then applied to the column and 5-ml fractions are collected automatically. The threshold volumes of some ketosteroids are given in Table 11.

Table 11. *Partition chromatogram of 17-ketosteroids* (J. JONES and STITCH 1953)

Compound	ml of effluent
4-Cholesten-3-one	24
3β-Chloro-5-androsten-17-one	33
6β-Hydroxy-3,5-cycloandrostan-17-one	90
3α-Hydroxyandrostan-17-one	90
3β-Hydroxyandrostan-17-one	130
3β-Hydroxy-5-androsten-17-one	170
3α-Hydroxyetiocholan-17-one	210
3α,11β-Dihydroxyetiocholan-17-one	210
4-Androstene-3,17-dione	270

Reprinted by permission of the authors and of Cambridge University Press from Biochemical Journal **53**, 681 (1953).

HAINES (1952, HAINES and KARNEMAAT 1954) has developed a method for the separation of adrenocortical steroids on silica gel columns impregnated with 1,2-ethanediol. The column was eluted with a series of cyclohexane-dichloromethane mixtures. The eluents were automatically changed by a device, which was regulated by the ultraviolet absorption of the eluates.

To separate 10 mg of each of the then-known six active adrenocortical hormones the mixture was absorbed into a few discs of filter paper. These were then placed on a chromatographic column, containing 30 g of silica gel and 15 ml of 1,2-ethanediol, which had previously been saturated with cyclohexane and dichloromethane. The flow rate was adjusted to 4 ml/min and the eluents, which had been saturated with 1,2-ethanediol, were passed through the column under pressure in succession (Table 12).

Non-aqueous organic solvent systems can also be used in such a way that the less polar solvent is kept stationary on a column while the more polar solvent is

Table 12. *Partition chromatogram of corticosteroids.* (Data of HAINES 1952)

Composition of the eluent	Vol., ml	Steroid eluted
Cyclo- : Dichloro- hexane methane		
10 : 1	600	21-Hydroxy-4-pregnene-3,20-dione
6 : 1	960	21-Hydroxy-4-pregnene-3,11,20-trione
4 : 1	720	11 β,21-Dihydroxy-4-pregnene-3,20-dione and 17 α,21-Dihydroxy-4-pregnene-3,20-dione
3 : 1	420	—
2 : 3	660	Cortisone
Dichloromethane	720	Cortisol

made the mobile phase. This so-called reversed-phase partition chromatography is a valuable method for the fractionation of very lipophilic steroids, but it has not been widely used in the analysis of adrenal steroids.

b) Stationary phase: Aqueous solvents

The method of MORRIS and WILLIAMS (1953) for the determination of individual adrenal hormones in blood illustrates the preliminary purification of extracts by reversed-phase partition chromatography, fractionation on partition columns with aqueous alcohol as the stationary phase (BUTT et al. 1951), and fractionation on partition columns with a non-aqueous polar stationary phase.

The reversed phase partition column is made of 4 g of Hyflo Super-cel, which has been previously treated with dichlorodialkylsilane to make it hydrophobic. It holds 2 ml of petroleum ether (b. r. 80—100°) stationary. The mobile phase is 50% (v/v) aqueous ethanol which has been equilibrated with the stationary phase. To prepare the column a slurry of Super-cel, stationary phase and mobile phase is poured into a chromatographic tube, 1 cm in diameter, and packed to about 4 cm height with the acid of a perforated tamper. An ethyl acetate extract of 20 ml of plasma, dissolved in 0.5 ml of the mobile phase, is then applied to the column and positive pressure (30 mm of Hg) is applied. After the first 0.5 ml has been forced into the column, the process is repeated with another 0.5-ml portion of mobile phase to complete the transfer. Three ml of the mobile phase is now added to the column and pressure applied. The total corticosteroids are contained in the first 4 ml of eluate, while practically all fats and phospholipides are retained on the stationary phase.

The extract is next fractionated on a column of 3 g of Hyflo Super-cel that holds 2 ml of 25% (v/v) aqueous ethanol stationary. The mobile phase is toluene, which has been equilibrated with the stationary phase. The chromatographic column is prepared as before, except that it measures 8×0.6 cm. The running rate of the column should be about 0.8 ml/hr under a positive pressure of 80—100 mm of Hg. After the mobile phase reaches a level of about 1 mm below the surface of the column, sufficient dry Super-cel is packed on top to form a layer 2—3 mm deep. The eluate from the first column, dissolved in 0.1 ml of the 25% ethanol phase is added to the dry Super-cel and forced down under 10—20 mm of Hg pressure. A solvent reservoir containing the toluene phase is now attached to the column and the mobile phase is allowed to descend under hydrostatic pressure of 80—100 mm of toluene overnight. One-ml fractions are collected by means of an automatic fraction collector. The first 4 ml of effluent contains a mixture of dehydrocorticosterone, corticosterone, and 17 α,21-dihydroxy-4-pregnene-3,20-dione, while cortisone is collected in fractions 7—8 and cortisol in fractions 10—11.

The third column serves to separate dehydrocorticosterone, corticosterone, and 17 α,21-dihydroxy-4-pregnene-3,20-dione. It is identical with the second one in all respects, except that the solvent system is composed of 1,2-ethanediol as the stationary phase and 20% (v/v) petroleum ether (b. r. 80—100°) in toluene as the mobile phase. The 4-ml forerun of the second column is freed of solvent and dissolved in 0.1 ml of the 1,2-ethanediol phase for transference to this column. Dehydrocorticosterone is collected in fractions 4—5, corticosterone in fractions 11—12, and 17 α,21-dihydroxy-4-pregnene-3,20-dione in fractions 17—23.

More recently MORRIS (1955) has reported that dehydrocorticosterone, corticosterone, cortisone and cortisol in plasma can be determined after separation on a single column, which holds 1,2-ethanediol stationary. The mobile phase is gradually changed from 25% petroleum ether in toluene to 1,2-dichloroethane. In 1954, HEFTMANN and JOHNSON described a gradient elution method for the determination of individual adrenocortical hormones on a partition column with water as the stationary phase, which was later (D. JOHNSON et al. 1956) applied to urine. The elution gradient is produced by mounting two separatory funnels on the chromatographic column, one on top of the other. As the bottom funnel, containing petroleum ether (b. r. 30–60°) empties into the column an equal amount of dichloromethane drops into it from the top funnel and the polarity of the eluent thus gradually and automatically increases. A detailed description of the method follows:

Silicic acid, dried for 3 hours at 90—100° is thoroughly mixed with half its weight of water by grinding in a mortar. The powder is then suspended in H_2O-saturated petroleum ether to make a mobile slurry, which is poured in batches into a chromatographic tube, 1.8 × 33 cm.

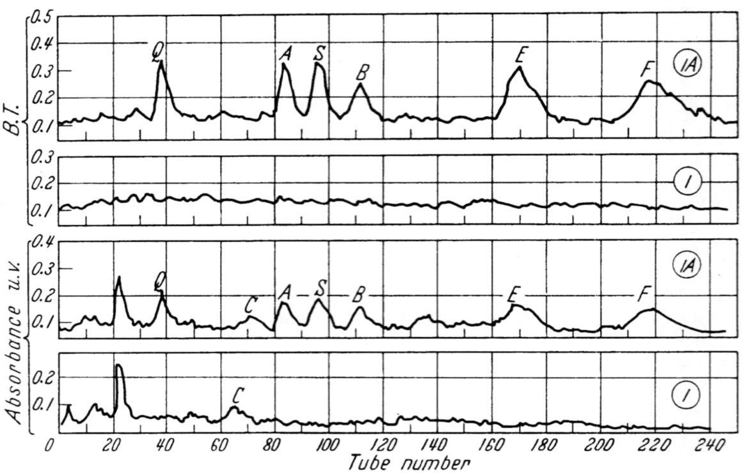

Fig. 92. Partition chromatogram of corticosteroids added to urine extract (D. JOHNSON et al. 1956). The elution patterns were obtained by plotting tube numbers against ultraviolet (U. V.) absorption at 240 mμ (two bottom curves) and absorption of reduced Blue Tetrazolium (B. T.) reagent (two top curves) before (1) and after (1A) addition of the steroids. Peaks in the U. V. and B. T. curves coincide where the hormones are located in the chromatogram. *Q* = deoxycorticosterone, *C* = caffeine, *A* = dehydrocorticosterone, *S* = 17α,21-dihydroxy-4-pregnene-3,20-dione, *B* = corticosterone, *E* = cortisone, and *F* = cortisol. Reproduced by permission of the editor from Acta Endocrinologica **23**, 346 (1956)

Each batch is packed down with a perforated tamper until a column, 20 cm high, is formed. The column is washed by irrigation with water-saturated petroleum ether overnight.

The mixture to be separated is absorbed on several filter paper disks, placed on the column, and covered with a 2-cm layer of silicic acid and then a 1-cm layer of sand. The solvent reservoirs are attached and eluate fractions are collected in a automatic fraction collector at a rate of 35 drops/min, 400 drops/fraction. Fig. 92 shows the position of the hormones in the chromatogram of a urine extract to which they were added. After spontaneous evaporation the fractions were analyzed by both ultraviolet absorption (U. V.) and reduction of Blue Tetrazolium (B. T.) reagent.

COOK et al. (1955) have also used silicic acid columns impregnated with water for the fractionation of corticosteroids in urine and adrenal extracts. A series of benzene-ether mixtures was used for elution. The method does not separate 17α,21-dihydroxy-4-pregnene-3,20-dione from dehydrocorticosterone, but the analysis of the fractions by both ultraviolet absorption at 240 mμ and by the Porter-Silber method permits quantitative differentiation between the two compounds.

A quantity of silicic acid calculated to contain 5.5 g of dry SiO_2 is mixed with water to give a total of 6.0 g of water. The powder is suspended in water-saturated benzene and poured into a chromatographic tube, 1.8×25 cm. When the benzene level approaches the top of the silicic acid, the mixture to be separated, dissolved in 2—5 ml of wet benzene, is added and the column is developed with a series of water-saturated solvents as follows (Table 13).

Ten-ml fractions are collected in an automatic fraction collector and evaporated to dryness at 90°. Figure 93 shows two chromatograms obtained by this method, one of a mixture of pure compounds (top) and one of an adrenal extract (bottom).

Table 13. *Solvent sequence for partition chromatography of corticosteroids.* (Data of Cook et al. 1955)

Vol., ml	Parts	
	Benzene	Ether
30	30	—
120	13	2
120	9	2
120	6	2
120	1	1
100	—	100

Chromatography on partition columns has been a very valuable tool in work on the isolation of aldosterone. Simpson et al. (1954a) used large kieselguhr columns (Hegedüs et al. 1953), containing water as the stationary phase, and a series of petroleum ether-benzene eluents, followed by benzene-chloroform mixtures for a preliminary fractionation of beef adrenal extract. This was followed by a second partition chromatogram on a cellulose column, which was developed in sequence with the organic solvent phases of 3 mixtures of benzene-petroleum ether-methanol-water of the following composition: first 1 : 1 :1 : 1, then 1.2 : 0.8 : 1 : 1, and finally 2 : 0 : 1 : 1. The last-mentioned solvent system was also

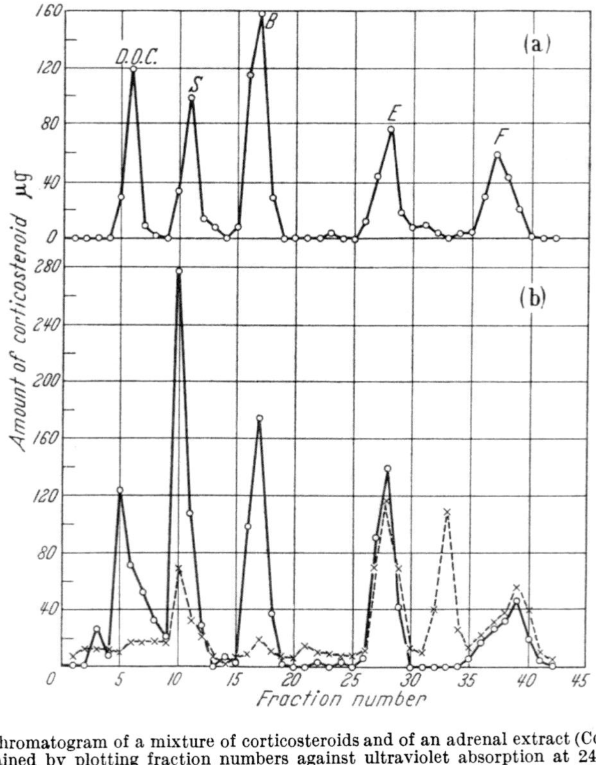

Fig. 93. Partition chromatogram of a mixture of corticosteroids and of an adrenal extract (Cook et al. 1955). The patterns were obtained by plotting fraction numbers against ultraviolet absorption at 240 mμ (full line) and absorption of Porter-Silber chromogens (broken line). The chromatogram of a mixture of corticosteroids (top) and of adrenal extract (bottom) are shown. For the meaning of letters, see Fig. 92, except that D.O.C. = deoxycorticosterone. Reproduced by permission of the authors and of the Society for Analytical Chemistry from The Analyst **80**, 217 (1955)

used by HARMAN et al. (1954) as the first step in the partition of beef adrenal extract. In the second step benzene containing 10% (v/v) of ethyl acetate was substituted for pure benzene in the solvent system. MATTOX and MASON (1956) have used powdered cellulose columns in their work on the isolation of aldosterone, since it was found that the side chain of corticosteroids is apparently oxidized by silica gel when a solvent system of chloroform-benzene (1 : 1), saturated with formamide, is used. It was also found that an abrupt change to a denser eluent (chloroform) distorts the chromatographic bands on the column. A gradual change from chloroform-benzene (1 : 1) to chloroform was obtained by use of the apparatus described by HEFTMANN and JOHNSON (1954).

More recently AYRES et al. (1957a) have described a method for the determination of aldosterone and other adrenal hormones in biological extracts by extensive purification of aldosterone diacetate on partition columns:

To a column, 60×1 cm, containing 26 g of Celite 545 with methanol-water (1 : 1) as the stationary and benzene-ethyl acetate (5.2 : 1) as the mobile phase, the purified extract is added in 2 ml of mobile phase. Fractions of 3 ml each are collected and a mixture of cortisol and aldosterone is obtained in Fractions 6—12. These fractions are pooled and acetylated. The acetates of aldosterone and cortisol are separated on another column, similar to but only half as long as the first, with methanol-water (4 : 1) as the stationary and petroleum ether (boiling range 80—100°)-benzene (3 : 2) as the mobile phase. Twelve 5-ml fractions containing aldosterone diacetate are collected, then the mobile phase is changed to petroleum ether-benzene (2 : 11) to elute cortisol acetate.

Additional applications of partition chromatography on columns to the separation of pregnane derivatives are the determination of progesterone in blood by BUTT et al. (1951) and the isolation of pregnane-3α,17α,20α-triol from urine by COX and MARRIAN (1953).

2. Paper

Partition chromatography on filter paper is based on the same principle as column chromatography and the same solvent systems are applicable to both techniques. Some of the advantages of paper chromatography over column chromatography have already been pointed out. Others, such as increased resolution by two-dimensional development and multiple color development after radial development, have hardly been exploited in the field of adrenal steroids.

ZAFFARONI et al. (1949) made the first attempt to apply paper chromatography to the identification of steroids. They prepared Girard's reagent T hydrazones of various ketosteroids and chromatographed them on filter paper with water-saturated butanol as the mobile phase. The hydrazones are not only more water-soluble than the ketosteroids, but also readily detectable on filter paper by use of various alkaloid reagents. Since the R_F values are mainly determined by the number of hydrazone groups in the molecule the separation of compounds having the same number of hydrazone groups is not very good. Moreover, steroids with two ketone groups frequently give rise to two spots, corresponding to the mono- and dihydrazone, respectively.

Another derivative of 17-ketosteroids, which can be separated in aqueous solvent systems (BUSH 1952), is that formed in the Zimmermann reaction. KELLIE and SMITH (1956) have suggested these derivatives, which absorb ultraviolet light and give a purple color on treatment with 10% ethanolic potassium hydroxide, for the characterization of 17-ketosteroids.

a) Stationary phase: Non-aqueous solvents

The application of non-aqueous solvent systems to the chromatographic separation of steroids was first described by ZAFFARONI et al. (1950). Details of this work have been reviewed by ZAFFARONI (1953), by HAINES and KARNEMAAT

(1954), and by SAVARD (1954). From a practical point of view, solvent systems with non-aqueous stationary phase are less convenient to use than aqueous solvent systems; they ordinarily require more time and are less reproducible, but they accommodate larger quantities of more complex mixtures and are better suited to the separation of the less polar steroids (BUSH 1954a). They have excellent resolving power, provided a suitable mobile phase is allowed to overflow the paper for a sufficiently long time. However, since the solvent runs off the paper, R_F values cannot be determined. The rate of movement is dependent on a

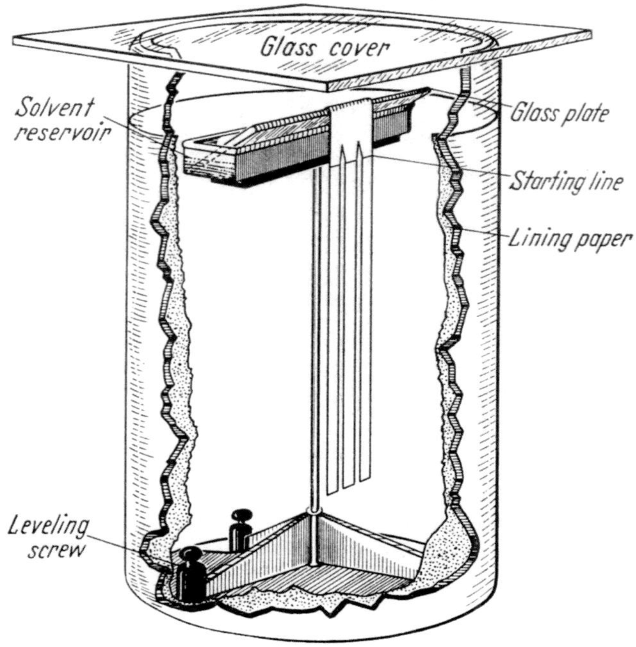

Fig. 94. Assembly for descending paper chromatography (BURTON et al. 1951). The atmosphere inside the chromatographic chamber is kept saturated with respect to the mobile phase with the aid of filter paper lining, which stands in a layer of solvent on the bottom of the jar. The lid is sealed to the jar with a starch-glycerol paste. The chromatographic paper is slotted to prevent sidewise diffusion of the steroids. Redrawn by permission of the authors and of the American Society of Biological Chemists, Inc., from the Journal of Biological Chemistry 188, 765 (1951)

number of factors, including the amount of stationary phase on the paper (DIRSCHERL and GERHARDS 1955). Diffusion of the zones occurs in the stationary phase, but it may be limited by cutting the filter paper into lanes. Diffusion of the spots increases with the amount of substance chromatographed and with the distance traveled. Esters give more diffuse zones than the corresponding alcohols. The chromatographic technique described by ZAFFARONI (1953) is as follows:

The chromatographic chamber is a battery jar 12″ in diameter and 18″ high, covered by plate glass (Fig. 94). The solvent reservoir is a rectangular tray, $8\times2^1/_2\times1^3/_8$″, held 14″ from the base on a tripod. The solvents are equilibrated in a separatory funnel. Whatman No. 1 filter paper, cut into strips 42 cm long and further subdivided to give several 1-cm wide limbs attached to a common head, is dipped into the stationary phase, which has been diluted with an equal volume of methanol. The methanol is allowed to evaporate, leaving the paper uniformly impregnated. The samples (in methanol solution) are applied to the starting line of the chromatograms by means of a medicine dropper, while the paper rests on a glass-plate. At this time the point of application is lifted away from the support by two glass rods and a stream of air is used intermittently to evaporate the methanol. Then the top edge of the paper is aligned with the long edge of a glass plate $7\times3^3/_8$″ and plate and paper are placed in the trough. The mobile solvent phase is added to the trough and any air bubbles between paper und glass plate are pressed out. After development, the paper is dried in a current of air.

Table 14 lists some of the solvent systems used in this type of chromatography and the type of steroids for which they are best suited.

Table 14. *Non-aqueous solvent systems for partition chromatography of steroids* (ZAFFARONI 1953)

Solvent systems							21-Monoacetates			Diacetates		Others
	$C_{19}O_2$	$C_{21}O_2$	$C_{21}O_3$	$C_{21}O_4$	$C_{21}O_5$	$C_{21}O_6$	$C_{21}O_3$	$C_{21}O_4$	$C_{21}O_5$	$C_{21}O_4$	$C_{21}O_5$	
n-Hexane/formamide	×	×	×				×			×		etioacids adrenosterone
n-Hexane/1,2-propanediol	×	×	×				×			×		
n-Hexane-benzene (1 : 1)/ formamide	×	×	×	×				×	×		×	
n-Hexane-benzene (1 : 1)/ 1,2-propanediol	×	×	×	×				×	×		×	
Benzene/formamide			×	×	×				×			
Benzene/1,2-propanediol			×	×	×				×			
Toluene/1,2-propanediol			×	×	×				×			
Chloroform/formamide				×	×	×						

Since complex mixtures of corticosteroids cannot be resolved by a single solvent system, ZAFFARONI (1953) has suggested a fractionation scheme, based on the development of a paper strip, impregnated with formamide with a series of mobile phases. Each mobile phase is saturated with respect to formamide and the paper strip is transferred from one chamber to another, allowing a few minutes for evaporation of the mobile phase between transfers and collecting the eluate in each case for 3 hours.

The mixture is applied to filter paper impregnated with formamide (50% in methanol) and developed with hexane. The eluate contains fats and cholesterol. The paper is transferred to a chamber with hexane-benzene (1 : 1) as the mobile phase and eluted. The second eluate contains $C_{19}O_2$ and $C_{21}O_2$ compounds as well as $C_{21}O_3$-monoacetates. Subsequent elution with benzene yields a fraction containing $C_{21}O_3$ compounds and $C_{21}O_4$-monoacetates. Final development with chloroform separates $C_{21}O_4$ compounds and $C_{21}O_5$-monoacetates in the eluate from $C_{21}O_5$ compounds on the paper strip.

The sequence in which steroids move in partition chromatograms depends primarily on the number and nature of their oxygen functions. Generalizations about the relation between their structure and chromatographic behavior are in the literature (ZAFFARONI 1953, SAVARD 1954), but many exceptions to these rules are known (AXELROD 1953a, REINEKE 1956), some of which have been used as confirmation of identity (PECHET 1955, NEHER and WETTSTEIN 1956b).

Table 15 lists the mobilities (units of distance per units of time) of various steroids in non-aqueous solvent systems. In this and the following table the steroids are arranged in the order: 1) increasing number of carbons, except that derivatives (esters, alkyl or halo derivatives, etc.) follow the parent compound, 2) increasing number of oxygens, 3) increasing number of hydroxyl groups, 4) increasing number of double bonds, 5) A/B *cis* compounds are followed by A/B *trans* compounds, 6) increasing number of position, 7) α configuration followed by β. Due to the difficulty in reproducing absolute mobilities no attempt has been made to correlate the figures in Table 15, which are collected from various papers. Most mobilities are expressed as ratios of the distance traveled by the steroid to the distance traveled by a reference steroid in the same chromatogram. The R_T values (SAVARD 1953) are corrected relative mobilities for the given steroids in cm/hr, calculated by the equation:

$$R_T = \frac{\text{distance in cm traveled by a steroid}}{\text{distance in cm traveled by the reference steroid}} \times K$$

Table 15. *Mobilities of steroids in non-aqueous solvent systems*

Compound	Solvent systems (see Legend)												
	1	2	3	4	5	6	7	8	9	10	11	12	13
3-Hydroxy-1,3,5(10)-estratrien-17-one	0.66	1.3	...	1.6	1.42	...	...	...	...	...	...	...	...
1,3,5(10)-Estratriene-3,17 β-diol	...	...	1.7	0.40	0.68	...	...	...	...	...	...	2.1	...
1,3,5(10)-Estratriene-3,17 α-diol	...	...	...	...	...	...	...	...	...	...	...	2.8	...
1,3,5(10)-Estratriene-3,16 α,17 β-triol	...	...	...	...	...	...	...	...	...	...	...	0.1	...
Androstan-3-one	...	...	...	...	...	...	...	...	...	...	6.25	...	...
Androstan-17-one	...	...	...	...	...	...	...	...	...	...	6.30	...	...
4-Androsten-3-one	...	...	...	...	...	...	...	...	...	4.81	...	...	...
3 β-Chloro-5-androsten-17-one	...	...	...	...	...	...	...	...	...	9.0	...	...	...
Androstan-3 α-ol	...	...	...	...	...	...	...	...	...	...	6.85	30	...
Androstan-3 β-ol	...	...	...	...	...	...	...	...	...	...	8.34	...	...
Androstan-17 α-ol	...	...	...	...	...	...	...	...	...	...	...	25	...
5-Androsten-3 β-ol	...	...	...	...	...	...	...	...	...	...	4.60	9	...
Etiocholane-3,17-dione	...	...	...	...	...	...	...	...	...	2.5	...	...	...
Androstane-3,17-dione	2.7	...	...	...	...	...	...	...	...	2.5	8.40	...	...
1-Androstene-3,17-dione	...	...	...	...	...	...	...	...	...	2.0	...	...	...
4-Androstene-3,17-dione	1.65	...	...	5.3	...	...	...	...	...	1.7	5.80	...	...
1,4-Androstadiene-3,17-dione	1.0	...	...	...	...	...	...	...	...	0.62	2.07	...	...
4,6-Androstadiene-3,17-dione	...	...	...	...	...	...	...	...	...	...	2.85	...	...
4,9(11)-Androstadiene-3,17-dione	1.65	...	...	...	...	...	...	...	...	...	...	...	...
3 α-Hydroxyetiocholan-17-one	...	...	...	...	...	...	...	...	...	0.70	1.91	...	...
3 β-Hydroxyetiocholan-17-one	...	...	...	...	...	...	...	...	...	1.00	...	...	...
17 β-Hydroxyetiocholan-3-one	...	...	...	...	...	...	...	...	...	0.70	...	...	...
3 α-Hydroxyandrostan-17-one	...	...	...	...	...	...	...	...	...	1.00	2.72	...	...
3 β-Hydroxyandrostan-17-one	...	...	...	...	...	...	...	...	...	0.70	1.66	...	...
17 β-Hydroxyandrostan-3-one	...	...	...	...	...	...	...	...	...	...	2.08	...	...
6 β-Hydroxy-3,5-cycloandrostan-17-one	...	...	...	...	...	...	...	...	...	1.5	...	...	...
3 β-Hydroxy-5-androsten-17-one	1.15	...	...	3.6	...	...	...	...	...	0.70	1.73	...	...
3 β-Acetoxy-5-androsten-17-one	5.5	...	...	...	...	...	...	...	...	...	...	...	...
17 α-Hydroxy-4-androsten-3-one	3.5	...	...	...	...	...	...	...	...	0.40	1.35	...	...
17 β-Hydroxy-4-androsten-3-one	0.89	1.50	...	2.0	1.44	...	...	...	...	0.33	1.00	...	...
17 β-Acetoxy-4-androsten-3-one	3.5	...	...	...	...	...	...	...	...	...	...	...	...
17 β-Hydroxy-1,4-androstadien-3-one	0.60	1.0	...	1.35	1.24	...	...	...	...	0.12	0.42	...	...
Etiocholane-3 α,17 β-diol	...	...	...	...	...	...	...	...	...	...	...	2.5	...
Androstane-3 α,17 β-diol	...	...	...	...	...	...	...	...	...	...	0.45	...	...
Androstane-3 β,17 α-diol	...	...	...	...	...	...	...	...	...	...	...	4.0	...

Table 15 (cont.)

Compound	Solvent systems (see Legend)												
	1	2	3	4	5	6	7	8	9	10	11	12	13
Androstane-3 β,17 β-diol	...	...	...	...	...	...	...	...	...	...	0.15	2.0	...
5-Androstene-3 β,17 α-diol	...	...	...	...	...	...	...	...	...	...	0.32	3.5	...
5-Androstene-3 β,17 β-diol	...	0.38	2.9	0.57	...	...	...	...	...	...	0.20	...	...
3,17-*seco*-13 α-Hydroxy-3-keto-4-androsten-17-oic acid lactone (13 → 17)	0.4	2.8	...	2.7	1.23	...	...	...	...	...	...	...	...
Androstane-3,7,17-trione	...	...	...	...	...	...	...	...	...	...	2.57	...	...
4-Androstene-3,6,17-trione	0.72	...	...	5.5	1.61	...	...	...	...	...	...	...	...
4-Androstene-3,11,17-trione	0.68	...	...	4.8	1.51	...	...	...	...	0.36	2.08	...	...
3 α-Hydroxyetiocholane-11,17-dione	...	...	...	...	...	...	...	...	...	0.26	...	...	...
3 α-Hydroxyandrostane-11,17-dione	...	...	...	...	...	...	...	...	...	0.30	...	...	...
11 β-Hydroxyandrostane-3,17-dione	...	...	...	...	...	...	...	...	...	0.17	...	...	...
6 α-Hydroxy-4-androstene-3,17-dione	...	...	...	...	...	...	...	...	...	0.024	0.11	...	...
6 β-Hydroxy-4-androstene-3,17-dione	...	1.0	8.4	0.89	0.74	...	...	...	...	0.080	0.33	...	...
11 α-Hydroxy-4-androstene-3,17-dione	...	0.65	4.6	0.49	0.45	...	...	...	...	...	...	...	...
11 β-Hydroxy-4-androstene-3,17-dione	...	1.50	...	1.4	1.0	...	...	...	...	0.10	...	...	...
14 α-Hydroxy-4-androstene-3,17-dione	...	0.90	5.9	0.79	0.55	...	...	...	...	0.039	...	...	...
15 α-Hydroxy-4-androstene-3,17-dione	...	...	4.0	0.40	0.41	...	...	...	...	...	...	...	...
17 β-Hydroxy-4-androstene-3,11-dione	...	0.82	5.3	0.64	0.65	...	...	...	...	...	...	...	...
3 α,11 β-Dihydroxyetiocholan-17-one	...	...	...	...	...	...	...	...	...	0.048	...	...	...
3 α,11 β-Dihydroxyandrostan-17-one	...	...	...	...	...	...	...	...	...	0.10	...	...	...
3 α,17 β-Dihydroxyandrostan-16-one	...	...	...	...	...	...	...	...	...	...	0.31	...	...
6 β,17 β-Dihydroxy-4-androsten-3-one	...	...	1.1	0.13	0.16	...	...	...	...	...	...	...	...
11 α,17 β-Dihydroxy-4-androsten-3-one	0.16	0.074	0.61	0.065	0.09	...	...	...	...	...	...	...	...
11 β,17 β-Dihydroxy-4-androsten-3-one	...	...	1.7	0.21	0.28	...	...	...	...	...	...	...	...
14 α,17 β-Dihydroxy-4-androsten-3-one	...	...	1.39	0.18	0.21	...	...	...	...	...	...	...	...
3 β,14 α-Dihydroxy-5-androsten-17-one	...	...	...	...	...	...	...	...	...	0.034	...	...	...
Pregnane-3,20-dione	3.5	...	...	...	...	...	...	...	...	3.0	...	...	...
Allopregnane-3,20-dione	3.5	...	...	...	...	...	...	...	...	3.0	...	...	...
4-Pregnene-3,20-dione	2.7	...	...	6.0	...	...	...	...	...	2.3	...	...	...
1,4-Pregnadiene-3,20-dione	1.65	...	...	...	...	...	...	...	...	...	...	...	...
4,6-Pregnadiene-3,20-dione	2.7	...	...	...	...	...	...	...	...	...	...	...	...
4,9(11)-Pregnadiene-3,20-dione	2.7	...	...	...	...	...	...	...	...	...	...	...	...
4,16-Pregnadiene-3,20-dione	2.7	...	...	...	...	...	...	...	...	...	...	...	...
3 α-Hydroxypregnan-20-one	1.95	...	...	...	...	...	...	...	...	1.30	...	...	...
3 α-Hydroxyallopregnan-20-one	...	...	...	...	...	...	...	...	...	1.40	...	...	...

3β-Hydroxyallopregnan-20-one	1.65	...	...	...	...	...	...	...	...	1.20	...	...	...
3β-Hydroxy-5-pregnen-20-one	1.65	...	...	...	...	...	...	...	...	1.20	...	...	...
20α-Hydroxy-4-pregnen-3-one	...	...	...	...	...	...	...	...	...	0.24	...	...	...
3β-Hydroxy-5,16-pregnadien-20-one	1.65	...	...	...	...	...	...	...	...	...	...	...	...
Pregnane-3α,20α-diol	...	...	...	...	...	...	...	...	...	...	...	1.9	...
Allopregnane-3α,20α-diol	...	...	...	...	...	...	...	...	...	...	...	1.7	...
Allopregnane-3β,20α-diol	...	...	...	...	...	...	...	...	...	...	...	3.0	...
Pregnane-3,11,20-trione	1.25	...	...	...	...	...	...	...	...	...	...	...	...
Allopregnane-3,11,20-trione	1.25	...	...	...	...	...	...	...	...	...	...	...	...
4-Pregnene-3,11,20-trione	1.00	...	...	4.0	...	...	...	...	...	...	...	...	...
3α-Hydroxypregnane-11,20-dione	0.79	1.6	...	2.1	...	...	...	...	...	...	...	...	...
3β-Hydroxypregnane-11,20-dione	0.79	1.6	...	2.1	...	...	...	...	...	...	...	...	...
11α-Hydroxypregnane-3,20-dione	0.79	1.6	...	2.1	...	...	...	...	...	...	...	...	...
11β-Hydroxypregnane-3,20-dione	1.0	...	...	3.3	...	...	...	...	...	...	...	...	...
3β-Hydroxyallopregnane-11,20-dione	0.64	1.6	...	1.9	...	...	...	...	...	...	...	...	...
11α-Hydroxyallopregnane-3,20-dione	0.64	1.6	...	1.9	...	...	...	...	...	...	...	...	...
21-Hydroxyallopregnane-3,20-dione	...	...	...	...	...	...	...	...	...	1.06	...	...	...
6α-Hydroxy-4-pregnene-3,20-dione	...	...	...	...	...	...	...	...	...	0.026	...	...	...
6β-Hydroxy-4-pregnene-3,20-dione	0.63	1.5	...	1.6	1.12	...	...	...	...	0.14	...	...	...
11α-Hydroxy-4-pregnene-3,20-dione	0.4	1.0	8.4	1.0	1.0	...	...	...	...	0.079	...	...	...
11α-Acetoxy-4-pregnene-3,20-dione	1.25	...	...	5.3		...	...	...	...	...	...	...	...
11β-Hydroxy-4-pregnene-3,20-dione	0.60	1.8	...	2.1	1.42	...	...	...	...	0.20	...	...	...
14α-Hydroxy-4-pregnene-3,20-dione	0.60	1.8	...	2.1	1.32	...	...	...	...	...	...	...	...
15α-Hydroxy-4-pregnene-3,20-dione	0.37	1.0	6.8	0.77	0.82	...	...	...	...	...	...	...	...
15β-Hydroxy-4-pregnene-3,20-dione	0.4	1.1	...	1.2	1.0	...	...	...	...	...	...	...	...
16α-Hydroxy-4-pregnene-3,20-dione	0.4	0.8	6.8	0.92	0.74	...	...	...	...	...	...	...	...
17α-Hydroxy-4-pregnene-3,20-dione	0.78	1.8	...	2.55	1.51	...	...	...	...	0.28	...	...	...
21-Hydroxy-4-pregnene-3,20-dione	0.86	...	...	4.0	1.5	...	...	...	1.20	0.45	...	...	1.000
21-Acetoxy-4-pregnene-3,20-dione	1.4	...	...	...		...	...	...	1.54	...	...	...	...
11α-Hydroxy-4,16-pregnadiene-3,20-dione	0.4	1.0	...	1.0	1.04	...	...	...	...	...	...	...	...
21-Acetoxy-4,16-pregnadiene-3,20-dione	1.15	...	...	...	...	...	...	...	...	...	...	...	...
3α,11α-Dihydroxypregnan-20-one	0.4	...	3.0	0.43	0.74	...	...	...	...	...	...	...	...
3α,17α-Dihydroxypregnan-20-one	...	...	...	...	...	...	...	...	...	0.090	...	...	...
3β,17α-Dihydroxypregnan-20-one	...	...	...	...	...	...	...	...	...	0.21	...	...	...
3β,11α-Dihydroxyallopregnan-20-one	0.24	...	2.3	0.29	0.63	...	...	...	...	...	...	...	...
3β,21-Dihydroxy-5-pregnen-20-one	...	...	...	...	...	...	...	...	...	0.34	...	...	...
Pregnane-3α,17α,20α-triol	...	...	...	...	...	...	...	...	...	...	...	1.0	...
17α-Hydroxypregnane-3,11,20-trione	0.64	1.5	...	1.8	...	...	...	...	...	...	...	...	...
6β-Hydroxy-4-pregnene-3,11,20-trione	...	0.91	...	0.57	0.58	...	...	...	...	...	...	...	...
17α-Hydroxy-4-pregnene-3,11,20-trione	0.4	1.0	...	1.0	0.89	...	...	...	...	0.061	...	...	...
21-Hydroxy-4-pregnene-3,11,20-trione	0.28	...	11.0	1.4	0.69	...	...	...	...	0.074	...	...	0.500

Table 15 (cont.)

Compound	Solvent systems (see Legend) 1	2	3	4	5	6	7	8	9	10	11	12	13
21-Acetoxy-4-pregnene-3,11,20-trione	0.47	...	...	4.7	1.6	...	...	...	1.39	...	...	...	...
3α,17α-Dihydroxypregnane-11,20-dione	0.35	...	1.9	0.27	0.48	...	...	...	...	...	...	...	...
3α,21-Dihydroxypregnane-11,20-dione	...	...	...	...	...	...	>6.00	3.75	...	...	...	...	0.140
6β,11α-Dihydroxypregnane-3,20-dione	...	...	1.0	...	0.067	...	...	...	...	...	...	...	...
3β,21-Dihydroxyallopregnane-11,20-dione	...	...	...	...	...	...	>6.00	3.71	...	...	...	...	...
6β,11α-Dihydroxy-4-pregnene-3,20-dione	...	0.074	0.61	0.04	0.05	...	...	...	...	...	...	...	...
6β,17α-Dihydroxy-4-pregnene-3,20-dione	...	...	1.7	0.19	0.21	...	...	...	...	...	...	...	...
6α,21-Dihydroxy-4-pregnene-3,20-dione	...	...	...	...	...	...	...	...	...	0.008	...	...	...
6β,21-Dihydroxy-4-pregnene-3,20-dione	...	...	3.5	0.28	0.19	...	...	...	...	0.016	...	...	...
11α,17α-Dihydroxy-4-pregnene-3,20-dione	...	0.14	1.0	0.09	0.13	...	...	...	...	...	...	...	...
11β,17α-Dihydroxy-4-pregnene-3,20-dione	...	0.38	2.9	0.29	0.39	...	...	...	...	...	...	...	...
11α,21-Dihydroxy-4-pregnene-3,20-dione	...	0.23	1.7	0.16	0.13	...	5.07	3.71	...	...	...	...	...
11β,21-Dihydroxy-4-pregnene-3,20-dione	0.22	0.63	4.6	0.49	0.39	...	...	3.51	...	0.025	...	...	0.204
21-Acetoxy-11β-hydroxy-4-pregnene-3,20-dione	0.4	2.6	...	2.1	1.36	...	...	...	1.00	...	...	...	...
14α,21-Dihydroxy-4-pregnene-3,20-dione	0.17	0.42	3.5	0.36	0.25	...	...	...	...	...	...	...	...
15α,21-Dihydroxy-4-pregnene-3,20-dione	...	...	1.36	0.12	0.086	...	...	...	...	...	...	...	...
21-Acetoxy-15α-hydroxy-4-pregnene-3,20-dione	...	0.92	...	0.73	0.79	...	...	...	...	...	...	...	...
17α,21-Dihydroxy-4-pregnene-3,20-dione	0.28	0.42	3.5	0.36	0.35	...	...	...	...	0.029	...	...	0.140
21-Acetoxy-17α-hydroxy-4-pregnene-3,20-dione	0.49	1.8	...	1.7	1.5	...	...	...	0.93	...	...	...	...
19,21-Dihydroxy-4-pregnene-3,20-dione	...	...	...	...	...	...	4.11	2.79	...	...	...	...	...
21-Acetoxy-19-hydroxy-4-pregnene-3,20-dione	...	...	...	...	...	...	...	...	0.56	...	...	...	...
19-Acetoxy-21-hydroxy-4-pregnene-3,20-dione	...	...	...	...	...	...	...	...	1.22	...	...	...	...
19,21-Diacetoxy-4-pregnene-3,20-dione	...	...	...	...	...	...	...	...	1.53	...	...	...	...
20β,21-Dihydroxy-4-pregnene-3,11-dione	...	...	...	...	...	...	2.41	2.94	...	...	...	...	...
3α,11β,21-Trihydroxypregnan-20-one	...	...	...	...	...	...	5.43	2.60	...	...	...	...	...
3α,17α,21-Trihydroxypregnan-20-one	...	...	...	...	...	...	2.80	2.79	...	...	...	...	...
3β,11β,21-Trihydroxyallopregnan-20-one	...	...	...	...	...	...	4.00	2.63	...	...	...	...	...
3β,17α,21-Trihydroxyallopregnan-20-one	...	...	...	...	...	...	2.34	2.50	...	...	...	...	...
Allopregnane-3β,17α,20β,21-tetrol	...	...	1.5	...	...	...	0.47	0.46	...	...	...	...	...
17α,21-Dihydroxypregnane-3,11,20-trione	0.21	...	9.5	0.11	0.11	...	5.31	2.84	...	...	...	...	...
21-Acetoxy-17α-hydroxypregnane-3,11,20-trione	0.4	1.5	...	...	1.0	...	...	...	...	...	...	...	...
17α,21-Dihydroxyallopregnane-3,11,20-trione	...	...	...	...	...	...	5.47	3.11	...	...	...	...	...
6β,21-Dihydroxy-4-pregnene-3,11,20-trione	...	...	...	...	...	...	1.87	1.62	...	...	...	...	...
6β,21-Diacetoxy-4-pregnene-3,11,20-trione	...	...	...	...	...	...	...	...	1.44	...	...	...	...
17α,21-Dihydroxy-4-pregnene-3,11,20-trione	0.16	...	1.0	0.085	0.055	1.95	2.51	2.39	...	...	...	...	...

21-Acetoxy-17α-hydroxy-4-pregnene-3,11,20-trione	0.22	0.91	6.0	0.64	0.65	...	...	...	0.42	...	...	...	...
17α,21-Dihydroxy-1,4-pregnadiene-3,11,20-trione	...	...	0.66	...	...	...	...	...	...	...	...	...	...
21-Acetoxy-17α-hydroxy-1,4-pregnadiene-3,11,20-trione	...	...	4.4	0.39	0.56	...	...	...	...	...	...	...	...
3α,17α,21-Trihydroxypregnane-11,20-dione	...	...	...	...	...	...	0.42	1.01	...	...	...	...	...
21-Acetoxy-3α,17α-dihydroxypregnane-11,20-dione	...	...	1.4	0.11	0.31	...	...	...	...	...	...	...	...
11β,17α,21-Trihydroxypregnane-3,20-dione	...	...	...	...	...	...	2.70	1.43	...	...	...	...	...
3β,17α,21-Trihydroxyallopregnane-11,20-dione	...	...	...	...	...	...	0.40	0.79	...	...	...	...	...
21-Acetoxy-3β,17α-dihydroxypregnane-11,20-dione	...	...	1.1	0.08	0.23	...	...	...	...	...	...	...	...
3β,21-Diacetoxy-17α-hydroxyallopregnane-11,20-dione	...	...	...	...	...	...	...	...	1.17	...	...	...	...
11β,17α,21-Trihydroxyallopregnane-3,20-dione	...	...	...	...	...	...	2.67	1.61	...	...	...	...	...
6β,11β,21-Trihydroxy-4-pregnene-3,20-dione	...	...	...	...	...	0.30	0.66	0.50	...	...	...	...	...
6β,21-Diacetoxy-11β-hydroxy-4-pregnene-3,20-dione	...	...	...	...	...	...	...	...	0.98	...	...	...	...
6β,17α,21-Trihydroxy-4-pregnene-3,20-dione	...	...	0.23	...	...	0.31	0.49	0.58	...	...	...	...	...
11α,17α,21-Trihydroxy-4-pregnene-3,20-dione	...	...	0.12	0.01	0.01	...	0.20	0.46	...	...	...	...	...
11β,17α,21-Trihydroxy-4-pregnene-3,20-dione	...	...	0.44	0.033	0.029	1.00	1.00	1.00	...	...	...	...	0.017
21-Acetoxy-11β,17α-dihydroxy-4-pregnene-3,20-dione	0.17	...	2.9	0.29	0.39	...	...	...	...	...	...	...	...
2-Methyl-11β,17α,21-trihydroxy-4-pregnene-3,20-dione	...	...	0.86	...	0.09	...	...	...	...	...	...	...	...
2-Methyl-21-acetoxy-11β,17α-dihydroxy-4-pregnene-3,20-dione	...	...	6.1	0.64	0.92	...	...	...	...	...	...	...	...
21-Acetoxy-9α-fluoro-11β,17α-dihydroxy-4-pregnene-3,20-dione	...	...	1.4	...	0.23	...	...	...	...	...	...	...	...
2-Methyl-9α-fluoro-11β,17α,21-trihydroxy-4-pregnene-3,20-dione	...	...	0.44	...	0.04	...	...	...	...	...	...	...	...
2-Methyl-21-acetoxy-9α-fluoro-11β,17α-dihydroxy-4-pregnene-3,20-dione	...	...	4.4	0.49	0.56	...	...	...	...	...	...	...	...
11β,19,21-Trihydroxy-4-pregnene-3,20-dione	...	...	...	...	...	0.60	0.69	1.37	...	...	...	...	...
19,21-Diacetoxy-11β-hydroxy-4-pregnene-3,20-dione	...	...	...	...	...	...	...	...	0.84	...	...	...	...
17α,19,21-Trihydroxy-4-pregnene-3,20-dione	...	...	...	...	...	0.13	0.19	0.36	...	...	...	...	...
19,21-Diacetoxy-17α-hydroxy-4-pregnene-3,20-dione	...	...	...	...	...	...	...	...	0.91	...	...	...	...
17α,20α,21-Trihydroxy-4-pregnene-3,11-dione	...	...	...	...	...	0.59	0.95	1.59	...	...	...	...	...
21-Acetoxy-17α,20α-dihydroxy-4-pregnene-3,11-dione	...	...	...	...	...	...	...	...	0.16	...	...	...	...
20α,21-Diacetoxy-17α-hydroxy-4-pregnene-3,11-dione	...	...	...	...	...	...	...	...	0.82	...	...	...	...

Table 15 (cont.)

Compound	Solvent systems (see Legend)												
	1	2	3	4	5	6	7	8	9	10	11	12	13
17 α,20 β,21-Trihydroxy-4-pregnene-3,11-dione	...	...	...	...	...	0.60	0.84	1.48	...	...	...	...	...
20 β,21-Diacetoxy-17 α-hydroxy-4-pregnene-3,11-dione	...	...	...	...	...	...	...	...	0.80	...	...	...	...
11 β,17 α,21-Trihydroxy-1,4-pregnadiene-3,20-dione	...	...	0.27	...	0.018	...	...	...	...	...	...	...	...
21-Acetoxy-11 β,17 α-dihydroxy-1,4-pregnadiene-3,20-dione	...	...	1.3	0.15	0.25	...	...	...	...	...	...	...	...
3 α,11 β,17 α,21-Tetrahydroxypregnan-20-one	...	...	...	...	...	...	0.29	0.28	...	...	...	...	...
3 α,11 β,17 α,21-Tetrahydroxyallopregnan-20-one	...	...	...	...	...	...	0.40	0.40	...	...	...	...	...
3 α,21-Diacetoxy-11 β,17 α-dihydroxyallopregnan-20-one	...	...	...	...	...	...	...	...	0.98	...	...	...	...
3 β,11 β,17 α,21-Tetrahydroxyallopregnan-20-one	...	...	...	...	...	...	0.20	0.22	...	...	...	...	...
3 β,21-Diacetoxy-11 β,17 α-dihydroxyallopregnan-20-one	...	...	...	...	...	...	...	...	0.98	...	...	...	...
11 β,17 α,20 α,21-Tetrahydroxy-4-pregnen-3-one	...	...	0.09	...	...	...	...	...	...	...	...	...	...
21-Acetoxy-11 β,17 α,20 α-trihydroxy-4-pregnen-3-one	...	...	1.0	...	0.055	...	...	...	...	...	...	...	...
11 β,17 α,20 β,21-Tetrahydroxy-4-pregnen-3-one	...	...	0.12	...	...	...	0.25	0.48	...	...	...	...	...
Allopregnane-3 β,11 β,17 α,20 β,21-pentol	...	...	...	...	...	...	0.04	0.04	...	...	...	...	...
11 β,21-Dihydroxy-3,20-diketo-4-pregnen-18-al	...	...	...	...	...	1.18	2.59	2.27	...	...	...	...	...
21-Acetoxy-11 β-hydroxy-3,20-diketo-4-pregnen-18-al	...	...	...	...	...	...	...	...	0.51 1.27	...	...	...	...
11 β,21-Diacetoxy-3,20-diketo-4-pregnen-18-al	...	...	...	...	...	...	...	...	...	...	...	...	...
20,20,21-Trihydroxy-3-keto-4-pregnen-18-oic acid lactone (18 → 20)	...	...	...	...	...	...	2.46	2.75	...	...	...	...	...
20,21-Diacetoxy-20-hydroxy-3-keto-4-pregnen-18-oic acid lactone (18 → 20)	...	...	...	...	...	...	...	...	1.39	...	...	...	...

Legend

No.	Mobile phase	Stationary phase	Mobility relative to	Reference
1	Methyl cyclohexane	Carbitol (diethylene glycol monoethyl ether)	4-Pregnene-3,11,20-trione	REINEKE (1956)
2	Toluene	Propylene glycol	11 α-Hydroxy-4-pregnene-3,20-dione	REINEKE (1956)
3	Toluene	Propylene glycol, diluted with methanol (1 : 1), dried at 37° for 10 min	Cortisone	REINEKE (1956)
4	Cyclohexane-benzene (1:1)	Propylene glycol, diluted with methanol (1 : 1), dried at 37° for 10 min	11 α-Hydroxy-4-pregnene 3,20-dione	REINEKE (1956)
5	Benzene	Formamide, diluted with methanol (1 : 1), dried at 37° for 10 min	4-Pregnene-3,11,20-trione	REINEKE (1956)
6	Benzene	Formamide (20% in acetone), dried 5—10 min at room temp.	Cortisol	NEHER and WETTSTEIN (1956b)
7	Toluene	Propylene glycol (20% in acetone), dried 5—10 min at room temp.	Cortisol	NEHER and WETTSTEIN (1956b)
8	Chloroform	Formamide (20% in acetone), dried 5—10 min at room temp.	Cortisol	NEHER and WETTSTEIN (1956b)
9	Toluene	Propylene glycol (20% in acetone), dried 5—10 min at room temp.	Corticosterone	NEHER and WETTSTEIN (1956b)
10	Ligroin (boiling range 65—115°)	Propylene glycol (50% in methanol), blotted	R_T[1]	SAVARD (1953)
11	Cyclohexane-benzene (1:1)	Propylene glycol (50% in methanol)	Testosterone	KOCHAKIAN and STIDWORTHY (1952)
12	Toluene	Propylene glycol (50% in methanol)	Pregnane-3α,17α,20α-triol	ROSENKRANTZ (1953a,b)
13	Toluene	Propylene glycol	Deoxycorticosterone	BURTON et al. (1951)

[1] Defined in the text.

where K = rate of movement (in cm/hr) for the reference steroid in the solvent system being used. It was found (SAVARD 1953) that the mobility of steroids is increased by admixture of steroids having a slightly lesser mobility. The "displaced R_T" values, i. e., R_T values in the presence of these steroids, may be found in SAVARD's paper.

A number of other non-aqueous solvent systems have been used for paper chromatography (e. g., AXELROD 1953a, ARROYAVE and AXELROD 1954). A solvent system of 2-phenoxyethanol (phenylcellosolve) as the stationary phase and heptane as the mobile phase, which was first proposed for the separation of relatively non-polar steroids by NEHER and WETTSTEIN (1952), is particularly suitable for 17-ketosteroids (SAVARD 1953, B. RUBIN et al. 1953a). Since 3α-hydroxyandrostan-17-one and 3α-hydroxyetiocholan-17-one do not separate from their respective $\Delta^{9(11)}$-unsaturated analogues, perbenzoic acid oxidation was used by B. RUBIN et al. (1953a) to convert the latter to the epoxides prior to chromatography.

Reversed phase partition chromatography on paper has also been described, notably by KRITCHEVSKY and TISELIUS (1951), by KUČERA et al. (1957) and by MARTIN (1957).

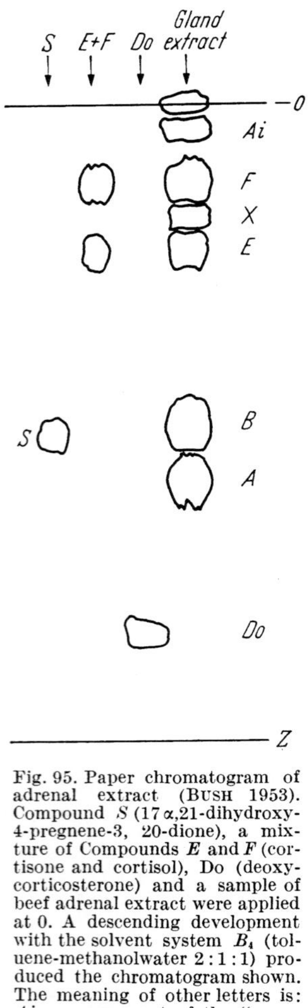

Fig. 95. Paper chromatogram of adrenal extract (BUSH 1953). Compound S (17α,21-dihydroxy-4-pregnene-3, 20-dione), a mixture of Compounds E and F (cortisone and cortisol), Do (deoxycorticosterone) and a sample of beef adrenal extract were applied at 0. A descending development with the solvent system B_4 (toluene-methanolwater 2:1:1) produced the chromatogram shown. The meaning of other letters is: Ai = component of the "amorphous fraction," X = unidentified component, B = corticosterone, A = 11-dehydrocorticosterone, and Z = solvent front. Reproduced by permission of the author and of Cambridge University Press from the Biochemical Journal **50**, 376 (1952)

b) Stationary phase: Aqueous solvents

When water-containing solvent systems are used in paper chromatography, it is not necessary, as a rule, to impregnate the filter paper with the stationary phase, since it already contains water. However, it has been reported (BUSH 1952) that streaking of the zones occurs unless the development is carried out at a slightly elevated temperature, usually between 30 and 38°. This inconvenience can be avoided by impregnating the paper with the stationary phase or 70% aqueous acetone (SHULL 1954). After the acetone has evaporated, the chromatogram can be developed at room temperature after only a short incubation in the chamber. EBERLEIN and BONGIOVANNI (1955) have used aqueous solvent systems at room temperature after equilibrating the paper in the chamber at 35–37° for 5–12 hours.

The advantages and disadvantages of aqueous solvent systems have been reviewed by BUSH (1954a) and some of them have been mentioned earlier. The convenience, rapidity, and reproducibility of paper chromatography in these solvent systems make this technique very useful, especially for the analysis of small quantities of relatively polar steroids. Examples given below will illustrate the flexibility of the method, which is achieved by changing the composition of the solvent systems. In all cases the components of the solvent system are equilibrated in a separatory funnel and the phase richer in organic solvents is used as the mobile phase. The atmosphere in the chromatographic chamber and the filter

Table 16. *R_F values of steroids in aqueous solvent systems*

Compound	Solvent systems (cf. Legend)														
	1	2	3	4	5	6	7	8	9	10	11[3]	12[3]	13[3]	14[3]	15
3-Hydroxy-1,3,5(10)-estratrien-17-one	0.14	0.74	0.52	. .	. .	. .	. .	. .	. .	. .	. .	. .	. .	. .	. .
1,3,5(10)-Estratriene-3,17 β-diol	. .	0.47	0.21	. .	. .	. .	. .	. .	. .	. .	. .	. .	. .	. .	. .
Androstane-3,17-dione	0.76	. .	. .	. .	. .	. .	. .	. .	. .	. .	. .	. .	. .	. .	. .
4-Androstene-3,17-dione	0.50	0.89	0.78	. .	. .	. .	. .	. .	. .	. .	. .	. .	. .	. .	. .
1,4-Androstadiene-3,17-dione	0.27	. .	0.78	. .	. .	. .	. .	. .	. .	. .	. .	. .	. .	. .	. .
4,9(11)-Androstadiene-3,17-dione	0.50	. .	0.78	. .	. .	. .	. .	. .	. .	. .	. .	. .	. .	. .	. .
3 β-Hydroxy-5-androsten-17-one	0.38	0.83	0.65	. .	. .	. .	. .	. .	. .	. .	. .	. .	. .	. .	. .
3 β-Acetoxy-5-androsten-17-one	0.85	0.99	0.99	. .	. .	. .	. .	. .	. .	. .	. .	. .	. .	. .	. .
17 β-Hydroxy-4-androsten-3-one	0.22	0.74	0.52	. .	. .	. .	. .	. .	. .	. .	. .	. .	. .	. .	. .
17 β-Acetoxy-4-androsten-3-one	0.83	. .		. .	. .	. .	. .	. .	. .	. .	. .	. .	. .	. .	. .
17 β-Hydroxy-1,4-androstadien-3-one	0.11	0.69	0.34	. .	. .	. .	. .	. .	. .	. .	. .	. .	. .	. .	. .
5-Androstene-3 β,17 β-diol	. .	0.62	. .	. .	. .	. .	. .	. .	. .	. .	. .	. .	. .	. .	. .
3,17-*seco*-13 α-Hydroxy-3-keto-4-androsten-17-oic acid lactone (13 → 17)	0.043	0.63	0.28	. .	. .	. .	. .	. .	. .	. .	. .	. .	. .	. .	. .
4-Androstene-3,6,17-trione	0.13	0.80	0.52	. .	. .	. .	. .	. .	. .	. .	. .	. .	. .	. .	. .
4-Androstene-3,11,17-trione	0.11	0.78	0.43	. .	. .	. .	. .	. .	. .	. .	. .	. .	. .	. .	. .
6 β-Hydroxy-4-androstene-3,17-dione	. .	0.43	0.19	0.90	. .	. .	. .	. .	. .	. .	. .	. .	. .	. .	. .
11 α-Hydroxy-4-androstene-3,17-dione	. .	0.25	0.06	0.82	. .	. .	. .	. .	. .	. .	. .	. .	. .	. .	. .
11 β-Hydroxy-4-androstene-3,17-dione	. .	0.53	0.26	. .	. .	. .	. .	. .	. .	. .	. .	. .	. .	. .	. .
14 α-Hydroxy-4-androstene-3,17-dione	. .	0.33	0.12	0.87	. .	. .	. .	. .	. .	. .	. .	. .	. .	. .	. .
15 α-Hydroxy-4-androstene-3,17-dione	. .	0.22	. .	0.82	. .	. .	. .	. .	. .	. .	. .	. .	. .	. .	. .
17 β-Hydroxy-4-androstene-3,11-dione	. .	0.31	0.13	. .	. .	. .	. .	. .	. .	. .	. .	. .	. .	. .	. .
6 β,17 β-Dihydroxy-4-androsten-3-one	. .	0.15	. .	0.67	. .	. .	. .	. .	. .	. .	. .	. .	. .	. .	. .
11 α,17 β-Dihydroxy-4-androsten-3-one	. .	0.07	. .	0.47	. .	. .	. .	. .	. .	. .	. .	. .	. .	. .	. .
11 β,17 β-Dihydroxy-4-androsten-3-one	. .	0.21	. .	0.74	. .	. .	. .	. .	. .	. .	. .	. .	. .	. .	. .
14 α,17 β-Dihydroxy-4-androsten-3-one	. .	0.12	. .	0.67	. .	. .	. .	. .	. .	. .	. .	. .	. .	. .	. .
Pregnane-3,20-dione	0.83	. .	. .	. .	. .	. .	. .	. .	. .	. .	. .	. .	. .	. .	. .
Allopregnane-3,20-dione	0.83	. .	. .	. .	. .	. .	. .	. .	. .	. .	. .	. .	. .	. .	. .
4-Pregnene-3,20-dione	0.76	0.95	0.96	. .	. .	. .	. .	. .	. .	. .	. .	. .	. .	. .	. .
1,4-Pregnadiene-3,20-dione	0.55	. .	0.78	. .	. .	. .	. .	. .	. .	. .	. .	. .	. .	. .	. .
4,6-Pregnadiene-3,20-dione	0.68	. .	0.96	. .	. .	. .	. .	. .	. .	. .	. .	. .	. .	. .	. .
4,9(11)-Pregnadiene-3,20-dione	0.68	. .	0.96	. .	. .	. .	. .	. .	. .	. .	. .	. .	. .	. .	. .
4,16-Pregnadiene-3,20-dione	0.76	. .	0.96	. .	. .	. .	. .	. .	. .	. .	. .	. .	. .	. .	. .
3 α-Hydroxypregnan-20-one	0.72	. .	0.93	. .	. .	. .	. .	. .	. .	. .	. .	. .	. .	. .	. .
3 β-Hydroxyallopregnan-20-one	0.72	. .	0.89	. .	. .	. .	. .	. .	. .	. .	. .	. .	. .	. .	. .

Table 16 (cont.)

Compound	Solvent systems (cf. Legend)														
	1	2	3	4	5	6	7	8	9	10	11[3]	12[3]	13[1]	14[3]	15
3β-Hydroxy-5-pregnen-20-one	0.64	. .	0.78	. .	. .	. .	. .	. .	. .	. .	. .	. .	. .	. .	. .
3β-Hydroxy-5,16-pregnadien-20-one	0.61	. .	0.78	. .	. .	. .	. .	. .	. .	. .	. .	. .	. .	. .	. .
Pregnane-3α,20α-diol	. .	. .	. .	. .	0.50	1.00	0.11	1.00	0.56	0.63	. .	. .	. .	. .	. .
Pregnane-3,11,20-trione	0.42	0.89	0.78	. .	. .	. .	. .	. .	. .	. .	. .	. .	. .	. .	. .
Allopregnane-3,11,20-trione	0.42	0.89	0.78	. .	. .	. .	. .	. .	. .	. .	. .	. .	. .	. .	. .
4-Pregnene-3,11,20-trione	0.25	0.82	0.65	. .	. .	. .	. .	. .	. .	. .	. .	. .	. .	. .	. .
3α-Hydroxypregnane-11,20-dione	0.14	. .	0.43	. .	. .	. .	. .	. .	. .	. .	. .	. .	. .	. .	. .
3β-Hydroxypregnane-11,20-dione	0.16	. .	0.43	. .	. .	. .	. .	. .	. .	. .	. .	. .	. .	. .	. .
11α-Hydroxypregnane-3,20-dione	0.12	. .	0.43	. .	. .	. .	. .	. .	. .	. .	. .	. .	. .	. .	. .
11β-Hydroxypregnane-3,20-dione	0.20	. .	0.65	. .	. .	. .	. .	. .	. .	. .	. .	. .	. .	. .	. .
3β-Hydroxyallopregnane-11,20-dione	0.11	. .	0.34	. .	. .	. .	. .	. .	. .	. .	. .	. .	. .	. .	. .
11α-Hydroxyallopregnane-3,20-dione	0.08	. .	0.34	. .	. .	. .	. .	. .	. .	. .	. .	. .	. .	. .	. .
6β-Hydroxy-4-pregnene-3,20-dione	0.097	0.69	0.34	. .	. .	. .	. .	. .	. .	. .	. .	. .	. .	. .	. .
11α-Hydroxy-4-pregnene-3,20-dione	0.034	0.47	0.19	. .	. .	. .	. .	. .	. .	. .	. .	. .	. .	. .	. .
11α-Acetoxy-4-pregnene-3,20-dione	0.25	. .	0.71	. .	. .	. .	. .	. .	. .	. .	. .	. .	. .	. .	. .
11β-Hydroxy-4-pregnene-3,20-dione	0.11	0.71	0.43	. .	. .	. .	. .	. .	. .	. .	. .	. .	. .	. .	. .
14α-Hydroxy-4-pregnene-3,20-dione	0.078	0.69	0.34	. .	. .	. .	. .	. .	. .	. .	. .	. .	. .	. .	. .
15α-Hydroxy-4-pregnene-3,20-dione	0.017	0.40	0.13	. .	. .	. .	. .	. .	. .	. .	. .	. .	. .	. .	. .
15β-Hydroxy-4-pregnene-3,20-dione	0.036	0.49	0.20	. .	. .	. .	. .	. .	. .	. .	. .	. .	. .	. .	. .
16α-Hydroxy-4-pregnene-3,20-dione	0.024	0.43	0.15	. .	. .	. .	. .	. .	. .	. .	. .	. .	. .	. .	. .
17α-Hydroxy-4-pregnene-3,20-dione	0.16	0.78	0.52	. .	0.60	1.00	0.07	1.00	0.40	0.77	. .	. .	. .	. .	. .
21-Hydroxy-4-pregnene-3,20-dione	0.23	0.81	0.52	. .	0.60	1.00	0.09	1.00	0.28	0.72	. .	. .	. .	. .	. .
21-Acetoxy-4-pregnene-3,20-dione	0.55	. .	0.89	. .	0.95	. .	0.24	1.00	0.49	1.00	0.70	. .	. .	. .	. .
11α-Hydroxy-4,16-pregnadiene-3,20-dione	0.04	0.47	0.19	. .	. .	. .	. .	. .	. .	. .	. .	. .	. .	. .	. .
21-Acetoxy-4,16-pregnadiene-3,20-dione	0.50	. .	0.89	. .	. .	. .	. .	. .	. .	. .	. .	. .	. .	. .	. .
3α,11α-Dihydroxypregnan-20-one	. .	0.41	. .	0.87	. .	. .	. .	. .	. .	. .	. .	. .	. .	. .	. .
3β-11α-Dihydroxyallopregnan-20-one	. .	0.29	. .	0.87	. .	. .	. .	. .	. .	. .	. .	. .	. .	. .	. .
Pregnane-3α,17α,20α-triol	. .	. .	. .	. .	0.10	0.43	0.02	1.00	0.30	0.19	. .	. .	. .	. .	. .
17αHydroxypregnane-3,11,20-trione	0.07	0.65	0.37	. .	. .	. .	. .	. .	. .	. .	. .	. .	. .	. .	. .
6β-Hydroxy-4-pregnene-3,11,20-trione	. .	0.28	. .	. .	. .	. .	. .	. .	. .	. .	. .	. .	. .	. .	. .
17α-Hydroxy-4-pregnene-3,11,20-trione	0.304	0.47	0.19	. .	. .	. .	. .	. .	. .	. .	. .	. .	. .	. .	. .
21-Hydroxy-4-pregnene-3,11,20-trione	. .	0.43	. .	0.92	0.14	0.69	0.01	0.63	0.06	0.17	0.28	. .	. .	. .	. .
21-Acetoxy-4-pregnene-3,11,20-trione	. .	0.75	0.43	. .	0.62	. .	0.02	1.00	0.16	. .	. .	. .	. .	. .	. .
3α,17α-Dihydroxypregnane-11,20-dione	. .	0.32	0.10	0.87	. .	. .	. .	. .	. .	. .	. .	. .	. .	. .	. .
3α,21-Dihydroxypregnane-11,20-dione	. .	. .	. .	. .	0.07	. .	. .	. .	. .	. .	0.87	3.71	2.28	1.92	2.18

3α,21-Dihydroxypregnane-11,20-dione	. .	. .	. .	. .	0.07	. .	. .	. .	. .	. .	0.87	3.71	2.28	1.92	2.18
6β,11α-Dihydroxypregnane-3,20-dione	. .	0.05	. .	0.47	. .	. .	. .	. .	. .	. .	. .	. .	. .	. .	. .
3β,21-Dihydroxyallopregnane-11,20-dione	. .	. .	. .	. .	. .	. .	. .	. .	. .	. .	. .	3.19	2.23	1.74	1.52
6β,11α-Dihydroxy-4-pregnene-3,20-dione	. .	0.04	. .	0.38	. .	. .	. .	. .	. .	. .	. .	. .	. .	. .	. .
6β,17α-Dihydroxy-4-pregnene-3,20-dione	. .	0.19	. .	0.75	. .	. .	. .	. .	. .	. .	. .	. .	. .	. .	. .
6β,21-Dihydroxy-4-pregnene-3,20-dione	. .	0.21	. .	0.75	. .	. .	. .	. .	. .	. .	. .	. .	. .	. .	. .
11α,17α-Dihydroxy-4-pregnene-3,20-dione	. .	0.09	. .	0.61	. .	. .	. .	. .	. .	. .	. .	. .	. .	. .	. .
11β,17α-Dihydroxy-4-pregnene-3,20-dione	. .	0.23	. .	0.83	. .	. .	. .	. .	. .	. .	. .	. .	. .	. .	. .
11α,21-Dihydroxy-4-pregnene-3,20-dione	. .	0.10	. .	0.61	. .	. .	. .	. .	. .	. .	. .	2.15	1.42	1.37	1.36
11β,21-Dihydroxy-4-pregnene-3,20-dione	. .	0.33	. .	0.87	0.08	0.47	0.00	0.79	0.10	0.18	0.40	. .	2.62	. .	. .
21-Acetoxy-11β-hydroxy-4-pregnene-3,20-dione	. .	0.71	0.38	. .	0.52	. .	0.03	1.00	0.23	0.52	. .	. .	. .	. .	. .
14α,21-Dihydroxy-4-pregnene-3,20-dione	. .	0.21	. .	0.77	. .	. .	. .	. .	. .	. .	. .	. .	. .	. .	. .
15α,21-Dihydroxy-4-pregnene-3,20-dione	. .	0.06	. .	0.53	. .	. .	. .	. .	. .	. .	. .	. .	. .	. .	. .
21-Acetoxy-15α-hydroxy-4-pregnene-3,20-dione	. .	0.34	0.10	. .	. .	. .	. .	. .	. .	. .	. .	. .	. .	. .	. .
17α,21-Dihydroxy-4-pregnene-3,20-dione	. .	0.33	. .	0.87	0.09	0.50	0.01	0.91	0.13	0.31	0.49	. .	. .	. .	. .
21-Acetoxy-17α-hydroxy-4-pregnene-3,20-dione	. .	0.73	0.43	. .	0.54	. .	0.03	1.00	0.29	. .	. .	. .	. .	. .	. .
19,21-Dihydroxy-4-pregnene-3,20-dione	. .	. .	. .	. .	. .	. .	. .	. .	. .	. .	. .	2.34	1.71	1.40	1.24
20β,21-Dihydroxy-4-pregnene-3,11-dione	. .	. .	. .	. .	. .	. .	. .	. .	. .	. .	. .	1.89	1.34	1.39	1.40
3α,11β,21-Trihydroxypregnan-20-one	. .	. .	. .	. .	0.04	. .	. .	0.88	. .	. .	0.42	3.24	2.21	2.01	1.62
3α,17α,21-Trihydroxypregnan-20-one	. .	. .	. .	. .	0.06	0.16	. .	1.00	0.31	. .	. .	2.81	1.99	1.81	2.71
3β,11β,21-Trihydroxyallopregnan-20-one	. .	. .	. .	. .	. .	. .	. .	. .	. .	. .	. .	2.64	1.87	1.67	1.61
3β-17α,21-Trihydroxyallopregnan-20-one	. .	. .	. .	. .	. .	. .	. .	. .	. .	. .	. .	2.87	2.02	1.63	2.51
Allopregnane-3β,17α,20β,21-tetrol	. .	. .	. .	. .	. .	. .	. .	. .	. .	. .	. .	. .	. .	. .	1.88
17α,21-Dihydroxypregnane-3,11,20-trione	. .	0.21	. .	0.75	0.04	0.16	0.00	0.78	0.10	0.05	0.33	2.83	2.04	1.73	1.66
21-Acetoxy-17α-hydroxypregnane-3,11,20-trione	. .	0.65	0.28	. .	0.33	. .	0.01	1.00	0.23	0.35	. .	. .	. .	. .	. .
17α,21-Dihydroxyallopregnane-3,11,20-trione	. .	. .	. .	. .	. .	. .	. .	. .	. .	. .	. .	2.80	2.07	1.71	1.60
6β,21-Dihydroxy-4-pregnene-3,11,20-trione	. .	. .	. .	. .	. .	. .	. .	. .	. .	. .	. .	0.91	0.70	0.75	0.49
17α,21-Dihydroxy-4-pregnene-3,11,20-trione	. .	0.09	. .	0.61	0.01	0.08	0.00	0.59	0.04	0.04	0.23	1.91	1.60	1.38	1.32
21-Acetoxy-17α-hydroxy-4-pregnene-3,11,20-trione	. .	0.44	. .	0.95	0.16	. .	0.01	1.00	0.12	0.27	. .	. .	. .	. .	. .
17α,21-Dihydroxy-1,4-pregnadiene-3,11,20-trione	. .	. .	0.15	0.57	. .	. .	. .	. .	. .	. .	. .	. .	. .	. .	. .
21-Acetoxy-17α-hydroxy-1,4-pregnadiene-3,11,20-trione	. .	0.40	0.12	0.90	. .	. .	. .	. .	. .	. .	. .	. .	. .	. .	. .
3α,17α,21-Trihydroxypregnane-11,20-dione	. .	. .	. .	. .	0.01	0.02	0.00	0.73	0.01	0.00	0.28	0.89	0.81	0.88	1.92
21-Acetoxy-3α,17α-dihydroxypregnane-11,20-dione	. .	0.28	. .	0.87	. .	. .	. .	. .	. .	. .	. .	. .	. .	. .	. .
3α,21-Diacetoxy-17α-hydroxypregnane-11,20-dione	. .	. .	. .	. .	0.77	. .	0.11	1.00	0.48	0.78	. .	. .	. .	. .	. .
11β,17α,21-Trihydroxypregnane-3,20-dione	. .	. .	. .	. .	0.02	0.09	0.00	0.70	0.08	0.10	0.31	1.99	1.63	1.51	1.49
21-Acetoxy-11β,17α-dihydroxypregnane-3,20-dione	. .	. .	. .	. .	0.22	. .	0.01	1.00	0.24	0.37	. .	. .	. .	. .	. .
3β,17α,21-Trihydroxyallopregnane-11,20-dione	. .	. .	. .	. .	. .	. .	. .	. .	. .	. .	. .	0.70	0.60	0.77	1.78
21-Acetoxy-3β,17α-dihydroxypregnane-11,20-dione	. .	0.21	. .	0.87	. .	. .	. .	. .	. .	. .	. .	. .	. .	. .	. .
3β,21-Diacetoxy-17α-hydroxyallopregnane-11,20-dione	. .	. .	. .	. .	. .	. .	. .	. .	. .	. .	. .	0.70	. .	. .	. .

Table 16 (cont.)

Compound	Solvent systems (cf. Legend)														
	1	2	3	4	5	6	7	8	9	10	11[3]	12[3]	13[3]	14[3]	15
11 β,17 α,21-Trihydroxyallopregnane-3,20-dione												1.86	1.58	1.43	1.40
6 β,11 β,21-Trihydroxy-4-pregnene-3,20-dione												0.31	0.35	0.40	0.41
6 β,17 α,21-Trihydroxy-4-pregnene-3,20-dione		0.024		0.22								0.44	0.53	0.59	0.90
11 α,17 α,21-Trihydroxy-4-pregnene-3,20-dione		0.013		0.12								0.24	0.25		
11 β,17 α,21-Trihydroxy-4-pregnene-3,20-dione		0.04		0.37	0.01	0.03	0.00	0.51	0.03	0.03	0.18	1.00	1.00	1.00	1.00
21-Acetoxy-11 β,17 α-dihydroxy-4-pregnene-3,20-dione		0.34	0.09	0.90	0.08		0.00	0.93	0.14	0.30					
2-Methyl-11 β,17 α,21-trihydroxy-4-pregnene-3,20-dione		0.10		0.69											
2-Methyl-21-acetoxy-11 β,17 α-dihydroxy-4-pregnene-3,20-dione		0.51	0.20	0.96											
2-Methyl-9 α-fluoro-11 β,17 α,21-trihydroxy-4-pregnene-3,20-dione		0.057		0.61											
2-Methyl-21-acetoxy-9 α-fluoro-11 β,17 α-dihydroxy-4-pregnene-3,20-dione		0.41	0.15	0.95											
21-Acetoxy-9 α-fluoro-11 β,17 α-dihydroxy-4-pregnene-3,20-dione		0.27		0.90											
11 β,19,21-Trihydroxy-4-pregnene-3,20-dione												0.80	0.64	0.77	0.83
17 α,19,21-Trihydroxy-4-pregnene-3,20-dione												0.20	0.31	0.35	0.64
17 α,20 α,21-Trihydroxy-4-pregnene-3,11-dione												0.72	0.46	0.58	0.72
17 α,20 β,21-Trihydroxy-4-pregnene-3,11-dione												0.67	0.50	0.58	0.79
11 β,17 α,21-Trihydroxy-1,4-pregnadiene-3,20-dione		0.025		0.27											
21-Acetoxy-11 β,17 α-dihydroxy-1,4-pregnadiene-3,20-dione		0.27	0.07	0.82											
3 α,11 β,17 α,21-Tetrahydroxypregnan-20-one					0.00	0.01	0.00	0.63	0.04	0.04	0.20	0.38	0.52	0.63	1.51
3 α,21-Diacetoxy-11 β,17 α-dihydroxypregnan-20-one					0.68		0.06	1.00	0.47	0.82					
3 α,11 β,17 α,21-Tetrahydroxyallopregnan-20-one												0.55	0.65	0.82	1.37
3 β,11 β,17 α,21-Tetrahydroxyallopregnan-20-one												0.33	0.42	0.57	1.12
11 β,17 α,20 α,21-Tetrahydroxy-4-pregnen-3-one				0.09											
21-Acetoxy-11 β,17 α,20 α-trihydroxy-4-pregnen-3-one		0.09		0.61											
11 β,17 α,20 β,21-Tetrahydroxy-4-pregnen-3-one				0.12								0.20	0.23	0.32	0.68
Allopregnane-3 β,11 β,17 α,20 β,21-pentol												0.00	0.10	0.13	0.67
11 β,21-Dihydroxy-3,20-diketo-4-pregnen-18-al								0.36			0.15	1.52	1.03	0.97	0.73
20,20,21-Trihydroxy-3-keto-4-pregnen-18-oic acid lactone (18 → 20)												2.76	1.81	1.67	1.39

Legend

No.	Solvent system	Temp.	Reference
1	Petroleum ether[1]-methanol-water (5:4:1)	34±1°	REINEKE (1956)
2	Petroleum ether-toluene-methanol-water (5:5:7:3)	34±1°	REINEKE (1956)
3	Petroleum ether-benzene-methanol-water (667:333:800:200)	34±1°	REINEKE (1956)
4	Benzene-methanol-water (2:1:1)	34±1°	REINEKE (1956)
5	Isoöctane[2]-toluene-methanol-water (5:5:7:3)	chamber preheated at 36±1°; chromatogram run at 25±3°	EBERLEIN and BONGIOVANNI (1955)
6	Isoöctane-toluene-methanol-water (33:67:60:40)		EBERLEIN and BONGIOVANNI (1955)
7	Isoöctane-methanol-water (10:9:1)		EBERLEIN and BONGIOVANNI (1955)
8	Isoöctane-*tert*-butanol-water (10:5:9)		EBERLEIN and BONGIOVANNI (1955)
9	Isoöctane-*tert*-butanol-methanol-water (20:9:9:2)		EBERLEIN and BONGIOVANNI (1955)
10	Isopropyl ether-*tert*-butanol-water (8:5:6)		EBERLEIN and BONGIOVANNI (1955)
11[3]	*n*-Hexane-*tert*-butanol-methanol-water (20:9:9:8)		EBERLEIN and BONGIOVANNI (1955)
12[3]	Benzene-methanol-water (2:1:1)	22±2°	NEHER and WETTSTEIN (1956b)
13[3]	Toluene-ethyl acetate-methanol-water (9:1:5:5)	22±2°	NEHER and WETTSTEIN (1956b)
14[3]	Toluene-ethyl acetate-methanol-water (9:1:5:5)	40±1°	NEHER and WETTSTEIN (1956b)
15	Isoöctane-*tert*-butanol-water (10:5:9)	40±1°	NEHER and WETTSTEIN (1956b)

[1] Boiling range 80—100°.
[2] 2,2,4-trimethylpentane.
[3] Values 11—14 are mobilities related to cortisol.

Solvent systems prepared by mixing solvents in above proportions by volume.

paper are saturated with respect to both the mobile and the stationary phases prior to chromatographic development. In addition to the descending technique, described earlier, the ascending method of paper chromatography may be used. The latter is considerably simpler technically, requiring only a closed chamber. The filter paper stands in the mobile phase, which rises by capillary attraction. The finished chromatograms dry rapidly at room temperature.

A sample chromatogram of beef adrenal extract is shown in Fig. 95. R_F values in various solvent systems are listed in Table 16.

Paper chromatography has been of great value in the study of aldosterone (SIMPSON and TAIT 1955) and various methods have been described for its separation from cortisone and cortisol (SCHMIDT and STAUDINGER 1954, PECHET 1955, VENNING et al. 1955, GORNALL and GWILLIAM 1957). NEHER and WETTSTEIN (1956a), for example, chromatographed urine extracts on filter paper in a chloroform/formamide solvent system for $2^1/_2$—3 hours, eluted the region occupied by a reference sample of cortisone and rechromatographed this eluate on a second sheet of filter paper in toluene-ethyl acetate-methanol-water (9 : 1 : 5 : 5). In this solvent system aldosterone separates readily from cortisone, but moves at the same rate as cortisol from which it has already been separated in the first chromatogram. NOWACZYNSKI et al. (1957a) applied urine extracts purified by adsorption chromatography on silica gel to filter paper impregnated with ethylene glycol and developed the paper chromatogram with toluene for 48 hours to separate aldosterone from cortisol (see also NOWACZYNSKI and KOIW 1957). Separation from cortisone and other contaminants was carried out in the solvent system of EBERLEIN and BONGIOVANNI (1955), isoöctane-*tert*-butanol-water (10 : 5 : 9).

Steroid glucosiduronates and sulfates, owing to their water-solubility, are readily separated by paper chromatography in aqueous solvent systems. Table 17 gives the R_F values of various steroid conjugates from the literature.

Table 17. *R_F values of steroid conjugates in aqueous solvent systems*

Compounds	Solvent systems (see Legend)						
	1	2	3	4	5	6	7
Estrone sulfate				0.63	0.51		
Estradiol monosulfate					0.39		
Estradiol disulfate				0.108	0.043		
Estriol monoglucosiduronate						0.57	0.55
3α-Hydroxyandrostan-17-one sulfate	0.38	0.47	0.45	0.51	0.27	0.77	
17β-Hydroxy-4-androsten-3-one sulfate				0.47	0.29	0.76	
3β-Hydroxy-5-androsten-17-one sulfate	0.25	0.34	0.42	0.53	0.25	0.74	0.79
3β-Hydroxy-5-androsten-17-one monoglucosiduronate	0.78	0.78	0.21				
5-Androstene-3β,17β-diol monosulfate				0.59	0.27		
5-Androstene-3β,17β-diol disulfate				0	0		
3β-Hydroxy-5-pregnen-20-one sulfate	0.40	0.52	0.52				
Pregnane-3α,20α-diol monoglucosiduronate	0.81	0.82	0.27	0.32	0.76	0.59	0.63
3α,17α,21-Trihydroxypregnane-11,20-dione monoglucosiduronate	0.39	0.42	0.13				
3α,17α,20β,21-Tetrahydroxypregnan-11-one monoglucosiduronate	0.14	0.16	0.08				

Legend

No.	Solvent system	Reference
1	*n*-Butyl acetate-*n*-butanol-10% acetic acid (4:1:5)	LEWBART and SCHNEIDER (1955)
2	*n*-Butyl acetate-*n*-butanol-10% formic acid (4:1:5)	
3	*n*-Butyl acetate-methanol-0.1 *M* barbiturate buffer in 50% aqueous methanol pH 8.2 (3:1:1)	
4	*n*-Butanol saturated with water	CAVINA (1955b)
5	Ethyl acetate-*n*-butanol-0.2 *N* acetic acid (9:1:10)	
6	Barbiturate buffer, pH 7	ANDREČ and KREČ (1956)
7	Barbiturate buffer, pH 9	

Values 1, 2, 3 obtained at 28.5—29.7° after equilibration at 29.7—32.0°.
Solvents 1, 2, 3, 5 prepared by mixing solvents in above proportions by volume.
Solvents 3, 6, 7 are "monophasic."

C. Identification

The examination of a steroid-containing material may have one of several objectives, most of which fall into one of the following categories:

(1) The quantitative determination of a certain class of compounds, as in the routine clinical assay of urinary 17-ketosteroids or reducing corticosteroids. For this purpose an analytical method that is usually based, for speed and convenience, on some form of spectrophotometry is required which can be applied to relatively crude extracts and which is sufficiently specific to distinguish steroids containing the functional group in question from interfering substances. The latter proviso has not always received sufficient attention. For instance, much if not most of

the materials measured in dog urine by the Zimmermann reaction were shown not to be 17-ketosteroids (GLENN and HEFTMANN 1951). Some representative group assay procedures are described in the following section on instrumental methods.

(2) Identification and quantitative estimation of the steroid constituents of a biological fluid. In this case the individual compounds must be separated and then determined by a suitable analytical method, again usually colorimetric, either on a paper chromatogram or an eluate fraction. Although extensive purification of steroids can be achieved by the methods described in the foregoing section, the amounts usually encountered preclude identification by crystallization, melting point determination, preparation of derivatives, etc., and quantitative determination by gravimetric methods, except in cases such as the analysis of pregnanediol in urine (WESTPHAL 1944). The qualitative characterization of the constituents can often be accomplished in the course of the chromatographic separation by reference to mobilities in different solvent systems and a variety of color reactions.

(3) Identification of a steroid not previously encountered or of uncertain constitution. For this end it is nearly always necessary to isolate the steroid in crystalline form, although it is equally important to establish the homogeneity of the substance by paper chromatography. A combination of color reactions and non-destructive physical measurements can furnish much information from an extremely small amount of material. The infrared spectrum may reveal identity with a compound of known structure, or suggest a structure which can be confirmed by synthesis and then comparison of spectra. If the information gained by these approaches is insufficient to permit positive identification, recourse must be made to the more elaborate chemical methods discussed in the earlier sections, such as the alteration or removal of functional groups and various degradative procedures which eventuate in a recognizable compound.

I. Crystalline properties

Implicit in the name is the fact that steroids are, with very few exceptions, crystalline compounds, usually of relatively high melting point. This property permitted systematic work on steroids at a time when distillation and crystallization were virtually the only physical methods available for the characterization of organic compounds. Thanks to the development of instrumentation for spectrophotometry, the crystalline form and melting point play a much less crucial role in the identification of steroids than was the case 20 years ago. It must be emphasized that crystallinity in itself is not a criterion of purity of steroids, nor is the reproducibility of the melting point on recrystallization. Crystallization is, nevertheless, for most steroid chemists a *sine qua non* in the description of a new compound.

Crystallization is an indispensable preliminary in removing traces of contaminants from a chromatographically homogeneous steroid prior to elemental analysis or the determination of spectra or rotation for publication. Many steroids, including rather highly oxygenated substances, can also be sublimed in high vacuum, and this sometimes furnishes a highly satisfactory alternative for final purification. In the majority of cases, recrystallization from scrupulously pure solvents, usually ether, ethanol, acetone and aliphatic hydrocarbons or mixtures of these, is carried out. Care must be taken to insure complete removal of solvent from the crystals. This can usually be accomplished by drying at 100° in vacuum, but solvent may be retained tenaciously. For example, cortisone crystallizes from chloroform as a 1 : 1 complex that is stable at 140°.

Fractional crystallization is seldom as efficient as chromatography for the resolution of mixtures of steroids. In the isolation of trace amounts of companion substances from a steroid, however, preliminary fractionation by crystallization can effect a considerable economy of labor and solvents (FIESER 1953b). The solubility of a crystalline steroid in a given solvent is almost invariably increased by the presence of small amounts of impurities; a procedure for the determination of purity of steroids based upon solubility equilibria has been developed by TARPLEY and YUDIS (1953) by which very small quantities of contaminants can be detected.

The melting point or range of a crystalline steroid is certainly the most readily determined and widely reported physical property. This temperature is usually recorded with a hot stage microscope (Kofler block) or less often in a sealed capillary tube, and should be corrected for thermometer exposure. The recrystallization solvent and drying should also be specified. It is probably safe to say that melting points of steroids are usually reproducible in different laboratories only to $\pm 2^\circ$. A number of instances of polymorphism, or the existence of two or more crystalline forms, with different melting points (and different infrared spectra in the solid state) have been reported with steroids. Double melting points are sometimes observed, in which the sample melts, completely or partially, resolidifies, and then melts again at a higher temperature. With many steroids having melting points above 200°, melting is accompanied by decomposition, and the melting point may vary considerably with the rate of heating. A significant depression is usually observed in the melting point of a mixture of two different compounds which have about the same melting point, and the lack of depression is one of the oldest criteria of identity. Many exceptions are known, however, as in the case of cortisone and 11-deoxycortisol (REICHSTEIN and SHOPPEE 1943), and this method of comparison has become less important with the advent of comparison of infrared spectra or X ray powder diagrams. Since the melting point determination of a mixture, however, is more rapidly performed than other comparisons, and since a mixture usually behaves like its components in paper chromatography, the former method still has much to recommend it over the latter.

The crystal structure of a compound is a unique property that can be used for both identification and structure elucidation. Although extensive crystallographic descriptions, including refractive indices, were often recorded for steroids in the older literature, these measurements are time-consuming and of limited usefulness, and are rarely made in contemporary work. More important are the diffraction effects obtained with X rays, which arise because the wave length of the radiation is of the order of magnitude of interatomic distances in the crystal lattice. The simplest application of X ray diffraction is the powder diagram which is obtained by directing a beam of X rays at randomly oriented crystals in a finely powdered sample. The diffraction pattern is recorded as a series of arcs on a strip of photographic film; the spacings and intensities of the lines, which correspond to spacings in the unit cell, are unique for any crystal. Comparison of X ray powder diagrams, which can readily be obtained with less than 1 mg of sample, is sometimes a more reliable test of the identity of steroids than frared spectral comparison (DJERASSI et al. 1955b). A catalogue of powder diagrams of steroids, including a number of corticosteroids, has been compiled by PARSONS et al. (1956) and their use in steroid identification has been discussed by BEHER et al. (1957). By the lengthy mathematical analysis of diffraction patterns from a properly oriented single crystal, the positions of all atoms in the crystal lattice can be determined (see monograph by J. ROBERTSON 1955). This furnishes the ultimate tool for structural clarification, since in principle all structural details, including exact bond lengths and absolute configuration, can be obtained without any chemical information whatever.

II. Color reactions

1. Reagents

a) Strong acids

The development of a color on treatment with 85–100% sulfuric acid is often a very characteristic and useful property of highly oxygenated steroids, particularly hydroxy compounds. The observation of a color and/or fluorescence has been used for qualitative identification and as a basis for quantitative assay methods in a number of instances.

The yellow color produced with pregnanediol (KOBER 1931) and many other steroids is not specific; 17α,21-dihydroxy-4-pregnene-3,20-dione gives a red color with sulfuric acid (WINTERSTEINER and PFIFFNER 1936); the other color reactions of corticosteroids are listed by REICHSTEIN and SHOPPEE (1943). Details of this test and references to earlier literature are given by DIRSCHERL and BREUER (1954) and by LINFORD (1957).

The color reaction may be modified by various additions to the sulfuric acid. Thus, the Liebermann-Burchard reaction for Δ^5-sterols with sulfuric acid in the presence of glacial acetic acid gives a blue color with testosterone (ROGERS and JAFFE 1948). Sulfuric acid-glacial acetic acid (3 : 2) gives a yellow color with cortisol and corticosterone, but not with cortisone and 11-dehydrocortisterone (BANES 1953, SZALKOWSKI et al. 1955). Testosterone shows a blue fluorescence on heating with sulfuric acid and thiocol (potassium guaiacolsulfonate) and a bright green color if copper or iron salts are also present (KOENIG et al. 1941). The test is also given by 4-androstene-3,17-dione.

MUNSON et al. (1948) found that the Pettenkofer reaction with sulfuric acid and furfural is given by a number of steroids with unsaturation or potential unsaturation in ring B and a hydroxyl group or double bond in ring A. Dehydroepiandrosterone and two of its hydrolysis artifacts, 3β-chloro-5-androsten-17-one and 3,5-androstadien-17-one, 5-androsten-3β-ol, and other steroids with additional functional groups in the molecule give a blue color (MUNSON and KENNY 1954).

Anthrone in sulfuric acid gives colored products with a variety of steroids (GRAFF et al. 1952a). The reaction has been used for the determination of 17-ketosteroids (GRAFF et al. 1952b) and cortisone (HOLECHEK and COLLINS 1953). Perchloric acid gives colored and fluorescent condensation products with a variety of steroids (TAUBER 1952, PONTIUS et al. 1955). Formic acid (BOSCOTT 1949) and *p*-toluenesulfonic acid (EPSTEIN et al. 1957) can also be used for the identification of steroids.

b) Zimmermann reaction and other group tests

m-Dinitrobenzene and potassium hydroxide give a purple color with all 17-ketosteroids (ZIMMERMANN 1935). 3-Ketosteroids, 20-ketosteroids, cardiac glycosides and various simpler organic compounds give a color which is unstable in alkali (ZIMMERMANN 1955, p. 54). The specificity of the test has been reviewed by MARLOW (1950). The potassium hydroxide may be dissolved in alcohol or water. Although aqueous potassium hydroxide is a more stable reagent than ethanolic potassium hydroxide, the latter has the advantages that the color produced follows Beer's law (FRIEDGOOD and WHIDDEN 1941, FRIEDGOOD and BERMAN 1941, MASON 1953), that it is the same for androsterone and dehydroepiandrosterone

(SAIER et al. 1941), and that a correction may be applied for urinary chromogens that interfere in the analysis (ENGSTROM and MASON 1943). Ethanolic potassium hydroxide may be preserved by addition of ascorbic acid (WILSON and CARTER 1947). Methanolic potassium hydroxide is less reactive, but stable for long periods (PFISTER 1952). The effect of alcohol and potassium hydroxide concentrations on the Zimmermann reaction was studied in detail by WILSON and NATHANSON (1945). The chromogenicity of various ketosteroids was also investigated by WILSON (1954). Similar studies have been reported by HANSEN et al. (1943). BONGIOVANNI et al. (1957) have suggested the use of N-benzyltrimethylammonium methoxide instead of potassium hydroxide in the Zimmermann reaction. It is claimed to be stable for several months and to be less prone to give atypical colors and high reagent blanks. In analogy to the Zimmermann reaction, 3,5-dinitrobenzoic acid and benzyltrimethylammonium hydroxide (TANSEY and CROSS 1950) give a purple color with ketosteroids. It is claimed that non-specific urinary chromogens interfere less with this reaction than with the Zimmermann test (S. COHEN and KALUSZYNER 1957).

Alcoholic steroids may be converted to 3,5-dinitrohemiphthalates, which give a red color with potassium hydroxide (L. ENGEL et al. 1950), or 3,5-dinitrobenzoates, which give a purple color with alkali (KELLIE et al. 1953). BAGGETT et al. (1955) have also proposed the determination of alcoholic steroids by acetylation and conversion of the steroid acetates into acetohydroxyamic acid. The latter gives a purple complex with ferric ion:

$$ROH + (CH_3CO)_2O \xrightarrow{\text{pyridine}} CH_3COOR + CH_3COOH$$

$$CH_3COOR + NH_2OH \xrightarrow{OH^-} CH_3CONHOH + ROH$$

$$3\,CH_3CONHOH + Fe^{+++} \longrightarrow (CH_3CONHO)_3Fe + 3\,H^+$$

PINCUS (1943) found that an intense blue color is developed by androsterone, 3β-hydroxyandrostan-17-one, 3α-hydroxyetiocholan-17-one, and 2-androsten-17-one after heating with concentrated antimony trichloride solution in glacial acetic acid, followed by dilution with glacial acetic acid. Dehydroepiandrosterone, androstane-3,17-diol and other steroids do not interfere, and vitamins A and D give the reaction only in chloroform solution. Antimony trichloride in acetic anhydride gives a blue color with dehydroepiandrosterone, but not with androsterone (CLARK and THOMPSON 1948).

The Δ^4-3-keto group present in many steroid hormones can be exploited for analytical purposes. In addition to ultraviolet absorption and fluorescence reactions, which will be discussed later, steroids containing this group give thiosemicarbazones with very high and characteristic ultraviolet absorption (PEARLMAN and CERCEO 1953). They also form hydrazones with isonicotinic acid hydrazide (UMBERGER 1955). The reaction with Δ^4-3-ketosteroids is quantitatively complete in an hour. The reactivity of carbonyl groups is discussed more fully on page 243. The yellow derivatives may be separated by adsorption chromatography and determined colorimetrically. Certain Δ^4-3-ketones react with 2,6-di-*tert*-butyl-*p*-cresol in alkali to give colored products which can be used for quantitative determination (SCHULZ and NEUSS 1957). Thus, cortisone gives a yellow-brown and cortisol a blue color. BURSTEIN (1953) has described a phosphomolybdic acid reagent, similar to the Folin-Wu reagent but without tungstic acid, which is reduced by compounds possessing conjugated double bonds, including the Δ^4-3-ketosteroids.

That enzymes can be used as very sensitive and specific reagents for steroid determination has been shown by HURLOCK and TALALAY (1957, 1958). By use of

α- and β-hydroxysteroid dehydrogenases they have estimated 3α, 3β-, and 17β-hydroxysteroids and 3- and 17-ketosteroids in mixtures. The method depends on changes in the extinction at 340 mμ, produced by changes in the concentration of DPNH.

c) Group tests of side chains

A number of reagents are available for the differentiation between the various side chains of C_{21} steroids of adrenal origin (L. Engel 1950, C. Brooks and Norymberski 1952) (Table 18). The α-ketols reduce alkaline solutions of silver diammine (Burton et al. 1951), cupric (Talbot et al. 1945) or ferricyanide (Broadbent et al. 1953) ions, phosphomolybdic acid (Heard and Sobel 1946),

Table 18. *Group reactions of C_{21} steroid side chains*

Side chain	Reducing	Formaldehydogenic	Acetaldehydogenic	17-Ketogenic				Phenylhydrazine
				IO_4^-	BiO_3^-	CrO_3	BH_4^- BiO_3^-	
$CH_2OHCHOHCOH<$	—	+	—	+	+	+	+	—
$CH_2OHCOCOH<$	+	+	—	—	+	+	+	+
$CH_3COCOH<$	—	—	—	—	—	+	+	—
$CH_3CHOHCOH<$	—	—	+	+	+	+	+	—
$CH_2OHCOCH<$	+	+	—	—	—	—	—	—
$CH_2OHCHOHCH<$	—	+	—	—	—	—	—	—

2,3,5-triphenyltetrazolium chloride (Burton et al. 1951) and other oxidizing agents. The reaction of reducing steroids with tetrazolium salts is illustrated by the reduction of 3,3'-dianisole-*bis*-4,4'-(3,5-diphenyl)tetrazolium chloride (Blue Tetrazolium) in alkaline medium (Fig. 96). The yellow ditetrazolium chloride

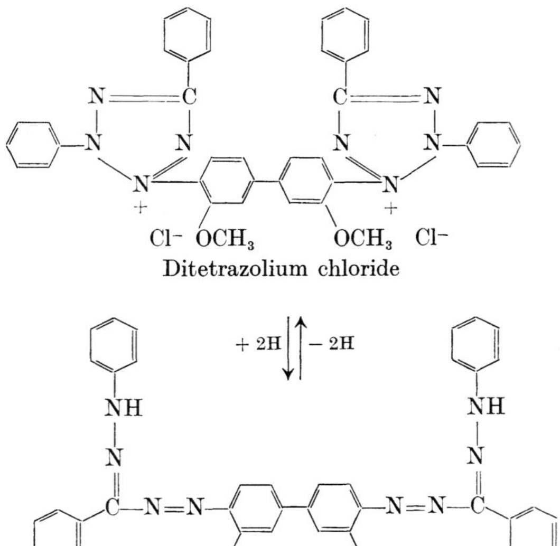

Fig. 96. Reduction of blue tetrazolium

yields a blue diformazan (RUTENBURG et al. 1950). The alkali may be tetramethylammonium chloride (MADER andBUCK 1952), sodium hydroxide (RECKNAGEL and LITTERIA 1956), or choline (CHEN et al. 1953). Other tetrazolium salts are also suitable (HENLY 1952, OERTEL 1954b).

The reducing power of α-ketols is modified by the presence of other groups in the molecule. Thus, HEARD and SOBEL (1946) found that the reducing power of α-ketols toward phosphomolybdic acid is diminished by a hydroxyl group at C-17 and by keto groups at C-11 or C-12. α,β-Unsaturated 3-ketones and α,β-diketones are also reducing and add to the reducing power of the α-ketol group if both are present in the molecule. The specificity of this method may be increased by determining the reduction of phosphomolybdic acid before and after destruction of the α-ketol side chain by boiling the steroid extract in 0.1 *N* sodium hydroxide for one hour (STAUDINGER and SCHMEISSER 1948, 1950). The reducing characteristics of various steroids toward the Blue Tetrazolium reagent have been investigated by A. MEYER and LINDBERG (1955) and by Izzo et al. (1957). Steroids with α-ketol groups reduce this compound more readily than steroids not containing this group. However, Δ^4-3-ketosteroids also give rise to the formazan, the rate and intensity of color formation depending on the location and configuration of other functional groups in the molecule.

Periodic acid oxidation yields one mole of formaldehyde with C-21 alcohols having either a carbonyl or a hydroxyl group in C-20 (FIESER et al. 1944, CORCORAN and PAGE 1948). Depending on the nature of the side chain, etio aldehydes, etio acids, hydroxyetio acids, or 17-ketosteroids (WILSON and FAIRBANKS 1955b) are formed (Fig. 97).

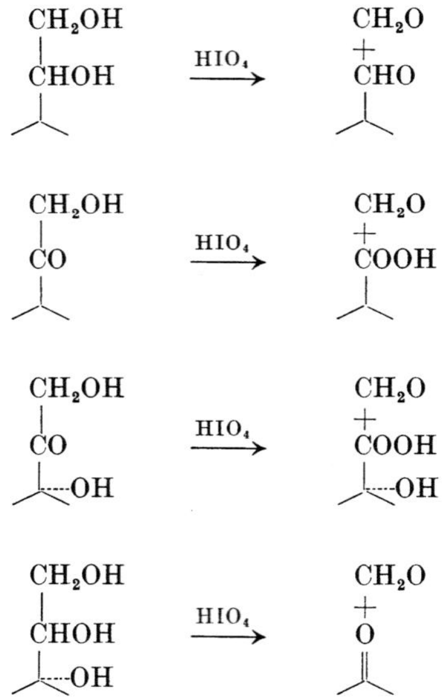

Fig. 97

The same steroids give formaldehyde with sodium bismuthate (EDWARDS and KELLIE 1954), but the 20-keto-17,21-diols are also converted to 17-ketosteroids (NORYMBERSKI 1952). Following the oxidation, tests for 17-ketosteroids or for formaldehyde may be used (C. BROOKS and NORYMBERSKI 1953). The periodate oxidation of pregnane-17,20-diols yields acetaldehyde (COX 1952) and affords a specific method for their determination.

Chromic acid oxidation converts only the 17-hydroxypregnanes (and 17-hydroxy C_{19} steroids) to the corresponding 17-ketosteroids, while the 17-unsubstituted pregnanes give the corresponding etio acids. A simple method for the determination of steroids giving 17-ketosteroids after chromic acid oxidation is described by WILSON and FAIRBANKS (1955a).

17-Hydroxypregnan-20-ones can be determined by measuring the 17-ketosteroids produced by bismuthate oxidation before and after the reduction to pregnane-17,20-diols with sodium borohydride (APPLEBY et al. 1955, APPLEBY and NORYMBERSKI 1955).

Phenylhydrazine in sulfuric acid gives a relatively specific (SILBER and BUSCH 1955) color test for 17,21-dihydroxy-20-ketosteroids (PORTER and SILBER 1950, SILBER and PORTER 1957). GORNALL and MACDONALD (1953) have utilized the 2,4-dinitrophenylhydrazones of corticosteroids, which give a deep red color in alkali, for quantitative assay. MOOLENAAR (1957) has recently applied this reaction to aldosterone-containing eluates from paper chromatograms.

2. Applications to paper chromatography

The methods of identification in the foregoing section are applicable to suitably purified extracts or fractions from biological material. Many of these methods may also be applied to paper chromatograms, but some reagents used in test tubes will corrode the paper, while some reagents used on paper are not suitable for test tube reactions.

Ultraviolet absorbing steroids, e. g., the adrenocortical hormones, may be detected or photographed with ultraviolet light (N. DRAKE et al. 1956). Other steroids are revealed by treating paper chromatograms with reagents.

Corrosive reagents may be used on glass-fiber paper (DIECKERT and REISER 1954). On paper chromatograms concentrated sulfuric acid (REICH et al. 1950) and fuming sulfuric acid (AXELROD 1953a) also give characteristic colors or fluorescence, but the filter paper soon disintegrates. The Liebermann-Burchard reagent is less destructive, but gives a fluorescence similar to the more convenient 15% phosphoric acid reagent (NEHER and WETTSTEIN 1951). Paper chromatograms drawn through this reagent and heated to 90° for 20 minutes show color reactions under the ultraviolet lamp, which are detailed in Table 19. The test can be made more sensitive by exposure of the finished chromatogram to iodine vapors (PROCHÁZKA 1953).

FINKELSTEIN and GOLDBERG (1957) have shown that $3\alpha,17\alpha,20\alpha$-trihydroxypregnan-11-one gives a blue fluorescence when the paper chromatogram is briefly dipped in 70% phosphoric acid and subsequently heated to 87° for 10 minutes and that the test may be applied to eluates from paper chromatograms for quantitative assay. A solution of 25 g of vanillin in 100 ml of 85% phosphoric acid gives an orange color with 17-hydroxypregnan-20-ones after heating at 70° for 20 minutes (MCALEER and KOZLOWSKI 1956). Pregnane-$3\alpha,17\alpha$,20-triols give a blue fluorescence when the chromatogram is sprayed with a 25% solution of trichloroacetic acid in ethanol and then heated at 75° for 10 minutes (DE COURCY 1956).

NEHER and WETTSTEIN (1951) applied the antimony trichloride reagent by drawing the paper chromatograms through a saturated solution of antimony trichloride in chloroform, followed by heating at 90° for 4 minutes. The color of the fluorescent spots is shown in Table 19. Many spots, e. g. of Δ^5-steroids, also appeared colored in daylight. Similar results were obtained by ROSENKRANTZ (1953a, 1953b), who used a solution of antimony trichloride in nitrobenzene

(Table 19). Generally, the test is not so sensitive as the phosphoric acid reaction mentioned above. The test has been modified by exposing the paper to an atmosphere of chlorine (SHULL et al. 1952) or by spraying it with chlorine in chloroform (PROCHÁZKA 1953) prior to spraying it with a solution of antimony trichloride in acetic anhydride. On addition of chlorine to antimony trichloride in chloroform the pentachloride is formed, which gives additional color tests (Table 19), but is more difficult to handle (KRITCHEVSKY and KIRK 1952b). Various colors and fluorescence are also produced when the paper chromatograms are sprayed with a solution of 20 g of anhydrous zinc chloride in 30 ml of glacial acetic acid and heated to 90° for 5 minutes, then drawn through a 50% solution of benzoyl chloride in chloroform and again briefly heated to 90° (NEHER and WETTSTEIN 1952).

Exposure of paper chromatograms to iodine vapors (KRITCHEVSKY and KIRK 1952) produces brown spots with many organic compounds, including steroids. The paper may also be suspended over a warm chloroform solution of bromine and subsequently sprayed with potassium iodide and starch solutions (MCMAHON et al. 1950) or drawn through a solution of iodine in either petroleum ether (BUSH 1950) or hydriodic acid (AXELROD 1953a). ZAFFARONI et al. (1950) found that 0.3% iodine in 5% potassium iodide gives a blue color with cortisone. Other steroids are known to give this test (SCHWARZ 1953b, MCALEER and KOZLOWSKI 1957).

Potassium permanganate (0.2% in 5% sodium carbonate) is reduced by many steroids (BURTON et al. 1951). Phosphomolybdic acid (10% in alcohol) treatment and heating at 110° for 1 minute gives blue spots on a greenish yellow background with many steroids (KRITCHEVSKY and KIRK 1952). The colors given by various steroids after spraying with 25% alcoholic solution of silicotungstic acid, followed by heating at 110° are indicated in Table 19 (KRITCHEVSKY and KIRK 1952). Color reactions and fluorescence produced by spraying paper chromatograms with anisaldehyde (0.5 ml in 50 ml glacial acetic acid and 1 ml of concentrated sulfuric acid) and heating at 90° for 2–5 minutes (NEHER and WETTSTEIN 1951) are also listed in Table 19.

The Zimmermann reaction is applied to paper chromatograms as follows (KOCHAKIAN and STIDWORTHY 1952):

The dried chromatogram is hung up in a stream of air at approximately 45°. The color reagent is prepared by mixing equal volumes of 2% *m*-dinitrobenzene in ethanol and 30% (w/v) potassium hydroxide in ethanol immediately before use. The reagent is sprayed on and the spots are outlined with pencil.

3-Ketosteroids immediately give blue spots, which fade on prolonged heating or after 2 days at room temperature. 17-Ketosteroids give a violet color, which persists. 16-Ketosteroids give tan spots.

The product of the Zimmermann reaction may be sprayed with 2,4-dinitrophenylhydrazine (1 g dissolved in 1 l of ethanol and 10 ml of hydrochloric acid) (KOCHAKIAN and STIDWORTHY 1952). The color will then disappear and the 2,4-dinitrophenylhydrazones of the ketosteroids will be formed. Steroids with α,β-unsaturated ketone groups react immediately and other ketosteroids more slowly to give yellow or orange spots with this reagent. When the excess reagent is washed out of the paper in running water the spots become more distinct.

Tollen's reagent is a very useful permanent staining reagent for reducing steroids (BURTON et al. 1951):

The reagent is prepared by mixing 10 ml of 0.1 *N* silver nitrate, 10 drops of concentrated ammonia, and 5 ml of 10% sodium hydroxide shortly before use. The chromatogram is immersed in the solution for an instant and then removed. When the black stain has reached maximum intensity the paper is washed, first in 5% sodium thiosulfate, then in water.

Other spot tests for reducing steroids include Nelson's arsenomolybdate reagent (SCHWARZ 1952, 1953a) and the tetrazolium salts (HOFMANN and STAUDINGER

Table 19. *Color reactions of steroids applicable to paper chromatograms*

Compound	Reagents (see Legend) 1	2	3	4	5	6	7	8	9
Cholesterol				blue violet	faint blue	brown	pink		
Cholesteryl acetate	faint violet	red violet	red violet					—	light blue
3-Hydroxy-1,3,5(10)-estratrien-17-one				pink-orange	orange	brown	orange		
1,3,5(10)-Estratriene-3,17 α-diol	yellow	green	orange	yellow-orange	faint red			light yellow	blue
1,3,5(10)-Estratriene-3,17 β-diol	green	green	yellow	red-orange	orange	brown	brown	light green	light blue
1,3,5(10)-Estratriene-3,16 α,17 β-triol				pink	orange				
Androstan-3 α-ol				pink-brown	orange				
Androstan-3 β-ol				pink-brown	orange				
Androstan-17 β-ol				pink-brown	orange				
5-Androsten-3 β-ol				yellow-green	dark violet				
Androstane-3,17-dione	violet	violet	—					yellow	blue
4-Androstene-3,17-dione	orange	violet	—	—	absorption			brown	dark brown
3 α-Hydroxyetiocholan-17-one				pink	red-orange				
3 β-Hydroxyetiocholan-17-one				blue-pink	red-orange				
3 α-Hydroxyandostan-17-one				pink	red-orange	gray	brown		
3 α-Acetoxyandrostan-17-one	blue	yellow-orange	—					—	—
3 β-Hydroxyandrostan-17-one				pink	red-orange				
17 β-Hydroxyandrostan-3-one				yellow-brown	yellow-orange				
3 β-Hydroxy-5-androsten-17-one				brown-red	red-orange	violet	brown		

Table 19 (cont.)

Compound	Reagents (see Legend)								
	1	2	3	4	5	6	7	8	9
3 β-Acetoxy-5-androsten-17-one	violet	red-brown	red			violet	brown	—	faint blue
17 β-Hydroxy-4-androsten-3-one	yellow	faint red	—	blue green	red-orange	—	pink	brown	brown
17 α-Hydroxy-4-androsten-3-one	red	red	blue	light blue	red-orange			blue	red
Etiocholane-3 α,17 β-diol				violet	yellow-orange				
Androstane-3 β,17 α-diol				orange-brown	yellow-orange				
Androstane-3 β,17 β-diol				orange-brown	yellow-orange				
5-Androstene-3 β,17 α-diol				blue-violet	red-orange				
3 α,11 β-Dihydroxyetiocholan-17-one				gray-blue	faint blue				
3 α,11 β-Dihydroxyandrostan-17-one				violet	faint blue				
4-Pregnene-3,20-dione	green	faint violet	—	—	absorption	light yellow	yellow	brown	brown
3 β-Hydroxy-5-pregnen-20-one	violet	red	red	pink-violet	red	brown	light brown	green	blue
Pregnane-3 α,20 α-diol				yellow-brown	red-orange	light brown	pink		
Allopregnane-3 α,20 α-diol				yellow-brown	red-orange				
Allopregnane-3 β,20 α-diol				yellow-brown	red-orange				
21-Hydroxy-4-pregnene-3,20-dione	red-violet	orange		faint red-brown	absorption	—	blue		
3 α,17 α-Dihydroxypregnan-20-one				orange-brown	red-orange				
Pregnane-3 α,17 α,20 α-triol				yellow	red				

21-Hydroxy-4-pregnene-3,11,20-trione	blue	faint gray-blue				—	—		
11 β,21-Dihydroxy-4-pregnene-3,20-dione	green	faint gray-blue		—	absorption	—	—		
17 α,21-Dihydroxy-4-pregnene-3,20-dione	orange-red	orange		faint light blue	absorption				
3 β,21-Diacetoxy-17 α-hydroxyallopregnan-20-one	faint red	yellow							
3 β,20 β,21-Triacetoxyallopregnan-17 α-ol	faint violet	—							
17 α,21-Dihydroxy-4-pregnene-3,11,20-trione	blue	faint gray-blue		—	absorption	—	—		
11 β,17 α,21-Trihydroxy-4-pregnene-3,20-dione	green-yellow	green		—	absorption	—	—		
3 β,21-Diacetoxy-17 α-hydroxyallopregnane-11,20-dione	blue	—							
20 β,21-Diacetoxy-17 α-hydroxy-4-pregnene-3,11-dione	faint-blue-violet	—							
20 β,21-Diacetoxy-11 β,17 α-dihydroxy-4-pregnen-3-one	yellow	red							

Legend

No.	Reagent	Result	Reference
1	15% Phosphoric acid	Fluorescence in ultraviolet light	NEHER and WETTSTEIN (1951)
2	Antimony trichloride in chloroform	Fluorescence in ultraviolet light	NEHER and WETTSTEIN (1951)
3	Antimony trichloride in chloroform	Color in daylight	NEHER and WETTSTEIN (1951)
4	Antimony trichloride in nitrobenzene	Color in daylight	ROSENKRANTZ (1953a, b)
5	Antimony trichloride in nitrobenzene	Fluorescence in ultraviolet light	ROSENKRANTZ (1953a, b)
6	Antimony pentachloride	Color in daylight	KRITCHEVSKY and KIRK (1952b)
7	Silicotungstic acid	Color in daylight	KRITCHEVSKY and KIRK (1952b)
8	Anisaldehyde-sulfuric acid	Color in daylight	NEHER and WETTSTEIN (1951)
9	Anisaldehyde-sulfuric acid	Fluorescence in ultraviolet light	NEHER and WETTSTEIN (1951)

1951, OERTEL 1954). Triphenyltetrazolium chloride (BURTON et al. 1951) may be applied by immersing the paper chromatogram in a freshly prepared mixture of 2 parts of 0.2% aqueous solution of the salt with 1 part of 10% sodium hydroxide.

AXELROD (1955) has applied various reactions to C_{21}-steroids on paper chromatograms, which permit differentiation between various side chains. When the paper is drawn through 0.1 *N* sodium hydroxide and heated on a glass plate at 100° for 3 minutes, 17,21-dihydroxypregnan-20-ones will no longer give the triphenyltetrazolium test, whereas 21-hydroxypregnan-20-ones are more resistant to alkali treatment (AXELROD 1953b). In a more elaborate scheme of differential analysis AXELROD (1955) subjected paper chromatograms to a series of reactions, including chromic acid oxidation, Oppenauer oxidation, lithium aluminium hydride reduction, the Zimmermann and the triphenyltetrazolium chloride test, etc., to determine the nature of the side chain in pregnane derivatives.

3-Hydroxy-Δ^5-steroids give a characteristic pink color test when the paper chromatogram is first sprayed with water-saturated phenol, then, after drying at room temperature for 1 hour, heated at 75° for 10 minutes and, finally, sprayed with a solution containing 1 g of ammonium molybdate and 2.5 ml of 60% perchloric acid in 100 ml of 0.1 *N* hydrochloric acid, and again heated at 75—80° for 8 minutes (WITTER and STONE 1957).

BUSH (1952) has introduced a very sensitive and specific (BUSH 1954a) test for Δ^4-3-ketosteroids: Two μg of these steroids per sq cm of paper give a primrose yellow fluorescence in ultraviolet light when the paper is sprayed with 10% (w/v) sodium hydroxide in 60% (v/v) aqueous methanol and dried at 60—100°. It was later found (BUSH 1954b) that the reaction is even more sensitive when aqueous instead of methanolic sodium hydroxide is used, to which a trace of triphenyltetrazolium chloride has been added to quench the background fluorescence of the paper. BODÁNSZKY and KOLLONITSCH (1955) have reported that *p*-phenylenediamine phthalate in alcohol gives yellow, orange, or brown spots selectively with Δ^4-3-ketosteroids. Osmium tetroxide vapors (MANARO and ZYGMUNTOWICZ 1951) are apparently also a specific reagent for Δ^4-3-ketosteroids, but the reagent is not so safe to use.

Qualitative or quantitative tests on eluates from paper chromatograms are treated elsewhere in this chapter (p. 362). Extractives from the filter paper may interfere with such analyses (LEWIS 1956), and non-volatile stationary phases may also contaminate eluates. Some workers prefer to stain the paper chromatograms first and then elute the stains for colorimetric assay. Thus, HOFMANN and STAUDINGER (1951) and TOUCHSTONE and HSU (1955) have proposed the analysis of corticosteroids after spraying paper chromatograms with tetrazolium salts. Paper chromatograms may also be evaluated without elution. In this case either the area of the spot is measured and related to the amount from a standard plot of area versus log of the amount or the paper strip is stained and drawn through a densitometer. Applications of the latter technique to 17-ketosteroids (OERTEL 1954a) and corticosteroids (OERTEL 1954b) have been described.

D. Instrumental methods

I. Spectrophotometry

When a beam of electromagnetic radiation is split up into its component frequencies and passed through a sample of an organic compound, the intensity of the transmitted radiation is diminished at certain frequencies. A plot of the intensity of the transmitted radiation, or the ratio of the incident and transmitted intensities versus the frequency or wave length of the radiation comprises an absorption spectrum. The absorption of energy from the radiation field by a molecule occurs when electrical work is performed in raising the molecule to a higher energy state; the frequency of the radiation corresponds to that of a certain intramolecular transition, either atomic or electronic. These frequencies, and the extent of absorption, are related to specific structural elements in the molecule, and absorption spectra can provide an great deal of information concerning the structure of the absorbing species.

Before we proceed to the different types of spectra and their uses and interpretation, a few important features of electromagnetic radiation and its measurement will be reviewed. The wave character of the radiation is expressed in terms of the frequency, $\tilde{\nu}$, in oscillations per unit time, the wave length, λ, and the velocity, c, by the equation $\lambda\tilde{\nu} = c$. Frequency is more conveniently and commonly expressed as the wave number, ν (or v) in oscillations per unit length, thus $\lambda\nu = 1$. Wave length is expressed in various units, depending upon the order of magnitude. In the infrared region of the electromagnetic spectrum, wave lengths are of the order of 10^{-3} to 10^{-4} cm, and are usually given in microns (μ), 1 $\mu = 10^{-6}$ m or 10^{-4} cm. In the visible and ultraviolet regions, the millimicron (mμ) ($= 10^{-7}$ cm) or Ångström unit (Å) ($= 10^{-8}$ cm) is used.

The process of absorption of electromagnetic energy by a molecule is described by quantum theory in terms of a "packet" of radiation, or photon, the energy of which is completely transferred to the molecule, when absorption occurs, as increased rotational, vibrational or electronic energy. The energy change in the molecule is given by the expression

$$\Delta E = h\tilde{\nu} = \frac{hc}{\lambda}$$

in which h is the Planck constant, 6.62×10^{-27} erg-sec per molecule. The energy is thus directly proportional to the frequency of the radiation, and inversely proportional to the wave length. In the conventional units of chemical energies, this fundamental frequency equation becomes

$$\Delta E = \frac{2.86 \times 10^{-3}}{\lambda} \text{ kcal/mole}$$

when the wave length is expressed in cm.

The amount of radiation absorbed by a sample can in most cases be expressed by the Beer-Lambert Law, which states that the absorption is proportional to the number of absorbing molecules in the light path. If the intensities of the incident and transmitted beams are denoted by I_o and I, respectively, the absorption by a solution of concentration c and length or thickness l is expressed by

$$\log_{10} \frac{I_o}{I} = kcl\,.$$

An assortment of conventions for expressing the quantities in this equation is encountered in the literature. The fractions I/I_o and $I_o—I/I_o$ are called *transmission* and *absorption*, respectively. The important quantity $\log_{10} I_o/I$ has been variously called *optical density* (D or $O.\,D.$), *extinction* (E) or, in currently preferred terminology, *absorbance* (A or A_s). The constant k is called the *extinction coefficient* (the term *absorption coefficient* has been used when the equation is expressed in natural logarithms) or the *absorbancy index* (a_s). With concentration expressed in molarity (moles per l.), the constant becomes the *molar extinction coefficient* (ε) or *molar absorbancy index* (a_M):

$$k = \frac{E}{cl} = \varepsilon \qquad a_s = \frac{A_s}{cl} = a_M$$

The value of ε or a_M, which is the extinction or absorbance of a 1 M solution in a 1 cm thickness, is a characteristic constant for any substance at a specified wave length.

1. Infrared spectrophotometry

The importance of infrared spectral measurements in the steroid field and the popularity of this tool among steroid chemists can be judged by the growing number of reviews of the subject. Atlases of steroid spectra have been compiled by Dobriner et al. (1953) and Roberts et al. (1958), and reviews have been presented by R. N. Jones and Dobriner (1949), A. Cole (1956), Rosenkrantz (1955), and

PLANTIN (1955). In addition to these surveys dealing primarily with steroid spectra, numerous treatises and reviews have been devoted to the theoretical aspects and interpretation of the infrared spectra of organic compounds, among which should be mentioned those of BELLAMY (1954) and of R. N. JONES and SANDORFY (1956). The present discussion makes no pretense to completeness, and is drawn largely from this wealth of authoritative summaries.

a) *Origin of infrared spectra*

The absorption spectra which are measured with standard infrared spectrophotometers extend over the region of 4000—600 cm^{-1} (2.5—15.5 μ) of the electromagnetic spectrum. The energies of radiation in this region are of the order of magnitude associated with atomic motions in the electronic ground state, specifically *vibrations* of the atoms joined by covalent bonds. The energies involved in *rotations* about a covalent bond are lower, and interactions of these transitions with radiation occur in the far infrared region (600—100 cm^{-1}).

The vibrations that give rise to absorption bands in the infrared consist primarily of longitudinal *stretching* along covalent bonds and transverse *bending* or deformation. The latter, which are in general of lower energy, can be classified for carbon-hydrogen bonds in methyl and methylene groups into several types denoted by the expressive terms, *rocking*, *wagging*, *twisting*, and *scissoring*, referring to the several symmetrical and unsymmetrical vibrational modes. The total number of fundamental vibrational modes in a molecule containing n atoms is given by the expression $3n—6$, and additional frequencies can arise from overtones and interactions of these. In the spectrum of a steroid, literally hundreds of bands might be anticipated, but in a typical spectrum (Fig. 98) there is a relatively small number of well-defined bands. The main reason for this lies in the fact that the frequencies associated with a given type of vibration, carbon-hydrogen stretching, for example, fall into a narrow range and appear as a single broad band. Furthermore, many bending vibrations give rise to low intensity bands that are not resolved in the usual spectrum.

The complete analysis of an infrared spectrum, i. e., the identification of every absorption band with a specific vibrational mode can be accomplished for very simple molecules by statistical mechanical treatment. An approximate location of the stretching frequency associated with a specific grouping can be obtained, however, by considering a covalent bond as a simple harmonic oscillator, the motion of which is described by the equation:

$$\nu = \frac{1}{2\pi c}\left(\frac{k}{\mu}\right)^{1/2} = 4.11\left(\frac{k}{\mu'}\right)^{1/2}$$

In this expression, the frequency, ν, of the stretching vibration is related to a *force constant*, k, in dynes per cm, and μ, the reduced mass of the atom pair. In the form on the right, the mass, μ', is expressed in atomic weight units:

$$\mu' = \frac{M_1 M_2}{M_1 + M_2}$$

The force constant is a function of the electronegativities of the atoms, the bond distance and the bond order (single, double or triple covalent). The values of k can be obtained from empirical expressions, and are of the order of 5×10^5 dynes per cm for a single bond between the lighter elements, and two or three times this value for a double or triple bond, respectively. From the above equations it is seen that the reduced mass for a C—H or O—H bond is slightly less than unity, while that for C—C or C—O bonds is 6—7 fold greater. Using a value of $k = 5\times10^5$, the frequencies given by the oscillator equation which are associated with C—H and O—H stretching vibrations are of the order of 3000—4000 cm^{-1}, and for C—C and C—O bonds, 1100—1200 cm^{-1}. The frequencies corresponding to C=C and C=O bonds are higher than those of the single bonds by approximately a factor of 2. The substitution of a deuterium atom for hydrogen in an O—H or C—H bond nearly doubles the reduced mass, and the frequency is reduced by a factor of $1/\sqrt{2}$.

b) *Measurement and presentation of spectra*

Descriptions of the design and optical systems of infrared spectrophotometers will be found in several of the reviews cited above. For the dispersion of infrared radiation, prisms prepared from large crystals of salts such as sodium chloride or calcium fluoride have the requisite transparency and resolving properties. For highly refined measurements, different prisms are used in the range of their highest dispersion. Thus lithium fluoride is superior for measurements at the high frequency end of the spectrum. For general purpose work, rock salt optics can be used over the region 4000—600 cm^{-1}.

Infrared spectrophotometers can be designed to scan and record the spectrum on a scale linear either in wave length or in frequency. The majority of instruments in use at present are linear in wave length, and for this reason it is logical to designate band positions in terms of this scale; this also follows the universal practice used in ultraviolet and visible spectra. A number of authors, however, including most spectroscopists, prefer a frequency or wave number scale for infrared spectra to facilitate comparisons with Raman spectra and because frequency is directly related to vibrational energy. In the extensive correlations of infrared spectral data on steroids published by R. N. JONES and his associates all band positions are expressed in frequency units (cm^{-1}). Both scales are encountered in the literature, and band positions given in the present discussion are expressed in both frequency and wave length.

Because of instrument design, curves recorded linearly in wave length are nearly always plotted against percent transmission 100 I/I_o with zero transmission at the bottom of the ordinate, while those recorded on a linear frequency scale are plotted against percent absorption (100 $(I_o—I)/I_o$ usually with zero at the bottom. The terms "absorption band" and "absorption maximum" are employed in either case, although in the former they are actually transmission minima. In curves which are plotted with linear frequency, the scale is changed at 2000 cm^{-1} to spread the bands more evenly across the spectrum. The curves obtained by recording the spectrum of the same sample by these different conventions differ somewhat in overall appearance, some bands being more compressed in one presentation than the other.

The description and presentation of infrared spectral data are vexing problems aside from the divergent conventions employed. The tremendous value of an infrared spectrum lies in the plurality of bands, usually with characteristic contours, and it is just this feature which prevents an adequate summarization of the absorption data. It is impossible to reproduce routinely complete spectral curves in the periodical literature. However, some authors include in the description of a new compound the wave length or frequency of the principal bands associated with important functional groups. Most of the complete infrared curves of steroids published for characterization purposes are plotted on a linear wave length scale and usually are obtained in the solid state (Fig. 98). In connection with the detailed interpretation of spectra obtained with solutions it is often convenient to reproduce only a portion of the curve; this is more often plotted on a linear frequency scale (Fig. 99).

c) *State of sample*

The detail and reproducibility of infrared spectra depend to a considerable extent on the preparation of samples, and an understanding of the factors involved is necessary for anyone using infrared data. Techniques are available for obtaining the infrared spectra of steroids in solution or as crystalline solids; each method has certain drawbacks and advantages. Melted films have occasionally been used to obtain liquid state spectra of steroids, but they are not generally applicable, especially for the high-melting corticosteroids (ROSENKRANTZ and ZABLOW 1953).

The two important advantages of solution techniques, as contrasted with methods using samples in the solid state, are the absence of effects due to scattering and reflection and to intermolecular forces in a crystal lattice, and the fact that concentrations and sample thickness are accurately known. The latter conditions are a prerequisite if quantitative measurements of the absorption intensity are to be made. The cells used to hold a solution sample are constructed of rock salt plates spaced about 1 mm apart and contain from 0.2 to 0.5 ml of solution; spectra can be obtained conveniently with 1—5 mg of sample.

The major difficulty in using solutions of steroids for infrared work is the limitation imposed in the choice of solvent. Even with double-beam spectrophotometers in which the absorption due to solvent is cancelled out by use of a blank cell, the spectrum of the solute in regions of strong solvent absorption is obscured. Since most common steroid solvents, containing C—O, C—H, O—H and sometimes C=O bonds, exhibit a full complement of absorption bands, non-polar solvents such as carbon disulfide, carbon tetrachloride or chloroform must be used. None of these compounds is completely transparent, but each is useful in certain regions of the spectrum. In the steroid curves compiled by DOBRINER et al. (1953), these solvents were used in the following regions:

carbon disulfide: 1800–1660, 1400–880 and 820–700 cm^{-1}
carbon tetrachloride: 1800–1600, and 1480–1300 cm^{-1}
chloroform: 1800–1600, 1150–940, and 920–850 cm^{-1}

Carbon disulfide and carbon tetrachloride are both suitable for the higher frequency region. A more serious drawback to these solvents is their very low solvent power for hydroxylated steroids. This has been circumvented by recording the spectra of acetates, but the fact remains that the spectra of many natural steroids, particularly the adrenocortical hormones, cannot be obtained in solution.

Several techniques for obtaining infrared spectra on crystalline samples are available. The simplest of these is the deposition of a crystalline film on a rock salt plate from a solution in a convenient solvent, but it is difficult to obtain reproducible spectra in this way. Another method comprises mulling the finely powdered steroid with mineral oil (Nujol) to form a homogeneous suspension in which scattering is minimized. Spectra obtained from such mulls contain the strong aliphatic C–H frequencies (3000–2700 cm^{-1} and 1500–1300 cm^{-1}) due to the oil. LABROSSE (1953) has described the preparation of mulls with microgram quantities of steroids from paper chromatographic eluates.

Probably the most generally satisfactory procedure for solid state samples is the potassium bromide disc method (STIMPSON and O'DONNELL 1952). The sample is thoroughly ground with a suitable amount of pure potassium bromide and the mixture is converted to a transparent pellet or wafer in a hydraulic press. This method is somewhat less demanding of experimental skill than is the mull procedure; sample thickness is more uniform, and the spectra are quite reproducible. The most important advantage lies in the fact that no extraneous bands appear in the spectrum. Potassium bromide pellets can be fashioned to conform to the size and shape of the spectrophotometer slit, permitting the use of 10–25 μg of sample (D. ANDERSON and WOODALL 1953, SCHIEDT and RESTLE 1954). Microscope attachments for micro scale spectra have been developed and are reviewed by A. COLE (1956).

In any procedure using samples in the solid state, significant shifts in the characteristic frequencies due to specific functional groups, notably carbonyl and hydroxyl, may be encountered. This is true also to a lesser extent of spectra obtained in different solvents. The changes are due to hydrogen bonding and dipole effects which are particularly important in a crystal. Other changes, caused by the grinding operation and the application of very high pressure, have been observed in the spectra of highly hydroxylated steroids in potassium bromide discs (ROBERTS 1957). The bulk of the literature on the correlation of specific group frequencies is based upon spectra obtained in solution, and for structural assignments or comparison of spectra, the same experimental technique must be employed for the reference and unknown samples.

For identification purposes, the sample should, of course, be of the highest purity obtainable. The spectrum of an impure compound is characterized by a lack of fine structure and general "smearing" of bands and, in some cases, the presence of very weak bands. SLAUNWHITE and NEELY (1957) have compared the spectra of urinary 17-ketosteroids obtained from samples of differing purity, and recommend charcoal treatment and protracted washing of the extracts with acid prior to obtaining the spectra. An indication of the presence of certain functional groups can be obtained on grossly impure samples, however, and this is sometimes a useful preliminary in choosing a method of purification.

d) *Interpretation of spectra*

The two *major functions* that can be filled by infrared spectra are the *detection of functional groups* in a compound of unknown structure and the *establishment* of *identity* or *non-identity* of two samples. It has been noted that the vibrational frequencies associated with certain bonds fall in a fairly narrow range, and a number of bands in the higher frequency end of the spectrum provide very reliable evidence for the presence of certain functional groups. There is, on the other hand, a sufficient number of bands associated with molecular "fine structure" to render the spectrum of every compound, excepting the case of enantiomorphs, unique and characteristic. These bands are found toward the lower frequency end of the curve, from 1350–650 cm^{-1}, and this portion of the spectrum is aptly called the

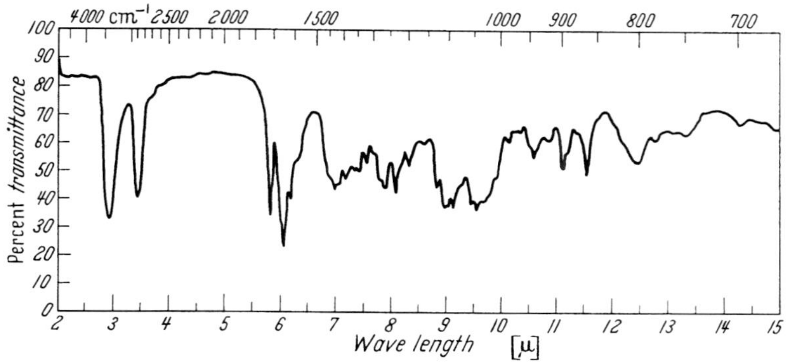

Fig. 98. Infrared spectrum of cortisol in KBr pellet (HAYDEN 1955). Reproduced from Analyt. Chem. **27**, 1486 (1955) with permission of the author and editor

"fingerprint" region. These features are discussed in detail in the following section; it will be useful to illustrate them first by reference to the spectrum of cortisol (Fig. 98).

From the left-hand side of the spectrum to 1500 cm^{-1} are found four strong bands, due to the stretching frequencies of the O–H, C–H, 20-keto and Δ^4-3-keto groupings. These bands would differ very little in the spectrum of any 17,21-dihydroxy-Δ^4-3,20-diketosteroid. In the region from 1450–1350 cm^{-1} is seen a complex of bands due to C–H bending frequencies. The bands from 1300–1200 cm^{-1} are probably associated with the Δ^4-3-keto grouping; the vibrational modes have not been specified. The absorption from 1150–1000 cm^{-1} can be attributed to vibrations of the C–O bonds. Specific assignments of many of the individual bands in the region from 1500–1000 cm^{-1} have been made by R. N. JONES and co-workers for the spectra of relatively simple steroids in solution, but these cannot be carried over to the spectrum of a highly substituted compound such as cortisol, particularly in the solid state. Below 1000 cm^{-1} is the fingerprint absorption due to skeletal deformations that are modified by the oxygen substituents; these are uniquely characteristic for cortisol in the solid state.

α) **Carbonyl groups.** The bands arising from carbonyl stretching frequencies have been investigated more extensively by far than any others in the infrared spectra of steroids. The carbonyl band is one of the strongest and most conspicuous ones in the spectrum of an aldehyde, ketone or ester, and the frequency of a band due to a specific carbonyl system varies very little over a range of compounds. R. N. JONES and HERLING (1954) have summarized the characteristic frequencies for some 130 different types of steroid carbonyl groups. A few of these values for

Table 20. *Characteristic frequencies of carbonyl bands*

CCl_4 or CS_2 Solution		Carbonyl type	A[2]	$CHCl_3$ solution	
ν, cm^{-1}	λ, μ			ν, cm^{-1}	λ, μ
1767—1764	5.66—5.67	Phenolic 3-acetate ester	2.75	1762—1753	5.67—5.70
1758—1754	5.69—5.70	*21-Acetoxy*-20-ketone[1]		1750—1745	5.71—5.73
1752	5.71	*12α-Acetoxy*-11-ketone[1]			
1752—1748	5.71—5.72	11,*17*-Diketone[1]		1742—1738	5.74—5.75
1749	5.72	16-Ketone	2.74	1742	5.74
1749	5.72	17β-Hydroxy-16-ketone		1742	5.74
1749	5.72	20,21-Diacetate ester		1739—1736	5.75—5.76
1749—1746	5.72—5.73	*6α-Acetoxy-Δ^4*-3-ketone[1]			
1745	5.73	*2-Acetoxy-Δ^4*-3-ketone[1]			
1745—1743	5.73—5.74	*6β-Acetoxy-Δ^4*-3-ketone[1]			
1745—1741	5.73—5.74	17-Ketone	2.69	1737—1733	5.76—5.77
1742—1737	5.74—5.76	17β-Acetate ester		1722—1720	5.81—5.82
1739—1736	5.75—5.76	17α-Hydroxy-20α or β-acetate ester			
1739—1733	5.75—5.77	3-Acetate ester	3.24	1728—1719	5.79—5.82
1736	5.76	*17α-Acetoxy*-20-ketone[1]			
1736—1733	5.76—5.77	20-Acetate ester			
1736—1724	5.76—5.80	21-Acetoxy-*20-ketone*[1]		1727—1720	5.79—5.81
1724—1717	5.80—5.82	3- or 17-Benzoate ester	3.55	1713—1710	5.84—5.85
1723—1718	5.80—5.82	3,6-Diketone			
1720	5.81	12α-Acetoxy-*11-ketone*[1]			
1719	5.82	17α-Acetoxy-*20-ketone*[1]			
1719—1713	5.82—5.84	3-Ketone	2.55	1709—1700	5.85—5.88
1719—1713	5.82—5.84	*11*,17-Diketone[1]		1711—1707	5.84—5.86
1718	5.82	10-Aldehyde			
1714—1713	5.83	6-Ketone (5α series)			
1710—1707	5.85—5.86	17α or β-Hydroxy-20-ketone			
1710—1706	5.85—5.86	20-Ketone	1.79	1707—1698	5.86—5.89
1710—1704	5.85—5.87	11-Ketone	2.21	1705—1698	5.86—5.89
1708—1706	5.85—5.86	6-Ketone (5β-series)			
1700	5.88	2-Acetoxy-*Δ^4-3-ketone*[1]			
1697—1690	5.89—5.92	17α or β-Hydroxy-20-ketone			
1692	5.91	Δ^4-3,6-Diketone			
1688	5.92	6β-Acetoxy-*Δ^4-3-ketone*[1]			
1686—1685	5.93—5.94	6α-Acetoxy-*Δ^4-3-ketone*[1]			
1684—1680	5.94—5.95	Δ^1-3-ketone			
1681—1677	5.95—5.96	Δ^4-3-ketone	3.65	1668—1660	5.99—6.02
1671—1663	5.98—6.01	$\Delta^{1,4}$-Diene-3-ketone	3.70	1666—1660	6.00—6.02
1669—1666	5.99—6.00	$\Delta^{4,6}$-Diene-3-ketone	4.11		

[1] These groups give rise to two carbonyl bands, both of which are listed; the italicized group is responsible for the frequency indicated.
[2] Integrated absorption intensity.

the groups most commonly encountered in corticosteroids and their derivatives are listed in Table 20.

It is evident from Table 20 that a precise knowledge of the carbonyl stretching frequency reveals considerable information on the nature and position of the group. The frequency is determined by a number of factors that influence the carbonyl resonance system:

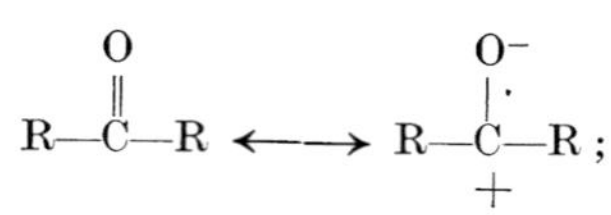

A few of these are summarized in the following generalizations.

Type of carbonyl. The frequencies of ester carbonyl bands are in general higher than those of ketones. In an enolic or phenolic ester, containing the grouping $-\overset{\overset{\displaystyle O}{\|}}{C}-O-\overset{|}{C}=\overset{|}{C}-$, the band is shifted to still higher frequencies.

Ring size. Steric strain in a cyclic ketone is reflected in a higher frequency of the carbonyl band. The carbonyl frequency of a cyclohexanone is very close to that of acyclic ketones; a cyclopentanone (ring D carbonyl group) absorbs some 30 cm^{-1} higher. This is a very general effect, and a band at 1745–1742 cm^{-1} is a very reliable indication of a five-membered ring ketone.

Adjacent functional groups. A substituent in the vicinity of a carbonyl group can alter the frequency of the carbonyl band by a simple mass or electronic effect or by vibrational interaction. The first type of influence is illustrated by the carbonyl frequencies of α-bromoketones. When the bromine substituent is axial the position of the band is not significantly changed from that of the parent ketone. An equatorial bromine atom, however, which is in the plane of the carbonyl bond and can thus affect the dipole, produces a shift of about 15 cm^{-1} to higher frequencies. A different effect is encountered in α-hydroxyketones, in which hydrogen bonding gives rise to a second, weaker band at slightly lower frequency (R. N. Jones et al. 1952b). More conspicuous effects may be observed when vibrational coupling can occur, as in an α-acetoxyketone. In the spectra of 21-acetoxy-20-ketones two bands are present at about 1756 and 1730 cm^{-1}, attributable to the acetate and ketone groups, respectively, but each is displaced about 20 cm^{-1} to higher frequencies from the "normal" position.

Conjugation. The effective extension of the carbonyl system that occurs in an α,β-unsaturated ketone and leads to intense ultraviolet absorption is reflected in a shift to lower frequency of about 40–50 cm^{-1} in the carbonyl stretching band and a significant increase in the intensity of the band. A similar effect is observed in the lower frequency of benzoate esters as compared to acetates.

The ranges indicated in Table 20 for the characteristic frequencies of carbonyl bands reflect the variation over a number of compounds; R. N. Jones and Dobriner (1949) suggest an uncertainty of $\pm$ 3 cm^{-1} in the measurement of the frequency with sodium chloride prisms. Carbonyl frequencies due to 5-membered ketones or esters, 6-membered or acyclic ketones and conjugated ketones are readily distinguished, but the several 6-membered cyclic carbonyl and C-20 carbonyl groups cannot be identified with certainty from the band position alone. Differences in the band intensity, and other bands associated with the carbonyl function, e. g., the C–O stretching band of esters or the methylene bending frequency of cyclic ketones, permit further delineation in certain cases, as in the distinction between 17-keto and acetate groups or C-3 and C-20 ketones. R. N. Jones et al. (1955) have drawn attention to several characteristic bands that occur in the spectra of ketones below 1350 cm^{-1}. These frequencies, presumably due to skeletal vibrations of carbon atoms adjacent to the carbonyl group, can be used with caution to confirm the position of the carbonyl group which is deduced from the stretching frequency.

The individual carbonyl bands in di- or polycarbonyl compounds can usually be discerned, at least as a bifurcation on a fused band, if the respective frequencies are separated by 10 cm^{-1} or more. As already mentioned, the two bands are both displaced to higher frequencies if interaction occurs. In the spectrum of a 3,11- or 3,20-diketone the carbonyl bands are not resolved and a single symmetrical band of high intensity is observed.

Values of the molecular extinction coefficient, E, for typical saturated carbonyl bands are of the order of 400—700, but the band intensity cannot be adequately expressed by a simple extinction coefficient. Because of the high ratio of the slit width to the width of the band, the shape of the band and thus the peak height will vary with the resolving power of the spectrophotometer. The intensity of the band is, therefore, measured as the *area* under the band and expressed as the *integrated absorption intensity* in units defined by the equation

$$A = \frac{10^{-4}}{l\,c} \int \ln \frac{I_o}{I}\, d\nu \,.$$

Methods for the determination of the true integrated absorption intensity (corrected for finite slit width) have been developed by RAMSAY (1952) and a spectrophotometer attachment to permit automatic measurement of this function has been described by WILLIAMS et al. (1955). The use of these values in the interpretation of the infrared spectra of steroidal ketones has been illustrated by R. N. JONES et al. (1952c) and SIMPSON et al. (1954a), among others. In the case of polycarbonyl compounds, the integrated absorption intensities are additive. Thus, in a 3,20-diketone, in which the separate bands are not resolved, the integrated intensity of the band, 4.45 units, corresponds well with the sum of the values for the individual bands (Table 20) of 4.34 units, and clearly establishes the presence of two carbonyl groups.

***β*) Hydroxyl groups.** The spectra of hydroxylated steroids in dilute solution contain a band in the region 3640—3600 cm^{-1} (2.75—2.78 μ) due to the O—H stretching frequency. This band, which has an extinction coefficient of about 50 for a monoalcohol, provides a reliable indication of the presence of a free hydroxyl group, which in the case of tertiary alcohols cannot be obtained by direct chemical methods. Correlation of the position of the band and the location of the hydroxyl group is not possible, but information on this point can be obtained from the spectra of the derived ketone or acetate. An important feature of the O—H band is the sensivity to hydrogen bonding, which is marked by the appearance of a second band at lower frequencies.

Intermolecular hydrogen bonding is observed in spectra of hydroxy compounds obtained in concentrated solutions and particularly in the crystalline state; the band due to bonded hydroxyl group (I) is very broad

(I)

and usually falls between 3550 and 3300 cm^{-1} (2.82—3.03 μ). In vicinal diols or α-hydroxyketones intramolecular hydrogen bonding may occur; this effect is distinguished from intermolecular association by its persistence in very dilute solutions. The band due to an internally bonded hydroxyl group is displaced to about 3500 cm^{-1} (2.86 μ) and can be very useful in the diagnosis of configurational relationships.

***γ*) Double bonds.** The stretching frequencies due to isolated olefinic linkages fall in a fairly narrow range between 1670—1620 cm^{-1} (5.98—6.7 μ); characteristic frequencies have been assigned to double bonds at a number of positions in the steroid skeleton (R. N. JONES and HERLING 1954). The C=C stretching band is of low intensity (ε 50) because of the symmetrical nature of the bond, and cannot be detected in the spectra of tetrasubstituted olefins such as $\Delta^{8(14)}$-steroids. In the spectra of non-conjugated unsaturated carbonyl compounds, the double bond stretching band is difficult to distinguish and may appear only as a shoulder. In conjugated dienes or unsaturated ketones the intensity of the band is increased and the position is shifted to lower frequency, just as in the case of the conjugated carbonyl band. The olefinic band in a Δ^4-3-ketone appears as a sharp band of medium intensity at 1620–1615 cm⁻¹ (6.17–6.21 μ). The stretching and bending frequencies of C–H bonds directly attached to an olefinic center provide more information concerning the environment and substitution of a double bond.

δ) Carbon-oxygen bonds. Strong bands (ε 100—250) attributed to the C—O stretching frequency are present in the spectra of alcohols, ethers and esters. The principle band in simple esters such as acetates appears at 1250–1230 cm^{-1} (8.00–8.13 μ) and is very prominent, usually obscuring any other absorption in this region. Acetates of equatorial C-3 alcohols are characterized by a single band at 1242 to 1236 cm^{-1} (8.07–8.12 μ) whereas axial 3-acetoxy compounds show a multiple band at the same or a slightly higher frequency (R. N. Jones et al. 1951).

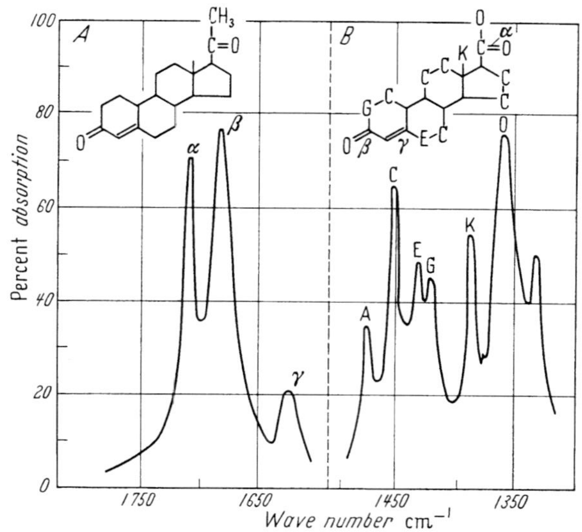

Fig. 99. Infrared spectrum of 19-nortestosterone in CCl_4 solution (R. N. Jones and Cole 1952). The letters on the bands correspond to the assignment to individual methyl, methylene and carbonyl groups as indicated in the formula. Reproduced from J. Amer. Chem. Soc. 74, 5648 (1952) with permission of the authors and editor

The C–OH stretching frequency is readily detected between 1060–1000 cm^{-1} (9.44–10.0 μ) in simple alcohols, and certain correlations of the position of this band and the position and configuration of the hydroxyl group have been made (A. Cole et al. 1952). Other functional groups, including the carbonyl groups in acetates and ketones, give rise to strong bands in this same region, however, and the band is of limited usefulness in polyfunctional compounds. Hayden (1955) has suggested that a band at 1096 cm^{-1} in the spectra of corticosteroids in KBr discs may be attributable to the 17-hydroxyl group.

ε) Carbon-hydrogen bonds. The spectra of all steroids exhibit an intense band at 2970–2930 cm^{-1} (3.37–3.39 μ) and a weaker band or shoulder at about 2870 cm^{-1} (3.48 μ). These bands arise from C–H *stretching* vibrations in methyl and methylene groups. With a lithium fluoride prism considerable fine structure can be seen in these bands, but assignments to individual groups have not been made. Hydrogen atoms attached to a doubly bonded carbon atom produce a very weak band at 3050–3000 cm^{-1} (3.22–3.33 μ), the frequency of which has been used to distinguish double bonds at various locations in the nucleus (Henbest et al. 1954).

Much more specific structural information is available from the C–H *bending* vibrations of methyl and methylene groups which occurs in the region from 1475–1300 cm^{-1} (6.77–7.69 μ). In most steroid spectra two groups of bands are observed at 1450–1410 cm^{-1} (6.89–7.14 μ) and 1390–1360 cm^{-1} (7.19–7.35 μ), corresponding to methylene and methyl absorption, respectively. These are of moderate intensity but are sometimes inconspicuous because the base-line absorption is high. The spectra of some 150 compounds in this region were examined by R. N. JONES and COLE (1952) using carbon tetrachloride solutions and a calcium fluoride prism. With these conditions it was found that the vibrations due to a number of individual methylene and methyl groups could be assigned characteristic frequencies, as indicated in Fig. 99. Several of these assignments were rigorously confirmed by the absence of the band due to selected groups when deuterium was substituted for hydrogen in the group in question (R. N. JONES et al. 1952a). The band positions of a few of the more important groups are given in Table 21.

Table 21. *Characteristic frequencies for methyl and methylene bending vibrations*

ν, cm^{-1}	λ, μ	Structural element
1470—1462	6.80—6.84	Aliphatic methylene group in side chain
1464—1446	6.83—6.92	Cyclic methylene group
1445—1432	6.92—6.98	Allylic methylene group, $-\overset{\mid}{C}=\overset{\mid}{C}-CH_2-$
1438—1426	6.95—7.01	Methylene group adjacent to keto group at C-4, -6, -7, 11, -12
1426—1415	7.01—7.07	Methylene group at C-2 or -4 adjacent to 3-keto group
1415—1412	7.07—7.08	C-21 methylene group in 21-acetoxy-20-ketone
1410	7.09	C-21 methylene group in 21-hydroxy-20-ketone (based on single compound)
1411—1404	7.09—7.12	Methylene group adjacent to keto group in five-membered ring
1392—1374	7.18—7.28	Angular methyl group, C-19, or terminal methyl group in sterol side chain
1383—1372	7.23—7.29	Angular methyl group, C-18
1382—1369	7.24—7.30	Acetate methyl group
1374—1360	7.28—7.35	Terminal *gem*-dimethyl group in side chain
1370—1360	7.30—7.35	Acetate methyl group
1356—1359	7.36—7.37	C-21 methyl group in CH_3CO-side chain

Measurements in this region of the spectrum are useful in confirming the position of carbonyl groups, and can be especially valuable in distinguishing between homologous steroids that differ only by a methylene group in the side chain (R. N. JONES and COLE 1952). In the latter case, the lower frequency ends of the two spectra, normally most significant in compound comparison, were nearly identical. It should be observed that the positive identification of individual methyl and methylene groups from this region of the infrared spectrum is by no means a routine matter. These bending vibrations, while surprisingly localized in many simple compounds, are nevertheless very sensitive to the presence of functional groups, and displacements and extra bands can make specific assignments impossible. Since analysis of this region has been largely restricted to compounds in carbon tetrachloride solution, detailed data on free corticosteroids are lacking.

Although detailed information on methyl and methylene groups can be adduced from stretching and bending vibrations in the infrared in favorable cases, nuclear magnetic resonance spectra (p. 376) provide a much more discerning and general method for the characterization of these structural elements.

ζ) Fingerprint region and compound comparison. Few correlations of band positions with structural elements have been made in the region below 1000 cm^{-1} (10.0 μ). From 860—730 cm^{-1} (11.6—13.7 μ) are found bands corresponding to aromatic C—H vibrations that are present in the spectra of estrogens and benzoate esters. Apart from these there are few conspicuous bands in this lower frequency region in most steroid spectra. The absorption is due to various skeletal vibrations which are very sensible to the presence of polar groups, and the introduction of oxygen substituents brings about a marked enrichment of the absorption. H. HIRSCHMANN (1953) has discussed the effect of spatially remote groups on the out-of-plane bending vibrations of hydrogen atoms attached to double bonds, especially in Δ^5-steroids, which are responsible for weak bands at 820—790 cm^{-1} (12.2—12.6 μ).

Because of the sensitivity to minor structural or stereochemical changes, this fingerprint region is the most significant part of the curve in establishing the identity or non-identity of two compounds. Since the spectral differences may be subtle ones, it is very important that the two curves being compared be run under identical conditions, preferably on the same spectrophotometer. If the spectra are truly superimposable, identity is assured. Small amounts of impurities, or, in the case of solid-phase spectra, differences in crystal form, may introduce sufficient changes to render the comparison equivocal, and the two samples should be subjected to the same purification and crystallization treatment if a rigorous test of identity is to be made.

Although the spectrum of a gross mixture of steroids is usually uninterpretable, slight deviations from the reference spectrum can furnish a clue to the presence in the sample of a small amount of a second component. This approach has been used on several occasions in the extensive studies of urinary steroids by the Sloan-Kettering group (DOBRINER et al. 1948a).

η) Quantitative applications. The quantitative determination of the concentration of a steroid in a sample by infrared spectrophotometry is completely feasible, but in practice this has been largely restricted to mixtures that cannot be analyzed by any other technique. Methods based upon infrared measurements are in general more time-consuming and complicated than those based upon visible or ultraviolet spectrophotometry. Probably the most serious limitation is the fact that for accurate quantitative work all components of the mixture must be known and their spectra available. On the other hand, it is evident that the uniqueness of infrared spectra furnishes a degree of discrimination unattainable in any other analytical method based upon spectrophotometry, making possible the analysis of mixtures of very closely related substances.

Semiquantitative estimation of the composition of a mixture of two steroids can be made by inspection of the spectra of synthetic mixtures of the two components and comparison with the unknown; the accuracy of this procedure is 5—10% (R. N. JONES and DOBRINER 1949). More refined analyses are accomplished by comparison of the intensities of selected bands in the spectra of the several components of the mixture. These are chosen, as far as possible, so that a band maximum in one curve corresponds to a position of minimum absorption in the others.

The composition can be obtained from the band intensities by means of a calibration curve, or by the base-line or ratio method (WRIGHT 1941, HEIGL et al. 1947). In this method a line is drawn through a point of minimum absorption near the peak to be measured and the absorbance at the band maximum is related to this line, avoiding the necessity of making absolute intensity measurements. An illustration of this method is furnished by the assay described by SLOMP et al. (1955) for mixtures of 4,22-stigmastadiene-3-one and the 5α- and 5β-3-ketones derived from it by catalytic hydrogenation. Other applications of infrared procedures to the analysis of steroid mixtures have dealt with mixtures of estrogens (CAROL et al. 1948) and plant sterols (J. JOHNSON et al. 1957).

2. Ultraviolet and visible spectrophotometry and fluorescence

a) Origin of absorption and fluorescence

The distinction between ultraviolet and visible light has a purely physiological basis; the two regions are contiguous, covering radiation of wave lenght 800—400 mμ in the visible and 400—200 mμ in the near ultraviolet. Absorption spectra in the region 800—200 mμ are measured with the same spectrophotometers, and the spectra have a common origin, namely the electronic transitions that are associated with unsaturated groupings. The reviews of BRAUDE (1955) and WEST (1956) are recommended for discussion of the theory of electronic spectra.

The energies involved in these transitions are some 20-fold greater than those associated with atomic vibrations in the infrared region, and it follows that any electronic transition must comprise a number of vibrational and rotational levels. This means that the absorption maxima in an ultraviolet curve are broad envelopes in contrast to the narrow bands in an infrared spectrum. A precise analysis of electronic spectra can be approached by molecular orbital theory, but it will suffice for the present discussion to state that the more labile the electrons, the lower the energy of the transition and the lower the frequency of the radiation absorbed. In terms of structural entities, *a system of unsaturation extending over several atoms and over which the unsaturation electrons can resonate will absorb radiation in the near ultraviolet* or visible regions. The electronic movements in an isolated double bond or a single bond are of much higher energy and absorb in the far ultraviolet region below 200 mμ.

Several paths are available for the stabilization of a molecule that has been excited by the absorption of visible or ultraviolet radiation. In the ultraviolet, the energy may be sufficient to cause dissociation, but more often the energy absorbed is dissipated by "radiationless" transitions and intermolecular collisions. In certain cases, however, return to the electronic ground state proceeds by a quantized transition of somewhat lower frequency (because of loss in vibrational energy), giving rise to the phenomenon of fluorescence. This "reradiation" or emission occurs when the molecule can remain in an electronically excited state for a relatively long time interval, of the order of 10^{-9} seconds, as compared to the 10^{-15} seconds required for the excitation. In exceptional cases, the decay time may be of the order of seconds, and phosphorescence is observed. The structural features which impart fluorescence to a molecule cannot be narrowly defined and the exact nature of the molecular species is frequently unknown.

The fluorescence phenomena that are useful for qualitative and quantitative work with corticosteroids do not usually result from the compound itself, but rather from some complex formed by the addition of a reagent, e. g., sulfuric acid, or, for the detection of Δ^4-3-ketones, strong base. For qualitative purposes, fluorescence is usually observed visually, under either visible or ultraviolet light, and a complete fluorescence spectrum is seldom obtained. The major importance of fluorescence measurements in quantitative work is the very high sensitivity, most fluorimetric procedures being designed for microgram quantities.

After treatment with sulfuric acid, cortisol and corticosterone, their 11-epimers and 11β, 17α, 20β, 21-tetrahydroxy-4-pregnen-3-one give an intense fluorescence. The only other pregnane derivative found to give moderately strong fluorescence under these conditions is 12, 21-dihydroxypregnane-3,20-dione (SWEAT 1954a).

b) Measurement and instrumentation

Spectrophotometric measurements in the visible and ultraviolet regions serve a variety of functions, and the instruments employed and the type of measurement depend on the information required. Ultraviolet curves provide a very convenient and sensitive means of detecting the presence of certain chromophoric groups in a steroid. A certain amount of information concerning the nature of substituents on the chromophoric system can also be adduced, and to this extent ultraviolet spectra complement infrared curves, but the latter supply vastly more structural information. The colorless natural steroids are, of course, transparent in the visible region, but absorption curves in the visible of colored derivatives or complexes are useful in identification. Perhaps the major usefulness of spectrophotometry in the ultraviolet and visible regions, however, lies in the accuracy and ease with which quantitative measurements of absorbance can be made, since this in turn leads to very convenient colorimetric assay procedures.

To obtain ultraviolet or visible absorption curves, a spectrophotometer with a monochromator and a photoelectric detector is used; several commercial instruments are available with automatic recorders. Detailed descriptions of these instruments and others mentioned below will be found in any treatise on instrumental analysis such as that of WEST (1949), KORTUM (1955), or WILLARD et al. (1957). Ultraviolet spectrophotometers are equipped with quartz optics and sample cells, providing a range of 200—800 mμ. A hydrogen lamp serves as the continuous source in the ultraviolet range, a tungsten filament in the visible. With automatic recording instruments, either transmission (I/I_o) or absorbance ($\log_{10}$ (I_o/I) can be directly measured.

Spectra are measured in very dilute solutions, usually in a cell of 1 cm path length. The solvent used for practically all steroid work is ethanol, which is transparent to 200 mμ. Many other solvents can be used; it should be noted that chloroform absorbs strongly below 250 mμ. Ultraviolet spectra are universally plotted on a wave length scale, the most commonly used unit being the millimicron. The intensity of absorption is preferably expressed as the molar extinction coefficient ε or a_m, or as $\log_{10}\varepsilon$. It is usually possible to describe an ultraviolet adsorption spectrum adequately by a statement of the wave length and intensity of each maximum, e. g. $\lambda^{\text{Ethanol}}_{\text{Max}}$ 241 mμ (ε = 16,500).

For routine assay work, in which the concentration of the absorbing species is determined by the absorbance at one or two wave lengths in the visible region, less complex and less expensive photometers or colorimeters are usually used. These instruments range from visible spectrophotometers with glass optics and photoelectric detectors to filter photometers and colorimeters in which light of the desired wave length is obtained by the use of glass filters. In the latter, comparison may be made visually by matching the intensity of a colored solution with a standard. Readings on these instruments are usually expressed as optical density or extinction, which is directly proportional to concentration. Binary mixtures can also be analyzed, provided the extinction of the individual components at suitable wavelengths are known. For the determination of mixtures of chromatographically inseparable pairs of steroids, e. g., dehydroepiandrosterone and its saturated analogue, or androsterone and its Δ^9-unsaturated analogue, BITMAN et al. (1957) measured the extinctions A_1 and A_2 of the mixture at the wavelengths λ_1 and λ_2, respectively, in sulfuric acid solution. The extinctions S_1 and S_2 of the saturated partner and U_1 and U_2 of the unsaturated partner at these two wavelengths were obtained from the expressions

$$A_1 = S_1 + U_1$$

$$A_2 = S_2 + U_2$$

by substitution in the equation $S_1 = A_1\,[K_s/(K_s - K_u)] - A_2\,[K_s K_u/(K_s - K_u)]$ where $K_s = S_1/S_2$ and $K_u = U_1/U_2$.

In fluorimetric measurements, the intensity of the emitted radiation is measured at right angles to the exciting beam; filters are inserted both before the sample and between the sample and the detector to avoid radiation from the source, which is usually a mercury vapor lamp. Several photometers can be adapted for fluorimetry by the use of attachments. The fluorescence is measured as a current from the detector, and concentration is obtained from a calibration curve.

c) *Interpretation of ultraviolet spectra*

The chromophoric groups encountered in corticosteroids and their metabolites that give rise to ultraviolet absorption are carbonyl groups, double bonds, conjugated dienes and, most important, unsaturated ketones. Additional chromophores are often introduced in the preparation of derivatives. An extensive summary of ultraviolet absorption data on steroids has been compiled by L. DORFMAN (1953).

α) Isolated carbonyl groups and double bonds. The ultraviolet maxima associated with simple carbonyl groups are due to "forbidden" transitions of low probability, and are of low intensity (ε 20—80). The maxima occur in the region 280—300 mμ and may appear as a shoulder on a band of high intensity. Since much more accurate information on carbonyl groups can be obtained in the infrared region, little effort has been made to use the low intensity ultraviolet absorption of carbonyl groups in structural correlations. In contrast to infrared vibrations, the ultraviolet carbonyl maximum in a 5-membered ring ketone occurs at slightly longer wave lengths than in a 6-membered ring ketone. The low intensity absorption of saturated carbonyl compounds is an extremely important factor in optical rotatory dispersion.

The ultraviolet absorption maxima due to isolated olefinic linkages occur at, wave lengths below 200 mμ. Highly substituted double bonds display relatively strong end absorption between 200 and 220 mμ, however, and absorption in this region has been used to differentiate tri- and tetra-substituted double bonds (BLADON et al. 1952). A. MEYER et al (1954) report that the presence of a Δ^4-3-keto group does not interfere with the end absorption due to isolated double bonds.

***β*) Conjugated chromophores.** The spectra of conjugated dienes and unsaturated carbonyl compounds are characterized by very intense absorption maxima (ε 5,000–20,000) in the range from 240–280 mμ and also weak bands from 300 to 330 mμ. The transitions responsible for these maxima are attributable to relatively low-energy excited states that arise from resonance interaction of the unsaturation (π) electrons in the two multiple bonds as represented in the hybrid:

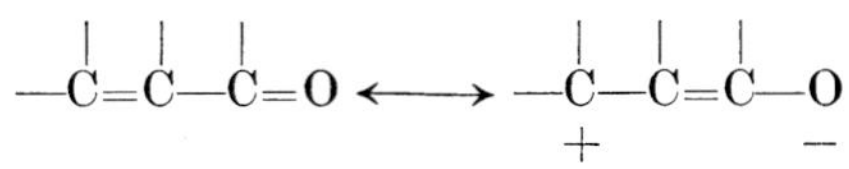

A necessary condition for such resonance is coplanarity of the atoms in the system and structural features that cause out-of-plane distortion of the chromophore markedly reduce the intensity of the absorption maximum.

Ultraviolet absorption spectra have been recorded for a wide range of steroidal dienes, polyenes and unsaturated ketones, and some very useful empirical correlations of the wave length of the principal absorption maximum and the location of the chromophoric system have been developed by WOODWARD (1942) and extended by FIESER and FIESER (1949). These generalizations are based on the progressive displacement of the maximum to longer wave lengths (bathochromic shift) that occurs with increasing number of substituents on the conjugated system.

Dienes are classified as heteroannular (I) or homoannular (II) and values of 214 mμ and 253 mμ, respectively, are assigned for hypothetical parent substances. An increment of 5 mμ in the wave length of the peak is caused by each substituent or ring residue and by each double bond which is exocyclic, e. g. a Δ^4-double bond (exocyclic to ring B). In the diene (I) there are three ring residues and one exocyclic double bond; the calculated value of 234 mμ ($214 + 4 \times 5$) accords well with that observed (235 mμ). The calculated and observed values for the diene (II) are 283 mμ and 280 mμ, respectively. An additional double bond extending the conjugation of a homocyclic diene, as in (III), produces a bathochromic shift of 30 mμ.

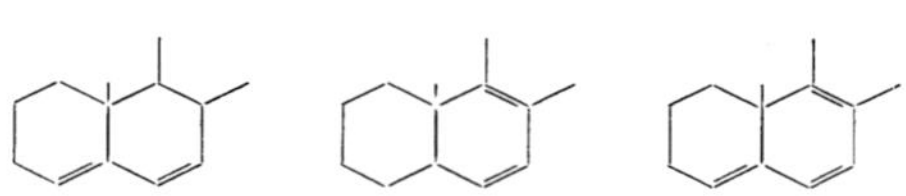

I II III

Similar displacements are observed in the spectra of unsaturated ketones and dienones; the effect of a substituent depends on the position relative to the carbonyl group. The values given by DORFMAN (1953) for the bathochromic shifts in the chromophoric systems (IV) and (V) are given in Table 22.

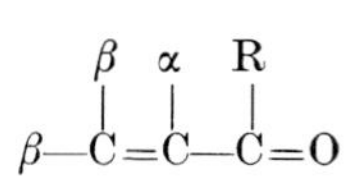

(IV)

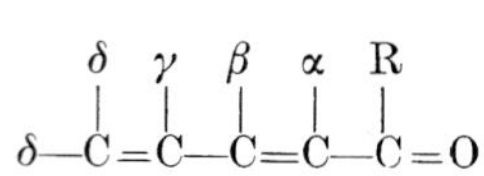

(V)

It is important to distinguish between extended and cross-conjugated dienones; in the latter case, exemplified by the $\Delta^{1,4}$-dien-3-one system (VI), the principal band appears at the position 236 mμ, calculated for the mono-unsaturated ketone. In the case of the $\Delta^{1,4,6}$-trien-3-one system three maxima are observed at about 223, 256 and 298 mμ; the middle band is of lowest intensity.

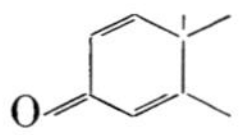
VI

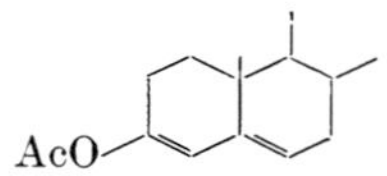
VII

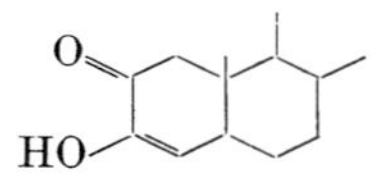
VIII

Substituents such as halogen atoms or acetoxy groups [e. g., enol acetates (VII) of unsaturated ketones] have approximately the same effect on the position of the absorption maximum as do alkyl substituents. Hydroxyl groups, as in the case of enolized α-diketones (VIII), produce a bathochromic shift of some 30 mμ. In this case the conjugation is extended by participation of the unshared electrons on oxygen.

Table 22. *Absorption maxima of unsaturated ketones*

Parent system (IV)		215 mμ
Increments for substituents (alkyl groups or ring residues)	α	10
	β	12
	γ or δ	18
Exocylic double bond		5
Each additional double bond extending conjugation		30
Homoannular diene component		38

The wave lengths calculated by these empirical rules agree, on the whole, quite satisfactorily with the spectra of a number of conjugated systems spread over the steroid nucleus. Disparities between predicted and observed values have permitted the clarification of structural uncertainties on several occasions. There are, of course, anomalies and exceptions to these generalizations, some of which can be explained on the basis of steric factors. In other cases a departure from the expected wave length is caused by a group apparently insulated from the chromophoric system, indicating interactions which may be mirrored in abnormal chemical behavior.

Of particular interest in the present discussion are the displacements observed in the absorption band of the Δ^4-3-ketone system in the corticosteroids and modified hormones. The "normal" position of the Δ^4-3-ketone maximum in compounds without substituents in rings A, B, or C is 240–242 mμ (ε 15,000–18,000). Substituents at most positions in these rings cause small but reproducible shifts in the position of the maximum which, taken together with other evidence, can be very useful in making structural assignments. Nearly all of these shifts, none of which is due to direct interaction of the substituent with the chromophoric system, are hypsochromic (to shorter wave lengths). In Table 23 are summarized the

Table 23. *Ultraviolet absorption maxima of substituted Δ^4-3-ketones*

Substituent	$\lambda_{max.}^{alc.}$, mμ	Reference	Substituent	$\lambda_{max.}^{alc.}$, mμ	Reference
2α-OH	242	a	11α-OH	242	e, f
2α-OAc	240	a	11α-OAc	240	e, f
2β-OAc	242	a	11β-OH	242	e
2α-CH_3	242	b	11-one	238	d, e
6α-OH	240	c	9α-F-11β-OH	238	g
6α-OAc	237	c	9α-F-11-one	234	g
6β-OH	235	c, d	19-OH	242	h
6β-OAc	235	c, d	19-OAc	239	h
6β-OH-11-one	232	d			

References. a) SONDHEIMER et al. 1953b. — b) HOGG et al. 1955b. — c) FLOREY and EHRENSTEIN 1954. — d) SONDHEIMER et al. 1954a. — e) ANTONUCCI et al. 1953. — f) ROMO et al. 1953. — g) FRIED and SABO 1957. — h) BARBER and EHRENSTEIN 1954.

average values of $\lambda_{max.}^{alc.}$ for some substituted Δ^4-3-ketones. The most prominent effects are caused by C-6 hydroxyl and 11-keto groups, and it will be noted that these shifts are additive, causing displacements of as much as 10 mμ.

γ) **Spectra in reactive media.** The ultraviolet and visible absorption curve of a neutral steroid in ethanol solution is usually slightly different from that in a hydrocarbon or other non-polar solvent insofar as hydrogen bonding or coordination may occur. Entirely different spectra can be expected, however, if the compound is dissolved in solvents that can give rise to new molecular species by ionization or complex formation. These changes are often observed qualitatively, as in the color reactions described in the foregoing section, but complete spectral measurements provide much more information.

The spectra of steroids in 90—100% sulfuric acid are due to highly conjugated chromophoric systems arising from protonation of weakly basic oxygen atoms by the strong acid, and further changes such as dehydration which may result. LINFORD (1957) has summarized some of the literature on the spectra of androgens and estrogens in strong acid and has speculated on the nature of the molecular species present. ZAFFARONI (1950) introduced the use of sulfuric acid chromogen spectra for the identification of microgram quantities of corticosteroids. NOWACZYNSKI and STEYERMARK (1955) determined the spectra of a number of steroids in 100% phosphoric acid.

Extensive data have been compiled by BERNSTEIN and LENHARD (1953,1954), who recorded the absorption maxima and extinction values of some 200 steroids in 97% sulfuric acid. Most polyfunctional steroids exhibit several maxima of decreasing intensity between 230 and 550 mμ, and these spectra thus take on some of the "fingerprint" character of infrared spectra. Certain bands can be ascribed to specific functional groups. Δ^4-3-Ketones usually exhibit an absorption maximum at 279—300 mμ, corresponding to a bathochromic shift of about 50 mμ. Hydroxysteroids usually exhibit at least one maximum in the region 230—278 mμ.

A. MEYER (1955b) has recorded the spectra of a number of Δ^4-3-ketones after exposure to 0.06 *N* ethanolic hydroxide ion. The curves of compounds containing an oxygen function at C-11 developed strong secondary bands at 377 mμ. Very characteristic spectral changes are also observed with 2- and 6-hydroxylated Δ^4-3-ketones, with new bands appearing at 355 mμ and 380 mμ, respectively. The changes that occur are not reversible, new spectra arising on acidification.

d) Quantitative applications — colorimetric methods

As already mentioned, the adaptability and great convenience of spectrophotometric measurements have led to their wide use in biochemical analysis. The basis for these colorimetric assay methods is the proportionality of concentration and light absorption expressed by the Beer-Lambert Law. It is unnecessary to evaluate the extinction coefficient of the absorbing species. In practice, the extinction (absorbance) at a suitable wave length, which should correspond to a maximum in the absorption spectrum, is measured and the concentration is then determined by reference to standard solutions or a calibration curve.

The preparation of samples is the most crucial operation in most colorimetric methods, the spectrophotometric measurements being a routine matter. The most pressing requirement in a colorimetric or fluorimetric assay procedure is the elimination of interfering substances that contribute to the absorption. Colorimetric methods for steroids are in most cases based upon a complex or product formed by a specific functional group with a color-forming reagent (p. 337), and the specificity and reproducibility of the color reaction is, of course, a major factor in the success of an assay procedure.

α) 17-Ketosteroids. For recent reviews, see L. ENGEL (1954), MUNSON and KENNY (1954), ZIMMERMANN (1955, p. 66), HENRY (1955), and KELLIE (1957).

The Zimmermann reaction is undoubtedly the most widely used test in steroid analysis. Although aqueous alkali is more convenient to use, alcoholic potassium hydroxide has been used extensively for the reasons already given. As an example of the latter technique, details of the method of CALLOW et al. (1938) are given below:

To 0.2 ml of extract in absolute ethanol are added 0.2 ml of 2% purified (ROSSI 1950, JANSSON and KYRKI 1952) *m*-dinitrobenzene in absolute ethanol and 0.2 ml of 2.5 *N* ethanolic potassium hydroxide. The mixture is maintained in the dark at 25° for 60 minutes. Then 10 ml of absolute ethanol is added and the extinction at 520 mμ is determined, using suitable blanks to correct for the color of the reagents and the extract.

In urine extracts not purified by use of Girard's reagent much of the extinction at 520 mμ may be due to substances other than 17-ketosteroids. Application of the correction method of GIBSON and EVELYN (1938) to such extracts gives a good approximation to the extinction of purified extracts (MASON and ENGSTROM 1950, CHANG and SLAUNWHITE 1955, SCHEDL et al. 1955) if non-aqueous potassium hydroxide is used in the Zimmermann reaction. The corrected extinction at 520 mμ is obtained by determining extinctions at 520 mμ and 420 mμ for the extract, a sample of the interfering impurities and a pure 17-ketosteroid, and substituting in the equation (FRASER et al. 1941, LANDAU 1949, TALBOT et al. 1942):

$$\text{Corrected extinction at 520 m}\mu = \frac{K_I E_G - E_V}{K_I - K_S}$$

where E_G = extinction at 520 mμ
E_V = extinction at 420 mμ
$K_I = E_V/E_G$ for a concentrate of interfering material
$K_S = E_V/E_G$ for a pure 17-ketosteroid

For routine use calculations may be eliminated by referring E_V and E_G readings to a nomogram (COOK and ROOKS 1952b). The elimination of interfering chromogens by solvent partition has also been suggested. The purple complex is extracted from the final solution by shaking it with chloroform (CAHEN and SALTER 1944), benzene (WALTER 1952), amyl acetate (MASUDA and THULINE 1953) or ether ZIMMERMANN et al. 1952, WERBIN and ONG 1954).

The methods based on the Zimmermann reaction with aqueous alkali may be exemplified by that of HOLTORFF and KOCH (1940), as modified by NATHANSON and WILSON (1943):

To the extract in 0.2 ml of ethanol are added 0.2 ml of 2% *m*-dinitrobenzene in 95% ethanol and 0.2 ml of 5 *N* aqueous potassium hydroxide. After it has stood in the dark at 25° for 105 minutes, the mixture is diluted with 10 ml of 60% ethanol and read at 520 mμ.

Fading of the purple color may be delayed by using 2-methoxyethanol (methyl cellosolve) for the final dilution instead of ethanol (KLENDSHOJ et al. 1953). WILSON (1954) found that less of the nonspecific brown color due to high alkalinity is produced when the color is allowed to develop at 0°.

A micro procedure for the determination of urinary 17-ketosteroids, described by ZIMMERMANN and PONTIUS (1954), is conveniently carried out in a single test tube:

To 10 ml of urine are added 1 ml of concentrated hydrochloric acid and 1 ml of 10% copper sulfate. The mixture is boiled under reflux for 20 minutes and then cooled. The 17-ketosteroids are extracted into 20 ml of ether by inverting the tubes mechanically for 5 minutes. The urine is siphoned off and the ether extract is washed in similar manner, first with 20 ml of 10% sodium hydroxide, then with two portions of water. After drying with sodium sulfate, the ether is evaporated. The residue is dissolved in 2 ml of ethanol and the solution is divided into two equal portions. To one 1-ml portion are added 1 ml of 2% alcoholic *m*-dinitrobenzene solution and 1 ml of aqueous 3 *N* potassium hydroxide solution. The other portion serves as a blank, in which alcohol is used instead of *m*-dinitrobenzene. The reagent blank contains alcohol instead of urine extract. Samples, blanks, and standards are kept at 25±0.2° for 90 minutes, then extracted with 4 ml of ether, which is then read in the colorimeter.

As an example of the various applications of the sulfuric acid reaction to the determination of dehydroepiandrosterone, etc., the method of W. M. ALLEN et al. (1950) is given in detail:

The dry extraction residue is dissolved in 2 ml of a mixture of 4 parts by volume of concentrated sulfuric acid and 1 part of 90% ethanol and warmed to 55° for 12 minutes. After cooling in ice water, 3 ml of 95% ethanol is added, the mixture is again cooled in ice water, and then read in the spectrophotometer at 600 mμ.

To correct for interfering chromogens, the extinction at 600 mμ and at two equidistant wavelengths, e. g., 560 and 640 mμ, may be read and substituted in the equation below to give a corrected extinction at 600 mμ (W. M. ALLEN 1950):

$$\text{Corrected extinction at 600 m}\mu = E_{600} - \frac{E_{560} + E_{640}}{2}$$

where E_{600}, E_{560}, and E_{640} are the observed extinctions at 600, 560, and 640 mμ, respectively.

The antimony trichloride reaction of PINCUS (1943) is applied as follows:

The reagent is prepared by dissolving 3.8 g of antimony trichloride in 1 ml of a 9:1 mixture of glacial acetic acid and acetic anhydride. The dry extraction residue is dissolved in 0.2 ml of this reagent and heated in a closed test tube on the water bath for 20 minutes. After cooling, the contents of the tube are suitably diluted with glacial acetic acid, allowed to stand at room temperature for 40—60 minutes and then read at 610 mμ.

The Pettenkofer reaction may be applied to urine extracts, using the method of MUNSON et al. (1948):

The extraction residue is dissolved in 0.5 ml of glacial acetic acid. Two ml of a 0.56% solution of furfural in 50% (w/w) acetic acid and 7.5 ml of 16 *N* sulfuric acid are added and the mixture is heated to 67±0.2° for 12 minutes. After the solution is cooled in ice, the extinction at 660 mμ is determined. Corrections for the color of the urine and furfural are made by blanks in which 50% acetic acid is substituted for either the furfural solution or the urine extract.

***β*) Corticosteroids.** For recent reviews, see GEMZELL (1955), BONGIOVANNI and EBERLEIN (1955a), HENRY (1955), BORTH (1956a and 1956b), DICZFALUSY et al. (1956), and GOLD (1957). Evaluations of current chemical assay methods have been published by WHEELER et al. (1953), DE COURCY and GRAY (1953), SPRECHLER (1955), HARWOOD and MASON (1956), and R. PETERSON et al. (1957).

The method of MADER and BUCK (1952) (see also NOWACZYNSKI et al. 1955) permits the determination of reducing steroids:

The extract is dissolved in 10 ml of 95% ethanol, and 1 ml each of 0.5% Blue Tetrazolium salt in 95% ethanol and 1% tetramethylammonium hydroxide in 95% ethanol are added. After allowing the mixture to stand for 20 minutes at room temperature, the extinction at 510 mμ is read against a reagent blank.

STAUDINGER and SCHMEISSER (1948, 1950) have modified the HEARD-SOBEL (1946) method by determining the reduction of phosphomolybdic acid before and after destruction of the α-ketol side chain with alkali:

Half of the extract is dissolved in 2 ml of 0.1 *N* sodium hydroxide and heated at 100° for 1 hour, then neutralized with 0.1 *N* acetic acid and evaporated. The residue and the other half of the extract are each taken up in 4 ml of glacial acetic acid. To each of the solutions is added 0.5 ml of phosphomolybdic acid reagent, prepared by mixing before use one part of a solution of 14 g of molybdenum oxide and 8 g of sodium hydroxide in 50 ml of water with four parts of a mixture of 32 ml of 85% phosphoric acid and 18 ml of 45% sulfuric acid. After being heated in a boiling water bath for at least an hour, the solutions are made up to 5 ml with glacial acetic acid, filtered if turbid, and then read at 720 mμ. The difference in the reduction between the two tubes is due to the α-ketols in half of the extract.

The C_{21} steroids giving formaldehyde on oxidation with periodic acid are determined by distilling off the formaldehyde (STIMMEL 1951a, CORCORAN et al. 1951) and allowing the latter to react with chromotropic acid (1,8-dihydroxy-naphthalene-3,6-disulfonic acid) to give a purple color. In the method of HOLLANDER et al. (1951) (see also BASSIL and HAIN 1950, WILSON 1953), the formaldehyde is allowed to diffuse directly into the chromotropic acid in Conway diffusion cells:

The dry sample, containing from 15—90 μg of corticosteroid is wet with 0.1 ml of glacial acetic acid, and after 10 minutes, 1.5 ml of 0.003 *M* periodic acid is added. The oxidation is allowed to proceed at room temperature for one hour and then stopped by addition of 0.4 ml of a solution of 2 g of stannous chloride in 5 ml of 0.1 *N* hydrochloric acid. A 0.4-ml aliquot of the reaction mixture is transferred to the outer well of a Conway cell and 0.8 ml of 0.2% chromotropic acid in 15 *M* sulfuric acid is placed in the inner well. Distillation is allowed to proceed at room temperature in the dark for about 15 hours. The contents of the inner well are then transferred to a small glass-stoppered test tube and heated in a boiling water bath for 30 minutes in the dark. After the tube is cooled in ice, the reading is taken at 570 mμ against a blank.

Urine contains material that binds formaldehyde (PATERSON and MARRIAN 1953) and prevents its diffusion. This material can be removed by ether extraction of the reaction mixture (WILSON 1953). The recovery of formaldehyde is increased if the diffusion is allowed to take place at 50° (HENLEY and POTTER 1952). A micro-technique that does not involve distillation or diffusion of formaldehyde was described by WILSON (1953) and later by BONGIOVANNI and EBERLEIN (1956). In this method the excess periodate is reduced by sodium bisulfite, which also binds the formaldehyde formed.

COX (1952) has described a method for the simultaneous formaldehyde and acetaldehyde determination after periodate oxidation of steroids.

The dry residue is dissolved in 0.5 ml of glacial acetic acid and 1 ml of a mixture of equal volumes of 0.12 *M* periodic acid in 0.4 *N* sulfuric acid and 1% glycine in 0.4 *N* sulfuric acid is added. The oxidation is allowed to proceed for 45 minutes, while a stream of dry air is passed through the mixture and into a trap, containing 2 ml of 1% sodium bisulfite solution. The formaldehyde is determined by reaction with chromotropic acid. For the determination of acetaldehyde the solution in the trap is then made up to 3 ml and a 1-ml aliquot is treated with 0.05 ml of 4% copper sulfate and 7 ml of concentrated sulfuric acid at 25°. Addition of 0.1 ml of 1.5% *p*-hydroxydiphenyl in 0.5% sodium hydroxide and boiling for 1.5 minutes then gives the color, which is read at 560 mμ.

WILSON and FAIRBANKS (1955a) have given the following directions for the chromic acid oxidation of 17-hydroxysteroids:

To the dry extraction residue containing up to 20 μg of steroid is added 0.02 ml of a freshly prepared 0.2% solution of chromic acid in glacial acetic acid. After heating at 40° for 1 hour, the 17-ketosteroid determination is carried out in the same test tube.

For the sodium bismuthate oxidation of 17-ketogenic steroids in urine NORYMBERSKI et al. (1953) give the following procedure:

The 24-hour urine is made up to 2 l with water. To 8-ml aliquots of urine in glass-stoppered centrifuge tubes are added 8 ml of glacial acetic acid and 2 μg of sodium bismuthate. The tubes are stoppered and shaken for one-half hour under exclusion of direct daylight and then centrifuged for 5 minutes. Ten-ml aliquots of the supernatant liquid are transferred to other centrifuge tubes and 5 drops of 5% sodium bisulfite solution and 10 ml of water are added to each. The mixtures are stirred, treated with 6 ml of concentrated hydrochloric acid, and stirred again. This is followed by a hydrolysis and extraction procedure for the determination of 17-ketosteroids, which is simultaneously carried out on 8-ml aliquots of urine that have not been oxidized by sodium bismuthate: the aliquots are boiled for 10 minutes with 2 ml of glacial acetic acid and 3 ml of concentrated hydrochloric acid. After cooling, they are extracted by shaking with 10 ml of 1,2-dichloroethane for 15 minutes. The tubes are centrifuged, the aqueous layers removed, and the dichloroethane extracts are purified by shaking with, first, 2.5 ml of water, and then 2.5 ml of 3 *N* sodium hydroxide, each wash being removed after centrifugation. The washed extracts are filtered through fluted filter paper and 5-ml aliquots (2.5-ml aliquots in the case of unoxidized urine) of the filtrates used for the Zimmermann test.

SCHILD (1954) found that the oxidized products are adsorbed on the sodium bismuthate and that higher values are obtained if the excess reagent is dissolved in hydrochloric acid. Glucosiduronates in the urine are oxidized to the corresponding 3α-formoxysteroids (NORYMBERSKI and SHERMIN 1953).

17-Hydroxypregnan-20-ones are also 17-ketogenic in this procedure if the urine has been treated overnight with 5 mg of sodium borohydride per ml of urine to reduce the C-20 carbonyl group (APPLEBY and NORYMBERSKI 1955).

The determination of 17,21-dihydroxypregnan-20-ones by the SILBER-PORTER (1957) method is carried out by heating the steroid extract with an ethanolic phenylhydrazine-sulfuric acid mixture. NELSON and SAMUELS (1952) recommend the following procedure for plasma extracts (see also SILBER and BUSCH 1956, RIONDEL et al. 1957):

The column eluate containing the 17-hydroxycorticosteroids (p. 314) is dissolved in 0.25 ml of methanol. To a 0.2-ml aliquot is added 0.3 ml of a reagent containing 16 mg of phenylhydrazine hydrochloride per 10 ml of sulfuric acid-water (31:19 v/v). After incubation at 60° for 1 hour, the tubes are cooled, centrifuged, and decanted into 0.5-ml cuvettes. The extinction is read at 370, 410 and 450 mμ and the corrected extinction at 410 mμ is calculated by W. M. ALLEN's method (1950).

For extracts of urine SILBER and PORTER (1954) use the following method:

To 10 ml of a chloroform extract, corresponding to about 4 ml of urine, is added 1 ml of a solution of 65 mg of phenylhydrazine hydrochloride in 150 ml of blank reagent. The blank reagent is prepared by mixing 50 ml of absolute ethanol with 100 ml of sulfuric acid-water (31:19 v/v). To a duplicate 10-ml chloroform extract, 1 ml of blank reagent is added. After shaking and centrifuging both tubes, the chloroform layers are removed. The 1-ml portions of reagent and blank reagent are left at about 25° for about 18 hours. The extinctions of the sample and its blank are measured at 380, 410, and 440 mμ. After subtracting the reading of the blank from that of the sample at each of these wavelengths, the correction is applied as before and the corrected extinction is compared with corrected extinction values similarly determined for standards.

γ) Δ^4-3-Ketosteroids. The intense absorption of the Δ^4-3-ketone system at 240 mμ has been utilized in the assay of Δ^4-3-keto corticosteroids, but the nonspecific background absorption in this region of the spectrum is a serious disadvantage in the use of this absorption maximum in quantitative analysis. Several alternative methods, including very sensitive fluorimetric procedures, have been developed for the assay of Δ^4-3-ketosteroids.

The application of the isonicotinic acid hydrazide reaction for Δ^4-3-ketosteroids to plasma extracts has recently been described by WEICHSELBAUM et al. (1957). An aliquot of the eluate from a Florisil chromatographic column, containing the 17-hydroxycorticosteroids (p. 312), is used for the assay:

Quadruplicate samples and quadruplicate solvent blank eluates are measured into centrifuge tubes and the solvents are evaporated. To one pair of each of the two residues is added 0.5 ml of a blank reagent containing 0.625 ml of concentrated hydrochloric acid per 1 of ethanol. To the other pair is added 0.5 ml of a solution of 25 mg of isonicotinic acid hydrazide in 50 ml of the blank reagent. After stoppering and mixing, the tubes are allowed to stand at room temperature for one hour. Readings are taken at 380 mμ and corrections are made for the reagent and solvent blanks.

The application of fluorimetry to purified corticosteroid fractions is illustrated by the procedure of SWEAT (1954a):

The sample, containing 0.05 to 5.0 μg of steroid, is brought to dryness in a cuvette and dissolved in 0.1 ml absolute ethanol. One ml of conc. sulfuric acid is added rapidly and the mixture is stirred thoroughly. After allowing the tube to stand for 20 minutes it is placed in a fluorimeter. Maximum fluorescence is excited by light of 475 mμ and the emitted fluorescence is measured at 540 mμ.

The fluorescence of Δ^4-3-ketosteroids produced by sodium hydroxide can be measured directly on the paper chromatograms by use of a fluorimeter attachment

(AYRES et al. 1957 b). ABELSON and BONDY (1955) have modified the soda fluorescence method by substituting potassium *tert*-butoxide in *tert*-butanol for sodium hydroxide. With larger amounts of steroid (above 10 μg/ml) this reagent gives a yellow color, which can be exploited for a colorimetric assay (ABELSON and BONDY 1956). The application of the fluorescence method to eluates from paper chromatograms has been described (BONDY and UPTON 1957, BONDY et al. 1957).

II. Isotope analysis

1. Nature and properties of isotopes

The several isotopes of an element consist of atoms having the same number of protons and electrons but different numbers of neutrons. For all but a few elements, two or more combinations of neutrons and protons form stable nuclei, and these isotopes contribute, usually very unequally, to the natural population of the element. For the lighter elements, the number of neutrons seldom exceeds the number of protons by more than two in stable nuclear configurations. A larger imbalance in the numbers of the nuclear particles gives rise to unstable isotopes which are radioactive, decaying in one of several ways with the emission of particles and energy. All of the isotopes of an element have the same atomic number and are thus chemically identical, but they may be distinguished by the difference in mass number or, in the case of unstable isotopes, by their characteristic radioactivity.

The natural isotopic composition of elements is, except for geological purposes, constant, so that a compound containing atoms with an isotopic content detectably different from the natural one is labeled, and can be traced through a series of chemical reactions provided that exchange of the tagged atom with the medium does not occur. In the past decade, the ready availability of suitable isotopes of a large number of elements, especially carbon, has made the use of tracer methods commonplace in biochemical and physiological studies, and numerous applications are discussed in subsequent chapters. In the present section methods for the measurement of the isotopic content of labeled compounds are briefly summarized. The recent reviews of KAMEN (1957) and of SCHMEISER (1957) are recommended for a fuller account.

Three isotopes are of commanding importance in tracer studies with steroids, H^2 (deuterium, D), H^3 (tritium, T) and C^{14}, and two others have been used occasionally for special purposes, C^{13} and O^{18}. In addition, the urinary sodium/potassium ratio assay procedure for mineralocorticoid activity of SIMPSON and TAIT (1952) is based upon tracer methods using Na^{24} and K^{42}. The properties of these isotopes are given in Table 24. The superscript on the chemical symbol is the mass number of the isotope and is indicated in a chemical formula (e. g. $C^{14}H_3C^{13}O_2H$) to designate enrichment in the isotope at a given position. For stable isotopes, the isotopic content of a compound is usually specified in terms of the excess amount over the natural abundance, or atom per cent excess, which is simply the excess in percentage of the labeling isotope. In steroid work, deuterium content is often expressed as the average number of D atoms per molecule. For radioactive isotopes, labeled content is stated as specific activity, which is the radioactivity per unit weight. Radioactivity is expressed as disintegrations per unit time, e. g., counts per minute, or by the curie (c) = 10^3 millicuries (mc) = 10^6 microcuries (μc). A curie, defined as 3.7×10^{10} distintegrations per second, corresponds to the activity of 1 g of radium, and is used to denote the amount of any radioactive substance having an equivalent radiation. The specific activities encountered in biochemical work are usually of the order of μc per millimole.

Table 24. *Properties of important isotopes* (KAMEN 1957)

Isotope	Natural abundance	Radiation characteristics		
		particle	energy	half-life
H^2(D)	0.02%	Stable		
H^3(T)	minute trace	β-	0.018	12.46 years
C^{13}	1.10%	Stable		
C^{14}	minute trace	β-	0.155	5570 years
O^{18}	0.204%	Stable		

The radioactivity of an unstable isotope is the result of a nuclear event which, in the cases of interest in the present discussion, results in the emission of particles or radiation and transformation of the atom into a new isotope having one higher atomic number. The radiation characteristics of any isotope are described by the *type* of particle, the *energy* of the particle and the *rate* of emission or decay of the element. Several entities have been observed in the decay of radioisotopes: α-particles, which are helium nuclei (He^{4+}), negative β-particles (sometimes called negatrons), which are electrons (β^- or e), positive β-particles, or positrons (β^+) and γ-rays, which are electromagnetic radiation of X-ray energy.

α-Particles are emitted only by fairly large nuclei. The radioisotopes of interest in steroid tracer work all emit β^--particles by reactions that can be illustrated with the case of C^{14}: ${}_6C^{14} \rightarrow {}_7N^{14} + \beta^-$. The average energy of the emitted β-particle is characteristic for each isotope; these energies are usually expressed in units of millions of electron volts (Mev $= 1.6 \times 10^{-6}$ ergs) and are called "hard" and "soft" to denote the relative order of energy.

The particles emitted by radioisotopes are ionizing radiation, their absorption by matter being accompanied by the production of pairs of ions. This property provides a means for detecting and measuring the emission, as described below, and is reponsible for radiation damage in tissue. The absorption of a beam of particles is proportional to the logarithm of the absorber thickness, analogous to the absorption of energy expressed by the Beer-Lambert Law. The penetrating power of radiation depends on the energy of the particles and is an important practical consideration, since absorption losses for soft radiation, especially that of H^3 and C^{14}, must be corrected for in radioactivity measurements.

The rate of emission is characteristic for each radioisotope. The decay process follows the universal rate expression, the rate being proportional to the number of nuclei present: $-\frac{dN}{dt} = \lambda N$. On integration, this equation becomes $N = N_0 e^{-\lambda t}$, in which λ is a disintegration constant characteristic of the isotope. It is convenient to express the decay rate in terms of the half-life, or time required for one-half of the nuclei present to decay. Thus $N = N_0/Z$, and we have $t_{1/2} = .69/\lambda$. The half-life of radioisotopes varies from fractions of a second to 10^6 years. In tracer work the decay rate is of importance only if the half-life is of the order of magnitude of the length of the experiment.

2. Analysis of isotopically labeled compounds

a) Stable isotopes

The stable isotopes of an element are distinguished by differences in mass and, in a number of cases, differences in nuclear spin. Methods for the measurement of isotopic composition in a compound labeled with a stable isotope have up to the present been based on mass difference. The recent development of nuclear magnetic resonance spectrometers (p. 376), however, may afford an important alternative based on nuclear spin signals for certain isotopes such as O^{17} (WEAVER et al. 1955).

Special methods are available for the analysis of deuterium because of the large mass ratio of 2:1. Determination of the density of water obtained by combustion provides a convenient and inexpensive method for deuterium assay since the density of pure D_2O is about 10% greater than of water. The density has been measured by the rate of fall of a drop of water through *o*-fluorotoluene (KESTON et al. 1937) or by the effect of temperature and/or pressure on the position of a float (Cartesian diver). Methods have also been devised for the determination of deuterium which are based on the difference in infrared spectral frequencies due to the increased mass of the deuterium atom.

The method generally used for the isotopic analysis of compounds containing stable tracer atoms is mass spectrometry. In a mass spectrometer, molecules of a gas are ionized by a beam of electrons from a hot filament, and the different ions produced are electromagnetically separated into a mass spectrum which reveals the ratios of ions of different masses. Mass spectrometric instruments and procedures are discussed in the text of WILLARD et al. (1957) and the monograph of A. ROBERTSON (1954).

For the analysis of deuterium in steroids, water from the combustion is reduced to hydrogen by zinc or lithium aluminum hydride, or by passing the steam over magnesium or magnesium amalgam. The ratio of the peaks corresponding to mass 2 (H_2) and 3 (HD) is then obtained in the mass spectrometer. In the determination of C^{13}, carbon dioxide from the combustion is used in the mass analysis. Acetic acid obtained by the drastic oxidation of a steroid has been analyzed directly in the mass spectrophotometer to determine the ratio of doubly-labeled compound, $C^{13}H_3C^{13}O_2H$ (mass 62) to singly labeled (mass 61) and unlabeled (mass 60) molecules; the results established the path by which the C-18 methyl group is introduced in biosynthesis (CORNFORTH et al. 1958).

b) Radioactive isotopes

α) Instrumentation. In the isotopic analysis of compounds labeled with radioisotopes, advantage is taken of one of the several types of interaction between ionizing radiation and matter. The most widely used method for the measurement of radioactivity in tracer work depends on the dissociation of atoms of a gas into electrons and positive ions that occurs in the absorption of radiation. Three instruments employing this effect have been developed and are in current use: the Geiger-Müller (G-M) tube, the proportional counter and the ionization chamber. Details of the design and characteristics of these radiation detectors will be found in the reviews cited above and in the recent monograph of PRICE (1958).

The basic principle in all of these instruments is the detection and measurement of an electrical charge or current developed by the electrons in a voltage gradient across two electrodes in the gas chamber. At high voltage the initial ion-pair is multiplied into an avalanche by subsequent ionizations in the gas, producing a relatively strong pulse in the circuit which is recorded by a mechanical counter; both the G-M tube and the proportional counter employ this pulse-type operation.

The *Geiger-Müller tube* consists of a metal cylinder which is the cathode and a coaxial wire anode maintained at a potential of about 1000 volts. The multiplication obtained by this voltage is so high that the pulse is independent of the initial energy of the ionizing radiation, i. e., any ionization event will trigger a pulse. When the electron charge has been collected at the anode, the slower-moving ions remain as an insulating sheath that must be dissipated before the next pulse can be detected. In order to reduce the time required for the discharge of these ions at the cathode, an organic vapor such as ethanol or ethyl formate is incorporated with the ionizing gas, which is usually argon or helium. The organic quenching gas is slowly consumed during operation and is a limiting factor in the lifetime of the tube. This quenching effect reduces but does not eliminate the "dead time" involved in restoration of the voltage gradient, during which no counts are registered, and coincidence corrections must be applied for any but very slow counting rates. While it is possible to place the sample inside the G-M tube to obtain maximum counting efficiency, it is more convenient to use a tube that is closed at one end by a thin window (end-window counters) and to mount the sample on a planchet of glass or metal held close to the window.

The *proportional counter* is similar to the G-M tube, but operates at a lower voltage, at which the pulse obtained is proportional to the radiation energy, and discrimination between pulses due to β-particles of different energy is possible. Also, since multiplication is less by a factor of about 10^3, quenching is much more rapid, permitting faster counting rates than are possible with a G-M tube. Most proportional detectors are operated as gas-flow counters, in which the counting gas mixture flows continuously through the tube, and no window is required.

In the *ionization chamber*, the current generated by the charged particles is measured by an electrometer without amplification by further ionizations. The major usefulness of this instrument in tracer work is in the measurement of the radioactivity of gaseous samples, which can be introduced directly into the sensitive portion of the chamber.

Another approach to the detection of radioactivity which is finding increasing use in tracer work is the *scintillation counter*, which is based on the excitation by radiation of atoms or molecules in the liquid or solid state. The fluorescence or phosphorescence thus produced is amplified by a photomultiplier tube. The pulse height obtained in a scintillation counter is proportional to the energy of the radiation, and it is possible to count the activities due to tritium and to C^{14} in the same sample.

Progress in instrumentation for the measurement of radioactivity is being made rapidly, and it would be unwarranted in this discussion to attempt an evaluation of the numerous modifications of these instruments which are available. Proportional counters and particularly ionization chambers which are suitable for counting samples in the gas phase are being increasingly used for soft radiations such as those for tritium and C^{14}. Thin-window counters are available (FULLER 1956) for location of C^{14}-labeled steroids in paper chromatograms, and automatic scanning of paper strips has been described (BERLINER et al. 1957b). Radioautographic techniques using X-ray film are also very useful for the detection and measurement of radioactive zones on paper chromatograms.

β) Preparation of samples and measurement. *Tritium.* The radiochemical assay of tritium poses some difficulties owing to the low energy of the radiation. Compounds containing this isotope are best burned to water, which is then reduced as indicated above for deuterium. The radioactivity of the tritium gas can then be determined in the ionization chamber, proportional counter or G-M counter. Alternatively, the water produced by combustion of the tritium-containing steroid may be allowed to react either with aluminum carbide or methyl magnesium iodide to give tritiomethane, or with butyl magnesium iodide to give tritiobutane; either of these hydrocarbons can be assayed in a proportional counter. The techniques involved in the counting of gas-phase samples are fully discussed in the monograph of GLASCOCK (1954). The manipulations necessary for the preparation of a gaseous sample are circumvented with the scintillation counter.

Carbon[14]. In most steroid tracer work with C^{14}, the carbon is converted to carbon dioxide or barium carbonate for the radioactivity measurement. In the degradation schemes developed for steroids, individual carbon atoms are usually separated as carbon dioxide either directly or indirectly by reactions such as the Schmidt degradation ($RCO_2H + HN_3 \rightarrow RNH_2 + CO_2 + N_2$). The total activity of a sample is determined by converting the carbon to carbon dioxide by a wet combustion technique. The carbon dioxide can be used for counting directly in an ionization chamber, or else precipitated as barium carbonate for counting as a solid with a G-M or proportional counter. A recent review of techniques for preparing samples for counting has been presented by D. Steinberg and Udenfriend (1957).

The film of barium carbonate deposited on the planchet must be evenly distributed (uniform sample geometry) to obtain reproducible counting rates. In counting a solid sample of $BaC^{14}O_3$, an apparent decrease in specific activity is observed with increasing thickness of the film until a limiting value is reached. This is due to self absorption of the soft β-radiation from lower layers of the film by molecules in the upper layers, and a correction must be applied by means of an absorption curve or else the samples may be made "infinitely thick" so that the limiting value is measured. Pearce et al. (1956) have prepared tissues for C^{14} assay by dissolving in warm formamide and then evaporating the solution directly in a planchet covered with lens paper.

In any radiochemical assay, a certain fraction of the disintegrations from the sample are lost due to various effects besides the coincidence losses due to the resolving time and self absorption which have already been mentioned. Absorption and scattering of the radiation by air and by the tube window are particularly serious when low-energy radiation is counted with solid samples placed outside the tube. These losses are taken into account by counting standard samples periodically. Any measured radioactivity must be corrected for background radiation in the atmosphere due to cosmic rays and other sources. This background, which is usually of the order of 20—40 counts per minute, is subtracted from the total counting rate. In order to obtain accurate activity measurements, a sufficient number of counts or sufficiently long counting times must be observed to average out random fluctuations.

γ) Analytical applications. The use of radioactive tracers in metabolic studies of adrenocortical hormones is discussed elsewhere in this volume, but certain important applications to chemical analysis will be mentioned in this section.

An important proviso in all tracer work is that fractionation of labeled and unlabeled species during separation processes, or discrimination in chemical (including enzymatic) reactions, must not occur. Isotope effects, or differences in physical properties and chemical reactivity in molecules of different isotopic composition may be very pronounced in compounds containing deuterium and tritium because of the large differences in mass and zero-point quantum energy between these isotopes and hydrogen, and this possibility should be taken into account with hydrogen tracers. Isotope effects involving C^{14} labeled compounds are usually negligible, however, and the fact that an isotopic label remains randomly distributed in a sample forms the basis of isotopic dilution analysis, which is a very powerful tool in both qualitative and quantitative analysis of complex mixtures.

The identity of a trace amount of a radioactive steroid can be confirmed by adding an excess of the non-radioactive species (normer), and isolation of the mixture of inactive carrier and radioactive material (radiomer). Confirmatory tests should then establish that the specific activity of the steroid is unchanged by a variety of fractionation procedures such as crystallization, paper chromatography, countercurrent distribution, column chromatography (coincidence of elution curve), etc. Further refinements of the technique may involve the preparation of derivatives, their crystallization, etc., and degradation of the steroid, followed by similar tests of radiochemical purity.

The isotope dilution method (review, Rosenblum 1957) permits the accurate determination of any component of a mixture without the need for quantitative isolation. However, the radiomer of this component must be available and a portion of the compound must be isolated in pure form. A small quantity w^* of the radiomer, having a known specific activity S^* is added to the mixture containing the unknown quantity w of the normer and a suitable portion of the steroid is isolated in pure form. Since the activity a of the radiomer is not changed by the dilution with the normer, a determination of the specific activity after dilution allows the calculation of the weight of normer w that was present in the original mixture.

$$S^* = \frac{a}{w^*}$$

$$S = \frac{a}{w^* + w}$$

$$S^* w^* = S(w^* + w)$$

$$w = \frac{w^*(S^* - S)}{S} = w^* \left(\frac{S^*}{S} - 1\right)$$

If the initial specific activity is much larger than the final one this expression becomes $w = w^* S^*/S$. If it is desired to measure the amount w^* of a radioactive steroid metabolite in a mixture, the inverse of the above procedure may be used. A known amount w of inactive carrier is added to the mixture and a suitable portion of the steroid is isolated in pure form. From the specific activity S^* and S before and after dilution w^* can be calculated from the above equation, solved for w^*.

$$S^* w^* = S(w^* + w)$$

$$w^* = \frac{wS}{S^* - S}$$

If the initial specific activity is much larger than the final one this expression becomes $w^* = wS/S^*$.

In the double dilution method the mixture containing w^* is divided into two, preferably equal, aliquots, unequal amounts w_1 and w_2 of normer are added to the aliquots, and, after purification, their respective specific activities S_1 and S_2 are determined. If the aliquots were equal w^* can be calculated as follows:

$$a = S^* w^* = S_1(w^* + w_1) = S_2(w^* + w_2)$$

$$w^* = \frac{w_2 S_2 - w_1 S_1}{S_1 - S_2}$$

Steroids may be labeled not only by the methods mentioned in an earlier section (p. 250), but also by the preparation of derivatives with radioactive adducts. Thus, RULIFFSON et al. (1953) have prepared labeled hydrazones of ketosteroids with acetylhydrazide-3-iodo^{131}-pyridinium bromide (I^{131}-Girard's reagent P). These hydrazones can be separated by paper chromatography and determined on the paper with an end window Geiger counter. AVIVI et al (1954) have used tritium-labeled acetic anhydride for the analysis of adrenocortical hormones by a double tracer technique.

4-C^{14}-Labeled cortisol was added to a blood extract, which was then fractionated on a partition column. The fractions showing radioactivity due to cortisol were pooled and acetylated with tritiated acetic anhydride. The cortisol acetate was purified by partition chromatography and the fractions analyzed by a mica-window Geiger-Müller counter and also by liquid scintillation counting. The amount w of cortisol present initially was calculated from the expression $w = (x/z)(r' - r)$, where x is the specific activity of the 4-C^{14}-labeled cortisol measured by the Geiger counter, z is the specific activity of the tritium-labeled cortisol measured by the scintillation counter, r is the sum of the radioactivity due to C^{14} and H^3 measured by the scintillation counter divided by the radioactivity due to C^{14} in the Geiger counter from the 4-C^{14}-labeled cortisol acetylated with H^3-acetic anhydride, and r' is the same ratio for the purified derivative of the acetylated extract.

Simpson and Tait (1953) have used the acetylation of aldosterone with 1-C^{14}-acetic anhydride to establish the presence of two easily acetylated hydroxyl groups in aldosterone. Later, the same group (Ayres et al. 1957a) published an isotope dilution method for the determination of adrenocortical hormones in biological extracts, in which aldosterone C^{14}-diacetate is used as a tracer for acetylated aldosterone, and 4-C^{14}-labeled radiomers for other adrenocortical hormones. R. Peterson (1957) has also used this method for the determination of corticosterone in blood. Berliner et al. (1956, 1957a, 1957b) have successfully combined isotope dilution and paper chromatographic techniques in their determination of radioactive steroids on paper chromatograms.

III. Other methods

1. Optical rotation

In the discussion of the stereochemistry of the steroids it was observed that the presence of asymmetric centers in the molecule gives rise to two enantiomeric forms (p. 189). It is this molecular asymmetry which is responsible for the optical activity of steroids, i. e., the property of rotating the plane of polarized light. Although the exact nature of the interaction between an asymmetric molecule and polarized light is imperfectly understood, the phenomenon of optical rotation has been an indispensable tool in the study of the stereochemistry of organic compounds for more than a century. A recent review of the measurement and use of optical rotations has been presented by Klyne (1955).

a) Measurement

Rotations are determined in a polarimeter in which polarized monochromatic light from a system of Nicol prisms is passed through a solution contained in a tube of accurately known length with parallel endplates of glass which must be free of strain. The beam then passes through a second Nicol prism; the angle of rotation is measured by rotation of this analyzing prism, mounted on a circular scale, until the sections of a split field are of equal intensity. Instruments with a photocell detector are available, but the matching is usually done visually in a dark room. The match point should be approached at least six times and the average value recorded. A similar series of readings for a solvent blank must always be made and subtracted from the observed rotation; the solution and solvent must be scrupulously free of dust. Subjective errors in matching the field cannot be entirely eliminated and may be rather serious for certain individuals. The reproducibility of readings is usually $\pm 0.02°$.

The optical rotatory power of a compound is expressed by the specific rotation $[\alpha]$, defined by the formula

$$[\alpha]_\lambda = \frac{100\,\alpha}{l\,c}$$

in which α is the observed angle of rotation, positive for counterclockwise rotation and negative for clockwise, l is the length of the solution path in decimeters and c the concentration in g per 100 ml. For the analysis of rotation data, values are expressed as molecular rotation, $[M]_\lambda$:

$$[M]_\lambda = \frac{[\alpha]_\lambda \cdot \text{Mol. Wt.}}{100}$$

The length of the polarimeter tube is most often 1.0 dm (occasionally 2 dm) and usually about 1 ml of solution is required to fill the tube. For most steroid work a concentration of 0.5—2 g per 100 ml, requiring about 10—20 mg of material, is used. Capillary tubes are available if rotations must be determined with a smaller quantity. Under these conditions the observed rotation for a typical steroid is of the order of 0.5—1.0°, and the accuracy of the measurement is about 2—3%. Since the specific rotation may vary from 0 to 300° it is more meaningful to speak of the absolute error; Klyne (1955) has suggested an uncertainty of 2—3° in replicate determinations of the specific rotation.

The specific rotation varies with concentration, temperature, solvent and wave length of the light. The first two effects are in general negligible for neutral steroids, and fluctuations in room temperature or concentration changes of 1—2 fold can usually be ignored. The effect of solvent can be quite pronounced, and the solvent used must be specified. Chloroform and dioxane are preferred solvents, and much of the data for steroid rotations is for values in chloroform, although ethanol has often been used for the highly oxygenated corticosteroids. Specific rotations in highly polar solvents such as acetone or alcohol are likely to be 5—10° more negative than values measured in chloroform, and the possibility of larger effects must be considered in comparing rotations determined in different solvents.

The vast majority of the rotations recorded in the literature are for light of wave length 5890 Å, the *D* line of sodium, thus the designation $[\alpha]_D$. This choice of wave length was dictated by the convenience of a sodium vapor source, and is in the region of optimum sensitivity of the eye, but, as discussed later, it is by no means the optimum wavelength for the measurement of optical rotations. Besides $[\alpha]_D$ values, a few data for rotations at 5460 Å, obtained with a mercury vapor lamp, have been recorded. The variation of rotation with wave length, which is called rotatory dispersion, is discussed in a later section.

b) Applications—molecular rotation differences

Optical rotatory data serve several functions in steroid chemistry. The specific rotation provides a characteristic physical constant, as does a melting point, and a report of the rotation is at least as important as the melting point in the description of a new steroid or derivative. In certain situations the specific rotation is a more sensitive index of purity than any other physical measurement, including the infrared spectrum. *The optical rotation is*, of course, *the only physical method that can be used to distinguish between or identify enantiomorphs*, or to measure the purity of a product obtained by resolution of a racemic mixture.

The most important application of optical rotations is in the investigation of structural and especially stereochemical problems. The correlation of molecular structure and rotation is based on the principle of optical superposition, which states that the rotation of a molecule is the sum of the rotational contributions of the asymmetric elements. The systematic analysis of rotational data for steroids is largely due to BARTON and co-workers, who have compiled extensive tables of molecular rotation differences due to the introduction of various substituents or the inversion of configuration of asymmetric centers. Summaries of these data have been presented by FIESER and FIESER (1949) and by KLYNE (1955); in addition, a catalogue of optical rotations of steroids covering the literature to 1953 has been compiled by MATHIEU and PETIT (1956).

Although space does not permit a tabulation of the extensive literature on molecular rotation differences, a few generalizations can be made concerning the rotatory contributions of representative functional groups and asymmetric centers. The saturated C_{19} and C_{21} steroid hydrocarbons have relatively small positive rotations, the asymmetric centers of the nucleus being nearly compensated. Skeletal distortions caused by modification of a ring junction may cause profound changes in the rotation, however, as in the case of an olefinic center at C-5. Whereas a change of configuration at C-5 from α to β causes a negligible rotation increment ($\Delta M_D + 6°$) in the saturated hydrocarbon, the introduction of a Δ^4 or Δ^5 double bond is attended by a ΔM_D value of $+159$ or -298, respectively. These changes are strongly reinforced on passing to the homoannular dienes, the ΔM_D values for the 2,4- and 5,7-dienes being $+531$ and -538, respectively.

The molecular rotation contributions of hydroxyl groups are in general not large; the ΔM_D values are often but not always of opposite sign for epimeric hydroxyl groups at a given position. The rotatory effect of a hydroxyl substituent is almost always accentuated by acetylation, and the ΔM_D values for acetylation of epimeric alcohols in several cases provide a very reliable tool for the diagnosis of configuration. Carbonyl groups frequently exhibit large rotatory contributions, as exemplified by the ΔM_D value of $+254$ for the 20-keto group in the pregnane series. The very large rotation change (ΔM_D + 450 to 700) attending inversion of the acetyl side chain from 17β to 17α (17-iso) furnishes a sensitive indication of the occurrence of isomerization at C-17 in reactions of 20-ketones (SHOPPEE 1949, FLOREY and EHRENSTEIN 1954).

The method of molecular rotation differences is purely empirical, and is valid only when sufficient data on known compounds are available, and when the rotation differences are significantly greater than the spread of values for the reference compounds. Another limitation that must not be ignored is the interaction of substituents, which may have a profound effect on the contribution of the individual groups to the total rotation. With due consideration of these requirements of adequate analogies and absence of vicinal effects, molecular rotation differences can provide a powerful tool in structural problems.

Correlations of molecular rotation differences with the position and configuration of substituents can be made in several ways. The most direct approach is the comparison of the molecular rotation difference between a substituted compound X and the parent unsubstituted compound Y with the ΔM_D values for the introduction of the substituent in question at various positions and orientations in Y.

A good illustration of this method is provided by the structure elucidation of a urinary metabolite by H. HIRSCHMANN and HIRSCHMANN (1950). Preliminary data indicated the compound to be a Δ^5-pregenetriol with two of the hydroxyl groups at C-3β and C-20α. The compound was reduced and acetylated, and the rotation of the product was compared to that of allopregnane-3β,20α-diol diacetate, revealing a ΔM_D contribution of -276 for the unknown acetoxy group. This large negative increment was consistent only with a 16α-acetoxy group, and the 5-pregnene-3β, 16α, 20α-triol structure was confirmed by synthesis.

The most noteworthy application of ΔM_D values for acetylation has been in the assignment of configuration of C-20 alcohols. Hydroyxl groups in both configurations at C-20 are encountered in many adrenal and urinary steroids, and are not distinguished by chemical behavior. FIESER and FIESER (1948) and SARETT (1949c) have demonstrated that the molecular rotation increments on acetylation ($M_D^{20-\mathrm{OAc}} - M_D^{20-\mathrm{OH}}$) for 20$\beta$-hydroxyl groups are invariably positive, and for 20α-hydroxyl groups negative. Although the rotations of the two hydroxy compounds may be too close to allow a distinction, the more dextrorotatory acetate can always be assigned the β-configuration. These rules are valid for monohydroxy compounds, diols and triols, notwithstanding the vicinal interaction of the adjacent groups, which leads to larger ΔM_D values for acetylation in the triol series.

c) Optical rotatory dispersion

The change in optical rotatory power with wave length has been of importance in theoretical investigations of the phenomenon of optical rotation for many years, but until the development of a commercially available spectropolarimeter (RUDOLF 1955) systematic studies of the optical rotatory spectra of organic compounds were quite limited. In the last six years the extensive studies of the rotatory dispersion of steroids by DJERASSI et al. (1955a, DJERASSI 1957) have firmly established the importance of these measurements. *Dispersion curves add* in essence *a further dimension to optical rotatory data*, and their increasing use in problems pertaining to molecular asymmetry seems assured. Since this tool is relatively new, and promises to be of considerable significance in the investigation of subtle stereochemical problems, it seems appropriate to include a brief discussion of dispersion curves even though their measurement has not yet become a routine procedure.

In very general terms, the optical rotation of a substance depends upon the influence of an asymmetric environment on the interaction of polarized light with a chromophoric group in the molecule. The shape of a rotatory dispersion curve, in which [α] or [*M*] is plotted against wave length λ, is therefore determined by the presence or absence of an absorption maximum in the wave length region studied. At wave lengths remote from the absorption band the equation for the curve for many compounds is given by the Drude equation:

$$[\alpha]_\lambda = \frac{K}{\lambda^2 - \lambda_0^2}$$

in which the rotation [α] at wave length λ is related to λ and λ_0, which is the wave length of an absorption band. The curve is thus hyperbolic, and with colorless steroids the absolute value of the rotation invariably *increases* with *decreasing* wave length in the visible region.

With compounds that do not have absorption bands in the region measured (usually 280–700 mμ) a "plain" dispersion curve (DJERASSI and KLYNE 1957), either positive or negative, is observed. When the *wave length approaches and coincides with an absorption maximum,* however, *the Cotton effect is observed,* in which the curve rises to a sharp peak and then falls precipitously to a trough (Fig. 100). More complex curves with several peaks and inflections are obtained in many cases, and it is in these Cotton-effect curves with characteristic shape that the great value of optical rotatory dispersion is realized.

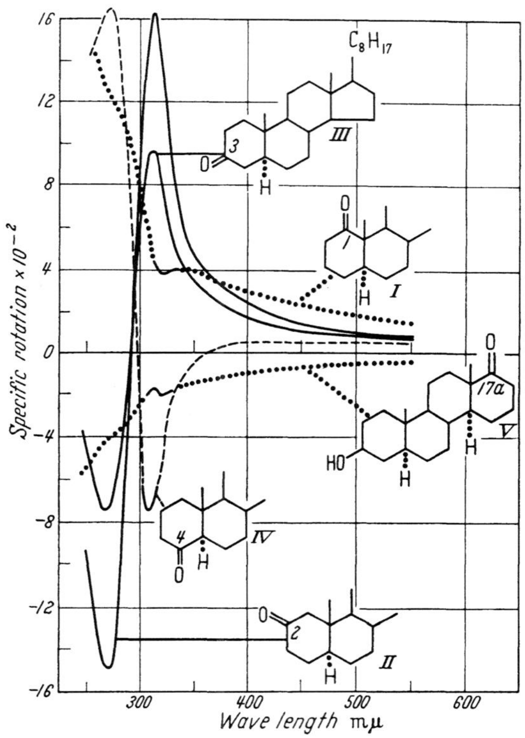

Fig. 100. Optical rotatory dispersion curves of steroidal ketones (DJERASSI 1957). Reproduced from Bull. Soc. Chim. (France) **1957**, 741 with permission of the author and editor

Because of instrumental limitations, rotations cannot be measured in regions of intense absorption, such as in unsaturated ketones at 240–250 mμ, but the low intensity absorption of unsaturated and saturated ketones in the 280–350 mμ region makes then particularly suitable for study. One of the most striking features of dispersion curves is the *extremely large specific rotation value at the peaks,* often of the order of 2000–3000, permitting measurements with fractional milligram quantities.

It is evident that a rotatory dispersion curve provides a much more complete picture of the effect of structural features on molecular shape and asymmetry than does a single measurement in the flat part of the curve. This fact is strikingly illustrated by the dispersion curves for saturated ketones (Fig. 100), in which the shape and peak intensities are highly characteristic for carbonyl groups at various positions. Although the dispersion curve characteristic for a carbonyl function at a given location is not significantly affected by the presence of hydroxyl groups or other optically inactive substituents, it is, as expected, very sensitive to stereochemical changes at a neighboring asymmetric center.

In the curves for some of the important adrenal hormones (Figs. 101 and 102), the individual contributions of the Δ^4-3-keto, 20-keto and 11-keto groups can be discerned. The double minima at 350—370 mμ are typical for Δ^4-3-ketones, and the very high peak at 300—310 mμ is due to the C-20 carbonyl. The curves of the 11-ketones (IX and VIII) are higher in the 350—370 region because of the contribution of the 11-keto group at this wave length.

The fundamental importance of rotatory dispersion measurements in steroids lies in the visualization of subtle features of the asymmetric "backbone" of the steroid skeleton.

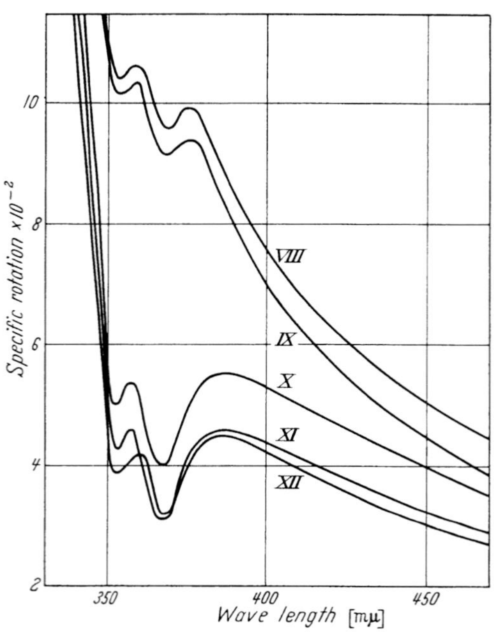

Fig. 101. Optical rotatory dispersion curves of 11-dehydrocorticosterone (VIII), cortisone (IX), corticosterone (X), cortisol (XI), and 11-deoxycorticosterone (XII) (FOLTZ et al. 1955). Reproduced from J. Amer. chem. Soc. **77**, 4361 (1955) with permission of the authors and editor

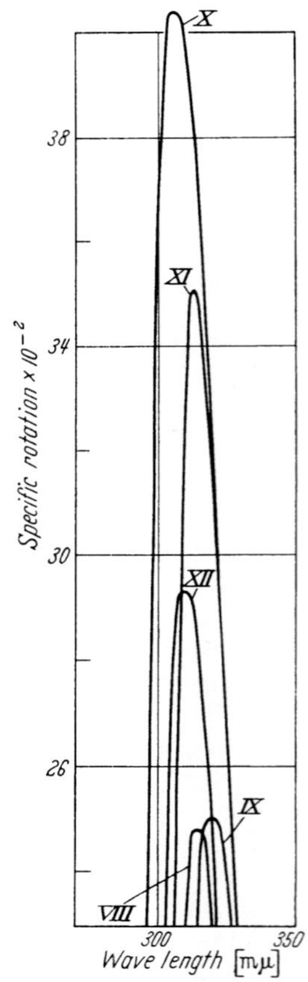

Fig. 102. Optical rotatory dispersion curves of corticosteroids (cf. Fig. 101) (FOLTZ et al. 1955). Reproduced from J. Amer. chem. Soc. **77**, 4359 (1955) with permission of the authors and editor

Conformational changes or distortions have a very pronounced effect on the rotatory spectrum (DJERASSI et al. 1956b), and a more precise understanding of conformation can be expected to emerge from further study of dispersion curves. Finally, it must be recalled that the investigation of optical rotatory dispersion in the steroid series is in a very early stage of development, and the full significance of the tool has scarcely been recognized.

2. Nuclear magnetic resonance

The most recent of the contributions of physics to the study of complex organic molecules is nuclear magnetic resonance spectrometry. Although the use of this tool in steroid chemistry has only begun, it promises to play a highly important role in certain structural problems, and a brief description of the method seems warranted. Reviews covering the theoretical basis of nuclear magnetic resonance and its applications in organic chemistry have been presented by GUTOWSKY (1956), SHOOLERY (1954) and GORDY (1956).

Certain atomic nuclei, among them the single proton in H^1, possess a magnetic moment because of nuclear spin; when placed in a permanent magnetic field such nuclei undergo a quantized precession. When an alternating magnetic field is imposed at right angles, energy is absorbed from the latter at a certain frequency which is in resonance with the precession. This angular frequency, ω, is determined by the field strength H of the permanent magnet and the magnetogyric ratio, γ, which has a characteristic value for a given nucleus: $\omega = \gamma H = 2\pi\tilde{\nu}$. For the proton, $\gamma = 2.675\ \text{sec}^{-1}\ \text{gauss}^{-1}$, and resonance occurs at a field strength of 9400 gauss with a radio frequency, $\tilde{\nu}$, of 40 Mc.

The resonance condition refers to the field strength at the nucleus, which differs by a very slight amount from the permanent field in the bulk sample because of *shielding of the nucleus by orbital electrons. The importance of nuclear magnetic resonance in organic chemistry lies in these changes in field strength, which are called chemical shifts.* The shift, δ, is defined by the relationship $\delta = \frac{H - H_R}{H_R} \cdot 10^6$, and is determined by measurement of ΔH for a reference compound and the resonance peak in question. Nuclear magnetic resonance is useful for the study of a number of nuclei with an odd number of protons, particularly those of H^1, F^{19} and P^{31}; the nuclei of C^{12} and O^{16} do not possess magnetic moments and are thus inactive.

A nuclear magnetic resonance spectrum consists essentially of an oscilloscope tracing of a sweep across a narrow range of field strengths, with resonance signals due to protons in different electronic environments appearing as very sharp peaks. The phenomenon of nuclear magnetic resonance in bulk samples depends on the difference in the number of nuclei in the two energy levels involved; this depends on the relaxation time for return to the lower level, which in turn is related to the interaction of the spinning nucleus with the molecular lattice and with other nuclei. In terms of experimental procedures, these factors lead to a requirement of liquid samples and solutions of low viscosity in order to obtain sharp peaks.

The experimental conditions for obtaining nuclear magnetic resonance spectra of steroids have been studied by SHOOLERY and RODGERS (1958), who have reported data on about fifty steroids, obtained at 40 Mc. and 9400 gauss. Most of the spectra were recorded with solutions of 50—75 mg of steroid in 0.4 ml of deuterated chloroform, $CDCl_3$, although useful spectra can be obtained with much less concentrated solutions. The reference standard used was benzene, contained in an annular cell surrounding the sample; samples were rotated during the measurement to average out inhomogeneities in the magnetic field.

The proton magnetic resonance spectra of simple organic molecules contain peaks due to each different "kind" of hydrogen atom. Thus in the spectrum of ethanol three peaks, due to the hydrogen atoms in the CH_3, CH_2 and OH groups, are present, with peak areas in the ratio of 3 : 2 : 1. With high resolution instruments these peaks are further split due to spin-spin interactions of the protons on adjacent atoms. The chemical shifts observed in the protons of a specific grouping, e. g., a terminal methyl group, with respect to a standard reference compound such as water or benzene, is the same, within fairly narrow limits, for a range of compounds. L. MEYER et al. (1953) have compiled a chart of characteristic shifts which range from $\delta = +4.5$[1] for $-CH_3$ to -7.0 for $-CO_2H$ with respect to water, with the highest (positive) values corresponding to maximum shielding.

As expected, the spectra of steroids are rather complex, and in general the signals for protons attached directly to a saturated skeletal carbon atom are not resolved. SHOOLERY and RODGERS (1958) suggest that different orders of relaxation times, and thus band widths, are involved for protons attached to the skeleton and those in substituents, which possess an additional degree of freedom. The resonance due to the skeletal protons forms a broad hump (Fig. 103) which has the fingerprint nature of the low-frequency region of an infrared spectrum, being characteristic for the molecule as a whole, and sensitive to minor structural changes.

[1] These δ values are as defined by the equation given. In the reference cited they are divided by 10 and of opposite sign.

The peaks due to the protons in substituents, including angular methyl groups, on the other hand, are very sharp and prominent, permitting a number of structural correlations. In Table 25 are summarized some of the shifts that have been identified by SHOOLERY and ROGERS (1958) for the resonance signals of specific protons.

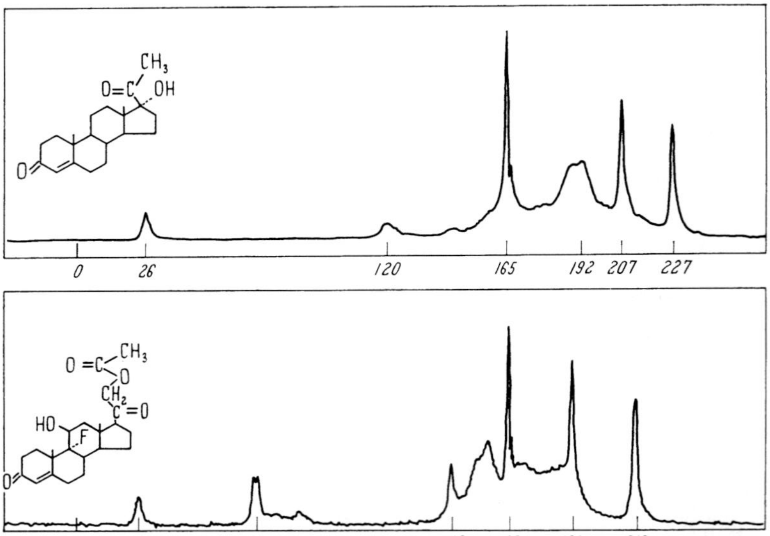

Fig. 103. Nuclear magnetic resonance spectra of 17α-hydroxyprogesterone (upper curve) and 9α-fluorocortisol acetate (lower curve) in $CDCl_3$ solution (SHOOLERY and ROGERS 1958). Reproduced from J. Amer. chem. Soc. **80**, 5121 (1958) with permission of the authors and editors

Table 25. *Nuclear magnetic resonance peaks associated with functional groups*

Group	Proton(s)	Shift[1]
Double bonds:		
Δ^1-3-Ketone	C-1	—54[4]
	C-2	19[4]
Δ^{16}-20-Ketone	C-16	—13[4]
Δ^4-3-Ketone	C-4	27 ± 3
Δ^5-3OH or OAc	C-6	42 ± 1
Hydroxyl groups:		
17β-Hydroxyl	OH	139, 151, 176[2]
3α- or β-Hydroxyl	OH	131, 164, 183[2]
17α-hydroxy-20-ketone	OH	120[2]
Methyl groups:		
20-Ketone	C-21 CH_3	171 ± 1
17α-Hydroxy-20-ketone	C-21 CH_3	166[1]
21-Acetoxy-20-ketone	Acetate CH_3	169
	C-21 CH_2	70 ± 1[3]
3- or 11-Acetoxy	Acetate CH_3	175 ± 2
3-Methoxy	OCH_3	122[4]
C-18	C-18 CH_3	218—234
C-19	C-19 CH_3	193—220

Since the chemical shift of a proton depends on the degree of electron shielding, any structural feature leading to decreased electron density at a proton will tend to lower the shift value, as seen for protons attached to a double bond. The peak for the C-4 hydrogen at 27 cps. is highly specific for the Δ^4-3-keto grouping as far as can be judged from the data available, and appears to provide a reliable indication of the presence of this system. It must be noted that as nuclear magnetic resonance data for additional functional groups become available, some

[1] Shifts in cycles per second with respect to benzene. δ Values are obtained by dividing by 40. All shifts positive except as noted.
[2] These shifts are temperature and concentration dependent. Measured in $> 1\ M$ solutions.
[3] Multiplet with very weak signals at 53 and 87 cps.
[4] Data on single compound.

of the assignments of characteristic shift values drawn by SHOOLERY and ROGERS may prove not to be unequivocal.

The shielding of the protons in hydroxyl groups is inversely proportional to the extent of hydrogen bonding, and, therefore, decreases with increasing concentration. The shift values are not generally useful for locating the position of hydroxyl groups in an unknown compound, but they can provide important information on hydrogen bonding. The highly hindered 11 β-hydroxyl group is not sufficiently hydrogen bonded in most cases for the proton signal to be shifted out of the non-specific skeletal resonance region.

A proton attached to a skeletal carbon atom which bears an oxygen substituent, e. g., the 3 β-proton in a 3 α-alcohol, is subject to a large change in shielding, and the signal is removed from the general skeletal resonance to about 80–120 cps. The shift for such protons is dependent on the conformation of the proton (opposite to that of the substituent) and is some 20 cps. lower for an equatorial than an axial proton. These peaks, due to a single proton which is subject to strong spin-spin coupling, are in general not sharp, and may be quite inconspicuous.

Probably of greatest significance in structural correlations with nuclear magnetic resonance spectra are the shifts due to protons in methyl groups. The relaxation times are smaller than for skeletal protons, and in all of the spectra recorded so far, *a sharp peak corresponding to each methyl group in the molecule is present*. The highest shielding is experienced by protons in methyl groups such as C-18 and C-19 when these are remote from electron-withdrawing groups, and the shifts become progressively lower as such groups are introduced. Protons on a carbon atom adjacent to a carbonyl, for example, are displaced some 40 cps. toward lower values, as seen in the case of acetates and 20-ketones. The protons on the C-21 methylene group in 20,21-ketols or the acetates are subject to reduced shielding by both the carbonyl group and the oxygen substituent, and the peaks are found at 70–90 cps.

Table 26. *Substituent effects in angular methyl proton shifts*

C-19	
Maximum shift:	220±1 cps.
Increment:	
3-Alcohol or acetate	0
3-Ketone	—5
Δ^4- or Δ^5-Double bond	—5
Δ^4-3-Ketone	—11
11-Ketone	—7
11 α-Alcohol	—4
11 β-Alcohol	—10
Δ^4-3,11-Diketone	—20
Δ^4-3-Keto-11 β-alcohol	—23
Δ^4-3-Keto-9 α-fluoro-11 β-alcohol	—27
C-18	
Maximum shift:	228—234
Increment:	
20-Ketone	0
11-Ketone	0
11 α-Alcohol	0
11 β-Alcohol	—10
17 β-Alcohol	—8
17α-Hydroxy-20-ketone	0
17-Ketone	—10
Δ^{16}-Double bond	—10
$\Delta^{17(20)}$-Double bond	—10

The effects of substituents on the chemical shifts of the protons in the angular methyl groups may arise in a number of ways. The displacement caused by unsaturation and substitution at neighboring centers are additive; SHOOLERY and ROGERS (1958) suggest that the most important influence is the magnetic anisotropy of electrons in the substituents. Base values (maximum shifts) can be assigned for the two methyl groups; substituent effects calculated from these values are summarized in Table 26.

From the data in Tables 25 and 26, it is evident that *nuclear magnetic resonance spectra furnish a unique means of visualizing the hydrogen atoms in a steroid*. As a catalogue of shift values is accumulated it should be possible to characterize a number of functional groups by reference to proton shifts at the skeletal center involved and the changes induced in angular methyl peaks. The structural data provided by infrared spectra and proton magnetic resonance spectra are complementary, and with both available a substantial amount of information on the structure of an unknown steroid can be obtained literally at a glance.

The full potentialities of nuclear magnetic resonance measurements in the steroid field are certainly not yet realized, and further advances in instrumentation are constantly being made. Apart from structural purposes, it can be anticipated that as the factors involved in nuclear shielding become better understood, proton magnetic resonance spectra may dispel much of the obscurity which still shrouds "long range" interactions of remote functional groups in steroids. Any new knowledge in this problem can in turn be expected to contribute significantly to an understanding of the effects of substituents on the biological activity of steroids.

3. Polarography

Polarography is a method for the electrolytic reduction of ions or organic compounds under special conditions, and provides a *very useful tool for the quantitative determination of reducible substances in dilute solution*. The application of polarography to the analysis of steroids has been somewhat limited, but several interesting applications have been described. A comprehensive discussion of the theory and instrumentation of polarography, and a review of applications of the method to organic compounds, including steroids, will be found in the treatise of Kolthoff and Lingane (1952).

A polarographic determination entails measurement of the current that flows during electrolysis at a polarized microelectrode as the applied potential is varied. This electrode, which is the cathode in polarographic reductions, is a drop of mercury that forms continually at a capillary tip and falls away; a reference calomel cell forms the anode. The current-voltage relationship is controlled entirely by the reduction occurring at the polarized cathode.

In a typical polarogram, the current begins to rise sharply at a certain applied voltage (negative with respect to the cathode) corresponding to the reduction potential of the reducible solute, and then levels off in a plateau. The current that flows in this region of the polarogram depends on diffusion of the reducible species from the concentration in the bulk solution through a boundary at the dropping mercury cathode, in which the concentration is depleted due to removal at the electrode. The difference in microamperes between the height of the plateau or inflection and the residual current, which is due to conductance of the solution in the absence of electrolysis, is called the *diffusion current*.

The current that is measured in the galvanometer of a polarograph is actually an oscillating one because of the periodic nature of the mercury drop; for this reason the curves obtained in an automatic-recording instrument have a "saw-tooth" profile. The diffusion current is a function of the concentration C and diffusion coefficient D of the reducible solute and of several parameters of the dropping mercury electrode. The average value of the current during the lifetime t of the drop having mass m for a reaction involving n electrons is given by the modified Ilkovic equation:

$$i_D = 607\, n D^{1/2} C m^{2/3} t^{1/6} \left(1 + \frac{39 D^{1/2}\, t^{1/6}}{m^{1/3}}\right)$$

This expression is derived from a consideration of Faraday's Law and the diffusion process at a spherical drop. It is evident that for a given reduction process using the same solvent and the same electrode and drop-time, the diffusion current is directly proportional to the concentration. For routine analytical purposes a calibration curve of concentration versus i_d is constructed; this plot is nearly linear.

The polarographic wave described by a cathodic reduction occurs at a voltage characteristic for the ion or compound undergoing reduction. This voltage is called the half-wave potential, $E_{1/2}$, and is taken as the voltage at the midpoint of the wave slope. This potential is related to the true reduction potential by the diffusion coefficients of the oxidized and reduced forms:

$$E_{1/2} = -E^\circ + \frac{0.0591}{n} \log \left(\frac{D_{red}}{D_{ox}}\right)^{1/2}$$

The half-wave potential for a reduction is usually strongly dependent on the p_H, especially for organic compounds, and a buffer system is often used as the supporting electrolyte. The practical voltage limit for polarographic reductions is determined by the decomposition potential of the supporting electrolyte or impurities in the solution; this limit is usually about —2.0 volts. Two or more waves can be detected in a polarogram and their heights measured, providing the $E_{1/2}$ values are separated by 0.15 to 0.2 volts, and it is thus possible to analyze simultaneously for several reducible components of a mixture.

Polarographic measurements on steroids have been largely restricted to the reduction of carbonyl groups or their derivatives. The reduction potentials of saturated ketones are in general too high to provide useful polarographic waves, but WOLFE et al. (1940) have shown that α-β-unsaturated 3-ketosteroids, dissolved in a mixture of *iso*-propanol and tetraethylammonium hydroxide or in aqueous ethanol (ZUMAN et al. 1954), have suitable half-wave potentials. Δ^1-3-Ketosteroids can be distinguished from Δ^4-3-ketosteroids, and Δ^5-3-hydroxysteroids can be determined after Oppenauer oxidation (HERSHBERG et al. 1941). 17-Ketosteroids and C-20 ketosteroids may be determined in the form of their Girard's T derivatives (see also BARNETT et al. 1946), but not the saturated 3-ketosteroids. BUTT et al. (1951) and MORRIS (1953, 1955) have described the polarographic determination of the Girard's T hydrazones of purified adrenocortical hormones from blood. The half-wave potentials due to the Δ^4-3-keto group occur in the range of 1.08—1.15 V and that due to the 20-keto group in the range of 1.35—1.37 V. The latter is absent in the polarogram of the 17-hydroxycorticosteroids.

Literature

ABELSON, D., and P. K. BONDY: Fluorometric determination of Δ^4-3-ketosteroids. Arch. Biochem. Biophys. **57**, 208—217 (1955).

— — Spectrophotometric determination of Δ^4-3-keto-steroids. Analyt. Chem. **28**, 1922—1924 (1956).

ADAMS, W. J., D. N. KIRK, D. K. PATEL, V. PETROW and I. A. STUART-WEBB: Ketals of *allo*-pregnane-12,20-diones. J. chem. Soc. **1954**, 2299—2304.

AGNELLO, E. J., B. L. BLOOM and G. D. LAUBACH: A new class of biologically active corticoids. J. Amer. chem. Soc. **77**, 4684—4685 (1955).

— and G. D. LAUBACH: A new method for dehydration of corticosteroids. J. Amer. chem. Soc. **79**, 1257—1258 (1957).

ALLEN, W. M.: A simple method for analyzing complicated absorption curves, of use in the colorimetric determination of urinary steroids. J. clin. Endocrin. **10**, 71—83 (1950).

— S. J. HAYWARD and A. PINTO: A color test for dehydroisoandrosterone and closely related steroids, of use in the diagnosis of adrenocortical tumors. J. clin. Endocrin. **10**, 54—70 (1950).

ALLEN, W. S., and S. BERNSTEIN: The preparation of Δ^{16}-steroids. J. Amer. chem. Soc. **77**, 1028—1032 (1955).

— — The preparation of 16α-hydroxyhydrocortisone and related compounds. J. Amer. chem. Soc. **78**, 1909—1913 (1956a).

— — The preparation of 16β,21-epoxido steroids. J. Amer. chem. Soc. **78**, 3223—3224 (1956b).

— — and R. LITTELL: The conversion of 11-*epi*-hydrocortisone into hydrocortisone. J. Amer. chem. Soc. **76**, 6116—6119 (1954).

AMENDOLLA, C., G. ROSENKRANZ and F. SONDHEIMER: Synthesis of steroidal hormone analogues hydroxylated at $C_{(6)}$. J. chem. Soc. **1954**, 1226—1233.

ANDERSON, A. J., and F. L. WARREN: Extraction of conjugated androgens from normal human urine. J. Endocrin. **7**, LXV (1951).

ANDERSON, D. H., and N. B. WOODALL: Infrared identification of materials in the fractional milligram range. Analyt. Chem. **25**, 1906—1909 (1953).

ANDERSON, H. V., E. R. GARRETT, F. H. LINCOLN, A. H. NATHAN and J. A. HOGG: Preparation and reactions of steroidal $\Delta^{17(20)}$-enol acetates. J. Amer. chem. Soc. **76**, 743—746 (1954).

ANDREČ, K., u. M. KREČ: Die Papierchromatographie konjugierter Steroide. Bull. sci., Conseil Acad. R. P. F. Yougoslav. **2**, 103—104 (1956).

— u. L. STRAUCH: Papierelektrophorese von konjugierten Steroiden. Bull. sci., Conseil Acad. R. P. F. Yougoslav. **2**, 101—102 (1956).

ANLIKER, R., O. ROHR u. M. MARTI: Über den Nachweis von Testosteron in einem virilisierenden Nebennierenrindentumor. Helv. chim. Acta **39**, 1100—1106 (1956).

ANNER, G., u. K. MIESCHER: Die Synthese des natürlichen Oestrons. Totalsynthesen in der Oestronreihe III. Helv. chim. Acta **31**, 2173 (1948).

ANTONUCCI, R., S. BERNSTEIN, M. HELLER, R. LENHARD, R. LITTEL and J. WILLIAMS: Hydrocortisone and related corticoids. J. org. Chem. **18**, 70—82 (1953).

APPLEBY, J. I., G. GIBSON, J. K. NORYMBERSKI and R. D. STUBBS: Indirect analysis of corticosteroids. I. The determination of 17-hydroxycorticosteroids. Biochem. J. **60**, 453—460 (1955).

— and J. K. NORYMBERSKI: Indirect analysis of corticosteroids. II. The determination and identification of urinary 17-hydroxy-20-oxysteroids unsubstituted at C_{21}. Biochem. J. **60**, 460—467 (1955).

ARROYAVE, G., and L. R. AXELROD: The paper chromatographic separation and qualitative identification of mixtures of $C_{19}O_3$ steroids. J. biol. Chem. **208**, 579—589 (1954).

ARTH, G. E., J. FRIED, D. B. R. JOHNSTON, D. R. HOFF, L. H. SARETT, R. H. SILBER, H. C. STOERK and C. A. WINTER: 16α-Methyl analogs of cortisone, a new group of anti-inflammatory steroids. 9α-Haloderivatives. J. Amer. chem. Soc. **80**, 3161—3163 (1958a).

— D. B. R. JOHNSTON, J. FRIED, W. W. SPOONCER, D. R. HOFF and L. H. SARETT: 16α-Methylated analogs of cortisone. A new group of anti-inflammatory steroids. J. Amer. chem. Soc. **80**, 3160—3161 (1958b).

ARX, E. VON, u. R. NEHER: Neue Hilfsmittel für die Papierchromatographie. Helv. chim. Acta **39**, 1664—1670 (1956).

ATTENBURROW, J., A. F. B. CAMERON, J. H. CHAPMAN, R. M. EVANS, B. A. HEMS, A. B. A. JANSEN and T. WALKER: A synthesis of vitamin A from cyclohexanone. J. chem. Soc. **1952**, 1094—1111.

ATWATER, N. W.: Direct monoalkylation of steroids at C-4. J. Amer. chem. Soc. **79**, 5315—5316 (1957).

AVIVI, P., S. A. SIMPSON, J. F. TAIT and J. K. WHITEHEAD: Use of hydrogen^3- and carbon^{14}-labeled acetic anhydride as analytical reagents in microbiochemistry. Proc. 2nd Radioisotope Conference, Oxford. I. Med. and Physiol. Application, 313—324 (1954).

AXELROD, L. R.: The separation and identification of mixtures of C_{21} and C_{19} steroids by paper chromatography. J. biol. Chem. **205**, 173—184 (1953a).

— A specific test differentiating between α-ketol and dihydroxy-acetone groups of C_{21}-steroids on paper chromatograms. J. Amer. chem. Soc. **75**, 4074 (1953b).

— Microchemical detection of characteristic functional groups in steroids. Side chains. Analyt. Chem. **27**, 1308—1311 (1955).

— and A. ZAFFARONI: The extraction of corticosteroids from blood and tissues by dialysis. Arch. Biochem. Biophys. **50**, 347—353 (1954).

AYRES, P. J., O. GARROD, S. A. SIMPSON and J. F. TAIT: A method for the determination of aldosterone, cortisol and corticosterone in biological extracts, particularly applied to human urine. Biochem. J. **65**, 639—651 (1957a).

— S. A. SIMPSON and J. F. TAIT: A fluorescence method for the microanalysis of Δ^4-3-oxo steroids on paper chromatograms. Biochem. J. **65**, 647—651 (1957b).

BACHMAN, W. E., W. COLE and A. R. WILDS: The total synthesis of the sex hormone equilenin. J. Amer. chem. Soc. **61**, 974—975 (1939).

BAGGETT, B., L. L. ENGEL and L. L. FIELDING: The estimation of hydroxysteroids. J. biol. Chem. **213**, 87—97 (1955).

BAKER, C. P., F. DOBSON and S. W. STROUD: Isolation of 17-hydroxycorticosterone and 17-hydroxy-11-dehydrocorticosterone from hog adrenals. Nature (Lond.) **168**, 114 (1951).

BANES, D.: The chemical assay of adrenal cortex extracts. J. Amer. pharm. Ass. **42**, 669—673 (1953).

BARAN, J. S.: 2-Hydroxy-$\Delta^{1,4}$-3-ketosteroids. J. Amer. chem. Soc. **80**, 1687—1691 (1958).

BARBER, G., and M. EHRENSTEIN: 19-Hydroxy-11-desoxycorticosterone and 19-hydroxyprogesterone. J. org. Chem. **19**, 1758—1765 (1954).

— — 19-Oxoprogesterone and 19-oxo-11-desoxycorticosterone. J. org. Chem. **20**, 1253—1259 (1955).

— — Synthese von 19-Nor-10ξ, 14β,17α-trioxyprogesteron und 19-Nor-11-desoxy-10ξ, 14β,17α-trioxycorticosteron aus Strophanthidin. Liebigs Ann. Chem. **603**, 89—109 (1957).

BARNETT, J., A. A. HENLEY and C. J. O. R. MORRIS: Polarographic estimation of steroid hormones. I. Polarography of neutral 17-ketosteroids in urinary extract. Biochem. J. **40**, 445—449 (1946).

BARTON, D. H. R.: The stereochemistry of cyclohexane derivatives. J. chem. Soc. **1953**, 1027—1040.

— and R. C. COOKSON: The principles of conformational analysis. Quart. Rev. chem. Soc. (Lond.) **10**, 44—82 (1956).

Barton, D. H. R., and W. C. Taylor: The photochemistry of prednisone acetate. Proc. chem. Soc. (Lond.) **1957**, 96.

— — The photochemistry of prednisone acetate in neutral solution. J. Amer. chem. Soc. **80**, 244—245 (1958).

— R. M. Evans, J. C. Hamlet, P. G. Jones and T. Walker: The preparation of 3β:17α-dihydroxyallopregnane-11:20-dione. J. chem. Soc. **1954**, 747—751.

— D. A. Ives and B. R. Thomas: A Wolff-Kishner reduction procedure for sterically hindered carbonyl groups. J. chem. Soc. **1955**, 2026.

Bassil, G. T., and A. M. Hain: Chemical estimation of urinary cortin. Nature (Lond.) **165**, 525 (1950).

Bauer, J., u. J. Karl: Der chemische und krystalline Nachweis von androgenen und corticoiden Wirkstoffen aus Urin und Geschwulstgewebe bei virilisierendem Ovarialtumor. Z. ges. exp. Med. **118**, 425—452 (1952).

Bayliss, R. I. S.: Some observations on the hydrolysis and extraction of formaldehydogenic corticosteroids. Biochem. J. **52**, 63—70 (1952).

Beher, W. T., J. Parsons and G. D. Baker: Comparative value of X-ray diffraction and infrared spectrophotometry in identifying certain sterols and their digitonides. Analyt. Chem. **29**, 1147—1151 (1957).

Bellamy, L. J.: Infrared spectra of complex molecules. London: Methuen 1954.

Berliner, D. L.: Microdetermination of acetylable steroids in plasma. Proc. Soc. exp. Biol. (N. Y.) **94**, 126—128 (1957).

— and H. A. Salhanick: Microchemical identification of radioactive and nonradioactive steroids. Analyt. Chem. **28**, 1608—1610 (1956).

— O. V. Dominguez and G. Westenskow: Determination of carbon-14 steroids on paper chromatograms. Analyt. Chem. **29**, 1797—1800 (1957).

Bernstein, S., and R. H. Lenhard: The absorption spectra of steroids in concentrated sulfuric acid. J. org. Chem. **18**, 1146—1165 (1953).

— — The absorption spectra of steroids in concentrated sulfuric acid. Structural correlations. J. org. Chem. **19**, 1269—1277 (1954).

— — The preparation of pregnane-5α,11β,17α,21-tetrol-3,20-dione. J. Amer. chem. Soc. **77**, 2233—2237 (1955).

— W. S. Allen, C. E. Linden and J. Clemente: The synthesis of 6β-hydroxyprogesterone. J. Amer. chem. Soc. **77**, 6612—6613 (1955).

— R. H. Lenhard, W. S. Allen, M. Heller, R. Littell, S. M. Stolar, L. I. Feldman and R. H. Blake: The synthesis of the 16α-hydroxy derivatives of 9α-halo steroids. J. Amer. chem. Soc. **78**, 5693—5694 (1956).

— M. Heller, R. Littell, S. M. Stolar, R. H. Lenhard, W. S. Allen and I. Ringler: The synthesis of the 16α-hydroxy derivatives of 2-methyl steroids. J. Amer. chem. Soc. **81**, 1696—1701 (1959a).

— M. Heller and S. M. Stolar: The preparation and epimerization of 16β-acetoxy-17α-hydroxy steroids. J. Amer. chem. Soc. **81**, 1256 (1959b).

Beyler, R. E., and F. Hoffman: Synthesis of 16β-chlorocortisone and related 16-substituted steroids. J. org. Chem. **21**, 572—574 (1956).

— R. Moriarty, F. Hoffman and L. H. Sarett: 17,20;20,21-Bismethylenedioxy steroids. J. Amer. chem. Soc. **80**, 1517—1518 (1958).

Bijvoet, J. M.: Determination of the absolute configuration of optical antipodes. Endeavour **14**, 71—77 (1955).

Birch, A. J.: The reduction of organic compounds by metal-ammonia solutions. Quart. Rev. chem. Soc. (Lond.) **4**, 69—93 (1950).

Bitman, J., and S. L. Cohen: Hydrolysis of urinary conjugated 17-ketosteroids by acetate buffer. J. biol. Chem. **191**, 351—363 (1951).

— J.-P. Rosselet, A. M. Reddy and S. Lieberman: A spectrophotometric method for the analysis of binary mixtures of urinary steroids. J. biol. Chem. **225**, 39—52 (1957).

Bladon, P., H. B. Henbest and G. W. Wood: Ultraviolet absorption spectra of ethylenic steroids. J. chem. Soc. **1952**, 2737—2743.

— — E. R. H. Jones, B. J. Lovell and G. F. Woods: Synthesis of cortisone acetate *via* 5α-hydroxysteroids. J. chem. Soc. **1954**, 125—130.

Blix, G., A. Tiselius and H. Svensson: Lipides and polysaccharides in electrophoretically separated blood-serum proteins. J. biol. Chem. **137**, 485—494 (1941).

Bloch, E., G. Pincus and R. I. Dorfman: The conversion of acetate to dehydroepiandrosterone by human adrenal gland slices. Arch. Biochem. **61**, 245—247 (1956).

Block, R. J., E. L. Durrum and G. Zweig: A manual of paper chromatography and paper electrophoresis. New York: Academic Press, Inc. 1955.

Bloom, B. M., and G. M. Shull: Epoxidation of unsaturated steroids by microorganisms. J. Amer. chem. Soc. **77**, 5767—5768 (1955).

BOCK, R. M., and N. S. LING: Devices for gradient elution in chromatography. Analyt. Chem. **26**, 1543—1546 (1954).
BODÁNSZKY, A., and J. KOLLONITSCH: Color reagent for paper chromatography of steroids. Nature (Lond.) **175**, 729 (1955).
BONDY, P. K., and G. V. UPTON: Simultaneous determination of cortisol and corticosterone in human plasma by quantitative paper chromatography. Proc. Soc. exp. Biol. (N. Y.) **94**, 585—589 (1957).
— D. ABELSON, J. SCHEUER, T. K. L. TSEU, V. UPTON, H. ABELSON and A. CLARKE-LEWIS: Determination of cortisol in human plasma by quantitative paper chromatography. J. biol. Chem. **224**, 47—55 (1957).
BONGIOVANNI, A. M., and W. R. EBERLEIN: Adrenocortical steroids in the peripheral blood of man. J. clin. Endocrin. **15**, 1524 (1955a).
— — Determination, recovery, identification, and renal clearance of conjugated adrenal corticoids in human peripheral blood. Proc. Soc. exp. Biol. (N. Y.) **89**, 281—285 (1955b).
— — Microformaldehydogenic method for the analysis of chromatographic fractions of adrenal steroids. J. Lab. clin. Med. **48**, 320—323 (1956).
— — M. M. GRUMBACH, J. J. VAN WYK and G. CLAYTON: Conjugates of adrenal corticoids in human plasma. Proc. Soc. exp. Biol. (N. Y.) **87**, 282—287 (1954).
— — and P. Z. THOMAS: Use of an organic base in the Zimmermann reaction. J. clin. Endocrin. **17**, 331—332 (1957).
BORGSTROM, E., and T. F. GALLAGHER: Partial synthesis of compounds related to adrenal cortical hormones. Reactions of ring C ketols in the preparation of an 11-keto bile acid. J. biol. Chem. **177**, 951—967 (1949).
BORTH, R.: Problems related to the chemical assay of corticosteroids in human body fluids. Acta endocr. (Kbh.) **22**, 125—135 (1956a).
— Chemical determination of urinary corticosteroids. Chimica **10**, 49—58, 81—90 (1956b).
BOSCOTT, R. J.: Significance and use of dehydration effects in the analysis of steroids. Nature (Lond.) **164**, 140 (1949).
BOWERS, A., and H. RINGOLD: 6α- and 6β-Fluorotestosterone and 6α- and 6β-fluoroprogesterone. Tetrahedron **3**, 14—27 (1958).
— — and R. DORFMAN: 19-Nordihydrotesterone derivatives. A potent class of anti-estrogenic compounds. J. Amer. chem. Soc. **79**, 4556—4557 (1957).
BOYD, G. S.: Estimation of serum lipoproteins. A micromethod based on zone electrophoresis and cholesterol estimations. Biochem. J. **58**, 680—685 (1954).
BRAUDE, E. A.: Ultraviolet and visible light absorption. In: E. A. BRAUDE and F. NACHOD, Eds., Determination of organic structures by physical methods, 131—186. New York: Academic Press 1955.
BRENNEISEN, K., C. TAMM u. T. REICHSTEIN: Synthese der L(—) β-Methoxy-adipinsäure aus L(—)-Äpfelsäure. Helv. chim. Acta **39**, 1233—1241 (1956).
BROADBENT, I. E., J. P. NEWHOUSE and W. KLYNE: The determination of reducing corticoids in urine. J. clin. Path. **6**, 190—194 (1953).
BROOKS, C. J. W., and J. K. NORYMBERSKI: Characterization of the corticosteroid side-chain. Chem. and Ind. **1952**, 804—805.
— — The oxidation of corticosteroids with sodium bismuthate. Biochem. J. **55**, 371—378 (1953).
BROOKS, R. V., W. KLYNE and E. MILLER: Separation of steroid alcohols by chromatography of their benzoates on alumina. Biochem. J. **54**, 212—217 (1953).
BUCHSCHACHER, P., J. KALVODA, D. ARIGONI and O. JEGER: Direct introduction of a nitrogen function at C-18 in a steroid. J. Amer. chem. Soc. **80**, 2905—2906 (1958).
BUEHLER, H. J., P. A. KATZMAN and E. A. DOISY: Hydrolysis of conjugates of neutral ketosteroids. Proc. Soc. exp. Biol. (N. Y.) **78**, 3—8 (1951).
BURSTEIN, S.: Reduction of phosphomolybdic acid by compounds possessing conjugated double bonds. Analyt. Chem. **25**, 422—424 (1953).
— Extraction of $C_{21}O_5$ and $C_{21}O_6$ steroids from aqueous media. Science **124**, 1030 (1956a).
— 2α-Hydroxylation of cortisol in the guinea pig. J. Amer. chem. Soc. **78**, 1769—1770 (1956b).
— and R. I. DORFMAN: Steroid metabolism in guinea pigs. J. biol. Chem. **213**, 581—595 (1955).
BURTON, R. B., A. ZAFFARONI and E. H. KEUTMANN: Paper chromatography of steroids. II. Corticosteroids and related compounds. J. biol. Chem. **188**, 763—771 (1951).
BUSH, I. E.: Chromatography of steroids on alumina-impregnated filter paper. Nature (Lond.) **166**, 445 (1950).
— Method of paper chromatography of steroids applicable to the study of steroids in mammalian blood and tissues. Biochem. J. **50**, 370—378 (1952).
— The possibilities and limitations of paper chromatography as a method of steroid analysis. Recent Progr. Hormone Res. **9**, 321—330 (1954a).
— Chromatography of steroids and sterols. Brit. med. Bull. **10**, 229—236 (1954b).

BUSH, I. E., and A. A. SANDBERG: Adrenocortical hormones in human plasma. J. biol. Chem. **205**, 783—793 (1953).
BUTT, W. R., A. A. HENLY and C. J. O. R. MORRIS: The polarographic estimation of steroid hormones. IV. Determination of 3α- and 3β-hydroxy-17-ketosteroids. Biochem. J. **42**, 447—452 (1948).
— P. MORRIS, C. J. O. R. MORRIS and D. C. WILLIAMS: The polarographic estimation of steroid hormones. V. Determination of progesterone in blood. Biochem. J. **49**, 434—438 (1951).
BUZZETTI, F., W. WICKI, J. KALVODA u. O. JEGER: Einführung einer Sauerstoffunktion an die quaternäre Methyl-Gruppe C-18 im intakten Steroid-Gerüst. Helv. chim. Acta **42**, 388—390 (1959).
CAHEN, R. L., and W. T. SALTER: Urinary 17-ketosteroids in metabolism. I. Standardized chemical estimation. J. biol. Chem. **152**, 489—499 (1944).
CAINELLI, G., M. L. MIHAILOVIC, D. ARIGONI u. O. JEGER: Direkte Einführung einer Sauerstoffunktion in die Methylgruppe C-18 im intakten Steroidgerüst. Helv. chim. Acta **42**, 1124—1127 (1959).
CALLOW, N. H., R. K. CALLOW and C. W. EMMENS: Colorimetric determination of substances containing the grouping —CH_2—CO in urine extracts as an indication of androgen content. Biochem. J. **32**, 1312—1331 (1938).
CALLOW, R. K., and V. H. T. JAMES: By-ways of synthesis of cortisone from hecogenin. 9:11-Dehydrohecogenin as intermediate in preparation of 11-oxygenated compounds (11-oxotigogenin and 3β-acetoxy-17α-hydroxyallopregnane-11:20-dione). J. chem. Soc. **1956**, 4739—4743.
CAMERINO, B., and A. VERCELLONE: Action of fermenting yeasts on 16,17-epoxy-20-ketosteroids (in Italian). Gazz. chim. ital. **86**, 260—265 (1956).
— B. PATELLI and A. VERCELLONE: Synthesis and anabolic activity of 4-substituted testosterone analogs. J. Amer. chem. Soc. **78**, 3540—3541 (1956).
CARDWELL, H. M., J. W. CORNFORTH, S. R. DUFF, H. HOLTERMAN and R. ROBINSON: Total synthesis of androgenic hormones. Chem. and Ind. **1951**, 389—390.
CAROL, J., J. C. MOLITOR and E. O. HAENNI: Determination of estrone, equilin and equilenin by infrared spectrophotometry. J. Amer. pharm. Ass. (Sci. ed.) **37**, 173—179 (1948).
CARSTENSEN, H.: Separation of corticosteroids by countercurrent distribution. Acta chem. scand. **9**, 1026—1027 (1955).
— Use of countercurrent distribution for the identification of aldosterone (Δ^4-pregnen-21-ol-3,20-dion-18-al). Arkiv Kemi **10**, 235—238 (1956a).
— Countercurrent distribution studies of adrenocortical steroids. Acta Soc. med. upsalien. **61**, 26—46 (1956b).
CASANOVA, R., C. W. SHOPPEE and G. SUMMERS: Partial synthesis of 9:11-anhydrocorticosterone actetate (21-acetoxypregna-4:9(11)-diene-3:20-dione). J. chem. Soc. **1957**, 2983—2992.
CASPI, E., G. ROSENFELD and R. I. DORFMAN: Degradation of cortisol-C^{14} biosynthesized from acetate-1-C^{14}. J. org. Chem. **21**, 814—815 (1956).
CASTELLS, J., and G. D. MEAKINS: A new method for locating double bonds in steroids. Chem. and Ind. **1956**, 248.
CAVINA, G.: Paper electrophoresis of sulfuric esters and glucuronosides of steroids (In Italian). Boll. Soc. ital. Biol. sper. **31**, 1666—1668 (1955a).
— Paper chromatography of some sulfuric esters and glucuronosides of steroids (in Italian). Boll. Soc. ital. Biol. sper. **31**, 1668—1670 (1955b).
— and L. TENTORI: Paper electrophoresis of sulfuric esters of hydroxy steroids: Application to the analysis of 3-hydroxy-17-oxygenated urinary steroids (in Italian). Boll. Soc. ital. Biol. sper. **32**, 1109—1111 (1956).
CHAMBERLIN, E. M., E. TRISTRAM, T. UTNE and J. M. CHEMERDA: The transformation of stigmasterol to 17α-hydroxypregnenolone and 17α-hydroxyprogesterone. J. Amer. chem. Soc. **79**, 456—459 (1957).
CHANG, E., and W. R. SLAUNWHITE: A comparison of urinary Zimmermann chromogens by Girard fractionation and by Allen's method of correction. J. clin. Endocrin **15**, 767—770 (1955).
CHAPMAN, J. H., J. ELKS, G. H. PHILLIPS and L. J. WYMAN: A new method of preparing 11-oxotigogenin. J. chem. Soc. **1956**, 4344-4350.
CHEN, C., J. WHEELER and H. E. TEWELL: Methods of estimation of adrenal cortical steroids with tetrazolium salts. J. Lab. clin. Med. **42**, 749—757 (1953).
— S. M. VOEGTLI, and S. FREEMAN: Estimation of total free reducing plasma steroids. J. biol. Chem. **217**, 709—719 (1955).
Ciba Foundation: Steroid nomenclature. Chem. and Ind. (**1951**) SN 1-SN 11 (June 23).
— —: — Helv. chim. Acta **34**, 1680—1695 (1951).

CLARK, L. C., JR., and H. THOMPSON: A new series of reagents for the colorimetric determination of steroids. Science **107**, 429—431 (1948).

CLARKE, R. L., K. DOBRINER, A. MOORADIAN and C. M. MARTINI: The reaction of lead tetraacetate with progesterone and testosterone. J. Amer. chem. Soc. **77**, 661—664 (1955).

CLINTON, R. O., R. G. CHRISTIANSEN, H. C. NEUMAN and S.C. LASKOWSKI: 17α-Ethinyl-D-homoetiocholane-3α,17a-diol-11-ones and related compounds. J. Amer. chem. Soc. **80**, 3389—3395 (1958a).

— H. C. NEUMAN, A. J. MANSON, S. C. LASKOWSKI and R. G. CHRISTIANSEN: The synthesis of D-homocortisone acetate and related compounds. J. Amer. chem. Soc. **80**, 3395—3402 (1958b).

COFFMAN, J. R.: Studies on colored esters of hydroxy steroids. J. biol. Chem. **140**, XXVIII (1941).

COHEN, S. L.: A simple, continuous, liquid-liquid extraction apparatus suitable for the removal of steroids at low temperatures. J. Lab. clin. Med. **36**, 769—772 (1950).

— The hydrolysis of steroid glucuronides with calf spleen glucuronidase. J. biol. Chem. **192**, 147—160 (1951).

— and A. KALUSZYNER: Color reaction between 17-ketosteroids and 3,5-dinitrobenzoic acid. Analyt. Chem. **29**, 161—164 (1957).

— and I. B. ONESON: The conjugated steroids. IV. The hydrolysis of ketosteroid sulfates. J. biol. Chem. **204**, 245—256 (1953).

COLE, A. R. H.: Infrared spectra of natural products. Fortschr. Chem. org. Naturst. **13**, 1—69 (1956).

— R. N. JONES and K. DOBRINER: A relationship between the stereochemical configuration of 3-hydroxysteroids and their infrared absorption spectra. J. Amer. chem. Soc. **74**, 5571—5575 (1952).

COLE, W., and P. JULIAN: Reduction of epoxy ketones by chromous salts. J. org. Chem. **19**, 131—138 (1954).

COLTON, F. G., W. R. NES, D. A. VON DORP, H. L. MASON and E. C. KENDALL: Introduction of the 17-hydroxy group in the partial synthesis of cortisone. J. biol. Chem. **194**, 235—245 (1952).

CONSTANTIN, J. M., and L. H. SARETT: Introduction of the 11-keto function in the steroids. J. Amer. chem. Soc. **74**, 3908—3910 (1952).

COOK, E. R., and M. E. ROOKS: An extractor and distillation head suitable for use in the determination of urinary cortin. Analyst **77**, 274—275 (1952a).

— — A nomogram for the calculation of urinary 17-ketosteroids. Analyst **77**, 525—529 (1952b).

— S. R. STITCH, A. E. HALL and M. P. FELDMAN: The fractionation of urinary neutral 17-ketosteroids by adsorption and partition chromatography. Analyst **79**, 24—29 (1954).

— B. DELL and D. J. WAREHAM: The fractionation and determination of corticosteroids in urine. Analyst **80**, 215—225 (1955).

COOLEY, G., B. ELLIS, F. HARTLEY and V. PETROW: Hydroxylation of pregn-16-en-20-ones. J. chem. Soc. **1955**, 4373—4383.

CORCORAN, A. C., and I. H. PAGE: Methods for the chemical determination of corticosteroids in urine and plasma. J. Lab. clin. Med. **33**, 1326—1333 (1948).

— H. P. DUSTAN and I. H. PAGE: Note on apparatus for determination of urinary formaldehydogenic corticoids. J. Lab. clin. Med. **38**, 780—782 (1951).

COREY, E. J., and W. E. HERTLER: The synthesis of dihydroconessine. A method for functionalizing steroids at C-18. J. Amer. chem. Soc. **80**, 2903—2904 (1958).

CORNFORTH, J. W.: Total synthesis of steroids. Progress in organic chemistry, Vol. 3, 1—43. New York: Academic Press 1955.

— I. YOUHOTSKY and G. POPJAK: Absolute configuration of cholesterol. Nature (Lond.) **173**, 536 (1954).

— R. H. CORNFORTH, A. PELTER, M. G. HORNING and G. POPJAK: Rearrangement of methyl groups in the enzymic cyclization of squalene to lanosterol. Proc. chem. Soc. (Lond.) **1958**, 112—113.

COX, R. I.: A method for the quantitative determination in urinary extracts of C_{21} 17:20-dihydroxy-20-methylsteroids. Biochem. J. **52**, 339—345 (1952).

— and G. F. MARRIAN: The isolation of pregnane-3α,17α,20α-triol from the urine of normal men. Biochem. J. **54**, 353—356 (1953).

CRAIG, L. C., D. CRAIG and E. G. SCHEIBEL: Extraction and purification. In: A. WEISSBERGER, Ed., Technique of organic chemistry, Ed. 2, Vol. III, 149—393. New York: Interscience Publ. 1956.

CRAWSHAW, A., H. B. HENBEST and E. R. H. JONES: Reactions of 11-hydroxy and 11-oxo-5α-steroids. J. chem. Soc. **1954**, 731—735.

Crépy, O., M.-F. Jayle et F. Meslin: Séparation chromatographique des sulfates et des glycuronides de stéroïdes urinaires. I. Application aux stéroïdes conjugués provenant de l'androstérone et de la dehydroepiandrostérone. Acta endocr. (Kbh.) **24**, 233—248 (1957).
Criegee, R., B. Marchand u. H. Wannowius: Zur Kenntnis der organischen Osmium-Verbindungen. Liebigs Ann. Chem. **550**, 99—133 (1942).
Dauben, H. J., B. Löken and H. J. Ringold: Preparation of ethylenedioxy derivatives of ketosteroids by exchange dioxolanation. An improved synthesis of testosterone from Δ^4-androstene-3,17-dione. J. Amer. chem. Soc. **76**, 1359—1363 (1954).
Dauben, W. G., and J. P. Eastham: The conversion of cholestanone to cholesterol. J. Amer. chem. Soc. **73**, 3260—3263 (1951a).
— — The reduction of cholestenone enol acetate by sodium borohydride. J. Amer. chem. Soc. **73**, 4463—4464 (1951b).
— R. A. Micheli and J. F. Eastham: The reduction of steroidal enol acetates with lithium aluminium hydride and sodium borohydride. J. Amer. chem. Soc. **74**, 3852—3855 (1952).
— D. Dickel, O. Jeger u. V. Prelog: Über die Anwendung der asymmetrischen Synthese zur Konfigurationsbestimmung bei Triterpenen und Steroiden. Helv. chim. Acta **36**, 325—335 (1953).
— E. J. Blanz, jr., J. Jiv and R. A. Micheli: The stereochemistry of the hydride reduction of some steroidal ketones. J. Amer. chem. Soc. **78**, 3752—3755 (1956a).
— G. J. Fonken and D. S. Noyce: The stereochemistry of hydride reductions. J. Amer. chem. Soc. **78**, 2579—2582 (1956b).
Daughaday, W. H., H. Jaffe and R. H. Williams: Chemical assay for "cortin": Determination of formaldehyde liberated on oxidation with periodic acid. J. clin. Endocrin. **8**, 166—174 (1948).
De Courcy, C.: A trichloracetic acid reagent for the detection of pregnane-3α,17α,20-triols on paper chromatograms. J. Endocrin. **14**, 164—170 (1956).
— and C. H. Gray: Studies on the determination of urinary corticosteroids. J. Endocrin. **9**, 391—400 (1953).
De Venuto, F., S. J. Mule and U. F. Westphal: Perextraction. A procedure to extract C_{21} steroids from tissues and blood. US Army Medical Research Lab. Fort Knox, Ky., Report No. 305 (1957).
Devis, R.: Les liaisons entre stéroïdes et proteines dans l'organisme animal. Bull. Soc. Chim. biol. (Paris) **37**, 563—586 (1955).
Diczfalusy, E., G. Birke u. L. O. Plantin: Corticosteroidbestimmungen und ihre klinische Bedeutung. Klin. Wschr. **34**, 225—231 (1956).
Dieckert, J. W., and R. Reiser: Glass-fiber paper impregnated with silicic acid as a new chromatographic tool. Science **120**, 678 (1954).
Dingemanse, E., L. G. Huis in't Veld and B. M. de Laat: Clinical method for the chromatographic-colorimetric determination of urinary 17-ketosteroids. I. J. clin. Endocrin. **6**, 535—548 (1946).
— — and S. L. Hartogh-Katz: Clinical method for the chromatographic-colorimetric determination of urinary 17-ketosteroids. II. J. clin. Endocrin. **12**, 66—85 (1952).
Dirscherl, W.: Katalytische Hydrierung von Follicular-hormon und seiner Acetyl-derivate. Hoppe-Seylers Z. Physiol. Chem. **239**, 53—66 (1936).
— u. H. Breuer: Über das Verhalten von Steroiden bei der Schwefelsäure-Wasser- und Schwefelsäure-Alkohol-Farbreaktion. Z. Vitamin-, Hormon- u. Fermentforsch. **6**, 287—294 (1954).
— and E. Gerhards: Paper chromatography of steroids. Acta endocr. (Kbh.) **19**, 233—243 (1955).
— W. Korus u. H. Schriefers: Trennung von Steroiden durch Verteilungschromatographie an Cellulosesäulen. Hoppe-Seylers Z. Physiol. Chem. Z. **305**, 116—122 (1956).
Djerassi, C.: Oppenauer Oxidation. Org. React. **6**, 207—272 (1951).
— Synthesis of cortisone and related steroids. Vitam. and Horm. **11**, 205—238 (1953).
— Structure et dispersion rotatoire. Bull. Soc. chim. France **1957**, 741—750.
— and W. Klyne: Recording and nomenclature of optical rotatory dispersion. Proc. chem. Soc., Lond. **1957**, 55—57.
— and C. T. Lenk: Reactions of enol acetates with N-iodosuccinimide. J. Amer. chem. Soc. **75**, 3493—3495 (1953).
— G. Rosenkranz, J. Romo, J. Pataki and S. Kaufmann: The dienone-phenol rearrangement in the steroid series. J. Amer. chem. Soc. **72**, 4540—4544 (1950).
— — J. Iriarte, J. Berlin and J. Romo: Aromatization experiments in the progesterone series. J. Amer. chem. Soc. **73**, 1523—1527 (1951).
— L. Miramontes and G. Rosenkranz: 19-Norprogesterone, a potent progestational hormone. J. Amer. chem. Soc. **75**, 4440—4442 (1953).
— — — and F. Sondheimer: Synthesis of 19-nor-17α-ethinyltestosterone and 19-nor-17α-methyltestosterone. J. Amer. chem. Soc. **76**, 4092—4094 (1954).

DJERASSI, C., E. W. FOLTZ and A. E. LIPPMAN: Optical rotatory dispersion studies in the androstane series. J. Amer. chem. Soc. **77**, 4354—4359 (1955a).
— L. B. HIGH, T. T. GROSSNICKLE, R. EHRLICH, J. A. MOORE and R. B. SCOTT: The configuration of the hydroxyl groups in gitogenin. Chem. and Ind. **1955**b, 474—476.
— R. ENGLE and A. BOWERS: The direct conversion of steroidal Δ^5-3β-alcohols to Δ^5 and Δ^4-3-ketones. J. org. Chem. **21**, 1547—1549 (1956a).
— A. LIPPMAN and J. GROSSMAN: 1-Methyl-19-norprogesterone and 1-methyl-19-nor-17α-hydroxyprogesterone. J. Amer. chem. Soc. **78**, 2479—2481 (1956b).
— R. RINIKER and B. RINIKER: Application of optical rotatory dispersion to problems of absolute configuration. J. Amer. chem. Soc. **78**, 6362—6377 (1956c).
— A. MANSON and H. BENDAS: The synthesis of 8-iso-progesterone and 8-iso-testosterone. Tetrahedron **1**, 22—27 (1957).
DOBRINER, K., S. LIEBERMAN and C. P. RHOADS: Studies in steroid metabolism. I. Methods for the isolation and quantitative estimation of neutral steroids present in human urine. J. biol. Chem. **172**, 241—261 (1948a).
— — — R. N. JONES, V. Z. WILLIAMS and R. B. BARNES: III. The application of infrared spectrometry to the fractionation of urinary ketosteroids. J. biol. Chem. **172**, 291—311 (1948b).
— E. R. KATZENELLENBOGEN and R. N. JONES: Infrared absorption spectra of steroids An Atlas. New York: Interscience 1953.
DOBSON, F., and S. S. RANDALL: An improved continuous extractor for liquids. Biochem. J. **49**, 399—400 (1951).
DODSON, R. M., P. SOLLMAN and B. RIEGEL: The preparation of D-homoprogesterone and D-homo-11-desoxycorticosterone. J. Amer. chem. Soc. **75**, 5132 (1953).
— A. H. GOLDKAMP and R. D. MUIR: Microbiological transformation of C_{19} steroids at C-1 and C-2. J. Amer. chem. Soc. **79**, 3921 (1957).
DORFMAN, L.: Ultraviolet absorption of steroids. Chem. Rev. **53**, 47—144 (1953).
DRAKE, B.: Theory of gradient elution analysis. Ark. Kemi **8**, 1—21 (1955).
DRAKE, N. A., W. J. HAINES, R. E. KNAUFF and E. D. NIELSON: Ultraviolet scanner-camera for paper chromatography. Analyt. Chem. **28**, 2036—2037 (1956).
DREIDING, A. C., W. J. PUMMER and A. J. TOMASEWSKI: The dienone-phenol rearrangement with mineral acids. J. Amer. chem. Soc. **75**, 3159—3161 (1953).
DREKTER, I. J., A. HEISLER, R. G. SCISM, S. STERN, S. PEARSON and T. H. MCGAVACK: The determination of urinary steroids. I. The preparation of pigment-free extracts and a simplified procedure for the determination of total 17-ketosteroids. J. clin. Endocrin. **12**, 55—65 (1952).
DULIN, W. E., B. J. BOWMAN and R. O. STAFFORD: Effect of 2-methylation on glucocorticoid activity of various C-21 steroids. Proc. Soc. exp. Biol. (N. Y.) **94**, 303—305 (1957).
DÜNNENBERGER, M., and M. EHRENSTEIN: The synthesis of 19-hydroxy-4-androstene-3,17-dione. J. org. Chem. **21**, 774—782 (1956).
DUTLER, H., H. BOSSHARD u. O. JEGER: Lichtkatalysierte Dienon-Phenol Umlagerung. Helv. chim. Acta **40**, 494—498 (1957).
DYRENFURTH, I., and E. H. VENNING: Preparation of urinary extracts for the aldosterone assay: Hydrolysis and extraction. Endocrinology **60**, 136—143 (1957).
EBERLEIN, W. R.: A compact apparatus for ascending paper chromatography. J. clin. Endocrin. **17**, 1476—1478 (1957).
— and A. M. BONGIOVANNI: New solvent systems for the resolution of corticosteroids by paper chromatography. Arch. Biochem. Biophys. **59**, 90—96 (1955).
EDWARDS, R. W. H., and A. E. KELLIE: Micro-estimation of "formalhydogenic steroids" by bismuthate oxydation. Biochem. J. **56**, 207—211 (1954).
— — Chromatographic separation of the glucuronides of urinary steroid ketones. Chem. and Ind. **1956**, 250—251.
— — and A. P. WADE: The extraction and oxidation of urinary steroid conjugates. Mem. Soc. Endocr. **2**, 53—63 (1953).
EHRENSTEIN, M., and T. STEVENS: New method of preparing 6(α)-acetoxyprogesterone. J. org. Chem. **6**, 908—919 (1941).
EIDINOFF, M. L., J. E. KNOLL, D. K. FUKUSHIMA and T. F. GALLAGHER: Isotopic fractionation factors in a catalytic hydrogenation. J. Amer. chem. Soc. **74**, 5280—5284 (1952).
EIK-NES, K.: Determination of 17,21-dihydroxy-20-ketosteroids in blood plasma. J. clin. Endocrin. **17**, 502—511 (1957).
— D. H. NELSON and L. T. SAMUELS: Determination of 17,21-hydroxycorticosteroids in plasma. J. clin. Endocrin. **13**, 1280—1282 (1953).
ELLIS, B., F. HARTLEY, J. PETROW and D. WEDLAKE: Hydroxylation of pregn-16-en-20-ones. J. chem. Soc. **1955**, 4383—4388.

ELY, R. S., E. R. HUGHES and V. C. KELLEY: Studies of adrenal corticosteroids I. Estimation of plasma corticosterone and cortisol. J. clin. Endocrin. **18**, 190—207 (1958).
ENGEL, C.: The synthesis of 11-dehydro-17α-methylcorticosterone. J. Amer. chem. Soc. **78**, 4727—4733 (1956).
— The synthesis of 17α-methylcorticosterone. Canad. J. Chem. **35**, 131—140 (1957).
— and G. JUST: The synthesis of 17α-methyl desoxycorticosterone. J. Amer. chem. Soc. **76**, 4909—4914 (1954).
ENGEL, L. L.: The chemical estimation of steroid hormone metabolites. Recent Progr. Hormone Res. **5**, 335—379 (1950).
— The assay of urinary neutral 17-ketosteroids. In: D. GLICK, Ed., Methods of Biochemical Analysis **1**, 479—509. New York: Interscience 1954.
— H. R. PATTERSON and H. WILSON: The colorimetric estimation of urinary neutral steroid alcohols. J. biol. Chem. **183**, 47—61 (1950).
— J. ALEXANDER, P. CARTER, J. ELLIOTT and M. WEBSTER: Relations between phase composition and partition coefficients for some neutral steroids. A nomographic approach. Analyt. Chem. **26**, 639—641 (1954).
— P. CARTER and L. L. FIELDING: Urinary metabolites of administered corticosterone. I. Steroids liberated by glucuronidase hydrolysis. J. biol. Chem. **213**, 99—106 (1955).
ENGSTROM, W. W., and H. L. MASON: A study of the colorimetric assay of urinary 17-ketosteroids. Endocrinology **33**, 229—236 (1943).
EPPSTEIN, S. H., P. D. MEISTER, H. C. MURRAY and D. H. PETERSON: Microbiological transformations of steroids and their applications to the synthesis of hormones. Vitam. and Horm. **14**, 359—432 (1956).
EPSTEIN, E., W. O. MADDOCK and A. J. BOYLE: p-Toluenesulfonic acid spot plate test for steroids. Analyt. Chem. **29**, 1548—1549 (1957).
EUW, J. v., u. T. REICHSTEIN: Über Bestandteile der Nebennierenrinde und verwandte Stoffe ,,Substanz V" und Konfigurationsbestimmungen in der $C_{21}O_5$ Gruppe. Helv. chim. Acta **25**, 988—1022 (1942).
— — Konfiguration der Cortico-Steroide. Helv. chim. acta **30**, 205—217 (1947).
— A. LARDON u. T. REICHSTEIN: Pregnandiol-(3α,11α)-on-(20) und Pregnandiol-(3β,11β)-on-(20). Helv. chim. Acta **27**, 821—839 (1944).
— R. NEHER u. T. REICHSTEIN: 11-Oxy-*d*-progesteron-18-säurelacton (18 → 11) und Teilsynthese von d-Aldosteron aus d-Corticosteron-18-säurelacton. Helv. chim. Acta **83**, (1423—1437 (1955).
— C. MEYSTRE, R. NEHER, T. REICHSTEIN u. A. WETTSTEIN: Isolierung von Substanz Y und 3α, 17,21-Trihydroxy-5α-pregnan-11,20-dion (,,11-Dehydro-C") aus Nebennierenrinde. Helv. chim. Acta **41**, 1516—1548 (1958).
EVANS, D. D., and C. W. SHOPPEE: Solvolysis of *epi* cholesteryl toluene-*p*-sulphonate and other steroid toluene-*p*-sulphonates. J. chem. Soc. **1953**, 540—547.
FIESER, L. F.: Partial dichromate oxidation [of cholesterol] to Δ4-cholestene-6β-ol-3-one. J. Amer. chem. Soc. **75**, 4377—4395 (1953a).
— Cholesterol and companions, III. Cholestanol, lathosterol and ketone 104. J. Amer. chem. Soc. **75**, 4395—4403 (1953b).
— Experiments in organic chemistry. Boston: Heath 1955.
— and M. FIESER: Cortical steroids; configurations at C-20 relative to C-17. Experientia (Basel) **4**, 285—295 (1948).
— — Natural products related to phenanthrene, Ed. 3. New York: Reinhold 1949.
— — Steroids. New York: Reinhold 1959.
— and S. RAJAGOPALAN: Selective oxidations and acylations in the bile acid series. J. Amer. chem. Soc. **72**, 5530—5536 (1950).
— and M. ROMERO: Rearrangement of 6β-bromo-Δ^4-cholestene-3-one to 2α-acetoxy-Δ^4-cholestene-3-one. J. Amer. chem. Soc. **75**, 4716—4719 (1953).
— M. FIELDS and S. LIEBERMAN: Concerning the characterization of possible cortical hormone metabolites in urine. J. biol. Chem. **156**, 191—201 (1944).
FILLERUP, D. L., and J. F. MEAD: Chromatographic separation of plasma lipides. Proc. Soc. exp. Biol. (N. Y.) **83**, 574—577 (1953).
FINKELSTEIN, M., and S. GOLDBERG: A test for qualitative and quantitative estimation of pregnane-3α,17α,20α-triol-11-one in urine, and its significance in adrenal disturbances. J. clin. Endocrin. **17**, 1063—1070 (1957).
FLOREY, K., and M. EHRENSTEIN: 6-Hydroxy derivatives of 17α-hydroxyprogesterone and Reichstein's compound S. Studies of 6β,11α-dihydroxyprogesterone. J. org. Chem. **19**, 1331—1350 (1954).
FOLTZ, E. W., A. LIPPMAN and C. DJERASSI: Optical rotatory dispersion studies on steroid hormones. J. Amer. chem. Soc. **77**, 4359—4364 (1955).

FONKEN, G. S., J. L. THOMPSON and R. H. LEVIN: The oxidation of 3-hydroxysteroids with *t*-butylhypochlorite. J. Amer. chem. Soc. **77**, 172 (1955).

FRAME, E. G.: A micro method for the separation of 17-ketosteroids into alpha and beta fractions. Endocrinology **34**, 175—180 (1944).

FRASER, R. W., A. P. FORBES, F. ALBRIGHT, H. SULKOWITCH and E. C. REIFENSTEIN, JR.: Colorimetric assay of 17-ketosteroids in urine. J. clin. Endocrin. **1**, 234—256 (1941).

FRIED, J., and E. SABO: Synthesis of 17α-hydroxycorticosterone and its 9α-halo derivatives from 11-epi-17α-hydroxycorticosterone. J. Amer. chem. Soc. **75**, 2273—2274 (1953).

—, — 9α-Fluoroderivatives of cortisone and hydrocortisone. J. Amer. chem. Soc. **76**, 1455—1456 (1954).

— — 9α-Halogen derivatives of cortisone and hydrocortisone. J. Amer. chem. Soc. **79**, 1130—1141 (1957).

— R. W. THOMA and A. KLINGSBERG: Oxidation of steroids by microorganisms. Side chain degradation, ring D cleavage, and dehydrogenation in ring A. J. Amer. chem. Soc. **75**, 5764—5765 (1953).

— K. FLOREY, E. SABO, J. HERZ, A. RESTIVO, A. BORMAN and F. SINGER: Synthesis and biological activity of 1- and 6-dehydro-9α-halocorticoids. J. Amer. chem. Soc. **77**, 4181—4182 (1955a).

— J. HERZ, E. SABO, A. BORMAN, F. SINGER and P. NUMEROF: 9α-Halo-11β-hydroxy- and 11-keto derivatives of progesterone, desoxycorticosterone and 17α-hydroxyprogesterone. J. Amer. chem. Soc. **77**, 1068—1069 (1955b).

— R. W. THOMA, D. PERLMAN, J. E. HERZ and A. BORMAN: The use of microorganisms in the synthesis of steroid hormones and hormone analogs. Recent Prog. Hormone Research **11**, 149—181 (1955c).

— A. BORMAN, W. B. KESSLER, P. GRABOWICH and E. F. SABO: Cyclic 16α,17α-ketals and acetals of 9α-fluoro-16α-hydroxycortisol and -prednisolone. J. Amer. chem. Soc. **80**, 2338—2339 (1958).

FRIEDGOOD, H. B., and R. A. BERMAN: Effect of water on the accuracy of *m*-dinitrobenzene reaction in the quantitative assay of urinary ketosteroids. Endocrinology **28**, 248—256 (1941).

— and H. L. WHIDDEN: Effect of water on the *m*-dinitrobenzene reaction in the determination of crystalline androgens by Evelyn photoelectric colorimeter. Endocrinology **28**, 237—247 (1941).

— E. H. TAYLOR and M. L. WRIGHT: Combined versus independent hydrolysis and extraction of urinary 17-ketosteroids, with special reference to the choice of solvents. J. clin. Endocrin. **3**, 638—647 (1943).

FUJIMOTO, G. I.: Labelling of steroids in the 4-position. J. Amer. chem. Soc. **73**, 1856 (1951).

FUKUSHIMA, D. K., and T. F. GALLAGHER: Studies on platinum-catalyzed exchange of hydrogen isotopes with steroids. J. biol. Chem. **198**, 871—884 (1952).

— — Identification of D-homo steroids in extracts of urine. J. biol. Chem. **220**, 951—956 (1956).

— — Metabolism of 21-deoxyhydrocortisone-4-C^{14} in man; 17-dehydroxylation, an artifact. J. biol. Chem. **226**, 725—733 (1957).

— T. H. KRITCHEVSKY, M. L. EIDINOFF and T. F. GALLAGHER: Synthesis of radioactive cortisone. J. Amer. chem. Soc. **74**, 487—490 (1952).

— S. DOBRINER, M. S. HAFFLER, T. H. KRITCHEVSKY, F. HERLING and G. ROBERTS: The D-homoannulation of 17α-hydroxy-20-ketosteroids. J. Amer. chem. Soc. **77**, 6585—6591 (1955).

— N. S. LEEDS, H. L. BRADLOW, T. H. KRITCHEVSKY, M. B. STOKEM and T. F. GALLAGHER: The characterization of four new metabolites of adrenocortical hormones. J. biol. chem. **212**, 449—460 (1956).

FULLER, R. C.: Modified end-window counting tube for paper chromatograms. Science **124**, 1253 (1956).

GALLAGHER, T. F.: The structure of the "3α,11-dihydroxy-12-ketocholanic acid" of Marker and Lawson and of the products obtained by Wolff-Kishner reduction. J. biol. Chem. **162**, 539—548 (1946).

— and T. H. KRITCHEVSKY: Perbenzoic acid oxidation of 20-keto steroids and the stereochemistry at C-17. J. Amer. chem. Soc. **72**, 882—885 (1950).

GARDNER, L. I.: Plasma neutral 17-ketosteroids. I. Techniques of estimation. J. clin. Endocrin. **13**, 941—947 (1953).

GAYLORD, N. G.: Reduction with Complex Metal Hydrides. New York: Interscience 1956.

GEMZELL, C. A.: Methods of estimating corticosteroids in plasma. Acta endocr. (Kbh.) **18**, 324—350 (1955).

GIBSON, J. G., 2nd., and K. A. EVELYN: Clinical studies of the blood volume. IV. Adaptation of the method to the photoelectric microcolorimeter. J. clin. Invest. **17**, 153—158 (1938).

GLASCOCK, R. F.: Isotopic gas analysis for biochemists. New York: Academic Press 1954.
GLENN, E. M., and E. HEFTMANN: Urinary excretion of neutral 17-ketosteroids by normal dogs. Proc. Soc. exp. Biol. (N. Y.) **77**, 147—149 (1951).
— and D. H. NELSON: Chemical method for the determination of 17-hydroxycorticosteroids and 17-ketosteroids in urine following hydrolysis with β-glucuronidase. J. clin. Endocrin. **13**, 911—921 (1953).
GOFMAN, J. W., F. T. LINDGREN and H. A. ELLIOTT: Ultracentrifugal studies of lipoproteins of human serum. J. biol. Chem. **179**, 973—979 (1949).
GOLD, J. J.: Blood corticoids: Their Measurement and significance — a review. J. clin. Endocrin. **17**, 296—316 (1957).
GORDY, W.: Microwave and radiofrequency spectroscopy. In: W. WEST, Ed., Chemical applications of spectroscopy(Technique of organic chemistry, Vol.IX), 71—185. New York: Interscience 1956.
GORNALL, A. G., and C. GWILLIAM: Studies on aldosterone. I. Isolation from adrenal cortical extract and a test for its identity. Canad. J. Biochem. **35**, 71—78 (1957).
— and P. M. MACDONALD: Quantitative determination of the steroid hormones with 2,4-dinitrophenylhydrazine. J. biol. Chem. **202**, 279—297 (1953).
GOULD, D., L. FINCKENOR, E. B. HERSHBERG, J. CASSIDY and P. I. PERLMAN: Long-acting testosterone esters. Some considerations on their biological utilization. J. Amer. chem. Soc. **79**, 4472—4475 (1957a).
— E. L. SHAPIRO, H. L. HERZOG, M. J. GENTLES, E. B. HERSHBERG, W. CHARNEY, M. GILMORE, S. TOLKSDORF, M. EISLER, P. L. PERLMAN and M. M. PECHET: Some new active corticosteroids. J. Amer. chem. Soc. **79**, 502—503 (1957b).
GRAFF, M. M., J. T. MCELROY and A. L. MOONEY: Chromogens produced by certain steroids with the anthrone reagent. J. biol. Chem. **195**, 351—356 (1952a).
— J. J. HOLECHEK and A. R. COLLINS: Determination of dehydroisoandrosterone and 17-ketosteroids. J. biol. chem. **197**, 741—749 (1952b).
GREENSPAN, G., C. P. SCHAFFNER, W. CHARNEY, H. L. HERZOG and E. B. HERSHBERG: Transformations of steroids by fungi. Indroduction of 1- and 2β-hydroxyl groups into Reichstein's Substance S. J. Amer. chem. Soc. **79**, 3922—3923 (1957).
GREUTER, F., J. KALVODA and O. JEGER: Thermal decomposition of 21-diazo-5α-pregnan-20-one. Proc. chem. Soc. 349 (1958).
GUT, M.: 3β-Hydroxypregn-4-en-20-one. J. org. Chem. **21**, 1327—1328 (1956).
GUTOWSKY, H. S.: Nuclear magnetic resonance. In: W. G. BERL, Ed., Physical methods in chemical analysis, Vol. III, 303—381. New York: Academic Press 1956.
HADD, H. E., and W. H. PERLOFF: An improved all-glass extractor. J. clin. Endocrin. **17**, 673—675 (1957).
HADLER, H. I.: The steric course of hydrogenation in the steroids. Experientia (Basel) **11**, 175—177 (1955).
HAINES, W. J.: Aspects of steroid hormone chemistry and physiology. Studies on the biosynthesis of adrenal cortex hormones. Recent Progr. Hormone. Res. **7**, 255—299 (1952).
— and J. N. KARNEMAAT: Chromatographic separation of the steroids of the adrenal gland. In: D. GLICK, Ed., Methods of biochemical analysis, Vol. I, 171—204. New York: Interscience Publishers 1954.
HAM, E. A., R. E. HARMAN, N. G. BRINK and L. H. SARETT: Studies on the chemistry of aldosterone. J. Amer. chem. Soc. **77**, 1637—1640 (1955).
HAMILTON, J. G., JR., and R. T. HOLMAN: Displacement analysis of lipides. VIII. Preliminary studies with steroids. Arch. Biochem. Biophys. **36**, 456—461 (1952).
HANSEN, L. P., A. CANTAROW, A. E. RAKOFF and K. E. PASCHKIS: Studies of the Zimmermann reaction: Factors affecting the color intensity: Relation of molecular structure to color production. Endocrinology **33**, 282—288 (1943).
HARMAN, R. E., E. A. HAM, J. J. DE YOUNG, N. G. BRINK and L. H. SARETT: Isolation of aldosterone (electrocortin). J. Amer. chem. Soc. **76**, 5035—5036 (1954).
HARWOOD, C. T., and J. W. MASON: A systematic evaluation of the Nelson-Samuels plasma 17-hydroxy-corticosteroid method. J. clin. Endocrin. **16**, 790—800 (1956).
HASLAM, R., and W. KLYNE: The precipitation of 3β-hydroxysteroids by digitonin. Biochem. J. **55**, 340—346 (1953).
HAYDEN, A. L.: Infrared spectra of adrenocortical hormones in potassium bromide disks. Analyt. Chem. **27**, 1486—1489 (1955).
HEARD, R. D. H.: Chemistry and metabolism of the adrenal cortical hormones. In: G. PINCUS and K. V. THIMANN, Eds., The hormones, Vol. 1, 549—629. New York: Academic Press. 1948.
— und H. SOBEL: Steroids. VIII. A colorimetric method for the estimation of reducing steroids. J. biol. Chem. **165**, 687—698 (1946).
— — and E. H. VENNING: The neutral lipide-soluble reducing substances of urine as an index of adrenal cortical function. J. biol. Chem. **165**, 699—710 (1946).

HECHTER, O., A. ZAFFARONI, R. P. JACOBSEN, H. LEVY, R. W. JEANLOZ, V. SCHENKER and G. PINCUS: The nature and the biogenesis of the adrenal secretory product. Recent Progr. Hormone Res. **6**, 215—241 (1951).

HECKER, E.: Verteilungsverfahren im Laboratorium. Monographien zur Angewandten Chemie und Chemie-Ingenieur-Technik. Weinheim/Bergstr.: Verlag Chemie GmbH. 1955.

HEER, J., u. A. WETTSTEIN: Eine vereinfachte Darstellung von $\Delta^{20,22;\,23}$-21-Bromocholadienen. Helv. chim Acta **36**, 891—895 (1953).

HEFTMANN, E.: Partition chromatography of steroids. Chem. Rev. **55**, 679—711 (1955).

— A simplified mathematical treatment of the theory of partition chromatography. Chromatog. Methods **2**, 5—8 (1957).

— and D. F. JOHNSON: Modified Hershberg-Wolfe extractor. Analyt. Chem. **25**, 1578 (1953).

— — Determination of adrenocortical steroids in mixtures. Analyt. Chem. **26**, 519—521 (1954).

— and E. MOSETTIG: Biochemistry of Steroids. New York: Reinhold 1960.

HEGEDÜS, H., C. TAMM u. T. REICHSTEIN: Die Glykoside von *Strophanthus intermedius* Pax 2. Trennung des Kristallisats Nr. 790. Helv. chim. Acta **36**, 357—369 (1953).

HEGNER, P., u. T. REICHSTEIN: Über Bestandteile der Nebennierenrinde und verwandte Stoffe. 11-Dehydro-progesteron. Helv. chim. Acta **26**, 715—721 (1943).

HEIGL, J. J., M. P. BELL and J. V. WHITE: Application of infrared spectroscopy to the analysis of liquid hydrocarbons. Analyt. Chem. **19**, 293—298 (1947).

HENBEST, H. B., G. D. MEAKINS and G. W. WOOD: Infrared absorption of disubstituted ethylenic centres. J. chem. Soc. **1954**, 800—805.

— E. R. H. JONES, A. A. WAGLAND and T. I. WRIGLEY: Bromination of 11-oxo-5α-steroids. J. chem. Soc. **1955**, 2477—2481.

HENLY, A. A.: Estimation of corticosteroids. Nature (Lond). **169**, 877 (1952).

— and M. POTTER: The estimation of formaldehydogenic steroids. Lancet **262**, 697—698 (1952).

HENRY, R.: Acquisitions récentes relative aux dosages et à l'identification des 17 céto-stéroïdes et des corticoïdes. Rec. Trav. chim. Pays-Bas. **74**, 442—466 (1955).

— M. THEVENET et P. JARRIGE: Etude de l'action hydrolysante de la glucuronidase et de la phénolsulfatase du suc digestif d'*Helix pomatia* L. sur les stéroïdes conjugués urinaires. Bull. Soc. Chim. biol. (Paris) **34**, 897—899 (1952).

HERR, M. E., and F. W. HEYL: "Enamine" derivatives of steroidal carbonyl compounds. J. Amer. chem. Soc. **74**, 3627—3630 (1952).

— — The synthesis of C_{11}-oxygenated testosterones. J. Amer. chem. Soc. **75**, 5927—5930 (1953).

HERRMANN, L. G., A. AMES and R. J. TAPKE: The lipocrit method for the determination of the lipoid content of the blood. J. Lab. clin. Med. **19**, 411—421 (1934).

HERSHBERG, E. B., and J. K. WOLFE: A rapid extractor for urinary steroids. J. biol. Chem. **133**, 667—673 (1940).

— — A rapid extractor for urinary steroids. II. Modifications for the simultaneous hydrolysis and extraction of urine with any solvent heavier than water. J. biol. Chem. **141**, 215—218 (1941).

— — and L. F. FIESER: Polarographic determination of dehydroisoandrosterone and other 3-hydroxy-Δ^5-steroids. J. biol. Chem. **140**, 215—232 (1941).

— M. RUBIN and E. SCHWENK: Synthesis of estrone from androstadienedione. J. org. Chem. **15**, 292—299 (1950).

— C. GEROLD and E. P. OLIVETO: Simultaneous oxidation and bromination of 21-acetoxy-pregnan-3α,17α-diol-11,20-dione. J. Amer. chem. Soc. **74**, 3849—3852 (1952).

HERZ, J., JOSEF FRIED, P. GRABOWICH and E. SABO: 21-Fluoro derivatives of 9α-fluoro- and 1-dehydrocorticosteroids. J. Amer. chem. Soc. **78**, 4812—4814 (1956a).

— — and E. SABO: 12α-Halo derivatives of 11β-hydroxyprogesterone. J. Amer. chem. Soc. **78**, 2017 (1956b).

HERZIG, P. T., and M. EHRENSTEIN: Studies on 6β-21-diacetoxyallopregnane-5-ol-3,20-dione. J. org. Chem. **15**, 1050—1063 (1951).

HERZOG, H. L., M. JEVNIK and E. B. HERSHBERG: The reduction of 11-carbonyl to 11α-hydroxyl in the etiocholane series. J. Amer. chem. Soc. **75**, 269—272 (1953).

— A. NOBILE, S. TOLKSDORF, W. CHARNEY, E. B. HERSHBERG, P. L. PERLMAN and M. M. PECHET: New antiarthritic steroids. Science **121**, 176 (1955a).

— C. C. PAYNE, M. A. JEVNIK, D. GOULD, F. L. SHAPIRO, E. P. OLIVETO and E. B. HERSHBERG: Synthesis and proof of structure of $\Delta^{1,4}$-pregnadiene-17α,21-diol-3,11,20-trione and $\Delta^{1,4}$-pregnadiene-11β,17α,21-triol-3,20-dione. J. Amer. chem. Soc. **77**, 4781—4784 (1955b).

— M. J. GENTLES, E. B. HERSBERG, F. CARVAJAL, D. SUTTER, W. CHARNEY and C. P. SCHAFFNER: Microbiological transformation of steroids. 2β-Hydroxylation. J. Amer. chem. Soc. **79**, 3921—3922 (1957a).

Herzog, H. L., C. C. Joyner, M. J. Gentles, M. T. Hughes, E. P. Oliveto, E. B. Hershberg and D. H. R. Barton: Wagner-Meerwein rearrangement of some 17α-hydroxysteroids. J. org. Chem. **22**, 1413—1417 (1957b).
Heusler, K., u. A. Wettstein: Herstellung und Reaktionen von 16-Oxy-pregnan-derivaten. Chem. Ber. **87**, 1301—1312 (1954).
Heusser, H., C. Engel, P. Herzig u. P. Plattner: Über 17-Methylprogesteron, ein hochaktives Gestagen, und seine Bereitung aus 21- bzw. 17-Halogenpregnenolon. Helv. chim. Acta **33**, 2229—2236 (1950a).
— M. Feuer, K. Eichenberger u. V. Prelog: Über 16,17α-Oxido-Verbindungen der Oestran- und Androstan-Reihe; ein Beitrag zur Stereochemie der isomeren Oestriole. Helv. chim. Acta **33**, 2243—2250 (1950b).
— P. T. Herzig, A. Fürst u. P. Plattner: D-Homo-dehydro-epi-androsteron und eine neue Synthese des D-Homo-testosterons. Helv. chim. Acta **33**, 1093—1105 (1950c).
— N. Wahba u. F. Winternitz: 17aα-Methyl-D-homotestosteron, ein hoch-aktives männliches Sexualhormon. Helv. chim. Acta **37**, 1052—1059 (1954).
Hill, B. R.: Extraction of neutral steroids for 17-ketosteroid determinations. Arch. Biochem. Biophys. **40**, 478 (1952).
Hilmer, P. E., and W. C. Hess: Spectrophotometric determination of androsterone and testosterone. Analyt. Chem. **21**, 822—823 (1949).
Hirschmann, H.: The Characterization of trisubstituted steroidal olefines by infrared spectroscopy. Ciba Found. Coll. Endocrin. **7**, 110—126 (1953).
— Chemistry of steroid hormones. In: G. Pincus and K. V. Thimann, Eds., The Hormones, Vol. III, 521—588. New York: Academic Press 1955.
— and F. B. Hirschmann: Steroid excretion in a case of adrenocortical carcinoma. J. biol. Chem. **167**, 7—25 (1947).
— — Steroid excretion in a case of adrenocortical carcinoma. Δ^5-Prenenetriol-3β,16α,20α. J. biol. Chem. **184**, 259—282 (1950).
— — and M. Daus: The configuration of the allopregnanetriol-3,16,20 of the urine of pregnant mares.. J. biol. Chem. **178**, 751—774 (1949).
— — — A partial synthesis of Δ^5-pregnenetriol-3β,16α,20α and of other16α-hydroxysteroids. J. Amer. chem. Soc. **74**, 539—542 (1952).
Hirschmann, R. F., R. Miller, R. E. Beyler, L. H. Sarett and M. Tishler: A new biologically potent steroid: 1-Dehydro-9α-fluorohydrocortisone acetate. J. Amer. chem. Soc. **77**, 3166—3167 (1955).
— G. Bailey, G. Poos, R. Walker and J. Chemerda: A contribution to the chemistry of steroid 17,21-oxides. J. Amer. chem. Soc. **78**, 4814 (1956a).
— — J. Wood and R. E. Jones: The preparation of 9α-fluoro-4-pregnene-11β, 17α,21-triol-3,20-dione-21-acetate and its 1-dehydro analog. J. Amer. chem. Soc. **78**, 4956—4959 (1956b).
Hoffman, F., R. E. Beyler and M. Tishler: Synthesis of 9α-methylhydrocortisone and 9α-methylprednisolone. J. Amer. chem. Soc. **80**, 5322—5323 (1958).
Hofmann, H., u. H. Staudinger: Papierchromatographische Trennung und Bestimmung der Nebennierenrindenhormone. Biochem. Z. **322**, 230—236 (1951).
Hogg, J. A., P. F. Beal, A. H. Nathan, F. H. Lincoln, W. F. Schneider, B. J. Magerlein, A. R. Hanze and R. W. Jackson: A "direct" synthesis of hydrocortisone acetate and cortisone acetate from 11α-hydroxyprogesterone. J. Amer. chem. Soc. **77**, 4436—4438 (1955a).
— F. H. Lincoln, R. W. Jackson and W. P. Schneider: Synthesis of 2-alkyl adrenal hormone analogs. J. Amer. chem. Soc. **77**, 6401—6402 (1955b).
— — A. H. Nathan, A. R. Hanze, W. P. Schneider, P. F. Beal and J. Korman: Synthesis of 1-dehydrocorticosteroid analogs. J. Amer. chem. Soc. **77**, 4438—4439 (1955c).
— G. B. Spero, J. L. Thompson, B. J. Magerlein, W. P. Schneider, D. H. Peterson, O. K. Sebek, H. C. Murray, J. C. Babcock, R. L. Pederson and J. A. Campbell: 6-Fluoro analogs of steroid hormones. Chem. and Ind. **1958**, 1002—1003.
Holechek, J. J., and A. R. Collins: Determination of 17-hydroxy-11-dehydrocorticosterone with anthrone reagent. Analyt. Chem. **25**, 991—992 (1953).
Hollander, V. P., S. DiMauro and O. H. Pearson: A diffusion method for the estimation of formaldehydrogenic steroids. Endocrinology **49**, 617—623 (1951).
Holtorff, A. G., and F. C. Koch: The colorimetric estimation of 17-ketosteroids and their application to urine extracts. J. biol. Chem. **135**, 377—392 (1940).
Holysz, R. P.: Preparation and dehydrohalogenation of 4-halo-3-keto-steroids. J. Amer. chem. Soc. **75**, 4432—4437 (1953).
Horeau, A., J. Jaques, J. P. Mathieu et A. Petit: A propos de la nomenclature des antipodes optiques dans les composés stéroïdes. Bull. Soc. chim. Fr. **22**, 1304 (1955).

HORWITT, B. N.: Bacterial β-glucuronidase as a reagent for the hydrolysis of urinary cortico steroids. J. Lab. clin. Med. **44**, 478—485 (1954).
HUANG-MINLON: Reduction of steroid ketones and other carbonyl compounds by modified Wolff-Kishner reduction. J. Amer. chem. Soc. **71**, 3301—3303 (1949).
— E. WILSON, N. L. WENDLER and M. TISHLER: Steroid 17α-acetates. J. Amer. chem. Soc. **74**, 5394—5396 (1952).
HUIS IN'T VELD, L.: Les 17-cétostéroïdes neutres dans l'urine. Leur signification pour le diagnostic clinique. Expos. ann. Biochem. med. **18**, 17—46 (1956).
HUNGER, A., u. T. REICHSTEIN: Die Konstitution von Gofrusid und Frugosid. Helv. chim. Acta **35**, 1073—1103 (1952).
HURLOCK, B., and P. TALALAY: Principles of the enzymatic measurement of steroids. J. biol. Chem. **227**, 37—52 (1957).
— — Enzymatic estimation of steroids in human urine. Endocrinology **62**, 201—215 (1958).
IZZO, A. J., E. H. KEUTMANN and R. B. BURTON: Special considerations in the use of Blue Tetrazolium for the measurement of alpha-ketols in mixtures with other steroids. J. clin. Endocrin. **17**, 889—901 (1957).
JANSSON, J. I., and J. KYRKI: Purification of *m*-dinitrobenzene for determination of 17-ketosteroids according to Zimmermann. Acta endocr. (Kbh.) **11**, 317—319 (1952).
JAYLE, M. F., et E. E. BAULIEU: Données expérimentales sur l'hydrolyse des 17-cétostéroïdes conjugués urinaires par les diastases du suc digestif d'*Helix pomatia* (L.). Bull. Soc. chim. biol. (Paris) **36**, 1391—1406 (1954).
— et O. CRÉPY: Technique du dosage du pregnandiol glycuronide (P. G.) et des glycuronides butylosoluble (G. B. S.). Bull. Soc. Chim. biol. (Paris) **34**, 435—443 (1952).
— — L'exploration polysteroïdique. Sem. Hôp. Paris **30**, 77—79 (1954).
JENSEN, C. C., and L. E. TÖTTERMAN: Hydrolysis of urinary neutral 17-ketosteroid conjugates. I. Comparison of various procedures. Acta endocr. (Kbh.) **10**, 221—232 (1952a).
— — II. Fractional hydrolytic procedure. Acta endocr. (Kbh.) **11**, 33—48 (1952b).
JOHNSON, D. F., E. HEFTMANN and A. L. HAYDEN: Determination of individual adrenocortical steroids in urine. Acta endocr. (Kbh.) **23**, 341—357 (1956).
JOHNSON, J. L., M. F. GROSTIC and A. O. JENSEN: Determination of stigmasterol in soybean sterols by infrared spectrophotometry. Analyt. Chem. **29**, 468—470 (1957)
— M. E. HERR, J. C. BABCOCK, A. E. FONKEN, J. E. STAFFORD and F. W. HEYL: "Enamine" derivatives of steroidal carbonyl compounds. Structural considerations. J. Amer. chem. Soc. **78**, 430—436 (1956).
JOHNSTON, C. D.: Identification and chromatography of adrogens as their 2,4-dinitrophenylhydrazones. Science **106**, 91 (1947).
JONES, J. K. N., and S. R. STITCH: The separation of ketosteroids by partition chromatography using nitromethane as stationary phase. Biochem. J. **53**, 679—684 (1953).
JONES, R. E., and S. A. ROBINSON: Derivatives at C-20 of 17-α-hydroxy-20-ketosteroid acylates. J. org. Chem. **21**, 586—587 (1956).
JONES, R. N., and A. R. H. COLE: The characterization of methyl and methylene groups in steroids by infrared spectrophotometry. Correlations of bending frequencies with molecular structure. J. Amer. chem. Soc. **74**, 5648—5661 (1952).
— and K. DOBRINER: Infrared spectrometry applied to steroid structure and metabolism. Vitam. and Horm. **7**, 293—363 (1949).
— and F. HERLING: Characteristic group frequencies in the infrared spectra of steroids. J. org. Chem. **19**, 1252—1268 (1954).
— and C. SANDORFY: The application of infrared and Raman spectrometry to the elucidation of molecular structure. In: W. WEST, Ed., Chemical applications of spectroscopy (Technique of Organic Chemistry, Vol. IX). 247—580. New York: Interscience Publishers 1956.
— P. HUMPHRIES, F. HERLING and K. DOBRINER: Steroid metabolism. X. The effects of stereochemical configuration at positions 3 and 5 on the infrared spectra of 3-acetoxy steroids. J. Amer. chem. Soc. **73**, 3215—3220 (1951).
— A. R. H. COLE and B. NOLIN: Methyl and methylene bending frequencies in steroids labelled with deuterium. J. Amer. chem. Soc. **74**, 5662—5668 (1952a).
— P. HUMPHRIES, F. HERLING and K. DOBRINER: XII. The determination of the structure of the side chains of C-21 steroids by infrared spectrometry. J. Amer. chem. Soc. **74**, 2820—2828 (1952b).
— D. A. RAMSAY, D. S. KEIR and K. DOBRINER: The intensities of carbonyl bands in the infrared spectra of steroids. J. Amer. chem. Soc. **74**, 80—88 (1952c).
— F. HERLING and E. KATZENELLENBOGEN: The infrared spectra of ketosteroids below 1350 cm^{-1}. J. Amer. chem. Soc. **77**, 651—661 (1955).
JULIA, S. A., P. PLATTNER u. H. HEUSSER: Über Derivate des 3-Keto-5-oxy-androstans, eine Gruppe von möglichen Zwischenprodukten in der Biosynthese von androgenen Hormonen. Helv. chim. Acta **35**, 665—672 (1952).

JULIAN, P. L., E. MEYER, W. KARPEL and I. RYDEN: 17α-Hydroxyprogesterone and 17α-hydroxy-11-desoxycorticosterone. J. Amer. chem. Soc. **71**, 3574—3575 (1949).
— — — and I. R. WALLER: 17α-Hydroxy-11-desoxycorticosterone (Reichstein's Substance S). J. Amer. chem. Soc. **72**, 5145—5147 (1950).
— — — and W. COLE: The partial synthesis of 4-pregnen-17α,20,21-triol-3-ones and Reichstein's Substance E. J. Amer. chem. Soc. **73**, 1982—1985 (1951).
— W. COLE, E. W. MEYER and B. M. REGAN: 16α-Hydroxy and 16α,17α-epoxyanalogs of cortisone. J. Amer. chem. Soc. **77**, 4601—4604 (1955).
JUNKMANN, K.: Corticoids of prolonged action (in German). Arch. exp. Path. Pharmak. **223**, 280—284 (1954).
KAHNT, F. W., R. NEHER u. A. WETTSTEIN: Bildung von 18- und 19-Oxycortexon durch Einwirkung von Nebennieren-Homogenaten auf Cortexon. Helv. chim. Acta **38**, 1237—1248 (1955).
KAMEN, M. D.: Isotopic tracers in biology. Ed. 3. New York: Academic Press 1957.
KAPPAS, A., and T. F. GALLAGHER: Studies in steroid metabolism. 28. The α-ketosteroid excretion pattern in normal females and their response to ACTH. J. clin. Invest. **34**, 1566—1572 (1955).
KATZENELLENBOGEN, E. R , K. DOBRINER and T. H. KRITCHEVSKY: Separation of steroid hormones and their metabolites by partition type chromatography J. biol. Chem. **207**, 315—321 (1954).
KAUFMAN, S., J. PATAKI, G. ROSENKRANZ, J. ROMO and C. DJERASSI: The bromination of steroidal 1,4-dien-3-ones. J. Amer. chem. Soc. **72**, 4531—4534 (1950).
KELLIE, A. E.: Determination of 17-ketosteroids and 17-ketogenic steroids. Analyst **82**, 722—734 (1957).
— and E. R. SMITH: Characterization of 17-oxosteroids by derivatives formed in the Zimmermann reaction. Nature (Lond.) **178**, 323 (1956).
— and A. P. WADE: Urinary steroids. II. Chromatography of 3,5-dinitrobenzoates. Application to the nonketonic alcoholic fraction. Biochem. J. **53**, 582—588 (1953).
— — Steroid conjugates. I. The separation of urinary 17-ketosteroid glucuronides and sulphates, and their composition in normal individuals. Acta endocr. (Kbh.) **23**, 357—370 (1956).
— — Analysis of urinary 17-oxosteroids by gradient elution. Biochem. J. **66**, 196—206 (1957).
— E. R. SMITH and A. P. WADE: Urinary steroids. I. Formation and determination of 3,5-: dinitrobenzoates of alcoholic steroids. Biochem. J. **53**, 578—582 (1953).
KEMP, A., A. KAPPAS, I. SALAMON, F. HERLING and T. GALLAGHER: Identification and characterization of 3β-hydroxy-$\Delta^{(11)}$-androstene-17-one and 3β,11β-dihydroxyandrostane-17-one from human urine. J. biol. Chem. **210**, 123—128 (1954).
KESTON, A. S., D. RITTENBERG and R. SCHOENHEIMER: Determination of deuterium in organic compounds. J. biol. Chem. **122**, 227—237 (1937).
KINSELLA, R. A., JR., and J. H. GLICK, JR.: Hydrolysis of conjugates of urinary corticoids with β-glucuronidase I. Neutral reducing lipides. J. biol. Chem. **203**, 1003—1011 (1953).
KIRK, D., D. PATEL and V. PETROW: Some steroidal 1,4-diene-3,11-diones and 1,4,6-triene-3,11-diones. J. chem. Soc. **1957**, 1046—1054.
KISSMAN, H. M., A. M. SMALL and M. J. WEISS: The synthesis of 2α-fluorohydrocortisone. J. Amer. chem. Soc. **81**, 1262 (1959).
KLASS, D. L., M. FIESER and L. F. FIESER: Digitogenin. J. Amer. chem. Soc. **77**, 3829—3833 (1955).
KLEIN, R., C. PAPADATOS, J. FORTUNATO and C. BYERS: Acid-hydrolyzable corticoids of serum. J. clin. Endocrin. **15**, 215—226 (1955).
KLENDSHOJ, N. C., M. FELDSTEIN and A. SPRAGUE: Determination of 17-ketosteroids in urine. J. clin. Endocrin. **13**, 922—927 (1953).
KLYNE, W.: The structure of uranediol. Nature (Lond.) **166**, 559—560 (1950).
— Some aspects of the stereochemistry of C-20. Ciba Found. Coll. Endocrin. **7**, 127—141 (1953).
— The conformations of six-membered ring systems. In: W. KLYNE, Ed., Progress in stereochemistry, Vol. 1, 36—89. London: Butterworths 1954.
— Optical rotation. In: E. A. BRAUDE and F. NACHOD, Eds., Determination of organic structures by physical methods, 73—130. New York: Academic Press 1955.
— The chemistry of the steroids. New York: Wiley 1957.
KNAUFF, R. E., E. D. NEILSON and W. J. HAINES: Studies on a mineralocorticoid from hog adrenal extract. J. Amer. chem. Soc. **75**, 4868—4869 (1953).
KOBER, S.: Eine kolorimetrische Bestimmung des Brunsthormons (Menformon). Biochem. Z. **239**, 209—212 (1931).

KOCHAKIAN, D. C., and G. STIDWORTHY: Paper chromatography of C_{19} steroids. J. biol. Chem. **199**, 607—612 (1952).
KOENIG, V. L., F. MELZER, C. M. SZEGO and L. T. SAMUELS: A colorimetric reaction for testosterone: J. biol. Chem. **141**, 487—493 (1941).
KOLTHOFF, I. M., and J. J. LINGANE: Polarography, Ed. 2 (2 Vols.). New York: Interscience 1952.
KORTUM, G.: Kolorimetrie, Photometrie and Spektrometrie, Band II. Berlin: Springer 1955.
KRITCHEVSKY, D. and M. R. KIRK: Paper chromatography of steroids. J. Amer. chem. Soc. **74**, 4484—4486 (1952a).
— — Detection of steroids in paper chromatography. Arch. Biochem. Biophys. **35**, 346—351 (1952b).
KRITCHEVSKY, T. H., and T. F. GALLAGHER: The preparation of 17α-hydroxyprogesterone and substances L and P. J. Amer. chem. Soc. **73**, 184—189 (1951).
— and A. TISELIUS: Reversed phase partition chromatography of steroids on silicone-treated paper. Science **114**, 299—300 (1951).
— D. GARMAISE and T. H. GALLAGHER: Preparation of cortisone and related compounds. J. Amer. chem. Soc. **74**, 483—486 (1952).
KUČERA, J., Ž. PROCHÁZKA u. K. VERES: Über Steroide XXVI. Die Papierchromatographie neutraler Steroidverbindungen. Coll. Czech. Chem. Comm. **22**, 1185—1191 (1957).
LABROSSE, E. H.: Micro method for the identification of substances on paper chromatograms. Fed. Proc. **12**, 471 (1953).
LAKSHMANAN, T. K., and S. LIEBERMAN: An improved method of gradient elution chromatography and its application to the separation of urinary ketosteroids. Arch. Biochem. Biophys. **53**, 258—281 (1954).
LANDAU, R. L.: Diagnostic significance and laboratory methods in determination of the 17 ketosteroids. Amer. J. clin. Path. **19**, 424—434 (1949).
LARDON, A., u. T. REICHSTEIN: Teilsynthese des 11-dehydro-corticosterons. Helv. chim. Acta **26**, 747—755 (1943).
— — Die Konfiguration der optisch aktiven β-Methoxyadipinsäuren und des Calciferols. Vorläufige Mitteilung. Helv. chim. Acta **32**, 2003—2009 (1949).
— — Partial synthesis of cortisone and related compounds from sarmentogenin (in German). Pharm. Acta Helvet. **27**, 287—302 (1952).
— — Synthesen von 11β-Oxy-corticoiden über die 11β-Formyloxy- und 11β-Trifluoracetoxy-Derivate. Helv. chim. Acta **37**, 443—450 (1954).
— O. SCHINDLER u. T. REICHSTEIN: Eine weitere Synthese von d,l-Aldosterone. Helv. chim. Acta **40**, 666—703 (1957).
LEANZA, W., J. CONBERE, E. F. ROGERS and K. PFISTER: Aldehydes derived from cortisone and hydrocortisone. J. Amer. chem. Soc. **76**, 1691—1694 (1954).
LEDERER, E., and M. LEDERER: Chromatography. A review of principles and applications. New York: Elsevier 1957.
LEDERER, M.: Introduction to paper electrophoresis and related methods. New York: Elsevier 1955.
LEEDS, N. S., D. K. FUKUSHIMA and T. F. GALLAGHER: Synthesis of estriol and androstane-3β,16α,17β-triol. J. Amer. chem. Soc. **76**, 2943—2948 (1954).
LETTRÉ, H. R., TSCHESCHE u. H. FERNHOLZ: Über Sterine, Gallensäuren und verwandte Naturstoffe, 2 Ed. Stuttgart: Enke. Vol. I, 1954; Vol. II, 1958.
LEVIN, B., and G. DAVIES: Separation of steroid conjugates by paper electrophoresis. Nature (Lond.) **178**, 918 (1956).
LEVIN, R. H., B. J. MAGERLEIN, A. V. MCINTOSH, A. R. HANZE, G. S. GONKEN, J. L. THOMPSON, A. M. SEARCY, M. A. SCHERI and E. S. GUTSELL: 17α-Hydroxycorticosterone. J. Amer. Chem. Soc. **76**, 546—552 (1954).
LEVY, H., and S. KUSHINSKY: Methods of isolating corticosteroids from blood. Recent Progr. Hormone Res. **9**, 357—377 (1954).
— — On the mineralocorticoid activity of 17α,19-dihydroxy-11-deoxycorticosterone. Arch. Biochem. **58**, 245—246 (1955).
LEWBART, M. L., and J. J. SCHNEIDER: Paper chromatography of steroid glucuronosides and sulfates. Nature (Lond.) **176**, 1175 (1955).
LEWIS, B.: Preparation of paper for quantitative chromatography of corticosteroids. Biochim. biophys. Acta **20**, 396 (1956).
LIEBERMAN, S., L. B. HARITON and D. K. FUKUSHIMA: Studies in steroid metabolism. V. The problem of Walden inversion in the reactions of steroid hydrogen sulfates and steroid sulfites. J. Amer. chem. Soc. **70**, 1427—1432 (1948a).
— K. DOBRINER, B. R. HILL, L. F. FIESER and C. P. RHOADS: II. Identification and characterization of ketosteroids isolated from urine of healthy and diseased persons. J. biol. Chem. **172**, 263—295 (1948b).

LIEBERMAN, S., D. K. FUKUSHIMA and K. DOBRINER: VII. Identification and characterization of additional ketosteroids isolated from urine of healthy and diseased persons. J. biol. Chem. **182**, 299—316 (1950).
— B. MOND and E. SMYLES: Hydrolysis of urinary ketosteroid conjugates. Recent Progr. Hormone Res. **9**, 113—130 (1954).
LINFORD, J. H.: Experiments on the use of silicic acid for steroid chromatography. Canad. J. Biochem. **34**, 1153—1168 (1956).
— Steroid color reactions in strong acid. Canad. J. Biochem. **35**, 299—325 (1957).
LITTEL, R., and S. BERNSTEIN: The preparation of 9α-hydroxyhydrocortisone. J. Amer. chem. Soc. **78**, 984—989 (1956).
LLOYD, C. W., and J. LOBOTSKY: Studies of urinary corticosteroids by a method permitting analysis of steroids poorly soluble in water. J. clin. Endocrin. **10**, 1559—1569 (1950).
LOGEMAN, W.: Über Ketole der 18α- und 18β-Glycyrrhetinsäure. Chem. Ber. **90**, 601—604 (1957).
LOGEMANN, W., and P. GIRALDI: The corticoid hormones and synthetic "models" of them. Raccolta pubbl. chim. biol. med. **1**, 391 (1953).
LÖKEN, B., S. KAUFMAN, G. ROSENKRANZ and F. SONDHEIMER: Synthesis and reactions of 16β-oxygenated pregnan-20-one derivatives. J. Amer. chem. Soc. **78**, 1738—1741 (1956).
LOMBARDO, M. E., P. H. MANN, T. A. VISCELLI and P. B. HUDSON: A method for the extraction of steroids from blood. J. biol. Chem. **212**, 345—351 (1955a).
— T. A. VISCELLI, A. MITTELMAN and P. B. HUDSON: The removal of non-steroidal pigments of urinary extracts by adsorption on charcoal. J. biol. Chem. **212**, 353 (1955b).
MACHEBOEUF, M. A.: Recherches sur les phosphoaminolipides et les stérides du sérum et du plasma sanguins. I. Entraînement des phospholipides, des stérols et des stérides par les diverses fractions au cours du fractionnement des protéïdes du sérum. Bull. Soc. Chim. biol. (Paris) **11**, 268—293 (1929).
MADAEVA, O. S., and F. A. LUR'I: Some properties of esters of p-toluenesulfonic acid with 17α- and β-hydroxysteroids (in Russian). Doklady Akad. Nauk S.S.S.R. **84**, 713—716 (1952); Chem. Abstr. **47**, 3326 (1953).
MADER, W. J., and R. R. BUCK: Colorimetric determination of cortisone and related ketol steroids. Analyt. Chem. **24**, 666—667 (1952).
MAGERLEIN, B. J., and J. A. HOGG: Preparation and reactions of 11-substituted 1,3,5(10)-estratrienes. Synthesis of 19-norhydrocortisone. J. Amer. chem. Soc. **80**, 2226—2229 (1958).
MAGERLEIN, B. M., and R. LEVIN: 3α,20β-Dihydroxypregnan-11-one 11-ethylene ketal. J. Amer. chem. Soc. **77**, 1904 (1955).
MAMOLI, L., u. A. VERCELLONE: Über die biochemische Dehydrierung in der Reihe der Keimdrüsenhormone: Bakterielle Oxydation von Dehydroandrosteron zu Androstendion. Chem. Ber. **71**, 1686—1687 (1938).
MANARO, J. M., and A. ZYGMUNTOWICZ: Use of osmium tetraoxide as a reagent for assaying corticosteroids. Endocrinology **48**, 114—115 (1951).
MANCERA, O., H. J. RINGOLD, C. DJERASSI, G. ROSENKRANZ and F. SONDHEIMER: A ten step conversion of progesterone to cortisone. The differential reduction of pregnane-3,20-diones with sodium borohydride. J. Amer. chem. Soc. **75**, 1286—1290 (1953).
MANN, K. M., F. R. HANSON, P. W. O'CONNEL, H. V. ANDERSON, M. P. BRUNNER and J. M. KARNEMAAT: The microbiological oxydation of steroids by Cunninghamella blakesleeana. The effect of alcohols and phenol. J. appl. Microbiol. **3**, 14—16 (1955).
MARIANI, A., et C. VICARI: Isolement des stéroides cétoniques et leur fractionnement par ionophorèse. Svensk farm. Tidskr. **60**, 806—809 (1956).
MARKER, R. E.: The conversion of the sapogenins to the pregnenolones. J. Amer. chem. Soc. **62**, 3350—3352 (1940).
— O. KAMM, E. L. WITTLE, T. S. OAKWOOD, E. J. LAWSON and J. F. LAUCIUS: The pregnandiols and pregnanolones. J. Amer. chem. Soc. **59**, 2291—2296 (1937).
MARLOW, H. W.: Groups involved in the Zimmermann and Kober reactions. J. biol. Chem. **183**, 167—171 (1950).
MARRIAN, G. F.: Observations on the determination of urinary formaldehydogenic substances. Recent Progr. Hormone Res. **9**, 303—310 (1954).
— and S. M. ATHERDEN: Urinary adrenocortical steroids. II. Formaldehydogenic and acetaldehydogenic substances in extracts of hot acid-hydrolyzed urine. Acta endocr. (Kbh.) **13**, 265—285 (1953).
— J. Y. F. PATERSON and S. M. ATHERDEN: Acid-labile and acid-stable formaldehydogenic substances in urine. Mem. Soc. Endoc. **2**, 4—8 (1953).
MARTI, M.: Die chromatographische Trennung und Bestimmung der 17-Ketosteroide im Urin. Methodik und klinisch-experimentelle Anwendung. Helv. med. Acta **18**, 215—240 (1951).
MARTIN, R. P.: Reversed phase paper chromatography and detection of steroids of the cholesterol class. Biochim. biophys. Acta **25**, 408—416 (1957).

MASON, H. L.: Isolation of adrenal cortical hormones from urine: 17-Hydroxycorticosterone and 17-hydroxy-11-dehydrocorticosterone. J. biol. Chem. **182**, 131—149 (1950).
— The determination of urinary 17-ketosteroids. J. clin. Endocrin. **13**, 1009—1010 (1953).
— Hydrolysis of conjugates and extraction of urinary corticosteroids. Recent Progr. Hormone Res. **9**, 267—278 (1954).
— and W. W. ENGSTROM: The 17-ketosteroids: Their origin, determination and significance. Physiol. Rev. **30**, 321—374 (1950).
— and E. C. KENDALL: Studies in the chemistry of the adrenal cortex. Conversion of compound E to the $C_{21}O_4$ series and to adrenosterone by the action of calcium hydroxide. Proc. Mayo Clin. **13**, 235—236 (1938).
MASUDA, M., and H. C. THULINE: An improved method for determination of urinary 17-ketosteroids. J. clin. Endocrin. **13**, 581—586 (1953).
MATHIEU, J. P., et A. PETIT: Constantes Sélectionées Pouvoir Rotatoire Naturel. I. Paris: Masson 1956.
MATTOX, V. R.: The formation of a glyoxal side chain from steroids with dihydroxyacetone and Δ^{16}-ketol side chains. J. Amer. chem. Soc. **74**, 4340—4347 (1952).
— Isolation of 19-hydroxy-11-deoxycorticosterone from beef adrenal extract. Proc. Mayo Clin. **30**, 180—182 (1955).
— and E. C. KENDALL: Adrenal cortical hormones; introduction of a double bond. J. biol. Chem. **188**, 287—297 (1951).
— and H. L. MASON: Chromatographic fractionation of beef adrenal extract. I. Isolation of aldosterone. J. biol. Chem. **223**, 215—225 (1956).
— E. L. WOROCH, G. A. FLEISCHER and E. C. KENDALL: Preparation of 6-dehydrocortisone. J. biol. Chem. **197**, 261—270 (1952).
— H. L. MASON, A. ALBERT and C. F. CODE: Properties of a sodium retaining principle from beef adrenal extract. J. Amer. chem. Soc. **75**, 4869—4870 (1953).
MCALEER, W. J., and M. A. KOZLOWSKI: Color test for 17-hydroxy-20-keto, 21-methyl steroids. Arch. Biochem. Biophys. **62**, 196—197 (1956).
— — Application of the I_2—KI color test to some oxygenated progesterones. Arch. Biochem. Biophys. **66**, 125—127 (1957).
MCDONALD, H. J.: Ionography. Electrophoresis in stabilized media. Chicago: Year Book Publ. 1955.
MCGUCKIN, W. F., and E. C. KENDALL: Introduction of the 4,5-double bond of cortisone. J. Amer. chem. Soc. **74**, 5811—5813 (1952).
— and H. L. MASON: Adrenal hormone analogs: 16α,17α-epoxy-11-dehydrocorticosterone acetate and Δ^{16}-11-dehydrocorticosterone. J. Amer. chem. Soc. **77**, 1822—1824 (1955).
MCKINLEY, W. P.: Paper electrophoresis of steroid derivatives. Science **121**, 139—140 (1955).
MCMAHON, J. M., R. B. DAVIS and G. KALNITSKY: Identification of sterols on filter paper and their separation by paper partition chromatography. Proc. Soc. exp. Biol. (N. Y.) **75**, 799—801 (1950).
Medical Research Council Committee on Clinical Endocrinology. Proposed standard method for 17-ketosteroid determination. Lancet **261**, 585—586 (1951).
MEYER, A. S.: Chemical transformations of steroids by adrenal perfusion. V. Blood incubations and perfusions with cortisone. J. biol. Chem. **203**, 469—487 (1953).
— 19-Hydroxylation of 4-androstene-3,17-dione and dehydroepiandrosterone by bovine adrenals. Experientia (Basel) **11**, 99—102 (1955a).
— Light absorption between 230—500 mμ of α,β-unsaturated and 3,6-diketo steroids in alkaline ethanol. J. org. Chem. **20**, 1240—1252 (1955b).
— and M. C. LINDBERG: Optical-analytical studies on steroids. Reducing characteristics of hydroxylated and ketonic steroids toward blue tetrazolium. Analyt. Chem. **27**, 813—817 (1955).
— H. ROSENKRANTZ and G. PINCUS: Optical analytical studies of steroids in the 200 mμ Region. Proc. 126th meeting. Amer. chem. Soc., 50 (1954).
MEYER, L. H., A. SAIKA and H. S. GUTOWSKY: The proton magnetic spectra of simple groups. J. Amer. chem. Soc. **75**, 4567—4573 (1953).
MEYSTRE, C., u. A. WETTSTEIN: Ein einfacher Abbau der Gallensäuren-Seitenkette XIII. Δ^{11}-Anhydro-corticosteron-acetat. Helv. chim. Acta **31**, 1890—1898 (1948).
— H. FREY, A. WETTSTEIN u. K. MIESCHER: Ein einfacher Abbau der Gallensäuren-Seitenkette zur Methylketonstufe. Helv. chim. Acta **27**, 1815—1824 (1944).
— E. TSCHOPP u. A. WETTSTEIN: Über 11-Dehydro-progesteron, das wirksamste Gestagen. Helv. chim. Acta **31**, 1463—1469 (1948).
— H. FREY, W. VOSER u. A. WETTSTEIN: Gewinnung von 1,4-Bisdehydro-3-oxo-steroiden. Helv. chim. Acta **39**, 734—742 (1956).
MIESCHER, K., u. H. KÄGI: Beitrag zur Farbreaktion auf 17α-Oxysteroide. Helv. chim. Acta **32**, 761—769 (1949).

MIESCHER, K., u. J. SCHMIDLIN: Ein einfacher Weg zur Bereitung von Corticosteroiden mit Dioxyaceton-Seitenkette, ausgehend von 17-Ketonen. Neue Teilsynthese von Reichsteins Substanz S. Helv. chim. Acta **33**, 1840—1847 (1950).
MILLER, A. M., and R. I. DORFMAN: Metabolism of the steroid hormones: Isolation of 13 steroid metabolites from a patient with adrenal hyperplasia. Endocrinology **46**, 514—525 (1950).
MILLER, W. R., C. W. TURNER, D. K. FUKUSHIMA and I. I. SALAMON: The identification of C_{19} steroids in bovine feces. J. biol. Chem. **220**, 221—225 (1956).
MOFFET, R. B., and J. H. HUNTER: The Wolff-Kishner reduction of 11-keto steroids. J. Amer. chem. Soc. **73**, 1973 (1951).
MOOLENAAR, A. J.: Colorimetric determination of aldosterone in urine. Acta endocr. (Kbh.) **25**, 161—172 (1957).
MORRIS, C. J. O. R.: The polarographic estimation of ketosteroids. Rec. Trav. chim.Pays-Bas **74**, 476—497 (1955).
— and D. C. WILLIAMS: The polarographic estimation of steroid hormones. VI. Determination of individual adrenocortical hormones in human peripheral blood. Biochem. J. **54**, 470—475 (1953).
MUNSON, P. L., M. E. JONES, P. J. MCCALL and T. F. GALLAGHER: A colorimetric method for estimation of dehydroisoandrosterone and its application to urine extracts. J. biol. Chem. **176**, 73—82 (1948).
— and A. D. KENNY: Colorimetric-analytical methods for neutral 17-ketosteroids of urine. Recent Progr. Hormone Res. **9**, 135—161 (1954).
NACE, H. R.: The reaction of α- and β-cholestanyl p-toluenesulfonates with methanol. J. Amer. chem. Soc. **74**, 5937—5938 (1952).
NADEL, E. M., and F. HIGHHOUSE: Tube-separatory funnel for rapid extraction. Chem. Anal. **43**, 25 (1954).
NATHANSON, I. T., and H. WILSON: Factors affecting colorimetric urinary 17-ketosteroid determinations. Endocrinology **33**, 189—203 (1943).
NAZAROW, I., and L. BERGEL'SON: Chemistry of steroid hormones (in Russian). Academy of Sciences, U.S.S.R. 1955.
NEHER, R., u. A. WETTSTEIN: Farbreaktionen mit Steroiden, insbesondere Corticosteroiden im Papierchromatogramm. Helv. chim. Acta **34**, 2278—2285 (1951).
— — Papierchromatographie von schwach polaren Steroiden. Helv. chim. Acta **35**, 276—283 (1952).
— — Physiochemical estimation of aldosterone in urine. J. clin. Invest. **35**, 800—805 (1956a).
— — Isolierung und Konstitutionsermittlung weiterer Pregnanverbindungen aus Nebennieren. Helv. chim. Acta **39**, 2062—2088 (1956b).
— P. DESAULLES, E. VISCHER, P. WIELAND and A. WETTSTEIN: Isolierung, Konstitution und Synthese eines neuen Steroids aus Nebennieren. Helv. chim. Acta **41**, 1667—1692 (1958).
— C. MEYSTRE u. A. WETTSTEIN: Neue 16α-Hydroxysteroide aus menschlichem Urin und aus Schweine-Nebennieren. Helv. chim. Acta **42**, 132—152 (1959).
NELSON, D. H., and L. T. SAMUELS: A method for the determination of 17-hydroxycorticosteroids in blood: 17-hydroxycorticosterone in peripheral circulation. J. clin. Endocrin. **12**, 519—526 (1952).
NEWHOUSE, J. P.: A continuous liquid-liquid extractor. J. clin. Path. **6**, 195—197 (1953).
NETERVAL, C. R.: A small apparatus for extracting urinary androgens. J. biol. Chem. **133**, 313—317 (1940).
NOBILE, A., W. CHARNEY, P. L. PERLMAN, H. L. HERZOG, C. C. PAYNE, M. E. TULLY, M. A. JEVNIK and E. B. HERSHBERG: $\Delta^{1,4}$-Diene-3-ketosteroids. J. Amer. chem. Soc. **77**, 4184 (1955).
NORYMBERSKI, J. K.: Determination of urinary corticosteroids. Nature (Lond.) **170**, 1074 (1952).
— The partial reduction of steroid hormones and related substances. J. chem. Soc. **1956**, 517—519.
— and A. T. SHERMIN: Oxidative fission of pregnanediol glucuronide. Biochem. J. **55**, 876—880 (1953).
— and R. D. STUBBS: Indirect analysis of corticosteroids; the determination of steroidal dihydroxyacetones. Biochem. J. **64**, 168—175 (1956).
— and G. F. WOODS: Partial reduction of steroid hormones and related substances. J. chem. Soc. **1955**, 3426—3430.
— R. D. STUBBS and H. F. WEST: Assessment of adrenocortical activity by assay of 17-ketogenic steroids in urine. Lancet **1953, i**, 1276.
NOWACZYNSKI, W., and J. GENEST: Recovery of aldosterone, hydrocortisone and cortisone following dialysis. J. clin. Endocrin. **15**, 1310—1312 (1955).

NOWACZYNSKI, W., and E. KOIW: Paper chromatographic system for the separation of polar corticosteroids. J. Lab. clin. Med. **49**, 815 (1957).
— and P. R. STEYERMARK: Absorption spectra of steroids in 100% phosphoric acid. Arch. Biochem. **58**, 453—460 (1955).
— M. GOLDNER and J. GENEST: Microdetermination of corticosteroids with tetrazolium derivatives. J. Lab. clin. Med. **45**, 818—821 (1955).
— E. KOIW, J. GENEST, R. TELLIER, I. MORIN, A. LAFLAMME and P. ROBINSON: Chemical method for the determination of urinary aldosterone. Canad. J. Biochem. **35**, 425 to 443 (1957a).
— T. SANDOR, E. KOIW and J. GENEST: Isolation of a new steroid from urine. Progr. 39th Meet. Endocrine Soc., 17—18 (1957b).
NUSSBAUM, A. L., and F. E. CARLON: The reactions of steroids with diazomethane. J. Amer. chem. Soc. **79**, 3831—3835 (1957).
O'CONNEL, P. W., K. M. MANN, E. D. NIELSON and F. R. HANSON: Microbiological oxidation of steroids by Cunninghamella blakesleeana. Medium design. J. appl. Microbiol. **3**, 16—20 (1955).
OERTEL, G.: Papierchromatographische Trennung und Bestimmung von 17-Ketosteroiden. Acta endocr. (Kbh.) **16**, 263—266 (1954a).
— Bestimmung von Corticoiden auf Papierchromatogrammen. Acta endocr. (Kbh.) **16**, 267—269 (1954b).
— Papierchromatographie von Steroiden. Naturwissenschaften **43**, 17 (1956a).
— Fraktionierte Papierchromatographie von Steroiden. Naturwissenschaften **43**, 18 (1956b).
— J. SCHMIDT-THOMÉ u. R. REUBER: Isolierung, Nachweis und Bestimmung der Steroide. Hoppe-Seyl./Thierfelder. Handbuch der physiologisch- und pathologisch-chemischen Analyse. 10. Ausg., Teil. III, S. 2, 1583—1647. Berlin: Springer 1955.
OLIVETO, E. P., and E. B. HERSHBERG: The selective reduction of steroidal 11,20-diketones with sodium borohydride. J. Amer. chem. Soc. **75**, 488—490 (1953).
— C. GEROLD, L. WEBER, H. JORGENSEN, R. RAUSSER and E. B. HERSHBERG: The acylation of 11 β-hydroxy steroids: The synthesis of compound F 11-acetate and related compounds. J. Amer. chem. Soc. **75**, 5486—5489 (1953).
— — R. RAUSSER and E. B. HEHRSBERG: The preparation of 17 α-hydroxycorticosterone 21-acetate (Kendall's Compound F acetate) via 11 β-formates. J. Amer. chem. Soc. **77**, 3564—3567 (1955).
— R. RAUSSER, L. WEBER, E. SHAPIRO, D. GOULD and E. B. HERSHBERG: The preparation of hydrocortisone from cortisone acetates. J. Amer. chem. Soc. **78**, 1736—1738 (1956a).
— H. Q. SMITH, C. GEROLD, R. RAUSSER and E. B. HERSHBERG: New syntheses of corticosterone. J. Amer. chem. Soc. **78**, 1414—1416 (1956b).
— — and E. B. HERSHBERG: Some transformation products of cortisone acetate. J. Amer. chem. Soc. **79**, 3596—1598 (1957a).
— — R. RAUSSER and E. B. HERSHBERG: D-Homo rearrangement of 11-oxygenated 17 α-hydroxy-20-ketosteroids. J. Amer. chem. Soc. **79**, 3594—3596 (1957b).
— R. RAUSSER, A. L. NUSSBAUM, W. GEBERT, E. B. HERSHBERG, S. TOLKSDORF, M. EISLER, P. L. PERLMAN and M. M. PECHET: 16 α-Methylprednisone and 16 β-methylprednisone. J. Amer. chem. Soc. **80**, 4428 (1958a).
— — H. L. HERZOG, E. B. HERSHBERG, S. TOLKSDORF, M. EISLER, P. L. PERLMAN and M. M. PECHET: 16 β-Methyl-9 α-fluoroprednisolone 21-acetate. J. Amer. chem. Soc. **80**, 6687—6688 (1958b).
ONCLEY, J. L., F. R. N. GURD and M. MELIN: Preparation and properties of serum and plasma proteins. XXV. Composition and properties of human serum β-lipoprotein. J. Amer. chem. Soc. **72**, 458—464 (1950).
OTT, A. C., M. F. MURRAY and R. L. PEDERSON: The reaction of dihydropyran with steroidal alcohols. Utility in the syntheses of testosterone acyl esters. J. Amer. chem. Soc. **74**, 1239—1241 (1952).
PAPPO, R.: Synthesis of 18-oxygenated progesterones. J. Amer. chem. Soc. **81**, 1010—1011 (1959).
PARSONS, J., W. T. BEHER and G. D. BAKER: Steroid X-ray diffraction powder data. Analyt. Chem. **28**, 1514—1518 (1956).
PATAKI, J., G. ROSENKRANZ and C. DJERASSI: Partial synthesis of allopregnane-3 β,11 β,21-triol-20-one and allopregnane-3 β,11 β,17 α,21-tetrol-20-one. J. biol. Chem. **195**, 751—754 (1952).
PATERSON, J. Y. F., and G. F. MARRIAN: Formaldehyde retention in urine extracts in the determination of urinary formaldehydogenic substances. Mem. Soc. Endocr. **2**, 1—3 (1953).
PAUL, M. H., and E. L. DURRUM: Ionophoresis in non-aqueous solvent systems. J. Amer. chem. Soc. **74**, 4721—4722 (1952).

PEARCE, E. M., F. DEVENUTO, W. M. FITCH, H. E. FIRSCHEIN and U. WESTPHAL: Rapid determination of radiocarbon in animal tissues. Analyt. Chem. **28**, 1762—1765 (1956).
PEARLMAN, W. H.: Recent experiences in the detection, estimation and isolation of progesterone and related C_{21} steroids. Recent Progr. Hormone Res. **9**, 27—39 (1954).
— and E. CERCEO: The isolation of progesterone from human placenta. J. biol. Chem. **198**, 79—83 (1952).
— — The estimation of saturated and α,β-unsaturated ketonic compounds in placental extracts. J. biol. Chem. **203**, 127—134 (1953).
PECHET, M. M.: Method for the chromatographic separation of very polar steroids. Science **121**, 39—40 (1955).
PEDERSON, R. L., J. A. CAMPBELL, J. C. BABCOCK, S. H. EPPSTEIN, H. C. MURRAY, A. WEINTRAUB, R. C. MEEKS, P. D. MEISTER, L. M. REINEKE and D. H. PETERSON: Preparation of a tertiary hydroxy-steroid, 10ξ-hydroxy-19-nor-testosterone by microbiological hydroxylation. J. Amer. chem. Soc. **78**, 1512—1513 (1956).
PETERSON, D. H., and H. C. MURRAY: Microbiological oxidation of steroids at carbon 11. J. Amer. chem. Soc. **74**, 1871—1872 (1952).
— — S. H. EPPSTEIN, L. M. REINEKE, A. WEINTRAUB, P. D. MEISTER and H. M. LEIGH: Introduction of oxygen at carbon-11 of progesterone. J. Amer. chem. Soc. **74**, 5933—5936 (1952).
— S. H. EPPSTEIN, P. D. MEISTER, H. C. MURRAY, H. M. LEIGH, A. WEINTRAUB and L. M. REINEKE: Microbiological transformations of steroids. IX. Degradation of C_{21} steroids to C_{19} ketones and testololactone. J. Amer. chem. Soc. **75**, 5768—5769 (1953).
PETERSON, R. E.: The identification of corticosterone in human plasma and its assay by isotope dilution. J. biol. Chem. **225**, 25—37 (1957).
— A. KARRER and S. L. GUERRA: Evaluation of Silber-Porter procedure for determination of plasma hydrocortisone. Analyt. Chem. **29**, 144—149 (1957).
PFEFFER, K. H., W. RUPPEL, HJ. STAUDINGER u. L. WEISSBECKER: Getrennte Bestimmung der 11-Oxy- und 11-Desoxycorticoide. Arch. exp. Path. Pharmak. **214**, 165—173 (1952).
PFIFFNER, J. J.: The adrenal cortical hormones. Adv. Enzymol. **2**, 325—356 (1942).
PFISTER, H.: Beitrag zur Methodik der 17-Ketosteroidbestimmung. Biochem. Z. **323**, 157—162 (1952).
PIERCE, C. E., and R. E. PETERSON: Liquid-liquid continuous extractor for solvents heavier than water. Analyt. Chem. **28**, 2029—2030 (1956).
PINCUS, G.: A new color reaction for certain urinary 17-ketosteroids. Endocrinology **32**, 176—184 (1943).
— and W. H. PEARLMAN: Fractionation of neutral urinary steroids. Endocrinology **29**, 413—424 (1941).
PLANTIN, L. O.: Infrared spectrographic analysis of corticosteroids. Acta endocr. (Kbh.) **18**, 351—355 (1955).
POND, M. H.: Hydrolysis of 17-ketosteroid conjugates in urine. J. clin. Endocrin. **13**, 1544—1546 (1953).
— 17-ketosteroid fractionation studies by a microchromatographic technique in endocrine disorders. J. Endocrin. **10**, 202—211 (1954).
PONTIUS, D., I. BECKMAN u. K. D. VOIGT: Untersuchungen über die Spezifität einzelner Farbreaktionen von Steroiden und ihre klinische Anwendung. Acta endocr. (Kbh.) **20**, 19—38 (1955).
POOS, G. I.: The C-20 epimers of 4-pregnen-11β,17α,20-triol-3-one. J. Amer. chem. Soc. **77**, 4932—4935 (1956).
— and L. H. SARETT: Synthesis of a tricyclic analog of cortisone. J. Amer. chem. Soc. **78**, 4100—4103 (1956).
— G. E. ARTH, R. E. BEYLER and L. H. SARETT: 4b-Methyl-7-ethylenedioxy-1,2,3,4,4aα, 4b,5,6,7,8,10,10aβ-dodecahydrophenanthrene-4β-ol-1-one and related tricyclic derivatives. J. Amer. chem. Soc. **75**. 422-429 (1953).
PORTER, C. C., and R. H. SILBER: A quantitative color reaction for cortisone and related 17,21-dihydroxy-20-keto steroids. J. biol. Chem. **185**, 201—207 (1950).
PRELOG, V., u. G. TSATSAS: Über die Konfiguration am Kohlenstoffatom 20 des Steroid-Gerüstes. Helv. chim. Acta **36**, 1178—1180 (1953).
PRICE, W. J.: Nuclear radiation detection. New York: McGraw-Hill 1958.
PRINS, D. A., u. T. REICHSTEIN: Über Bestandteile der Nebennierenrinde und verwandte Stoffe. Das vierte isomere Allo-pregnan-triol-(3β,17,20). Helv. chim. Acta **23**, 1490—1501 (1940).
PROCHÁZKA, Ž.: On steroids, VIII. Paper chromatography of steroid acids (In Czech.). Chem. Listy **47**, 718—725 (1953).
RAMSAY, D. A.: Intensities and shapes of infrared absorption bands of substances in the liquid phase. J. Amer. chem. Soc. **74**, 72—80 (1952).

REBER, F., A. LARDON u. T. REICHSTEIN: Teilsynthese von 11-Epicorticosteron. Helv. chim. Acta 37, 45—58 (1954).
RECKNAGEL, R. O., and M. LITTERIA: A simplified blue tetrazolium reaction. J. Lab. clin. Med. 48, 463—468 (1956).
REDDY, W. J.: Modification of the Reddy-Jenkins-Thorn method for the estimation of 17-hydroxycorticoids in urine. Metabolism 3, 489—492 (1954).
— D. JENKINS and G. W. THORN: Estimation of 17-hydroxycorticoids in urine. Metabolism 1, 511—527 (1952).
— N. A. HAYDAR, J. C. LAIDLAW, A. RENOLD and G. W. THORN: Determination of total 17-hydroxycorticoids in plasma. J. clin. Endocrin. 16, 380—390 (1956).
REICH, H., D. H. NELSON and A. ZAFFARONI: Isolation of 17-hydroxycorticosterone from blood obtained from adrenal veins of dogs. J. biol. Chem. 187, 411—417 (1950).
— and B. K. SAMUELS: The reaction of 21-hydroxypregnenolone and 5-pregnene-3β, 17α,21-triol-20-one with 2,4-dinitrophenylhydrazine. J. org. Chem. 19, 1041—1045 (1954).
— — The reactions of adrenal cortical hormones with 2,4-dinitrophenylhydrazine. J. org. Chem. 21, 65—67 (1956).
— u. T. REICHSTEIN: Über Gallensäuren und verwandte Stoffe. 11-Keto- und 11α-Oxy-cholansäure. Helv. chim. Acta 26, 562—585 (1943).
— S. J. SANFILIPPO and K. F. CRANE: Separation and quantitative determination of progesterone and pregnenolone. J. biol. Chem. 198, 713—719 (1952).
REICHSTEIN, T.: Über die Bestandteile der Nebennieren-Rinde V. Chemischer Nachweis des Androstan-skelettes. Helv. chim. Acta 19, 979—984 (1936).
— Chemie des Cortins und seiner Begleitstoffe. In: L. RUZICKA and W. STEPP, Herausgeb. Ergebnisse der Vitamin- und Hormonforschung. Band 1, 334—368. Leipzig: Akademische Verlagsgesellschaft 1938.
— u. J. v. EUW: Über Bestandteile der Nebennierenrinde. Die Verseifung von Estern reduzierender Ketale mit schwachen Alkalien. Helv. chim. Acta 21, 1181—1185 (1938).
— and C. W. SHOPPEE: The hormones of the adrenal cortex. Vitam. and Horm., 1, 346—413 (1943).
— — Chromatography of steroids and other colorless substances by the method of fractional elution. Disc. Faraday Soc. 7, 305—311 (1949).
REINEKE, L. M.: Paper chromatography in steroid determination. Analyt. Chem. 28, 1853—1858 (1956).
REPETTO, O. M., and A. A. DREWES: Determination of neutral 17-ketosteroids. A new apparatus for extraction of urine. An. Asoc. quím. argent. 35, 5—20 (1947).
REUBER, R., R. TSCHESCHE, J. SCHMIDT-THOME, G. OERTEL u. H. A. OFFE: Steroide. Hoppe-Seyler/Thierfelder. Handbuch der physiologisch und pathologisch-chemischen Analyse. 10 Aufl., Bd. III, Teil 2, S. 1373—1647. Berlin, Göttingen, Heidelberg: Springer 1955.
RINGOLD, H. J., and G. STORK: Introduction of the cortical hormone side-chain. J. Amer. chem Soc. 80, 250 (1958).
— E. BATRES, O. MANCERA and G. ROSENKRANZ: Synthesis of 4-halo hormone analogs. J. org. Chem. 21, 1432—1435 (1956a).
— G. ROSENKRANZ and F. SONDHEIMER: Synthesis of 4-pregnen-17α,21-diol-3,20-dione. J. Amer. chem. Soc. 78, 820—825 (1956b).
— — — 1-Methyl-19-nortestosterone and 1-methyl-17α-ethinyl-19-nortestosterone. J. Amer. chem. Soc. 78, 2477—2479 (1956c).
RINIKER, B., D. ARIGONI u. O. JEGER: Über die direkte konfigurative Verknüpfung der Steroide mit dem Citronellal, ein Beitrag zur Bestimmung der absoluten Konfiguration der Steroide. Helv. chim. Acta 37, 546—552 (1954).
RIONDEL, A. M., M. C. SANZ and A. F. MULLER: Estimation of 17,21-dihydroxy-20-ketosteroids in plasma by a micro-modification of the Silber-Porter assay. Acta endocr. (Kbh.) 26, 57—64 (1957).
RIVOIRE, R., et M. PASZKOWSKI: Le dosage des corticoïdes urinaires. Press méd. 60, 214—216 (1952).
ROBBIE, W. A., and R. B. GIBSON: Rapid clinical determination of urinary 17-ketosteroids. J. clin. Endocrin. 3, 200—205 (1943).
ROBERTS, G.: Alteration of infrared spectra of steroids in potassium bromide. Analyt. Chem. 29, 911—916 (1757).
— B. S. GALLAGHER and R. N. JONES: Infrared absorption spectra of steroids. An Atlas. Vol. II. New York: Interscience 1958.
ROBERTSON, A. J. B.: Mass spectrometry. New York: Wiley 1954.
ROBERTSON, J. M.: X ray diffraction. In: E. A. BRAUDE and F. NACHOD, Eds., Determination of organic structures by physical methods 463—502. New York: Academic Press 1955.

Robinson, A. M.: The uses and limitations of adsorption chromatography for the separation of urinary ketosteroids. Recent Progr. Hormone Res. **9**, 163—176 (1954).
— and J. M. Norton: Estimation of corticosteroid-like substances in human urine. J. Endocrin. **7**, 321—329 (1951).
Rogers, W. F., jr., and H. Jaffe: A simple test of clinical value for the differentiation of testosterone and desoxycorticosterone in pellet form. J. clin. Endocrin. **8**, 1099 (1948).
Romanoff, L. P., and R. S. Wolf: The fractionation of corticosteroid metabolites in human urine. Recent Progr. Hormone Res. **9**, 337—352 (1954).
Romo, J., and A. R. de Vivar: Synthesis of 16α,17α-dihydroxy-steroids and related compounds. J. org. Chem. **21**, 902—909 (1956).
— G. Rosenkranz, C. Djerassi and F. Sondheimer: Synthesis of 11α,17α-dihydroxyprogesterone and 11α,17α,21-trihydroxyprogesterone. J. Amer. chem. Soc. **75**, 1277—1282 (1953).
— — and F. Sondheimer: Synthesis of allopregnane-3β,6β,21-triol-20-one triacetate. J. Amer. chem. Soc. **76**, 5169—5171 (1954).
Rosenblum, C.: Principles of isotope dilution assays. Analyt. Chem. **29**, 1741—1744 (1957).
Rosenfeld, R. S.: Tertiary hydroxyl group elimination in steroid ketols. J. Amer. chem. Soc. **79**, 5540—5542 (1957).
— and T. H. Gallagher: Deacetoxylation of steroid Ring C ketol acetates. J. Amer. chem. Soc. **77**, 4367—4370 (1955).
Rosenkrantz, H.: An antimony trichloride reagent suitable for the detection and estimation of nonketonic steroids. Arch. Biochem. Biophys. **44**, 1—7 (1953a).
— Erratum: An antimony trichloride reagent suitable for the detection and estimation of nonketonic steroids. Arch. Biochem. Biophys. **46**, 260 (1953b).
— Analysis of steroids by infrared spectrometry. In: D. Glick, Ed., Methods of biochemical Analysis, Vol. II, 1—56. New York: Interscience 1955.
— and L. Zablow: Comparison of infrared absorption spectra of steroids obtained on solid films, mulls and solutions. Analyt. Chem. **25**, 1025—1028 (1953).
Rosenkranz, G., O. Mancera, J. Gatica and C. Djerassi: A method for the conversion of allosteroids into Δ^4-3-ketosteroids. J. Amer. chem. Soc. **72**, 4077—4080 (1950a).
— J. Pataki, S. Kaufmann, J. Berlin and C. Djerassi: The synthesis of adrenal steroids and analogs from allopregnan-3β-ol-20-one. J. Amer. chem. Soc. **72**, 4081—4085 (1950b).
— and F. Sondheimer: Syntheses of cortisone. Fortschr. Chem. org. Naturst. **10**, 274—389 (1953).
— O. Mancera and F. Sondheimer: Synthesis of C-2 oxygenated derivatives of Reichstein's substance S and of cortisone. J. Amer. chem. Soc. **77**, 145—148 (1955).
Rosselet, J. P., J. W. Jailer and S. Lieberman: The metabolism of 21-deoxysteroids. J. biol. Chem. **225**, 977—993 (1957).
Rossi, M.: New method for the purification of *m*-dinitrobenzene for use in the color testing of 17-ketosteroids. Riv. Ist. sieroter. ital., Sez. prima **25**, 251—254 (1950).
Rubin, B. L., R. I. Dorfman and G. Pincus: A method for the quantitative analysis of 17-ketosteroid mixtures. J. biol. Chem. **203**, 629—645 (1953a).
— H. Rosenkrantz, R. I. Dorfman and G. Pincus: Separation of 17-ketosteroids in a 24-hour urine sample. J. clin. Endocrin. **13**, 578—580 (1953b).
Rubin, M., H. Wishinsky and F. Bompard: Bromination of allopregnane-3,20-dione. J. Amer. chem. Soc. **73**, 2338—2340 (1951).
Rudolph, H.: A photoelectric polarimeter attachment. J. opt. Soc. Amer. **45**, 50—59 (1955).
Ruliffson, W. S., H. M. Lang and J. P. Hummel: Radiochemical determination of ketosteroids. J. biol. Chem. **201**, 839—846 (1953).
Ruschig, H.: Zur Darstellung von 21-Estern des Pregnen-5-diol-3β,21-ons-20 aus Pregnen-5-ol-3β-on-20. Chem. Ber. **88**, 878—883 (1955).
Rutenburg, A. M., R. Gofstein and A. M. Seligman: Preparation of a new tetrazolium salt which yields a blue pigment on reduction and its use in the demonstration of enzymes in normal and neoplastic tissues. Cancer Res. **10**, 113—121 (1950).
Ruzicka, L., u. A. Wettstein: Künstliche Herstellung des männlichen Sexualhormons *trans*-Dehydroandrosteron. Helv. chim. Acta **18**, 986—994 (1935).
— N. Wahba, P. Herzig u. H. Heusser: 17aβ-Methyl-D-homotestosteron; ein Beitrag zur Bestimmung des sterischen Verlaufes von Reaktionen am Kohlenstoff 17a der D-Homosteroide. Chem. Ber. **85**, 491—497 (1952).
Saier, E., M. Warga and R. C. Grauber: Differences in calibration curves of androsterone and dehydroisoandrosterone. J. biol. Chem. **137**, 317—323 (1941).
Sakal, E. H., and E. J. Merrill: Paper chromatography of corticosteroids. Science **117**, 451—452 (1953).
Sandberg, A. A., W. R. Slaunwhite, jr., and H. N. Antoniades: The binding of steroids and steroid conjugates to human plasma proteins. Recent Progr. Hormone Res. **13**, 209—260 (1957).

SANDOVAL, A., G. H. THOMAS, C. DJERASSI, G. ROSENKRANZ and F. SONDHEIMER: Synthesis of 19-nor-11-desoxycorticosterone, a potent mineralocorticoid hormone. J. Amer. chem. Soc. **77**, 148—151 (1955).

SARETT, L. H.: Partial synthesis of dehydrocortisone acetate. J. Amer. chem. Soc. **68**, 2478—2483 (1946a).

— Partial synthesis of pregnene-4-triol-17 β,20 β,21-dione-3,11 and pregnene-4-diol-17 β, 21-trione-3,11,20 monoacetate. J. biol. Chem. **162**, 601—631 (1946b).

SARETT, L. H.: A new method for the preparation of 17(α)-hydroxy-20-ketopregnanes. J. Amer. chem. Soc. **70**, 1454—1458 (1948).

— Stereoisomeric substituted 11-keto-20-hydroxypregnanes. I. J. Amer. chem. Soc. **71**, 1165—1169 (1949a).

— Stereoisomeric substituted 11-keto-20-hydropregnanes. II. J. Amer. chem. Soc. **71**, 1169—1174 (1949b).

— Stereoisomeric substituted 11-keto-20-hydroxypregnanes. III. J. Amer. chem. Soc. **71**, 1174—1184 (1949c).

— G. E. ARTH, R. M. LUKES, R. E. BEYLER, G. I. POOS, W. F. JOHNS and J. M. CONSTANTIN: Stereospecific total synthesis of cortisone. J. Amer. chem. Soc. **74**, 4974—4976 (1952).

SATO, Y., and W. A. JACOBS: The structure o fisorubijervine. J. biol. Chem. **191**, 63—69 (1951).

SAVARD, K.: Paper paritition chromatography of C_{19}- and C_{21}-ketosteroids. J. biol. Chem. **202**, 457—477 (1953).

— Some theoretical and some practical aspects of partition chromatography of ketosteroids. Recent Progr. Hormone Res. **9**, 185—208 (1954).

— S. BURSTEIN, H. ROSENKRANTZ and R. I. DORFMAN: The metabolism of adrenosterone in vivo. J. biol. Chem. **202**, 717—725 (1953).

SCHEDL, H. P., W. B. BEAN, B. M. STEVENSON and E. R. SCHUMACHER: Correction for color difference between standards and urine extracts and their ketonic fractions in the Callow-Zimmermann reaction. J. Lab. clin. Med. **45**, 191—214 (1955).

SCHIEDT, U., u. H. RESTLE: Eine einfache Methode zur Bestimmung der Kettenlänge von Peptiden. Z. Naturf. **9b**, 182—186 (1954).

SCHILD, W.: Eine Modification der Urin-Corticoidbestimmung nach J. K. NORYMBERSKI, R. D. STUBBS, and H. F. WEST. Klin. Wschr. **32**, 326—327 (1954).

SCHMEISER, K.: Radioaktive Isotope, ihre Herstellung und Anwendung. Berlin: Springer 1957.

SCHMID, O. J., and K. D. VOIGT: Solubility and electrophoretic behavoir of steroid-detergent mixtures in aqueous buffers. Nature (Lond.) **174**, 129 (1954).

SCHMIDLIN, J., G. ANNER, J.-R. BILLETER u. A. WETTSTEIN: Totalsynthese des racemischen Aldosterons. Experientia (Basel) **11**, 365—368 (1955).

— — — K. HEUSLER, H. UEBERWASSER, P. WIELAND u. A. WETTSTEIN: Racemisches Aldosteron und die beiden Enantiomeren des entsprechenden 18→11 Lactons. Helv. chim Acta **40**, 2291—2321 (1957).

SCHMIDT, H., u. H. STAUDINGER: Papierchromatographischer Nachweis der Aldosterons. Angew. Chem. **66**, 711—712 (1954).

SCHMIDT-THOMÉ, J.: Beckmannsche Umlagerung von 17α-Hydroxy-20-ketosteroidoximen. Liebigs Ann. Chem. **603**, 43—45 (1957).

SCHREIBER, J., u. A. ESCHENMOSER: Über die relative Geschwindigkeit der Chromsäureoxydation sekundärer, alicyclischer Alkohole. Helv. chim. Acta **38**, 1529—1936 (1955).

SCHROEDER, W., u. K. D. VOIGT: Über die Möglichkeit der Trennung von Steroiden mittels Papier-Elektrophorese und Gegenstromverteilung. Rec. Trav. chim. Pays-Bas **74**, 603 bis 612 (1955).

SCHROEDER, W., K. D. VOIGT u. I. BECKMANN: Über die kombinierte Anwendung von Craig-Verteilung und Papierelektrophorese bei der Trennung von Steroidgemischen. Hoppe-Seylers Z. physiol. Chem. **302**, 63—67 (1955).

SCHUBERT, A., G. LANGBEIN u. R. SIEBERT: Mikrobielle Hydroxylierung von Steroiden in 12- und 15-Stellung. Chem. Ber. **90**, 2576—2582 (1957).

SCHULZ, E. P., and J. D. NEUSS: Colorimetric assay for cortisone, hydrocortisone and related steroids. Analyt. Chem. **29**, 1662—1665 (1957).

SCHWARZ, V.: A rapid-semi-quantitative spot test for reducing steroids. Nature (Lond.) **169**, 506 (1952).

— Quantitative paper chromatography of reducing steroids of the adrenal cortex. Biochem. J. **53**, 148—152 (1953a).

— Qualitative and quantitative examination of corticoid mixture by chromatographic methods. Mem. Soc. Endocr. **2**, 75—82 (1953b).

SCHWENK, E., and N. T. WERTHESSEN: Purification of ^{14}C-cholesterol from perfusion with liver and other organs. Arch. Biochem. **40**, 334—341 (1952).

— M. GUT and J. BELISLE: Preparation of C^{14}-cholesterol from C^{14}-cholestenone. Arch. Biochem. **31**, 456—457 (1951).

Shapiro, E. L., D. Gould and E. B. Hershberg: Reduction of a steroidal Δ^{16}-20-ketone with lithium hydride. J. Amer. chem. Soc. **77**, 2912—2913 (1955).
Shoolery, J. N.: Nuclear magnetic resonance spectroscopy. Analyt. Chem. **26**, 1400—1403 (1954).
— and M. T. Rogers: Nuclear magnetic resonance spectra of steroids. J. Amer. chem. Soc. **80**, 5121—5135 (1958).
Shoppee, C. W.: The Serini reaction. J. chem. Soc. **1949**, 1671.
— The chemistry of cortisone. Ann. Rev. Biochem. **22**, 261—298 (1953).
— and G. R. Summers: Reduction of cholest-5-en-3-one with lithium aluminium hydride. J. chem. Soc. **1950**, 687—689.
— — The stereochemistry and the mechanism of the *i*-steroid rearrangement. J. chem. Soc. **1952**, 3361—3374.
Shull, G. M.: Paper chromatography of steroid fermentation products. Proc. 126th Meet. Amer. chem. Soc., 9 A (1954).
— Transformations of steroids by molds. Trans. N. Y. Acad. Sci., Ser. II, **19**, 147—172 (1956).
— J. L. Sardinas and R. C. Nubel: Paper chromatography of steroid compounds. Arch. Biochem. Biophys. **37**, 186—198 (1952).
Silber, R. H., and R. D. Busch: The specificity of the reaction of phenylhydrazine with 17,21-dihydroxy-20-ketosteroids. J. clin. Endocrin. **15**, 505—507 (1955).
— — An improved procedure for the determination of hydrocortisone in human plasma. J. clin. Endocrin. **16**, 1333—1336 (1956).
— and C. C. Porter: The determination of 17,21-dihydroxy-20-ketosteroids in urine and plasma. J. biol. Chem. **210**, 923—932 (1954).
— — Determination of 17,21-dihydroxy-20-ketosteroids in urine and plasma. In: D. Glick, Ed., Methods of Biochemical Analysis, **4**, 139—169. New York: Interscience Publishers 1957.
Simpson, S. A., and J. F. Tait: A quantitative method for the bioassay of the effect of adrenal cortical steroids on mineral metabolism. Endocrinology **50**, 150—161 (1952).
— — Physico-chemical methods of detection of a previously unidentified adrenal hormone. Mem. Soc. Endocr. **2**, 9—24 (1953).
— — Recent progress in methods of isolation, chemistry and physiology of aldosterone. Recent Progr. Hormone Res. **11**, 183—210 (1955).
— — A. Wettstein, R. Neher, J. v. Euw and T. Reichstein: Isolation of a new crystalline compound from beef adrenal extracts with especially high activity on mineral metabolism. Experientia (Basel) **9**, 333—335 (1953).
— — — — — O. Schindler u. T. Reichstein: Über Bestandteile der Nebennierenrinde und verwandte Stoffe 91. Aldosteron. Isolierung und Eigenschaften. Helv. chim. Acta **37**, 1163—1200 (1954a).
— — — — — — — Die Konstitution des Aldosterons. Über Bestandteile der Nebennierenrinde und verwandte Stoffe. Helv. chim. Acta **37**, 1200—1223 (1954b).
Slaunwhite, W. R., jr., and L. Neely: Purification of urinary 17-ketosteroid extracts for infrared analysis. Analyt. Chem. **29**, 1614—1616 (1957).
Slomp, G., Y. F. Shealy, J. L. Johnson, R. A. Donia, B. A. Johnson, R. P. Holysz, R. L. Pederson, A. O. Jensen and A. C. Ott: A synthesis of pregnane-3,20-dione from stigmasterol and ergosterol. J. Amer. chem. Soc. **77**, 1216—1221 (1955).
Sneeden, R. P. A., and R. B. Turner: Ouabagenin. The hydroxyl groups of the A/B ring systems. J. Amer. chem. Soc. **77**, 130—134 (1955a).
— — Catalytic dehydrogenation of primary and secondary alcohols with platinum and oxygen: Selective dehydrogenation in the steroid series. J. Amer. chem. Soc. **77**, 190—191 (1955b).
Soloway, A. H., W. J. Considine, D. K. Fukushima and T. F. Gallagher: Some reactions of epoxides of steroid enol acetates. J. Amer. chem. Soc. **76**, 2941—2943 (1954).
Sondheimer, F., C. Amendolla and G. Rosenkranz: $\Delta^{4,6}$-Dien-3-ones. J. Amer. chem. Soc. **75**, 3932—5935 (1953a).
— S. Kaufman, J. Romo, H. Martinez and G. Rosenkranz: Synthesis of steroidal hormone analogs hydroxylated at C-2. J. Amer. chem. Soc. **75**, 4712—4715 (1953b).
— O. Mancera, G. Rosenkranz and C. Djerassi: The chemical reduction of Δ^{8}-11-keto and of saturated 11-keto steroids. A new approach to 11-keto and 11α-hydroxy steroids. J. Amer. chem. Soc. **75**, 1281—1285 (1953c).
— — — C-6 oxygenated derivatives of cortical hormones. J. Amer. chem. Soc. **76**, 5020 to 5023 (1954a).
— M. Velasco, E. Batres and G. Rosenkranz: Reduction of Δ^1 and Δ^4 double bonds with $NaBH_4$. Chem. and Ind. **1954**, 1482—1483 (b).

SONDHEIMER, F., O. MANCERA, M. URQUIZA and G. ROSENKRANZ: The decarboxylation of unsaturated steroidal acids. J. Amer. chem. Soc. **77**, 4145—4149 (1955).
— G. ROSENKRANZ, O. MANCERA and C. DJERASSI: Transformation of sapogenins to androgens and estrogens. Beckmann rearrangement of Δ^{16}-20-ketosteroids. J. org. Chem. **21**, 520—522 (1956).
— and Y. MAZUR: Synthesis of 4-methylated steroids. J. Amer. chem. Soc. **79**, 2906—2910 (1957).
SORKIN, M., u. T. REICHSTEIN: Berichtigung zur Stereochemie von Sterinen und Steroiden bezüglich der Substituenten an den Ringen C and D. Helv. chim. Acta **29**, 1218—1229 (1946).
SPERO, G. B., A. MCINTOSH, JR., and R. H. LEVIN: Desulfurization of thiol esters of desoxycholic acid. J. Amer. chem. Soc. **70**, 1907—1910 (1948).
— J. L. THOMPSON, B. J. MAGERLEIN, A. R. HANZE, H. C. MURRAY, O. K. SEBEK and J. A. HOGG: 6-Methyl steroids. J. Amer. chem. Soc. **78**, 6213—6214 (1956).
— — F. H. LINCOLN, W. P. SCHNEIDER and J. A. HOGG: Fluorinated 6-methyl steroids. J. Amer. chem. Soc. **75**, 1515—1516 (1957).
SPERRY, W. M.: Lipide analysis. In: D. GLICK, Ed., Methods of Biochemical Analysis, Vol. 2, 83—111. New York: Interscience Publ. 1955.
SPRECHLER, M.: Comparison of various methods for the chemical assay of corticoids in urine. Acta endocr. (Kbh.) **18**, 374—378 (1955).
STAUDINGER, H., and V. BAUER: Zur Bestimmung der an Glucuronsäure gebundenen Corticoide im Harn. Klin. Wschr. **32**, 330—331 (1954).
— u. M. SCHMEISSER: Quantitative chemische Bestimmung der Nebennierenrindenhormone. Hoppe-Seylers Z. physiol. Chem. **283**, 54—63 (1948).
— — Bestimmung der Nebennierenrindenhormone im Harn. Biochem. Z. **321**, 83—92 (1950).
STEIGER, M., u. T. REICHSTEIN: Δ^5-3-Oxy-ätio-cholensäure und einige ihrer Umwandlungsprodukte. Helv. chim. Acta **20**, 1040—1054 (1937).
— — Über Bestandteile der Nebennieren-Rinde. Die Substanzen J, K, N und O. Helv. chim. Acta **21**, 546—564 (1938).
STEINBERG, D., and S. UDENFRIEND: The measurement of radioisotopes. In: S. P. COLOWICK and N. O. KAPLAN, Eds., Methods in Enzymology. Vol. IV, New York: Academic Press 1957.
STEINBERG, N. G., R. HIRSCHMANN and J. M. CHEMERDA: Partial synthesis of 4-methylhydroxycortisone-4-C^{14} and related compounds. Chem. and Ind. **1958**, 975—976.
STIMMEL, B. F.: A dual semimicrodistillation unit for simultaneous distillation of formaldehyde from oxidized and unoxidized aliquots of urinary corticoid residues. J. Lab. clin. Med. **37**, 948—949 (1951a).
— Effect of zinc-hydrochloric acid hydrolysis on formaldehydogenic steroids in human urine. Fed. Proc. **10**, 254 (1951b).
STIMPSON, M. M., and M. J. O'DONNELL: The infrared and ultraviolet absorption spectra of cytosine and isocytosine in the solid state. J. Amer. chem. Soc. **74**, 1805—1808 (1952).
STITCH, S. R., and I. D. K. HALKERSTON: The enzymic hydrolysis of steroid conjugates. II. Hydrolysis of steroid conjugates in urine. Biochem. J. **63**, 710—715 (1956).
— — and J. HILLMAN: I. Sulphatase and β-glucuronidase activity of molluscan extracts. Biochem. J. **63**, 705—710 (1956).
STOKES, W. M., W. A. FISH and F. C. HICKEY: Chromatographic separation and quantitative estimation of iodine-131-labeled derivatives of sterols, amines, acids and aldehydes. Analyt. Chem. **27**, 1895—1898 (1955).
SWEAT, M. L.: Sulfuric acid-induced fluorescence of corticosteroids. Analyt. Chem. **26**, 773—776 (1954a).
— Silica gel microcolumn for chromatographic resolution of cortical steroids. Analyt. Chem. **26**, 1964—1967 (1954b).
SWERN, D.: Epoxidation and hydroxylation of ethylenic compounds with organic peracids. Org. Reac. **7**, 378—433 (1953).
SZALKOWSKI, C. R., M. G. O'BRIEN and W. J. MADER: Determination of hydrocortisone. Analyt. Chem. **27**, 944—946 (1955).
TALALAY, P., F. A. LOEWUS and B. VENNESLAND: The enzymatic transfer of hydrogen. IV. The reaction catalyzed by α,β-hydroxysteroid dehydrogenase. J. biol. Chem. **212**, 801—809 (1955).
TALBOT, N. B., and G. O. LANGSTROTH: A rapid extractor for urinary androgens. Factor to be considered in the preparation of extracts for colorimetric assay. Endocrinology **25**, 729—736 (1939).
— R. A. BERMAN and E. A. MACLACHLAN: Elimination of errors in the colorimetric assay of neutral urinary 17-ketosteroids by means of a color correction equation. J. biol. Chem. **143**, 211—218 (1942).

TALBOT, N. B., A. H. SALTZMAN, R. L. WIXOM and J. K. WOLFE: The colorimetric assay of urinary corticosteroid-like substances. J. biol. Chem. **160**, 535—546 (1945).
— S. ULICK, A. KOUPREIANOW and A. ZYGMUNTOWICZ: Countercurrent distribution procedure for the quantitative-qualitative analysis of small amounts of urinary corticosteroids. J. clin. Endocrin. **15**, 301—316 (1955).
TAMM, C., G. VOLPP u. G. BAUMGARTNER: Beweis des Steringerüstes von Ouabagenin. Helv. chim. Acta **40**, 1469—1496 (1957).
TAMM, J., I. BECKMANN and K. D. VOIGT: Determination of neutral steroids in human blood. Acta endocr. (Kbh.) **31**, 219—225 (1957).
TANNHAUSER, P., R. J. PRATT and E. V. JENSEN: The preparation of 21-fluorosteroids. J. Amer. chem. Soc. **78**, 2658—2659 (1956).
TANSEY, R. P., and J. M. CROSS: A quantitative colorimetric method for certain ketosteroids with a modification of the Zimmermann reaction. J. Amer. pharm. Ass. **39**, 660—663 (1950).
TARPLEY, W., and M. YUDIS: Determination of purity of steroids by solubility analysis. Analyt. Chem. **25**, 121—127 (1953).
TAUB, D., R. H. PETTEBONE, N. L. WENDLER and M. TISHLER: The transformation of cortisone and hydrocortisone to cortisol. J. Amer. chem. Soc. **76**, 4094—4097 (1954).
— R. D. HOFSOMMER and N. L. WENDLER: Cortical steroids substituted at C-12. J. Amer. chem. Soc. **79**, 452—456 (1957).
— — H. L. SLATES and N. L. WENDLER: 16β-Methyl cortical steroids. J. Amer. chem. Soc. **80**, 4435 (1958).
— R. D. HOFFSOMMER and N. L. WENDLER: Neighboring group effects in the cortical side chain. J. Amer. chem. Soc. **81**, 3291—3259 (1959).
TAUBER, H.: New color reaction for steroids with perchloric acid. Analyt. Chem. **24**, 1494 (1952).
TENNENT, D., J. B. WHITLA and K. FLOREY: Two techniques in paper chromatography. Application to ketosteroids. Analyt. Chem. **23**, 1748—1749 (1951).
THOMA, R. W., J. FRIED, S. BONANNO and P. GRABOWICH: 16α-Hydroxylation of 9α-fluorohydroxycortisone. J. Amer. chem. Soc. **79**, 4818 (1957).
TISHLER, M.: Developments in the synthesis of cortical hormones. Rec. chem. Progr. **13**, 161—177 (1952).
TOMPSETT, S. L.: An investigation into the determination of corticosteroids in urine. J. clin. Path. **6**, 74—77 (1953).
— and D. C. SMITH: Determination of 17-ketogenic steroids in urine. Acta endocr. (Kbh.) **20**, 303—309 (1955a).
— — Extraction of polyhydroxy steroids by chloroform. Acta endocr. (Kbh.) **20**, 319—321 (1955b).
TOUCHSTONE, J. C., and C.-T. HSU: Determination of alpha-ketolic substances in urinary extracts and paper chromatograms. Analyt. Chem. **27**, 1517—1519 (1955).
TURNER, R. B.: Stereochemistry of the steroids. In: L. FIESER and M. FIESER, Natural products related to phenanthrene, Ed. 3 p. 620 New York: Reinhold 1949.
— Steroids labelled with isotopic carbon: Cholestenone and testosterone: J. Amer. chem. Soc. **72**, 579—585 (1950a).
— Stereochemistry of peracid oxidation of ketones. J. Amer. chem. Soc. **72**, 878—882 (1950b).
— Structural evidence for the reaction product of cholesterol-α-oxide and methylmagnesium iodide. J. Amer. chem. Soc. **74**, 5362—5364 (1952).
— Stereospecificity in the rearrangement of 17-hydroxy-20-ketosteroids. J. Amer. chem. Soc. **75**, 3484—3488 (1953a).
— Acylation of 17-hydroxy-20-ketosteroids. J. Amer. chem. Soc. **75**, 3489—3492 (1953b).
— V. MATTOX, L. ENGEL, B. MCKENZIE and E. KENDALL: Steroids derived from bile acids. Introduction of oxygen at C_{11}. J. biol. Chem. **166**, 345—365 (1946).
UMBERGER, E. J.: Isonicotinic acid hydrazide as a reagent for determination of Δ^4-3-ketosteroids. Determination of progesterone and testosterone propionate in oil solutions. Analyt. Chem. **27**, 768—773 (1955).
ULICK, S., and S. LIEBERMAN: The occurrence of a metabolite of aldosterone in urine. J. Amer. chem. Soc. **79**, 6567—6568 (1957).
VANEK, R., et R. DEVIS: A propos de l'hydrolyse et de l'extraction des corticoïdes urinaires. Ann. Endocr. (Paris) **15**, 201—212 (1954).
VELLUZ, L., A. PETIT, M. PESEZ et R. BERRET: Saponification de deoxycorticostérone. Bull. Soc. chim. Fr. (**1947**) 123—125.
VENNING, E. H., I. DYRENFURTH and V. E. KAZMIN: Hydrolysis and extraction of corticoids and 17-ketosteroids from body fluids. Recent Progr. Hormone Res. **8**, 27—50 (1953).
— C. J. P. GIROUD, I. DYRENFURTH and J. C. BECK: Studies with aldosterone. Canad. J. Biochem. **33**, 605—614 (1955).

VISCHER, E., and A. WETTSTEIN: Side-chain degradation of steroids by fungi. Experientia (Basel) **9**, 371—372 (1953).
— C. MEYSTRE u. A. WETTSTEIN: Herstellung weiterer 1-Dehydrosteroide auf mikrobiologischem Weg. Helv. chim. Acta **38**, 1502—1507 (1955).
— J. SCHMIDLIN u. A. WETTSTEIN: Mikrobiologische Spaltung racemischer Steroide; Synthese von d-Aldosteron. Experientia (Basel) **12**, 50—52 (1956).
VISCONTINI, M., u. P. MIGLIORETTO: Die optische Konfiguration der beiden β-Methoxyadipinsäuren und der Steroide. Helv. chim. Acta **38**, 930—935 (1955).
VOIGT, K.-D., u. I. BECKMANN: Papierelectrophorese von Steroiden. I. Acta Endocr. (Kbh.) **13**, 19—23 (1953).
— — Papierelektrophorese von Steroiden. II. Fraktionierung der neutralen Ketosteroide des Urins unter normalen und pathologischen Bedingungen. Acta endocr. (Kbh.) **15**, 251—264 (1954).
— W. SCHROEDER and I. BECKMANN: Chromatographic and electrophoretic determination of corticosteroids. Acta endocr. (Kbh.) **18**, 325—341 (1955).
WAGNER, R. B., and J. A. MOORE: A new synthetic route to the cortical side chain. J. Amer. chem. Soc. **72**, 5301—5305 (1950).
WALTER, K.: Bestimmung von 17-Ketosteroiden im Harn. Klin. Wschr. **30**, 474—475 (1952).
WEAVER, H. E., B. M. TOLBERT and R. C. LAFORCE: Observation of chemical shifts of O^{17} nuclei in various chemical environments. J. chem. Phys. **23**, 1956—1957 (1955).
WEICHSELBAUM, T. E., H. W. MARGRAF and D. C. KING: Determination of Δ^4-3-keto-corticosteroids in human peripheral plasma by the isonicotinic acid hydrazide (INH) reaction. J. clin. Endocrin. **17**, 959—965 (1957).
WEISENBORN, F., and D. BURN: The structure of isorubijervine. J. Amer. chem. Soc. **75**, 259—262 (1951).
WEISSBECKER, L., u. H. STAUDINGER: Trennung der C_{11}-Oxy- bzw. Oxo- von den C_{11}-Desoxy- bzw. Desoxocorticoiden und deren quantitative Bestimmung. Klin. Wschr. **29**, 59—60 (1951).
WEISSBERGER, A., E. S. PROSKAUER, J. A. RIDDICK and E. E. TOOPS, JR.: Organic solvents. Physical properties and methods of purification. In: A. WEISSBERGER, Ed., Technique of organic chemistry, vol. VII. New York: Interscience Publ. 1955.
WENDLER, N. L., and R. P. GRABER: Alkaline degradation of the cortical side chain. Chem. and Ind. **1956**, 549.
— and D. TAUB: D-Homo rearrangement of cortical steroids. Chem. and Ind. **1957** a, 822—823.
— — D-Homoannulation of 16α,17α-dihydroxy-20-keto steroids. Chem. and Ind. **1957** b, 1237—1238.
— — and H. L. SLATES: The structure of steroid D-Ring lactones. J. Amer. chem. Soc. **77**, 3559—3561 (1955).
— — S. DOBRINER and D. FUKUSHIMA: The structure and synthesis of D-homosteroids. J. Amer. chem. Soc. **78**, 5027—5032 (1956).
— R. P. GRABER, C. S. SNODDY, JR., and F. BOLLINGER: Transannular hydrogen transfer in the steroid series. J. Amer. chem. Soc. **79**, 4476—4487 (1957).
WERBIN, H., and S. ONG: Improved colorimetric determination of urinary 17-ketosteroids. Analyt. Chem. **26**, 762—764 (1954).
WEST, W.: Colorimetry, photometric analysis, fluorimetry and turbidimetry. In: A. WEISSBERGER, Ed., Physical methods of organic chemistry, 1399—1490. New York: Interscience Publ. 1949.
— Introductory survey of molecular spectra. In: W. WEST, Ed., Chemical applications of spectroscopy (Technique of organic chemistry, vol. IX), 1—70. New York: Interscience Publ. 1956.
WESTPHAL, U.: Über den Nachweis von Pregnandiolglukonid im Harn normaler Frauen und Männer. Hoppe-Seylers Z. physiol. Chem. **281**, 14—24 (1944).
— H. E. FIRSCHEIN and E. M. PEARCE: Transport of cortisol-4-C^{14} and progesterone -4-C^{14} by serum proteins as demonstrated by paper electrophoresis and paper chromatography. Amer. J. Physiol. **185**, 54—60 (1956).
WETTSTEIN, A.: Über $\Delta^{4,5}$-3-Ketone der Androstan- und Pregnan-Reihe. Helv. chim. Acta **23**, 388—399 (1940).
— Conversion of steroids by microorganisms. Experientia (Basel) **11**, 465—479 (1955).
— Aldosteron und andere Nebennierenrinden Hormone. Verh. Schweiz. naturf. Ges. **136**, 22—41 (1956).
— L'Aldostérone. Exp. Ann. Biochim. Méd. **19**, 171—191 (1957).
— P. DESAULLES, K. HEUSLER, R. NEHER, J. SCHMIDLIN, H. ÜBERWASSER u. P. WIELAND: Substanzen mit Wirkung auf die Elektrolytausscheidung. Angew. Chem. **69**, 689 (1957).
— R. NEHER u. H. J. URECH: Über Umwandlungen von Progesteron und 16α-Hydroxyprogesteron durch Enzyme tierischer Gewebe. Helv. chim. Acta **42**, 956—971 (1959).

WHEELER, J., S. FREEMAN and C. CHEN: A comparison of the tetrazolium and other methods of steroid estimation applied to urine. J. Lab. clin. Med. **42**, 758—765 (1953).
WIELAND, P., u. K. MIESCHER: Abbau der Oxy-bisnor-cholensäure III. Beitrag zur Asymmetrie am Kohlenstoffatom 20. Helv. chim. Acta **32**, 1922—1932 (1949).
WILDS, A., J. A. JOHNSON and R. E. SUTTON: The stereochemical course of hydrogenation of Ring D-unsaturated equilenane derivatives. J. Amer. chem. Soc. **72**, 5524—5529 (1950).
WILDS, A. L., and N. A. NELSON: A superior method of reducing phenol ethers to dihydro derivatives and unsaturated ketones. J. Amer. chem. Soc. **75**, 5360—5365 (1953a).
— — The facile synthesis of 19-nortestosterone and 19-norandrostenedione from estrone. J. Amer. chem. Soc. **75**, 5366—5369 (1953b).
WILLARD, H. H., L. L. MERRITT, JR., and J. A. DEAN: Instrumental methods of analysis, Ed. 3, Princeton: Van Nostrand 1957.
WILLIAMS, V. Z., V. J. COATES and F. GAARDE: A Device for automatic measurement of integrated absorbance. Analyt. Chem. **27**, 2017 (1955).
WILSON, D., R. C. GRAUER and E. SAIER: A simplified continuous extractor for estrogens and androgens. J. Lab. clin. Med. **26**, 581—585 (1940).
WILSON, H.: Methods for the determination of formaldehydogenic steroids by diffusion and by direct reaction. J. clin. Endocr. **13**, 1465—1479 (1953).
— Chromogenic values of various ketosteroids in a micro modification of the Zimmermann reaction: Comparison with the macro procedure. Arch. Biochem. Biophys. **52**, 217—235 (1954).
— and P. CARTER: Stabilization of the alcoholic potassium hydroxide in colorimetric 17-ketosteroid determinations. Endocrinology **41**, 417—421 (1947).
— and R. FAIRBANKS: Chromogenic values of 17-hydroxy corticosteroids in a modified Porter-Silber reaction. Arch. Biochem. Biophys. **53**, 71—76 (1954).
— — A rapid micromethod for the determination of 17-hydroxy and 17-ketosteroids. Arch. Biochem. Biophys. **54**, 440—456 (1955a).
— — A micromethod for the detection and assay of steroid C_{21}-17-hydroxy-α-glycols. Arch. Biochem. Biophys. **54**, 457—466 (1955b).
— and I. T. NATHANSON: The effect of alcohol and potassium hydroxide concentration on the reaction of 17-ketosteroids with *m*-dinitrobenzene. Endocrinology **37**, 208—216 (1945).
WINTERSTEINER, O., and J. J. PFIFFNER: Chemical studies on the adrenal cortex. III. Isolation of two physiologically inactive compounds. J. biol. Chem. **116**, 291—305 (1936).
WITTER, R. F., and S. STONE: Spot test for 3-hydroxy delta-5-steroids. Analyt. Chem. **29**, 156—157 (1957).
WOLFE, J. K., E. B. HERSHBERG and L. F. FIESER: The polarographic determination of ketonic steroids. J. biol. Chem. **136**, 653—687 (1940).
WOODWARD, R. B.: Structure and absorption spectra. J. Amer. chem. Soc. **64**, 72—77 (1942).
— SONDHEIMER, F., D. TAUB, K. HEUSLER and W. MCLAMORE: The total synthesis of a steroid. J. Amer. chem. Soc. **73**, 2403—2404 (1951).
— — — — — The total synthesis of steroids. J. Amer. chem. Soc. **74**, 4223—4251 (1952).
— H. INHOFFEN, H. LARSON u. K. MENZEL: Synthese des 3,8-Dimethyl-1,2-cyclopentenophenanthrens. Chem. Ber. **86**, 594—601 (1953).
WRIGHT, N.: Application of infrared spectroscopy to industrial research. Industr. Engng. Chem. (Anal.) **13**, 1—8 (1941).
WYLIE, H. B., and M. J. SCHMULOVITZ: A simple combination of continuous extractor and extract washer. Science **83**, 398 (1936).
ZAFFARONI, A.: Absorption spectra of sulfuric acid chromogens obtained from adrenal steroids and related compounds. J. Amer. chem. Soc. **72**, 3828 (1950).
— Micromethods for the analysis of adrenocortical steroids. Recent Progr. Hormone Res. **8**, 51—83 (1953).
— R. B. BURTON and E. H. KEUTMANN: The application of paper partition chromatography to steroid analysis. I. Ketosteroids. J. biol. Chem. **177**, 109—116 (1949).
— — — Adrenal cortical hormones: Analysis by paper partition chromatography and occurrence in the urine of normal persons. Science **111**, 6—8 (1950).
— H. J. RINGOLD, G. ROSENKRANZ, F. SONDHEIMER, G. H. THOMAS and C. DJERASSI: Synthesis of 19-norhydrocortisone and related 19-nor-adrenal hormones. J. Amer. chem. Soc. **76**, 6210—6211 (1954).
— V. TRONCOSO and M. GARCIA: C-19 hydroxylation of 11-deoxycorticosterone by the adrenal gland. Chem. and Ind. **1955**, 634.
ZIMMERMANN, W.: Eine Farbreaktion der Sexualhormone und ihre Anwendung zur quantitativen colorimetrischen Bestimmung. Hoppe-Seylers Z. physiol. Chem. **233**, 257—264 (1935).
— Chemische Bestimmungsmethoden von Steroidhormonen in Körperflüssigkeiten. Berlin: Springer 1955.

ZIMMERMANN, W., u. D. PONTIUS: Zur Mikrobestimmung der neutralen 17-Ketosteroide. Hoppe-Seylers Z. physiol. Chem. **297**, 157—163 (1954).
— H. V. ANTON u. D. PONTIUS: Die Eliminierung störender Chromogene bei der Bestimmung der 17-Ketosteroide mit *m*-Dinitrobenzol. Hoppe-Seylers Z. physiol. Chem. **289**, 91—102 (1952).
ZUMAN, P., J. TENYGL u. M. BŘEZINA: Polarographie der Steroide. I. Direkt reduzierbare Ketosteroide: Δ^4-Cholestenon, Methyltestosteron, Testosteron, Progesteron und Desoxycorticosteron. Coll. Czech. Chem. Comm. **19**, 46—55 (1954).
ZWINGELSTEIN, G., H. PACHECO and J. JOUANNETEAU: Chromatography of ketosteroids as 2,4-dinitrophenylhydrazones on alumina and on paper. C. R. Acad. Sci. (Paris) **236**, 1561—1563 (1953).
ZYGMUNTOWICZ, A. S., M. WOOD, E. CHRISTO and N. B. TALBOT: Studies of 17-ketosteroid excretion by means of a new micro-chromatographic fractionation procedure. J. clin. Endocr. **11**, 578—596 (1951).

Chapter 3

Biochemistry of the adrenocortical hormones

By

RALPH I. DORFMAN*

With 37 Figures

The adrenal cortex produces a variety of steroids, some biologically active and others, by present bioassay methods, inactive. These compounds possess from 18 to 27 carbon atoms. Some of these adrenocortical steroids as well as their metabolites are present in the blood and urine, others are represented in these fluids only by their metabolites. During the past 20 years, a wealth of information gathered from isolation studies and from metabolism studies (on biosynthesis and catabolism) has so enriched our knowledge that it is now possible to draw a reasonable balance sheet between steroid production in the adrenal gland and the levels of related steroids in blood and urine. This review will be concerned with (1) the biosynthesis of adrenocortical steroids, (2) factors relating to their metabolism, and (3) the enzymes that regulate the various reactions.

Isolation of steroids and sterols

A. Adrenal cortex

The first isolations of steroids from adrenal tissue were accomplished by REICHSTEIN, KENDALL, WINTERSTEINER and their co-workers in studies begun in about 1932. More recently several additional laboratories, notably that of WETTSTEIN, have been concerned with this important problem. The older as well as the more recent work will be discussed in the following sections. Further information is presented in Chapter 2.

I. C_{18} steroids

The presence of C_{18} steroids, the estrogens, in adrenal tissue has been demonstrated by indirect and direct methods. Indirect evidence for estrogen production by adrenals includes the increased excretion of estrogens by patients suffering from adrenocortical hyperactivity, the continued excretion of estrogens after gonadectomy, and the increased excretion of estrogens after treatment with ACTH. Direct evidence for adrenocortical production of estrogens includes the definitive isolation and identification of estrone from the gland and the biosynthesis of estrogens by adrenal tissue *in vitro*.

Patients with adrenocortical hyperactivity frequently show increased excretion of estrogens in urine, although biological manifestations may not be apparent.

* The author is especially grateful to Dr. BETTY L. RUBIN and the sub-editor for the many helpful suggestions during the preparation of this chapter for press. The chapter was completed late in 1958.

(Possibly this absence of feminization may be due to excessive secretion of androgens that counteract the physiological action of the estrogens.) MASON and KEPLER (1945) isolated estrone in milligram quantities per day from the urine of a patient with an adrenal cancer. Other patients with adrenocortical tumors have in some instances shown enormous values for urinary estrogens (HUIS IN'T VELD and DINGEMANSE 1948, WALLACH et al. 1957). In such cases values up to 12,000 I. U. of estrone equivalent (i.e., 1–2 mg) per day have been observed. This is 40 to 80 times the amount found in the urine of normal individuals.

True feminization due to adrenal tumors has been observed in men. SOFFER (1956) and WALLACH et al. (1957) reviewed some of these cases, which illustrate that estrogen production by the tumor may be sufficiently great to produce breast enlargement, testicular atrophy, loss of libido and partial or complete impotence. Surgical removal of these tumors reverses these symptoms. Estrogen determinations have been reported by a number of workers, and DICZFALUSY and LUFT (1956) have made a detailed study of the urine of one such patient, using countercurrent distribution methods. Excessive concentrations of estrone, estradiol-17β, and estriol were demonstrated.

High estrogen production may be associated with adrenal hyperplasia and with the adrenogenital syndrome.

Gonadectomy decreases but does not eliminate completely the excretion of estrone. In the urine of women whose ovaries have been removed, a significant excretion of estrone and estriol has been reported (WEST et al. 1958).

Indirect evidence for estrogen production from adrenals also comes from the studies of WEST et al. (1958), who administered ACTH to gonadectomized women and showed that increased levels of estrogens were present in the urine. These studies included identification of the increased urinary estrogens as estrone, estradiol, and estriol.

A word of caution in interpreting all of this indirect evidence for adrenal estrogen production is in order. A serious point to consider is the fact that estrogens may be derived from androgens by peripheral (probably hepatic) metabolism, and it is possible that the increment in estrogens under conditions of excessive androgen production may in fact only reflect these peripheral mechanisms.

Direct evidence for estrogen formation consists of the isolation of estrone and the detection of equilenin (Table 1 and Fig. 1), as well as evidence for the biosynthesis of estrogens by normal and abnormal adrenal tissue (MEYER 1955, BAGGETT et al. 1958).

Table 1. *C_{18} steroids of adrenal origin*

Steroid demonstrated	Species	Reference
Estrone	Ox	BEALL (1939) BEALL (1940)
Equilenin	Man	SALHANICK and BERLINER (1957)

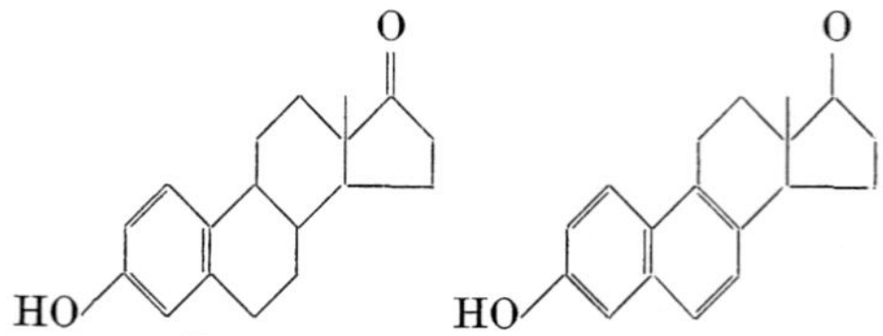

Estrone Equilenin

Fig. 1. C_{18} steroids detected in adrenocortical tissue

An important but as yet unresolved problem pertains to the continued excretion of estrogens after both ovariectomy and adrenalectomy (GREENWOOD and BULBROOK 1956, STRONG et al. 1956, BULBROOK and GREENWOOD 1957, BULBROOK et al. 1958, DICZFALUSY et al. 1958, WEST et al. 1958). Estrogens have been detected by chemical assay as well as by bioassay procedures in the urine of some doubly-operated patients, and in the studies of WEST et al. (1958) estriol was characterized by countercurrent distribution. This was a breast cancer patient who was hypophysectomized in addition to being ovariectomized and adrenalectomized.

Two types of C_{18} neutral steroids may be expected to occur in adrenal tissue on the basis of the existence of C_{18} and C_{19} oxygenated compounds.

MEYER (1955b) has demonstrated that the adrenal androgen Δ^4-androstene-3,17-dione may be hydroxylated at carbon 19 by beef adrenal homogenates, and KAHNT and co-workers (1955a) have indicated the presence of 18-oxygenated steroids. It might therefore be speculated that, after hydroxylation to form primary alcohol groups, the corresponding aldehydes and carboxylic acid derivatives could be formed. Decarboxylation would produce the nor compounds. Thus 18- and 19-nor-Δ^4-androstene-3,17-dione may be adrenal steroids, and to these could be added 18,19-bis-nor-Δ^4-androstene-3,17-dione (Fig. 2).

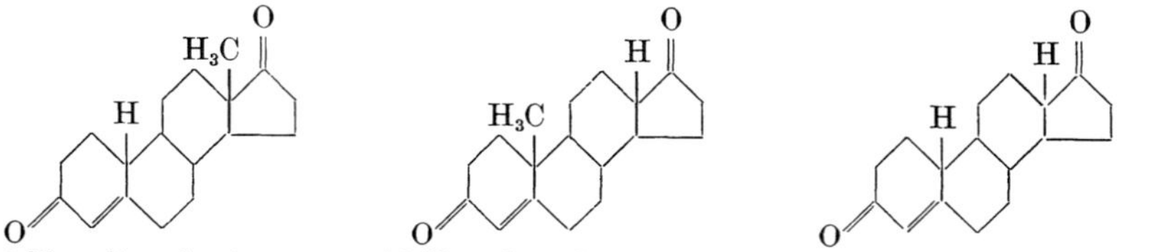

Fig. 2. C_{17} and C_{18} steroids that may be produced by adrenal tissue

II. C_{19} steroids

The presence of C_{19} steroids in adrenal tissue was noted during the early studies of REICHSTEIN and co-workers. This group not only noted their presence but quickly realized that the biologically active steroids of the corticoid series (C_{21}) could be easily converted to the C_{19} group. The fact that some C_{19} steroids were androgenic aided in the elucidation of the structure of the adrenocortical hormones.

Nine C_{19} steroids have now been isolated from adrenal tissues under conditions that point to their presence as physiologically important constituents (Table 2 and Fig. 3). In addition to these 9 steroids, certain other C_{19} steroids may also be present in adrenal tissues, since both their immediate precursors and the necessary enzyme systems are known to be present. Included in the group are a variety of 6-,18-, and 19-hydroxylated C_{19} steroids (Fig. 4).

Among the C_{19} steroids, 11β-hydroxy-Δ^4-androstene-3,17-dione and adrenosterone appear to be produced exclusively in the adrenal cortex. This is because of the absence of 11β-hydroxylase in other steroid-producing tissues. In a recent study 11β-hydroxylase activity was, however, detected in slices from what was judged to be an interstitial-cell tumor of the testis from a young boy (SAVARD et al. 1956). No such activity could be detected in normal testicular tissue, and one cannot exclude with certainty the possibility that adrenal tissue may actually have been present in this tumor.

Dehydroepiandrosterone seems to be produced in the adrenal but not in gonadal or placental tissue. OERTEL and EIK-NES (1958a) have reported the peripheral formation of dehydroepiandrosterone when pregnenolone is perfused through muscle tissue. If substantiated, this would indicate a mechanism involving the adrenal secretion of pregnenolone and its peripheral metabolism to this C_{19} steroid.

Attempts to isolate dehydroepiandrosterone from normal adrenal tissue have failed, probably due to the high concentration of 3β-ol-dehydrogenase, which could convert any dehydroepiandrosterone that may be present to Δ^4-androstene-3,17-dione. Recently, however, the presence of this Δ^5-3β-hydroxy androgen in adrenal tissue has been firmly established by 3 different studies. Two involved incubations with labeled acetate of human adrenal tissue from patients suffering from the virilizing adrenogenital syndrome and Cushing's disease, respectively,

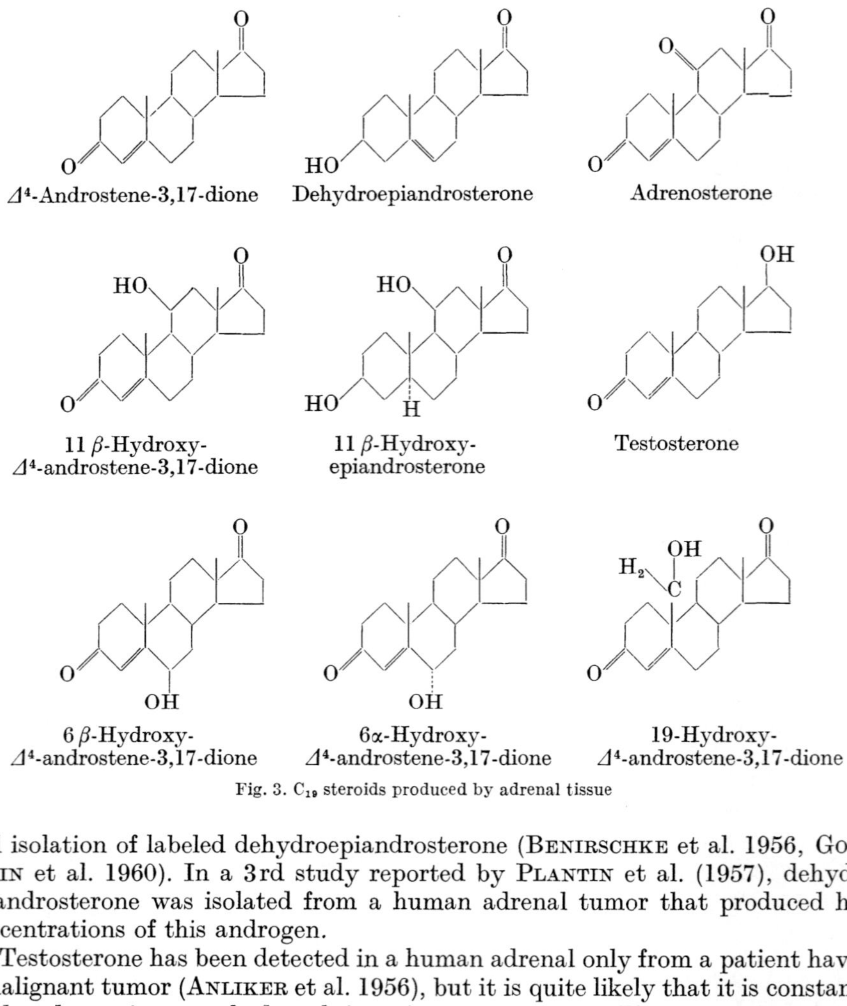

Fig. 3. C_{19} steroids produced by adrenal tissue

and isolation of labeled dehydroepiandrosterone (Benirschke et al. 1956, Goldstein et al. 1960). In a 3rd study reported by Plantin et al. (1957), dehydroepiandrosterone was isolated from a human adrenal tumor that produced high concentrations of this androgen.

Testosterone has been detected in a human adrenal only from a patient having a malignant tumor (Anliker et al. 1956), but it is quite likely that it is constantly produced even in normal adrenal tissue in trace amounts. The companion steroid, Δ^4-androstene-3,17-dione, is present in the adrenals of all species thus far studied (Table 2). It is also produced by the gonads and the placenta, and may arise from certain C_{21} steroids by peripheral metabolism.

19-Hydroxy-Δ^4-androstene-3,17-dione is an intermediate in the biosynthesis of estrogens, and two 6-oxygenated C_{19} steroids are known, 6α- and 6β-hydroxy-Δ^4-androstene-3,17-dione.

Six theoretically possible C_{19} adrenal steroids are represented in Fig. 4. These are only a few of the many that may be expected to be present but have as yet not been isolated. The compound 6β,11β-dihydroxy-Δ^4-androstene-3,17-dione has actually been isolated from guinea pig urine (Peron and Dorfman 1951) but this finding

Table 2. *C_{19} steroids of adrenal origin*

Steroid demonstrated	Species	Reference
Δ^4-Androstene-3,17-dione	Ox	v. EUW and REICHSTEIN (1941), E. BLOCH et al. (1954)
	Man	E. BLOCH et al. (1956a)
	Man (Fetal)	E. BLOCH et al. (1955), BENIRSCHKE et al. (1956)
Dehydroepiandrosterone	Man	E. BLOCH et al. (1956a)
	Man (Tumor)	PLANTIN et al. (1957)
Testosterone	Man (Cancer—adrenogenital and Cushing's)	ANLIKER et al. (1956)
Adrenosterone	Ox	REICHSTEIN (1936a, b), E. BLOCH et al. (1954)
11β-Hydroxy-Δ^4-androstene-3,17-dione	Man (Fetal)	E. BLOCH et al. (1956a)
	Ox	E. BLOCH et al. (1954)
	Pig	WETTSTEIN and ANNER (1954)
11β-Hydroxyepiandrosterone	Ox	REICHSTEIN and v. EUW (1938), v. EUW and REICHSTEIN (1941)
6β-Hydroxy-Δ^4-androstene-3,17-dione	Ox	MEYER et al. (1955)
6α-Hydroxy-Δ^4-androstene-3,17-dione	Ox	MEYER et al. (1955)
19-Hydroxy-Δ^4-androstene-3,17-dione	Ox	MEYER (1955b)

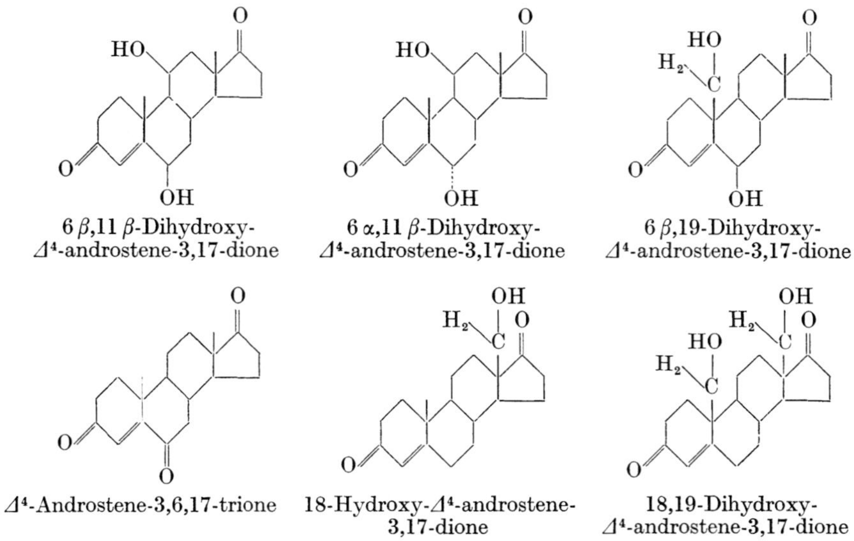

Fig. 4. Some possible C_{19} adrenocortical steroids

does not necessarily indicate that the compound was produced by the adrenal gland; it may have been a peripheral metabolite of 6β-hydroxycortisol. The 6-keto compound could be expected to be present on the basis of the finding of both the 6α- and 6β-hydroxy forms, with the possibility that 6β-hydroxylation of a C_{19} substrate takes place, that the product formed is oxidized to the 6-keto derivative, and that this latter compound is then reduced to the 6α-hydroxy form. It is quite likely that the 16-hydroxylated C_{19} steroids are also present.

III. C_{21} and C_{27} steroids

C_{27} steroids isolated: R =

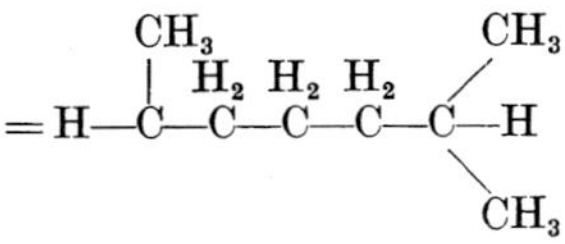

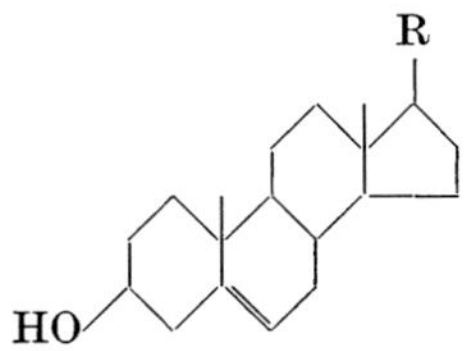

Cholesterol

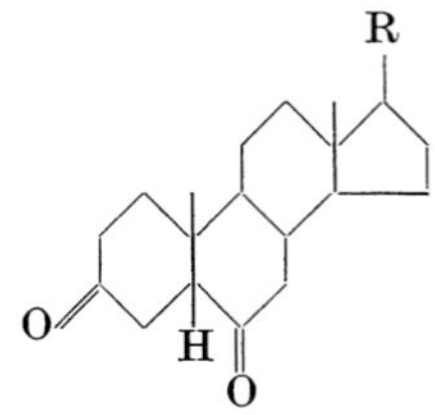

Cholestane-3,6-dione

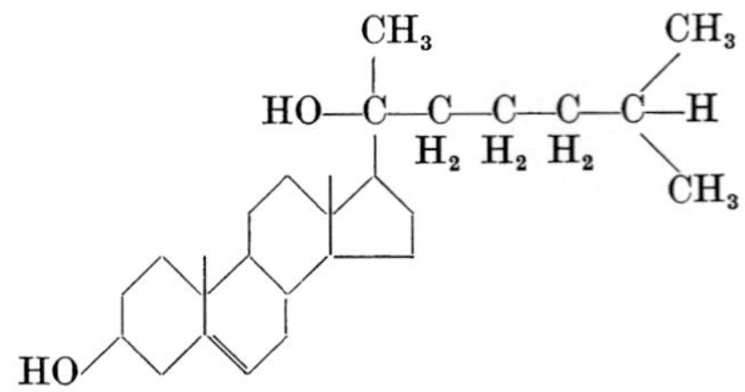

20 β-Hydroxycholesterol

Possible C_{27} steroids:

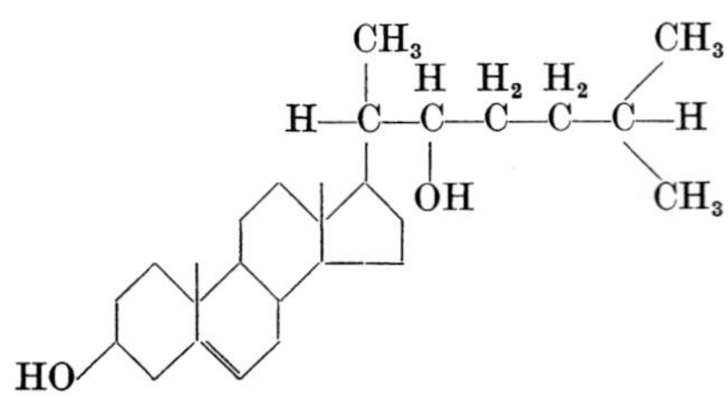

22-Hydroxycholesterol

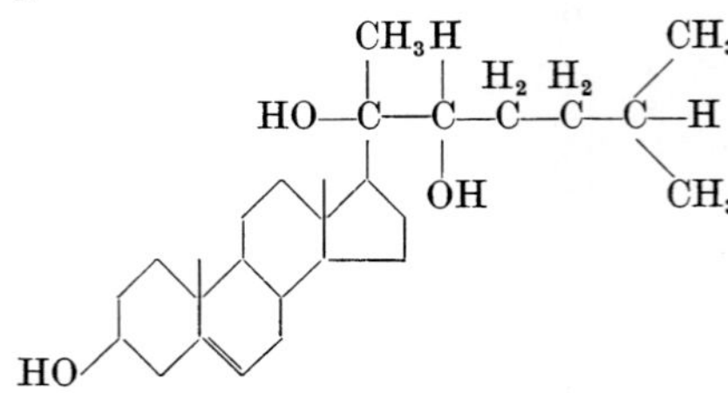

20 β, 22-Dihydroxycholesterol

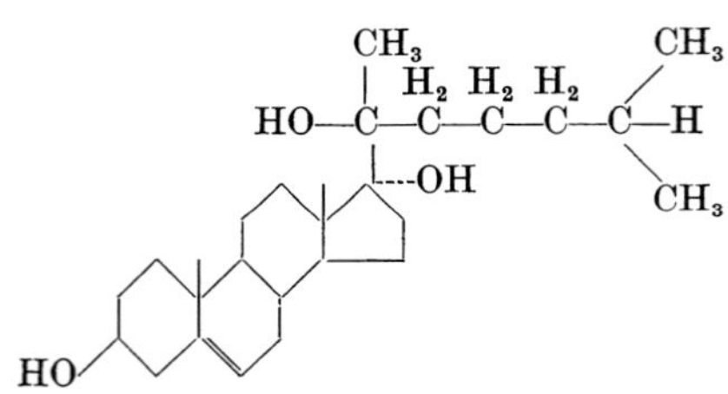

17 α,20 β-Dihydroxycholesterol

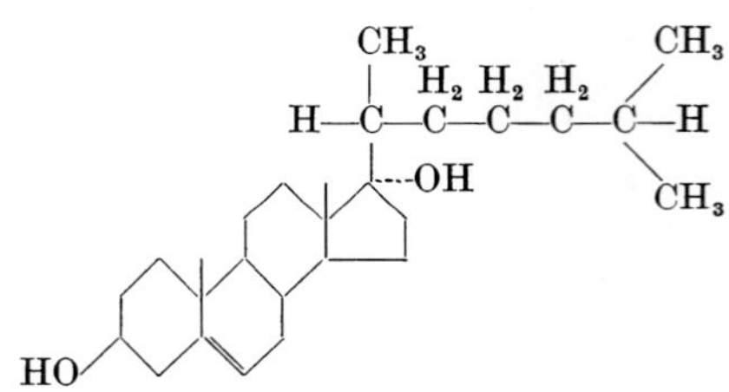

17 α-Hydroxycholesterol

Fig. 5. C_{27} steroids in adrenal tissue — those demonstrated and those possible

Two C_{27} steroids as well as the ever-present cholesterol (Fig. 5) have been isolated from adrenocortical tissues. Of these, cholestane-3,6-dione has been detected, but no specific function is apparent. On the other hand, 20 β-hydroxycholesterol is a possible intermediate in the biosynthesis of the steroid hormones from cholesterol. Other likely derivatives of cholesterol from the viewpoint of biosynthesis include the four hydroxylated compounds pictured in Fig. 5. These derivatives are logical substrates for the action of a suitable 17,20-desmolase to form C_{19} steroids and for the action of a suitable 20,22-desmolase to form C_{21} steroids.

In the C_{21} series of adrenocortical steroids, there are 2-, 3-, 4-, and 5-oxygenated representatives, and it is not altogether unlikely that 6- and 7-oxygenated representatives are produced by the adrenal. Most of the isolation studies have been made on bovine and porcine tissue, with a limited number of studies reported for human and other animal tissues. Recently CHESTER JONES et al. (1959) have detected.

with reasonable certainty, corticosterone, cortisol, and aldosterone in the adrenal effluent of lower vertebrates ranging from fishes to birds (see Chapter 1, Table 15, p. 75). These studies deserve considerable extension in the interest of comparative endocrinology and biochemistry. Figs. 6, 7, 8, and 9 represent the C_{21} steroids that have been isolated, and Table 3 lists the specific isolation studies. Further data are presented in Chapter 2.

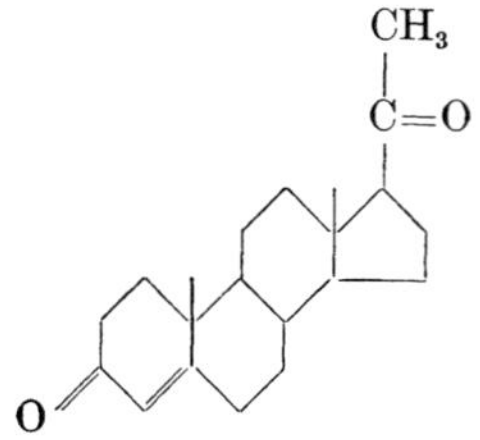

Progesterone

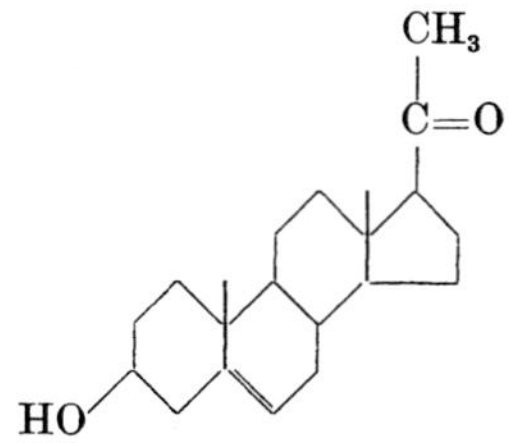

Pregnenolone

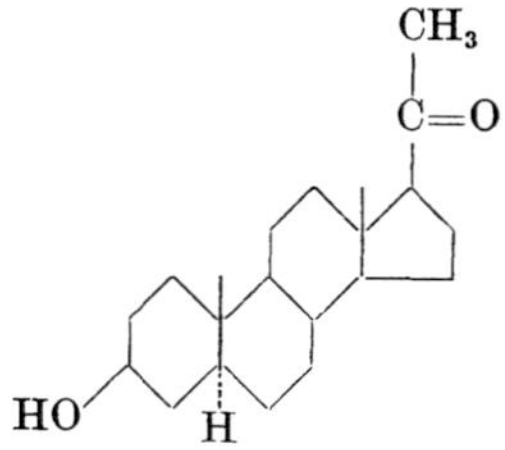

3 β-Hydroxyallopregnan-20-one

Fig. 6. $C_{21}O_2$ steroids of adrenal tissue

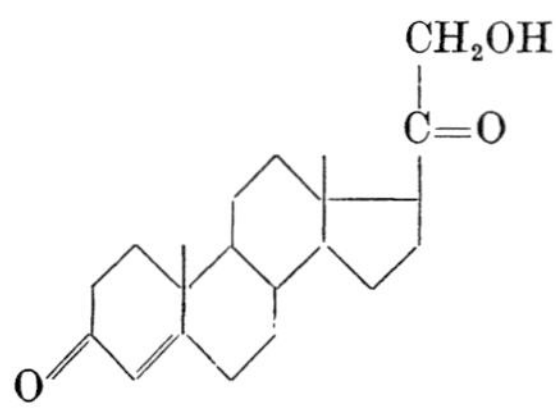

Deoxycorticosterone

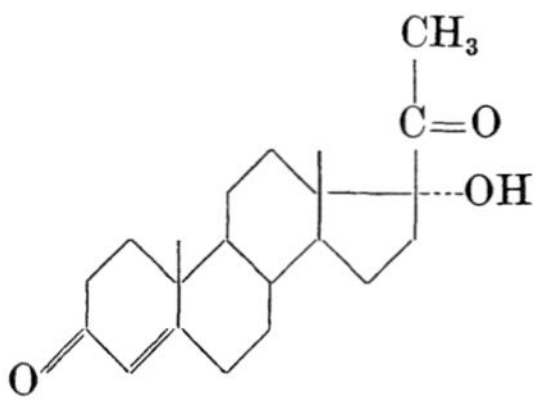

17 α-Hydroxyprogesterone

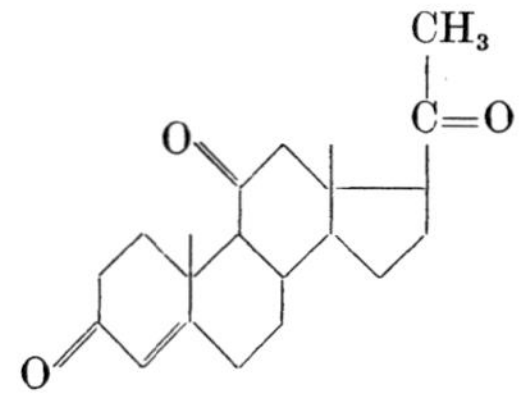

11-Ketoprogesterone

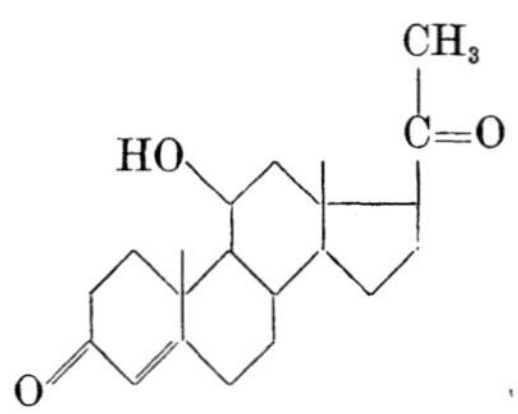

11 β-Hydroxyprogesterone

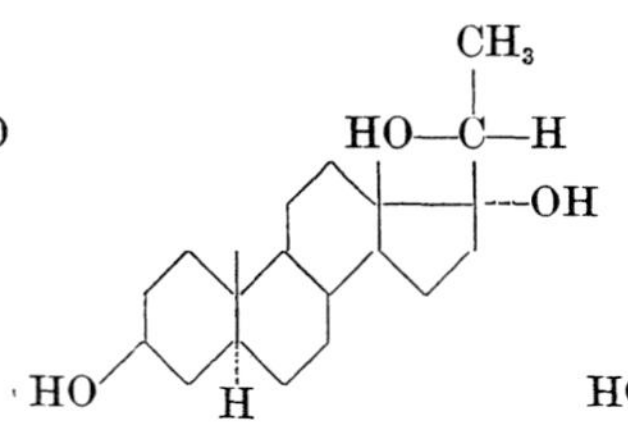

Allopregnane-3 β,17 α,20 β-triol

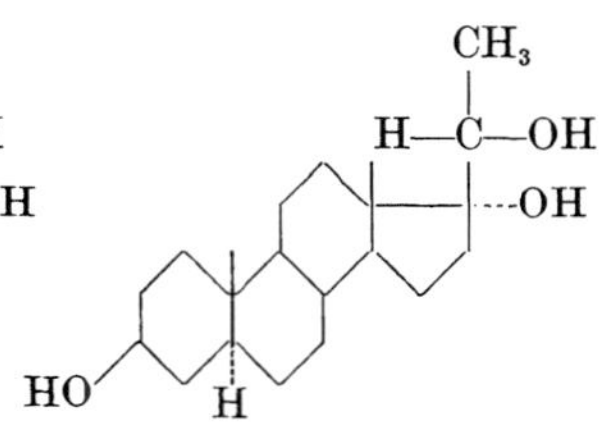

Allopregnane-3 β,17α,20 α-triol

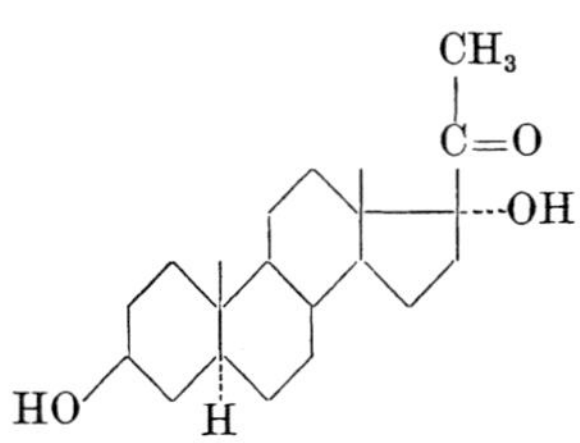

3 β,17α-Dihydroxyallopregnan-20-one

Fig. 7. $C_{21}O_3$ steroids of adrenal tissue

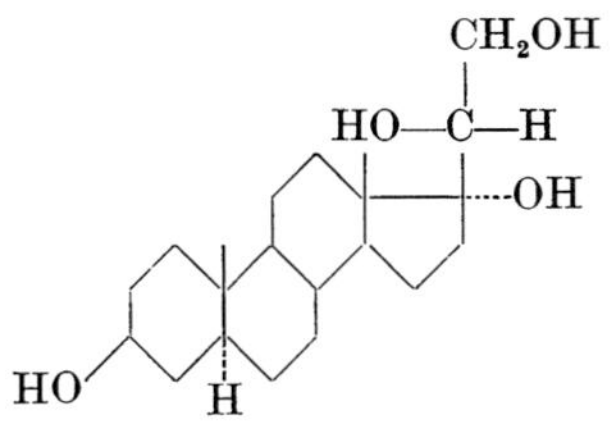

Allopregnane-
3 β,17α,20 β,21-tetrol

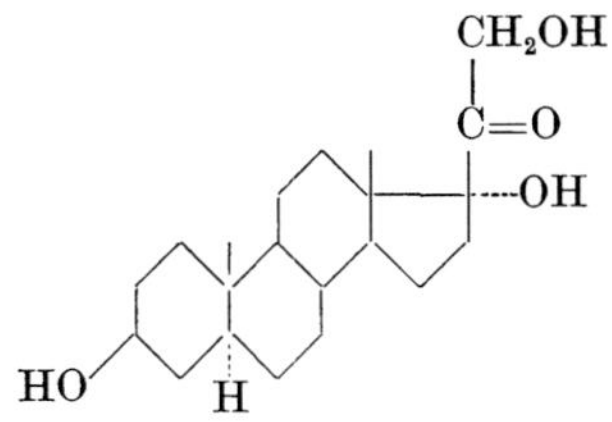

3 β,17α,21-Trihydroxyallo-
pregnan-20-one

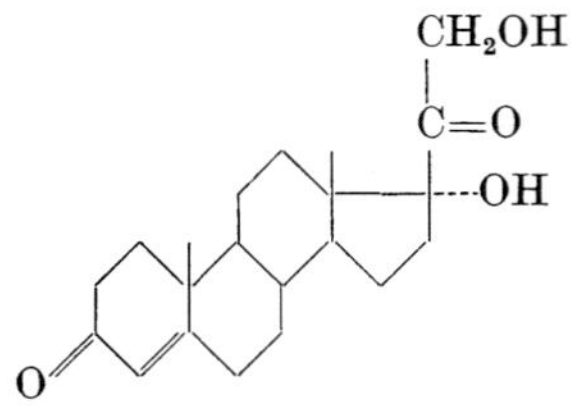

11-Deoxycortisol

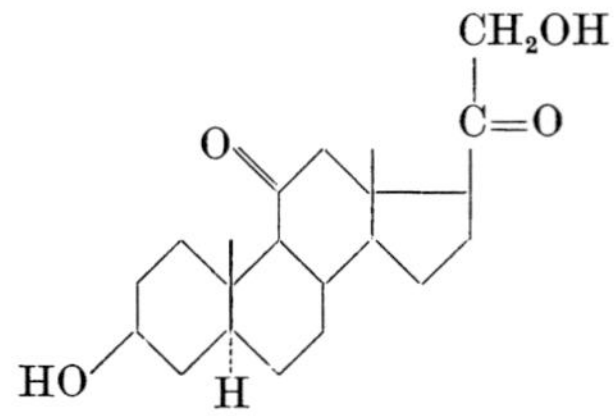

3 β,21-Dihydroxyallopregnane-
11,20-dione

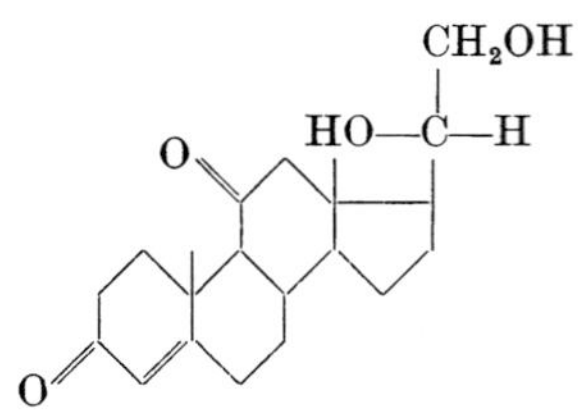

20 β,21-Dihydroxy-
Δ^4-pregnene-3,11-dione

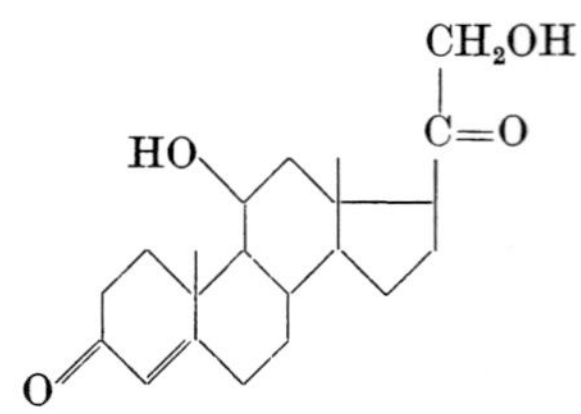

Corticosterone

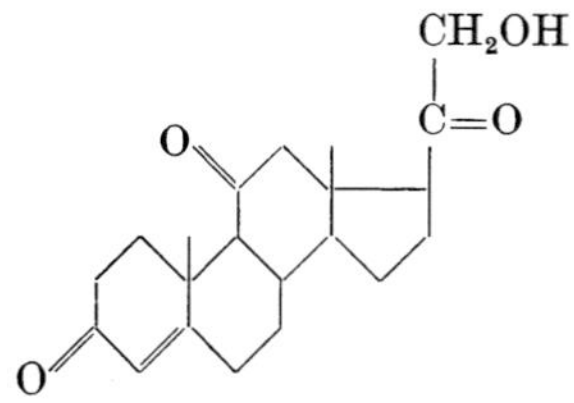

11-Dehydrocorticosterone

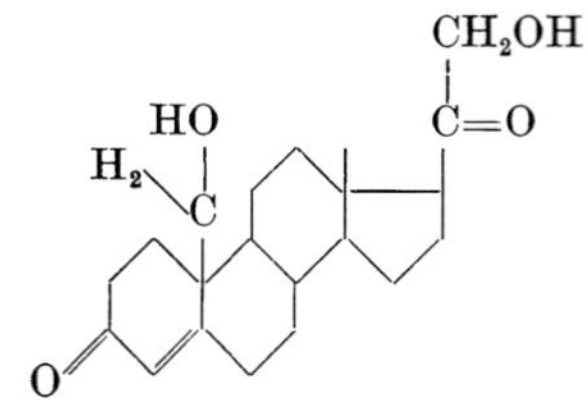

19-Hydroxydeoxycorticosterone

Fig. 8. $C_{21}O_4$ steroids of adrenal tissue

IV. Nitrogen-containing steroids

Nitrogen-containing steroids may be present in adrenal tissue (Voigt and Schroeder 1953, 1955, 1956, Voigt et al. 1953, Schroeder et al. 1953). Evidence for urinary compounds of a similar type has also been reported (Eades et al. 1954, Hudson and Lombardo 1955). Although the crude extracts containing these N components showed activity in a life-maintenance test, none of the 6 components reported by Voigt and Schroeder (1955) was active.

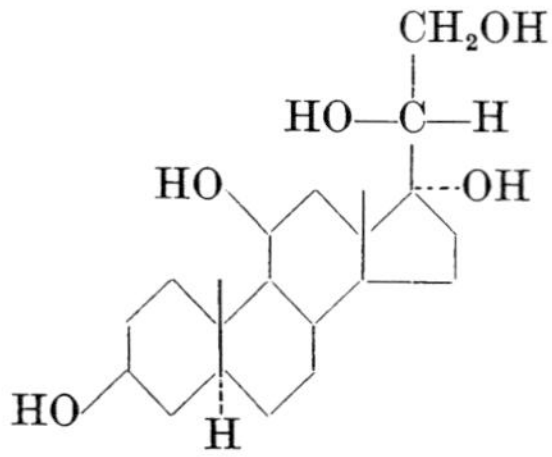

Allopregnane-
3 β,11 β,17 α,20 β,21-pentol

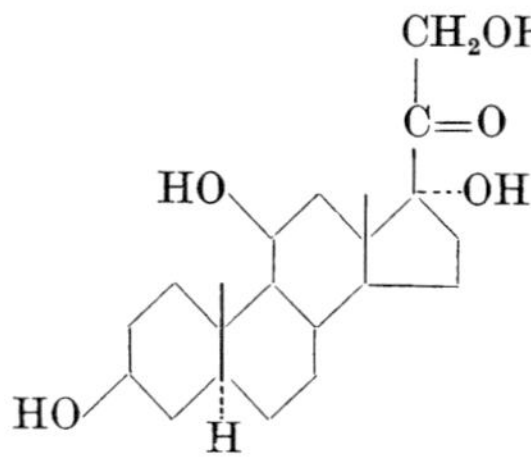

3 β,11 β,17 α,21-
Tetrahydroxyallopregnan-
20-one

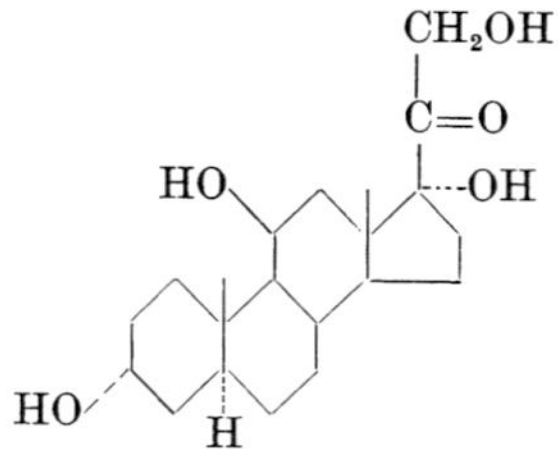

3α,11 β,17α,21-
Tetrahydroxy-
allopregnan-20-one
(Urocortisol)

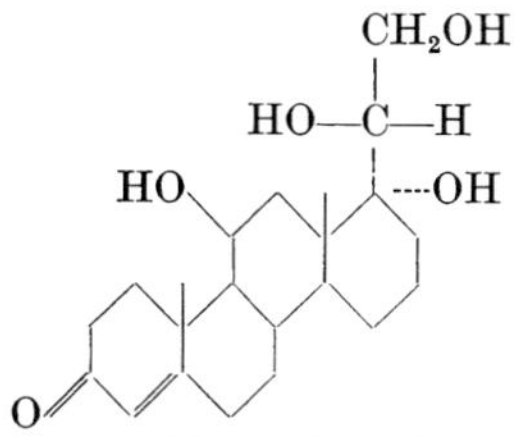

11 β,17 α,20 β,21-Tetrahydroxy-
Δ⁴-pregnen-3-one

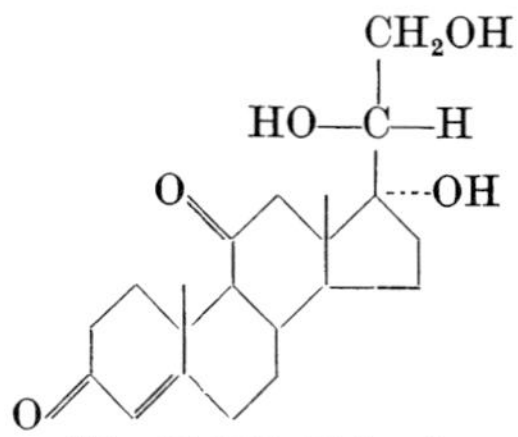

17 α,20 β,21-Trihydroxy-
Δ⁴-pregnene-3,11-dione

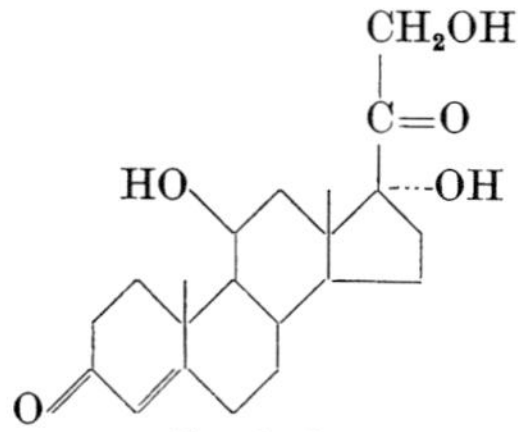

Cortisol

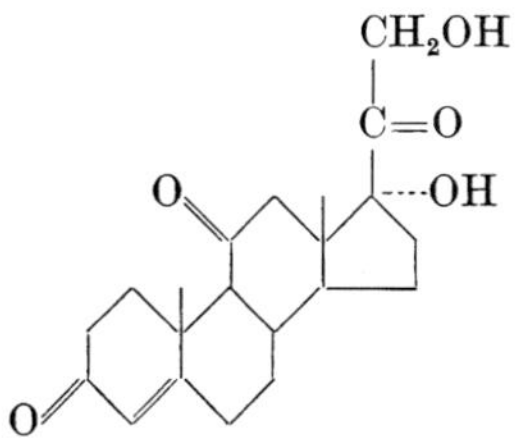

Cortisone

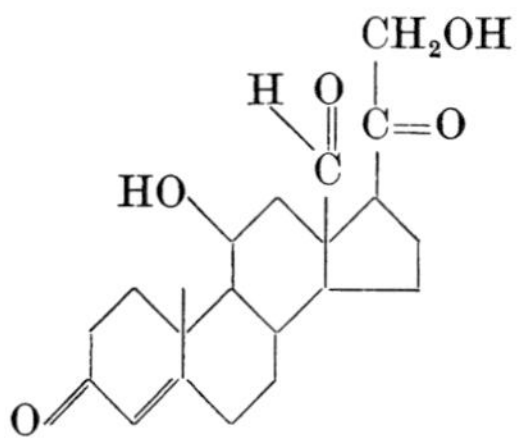

Aldosterone

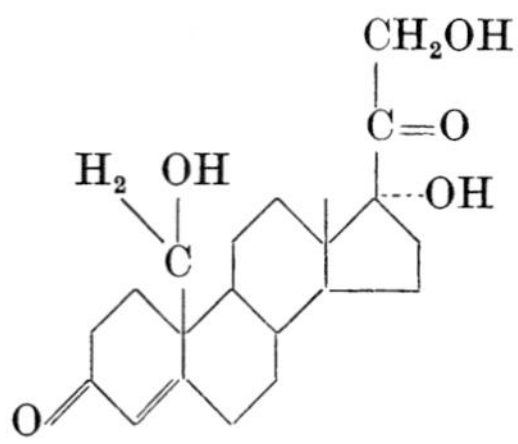

17 α,19,21-Trihydroxy-
Δ⁴-pregnene-3,20-dione

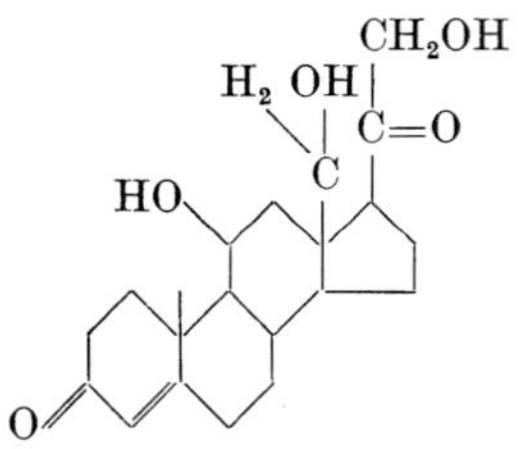

18-Hydroxycorticosterone

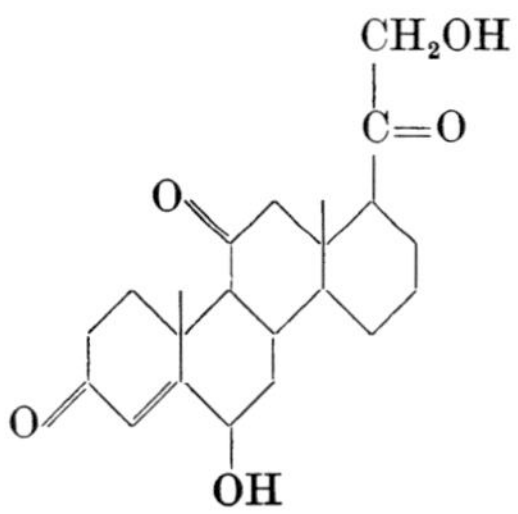

6β,21-Dihydroxy-Δ⁴-pregnene-
3,11,20-trione

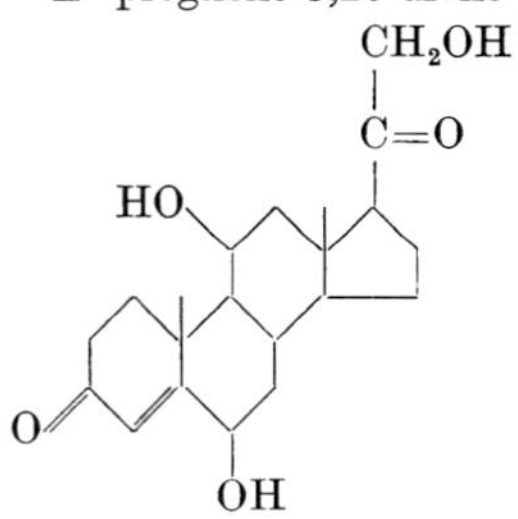

6 β,11 β,21-Trihydroxy-
Δ⁴-pregnene-3,20-dione

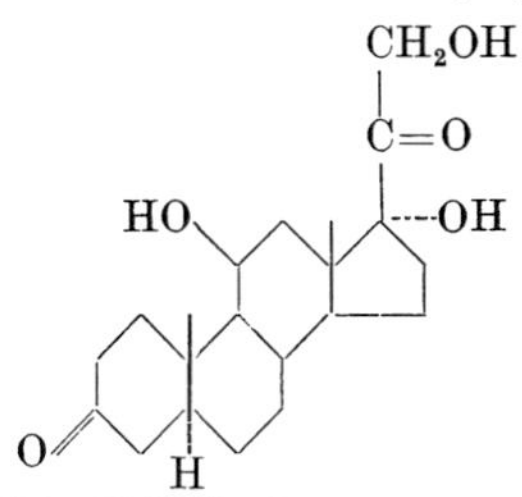

11 β,17 α,21-Trihydroxy-
allopregnane-3,20-dione

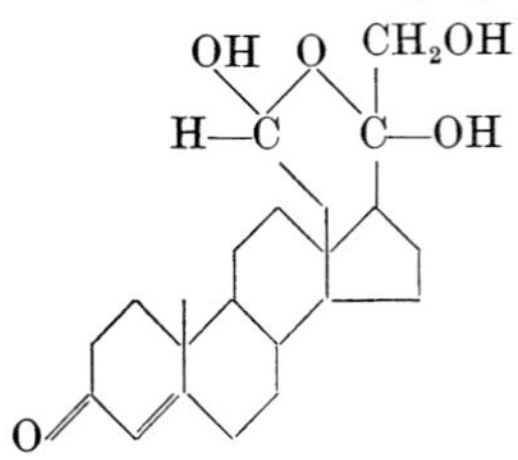

20,20,21-Trihydroxy-18-hemiacetal
(18→20)-Δ⁴-pregnen-3-one

Fig. 9. $C_{21}O_5$ steroids of adrenal tissue

Table 3. *C_{21} steroids of adrenal origin*

Group	Steroid	Species	Reference
$C_{21}O_2$	Δ^4-Pregnene-3,20-dione (Progesterone)	Rat and Pig	ANLIKER et al. (1956)
		Ox	BEALL and REICHSTEIN (1938), BEALL (1938), SABA and HECHTER (1955), HAYANO et al. (1956)
		Man (Tumor)	SALHANICK and BERLINER (1957)
	3β-Hydroxyallopregnan-20-one	Ox	SABA and HECHTER (1955),
	3β-Hydroxy-Δ_5-pregnen-20-one (Pregnenolone)		STAPLE et al. (1956), HAYANO et al. (1956)
$C_{21}O_3$	21-Hydroxy-Δ^4-pregnene-3,20-dione (Deoxycorticosterone, Cortexone)	Ox	REICHSTEIN and v. EUW (1939)
	17α-Hydroxy-Δ^4-pregnene-3,20-dione (17α-Hydroxyprogesterone)	Ox	PFIFFNER and NORTH (1940, 1941), v. EUW and REICHSTEIN (1941)
	Allopregnane-3β-, 17α, 20β-triol	Ox	STEIGER and REICHSTEIN (1938)
	Allopregnane-3β, 17α, 20α-triol	Ox	REICHSTEIN (1936c)
	3β, 17α-Dihydroxyallopregnan-20-one	Ox	REICHSTEIN (1936c), WINTERSTEINER and PFIFFNER (1936)
	17α-Hydroxyprogesterone	Pig	HEARD et al. (1956)
	11-Ketoprogesterone	Pig	HEARD et al. (1956)
	11β-Hydroxyprogesterone	Ox	HAYANO et al. (1956)
$C_{21}O_4$	Allopregnane-3β, 17α, 20β, 21-tetrol	Ox	STEIGER and REICHSTEIN (1938)
	3β, 17α, 21-Trihydroxyallopregnan-20-one	Ox	REICHSTEIN and GATZI (1938b)
	17α, 21-Dihidroxy-Δ^4-pregnene-3,20-dione (11-Deoxycortisol)	Ox	REICHSTEIN and GATZI (1938b), REICHSTEIN and v. EUW (1938)
		Pig	DOBRINER et al. (1954)
	3β,21-Dihydroxyallopregnane-11,20-dione	Ox	MASON et al. (1937), STEIGER and REICHSTEIN (1938)
	20β,21-Dihydroxy-Δ^4-pregnene-3,11-dione	Ox	REICHSTEIN and v. EUW (1939)
	11β,21-Dihydroxy-Δ^4-pregnene-3,20-dione (Corticosterone)	Ox	REICHSTEIN (1937), MASON et al. (1937), HAYANO et al. (1956), SABA and HECHTER (1955)
		Rat	ELLIOTT and SCHALLY (1955)
		Pig	KUIZENGA and CARTLAND (1940), DOBRINER et al. (1954)
		Dog	REICH and LEHNINGER (1955)
	21-Hydroxy-Δ^4-pregnene-3,11,20-trione (11-Dehydrocorticosterone)	Ox	MASON et al. (1937), REICHSTEIN and v. EUW (1938), KUIZENGA and CARTLAND (1939), HEARD et al. (1956)
		Pig	DOBRINER et al. (1954)
	19,21-Dihydroxy-Δ^4-pregnene-3,20-dione	Ox	MATTOX (1955), NEHER and WETTSTEIN (1956), MATTOX and MASON (1956)
		Pig	NEHER and WETTSTEIN (1956)
$C_{21}O_5$	Allopregnane-3β,11β,17α,20β,-21-pentol	Ox	WINTERSTEINER and PFIFFNER (1935), STEIGER and REICHSTEIN (1938), REICHSTEIN and GATZI (1938a)
	3β,11β,17α,21-Tetrahydroxy-allopregnan-20-one	Ox	v. EUW and REICHSTEIN (1942)

Table 3 (cont.)

Group	Steroid	Species	Reference
$C_{21}O_5$ (cont.)	3α,11β,17α,21-Tetrahydroxy-allopregnan-20-one (urocortisol)	Ox	REICHSTEIN (1936a), KUIZENGA and CARTLAND (1939), MASON et al. (1936a), WINTERSTEINER and PFIFFNER (1935)
	11β,17α,20β,21-Tetrahydroxy-Δ⁴-pregnen-3-one	Ox	REICHSTEIN (1936a), REICHSTEIN (1937)
	17α,20β,21-Trihydroxy-Δ⁴-pregnene-3,11-dione	Ox	REICHSTEIN and v. EUW (1938)
	11β,17α,21-Trihydroxy-Δ⁴-pregnene-3,20-dione (Cortisol)	Ox	REICHSTEIN (1937), MASON et al. (1938), KUIZENGA and CARTLAND (1939), CASPI et al. (1956a), SABA and HECHTER (1955)
		Rat	ELLIOTT and SCHALLY (1955), APRILE et al. (1956)
		Pig	DOBRINER et al. (1954), HEARD et al. (1956), APRILE et al. (1956)
		Man (Newborn)	GARDNER and TICE (1957)
		Guinea pig	BURSTEIN and NADEL (1956)
		Dog	REICH and LEHNINGER (1955)
	17α,21-Dihydroxy-Δ⁴-pregnene-3,11,20-trione (Cortisone)	Pig	DOBRINER et al. (1954)
		Ox	MASON et al. (1936), REICHSTEIN and v. EUW (1938), KUIZENGA and CARTLAND (1939)
	Aldosterone	Ox	SIMPSON et al. (1954), MATTOX and MASON (1956), HARMAN et al. (1954)
	19,17α,21-Trihydroxy-Δ⁴-pregnene-3,20-dione	Pig and Ox	NEHER and WETTSTEIN (1956)
	11β,19,21-Trihydroxy-Δ⁴-pregnene-3,20-dione	Ox and Pig	NEHER and WETTSTEIN (1956)
	6β,21-Dihydroxy-Δ⁴-pregnene-3,11,20-trione	Ox	NEHER and WETTSTEIN (1956)
	6β,11β,21-Trihydroxy-Δ⁴-pregnene-3,20-dione	Ox and Pig	NEHER and WETTSTEIN (1956)
	11β,17α,21-Trihydroxyallo-pregnane-3,20-dione	Ox and Pig	NEHER and WETTSTEIN (1956)
	20,20,21-Trihydroxy-18-hemi-acetal (18→20)-Δ⁴-pregnen-3-one	Pig	NEHER and WETTSTEIN (1956)
$C_{27}O_2$	Cholestane-3,6-dione	Man (Adenoma)	KELLER (1956)
	20β-Hydroxycholesterol	Ox	SOLOMON et al. (1956)

B. Adrenocortical steroids and their derivatives in other tissues

The placenta has been implicated as an organ capable of synthesizing corticoids. Extracts from human placenta can promote liver glycogen deposition in adrenalectomized rats (R. JOHNSON and HAINES 1952), a test that is considered to be highly specific for glucocorticoids (Chapter 7). Also, two biologically-active corticoids, cortisol and cortisone have been isolated in addition to the inactive metabolite of cortisone, 3α,17,21-trihydroxy pregnane-11,20-dione (urocortisone), as indicated in Table 4. When placentas were incubated in Krebs-Ringer solution and maternal plasma, the medium contained a significantly increased (2.5-fold) amount of 17-hydroxycorticoids as measured by the Nelson-Samuels procedures

(LITTLE and ROSSI 1957). All these experiments point to the possibility of a biosynthetic mechanism for corticoid production in the placenta, but certainly additional evidence is needed.

Table 4. *Adrenocortical steroids in various tissues*

Steroid	Species	Tissue	Reference
Cortisol, Cortisone and Urocortisone	Man	Placenta	BERLINER et al. (1956)
Cortisol (Paper chromatographic evidence)	Man	Synovial fluid	COPE and SEWELL (1956)
Glycogenic material	Pig	Spleen	FAZEKAS (1957)

Paper chromatographic evidence suggests the presence of cortisol in synovial fluid, and a rather high amount of glycogenic material has been reported present in spleen extracts (Table 4).

C. Adrenocortical steroids and their derivatives in bile

The two reduced corticoids quantitatively important in urine, $3\alpha,11\beta,17,21$-tetrahydroxypregnan-20-one (urocortisol) and urocortisone, have been identified in extracts of bovine bile. Guinea pig bile contains a reasonable concentration of cortisol (Table 5).

Table 5. *Adrenocortical steroids in bile*

Steroid	Species	Reference
Urocortisol and Urocortisone	Ox	GLICK (1957)
Cortisol	Guinea pig	EIK-NES et al. (1957)

D. Adrenocortical steroids and their derivatives in blood

The blood contains most of the steroids present in the adrenal as well as many of the metabolites found in urine, as indicated in Tables 6 and 7.

Table 6. *C_{19} adrenocortical steroids and their metabolites in blood*

Steroid	Species	Reference
Dehydroepiandrosterone	Man	MIGEON and PLAGER (1954), TAMM et al. (1958)
Δ^4-Androstene-3,17-dione	Man (Adrenal venous)	E. ROMANOFF et al. (1953)
Androsterone	Man	MIGEON (1956), TAMM et al. (1958)
Etiocholanolone	Man	TAMM et al. (1958)
11β-Hydroxy-Δ^4-androstene-3,17-dione	Man (Adrenal venous)	E. ROMANOFF et al. (1953), PINCUS and ROMANOFF (1955)
	Man (Peripheral)	COOPER et al. (1958), TAMM et al. (1958)
	Rat (Adrenal venous)	LONGWELL et al. (1956)
11β-Hydroxyetiocholanolone	Man	TAMM et al. (1958), SAVARD (1957)
11-Ketoetiocholanolone	Man	TAMM et al. (1958)
11β-Hydroxyandrosterone	Man	SAVARD (1957)

Table 7. C_{21} *adrenocortical steroids and their metabolites in blood*

Steroid	Species	Reference
Progesterone	Man	ZANDER and SUMMER (1954)
	Goat	SHORT (1957)
	Ox	SHORT (1957)
	Pig	SHORT (1957)
	Sheep	SHORT (1957)
	(Adrenal venous)	
	Rat	WIEST (1958)
20β-Hydroxy-Δ^4-pregnen-3-one	Rat	WIEST (1958)
20α-Hydroxy-Δ^4-pregnen-3-one	Man	FORBES and ZANDER (1956)
	Sheep	SHORT (1957)
Deoxycorticosterone	Dog (Adrenal venous)	FARRELL et al. (1954)
11-Deoxycortisol	Man (Congenital adrenal hyperplasia with hypertension)	EBERLEIN and BONGIOVANNI (1956)
	Dog (Adrenal venous)	CARSTENSEN (1956)
	Man (Adrenal venous)	TOUCHSTONE (1958)
11-Dehydrocorticosterone	Man	SALVADORI et al. (1956)
Corticosterone	Dog (Adrenal venous)	ZAFFARONI and BURTON (1953)
	Rat (Adrenal venous)	LONGWELL et al. (1955, 1956), VOGT (1957)
	Man (Peripheral)	SALVADORI et al. (1956), AYRES et al. (1957), COOPER et al. (1958), TAMM et al. (1958)
	Man (Adrenal venous)	PINCUS and ROMANOFF (1955)
	Mouse	SOUTHCOTT et al. (1956a, 1956b)
11β,20α,21-Trihydroxy-Δ^4-pregnen-3-one	Mouse	SOUTHCOTT et al. (1956a, 1956b)
Cortisone	Man	DE HACHEN (1956), SALVADORI et al. (1956), SAVARD (1957), COOPER et al. (1958)
Cortisol	Man (Adrenal venous)	PINCUS and ROMANOFF (1955)
	Man (Peripheral)	DE HACHEN (1956), AYRES et al. (1957), COOPER et al. (1958), TAMM et al. (1958)
	Dog (Adrenal venous)	ZAFFARONI and BURTON (1953)
Aldosterone	Monkey (Adrenal venous)	SIMPSON et al. (1952)
	Dog (Adrenal venous)	FARRELL et al. (1954)
	Man	AYRES et al. (1957)
Urocortisol glucosiduronate	Man	VERMEULEN (1956), SAVARD (1957)
Urocortisone glucosiduronate	Man	VERMEULEN (1956), SAVARD (1957)
Urocortisol	Man	REDDY et al. (1955), DE HACHEN (1956), VERMEULEN (1956c), TAMM et al. (1958)
Urocortisone	Man	VERMEULEN (1956c), KLEIN et al. (1957), TAMM et al. (1958)
3α,17α,21-Trihydroxypregnan-20-one	Man (Congenital adrenal hyperplasia with hypertension)	EBERLEIN and BONGIOVANNI (1956)

The adrenal androgens, 11 β-hydroxy-Δ^4-androstene-3,17-dione, Δ^4-androstene-3,17-dione, and dehydroepiandrosterone, are present in blood together with the metabolites androsterone and 3 α-hydroxyetiocholan-17-one (etiocholanolone) (Table 6). Some of these androgens are present in free form, while others seem to be conjugated with glucuronic acid, as indicated by the fact that hydrolysis with β-glucuronidase releases a significantly greater quantity of 17-ketosteroids.

A number of C_{21} adrenocortical steroids have been identified in human and animal blood (Table 7). Cortisol and corticosterone are consistently present in the blood of all species, and here an interesting variation has been noted by Bush (1953) and elaborated by others. Bush found that the proportion of plasma cortisol to corticosterone varies in mammalian species.

There is a characteristic value for any given species. A striking contrast exists between the rat and the guinea pig. Cortisol still has not been conclusively demonstrated in either blood or adrenal of the rat, while corticosterone is easily detected. On the other hand, corticosterone has not been demonstrated in the guinea pig adrenal or in the animal's blood. In an *in vivo* experiment on guinea pigs with acetate-1-C^{14} (Burstein and Nadel 1956), it was possible to demonstrate the formation of cortisol-C^{14} but not corticosterone-C^{14} by the reverse isotope dilution technique. Other species secrete varying amounts of the two steroids. Man, monkey, and sheep exhibit high ratios of cortisol to corticosterone, 10, 20, and 18, respectively; cat, dog, and ferret have intermediate ratios of 5, 3, and 2; while the rabbit has a rather low ratio of approximately 0.05 (Bush 1953).

In some instances a peripheral metabolite may well be identical to a product secreted by the gland. Specifically, it is hard to decide whether the 20 α and 20 β reduced products of progesterone may be secreted or be peripherally-reduced metabolites of progesterone (Table 7). Similarly, Δ^4-androstene-3,17-dione may be secreted by the adrenal and may also enter the blood as a peripheral metabolite of 17 α-hydroxyprogesterone.

Corticosterone is found in peripheral human blood in amounts ranging from 0.5 to 2.0 μg-%, with a mean of 1.1 $\pm$ 0.3 μg-%, making a ratio of cortisol to corticosterone of 15/1. The blood corticosterone content is increased by ACTH and decreased by the synthetic glucocorticoid, 9 α-fluoroprednisolone (= inhibition of ACTH). As expected, the production of corticosterone is low in Addison's disease and in the virilizing adrenogenital syndrome, and increased in the blood of patients with adrenocortical carcinoma (see Chapters 8 and 9).

I. Steroid conjugates in blood

The presence of conjugated corticoids in blood was demonstrated first by Savard et al. (1952), who obtained good evidence for corticosteroids conjugated with glucuronic acid in a dialyzate obtained from dog blood. Similar findings were reported by Vliers et al. (1955), who demonstrated that two-thirds of the total corticoids of plasma could be liberated by incubation of the plasma for 24 hours with either spleen β-glucuronidase or an extract prepared from the snail, *Helix pomatia*.

Corticoid conjugates in blood have been studied by Bongiovanni et al. (1954) in normal individuals as well as those under stress and those suffering from liver disease. Under normal conditions about one-half of the blood corticoids were found to be conjugated as glucosiduronates, as shown by the hydrolysis effected by β-glucuronidase. In Cushing's disease the conjugated fraction was significantly greater than the free fraction. Stress, ACTH, or administration of cortisone or cortisol led to an elevation in both free and conjugated corticoids, while the rate of conjugation was significantly decreased in patients with liver disease. Other reports of corticoid conjugates in blood include those of KLEIN et al. (1955a, b), HAYDAR et al. (1955), WEICHSELBAUM and MARGRAF (1955), BONGIOVANNI and EBERLEIN (1955), and SOUTHCOTT et al. (1956a, b). BONGIOVANNI and EBERLEIN (1955) demonstrated that after the administration of free urocortisone it could be detected as the glucosiduronate in blood, and VERMEULEN (1956) actually demonstrated the presence of both urocortisone and urocortisol glucosiduronates in blood after the administration of cortisol or ACTH to human patients (Table 7).

The presence of phosphate conjugates in human plasma has been claimed by OERTEL and EIK-NES (1958), who reported that the incubation of human plasma with acid or alkaline phosphatase liberated dehydroepiandrosterone.

E. Adrenocortical steroids and related compounds in urine

Since the adrenal produces all four types of steroid hormones — corticoids, androgens, estrogens, and progestational material — the urine should contain these steroids and/or their representative metabolites. Those urinary steroids derived all or in part from adrenocortical precursors are indeed numerous and are here recorded in a series of 3 tables classified on the basis of the number of carbon atoms contained in the molecule. The structures are illustrated in corresponding figures.

I. C_{18} steroids

Nine urinary estrogens that may originate directly or indirectly from the adrenal are illustrated in Table 8 and Fig. 10.

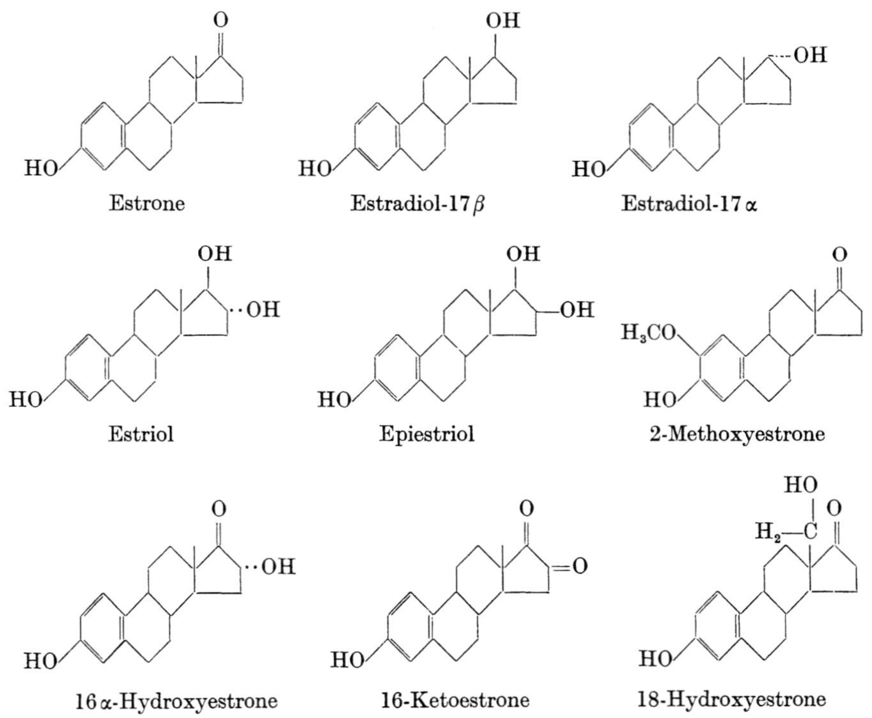

Fig. 10. C_{18} urinary steroids that may be derived directly or indirectly from the adrenal

Table 8. C_{18} *adrenocortical steroids and related compounds in urine*

Class	Urinary steroid	Precursor in adrenal	Species	Reference
$C_{18}O_2$	Estrone	Estrone or Estradiol-17β	Man	DOISY et al. (1929), BUTENANDT (1929), DINGEMANSE et al. (1930, 1938), ENGEL et al. (1952)
			Horse	DE JONGH et al. (1931), DEULOFEU and FERRARI (1934), HAUSSLER (1934)
			Ox	MARKER (1939a, b)
	Estradiol-17β	Estrone or Estradiol-17β	Man	G. SMITH et al. (1939), ENGEL et al. (1952)
			Horse	WINTERSTEINER et al. (1935), LEVIN (1945)
	Estradiol-17α	Estrone or Estradiol-17β	Horse	HIRSCHMANN and WINTERSTEINER (1937/38)
$C_{18}O_3$	Estriol	Estrone or Estradiol-17β	Man	MARRIAN (1930), DOISY et al. (1930), ENGEL et al. (1952)
	18-Hydroxyestrone	18-Hydroxyestrone (?) or 18-Hydroxyestradiol-17β (?)	Man	LAKE and MARRIAN (1958)
	16α-Hydroxyestrone	Estrone or Estradiol-17β	Man	WATSON and MARRIAN (1955), MARRIAN et al. (1957)
	2-Methoxyestrone	Estrone or Estradiol-17β	Man	LAKE and MARRIAN (1958)
	16-Ketoestrone	Estrone or Estradiol-17β	Man	SERCHI (1953)
	16-Epiestriol	Estrone or Estradiol-17β	Man	MARRIAN and BAULD (1955), WATSON and MARRIAN (1956)

II. C_{19} steroids

The $C_{19}O$, $C_{19}O_2$, $C_{19}O_3$, and $C_{19}O_4$ adrenocortical steroids and related compounds in urine are summarized in Table 9. One $C_{19}O$ compound, Δ^{16}-androsten-3α-ol (Fig. 11), is present in urine of normal individuals and is increased in patients with adrenocortical hyperactivity.

Three types of $C_{19}O_2$ steroids have been isolated from the urine (Fig. 11), including those of the Δ^5-3β-hydroxy type, the reduced ring A type, and a limited amount of a Δ^4-3-ketone. In the last named category only one $C_{19}O_2$ has been found, i.e., Δ^4-androstene-3,17-dione, and then only in conditions in which greatly increased quantities are present in the blood. Of the $C_{19}O_2$ group of urinary steroids, androsterone and etiocholanolone make up the bulk of the urinary 17-ketosteroids in human urine.

The C-3, C-16 oxygenated C_{19} steroids are included in the list (Table 9), but there is no real certainty that they are all of adrenal origin. Three steroids of this group have been isolated that have the 3β-hydroxy grouping in common, and on this basis they appear to be closely related and most likely of common origin. Their structures are unique in lacking oxygen at C-17. This fact poses a most difficult problem for postulating the biosynthetic route.

The $C_{19}O_3$ steroids (Fig. 12) having a C-11 oxygen function are clearly of adrenal origin. Those of this group lacking this function may or may not be of adrenal origin, since 6- and 16-hydroxylation can occur not only in the adrenal but in other steroid-producing organs and in the metabolism of C_{21} steroids by peripheral tissues.

Table 9. *C_{19} adrenocortical steroids and related compounds in urine*

Class	Urinary steroid	Precursor in adrenal	Species	Reference
$C_{19}O$	Δ^{16}-Androsten-3α-ol	Testosterone (?)	Man	BROOKSBANK and HASLEWOOD (1949), MASON and SCHNEIDER (1950), A. MILLER et al. (1953), ENGEL et al. (1953)
$C_{19}O_2$	Δ^4-Androstene-3,17-dione	Δ^4-Androstene-3,17-dione or Dehydroepiandrosterone	Man	A. MILLER et al. (1953), LIEBERMAN et al. (1953)
	Dehydroepiandrosterone	Dehydroepiandrosterone	Man	BUTENANDT (1931), CALLOW and CALLOW (1939, 1940), ENGEL et al. (1941), HIRSCHMANN (1940)
			Ox	MARKER (1939a, b), HOLTZ (1954)
			Horse	OPPENAUER (1941)
	Δ^5-Androstene-3β,17β-diol	Dehydroepiandrosterone	Man	HIRSCHMANN and HIRSCHMANN (1945), MASON and KEPLER (1945b), SCHILLER et al. (1945), UNGAR and DORFMAN (1953)
	3α(?)-Hydroxy-Δ^5-androsten-17-one	?	Man	KELLER (1956)
	Androsterone	Major: Dehydroepiandrosterone and Δ^4-Androstene-3,17-dione; Minor: 11-Deoxycortisol-17α,	Man	BUTENANDT (1932), BUTENANDT and WESTPHAL (1934), ENGEL et al. (1941), CALLOW (1939), CALLOW and CALLOW (1939), MASON (1948)
		Hydroxyprogesterone	Ox	MARKER (1939a, b)
	Epiandrosterone	Dehydroepiandrosterone, Δ^4-Androstene-3,17-dione	Man Horse	PEARLMAN (1942) OPPENAUER (1941)
	Etiocholanolone	Major: Dehydroepiandrosterone and Δ^4-Androstene-3,17-dione; Minor:	Man	CALLOW (1939), CALLOW and CALLOW (1939, 1940), HIRSCHMANN (1940), EBERLEIN and BONGIOVANNI (1956)
		Deoxycortisol,17α, Hydroxy-progesterone	Guinea pig Ox	PERON and DORFMAN (1956) HOLTZ (1954)
	3β-Hydroxyetiocholane-17-one	Dehydroepiandrosterone and Δ^4-Androstene-3,17-dione	Man	DOBRINER and LIEBERMAN (1952), FUKUSHIMA and GALLAGHER (1957)
	Etiocholane-3,17-dione	Dehydroepiandrosterone and Δ^4-Androstene-3,17-dione	Man	LIEBERMAN et al. (1948)
	Androstane-3,17-dione	Dehydroepiandrosterone and Δ^4-Androstene-3,17-dione	Man	LIEBERMAN et al. (1948)
	Etiocholane-3α,17β-diol	Dehydroepiandrosterone and Δ^4-Androstene-3,17-dione	Man	A. MILLER and DORFMAN (1949), LIEBERMAN et al. (1953), FUKUSHIMA and GALLAGHER (1957)

Table 9 (cont.)

Class	Urinary steroid	Precursor in adrenal	Species	Reference
$C_{19}O_2$ (cont.)	Androstane-3α,17β-diol	Dehydroepiandrosterone and Δ⁴-Androstene-3,17-dione	Man	LIEBERMAN et al. (1953)
	Δ¹-Androstene-3,17-dione (5α)	?	Man	LIEBERMAN et al. (1948)
	3β-Hydroxyandrostan-16-one	?	Horse	HEARD and McKAY (1939), OPPENAUER (1941), HOFFMAN and LOTT (1951)
	Androstane-3β,16β-diol	?	Horse	BROOKS and KLYNE (1956)
	Androstane-3β,16α-diol	?	Horse	BROOKS and KLYNE (1956)
$C_{19}O_3$	11β-Hydroxy-etiocholanolone	11β-Hydroxy-Δ⁴-androstene-3,17-dione, Cortisol, 21-Deoxycortisol	Guinea pig Man	PERON and DORFMAN (1956) LIEBERMAN and DOBRINER (1948), PLANTIN and BIRKE (1955)
	11-Ketoetio-cholanolone	11β-Hydroxy-Δ⁴-androstene-3,17-dione, Cortisol, 21-Deoxycortisol	Guinea pig Man	PERON and DORFMAN (1956), BURSTEIN and DORFMAN (1958) LIEBERMAN and DOBRINER (1946, 1948)
	11β-Hydroxy-androsterone	11β-Hydroxy-Δ⁴-androstene-3,17-dione, Cortisol, 21-Deoxycortisol	Guinea pig Man	PERON and DORFMAN (1956) MILLER et al. (1946), MASON (1948), LIEBERMAN and DOBRINER (1948)
	Etiocholane-3,11,17-trione	11β-Hydroxy-Δ⁴-androstene-3,17-dione, Cortisol, 21-Deoxycortisol	Man	FUKUSHIMA and GALLAGHER (1957b)
	3β-Hydroxy-Δ⁵-androstene-7,17-dione	Dehydroepiandrosterone	Man	FUKUSHIMA and GALLAGHER (1957b)
	3β-Hydroxyetio-cholane-11,17-dione	11β-Hydroxy-Δ⁴-androstene-3,17-dione, Cortisol, 21-Deoxycortisol	Man	FUKUSHIMA and GALLAGHER (1957b)
	11β-Hydroxy-etiocholane-3,17-dione	11β-Hydroxy-Δ⁴-androstene-3,17-dione, Cortisol, 21-Deoxycortisol	Man	FUKUSHIMA and GALLAGHER (1957b)
	Etiocholane-3α,16α,17β-triol	?	Man	LIEBERMAN et al. (1953), LIEBERMAN and DOBRINER (1950)
	Androstane-3α,16α,17β-triol	?	Man	LIEBERMAN and DOBRINER (1950), LIEBERMAN et al. (1953)
	Δ⁵-Androstene-3β,16α,17β-triol	Δ⁵-Androstene-3β,16α,17β-triol or Dehydroepiandrosterone	Man	HIRSCHMANN (1943), MARRIAN (1944), FINKELSTEIN et al. (1953), FOTHERBY et al. (1957)
	11β-Hydroxyepi-androsterone	Cortisol and 11β-Hydroxy-Δ⁴-androstene-3,17-dione	Man	KEMP et al. (1954)
	11-Ketoandro-sterone	Cortisol and 11β-Hydroxy-Δ⁴-androstene-3,17-dione	Man	LIEBERMAN et al. (1950)

Table 9 (cont.)

Class	Urinary steroid	Precursor in adrenal	Species	Reference
$C_{19}O_3$ (cont.)	11β-Hydroxy-Δ^4-androstene-3,17-dione	11β-Hydroxy-Δ^4-androstene-3,17-dione and Cortisol	Man Ox	SALAMON and DOBRINER (1952) HOLTZ (1954)
	Δ^5-Androstene-3β,16β,17β-triol	Dehydroepiandrosterone (?)	Man	FOTHERBY (1957)
	3β,16α-Dihydroxy-Δ^5-androsten-17-one	Dehydroepiandrosterone	Man	FOTHERBY et al. (1957)
$C_{19}O_4$	6β,11β-Dihydroxy-Δ^4-androstene-3,17-dione	11β-Hydroxy-Δ^4-androstene-3,17-dione and Cortisol	Guinea pig	PERON and DORFMAN (1957)

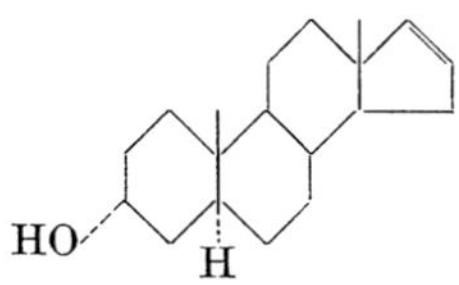
Δ^{16}-Androsten-3α-ol

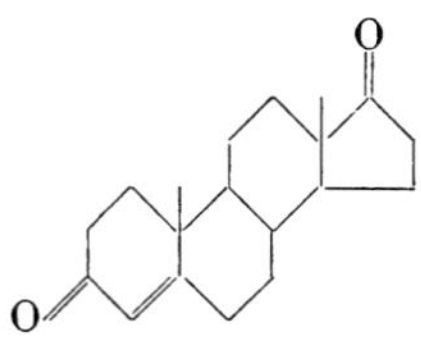
Δ^4-Androstene-3,17-dione

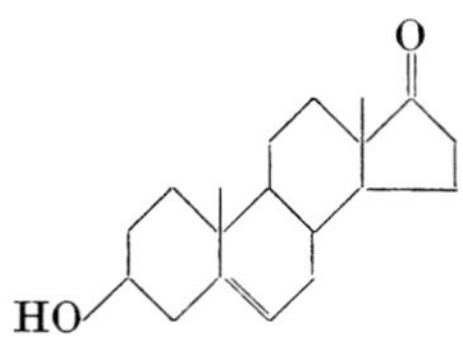
Dehydroepiandrosterone

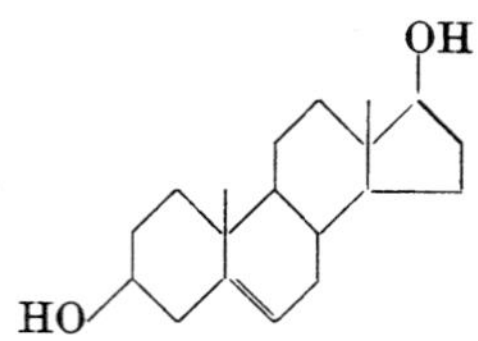
Δ^5-Androstene-3β,17β-diol

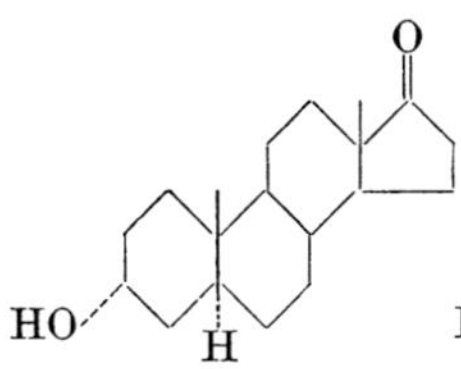
Androsterone

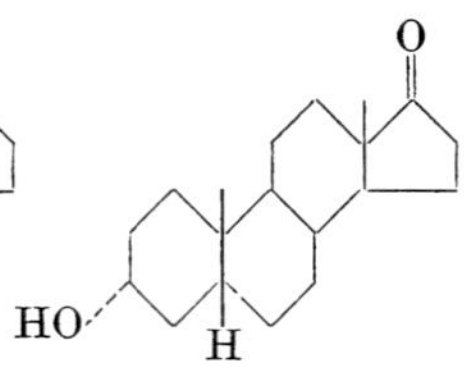
Etiocholanolone

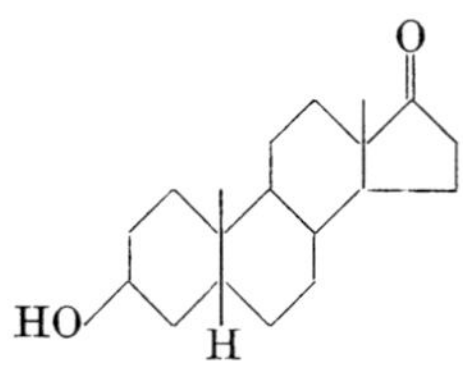
Epietiocholanolone

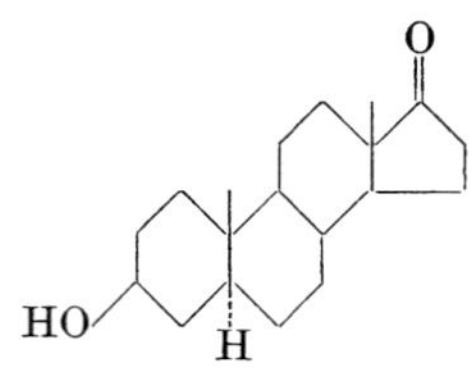
Epiandrosterone

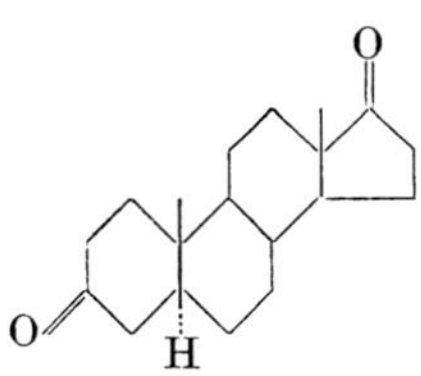
Androstane-3,17-dione

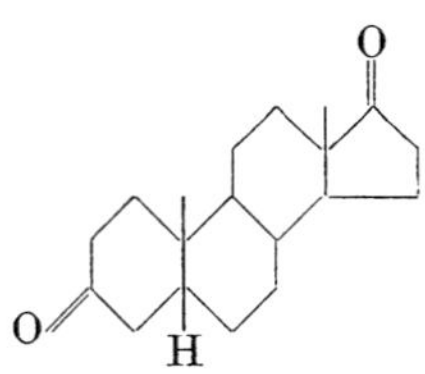
Etiocholane-3,17-dione

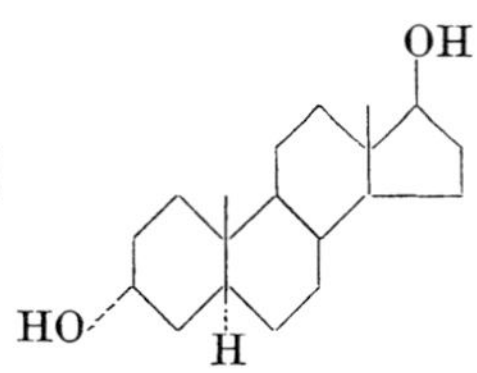
Androstane-3α,17β-diol

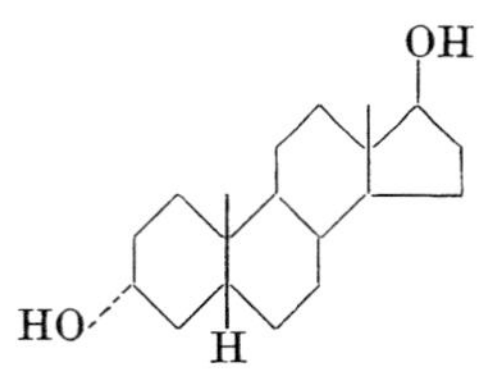
Etiocholane-3α,17β-diol

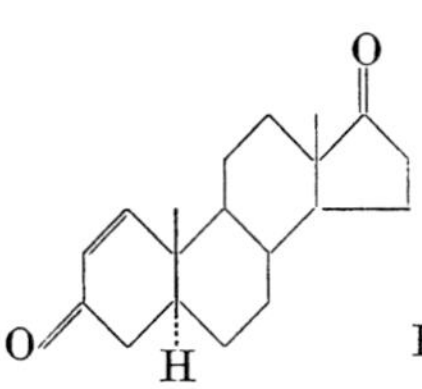
Δ^1-Androstene-3,17-dione

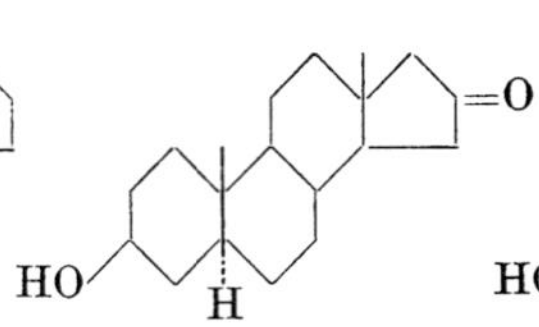
3β-Hydroxy-androstan-16-one

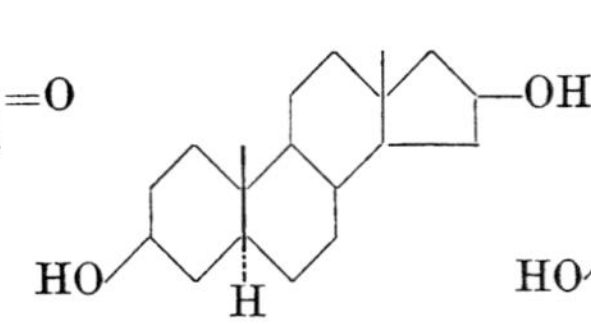
Androstane-3β,16β-diol

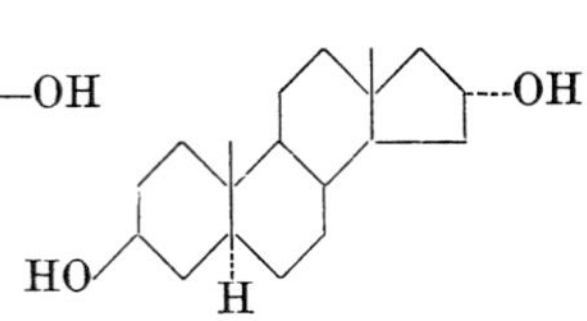
Androstane-3β,16α-diol

Fig. 11. $C_{19}O$ and $C_{19}O_2$ urinary steroids that may arise directly or indirectly from the adrenal

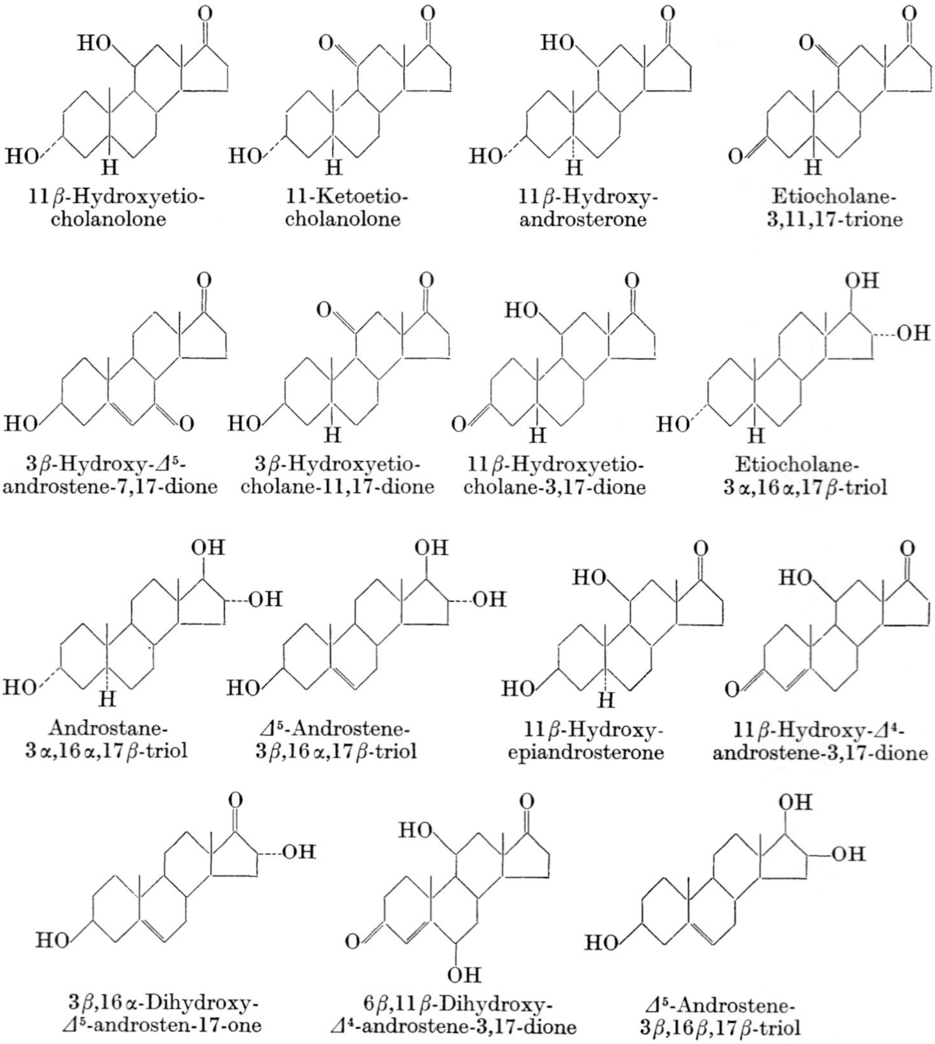

Fig. 12. $C_{19}O_3$ and $C_{19}O_4$ urinary steroids that may arise directly or indirectly from the adrenal

The $C_{19}O_4$ steroid, 6β,11β-dihydroxy-Δ^4-androstene-3,17-dione (Fig. 12), has been observed only in guinea pig urine, and it is not clear whether this steroid is secreted by the adrenal or whether the secreted compound is actually 11β-hydroxy-Δ^4-androstene-3,17-dione, which becomes 6β-hydroxylated during passage through the liver. Still other alternatives include the possibility that 6β-hydroxycortisol is secreted and converted peripherally to the $C_{19}O_4$ compound, or that the cortisol secreted by the adrenal is converted peripherally to 6β-hydroxycortisol, which then may be transformed to the $C_{19}O_4$ steroid.

III. C_{21} steroids

The C_{21} urinary steroids of possible adrenal origin vary from those with 2 oxygen functions to those carrying 6 oxygen groups (Table 10). One C_{21} steroid with a

Table 10. *C_{21} adrenocortical steroids and related compounds in urine*

Class	Urinary steroid	Precursor in adrenal	Species	Reference
$C_{21}O_2$	3α-Hydroxy-pregnan-20-one	Progesterone	Pig	Marker and Rohrmann (1939b)
			Man	Marker and Kamm (1937), Dobriner et al. (1944), Dobriner (1951)
	3α-Hydroxyallo-pregnan-20-one	Progesterone	Man	Marker and Kamm (1937), Dobriner et al. (1944), Lieberman et al. (1948)
	3β-Hydroxyallo-pregnan-20-one	Progesterone	Man	Pearlman et al. (1942)
			Horse	Marker et al. (1938), Savard et al. (1958)
			Pig	Marker and Rohrmann (1939b)
	Pregnane-3α,20α-diol	Progesterone	Man	Engel et al. (1941), Heard and Hoffman (1941a), Eberlein and Bongiovanni (1956), Fukushima and Gallagher (1957), Marrian (1929), Venning and Browne (1937), Butler and Marrian (1937), Hirschmann (1940), Mason and Kepler (1945b)
	Allopregnane-3α,20α-diol	Progesterone	Man	Hartmann and Locher (1935), Marker and Kamm (1937), Lieberman et al. (1953)
			Ox	Marker (1938a, b)
			Horse	Marker et al. (1937), Savard et al. (1958)
	Δ^5-Pregnene-3β,20α-diol	Pregnenolone	Man	Schiller et al. (1945), Mason and Kepler (1945b), Hirschmann and Hirschmann (1945)
			Horse	Marker and Rohrmann (1938)
	Allopregnane-3β,20β-diol	Progesterone	Horse	Bauld and Heard (1948), Savard et al. (1958)
	3α-Hydroxy-Δ^{16}-pregnen-20-one		Man	Fukushima et al. (1954c)
$C_{21}O_3$	3α,17α-Dihydroxy-pregnan-20-one	17α-Hydroxy-progesterone	Man	Mason and Strickler (1947), Eberlein and Bongiovanni (1956)
	Pregnane-3α,17α,20α-triol	17α-Hydroxy-progesterone	Man	Cox and Marrian (1953), Eberlein and Bongiovanni (1956), Fukushima and Gallagher (1957)
	3α,17α-Dihydroxy-allopregnan-20-one	17α-Hydroxy-progesterone	Man	Fukushima et al. (1954c)
	Allopregnane-3α,17α,20α-triol	17α-Hydroxy-progesterone	Man	Fukushima et al. (1956), Fukushima and Gallagher (1957)
	Allopregnane-3α,17α,20β-triol	17α-Hydroxy-progesterone	Man	Fukushima and Gallagher (1957)
	Pregnane-3α,11β,20α-triol	11β-Hydroxy-progesterone	Man	Fukushima et al. (1954c)
	3α,11β-Dihydroxy-pregnan-20-one	11β-Hydroxy-progesterone	Man	Fukushima et al. (1954c)

Table 10 (cont.)

Class	Urinary steroid	Precursor in adrenal	Species	Reference
$C_{21}O_3$ (cont.)	3α,20α-Dihydroxy-pregnan-11-one	11β-Hydroxy-progesterone	Man	LIEBERMAN et al. (1953)
	11-Ketopregnane-3α,20α-diol	11β-Hydroxy-progesterone	Man	FUKUSHIMA et al. (1956), FUKUSHIMA and GALLAGHER (1957)
	Δ^5-Pregnene-3β,17α,20α-triol	17α-Hydroxy-pregnenolone	Man	HIRSCHMANN and HIRSCHMANN (1950)
	Δ^5-Pregnene-3β-16α,17α-triol	Pregnenolone	Man	HIRSCHMANN and HIRSCHMANN (1949)
	3β-17α-Dihydroxy-Δ^5-pregnen-20-one	17α-Hydroxy-pregnenolone	Man	HIRSCHMANN and HIRSCHMANN (1947)
	3α,21-Dihydroxy-pregnan-20-one	Deoxycortico-sterone	Man	RICHARDSON et al. (1955), EBERLEIN and BONGIOVANNI (1956)
$C_{21}O_4$	Pregnane-3α,11β,17α,20α-tetrol	21-Deoxycortisol or 21-Deoxy-cortisone	Man	FUKUSHIMA et al. (1956), FUKUSHIMA and GALLAGHER (1957)
	3α,17α,20α-Tri-hydroxypregnan-11-one	21-Deoxycortisol or 21-Deoxy-cortisone	Man	FINKELSTEIN et al. (1953), FUKUSHIMA et al. (1956)
	3α,17α,21-Tri-hydroxypregnane-20-one	11-Deoxycortisol	Man	ROSSELET et al. (1954), BONGIOVANNI and EBERLEIN (1955b), RICHARDSON et al. (1955), DOHAN et al. (1955), UNGAR and DORFMAN (1956), EBERLEIN and BONGIOVANNI (1956), TOUCHSTONE et al. (1957)
	Corticosterone	Corticosterone	Man	DOHAN et al. (1955)
	11-Dehydro-corticosterone	11-Dehydro-corticosterone	Man	DOHAN et al. (1955)
	3α,11β,21-Tri-hydroxypregnan-20-one	Corticosterone or 11-Dehydro-corticosterone	Man	DOHAN et al. (1955)
	3α,11β,21-Tri-hydroxyallo-pregnan-20-one	Corticosterone or 11-Dehydro-corticosterone	Man	DOHAN et al. (1955)
	3α,21-Dihydroxy-pregnane-11,20-dione	Corticosterone or 11-Dehydro-corticosterone	Man	DOHAN et al. (1955)
$C_{21}O_5$	Cortisone	Cortisol	Man	SCHNEIDER (1952), DOHAN et al. (1955), HOLNESS et al. (1956)
	Cortisol	Cortisol	Man	SCHNEIDER (1952), HOLNESS et al. (1956)
			Guinea Pig	NADEL and BURSTEIN (1956)
	11β,17α,21-Tri-hydroxypregnane-3,20-dione	Cortisol	Man	SCHNEIDER (1952)
	11β,17α,20α,21-Tetrahydroxy-Δ^4-pregnen-3-one	Cortisol	Man	R. PETERSON et al. (1958)
	Urocortisone	Cortisol	Man	SCHNEIDER (1952), DOBRINER and LIEBERMAN (1952), DOHAN et al. (1955), HOLNESS et al. (1956)

Table 10 (cont.)

Class	Urinary steroid	Precursor in adrenal	Species	Reference
$C_{21}O_5$ (cont.)	Urocortisol	Cortisol	Man	DOBRINER and LIEBERMAN (1952), DOHAN et al. (1955), HOLNESS et al. (1956), BUSH and WILLOUGHBY (1957)
	Cortol	Cortisol	Man	FUKUSHIMA et al. (1955)
	β-Cortol	Cortisol	Man	FUKUSHIMA et al. (1955)
	17α,20β,21-Trihydroxy-Δ^4-pregnene-3,11-dione	Cortisol	Man	HOLNESS et al. (1956)
	Cortolone	Cortisol	Man	FUKUSHIMA et al. (1955)
	β-Cortolone	Cortisol	Man	FUKUSHIMA et al. (1955)
	Aldosterone	Aldosterone	Man	MAJNARICH and DILLON (1954), LUETSCHER et al. (1954, 1956), LUETSCHER (1956), WETTSTEIN et al. (1956)
	17α,21-Dihydroxy-pregnane-3,11,20-trione	Cortisol	Man	SCHNEIDER (1952)
	3α,11β,17α,21-Tetrahydroxyallo-pregnan-20-one	Cortisol	Man	BUSH (1957)
$C_{21}O_6$	6β-Hydroxy-cortisol	Cortisol	Man Guinea pig	NADEL et al. (1956) BURSTEIN and DORFMAN (1955), NADEL and BURSTEIN (1956)
	2α-Hydroxy-cortisol	Cortisol	Guinea pig	BURSTEIN (1956), PERON and DORFMAN (1956), NADEL and BURSTEIN (1956)

single oxygen function was reported by MARKER and LAWSON (1938) in human pregnancy urine. The origin of this steroid is uncertain, and no confirmatory isolation has been reported. In the $C_{21}O_2$ class, 6 metabolites of progesterone have been isolated and Δ^5-pregnene-3β,20α-diol, most likely a reduction product of pregnenolone, has been demonstrated. The tissue precursor of 3α-hydroxy-Δ^{16}-pregnen-20-one is unknown.

The 11 $C_{21}O_3$ urinary steroids isolated from human urine appear to be metabolites of 17α-hydroxyprogesterone, 11β-hydroxyprogesterone, and deoxycorticosterone. Other urinary metabolites of the $C_{21}O_3$ series that probably are present but have not as yet been isolated include reduced derivatives of 6-, 16-, 18-, and 19-hydroxylated derivatives of progesterone.

The urine does not seem to contain isolatable derivatives of either $C_{21}O_2$ or $C_{21}O_3$ steroids with the Δ^4-3-keto grouping, but in the $C_{21}O_4$, $C_{21}O_5$, and $C_{21}O_6$ series these derivatives have been clearly demonstrated. In the $C_{21}O_4$ series, corticosterone and 11-dehydrocorticosterone have been isolated. Other urinary $C_{21}O_4$ steroids are reduced products of these adrenal steroids, 11-deoxycortisol and 21-deoxycortisol (Table 10).

Of the $C_{21}O_5$ steroids, cortisol and its reduced products make up the bulk of urinary steroids, with aldosterone being a constant but quantitatively smaller constituent. ULICK and LIEBERMAN (1957) have published evidence for the presence of a ring A-reduced metabolite of aldosterone, but the compound has yet to be rigorously identified. Only two $C_{21}O_6$ urinary steroids are now known, the 2α- and the 6β-hydroxy derivatives of cortisol (Table 10).

IV. Urinary steroids as conjugates of sulfuric and glucuronic acid

Some steroids are present in urine as free compounds, and these are mainly compounds with the Δ^4-3-keto grouping. By far the greatest amounts of urinary steroids are present as conjugates of sulfuric and glucuronic acid. Those that are known to be conjugated with glucuronic acid include pregnane-3α,20α-diol (VENNING and BROWNE 1936, HEARD et al. 1944), pregnane-3α,17α,20α-triol (MASON and KEPLER 1945b), 3α-hydroxypregnan-20-one (SUTHERLAND and MARRIAN 1947), urocortisol, urocortisone, cortol, β-cortol, cortolone, and β-cortolone. All these conjugates can be conveniently hydrolyzed with β-glucuronidase, and procedures of this type are extensively employed for quantitative determination of these components.

Steroid sulfate esters were demonstrated in urine with the isolation of androsterone sulfate (VENNING et al. 1942), dehydroepiandrosterone sulfate (MUNSON et al. 1944), and estrone sulfate (SCHACTER and MARRIAN 1938). Aryl and aliphatic steroid sulfatases are known to be present in certain mollusks. Extracts of 3 species, namely *Patella vulgata* (STITCH and HALKERSTON 1953, ROY 1954, 1956), *Otala punctata* (SAVARD et al. 1954), and *Helix pomatia* (HENRY et al. 1952), are especially efficient in hydrolyzing 3β-sulfate esters of the 5α and Δ^5 groups of steroids. On this basis, ROY (1956) has suggested the name 3β-steroid sulfatase for the enzyme. Preparations of both *Otala punctata* and *Patella vulgata* hydrolyze cortisone-21-sulfate, but in the opinion of ROY (1956) this action is probably due to the glycosulfatase contained in the extract. The 3β-steroid sulfatase from *Patella vulgata* was found to have a p_H optimum at 4.5 in acetate buffer and to be inhibited by Cl^-, $SO_4^=$ and $H_2PO_4^-$ ions.

JAYLE and co-workers have studied the hydrolysis of urinary conjugates using glucuronidase and aliphatic sulfatase preparations from *Helix pomatia*. In a series of papers, they were able to show that dehydroepiandrosterone, Δ^5-androstene-3β,17β-diol, and Δ^5-androstene-3β,16α,17β-triol were present in human urine almost exclusively as sulfates (MALASSIS et al. 1957, WEINMANN et al. 1957a, b) while 11-ketoetiocholanolone was present only in the form of the glucosiduronate and 11-dehydrocorticosterone, cortisone, 11β-hydroxyetiocholanolone, and 11β-hydroxyandrosterone were present partly as glucosiduronates and partly as sulfates (WEINMANN et al. 1957a, CHEMAMA et al. 1957).

Biosynthesis of adrenocortical hormones

With the elucidation of the structures of cholesterol, estrogens, progesterone, and androgens, a number of investigators, including BUTENANDT (1935), FIESER (1937), KOCH (1937), and KARRER (1938), suggested that the steroid hormones are probably derived from cholesterol. These suggestions were prophetic since, except for minor details and the inclusion of the adrenocortical hormones, the biosynthetic pathways predicted by these investigators are those that are now considered to be the best representation of the facts.

The advent of isotopic research in biochemistry, coupled with refined technics of chromatography, opened the way to effective research in the biosynthesis of the hormones. K. BLOCH (1945) provided the first experiment in the field when he administered deuterated cholesterol to a pregnant woman and isolated deuterated pregnane-3α,20α-diol from the urine. This indicated that the cholesterol-H^2 was converted to progesterone-H^2, which in turn was reduced to and excreted as pregnane-3α,20α-diol-H^2. Since this experiment was reported, a great number of additional studies have been recorded, some using *in vitro* methods and others using *in vivo* methods, all indicating a conversion of cholesterol to steroid hormones (Table 11).

Table 11. *Cholesterol as a precursor of steroid hormones*

Species	Conditions of experiment	Product	References
Man	*In vivo*, Pregnant women	Pregnane-3α,20α-diol-H^2 from urine	K. BLOCH (1945)
Ox	Adrenal perfusion, Cholesterol-C^{14}	Cortisol-C^{14}, Corticosterone-C^{14}	ZAFFARONI et al. (1951)
Man	*In vivo*, Patient with adrenal cancer, Cholesterol-C^{14}	Androsterone-C^{14}, Etiocholanolone-C^{14}, Dehydroepiandrosterone-C^{14}, Δ^5-Androstene-3β,17β-diol-C^{14}, Δ^5-Pregnene-3β,20α-diol-C^{14}	UNGAR and DORFMAN (1953)
Ox	*In vitro*, Adrenal homogenate, Cholesterol-C^{14}	Pregnenolone-C^{14}	SABA et al. (1954)
Ox and Dog	*In vitro*, Adrenal homogenate, Cholesterol	Cortisol, Corticosterone	REICH and LEHNINGER (1955)
Ox	Adrenal homogenate supernatant	Pregnenolone, Isocaproic acid	STAPLE et al. (1956)
Ox	*In vitro*, Adrenal homogenate, Cholesterol-C^{14}	Pregnenolone, Progesterone, Cortisol, Corticosterone	STAPLE et al. (1956)
Ox	*In vitro*, Adrenal mitochondrial preparation, Cholesterol-C^{14}	Pregnenolone, Progesterone, 11β-Hydroxyprogesterone, Corticosterone, Cortisol	HAYANO et al. (1956)
Man	*In vivo*, Cholesterol-C^{14}-H^3	Urocortisone-C^{14}-H^3, Urocortisol-C^{14}-H^3, Androsterone-C^{14}-H^3, Etiocholanolone-C^{14}-H^3, 11-Ketoetiocholanolone-C^{14}-H^3	WERBIN and LEROY (1955), WERBIN and LEROY (1954), WERBIN et al. (1957a, b)
Ox	*In vitro*, Adrenal homogenate, Cholesterol-C^{14}	Pregnenolone, 20β-Hydroxycholesterol	SOLOMON et al. (1956)
Ox	Adrenal perfusion	Progesterone, Corticosterone, Cortisol	SOLOMON et al. (1954)
Man	*In vivo*, Cholesterol	Estrogens	WERBIN et al. (1957)
Man	Placental perfusion	Progesterone	SOLOMON et al. (1954)
Guinea pig	*In vivo*, Cholesterol-H^3	Cortisol-H^3	WERBIN et al. (1958)
Ox	*In vitro*, Adrenal, Cholesterol-21-C^{14}	Cortisol-21-C^{14}, Corticosterone-21-C^{14}	GANIS and RADAKOVICH (1958)

Particularly informative have been a series of experiments in which doubly-labeled cholesterol (C^{14} and H^3) was administered and the ratio C^{14}/H^3 of the isolated urinary product shown to be identical to that of the administered cholesterol. In these experiments WERBIN and LEROY (1954, 1955) isolated the urinary compounds urocortisol and urocortisone, the metabolites of cortisol, with the expected C^{14}/H^3 ratio, thus demonstrating that cholesterol was converted to cortisol without prior breakdown to small particles and ring-system resynthesis. Additional experiments with doubly-labeled cholesterol indicated that the same was true for androsterone, etiocholanolone, 11-ketoetiocholanolone and estrogens (WERBIN et al. 1957).

Another direct approach to the problem was made by CASPI et al. (1956a, 1957), who examined carbon atoms 20 and 21 of cortisol, corticosterone, and 3α,17α,21-trihydroxypregnan-20-one for radioactivity when synthesized from acetate-1-C^{14}. The trihydroxypregnanone was produced by a patient with Cushing's disease and is the urinary metabolite of the adrenal steroid, 11-deoxycortisol. The labeled corticosterone and cortisol were synthesized by perfusion of the calf adrenal with acetate-1-C^{14}. In each case carbon 21 was devoid of radioactivity, while carbon 20 was labeled with C^{14}. This distribution would be expected if the corticoids are derived from cholesterol, assuming that the scheme postulated by WOODWARD and BLOCH (1953) is correct.

A. Biosynthesis of cholesterol

On the basis of detailed information documented in other sections of this review, cholesterol appears to be an important precursor of the steroid hormones. This section contains a summary of the biosynthetic steps to cholesterol which in effect are also steps in the formation of steroid hormones.

The clue that acetate is an important precursor for sterols came from the studies of SONDERHOFF and THOMAS (1937), who isolated deuterated ergosterol after the incubation of yeast

Table 12. *Conversion of various compounds to cholesterol*

Substrate	References
Acetate	K. BLOCH and RITTENBERG (1942), BUCHER (1953), RABINOWITZ and GURIN (1953), ZABIN and BARKER (1953), EIDINOFF et al. (1954), FRANTZ and BUCHER (1954), HELLMAN et al. (1954a), R. ROSENFELD et al. (1954), BUCHER and MCGARRAHAN (1956)
Pyruvate	RITTENBERG and SCHOENHEIMER (1937), K. BLOCH and RITTENBERG (1944), BRADY and GURIN (1950), POPJAK et al. (1953)
Ethanol	RITTENBERG and SCHOENHEIMER (1937), K. BLOCH and RITTENBERG (1944), K. BLOCH et al. (1945), CURRAN and RITTENBERG (1951)
Butyric acid	RITTENBERG and SCHOENHEIMER (1937), K. BLOCH (1944), BRADY and GURIN (1950), K. BLOCH (1951), ZABIN and BLOCH (1951a)
Valeric acid	RITTENBERG and SCHOENHEIMER (1937), K. BLOCH and RITTENBERG (1942)
Hexanoic acid	BRADY and GURIN (1950)
Acetaldehyde	K. BLOCH et al. (1943), BRADY and GURIN (1951)
Acetone	BORCK and RITTENBERG (1949), PRICE and RITTENBERG (1950), ZABIN and BLOCH (1950a, b), BRADY and GURIN (1951)
Acetoacetic acid	ZABIN and BLOCH (1950a, b), BRADY and GURIN (1951), CURRAN (1951), BLECHER and GURIN (1954)
Valine	K. BLOCH (1951)
Leucine	RITTENBERG and SCHOENHEIMER (1937), K. BLOCH (1944)
Isovaleric acid	RITTENBERG and SCHOENHEIMER (1937), K. BLOCH (1944), ZABIN and BLOCH (1950a, b, 1951a, b), K. BLOCH (1951), BRADY and GURIN (1951)
Mevalonic acid	TAVORMINA et al. (1956), GOULD and POPJAK (1957)
Octanoic acid	LYON et al. (1956)
Isobutyric acid	K. BLOCH (1944), KRITCHEVSKY and GRAY (1951)
Glucose	POPJAK et al. (1953)
β-Hydroxy-β-methylglutaric acid	RABINOWITZ and GURIN (1954b)
Dimethylacrylic acid	K. BLOCH et al. (1954)
Squalene	LANGDON and BLOCH (1952, 1953a), SCHWENK et al. (1955b), NICOLAIDES et al. (1955), DITURI et al. (1956)
Cholestenone	ANKER and BLOCH (1949)
Lanosterol	R. B. CLAYTON and BLOCH (1956b), OLSON and BLOCH (1956)
Zymosterol	SCHWENK et al. (1955b), JOHNSTON and BLOCH (1957)
Lathosterol (Δ^7-Cholestenol)	BIGGS et al. (1954), DAVIDSON et al. (1957)
Δ^7-Dehydrocholesterol	COOK et al. (1954)

with acetate-H^2. Similar experiments were reported by K. BLOCH and RITTENBERG (1942), who administered acetate-H^2 to mice and rats and isolated cholesterol-H^2. Although acetate is an important precursor of cholesterol as demonstrated by *in vivo* (K. BLOCH and RITTENBERG 1942) and *in vitro* methods (K. BLOCH et al. 1946, BUCHER 1953), a variety of other small molecules including acetaldehyde, pyruvate, acetone, glucose, and the amino acids valine and leucine are also precursors (Table 12) after preliminary conversion to acetate, probably as acetyl-coenzyme A (GOULD 1938).

Cholesterol biosynthesized from methyl- and carboxyl-labeled acetate has been subjected to carbon by carbon degradation with the clear result that each carbon atom of the molecule is derived from acetate (Table 13).

Table 13. *Origin of carbon atoms in cholesterol synthesized from acetate*

Cholesterol carbon atom number	From methyl carbon of acetate	From carboxyl carbon of acetate	References
1	+		CORNFORTH et al. (1953) (carbons 1–6)
2		+	
3	+		
4		+	
5	+		
6		+	
7	+		K. BLOCH (1953), DAUBEN and TAKEMURA (1953)
8		+	CORNFORTH et al. (1956, 1957a)
9	+		
10		+	CORNFORTH et al. (1953)
11		+*	CORNFORTH et al. (1956, 1957a)
12		+*	
13	+		WOODWARD and BLOCH (1953), CORNFORTH and POPJAK (1954)
14		+	CORNFORTH et al. (1956, 1957a)
15	+		
16		+	
17	+		WUERSCH et al. (1952)
18	+		
19	+		CORNFORTH et al. (1953)
20		+	WUERSCH et al. (1952) (carbons 20–27)
21	+		
22	+		
23		+	
24	+		
25		+	
26	+		
27	+		

* DAUBEN and HUTTON (1956) demonstrated that these two carbon atoms originated from the carboxyl carbon of acetate, which was used to biosynthesize ergosterol.

Since the feeding of squalene in the rat also causes increased concentrations of cholesterol in the liver (CHANNON 1926), K. BLOCH (1954) has proposed that biosynthesis of cholesterol probably involves the pathway acetate → "isoprenoid unit" → squalene → cholesterol.

The role of squalene as an intermediate was established by conversion of squalene-C^{14} to cholesterol-C^{14} (LANGDON and BLOCH 1953a). The direct relationship of acetate to squalene was rigorously established by the incubation of rat liver mince with acetate-2-C^{14}, the isolation of the formed squalene-C^{14}, and the carbon-by-carbon analysis of the molecule for C^{14} content. The results of this analysis were consistent with the hypothesis that acetate is converted to an isoprenoid unit (C_5) and that six of these units are polymerized to squalene (CORNFORTH and POPJAK 1954).

In the biosynthetic pathway between acetate and squalene, mevalonic acid is either an important direct intermediate or easily converted to such. The incubation of this C_6 compound with a rat liver homogenate preparation produces cholesterol in high yield (TAVORMINA et al. 1956). Other experiments show that when mevalonic acid is incorporated into cholesterol the carboxyl carbon is lost, since C^{14} could not be detected in the product when mevalonic acid-1-C^{14} was used (TAVORMINA and GIBBS 1956). In a second experiment CORNFORTH et al. (1957b) labeled mevalonic acid in position 2 and degraded the squalene that was formed. The results were consistent with the idea that the carboxyl group is eliminated, that water is removed by elimination of the hydroxyl group at carbon 3, and that the polymer is formed between carbon atom 2 of one isoprenoid unit and carbon 5 of another. This is repeated so that a total of three such units are attached to form a C_{15} unit. Squalene is finally formed by two C_{15} units condensing head to tail.

In addition to mevalonic acid, a variety of C_5 and C_6 compounds have been shown to be derived from acetate and to be precursors of cholesterol in biological systems (Table 14). The biosynthetic yields from these compounds, however, have been remarkably low as compared to that found for mevalonic acid and are in general of minor importance.

Table 14. *Biosynthesis of direct and indirect intermediates of cholesterol*

Substrate	Product	References
Acetate-2-C^{14}	Squalene-C^{14}	EIDINOFF et al. (1954)
Acetate-1-C^{14}	Dimethylacrylic acid-C^{14}, Senecioic acid-C^{14}	RABINOWITZ (1954)
Acetate-C^{14}, Pyruvate-C^{14}	β-Hydroxy-β-methyl-glutaric acid-C^{14}	RABINOWITZ and GURIN (1954a)
β-Hydroxy-β-methylglutaric acid-C^{14}	β-Methylcrotonic acid-C^{14}, β-Methylglutaconic acid-C^{14}	RABINOWITZ and GURIN (1954b)
Acetate-C^{14}	Dimethylacrylic acid-C^{14}	RUDNEY (1954)
Squalene	Lanosterol, Agnosterol	R. B. CLAYTON and BLOCH (1956a)
Acetate-C^{14}	Lanosterol-C^{14}	R. B. CLAYTON and BLOCH (1955, 1956b)
Acetate-1-C^{14}	Squalene-C^{14} Zymosterol	SCHWENK et al. (1955a, b)
Acetate-C^{14}, β-Hydroxy-β-methylglutaric acid-C^{14}	Squalene-C^{14}	DITURI et al. (1955), RABINOWITZ et al. (1955)
Acetate-C^{14}	Squalene	LANGDON and BLOCH (1953b), POPJAK (1954), CORNFORTH and POPJAK (1954)
β-Hydroxy-β-methylglutaric acid-C^{14}	Farnesinic acid-C^{14} (no farnesol-C^{14} found)	RABINOWITZ et al. (1955), DITURI et al. (1956)
β-Hydroxyisovaleric acid-C^{14}	Squalene-C^{14}	RABINOWITZ et al. (1955), DITURI et al. (1956)
Mevalonic acid	Squalene	AMDUR et al. (1957), CORNFORTH et al. (1958).

The conversion of squalene to cholesterol involves a cyclization of the latter compound. WOODWARD and BLOCH (1953) proposed what seems to be the correct arrangement, as indicated in Figure 13, and this formulation has the features of accounting for the origin from acetate of all the carbons through the intermediate squalene. This scheme has now been experimentally verified by the careful degradation of cholesterol-C^{14} biosynthesized from labeled acetate, as summarized in Table 13.

The first steroid that is known to arise from cyclization of squalene is lanosterol. Lanosterol is synthesized from acetate by rat liver preparations and lanosterol in turn may be converted into cholesterol (R. B. CLAYTON and BLOCH 1956). During this process a key reaction appears to be the hydroxylation reaction at carbon 3, with the reaction involving O_2 in a manner similar to that shown earlier for hydroxylations at various positions on the C_{21}-steroid molecule (TCHEN and BLOCH 1957a, b). Lanosterol appears to be converted to zymosterol, on the basis of the latters being present in trace amounts in pig liver (ALEXANDER and SCHWENK 1957) and the fact that it may be converted to cholesterol *in vivo* (SCHWENK et al. 1955a) and *in vitro* (JOHNSTON and BLOCH 1957). For this conversion, lanosterol to zymosterol, the 4,4'-dimethyl groups must be lost as well as the 14 methyl group, and a mechanism for these reactions has been indicated. These carbon atoms are probably lost as CO_2, since lanosterol derived from acetate-2-C^{14} yielded 3 moles of C^{14}-labeled CO_2 when converted to cholesterol (OLSON and BLOCH 1956).

On a theoretical basis a rather large number of intermediates are possible between lanosterol and zymosterol. These include the dimethyl derivatives 4,4′ and 4,14, the monmethyl 4 and 14 (Figure 13). Among these possibilities may well be an intermediate detected by GAUTSCHI and BLOCH (1957) which appears to be 4,4′-dimethylcholestadienol.

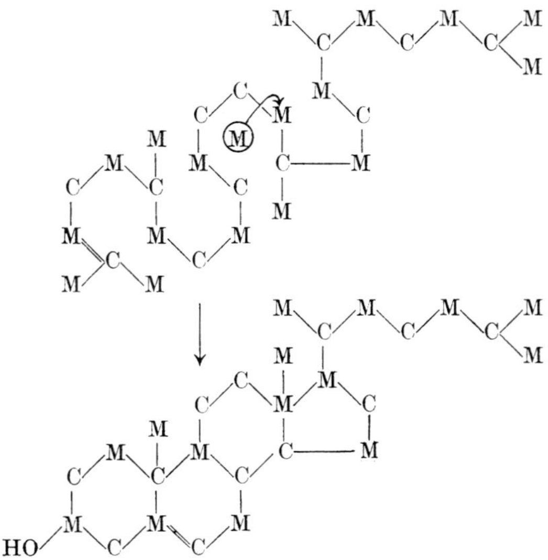

Fig. 13. Biosynthesis of cholesterol from acetate and squalene (M = methyl C of acetate; C = carboxyl C of acetate)

B. Biosynthesis of corticoids

The fact that the corticoids — corticosterone, cortisol and related C_{21} compounds — may originate from cholesterol has already been indicated. Table 15 presents a more extensive list of biosynthetic reactions, observed by *in vivo* and *in vitro* methods, leading to the corticoids. These reactions have been observed in man, ox, dog, guinea pig, rat and pig, and essentially no differences in the biosynthetic pathways are observable, except for the lack of cortisol formation in rats and of corticosterone formation in the guinea pig. The over-all pattern is represented in Figs. 14 and 15.

The reactions from cholesterol to pregnenolone have not been completely elucidated, but two important observations have been made. The first involves the over-all reaction of cholesterol to pregnenolone and isocaproic acid. The isolation of isocaproic acid was reported by LYNN et al. (1955) and STAPLE et al. (1956), while SABA et al. (1954), SOLOMON et al. (1954), SABA and HECHTER (1955), and STAPLE et al. (1956) have isolated the steroid product of the reaction, pregnenolone. The second important observation in this area was the demonstration by SOLOMON et al. (1956a) that cholesterol incubated with cow adrenal homogenates formed 20β-hydroxycholesterol. DORFMAN (1957) suggested that the immediate precursor of pregnenolone may be 20β,22ξ-dihydroxycholesterol and that the C-20, C-22 bond may be split by a proper C-20,22 desmolase (Fig. 14).

The biosynthetic route from pregnenolone to corticosterone and cortisol involves an oxidation of carbon 3 to a ketone and rearrangement of the double bond from carbon atoms 5,6 to 4,5 and a series of hydroxylating reactions at specific points on the molecule (Fig. 15). The first reaction, pregnenolone to progesterone, is catalyzed by the enzyme Δ^5-3β-hydroxydehydrogenase (SAMUELS 1953,

BEYER and SAMUELS 1956, HAYANO et al. 1956a, RUBIN and DORFMAN 1957). The hydroxylating enzymes include the 11β-, 17α-, and 21-hydroxylases and good evidence exists for the formation of all seven hydroxylated derivatives of progesterone (Table 15 and Fig. 15). The arrows in Fig. 15 indicate the biosynthetic routes that have been demonstrated.

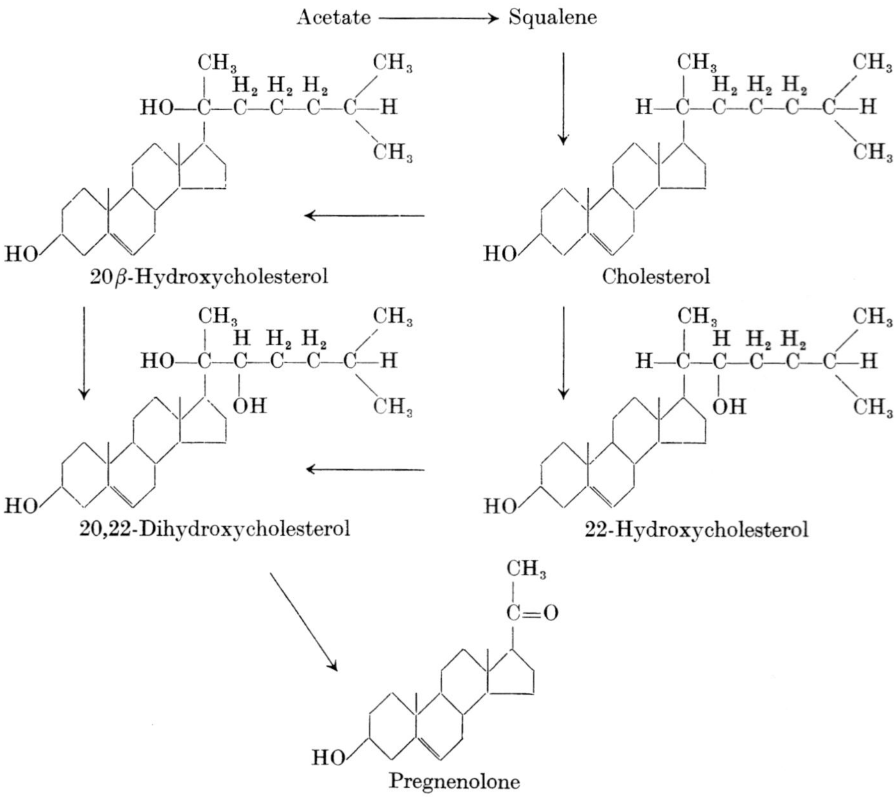

Fig. 14. Biosynthesis of pregnenolone

In earlier reviews of biosynthesis of cortisol, a pathway involving the formation of this corticoid from corticosterone was suggested (DORFMAN 1955, 1956). This reaction would involve the 17α-hydroxylation of a steroid previously 11β- and 21-hydroxylated. This suggestion was originally based on the observation that deoxycorticosterone could give rise to cortisol (DORFMAN et al. 1953). Although this view appeared to be strengthened by the findings of HEARD et al. (1956), who reported that deoxycorticosterone-C^{14} was converted to cortisol-C^{14} by rat, rabbit, porcine and bovine adrenals, studies by EICHHORN and HECHTER (1957) have not confirmed these observations. These latter workers showed that incubation with deoxycorticosterone produced a substance with a mobility in a paper chromatographic system like that of cortisol, but after extensive purification this material was shown to be distinctly different and separable from cortisol.

The corticoids in Fig. 15 do not represent all the C_{21} steroids formed in the adrenal. Compounds with nuclear reduction in ring A and at carbon atoms 3 and 20 are probably produced by the adrenal, but only after the formation of the

Table 15. *Biosynthesis of corticoids*

Species	Test system	Substrate	Product	Reference
Ox	Adrenal homogenate	Cholesterol-C^{14}	20β-Hydroxycholesterol	SOLOMON et al. (1954)
Ox	Adrenal perfusion	Cholesterol-C^{14}	Cortisol-C^{14}, Corticosterone-C^{14}	ZAFFARONI et al. (1951)
Man	*In vivo*, adrenal cancer	Cholesterol-C^{14}	Androsterone-C^{14}, Etiocholanolone-C^{14}, Dehydroepiandrosterone-C^{14}, Δ^5-Androstene-3β,17β-diol-C^{14}, Δ^5-Pregnene-3β,20α-diol-C^{14}	UNGAR and DORFMAN (1953)
Ox	*In vitro*	Cholesterol-C^{14}	Pregnenolone-C^{14}	SABA et al. (1954)
Ox and Dog	*In vitro*, adrenal homogenate	Cholesterol	Cortisol	REICH and LEHNINGER (1955)
Ox	*In vitro*, adrenal homogenate supernatant (100000 × g, 30 min.)	Cholesterol	Pregnenolone, Progesterone, Cortisol, Corticosterone	STAPLE et al. (1956)
Ox	*In vitro*, adrenal mitochondrial preparation	Cholesterol-C^{14}	Pregnenolone, Progesterone, 11β-Hydroxyprogesterone, Corticosterone Cortisol	HAYANO et al. (1956)
Ox	Adrenal perfusion	11β-Hydroxyprogesterone, Progesterone	Cortisol, Corticosterone	EICHHORN and HECHTER (1958)
		Deoxycorticosterone	Corticosterone	
Man	*In vivo*	Cholesterol-C^{14},H^3	Urocortisone-C^{14},H^3, Urocortisol	WERBIN and LEROY (1955), WERBIN and LEROY (1954), WERBIN et al. (1957a, b)
Ox	Adrenal perfusion	Cholesterol-C^{14}	Progesterone, Corticosterone, Cortisol	SOLOMON et al. (1954)
Ox	Adrenal perfusion	Acetate	Cortisol, Corticosterone	CASPI et al. (1956)
Man	*In vivo*, adrenal cancer	Acetate	3α,17α,21-Trihydroxypregnan-2-one	CASPI et al. (1957)
Guinea pig	Adrenal (normal and scorbutic)	Acetate	Cortisol	BURSTEIN and NADEL (1956)
Ox	Adrenal perfusion	Progesterone	17α-Hydroxyprogesterone, 11β-Hydroxyprogesterone, Corticosterone, Cortisol, 11-Deoxycortisol	H. LEVY et al. (1954), HECHTER et al. (1951), PINCUS (1954)
		17α-Hydroxyprogesterone	Cortisol	H. LEVY et al. (1950, 1954), HECHTER et al. (1951)
		Pregnenolone	Progesterone, Corticosterone, Cortisol	H. LEVY et al. (1950, 1954), HECHTER et al. (1951)

Table 15 (cont.)

Species	Test system	Substrate	Product	Reference
Ox (cont.)	Adrenal perfusion (cont.)	Progesterone	Progesterone	KUSHINSKY (1955)
			20β-Hydroxy-Δ^4-pregnen-3-one, Allopregnane-3,20-dione, 3β-Hydroxypregnan-20-one, 17α-Hydroxyallopregnane-3,20-dione, 3β,17α-Dihydroxyallopregnan-20-one, 17α-Hydroxyprogesterone, 11β-Hydroxyprogesterone, 6β-Hydroxyprogesterone, 11-Deoxycortisol, Corticosterone, 19-Hydroxydeoxycorticosterone, Cortisol 3α,17α,11β,21-Tetrahydroxyallopregnan-20-one, 6β,17α,21-Trihydroxy-Δ^4-pregnene-3,20-dione, 3β,17α,11β,21-Tetrahydroxyallopregnan-20-one, 17α,19,21-Trihydroxy-Δ^4-pregnene-3,20-dione	H. LEVY and KUSHINSKY (1955)
Rat and Pig	Adrenal slices and cell free preparations	Acetate and β,β-Dimethylacrylate	Progesterone	APRILE et al. (1956)
Ox	Adrenal slices	Acetate-1-C^{14}	Cortisol-C^{14}	HAYNES et al. (1952, 1954)
Pig	Adrenal slices	Acetate-1-C^{14}	Cortisol-C^{14}, Cortisone-C^{14}, Corticosterone-C^{14}, 11-Deoxycortisol-C^{14}	HAINES (1952)
		11-Deoxycortisol	Cortisol, Cortisone	
		Deoxycorticosterone	Corticosterone, 11-Dehydrocorticosterone, 6β-Hydroxydeoxycorticosterone	
Ox	Adrenal perperfusion	Progesterone	6β-Hydroxyprogesterone	H. LEVY et al. (1954)
Ox, Man, Pig	Adrenal perfusion	Deoxycorticosterone	Corticosterone	HECHTER et al. (1949), HECHTER et al. (1951), H. LEVY et al. (1953), UNGAR et al. (1958)
Ox	Adrenal perfusion	11-Deoxycortisol	Cortisol	HECHTER et al. (1950, 1951)
Ox	Adrenal perfusion	21-Deoxycortisol	Cortisone	MEYER et al. (1954)
Ox and Pig	Adrenal brei	Deoxycorticosterone	6β,17β,21-Trihydroxy-Δ^4-pregnene-3,20-dione	NEHER and WETTSTEIN (1956)
Pig	Adrenal brei	Deoxycortico-	19,21-Dihydroxy-Δ^4-pregnene-3,20-dione	NEHER and WETTSTEIN (1956)
Pig and Ox	Adrenal	Deoxycorticosterone	6β,11β,21-Trihydroxy-Δ^4-pregnene-3,20-dione	NEHER and WETTSTEIN (1956)

Table 15 (cont.)

Species	Test system	Substrate	Product	Reference
Pig and Ox	Adrenal brei	Deoxycorticosterone	11β,19,21-Trihydroxy-Δ⁴-pregnene-3,20-dione	NEHER and WETTSTEIN (1956)
Man	Adrenal slices	Progesterone 17α-Hydroxyprogesterone	Cortisol	LOMBARDO et al. (1956)
Ox	Adrenal homogenate	Deoxycorticosterone plus O_2^{18}	Corticosterone-O^{18}	HAYANO et al. (1955)
Ox	Adrenal mitochondria	Deoxycorticosterone-O_2^{18}	Corticosterone-O^{18}	SWEAT et al. (1956)
Ox (Calf)	Adrenal perfusion	11-Deoxycortisol 21-Deoxycortisol Deoxycorticosterone Progesterone	Cortisol (?) Cortisol Corticosterone Cortisol	G. ROSENFELD et al. (1955)
Ox	Adrenal mitochondria	Progesterone	11β-Hydroxyprogesterone	BROWNIE et al. (1954), SABA and HECHTER (1955)
Ox	Adrenal homogenate, Supernatant plus microsomal system	Progesterone 17α-Hydroxyprogesterone 21-Deoxycortisol	11-Deoxycorticosterone 11-Deoxycortisol Cortisol	RYAN (1956) RYAN and ENGEL (1956)
Ox	Adrenal homogenate	Deoxycorticosterone	18-Hydroxy-11-deoxycorticosterone, 19-Hydroxy-11-deoxycorticosterone	KAHNT et al. (1955)
Ox	Washed residue of adrenal homogenate (5000 × g)	Deoxycorticosterone	19-Hydroxy-11-deoxycorticosterone	HAYANO and DORFMAN (1955)
Ox	Adrenal homogenate supernatant	Progesterone	11-Deoxycortisol	PLAGER and SAMUELS (1954)
Ox	Adrenal homogenate	Pregnenolone Progesterone	Progesterone Cortisol, Corticosterone	PLAGER and SAMUELS (1952)
Pig	Adrenal brei (?)	Deoxycorticosterone	6β-21-Dihydroxy-Δ⁴-pregnene-3,20-dione	HAINES (1952)
Ox	Adrenal homogenate	11-Deoxycortisol Deoxycorticosterone	Cortisol Corticosterone	HAYANO and DORFMAN (1953)
Ox	Adrenal homogenate	17α-Hydroxyprogesterone	Cortisol	HAYANO and DORFMAN (1952)
Ox and Pig	Adrenal homogenate or slices	11-Deoxycortisol	Cortisol	MCGINTY et al. (1950), SAVARD et al. (1950), HAINES (1952)
Pig	Adrenal cell-free extract	Progesterone	Cortisol, Cortisone, Aldosterone, 11-Deoxycortisol, Corticosterone, 16α-Hydroxyprogesterone, 11-Dehydrocorticosterone, 11β-Hydroxyprogesterone, 17α-Hydroxyprogesterone, 11-Ketoprogesterone	RAO and HEARD (1957)

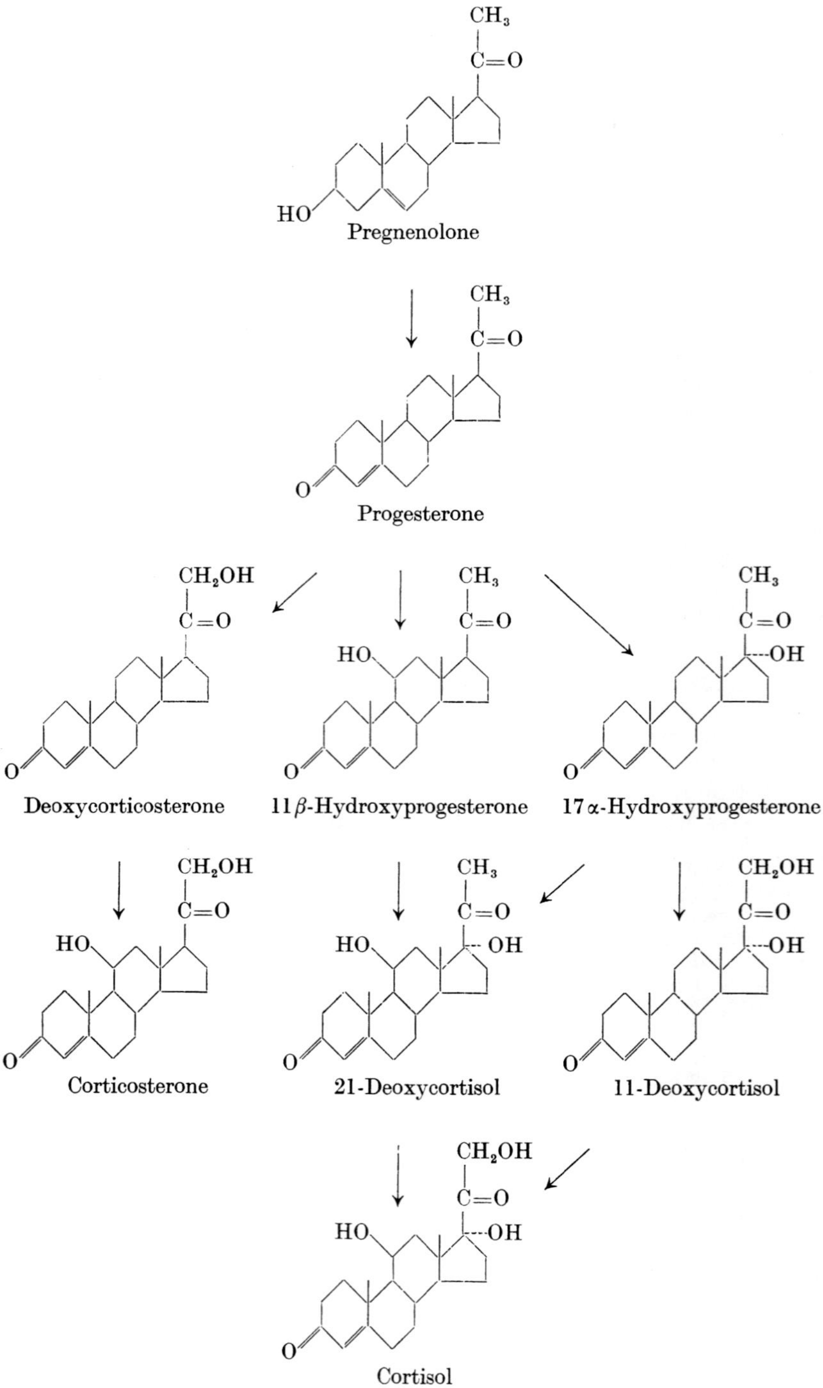

Fig. 15. Biosynthesis of cortisol and corticosterone

Δ^4-3,20-diketones. Another group of corticoids are those C_{21} steroids that have oxygen functions at carbon atoms 6(β), 16(α), 18 (excluding aldosterone), and 19, in addition to oxygen at positions 3, 11, 17, 20, and 21. These compounds are listed in Fig. 16. The finding of 2α-hydroxycortisol in guinea pig urine points to the possibility that this steroid is produced in the adrenal gland (BURSTEIN 1956, PERON and DORFMAN 1956), as well as other 2α-hydroxy corticoid derivatives.

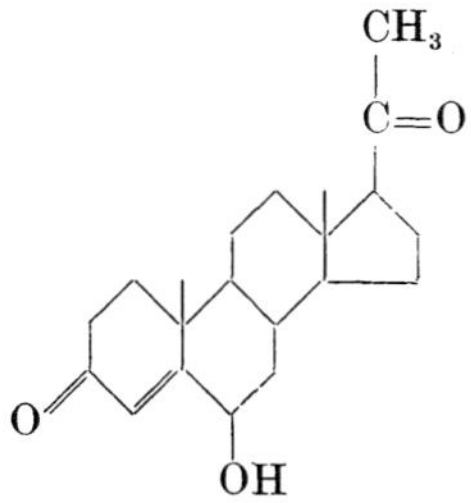

6β-Hydroxyprogesterone

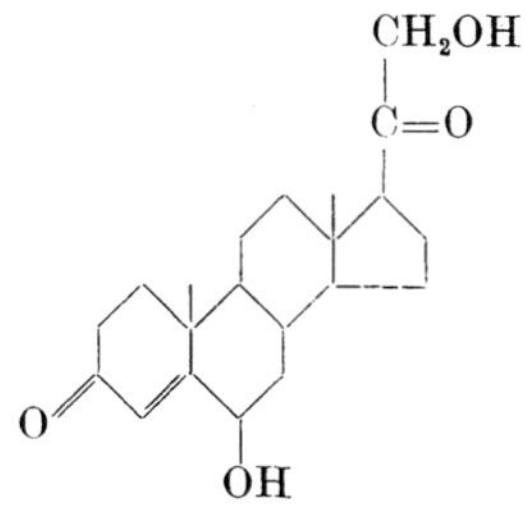

6β,21-Dihydroxy-Δ^4-pregnene-3,20-dione

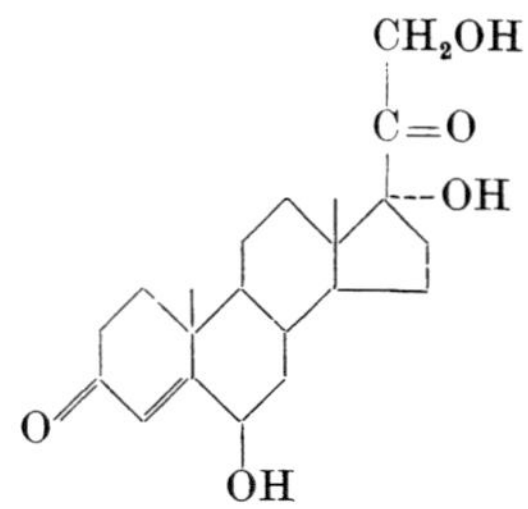

6β,17α,21-Trihydroxy-Δ^4-pregnene-3,20-dione

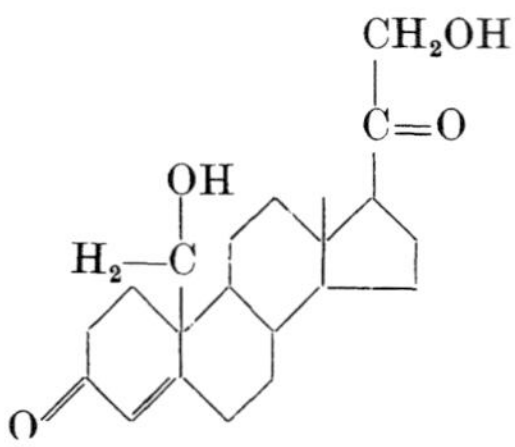

19,21-Dihydroxy-Δ^4-pregnene-3,20-dione

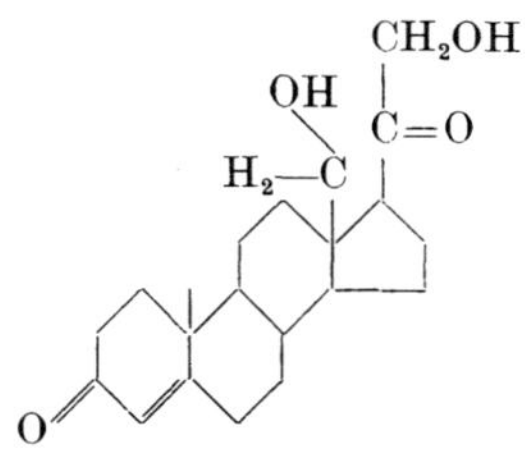

18,21-Dihydroxy-Δ^4-pregnene-3,20-dione

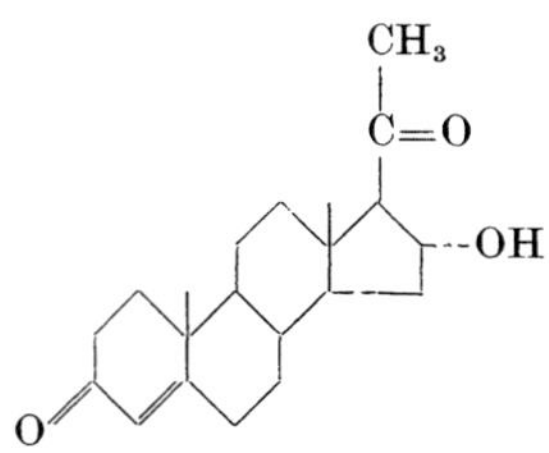

16α-Hydroxyprogesterone

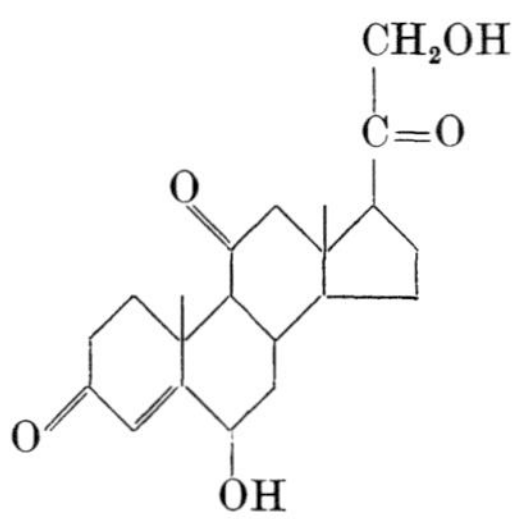

6β,21-Dihydroxy-Δ^4-pregnene-3,11,20-trione

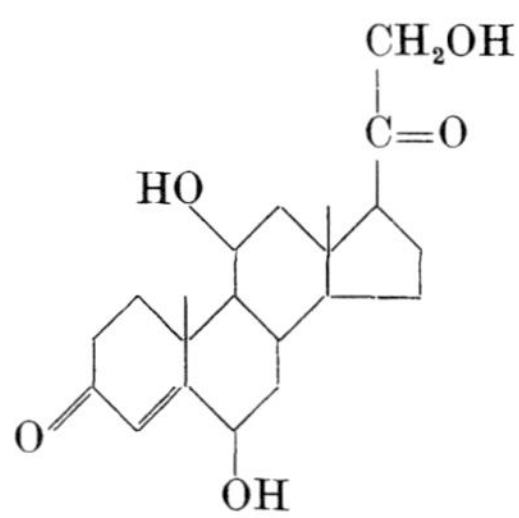

6β-Hydroxycorticosterone

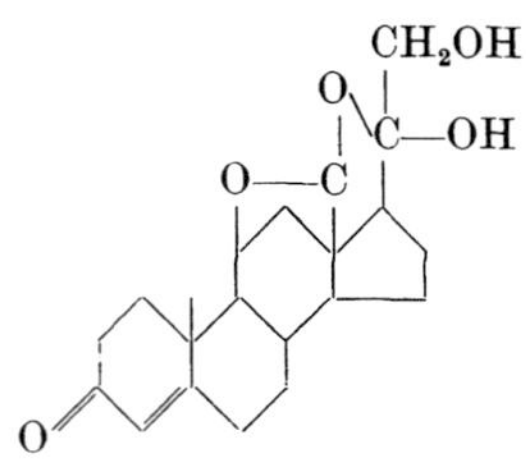

11β, 20,20,21-Tetrahydroxy-(18 → 11,18→20)-acetal-20-hemiketal-Δ^4-pregnen-3-one

Fig. 16. 6β,16α,18,19-Hydroxylated steroids of the adrenal gland

On theoretical grounds, HIRSCHMANN and HIRSCHMANN (1947) suggested that C_{21} steroids may be biosynthesized by the condensation of C_2 units with C_{19} steroids, and HECHTER (1953) reported some preliminary results consistent with this idea. A more extensive study has now appeared by UDAEV and DRUZHINIA (1958), claiming the conversion of androsterone, Δ^4-androstene-3,17-dione, and estrone to cortisol and the conversion of adrenosterone to cortisone when the substrates were incubated with guinea pig adrenal slices in a Krebs-Ringer bicarbonate medium in a 95% oxygen atmosphere at 37° C. Identity of the products was based on paper chromatographic mobility, reactions with the Porter-Silber reagent, ultraviolet absorption at 240 mμ, and sulfuric acid spectrum of the rechromatographed product. Quantitative yields were assessed by ultraviolet absorption determinations at 240 mμ and by the Porter-Silber method.

Administration of dehydroepiandrosterone to normal individuals is reported to increase the urinary corticoid excretion (LAMEDICA and MISSALE 1958). The authors interpreted this finding as meaning that dehydroepiandrosterone is utilized for the biosynthesis of corticoids. An alternative explanation may involve the peripheral formation of a 16-oxygen derivative of dehydroepiandrosterone with a 16,17-α-ketol configuration, which reacts with the corticoid reagent. The presence of 16,17-α-ketols in human urine is known from the work of FOTHERBY et al. (1957), who isolated 3β,16α-dihydroxy-Δ^5-androsten-17-one.

I. A non-cholesterol pathway to corticoids

In various experiments involving the conversion of acetate-C^{14} to corticoids, suggestive evidence has been reported for the idea of a biosynthetic pathway that does not involve cholesterol (HECHTER 1952, 1953, STONE and HECHTER 1954, HEARD et al. 1956). The experiments of HEARD et al. (1956) were particularly striking, since these workers isolated C^{14}-containing corticoids from an adrenal homogenate incubated with acetate-1-C^{14} when no C^{14} could be detected in the isolated cholesterol. HECHTER (1958) has reviewed the difficulties in interpreting these experiments because of lack of kinetic data on the relative specific activities of cholesterol and corticoids and therefore the inability to evaluate the possible existence of diverse metabolic pools.

HECHTER (1958) has speculated on the possibility of corticoid biosynthesis proceeding by the condensation of 4 isoprene units to yield a C_{20} unit that in turn condenses with two C_2 units to yield, after proper migration of one carbon to form the C-18 angular methyl group, a 4,4′,14-trimethyl derivative of a C_{21} steroid, which could have the same arrangement of methyl and carboxyl carbons (from acetate) as has been demonstrated for the first 21 carbon atoms of cholesterol originating also from acetate. This pathway is suggested as an alternative to the cholesterol→ corticoid pathway so well established by other experiments. Further, it is to be mentioned that HECHTER's theoretical scheme is only one of many that could be suggested. Others could include the formation of a C_{25} isoprenoid unit, with the loss of four methyl groups (possibly 4,4′,14 and 20) to yield perhaps pregnenolone.

C. Biosynthesis of aldosterone

The principal electrolyte-regulating factor of the adrenal appears to be aldosterone, which has an action promoting retention of sodium and excretion or potassium. Although aldosterone has only recently been isolated and identified (SIMPSON et al. 1954, HARMAN et al. 1954) considerable progress has already been made toward elucidating the pathways of biosynthesis.

Experimental evidence (Table 16) now clearly indicates that progesterone, deoxycorticosterone, 11β-hydroxyprogesterone, corticosterone, and cholesterol are precursors of aldosterone. The first clear demonstration of aldosterone biosynthesis from deoxycorticosterone resulted from the studies of KAHNT et al. (1955 b), who incubated the C^{14}-labeled substrate with beef adrenal homogenate and isolated labeled aldosterone. In another series of experiments this group of investigators observed a decreased aldosterone production when corticosterone or progesterone was incubated with a bovine adrenal homogenate. ROSEMBERG et al. (1956) were able to demonstrate a conversion of progesterone to aldosterone by perfusing the substrate through a calf adrenal, but neither corticosterone nor deoxycorticosterone increased the aldosterone content of the perfusate.

Table 16. *Biosynthesis of aldosterone*

Species	Test system	Substrate	Reference
Ox	Adrenal homogenate	Deoxycorticosterone	KAHNT et al. (1955b)
Ox	Adrenal perfusion	Progesterone	ROSEMBERG et al. (1956)
Ox	Adrenal capsule strippings, mainly zona glomerulosa	Progesterone	AYRES et al. (1956)
Ox	Adrenal perfusion	Corticosterone, Progesterone	CHEN et al. (1958)
Ox	Adrenal capsule strippings	Progesterone Deoxycorticosterone Corticosterone, Cholesterol	AYRES et al. (1957a, b) HECHTER (1958)
Ox	Zona glomerulosa	Progesterone, Deoxycorticosterone, Corticosterone, 11β-Hydroxyprogesterone	GIROUD et al. (1958)
Ox	Adrenal capsule strippings	Progesterone, Corticosterone	TRAVIS and FARRELL (1958)
Pig	Adrenal cell-free extract	Progesterone	RAO and HEARD (1957)

Studies of the biosynthesis of aldosterone in discrete adrenocortical zones have yielded clear cut results that strongly support the views of DEANE and co-workers concerning the secretion of electrolyte regulating hormone by the zona glomerulosa (Chapter 1). Aldosterone was shown to be produced by this zone by AYRES et al. (1956) using bovine tissue and by GIROUD et al. (1956) using rat tissue. Of further importance is the observation that aldosterone is produced almost exclusively in the zona glomerulosa. On the other hand, the inner zones (fasciculata and reticularies) of the bovine adrenal seem to be the exclusive site for cortisol production (GIROUD et al. 1958). Corticosterone is produced by all three layers of the adrenal cortex. From this it is possible to conclude that in the zona glomerulosa the 17α-hydroxylating reaction is absent, while the 18-hydroxylase is not present in the fasciculata-reticularis.

Possible pathways for aldosterone biosynthesis are illustrated in Fig. 17. Not all of the reactions have been demonstrated, nor have all the compounds been shown to be present in the adrenal gland. Those, however, that are named and numbered have either been isolated or have been biosynthesized from naturally-occurring substrates. Five different steroids have been converted to aldosterone, and an important need exists for testing the other possible intermediates that are indicated. It is not possible to speak of an aldosterone biosynthetic pathway on the basis of existing information.

The incubation of 11-dehydrocorticosterone with beef zona glomerulosa by GIROUD et al. (1958) has resulted in the formation of a steroid with a mobility similar to that of aldosterone in three different chromatographic systems. This product could not be produced by incubation of 11-dehydrocorticosterone with

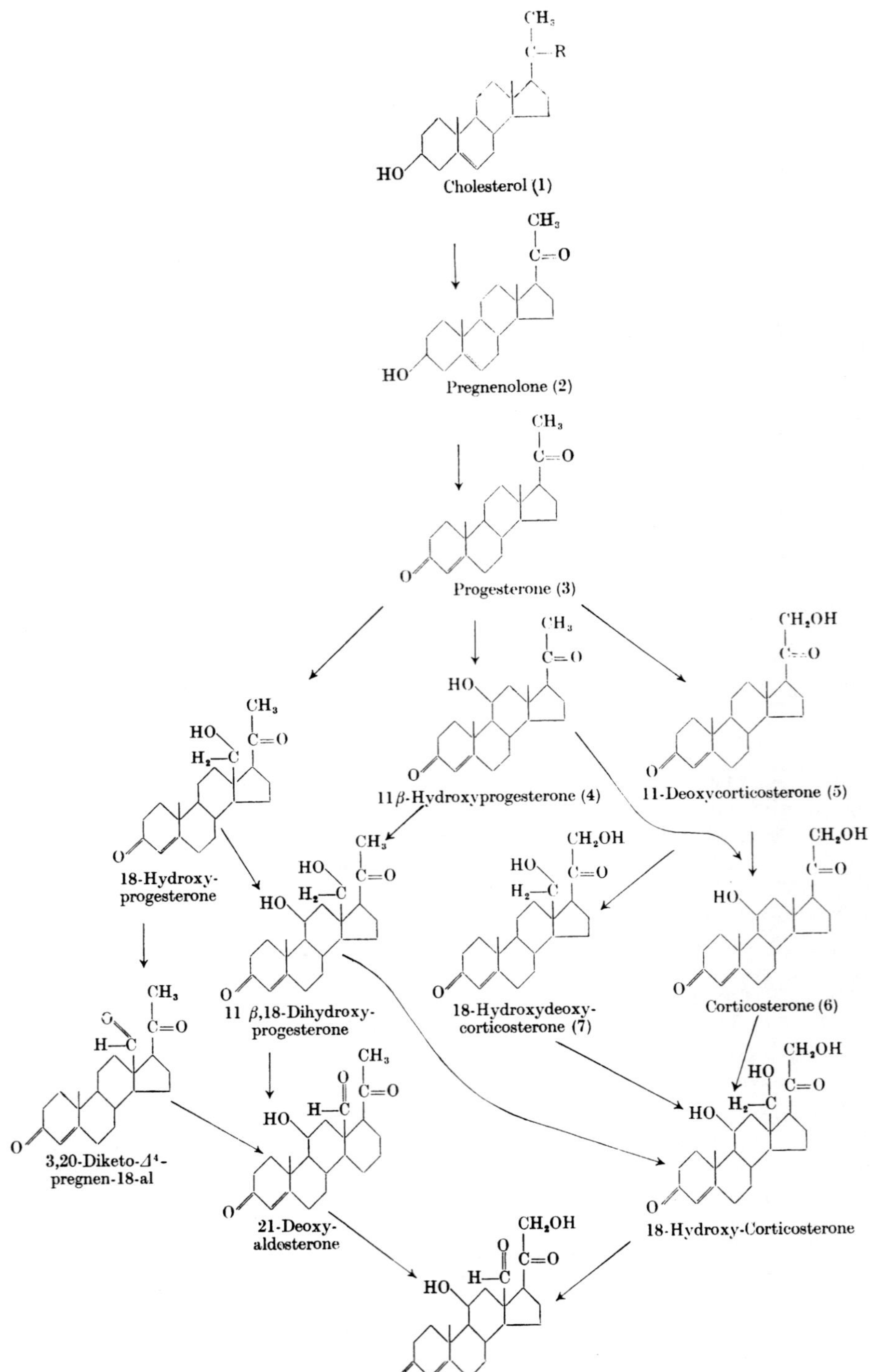

Fig. 17. Possible pathways for biosynthesis of aldosterone (numbered compounds are known to be produced by adrenal). Biosynthetic conversions that have been demonstrated: 1, 3, 4, 5, and 6 → 8

fasciculata-reticularis tissue. This indicates the possible 18-hydroxylation of the substrate, but the compound (tentatively called Compound III) does not cause sodium retention by bioassay.

D. Biosynthesis of adrenal androgens

Androgen production in adrenal tissue probably occurs by two different pathways, one of which has been carefully documented, while a second is indicated but not proved. It is now a common observation that administration of ACTH results in an increase in androgens and related steroids. These observations, however, must be considered to be only *indirect proof* of androgen production in the gland, since it is possible that ACTH only stimulates the production of C_{21} steroids, which are converted to C_{19} androgens by catabolic reactions in tissues such as the liver. For this reason, it is important to examine certain more direct evidence that ACTH actually increases androgen production. Such an experiment is that of E. BLOCH et al. (1957), who showed that C_{19} steroids were actually increased when adrenal slices from a patient with virilizing adrenogenital syndrome were

Table 17. *Biosynthesis of adrenal androgens*

Species	Test system	Substrate	Product	Reference
Man	*In vivo*, adrenal carcinoma	Acetate-1-C^{14}, Cholesterol-C^{14}	Androsterone, Etiocholanolone, Dehydroepiandrosterone, Δ^5-Androstene-3β-17β-diol	UNGAR and DORFMAN (1953)
Ox	Adrenal perfusion	Δ^4-Androstene-3,17-dione	11β-Hydroxy-Δ^4-androstene-3,17-dione	JEANLOZ et al. (1950, 1953)
Ox	Adrenal perfusion	Progesterone-C^{14}	11β-Hydroxy-Δ^4-androstene-3,17-dione	KUSHINSKY (1955)
Man (Androgenital syndrome)	Adrenal slices	Acetate-1-C^{14}	Δ^4-Androstene-3,17-dione, Dehydroepiandrosterone, 11β-Hydroxy-Δ^4-androstene-3,17-dione	E. BLOCH et al. (1956a, b)
Man	*In vivo*	Cholesterol-C^{14},H^3	Androsterone, Etiocholanolone, 11-Keto-etiocholanolone	WERBIN and LEROY (1955), WERBIN et al. (1957a, b)
Man (Hyperactive adrenals)	Adrenal slices	—	11β-Hydroxy-Δ^4-androstene-3,17-dione	COOPER et al. (1955), TOUCHSTONE et al. (1955)
Pig	Adrenal homogenate	Acetate	Δ^4-Androstene-3,17-dione	BLIGH et al. (1955), HEARD et al. (1956)
Calf	Adrenal perfusion	Dehydroepiandrosterone	11β-Hydroxy-Δ^4-androstene-3,17-dione	G. ROSENFELD et al. (1955)
Ox	Adrenal homogenate	Δ^4-Androstene-3,17-dione	19-Hydroxy-Δ^4-androstene-3,17-dione	MEYER (1955b)
Ox	Washed residue from adrenal homogenate (5000 × g)	Δ^4-Androstene-3,17-dione	6α-Hydroxy-Δ^4-androstene-3,17-dione, 6β,11β-Dihydroxy-Δ^4-androstene-3,17-dione	MEYER et al. (1955)
Ox	Adrenal homogenate	Δ^4-Androstene-3,17-dione	11β-Hydroxy-Δ^4-androstene-3,17-dione	HAYANO and DORFMAN (1953)
Ox	Adrenal homogenate	Dehydroepiandrosterone, Δ^4-Androstene-3,17-dione	11β-Hydroxy-Δ^4-androstene-3,17-dione, Adrenosterone	MEYER et al. (1955)
Pig	Adrenal cell-free extract	Progesterone	Δ^4-Androstene-3,17-dione	RAO and HEARD (1957)

incubated with acetate-1-C^{14} and ACTH. A 5-fold increase (over the amount obtained without ACTH) in the formation of 11 β-hydroxy-Δ^4-androstene-3,17-dione was observed, as well as an increase in androstane-3,17-dione. On the basis of C^{14} content, the increments in Δ^4-androstene-3,17-dione and dehydroepiandrosterone were not significant.

The experiments summarized in Table 17 provide direct evidence for adrenal androgen biosynthesis from acetate, cholesterol, progesterone, and 17 α-hydroxyprogesterone. The active materials secreted into the blood may be considered to be three main components, namely, dehydroepiandrosterone, Δ^4-androstene-3,17-dione, and 11 β-hydroxy-Δ^4-androstene-3,17-dione.

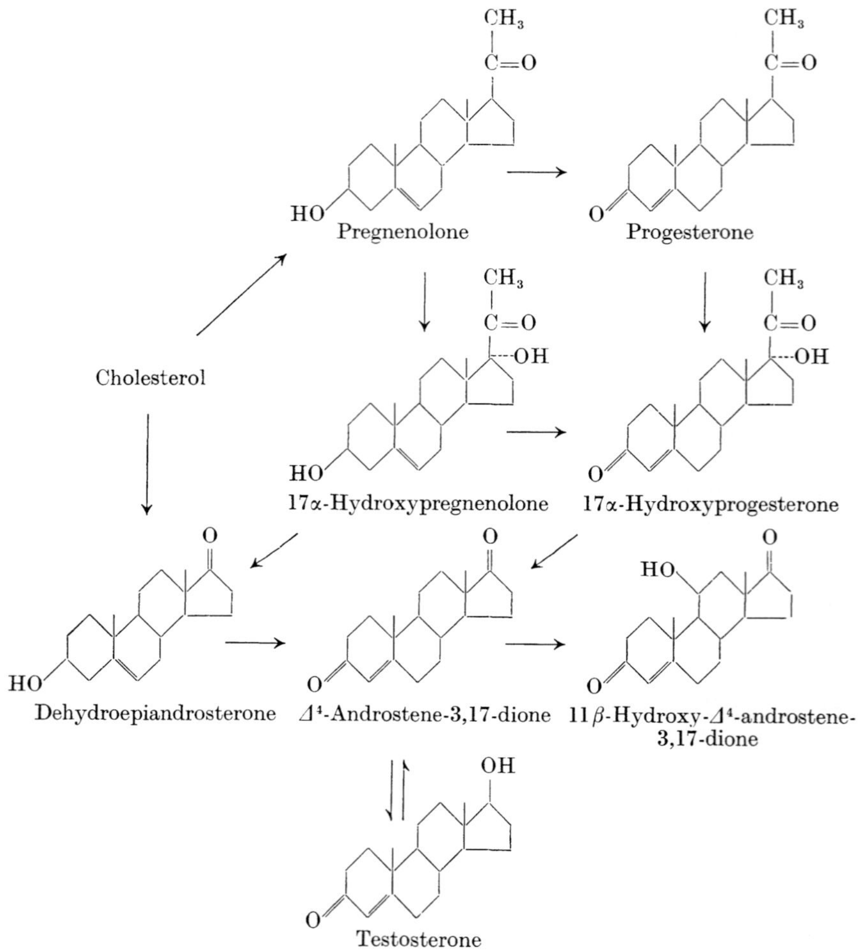

Fig. 18. Biosynthesis of adrenocortical androgens

The results are consistent with a pathway illustrated in Fig. 18, involving the sequence acetate → cholesterol → pregnenolone → progesterone → 17 α-hydroxyprogesterone → Δ^4-androstene-3,17-dione → 11 β-hydroxy-Δ^4-androstene-3,17-dione. On the basis of indirect evidence, one would expect a second biosynthetic pathway, involving the direct formation of dehydroepiandrosterone from cholesterol without going through a C_{21} intermediate. This pathway of cholesterol → dehydroepiandrosterone → Δ^4-androstene-3,17-dione → 11 β-hydroxy-Δ^4-androstene-3,17-dione is also represented in Fig. 18. Still a third pathway for possible formation of adrenal androgens must be considered. That is the 17 α-hydroxylation of pregenolone and the formation of dehydroepiandrosterone by desmolase action (Fig. 18).

E. Biosynthesis of adrenal estrogens

Adrenal tissue produces estrogens from acetate. This has been established by a series of experiments involving conversion of acetate to the androgen, Δ^4-androstene-3,17-dione, and the conversion of the latter compound to estrone. That Δ^4-androstene-3, 17-dione can arise from acetate has been established by the incubation of acetate-1-C^{14} with adrenal slices from a patient with the adrenogenital syndrome (E. Bloch et al. 1956b, 1957) and by the incubation of the same substrate with hog adrenal homogenates (Bligh et al. 1955, Heard et al. 1956). The second series of reactions was demonstrated by Meyer (1955), who incubated Δ^4-androstene-3,17-dione with an adrenal homogenate and isolated estrone.

An adrenal tumor of a 66-year-old male was shown to convert testosterone to estrogens. This individual excreted a markedly elevated quantity of estrogens which, on a daily basis, amounted to 187 μg of estrone, 47 μg of estradiol-17 β, and 170 μg of estriol (Engel et al. 1950). Slices of the 2 kg adrenal tumor were incubated with testosterone-3-C^{14}. Radioactivity was found associated with estrone and estradiol-17 β but not with estriol (Baggett et al. 1959).

Salhanick and Berliner (1957) have isolated estrone and equilenin from adrenal tumor tissue. This is the first demonstration of equilenin in human tissues or body fluids, all other studies on this estrogen being confined to the urine of the pregnant mare. Information about the biosynthesis of this ring B-unsaturated estrogen is meager, but certain facts are available to suggest that equilenin is synthesized by a pathway independent of cholesterol and androgens (Dorfman 1956). These facts may be listed as (a) estrone and the ring B-unsaturated estrogens, equilin and equilenin, biosynthesized from acetate-1-C^{14} in the same experiment, have significantly different specific activities (Heard et al. 1956), (b) estrone is not converted to ring B-unsaturated estrogens by the pregnant mare, (c) and testosterone is converted to estrone but not to equilenin (Heard et al. 1955).

A possible pathway for the biosynthesis of equilenin that accounts for the experimental facts now available includes the condensation of "an ethyltoluene residue" (which may be derived from the diet) plus a C_{10} isoprenoid unit. This condensation plus hydroxylation at steroid positions 3 and 17 and demethylation at carbon 14 would yield dihydroequilenin (Dorfman 1956, Savard et al. 1958). It is further possible that dihydroequilenin could be reduced in ring B to yield estradiol-17β, which would be an alternate pathway to formation of the usual ring A-aromatic estrogens.

This alternative pathway for estrogen biosynthesis is intriguing, since it could possibly explain the presence of two steroids isolated from pregnant mare's urine, which contains as one important constituent equilenin. The first steroid, 3-deoxyequilenin (Prelog and Führer 1945), could represent an incomplete hydroxylated product, while 3β-hydroxy-$\Delta^{5,7,9}$-estratriene-17-one (Heard and Hoffman 1941) could represent a ring A-reduction product of equilenin.

18-Hydroxyestrone (Loke and Marrian 1958) may also be formed in the adrenal, since other C-18 oxygenated steroids are produced by that gland. In view of the demonstration by Ayres et al. (1956) and Giroud et al. (1958) that aldosterone biosynthesis, which requires 18-hydroxylation, occurs exclusively in the zona glomerulosa, it is likely that 18-hydroxyestrone is not only a specific adrenal product but the product of a specific portion of this gland.

F. Biosynthesis of steroid conjugates

Steroids of blood and urine are, at least in part, conjugated with sulfuric or glucuronic acid. The formation of both these conjugates by tissue preparations has now been accomplished.

Conjugation of dehydroepiandrosterone with sulfuric acid can be accomplished by a system composed of a soluble enzyme preparation from rat liver, in the presence of ATP and magnesium (DeMeio and Lewycka 1955). In a comprehensive study, Schneider and Lewbart (1956) used a similar liver preparation from rabbits and incubated the substrates at 37° in air for periods from 3 to 4 hours. Steroid sulfates were formed from the following compounds: dehydroepiandrosterone, androsterone, epiandrosterone, androstane-3α, 17α-diol, 17α-methyl-Δ^5-androstene-3β,17β-diol, androstane-3α,16β,17α-triol, Δ^5-androstene-3β,16β,17α-triol, testosterone, deoxycorticosterone, 3β-hydroxy-Δ^5-pregnen-20-one, 3β-hydroxy-$\Delta^{5,16}$-pregnadien-20-one, 3β,21-dihydroxy-Δ^5-pregnen-20-one, allopregnane-3β, 20,21-triol and Δ^5-pregnene-3β,17α,20-triol. The enzyme system employed, however, did not yield sulfate esters of the following steroids: cholesterol, Δ^{16}-androsten-3α-ol, pregnan-3α-ol, 3α-hydroxypregnan-20-one, cortisol, cortisone, and urocortisone.

Schneider and Lewbart (1956) were able to make certain interesting generalizations on the basis of their experimental data. In the C_{21} series, the yield of sulfate esters was greater for steroids possessing the Δ^5-3β-hydroxy grouping than the related fully saturated representatives. Steroids belonging to the 5α (allo-A/B ring fusion trans) were more readily conjugated than those of the 5β series.

The facts that glucosiduronide formation can take place in liver and that this tissue is the primary site for this reaction were demonstrated by Lipschitz and Bueding (1939). These observations were confirmed and extended by subsequent investigators, who found that glucose, without prior metabolism to smaller units, was utilized for glucosiduronide formation.

Douglas and King (1952), for example, administered C-1 and C-6-labeled glucose to guinea pigs together with borneol and isolated the properly tagged glucosiduronide. Similar results were reported by Eisenberg and Gurin (1952) with menthol in the rabbit. Of further importance was the fact that C^{14}-labeled glucuronic acid administered to guinea pigs only appeared in trace amounts in borneol glucuronide (Douglas and King 1952). Studies on liver slices and homogenates revealed that uridine diphosphate glucuronic acid was the key to the formation of the glucuronides (Smith et al. 1953, Smith and Mills 1954, Mills et al. 1954 Dutton and Storey 1954, Storey and Dutton 1955).

The synthesis of a steroid conjugate, urocortisone glucosiduronide, by liver enzymes is credited to Isselbacher and Axelrod (1955) and Isselbacher (1956). The mechanism involves glucose oxidation and attachment to uridine diphosphate to form uridine diphosphate glucuronic acid. The synthesis of the conjugate involves a glucuronosyl transfer to urocortisone. The enzyme responsible for this reaction is found in the microsome fraction of mammalian liver homogenates.

G. Factors influencing biosynthesis of adrenocortical steroids

The production of adrenocortical steroids may be influenced in a variety of ways, ranging from physiological mechanisms to pharmacological agents. The prime physiological agent is ACTH, and the fact that this pituitary hormone can stimulate the production of corticoids, androgens, and estrogen is fully documented. The manner in which ACTH produces this effect is still open to considerable question in spite of significant and real advances.

Thus far ACTH is known only to act on adrenal tissue with intact cells, such as the gland *in situ*, whole isolated adrenal, slices, and brei. Some effects of ACTH on cell-free systems have been reported but these effects, thus far, have not been consistently reproduced.

ACTH acts in two ways: (a) it controls the growth and differentiation of the adrenal gland, which is an action that takes hours to days (Chapter 1, p. 82), and (b) it stimulates hormone production, which takes seconds to minutes. It is the latter action that concerns us here.

Among other possible actions of ACTH, the conversion of cholesterol to pregnenolone has been considered to be the principal point of action. This idea has been advanced mainly because (a) adrenal tissue, either when perfused, in slices, or in homogenates, produces corticoids from pregnenolone or progesterone *without* the need for ACTH, but (b) corticoid production from cholesterol *is* stimulated by ACTH. Studies by HAYNES and BERTHET (1957) have indicated that one mechanism of ACTH involves the effective increase in phosphorylase, leading to increases in glucose-1-PO_4 and glucose-6-PO_4. This latter compound, when oxidized, reduces TPN to TPNH. The active component formed in this system seems to be the TPNH. The mechanism by which ACTH effectively increases phosphorylase is through increasing the amount of cyclo adenine nucleotide (HAYNES 1958). If this chain of events is triggered by ACTH, then the addition of the cyclo adenine nucleotide to adrenal tissue should stimulate corticoid production. This stimulus has been demonstrated by HAYNES et al. (1959) in quartered rat adrenals.

Since TPNH, the level of which is governed in part by the action of ACTH, is an obligatory coenzyme for the hydroxylation reaction, possibly the increased conversion of cholesterol to pregnenolone is really a result of making TPNH available and facilitating the hydroxylation of cholesterol prior to rupture between C-20 and C-22. If this is true however, why does not ACTH stimulate the hydroxylation of progesterone at 11β, 17α, and 21 when these enzyme systems also require TPNH as an obligatory coenzyme? HEARD et al. (1956) did report that progesterone and deoxycorticosterone incubated with rat adrenal slices formed increased amounts of corticoids in the presence of ACTH, but this observation lacks confirmation at this time.

KORITZ and PERON (1958) have recently confirmed and extended the findings of HAYNES and BERTHET (1957). They showed that rat adrenal quarters produced corticosterone more efficiently if TPNH was added to the incubation medium, but, after a maximum stimulation due to TPNH, the addition of ACTH caused a further increment in corticoid production. In another set of experiments, rat adrenal slices were maximally stimulated with ACTH, that is, to a point where the further addition of ACTH caused no further increment in corticoid production. At this point the addition of a TPNH-generating system caused a further rise in corticoid production. These data together with the observations of HAYNES and BERTHET (1957) suggest that ACTH perhaps has two actions in corticoid production, (a) the first to facilitate an adequate concentration of TPNH, and (b) the second to make available a critical limiting substrate. The second action may be the liberation of a substrate from an enzymatically-unavailable, bound form.

The quartered-rat-adrenal technique of SAFFRAN and BAYLISS (1953) has served as an important tool in dissecting the factors relating to corticoid production. Using this technique, BIRMINGHAM and KURLENTS (1958) demonstrated that rat adrenals removed ACTH from the medium within 5 minutes and that the quantity of hormones produced was dependent upon the amount of tissue present and independent of the ACTH content of the medium. Addition of cortisol or corticosterone to the medium at a concentration of 10^{-5} *M* inhibited corticoid production.

The requirements for ACTH stimulation of adrenal tissue *in vitro* have been investigated in some detail. Glucose (SCHÖNBAUM et al. 1956, PERON and KORITZ 1958), calcium (BIRMINGHAM et al. 1953, PERON and KORITZ 1958), and potassium (PERON and KORITZ 1958) are required for the stimulation. The studies of PERON and KORITZ (1958) further indicated that rubidium can replace the potassium and that the calcium requirement can be satisfied by strontium but not by magnesium, manganese or barium. Corticoid production as a result of ACTH is independent of adrenal protein synthesis (KORITZ et al. 1957).

I. Role of ascorbic acid

Ascorbic acid in rat adrenals decreases rapidly after ACTH stimulation or whenever the animal is subjected to a procedure which would be expected to cause an increase in circulating ACTH (SAYERS et al. 1946, LONG 1947). This phenomenon has been utilized as an indirect method for the assay of micro quantities of ACTH and has led to considerable speculation concerning the relationship of ascorbic acid and corticoid production.

SMIDT and STAUDINGER (1956) have claimed that corticoid production by porcine adrenals may be stimulated by ascorbic acid, a view shared by relatively few workers. Actually the greater weight of evidence indicates that corticoid production is inversely proportional to the concentration of ascorbic acid. In the scorbutic guinea pig, cortisol production is markedly increased as measured by urinary levels (NADEL and SCHNEIDER 1952, BURSTEIN et al. 1954a, R. E. CLAYTON et al. 1957). Direct analysis of the biosynthetic capacity of the adrenal indicates that guinea pig adrenals in late scurvy actually produce an excessive amount of cortisol (BURSTEIN and NADEL 1956). This increased cortisol production is associated with an increased serum ACTH, which could be decreased by the addition of ascorbic acid to the diet. This decrease in blood ACTH produced by ascorbic acid in late scurvy correlates with a decrease in cortisol concentrations in plasma (DONE et al. 1953) and a decrease in urinary titers of corticoids (NADEL and SCHNEIDER 1952). The plasma ACTH levels can be decreased by cortisol, cortisone and deoxycorticosterone but not by progesterone, estradiol-17β, or testosterone (R. E. CLAYTON et al. 1957).

These facts argue against the idea that ascorbic acid is utilized for corticoid formation. Rather they seem to suggest that ascorbic acid may actually be an inhibitory factor. In line with the idea that ascorbic acid acts as a "brake" to corticoid biosynthesis is the observation that the ascorbic acid disappearing from the adrenal glands in normal rats subjected to the stress of laporatomy and in hypophysectomized rats treated with ACTH can be accounted for in the adrenal venous effluent (SLUSHER and ROBERTS 1957). BRIGGS and TOEPEL (1958) agree with this observation for ACTH-treated animals but find more ascorbic acid disappearing under stress than is recoverable from the adrenal venous blood. To these facts may be added the report that at least one important biosynthetic reaction, 11β-hydroxylation, can be inhibited *in vitro* with ascorbic acid (HAYANO et al. 1956a)*.

II. Chorionic gonadotrophin

Whether human chorionic gonadotrophin can stimulate adrenocortical steroid production is a subject of some controversy. As seen in Table 18, some increases in 17-ketosteroid excretion in ovariectomized women have been reported, but more detailed information is needed.

* In many species, ascorbic acid concentration in the adrenal glands is low; in others, the level does not change with stimulation. See Chapter 1, p. 84 ff. sub-ed.

Table 18. *Influence of human chorionic gonadotrophin (HCG) on 17-ketosteroid excretion of ovariectomized women*

No. of subjects	Total dose of HCG, I. U.	Duration of treatment, days	Response	Reference
6	5,000	1	Increase	Gomez-Maestro (1951)
5	25,000	5	0	Borth et al. (1953)
1	50,000			
18	50,000	1	Increase	Borell (1954)
2	60,000	3	Increase	Plate (1952)

III. Δ^4-Cholestenone

When rats are fed a diet containing 1% of Δ^4-cholestenone, decreased synthesis of cholesterol and hypertrophy of the adrenal are observed (Tomkins et al. 1953, Steinberg and Fredrickson 1956). This increased adrenal size is probably due to an increased ACTH production resulting from decreased corticosterone production. Fredrickson et al. (1958) showed that even when the adrenals were hypertrophied 6 to 8 times the corticosterone and aldosterone contents of the adrenal venous effluent were decreased. Since cholesterol biosynthesis is known to be inhibited by Δ^4-cholestenone, it has been suggested that corticoid production is low because of a relative deficiency of cholesterol in the adrenal (Steinberg et al. 1958), or in other words an inadequate supply of necessary substrate.

IV. Estrogens

Estrogens cause hypertrophy and hyperplasia of the adrenal cortex in the rat (Korenchevsky and Dennison 1935, Selye et al. 1935). This hypertrophy, when produced by estradiol-17β, appears to be due to a true increase in circulating ACTH (Selye and Collip 1936, Gemzell 1952). Further, the influence of estrogens can only be demonstrated in the presence of the pituitary (Selye and Collip 1936) (Chapter 1).

The reason for the ACTH increase as a result of hexestrol treatment has been elucidated by Vogt (1955), who showed that the actual rate of corticosterone secretion by the rat is decreased 70% after 3 days of treatment with this synthetic estrogen and that the decreased secretion is not preceded by an earlier period of excessive corticosterone secretion. In the rat, other estrogens, including estradiol-17β, chlorotrianisene, and ethinyl estradiol-17β, inhibited corticoid synthesis in maximally-stimulated adrenals (Holzbauer 1957). Interference with cholesterol biosynthesis has been suggested as a possible mechanism of action.

Rabbits did not show the same effect (Zondek and Burstein 1952), since both estrone and stilbestrol caused an immediate increase in urinary corticoids as assayed by a formaldehydogenic method (Burstein 1952).

Estrogens cause an increased blood concentration of adrenocortical steroids in man. Halkerston et al. (1956) found an increased excretion of dehydroepiandrosterone after estradiol-17β treatment, and Taliaferro et al. (1956) reported an increased plasma corticoid titer in men treated with stilbestrol but not in an Addisonian patient similarly treated. These results were essentially confirmed and extended by Wallace et al. (1957), who found maked plasma corticoid elevations after ethinyl estradiol treatment in post-menopausal and castrate women but not in an Addisonian man. Two other facts were uncovered by these investigators. Ethinyl estradiol appeared to cause an increased responsiveness of the adrenal to ACTH and a delayed rate of plasma clearance of administered cortisol. Both of these effects could be related to the one fundamental interference with cortisol inactivation.

HOLZBAUER and VOGT (1957) have described but did not identify a material in rat plasma that restored the corticosterone secretion by hexestrol-inhibited rat adrenals. Amphenone (0.2 g/kg -by mouth) and an intravenous suspension of cholesterol did the same.

V. Progestational agents

Progesterone given to guinea pigs even at a high dose level (10 mg) did not change the excretion of corticoids (ZONDEK and BURSTEIN 1952), nor did it produce any change in the corticoid level in the adrenal venous effluent in the rat (VOGT 1955). On the other hand, significant lowering of urinary corticoids and 17-ketosteroids has been found with 19-nor-17-ethyltestosterone (BROOKS and PRUNTY 1957).

VI. Acetylcholine, physostigmine and atropine

Acetylcholine exerted a stimulative effect on the isolated calf adrenal perfused with an artificial medium (G. ROSENFELD 1955b). The concentration of added stimulant was of the order of 2.7×10^{-7} *M*. Physostigmine sensitized, whereas atropine prevented, the action of acetylcholine. These studies suggest that acetylcholine transmits the effect of nerve excitation and is the direct stimulant of the adrenal cortex in those animals in which electrical stimulation of the splanchnic nerves results in increased corticoid production (VOGT 1944, OKINAKA et al. 1952).

VII. Amphenone

HERTZ et al. (1951, 1955) reported that amphenone (1,1-bis-(*p*-aminophenyl)-2-methyl propanone-1-dichloride) (Fig. 19) causes enlargement of the adrenal glands of rats. This was interpreted to mean that corticosteroid production was blocked, causing decreased corticoids in circulation and a compensatory increase in the production of ACTH. The increased production of ACTH caused an enlarged gland. Studies by TULLNER et al. (1956) and by HUME and NELSON (1958) indicated

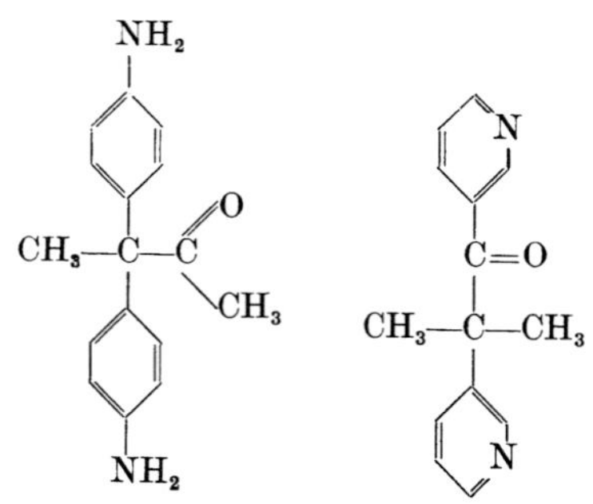

Amphenone SU 4885

Fig. 19. Structure of Amphenone and of SU 4885

that the rate of corticoid output in hypophysectomized dogs stimulated with ACTH was suppressed with amphenone. In the intact dog, after initial suppression there was a rapid return to pretreatment level of production.

GALLAGHER et al. (1956) and GALLAGHER (1957) studied the influence of amphenone on the excretion of 17-ketosteroids in patients with adrenal cancer. Dehydroepiandrosterone was decreased dramatically; androsterone, etiocholanolone and their 11-oxygenated derivatives were decreased less. HERTZ et al. (1956)

found irregular decreases in plasma and urinary corticoids but essentially no change in 17-ketosteroids in patients with breast carcinoma and adrenal hyperplasia. Similar data were reported by THORN et al. (1956).

In a definitive study, R. PETERSON et al. (1957a) demonstrated that amphenone causes an inhibition of glucocorticoid biosynthesis and that this was most marked with adrenal tissue relatively unresponsive to ACTH. This compound, when administered to a patient with adrenal carcinoma and to normal subjects, produced a fall in plasma cortisol and corticosterone as well as in urinary corticoids, but no striking decrease in 17-ketosteroid excretion. Amphenone did not influence the rate of cortisol, corticosterone, or urocortisol metabolism. In an amphenone-treated patient suffering from congenital adrenal hyperplasia, pregnanetriol and corticoid excretion were suppressed, but no change was observed in the 17-ketosteroids (J. JOHNSON et al. 1957). In still another study (MCCULLAGH and TRETBAR 1958), intravenous administration of amphenone to a Cushing's patient produced a significant decrease in urinary glucocorticoids but the 17-ketosteroids were variable.

Amphenone also affects aldosterone biosynthesis. The studies of RENOLD et al. (1957) and HERNANDO et al. (1957) indicated that aldosterone biosynthesis was decreased by amphenone, at a dose level so small that the cortisol biosynthesis was not influenced. In keeping with the decreased aldosterone excretion was an increased excretion of sodium. Complete disappearance of aldosterone from the urine was found by MACH and MULLER (1957) after the administration of 4 to 6 g of amphenone per day for 3 days.

Amphenone perfused through a calf adrenal reduced with corticoid production (G. ROSENFELD and BASCOM 1956). Amphenone, therefore, interferes with corticoid biosynthesis in the dog, man, and calf. Such does not seem to be the case for the golden hamster, where amphenone administration did not reduce urinary adrenocortical steroid metabolites or induce adrenal enlargement (MARKS et al. 1958). These investigators suggested that the hamster adrenal may not utilize cholesterol as a steroid precursor, but rather may produce its steroids from a pathway involving the condensation of small units directly to the C_{21} steroids.

VIII. 2-Methyl-1,2-bis-(3-pyridyl)-1-propanone

This analogue of amphenone, SU 4885 (Fig. 19), inhibited the secretion of corticoids by dog adrenals but was apparently inactive in rats (CHART et al. 1958). *In vitro* studies indicated suppression of hormone synthesis by rat, guinea pig, and dog adrenal slices. The compound was inactive as a progestational agent, androgen, or estrogen. (See also Chapters 1, 4.)

IX. Phenylethylacetic and diphenylethylacetic acids

Diphenylethylacetic acid caused a decrease in 17-ketosteroid excretion but phenylethylacetic acid was inactive. The inactive acid, however, intensified the action of diphenylethylacetic acid (FIDANZA et al. 1957a).

X. 1,1-Dichloro-2,2-bis(chlorophenyl)ethane

The insecticide DDD was found to cause severe adrenocortical atrophy in the dog (NELSON and WOODWARD 1949). Atrophy of the gland was accompanied by a decreased production of corticoids as indicated by the animals' becoming sensitive to insulin; the compound also caused amelioration of alloxan diabetes (NICHOLS and GARDNER 1951, NICHOLS and SHEEHAN 1952, NICHOLS et al. 1953). Direct studies of corticoid production by dogs treated with DDD showed a significant drop in urinary Porter-Silber chromogens at a time when the adrenal was undergoing necrosis.

The hypertrophy produced by amphenone is prevented by the simultaneous administration of DDD in the dog (NICHOLS and HENNIGAR 1957).

Attempts have been made to produce adrenal effects with DDD in man. SHEEHAN et al. (1953) found no significant clinical, biochemical, or histological effects with the compound in a patient suffering from Cushing's syndrome. ZIMMERMAN et al. (1956) administered DDD to 10 patients with carcinoma and found significant decreases in 17-ketosteroid excretion; other steroid excretion titers were unaffected.

The relation of structural modifications in the molecule to biological activity has been studied by LARSON et al. (1955) and by BLEIBERG and LARSON (1957). Both phenyl groups proved to be essential for the adrenal effect but the *para* phenyl substitution was not obligatory. Other changes in the molecule, including the number of chlorine substitutions and desaturation or removal of the hydroxy group, produced inactive molecules. *Para* methoxy and nitro derivatives were inactive, but activity was retained by the methyl derivatives. (See also Chapters 1,4.)

XI. Pantothenic acid

The need for panthothenic acid (a constituent of coenzyme A) in the biosynthesis of adrenal cholesterol has been demonstrated (DUMM et al. 1953). A severe decrease in this factor caused decreased steroid-hormone biosynthesis in rats (WINTERS et al. 1952a, b, DUMM et al. 1953, EISENSTEIN 1957). When EISENSTEIN and WENNEKER (1958) administered omega-methyl panthothenic acid, a vitamin antagonist, to guinea pigs, corticoid excretion was significantly decreased as compared to pair-fed controls. Consistent with these findings was the observation that no stimulation of corticoid output by ACTH could be produced in pantothenic acid-deficient rats. In humans panthotenic acid deficiency caused a 50% decrease in 17-ketosteroid output, but the question of a nonspecific malnutrition effect must be considered (FIDANZA et al. 1957b).

XII. Pyrogenic bacterial polysaccharides

Purified, pyrogenic bacterial polysaccharides derived from *Pseudomonas aeruginosa*, *Escherichia coli*, and *Streptococcus marcescens* inhibited the biosynthesis of adrenocortical steroids in an adrenal perfusion study when homologous whole blood was used, but when an artificial medium was employed the same pyrogens were ineffective as inhibitors. G. ROSENFELD (1955a), who reported these findings suggested that this phenomenon may be related to the relative adrenocortical insufficiency reported in the bacteremia and septicemia associated with fulminating infectious disease. The facts that the pyrogens were ineffective as inbitors when the artificial medium was employed but effective when blood was used as the perfusing medium suggest that a reaction between the pyrogen and some cellular or humoral blood constituent may produce the effective agent.

Catabolism of adrenocortical steroids

A. C_{21} steroids

The catabolic reactions of corticoids are presented in Table 19 for *in vivo* studies and Table 20 for *in vitro* studies. By far the greatest number of *in vivo* studies have been done in human beings, while *in vitro* studies have been confined

Table 19. *In vivo metabolism of corticoids. Product isolated from urine unless otherwise noted*

Compound	Species	Product	Reference
Deoxycorticosterone	Chimpanzee (ovariectomized)	Pregnanediol	W. FISH et al. (1943)
	Man	Pregnanediol	HORWITT et al. (1944)
	Rabbit	Pregnanediol	WESTPHAL (1942), HOFFMAN et al. (1943)
	Man	3α,21-Dihydroxypregnan-20-one	RICHARDSON et al. (1955)
21-Deoxycortisone	Man	11β-Hydroxyetiocholanolone, 11-Ketoetiocholanolone, 3α,17α-Dihydroxypregnane-11,20-dione	BURSTEIN et al. (1953b)
	Man	11β-Hydroxyandrosterone, 11-Ketoetiocholanolone, 3α,17α-Dihydroxypregnane-11,20-dione, 11-Ketopregnanolone	ROSSELET et al. (1954a, 1955)
	Man	3α,17α,20α-Trihydroxy-pregnan-11-one	FUKUSHIMA and GALLAGHER (1957a)
17α-Hydroxy-progesterone	Man	Androsterone, Etiocholanolone	DORFMAN (1954)
	Man	Pregnanolone	ROSSELET et al. (1954a)
11-Dehydrocorticosterone	Man	11-Ketopregnanediol	MASON (1948)
		3α,11β,21-Trihydroxypregnan-20-one, 3α,21-Dihydroxy-pregnane-11,20-dione	RICHARDSON et al. (1955)
		3α,21-Dihydroxypregnane-11,20-dione	GOMEZ-MONT and BERLINER (1953)
		11β,20β,21-Trihydroxy-Δ^4-pregnen-3-one	RICHARDSON et al. (1958)
Cortisone	Man	11-Ketoetiocholanolone, 11β-Hydroxyetiocholanolone	BIRKE and PLANTIN (1953)
		11-Ketoetiocholanolone, 11β-Hydroxyetiocholanolone, 3α,17α,20α-Trihydroxy-pregnan-11-one, Cortisol	BURSTEIN et al. (1953b)
		11-Ketoetiocholanolone, 11-Hydroxyetiocholanolone, 3α,17α,20α-Trihydroxy-pregnan-11-one	DOBRINER and LIEBERMAN (1952)
		Cortisone, Cortisol, Urocortisol, Urocortisone	BURTON et al. (1953)
		Cortisol, Urocortisone, 11β-Hydroxyetiocholanolone, 11-Ketoetiocholanolone, 17α,20β,21-Trihydroxy-Δ^4-pregnene-3,11-dione	LOMBARDO and HUDSON (1957)
		Cortolone, β-Cortolone	FUKUSHIMA et al. (1955)
		Cortisol, Urocortisol, Urocortisone	COPE and HURLOCK (1954)
		Urocortisol, Urocortisone	RICHARDSON et al. (1955)
	Rat	Cortisol, 11β,17α,20β,21-Tetrahydroxy-Δ^4-pregnen-3-one	CASPI and BERGEN (1955)

Table 19 (cont.)

Compound	Species	Product	Reference
Cortisol	Man	11β-Hydroxyepiandrosterone	KEMP et al. (1954)
	Guinea pig	11β,17α,20α,21-Tetrahydroxy-Δ4-pregnen-3-one, 11β,17α,20β,21-Tetrahydroxy-Δ4-pregnen-3-one, 6β-Hydroxycortisol	BURSTEIN and DORFMAN (1955)
	Rat	17β,17α,20β,21-Tetrahydroxy-Δ4-pregnen-3-one (Plasma)	ABELSON et al. (1955)
	Man	Urocortisone, Urocortisol	RICHARDSON et al. (1955)
	Man	6β-Hydroxycortisol	BURSTEIN et al. (1954)
	Mouse	Urocortisone, Urocortisol (Blood and connective tissue)	DOUGHERTY et al. (1958)
	Man	Urocortisone	COPE and HURLOCK (1954)
	Man	Cortol, β-Cortol, Cortolone, β-Cortolone	FUKUSHIMA et al. (1955)
	Man	Urocortisone glucosiduronate (Blood) Urocortisol glucosiduronate (Blood)	VERMEULEN (1956c)
	Man	11β-Hydroxyandrosterone, 11β-Hydroxyetiocholanolone, 11-Ketoetiocholanolone, 11-Ketoandrosterone	GALLAGHER (1957)
	Man	11β,17α,20α,21-Tetrahydroxy-pregnan-3-one	R. PETERSON et al. (1957b)
Corticosterone	Man	3α,11β,21-Trihydroxypregnan-20-one, 3α,21-Dihydroxy-pregnane-11,20-dione	RICHARDSON et al. (1955)
	Man	Pregnane-3α,11β-20-triol, 3α,11β-Dihydroxypregnan-20-one	FUKUSHIMA et al. (1954c)
	Man	3α,21-Dihydroxypregnane-11,20-dione, 3α,11β,21-Tri-hydroxyallopregnan-20-one, 3α,11β,21-Trihydroxypregnan-20-one, 3α,20α-Dihydroxypreg-nan-11-one	ENGEL et al. (1955)
	Man	11β,20β,21-Trihydroxy-Δ4-pregnen-3-one	RICHARDSON et al. (1958)
11-Deoxycortisol	Man	3α,17α,21-Trihydroxypregnan-20-one	RICHARDSON et al. (1954)
	Man	Androsterone, Etiocholanolone	DORFMAN (1954)
Urocortisone	Man	β-Cortolone, Urocortisone (as glucosiduronate)	SCHNEIDER et al. (1955)
	Man	Urocortisone conjugate (Blood) Cortolone or β-Cortolone	BONGIOVANNI and EBERLEIN (1955a)
11-Ketoprogesterone	Man	11-Ketopregnanediol	SARVARD and DORFMAN (1954a)
17α,21-Dihydroxy-pregnane-3,20-dione	Man	Etiocholanolone, Pregnanetriol, 3α,17α,21-Trihydroxypregnan-20-one	UNGAR et al. (1954a)
11β-17α,21-Tri-hydroxypregnane-3,20-dione and Urocortisol	Man	Urocortisone, 3α,17α,21-Tri-hydroxypregnane-11,20-dione, 11β-Hydroxyetiocholanolone, 11-Ketoetiocholanolone	SAVARD and GOLD-FADEN (1954)

Table 20. *In vitro metabolism of corticoids* (In tissues other than adrenal)

Substrate	Species	Test system	Product	Reference
Deoxycorticosterone	Rat	Liver slices	3β,21-Dihydroxyallopregnan-20-one	Schneider and Horstmann (1951)
	Dog	Liver slices	6β,21-Dihydroxy-Δ^4-pregnene-3,20-dione	Axelrod and Miller (1954)
	Ox	Corpus luteum homogenate	20β-21,Dihydroxy-Δ^4-pregnen-3-one	Hayano et al. (1954)
	Ox	Liver perfusion	Δ^4-Etiocholanic acid-3-one	Picha et al. (1952)
	Pig	Liver brei	20α,21-Dihydroxy-Δ^4-pregnen-3-one	Caspi et al. (1956b)
	Rabbit	Liver homogenate	3β,21-Dihydroxyallopregnan-20-one, 3α,21-Dihydroxypregnan-20-one	Taylor (1957)
	Rabbit	Kidney homogenate	Progesterone	Ishizuka et al. (1957, 1958)
Urocortisone	Rat	Liver homogenate	β-Cortolone	DeCourcy and Schneider (1956)
	Rat	Liver homogenate	Cortolone	Hübener (1956), Hübener et al. (1956)
Urocortisol	Rat	Liver homogenate	β-Cortol	Hübener et al. (1956, Hübener (1956)
	Guinea Pig	Liver homogenate microsomes	Urocortisone glucosiduronate	Isselbacher and Axelrod (1955),
	Rat	Liver microsomes	Urocortisol glucosiduronate	Isselbacher (1956)
Cortisol	Rat	Loose connective tissue (90% mast cells)	Cortisone, 11β,17α,21-Trihydroxypregnane-3,20-dione, Corticosterone, 11β-Hydroxy-Δ^4-androstene-3,17-dione	Berliner and Dougherty (1958b)
	Ox	Kidney mince	Adrenosterone, 11β-Hydroxy-Δ^4-androstene-3,17-dione, Cortisone, 11β,17α,20β,21-Tetrahydroxy-Δ^4-pregnen-3-one	Ganis et al. (1956)
	Dog	Perfused liver	(Presumptive evidence) Cortolone, Urocortisone, Urocortisol, 11-Epicortisol, Cortisone, 11β-Hydroxy-Δ^4-androstene-3,17-dione	Axelrod and Miller (1956)
	Rat	Perfused liver	17α,20β,21-Trihydroxy-Δ^4-pregnene,3-11-dione Allopregnane-3α,11β,17α,20β,21-pentol, 3β,11β,17α-21-Tetrahydroxyallopregnan-20-one, 11β-Hydroxy-Δ^4-androstene, 3,17-dione, 3β,11β-Dihydroxyandrostan-17-one	Caspi (1955), Caspi and Hechter (1956)
	Rat	Liver extract	Urocortisol	Tomkins (1956b)
	Rat	Liver homogenate	β-Cortol	Hübener et al. (1956), Hübener (1956)

Table 20 (cont.)

Substrate	Species	Test system	Product	Reference
Cortisol (cont.)	Rat	Connective tissue (fibroblasts 90%)	Cortisone, 17α,20α,21-Trihydroxy-Δ^4-pregnen-3-one, Corticosterone, 11β-Hydroxy-Δ^4-androstene-3,17-dione, 11β,17α,21-Trihydroxypregnane-3,20-dione	BERLINER and DOUGHERTY (1958a)
11-Deoxycortisol	Pig	Liver brei	17α,20α,21-Trihydroxy-Δ^4-pregnen-3-one	CASPI et al. (1956b)
	Rat	Liver homogenate	17α,20α,21-Trihydroxy-Δ^4-pregnen-3-one	HÜBENER et al. (1956), HÜBENER (1956)
	Rat	Liver homogenate	17α,21-Dihydroxyallopregnane-3,20-dione, 3α,17α,21-Trihydroxyallopregnan-20-one, 3β,17α,21-Trihydroxyallopregnan-20-one, 3α,17α,21-Trihydroxypregnan-20-one, 3β-17α,21-Trihydroxypregnan-20-one, 3β,17α,20β,21-Allopregnanetetrol, 17α,20β,21-Trihydroxy-Δ^4-pregnen-3-one, 17α,20α,21-Trihydroxy-Δ^4-pregnen-3-one, 6β,17α,21-Trihydroxy-Δ^4-pregnene,3-20-dione, Androsterone, Androstane-3,17-dione, Δ^4-Androstene-3,17-dione, Etiocholanolone (Tentatively identified)	FORCHIELLI et al. (1955), FORCHIELLI (1956), FORCHIELLI and DORFMAN (1956)
Cortisone	Rat	Perfused liver	Adrenosterone, 11-Ketoandrosterone, 3β,17α,20β,21-Tetrahydroxyallopregnan-11-one, 11β,17α,20β,21-Tetrahydroxy-Δ^4-pregnen-3-one, 17α,20β,21-Trihydroxy-Δ^4-pregnene-3,20-dione, Cortisol, 3α,17α,21-Trihydroxyallopregnane-11,20-dione, 3β,17α,21-Trihydroxyallopregnane-11,20-dione, 3α,11β,17α,21-Tetrahydroxyallopregnan-20-one, 3β,11β,17α,21-Tetrahydroxyallopregnan-20-one	CASPI and HECHTER (1954), CASPI (1955)
	Rat	Perfused liver (normal)	Cortisol, 3α,17α,21-Trihydroxyallopregnane-11,20-dione, 3β,11β,17α,21-Tetrahydroxyallopregnan-20-one, 3α,11β,17α,21-Tetrahydroxyallopregnan-20-one	MILLER and AXELROD (1954)
		Perfused liver (cirrhotic)	6β,21-Dihydroxy-Δ^4-pregnene-3,20-dione, 17α,21-Dihydroxyallopregnane-3,11,20-trione	

Table 20 (cont.)

Substrate	Species	Test system	Product	Reference
Cortisone (cont.)	Rat	Liver extract	Urocortisone	TOMKINS (1956b)
	Rat	Liver homogenate	Cortisol	HÜBENER et al. (1956), HÜBENER (1956)
	Rat and Pig	Liver microsomes	Cortisol	AMELUNG et al. (1953)
	Rat	Liver slices	Cortisol	EISENSTEIN (1953)
	Rat, Ox, and Pig	Liver homogenate	Cortisol	C. FISH et al. (1953)
Urocortisone	Rat	Liver microsomes	Urocortisone glucosiduronate	ISSELBACHER and AXELROD (1955)
17α,21-Dihydroxypregnane-3,20-dione	Rat, Rabbit	Extract of liver homogenate powder	3α,17α,21-Trihydroxypregnan-20-one	UNGAR and DORFMAN (1954)
	Rabbit	Kidney homogenate	3α,17α,21-Trihydroxypregnan-20-one, 3β,17α,21-Trihydroxypregnan-20-one	UNGAR and DORFMAN (1954)
11β,17α,21-Trihydroxypregnane-3,20-dione	Rat	Liver homogenate	Urocortisol	TOMKINS (1956a)
17α,21-Dihydroxypregnane-3,11,20-trione	Rat	Liver homogenate	Urocortisone	TOMKINS (1956a)
17α-Hydroxypregnenolone	Dog	Perfused hind limb	Dehydroepiandrosterone	OERTEL and EIKNES (1958)

almost exclusively to animal tissues. The catabolism of the major and most of the minor adrenocortical steroids will be discussed separately and illustrated in a series of Figures (20 to 34).

I. Pregnenolone and progesterone

There is a strong possibility that both pregnenolone and progesterone are secreted by the adrenal gland. This is suggested by the fact that pregnane-3α,20α-diol, the major urinary metabolite of progesterone and pregnenolone, is present in men's urine at the level of approximately 0.9 mg per day, in post-menopausal urine at 0.63 mg per day, and in women in the follicular phase of the cycle at 1.12 mg per day (KLOPPER et al. 1955). Intravenous infusion of ACTH into a normal male increased the pregnanediol-3α,20α-diol excretion 4 to 5 fold. This procedure has been suggested as a test for evaluating adrenal function in man (KLOPPER et al. 1957).

Fig. 20 lists the many products of progesterone metabolism, but the only quantitatively-important urinary constituent is pregnane-3α,20α-diol. Catabolic reactions of progesterone, which destroy biological activity, are reductive in nature and are characteristic of the metabolic changes seen in all neutral steroid hormones, with the addition of some hydroxylation reactions. For detailed references on progesterone metabolism the reader is referred to DORFMAN and UNGAR (1953) and DORFMAN (1957b).

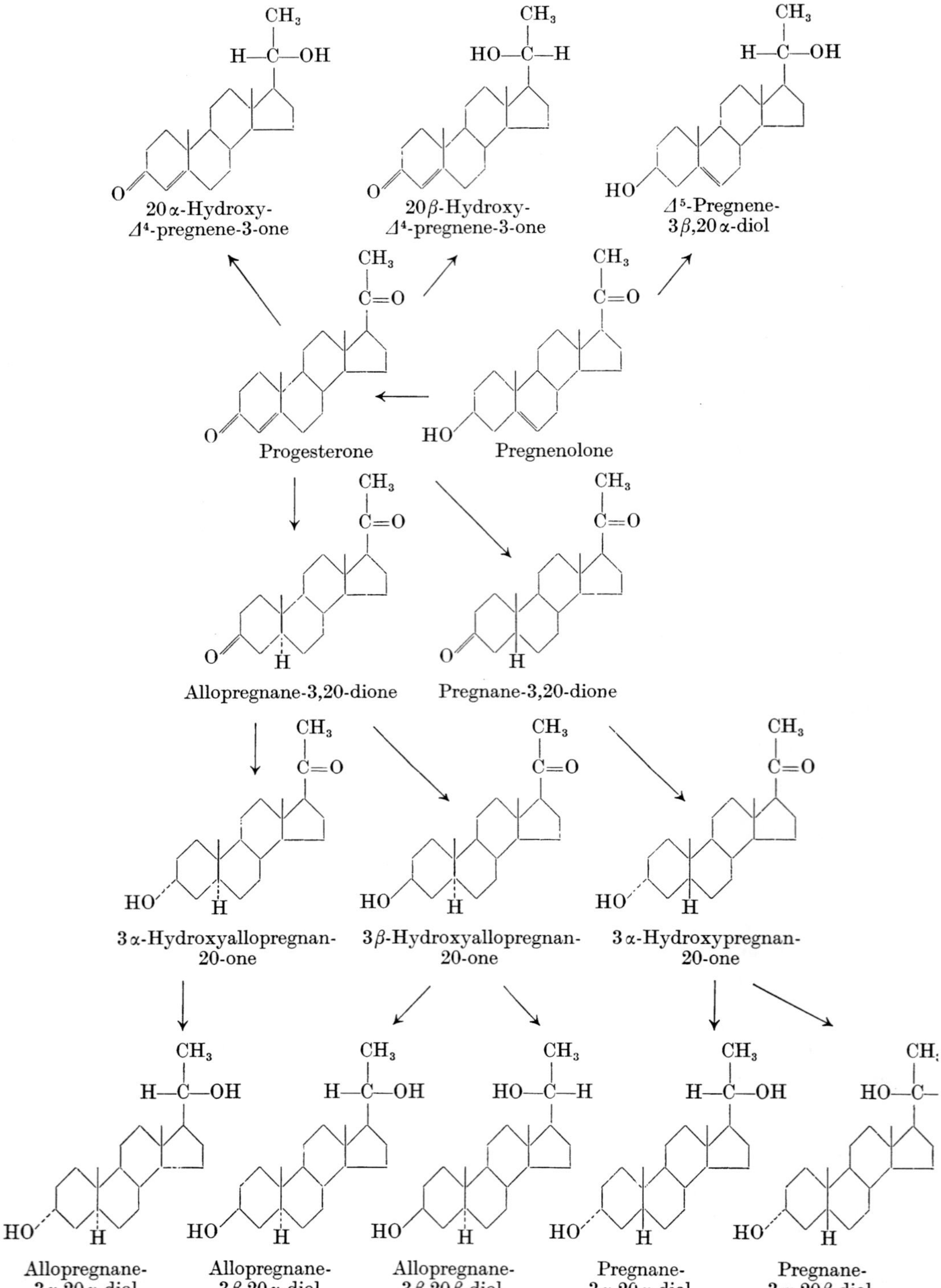

Fig. 20. Catabolism of $C_{21}O_2$ adrenocortical steroids

II. 16α-Hydroxyprogesterone

The 16α-hydroxy derivative of progesterone has been isolated from adrenal incubated with progesterone (RAO and HEARD 1957), and its secretion is consistent with the presence of three $C_{21}O_3$ 16α-hydroxylated products in urine — allo-pregnane-3α,16α,20α-triol, allopregnane-3β,16α,20α-triol, and Δ^5-pregnane-3β, 16α,20α-triol. The last-named compound is of particular interest since is probably derived from pregnenolone by C-20 reduction and 16α-hydroxylation (Fig. 21).

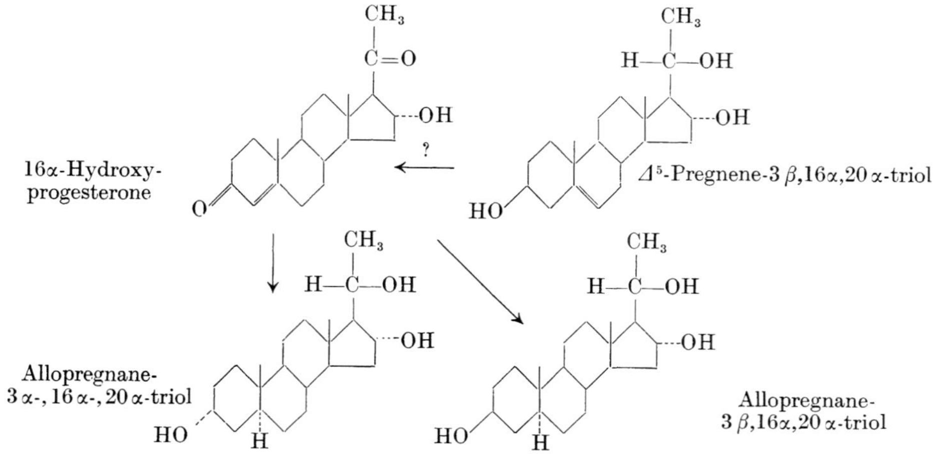

Fig. 21. Catabolism of 16α-hydroxyprogesterone

III. 17α-Hydroxyprogesterone

The presence of this steroid in adrenal tissue was detected in the first isolation studies; it was considered to be an androgen. Synthetic samples have failed to verify these early claims of androgenicity. 17α-Hydroxyprogesterone is not an androgen on the basis of the usual mammalian end-points, and chick-comb assays have been inconclusive (see Chapter 5). The compound, however, is an important intermediate in corticoid biosynthesis, particularly on the pathway to cortisol formation (see Fig. 15, p. 444).

Catabolic reactions of 17α-hydroxyprogesterone are summarized in Fig. 22. In addition to the usual reductive changes, entirely similar to those seen for progesterone (Fig. 20), a reaction involving the rupture of the C-17, C-20 bond has been demonstrated. Since at least in limited studies the principal 17-keto-steroid product of 17α-hydroxyprogesterone metabolism *in vivo* is etiocholanolone, it is quite likely that the oxidative removal of the side chain proceeds, in man, after ring A reduction.

The principal C_{21} urinary metabolite or 17α-hydroxy-progesterone is pregnane-triol. This compound is elevated remarkably in patients with the adrenogenital syndrome. After ACTH treatment of such patients, urinary titers rise to comparatively enormous values, apparently reflecting the high adrenal secretion of 17α-hydroxyprogesterone. During the past few years an *in vivo* metabolic reaction involving 17-dehydroxylation has been reported (ROSSELET et al. 1954, JAILER et al. 1955). This has now been shown to be due to artifact formation (during acid hydrolysis of the urine) by the group making the original claim (ROSSELET et al. 1957) and by FUKUSHIMA and GALLAGHER (1957). These latter workers demonstrated that the metabolism of 21-deoxycorticosterone-4-C^{14} proceeded to 3α,17α,

20α-trihydroxypregnan-11-one to the extent of 44%, that 3α-hydroxypregnane-11,20-dione was not a metabolite, but that on treatment of 3α,17α,20-trihydroxy-pregnan-11-one with mineral acid and heat the artifact 3α-hydroxypregnane-11,20-dione was formed.

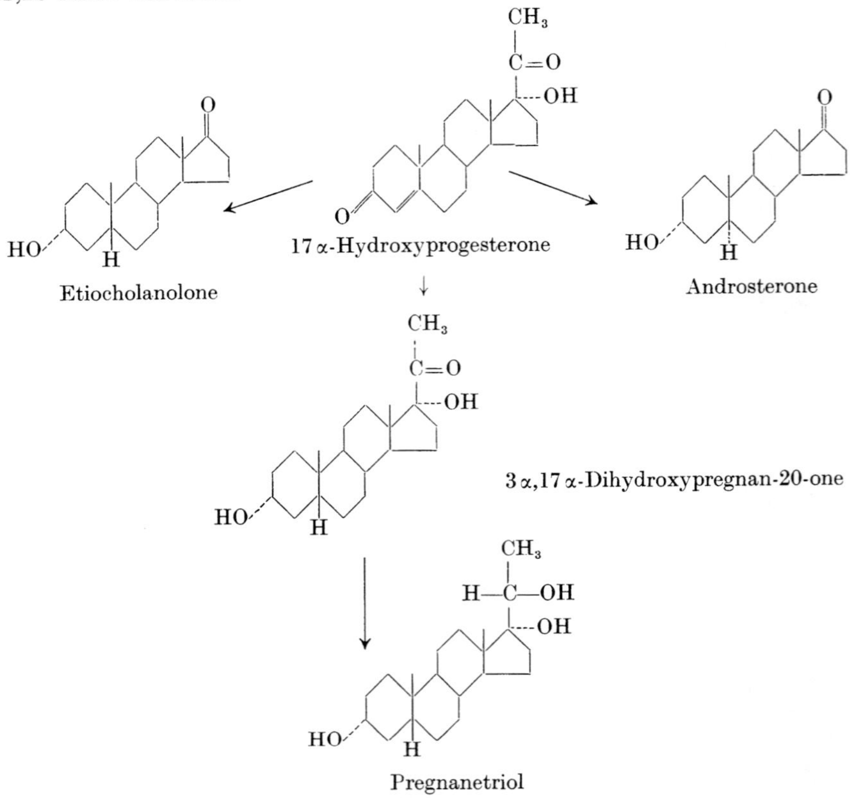

Fig. 22. Catabolism of 17α-hydroxyprogesterone

IV. 11β-Hydroxyprogesterone and 11-ketoprogesterone

11-Ketoprogesterone has been detected in incubation studies, and the corresponding 11β-hydroxyprogesterone has been found both as a product of progesterone incubation with adrenal tissue and of perfusion of progesterone through bovine adrenal glands (Table 15, p. 44). It is certainly quite likely that both 11β-hydroxyprogesterone and the 11-keto derivative are interconvertible in metabolism. The catabolic reactions are of the expected ring-A reductive type. Due to the absence of a 17α-hydroxy function, it is to be expected that neither of these two steroids would yield any C_{19} metabolites directly (Fig. 23). Known metabolites include 11-ketopregnanediol and 11-ketopregnanolone.

V. Deoxycorticosterone

Deoxycorticosterone, a minor adrenal corticoid, both from the quantity present in blood and tissues and from over-all importance in electrolyte regulation, undergoes reduction at the 4,5-nuclear double bond and the C-3 and C-20 carbonyl group, and in addition reduction of the C-21 alcohol grouping to the C-21 hydrocarbon (Fig. 24).

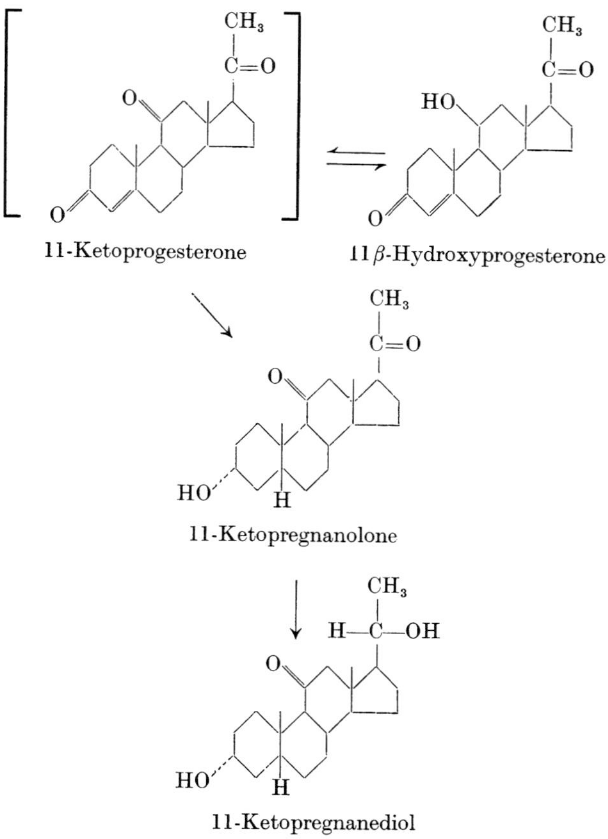

Fig. 23. Catabolism of 11-keto and 11β-hydroxyprogesterone

VI. 11-Deoxycortisol

The catabolism of 11-deoxycortisol proceeds by two routes, (a) a minor one involving carbon-carbon bond fission with the formation of C_{19} steroids and (b), what appears to be the major pathway, that involving reduction at carbons 4, 5, 3, and 20. These reactions, which occur *in vivo* and *in vitro*, are summarized n Fig. 25.

VII. Corticosterone and 11-dehydrocorticosterone

Corticosterone is present in human blood but in amounts relatively small as compared to the concentration of cortisol. In a species such as the rat, however, this corticoid is present as the major hormone component in the adrenal and the blood. As indicated in Fig. 26, ring-A reduction and C-20 reduction reactions are known, as well as C-21 reduction. No C_{19} compounds have been detected as metabolic products nor are they expected, owing to lack of a 17α-hydroxy group in the molecule.

The catabolic reactions for 11-dehydrocorticosterone are altogether similar to those found for corticosterone (Fig. 27), and the 11-keto compound is considered to be a secondary product derived from corticosterone. The separate presentation

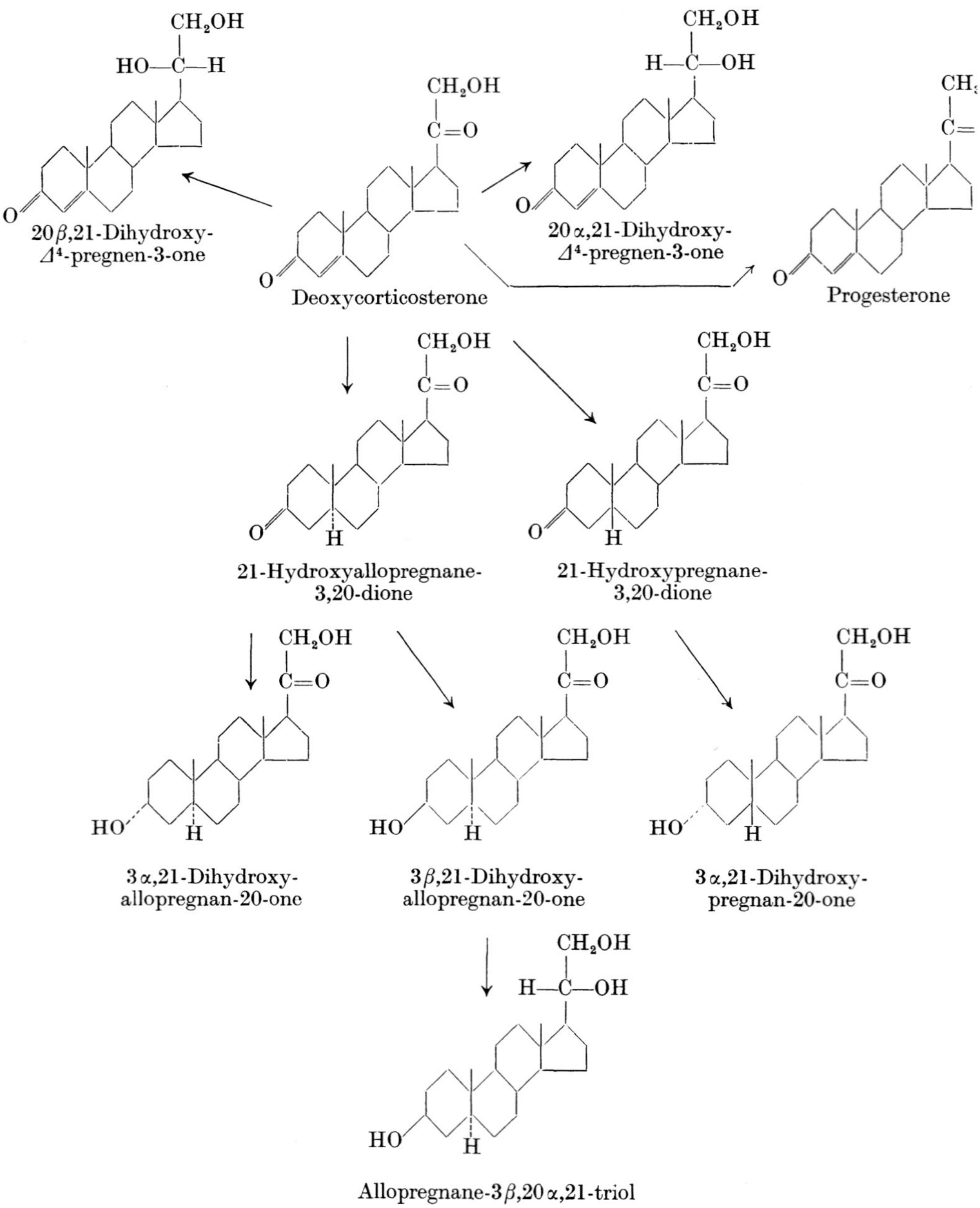

Fig. 24. Catabolism of deoxycorticosterone

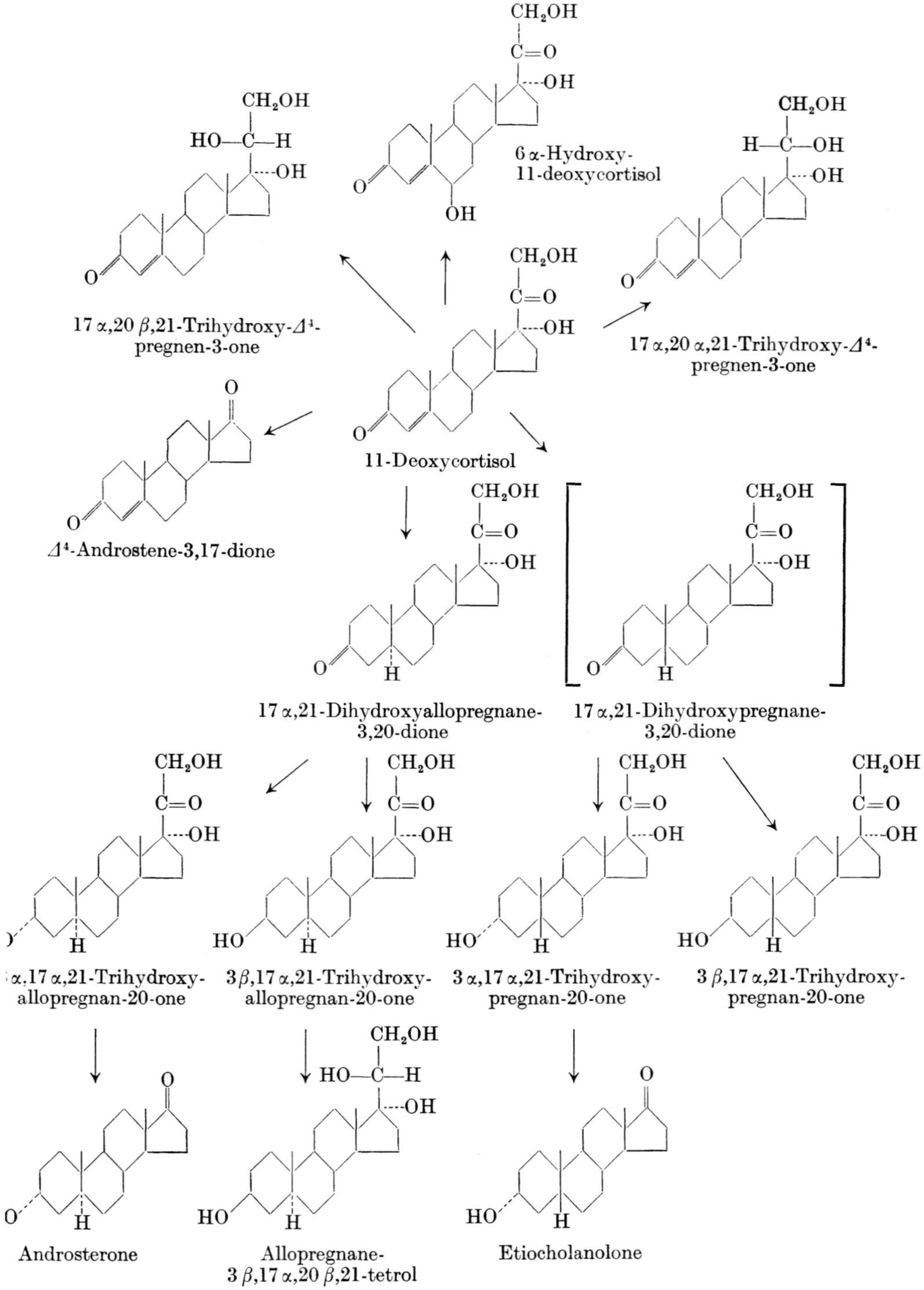

Fig. 25. Catabolism of 11-deoxycortisol

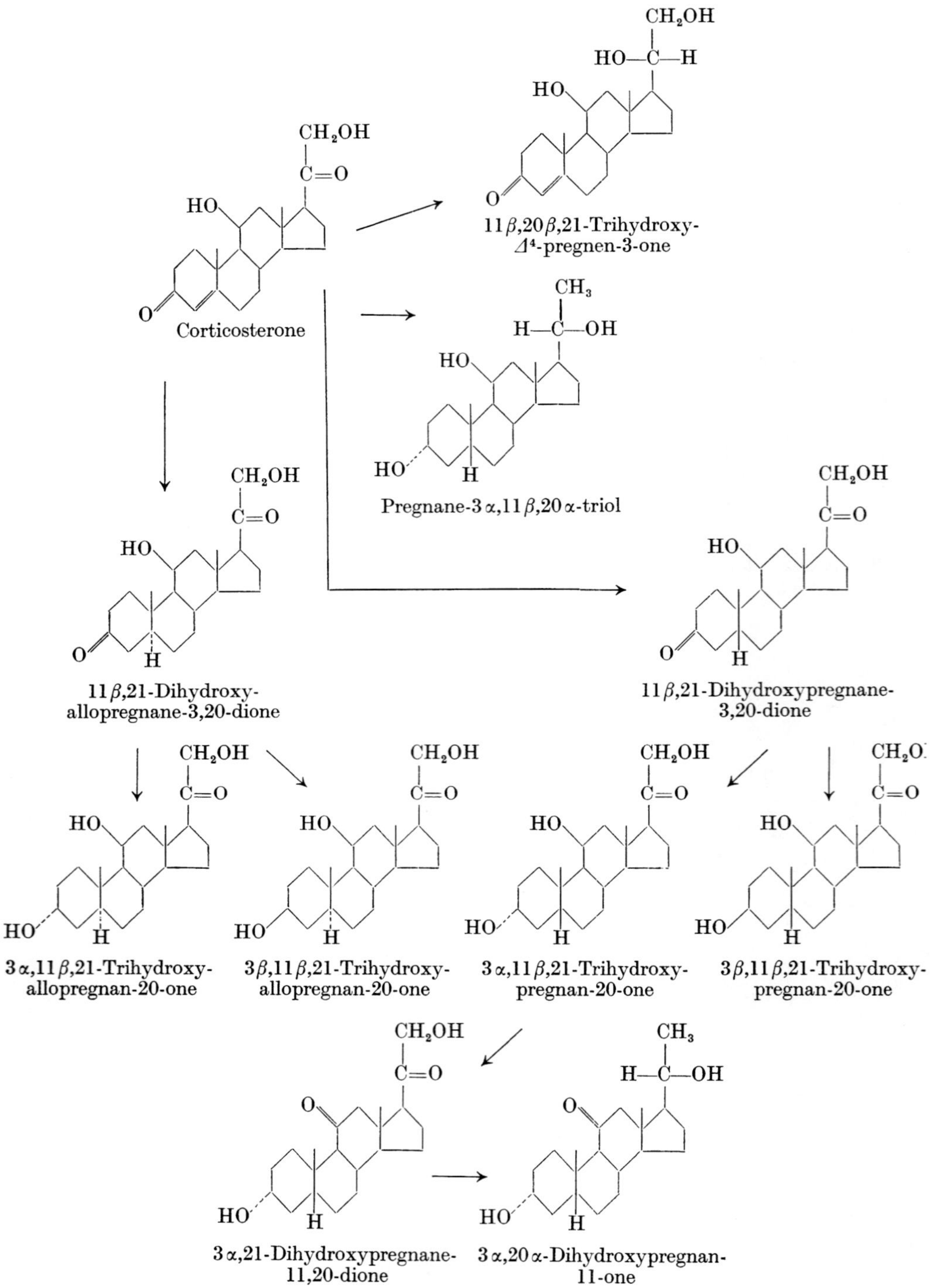

Fig. 26. Catabolism of corticosterone

of the 11-keto and 11 β-hydroxy compounds is for convenience rather than for the purpose of implying physiological significance.

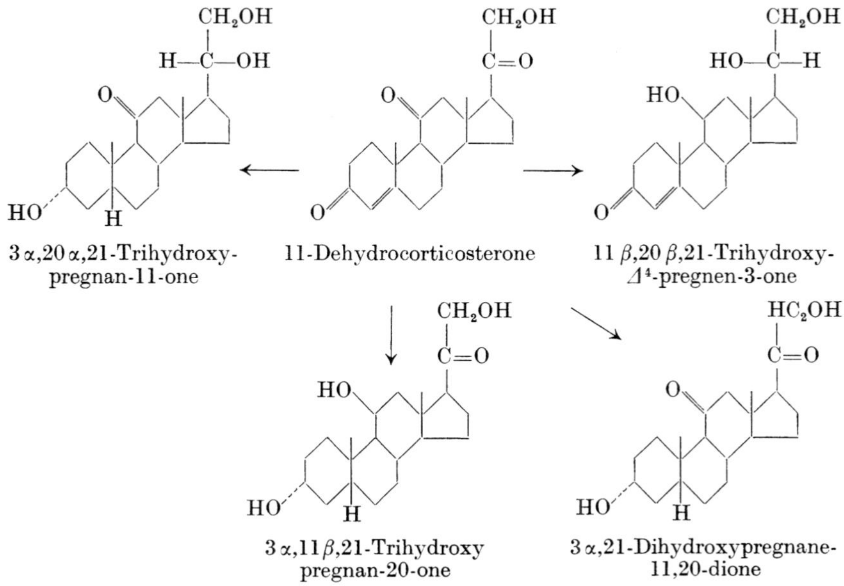

Fig. 27. Catabolism of 11-dehydrocorticosterone

VIII. Cortisol and cortisone

Of these two corticoids, cortisol appears to be the primary biosynthetic product, and the two compounds are interchangeable in metabolism. Both steroids are probably secreted into the blood.

The half-life of cortisol, in man, ranges from 90 to 115 minutes. DONE et al. (1955) reported a value of 90 ± 7.4 minutes, H. BROWN et al. (1954) found 111 ± 5,3. and R. PETERSON and WYNGAARDEN (1955) a value of 115. [These values for intravenously-administered cortisol are in contrast to that for prednisolone, which was found to have a half-life of 192 ± 10 minutes (ELY et al. 1956). The suggestion has been advanced that the increased biological activity for the Δ^1 steroid can be explained at least in part as being due to its longer survival in the body.]

Using trace quantities of intravenously-administered cortisol-4-C^{14}, R. PETERSON and WYNGAARDEN (1956) found the miscible pool of cortisol to be 1.1 to 2.4 mg in an apparent volume of 8—17 liters. After maximum ACTH stimulation the pool size was increased to 10 mg. The miscible pool could be effectively decreased by prednisone administration.

Cortisol metabolism is altered during pregnancy. Near term, the half-life is prolonged 100%. This has been established using C^{14}-labeled hormone. The increased plasma content suggests a slower rate of catabolism (MIGEON et al. 1957).

The over-all fate of cortisol-4-C^{14} has been studied in man. After intravenous administration of the labeled compound, 90% of the C^{14} was detected in the urine and feces within a 72 hour period. No significant amount of $C^{14}O_2$ could be detected in the expired air, nor could any activity be detected in the spinal fluid within the first 8 hours after infusion. The rate of excretion was independent of the amount injected (HELLMAN et al. 1954).

The catabolic fates of cortisol and cortisone are considered in Fig. 28, 29 and 30. These corticoids are converted peripherally to at least 7 *17-ketosteroids* of the 11-oxygenated type, including 11-keto and 11 β-hydroxy derivatives (Fig. 28.) Burstein et al. (1953 a) demonstrated, and others have confirmed, that the 11-oxygen function is not removed either in C_{19} or C_{21} metabolites. However, in a single, unconfirmed experiment reported by Miller and Axelrod (1954), perfusion of cortisone through liver from a cirrhotic rat resulted in the formation of the 11-deoxy steroid, 6 β,21-dihydroxy-Δ^4-pregnene-3,20-dione.

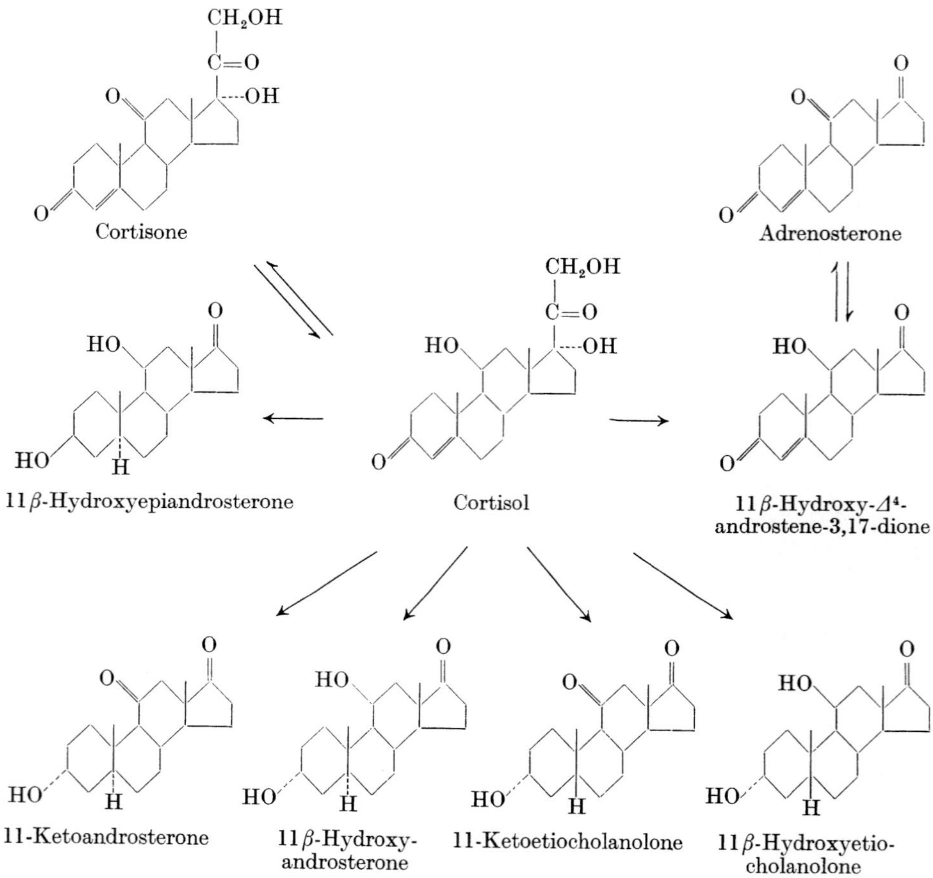

Fig. 28. Conversion of cortisol and cortisone to 17-ketosteroids

The *reduced* $C_{21}O_5$ *metabolites* having an 11 β-hydroxyl or 11-carbonyl function are presented separately in Figs. 29 and 30 for convenience only, since it has been amply established that these groups are interconvertible.

IX. Daily production of cortisol and corticosterone

Various methods have been employed for the estimation of the daily cortisol production in man and the reported results are reasonably consistent. Dorfman (1954 a) employed a method involving the quantification of 11-oxy 17-ketosteroids of both 5 α and 5 β forms together with the percentage recovery values of these

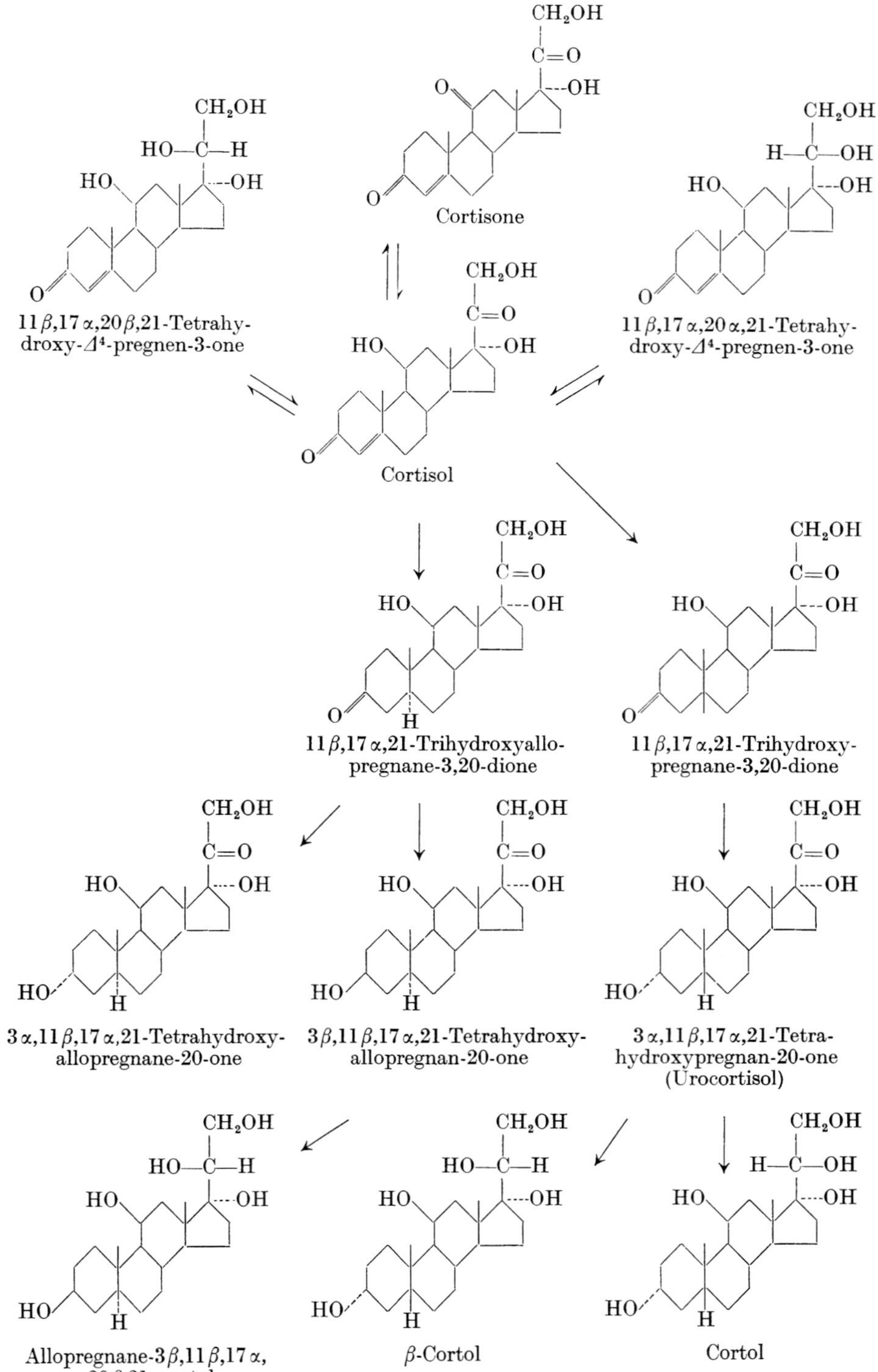

Fig. 29. Catabolic reactions of cortisol to C_{21} metabolites

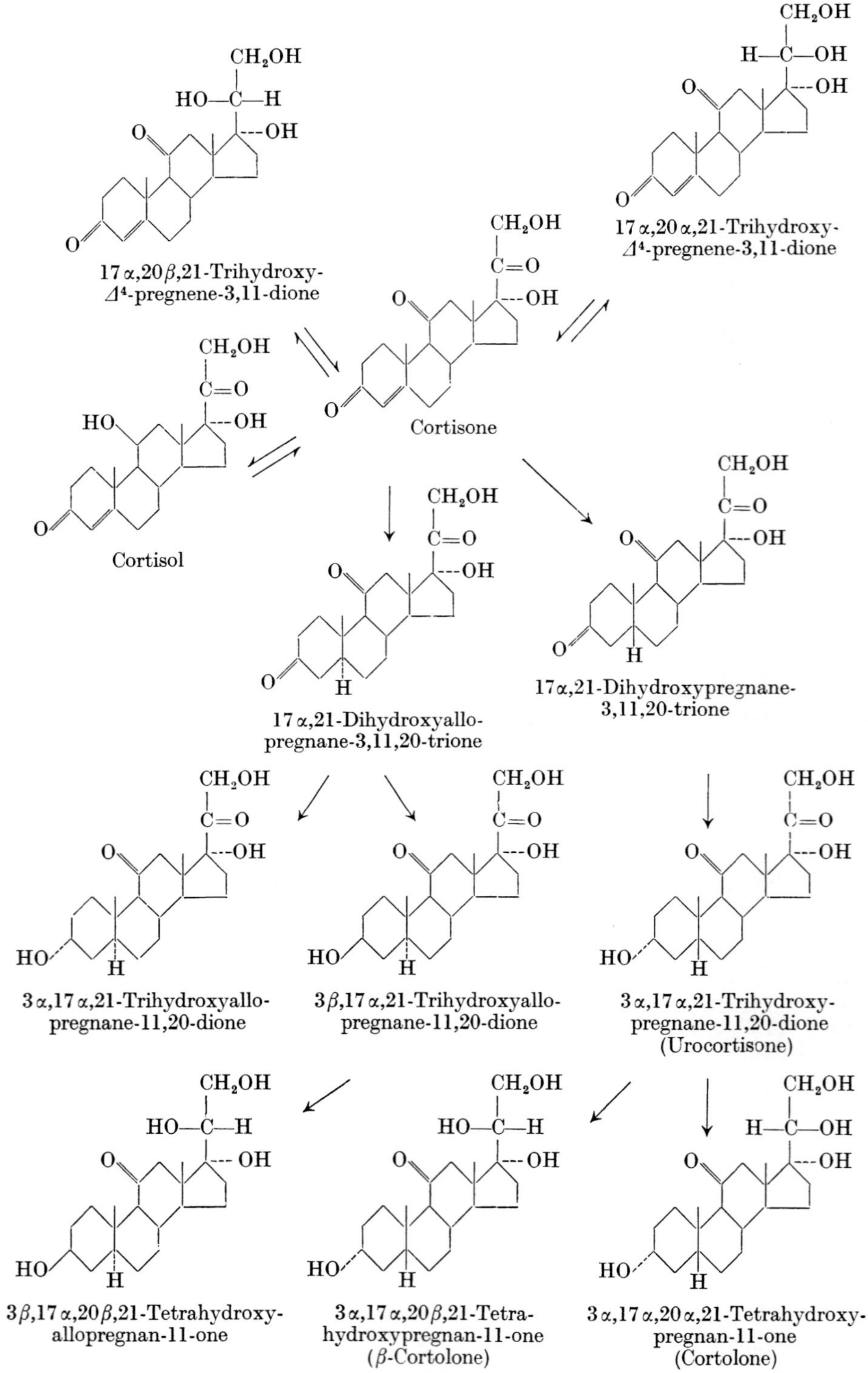

Fig 30. Catabolic reactions of cortisone to C_{21} metabolites

metabolites following the administration of cortisol. By this technique, an estimate of secretion of 21 and 18 mg of cortisol per day for men and women, respectively, was made. BONDY and ALTROCK (1953) arrived at a figure of 21 mg per day, on the basis of renal vein catheterization studies. This value may well be the result of compensating errors — the method giving low values and the stress of the procedure causing an elevated cortisol production. SILBER (1955) estimated the daily production of cortisol to vary from 30—60 mg per day on the basis of the fraction of a test dose excreted as Silber-Porter chromogens in urine. R. PETERSON and WYNGAARDEN (1956), using cortisol-4-C^{14}, calculated a production of 17—29 mg per day in 9 normal subjects. ACTH treatment increased the production to 160 mg per day.

Other estimates of daily production of cortisol in mg include: MOXHAM and NABARRO (1956) — men 9 to 49, women 6 to 40; SAMUELS et al. (1957) — normal 36, ACTH stimulation 250; HARDY and TURNER (1957) — stress 34; COPE and BLACK (1958a, b) — resting convalescent patients 4.2 to 24, normals 5 to 25, ACTH stimulation up to 257, severe disease 10 to 32, Addison's disease 0.6 and 1.2, 9α-fluorocortisol treatment 2.3, a pseudohermaphrodite patient 1; and GEYER (1956) — normal adults 15 to 57. Finally, the maintenance dose for adrenalectomized subjects has been set at 25 to 35 mg per day (KNOWLTON 1952).

Studies on the *diurnal variation of cortisol production* in man indicate a maximal production in mid-morning, relative stability in the forenoon, a gradual decline throughout the afternoon and evening, and minimal activity in late evening and early morning. This variation has been shown in plasma levels (R. PETERSON and WYNGAARDEN 1956) and in urinary levels (SANDBERG et al. 1953).

The question of *altered metabolism of cortisol* in various conditions has been studied for some time, but few published results seem to be precise and meaningful. One of the most informative papers is that of GOLD et al. (1958). Earlier studies by COPE and HURLOCK (1954) had suggested that urocortisol excretion increased more than urocortisone excretion after surgery. GOLD et al. (1958) have demonstrated that the ratio of urocortisol to urocortisone in urine is significantly increased in 4 different conditions, all characterized by an increase in circulating cortisol: (1) during and after surgery, (2) after ACTH treatment of normal subjects, (3) after administration of cortisol to a normal subject, and (4) in patients with severe burns. Other studies (SANDBERG et al. 1954, TYLER et al. 1954, FRANKSSON et al. 1954, STEENBURG et al. 1954) have indicated a decreased rate of metabolism of cortisol in surgical patients.

X. Aldosterone

Aldosterone, the newest member of the corticoid series, and a steroid that up to now has been available only in microgram quantities, has been studied only to a limited extent. VENNING et al. (1955) studied the over-all metabolism of aldosterone in a bilaterally-adrenalectomized human patient. A dose of 600 μg was administered by continuous intravenous administration over a 24 hour period to a patient who was maintained on a daily dose of 15 mg of cortisone. Thirteen per cent of the administered aldosterone was recovered from the urine during the treatment period and the following 48 hours. The material in the urine was present both in the free and conjugated forms. Not all the conjugated material could be released by liver β-glucuronidase, but more could be released by acid hydrolysis at p_H 1.5 at room temperature for 48 hours. Since the active material was chromatographed by the Bush B-5 system (see Chapter 2), the most likely possibility is that the salt-retaining activity was actually aldosterone. The fact that aldosterone is excreted as a conjugate is unusual, since other corticoids possessing the Δ^4-3-keto configuration, such as cortisone and cortisol, are excreted almost entirely in the free form.

ULICK and LIEBERMAN (1957) have presented evidence for a ring-A reduced metabolite of aldosterone, which appears to be a $C_{21}O_5$-pregnane compound with hydroxyl groups at C_3, C_{18}, and C_{21} and carbonyl groups at C_{11} and C_{20}.

XI. Prednisone and prednisolone

Since the introduction of prednisone and prednisolone as therapeutic agents, catabolism of these steroids has been studied by *in vivo* and *in vitro* methods (Table 21 and Fig. 31). These Δ^1 dehydro corticoids show essentially the same

Table 21. *Metabolism of prednisone and prednisolone*

Compound	Test system	Species	Product	Reference
Prednisone	*In vivo*	Man	Prednisone, Prednisolone (?)	GRAY et al. (1955)
Prednisolone	*In vivo*	Man	$\Delta^{1,4}$-Androstadiene-3,11,17-trione, 11β-Hydroxy-$\Delta^{1,4}$-androstadiene-3,17-dione	VERMEULEN (1957)
Prednisone	*In vivo*	Man	Prednisolone	VERMEULEN (1956a)
Prednisolone	*In vivo*	Man	Prednisolone, Urocortisol, Δ^1-Dehydrourocortisol, 3α,11β,17α,21-Tetrahydroxyallopregnan-20-one, 11β-Hydroxyetiocholanolone, 11-Ketoetiocholanolone	CASPI and PECHET (1957)
Prednisone	*In vivo*	Man	Prednisolone	D. JOHNSON et al. (1957)
Prednisolone	*In vivo*	Man	Prednisone	D. JOHNSON et al. (1957)
Prednisone	*In vitro*	Rat	Prednisolone	SHRIEFERS et al. (1957)
Prednisolone	*In vivo*	Man	Prednisone, 11β,17α,20β,21-Tetrahydroxy-$\Delta^{1,4}$-pregnadien-3-one, 17α,20α,21-Trihydroxy-$\Delta^{1,4}$-pregnadiene-3,11-dione	GRAY et al. (1956)
Prednisone	*In vivo*	Man	Prednisolone, 17α,20β,21-Trihydroxy-$\Delta^{1,4}$-pregnadiene-3,11-dione	GRAY et al. (1956)
Prednisolone	*In vitro*	Rat	Cortisol, 11β,17α,21-Trihydroxyallopregnane-3,20-dione, 11β,17α,20β,21-Tetrahydroxy-Δ^4-pregnen-3-one	VERMEULEN and CASPI (1958)
Prednisone	*In vivo*	Man	Cortisone, Cortisol, 17α,20β,21-Trihydroxy-Δ^4-pregnene-11,20-dione	LAMEDICA et al. (1957)
Prednisolone	*In vivo*	Man	3α,11β,17α,21-Tetrahydroxypregnan-20-one, 11β,17α,21-Trihydroxy-$\Delta^{1,4}$-pregnadiene,3-20-dione, 17α,21-Dihydroxy-Δ^1-pregnene-3,11,20-trione, 11β-Hydroxyetiocholanolone, 11-Ketoetiocholanolone	CASPI and PECHET (1958)
Prednisone	*In vitro*	Rat (liver)	Urocortisone	TOMKINS (1956b)

catabolic reactions as cortisone and cortisol, to which they may be converted by Δ^1 saturation, in addition to modifications that arise from the presence of the Δ^1 bond. Saturation of the Δ^1 nuclear bond was expected, since such a reaction had previously been demonstrated in the C_{19} series when Δ^1-androstene-3,17-dione was converted to androsterone (UNGAR and DORFMAN 1952).

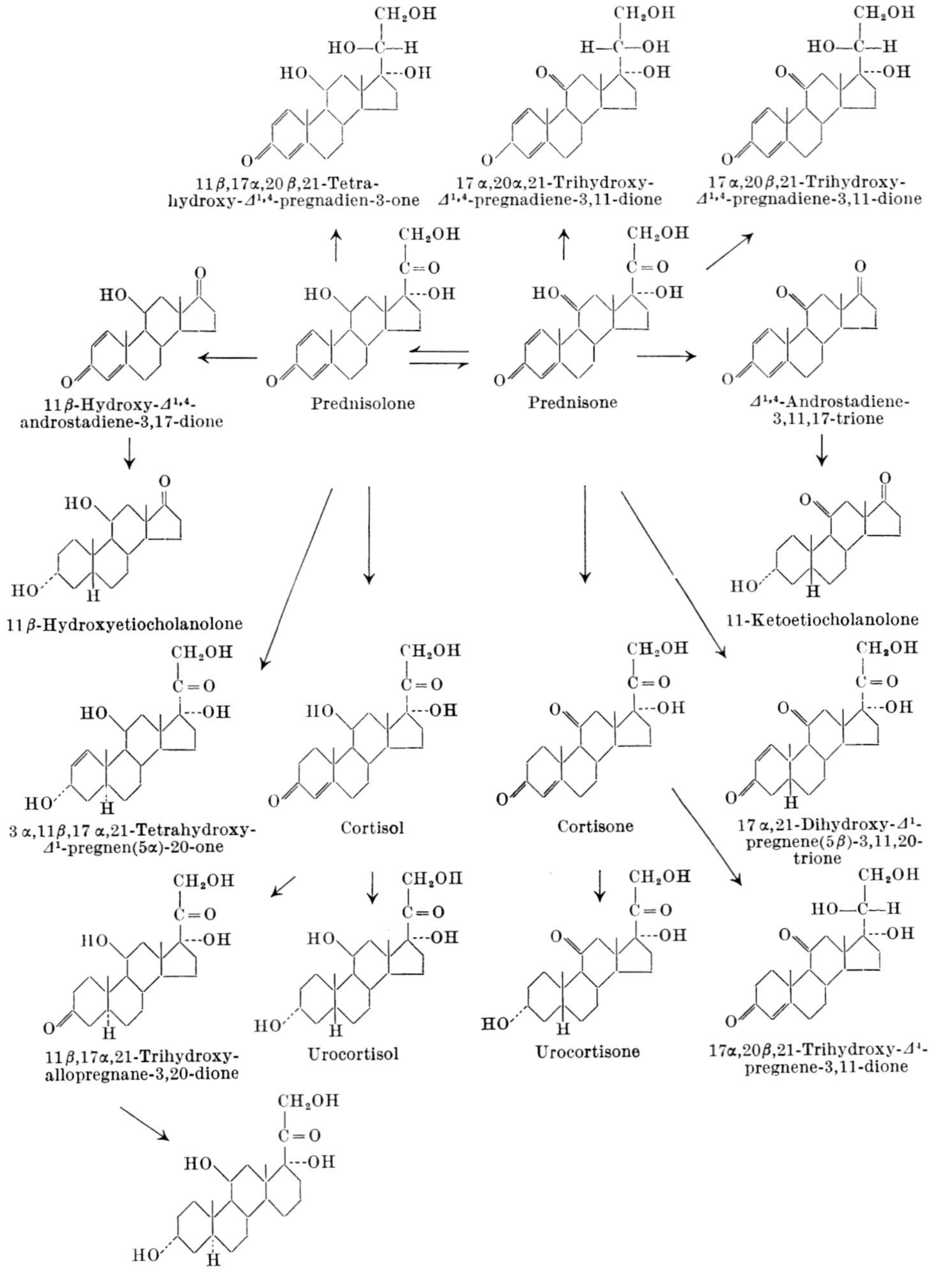

Fig. 31. Metabolism of prednisone and prednisolone

B. Adrenal androgens

I. $C_{19}O_2$ compounds

The recent finding by ANLIKER et al. (1957) that testosterone is present in adrenal tissue provides evidence that the adrenal produces all of the androgens synthesized by the testis and the ovary. A number of metabolism studies employing testosterone-C^{14} have now appeared. These have confirmed and extended the initial findings of DORFMAN et al. (1939) and CALLOW (1939) that androsterone and etiocholanolone are the principal metabolites. With labeled testosterone

Table 22. *In vivo metabolism of adrenal androgens*

Compound	Species	Urinary products	Reference
Dehydroepiandrosterone	Man	Androsterone, Etiocholanolone, Δ^5-Androstene-$3\beta,17\beta$-diol, Etiocholane-$3\alpha,17\beta$-diol	MASON and KEPLER (1954a, b), A. MILLER et al. (1950)
		Δ^5-Androstene-$3\beta,16\alpha,17\beta$-triol	MASON and KEPLER (1945a, 1947)
		16α-Hydroxydehydroepiandrosterone	FOTHERBY et al. (1957)
Δ^4-Androstene-3,17-dione	Man	Androsterone, Etiocholanolone	DORFMAN and HAMILTON (1940), DORFMAN et al. (1950), GALLAGHER et al. (1951), DORFMAN (1954b)
Adrenosterone	Man	11-Ketoandrosterone, 11-Ketoetiocholanolone, 11-Hydroxyetiocholanole	SAVARD et al. (1953)
11β-Hydroxy-Δ^4-androstene-3,17-dione	Man	11-Ketoandrosterone, 11-Ketoetiocholanolone, 11β-Hydroxyandrosterone, 11β-Hydroxyetiocholanolone	BRADLOW and GALLAGHER (1957)

Table 23. *In vitro metabolism of adrenal androgens*

Substrate	Species	Test system	Products	Reference
Dehydroepiandrosterone	Rat	Perfused liver	Δ^4-Androstene-3,17-dione, Δ^5-Androstene-$3\beta,17\beta$-diol	UNGAR et al. (1954b)
	Rabbit	Liver homogenate	Δ^5-Androstene-$3\beta,17\alpha$-diol	ROSENKRANZ and DORFMAN (1953)
	Rabbit	Liver slices	Δ^5-Androstene-$3\beta,16\alpha,7\beta$-triol	SCHNEIDER and MASON (1948)
	Rat and Ox	Liver homogenate	Dehydroepiandrosterone sulfate	DEMEIO and LEWYCKA (1955), DEMEIO et al. (1956)
	Rabbit	Liver homogenate	Dehydroepiandrosterone sulfate	SCHNEIDER and LEWBART (1956)
	Rat	Testis suspension	Δ^4-Androstene-3,17-dione	SAVARD and DORFMAN (1954b)
	Rabbit	Liver slices	Androstene-$3\beta,17\beta$-diol	SCHNEIDER and MASON (1948), UNGAR et al. (1954b)
Δ^4-Androstene-3,17-dione	Rat	Perfused liver	6β-Hydroxy-Δ^4-androstene-3,17-dione	AXELROD and MILLER (1954)
	Rat	Liver homogenate	Androsterone, Epiandrosterone	RUBIN (1957)
	Rabbit	Liver brei	Testosterone, Epitestosterone	CLARK et al. (1947)

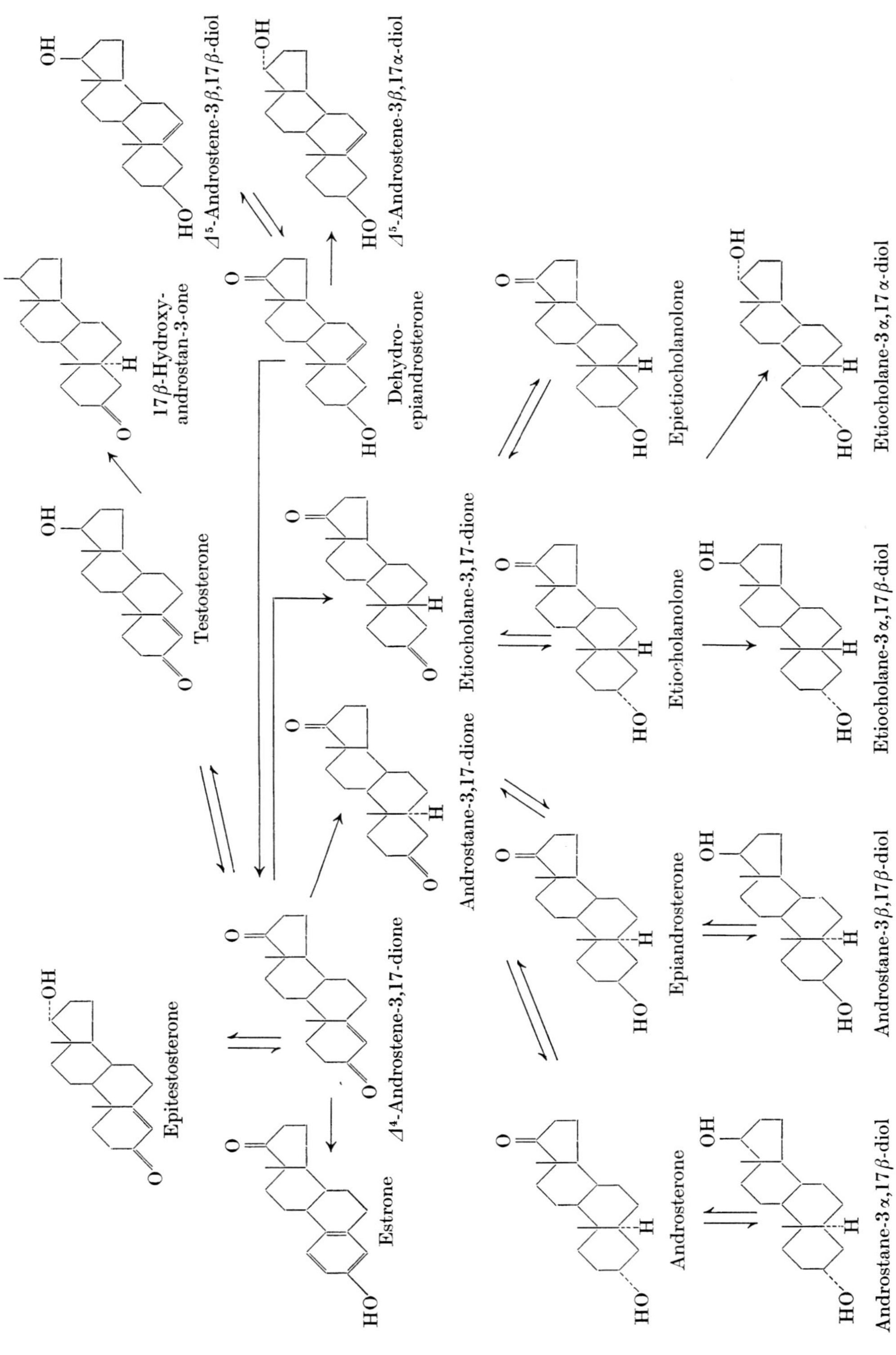

Fig. 32 Catabolism. of $C_{19}O_2$ steroids

it has been possible to show that in the human only trace amounts are in the feces, while by far the greatest amount occurs in the urine (FUKUSHIMA et al. 1954a, b). In other species such as the dog, mice, and rats, however, a good portion of this hormone appears in the bile and feces (ASHMORE et al. 1953, BARRY et al. 1952, PASCHKIS et al. 1954). Excretion routes are not consistent even in rodents, since the guinea pig, unlike the rat and mouse, excretes the bulk of the metabolites by way of the urine (BURSTEIN et al. 1955).

The catabolic reactions of all adrenal androgens ($C_{19}O_2$, $C_{19}O_3$) are summarized in Tables 22 and 23 and Figs. 32 and 33. The $C_{19}O_2$ adrenal androgens include major amounts of Δ^4-androstene-3,17-dione and dehydroepiandrosterone and a minor quantity of testosterone. The interrelations among this group of steroids (Fig. 32) have been documented in detail.

SANDBERG and SLAUNWHITE (1956) demonstrated that intravenous administration of testosterone-4-C^{14} to a human subject resulted within 4 hours in a urinary excretion of C^{14} products representing 50% of the dose and that by 48 hours the material was excreted almost quantitatively in the urine. The clearance from the plasma followed two different rates, one having a half-life of 11 minutes and the second of 100 minutes. The concentration of conjugated metabolites reached a peak in the plasma at 15 minutes. About 6% could be detected in the feces, and from 12 to 14% could be detected in direct studies of bile in patients with T tubes. The studies of SANDBERG and SLAUNWHITE indicated that metabolites in bile were conjugated, while the material in the stool was in free form.

II. Adrenosterone and 11β-hydroxy-Δ^4-androstene-3,17-dione

SAVARD et al. (1953) administered adrenosterone to a patient and isolated a group of 11-oxygenated 17-ketosteroids including 11β-hydroxyandrosterone, 11-ketoandrosterone, 11β-hydroxyetiocholanolone, and 11-ketoetiocholanolone (Fig. 33). This study clearly established that the 11-keto group was reduced to the 11β-hydroxy group, that oxygen was not removed from carbon 11, and that the principal metalboites were of the 5α series.

Metabolites of 11β-hydroxy-Δ^4-androstene-3,17-dione marked with C^{14} at carbon 4 were excreted almost quantitatively within 8 to 12 hours when this compound was administered intravenously (SANDBERG and SLAUNWHITE 1957). The bile contained negligible amounts of C^{14}. Most of the radioactivity in the plasma was conjugated to glucuronic acid, and this was in evidence 45 minutes after administration. The free labeled steroids were cleared from the plasma at apparently 2 different rates, one with a half-life of 30 minutes and the other with a slower rate of 80 minutes.

The metabolism of 11β-hydroxy-Δ^4-androstene-3,17-dione-C^{14} has been studied in human subjects by precise methods (BRADLOW and GALLAGHER 1957). The rate of excretion of metabolic products was found to be faster than that found for cortisol, testosterone, and estradiol-17β. A total of 10 steroids were sought by reverse isotopic dilution analysis, and 4 of these components were found to be present in significant amounts, accounting for 80% of the neutral urinary metabolites. These included 11β-hydroxyandrosterone, 11β-hydroxyetiocholanolone, 11-ketoandrosterone, and 11-ketoetiocholanolone (Fig. 33). Six other compounds were virtually devoid of radioactivity and therefore are not considered to be metabolites of this androgen. These included 3β,11β-dihydroxyandrostan-17-one, 3β,11β-dihydroxyetiocholan-17-one, androstane-3β,11β,17β-triol, etiocholane-3α,11α,17β-triol, 3α-hydroxyandrostan-17-one, and 3α-hydroxyetiocholan-17-one (BRADLOW and GALLAGHER 1957).

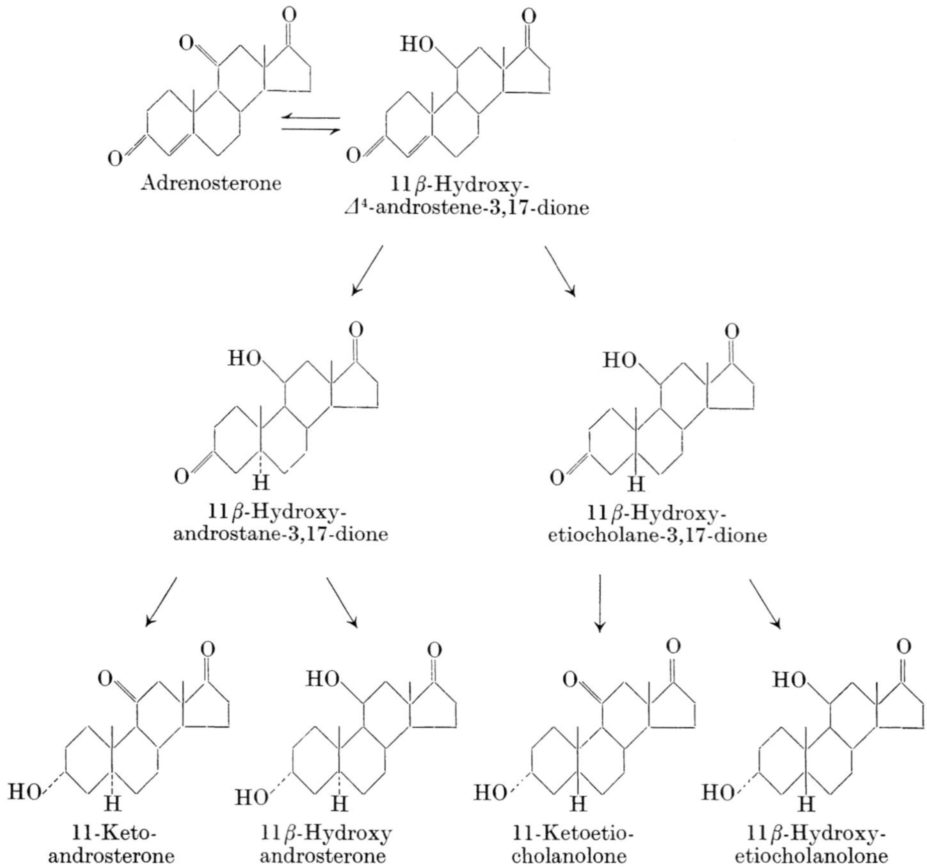

Fig. 33. Catabolism of $C_{19}O_3$ steroids

C. Generalizations on steroid metabolism

DORFMAN (1954a, b) suggested that certain tentative generalizations could be made with respect to *in vivo* steroid hormone metabolism. This led to a fuller meaning of the quantitative information on urinary constituents and a practical method for the calculation of the daily production of individual components produced by the gonads and adrenals, and it finally assisted in the understanding of the biosynthetic changes associated with hypo- and hypersecretion in various disease states. Since these generalizations were suggested, new data have appeared that essentially verify and strengthen the original general statements.

The generalizations are concerned with the orientation of rings A and B of the steroids after reduction of the nuclear double bond (Δ^4), that is, whether the resulting steroid metabolite is 5α(allo) or 5β (normal). The generalizations may be listed as follows:

(1) C_{19} steroids possessing the Δ^4-3-ketone grouping in ring A and an additional oxygen atom at carbon 17 are reduced at carbons 4 and 5 with the formation of slightly more 5β- than 5α-stereoisomer, with the ratio of $5\beta/5\alpha$ being between 1.5 and 2.0.

(2) C_{19} steroids possessing the Δ^4-3-ketone grouping in ring A and additional oxygen atoms at carbons 11 and 17 are reduced at carbons 4 and 5, with the formation of significantly more of the 5α-stereoisomer than of the 5β-stereoisomer.

(3) C_{21} steroids possessing the Δ^4-3-ketone grouping in ring A, with or without oxygen at carbons 11, 17, and 21, are reduced at carbons 4 and 5 to form the 5β-steroisomers, with only minor amounts of the 5α-steroisomers.

(4) C_{21} steroids possessing the Δ^4-3-ketone grouping in ring A and oxygen at carbons 17 and 20 are converted in part to C_{19} saturated 17-ketosteroids. The conversion of these C_{21} steroids to the C_{19} saturated 17-ketosteroids probably proceeds by two pathways — the major one by way of a saturated C_{21} intermediate and the minor by way of a Δ^4-3-keto-C_{19} steroid.

D. Rate of production of 17-ketosteroid precursors

A method for the calculation of endogenous production of 17-ketosteroid precursors was described by DORFMAN (1954a) on the basis of the above generalizations and quantitative and qualitative knowledge of *in vivo* metabolism of various adrenal and gonadal steroids. It was calculated (a) that normal men secrete a total of 18 mg of the three C_{19} steroids — testosterone, dehydroepiandrosterone, and Δ^4-androstene-3,17-dione; (b) that the two principal C_{21} donors of 11-deoxy-17-ketosteroids (11-deoxycortisol and 17α-hydroxyprogesterone) are secreted in amounts of less than 2.6 mg per day; (c) that 11β-hydroxy-Δ^4-androstene-3,17-dione is produced at the level of 7.4 mg per day; and (d) that cortisol is produced in an amount of approximately 21 mg per day.

Since these original findings were published, various parts of this work have been repeated, confirmed, and extended. FUKUSHIMA and co-workers (1954b) have estimated the secretion of androsterone and etiocholanolone precursors and placed the figure at not more than 36 mg and not less than 17 mg.

In previous reports from our laboratory it was stated that 11-oxygenated C_{19} steroids were reduced during *in vivo* metabolism so that the ratio of $5\beta/5\alpha$ metabolites was 0.19 (SAVARD et al. 1953, DORFMAN 1954a). These studies have recently been repeated by BRADLOW and GALLAGHER (1957), using 11β-hydroxy-Δ^4-androstene-3,17-dione-4-C^{14}. The ratio of $5\beta/5\alpha$-reduced metabolites was 0.32 and 0.21 for two patients.

As a result of various studies on the metabolism of cortisone and cortical in the intact human being, it was concluded that the 17-ketosteroids were predominantly of the 11-oxygenated 5β type and that the ratio of $5\beta/5\alpha$ was approximately 9 (DORFMAN 1954a). More recently, studies by GALLAGHER (1957) have indicated that cortisol was metabolized to both 5β and 5α 11-oxygenated 17-ketosteroids and that the ratio of $5\beta/5\alpha$ was 8.5.

BUSH (1956a) obtained adrenal venous blood at operation for adrenalectomy in a woman with virilism and collected urine preoperatively and postoperatively. Using percentage conversion values similar to DORFMAN's (1954a), he calculated the secretory ratio of cortisol to 11β-hydroxy-Δ^4-androstene-3,17-dione as 2.2–3.35, as compared to an observed value of 2.4. Administration of adrenosterone to a healthy young man led to products the ratios of which agreed with previous ratios, but in a castrated-adrenalectomized man a completely reversed ratio was found (BUSH 1956b).

BUSH also has suggested that certain modifications are necessary on the basis of the recent finding of $3\alpha,11\beta,17\alpha,21$-tetrahydroxyallopregnan-20-one in human urine (BUSH 1956b, ROMANOFF et al. 1957). The finding of this constituent does not change the generalizations about 17-ketosteroids. With respect to C_{21} metabolites, the few milligrams of 5α constituent are more than balanced by the amounts of the recently-described 5β metabolites of cortisol: cortolone, β-cortol and β-cortolone (FUKUSHIMA et al. 1955) (see Figs. 29 and 30, p. 473 and p. 474).

BRADLOW et al. (1958) have reported the important result that, whereas normal individuals metabolize testosterone to etiocholanolone (5β) and androsterone (5α) in a ratio of 1/1, myxedemic patients convert this hormone to these metabolites in the greatly altered ratio of 8/1. Administration of triiodothyronine modified the ratio toward normal.

E. Adrenal estrogens

A scheme for the catabolism of estradiol-17β and estrone formed in the adrenal is presented in Fig. 34, and the *in vitro* and *in vivo* data and references are given in Tables 24 and 25. The catabolism of 18-hydroxyestrone, probably the one characteristic estrogen produced by the adrenal, has not as yet been elucidated.

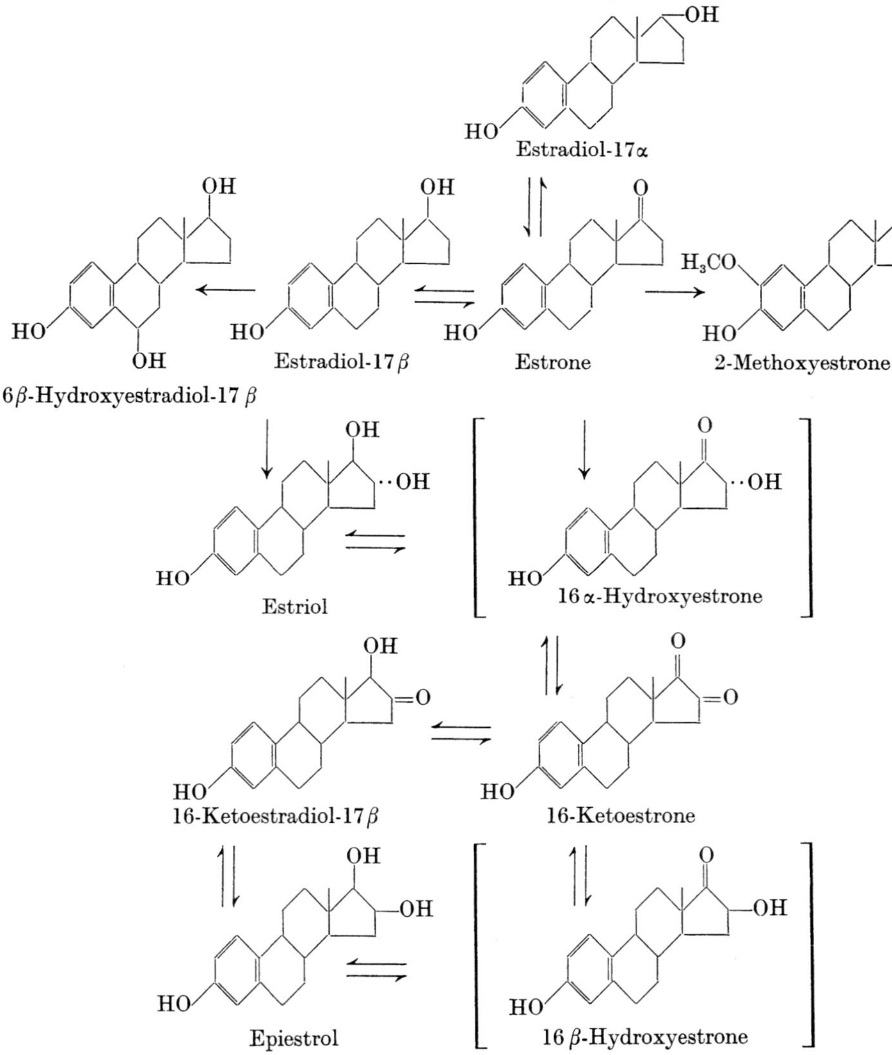

Fig. 34. Catabolism of estrone and estradiol-17β

Table 24. *In vivo metabolism of estrogens*

Compound	Species	Urinary product	Reference
Estrone	Rabbit	Estradiol-17α	STROUD (1939)
	Man	Estriol	PEARLMAN and PINCUS (1942)
		Estrone, Estriol	BEER and GALLAGHER (1955)
		16-Ketoestrone	SLAUNWHITE and SANDBERG (1956)
	Dog	Estradiol-17α (bile)	PEARLMAN et al. (1948)
Estradiol-17β	Man	Estrone, Estradiol-17β, Estriol, 2-Methoxyestrone	ENGEL et al. (1957)
	Rabbit (Ovariectomized-hysterectomized)	Estrone, Estradiol-17α	W. FISH and DORFMAN (1942)
	Guinea pig	Estrone	W. FISH and DORFMAN (1940a, 1941)
	Rabbit	Estrone, Estradiol-17α	HEARD et al. (1941)
	Man	Estrone	HEARD and HOFFMAN (1941b)
	Monkey	Estrone	DORFMAN et al. (1945)
	Dog	Estrone, Estradiol-17β, Estriol	MAYER (1952)
	Man	2-Methoxyestrone	KRAYCHY and GALLAGHER (1957a, b)
		16-Ketoestradiol-17β	LEVITZ et al. (1956)
16-Ketoestrone	Man	16-Ketoestradiol-17β, Estriol, Epiestriol	STIMMEL (1958)
16α-Hydroxyestrone	Man	Estriol	J. B. BROWN and MARRIAN (1957)

Table 25. *In vitro metabolism of adrenal estrogens*

Substrate	Species	Test system	Product	Reference
Estrone	Rat	Liver	Estradiol-17β	RYAN and ENGEL (1953a)
	Man	Various tissues	Estradiol-17β	RYAN and ENGEL (1953b)
	Rabbit Man	Red blood cells	Estradiol-17β	GRAY and BISCHOFF (1955)
	Ox	Blood	Estradiol-17α	L. LEVY et al. (1958)
Estradiol-17β	Man	Liver and ovary	Estrone and Estriol	DOWBEN and RABINOWITZ (1956)
		Testis, kidney	Estrone	
	Mouse and Rat	Liver microsomes	6β-Hydroxyestradiol-17β	RUMNEY (1956), MUELLER and RUMNEY (1957)
	Rat	Liver	Estriol	HAGOPIAN (1955)
	Man	Placenta, liver etc.	Estrone	RYAN and ENGEL (1953a)
	Rat	Liver	Estrone	PEARLMAN and DEMEIO (1949), RYAN and ENGEL (1953a)
	Ox	Liver powder	Estrone	LEDOGAR and JONES (1953)

Enzymes concerned with steroid metabolism

A. Hydroxylating enzymes

Hydroxylation reactions play extremely important roles in biosynthesis and catabolism of steroids. The oxygen for this reaction is derived from molecular oxygen (HAYANO et al. 1955, 1956a, SWEAT et al. 1956), and the stereospecific introduction of oxygen occurs as a direct substitution of the hydrogen by an hydroxyl group without inversion (HAYANO et al. 1958). HAYANO and DORFMAN (1954) observed that no deuterium entered the steroid molecule in a stable position when the 11 β-hydroxylation reaction was carried out with an adrenal preparation in D_2O, nor was O^{18} present in the reaction product when H_2O^{18} was employed. However, when the gas phase contained O_2^{18}, 92% of the theoretical quantity of O_2^{18} was contained in the hydroxylated steroid (Fig. 35). A similar mechanism was found for the following hydroxylations, using microbiological systems: 11 α, 11 β, 6 β, 17 α and 21 (HAYANO et al. 1956b).

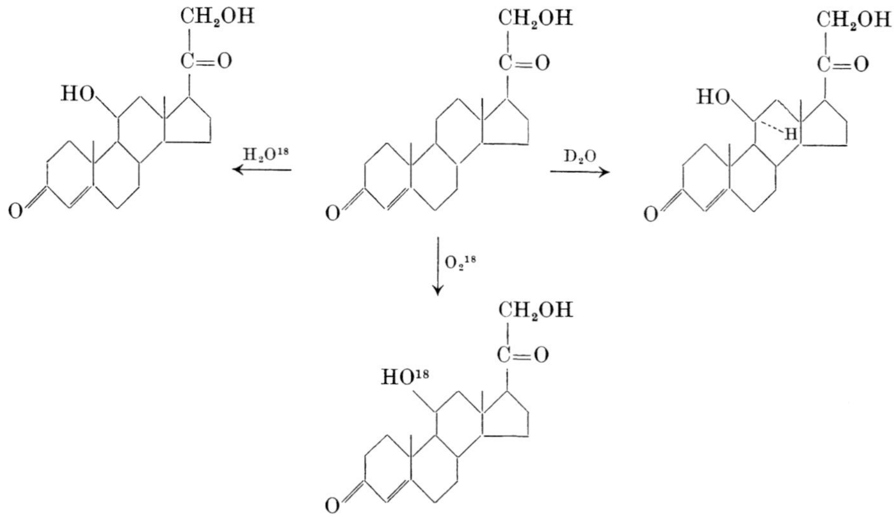

Fig. 35. Mechanism of 11 β-hydroxylation

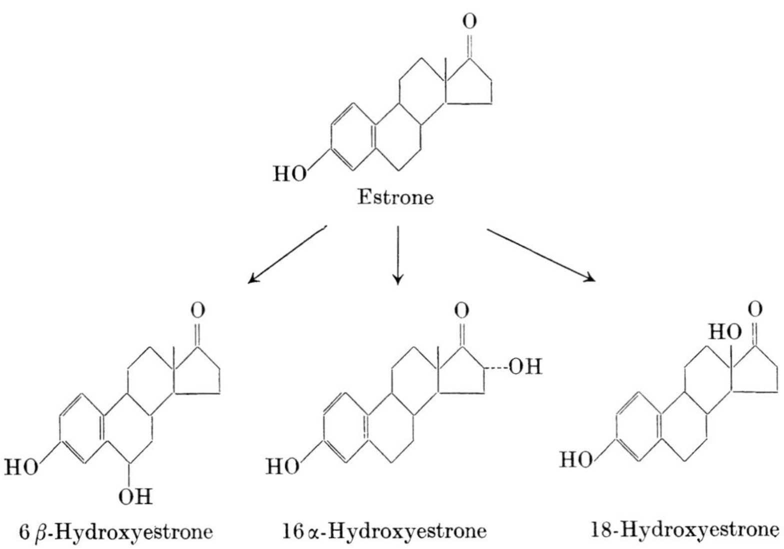

Fig. 36. Hydroxylase reactions of C_{18} steroids

I. Hydroxylases of C_{18} steroids

If we assume that estrone and estradiol-17 β are the primary estrogens formed in the adrenal from Δ^4-androstene-3,17-dione and testosterone, respectively, then three hydroxylation definitely occur, while a fourth may take place. The three hydroxylases that have been demonstrated include 6 β, 16 α, and 18 (Fig. 36). The fourth reaction leads to the formation of a 2-methoxy function, but the mechanism is not known.

6 β-Hydroxylation of estradiol-17 β takes place in the liver (MUELLER and RUMNEY 1957), and 16 α-hydroxylation of the same substrate has also been

Table 26. *Hydroxylation of C_{19} substrates by liver*

Position	Tissue preparation	Substrate	Product	Reference
2β	Dog liver, perfusion	Testosterone	2β-Hydroxytestosterone	AXELROD et al. (1956)
6β	Rat liver, perfusion	Testosterone or Δ^4-Androstene-3,17-dione	6β-Hydroxy-Δ^4-androstene-3,17-dione	AXELROD and MILLER (1954)
	Dog liver, perfusion	Testosterone	6β-Hydroxytestosterone, 6β-Hydroxy-Δ^4-androstene-3,17-dione	AXELROD et al. (1956)
16α	Dog liver, perfusion	Testosterone	16α-Hydroxytestosterone	AXELROD et al. (1956)
	Rabbit liver	Dehydroepiandrosterone	Δ^5-Androstene-$3\beta,16\alpha,17\beta$-triol	SCHNEIDER and MASON (1948)

Table 27. *Hydroxylations of C_{19} substrates by adrenal*

Position	Tissue preparation	Substrate	Product	References
6α	Bovine adrenal homogenate, washed residue ($5000 \times g$)	Δ^4-Androstene-3,17-dione	6α-Hydroxy-Δ^4-androstene-3,17-dione	MEYER et al. (1955)
6β	Bovine adrenal homogenate, washed residue ($5000 \times g$)	Δ^4-Androstene-3,17-dione	6β-Hydroxy-Δ^4-androstene-3,17-dione	MEYER et al. (1955)
11β	Bovine adrenal homogenate, washed residue ($5000 \times g$)	Δ^4-Androstene-3,17-dione	11β-Hydroxy-Δ^4-androstene-3,17-dione	HAYANO and DORFMAN (1953)
	Adrenal	Testosterone	11β-Hydroxytestosterone	AXELROD and ARRAYAVE (1953)
	Bovine adrenal perfusion	Androsterone	11β-Hydroxyandrosterone	HECHTER et al. (1951)
		Δ^4-Androstene-3,17-dione	11β-Hydroxy-Δ^4-androstene-3,17-dione, 11β-Hydroxyandrostene-3,17-dione	JEANLOZ et al. (1950, 1953)
		Epiandrosterone	11β-Hydroxyandrostane-3,17-dione, 11β-Hydroxyepiandrosterone	MEYER et al. (1953)
		Dehydroepiandrosterone	11β-Hydroxy-Δ^4-androstene-3,17-dione	UNGAR et al. (1954b)
19	Bovine adrenal homogenate	Δ^4-Androstene-3,17-dione	19-Hydroxy-Δ^4-androstene-3,17-dione	MEYER (1955b)

demonstrated in liver (HAGOPIAN 1955, HAGOPIAN and LEVY 1958). The 18-hydroxylation of estrone probably takes place exclusively in the adrenal.

The presence of 16β-hydroxy derivatives may result from reduction of the 16-ketone group and thus indirectly from the 16α-hydroxy function.

II. Hydroxylases of C_{19} steroids

Hydroxylation studies in the C_{19} series are here reviewed in Tables 26, 27, and 28 and 29. These include hydroxylating reactions in liver, adrenal and microbial systems. Liver tissue has been shown to contain three distinct hydroxylating systems for C_{19} steroids, including 2β, 6β, and 16α (Table 26). Adrenal perfusion and adrenal tissue incubation studies have demonstrated hydroxylating reactions for C_{19} steroids, at 6α, 6β, 11β and 19 (Table 27).

Table 28. *Hydroxylation of C_{19} substrates (including 19-nor and 17-methyl steroids) by microorganisms*

Reaction	Substrate	Product	Microorganism	Reference
1α	Δ^4-Androstene-3,17-dione	1α-Hydroxy-Δ^4-androstene-3,17-dione	*Penicillium*	DOBSON et al. (1937)
	Dehydroepiandrosterone	1α-Hydroxydehydroepiandrosterone		
2β	Δ^4-Androstene-3,17-dione	2β-Hydroxy-Δ^4-androstene-3,17-dione	*Penicillium*	DOBSON et al. (1937)
6β	Δ^4-Androstene-3,17-dione	6β-Hydroxy-Δ^4-androstene-3,17-dione	*Cunninghamella blakesleeana*	MURRAY and PETERSON (1952)
	19-Nortestosterone	6β-Hydroxy-19-nortestosterone	*Rhizopus nigricans*	PEDERSON et al. (1956)
	Testosterone or Δ^4-Androstene-3,17-dione	6β-Hydroxy-Δ^4-androstene-3,17-dione	*Rhizopus nigricans*	DJERASSI et al. (1955)
	Testosterone	6β-Hydroxytestosterone	*Rhizopus* sp.	EPPSTEIN et al. (1954)
7α	Testosterone	7α-Hydroxytestosterone	*Coniothyrium helleborine*	THOMA and FRIED (1957)
10ξ	19-Nortestosterone	10ξ-Hydroxy-19-nortestosterone	*Rhizopus nigricans*	PEDERSON et al. (1956)
11α	Δ^4-Androstene-3,17-dione	11α-Hydroxy-Δ^4-androstene-3,17-dione	*Rhizopus nigricans, Rhizopus arrhizus*	MURRAY and PETERSON (1952)
	19-Nortestosterone	11α-Hydroxy-19-nortestosterone	*Rhizopus nigricans*	PEDERSON et al. (1956)
	Δ^4-Androstene-3,17-dione	11α-Hydroxy-Δ^4-androstene-3,17-dione	*Rhizopus nigricans*	EPPSTEIN et al. (1954)
	Testosterone	11α-Hydroxytestosterone	*Rhizopus* sp.	
	Methyltestosterone	11α-Hydroxymethyltestosterone		
14α	Δ^4-Androstene-3,17-dione	14α-Hydroxy-Δ^4-androstene-3,17-dione	*Helicostylum piriforme*	MEISTER et al. (1953a)
	Testosterone	14α-Hydroxytestosterone		
15α	Δ^4-Androstene-3,17-dione, Testosterone	15α-Hydroxy-Δ^4-androstene-3,17-dione	*Fusarium* sp.	EPPSTEIN et al. (1956)
	Testosterone	15α-Hydroxytestosterone		
16α	Δ^4-Androstene-3,17-dione	16α-Hydroxy-Δ^4-androstene-3,17-dione	*Streptomyces roseochromogenus*	FRIED et al. (1955)

11 β-Hydroxylation of C_{19} steroids by normal gonadal or placental tissue has not been demonstrated, but this reaction on testosterone has been observed with tissue slices from an interstitial-cell tumor of the testis of a 5-year-old boy. The products were 11 β-hydroxytestosterone, 11 β-hydroxy-Δ^4-androstene-3,17-dione, and adrenosterone (SAVARD et al. 1956a, b).

Table 29. *Comparison of positions hydroxylated on the C_{19} steroid molecule by mammalian tissues and microorganisms*

Position	Liver	Adrenal	Microorganisms
1 α			+
2 β	+		+
6 α		+	
6 β	+	+	+
7 α			+
10 ξ			+
11 α			+
11 β		+	
14 α			+
15 α			+
16 α	+		
19		+	

Table 28 summarizes the many types of microbial hydroxylations that have been observed with C_{19} steroid substrates. Nine different positions may be attacked, depending upon the organism, the substrate, and the conditions. Table 29 compares the positions hydroxylated by liver and adrenal with those hydroxylated by microorganisms. The differences indicated are probably not very significant since more and more similarities between mammalian and microorganism reactions are found, as these studies are extended.

III. Hydroxylases of C_{21} steroids

The bulk of the detailed work on hydroxylases of mammalian origin has been done with C_{21} steroids as substrates and the 11 β- and the 21-hydroxylases have been studied most thoroughly. For 21-hydroxylation, in addition to O_2 and reduced TPN, both 105,000 $\times$ g supernatant of adrenal homogenate and adrenal microsomes were required (RYAN 1956, RYAN and ENGEL 1956). Either fraction was inactive alone. The 105,000 $\times$ g supernatant fraction from the adrenal homogenate could be replaced by a similar supernatant from rat liver, but the microsomal fraction of rat liver could not replace this fraction from adrenal. TOMKINS (1956c) obtained results with the 11 β-hydroxylase system. One of the two cellular fractions required enzymes could be replaced by a heat-labile factor from rat liver; the second factor was reported to be exclusively present in the adrenal. More recently TOMKINS et al. (1958) have reported the presence of still another heat-labile 11 β-hydroxylase-system factor.

Evidence for a multiplicity of substrate-specific 11 β-hydroxylases comes from the work of Tomkins. In crude preparations many substrates can be hydroxylated, but TOMKINS et al. (1957) have reported the separation of a system which 11 β-hydroxylates deoxycorticosterone from a preparation which 11 β-hydroxylates 11-deoxycortisol. If this point is clearly established, the complexity of the problem of biochemical synthetic mechanisms will be greatly increased.

Hydroxylase systems in mammalian tissues influencing C_{21} steroids include 2 α, 6 β, 11 β, 17 α, 18, 19, and 21. Of these only a 6 β-hydroxylating system has been observed in liver, the substrates being deoxycorticosterone (AXELROD and

Table 30. *Hydroxylations of C_{21} substrates by adrenal*

Position	Tissue preparation	Substrate	Product	References
6 β	Adrenal perfusion	Progesterone	6 β,17 α-Dihydroxy-progresterone	KUSHINSKY (1955)
	Adrenal perfusion	Deoxycortico-sterone	6 β-Hydroxyde-oxycorticosterone	HAINES (1952)

Table 30 (cont.)

Position	Tissue preparation	Substrate	Product	References
11 β	Adrenal slices and homogenate	Deoxycorticosterone	Corticosterone	McGinty et al. (1950), Savard et al. (1950), Haines (1952), Hayano and Dorfman (1953)
		11-Deoxycortisol	Cortisol	
		21-Hydroxyallopregnane-3,20-dione	11 β,21-Dihydroxyallopregnane-3,20-dione	Hayano and Dorfman (1953)
	Adrenal perfusion	Progesterone	11 β-Hydroxyprogesterone, Corticosterone, Cortisol	H. Levy et al. (1950), Hechter et al (1951)
		Androsterone	11 β-Hydroxyandrosterone	Hechter et al. (1951)
		Δ^4-Androstene-3,17-dione	11 β-Hydroxy-Δ^4-androstene-3,17-dione, 11 β-Hydroxyandrostane-3,17-dione	Jeanloz et al. (1950, 1953)
		11-Deoxycortisol	Cortisol	Hechter et al. (1950, 1951)
		Deoxycorticosterone	Corticosterone	Hechter et al. (1949, 1951), Ungar et al. (1958), H. Levy et al. (1953)
		Epiandrosterone	11 β-Hydroxyandrostane-3,17-dione, 11 β-Hydroxyepiandrosterone	Meyer et al. (1953)
		21-Hydroxyallopregnane-3,20-dione	11 β,21-Dihydroxyallopregnane-3,20-dione	Meyer et al. (1953)
		Pregnenolone	Corticosterone Cortisol	H. Levy et al. (1950), Hechter et al. (1951)
		17 α-Hydroxyprogesterone	Cortisol	H. Levy et al. (1950), Hechter et al. (1951)
	Adrenal mitochondria	Progesterone	11 β-Hydroxyprogesterone	Saba and Hechter (1955), Brownie et al (1954)
	Adrenal mitochondria (O^{18})	Deoxycorticosterone	Corticosterone-O^{18}	Sweat et al. (1956)
	Adrenal homogenate residue (O^{18})	Deoxycorticosterone	Corticosterone-O^{18}	Hayano et al. (1955)
	Adrenal perfusion	Pregnenolone	Cortisol, Corticosterone	H. Levy et al. (1954)
		Progesterone	17 α-Hydroxyprogesterone, Cortisol, Corticosterone	
		11-Deoxycortisol Progesterone	Cortisol	G. Rosenfeld et al (1955), Tomkins (1956b)
		Deoxycorticosterone	Corticosterone	
17 α	Adrenal perfusion	Progesterone	11-Deoxycortisol	H. Levy et al. (1954)

Table 30 (cont.)

Position	Tissue preparation	Substrate	Product	References
17α (cont.)	Adrenal perfasion, cont.	Pregnenolone	Cortisol	H. Levy et al. (1950), Hechter et al. (1951)
		Progesterone	17α-Hydroxyprogesterone, Cortisol	H. Levy et al. (1950), Pincus (1954), Hechter et al. (1951)
			3β,17α-Dihydroxyallopregnan-20-one, 17α-Hydroxyallopregnane-3,20-dione	H. Levy et al. (1954), Pincus (1954)
	Adrenal residue and homogenate	Progesterone	Cortisol	Plagner and Samuels (1952), Hayano and Dorfman (1953)
	Adrenal homogenate supernatant	Progesterone	11-Deoxycortisol	Plager and Samuels (1954)
	Adrenal slices (human)	Progesterone	Cortisol	Lombardo et al. (1956)
	Adrenal perfusion	Progesterone	Cortisol	G. Rosenfeld et al. (1955)
	Adrenal perfusion	Pregnenolone	Cortisol	H. Levy et al. (1954)
		Progesterone	17α-Hydroxyprogesterone, Cortisol	
18	Adrenal homogenates	Deoxycorticosterone	18-Hydroxydeoxycorticosterone	Kahnt et al. (1955a)
19	Adrenal perfusion	Progesterone	19-Hydroxy-11-deoxycortisol 19-Hydroxydeoxycorticosterone	H. Levy and Kushinsky (1955)
	Adrenal brei	Deoxycorticosterone	19-Hydroxydeoxycorticosterone	H. Levy and Kushinsky (1955)
	Adrenal homogenate residue	Deoxycorticosterone	19-Hydroxydeoxycorticosterone	Hayano and Dorfman (1955)
	Adrenal homogenate	Deoxycorticosterone	19-Hydroxydeoxycorticosterone	Kahnt et al. (1955a)
21	Adrenal homogenate supernatant	Progesterone	11-Deoxycortisol	Plager and Samuels (1954)
	Adrenal perfusion	Pregnenolone Progesterone	Cortisol Corticosterone	H. Levy et al. (1954)
	Adrenal homogenate supernatant	Progesterone	Deoxycorticosterone	Ryan (1956), Ryan and Engel (1956)
		17α-Hydroxyprogesterone	11-Deoxycortisol	
		21-Deoxycortisol	Cortisol	
	Adrenal perfusion	Progesterone	Cortisol	G. Rosenfeld et al. (1955)
		21-Deoxycortisone	Cortisone	
		Progesterone	Corticosterone Cortisol	H. Levy et al. (1950), Hechter et al. (1951)
		17α-Hydroxyprogesterone	Cortisol	
		Pregnenolone	Cortisol Corticosterone	
		21-Deoxycortisone	Cortisone, 17α-Hydroxyallopregnane-3,11,20-trione	Meyer et al. (1954)

Table 31. *Microbiological hydroxylations of C_{21} steroids*

Position	Substrate	Product	Microorganism	Reference
6β	Deoxycorticosterone	6β-Hydroxydeoxycorticosterone	*Trichothecium roseum*	MEYSTRE et al. (1955)
			Lenzites abientina	WETTSTEIN (1955)
			Rhizopus arrhizus	HAYANO et al. (1956b)
		6β,21-Dihydroxy-Δ^4-pregnene-3,20-dione	*Rhizopus* sp.	EPPSTEIN et al. (1953)
			Trichothecium roseum	MEYSTRE et al. (1954)
		6β,17α,21-Trihydroxy-Δ^4-pregnene-3,20-dione	*Cephalothecium roseum*	MEISTER et al. (1954)
	Progesterone	6β-Hydroxyprogesterone	*Streptomyces aureofaciens*	INGLE et al. (1953)
			Penicillum urticae	WETTSTEIN (1954), MURRAY and PETERSON (1953a)
		6β,11α-Dihydroxyprogesterone	*Rhizopus* sp.	D. PETERSON et al. (1952)
			Aspergillus niger	FRIED et al. (1952)
	16α-Hydroxyprogesterone	6β,16α-Dihydroxyprogesterone	*Aspergillus nidulans*	FRIED et al. (1955)
7α	Progesterone	7α-Hydroxyprogesterone	*Physomyces blakesleeanus*	FRIED and SABO (1953), FRIED et al. (1955)
	Deoxycorticosterone	7α-Hydroxydeoxycorticosterone	Curvulariae: *Peziza* sp.	MEYSTRE et al. (1955)
7β	3β-Hydroxyallopregnan-20-one	3β,7β-Dihydroxyallopregnan-20-one	*Rhizopus arrhizus*	EPPSTEIN et al. (1956), MURRAY and PETERSON (1952)
	Pregnenolone	3β,7β,11α-Trihydroxy-Δ^5-pregnen-20-one	*Rhizopus arrhizus*	EPPSTEIN et al. (1956), MURRAY and PETERSON (1955a)
	3β,21-Dihydroxyallopregnan-20-one	3β,7β,21-Trihydroxyallopregnan-20-one	*Rhizopus* sp.	KAHNT et al. (1952)
8ξ or 9ξ	Progesterone	8ξ- or 9ξ-Hydroxyprogesterone	*Streptomyces aureofaciens*	FRIED et al. (1955)
	Deoxycorticosterone	8ξ- or 9ξ-Hydroxydeoxycorticosterone	*Mucor* sp., *Helicostylum piriforme*	EPPSTEIN et al. (1956), MURRAY and PETERSON (1955c)
			Neurospora crassa	STONE et al. (1955)
			Curvularia pallescens	WETTSTEIN (1955)
	11-Deoxycortisol	8ξ or 9ξ,17α,21-Trihydroxy-Δ^4-pregnene-3,20-dione	*Mucor* sp., *Helicostylum piriforme*	MURRAY and PETERSON (1952), EPPSTEIN et al. (1956)
11α	Deoxycorticosterone	Epicorticosterone	*Rhizopus nigricans*	HAYANO et al. (1956b)
			Aspergillus sp.	EPPSTEIN et al. (1956)
		Epicortisol	*Cephalothecium roseum*	MEISTER et al. (1954)
	16,17-Oxidoprogesterone	11α-Hydroxy-16,17-oxidoprogesterone	*Rhizopus* sp.	D. PETERSON et al. (1955)

Table 31 (cont.)

Position	Substrate	Product	Microorganism	Reference
11α (cont.)	Progesterone	11α-Hydroxy-progesterone	*Pestalotia foedans*	SHULL et al. (1954)
	11-Deoxycortisol	Epicortisol	*Aspergillus nidulans*	DJERASSI et al. (1955)
			Helicostylum piriforme	EPPSTEIN et al. (1956)
	3β-Hydroxyallo-pregnan-20-one	3β,11α-Dihydroxy-allopregnan-20-one	*Rhizopus nigricans*	EPPSTEIN et al. (1956)
	Progesterone	11α-Hydroxy-progesterone	*Aspergillus ochraceus*	DULANEY et al. (1955)
	Δ^6-Dehydro-progesterone	11α-Hydroxy-Δ^6-dehydro-progesterone	*Rhizopus nigricans*	D. PETERSON et al. (1953)
	Progesterone	11α-Hydroxy-progesterone	*Rhizopus* sp.	PETERSON and MURRAY (1952), PETERSON et al. (1952), KAHNT et al. (1952), MANCERA et al. (1952)
			Aspergillus sp.	FRIED et al. (1952), MURRAY and PETERSON (1953b)
	Δ^{16}-Dehydro-progesterone	11α-Hydroxy-17α-progesterone	*Rhizopus nigricans*	MEISTER et al. (1953b)
	Pregnenolone	3β,11α-Dihydroxy-Δ^5-pregnene-7,20-dione	*Rhizopus arrhizus*	EPPSTEIN et al. (1956), MURRAY and PETERSON (1955b)
	Deoxycortico-sterone	11α,21-Dihydroxy-Δ^4-pregnene-3,20-dione	*Rhizopus* sp.	EPPSTEIN et al. (1953)
			Aspergillus sp.	FRIED et al. (1952) MURRAY and PETERSON (1953b) EPPSTEIN et al. (1956)
	Progesterone	6β,11α-Dihydroxy-progesterone	*Rhizopus* sp.	D. PETERSON et al. (1952)
			Aspergillus niger	FRIED et al. (1952)
	16α-Hydroxy-progesterone	11α,16α-Dihydroxyprogesterone	*Aspergillus niger*	FRIED et al. (1955)
	Progesterone	11α,17α-Dihydroxyprogesterone	*Rhizopus* sp.	MEISTER et al. (1953c)
			Aspergillus niger	FRIED et al. (1952)
	17α-Hydroxy-progesterone	11α,17α-Dihydroxyprogesterone	*Cephalothecium roseum*	MEISTER et al. (1954)
	Deoxycortico-sterone	11α,17α,21-Trihydroxy-Δ^4-pregnene-3,20-dione		
	Allopregnane-3,20-dione	11α-Hydroxyallo-pregnane-3,20-dione	*Rhizopus nigricans*	EPPSTEIN et al. (1953b)
	Pregnane-3,20-dione	11α-Hydroxy-pregnane-3,20-dione	*Rhizopus nigricans*	EPPSTEIN et al. (1953b)
	Pregnenolone	3β,7β,17α-Trihydroxy-Δ^5-pregnen-20-one	*Rhizopus arrhizus*	EPPSTEIN et al. (1956), MURRAY and PETERSON (1955a)

Table 31 (cont.)

Position	Substrate	Product	Microorganism	Reference
11α (cont.)	11-Deoxycortisol or 17α,21-Dihydroxypregnane-3,20-dione	11α,17α,21-Trihydroxypregnane-3,20-dione	*Rhizopus nigricans*	D. PETERSON et al. (1953a)
	11-Deoxycortisol	Epicortisol	*Corticum sasaki*	NISHIKAWA and HAGIWARA (1958)
11β	Progesterone	11β-Hydroxyprogesterone	Curvulariae	SHULL and KITA (1955)
			Cunninghamella blakesleeana	EPPSTEIN et al. (1956)
	17α-Hydroxyprogesterone	21-Deoxycortisol	*Curvularia lunata*	SHULL et al. (1955) SHULL and KITA (1955)
	Deoxycorticosterone	Corticosterone		
	11-Deoxycortisol	Cortisol		
	11-Deoxycortisol	Cortisol	*Cunninghamella blakesleeana*	MANN et al. (1955a, b)
		Cortisone		O'DONNELL et al. (1955)
	17α-Hydroxyprogesterone	11β,14α,17α-Trihydroxy-Δ^4-pregnene-3,20-dione	*Curvularia lunata*	SHULL et al. (1955)
	11-Deoxycortisol	Cortisol	*Corticum sasaki*	NISHIKAWA and HAGIWARA (1958)
14α	Progesterone	14α-Hydroxyprogesterone	*Bacillus cereus*	FRIED et al. (1955)
			Helicostylum piriforme	EPPSTEIN et al. (1956)
	Deoxycorticosterone	14α-Hydroxydeoxycorticosterone	*Cunninghamella blakesleeana*	MANN et al. (1955a, b)
			Helicostylum piriforme	MURRAY and PETERSON (1955c)
	17α-Hydroxyprogesterone	11β,14α,17α-Trihydroxy-Δ^4-pregnene-3,20-dione	*Curvularia lunata*	SHULL et al. (1955)
	11-Deoxycortisol	14α,17α,21-Trihydroxy-Δ^4-pregnene-3,20-dione	Mucors, *Helicostylum piriforme*	MEISTER et al. (1953a), EPPSTEIN et al. (1956)
			Cunninghamella blakesleeana	EPPSTEIN et al. (1956)
15α	Progesterone	15α-Hydroxyprogesterone	*Colletotrichum antirrhini*	FRIED et al. (1955)
			Gibberellae and *Fusaria*	EPPSTEIN et al. (1956)
			Penicillum urticae	EPPSTEIN et al. (1956)
			Phycomyces blakesleeanus	FRIED et al. (1955)
	Deoxycorticosterone	15α,21-Dihydroxy-Δ^4-pregnene-3,20-dione	Gibberellae	MEYSTRE et al. (1955), EPPSTEIN et al. (1956)
			Fusaria	EPPSTEIN et al. (1956)
15β	Deoxycorticosterone	15β,21-Dihydroxy-Δ^4-pregnene-3,20-dione	*Lenzites abietina*	DJERASSI et al. (1955), MEYSTRE et al. (1955), EPPSTEIN et al. (1956)
	Progesterone	15β-Hydroxyprogesterone	*Phycomyces blakesleeanus*	FRIED et al. (1955)

Table 31 (cont.)

Position	Substrate	Product	Microorganism	Reference
16α	Progesterone	16α-Hydroxypregnane-3,20-dione	*Streptomyces* sp.	PERLMAN et al. (1955)
	Deoxycorticosterone	16α,21-Dihydroxy-Δ^4-pregnene-3,20-dione	*Streptomyces* sp.	VISCHER et al. (1954), FRIED et al. (1955)
			Didymella vodakii	WETTSTEIN (1955)
	Progesterone	16α-Hydroxyprogesterone	Actinomycetes	PERLMAN et al. (1952)
17α	Deoxycorticosterone	11-Deoxycortisol	*Cephalothecium roseum*	PETERSON (1955)
			Trichothecium roseum	MEYSTRE et al. (1954)
	Corticosterone	Cortisol	*Cephalothecium roseum*	MEISTER et al. (1954)
			Trichothecium roseum	MEYSTRE et al. (1954)
	Deoxycorticosterone	6β,17α,21-Trihydroxy-Δ^4-pregnene-3,20-dione 11α,17α,21-Trihydroxy-Δ^4-pregnene-3,20-dione	*Cephalothecium roseum*	MEISTER et al. (1954)
	Progesterone	17α-Hydroxyprogesterone	*Cephalothecium roseum*	HAYANO et al. (1956b)
	16-Dehydroprogesterone	11α,17α-Dihydroxyprogesterone	*Rhizopus nigricans*	HERZOG et al. (1955)
	Deoxycorticosterone	11-Deoxycortisol	*Cephalothecium roseum*	EPPSTEIN et al. (1956)
			Trichothecium roseum	MEYSTRE et al. (1954).
	Progesterone	11α,17α-Dihydroxyprogesterone	*Cephalothecium roseum*	MEISTER et al. (1954)
	Corticosterone	Cortisol	*Cephalothecium roseum*	MEISTER et al. (1954)
			Trichothecium roseum	MEYSTRE et al. (1954)
	Deoxycorticosterone	6β,17α,21-Trihydroxy-Δ^4-pregnene-3,20-dione, 11α,17α,21-Trihydroxy-Δ^4-pregnene-3,20-dione	*Cephalothecium roseum*	MEISTER et al. (1954)
19	11-Deoxycortisol	19-Hydroxy-11-deoxycortisol	*Corticum sasaki*	NISHIKAWA and HAGAWARA (1958)
21	Progesterone	Deoxycorticosterone	*Ophiobolus herbotrichus*	MEYSTRE et al. (1954)
	11-Ketoprogesterone	11-Dehydrocorticosterone	*Aspergillus niger*	ZAFFARONI et al. (1955)
	11β-Hydroxyprogesterone	Corticosterone	*Aspergillus niger*	ZAFFARONI et al. (1955)
	17α-Hydroxyprogesterone	11-Deoxycortisol	*Ophiobolus herbotrichus*	MEYSTRE et al. (1954)
	19-Norprogesterone	19-Nordeoxycorticosterone		
	Progesterone	Deoxycorticosterone	*Cephocethium roseum*	HAYANO et al. (1956b)
	6β-Hydroxyprogesterone	6β-Hydroxydeoxycorticosterone	*Aspergillus niger*	ZAFFARONI et al. (1955)

Miller 1954) and 11-deoxycortisol (Forchielli 1956, Forchielli and Dorfman 1956). The presence of 6β-hydroxycortisol in human and guinea pig urine may well indicate 6β-hydroxylation of cortisol by liver tissue.

Six different hydroxylation reactions have been reported for adrenal tissue and these are summarized in Table 30.

The hydroxylations of C_{21} steroids by microbial systems have been studied extensively and are of commercial importance, particularly in the preparation of anti-inflammatory corticoids (Table 31). The activities of mammalian and microbial enzymes are compared in Table 32.

Organisms such as *Curvalaria lunata* and *Cunninghamella blakesleeana*, which possess 11β-hydroxylating systems, can form 9β,11β-epoxido-17α,21-dihydroxy-Δ^4-pregnene-3,20-dione from the corresponding $\Delta^{9(11)}$steroid. With the same substrate and 11α-hydroxylating organisms such as *Rhizopus nigricans*, *Rhizopus arrhizus*, and *Pestalotia foedans*, no epoxide could be formed (Bloom and Shull 1955). When $\Delta^{9(11)}$substrate was incubated with the washed residue of a bovine adrenal homogenate, the 9β,11β-epoxide was tentatively identified (Bloom et al. 1956).

Table 32. *Comparison between enzyme systems hydroxylating* C_{21} *steroids in mammalian tissues and microorganisms*

Position	Present in mammalian tissues	Present in microorganisms
1		
2α	+	
6α		
6β	+	+
7α		+
7β		+
8ξ or 9ξ		+
11α		+
11β	+	+
14α		+
15α		+
15β		+
16α	+	+
17α	+	+
18	+	
19	+	+
21	+	+

B. Dehydrogenases and reductases

I. 3α-Dehydrogenases and α-orienting 3-keto reductases

A variety of steroid dehydrogenases have been detected in tissues and microorganisms. Ungar and Dorfman (1954) reported the presence of a 3α-dehydrogenase in rabbit liver and kidney associated with the supernant of a homogenate. This activity was found also in an extract of an acetone powder of rat liver. Detailed reports on this enzyme have been published by Tomkins and Isselbacher (1954) and Tomkins (1956a, b). The suggestion is made that a single enzyme catalyzes the oxidation-reduction at carbon 3, regardless of whether the steroid substrate is a C_{19} or C_{21} compound. Reactions were observed with the following compounds, for which the relative rates are noted in parentheses: 17α,21-dihydroxypregnane-3,11,20-trione (100); 11β,17α,21-trihydroxypregnane-3,20-dione (36); etiocholane-3,17-dione (110); androstane-3,17-dione (22); and 17β-hydroxyandrostan-3-one (38). The specificity for steroid substrates was inferred from the fact that no reaction was observed with cyclohexanone or acetone. The 3β-hydroxy steroid, 3β-hydroxyandrostan-17-one, was not oxidized nor were other steroids reduced such as cholestan-3-one, coprostan-3-one, cortisone, and Δ^1-androstene-3,17-dione. DPN and TPN act as hydrogen acceptors. The enzyme was detected in liver, kidney and testis, but not in lung, spleen, muscle and brain.

Rubin (1957) also described a 3α-dehydrogenase in rat liver homogenate, which reduced androstane-3,17-dione, etiocholanedione, and 17,21-dihydropregnane-3,20-dione. There appeared to be an androgen-dependent sex difference in the amount of product obtainable, but recent studies (Rubin, personal communication) have indicated that, in the presence of adequate amounts of substrate, the 3α-reducing capacities of male and female rat livers are of the same order of magnitude.

Pseudomonas testosteroni, an organism capable of utilizing C_{19} steroids as the sole source of organic carbon, has been grown on a medium containing testosterone. This procedure induces development of a 3α-dehydrogenase. This enzyme has been prepared in cell-free extracts and shown to be a DPN-linked steroid dehydrogenase (Talalay et al. 1952, Marcus and Talalay 1956). In keeping with the observations that rat liver 3α-dehydrogenase can act on C_{19} and C_{21} substrates, Talalay and Marcus (1956) demonstrated that the induced *Pseudomonas* enzyme oxidized the following substrates: androsterone, androstane-3α,17β-diol, etiocholanolone, 11β-hydroxyetiocholanolone, urocortisone, and androstan-3α-ol. Deoxycholic acid was also oxidized.

Both rat and chick livers contain a 3α-dehydrogenase that can oxidize Δ^4-androstene-3α,17β-diol to Δ^4-androstene-3,17-dione. This 3α-dehydrogenase is present in the soluble fraction following centrifugation at 78,000 × g. The activity can be precipitated with ammonium sulfate between 0.40 and 0.55 saturation. Both DPN and TPN are equally effective. This enzyme preparation is equally effective on a C_{21} steroid with the same configuration in ring A (Ungar et al. 1957, Ungar and Bloom 1957).

II. 3β-Dehydrogenases and β-orienting 3-keto reductases

Rat liver contains enzymes that reduce the 3-keto group of both C_{19} and C_{21} steroids (Ungar and Dorfman 1954) to the 3β-hydroxyl group. Rat and chick liver contain, in the soluble fractions (78,000 × g) of liver, a DPN- or TPN-linked 3β-dehydrogenase which acts on C_{19} and C_{21} steroids having the Δ^4-3β-hydroxy function in ring A (Ungar et al. 1957; Ungar and Bloom 1957).

Rubin (1957) described apparently irreversible conversion of androstane-3,17-dione to epiandrosterone by homogenates of livers of rats, rabbits, mice, hamsters, and guinea pigs. In all these species except guinea pigs, there was greater activity in the homogenates of male livers than of female livers, as demonstrated by differences in the ratio of epiandrosterone to androsterone produced under conditions of competition for substrate. Recently Rubin (personal communication) has shown that when substrate is not a limiting factor, livers of male rats contain significantly more 3β-orienting reductase than livers of female rats, and there is evidence that the liver of male rats contains at least two different 3β-reducing enzymes.

The β-dehydrogenase prepared from *Pseudomonas testosteroni* grown on testosterone as the only source of organic carbon catalyzes oxidation of 3β,16β- and 17β-hydroxy steroids, including ring A aromatic compounds. The enzyme is DPN-linked (Talalay and Marcus 1956).

Steroid-producing tissues such as ovary, testis, adrenal, and placenta contain enzyme systems that catalyze the oxidation of the Δ^5-3β-hydroxyl group of both C_{19} and C_{21} steroids to form the Δ^4-3-keto group (Samuels et al. 1951). This enzyme has been referred to as steroid 3β-ol-dehydrogenase. The cofactor is DPN. Detailed studies with adrenal tissue indicated that the highest activity is associated with the microsomal fraction (Beyer and Samuels 1956).

Testicular 3β-ol-dehydrogenase may be doubled when the animal is treated with chorionic gonadotrophin. In hypophysectomized rats, after a temporary increase in activity, this enzyme significantly decreases in the testis (Samuels and Helmreich 1956), when expressed in terms of activity per gland. A quantitative method for the assay of this enzyme in adrenal tissue has been described by Rubin and Dorfman (1957).

The studies of TALALAY (1957) have indicated that the 3β-ol-dehydrogenase reaction may be separated into two distinct steps. The first involves the oxidation of the 3β-hydroxy function to the 3-ketone, with the formation of the Δ^5-3-keto group. The second reaction involves the rearrangement of the Δ^5 group to the Δ^4 group. This second enzyme system, an isomerase, is demonstrable in both *Pseudomonas testosteroni* and mammalian tissues. Rat adrenal, testis, ovary, and liver are rich sources of this enzyme. The steroid isomerase can convert the Δ^5 and the $\Delta^{5(10)}$ group to the Δ^4 group in both C_{19} and C_{21} steroids.

The liver isomerase is present in the supernatant fraction after centrifuging at 100,000 × g. On the basis of H_2^2O and H_2^3O experiments, the suggested mechanism concerns the direct transfer of hydrogen from C_4 to C_6. The isomerase reaction is considered to be reversible.

III. 11β-Dehydrogenases and β-orienting 11-keto reductases

The presence of 11β-dehydrogenase in liver tissue has been demonstrated by various investigators. The reduction of the 11-keto group is dependent upon the presence of a Δ^4-3-ketone group in ring A (HÜBENER et al. 1956). Better than 50% reduction of cortisone and $17\alpha,20\beta,21$-trihydroxy-Δ^4-pregnene-3,11-dione to the corresponding 11β-hydroxy compounds occurred with a rat liver preparation, while urocortisone, cortolone, β-cortolone and 11-ketoetiocholanolone were unaltered. The enzyme seemed to be independent of TPN, DPN, and cytochrome *c* (AMELUNG et al. 1953).

The *in vitro* conversion of cortisone to cortisol has been demonstrated in porcine, bovine, and rat tissues by C. FISH et al. (1953). These investigators reported conversions with whole liver homogenates, a supernatant fraction after centrifugation at 5000 × g and acetone powder of whole homogenate supplemented with DPN.

IV. 20-Keto reductases

Rat liver supernatant from a 17,500 × g centrifugate contained both the 20α- and 20β-enzymes. Urocortisol and cortisol were reduced primarily to 20β-hydroxy derivatives, while urocortisone and 11-deoxycortisol were reduced to 20α-hydroxy derivatives (HÜBENER et al. 1956). In another study 11-deoxycortisol incubated with rat liver homogenates yielded both 20α- and 20β-hydroxy derivatives (FORCHIELLI 1956, FORCHIELLI and DORFMAN 1956, FORCHIELLI et al. 1955). Both 20α- and 20β-reduction was observed in still another study when liver microsomes plus supernatant were used (DECOURCY and SCHNEIDER 1956).

A sex difference in the reduction of the C-20 keto group of cortisone (stereoisomer not specified) is indicated from the work of TROOP (1958, 1959). The homogenates of male rat livers produced from 3 to 6 times the 20-keto reduction found for similar preparations from female rats. This sex difference was not found in rats before 25 days of age. Androgens were associated with increased amounts of the C-20 keto reducing enzyme, while estrogens tended to produce decreased conversions.

C. Hydrogenases

I. Δ^1-Hydrogenases

In *in vivo* experiments in the human beings UNGAR and DORFMAN (1952, 1953) demonstrated the reduction of the Δ^1 double bond in C_{19} steroids. *In vitro* experiments have been reported by TOMKINS (1956b) and by VERMEULEN and CASPI (1958).

TOMKINS' enzyme preparation reduced the Δ^1 group of both C_{19} ($\Delta^{1,4}$-androstadiene-3,17-dione) and C_{21} (prednisone) steroids. In the presence of TPNH, both the Δ^1 and Δ^4 groups were reduced. This investigator suggested that the initial reduction must be a reduction of the Δ^1 group in compounds having both Δ^1 and Δ^4 unsaturation and that separate enzyme systems exist for each reduction. This is contrary, however, to the report of CASPI and PECHET (1957), who isolated 11β,17α,21-trihydroxy-Δ^1-5α-pregnane-3,20-dione after the administration of prednisolone, indicating that the Δ^4 double bond had been attacked without saturation of the Δ^1 nuclear double bond.

II. Δ^4-5β-Hydrogenases

These hydrogenases have been prepared in soluble form, from rat liver, and subjected to ammonium sulfate precipitation and further purification on calcium phosphate gel (TOMKINS and ISSELBACHER 1954, TOMKINS 1956b). The cofactor for these reactions was TPNH, DPNH being completely inactive. Although the crude liver extract showed activity of about the same order of magnitude for a wide variety of substrates, ranging from cholestenone to 9α-fluorocortisol, on further purification the relative activities of the enzyme preparation with respect to 3 substrates, cortisone, cortisol, and adrenosterone, changed. Therefore it appears that at least 3 different enzymes are actually present. Sulfhydryl groups appear to be involved, since addition of *p*-chloromercuribenzoate inhibited the reduction of cortisone.

FORCHIELLI and DORFMAN (1956) demonstrated that a Δ^4-5β-hydrogenase enzyme preparation for 11-deoxycortisol could be obtained from a rat liver homogenate, supernatant obtained by centrifugation at 78,000 × g.

III. Δ^4-5α-Hydrogenase

Perfusion of corticoids through isolated livers, or incubation of these substrates with slices or whole homogenates, yields essentially only 5α-reduced steroids. More careful studies have now demonstrated that Δ^4-5α-hydrogenase, at least for 11-deoxycortisol, is present in the microsomal fraction of liver and requires TPNH (FORCHIELLI and DORFMAN 1956).

IV. Factors influencing Δ^4-5α- and Δ^4-5β-hydrogenases

There are some interesting variations in effective enzyme concentrations dependent on the physiological status of the animal. FORCHIELLI et al. (1958) found that the rate of Δ^4-reduction of 11-deoxycortisol was 3 to 4 fold greater in whole homogenates of female rat livers than in male livers. The same sex difference was found in microsomal fractions containing the Δ^4-5α-hydrogenase. Evidence was found for only one Δ^4-hydrogenase (5α) in female rat liver, in contrast to the male liver, which contains a soluble Δ^4-5β-hydrogenase as well. Ovariectomy of the adult female rats did not produce any marked change in enzyme titer. Hypophysectomy resulted in a sharp drop, which could be reversed in part by ACTH, growth hormone, or pregnant-mare's-serum gonadotrophin. Rapid increases in the titer of Δ^4-5α-hydrogenase in the liver of the female rat occurred prior to sexual maturation.

Various factors influencing reduction of Δ^4-3-ketosteroids by rat liver have been studied in detail by YATES et al. (1958). A higher effective concentration of Δ^4-hydrogenases in female livers, as compared to male livers, was found for each of the following substrates: deoyxcorticosterone, cortisone, cortisol, aldosterone, corticosterone, testosterone, and progesterone. In keeping with the observations of FORCHIELLI et al. (1958), female liver had 3 to 10 times the activity of male liver. A representation of the factors involved is portrayed in Fig. 37. During the age period from approximately 35 to 60 days of age, gonadal influences predominate, with estrogens increasing the Δ^4-hydrogenase content of liver and androgens producing the opposite effect. The estrogen effect becomes insignificant at

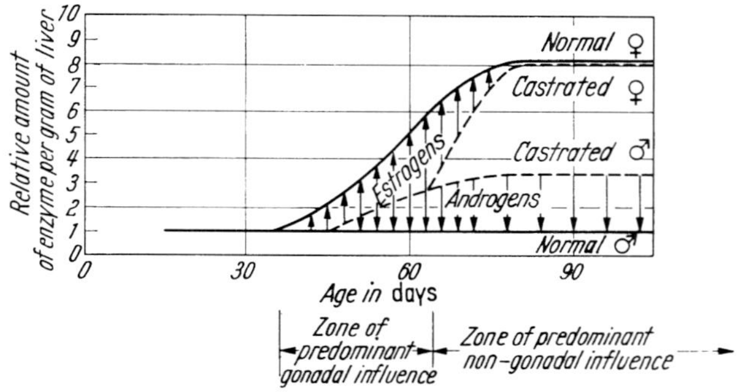

Fig. 37. Change in rat liver Δ^4-hydrogenase concentration with age. Reproduced with permission of authors and publisher from YATES et al., Endocrinology **63**, 887 (1958)

about 75 days of age in the female, whereas androgen influence continues to be an important factor in the older males. Beyond 60 days of age a non-gonadal process maintains the high enzyme titer in female liver. Adrenocortical secretions do not seem to account for this non-gonadal control.

D. Miscellaneous enzyme systems

A liver acetone powder contained an enzyme which reduced the aldehyde group of 11 β,17 α-dihydroxy-21-oxo-Δ^4-pregnene-3,20-dione to the corresponding alcohol function (SCHNEIDER 1953). In another study, both this substrate and the corresponding 11-keto steroid were reduced. The cofactor requirement for this system is DPNH or TPNH. Corresponding enzyme systems are indicated for the oxidation and reduction of carbon atoms 18 and 19. In addition the isolation of C-18 carboxylic acid derivatives (NEHER and WETTSTEIN 1956) indicates the presence of an enzyme system in the adrenal gland which oxidizes an 18-oxo group to the corresponding acid function.

Few enzymological studies have been reported on a series of desmolase reactions that have been established. These reactions include the conversion of cholesterol to pregnenolone and isocaproic acid (LYNN et al. 1954, SABA et al. 1954). The enzyme system, C-20,22-desmolase, for this reaction seems to be present in the adrenals, ovaries and testes. A second possible desmolase reaction involving cholesterol is C-17,20-desmolase, which would be concerned with the complete removal of the sidechain, resulting in the direct formation of dehydroepiandrosterone.

Steroids of the C_{21} series having oxygen functions at C-17 and C-20 are known to be converted *in vivo* and *in vitro* to 17-ketosteroids. The enzyme system for this type of transformation would be another desmolase enzyme system.

Literature

ABELSON, D., F. ULRICH and C. N. H. LONG: Proc. Soc. exp. Biol. (N. Y.) **89**, 386 (1955).
ALEXANDER, G. J., and E. SCHWENK: Arch. Biochem. **66**, 381 (1957).
AMDUR, B. H., H. RILLING and K. BLOCH: J. Amer. chem. Soc. **79**, 2646 (1957).

AMELUNG, D., H. J. HÜBENER, L. ROKA u. G. MEYERHEIM: Klin. Wschr. **15**, 386 (1953).
— — — Hoppe-Seylers Z. physiol. Chem. **294**, 36 (1953).
ANKER, H. S., and K. BLOCH: J. biol. Chem. **178**, 971 (1949).
ANLIKER, R., O. ROHR u. M. MARTI: Helv. chim. Acta **39**, 1100 (1956).
— — u. L. RUZICKA: Liebigs Ann. Chem. **603**, 109 (1957).
APRILE, M., E. G. BLIGH, J. L. WEBB and R. D. H. HEARD: Rev. canad. Biol. **15**, 232 (1956 bis 1957).
ASHMORE, J., W. H. ELLIOTT, E. A. DOISY, JR., and E. A. DOISY: J. biol. Chem. **200**, 661 (1953).
AXELROD, L. R., and G. ARRAYAVE: J. Amer. chem. Soc. **75**, 5729 (1953).
— and L. L. MILLER: Arch. Biochem. **49**, 248 (1954).
— — Arch. Biochem. **60**, 373 (1956).
— — and F. HERLING: J. biol. Chem. **219**, 455 (1956).
AYRES, P. J., R. P. GOULD, S. A. S. SIMPSON and J. F. TAIT: Biochem. J. **63**, 19 (1956).
— O. HECHTER, N. SABA, S. A. SIMPSON and J. F. TAIT: Biochem. J. **65**, 22P (1957a).
— O. GARROD, S. A. S. TAIT, J. F. TAIT, G. WALKER and W. H. PEARLMAN: Ciba Found. Coll. Endocrin. **11**, 309 (1957b).
BAGGETT, B., L. L. ENGEL, L. BALDERAS, G. LANSAN, K. SAVARD and R. I. DORFMAN: Endocrinology **64**, 600 (1959).
BARRY, M. C., M. L. EIDINOFF, K. DOBRINER and T. F. GALLAGHER: Endocrinology **50**, 587 (1952).
BARTTER, F. C.: Recent Progr. Hormone Res. **12**, 187 (1956).
BAULD, W. L., and R. D. H. HEARD: In: G. PINCUS and K. V. THIMANN, Eds., The hormones, Vol. I, 620. New York: Academic Press 1948.
BEALL, D.: Nature (Lond.) **144**, 76 (1939).
— J. Endocr. **2**, 81 (1940).
— and T. REICHSTEIN: Nature (Lond.) **142**, 479 (1938).
BEER, C. T., and T. F. GALLAGHER: J. biol. Chem. **214**, 335 (1955).
BENIRSCHKE, K., E. BLOCH and A. T. HERTIG: Endocrinology **58**, 598 (1956).
BERGSTRÖM, S., and A. NORMAN: Proc. Soc. exp. Biol. (N. Y.) **83**, 71 (1953).
— and J. SJÖVALL: Acta chem. scand. **8**, 611 (1954).
— M. ROTTENBERG u. J. SJÖVALL: Hoppe-Seylers Z. physiol. Chem. **295**, 278 (1953).
— J. SJÖVALL and J. VOLTZ: Acta physiol. scand. **30**, 22 (1953).
— K. PÄÄBO, K. and J. A. RUMPF: Acta chem. scand. **8**, 1109 (1954).
BERLINER, D. L., and T. F. DOUGHERTY: Fed. Proc. **17**, 189 (1958a).
— — Proc. Soc. exp. Biol. (N. Y.) **98**, 3 (1958b).
— J. E. JONES and H. A. SALHANICK: J. biol. Chem. **223**, 1043 (1956).
BEYER, K. F., and L. T. SAMUELS: J. biol. Chem. **219**, 69 (1956).
BIGGS, M. W., R. M. LEMMON and F. T. PIERCE, JR.: Arch. Biochem. **51**, 155 (1954).
BIRKE, G., and L. O. PLANTIN: Acta med. scand. **146**, 184 (1953).
BIRMINGHAM, M. K., and E. KURLENTS: Endocrinology **62**, 47 (1958).
— F. H. ELLIOT and P. H. L. VALERE: Endocrinology **53**, 6867 (1953).
BLECHER, M., and S. GURIN: J. biol. Chem. **209**, 953 (1954).
BLEIBERG, M. J., and P. S. LARSON: J. Pharmacol. **121**, 421 (1957).
BLIGH, E. G., R. D. H. HEARD, V. J. O'DONNELL, J. L. WEBB, M. SAFFRAN and E. SCHONBAUM: Arch. Biochem. **58**, 249 (1955).
BLOCH, E., R. I. DORFMAN and G. PINCUS: Proc. Soc. exp. Biol. (N. Y.) **85**, 106 (1954).
— K. BENIRSCHKE and R. I. DORFMAN: J. clin. Endocr. **15**, 379 (1955).
— R. I. DORFMAN and G. PINCUS: Arch. Biochem. **61**, 245 (1956a).
— — — J. biol. Chem. **224**, 737 (1957).
— G. PINCUS and R. I. DORFMAN: Fed. Proc. **15**, 222 (1956b).
BLOCH, K.: J. biol. Chem. **155**, 255 (1944).
— J. biol. Chem. **157**, 661 (1945).
— Recent Progr. Hormone Res. **6**, 111 (1951).
— Helv. chim. Acta **36**, 1611 (1953).
— Harvey Lect. **48**, 68 (1954).
— Vitam. and Horm. **15**, 119 (1957).
— and D. RITTENBERG: J. biol. Chem. **145**, 625 (1942).
— — J. biol. Chem. **155**, 243 (1944).
— B. N. BERG and D. RITTENBERG: J. biol. Chem. **149**, 511 (1943).
— — — J. biol. Chem. **162**, 441 (1945).
— E. BOREK and D. RITTENBERG: J. biol. Chem. **162**, 441 (1946).
— C. CLARK and I. HARARY: J. biol. Chem. **211**, 687 (1954).
BLOOM, B. M., and G. M. SHULL: J. Amer. chem. Soc. **77**, 5767 (1955).
— M. HAYANO, A. SAITO, D. STONE and R. I. DORFMAN: Fed. Proc. **15**, 222 (1956).
BONDY, P. K., and J. R. ALTROCK: J. clin. Invest. **32**, 703 (1953).

BONGIOVANNI, A. M., and W. R. EBERLEIN: Proc. Soc. exp. Biol. (N. Y.) **89**, 281 (1955a).
— — Pediatrics **16**, 628 (1955b).
— — M. M. GRUMBACH, J. J. VAN WYK and G. CLAYTON: Proc. Soc. exp. Biol. (N. Y.) **87**, 282 (1954).
BONNER, J., and B. ARREGUIN: Arch. Biochem. **21**, 109 (1949).
BORELL, U.: Acta endocrin. (Kbh.) **17**, 13 (1954).
BORCK, E., and D. RITTENBERG: J. biol. Chem. **179**, 843 (1949).
BORTH, R., M. GESELL and H. DE WATTEVILLE: Acta endocrin. (Kbh.) **14**, 316 (1953).
BRADLOW, H. L., and T. F. GALLAGHER: J. biol. Chem. **229**, 505 (1957).
— L. HELLMAN, B. ZUMOFF and T. F. GALLAGHER: Science **124**, 1206 (1956).
— — — — Proc. 40th Meeting Endocrine Society, p. 23, (1958).
BRADY, R. O., and S. GURIN: J. biol. Chem. **186**, 461 (1950).
— — J. biol. Chem. **189**, 371 (1951).
BRIGGS, F. N., and W. TOEPEL: Endocrinology **62**, 24 (1958).
BROOKS, R. V., and W. KLYNE: Biochem. J. **62**, 21P (1956).
— and F. T. G. PRUNTY: J. Endocr. **15**, 385 (1951).
BROOKSBANK, B. W. L., and G. A. D. HASLEWOOD: Biochem. J. **44**, 111 (1949).
BROWN, H., D. C. WILLARDSON, L. T. SAMUELS and F. H. TYLER: J. clin. Invest. **33**, 1524 (1954).
BROWN, J. B., and G. F. MARRIAN: J. Endocr. **15**, 307 (1957).
BROWN, J. H. U., J. GRIFFIN and R. B. SMITH: Metabolism **4**, 542 (1955).
BROWNIE, A. C., J. K. GRANT and D. W. DAVIDSON: Biochem. J. **58**, 218 (1954).
BUCHER, N. L. R.: J. Amer. chem. Soc. **75**, 498 (1953).
— and K. MCGARRAHAN: J. biol. Chem. **222**, 1 (1956).
BULBROOK, R. D., and F. C. GREENWOOD: Brit. med. J. **1957**, 662.
— — and P. C. WILLIAMS: In: Endocrine aspects of breast cancer, p. 181. London: E. & S. Livingstone 1958.
BURSTEIN, S.: Endocrinology **50**, 412 (1952).
— J. Amer. chem. Soc. **78**, 1769 (1956).
— and R. I. DORFMAN: J. biol. Chem. **213**, 581 (1955).
— — Unpublished (1958).
— and E. M. NADEL: Fed. Proc. **15**, 228 (1956).
— K. SAVARD and R. I. DORFMAN: Endocrinology **52**, 448 (1953a).
— — — Endocrinology **53**, 267 (1953b).
— R. I. DORFMAN and E. M. NADEL: Fed. Proc. **13**, 188 (1954a).
— — — Arch. Biochem. **53**, 307 (1954b).
— F. UNGAR, M. GUT and R. I. DORFMAN: Endocrinology **56**, 267 (1955).
BURTON, R. B., E. H. KEUTMANN and C. WATERHOUSE: J. clin. Endocr. **13**, 48 (1953).
BUSH, I. E.: Ciba Found. Coll. Endocrin. **7**, 210 (1953a).
— J. Endocr. **9**, 95 (1953b).
— Biochem. J. **62**, 16P (1956a).
— Experientia (Basel) **12**, 325 (1956b).
— Biochem. J. **66**, 28 (1957).
— and M. WILLOUGHBY: Biochem. J. **67**, 689 (1957).
BUTENANDT, A.: Naturwissenschaften **17**, 878 (1929).
— Z. angew. Chem. **44**, 905 (1931).
— Z. angew. Chem. **45**, 655 (1932).
— Dtsch. med. Wschr. **61**, 823 (1935).
— and U. WESTPHAL: Hoppe-Seylers Z. physiol. Chem. **67**, 1140 (1934).
BUTLER, G. C., and G. F. MARRIAN: J. biol. Chem. **119**, 565 (1937).
CALLOW, N. H.: Biochem. J. **33**, 559 (1939).
— and R. K. CALLOW: Biochem. J. **33**, 931 (1939).
— — Biochem. J. **34**, 276 (1940).
CARSTENSEN, H.: Acta chem. scand. **10**, 474 (1956).
CASPI, E.: Transformation products resulting from cortisone and cortisol perfusion through rat livers. Doctoral Thesis, Clark Univ., Worcester, Mass. (1955).
— and J. R. BERGEN: Arch. Biochem. **59**, 207 (1955).
— and O. HECHTER: Arch. Biochem. **52**, 478 (1954).
— — Arch. Biochem. **61**, 299 (1956).
— and M. M. PECHET: Arch. Biochem. **68**, 236 (1957).
— — J. biol. Chem. **230**, 843 (1958).
— G. ROSENFELD and R. I. DORFMAN: J. org. Chem. **21**, 814 (1956a).
— M. C. LINDBERG, M. HAYANO, J. L. COHEN, M. MATSUBA, H. ROSENKRANTZ and R. I. DORFMAN: Arch. Biochem. **61**, 267 (1956b).
— F. UNGAR and R. I. DORFMAN: J. org. Chem. **22**, 326 (1957).

CHANNON, H. J.: Biochem. J. **20**, 400 (1926).
CHART, J. J., H. SHEPPARD, M. J. ALLEN, W. L. BENCZE and R. GAUNT: Experientia (Basel) **14**, 151 (1958).
CHEMAMA, R., E. E. BAULIEU, O. CREPY and M. F. JAYLE: C. R. Soc. Biol. (Paris) **151**, 454 (1957).
CHEN, P. S., jr., H. P. SCHEDL, G. ROSENFELD and F. C. BARTTER: Proc. Soc. exp. Biol. (N. Y.) **97**, 683 (1958).
CHESTER JONES, I., J. G. PHILLIPS and W. N. HOLMES: In: A. GORBMAN, Ed., Comparative endocrinology, p. 582. New York: Wiley 1959.
CLARK, L. C., JR., C. D. KOCHAKIAN and J. LOBOTSKY: J. biol. Chem. **171**, 493 (1947).
CLAYTON, R. B., and K. BLOCH: Fed. Proc. **14**, 194 (1955).
— — J. biol. Chem. **218**, 305 (1956a).
— — J. biol. Chem. **218**, 319 (1956b).
CLAYTON, R. E., J. E. HAMMANT and P. ARMITAGE: J. Endocr. **15**, 284 (1957).
COOK, R. P., A. KLIMAN and L. F. FIESER: Arch. Biochem. **52**, 439 (1954).
COOPER, D. Y., J. C. TOUCHSTONE and J. M. ROBERTS: Fed. Proc. **14**, 196 (1955).
— M. KASPAROW, W. S. BLAKEMORE and O. ROSENTHAL: Fed. Proc. **17**, 205 (1958).
COPE, C. L., and E. G. BLACK: Brit. med. J. **1958 i**, 1020.
— — Clin. Sci. **17**, 147 (1958b).
— and B. HURLOCK: Clin. Sci. **13**, 69 (1954).
— and C. E. SEWELL: Ann. rheum. Dis. **14**, 392 (1956).
CORNFORTH, J. W., and G. POPJAK: Biochem. J. **58**, 403 (1954).
— G. D. HUNTER and G. POPJAK: Biochem. J. **54**, 590, 597 (1953).
— G. POPJAK and I. Y. GORE: In: Biochemical problems of lipids, p. 216. London: Butterworth's Scientific Publications 1956.
— I. Y. GORE and G. POPJAK: Biochem. J. **65**, 94 (1957a).
— R. H. CORNFORTH, G. POPJAK and I. YOUHOTSKY-GORE: Biochem. J. **66**, 10P (1957b).
— — — — Biochem. J. **69**, 146 (1958).
COX, R. I., and G. F. MARRIAN: Biochem. J. **54**, 353 (1953).
CURRAN, G. L.: J. biol. Chem. **191**, 775 (1951).
— and D. RITTENBERG: J. biol. Chem. **190**, 17 (1951).
DAUBEN, W. G., and T. W. HUTTON: J. Amer. chem. Soc. **78**, 2647 (1956).
— and K. H. TAKEMURA: J. Amer. chem. Soc. **75**, 6302 (1953).
DAVIDSON, A. G., E. G. DULIT and I. D. FRANTZ, JR.: Fed. Proc. **16**, 169 (1957).
DECOURCY, D., and J. J. SCHNEIDER: Chem. and Ind. **1956**, 315.
DE MEIO, R. H., and C. LEWYCKA: Endocrinology **56**, 489 (1955).
— — and M. WIZERKANIUK: Fed. Proc. **15**, 241 (1956).
DEULOFEU, V., and J. FERRARI: Z. physiol. Chem. **226**, 192 (1934).
DICZFALUSY, E., and R. LUFT: Acta endocr. (Kbh.) **21**, 321 (1956).
— G. BIRKE, C. FRANKSSON, J. HELLSTROM, S. HULTBERG, L. O. PLANTIN and A. WESTMAN: In: Endocrine aspects of breast cancer, p. 186. London: E. & S. Livingstone, Ltd. 1958.
DINGEMANSE, E., S. E. DEJONGH, S. KOBER and E. LAQUEUR: Dtsch. med. Wschr. **56**, 301 (1930).
— E. LAQUEUR and O. MUHLBOCH: Nature (Lond.) **141**, 927 (1938).
DITURI, F., F. COBEY, J. V. B. WARMS and S. GURIN: Fed. Proc. **14**, 203 (1955).
— — — — J. biol. Chem. **221**, 181 (1956).
DJERASSI, C., L. B. HIGH, J. FRIED and E. F. SABO: J. Amer. chem. Soc. **77**, 3673 (1955).
DOBRINER, K.: Steroids in experimental and clinical practice, p. 130. Philadelphia: Blakiston 1951.
— and S. LIEBERMAN: Ciba Found. Coll. Endocrin. **2**, 381 (1952).
— C. P. RHOADS, S. LIEBERMAN, B. R. HILL and L. F. FIESER: Science **99**, 494 (1944).
— E. R. KATZENELLENBOGEN and R. SCHNEIDER: Arch. Biochem. **48**, 167 (1954).
DOBSON, R. M., A. H. GOLDKAMP and R. D. MUIR: J. Amer. chem. Soc. **79**, 3921 (1937).
DOHAN, F. C., J. C. TOUCHSTONE and E. M. RICHARDSON: J. clin. Invest. **34**, 485 (1955).
DOISY, E. A., C. D. VELER and S. A. THAYER: Amer. J. Physiol. **90**, 329 (1929).
— S. A. THAYER, L. LEVIN and J. M. CURTIS: Proc. Soc. exp. Biol. (N. Y.) **28**, 88 (1930).
DONE, A. K., R. S. ELY, L. R. HEISELT and V. C. KELLEY: Proc. Soc. exp. Biol. (N. Y.) **83**, 722 (1953).
— — L. J. OLSEN and V. C. KELLEY: Metabolism **4**, 416 (1955).
DORFMAN, R. I.: Recent Progr. Hormone Res. **9**, 5 (1954a).
— J. clin. Endocr. **14**, 318 (1954b).
— In: G. PINCUS and K. V. THIMANN, Eds. The hormones, Vol. III, p. 589. New York: Academic Press 1955.
— Amer. J. Med. **21**, 679 (1956).
— Cancer Res. **17**, 535 (1957a).
— Ann. Rev. Biochem. **26**, 523 (1957b).

DORFMAN, R. I., and J. B. HAMILTON: J. biol. Chem. **133**, 753 (1940).
— and F. UNGAR: Metabolism of steroid hormones. Minneapolis: Burgess 1953.
— J. W. COOK and J. B. HAMILTON: J. biol. Chem. **130**, 285 (1939).
— J. WISE and G. VAN WAGENEN: Endocrinology **36**, 347 (1945).
— — and R. A. SHIPLEY: Endocrinology **42**, 81 (1948).
— — — Endocrinology **46**, 127 (1950).
— M. HAYANO, R. HAYNES and K. SAVARD: Ciba Found. Coll. Endocrin. **7**, 191 (1953).
DOUGHERTY, T. F., H. E. BROWN and D. L. BERLINER: Endocrinology **62**, 455 (1958).
DOUGLAS, J. F., and C. G. KING: J. biol. Chem. **198**, 187 (1952).
DOWBEN, R. M., and J. RABINOWITZ: Nature (Lond.) **178**, 696 (1956).
DULANEY, E. L., E. O. STAPLEY and C. HLAVAC: Mycologia **47**, 464 (1955).
DUMM, M. E., H. GERSCHBERG, E. M. BECK and E. P. RALLI: Proc. Soc. exp. Biol. (N. Y.) **82** 659 (1953).
DUTTON, G. J., and I. D. E. STOREY: Biochem. J. **57**, 275 (1954).
EADES, H. C., jr., R. L. POLLACK and J. S. KING, JR.: Fed. Proc. **13**, 201 (1954).
EBERLEIN, W. R., and A. M. BONGIOVANNI: J. biol. Chem. **223**, 85 (1956).
EICHHORN, J., and O. HECHTER: Proc. Soc. exp. Biol. (N. Y.) **95**, 311 (1957).
— — Proc. Soc. exp. Biol. (N. Y.) **97**, 614 (1958).
EIDINOFF, M. L., R. S. ROSENFELD, J. E. KNOLL, B. J. MARANO and L. HELLMAN: J. clin. Invest. **33**, 333 (1954).
EIK-NES, K., J. A. SCHELLMANN, R. LUMRY and L. T. SAMUELS: J. biol. Chem. **206**, 411 (1954).
— D. A. DEMETRIOU, Y. C. MAYNE and R. S. JONES: Proc. Soc. exp. Biol. (N. Y.) **96**, 409 (1957).
EISENBERG, F., and S. GURIN: J. biol. Chem. **195**, 317 (1952).
EISENSTEIN, A. B.: Proc. Soc. exp. Biol. (N. Y.) **83**, 27 (1953).
— Endocrinology **60**, 298 (1957).
— and A. S. WENNEKER: Proc. Soc. exp. Biol. (N. Y.) **97**, 844 (1958).
ELLIOTT, F. H., and A. V. SCHALLY: Canad. J. Biochem. **33**, 174 (1955).
ELY, R. S., A. K. DONE and V. C. KELLEY: Proc. Soc. exp. Biol. (N. Y.) **91**, 503 (1956).
ENGEL, L. L., G. W. THORN and R. A. LEWIS: J. biol. Chem. **137**, 205 (1941).
— W. R. SLAUNWHITE, JR., P. CARTER and I. T. NATHANSON: J. biol. Chem. **185**, 255 (1950).
— R. WILSON, W. R. SLAUNWHITE, JR., P. CARTER, P. C. OLMSTED and I. T. NATHANSON: Ciba Found. Coll. Endocrin. **2**, 104 (1952).
— R. I. DORFMAN and A. R. ABARBANEL: J. clin. Endocr. 8, 903 (1953).
— P. CARTER and L. L. FIELDING: J. biol. Chem. **213**, 99 (1955).
— B. BAGGETT and P. CARTER: Endocrinology **61**, 113 (1957).
EPPSTEIN, S. H., P. D. MEISTER, D. H. PETERSON, H. C. MURRAY, H. M. LEIGH OSBORNE, D. A. LYTTLE, L. M. REINEKE and A. WEINTRAUB: J. Amer. chem. Soc. **75**, 408 (1953a).
— D. H. PETERSON, H. M. LEIGH OSBORNE, H. C. MURRAY, A. WEINTRAUB, L. M. REINEKE and P. D. MEISTER: J. Amer. chem. Soc. **75**, 421 (1953b).
— P. D. MEISTER, D. H. PETERSON, H. C. MURRAY, L. M. REINEKE and A. WEINTRAUB: J. Amer. chem. Soc. **76**, 3174 (1954).
— — H. C. MURRAY and D. H. PETERSON: Microbiological transformations of steroids and their applications to the synthesis of hormones. Vitam. and Horm. **14**, 359 (1956).
EUW, J. v., and T. REICHSTEIN: Helv. chim. Acta **24**, 879 (1941).
— — Helv. chim. Acta **25**, 988 (1942).
FARRELL, G. L., E. W. RAUSCHKOLB, P. C. ROYCE and H. HIRSCHMANN: Proc. Soc. exp. Biol. (N. Y.) **87**, 587 (1954).
— — — Amer. J. Physiol. **182**, 269 (1955).
— R. C. BANKS and S. KOLETSKY: Endocrinology **58**, 104 (1956).
FAZEKAS, I. G.: J. Endocrin. **15**, 440 (1957).
FIDANZA, A., A. DE CICCO and L. EPOSITO: Boll. Soc. ital. Biol. sper. **33**, 940 (1957a).
— G. CARVINA and A. DE CICCO: Bull. Soc. ital. Biol. sper. **33**, 942 (1957b).
FIESER, L. F.: The chemistry of natural products related to phenanthrene, New York: Reinhold 2nd Ed. 1937.
FINKELSTEIN, M., J. VON EUW and T. REICHSTEIN: Helv. chim. Acta **36**, 1266 (1953).
FISH, C. A., M. HAYANO and G. PINCUS: Arch. Biochem. **42**, 480 (1953).
FISH, W. R., and R. I. DORFMAN: Science **91**, 388 (1940a).
— — J. biol. Chem. **135**, 349 (1940b).
— — J. biol. Chem. **140**, 83 (1941).
— — J. biol. Chem. **143**, 15 (1942).
— B. N. HORWITT and R. I. DORFMAN: Science **97**, 227 (1943).
FORBES, T. R., and J. ZANDER: Recent Progr. Hormone Res. **13**, 379 (1956).
FORCHIELLI, E.: The in vitro metabolism of a typical adrenocortical steroid (11-desoxycortisol). Doctoral Thesis, Boston Univ., Boston, Mass. (1956).

Forchielli, E., and R. I. Dorfman: J. biol. Chem. **223**, 443 (1956).
— H. Rosenkrantz and R. I. Dorfman: J. biol. Chem. **215**, 713 (1955).
— K. Brown-Grant and R. I. Dorfman: Steroid Δ^4-hydrogenases of rat liver. Proc. Soc. exp. Biol. (N. Y.) **99**, 594 (1958).
Fotherby, K.: Biochem. J. **67**, 705 (1957).
— A. Colas, S. M. Atherden and G. F. Marrian: Biochem. J. **66**, 664 (1957).
Franksson, C., C. A. Gemzell and U. S. von Euler: J. clin. Endocr. **14**, 608 (1954).
Frantz, I. D., jr., and N. L. R. Bucher: J. biol. Chem. **206**, 471 (1954).
Fredrickson, D. S., R. E. Peterson and D. Steinberg: Science **127**, 704 (1958).
Fried, J., and E. F. Sabo: J. Amer. chem. Soc. **75**, 2273 (1953).
— R. W. Thoma, J. R. Gerke, J. E. Herz, M. N. Donin and D. Perlman: J. Amer. chem. Soc. **74**, 3962 (1952).
— — D. Perlman, J. E. Herz and A. Borman: Recent Progr. Hormone Res. **11**, 157 (1955).
Fukushima, D. K., and T. F. Gallagher: J. biol. Chem. **226**, 725 (1957a).
— — J. biol. Chem. **229**, 85 (1957b).
— K. Dobriner and T. F. Gallagher: J. biol. Chem. **206**, 845 (1954a).
— H. L. Bradlow, K. Dobriner and T. F. Gallagher: J. biol. Chem. **206**, 863 (1954b).
— A. D. Kemp, R. Schneider, M. B. Stokem and T. F. Gallagher: J. biol. Chem. **210**, 129 (1954c).
— N. S. Leeds, H. L. Bradlow, T. H. Kritchevsky, M. B. Stokem and T. F. Gallagher: J. biol. Chem. **212**, 449 (1955).
— C. D. Meyer, E. Ashworth and T. F. Gallagher: Fed. Proc. **15**, 257 (1956).
Gallagher, T. F.: Cancer Res. **17**, 520 (1957).
— D. K. Fukushima, M. C. Barry and K. Dobriner: Recent Progr. Hormone Res. **6**, 131 (1951).
— A. Kappas, H. Spencer and D. Lazlo: Science **124**, 487 (1956).
Ganis, F. M., and M. Radakovich: Fed. Proc. **17**, 226 (1958).
— L. R. Axelrod and L. L. Miller: J. biol. Chem. **218**, 841 (1956).
Gardner, L. I., and A. A. Tice: Helv. paediat. Acta **12**, 147 (1957).
Gautschi, F., and K. Bloch: J. Amer. chem. Soc. **79**, 684 (1957).
Gemzell, C. A.: Acta endocr. (Kbh.) **11**, 221 (1952).
Geyer, G.: Verh. d. Ges. inn. Med. **62**, 476 (1956).
Giroud, C. J. P., M. Saffran, A. V. Schally, J. Stachenko and E. H. Venning: Proc. Soc. exp. Biol. (N. Y.) **92**, 855 (1956).
— J. Stachenko and P. Piletta: In: A. F. Muller and C. M. O'Conner, Eds., Aldosterone. Boston: Little, Brown 1958.
Glick, J. H., jr.: Endocrinology **60**, 368 (1957).
Gold, N. I., E. Singleton, D. A. Macfarlane and F. D. Moore: J. clin. Invest. **37**, 813 (1958).
Goldstein, M., M. Gut and R. I. Dorfman: Biochim. biophys. Acta **38**, 190 (1960).
Gomez-Maestro, D. V.: Acta ginec. (Mad.) **2**, 573 (1951).
Gomez-Mont, F., and D. Berliner: Gac. méd. Méx. **83**, 413 (1953).
Gould, R. G.: In: R. P. Cook, Ed., Cholesterol, chemistry, biochemistry and pathology, 209—235. New York: Academic Press 1958.
— and G. Popjak: Biochem. J. **66**, 51 P (1957).
Gray, C. L., and F. Bischoff: Amer. J. Physiol. **180**, 279 (1955).
— M. A. S. Green, J. J. Holness and J. B. Lunnon: Lancet **1955 II**, 1067.
— — — — J. Endocr. **14**, 146 (1956).
Greenwood, F. C., and R. D. Bulbrook: J. Endocr. **13**, XXXIII (1956).
de Hachen, D. P.C.: Rev. Soc. argent. Biol. **32**, 23 (1956).
Hagopian, M.: An in vitro study of estrogen metabolism. Master of Science Thesis, Clark University, Worcester, Mass. (1955).
— and L. Levy: Biochim. biophys. Acta. **30**, 641 (1958).
Haines, W. J.: Recent Progr. Hormone Res. **7**, 255 (1952).
Halkerston, I. D. K., J. Hillman, D. Palmer and A. Rundle: J. Endocr. **13**, 433 (1956).
Hardy, J. D., and M. D. Turner: Amer. J. Med. **22**, 967 (1957).
Harman, R. E., E. A. Ham, J. DeYoung, N. G. Brink and L. H. Sarett: J. Amer. chem. Soc. **76**, 5035 (1954).
Hartmann, M., and F. Locher: Helv. chim. Acta **18**, 160 (1935).
Haussler, E. P.: Helv. chim. Acta **17**, 531 (1934).
Hayano, M., and R. I. Dorfman: Arch. Biochem. **36**, 237 (1952).
— — J. biol. Chem. **201**, 175 (1953).
— — Arch. Biochem. **50**, 218 (1954).
— — Arch. Biochem. **55**, 289 (1955).

HAYANO, M., M. C. LINDBERG, M. WIENER, H. ROSENKRANTZ and R. I. DORFMAN: Endocrinology **55**, 326 (1954).
— — R. I. DORFMAN, J. E. H. HANCOCK and W. VON E. DOERING: Arch. Biochem. **59**, 529 (1955).
— N. SABA, R. I. DORFMAN and O. HECHTER: Recent Progr. Hormone Res. **12**, 79 (1956a).
— A. SAITO, D. STONE and R. I. DORFMAN: Biochim. biophys. Acta **21**, 380 (1956b).
— M. GUT, R. I. DORFMAN, O. K. SEBEK and D. H. PETERSON: J. Amer. chem. Soc. **80**, 2336 (1958).
HAYDAR, N. A., J. C. LAIDLAW, W. J. REDDY and G. W. THORN: J. clin. Endocr. **15**, 858 (1955).
HAYNES, R. C., JR.: Fed. Proc. **17**, 376 (1958).
— and L. BERTHET: J. biol. Chem. **225**, 115 (1957).
— K. SAVARD and R. I. DORFMAN: J. clin. Endocr. **12**, 972 (1952).
— — — J. biol. Chem. **207**, 925 (1954).
— S. KORITZ and F. PERON: J. biol. Chem. **234**, 1421 (1959).
HEARD, R. D. H., and M. M. HOFFMAN: J. biol. Chem. **138**, 651 (1941a).
— — J. biol. Chem. **141**, 329 (1941b).
— and A. F. MCKAY: J. biol. Chem. **131**, 371 (1939).
— W. S. BAULD and M. M. HOFFMAN: J. biol. Chem. **141**, 709 (1941).
— M. M. HOFFMAN and G. E. MACK: J. biol. Chem. **155**, 607 (1944).
— P. H. JELLINCK and V. J. O'DONNELL: Endocrinology **57**, 200 (1955).
— E. G. BLIGH, M. C. CANN, P. H. JELLINCK, V. J. O'DONNELL, B. G. RAO and J. L. WEBB: Recent Progr. Hormone Res. **12**, 45 (1956).
HECHTER, O.: Biogenesis of adrenal cortical hormones. In: E. RALLI, Ed., Trans. 3rd Macy Conference on Adrenal Cortex, 115. 1951.
— Ciba Found. Coll. Endocr. **7**, 161 (1953).
— In: R. P. COOK, Ed., Cholesterol, chemistry, biochemistry and pathology, 309—347. New York: Academic Press 1958.
— R. P. JACOBSEN, R. W. JEANLOZ, H. LEVY, C. W. MARSHALL, G. PINCUS and V. J. SCHENKER: J. Amer. chem. Soc. **71**, 3261 (1949).
— V. SCHENKER and G. PINCUS: Prog. 118th Meet. Amer. chem. Soc., 29C (1950).
— A. ZAFFARONI, R. P. JACOBSEN, H. LEVY, R. W. JEANLOZ, V. SCHENKER and G. PINCUS: Recent Progr. Hormone Res. **6**, 215 (1951).
HELLMAN, L., R. S. ROSENFELD and T. F. GALLAGHER: J. clin. Invest. **33**, 142 (1954a).
— H. L. BRADLOW, J. ADESMAN, D. K. FUKUSHIMA, J. L. KULP and T. F. GALLAGHER: J. clin. Invest. **33**, 1106 (1954b).
HENRY, R., M. THERENET and P. JARRIGE: Bull. Soc. chim. Biol. **34**, 897 (1952).
HERNANDO, L., J. CRABBÉ, E. J. ROSS, W. J. REDDY, A. E. RENOLD, D. H. NELSON and G. W. THORN: Metabolism **6**, 518 (1957).
HERTZ, R., W. TULLNER and M. J. ALLEN: Proc. Soc. exp. Biol. (N. Y.) **77**, 480 (1951).
— — J. A. SHRICKER, F. G. DHYSE and L. F. HALLMAN: Recent Progr. Hormone Res. **11**, 119 (1955).
— J. A. PITTMAN and M. M. GRAFF: J. clin. Endocr. **16**, 705 (1956).
HERZOG, H. L., A. NOBILE, S. TOLKSDORF, W. CHARNEY, E. B. HERSHBERG, P. L. PERLMAN and M. M. PECHET: Science **121**, 176 (1955).
HIRSCHMANN, H.: J. biol. Chem. **136**, 483 (1940).
— J. biol. Chem. **150**, 363 (1943).
— and F. HIRSCHMANN: J. biol. Chem. **157**, 601 (1945).
— — J. biol. Chem. **167**, 7 (1947).
— — Fed. Proc. **8**, 206 (1949).
— — J. biol. Chem. **187**, 137 (1950).
— and O. WINTERSTEINER: J. biol. Chem. **122**, 303 (1937/1938).
HOFFMAN, M. M., V. E. KAZMIN and J. S. L. BROWNE: J. biol. Chem. **147**, 259 (1943).
HOFFMAN, M. N., and M. H. LOTT: J. Amer. chem. Soc. **73**, 878 (1951).
HOLNESS, N. J., J. B. LUNNON and C. H. GRAY: J. Endocr. **14**, 138 (1956).
HOLTZ, A. H.: Nature (Lond.) **174**, 316 (1954).
HOLZBAUER, M.: J. Physiol. (Lond.) **139**, 306 (1957).
— and M. VOGT: J. Physiol. (Lond.) **138**, 460 (1957).
HORWITT, B. N., R. I. DORFMAN, R. A. SHIPLEY and W. R. FISH: J. biol. Chem. **155**, 213 (1944).
HÜBENER, H. J.: Recent Progr. Hormone Res. **12**, 146 (1956).
— D. K. FUKUSHIMA and T. F. GALLAGHER: J. biol. Chem. **220**, 499 (1956).
HUDSON, P. B., and M. E. LOMBARDO: J. clin. Endocr. **15**, 324 (1955).
HUIS IN'T VELD, L. G., and E. DINGEMANSE: Acta brev. neerl. Physiol. **16**, 9 (1948).
HUME, D. M., and D. H. NELSON: Personal communication (1958).

INGLE, D. J., D. F. BEARY and A. PURMALIS: Endocrinology **53**, 221 (1953).
ISHIZUKA, N., M. MAEYAMA, T. IDE and M. KAWATA: Med. J. Osaka Univ. **7**, 953 (1957); Chem. Abstr. **52**, 2116 (1958).
ISSELBACHER, K. J.: Recent Progr. Hormone Res. **12**, 134 (1956).
— and J. AXELROD: J. Amer. chem. Soc. **77**, 1070 (1955).
JAILER, J. W., J. J. GOLD, R. VANDE WIELE and S. LIEBERMAN: J. clin. Invest. **34**, 1639 (1955).
JAYLE, M. F., L. G. HUIS IN'T VELD, E. BAULIEU and O. CREPY: Acta endocr. (Kbh.) **21**, 115 (1956).
JEANLOZ, R. W., H. LEVY, R. P. JACOBSEN, O. HECHTER, V. SCHENKER and G. PINCUS: The chemical transformations of progesterone by adrenal perfusion. Abst. 118th Meeting Amer. chem. Soc., Chicago, Ill. p. 29 C (1950).
— — — — — — J. biol. Chem. **203**, 453 (1953).
JOHNSON, B. B., A. H. LIEBERMANN and P. J. MULROW: J. clin. Invest. **36**, 757 (1957).
JOHNSON, D. F., E. HEFTMANN and J. J. BUNIM: Proc. Soc. exp. Biol. (N. Y.) **94**, 291 (1957).
JOHNSON, J. E., jr., A. R. REMMERS, G. W. CLAYTON and R. GREGORY: Texas Rep. Biol. Med. **15**, 381 (1957).
JOHNSON, R. H., and W. J. HAINES: Science **116**, 456 (1952).
JOHNSTON, J. D., and K. BLOCH: J. Amer. chem. Soc. **79**, 1145 (1957).
DE JONGH, S. E., S. KOBER and E. LAQUEUR: Biochem. Z. **240**, 247 (1931).
KAHNT, F. W., C. MEYSTRE, R. NEHER, E. VISCHER and A. WETTSTEIN: Experientia (Basel) **8**, 422 (1952).
— R. NEHER and A. WETTSTEIN: Helv. chim. Acta **38**, 1237 (1955a).
— — — Experientia (Basel) **11**, 446 (1955b).
KARRER, P.: Organic chemistry. Amsterdam: Elsevier 1938.
KELLER, M.: J. clin. Endocr. **16**, 1075 (1956).
KEMP, A. D., A. KAPPAS, I. I. SALAMON, F. HERLING and T. F. GALLAGHER: J. biol. Chem. **210**, 123 (1954).
KLEIN, R., C. PAPADATOS, J. FURTUNATO and C. BYERS: J. clin. Endocr. **15**, 215 (1955a).
— — — — J. clin. Endocr. **15**, 943 (1955b).
— J. FORTUNATO, Z. LARON and C. PAPADATOS: J. clin. Endocr. **17**, 256 (1957).
KLOPPER, A., J. A. STRONG and L. R. COOK: J. Endocr. **15**, 180 (1957).
KNOWLTON, A. I.: Med. Clin. N. Amer. **1952**, 721.
KOCH, F. C.: Physiol. Rev. **17**, 153 (1937).
KORENCHEVSKY, V., and M. DENNISON: J. Path. Bact. **41**, 323 (1935).
KORITZ, S. B., and F. G. PERON: J. biol. Chem. **230**, 343 (1958).
— — and R. I. DORFMAN: J. biol. Chem. **226**, 643 (1957).
KRAYCHY, S., and T. F. GALLAGHER: J. Amer. chem. Soc. **79**, 754 (1957a).
— — J. biol. Chem. **229**, 519 (1957b).
KRITCHEVSKY, D., and I. GRAY: Experientia (Basel) **7**, 183 (1951).
KUIZENGA, M. H., and G. F. CARTLAND: Endocrinology **24**, 526 (1939).
— — Endocrinology **27**, 647 (1940).
KUSHINSKY, S.: The isolation of products from bovine adrenal perfusion with progesterone and progesterone-4-C^{14}. Doctoral Thesis, Boston University, Boston, Mass. (1955).
LAMEDICA, G., and G. MISSALE: Arch. E. Maragliano Pat. Clin. **13**, 257 (1957); Chem. Abstr. **51**, **13**, 226 (1957).
— L. CORREALE and R. BONANNI: Arch. Stud. Fisiopat. Ricambio **21**, 41 (1957).
LANGDON, R. G., and K. BLOCH: J. Amer. chem. Soc. **74**, 1869 (1952).
— — J. biol. Chem. **200**, 135 (1953a).
— — J. biol. Chem. **200**, 129 (1953b).
LARSON, P. S., G. R. HENNIGAR, J. K. FINNEGAN, R. B. J. SMITH and H. B. HAAG: J. Pharmacol. exp. Ther. **115**, 408 (1955).
LEDOGAR, J. A., and H. W. JONES, JR.: Science **112**, 536 (1953).
LEVIN, L.: J. biol. Chem. **158**, 725 (1945).
LEVITZ, M., J. R. SPITZER and G. H. TWOMBLY: J. biol. Chem. **222**, 981 (1956).
LEVY, H., and S. KUSHINSKY: Arch. Biochem. **55**, 290 (1955).
— R. W. JEANLOZ, R. P. JACOBSEN, O. HECHTER, V. SCHENKER and G. PINCUS: Proc. 118th Meeting Amer. chem. Soc. p. 29C (1950).
— — C. W. MARSHALL, R. P. JACOBSEN, O. HECHTER, V. SCHENKER and G. PINCUS: J. biol. Chem. **203**, 433 (1953).
— — R. P. JACOBSEN, O. HECHTER, V. SCHENKER and G. PINCUS: J. biol. Chem. **211**, 867 (1954).
LEVY, L. K., M. HAGOPIAN and C. A. FISH: Private communication (1958).
LIDDLE, G. W., F. C. BARTTER, L. E. DUNCAN, JR., J. K. BARBER and C. DELEA: J. clin. Invest. **34**, 949 (1955).

LIEBERMAN, S., and K. DOBRINER: J. biol. Chem. **166**, 773 (1946).
— — Recent Progr. Hormone Res. **3**, 71 (1948).
— — Proc. Amer. chem. Soc. 117th Meeting, p. 19C (1950).
— — B. R. HILL, L. L. FIESER, and C. P. RHOADS: J. biol. Chem. **172**, 263 (1948).
— D. K. FUKUSHIMA and K. DOBRINER: J. biol. Chem. **182**, 299 (1950).
— B. PRAETZ, P. HUMPHRIES and K. DOBRINER: J. biol. Chem. **204**, 491 (1953).
LIPSCHITZ, W. L., and E. BUEDING: J. biol. Chem. **129**, 333 (1939).
LITTLE, B., and E. ROSSI: J. clin. Endocr. **61**, 109 (1957).
LLAURADO, J. G., R. NEHER and A. WETTSTEIN: Clin. chim. Acta **1**, 236 (1956).
LOKE, K. H., and G. F. MARRIAN: Biochim. biophys. Acta **27**, 213 (1958).
LOMBARDO, M. E., and P. B. HUDSON: J. biol. Chem. **229**, 181 (1957).
— E. ROITMAN and P. B. HUDSON: J. clin. Endocr. **16**, 1283 (1956).
LONG, C. N. H.: Recent Progr. Hormone Res. **1**, 99 (1947).
LONGWELL, B. B., A. E. REIF, E. HANSBURY and H. S. BAKER: Fed. Proc. **14**, 246 (1955).
— — — Fed. Proc. **15**, 303 (1956).
LUETSCHER, J. A., JR.: Recent Progr. Hormone Res. **12**, 175 (1956).
— and B. J. AXELRAD: Proc. Soc. exp. Biol. (N. Y.) **87**, 650 (1954).
— R. NEHER and A. WETTSTEIN: Experientia (Basel) **10**, 456 (1954).
— — — Experientia (Basel) **12**, 22 (1956).
LYNN, W. S., JR.: Fed. Proc. **15**, 305 (1956).
— and R. BROWN: Biochim. biophys. Acta **21**, 403 (1956).
— E. STAPLE and S. GURIN: J. Amer. chem. Soc. **76**, 4048 (1954).
— — — Fed. Proc. **14**, 783 (1955).
LYON, I., R. P. GEYER and L. D. MARSHALL: J. biol. Chem. **217**, 757 (1956).
MACH, R. S., and A. F. MULLER: Schweiz. med. Wschr. **87**, 406 (1957).
MAJNARICH, J. J., and R. N. DILLON: Arch. Biochem **49**, 247 (1954).
MALLASSIS, D., E. E. BAULIEU, S. WEINMANN, O. CREPY and M. F. JAYLE: C. R. Soc. Biol. (Paris) **151**, 447 (1957).
MANCERA, O., A. ZAFFARONI, B. A. RUBIN, F. SONDHEIMER, G. ROSENKRANZ and C. DJERASSI: J. Amer chem. Soc. **74**, 3711 (1952).
MANN, K. M., F. R. HANSON and P. W. O'CONNELL: Fed. Proc. **14**, 251 (1955a).
— — — H. ANDERSON, M. P. BRUNNER and J. N. KARNEMAAT: Appl. Microbiol. **3**, 14 (1955b).
MARCUS, P. I., and P. TALALAY: J. biol. Chem. **218**, 661 (1956).
MARKER, R. E.: J. Amer. chem. Soc. **60**, 2442 (1938a).
— J. Amer. chem. Soc. **60**, 2931 (1938b).
— J. Amer. chem. Soc. **61**, 944 (1939a).
— J. Amer. chem. Soc. **61**, 1287 (1939b).
— and O. KAMM: J. Amer. chem. Soc. **59**, 1373 (1937).
— and E. J. LAWSON: J. Amer. chem. Soc. **60**, 2928 (1938).
— and E. ROHRMANN: J. Amer. chem. Soc. **60**, 1565 (1938).
— — J. Amer. chem. Soc. **61**, 2719 (1939a).
— — J. Amer. chem. Soc. **61**, 3476 (1939b).
— O. KAMM and R. V. MCGREW: J. Amer. chem. Soc. **59**, 616 (1937).
— E. J. LAWSON, E. L. WITTLE and H. M. CROOKS: J. Amer. chem. Soc. **60**, 1559 (1938).
MARKS, B. H., M. ALPERT and F. A. KRUGER: Endocrinology **63**, 75 (1958).
MARRIAN, G. F.: Biochem. J. **23**, 1090 (1929).
— J. Soc. chem. Ind. (Lond.) **49**, 515 (1930).
— Nature (Lond.) **154**, 19 (1944).
— and W. S. BAULD: Biochem. J. **59**, 136 (1955).
— K. H. LAKE, E. J. D. WATSON and M. PANATTONI: Biochem. J. **66**, 60 (1957).
MASON, H. L.: J. biol. Chem. **172**, 783 (1948).
— and E. J. KEPLER: J. biol. Chem. **160**, 255 (1945a).
— — J. biol. Chem. **161**, 235 (1945b).
— — J. biol. Chem. **167**, 73 (1947).
— and J. J. SCHNEIDER: J. biol. Chem. **184**, 593 (1950).
— and H. S. STRICKLER: J. biol. Chem. **171**, 543 (1947).
— C. S. MEYERS and E. C. KENDALL: J. biol. Chem. **114**, 613 (1936).
— W. M. HOEHN, B. F. MCKENZIE and E. C. KENDALL: J. biol. Chem. **120**, 719 (1937).
— — and E. C. KENDALL: J. biol. Chem. **124**, 459 (1938).
MATTOX, V. R.: Proc. Mayo Clin. **30**, 180 (1955).
— and H. L. MASON: J. biol. Chem. **223**, 215 (1956).
MAYER, C.: Ann. Endocr. (Paris) **13**, 675 (1952).
MCCULLAGH, E. P., and H. A. TRETBAR: J. clin. Endocr. **18**, 134 (1958).
MCGINTY, D. A., G. N. SMITH, M. L. WILSON and C. S. WORREL: Science **112**, 506 (1950).

MEISTER, P. D., S. H. EPPSTEIN, D. H. PETERSON, H. C. MURRAY, H. M. LEIGH, A. WEINTRAUB and L. M. REINEKE: Abst. 123rd Meeting Amer. Chem. Soc., Los Angeles, Calif., p. 5C (1953a).
— D. H. PETERSON, H. C. MURRAY, S. H. EPPSTEIN, L. M. REINEKE, A. WEINTRAUB and H. M. LEIGH: J. Amer. chem. Soc. **75**, 55 (1953b).
— — — G. B. SPERO, S. H. EPPSTEIN, A. WEINTRAUB, L. M. REINEKE and H. M. LEIGH: J. Amer. chem. Soc. **75**, 416 (1953c).
— L. M. REINEKE, R. C. MEEKS, H. C. MURRAY, S. H. EPPSTEIN, H. M. L. OSBORN, A. WEINTRAUB and D. H. PETERSON: J. Amer. chem. Soc. **76**, 4050 (1954).
MEYER, A. S.: Private communication (1952).
— Biochim. biophys. Acta **24**, 1435 (1955a).
— Experientia (Basel) **11**, 99 (1955b).
— R. W. JEANLOZ and G. PINCUS: J. biol. Chem. **203**, 463 (1953).
— O. G. RODGERS and G. PINCUS: Acta endocr. (Kbh.) **16**, 293 (1954).
— M. HAYANO, M. C. LINDBERG, M. GUT and O. G. RODGERS: Acta endocr. (Kbh.) **18**, 148 (1955).
MEYSTRE, C., E. VISCHER and A. WETTSTEIN: Helv. chim. Acta **37**, 1548 (1954).
— — — Helv. chim. Acta **38**, 381 (1955).
MIGEON, C. J.: J. biol. Chem. **218**, 941 (1956).
— and J. E. PLAGER: J. biol. Chem. **209**, 767 (1954).
— J. BERTRAND, P. E. WALL, R. S. STEMPFEL and H. PRYSTOWSKY: Ciba Found. Coll. Endocrin. **11**, 338 (1957).
MILLER, A. M., and R. I. DORFMAN: Endocrinology **42**, 174 (1949).
— — and M. MILLER: Endocrinology **46**, 105 (1950).
— — and E. H. SEVRINGHAUS: Endocrinology **38**, 19 (1946).
— H. ROSENKRANTZ and R. I. DORFMAN: Endocrinology **53**, 238 (1953).
MILLER, L. L., and L. R. AXELROD: Metabolism **3**, 438 (1954).
MILLS, G. T., R. ONDARZA and E. E. B. SMITH: Biochim. biophys. Acta **14**, 159 (1954).
MOXHAM, A., and J. D. N. NABARRO: J. clin. Pathol. **9**, 351 (1956).
MUELLER, G. C., and G. RUMNEY: J. Amer. chem. Soc. **79**, 1004 (1957).
MUNSON, P. L., T. F. GALLAGHER and F. C. KOCH: J. biol. Chem. **152**, 67 (1944).
MURRAY, H. C., and D. H. PETERSON: U. S. Patent 2,602,769 July 8 (1952).
— — U. S. Patent 2,649,400 (1953a).
— — U. S. Patent 2,649,402 (1953b).
— — U. S. Patent 2,702,809 (1955a).
— — U. S. Patent 2,703,326 (1955b).
— — U. S. Patent 2,703,806 (1955c).
NADEL, E. M., and S. BURSTEIN: J. nat. Cancer Inst. **17**, 213 (1956).
— and J. J. SCHNEIDER: Endocrinology **51**, 5 (1952).
— S. BURSTEIN and R. I. DORFMAN: Arch. Biochem. **61**, 144 (1956).
NEHER, R., and A. WETTSTEIN: Helv. chim. Acta **39**, 2062 (1956).
NELSON, A. A., and G. WOODWARD: Arch. Path. (Chicago) **48**, 387 (1949).
NICHOLS, J., and L. I. GARDNER: J. Lab. clin. Med. **37**, 229 (1951).
— and G. HENNIGAR: Endocrinology **61**, 226 (1957).
— and H. L. SHEEHAN: Endocrinology **51**, 362 (1952).
— C. DAVIS and H. D. GREEN: Endocrinology **53**, 541 (1953).
NICOLAIDES, N., O. K. REESS and R. G. LANGDON: J. Amer. chem. Soc. **77**, 1535 (1955).
NISHIKAWA, M., and H. HAGIWARA: Chem. pharm. Bull. **6**, 226 (1958).
O'DONNELL, P. W., K. M. MANN, E. D. NIELSON and F. R. HANSON: Appl. Microbiol. **3**, 16 (1955).
OERTEL, G. W., and K. EIK-NES: Fed. Proc. **17**, 400 (1958a).
— — Acta endocr. (Kbh.) **28**, 293 (1958b).
OKINAKA, S., K. NAKAO, M. NISHIKAWA, M. IKEDA and H. IBAYASHI: Tôhoku J. exp. Med. **56** 153 (1952).
OLSON, J. A., and K. BLOCH: Fed. Proc. **15**, 323 (1956).
OPPENAUER, R.: Z. physiol. Chem. **270**, 97 (1941).
PASCHKIS, K. E., A. CANTAROW, A. E. RAKOFF, L. HANSEN and A. A. WALKING: Proc. Soc. exp. Biol. (N. Y.) **55**, 127 (1954).
PEARLMAN, W. H.: Endocrinology **30**, 270 (1942).
— and R. H. DE MEIO: J. biol. Chem. **179**, 1141 (1949).
— and G. PINCUS: J. biol. Chem. **144**, 569 (1942).
— — and N. T. WERTHESSEN: J. biol. Chem. **142**, 649 (1942).
— A. E. RAKOFF, K. E. PASCHKIS, A. CANTAROW and A. A. WALKLING: J. biol. Chem. **173**, 175 (1948).

PEDERSON, R. L., J. A. CAMPBELL, J. C. BABCOCK, S. H. EPPSTEIN, H. C. MURRAY, A. WEINTRAUB, R. C. MEEKS, L. M. MEISTER, L. M. REINEKE and D. H. PETERSON: J. Amer. chem. Soc. **78**, 1512 (1956).
PERLMAN, D., E. TITUS and J. FRIED: J. Amer. chem. Soc. **74**, 2126 (1952).
— E. O'BRIEN, A. P. BAYAN and R. B. GREENFIELD, JR.: J. Bact. **69**, 347 (1955).
PERON, F., and R. I. DORFMAN: J. biol. Chem. **223**, 877 (1956).
— — Arch. Biochem. **67**, 490 (1957).
— and S. B. KORITZ: J. biol. Chem. **223**, 256 (1958).
PETERSON, D. H.: In: S. A. WAKSMAN, Ed., Perspectives and horizons in microbiology, p. 121. New Brunswick, N. J.: Rutgers Univ. Press 1955.
— and H. C. MURRAY: J. Amer. chem. Soc. **74**, 1871 (1952).
— — S. H. EPPSTEIN, L. M. REINEKE, A. WEINTRAUB, P. D. MEISTER and H. M. LEIGH: J. Amer. chem. Soc. **74**, 5933 (1952).
— S. H. EPPSTEIN, P. D. MEISTER, B. J. MAGERLEIN, H. C. MURRAY, H. M. LEIGH, A. WEINTRAUB and L. M. REINEKE: J. Amer. chem. Soc. **75**, 412 (1953a).
— A. H. NATHAN, P. D. MEISTER, S. H. EPPSTEIN, H. C. MURRAY, A. WEINTRAUB, L. M. REINEKE and H. M. LEIGH: J. Amer. chem. Soc. **75**, 419 (1953b).
— P. D. MEISTER, A. WEINTRAUB, L. M. REINEKE, S. H. EPPSTEIN, H. C. MURRAY and H. M. L. OSBORN: J. Amer. chem. Soc. **77**, 4428 (1955).
PETERSON, R. E., and J. B. WYNGAARDEN: Ann. N. Y. Acad. Sci. **61**, 297 (1955).
— — J. clin. Invest. **35**, 552 (1956).
— R. HERTZ and H. A. LUBS: Proc. Soc. exp. Biol. (N. Y.) **94**, 421 (1957a).
— C. E. PIERCE and B. KLIMAN: Arch. Biochem. **70**, 614 (1957b).
— — — Personal communication (1958).
PFIFFNER, J. J., and H. B. NORTH: J. biol. Chem. **132**, 459 (1940).
— — J. biol. Chem. **139**, 855 (1941).
PICHA, G. M., F. J. SAUNDERS and D. M. GREEN: Science **115**, 704 (1952).
PINCUS, G.: The biosynthesis of adrenal steroids. In: P. KALLOS, Ed., Progress in allergy, Vol. IV, 199—226. Basel-New York: Karrer 1954.
— and E. B. ROMANOFF: Ciba Found. Coll. Endocrin. **8**, 97 (1955).
PLAGER, J. E., and L. T. SAMUELS: Fed. Proc. **11**, 383 (1952).
— — J. biol. Chem. **211**, 21 (1954).
PLANTIN, L.-O., and G. BIRKE: Acta Endocr. **19**, 8 (1955).
— E. DICZFALUSY and G. BIRKE: Nature (Lond.) **179**, 421 (1957).
PLATE, W. P.: Acta Endocr. **11**, 119 (1952).
POPJAK, G.: Arch. Biochem. **48**, 102 (1954).
— Ann. Rev. Biochem. **27**, 533 (1958).
— G. D. HUNTER, and T. H. FRENCH: Biochem. J. **54**, 238 (1953).
PRELOG, V., and J. FÜHRER: Helv. chim. Acta **28**, 583 (1945).
PRICE, T. D., and D. RITTENBERG: J. biol. Chem. **185**, 449 (1950).
RABINOWITZ, J. L.: J. Amer chem. Soc. **76**, 3037 (1954).
— and S. GURIN: Biochim. biophys. Acta **10**, 345 (1953).
— — J. biol. Chem. **208**, 307 (1954a).
— — J. Amer. chem. Soc. **76**, 5168 (1954b).
— F. DITURI, F. COBEY and S. GURIN: Fed. Proc. **14**, 760 (1955).
RAO, B. G., and R. D. H. HEARD: Arch. Biochem. **66**, 504 (1957).
REDDY, W. J., N. A. HAYDAR and J. C. LAIDLAW: J. clin. Endocr. **15**, 891 (1955).
REICH, E., and A. L. LEHNINGER: Biochim. biophys. Acta **17**, 136 (1955).
REICHSTEIN, T.: Helv. chim. Acta **19**, 29 (1936a).
— Helv. chim. Acta **19**, 223 (1936b).
— Helv. chim. Acta **19**, 1107 (1936c).
— Helv. chim. Acta **20**, 953 (1937).
— and J. v. EUW: Helv. chim. Acta **21**, 1197 (1938).
— — Helv. chim. Acta **22**, 1222 (1939).
— and K. GATZI: Helv. chim. Acta **21**, 1185 (1938a).
— — Helv. chim. Acta **21**, 1490 (1938b).
RENOLD, A. E., J. CRABBÉ, L. HERMANDO-AVENDANO, D. H. NELSON, E. J. ROSS, K. EMERSON, JR., and G. W. THORN: New Engl. J. Med. **256**, 16 (1957).
RICHARDSON, E. M., J. C. TOUCHSTONE and F. C. DOHAN: Fed. Proc. **13**, 118 (1954).
— — — J. clin. Invest. **34**, 285 (1955).
— H. BULASCHENKO and H. C. DOHAN: Fed. Proc. **17**, 132 (1958).
RITTENBERG, D., and R. SCHOENHEIMER: J. biol. Chem. **121**, 235 (1937).
ROMANOFF, E. B., P. HUDSON and G. PINCUS: J. clin. Endocr. **13**, 1546 (1953).
ROMANOFF, L. P., J. SEELYE, R. RODRIGUEZ and G. PINCUS: J. clin. Endocr. **17**, 434 (1957).
ROSEMBERG, E., G. ROSENFELD, F. UNGAR and R. I. DORFMAN: Endocrinology **58**, 708 (1956).

ROSENFELD, G.: Amer. J. Physiol. **182**, 57 (1955a).
— Amer. J. Physiol. **183**, 272 (1955b).
— and W. D. BASCOM: J. biol. Chem. **222**, 565 (1956).
— F. UNGAR, R. I. DORFMAN and G. PINCUS: Endocrinology **56**, 24 (1955).
— E. ROSEMBERG, F. UNGAR and R. I. DORFMAN: Endocrinology **58**, 255 (1956).
ROSENFELD, R. S., L. HELLMAN, W. J. CONSIDINE and T. F. GALLAGHER: J. biol. Chem. **208**, 73 (1954).
ROSENKRANTZ, H., and R. I. DORFMAN: Proc. 19th Int. physiol. Congress (Montreal), p. 713 (1953).
ROSSELET, J.-P., M. FURMAN, S. LIEBERMAN and J. W. JAILER: Science **120**, 788 (1954a).
— L. OVERLAND, J. W. JAILER and S. LIEBERMAN: Helv. chim. Acta **37**, 1933 (1954b).
— J. W. JAILER and S. LIEBERMAN: Fed. Proc. **14**, 272 (1955).
— — — J. biol. Chem. **225**, 977 (1957).
ROY, A. B.: Biochim. biophys. Acta **15**, 300 (1954).
— Biochem. J. **62**, 41 (1956).
RUBIN, B. L.: J. biol. Chem. **227**, 917 (1957).
— and R. I. DORFMAN: Proc. Soc. exp. Biol. (N. Y.) **91**, 585 (1956).
— — Endocrinology **61**, 601 (1957).
— — and M. MILLER: Endocrinology **51**, 463 (1952).
RUDNEY, H.: Fed. Proc. **13**, 286 (1954).
— and T. G. FARKAS: Fed. Proc. **14**, 757 (1955).
RUMNEY, G.: Fed. Proc. **15**, 343 (1956).
RYAN, K. J.: Fed. Proc. **15**, 344 (1956).
— and L. L. ENGEL: Endocrinology **52**, 277 (1953a).
— — Endocrinology **52**, 287 (1953b).
— — J. Amer. chem. Soc. **78**, 2654 (1956).
SABA, N., and O. HECHTER: Fed. Proc. **14**, 775 (1955).
— — and D. STONE: J. Amer. chem. Soc. **76**, 3862 (1954).
SAFFRAN, M., and M. J. BAYLISS: Endocrinology **52**, 140 (1953).
SALAMON, I. I., and K. DOBRINER: J. clin. Endocr. **12**, 967 (1952).
SALHANICK, H. A., and D. L. BERLINER: J. biol. Chem. **227**, 583 (1957).
SALVADORI, B., U. LEONE e I. CASSANG: Quad. Clin. ostet. ginec. **11**, 261 (1956).
SAMUELS, L. T.: Ciba Found. Coll. Endocrin. **7**, 176 (1953).
— and M. L. HELMREICH: Endocrinology **58**, 435 (1956).
— — M. B. LASATER and H. REICH: Science **113**, 490 (1951).
— H. BROWN, K. EIK-NES, F. H. TYLER and O. V. DOMINGUEZ: Ciba Found. Coll. Endocrin. **11**, 221 (1957).
SANDBERG, A. A., and W. R. SLAUNWHITE, JR.: J. clin. Invest. **35**, 1331 (1956).
— — Proc. Soc. exp. Biol. (N. Y.) **96**, 658 (1957).
— D. H. NELSON, E. M. GLENN, F. H. TYLER and L. T. SAMUELS: J. clin. Endocr. **13**, 1445 (1953).
— K. EIK-NES, L. T. SAMUELS and F. H. TYLER: J. clin. Invest. **33**, 1509 (1954).
— W. R. SLAUNWHITE, JR., and H. N. ANTONIADES: Recent Progr. Hormone Res. **13**, 209 (1957).
SANDERMANN, W., and H. STOCKMANN: Naturwissenschaften **43**, 581 (1956).
SAVARD, K.: Ciba Found. Coll. Endocrin. **11**, 252 (1957).
— and R. I. DORFMAN: Rev. canad. Biol. **13**, 495 (1954).
— and S. H. GOLDFADEN: Fed. Proc. **13**, 288 (1954).
— A. A. GREEN and L. A. LEWIS: Endocrinology **47**, 418 (1950).
— W. KOLFF and A. C. CORCORAN: Endocrinology **50**, 366 (1952).
— S. BURSTEIN, H. ROSENKRANTZ and R. I. DORFMAN: J. biol. Chem. **202**, 717 (1953).
— E. BAGNOLI and R. I. DORFMAN: Fed. Proc. **13**, 289 (1954).
— R. I. DORFMAN, B. BAGGETT and L. L. ENGEL: Fed. Proc. **15**, 346 (1956a).
— — — — L. M. LISTER and F. L. ENGEL: J. clin. Endocr. **16**, 970 (1956b).
— — — — J. clin. Endocr. **16**, 1629 (1956c).
— K. ANDREC, B. W. L. BROOKSBANK, C. REYNERI and R. I. DORFMAN: J. biol. Chem. **231**, 765 (1958).
SAYERS, G., M. A. SAYERS, T. Y. LIANG and C. N. H. LONG: Endocrinology **38**, 1 (1946).
SCHACTER, B., and G. F. MARRIAN: J. biol. Chem. **126**, 663 (1938).
SCHILLER, S., A. M. MILLER, R. I. DORFMAN, E. H. SEVRINGHAUS and E. P. MCCULLOUGH: Endocrinology **27**, 262 (1945).
SCHNEIDER, J. J.: J. biol. Chem. **194**, 337 (1952).
— J. Amer. chem. Soc. **75**, 2024 (1953).
— and P. M. HORSTMANN: J. biol. Chem. **191**, 327 (1951).
— and M. L. LEWBART: J. biol. Chem. **222**, 787 (1956).

SCHNEIDER, J. J., and H. L. MASON: J. biol. Chem. **172**, 771 (1948).
— M. LEWBART, P. LEVITAN and S. LIEBERMAN: J. Amer. chem. Soc. **77**, 4184 (1955).
SCHÖNBAUM, E., M. K. BIRMINGHAM and M. SAFFRAN: Canad. J. Biochem. **34**, 527 (1956).
SCHROEDER, W., K. D. VOIGT, H. VAN DER WERTH and I. BECKMANN: Acta endocr. (Kbh.) **14**, 12 (1953).
SCHWENK, E., G. J. ALEXANDER, C. A. FISH and T. H. STOUDT: Fed. Proc. **14**, 752 (1955a).
— — T. H. STOUDT and C. A. FISH: Arch. Biochem. **55**, 274 (1955b).
SELYE, H., and J. B. COLLIP: Endocrinology **20**, 667 (1936).
— — and D. L. THOMSON: Proc. Soc. exp. Biol. (N. Y.) **32**, 1377 (1935).
SERCHI, G.: Chimica (Milan) 8, 9 (1953).
SHEEHAN, H. L., V. K. SUMMERS and J. NICHOLS: Lancet **1953 I**, 312.
SHORT, R. V.: Ciba Found. Coll. Endocrin. **11**, 362 (1957).
SHRIEFERS, H., W. KORUS and W. DIRSCHERL: Acta endocr. (Kbh.) **26**, 331 (1957).
SHULL, G. M., and D. A. KITA: J. Amer. chem. Soc. **77**, 763 (1955).
— J. L. SARDINAS and J. B. ROUTIEN: Can. Patent 507,009 (1954).
— — and J. W. DAVISSON: U. S. Patent 2,702,812 (1955).
SILBER, R. H.: Clin. Chem. **1**, 234 (1955).
SIMPSON, S. A., J. F. TAIT and I. E. BUSH: Lancet **1952 ii**, 226.
— — A. WETTSTEIN, R. NEHER, J. v. EUW, O. SCHINDLER and T. REICHSTEIN: Helv. chim. Acta **37**, 1163 (1954).
SIPERSTEIN, M. D., and I. L. CHAIKOFF: Fed. Proc. **14**, 767 (1955).
SLAUNWHITE, W. R., jr., and A. A. SANDBERG: Arch. Biochem. **63**, 478 (1956).
— — J. clin. Endocr. **17**, 395 (1957).
SLUSHER, M. A., and S. ROBERTS: Endocrinology **61**, 98 (1957).
SMIDT, H., and H. STAUDINGER: Biochem. Z. **328**, 24 (1956).
SMITH, E. E. B., and G. T. MILLS: Biochim. biophys. Acta **13**, 386 (1954).
— A. MUNCH-PETERSEN and G. T. MILLS: Nature (Lond.) **172**, 1038 (1953).
SMITH, G. V. S., O. W. SMITH, M. N. HUFFMAN, D. W. MACCORQUODALE, S. A. THAYER and E. A. DOISY: J. biol. Chem. **130**, 431 (1939).
SOFFER, L. J.: Diseases of the endocrine glands, Ed. 2, p. 375. Philadelphia: Lea and Febiger 1956.
SOLOMON, S., A. L. LENZ, R. VANDE WIELE and S. LIEBERMAN: Abst. Am. Chem. Soc. 126th Meeting, New York, p. 29C (1954).
— P. LEVITAN and S. LIEBERMAN: Rev. canad. Biol. **15**, 282 (1956—1957).
SONDERHOFF, R., and H. THOMAS: Liebigs Ann. Chem. **530**, 195 (1957).
SOUTHCOTT, C. M., S. K. GANDOSSI, A. D. BARKER, H. E. BANDY, H. MCINTOSH and M. DARRACH: Canad. J. Biochem. **34**, 146 (1956a).
— H. E. BANDY, S. E. NEWSOM and M. DARRACH: Canad. J. Biochem. **34**, 913 (1956b).
STAPLE, E., W. S. LYNN, JR., and S. GURIN: J. biol. Chem. **219**, 845 (1956).
STEENBURG, R. W., R. LENNIHAM and F. D. MOORE: Ann. Surg. **143**, 180 (1956).
STEIGER, M., and T. REICHSTEIN: Helv. chim. Acta **21**, 546 (1938).
STEINBERG, D., and D. S. FREDRICKSON: Ann. N. Y. Acad. Sci. **64**, 579 (1956).
— — and J. AVIGAN: Proc. Soc. exp. Biol. (N. Y.) **97**, 784 (1958).
STIMMEL, B. F.: Fed. Proc. **17**, 317 (1958).
STITCH, S. R., and I. D. HALKERSTON: Nature (Lond.) **172**, 398 (1953).
STOKES, W. M., W. A. FISH and F. C. HICKEY: J. biol. Chem. **220**, 415 (1956).
STONE, D., and O. HECHTER: Arch. Biochem. **51**, 457 (1954).
— M. HAYANO, R. I. DORFMAN, O. HECHTER, C. R. ROBINSON and C. DJERASSI: J. Amer. chem. Soc. **77**, 3926 (1955).
STOREY, I. D. E., and G. J. DUTTON: Biochem. J. **59**, 279 (1955).
STRONG, J. A., J. B. BROWN, J. BRUCE, M. DOUGLAS, A. J. KLOPPER and J. A. LORAINE: Lancet **1956 ii**, 955.
STROUD, S. W.: J. Endocr. **1**, 201 (1939).
SUTHERLAND, E. S., and G. F. MARRIAN: Biochem. J. **41**, 193 (1947).
SWEAT, M. L., R. A. ALDRICH, C. H. DE BRUIN, W. L. FOWLKS, L. R. HEISELT and H. S. MASON: Fed. Proc. **15**, 367 (1956).
TALALAY, P.: Physiol. Rev. **37**, 362 (1957).
— and P. I. MARCUS: Nature (Lond.) **173**, 1189 (1954).
— — J. biol. Chem. **218**, 675 (1956).
— M. M. DOBSON, and D. F. TAPLEY: Nature (Lond.) **170**, 620 (1952).
TALIAFERRO, I., and L. LEONE: New Engl. J. Med. **257**, 855 (1957).
— F. COBEY and L. LEONE: Proc. Soc. exp. Biol. (N. Y.) **92**, 742 (1956).
TAMM, J., I. BECKMANN and K. D. VOIGHT: Acta endocr. (Kbh.) **27**, 403 (1958).
TAVORMINA, P. A., and M. H. GIBBS: J. Amer. chem. Soc. **78**, 6210 (1956).
— — and J. W. HUFF: J. Amer. chem. Soc. **78**, 4498 (1956).

Taylor, W.: Biochem. J. **66**, 58P (1957).
Tchen, T. T., and K. Bloch: J. biol. Chem. **226**, 921 (1957a).
— — J. biol. Chem. **226**, 931 (1957b).
Thoma, R. W., and J. Fried: U. S. Patent 2,801,251 July 30 (1957).
Thorn, G. W., A. E. Renold, A. Goldfien, D. H. Nelson, W. J. Reddy and R. Hertz: New Engl. J. Med. **254**, 547 (1956).
Tomkins, G. M.: J. biol. Chem. **218**, 437 (1956a).
— Recent Progr. Hormone Res. **12**, 125 (1956b).
— Abst. Amer. chem. Soc. 130th Meeting, p. 57C (1956c)
— Abst. Amer. chem. Soc. 131st Meeting, p. 4C (1957).
— and K. J. Isselbacher: J. Amer. chem. Soc. **76**, 3100 (1954).
— H. H. Sheppard and I. L. Chaikoff: J. biol. Chem. **203**, 701 (1953).
— P. J. Michael and J. F. Curran: Biochim. biophys. Acta **23**, 655 (1957).
— J. F. Curran and P. J. Michael: Fed. Proc. **17**, 323 (1958).
Touchstone, J. C., L. Glazer, D. Y. Cooper and J. M. Roberts: J. clin. Endocr. **15**, 382 (1955).
— H. Bulaschenko, E. M. Richardson and F. C. Dohan: J. clin. Endocr. **17**, 250 (1957).
Travis, R. H., and G. L. Farrell: Fed. Proc. **17**, 324 (1958).
Troop, R. C.: Fed. Proc. **17**, 415 (1958).
— Influence of gonadal hormones on the metabolism of cortisone. Endocrinology **64**, 671 (1959).
Tullner, W. W., M. M. Graff and R. Hertz: Endocrinology 58, 802 (1956).
Tyler, F. H., C. D. Schmidt, K. Eik-Nes, H. Brown and L. T. Samuels: J. clin. Invest. **33**, 1517 (1954).
Udaev, N. A., and K. V. Druzhinia: Probl. Endocr. Horm. Ther. (U.S.S.R.) **4**, 21 (1958).
Ulick, S., and S. Lieberman: J. Amer. chem. Soc. **79**, 6567 (1957).
Ungar, F., and B. R. Bloom: Biochim. biophys. Acta **24**, 431 (1957).
— and R. I. Dorfman: Science **115**, 115 (1952).
— — J. biol. Chem. **205**, 125 (1953).
— — J. Amer. chem. Soc. **76**, 1197 (1954).
— — J. clin. Endocr. **16**, 920 (1956).
— J. W. Davis, H. Rosenkrantz and R. I. Dorfman: J. biol. Chem. **207**, 375 (1954a).
— A. M. Miller and R. I. Dorfman: J. biol. Chem. **206**, 597 (1954b).
— M. Gut and R. I. Dorfman: J. biol. Chem. **224**, 191 (1957).
— E. B. Romanoff, R. I. Dorfman and G. Pincus: Unpublished (1958).
Venning, E. H.: Recent Progr. Hormone Res. **12**, 185 (1956).
— and J. S. L. Browne: Proc. Soc. exp. Biol. (N. Y.) **34**, 792 (1936).
— — Endocrinology **21**, 711 (1937).
— M. M. Hoffman and J. S. L. Browne: J. biol. Chem. **146**, 369 (1942).
— C. J. P. Giroud, I. Dyrenfurth and J. C. Bell: Canad. J. Biochem. **33**, 605 (1955).
Vermeulen, A.: Acta endocr. (Kbh.) **23**, 113 (1956a).
— J. clin. Endocr. **16**, 163 (1956b).
— Acta endocr. (Kbh.) **22**, 115 (1956c).
— Lancet **273**, 79 (1957).
— and E. Caspi: J. biol. Chem. **233**, 54 (1958).
Vischer, E., J. Scmidlin and A. Wettstein: Helv. chim. Acta **37**, 321 (1954).
Vliers, M., L. Demey, C. Heusghem and H. van Cauwenberge: C. R. Soc. Biol. (Paris) **149**, 2271 (1955).
Vogt, M.: J. Physiol. **103**, 317 (1944).
— J. Physiol. **130**, 601 (1955).
— Ciba Found. Coll. Endocrin. **11**, 193 (1957).
Voigt, K. D., and W. Schroeder: Naturwissenschaften **40**, 485 (1953).
— — Nature (Lond.) **176**, 599 (1955).
— — Acta endocr. (Kbh.) **21**, 343 (1956).
— — I. Beckmann and H. van der Werth: Acta endocr. (Kbh.) **14**, 1 (1953).
Wallace, E. Z., H. I. Wilverberg, A. C. Carter and L. S. Hersch: Proc. Soc. exp. Biol. (N. Y.) **95**, 805 (1957).
Wallach, S., H. Brown, E. Englert and K. Eik-Nes: J. clin. Endocr. **17**, 945 (1957).
Watson, E. J. D., and G. F. Marrian: Biochem. J. **61**, 24 (1955).
— — Biochem. J. **63**, 64 (1956).
Weichselbaum, T. E., and H. W. Margraf: J. clin. Endocr. **15**, 970 (1955).
Weinmann, S., E. E. Baulieu, A. Alfsen, B. Lisboa and M. F. Jayle: C. R. Soc. Biol. **151**, 454 (1957a).
— F. F. Demoisson, E. E. Baulieu and M. F. Jayle: C. R. Soc. Biol. **151**, 518 (1957b).
Werbin, H., and G. V. LeRoy: J. Amer. chem. Soc. **76**, 5260 (1954).
— — Fed. Proc. **14**, 303 (1955).

WERBIN, H., J. PLOTZ, G. V. LEROY and E. DAVIS: Fed. Proc. **16**, 346 (1957a).
— D. M. BERGENSTAL, R. G. GOULD and G. V. LEROY: J. clin. Endocr. **17**, 337 (1957b).
— I. L. CHAIKOFF and E. JONES: Fed. Proc. **17**, 333 (1958).
WEST, C. D., B. DAMAST and O. H. PEARSON: J. clin. Invest. **37**, 341 (1958).
WESTPHAL, U.: Hoppe-Seylers Z. physiol. Chem. **273**, 13 (1942).
WETTSTEIN, A.: Experientia (Basel) **10**, 397 (1954).
— Experientia (Basel) **11**, 465 (1955).
— and G. ANNER: Experientia (Basel) **10**, 397 (1954).
— H. NEHER and J. G. LLAURADO: Proc. Univ. Otago med. Sch. **34**, 7 (1956).
WIEST, W. G.: Fed. Proc. **17**, 335 (1958).
WINTERS, R. W., R. B. SCHULTZ and W. A. KREHL: Endocrinology **50**, 377 (1952a).
— — — Endocrinology **50**, 388 (1952b).
WINTERSTEINER, O., and J. J. PFIFFNER: J. biol. Chem. **111**, 599 (1935).
— — J. biol. Chem. **116**, 291 (1936).
— E. SCHWENK and B. WHITMAN: Proc. Soc. exp. Biol. (N. Y.) **32**, 1087 (1935).
WOODWARD, R., and K. BLOCH: J. Amer. chem. Soc. **75**, 2023 (1953).
WUERSCH, J., R. L. HUANG and K. BLOCH: J. biol. Chem. **195**, 439 (1952).
YATES, F. E., A. L. HERBST and J. URQUHART: Sex difference in rate of ring A reduction of Δ^4-3-keto-steroids in vitro by rat liver. Endocrinology **63**, 887—902 (1958).
ZABIN, I., and W. F. BARKER: J. biol. Chem. **205**, 633 (1953).
— and K. BLOCH: J. biol. Chem. **185**, 117 (1950a).
— — J. biol. Chem. **185**, 131 (1950b).
— — J. biol. Chem. **192**, 261 (1951a).
— — J. biol. Chem. **192**, 267 (1951b).
ZAFFARONI, A., and R. B. BURTON: Arch. Biochem. **42**, 1 (1953).
— O. HECHTER and G. PINCUS: J. Amer. chem. Soc. **73**, 1390 (1951).
— C. C. CAMPILLO, F. CARDOBA and G. ROSENKRANZ: Experientia (Basel) **11**, 219 (1955).
ZANDER, J., and H. SUMMER: Klin. Wschr. **32**, 529 (1954).
ZIMMERMAN, N. B., H. S. BLOCH, W. L. WILLIAMS, C. R. HITCHCOCK and B. HOELSCHER: Cancer (N.Y.) **9**, 940 (1956).
ZONDEK, B., and S. BURSTEIN: Endocrinology **50**, 419 (1952).

Chapter 4

Mineralocorticoid action of adrenocortical hormones*

By

ROBERT GAUNT and J. J. CHART

Introduction

All of the physiologically active steroids of the adrenal cortex affect the metabolism of water and electrolytes. The term "mineralocorticoids" has been used conventionally to designate those steroids such as deoxycorticosterone (DOC) and aldosterone whose known major effects *at low dosage* are on electrolyte and water metabolism. Such steroids, however, cannot be considered from a physiological standpoint without reference to the related actions of the "glucocorticoids," i.e., those compounds whose major recognized actions *at low doses* are on some aspect of organic metabolism. Therefore, this chapter will deal with mineralocorticoid actions of the adrenal steroids in general.

Any precise and final definition of the role of adrenal hormones in fluid and electrolyte metabolism is impossible at the present time and is unlikely to be achieved in the near future. The reason for this pessimistic outlook lies in the complexity of the mechanisms involved. All protoplasmic activity is conducted in an aqueous electrolyte medium. All metabolic activities affect and are affected by this medium. Therefore, the regulation of electrolyte and water metabolism is essentially as complex as that of life itself. If, for instance, one simplifies the problem greatly by considering only those factors affecting sodium and water excretion by the mammalian kidney, a huge number of interacting variables can be and have been demonstrated (H. SMITH 1957). These factors include such things as diet, body position, liver function, blood pressure, emotions, various specific and localized renal functions, as well as hormones.

The hormones involved include those of practically all endocrine glands: pancreas, thyroid, parathyroid, anterior and posterior pituitary, hypothalamus and gonads, as well as the several active steroids of the adrenal cortex. The different corticoids may often have antagonistic actions and thereby confuse the analysis of gross over-all adrenal effects. Thus the adrenal hormones constitute one important but not generally dominant element in this maze of homeostatic complexity. In this field the ideal experiment of allowing one influence to vary while keeping all others constant is difficult to achieve for the simple reason that the one variable will induce others to occur and often these may be unrecognized. Realization of the facts should promote humility in all investigators concerned.

Nevertheless, recognition of difficulties should not paralyze initiative. Despite the self-evident obstacles, the scientific endeavor of the last quarter century has

* This review was completed in 1959; a few selected references were added to galley proof in the spring of 1961.

added a great and useful body of knowledge concerning the action of mineralo-corticoids. This knowledge has permitted the development of practical therapies and comprehensible, although limited, physiological theory.

History

Anyone who approaches problems of adrenal physiology today must inevitably be impressed, if not overwhelmed, by the seemingly endless number of measurable effects of the cortical steroids. It is perhaps difficult for any beginner to understand that the 75 years of research following the pioneer work of ADDISON and BROWN-SÉQUARD did not establish definitely a single function of the adrenal cortex. STEWART (1924) wrote that the function of the adrenal cortex was "utterly unknown"; in 1930, despite the availability of much more bibliographic ammunition, BRITTON (1930) summarized the problem with the discouraging conclusion that "the very meagre knowledge of corticoadrenal function does not lead itself at the present time even to rational theorizing." The known and accepted facts by then were that the adrenal cortex was essential for life and that in view of the known cases of the adrenogenital syndrome, it must have some relation to reproductive processes. The many attempts that had been made to measure specific deficiencies after adrenalectomy resulted in a plethora of data but little enlightenment. Many measurable changes occurred in adrenal insufficiency, but rational correlation and interpretation of these events were not forthcoming.

These and related historical problems are considered in more detail elsewhere (GAUNT and EVERSOLE 1949), but suffice it to say here that subsequent progress depended upon the preparation in 1929 and 1930 of potent active extracts of adrenocortical tissue. This led in due time, and perhaps after undue confusion and debate, first to some comprehensible outlines of the spheres of activity of cortical hormones, and secondly to the isolation of specific adrenal steroids. With that background of accomplishment, subsequent progress was rapid.

Before 1930, however, *various laboratories provided data that serve as the historical basis of knowledge of the relation of the adrenal cortex to electrolyte metabolism.* Unfortunately, like much other pioneer work, this was ahead of its day and there was not adequate recognition of its significance. In 1913 GRADINESCO and SANTA (see 1937) saw a loss of plasma into tissues in adrenal insufficiency, and at later dates other evidences of hemoconcentration and failing renal functions were noted: increases in blood urea (GAILLARD 1914, PORAK and CHEBANIER 1914, MARSHALL and DAVIS 1916); rise in viscosity and concentration of blood (KELLAWAY and COWELL 1923, LUCAS 1926); rise in protein and total nitrogen (SWINGLE 1927); elevated total solids, potassium and magnesium (BAUMANN and KURLAND 1927). Specific alterations in renal function after adrenalectomy were implied by GAILLARD (1914) and definitely established by the critical work of MARSHALL and DAVIS (1916).

LUCAS (1926) and ROGOFF and STEWART (1926) found that blood chlorides were reduced after adrenalectomy, and SWINGLE and EISENMANN (1927) found the same to be true of sodium, p_H, CO_2 tension and serum bicarbonate. The work of KELLAWAY and COWELL (1923) and particularly of WYMAN and TUM SUDEN (1930) showed that low blood volume was a consequence of adrenal insufficiency.

The highly significant fact that replacement therapy with sodium salts was beneficial in adrenal insufficiency was established by STEWART and ROGOFF (1925, ROGOFF and STEWART 1928a). They extended the lives of adrenalectomized dogs by periodic intravenous infusions of Ringer's solution but interpreted the beneficial effects as being probably due to a flushing-out of toxins. The accumulation-of-toxin theory, which could always serve as a working hypothesis to explain the cause of death from any glandular deficiency, did little service in this as in other instances.

The work of MARINE and BAUMANN (1927) *should be recognized as a landmark in adrenal history*, but again it was relatively uninfluential. *They showed in systematic studies that the administration of electrolyte solutions tended to prevent the development of adrenal insufficiency and* specifically demonstrated *that the sodium ion was the responsible salutary factor.*

In the light of present knowledge, it is easy enough to take the sum total of the above-mentioned reports and piece together a syndrome consisting of *renal dysfunction, loss of sodium salts, accumulation of potassium salts, hemoconcentration, shock and death*, which can be prevented or ameliorated by sodium therapy. This *is the typical picture of adrenal insufficiency* in animals in which death is not prematurely precipitated by failures of carbohydrate metabolism (as in fasting) or by non-specific stress.

The syndrome was, however, not recognized in its entirety. This was due to the fact that knowledge of the nature and importance of electrolyte metabolism and the genesis and nature of shock-like conditions was in an immature state, that there was a general lack of interest in the problem, and also that there was a lack of widely accepted and convincing evidence that the deficiencies resulting from adrenal removal were specifically due to lack of adrenocortical hormones secreted into the blood. The fact that the adrenal *cortex*, and not the *medulla*, was primarily involved was not widely appreciated before 1930, although critical evidence on the point was available prior to that time (e. g., WHEELER and VINCENT 1917, HOUSSAY and LEWIS 1923). Furthermore, experimental technique in the field was not ideal in many centers — a fact to which STEWART (1924) called attention in his eloquent but caustic evaluations or devaluations of contemporary investigations.

Modern adrenocortical history opened in 1930. In that year, after previous efforts with less clear success by ROGOFF and by HARTMAN (reviewed by BRITTON 1930), adrenal extracts of unquestioned efficacy in adrenal substitution therapy were prepared in some quantity and made available for extensive testing by SWINGLE and PFIFFNER (1930) and by HARTMAN et al. (1930). The availability of such extracts opened new techniques of study, increased interest in the field and led to chemical identification and synthesis of the adrenal hormones.

Profiles of major corticoids having mineralocorticoid action

In subsequent sections the effects of different corticoids on different functions and processes will be considered. As an introduction, we list and discuss in this section the major known steroids involved and try to relate them in a general way, particularly from the historical point of view, to the problem of the control of electrolyte and water metabolism. The listing is mainly chronological rather than in terms of potency. The formulas of the active compounds are given on p. 205, Chapter 2.

A. Extracts of the adrenal cortex

Early studies on adrenal insufficiency and with whole lipide extracts of the adrenal cortex indicated that some component of the adrenal secretions tended to cause sodium retention and potassium excretion, to affect the distribution of body fluids and to prevent the development of circulatory collapse (cf. p. 528). This seemed to be related to the fact that life could also be maintained, albeit in a precarious manner, by giving a high sodium-low potassium diet. When individual steroids were separated from adrenal extracts, it became apparent that these affected mineral and water metabolism in different ways both from a quantitative and qualitative point of view.

B. Deoxycorticosterone (cortexone)

The compound which probably still best fits the vague definition of a "mineralo-corticoid" is deoxycorticosterone (DOC). STEIGER and REICHSTEIN (1937) first prepared it by partial synthesis; later REICHSTEIN and v. EUW (1938) found it in trace amounts in adrenal tissue. The failure of others for some time to confirm its presence in adrenal extracts led to widespread doubts that the substance was a hormone or even a natural product. More recently its presence has been demonstrated unequivocally in adrenal extracts (ZAFFARONI and BURTON 1951, MATTOX pers. comm.) and adrenal perfusates *in vitro* (HECHTER et al. 1951). FARRELL et al. (1954) have added the information that DOC is released into adrenal vein blood of dogs in amounts that are reduced by hypophysectomy and increased by ACTH. The amounts secreted are so small, however, that it is unlikely that they are able to induce any significant metabolic effects. It appears, therefore, that DOC is not a hormone in the general sense but that it is a natural product. What then is its function? Presumably it is a metabolic intermediate in the biosynthesis of other corticoids in adrenal tissue (see Fig. 15, p. 444); presumably also, some of this DOC leaks into the adrenal blood stream. KAHNT et al. (1955) and STACHENKO and GIROUD (1959a, b) have provided suggestive evidence that DOC may be a precursor of corticosterone and aldosterone. The latter authors report that the conversion is limited to the zona glomerulosa of the adrenal.

Unless some unanticipated significance is discovered to renew interest in DOC, it will have a waning role in adrenal studies because of the now obvious greater physiological importance of its biological counterpart, aldosterone, and because of the discovery of new synthetic corticoids of even greater potency. From the historical standpoint, however, it was of the utmost importance. For a decade, during which time whole cortical extracts were rare and expensive (as they still are), DOC was the only active adrenal steroid available cheaply and in unlimited quantities by synthetic methods. It served as a life-maintaining therapy in Addison's disease and as a major tool in virtually all adrenal biological studies. Later its properties were studied particularly in contrast to the glucocorticoids, cortisone and cortisol, when they became available.

Under most conditions, DOC is a highly active agent in causing sodium retention and potassium excretion (see Table 1, p. 524). *It generally also causes water retention* (exceptions discussed on p. 531ff. and 542ff.). It will not relieve effectively the deficiencies of water excretion in adrenal insufficiency. Its glucocorticoid actions are either weak or lacking. Use of DOC led to discovery of some of the important corticoid-overdosage syndromes (see p. 534; p. 547). These consisted of fluid retention and edema when cortexone was given in the absence of other corticoids; the diabetes insipidus-like syndrome in susceptible species, notably the dog; and the "DOC-hypertensive" syndrome in the rat.

C. Cortisol (hydrocortisone)

In man and most other mammalian species (the mouse, rat and rabbit are exceptions), cortisol is the steroid secreted in greatest amounts by adrenal glands (BUSH 1953). It is the major glucocorticoid and accounts for most of the adrenal effects on carbohydrate and protein metabolism, eosinophil count, suppression of ACTH release, etc. It is the natural prototype of the anti-inflammatory steroids.

Its effects on mineral metabolism are weak in relation to those of deoxycorticosterone or aldosterone, and are in part probably secondary to other major actions (see Table 1, p. 524). These effects are, furthermore, difficult to define because they are notably reversible depending upon physiological conditions and the species in which they are tested. *In chronic use in human therapy, cortisol and its esters undoubtedly cause sodium and water retention* with a concomitant and potentially dangerous renal loss of potassium (see p. 529). In some acute tests in animals, similar actions are sometimes obtained (SIMPSON and TAIT 1952). On the other hand, *in most acute tests* and in some human instances, *cortisol is a potent natriuretic agent, which can antagonize the opposite actions of deoxycorticosterone and aldosterone* (DESAULLES et al. 1955, GAUNT et al. 1957). In all circumstances it tends to enhance potassium excretion qualitatively like the classical mineralocorticoids. *It is the most potent of the natural corticoids in stimulating the rate of water diuresis and in repairing deficiencies of handling water loads in adrenal insufficiency* (CHART et al. 1956).

D. Cortisone

The effects of cortisone on water and mineral metabolism and other metabolic functions are essentially like those of cortisol. The differences when existent are largely or wholly quantitative, with cortisol generally being the more potent (LAIDLAW et al. 1955).

E. Corticosterone

The physiological role of corticosterone is somewhat enigmatic despite the fact that it was the first adrenal steroid for which the structural formula was established. Next to cortisol it is the one produced in greatest quantity by the adrenal of man and most other mammalian species. In the rat, rabbit and mouse, however, it is the major corticoid (BUSH 1953, HOFMANN 1956). *It combines in weaker form the biological properties of the most active glucocorticoids such as cortisol and of the mineralocorticoids such as aldosterone.* In man it has distinct but weak effects on carbohydrate metabolism and somewhat more potent effects on electrolyte metabolism (CONN et al. 1951, THORN et al. 1953). In the rat it exerts definite glucocorticoid effects but is weaker than cortisol or cortisone (KUIZENGA 1944); its effect on electrolyte metabolism is weaker than that of DOC (see Table 1, p. 524). It is a curious fact that the rat, one of the most rugged and adaptable mammals, apparently depends upon one of the weaker anti-stress steroids to support its highly successful struggle for existence.

F. 11-Dehydrocorticosterone

This compound, produced by the adrenals of most species in small amounts, has not been extensively studied. Its biological properties resemble those of corticosterone (THORN et al. 1953).

G. 11-Deoxycortisol (Reichstein's Compound S)

11-Deoxycortisol is secreted into adrenal-vein blood in relatively small quantity. It is a weak steroid exerting primarily mineralocorticoid effects (see Table 1, p. 524). There is no reason to believe from present evidence that it contributes significantly to the metabolic effects of the total adrenal secretion in normal individuals.

H. Aldosterone

Aldosterone is the major mineralocorticoid secreted by the adrenal gland. This fact makes the long and devious history of its recognition and identification a topic of particular interest.

During the two decades when most of the definitive work on identification and biological categorization of other adrenal hormones was in progress, the hormone we now call aldosterone was dimly recognized in a shadowy background. There was good reason to suspect its existence, but in tangible form it eluded all investigators. Recognition of its potential activities, without definite knowledge concerning its nature or certainty of its existence, was a disconcerting influence on those who tried to develop comprehensive theory concerning adrenal function.

In 1934 Wintersteiner et al. (also, Pfiffner et al. 1935), in attempting to isolate from adrenal cortical extracts the supposed one adrenocortical hormone, obtained several crystalline steroids, one of which was cortisone. They tested for the ability of the steroids to maintain life and well-being in adrenalectomized dogs on the basis of the Swingle dog test (see Table 3, page 527). The residue in their non-crystalline mother liquors — "the amorphous fraction" — was highly active: 200 to 400 dog units per mg. The activity was probably due largely to aldosterone. Since by the test used the crystalline derivatives were inactive at relatively high doses, they were regarded as being of no probable significance. Soon after, H. Mason et al. (1936), using Ingle's muscle work test, recognized the activity of cortisone. The comparison may be seen in Table 2, p. 527, and Table 3, p. 527.

All of the leading chemical groups working in the field subsequently obtained evidence that some activity was present in adrenal extracts not readily explained by the known compounds. In fact, it is yet to be demonstrated that any combination of known steroids will have all of the effects of whole cortical extracts (Ingle et al. 1952, Rowntree 1955). While such work was in progress, DOC had been synthesized and was recognized to exert actions, particularly on electrolyte metabolism, different from those of other adrenal steroids. It was thought by many to be a prototype of the electrolyte-regulating, life-maintaining substance in the *amorphous fraction* of adrenal extracts.

Most of the work with the *amorphous fraction* was done with preparations made and distributed by Kendall. Hartman and Spoor (1940) obtained a fraction from adrenal extracts, still chemically uncharacterized, which was designated the *sodium factor* because of its high and apparently specific sodium-retaining actions.

Work with amorphous fractions indicated that their activity in life-maintenance was greater than that of DOC and not attributable to other known steroids (Kendall et al. 1937, Cartland and Kuizenga 1939, Kuizenga 1944). The preparations of Kendall, Mason and associates contained about half of the life-maintaining activity found in whole adrenal extracts and maintained adrenalectomized dogs in doses of 1—2 μg/kg/day (H. Mason 1939). The amorphous fraction caused less growth inhibition and thymic and adrenal atrophy than corticosterone and its action on carbohydrate metabolism was weak (B. Wells and Kendall 1940a, b). It provided protection against the stress of cold (Kendall 1940) and water intoxication (Eversole et al. 1942), and supported lactation (Gaunt et al. 1942) in adrenalectomized animals. In doses of 5 μg/day it sustained weight gains and repaired deficient muscle work performance to some degree in adrenalectomized rats (Ingle and Kendall 1940).

In the meantime, Luetscher and his associates first demonstrated in 1950 (Luetscher 1956a) that there was some steroid-like substance accumulated in urine of patients with congestive heart failure or the nephrotic syndrome that induced sodium retention in assay animals. The existence of this substance was confirmed and its nature elaborated by others (CHART et al. 1951, CHART and SHIPLEY 1953, GORDON et al. 1954, SINGER and VENNING 1953, SINGER and WENER 1953, VENNING et al. 1954). It was eventually identified as aldosterone (LUETSCHER et al. 1955a, LLAURADO et al. 1956).

In 1950 SIMPSON and TAIT (review, 1955) began their classic studies that ended with the recognition and identification of aldosterone. They took advantage of the greatly expanded knowledge of the chemical and biological properties of steroids then known, including chromatographic methods of separation and identification, and of their own newly-devised sensitive bioassay method for the activity of mineralocorticoids. This assay was based on the ratio of urinary Na^{24} and K^{42} after administration of these substances to adrenalectomized rats (SIMPSON and TAIT, 1952) (see Table 1, p. 524). Briefly, the ratio is high in adrenalectomized animals, lower in intact animals or experimental animals treated with a mineralocorticoid. With this method they detected the fact that something appeared in their paper chromatograms along with cortisone which could not be cortisone itself. With this information, subsequent work led to the separation of a chemically-uncharacterized substance called electrocortin, the most potent mineralocorticoid then known.

At this point the work was carried on by two of the well-known steroid laboratories of the world: that led by REICHSTEIN at the University of Basle and by WETTSTEIN of CIBA Ltd., Basle. In collaboration with SIMPSON and TAIT, these groups soon succeeded in crystallizing electrocortin (SIMPSON et al. 1953), as did others (MATTOX et al. 1953), and establishing its formula (SIMPSON et al. 1954). It was then given the generic name *aldosterone*. Somewhat later its total synthesis was accomplished by SCHMIDLIN et al. (1955).

With these chemical developments, plus the previously-mentioned demonstration that the sodium-retaining substance excreted in excess amounts in numerous diseases was aldosterone, the advancing fronts of chemistry and biology were united. This resulted in a flood-tide of new work which is clarifying knowledge of the patho-physiology of mineralocorticoids (GROSS 1956a, LUETSCHER 1956b, MULLER and O'CONNOR 1958). What information is available on the secretory rate of aldosterone and of the major glucocorticoids for various species is summarized in Chapter 1, Table 14, p. 74. It should be reemphasized that these are probably maximal rates.

I. Synthetic corticoids

It is now apparent that alterations of the natural steroid hormone molecules can produce important and therapeutically useful changes in their biological properties. Current work in this field was initiated by the discovery that the 9α-halo-derivatives of cortisol were much more potent in certain respects than the parent steroid (FRIED et al. 1955). (See Chapter 2, p. 264ff.), The most widely used of these, 9α-fluorocortisol, is more potent in all respects than cortisol itself but its predominant characteristic is a high salt-retaining activity (Table 3, p. 527).

Subsequently, the 1-dehydro derivatives of cortisol and cortisone, namely, prednisolone and prednisone, were shown to have increased anti-inflammatory and reduced sodium-retaining properties, thus making them highly useful therapeutic tools (BUNIM et al. 1955). It is interesting that the 1-dehydro compounds, while failing to cause the sodium retention produced by the parent substances, do show enhanced natriuretic activity when tested under circumstances in which the glucocorticoids show such action (CHART et al. 1956).

Further chemical work has led to the production of mineralocorticoids more active than aldosterone itself, e. g., 2-methyl-9 α-fluorocortisol (BYRNES et al. 1956, LIDDLE et al. 1956). A rapidly-growing literature records the energetic efforts now being made to modify further the mineralocorticoid and other activities by chemical manipulation of natural steroid molecules (FRIED 1955, STAFFORD et al. 1955, LIDDLE et al. 1956, SWINGLE et al. 1956, FRIED 1957, KAGAWA and VAN ARMAN 1957, LYSTER et al. 1957, PERRINE et al. 1958).

J. A sodium-excreting hormone (?)

CRIGLER and co-workers (1952) suggested the possibility that the adrenal secreted a hormone which had the function of stimulating sodium excretion and thus sought to explain the adrenogenital salt-losing syndrome of infants (see Table 2, p. 584, Chapter 5). This hypothesis was re-explored and supported by PRADER et al. (1955). In an attempt to identify the substance concerned, NEHER et al. (1958) isolated, crystallized and synthesized 3 β,16 α-dihydroxyallopregnan-20-one from adrenal tissue. The biological actions of this compound are weak and inconstant (DESAULLES 1959, CHART, RENZI and GAUNT, unpublished) and not such as to suggest that it is an important hormone in the regulation of sodium metabolism.

Simultaneously, KLEIN et al. (1957, 1958) made a series of studies on a sodium-excreting substance in the urine of human patients. As yet, the active material involved is unidentified and its physiological significance uncertain. The problem is a difficult one because in test animals almost any of the glucocorticoids may enhance sodium excretion. Therefore assessment of the role and importance of any new substance is dependent upon chemical identification or clear-cut uniqueness of biological properties.

Adrenocortical inhibitors

A. Inhibition of the adrenal by corticoids

Control of adrenocortical function is achieved partly by the so-called "thermostat" or "feed-back" mechanism: an excess of corticoid suppresses the secretion of ACTH, and a deficit of corticoid allows the release of ACTH (see Chapter 10). In this respect, cortisol is the most potent natural corticoid, but derived compounds, such as 9 α-fluorocortisol, are more active still (RENOLD et al. 1955, LIDDLE et al. 1956). Practical use of this physiological mechanism is made with great success by using steroids to treat cases of the adrenogenital syndrome and in other conditions (see Chapters 5, 9). Aldosterone in large (and probably unphysiological) doses will inhibit ACTH release (RENZI et al. 1954). Attempts have been made to show the converse, namely, that large doses of potent ACTH-inhibiting cortisol derivatives will inhibit aldosterone secretion. In direct tests these attempts were unsuccessful in dogs (FARRELL et al. 1956). This might have been expected since it is now generally recognized that the secretion of aldosterone is largely but not completely independent of ACTH (see Chapter 1, p. 109 ff.). Nevertheless, while direct suppression of aldosterone by such means may be both unexpected and impossible, in certain hydropic diseases, characterized by secondary aldosteronism, steroids such as cortisone and prednisolone may be able to achieve a reduction in aldosterone output by some indirect means, presumably through an action on electrolyte and water balance (LUETSCHER 1956 a, VESIN et al. 1957 b).

B. Pharmacological inhibition — Miscellaneous drugs

Pharmacological inhibition of endocrine function, as a research tool or for therapeusis, has never been widely used except as applied to the thyroid and adrenal medulla. This is in contrast to the widespread use of drugs to control cardiovascular, neurological, and many other phenomena. Recent work has revealed, however, that there are a number of means by which some degree of pharmacological inhibition of the adrenal cortex may be achieved. An effective and non-toxic inhibitor of aldosterone secretion might be especially useful. As yet, nothing approaching an ideal drug has been discovered, but enough is known to suggest that systematic studies in the field may be fruitful. This subject is considered from other points of view in Chapter 1, p. 135ff., and Chapter 3, p. 455ff.

I. ACTH inhibitors

One approach to the problem is that of finding drugs that inhibit the production or release of ACTH (see Chapter 10). To a certain degree and under certain conditions, this can be accomplished, as, for instance, by the barbiturates and morphine (MUNSON and BRIGGS 1955), reserpine (H. WELLS et al. 1956), and chlorpromazine (HAMBURGER 1955). The conditions under which such demonstrations can be made are complex and the practical consequences are obscure. In general such drugs appear to inhibit, presumably at a central nervous system site, the neurohumoral mechanisms which cause stress-induced release of ACTH. They cannot be used to produce adrenal insufficiency but only to dampen the excess outpourings of corticoids that follow numerous stressful procedures. An especially clear-cut example is that in which reserpine has been shown to prevent an anxiety-induced rise of cortisol levels in the blood of monkeys (J. MASON and BRADY 1956). The reaction to such drugs may be biphasic. An initial injection of either morphine (MUNSON and BRIGGS 1955) or reserpine (H. WELLS et al. 1956, EGDAHL et al. 1956) will cause a large outpouring of ACTH. Continued administration, however, inhibits the ACTH-releasing effects of other stimuli. From a practical standpoint it is not clear whether such inhibition of cortisol output is beneficial to the organism or not.

The dual control centers for the regulation of aldosterone, the anterior pituitary and posterior commissure-pineal area of the brain stem, as described by FARRELL (1959a), make the inhibition of secretion of this potent sodium-retaining corticoid by suppression of the trophic hormones a more complex problem. Presumably it would be necessary to suppress both ACTH and "glomerulotropin" in order to achieve a significant depression in aldosterone secretion.

II. Inhibitors of corticoid synthesis

1. Amphenone B

The synthesis of amphenone B by ALLEN and CORWIN (1950) and its subsequent biological study by HERTZ et al. (1955) and others (cited by CHART et al. 1958) opened a new field. This compound reduces corticoid secretion apparently by inhibiting the enzymes responsible for the biosynthesis of the cortical hormones (ROSENFELD and BASCOM 1956, PETERSON et al. 1957). There is evidence that it acts especially to inhibit aldosterone secretion (RENOLD et al. 1957, MACH and MULLER 1957). Unfortunately, however, the substance is somewhat toxic, must be given in large doses, and has many other actions, including a thiouracil-like inhibition of the thyroid and anesthetic effects. Hence, it has only limited value as a therapeutic agent.

2. Pinacolones

Preliminary investigation has shown that numerous compounds can be made which have amphenone-like effects on the adrenal but do not necessarily have all of the effects of amphenone. One, a pinacolone, 2-methyl-1,2-bis-(3-pyridyl)-propanone-1 (Su 4885), has been studied in animals (CHART et al. 1958, JENKINS et al. 1958) and is known to be active in man (LIDDLE et al. 1958a, b, JENKINS et al. 1959). One complication in working on such compounds is that there is a high degree of species specificity in response to them. Despite early debate on the point (HERTZ et al. 1951, HEMING et al. 1952, HOGNESS et al. 1953), the direct evidence available indicates that amphenone B is an adrenal stimulant rather than a suppressant in the rat (VOGT 1956), whereas it is clearly a suppressant in the dog and man. The pinacolones lack many amphenone-like effects in the rat but have such effects in dogs when given in comparable dosage. They do, nevertheless, suppress corticoid production in rat as well as guinea pig and dog adrenals *in vitro* (SHEPPARD et al. 1958). Interestingly enough, glucocorticoid production seems to be inhibited by amphenone B in amphibia (SCHOTTÉ and CHRISTIANSEN 1957).

The major difference between the pinacolone (Su 4885) and amphenone, however, is that the former given in moderate dosage to dogs or men selectively inhibits the production of 11-hydroxylated steroids, e. g., cortisol, corticosterone and aldosterone. As a result, due to excess release of ACTH, there is a compensatory increase of 11-deoxycortisol and DOC (LIDDLE et al. 1958a, b, JENKINS et al. 1958, 1959). The latter two compounds have sodium-retaining actions and their presence may prevent sodium loss or even cause sodium retention despite suppression of aldosterone secretion. If, however, Su 4885 is given simultaneously with an ACTH-inhibiting steroid, then the excess production of 11-deoxycortisol and deoxycorticosterone is prevented and sodium diuresis ensues (COPPAGE et al. 1959).

3. DDD

It has been known for some time that large doses of the insecticide DDD can cause adrenal atrophy (NICHOLS and SHEEHAN 1952) and hypofunction (NICHOLS et al. 1953, J. BROWN et al. 1955) in dogs. It does not cause adrenal atrophy in rats and there is only indirect evidence (J. BROWN 1953) that it inhibits adrenal function in this species. Interest in DDD has recently been accelerated by the discovery made independently by NICHOLS and HENNIGAR (1957b) and CUETO and BROWN (1958) that the active adrenal inhibitor may be a contaminant of crude DDD, such as o,p'-DDD. The latter compound has definite adrenal-inhibiting activity in man, including an inhibition of aldosterone secretion (BERGENSTAL et al. 1960).

The mechanism of action of the DDD-like substances (see Chapter 1, p. 135) must differ fundamentally from that of the amphenone analogues. The DDD derivatives, for instance, cause adrenal atrophy, involving some block in response to ACTH (NICHOLS and HENNIGAR 1957a), whereas amphenone causes adrenal hypertrophy. J. BROWN (1960) has recently reviewed the whole problem.

4. Hexestrol

HOLZBAUER (1957) and HOLZBAUER and VOGT (1957) have shown that the administration of hexestrol and ethinyl estradiol to rats decreases both the adrenal stores of corticosterone and the ability to secrete this major steroid of the rat adrenal. This action is attributed to an inhibition of cholesterol synthesis by these estrogens, thus making unavailable the precursors for the synthesis of corticoids.

It is of interest to speculate whether the increased size of the adrenal in females of various mammalian species is a compensatory reaction to such action of estrogens (Chapter 1, p. 106; Chapter 3, p. 455). At least, the adrenal hypertrophy that follows hexestrol administration is attributed to low blood levels of corticoids (VOGT 1957). Amphenone B, itself an adrenal inhibitor in some species, antagonizes the effects of hexestrol in the rat (VOGT 1957).

III. Aldosterone antagonists

KAGAWA et al. (1957) and LIDDLE (1957) first reported that some steroidal spirolactones act as true aldosterone antagonists: they are inactive when given alone but antagonize the sodium-retaining action of aldosterone or DOC when the latter are present. The consequence of this action, i. e., the production of a sodium diuresis, can be manifested in clinical states of secondary aldosteronism. More recently, another compound of this series, spironolactone, has been introduced as a diuretic and given extensive clinical study. It seems to have distinct utility in certain types of renal, hepatic and cardiovascular disease.

Relation of mineralocorticoids to the life-maintaining function of the adrenal cortex

The adrenals are essential for life. The mineralocorticoids, and aldosterone in particular, are much the most potent of the adrenal steroids in maintaining life after adrenalectomy. DOC, for instance, given to adrenalectomized animals in small dosage under optimal conditions of diet and care, maintains health, vigor, growth, sleekness, reproductive capacity, and every superficial appearance of well-being in most laboratory animals. Evidence from short-term experiments suggests that the same is true of aldosterone (GROSS and MEIER 1951, SWINGLE et al. 1956). It is abundantly clear, however, that an animal maintained on minimal-effective doses of DOC is not normal, although superficially it may appear so. Deficiencies of many kinds, notably in response to stresses or in other situations

Table 1. *Comparison of the effects of cortical steroids on sodium retention, potassium excretion maintenance of life, and glycogen storage in adrenalectomized animals*.* (Compiled by GROSS 1956b)

	Na^{24}/K^{42} in urine, adx. rats[1]	Adx. rats[2]		Maintenance of life, adx. dogs[3]	Liver gluconeogenesis, adx. mice[4]
		Na^+ retention	K^+ excretion		
11-Deoxycorticosterone (cortexone)	100	100	100	100	1
11-Deoxycortisol .	8	3	1.5	7	10
Corticosterone . .	14	0.5	15	14	50
11-Dehydrocorticosterone	7	10.3	10	14	50
Cortisone.	6	0.01	14	2.5	100
Cortisol	8	0.03	24	10	155
Aldosterone . . .	10,000	2,500	500	2,500	30

* For mineralocorticoid activity, DOC is given the arbitrary action of 100; for glucocorticoid activity, cortisone serves as the standard.

[1] From TAIT et al. 1952.

[2] From DESAULLES and MEIER (see GROSS 1956b).

[3] From GROSS and MEIER 1951; GROSS and GYSEL 1954.

[4] From SCHULER et al. 1954.

in which the maintenance of homeostasis requires glucocorticoids, can be demonstrated readily by appropriate experimentation. The same is probably true in animals maintained chronically on minimal-effective doses of aldosterone but has not been demonstrated. It would seem to be a simple matter to define what these "life-maintaining" steroids do and do not do that accounts for their distinctive but limited actions. Unfortunately this is not the case, largely perhaps because there are few absolute qualitative distinctions between the biological actions of the adrenal steroids; the differences are generally but not invariably quantitative.

The one primary action of the mineralocorticoids with which their life-maintaining action can be most closely associated is that of regulating the metabolism of sodium and potassium. This is clearly illustrated in work tabulated from various sources by GROSS (1956b) in Table 1. The most important effector organ involved, but certainly not the only one, is the kidney. With some exceptions noted below, *adrenal insufficiency in various species is generally associated with a negative sodium and positive potassium balance that is partially preventable by a high sodium or a high sodium-low potassium diet.*

A. Survival after adrenalectomy — General

Adrenal insufficiency may follow two general patterns or courses to a fatal end. First, there may be at any time, including immediate post-operative stages after adrenalectomy, a sudden collapse associated with drastic fall in blood pressure and possible hypoglycemia, leading to a rapid death. Such events are not necessarily associated with any marked over-all electrolyte loss or accumulation and can be reasonably attributed primarily to a susceptibility to shock due to lack of glucocorticoids — and/or hypoglycemia due to self-imposed fasting (HARROP et al. 1935).

The second and more usual picture, under conditions of optimal surgical, post-operative and laboratory care, is for animals to recover after adrenalectomy to a fairly normal state and then gradually deteriorate over a period of one to two weeks.

The actual length of survival after adrenalectomy varies with species and from individual to individual within any species group. The great effects of the investigator's care and experience, in the absence of any specific therapy, on the survival of adrenalectomized animals is well illustrated by the results of pioneer studies by STEWART and ROGOFF in dogs and SWINGLE and co-workes in cats. During the 1920s these investigators studied many aspects of the adrenal insufficiency syndrome. In the course of their work they observed the survival of untreated "control" animals. Over a period of a few years, the survival time of their control animals increased greatly, as can be seen from the following compilation, made partly by BRITTON (1930) and added to by us:

Survival — Cats		*Survival — Dogs*	
Year	*Days*	*Year*	*Days*
1925/26	3—4	1925	5.8
1929	7	1928	9.9
1930	9.8	1929	11.9

It can be assumed that the evolution observed here was in the technique of the investigators and not in the biology of the species concerned! In general terms, the basis of such findings is now clear: adrenalectomized animals are highly susceptible to any sort of stress, including self-induced fasting, and an investigator who wishes to maintain his animals will learn, consciously or unconsciously, to avoid such stresses. These problems are of less consequence now that effective corticoids are available to administer after adrenalectomy for the prevention of post-surgical collapse. Subsequent withdrawal of replacement therapy produces an uncomplicated adrenal insufficiency. (See also Chapter 8.)

B. Adrenalectomy in the rat

At the present time the use of the rat in adrenal problems of many kinds is commonplace. The introduction of this species into adrenal experimentation, however, involved considerable misgiving during the 1920s and 1930s, and distinguished authorities decried the use of rats and questioned the validity of any results obtained with them. The argument revolved around the simple point of whether a rat lived or died after adrenalectomy. Recorded results were available to support all points of view (STEWART 1924, BRITTON 1930, GAUNT 1933).

While at the present time the usefulness of the rat in many types of investigations is universally accepted, as is the fact that adrenalectomy is generally followed by a fatal adrenal insufficiency, some of the problems which were at the root of the original arguments remain unsolved. Unless the usual diets are fortified with sodium, most young rats die within 5—10 days after adrenalectomy. If, however, older animals are used or if younger ones are maintained on and then withdrawn from some supportive therapy, survival may be prolonged indefinitely under optimal living conditions. Such instances of survival are generally attributed to the development of accessory adrenal cortices. In numerous instances, however, accessory tissue cannot be located. One clear fact is that there are hereditary differences among colonies of rats of such order that in some colonies most adults survive adrenalectomy and in others most do not (GAUNT 1933, GAUNT et al. 1935). These surviving animals, however, are fragile and do not withstand stress well, but will grow and maintain an appearance of well-being. There seem to be two alternative explanations for this phenomenon. Either the rat, given an optimal diet and living conditions, with its inherent great adaptability, can sometimes survive indefinitely without cortical hormones, or there is some as yet unidentified source of corticoids available to them. It has been suggested on the basis of equivocal evidence that previously undifferentiated peritoneal tissues in the adrenal region can form microscopic nests of corticoid-producing tissue if adequate time is available after adrenalectomy (MACFARLAND 1945). This possibility should be explored further by modern histochemical techniques. In unpublished studies made in 1934, the senior author saw tissues which might serve this function and hence he would be sympathetic to the working hypothesis that they exist.

C. Adrenalectomy in other mammalian species

The effects of adrenalectomy have been studied in numerous other mammalian species and the results in sloths, ground squirrels, mice, rabbits, marmots, guinea pigs and monkeys reviewed by HARTMAN and BROWNELL (1949). Elsewhere data are recorded on ferrets (GAUNT and HAYS 1938) and hamsters (SNYDER and WYMAN 1951). In general the pattern of events in these various species is similar. There are, however, exceptions to this rule. For instance, in the marmot and opossum, the changes in plasma sodium and potassium, generally characteristic of adrenal insufficiency in mammals, may not occur (SILVETTE and BRITTON 1936, HARTMAN et al. 1943). The ground squirrel shows an unusual ability to survive adrenalectomy if life is maintained with salt for a period after operation, due to the development of adrenal-like tissue in the ovary (GROAT 1944). This is reminiscent of the ability of luteinized ovaries to provide life-maintaining hormones after adrenalectomy in dogs (ROGOFF and STEWART 1928b, SWINGLE et al. 1937c) and ferrets (GAUNT and HAYS 1938).

The results of adrenal insufficiency in man as well as animals are surveyed in Chapter 8.

D. Effects of different corticoids on survival

The potency of any particular steroid in maintaining life in adrenalectomized animals will vary with the species and the conditions of care, strain, and diets used in any particular laboratory. Furthermore, results will differ greatly if one adds the criterion of growth to that of survival as a measure of effectiveness. All active corticoids will probably maintain life after adrenalectomy — with the mineralocorticoids being the most effective. In addition, it is common experience that the mineralocorticoids support essentially normal growth. On the other hand, the strong glucocorticoids, as one manifestation of their catabolic action, fail to support normal growth, or even may cause weight loss while sustaining life (GAUNT 1954). No comprehensive quantitative data are available concerning the effects of aldosterone and the new derived steroids on survival of adrenalectomized rats. KUIZENGA (1944), however, has summarized data on survival and muscle work performance of longer-known corticoids in adrenalectomized rats (Table 2).

Table 2. *Companion of activities of corticoids on survival and work capacity in adrenalectomized rats* (KUIZENGA 1944)

Compound	Units per mg	
	Survival test	Work test
Deoxycorticosterone	40	0.1
Cortisol	10	6.3
Cortisone	10	5.0
Corticosterone	8	2.4
11-Deoxycortisol	8	0.1
11-Dehydrocorticosterone	6	1.7

Excellent data on a number of steroids are available in dogs. It is clear that aldosterone is the most potent life-maintainer of the natural steroids (GROSS and MEIER 1951, SWINGLE et al. 1956), and that 2-methyl-9α-fluorocortisol is the most potent of the known derived steroids. The relative potencies of these and other steroids, according to the data of SWINGLE et al. (1956, 1957), are given in Table 3.

Table 3. *Relative potency of some naturally-occurring and chemically-modified adrenal steroids in the maintenance of adrenalectomized dogs.* Compiled from SWINGLE et al. (1956) and SWINGLE et al. (1957)

Compound	No. dogs tested	MM dose/day[1]	Potency, DOC = 1
2-Methyl-9α-fluorocortisol	3	1.5 μg	166.0
Aldosterone	4	6.25—12.5 μg	20—40.0
1-Dehydro-9α-fluorocortisol	4	6.2—12.5 μg	20—40.0
Fluoro-corticosterone acetate	4	6.8—13.7 μg	18—36.0
Fluoro-cortisol acetate	3	27.5 μg	9.1
Chloro-cortisol acetate	5	55.0—110.0 μg	2.3—4.5
Fluoro-11β-hydroxyprogesterone	2	150.0 μg	1.67
Deoxycorticosterone (DOC)	4	125.00—250.0 μg	1.0
1-Dehydrodeoxycorticosterone	4	250.0 μg	1.0
Fluoro-11β, 17α-dihydroxyprogesterone	2	300.0 μg	0.83
Cortisone	4	5.0 mg	0.05
Cortisol	4	5.0 mg	0.05
1-Dehydrocortisone	4	20.0 mg	0.0125
1-Dehydrocortisol	4	20.0 mg	0.0125

[1] MM = minimum maintenance dose — amount of steroid, given daily, necessary to maintain dogs free of gross signs of adrenal insufficiency and with normal blood pressure and blood chemistry.

When adrenalectomized dogs were kept on an intake of 19 meq. of sodium daily, LICHTLEN (1958) found that maintenance was achieved with 10 mg cortisol, 40 mg cortisone or 60 mg prednisone. Such maintenance was associated, however, particularly when prednisone was used, with the development of a diabetes insipidus-like state (cf. p. 534).

Less systematic studies in human subjects suggest that such data are directly applicable to man. This does not mean, however, that the most potent steroid is the *best* therapy in adrenal insufficiency. Life-maintaining potency is clearly associated directly with sodium-retaining potency, but survival with mineralocorticoids may be effected without obtaining simultaneous glucocorticoid actions which are desirable or even essential in meeting the stressful exigencies of life. There is reason to believe that the best replacement therapy in adrenal insufficiency is still whole adrenal cortical extract (INGLE et al. 1952, ROWNTREE 1955). This suggests that a complex mixture of crystalline corticoids might be therapeutically desirable, but to our knowledge it has not been tried.

E. Maintenance with sodium salts

In most mammalian species, the only recognized effective substitute for steroids in the maintenance of life after adrenalectomy is the administration of sodium salts. This subject is discussed subsequently (p. 530).

The adrenal cortex and mineral metabolism

The work of LOEB (1932, 1933, LOEB et al. 1933), while anticipated to some degree by that of MARINE and BAUMANN (1927), ushered in the modern study of the action of the adrenals on mineral metabolism and established for the scientific record and in public consciousness its physiological importance. They showed *in an Addisonian patient*, supplemented by data on adrenalectomized dogs, that *reduced blood levels of sodium, chloride and bicarbonate were associated with renal loss of these ions, and that replacement therapy with sodium chloride sustained life.* In contrast, *the blood levels of potassium were elevated in adrenal insufficiency.* At the same time HARROP and co-workers were studying independently electrolyte and fluid balance in adrenalectomized dogs and confirmed and extended the work of LOEB's group. They, too, recognized that an important feature of the adrenal insufficiency syndrome was the shock-like state resulting from a renal loss of sodium and, also, that sodium replacement therapy was beneficial (HARROP et al. 1933a, 1933b).

By such observations it was established and widely-recognized that at least one adrenal function was to regulate electrolyte metabolism. Indeed, some subsequent authors felt that the adrenal problem had in essence been solved: It was supposed that there was only one cortical hormone, the action of which on the renal handling of sodium salts accounted for all adrenocortical functions. Such a simple concept was not long to endure. In fact, it had its vigorous dissenters from the beginning. BRITTON and SILVETTE (1932) were detecting the now well-known effects of cortical hormones on carbohydrate metabolism and insisting that such effects constituted the "prepotent function" of the adrenal cortex. They and others pointed to the salient fact that the alterations in carbohydrate metabolism and other aspects of the adrenal insufficiency syndrome were ill-explained by renal loss of electrolytes. The chemical work identifying more than one adrenal hormone and the demonstration that these new hormones had different and in some instances antagonistic effects led to the demise of any idea that all adrenal function could be explained in terms of any single, simple action of one substance.

A. Adrenalectomy and electrolyte balance

When adrenals are removed or replacement therapy is withdrawn in adrenalectomized dogs, a negative sodium, and generally, but less consistently, a positive potassium balance appear (LOEB et al. 1933, HARROP et al. 1933a, ANDERSON and JOSEPH 1939). At death, the levels of plasma sodium may be depleted markedly. Both in adrenal insufficiency (LOEB et al. 1933, HARROP et al. 1933a) and after withdrawal of replacement therapies (LUFT et al. 1954, SWINGLE et al. 1958), the corticoids generally have greater quantitative effects on sodium than chloride levels in body fluids, although usually the changes are in the same direction. PULLMAN and MCCLURE (1954) found that DOC did not influence chloride excretion in man unless the sodium intake was high enough for a measurable sodium retention to be induced. Sodium excretion was responsive to this steroid, however, when chloride intake was low.

The extent of the elevation of potassium is variable but may reach double normal plasma values. After adrenalectomy, rats may differ somewhat from the dog and man and show negative balances of potassium, calcium, magnesium, phosphorus, chloride and nitrogen as well as sodium (RUBIN and KRICK 1936); this potassium loss may be the basis for the originally unexpected observation that after adrenalectomy rats show an increased appetite for potassium as well as sodium and other ions (C. RICHTER and ECKERT 1938). The changing plasma levels of sodium under the influence of adrenocortical hormones is in part the result of altered renal functions but, as discussed in Subsection J (p. 535ff.), also the result of altered internal distribution of this ion.

In general, magnesium behaves like potassium in response to varying levels of corticoids, since plasma levels rise in adrenal insufficiency (BAUMANN and KURLAND 1927) and are reduced by excess corticoids, e. g., in primary aldosteronism in man (MADER and ISERI 1955) or by corticoid administration in animals (WOODBURY et al. 1950). The hypothesis has been advanced that these changes in magnesium metabolism are causally related to the effects of corticoids on carbohydrate metabolism (HINGERTY 1957).

Associated with the natriuresis of adrenal insufficiency there may be a relatively large urine volume, hemoconcentration and alterations in extra- and intracellular water and electrolytes, varying in different tissues and species (WYMAN and TUM SUDEN 1930, SWINGLE et al. 1933, HARRISON and DARROW 1938, DARROW et al. 1939).

Potassium. When it became evident that the whole syndrome of adrenal insufficiency could not be explained simply in terms of the role of the cortical hormones on sodium and water metabolism, attention was focused on the significance of the concurrent changes in potassium metabolism. HASTINGS and COMPERE (1931) had shown that serum potassium of the dog was elevated after adrenalectomy, as did later workers. HARROP (1936a) studied systematically in dogs the relation between sodium and potassium in cycles of adrenal insufficiency and in therapy with cortical extract and sodium salts. He noted the marked susceptibility of adrenalectomized animals to the toxic action of potassium. Subsequently ZWEMER's group (ZWEMER and TRUSZKOWSKI 1937) advanced the hypothesis that the primary function of the adrenal hormones was to regulate potassium metabolism. Since that time, the detailed relationship between the corticoids and potassium and sodium have been extensively studied and certain general conclusions have emerged, the basis of which are considered in more detail later:

(1) The susceptibility of adrenalectomized animals (TRUSZKOWSKI and DUSZYNSKA 1940, LOWENSTEIN and ZWEMER 1943) and Addisonian patients to excess potassium is well established and can serve as the basis of testing procedures (COLLINS 1956).

(2) Potassium may accumulate in the blood in adrenal insufficiency but this is not invariable, and the fatal consequences of adrenalectomy are not necessarily dependent on it. When it does occur, however, it may be a primary or contributory cause of specific signs of adrenal insufficiency and death (SWINGLE and REMINGTON 1944).

(3) The alterations in sodium and potassium are apparently to some degree causally reciprocal but in many instances these ions seem to behave independently of each other.

B. Electrolyte replacement therapy

Early work showing that a high-sodium diet can ameliorate adrenal insufficiency has been cited above. Most of the early investigators who used *adrenalectomized dogs* found that extra sodium would maintain life for variable but not indefinite periods. Later, more elegant studies were conducted in which electrolyte balance was followed daily and the toxic effects of potassium were taken into account. Thus, diets were devised with combinations of ions designed to compensate for daily urinary loss which could maintain life more or less indefinitely in adrenalectomized dogs (HARROP et al. 1935, ALLERS and KENDALL 1937).

Adrenalectomized rats are easier to maintain than dogs on salt alone. After much experimentation, in which combinations of electrolytes and different concentrations of sodium chloride have been given by different routes, one standard and simple procedure for the maintenance with salt of adrenalectomized rats became almost universally adopted: they are given 1% sodium chloride as the sole drinking fluid. Higher concentrations of sodium are not helpful and may be deleterious. On ordinary commercial diets, most adult adrenalectomized rats survive indefinitely on this treatment. We have less than 100% success if immature animals are used. It is interesting that the adrenalectomized rat, if allowed choice, prefers to drink 0.15 *M* sodium chloride, a concentration very close to that routinely used for maintenance (FREGLY 1958). Salt-maintained animals are susceptible to environmental stresses (temperature fluctuations, etc.) and show deficiencies in any situation requiring glucocorticoids, e.g., water diuresis tests. Under optimal conditions, however, they will behave, grow and reproduce in a near-normal fashion.

C. Steroid replacement therapy and overdosage

When aldosterone or deoxycorticosterone is administered acutely to normal or adrenalectomized animals, there is generally a very prompt reduction in the excretion of sodium which may or may not be associated with a parallel and proportional retention of water. Aldosterone is from 25 to more than 100 times as potent as DOC in this respect, the differences in estimates depending upon the test used (GAUNT et al. 1955, GROSS 1956b). (See Table 1, p. 524.) This variation in relative potency highlights the fact, which has been often and properly emphasized, that the qualitative and quantitative responses to steroids vary with the physiological setting under which their action is observed (GAUNT et al. 1949). In fact, the sodium-retaining action of mineralocorticoids may be absent or reversed.

In conditions of sodium deficiency, in which little sodium is being excreted, *deoxycorticosterone* has virtually no effects on renal functions (WELT et al. 1951);

likewise it fails to cause pathological changes (Knowlton et al. 1947, Selye and Stone 1949). In the condition of Cushing's syndrome (Soffer et al. 1944) or of intense water loading, or vasopressin administration, DOC may actually cause natriuresis (Gaunt et al. 1949, Bourguet and Debauche 1957). Also the rebound response to a DOC- or cortisone-induced sodium retention may result in net sodium loss (Perera 1948, Bartter et al. 1950, Ingle et al. 1950, Simpson and Tait 1950, Luft and Sjögren 1952, Luft et al. 1954). Very small doses of DOC, in contrast to larger ones, are said to cause increased sodium excretion (Forsyth 1953). Natriuresis has also been seen as a response to DOC in myotonic dystrophy (Zierler and Lilienthal 1948).

Aldosterone has not been available to study in the variety of conditions in which cortexone has been used. Suggestive evidence is available, however, to indicate that its effects, too, vary in different physiological circumstances (Dettbarn and Gross 1955, Renzi et al. 1956).

Another important variable that determines the response to exogenous mineralocorticoids is the status of endogenous adrenal function at the time they are given. In adrenalectomized animals or Addisonian patients, the sodium-retaining effect of DOC given chronically may be a long-continued one leading to edema (Ferrebee et al. 1939). Other evidences indicate that adrenalectomy leads to hypersensitivity to DOC (Martin and White 1952). In normal individuals, continued administration of DOC leads to various compensatory adjustments, the effect of which often is a reversal or negation of action, the end result being to prevent massive sodium and water retention (Relman and Schwartz 1952, Sjögren 1952, Soffer et al. 1957). In general, new homeostatic levels seem to be set at which sodium balance is achieved, perhaps at higher levels of total body sodium, but without gross edema. One bizarre adjustment to chronic DOC administration is the development of the "diabetes insipidus-like syndrome" in the dog, as discussed below (Subsection I, p. 534).

Another example of these "homeostatic adjustments" is described by Barger et al. (1959). Intrarenal infusion of aldosterone in the normal dog induced kaliuresis, but sodium retention was not observed. On the other hand, in adrenalectomized dogs or in dogs with cardiovalvular lesions, a prompt retention of sodium was observed following intrarenal aldosterone.

No one has yet given aldosterone in large quantities on a chronic basis in any species. In man, however, the condition of primary aldosteronism [which may not be *pure* aldosteronism (Editorial, Lancet 1956)], while causing some chronic sodium retention, does not usually cause edema, although exceptions have been noted (Mach et al. 1955, Hernando et al. 1957). It does, however, lead to hypokalemia, polyuria, metabolic alkalosis and a probable human counterpart of the diabetes insipidus-like syndrome (Conn and Louis 1956). This leads to the suggestion that DOC itself would produce such a syndrome in human subjects if given in chronic toxic overdosage.

DOC is without effect on plasma calcium but reduces magnesium markedly in rats (Woodbury et al. 1950).

D. Experimental heart failure

A striking example of the varying roles of different corticoids in pathological states is seen in the studies of *experimental heart failure* or thoracic caval constriction in dogs (Davis and Howell 1953, Hamilton et al. 1954, Davis et al. 1956). In these states, a severely reduced sodium excretion and ascites seem largely adrenal-dependent. The condition is relieved by adrenalectomy, is restored by DOC but not by cortisone, and is usually associated with an increased excretion of aldosterone in intact animals. In other words, it seems to replicate rather well the clinical states of secondary aldosteronism as studied extensively by WOLFF and collaborators (WOLFF et al. 1956, 1957) and others. The exact role of the corticoids in such cases is not known but the example illustrates again the fact that a corticoid can manifest chameleon-like changes in action when confronted with an altered physiological state. Thus, in this case, the probable salient feature of which is an elevated portal venous pressure, DOC and cortisone have opposite effects on sodium excretion, although under ordinary conditions of clinical use they both may cause sodium retention.

An experimental nephrotic syndrome in rats is also associated with increased aldosterone output (SINGER 1957).

E. Eclampsia-like syndrome

MASSON et al. (1951) showed that the action of renin could be markedly altered by an excess of DOC. They found that in rats sensitized by DOC, renin caused widespread vascular damage, massive edema, and renal damage and suppression. These effects were not observed when renin was given to normal animals. The antihypertensive drug, hydralazine hydrochloride, was strikingly effective in blocking this interaction between renin and DOC (RENZI and GAUNT 1953).

F. Corticoids and potassium metabolism

All of the active corticosteroids investigated enhance the excretion of potassium (Table 1, p. 524, Table 3, p. 527), and apparently deplete its levels in both intra- and extracellular sites. The most potent natural substance in this respect is aldosterone (DESAULLES et al. 1953). In general, the most potent sodium-retaining steroids are also the more potent kaliuretics. This is, however, not a strict rule. For instance, aldosterone is more potent relative to DOC in causing sodium retention than it is in causing kaliuresis in the rat (DESAULLES et al. 1953). Furthermore, hormones such as cortisol, particularly in acute experiments, frequently cause a simultaneous natriuresis and kaliuresis (DESAULLES et al. 1955). Conversely, evidence has been presented that DOC does not cause kaliuresis unless it is at the same time causing some sodium retention (SELDIN et al. 1956), the presumption being that renal tubules must exchange Na^+ for H^+ and K^+ ions. When potassium is depleted in the presence of retained sodium, the sodium replaces potassium in cells in large amounts (H. MILLER and DARROW 1940, FERREBEE et al. 1941). A state of hypokalemic alkalosis ensues, one example of which is seen in primary aldosteronism. DOC prevents accumulation of potassium in muscle even in the presence of a potassium plethora (COLE 1953). The possible interactions between potassium and sodium as related to the action of corticoids on renal tubules and other responsive cellular systems have been discussed recently in terms of unifying concepts by CONWAY (1957). WOODBURY and KOCH (1957) have reported different actions of aldosterone and DOC in this regard. The effects of corticoids cannot be explained as consequences of their anabolic or catabolic action (BARTTER and FOURMAN 1951), as was implied in the older literature.

Whatever may be the dynamics of the physiology involved, it is certain that potassium depletion is responsible for many aspects of the corticoid-overdosage syndromes, as is potassium accumulation for the symptoms of adrenal insufficiency. It is, for instance, well established that prolonged excess corticoids may cause cardiac dysfunction and skeletal muscle paralysis as a result of potassium depletion (RAGAN et al. 1940, VAN BUCHEM 1957). This is perhaps best seen in the rabbit, in which, according to GROSS and SCHMIDT (1958a), high doses of DOC, salt and fluid cause fatal hypokalemia with paralysis, etc., within 2 weeks. Aldosterone, despite its greater sodium-retaining action, is somewhat less active than DOC in causing these overdosage phenomena (GROSS and SCHMIDT 1958b).

G. Effects of cortisol, cortisone and derivatives

The effects of the glucocorticoid hormones on electrolyte metabolism are much more difficult to define than those of the mineralocorticoids — largely because they are more variable in different conditions. One problem in the chronic use of these steroids *in the dosage required to control inflammatory diseases* is that *they may cause sodium retention and edema* of such order as to oblige the termination of therapy. This is a common clinical observation, and the pattern of response suggests that the glucocorticoids should properly be classified as weak mineralocorticoids. In fact, the renal response to DOC and cortisone in dogs (ROBERTS and RANDALL 1955) under some conditions is essentially identical as far as causing sodium retention is concerned.

On the other hand, *the glucocorticoids can serve as powerful natriuretic agents.* For instance, in the rat given either a saline or water load, they may enhance sodium excretion by several fold, as seen by many investigators (e.g., DESAULLES et al. 1955). Similar results have been observed frequently in man (e.g., LAIDLAW et al. 1955).

In this respect, the 1-dehydro-corticoids, prednisolone and prednisone, are more potent as natriuretic agents (CHART et al. 1956) and less likely under any circumstances to cause sodium retention. In fact, they can be used therapeutically in some cases as diuretics and natriuretic agents (see p. 546). The relative potencies and variable activities of a number of different natural and derived corticoids are summarized in Table 4 from the work of LIDDLE et al. (1956), who used acute tests in adrenalectomized dogs as a basis for their data.

Table 4. *Relative effectiveness of natural and synthetic steroids on excretion of sodium and potassium in the adrenalectomized dog, using deoxycorticosterone as a standard.* (From LIDDLE et al. 1956)

Steroid	Sodium excretion		Potassium excretion	
	Effect	Potency	Effect	Potency
Deoxycorticosterone	Retention	1.0	Loss	1.0
Aldosterone	Retention	39.0 (26—61)	Loss	29.0 (20—41)
Cortisol	Loss		Loss	0.04 (0.026—0.063)
2-Methylcortisol	Variable		Loss	1.0 (0.5—2.1)
9α-Fluorocortisol	Variable		Loss	5.5 (3.1—10.7)
2-Methyl-9α-fluorocortisol	Retention	49.0 (19—100)	Loss	155.0 (75—353)
11β-Hydroxyprogesterone	Variable		Loss	0.018 (0.011—0.028)
2-Methyl-11β-hydroxy-progesterone	Variable		Loss	0.069 (0.043—0.111)
Cortisone	Variable		Loss	0.021 (0.012—0.039)
2-Methylcortisone	Variable		Loss	0.009 (0.004—0.020)
Deoxycorticosterone acetate	Retention	0.91 (0.4—1.8)	Loss	0.92 (0.6—1.5)
2-Methyldeoxycortico-sterone acetate	Retention	0.08 (0.03—0.17)	Loss	0.034 (0.017—0.067)

The figures in parentheses represent 95% confidence limits. Ratios of potency are adjusted for molecular weights so that the steroids are compared on an equimolar basis.

H. Corticoids and brain functions

Deoxycorticosterone antagonizes the seizure-provoking effects of vasopressin in epileptic subjects and decreases the tendency for grand mal seizures (Mc Quarrie et al. 1942). Woodbury (1954) in extensive studies found that DOC elevated the electrical seizure threshold in rats, an effect weakly mimicked by ACTH but not by adrenocortical extract. Adrenocortical extract, furthermore, antagonized the effects of DOC, but ACTH did not. The actions of DOC could be directly correlated with its action in elevating plasma sodium, effects which also were antagonized by adrenocortical extract (Woodbury et al. 1950). DOC synergized the actions of drugs elevating electroshock threshold (phenobarbital and diphenylhydantoin) and was itself potentiated by trimethadione — a drug inactive by itself. These results were not correlated with changes in measured parameters of electrolyte metabolism other than those attributable to DOC alone (Woodbury 1952a). A further study, comparing various corticoids, showed that 11-deoxycortisol acetate had a weak DOC-like effect in increasing electroshock thresholds; corticosterone acetate was inactive; 11-dehydrocorticosterone acetate decreased electroshock thresholds slightly; cortisol acetate caused a considerable decrease, and cortisone acetate had a similar but greater effect, which could be antagonized by DOC (Woodbury 1952b).

These phenomena appear to be related to those of *water intoxication*, a state in which electroencephalographic and other phenomena resemble those of epilepsy (Gellhorn and Ballin 1946), and one in which the corticoids provide a protection that is not entirely due to their effects on electrolyte excretion or electrolyte balance (see p. 546ff.).

Hoagland (1954) has reported metabolic and electroencephalogram alterations in the brains of adrenalectomized rats, associated with a reduction in blood flow through the brain. These abnormalities were corrected by glucocorticoids but not DOC. The significance of concomitant changes in the turnover rate (but not total content) of brain potassium is obscure. Gordan (1956) found that in certain conditions DOC, like numerous other steroids, inhibited oxygen uptake by brain tissues. Presumably, however, the effects he observed are not primarily related, as in Woodbury's experiments, to alterations in electrolyte metabolism, since some of his most active compounds are generally considered to be without electrolyte effects, e. g., 21-hydroxypregnane-3,20-dione sodium succinate.

I. Diabetes insipidus-like syndrome

In 1939 Kuhlman et al. first reported that chronic treatment with large doses of DOC in dogs caused the development of a *diabetes insipidus-like syndrome,* characterized by slight hypernatremia, striking hypokalemia, periodic weakness and paralysis, and a vasopressin-resistant polyuria (Ragan et al. 1940). There was a replacement of intracellular potassium by sodium. The polyuria was associated with hypernatremia and was salt-dependent (Ferrebee et al. 1941). The critical effects were paralysis and cardiac complications, which were apparently due to potassium loss and were responsive to potassium replacement therapy.

The causal chain of events resulting in this diabetes-insipidus-like syndrome is still obscure but is probably complex, involving problems of thirst mechanisms, renal actions, etc. (Gaunt 1951). The more recent concept is that the urine-concentrating mechanism is deficient in states of potassium deficiency and may be the primary mechanism (Hollander et al. 1957, H. Richter 1958). Other species may respond to corticoid overdosage differently from the dog. DOC causes diabetes insipidus-like polyuria in the rat only in the presence of a high salt intake and, even so, gross manifestations of potassium deficiency are not evident. In man, however, excess endogenous aldosterone production (primary aldosteronism) (CONN and LOUIS 1956) appears to cause effects much like those resulting from excess DOC in dogs.

Be that as it may, the production of this syndrome is not limited to the mineralocorticoids as such. Cortisone (MUSHETT et al. 1951, SIREK and BEST 1952, SWINGLE et al. 1953) and particularly some of the synthetic steroids, for instance prednisone (PERLMAN et al. 1956), 9α-fluorocortisol (POUTSIAKA et al. 1957), and 1-dehydro-9α-fluorocortisol (SWINGLE et al. 1956), are even more potent in dogs. Since one factor common to the action of all of these steroids is the production of kaliuresis, the concept stated above that potassium deficiency may contribute to the action of DOC is equally applicable to other steroids and becomes more impressive as a possible causal factor.

J. Extra-renal functions of the mineralocorticoids

While it is clear that a major site of action of the mineralocorticoids is on the kidney, it is also clear that perhaps equally important actions are exerted at extra-renal sites. *The mineralocorticoids tend to put a dam in front of all the major paths of loss of sodium from the body:* sweat glands, salivary glands, glands of the lower digestive tract, as well as the kidneys. *In addition they have profound effects on the internal distribution of electrolytes and water.* In many experimental and clinical situations, however, a separation of renal and extra-renal sites of action is all but impossible to define because of the inherent difficulty in establishing causal primacy in events that seem to occur concomitantly. The following summary, therefore, is more descriptive than analytical. (See also Chapter 8.)

I. Balance studies

Since the early 1930s (SWINGLE et al. 1933), studies have continued concerning the role of steroids in the internal distribution of fluid and electrolyte. STAHL et al. (1936) noted that symptomatic recovery from adrenal insufficiency in dogs occurred before any consistent change in serum sodium, potassium, urea or water. MUNTWYLER et al. (1940a) made similar observations.

HARROP et al. (1933a) stated that urinary loss of sodium and chloride is much greater in the early stages of adrenal insufficiency than can be accounted for by fall in plasma electrolytes; some of the electrolytes must come from tissues. Conversely, the water loss from extracellular fluid cannot be accounted for by urine volume and hence fluid must be shifted into cells (HARROP 1936b).

HEGNAUER and ROBINSON (1936) found that these changes were not associated with changes in the osmotic pressure of plasma but were associated with intracellular hydration. INGLE et al. (1937), using nephrectomized-adrenalectomized rats, found distinct extra-renal effects of adrenocortical extract, particularly on potassium distribution. KNIGHT et al. (1955) concluded that one of the first major effects of cortisol on electrolyte metabolism in man was to affect the distribution of cations between the intracellular and extracellular fluid. LAIDLAW et al. (1955) found in human adrenal insufficiency that the total exchangeable sodium and potassium of the body were not changed at the time serum sodium was depleted and potassium elevated. They, too, concluded that extra-renal effects of cortisol and cortisone on electrolyte distribution were factors of major importance in their action.

SWINGLE et al. (1937b) demonstrated that the adrenalectomized dog could be maintained on a "salt-free" diet when adequate amounts of adrenocortical extract were given. When replacement therapy was withdrawn, typical adrenal insufficiency with hemoconcentration occurred without further significant loss of sodium through the kidney. Reinstitution of extract therapy brought about revival and hemodilution, which could not be accounted for by alteration in external sodium balance. More recently this same type of experiment has been elaborated with more dramatic results (SWINGLE et al. 1958a). Adrenal insufficiency was allowed to develop. Recovery and subsequent temporary maintenance were effected with 2-methyl-9α-fluorocortisol without administration of food, salt or water. A huge sodium diuresis occurred during the recovery process, despite which blood volume and blood sodium were restored to above-normal levels. DOC was weak or inactive in accomplishing such results and the same was later found to be true of aldosterone (SWINGLE et al. 1958b). This certainly means that some store of sodium is available for mobilization under the influence of glucocorticoids, which can bring about the same results as administered sodium.

One interesting and presumably significant difference between the active glucocorticoids and the relatively inactive DOC in this experimental situation *was that the former restored blood pressure effectively and the latter did not* (see p. 547ff.). It can be presumed that the restoration of vigor was due in part to this restoration of circulation, but the relation of the circulatory effects to concomitant electrolyte phenomena is uncertain.

Table 5. *Sodium and chloride balance during* DOC *administration, acute adrenal insufficiency and recovery in dogs.* (From FLANAGAN et al. 1950)

	Na^+	Cl^+
Total "extracellular" ions		
1. Adrenalectomized on DOC (1.5 mg/day for 18 days)	627 m*M*	468 m*M*
2. Acute adrenal insufficiency	368 m*M*	275 m*M*
Total "extracellular" loss	259 m*M*	193 m*M*
3. Acute adrenal insufficiency	368 m*M*	275 m*M*
4. 60th hour of recovery on DOC (1.5 mg/day)	583 m*M*	420 m*M*
Total "extracellular" gain	215 m*M*	145 m*M*
External ionic balance		
1. Acute adrenal insufficiency		
Intake	377 m*M*	331 m*M*
Output	472 m*M*	429 m*M*
Balance	—95 m*M*	—98 m*M*
Total "extracellular" loss insufficiency	259 m*M*	193 m*M*
External balance (intake-output)	—95 m*M*	—98 m*M*
"Extracellular" ionic loss unaccounted for	164 m*M*	95 m*M*
2. Recovery on DOC		
Intake	68 m*M*	66 m*M*
Output	19 m*M*	14 m*M*
Balance	+49 m*M*	+52 m*M*
Total "extracellular" gain on recovery	215 m*M*	145 m*M*
External balance	49 m*M*	52 m*M*
"Extracellular" ionic gain unaccounted for	166 m*M*	93 m*M*

Flanagan et al. (1950) have studied systematically in dogs the problem of the disposition of the sodium, other than that lost through the kidney, which disappears from the extracellular fluid in adrenal insufficiency, but which can "return" under the influence of corticoids. The question has not been satisfactorily answered. Balance data which illustrate the nature of the problem are shown in Table 5. The quantitative relations are such that it would seem that the only available storage site would be bone, but attempts have not succeeded in demonstrating that there are actually labile sodium stores in bone, under control of corticoids, adequate to explain the phenomena observed (Overman et al. 1951, White and Rolf 1954, Sweet et al. 1954). While such events are related to concomitant changes in renal function, experiments of various types, including the use of nephrectomized animals, show that internal electrolyte changes of potassium as well as sodium may be independent of kidney alterations (Birnie et al. 1948, Levitt and Bader 1951, Woodbury 1951, Luft and Sjögren 1952, Hays 1952, Frost and Talmage 1952).

II. Role of sodium in adrenal insufficiency

Results such as those cited above raise interesting questions concerning the role of sodium depletion in the syndrome of adrenal insufficiency. *A priori*, it would seem clear from the well-known effectiveness of sodium as a replacement therapy in adrenal insufficiency that its loss must be a major precipitating factor in that syndrome. On the other hand, it is also clear that the *total* body stores of sodium are not depleted in adrenal insufficiency in amounts adequate to account for the shock-like collapse of that state. Furthermore, recovery from adrenal insufficiency under influence of the most active steroids may be associated with renal wastage of sodium, even in absence of dietary intake of that ion. It must be, therefore, that *adrenal insufficiency is a state in which the glucocorticoids can effect an internal redistribution of electrolytes adequate to maintain circulatory function and prevent shock*, but *in the absence of such steroids a plethora of sodium or a conservation of sodium by mineralocorticoids can exert similar effects.*

III. Corticoids and fluid compartments

Early evidence from many sources indicated that *adrenal insufficiency was characterized by a hemoconcentration and evidences of a declining blood volume*, an effect relieved by the administration of adrenalocortical extract (Swingle and Remington 1944) or DOC (Thorn and Eisenberg 1939). Clinton et al. (1942) found that DOC given to normal dogs could increase plasma volume above normal levels — an effect associated with sodium retention and decreased urine volume. In human subjects, adrenocortical extract prevented the fall in blood volume characteristically resulting from etherization (McAllister and Thorn 1937), as did DOC (Ragan et al. 1939). Harkness et al. (1942) found that the increased plasma volume was associated with an increased extracellular volume. In further studies typified by those of Gaudino and Levitt (1949), sharp distinctions could be drawn between the effects of adrenocortical extract (probably the dominant constituents of which were cortisol and aldosterone) and DOC. In severe adrenal insufficiency, total body water (measured with D_2O) was variably affected, but intracellular water increased 15—30% while the extracellular water (inulin space) declined about 60%. The *effect of DOC* either in adrenalectomized or intact dogs *was to increase extracellular fluid at the expense of the intracellular fluid.* There was an increase in total body sodium but not of total body water. *Adrenocortical extract*, on the other hand, *increased intracellular volume without affecting extracellular volume*, and increased total body water without changing plasma sodium levels. All of these alterations tended to revert toward normal while treatment was still continued and were not wholly explicable on any evident osmotic basis. Ziff et al. (1952) noticed a consistent increase of total body water in patients treated with ACTH, which in some cases (associated with edema) was primarily extracellular, while in others it was primarily intracellular. Levitt and Bader (1951), using patients on low-salt diets in order to minimize the effect of primary sodium retention, found that ACTH and cortisone both caused a transient (regressed after 8—9 days on therapy) shift of body water to the intracellular space, associated with an increase in glomerular filtration rate.

Luft et al. (1954) in careful human balance studies, found DOC to cause an increase in exchangeable sodium even in the face of a negative sodium balance (which followed initial sodium retention) and increased intracellular sodium. There was also an increased urinary loss of calcium with negative balance. Only minimal effects, however, were obtained when the subjects were on a restricted sodium intake. The effects of ACTH were much like those of DOC. Cortisone administration, over a period of 74 days, caused a net loss of sodium and a greater loss of chloride without change in exchangeable sodium or plasma sodium and chloride concentrations. Plasma potassium concentration decreased. Extracellular volume (inulin space) decreased by one liter.

In relatively short-term experiments and using small doses of steroids, Woodbury and Koch (1957) found distinct differences between the effects of DOC and aldosterone on the transport of sodium and potassium between the intracellular and extracellular sites in mice. They concluded that aldosterone appears to be capable of enhancing electrolyte transport at all sites which have been studied, whereas DOC has more limited actions.

Such results emphasize again the semantic difficulties in using the traditional terms "mineralocorticoid" and "glucocorticoid." As indicated in the Introduction, all corticoids are mineralocorticoids, but they may act in different ways.

IV. Corticoids and tissue electrolytes

A huge and confused literature records the effects of adrenalectomy on cellular, tissue and organ contents of electrolytes (Darrow et al. 1939, Muntwyler et al. 1940). No general summary of this work is possible because of the large number of variables in different experimental conditions, differences between different tissues and organs and between the effects of different corticoids. *Skeletal muscle is one of the most responsive tissues in showing variations in electrolyte content after administration of various corticoids.* Generally, the mineralocorticoids cause an increase in intracellular sodium and decrease in potassium; adrenal insufficiency results in opposite changes (Swingle and Remington 1944, Efron 1957). An excess of glucocorticoids may have different effects. Even aldosterone and DOC have somewhat different actions (Woodbury and Koch 1957). Each tissue seems to follow its own rules and *brain is notably resistant to corticoid-induced change* (Woodbury 1952a). The corticoids may affect the dynamics of electrolyte turnover *across capillary membranes*, even in the absence of effects on total organ contents (Overman et al. 1951). Generally there is a decrease in potassium turnover and an increase in sodium turnover in adrenal insufficiency (Efron 1957). Analogous phenomena have been observed in the movement of cations across the membranes of erythrocytes and other cells (STREETEN and SOLOMON 1954, SCHATZMANN 1954, JONES 1955). The tabular data in Tables 6 and 7, selected for presentation because of their comprehensiveness, show typical results concerning the effects of corticoids on tissue electrolytes and illustrate particularly tissue variability.

Table 6. *Influences of cortisone and deoxycorticosterone acetate on Na, K, Cl, and H_2O content of tissues of dogs.* (From DAVIS et al. 1951)

Tissue	Na^+, mM/kg wet tissue Av.	Range	K^+, mM/kg wet tissue Av.	Range	Cl^-, mM/kg wet tissue Av.	Range	H_2O, ml/kg wet tissue Av.	Range
Liver								
N	48	40—61	76	64—88	38	33—44	739	701—810
N + CA .	39	37—42	77	67—83	38	28—48	716	698—733
A + CA	34	29—38	70	69—73	30	25—39	706	694—714
N + DOC	62	52—85	80	71—87	41	36—46	740	730—752
A + DOC	52	44—56	73	55—83	44	38—58	741	713—772
Muscle								
N	25	18—32	100	87—117	17	11—21	752	719—785
N + CA .	38	31—43	91	84—95	26	18—32	757	738—773
A + CA .	29	21—40	96	88—110	19	12—31	755	741—769
N + DOC	32	26—39	93	78—103	18	16—21	738	701—753
A + DOC	27	22—39	93	80—100	17	15—20	747	732—757
Heart								
N	40	33—49	80	67—89	29	22—33	779	760—820
N + CA .	43	40—48	81	79—84	29	24—42	779	769—786
A + CA .	38	34—42	84	83—86	26	23—32	776	766—781
N + DOC	47	44—53	79	76—81	30	29—34	775	772—780
A + DOC	46	43—48	75	64—84	33	28—40	786	775—798
Ileum								
N	50	45—55	90	77—100	49	40—54	783	758—840
N + CA .	51	48—54	89	77—96	47	42—51	784	766—796
A + CA .	53	38—61	96	89 —102	45	39—56	781	751—803
N + DOC	57	52—64	95	91—98	51	48—55	782	773—790
A + DOC	52	43—59	91	83—98	50	47—53	784	777—792
Spleen								
N	64	48—83	71	50—89	50	45—55	761	740—788
N + CA .	59	44—66	82	72—99	49	45—54	763	738—791
A + CA .	73	66—82	55	48—70	43	42—45	742	728—762
A + DOC	63	48—81	69	51—94	44	36—48	763	744—808
Lung								
N	88	79—91	53	49—66	71	65—80	788	781—796
N + CA .	90	79—102	52	43—59	67	62—75	790	770—831
A + CA .	81	72—87	56	53—60	65	60—70	777	770—781
A + DOC	91	80—98	55	46—67	72	71—76	815	800—821
Skin								
N	73	53—93	17	12—23	68	51—96	538	409—684
N + CA .	78	70—92	16	13—19	74	64—86	581	562—604
A + CA .	80	71—87	21	16—24	74	58—98	629	615—636
N + DOC	82	79—86	15	13—18	72	69—79	587	541—641
A + DOC	81	68—97	19	13—22	74	55—82	632	511—692
Brain								
N	57	49—63	94	81—109	39	30—47	778	751—810
N + CA .	55	43—61	87	65—95	35	25—41	779	763—798
A + CA .	55	52—58	92	89—97	37	35—40	768	729—780
N + DOC	57	55—60	95	92—101	38	36—41	767	747—795
A + DOC	56	52—58	90	81—95	39	36—42	780	753—789

Abbreviations: N = normal dogs; A = bilaterally adrenalectomized dogs; DOC = deoxycorticosterone acetate; CA = cortisone acetate. The dose of DOC was 0.5 mg/kg/day for a week; that of CA 5 mg/kg/day for a week.

Table 7. *Sodium (Na^{24}) and potassium (K^{42}) content and turnover in various organs of normal and adrenalectomized rats. 12 animals per group.* (Condensed from two tables in EFRON 1957)

Organ	mg K+/100 g fresh tissue ± standard error		c.p.m./μg K+			mg Na+/100 g fresh tissue ± standard error		c.p.m./μg Na+		
	Normal	Adx.[1]	Normal	Adx.	Difference	Normal	Adx.	Normal	Adx.	Diffe-enc
Aorta	61±6	102±19	117.2	54.0	—63.2	260±42	176±9	55.3	82.8	+2
Diaphragm	309±7	333±5	44.7	40.9	—3.8	87±9	76±1	68.5	84.3	+1
M. sartorius	394±15	388±20	22.3	9.8	—12.5	72±17	49±6	54.3	59.8	+
Sciatic nerve	157±11	171±11	11.4	12.8	+1 4	168±10	135±11	65.9	82.8	+1
Brain	345±21	336±7	3.5	3.8	+0.3	114±80	106±12	44.5	49.0	+
Small intestine	333±9	356±8	45.0	41.0	—4.0	110±3	97±13	75.3	78.5	+
Large intestine	307±13	327±9	42.3	39.6	—3.7	92±5	91±4	81.4	87.4	+
Liver	255±13	272±17	54.0	50.6	—3.4	106±6	109±6	41.0	43.2	+
Kidney	141±10	165±10	72.4	72.8	+0.4	141±8	142±5	60.1	60.3	+
Bone	68±6	81±5	34.0	30.3	—3.7	261±20	275±14	52.2	53.8	+

[1] Rats adrenalectomized 6—8 days.

V. Gastrointestinal functions

The absorption of electrolytes from the gastrointestinal tract is altered and *generally delayed in adrenal insufficiency* (CLARK 1939, DENNIS and WOOD 1940, STEIN and WERTHEIMER 1941, GAUNT 1944). This fact has less to do with the delayed excretion of water and electrolyte following the acute administration of these substances than might be anticipated (GAUNT 1944). Because of the many abnormalities in adrenal insufficiency, the specificity of absorption defects could be questioned. The work of BERGER et al. (1951) points more strongly, however, to a specific role of the cortical hormones on such functions. They found that DOC clearly inhibited the fecal elimination of sodium in the intact rat and reduced the capacity of ion exchange resin to increase fecal sodium in man. Similar results have also been obtained by THORN et al. (1953), who suggested that the typical gastrointestinal dysfunction of Addison's disease may be related to the inability to absorb electrolytes normally.

In dogs, 9α-fluorocortisol reduced the fecal elimination of sodium and increased that of potassium by several fold even under conditions in which sodium was not being retained by the kidney (POUTSIAKA et al. 1957).

Various corticoids clearly increase the secretions of gastric acid and enzymes (GRAY and RAMSEY 1957, HIRSCHOWITZ 1957), at least under certain conditions. Generally the glucocorticoids are most potent in this respect, but HIRSCHOWITZ (1957) found that aldosterone increased pepsin secretion in man more effectively than any other steroid.

VI. Sweat

CONN (1949) *demonstrated an unequivocal effect of corticoids on the secretion of electrolytes by the sweat glands of man.* He found the concentration of sodium and chloride in sweat to be reduced by DOC, adrenocortical extract or ACTH; reduced in patients with adrenocortical carcinoma, Cushing's syndrome or adrenogenital syndrome, and increased in Addison's disease. Cortisone was weak or inactive in affecting sweat electrolyte content (LOCKE et al. 1951). The continued administration of DOC does not result in an "escape" from its effects on sweat glands, such as that which occurs in kidneys of subjects with normal adrenal function (THORN et al. 1953).

As would be expected, factors other than the levels of corticoids also affect the electrolyte composition of sweat (LOCKE et al. 1951), but under adequately-controlled conditions, electrolyte values become useful indices of adrenal function.

VII. Saliva

FRAWLEY and THORN (1951) showed that the electrolytes of saliva were to some degree under corticoid control. GRAD (1952) made similar observations. FRAWLEY and THORN (1951) reported that the Na^+/K^+ ratio of saliva was markedly elevated in Addison's disease and lowered in Cushing's syndrome. It was also reduced by ACTH and DOC. GRAD (1952) found, too, that the Na^+/K^+ ratio was reduced by ACTH and suggested that the diurnal variations in the salivary Na^+/K^+ ratios may be causally related to similar variations in adrenocortical function. BONGIOVANNI (see discussion, FRAWLEY and THORN 1951) found the salivary Na^+/K^+ ratio reduced in patients with cirrhosis and fluid retention — conditions now known to be associated with secondary aldosteronism. These results would be expected in view of the subsequent demonstration by SIMPSON and TAIT (1955) that *aldosterone in doses of 10—80 μg similary lowered the salivary* Na^+/K^+ *ratio*, according to a linear dose-response pattern. In fact, *this is the most sensitive response yet recorded of the human being to aldosterone.*

WARMING-LARSEN et al. (1952) used human subjects in whom salivary flow was stimulated by pilocarpine and showed in this test situation that ACTH had its typical effects in reducing salivary secretion of sodium.

The phenomena involved are not limited to the human species. In sheep, the Na^+/K^+ ratio of the secretion of cannulated parotid glands was raised by adrenalectomy and reduced by DOC (and probably less so, if at all, by cortisone) (GODING and DENTON 1956). Alterations in electrolyte balance modified but did not eliminate the corticoid effects.

K. Concluding comment

It is clear from functional studies of various organs in which sodium and potassium are involved in secretory or excretory processes that there is probably no instance in which the movement of these ions is not influenced by corticoids. The effect is nearly always that of conserving sodium for the body and eliminating potassium. The kidney responds to corticoids less consistently by sodium retention than do the other portals of potential sodium loss. The kidney may either eliminate or retain sodium under the influence of corticoids, depending upon the nature of the corticoid or other physiological conditions. Perhaps this is not surprising in view of the fact that the over-all function of the kidney is an algebraic summation of many different actions, including vascular functions, filtration, reabsorption and secretion, which occur at different sites and may be differently affected by different corticoids. Although details are less easy to define and measure, it is probable that the membranes of cells respond to corticoids, as far as electrolyte transfer is concerned, in a manner analogous to that seen more dramatically in cells with obvious secretory and excretory functions. The behavior of potassium in response to corticoids is more consistent than that of sodium. In virtually every situation examined in which corticoids influence potassium, they have the general effect of getting it out of cells and out of the body.

From a phylogenetic and teleological standpoint, there is at least some rationale to such processes. As the ancestral vertebrate emerged from salt water to fresh water and onto the land, it had the difficult problem of maintaining an extracellular fluid in which sodium was the major cation, under circumstances in which exogenous sodium and even water itself (on land) were not readily available. Furthermore, the only diets available were rich in potassium. While potassium is essential for physiological processes, it is toxic in excess. In broad terms, may not the physiology of mineralocorticoids be thought of as phenomena adapted to these problems?

Adrenal steroids and water metabolism

The adrenal steroids apparently affect the excretion of water in a variety of ways. The first and most obvious means is that in which the excretion of water is the indirect result of steroid effects on sodium excretion, i.e., *the corticoids cause retention of sodium and a consequent osmotic retention of water*. Analagous factors presumably operate in the internal distribution of water. Insofar as these factors are predominant, the corticoids are diuretic or antidiuretic, depending on whether they are natriuretic or antinatriuretic.

Despite these influences, it is clear that other mechanisms are involved in the now well-recognized facts: (a) that *in the absence of glucocorticoids, it is impossible for an animal or man to excrete a water excess at anything approaching a normal rate*, and (b) that the corticoids may serve as water diuretic agents.

A. Water diuresis in adrenal insufficiency and replacement therapy

The inability to excrete water at a rapid rate in adrenal insufficiency was first noted in Addisonian patients by ROWNTREE and SNELL (1931). Subsequently and probably independently, SILVETTE and BRITTON (1933) observed essentially the same phenomenon in adrenalectomized cats, as did RIGLER (1935) in adrenalectomized mice. SWINGLE et al. (1937b), working from other points of view, first pointed out in work on adrenalectomized dogs that there was a marked susceptibility to water intoxication. The significance of these preliminary observations, which indicated that the adrenal hormones were essential for the regulation of water excretion, was not at first widely appreciated, due to preoccupation with the idea that as far as the cortical hormones were concerned the movements of water were secondary to those of salt and that the primary action of adrenal hormones was to cause salt and hence water retention.

Attention to the problem of water excretion as such was stimulated by the description of a "water test" or "Kepler's test" for Addison's disease by ROBINSON et al. (1941) and by an extensive series of animal experiments describing and to some degree analyzing the phenomena involved (GAUNT et al. 1949, FROST and TALMAGE 1951, SLESSOR 1951, LLOYD 1952, GROSS 1956a). Similarly, the rate of water excretion after water-loading can serve as a rough index of the functional capacity of the animals under various experimental conditions (EVERSOLE et al. 1940, OLEESKY and MANC 1953, BEATTY et al. 1954).

Adrenalectomized animals allowed to develop adrenal insufficiency may exhibit a negative water balance and have urine volumes higher than normal, particularly in the early stages before extracellular fluid loss becomes severe (GAUNT et al. 1938). This is associated with a negative sodium balance and presumably is a consequence of it. *The specific deficiency of greatest interest is the inability to make the usual homeostatic adjustments necessary to excrete a water load rapidly.* It has been common experience that the ability to maintain a maximal rate of water diuresis is not restored in adrenalectomized animals or Addisonian patients by maintenance with salt and is restored only partially, if at all, by DOC (SOFFER et al. 1957).

Aldosterone is weak in repairing the deficiency, relative to its potent action on electrolyte excretion (RENZI et al. 1956). *Effective replacement is best and readily obtained with the glucocorticoids,* the action of which can be augmented by DOC (MOREL 1952).

Not only do *the corticoids* repair the deficiencies of adrenalectomized animals but *in high doses* they *increase water diuresis to above-normal levels* in either adrenalectomized or intact animals (GAUNT et al. 1949). In this respect, cortisol is probably the most potent natural corticoid, and prednisolone the most active of all presently-known steroids (CHART et al. 1956).

The mechanism of the deficiency in handling water has been widely studied, but all factors involved have not been satisfactorily analyzed. One reason is that the phenomenon almost certainly is due to multiple interrelated influences (SOFFER et al. 1957).

One of the most obvious factors that could contribute to an inhibition of rapid water excretion is the *decreased glomerular filtration rate* characteristic of adrenal insufficiency in all species. When present it undoubtedly contributes to the phenomena observed. It is clear, however, that when adrenalectomized animals are maintained with salt or DOC, *a deficiency in handling water is* still *present even when the glomerular filtration rate is normal* within the limits of available techniques of measurement. Similarly, while the glucocorticoids can and often do cause a marked stimulation of the glomerular filtration rate — a fact which could cause or contribute to a diuretic action when present — much evidence suggests that they can have diuretic actions in the absence of measurable elevation of the glomerular filtration rate (NABARRO 1956). Since the diuretic action of corticoids can be manifested acutely (i.e., in less than one hour), it cannot be due exclusively to hypokalemia and a consequent unresponsiveness to antidiuretic hormone and inability to concentrate urine due to potassium deficiency, as may be the case after chronic treatment (HOLLANDER et al. 1957, H. RICHTER 1958). Although the water diuresis induced by corticoids is oftentimes associated with natriuresis, it is not necessarily so. If all this be true, it is clear that the glucocorticoids by some indirect or direct mechanism have the over-all effect of inhibiting the tubular reabsorption of water, a phenomenon which itself could have varied and multiple causes.

B. The adrenal-neurohypophyseal interrelationship

The set of facts listed above would and did set the stage for *the working hypothesis that some vasopressin-like humoral agent* was present which *inhibited the excretion of water in the absence of the proper corticoids* (GAUNT et al. 1949). The search for and identity of such a substance has led to a literature of remarkable confusion. Early reports that there was an excess of antidiuretic-hormone-like material in the blood of adrenalectomized animals and Addisonian patients have been justly criticized on the basis of techniques employed. Nevertheless, such reports, with various technical improvements, continue to appear with frequency. On the other hand, a large body of evidence assails the idea that any such substance exists (reviews, HELLER 1952, MONTASTRUC 1954, MOREL 1955, GAUNT et al. 1957; see p. 545). The problem can only be resolved finally when methods of measuring antidiuretic hormone in body fluids have reached a greater sensitivity and specificity than is now the case. It is still the authors' working hypothesis — more a statement of faith than fact — that *there must be some antidiuretic-hormone-like humoral influence, either present in excess or acting unantagonized, that acts on the renal tubule to increase the reabsorption of water in the absence of glucocorticoids.*

The first question to ask is whether this is or can be an excess of antidiuretic hormone itself. As indicated above, no conclusive evidence has been presented that such is the case. Furthermore, it has been pointed out frequently that the sensitive homeostatic adjustments regulating the release of antidiuretic hormone should preclude its accumulation when not needed, i.e., a decreasing tonicity of extracellular fluids should inhibit antidiuretic hormone release. On the other hand, it is clear that influences other than the osmotic pressure of body fluid affect the release of antidiuretic hormone, e.g., extracellular fluid volume, drugs, psychic stimuli, etc. Humoral or other influences might be present in the absence of glucocorticoids that either affect tubular function directly or the release or rate of destruction of antidiuretic hormone. Some of the better established facts regarding the possible interrelationships between the adrenal cortex and neurohypophysis in this connection are outlined below.

(1) Hormones from the adrenal cortex support the mechanisms, possibly enzymatic in nature, which account for the rapid inactivation of vasopressin. This was seen in both *in vitro* (Birnie 1953) and *in vivo* (Ginsberg 1954) experiments. *The half-life of circulating exogenous vasopressin is doubled after adrenalectomy*, and repair of the deficiency is probably best effected by the glucocorticoids. The physiological significance of a retarded inactivation of vasopressin, however, must remain obscure unless accompanied by data on the simultaneous secretory rate of the hormone. Such data are not available.

(2) Some humoral substance from hypothalamic neurosecretory cells apparently traverses the hypothalmico-hypophyseal portal blood vessels after appropriate stimulation to cause release of ACTH from the anterior pituitary (Harris 1955). (See Chapter 10.) [In addition some substance from a cranial but non-pituitary source is probably released into the blood stream to stimulate aldosterone secretion (Rauschkolb and Farrell 1956, Farrell 1959b, c).] The chemical nature of the former substance is unknown, but much work implies that the ACTH-releasing material is vasopressin-like (review, Fortier 1957). If so, the interrelationship of the neurohypophyseal and adrenocortical systems may have therein a connecting link.

(3) From a pharmacological standpoint, if vasopressin and the various corticoids are administered concomitantly, their actions on electrolyte and water excretion may be antagonistic, additive or independent, depending upon the variable measured or the steroid used (Gaunt et al. 1957). Such observations give no clear clues as to what would be expected to happen when the effects of variable amounts of the neurohypophyseal, anterior pituitary and adrenocortical principles are "blended" in the blood stream under physiological conditions. The certain fact is that an experimental dissection of the separate roles of these interacting hormones will be a discouragingly difficult task.

(4) The corticoids have an effect on the amount of vasopressin present in the posterior pituitary and the hypothalamus itself. This was first shown by Cavallero et al. (1954), who found the pituitary content of antidiuretic hormone reduced by adrenalectomy and restored to some degree by salt and DOC but more effectively by cortisone. Lloyd (see Gaunt et al. 1957), studying the effects of various influences on the pituitary and hypothalamic content of antidiuretic hormone-like material, found that cortisol always increased the hormone titer in the posterior lobe and generally in the hypothalamus. Similarly, cortisol tends to prevent the depletion of the stainable neurosecretory substance of the posterior pituitary that is characteristically induced by dehydration and, conversely, depletion was facilitated by adrenalectomy (review, Gaunt et al. 1957). This type of result could be explained either on the basis that *the corticoids inhibit release of anti-*

diuretic hormone and its visible neurosecretory substrate *or stimulate its synthesis*. If the former were correct, it would presumably help explain the diuretic action of corticoids, and this viewpoint has been supported recently by the direct evidence of McCann et al. (1958). At present no final choice can be made between these and other possible alternatives. Furthermore, physiological interpretation of the experiments cited must be cautions since (a) the dosage of corticoid required to affect the hormone content of the neurohypophysis is high and possibly unphysiological, and (b) it is difficult to exclude the possibility that the direct effect of the corticoids is on internal fluid and electrolyte distribution or external balance and that by these means the neurohypophysis is affected indirectly.

(5) The finding of Ginsburg (1954) that adrenalectomy in animals facilitates the release of antidiuretic hormone is consistent with the studies on the effects of corticoids on the neurohypophyseal content of this hormone. It is probable that the increased amounts of antidiuretic hormone-like material found by Birnie et al. (1950) in the blood of adrenalectomized rats may have been the result of a facilitated release of the hormone following bleeding, although other interpretations were made at the time. Dingman and associates (review, Dingman 1958) reported that nicotine-induced release of antidiuretic hormone was blocked by glucocorticoids in man as judged by effects on free water clearance. The effects of other stimuli, e.g., hypertonic saline, were not similarly blocked. This and much additional evidence has been well summarized by Dingman (1958), who points to the varied routes by which antidiuretic hormone release may be achieved, including a possible direct action of corticoids.

Elsewhere the authors (Gaunt et al. 1957) concluded a summary of work in this field as follows:

"1. Different types of adrenal hormones have different effects on the responses to vasopressin. Vasopressin actions may be antagonized, augmented or unaffected by simultaneous steroid administration, depending upon the nature of the steroid, its dose and the variable measured. Therefore, generalization in this regard about 'adrenal' influences as such are not appropriate.

"2. The cortisol-like adrenal steroids, under experimental conditions at least, either inhibit the release or facilitate the synthesis of antidiuretic hormone and its visible neurosecretory substrate. The former interpretation seems to be the preferred one, although a combination of the two is not excluded. A distinct difference exists, possibly qualitative in nature, between the effects of cortisol and cortexone in this respect.

"3. The maintenance of the mechanism in various tissues which accounts for the rapid inactivation of antidiuretic hormone is functionally dependent on cortisol-like adrenal steroids.

"4. The postulated neurohumoral substance of hypothalamic origin which causes release of adrenocorticotrophic hormone is probably closely related chemically to the known neurohypophyseal hormones.

"Such findings suggest that the functional 'interrelationship' between the adrenal cortex and neurohypophysis is a many-faceted one. The corticoids apparently affect neurohypophyseal activity *per se* but in addition affect the reponse to neurohypophyseal hormones and such related matters as their rate of inactivation. A reciprocal aspect is that adrenocorticotrophic hormone release, and hence certain aspects of adrenal function, is probably under the control of hypothalamic neurohumors closely related to the known neurohypophyseal hormones. If all of this is the case, the relationship involved cannot be reduced to terms of simple definition and generalization analogous to those between the trophic hormones and their target glands."

C. The adrenals and water intoxication

Rigler (1935), working with mice, and Swingle et al. (1937a), using dogs, first reported *the susceptibility of adrenalectomized animals to water intoxication*. From what has been said above about the necessity of corticoids for the rapid excretion of water, this susceptibility would be anticipated; at the time, however, it was not.

All known corticoids that have been studied by us (adrenocortical extract, DOC, aldosterone, cortisone, cortisol, corticosterone, prednisolone) protect adrenalectomized animals against water intoxication, but, as *in the maintenance of simple water diuresis, the glucocorticoids are most effective.* In addition, the corticoids protect normal animals dramatically from excess water (GAUNT 1943). In fact, this is one of the few and first-discovered instances in which high doses of adrenal hormones protect normal animals against an otherwise lethal stress. It has long been supposed, and probably correctly, that *the disturbances of water intoxication are primarily due to an induced hypotonicity of extracellular fluids,* allowing cellular hydration. Sodium chloride itself is protective, as might be expected, whereas a substance such as urea, not limited in distribution to extracellular fluid, is not. It is of interest, however, that the protective adrenal hormones do not act by conserving salt: they cause increased salt as well as water excretion under conditions of water intoxication. The situation is presumably analogous to that described on p. 536, in which corticoids mobilized salt to the extracellular fluid from internal sources under conditions of water deprivation; probably their salutary effects in water intoxication are due in some measure to this phenomenon.

It is also clear from other evidence that the corticoids act other than on the kidney in water intoxication. Adrenalectomized animals show signs of water intoxication at lower water loads than normal ones (SWINGLE et al. 1941, LILING and GAUNT 1945, HAYS and MATHIESON 1945), and the life-span of water-loaded adrenalectomized (but not intact) animals can be prolonged by corticoids after nephrectomy (BIRNIE et al. 1948). The protection involved is probably not associated with prevention of circulatory collapse as in many other forms of stress in adrenalectomized animals, since blood pressure is relatively well maintained even in untreated adrenalectomized animals during water intoxication (GAUNT 1944).

Wide clinical experience shows that the problems of handling excess water, as they relate to adrenal function, are essentially the same in man as in experimental animals. The danger of water intoxication is real whenever fluid low in sodium content is forced on a patient in whom adrenal hypofunction is present, due either to adrenal disease, pituitary disease or to adrenal atrophy following withdrawal of corticoid therapy.

D. Use of corticoids as diuretics

The experimental physiologist, working with facts discussed above, has been led inevitably to this question: If, in a variety of situations in animals, corticoids enhance sodium and water excretion dramatically, are there not some clinical situations in man in which the same phenomenon could be observed and usefully applied? Until recently, except in adrenal insufficiency itself, the results have been generally negative. LUETSCHER et al. (1955b) and numerous others, however, have used ACTH and glucocorticoids in the treatment of nephroses, but the salutary effects oftentimes occurred after rather then during treatment. SCHEMM and CAMARA (1954, HEIDORN and SCHEMM 1955, HEIDORN et al. 1955) reported the successful use of ACTH in cases of myocardial insufficiency refractory to other forms of therapy. More recently, particularly in Europe, a number of investigators have found the 1-dehydrocorticoids useful diuretics in a variety of diseases characterized by edema (BARIETY et al. 1956, CATTAN and VESIN 1956, RIEMER 1956, FRAISSE et al. 1957, GENNES et al. 1957, GUTNER et al. 1957, VESIN et al. 1957). Cortisone and prednisolone have been found to promote diuresis and decrease total body water in experimental nutritional edema — a situation in which conventional diuretics are relatively ineffective (HELLER et al. 1958). The extent to which the beneficial effects of the 1-dehydrocorticoids may be due to some indirect inhibition of aldosterone secretion in secondary aldosteronism is an interesting problem that has not been finally resolved (VESIN et al. 1957a).

Adrenal cortex and hypertension

Extensive studies have been made both in animal and human subjects on the possible relation of the corticoids in general, and the mineralocorticoids in particular, to hypertensive disease. Only a brief summary of this work is included here. The subject is also considered in Chapter 9 on adrenocortical hyperactivity and in Chapter 8 on adrenocortical hypoactivity. Various aspects of the problem have been reviewed recently elsewhere (MOYER 1959).

Most considerations of the subject have been based on the following basic facts, which also have been reviewed in greater detail in another volume (GAUNT et al. 1959):

(1) *Adrenal hormones are necessary for the maintenance of normal blood pressure*; hypotension is an all but invariable consequence of adrenal insufficiency. Under ordinary conditions the mineralocorticoids in small dosage provide effective replacement therapy and maintain blood pressure after adrenalectomy. In stressful conditions, however, the glucocorticoids are much more active or even essential in preventing circulatory collapse (SWINGLE et al. 1958b). Thus, the suggestion is evident that the two types of corticoids act by different means, a suggestion supported by other evidence cited below.

(2) *Overdosages of most if not all of the corticoids cause hypertension* in certain species of animals, *notably in the rat*. The hypertensive syndrome produced by mineralocorticoids such as DOC differs from that produced by the glucocorticoids (e. g., cortisol) in numerous ways (KNOWLTON et al. 1952). For instance, *the hypertensive action of DOC is dependent upon dietary sodium*, while *that of cortisol may*, if anything, *be intensified by a low sodium intake*; DOC causes widespread renal and cardiovascular pathology, while cortisol does not; uninephrectomy does not increase the hypertensive action of cortisol as it does that of DOC; in severe potassium depletion the glucocorticoids will cause hypertension while DOC will not.

In man, diseases characterized by the excess production of corticoids are often associated with hypertension, e. g., primary aldosteronism (CONN and LOUIS 1956).

(3) In different types of experimental hypertension the maintenance of a hypertensive state is to varying degrees dependent on the presence of corticoids. The results suggest at least a permissive if not a causal action of corticoids in numerous cases (GAUNT et al. 1959).

(4) The action of pressor drugs and other pressor influences is in some but not all cases exaggerated by excess corticoids and diminished in their absence (GROSS and LICHTLEN 1958).

(5) Complex studies concerning the effect of corticoids on the renin content of the kidney and degranulation of the juxtaglomerular apparatus have led to the suggestion from different sources that renin may stimulate the secretion of aldosterone (DEANE and MASSON 1951, GROSS 1958).

From all such work no clear picture emerges as to the exact role of the adrenal cortex in hypertensive disease. It seems certain that the corticoids have a supporting or permissive role in most types of hypertension, but only in those instances of demonstrable excess corticoid secretion can the adrenal steroids be said with confidence to be a direct and primary cause of hypertension. The regulation of normal blood pressure is a highly complex affair; the causes of hypertension are no less so.

One disconcerting note, in attempts to arrive at rational theory concerning the role of corticoids in hypertension, was the discovery by SKELTON (1959) of the experimental syndrome called *adrenal regeneration hypertension*. This condition was produced by removing one adrenal in rats and enucleating the other. While the remaining adrenal was regenerating, hypertension developed if one kidney was removed and the animals were given a high salt diet. Considerable evidence has accumulated to indicate that the abnormal adrenal function involved consists of a distinct *hypofunction*, suggesting that deficient adrenal function may sensitize the animal to the hypertensive influence of a high sodium intake and uninephrectomy (reviews, GAUNT et al. 1959, SKELTON 1959). When the adrenals were completely removed no hypertension developed. If adrenal hypofunction as well as hyperfunction predisposes to hypertension, then indeed the analysis of the role of corticoids in hypertensive disease becomes a formidable problem.

Mineralocorticoids in lower vertebrates

As is apparent from other sections of this monograph, a great amount of work has been spent on the study of adrenal function in mammals. In contrast, very little study has been devoted to the adrenal or any other aspect of the endocrinology of lower vertebrates. The efforts of such people as CHESTER JONES (1957), PICKFORD and ATZ (1957) and GORBMAN (1959) to collect available information in symposia and monographs will serve, it is hoped, to stimulate work in comparative endocrinology; at least, they indicate well the immensity of the unexplored field. (See, also, Chapter 1, p. 141 ff.)

A. Cyclostomata

Adrenocortical tissue was identified in the Cyclostomata by GIACOMINI in 1902. Scattered islets, often consisting of only two or three cells, were found lying in the tissue surrounding the cardinal veins. It is understandable that no physiological data are available concerning the function of these scattered bits of tissue. The usual techniques of adrenalectomy are of course impossible. There also appear to be no recorded investigations involving the use of mammalian cortical hormones in this class.

B. Elasmobranchii

In elasmobranchs the separate interrenal body, not anatomically associated with chromaffin tissue, is homologous to the adrenal cortex of higher forms. Various investigators have performed interrenalectomy in the skate, but only HARTMAN et al. (1944) have attempted to study the effects of this operation on electrolyte metabolism in *Raja erinacea*. These investigators concluded that the interrenal body played no role in the control of electrolyte balance. CHESTER JONES (1956), however, has reviewed the data from these investigations and is of the opinion that *there is a sufficient rise in plasma potassium in the interrenal-ectomized animals to indicate some control of electrolytes by the interrenal body.* Further investigation of this subject is obviously needed.

C. Teleostei

In teleosts the interrenal body is also homologous to the adrenal cortex. Interrenalectomy, however, is impossible because the tissue is scattered in the cephalic kidney and around the walls of the anterior and posterior cardinal veins. It is impossible, therefore, to obtain by ablation experiments direct evidence that this organ controls electrolyte metabolism.

The relationship of the pituitary-adrenal system to electrolyte metabolism has been studied by various investigators. FONTAINE and HATEY (1953) observed adrenal atrophy in the fresh water eel *(Anguilla anguilla)* after hypophysectomy. The effects of this procedure on electrolyte tolerance have been studied by CHAVIN (1954, 1956). Hypophysectomized goldfish *(Carassius auratus)* survive indefinitely in either 0.7% saline or fresh water. Other fresh water forms such as *Phoxinus phoxinus* and *Ameiuris nebulosus* also generally survive hypophysectomy with no evidence that fatalities are due to failure of osmoregulation. The marine species, *Gobius pagenellus*, survives in salt water after hypophysectomy. The salt-regulatory function of the adrenal cortex in fish (if such exists) therefore appears to resemble that of mammals, i. e., it is relatively independent of pituitary control.

Changes in adrenal function have been observed in response to changes in salinity of the habitat. PICKFORD (1953a, 1953b) has found that the normal killifish *(Fundulus heteroclitus)* in salt water has adrenal glands which, by histological criteria, appear to be in a state of quiescence similar to that obtained after hypophysectomy. FONTAINE (1956), using the Nelson-Samuels technique for corticosteroid analysis of the plasma of salmon *(Salmo salar)*, has found the highest concentration of corticoid-like material occurs at the smolt stage. This is the stage at which seaward migration takes place, and FONTAINE postulates that a change in metabolism in preparation for a marine environment may be involved. These observations suggest the hypothesis, as stated by OLIVEREAU and FROMENTIN (1954), that high salinities inhibit interrenal function and low salinities stimulate it. The data on which this hypothesis is based certainly are not definitive, however, and this subject is currently under investigation by FONTAINE and co-workers.

CALLAMAND (1943) has made a study of European eels *(Anguilla anguilla)* during their migrations. It was found that the immature yellow eel is completely amphibiotic. Mature silver eels, however, at the stage of reproductive migration cannot tolerate fresh water. In this medium they suffered a loss of chloride which resulted in asthenia and death. The serum Na^+/K^+ ratio still remained high and treatment with DOC acetate did not alter this nor did it prevent chloride loss and death. This would suggest that the adrenal may not be involved at this stage. Young elvers, however, become highly halophobic at the time when they move to fresh water. At this time, as shown by FONTAINE and MOTAIS (1956), DOC glucoside was effective in abolishing this condition and could prolong the survival of elvers in 37% sea water. DRILHON (1956) also observed in carp *(Cyprinus)* that cortisone increased the resistance to higher saline concentrations. D. SMITH (1956), however, found that adrenocortical extract given at a dose of 0.1 ml per fish per day showed no effect on the tolerance to saline of trout *(Salmo trutta)*. This dose, however, may have been too low to elicit any effects.

With DOC, cortisone, and ACTH, SPALDING (cited by CHESTER JONES 1957) was able to induce an increase in muscle sodium and a reduction of muscle potassium in trout *(Salmo trutta)*. In the goldfish, SEXTON (1955) found that DOC inhibited the uptake of radioactive sodium by the gills but did not affect sodium loss. No effect on muscle sodium was seen, but a decrease occurred in muscle potassium.

Recent advances in analytical techniques have resulted in increased knowledge concerning the nature and function of the adrenal cortex in teleosts.

EDELMAN et al. (1960) showed that cortisol, 9α-fluorocortisol and D,L-aldosterone were not effective in preventing salt loss and prolonging life in various stenohaline and euryhaline marine teleosts placed in fresh water. Under somewhat different conditions HOLMES (1959) observed that both DOC acetate and cortisol caused a net sodium loss in trout (*Salmo gairdneri* Kamloops, Richardson) after an intraperitoneal sodium load.

Interrenal tissue of the sable fish *(Anoplopoma fimbria)* incubated *in vitro* produced cortisol, tissue from the mullet *(Mugil cephalus)* produced cortisol and cortisone, and cortisone was produced by the cichlid *(Tilapia mossambica)* (NANDI and BERN 1960). *In vitro* studies by PHILLIPS and MULROW (1959) have shown what interrenal tissue from *Fundulus heteroclitus* can convert progesterone to cortisol, cortisone and aldosterone.

Blood corticoids in the Fraser River sockeye salmon increase markedly during the spawning run (IDLER et al. 1959). Cortisone, cortisol and 17-hydroxyprogesterone were detected and the combined levels of cortisol and cortisone reached a concentration seventeen times that normally found in man. PHILLIPS et al. (1959) identified cortisol, cortisone, corticosterone and aldosterone in salmon blood *(Oncorhynchus nerka)* and BONDY et al. (1957) have found cortisol in the carp *(Cyprinus carpio)* and the black headed mullet *(Pseudopleuronectes americanus)*.

These investigations, conducted mainly during the two-year period 1959 to 1960, provide interesting new information concerning adrenocortical function in teleosts. Many of the biologically active corticoids found in man have also been found in fish. The use of the terms "mineralocorticoid" and "glucocorticoid" with reference to the activity of specific corticoids in man have their shortcomings but do serve to delineate major activities. In the teleosts the data available as of now would indicate that these same corticoids have a much broader scope of activity and cannot be so readily categorized as either "mineralocorticoids" or "glucocorticoids."

D. Amphibia

The study of adrenal function in amphibians has been more extensive than that in fishes. In this class, surgical adrenalectomy can be accomplished in a reasonably complete fashion. Thus, deficiency and overdosage phenomena can and have been studied. The discouraging fact that conflicting results prevent comprehensive conclusions will undoubtedly be changed in time when technical details are mastered.

I. Effects of exogenous cortical hormones

The effects of adrenocortical hormones on the intact animal have been studied by various investigators. DOW and ZUCKERMAN (1939) observed a decrease in weight in the axolotl *(Ambystoma)* after administration of aqueous adrenocortical extract or DOC. JOHNSON and USSING (1949) found a tremendous increase in sodium uptake that lasted for 5 days following administration of ACTH, but they observed no such effects from DOC treatment in this species.

In the frog *(Rana pipiens)*, HSIEH (1950) has reported that DOC causes a questionable potentiation of water retention by Pitressin. Studies by SAWYER et al. (1950) indicated no effect on water balance in the frog following administration of DOC glucoside, aqueous adrenocortical extract, DOC acetate, cortisone or ACTH. SAWYER (1956), however, did see a consistently positive chloride balance following these treatments.

Suggestive evidence that exogenous DOC may be replacing a normal corticoid function in amphibia is supplied by histological studies of the adrenals of animals given 1 to 2 mg daily. M. MILLER (1953) observed a general depression of adrenal cell volume and degeneration of nuclei in the nonbreeding newt *(Taricha torosa)*. FOWLER (cited by CHESTER JONES 1957) saw much the same response to DOC in "summer" frogs *(Rana temporaria)*.

In view of these somewhat conflicting observations, SAWYER (1956) concluded that *the amphibian adrenal may be essential to prevent overhydration and may contribute to sodium and chloride balance* but that satisfactory evidence on the point is still lacking.

II. Adrenalectomy in amphibians

Adrenalectomy can be performed in frogs and the consequences were first observed by ABELOUS and LANGLOIS (1891 a, b). Relative to other members of the lower vertebrates, with the possible exception of elasmobranchs, the ease with which this may be accomplished is probably responsible for many of the early studies, as reviewed by CHESTER JONES (1957). The literature contains what appear to be conflicting reports regarding survival after adrenalectomy. ANGERER and ANGERER (1942, 1949) reported that adrenalectomized frogs *(Rana pipiens)* showed an increase in body water. FOWLER and CHESTER JONES (1955) observed that adrenalectomized "winter" frogs *(Rana temporaria)* live for extended periods, are rather quiescent, show no great uptake of water or change in body weight, and after 14 days muscle sodium is decreased only slightly. In contrast, "summer" frogs generally die within 48 hours following adrenalectomy. The usual symptoms are edema and asthenia. These symptoms are also described for *Rana esculenta* and *Rana temporaria* by MAES (1937) and others (see CHESTER JONES 1956). One disturbing observation in the FOWLER and CHESTER JONES study was that sham-operated controls also died within this same period. The duration of survival, being longer in winter, appears to vary with the season. This is probably a reflection of the metabolic rate, the rate of corticoid utilization in the "summer" frog being such that secretion cannot keep up with the demand after the stress of surgery.

This postulate receives support from further studies (CHESTER JONES 1957) in which both adrenalectomized and sham-operated "summer" frogs were maintained in isotonic saline following the operation. Several days later, upon removal to tap water, the adrenalectomized animals died while the sham-operated did not. These results would also indicate that the adrenal is in some way directly involved in salt metabolism.

Direct evidence for adrenal control of serum electrolytes in amphibians is supplied by the work of MARENZI and FUSTINONI (1938), who observed a 24% drop in serum sodium following adrenalectomy in toads *(Bufo arenarum)*. This loss of sodium has also been investigated by FOWLER (cited by CHESTER JONES 1957) in adrenalectomized frogs *(Rana temporaria)*. Excretion studies showed that these animals lose body sodium via the kidney more rapidly than normal. Also, a sodium load was excreted more rapidly. Loss of sodium via the skin is also increased in hypophysectomized frogs, as shown by the studies of JØRGENSEN (1947). He postulated that this increase in permeability may be a result of adrenal insufficiency.

Cicardo (1947), using adrenalectomized toads *(Bufo arenarum)*, observed that muscle potassium was lost more quickly than normal when tissues were perfused with potassium-free Ringer's solution.

Replacement therapy in adrenalectomized anurans with known corticoid hormones has not been thoroughly investigated and the results are somewhat inconclusive. Fustinoni (1938) was not able to increase the survival time of adrenalectomized toads *(Bufo arenarum)* with adrenocortical extract. The results of Clark et al. (1944) show that DOC is only slightly effective in increasing survival time in adrenalectomized toads *(Bufo marinus)*. Cicardo (1947), however, observed that pretreatment of adrenalectomized toads with DOC reduced the loss of muscle potassium in his muscle-perfusion studies. Also, using perfused frog hind-limb preparations, Hyman and Chambers (1943) observed that the rate of edema formation was reduced by 11-dehydrocorticosterone, DOC and cortisone, in decreasing order of efficacy.

In conclusion, it is seen that some of the same problems mentioned in work on fishes are apparent in amphibian studies. The failure to obtain a satisfactory replacement therapy for long periods of time after adrenalectomy adds to the difficulty of satisfactory experimentation. It seems clear, however, that the corticoids have some role in regulating fluid and electrolyte metabolism in amphibia roughly analagous to that demonstrable in mammals.

E. Reptilia

Knowledge of adrenocortical function in reptiles is very meager. We do know that the adrenal cortex of the reptile is pituitary-dependent as shown by Schaeffer (1933) in *Thamnophis sirtalis* and *T. radix*, by M. Miller (1952) in *Xantusia vigilis*, and by Wright and Chester Jones (1957) in *Agama* and *Natrix*. The adrenal cortex in hypophysectomized animals becomes shrunken, can be restored to normal weight and appearance by ACTH or by implantation of hypophyses, and will also hypertrophy following unilateral adrenalectomy.

Cortisone in the intact *Xantusia vigilis* causes a loss of lipide from and an over-all shrinkage of cortical tissue with degeneration of the cells (M. Miller 1953). In this same study it was shown that DOC produces a loss of lipide, a shrinkage of cortical tissue, but little or no cellular degeneration. From such evidence it seems probable that the reciprocal control of pituitary-adrenal functions in reptiles follows a pattern closely resembling that known to occur in mammals in function and perhaps in secretory products.

Only a few reports are recorded concerning the effects of adrenalectomy in reptiles. In the lizard *Anolis*, Adams and Harland (1929) and Kleinholz (1938) found that survival in the adrenalectomized animal varied from a minimum of 8 hours to a maximum of 6 days. Do Valle (1945) reported that the snake *Philodryas* survives adrenalectomy for a period of 7 days. Wright (cited by Chester Jones 1956) observed that the lizard *Lacerta viridis* survives adrenalectomy for a period of 3 days and the snake *Natrix natrix* may survive for 15 days. The muscle sodium of this snake falls after adrenalectomy.

These studies indicate that, as in mammals, *the adrenal gland in snakes is necessary for survival and is very probably involved in electrolyte control.* Final confirmation of this latter point must await metabolic studies conducted after adrenalectomy and replacement therapy.

F. Aves

I. Pituitary-adrenal relationship

The bird adrenal glands, although discrete bodies in themselves, are a rather heterogenous composite of cords of cortical tissue intermingling with chromaffin tissue. Various species have been studied in an attempt to find some semblance of zonation, and it appears that the adrenal of the brown pelican *(Pelecanus occidentalis)* most closely approaches cortical zoning. This, at best, is only a poor facsimile of the zonation found in the typical mammalian gland. (Cf. p. 15.)

The adrenal gland has been studied extensively in several species of birds. Most of these investigations, however, have been largely morphological. Physiological studies so far have been few, perhaps with good reason. The contiguous anatomical relationship between the adrenals and the posterior vena cava is such that adrenalectomy in the bird or sampling of adrenal vein blood is either extremely difficult or impossible. Also, the processes by which urinary waste is dumped into the cloaca, water is reabsorbed, and the solids are then excreted in admixture with feces has made study of electrolyte, carbohydrate and nitrogenous wastes very difficult. Some of these obstacles have been overcome, as will be pointed out later, and the appearance of physiological studies of adrenal function in birds should become increasingly more frequent.

Proof that the adrenal gland of birds is under control of the pituitary gland has been derived from various sources. R. MILLER and RIDDLE (1942) found that hypophysectomy in Carneau pigeons *(Columba livia)* was followed by a 25% loss of adrenal weight at 10 days. This weight loss was reflected by a decrease in the percentage of cortical tissue in the gland, which became even more pronounced with time. SCHOOLEY (1939) has reported that both pigeons and ducks *(Anas)* survive hypophysectomy well. Reports of hypophysectomy in the chicken *(Gallus)* by MITCHELL (1929) and HILL and PARKES (1934), however, indicated that survival was rather poor. It is probable that death was due to difficulties other than adrenal dysfunction, and these problems have now largely been solved by the work of NALBANDOV and CARD (1942) and BAUM and MEYER (1956), who obtained a high percentage of survivors in hypophysectomized chickens.

From these studies it is assumed that, as in mammals, *the life-maintaining principle of the adrenal cortex in birds is only partially under the control of the pituitary*, if at all.

Using ACTH preparations obtained from beef pituitaries, BATES et al. (1940) and R. MILLER and RIDDLE (1942) showed that adrenal hypertrophy occurred following injection of this material. These authors postulated that the bird pituitary secretes a trophic hormone of similar nature, because a compensatory hypertrophy was noted following unilateral adrenalectomy. K. BROWN (1956) and K. BROWN et al. (1958) have also shown by physiological measurements that an intact pituitary-adrenal axis is necessary for an adequate response to stressful stimuli. Following an operation that exteriorizes the ureters of the chicken, K. BROWN (1956) observed adrenal hypertrophy. This operation also facilitated the collection of 24-hour urine samples that had not been subjected to water reabsorption in the cloaca or contamination by feces. Urine analysis following the operation showed a progressive increase in the excretion of sodium, potassium and nitrogenous waste for a period of 4 days and then a decline to a stable level. In hypophysectomized birds these changes in urinary constituents did not occur in response to exteriorizing the ureters, nor was any hypertrophy of the adrenals noted.

II. Adrenalectomy in birds

Early attempts at adrenalectomy in the pigeon (GOURFEIN 1896), and in the duck and chicken (PARKES and SELYE 1936, BÜLBRING 1937a, 1937b, 1940, and HERRICK and TORSTVEIT 1938) usually resulted in very short survival ranging from 5 to 24 hours. PARKINS (1931) and K. BROWN (1956), however, who did two-stage adrenalectomies in the chicken, found survivals ranging up to 6 days and 2.5 days respectively.

Salt therapy has been employed in adrenalectomized birds but the results are inconclusive. BÜLBRING (1937a, 1937b) found it to be of little help for survival in the adrenalectomized drake *(Anas)*. In contrast, HERRICK and TORSTVEIT (1938) observed adrenalectomized chickens *(Gallus)* surviving up to 82 days on salt therapy. In 1950 adrenalectomized Carneau pigeons *(Columba livia)*, RIDDLE et al. (1944) found that sodium chloride and sodium bicarbonate therapy prolonged life to an average of 9 days. Further study on the efficacy of salt therapy appears to be in order. K. BROWN (1956) and K. BROWN et al. (1958) have shown that chickens in which sodium intake was controlled by force-feeding a complete synthetic diet responded to adrenalectomy by reducing water intake and output of urinary sodium, uric acid, and total nitrogen. Serum sodium was also slightly decreased. On the basis of this evidence sodium must have been either passing into the tissues or being excreted in the feces.

III. Effects of exogenous cortical hormones

The effects of known adrenocortical hormones in birds have been described by various investigators. R. MILLER and RIDDLE (1942) found that implantation of DOC pellets in either intact or hypophysectomized pigeons resulted in adrenal atrophy. K. BROWN (1956) and K. BROWN et al. (1958) injected 4 mg per kg of DOC acetate into intact chickens and observed polyuria, polydipsia, sodium retention and possible potassium retention. Fifteen mg per kg of cortisone acetate daily also induced polyuria but caused marked increases in the excretion of sodium, potassium, uric acid and total nitrogen. No change in serum sodium or potassium was noted with either steroid.

Using corticoids in adrenalectomized birds, R. MILLER and RIDDLE (1943) reported that 2 mg of DOC per day maintained normal growth in adrenalectomized pigeons. In adrenalectomized drakes, BÜLBRING (1940) and LEROY and BENOIT (1954) found that daily doses of 5 mg of DOC or 50 mg of cortisone, respectively, would maintain survival. Four mg DOC per kg per day is also the life-maintenance dose reported by K. BROWN (1956) and K. BROWN et al. (1958) for adrenalectomized chickens.

Urinary excretion in the adrenalectomized chicken under hormone therapy has also been investigated by K. BROWN (1956) and K. BROWN et al. (1958). DOC at 4 mg per kg per day raised the water intake and urinary total nitrogen excretion to normal. Urinary sodium excretion was still low but the serum levels of sodium and potassium were returned to normal. The low blood glucose which K. BROWN has reported to be characteristic of adrenalectomized chickens was also raised to normal levels by DOC. He found, in addition, that cortisone acetate, 10 mg per kg per day, returned most measured functions in the adrenalectomized chicken to nearly normal. Serum and urinary sodium and potassium, water intake, and urine volume were essentially at pre-operative levels. At this dosage there was a rather marked daily weight loss, presumably due to the catabolic effects of the cortisone.

Various aspects of adrenal physiology of the bird, based on the work of K. BROWN (1956) and K. BROWN et al. (1958), are intriguing because of *the differences noted between the response of chickens and those usually seen in mammals to various corticoids and to adrenalectomy.* In the bird, (a) adrenalectomy is followed by sodium retention while potassium excretion remains normal; (b) DOC causes retention of both sodium and potassium; (c) cortisone causes marked excretion of both sodium and potassium; (d) DOC raises blood sugar markedly. Further work will be necessary to assess the mechanism and significance of these apparent class differences.

G. Chemical nature of corticoids in lower vertebrates

One marked and significant advance in knowledge of the adrenals of lower vertebrates has been provided recently by PHILLIPS and CHESTER JONES (1957). They employed ingenious methods to collect adrenal blood samples prior to systemic dilution in several species and identified chromatographically some of the corticoids being secreted. In all species either corticosterone, cortisol, cortisone, or all three were present (Table 8). Aldosterone was detectable in the chicken. More recent elaboration of this work is further described by CHESTER JONES et al. (1959). PHILLIPS and MULROW (1959) have also obtained results that strongly suggest that the head kidney in the killifish *(Fundulus heteroclitus)*, which contains interrenal tissue, is capable of producing cortisol, cortisone and aldosterone *in vitro*. CARSTENSEN et al. (1959) have also shown that adrenal cortical tissue of the bullfrog, *Rana catesbiana*, produces *in vitro* relatively large amount of aldosterone in response to mammalian ACTH.

Table 8. *The identity of adrenocortical secretions in adrenal-venous blood of lower vertebrates.* (Compiled from PHILLIPS and CHESTER JONES 1957)

Species	Compound identified	Conc., μg %
Ray	Corticosterone	8
Dogfish	Cortisol	2—3
Cod	Cortisol	1
Lungfish	Cortisol	15
Xenopus	Cortisol	7
Snake (1 hour after 25 I.U. ACTH)	Corticosterone	156
	Cortisol	8.5
	Cortisone	2
Capon (untreated)	Corticosterone	312
	Cortisol	3
	Cortisone	Trace
Capon No. 1, (25 I.U. ACTH per day for 3 days)	Corticosterone	68.5
	Cortisol	3
	Cortisone	Trace
Capon No. 2, (25 I.U. ACTH per day for 3 days)	Corticosterone	65.5
	Cortisol	3.5
	Cortisone	2.5
	Aldosterone	1.0

These results do not indicate the amounts of these steroids secreted nor prove that they are the only active ones produced, but they do show that the lower vertebrates secrete in part the same adrenal hormones as do mammals. This information adds much in confidence and significance to experimentation in which mammalian steroids are injected into lower vertebrates. It means that the results not do have to carry the burden of suspicion that they are entirely pharmacological.

With such information accumulating, it can be anticipated that within the near future a much more informative chapter can be written on the comparative endocrinology of the adrenal cortex and its homologues.

Literature

ABELOUS, J.-E., Note sur les fonctions des capsules surrénales chez la grenouille. C. R. Soc. Biol. (Paris) **43**, 792—789 (1891a).

— — and P. LANGLOIS: Le mort des grenouilles après la destruction des deux capsules surrénales. C. R. Soc. Biol. (Paris) **43**, 855—857 (1891b).

ADAMS, A. E., and M. HARLAND: The effects of adrenalectomy in the lizard. Anat. Rec. **41**, 42—43 (1929).

ALLEN, M. J., and A. H. CORWIN: Chemical studies on pinacols obtained from *p*-aminoacetophenone and *p*-aminopropiophenone. J. Amer. chem. Soc. **72**, 117—121 (1950).

ALLERS, W. D., and E. C. KENDALL: Maintenance of adrenalectomized dogs without cortin, through control of the mineral constituents of the diet. Amer. J. Physiol. **118**, 87—94 (1937).

ANDERSON, E., and M. JOSEPH: Urinary excretion of radioactive Na and K in adrenalectomized rats, with and without salt. Proc. Soc. exp. Biol. (N. Y.) **40**, 347—350 (1939).

ANGERER, C. A., and H. ANGERER: The effect of dehydration on the viability of adrenal-insufficient frogs. Fed. Proc. **1**, 3 (1942).

— — The rate and total loss of body water on the survival time of adrenalectomized frogs. Proc. Soc. exp. Biol. (N. Y.) **71**, 661—665 (1949).

BARGER, A. C., F. P. MULDOWNEY and M. R. LIEBOWITZ: Role of the kidney in the pathogenesis of congestive heart failure. Circulation **20**, 273—285 (1959).

BARIETY, M., A. GAJDOS, C. COURY et M. KOVACEVIC: Action comparée de la métacortandracine sur la diurèse et le métabolisme du sodium et du potassium chez le sujet normal et au cours de la cirrhose hépatique. Presse méd. **64**, 1699—1701 (1956).

BARTTER, F. C., and P. FOURMAN: The loss and the retention of potassium which occur immediately after starting and stopping adrenocorticotropic hormone. J. clin. Invest. **30**, 627—628 (1951).

— — F. ALBRIGHT, A. P. FORBES, W. K. MCK. JEFFERIES, G. GRISWOLD, E. DEMPSEY, D. BRYANT and E. CARROSS: The effect of adrenocorticotropic hormone in panhypopituitarism. J. clin. Invest. **29**, 950—971 (1950).

BATES, R. W., O. RIDDLE and R. A. MILLER: Preparation of adrenotropic extracts and their assay on two-day chicks. Endocrinology **27**, 781—792 (1940).

BAUM, G. J., and R. K. MEYER: Influence of diethylstilbestrol on lipids in intact and hypophysectomized cockerels. Endocrinology **58**, 338—346 (1956).

BAUMANN, E. J., and S. KURLAND: Changes in inorganic constituents of blood in suprarenalectomized cats and rabbits. J. biol. Chem. **71**, 281—302 (1927).

BEATTY, C. H., R. M. BOCEK and R. D. PETERSON: Water tolerance test for functioning adrenal cortical tissue in the rat. Amer. J. Physiol. **177**, 287—291 (1954).

BERGENSTAL, D. M., R. HERTZ, M. B. LIPSETT and R. H. MOY: Chemotherapy of adrenocortical cancer with o,p′-DDD. Ann. Intern. Med. **53**, 672—682 (1960).

BERGER, E. Y., G. P. QUINN and M. A. HOMER: Effect of desoxycorticosterone on the colon: its relation to the action of cation exchange resins in man. Proc. Soc. exp. Biol. (N. Y.) **76**, 601—604 (1951).

BIRNIE, J. H.: The inactivation of posterior pituitary antidiuretic hormone by liver extracts. Endocrinology **42**, 33—38 (1953).

— W. J. EVERSOLE and R. GAUNT: The extra-renal action of desoxycorticosterone: survival and water intoxication studies. Endocrinology **42**, 412—420 (1948).

— — W. R. BOSS, C. M. OSBORN and R. GAUNT: An antidiuretic substance in the blood of normal and adrenalectomized rats. Endocrinology **47**, 1—12 (1950).

BONDY, P. K., G. V. UPTON and G. E. PICKFORD: Demonstration of cortisol in fish blood. Nature **179**, 1354—1355 (1957).

BOURGUET, J., et R. DEBAUCHE: Influence de la désoxycorticostérone sur l'action antidiurétique de la pituitrine. C. R. Soc. Biol. (Paris) **151**, 402—405 (1957).

BRITTON, S. W.: Adrenal insufficiency and related considerations. Physiol. Rev. **10**, 617—682 (1930).

— and H. SILVETTE: The apparent prepotent function of the adrenal glands. Amer. J. Physiol. **100**, 701—713 (1932).

BROWN, J. H. U.: Influence of the drug DDD on adrenal cortical function in adult rats. Proc. Soc. exp. Biol. (N. Y.) **83**, 59—62 (1953).

— Specific inhibitors of the adrenal cortex. Nature **187**, 985—988 (1960).

— J. GRIFFIN and R. B. SMITH, III.: Excretion of urinary 17-hydroxycorticoids in dogs fed DDD. Metabolism **4**, 542—544 (1955).

BROWN, K. I.: Electrolyte and water metabolism in male chickens. Ph. D. Thesis, Univ. Wisconsin (1956).

— D. J. BROWN and R. K. MEYER: Effect of surgical trauma, ACTH, and adrenal cortical hormones on electrolytes, water balance, and gluconeogenesis in male chickens. Amer. J. Physiol. **192**, 43—50 (1958).

BUCHEM, F. S. P. VAN: The electrocardiogram and potassium metabolism. Amer. J. Med. **23**, 376—384 (1957).

BÜLBRING, E.: Relation between size of testes and requirement of cortical extract in adrenalectomized drakes. J. Physiol. (Lond.) **91**, 18P (1937a).

BÜLBRING, E.: The standardization of cortical extract by the use of drakes. J. Physiol. (Lond.) **89**, 64—80 (1937b).
— The relation between cortical hormone and the size of the testis in the drake, with some observations on the effect of different oils as solvents and on desoxycorticosterone acetate. J. Pharmacol. **69**, 52—63 (1940).
BUNIM, J. J., M. M. PECHET and A. J. BOLLET: Studies on metacortandralone and metacortandracin in rheumatoid arthritis. J. Amer. med. Ass. **157**, 311—318 (1955).
BUSH, I. E.: Species differences in adrenocortical secretion. J. Endocr. **9**, 95—100 (1953).
BYRNES, W. W., L. F. BARNES, B. J. BOWMAN, W. E. DULIN, E. H. MORLEY and R. O. STAFFORD: Adrenal corticoid activities of 2-methylhydrocortisone acetate and 2-methyl-9α-fluorohydrocortisone acetate. Proc. Soc. exp. Biol. (N. Y.) **91**, 67—70 (1956).
CALLAMAND, O.: L'anguille européenne (Anguilla anguilla L.). Les bases physiologiques de sa migration. Ann. Inst. océanogr. Monaco **21**, 361—440 (1943).
CARSTENSEN, H., A. C. J. BURGERS and C. H. LI: Isolation of aldosterone from incubates of adrenals of the American bullfrog and stimulation of its production by mammalian adrenocorticotrophin. J. Amer. chem Soc. **81**, 4109—4110 (1959).
CARTLAND, G. F., and KUIZENGA, M. H.: Fractionation studies on adrenal cortex extract with notes on the distribution of biological activity among the crystalline and amorphous fractions. Endocrinology **24**, 526—535 (1939).
CATTAN, R., et P. VESIN: Bases cliniques et physio-pathologiques du traitement des syndromes oedémateux par la delta-cortisone (métacortandracine). Sem. Hôp. Paris **32**, 712—719 (1956).
CAVALLERO, C., E. DOVA and L. ROSSI: Antidiuretic activity in the neurohypophysis of rats after adrenalectomy and replacement therapy. J. Endocr. **10**, 228—237 (1954).
CHART, J. J., and E. S. SHIPLEY: The mechanism of sodium retention in cirrhosis of the liver. J. clin. Invest. **32**, 560 (1953).
— — and E. S. GORDON: Evidence for a sodium retaining factor in toxemia of pregnancy. Proc. Soc. exp. Biol. (N. Y.) **78**, 244—246 (1951).
— N. HETZEL and R. GAUNT: Effect of new adrenal steroids on electrolyte and water excretion. Proc. Soc. exp. Biol. (N. Y.) **91**, 73—75 (1956).
— H. SHEPPARD, M. J. ALLEN, W. L. BENCZE and R. GAUNT: New amphenone analogs as adrenocortical inhibitors. Experentia (Basel) **14**, 151 (1958).
CHAVIN, W.: The role of the pituitary-adrenal mechanism in the reappearance of melanin and melanophores in the goldfish, *Carassius auratus* L. Ph. D. Dissertation, New York Univ. (1954).
— Pituitary-adrenal control of melanization in xanthic goldfish, *Carassius auratus* L. J. exp. Zool. **133**, 1—45 (1956).
CHESTER JONES, I.: The role of the adrenal in the control of salt-electrolyte metabolism in vertebrates. Mem. Soc. Endocr. **5**, 120—124 (1956).
— The Adrenal Cortex. London: Cambridge Univ. Press 1957.
— J. G. PHILLIPS and W. N. HOLMES: Comparative physiology of the adrenal cortex. In: A. GORBMAN, Ed., Comparative Endocrinology, 582—612. New York: Wiley 1959.
CICARDO, V.-H.: Biological importance of potassium (in Spanish). Buenos Aires: El Ateneo 1947.
CLARK, W. G.: Effect of adrenalectomy upon intestinal absorption of sodium chloride. Proc. Soc. exp. Biol. (N. Y.) **40**, 468—470 (1939).
— E. L. BRACKNEY and R. A. MILINER: Adrenalectomy in frogs and toads. Proc. Soc. exp. Biol. (N. Y.) **57**, 222—224 (1944).
CLINTON, M., JR., G. W. THORN, H. EISENBERG and K. E. STEIN: Effect of synthetic desoxycorticosterone acetate therapy on plasma volume and electrolyte balance in normal dogs. Endocrinology **31**, 578—581 (1942).
COLE, D. F.: The action of deoxycorticosterone acetate and of dietary sodium and potassium on skeletal muscle electrolyte in adrenalectomized rats. Acta endocr. (Kbh.) **14**, 245—253 (1953).
COLLINS, E. J.: Steroid-induced adrenal-pituitary hypofunction. Endocrinology **58**, 777—780 (1956).
CONN, J. W.: Electrolyte composition of sweat. Arch. intern. Med. **83**, 416—428 (1949).
— and L. H. LOUIS: Primary aldosteronism, a new clinical entity. Ann. intern. Med. **44**, 1—15 (1956).
— S. S. FAJANS and L. H. LOUIS: Metabolic effects in man of orally and parenterally administered corticosterone (compound B). Trans. Ass. Amer. Phycns **64**, 269—278 (1951).
CONWAY, E. J.: Electrolyte physiology. Nature (Lond.) **180**, 1017—1020 (1957).
COPPAGE, W. S., D. ISLAND, M. SMITH and G. W. LIDDLE: Inhibition of aldosterone secretion and modification of electrolyte excretion in man by a chemical inhibitor of 11β-hydroxylation. J. clin. Invest. **38**, 2101—2110 (1959).

CRIGLER, J. F., JR., S. H. SILVERMAN and L. WILKINS: Further studies on the treatment of congenital adrenal hyperplasia with cortisone. IV. Effect of cortisone and Compound B in infants with disturbed electrolyte metabolism. Pediatrics **10**, 397—413 (1952).

CUETO, C., and J. H. U. BROWN: Biological studies on an adrenocorticolytic agent and the isolation of the active components. Endocrinology **62**, 334—339 (1958).

DARROW, D. C., H. E. HARRISON and M. TAFFEL: Tissue electrolytes in adrenal insufficiency. J. biol. Chem. **130**, 487—502 (1939).

DAVIS, A. K., A. C. BASS and R. R. OVERMAN: Comparative effects of cortisone and DCA on ionic balance and fluid volumes of normal and adrenalectomized dogs. Amer. J. Physiol. **166**, 493—503 (1951).

DAVIS, J. O., and D. S. HOWELL: Diuresis during maintenance of adrenalectomized dogs with ascites on cortisone. Endocrinology **53**, 653—660 (1953).

— M. J. GOODKIND, M. M. PECHET and W. C. BALL, JR.: Increased excretion of aldosterone in urine from dogs with rightsided congestive heart failure and from dogs with thoracic inferior vena cava constriction. Amer. J. Physiol. **187**, 45—50 (1956).

DEANE, H. W., and G. M. C. MASSON: Adrenal cortical changes in rats with various types of experimental hypertension. J. clin. Endocr. **11**, 193—208 (1951).

DENNIS, C., and E. H. WOOD: Intestinal absorption in the adrenalectomized dog. Amer. J. Physiol. **129**, 182—190 (1940).

DESAULLES, P. A.: Preliminary note on certain pharmacological properties of 3β,16α-dihydroxyallopregnan-20-one. Experientia (Basel) **15**, 301—303 (1959).

— J. TRIPOD u. W. SCHULER: Wirkung von Elektrocortin auf die Elektrolyt- und Wasserausscheidung im Vergleich zu Desoxycorticosteron. Schweiz. med. Wschr. **83**, 1088—1089 (1953).

— W. SCHULER u. R. MEIER: Wirkungen von Kombinationen von Nebennierenrindenhormonen auf Elektrolyt-Wasser-Stoffwechsel und mesenchymale Reaktionen. Schweiz. med. Wschr. **85**, 662 (1955).

DETTBARN, W., u. F. GROSS: Wasser- und Kochsalzbelastungen an nebennierenlosen Hunden unter Cortexon, Aldosteron und 9-α-Fluorocortisol. Schweiz. med. Wschr. **85**, 662 (1955).

DINGMAN, J. F.: Hypothalamus and the endocrine control of sodium and water metabolism in man. Amer. J. med. Sci. **235**, 79—99 (1958).

DO VALLE, J. R.: Survival of snakes (*Philodryas* sp.) following adrenalectomy (in Portuguese). Mem. Inst. Butantan (S. Paulo) **18**, 237—240 (1945).

DOW, D., and S. ZUCKERMAN: The effect of vasopressin, sex hormones and adrenal cortical hormone on body water in axolotls. J. Endocr. **1**, 387—398 (1939).

DRILHON, A.: Modifications des protides sériques (globulines-euglobulines) chez la carpe soumise à la cortisone. C. R. Soc. Biol. (Paris) **150**, 476—479 (1956).

EDELMAN, I. S., H. L. YOUNG and J. B. HARRIS: Effects of corticosteroids on electrolyte metabolism during osmoregulation in teleosts. Amer. J. Physiol. **199**, 666—670 (1960).

Editorial: Aldosterone. Lancet **1956 i**, 796.

EFRON, D. H.: The effect of adrenalectomy on the content and turnover of sodium and potassium in various organs. Acta endocr. (Kbh.) **26**, 209—214 (1957).

EGDAHL, R. H., J. B. RICHARDS and D. M. HUME: Effect of reserpine on adrenocortical function in unanesthetized dogs. Science **123**, 418 (1956).

EVERSOLE, W. J., A. EDELMANN and R. GAUNT: Effect of adrenal cortical transplants on life-maintenance and water intoxication. Anat. Rec. **76**, 271—281 (1940).

— R. GAUNT and E. C. KENDALL: The effect of adrenal steroids in water intoxication. Amer. J. Physiol. **135**, 378—382 (1942).

FARRELL, G. (L.): The physiological factors which influence the secretion of aldosterone. Recent Progr. Hormone Res. **15**, 275—298 (1959a).

— Steroidogenic properties of extracts of beef diencephalon. Endocrinology **65**, 29—33 (1959b).

— Glomerulotropic activity of an acetone extract of pineal tissue. Endocrinology **65**, 239—241 (1959c).

— Isolation of desoxycorticosterone from adrenal venous blood of the dog; effect of hypophysectomy and ACTH. Proc. Soc. exp. Biol. (N. Y.) **87**, 587—590 (1954).

— R. C. BANKS and S. KOLETSKY: The effect of corticosteroid injection on aldosterone injection. Endocrinology **58**, 104—108 (1956).

FERREBEE, J. W., C. RAGAN, D. W. ATCHLEY and R. F. LOEB: Desoxycorticosterone esters. Certain effects in the treatment of Addison's disease. J. Amer. med. Ass. **113**, 1725—1731 (1939).

— D. PARKER, W. H. CARNES, M. K. GERITY, D. W. ATCHLEY and R. F. LOEB: Certain effects of desoxycorticosterone: The development of "diabetes insipidus" and the replacement of muscle potassium by sodium in normal dogs. Amer. J. Physiol. **135**, 230—237 (1941).

FLANAGAN, J. B., A. K. DAVIS and R. R. OVERMAN: Mechanism of extracellular sodium and chloride depletion in the adrenalectomized dog. Amer. J. Physiol. **160**, 89—102 (1950).

FONTAINE, M.: The hormonal control of water and salt-electrolyte metabolism in fish. Mem. Soc. Endocr. **5**, 69—82 (1956).
— et J. HATEY: Recherches sur le contrôle hypophysaire de l'interrénal antérieur d'un poisson téléostien, l'anguille *(Anguilla anguilla* L.*)*. I. Variations pondérales de l'interrénal antérieur. C. R. Soc. Biol. (Paris) **147**, 217—220 (1953).
— et R. MOTAIS: Influence du glucoside de desoxycorticostérone sur l'halotropisme de la jeune anguille (civelle) et sa résistance aux eaux de mer sursalées. C. R. Acad. Sci. (Paris) **242**, 1359—1361 (1956).
FORSYTH, B. T.: A sodium-excreting effect of desoxycorticosterone in adrenalectomized mice. Endocrinology **52**, 65—72 (1953).
FORTIER, C.: Neuroendocrine relationships. Progr. Neurol. Psychiat. **12**, 252—261 (1957).
FOWLER, M. A., and I. CHESTER JONES: The adrenal cortex in the frog *Rana temporaria* and its relation to water and salt electrolyte metabolism. J. Endocr. **13**, 6—7 (1955).
FRAISSE, PLAN et BRONDEL: Action de la delta-cortisone dans les cirrhoses ethyliques avec ascité. J. Méd. Lyon **38**, 255—280 (1957).
FRAWLEY, T. F., and G. W. THORN: The relation of the salivary sodium-potassium ratio to adrenal cortical activity. In: J. R. MOTE, Ed., Proc. 2nd Clinical ACTH Conference, Vol. 1, 115—122. New York: Blakiston 1951.
FREGLY, M. J.: NaCl appetite of adrenalectomized rats. Proc. Soc. exp. Biol. (N. Y.) **97**, 144—148 (1958).
FRIED, J.: Biological effects of 9α-fluorohydrocortisone and related halogenated steroids in animals. Ann. N. Y. Acad. Sci. **61**, 573—581 (1955).
— Structure-activity relationships in the field of halogenated steroids. Cancer (N. Y.) **10**, 752—756 (1957).
— R. W. THOMA, D. PERLMAN, J. E. HERZ and A. BORMAN: The use of microorganisms in the synthesis of steroid hormones and hormone analogues. Recent Progr. Hormone Res. **11**, 149—177 (1955).
FROST, R. C., and R. V. TALMAGE: Fluid and electrolyte shifts in the hydrated adrenalectomized rat. Endocrinology **49**, 606—612 (1951).
— — Fluid and electrolyte shifts in the hydrated adrenalectomized rat after nephrectomy. Proc. Soc. exp. Biol. (N. Y.) **80**, 171—175 (1952).
FUSTINONI, O.: L'asthénic des insuffisances hypophysaire et surrénale du crapaud. C. R. Soc. Biol. (Paris) **129**, 1252—1253 (1938).
GAILLARD, L.: Insuffisance surrénale et azotémie. Bull. Soc. Méd. Paris **37**, 272—276 (1914).
GAUDINO, M., and M. F. LEVITT: Influence of the adrenal cortex on body water distribution and renal function. J. clin. Invest. **28**, 1487—1497 (1949).
GAUNT, R.: Adrenalectomy in the rat. Amer. J. Physiol. **103**, 494—510 (1933).
— Protection of normal rats against death from water intoxication with adrenal cortical substances. Proc. Soc. exp. Biol. (N. Y.) **54**, 19—21 (1943).
— Water diuresis and water intoxication in relation to the adrenal cortex. Endocrinology **34**, 400—415 (1944).
— The adrenal cortex in salt and water metabolism. Recent Progr. Hormone Res. **6**, 247—268 (1951).
— Chemical control of growth in animals. In: E. J. BOELL, Ed., Dynamics of Growth Processes, 183—211. Princeton, N. J.: Princeton University Press 1954.
— and W. J. EVERSOLE: Notes on the history of the adrenal cortical problem. Ann. N. Y. Acad. Sci. **50**, 511—521 (1949).
— and H. W. HAYS: Role of progesterone and other hormones in survival of pseudopregnant andrenalectomized ferrets. Amer. J. Physiol. **124**, 767—773 (1938).
— J. H. GAUNT and C. E. TOBIN: Colony differences in survival of adrenalectomized rats. Proc. Soc. exp. Biol. (N. Y.) **32**, 888—892 (1935).
— H. E. POTTS and E. LOOMIS: Post-adrenalectomy diuresis: effects of cortical extracts, salts and estrone. Endocrinology **23**, 216—222 (1938).
— W. J. EVERSOLE and E. C. KENDALL: Influence of some steroid hormones on lactation in adrenalectomized rats. Endocrinology **31**, 84—88 (1942).
— J. H. BIRNIE and W. J. EVERSOLE: Adrenal cortex in water metabolism. Physiol. Rev. **29**, 281—310 (1949).
— A. A. RENZI and J. J. CHART: Aldosterone, a review. J. clin. Endocr. **15**, 621—646 (1955).
— C. W. LLOYD and J. J. CHART: The adrenal-neurohypophysial interrelationship. In: H. S. HELLER, Ed., Proceedings Eighth Symposium Colsten Research Society, 233—249. London: Butterworths Scientific Publications 1957.
— F. GROSS, A. A. RENZI and J. J. CHART: The adrenal cortex in hypertension (with particular reference to adrenal regeneration hypertension). In: J. MOYER, Ed., Hypertension, 219—232. Philadelphia: W. B. Saunders 1959.

GELLHORN, E., and H. M. BALLIN: Water intoxication and the electroencephalogram. Amer. J. Physiol. **146**, 559—566 (1946).
GENNES, L. de, B. M. DE MOREAU, H. BRICAIRE, R. TOURNEUR et JANSON: La delta-cortisone dans le traitement des cirrhoses ascitiques. Pr. méd. **65**, 709—712 (1957).
GIACOMINI, E.: Contribution to the knowledge of the adrenal capsule in the Cyclostomata. On the adrenal capsule in Petromyzon (in Italian). Monit. zool. ital. **13**, 143—162 (1902).
GINSBURG, M.: The secretion of antidiuretic hormone in response to haemorrhage and the fate of vasopressin in adrenalectomized rats. J. Endocr. **11**, 165—176 (1954).
GODING, J. R., and D. A. DENTON: Adrenal cortex and the parotid secretion of sodium-depleted sheep. Science **123**, 986—987 (1956).
GORBMAN, A., Ed.: Comparative Endocrinology. New York: Wiley 1959.
GORDAN, G. S.: Influence of steroids on cerebral metabolism in man. Recent Progr. Hormone Res. **12**, 153—170 (1956).
GORDON, E. S., J. J. CHART, D. HAGEDORN and E. G. SHIPLEY: Mechanism of sodium retention in preeclamptic toxemia. Obstet. and Gynec. **4**, 39—50 (1954).
GOURFEIN, D.: Recherches physiologiques sur la fonction des glandes surrénales. Rev. méd. Suisse rom. **16**, 113—142 (1896).
GRAD, B.: The influence of ACTH on the sodium and potassium concentrations of human mixed saliva. J. clin. Endocr. **12**, 708—718 (1952).
GRADINESCO, A., et N. SANTA: Sur le mécanisme régulateur du volume sanguin chez les animaux surrénalectomisés. C. R. Soc. Biol. (Paris) **125**, 195—187 (1937).
GRAY, S. J., and C. G. RAMSEY: Adrenal influences upon the stomach and the gastric responses to stress. Recent Progr. Hormone Res. **13**, 583—613 (1957).
GREEN, D. M., A. FARAH and W. W. KLEMPERER: Mechanisms of desoxycorticosterone action: renal effects of the water-soluble glycoside. Endocrinology **47**, 281—289 (1950).
GROAT, R. A.: Formation and growth of adrenocortical-like tissue in ovaries of the adrenalectomized ground squirrel. Anat. Rec. **89**, 33—41 (1944).
GROSS, F.: Experimental analysis of the effects of aldosterone and other corticosteroids on salt and water metabolism. Rev. ibér. Endocr. **3**, 603—605 (1956a).
— Nebennierenrinde und Wasser-Salzstoffwechsel unter besonderer Berücksichtigung von Aldosteron. Klin. Wschr. **34**, 929—941 (1956b).
— Renin und Hypertensin, physiologische oder pathologische Wirkstoffe? Klin. Wschr. **36**, 693—706 (1958).
— and H. GYSEL: The action of electrocortin in the adrenalectomized dog. Acta endocr. (Kbh.) **15**, 199—209 (1954).
— u. P. LICHTLEN: Verstärkung der pressorischen Wirkung von Renin, drucksteigernden Peptiden und adrenergischen Wirkstoffen nach Nephrektomie und bei Überdosierung von Cortexon und Kochsalz an der Ratte. Naunyn-Schmiedebergs Arch. exp. Path. Pharmak. **233**, 323—337 (1958).
— u. R. MEIER: Versuch zur Differenzierung der Wirkungen verschiedener Nebennierenrinden-Steroide auf den Wasser- und Elektrolytstoffwechsel. Schweiz. med. Wschr. **81**, 1013—1018 (1951).
— u. H. SCHMIDT: Überdosierung von Cortexon (DOC) und Kochsalz beim Kaninchen. Naunyn-Schmiedebergs Arch. exp. Path. Pharmak. **232**, 408—418 (1958a).
— — Aldosterone overdosage in the rabbit. Acta endocr. (Kbh.) **28**, 467—478 (1958b).
GUTNER, L. B., J. B. MOSES, S. DANN and H. S. KUPPERMAN: The use of prednisone in congestive heart failure. Amer. J. med. Sci. **234**, 281—286 (1957).
HAMBURGER, C.: Substitution of hypophysectomy by the administration of chlorpromazine in the assay of corticotrophin. Acta endocr. (Kbh.) **20**, 383—390 (1955).
HAMILTON, W. F., R. G. ELLISON, R. W. PICKERING, E. E. HAGUE and J. T. TUCKER: Hemodynamic and endocrine responses to experimental mitral stenosis. Amer. J. Physiol. **176**, 445—451 (1954).
HARKNESS, D. M., E. MUNTWYLER, F. R. MAUTZ and R. C. MELLORS: Electrolyte and water exchange between skeletal muscle, "available (thiocyanate) fluid," and plasma in the dog following the administration of DCA. J. Lab. clin. Med. **28**, 307—313 (1942).
HARRIS, G. W.: Neural control of the pituitary gland. London: Edward Arnold 1955.
HARRISON, H. E., and D. C. DARROW: Distribution of body water and electrolytes in adrenal insufficiency. J. clin. Invest. **17**, 77—86 (1938).
HARROP, G. A.: The water and salt hormone of the adrenal cortex. Johns Hopk. Hosp. Bull. **59**, 25—34 (1936a).
— Influence of the adrenal cortex upon the distribution of body water. Johns Hopk. Hosp. Bull. **59**, 11—24 (1936b).
— L. J. SOFFER, R. ELLSWORTH and J. H. TRESCHER: Studies on suprarenal cortex. III. Plasma electrolytes and electrolyte excretion during suprarenal insufficiency in the dog. J. exp. Med. **58**, 17—38 (1933a).

HARROP, G. A., A. WEINSTEIN, L. J. SOFFER and J. H. TRESCHER: II. Metabolism, circulation and blood concentration during suprarenal insufficiency in the dog. J. exp. Med. **58**, 1—16 (1933b).
— L. J. SOFFER, W. M. NICHOLSON and M. STRAUSS: IV. The effect of sodium salts in sustaining the suprarenalectomized dog. J. exp. Med. **61**, 839—860 (1935).
HARTMAN, F. A., and K. A. BROWNELL: The adrenal gland. Philadelphia: Lea and Febiger 1949.
— and H. J. SPOOR: Cortin and the Na factor of the adrenal. Endocrinology **16**, 871—878 (1940).
— K. A. BROWNELL and W. E. HARTMAN: A further study of the hormone of the adrenal cortex. Amer. J. Physiol. **95**, 670—680 (1930).
— D. E. SMITH and L. A. LEWIS: Adrenal functions in the opossum. Endocrinology **32**, 340—344 (1943).
— L. A. LEWIS, K. A. BROWNELL, C. A. ANGERER and F. F. SHELDEN: Effect of interrenalectomy on some blood constituents in the skate. Physiol. Zool. **17**, 228—238 (1944).
HASTINGS, A. B., and E. L. COMPERE: Effect of bilateral suprarenalectomy on certain constituents of blood of dogs. Proc. Soc. exp. Biol. (N. Y.) **28**, 376—378 (1931).
HAYS, H. W.: The influence of sodium chloride on water diuresis and its relation to the adrenal. Ciba Found. Coll. Endocrin. **4**, 481—493 (1952).
— and D. R. MATHIESON: Studies on water intoxication in adrenalectomized rats and the influence of desoxycorticosterone acetate and epinephrine in water diuresis. Endocrinology **37**, 147—156 (1945).
HECHTER, O., A. ZAFFARONI, R. P. JACOBSEN, H. LEVY, R. W. JEANLOZ, V. SCHENKER and G. PINCUS: The nature and the biogenesis of the adrenal secretory product. Recent Progr. Hormone Res. **6**, 215—241 (1951).
HEGNAUER. A. H., and E. J. ROBINSON: The water and electrolyte distribution among plasma, red blood cells, and muscle after adrenalectomy. J. biol. Chem. **116**, 769—778 (1936).
HEIDORN, G. H., and F. R. SCHEMM: The clinical use of corticotropin (ACTH) and adrenal corticosteroids in the therapy of intractable edema. Amer. J. med. Sci. **229**, 621—631 (1955).
— — and J. A. LAYNE: The varied patterns of water and sodium during corticotropin (ACTH) therapy of the nephrotic syndrome. Amer. J. med. Sci. **229**, 180—187 (1955).
HELLER, H.: The influence of the suprarenal cortex on mineral and water metabolism. In: J. M. YOFFEY, Ed., The Suprarenal Cortex, 187—195. London: Butterworths 1953.
— T. F. HEWER, J. M. HUGHES and H. SCHNIEDEN: The effect of diuretics, cortisone and prednisolone on weanling rats with oedema produced by an African plantain diet. Brit. J. Nutr. **12**, 52—64 (1958).
HEMING, A. E., D. E. HOLTKAMP, J. F. KERWIN, L. F. MANSOR and J. H. DACANAY: Effect of amphenone "B" on adrenal and thyroid function of adult ovariectomized rats. Proc. Soc. exp. Biol. (N. Y.) **80**, 154—157 (1952).
HERNANDO, L., J. CRABBÉ, E. J. ROSS, W. J. REDDY, A. E. RENOLD, D. H. NELSON and G. W. THORN: Clinical experience with a physicochemical method for estimation of aldosterone in urine. Metabolism **6**, 518—543 (1957).
HERRICK, E. H., and O. TORSTVEIT: Some effects of adrenalectomy in fowls. Endocrinology **22**, 469—473 (1938).
HERTZ, R., W. W. TULLNER and M. J. ALLEN: Selective inhibition of adrenal and thyroid stimulating effect of Amphenone B by cortisone and thyroxine. Proc. Soc. exp. Biol. (N. Y.) **77**, 480—481 (1951).
— — J. A. SCHRICKER, F. G. DHYSE and L. F. HALLMAN: Studies on amphenone and related compounds. Recent Progr. Hormone Res. **11**, 119—147 (1955).
HILL, R. T., and A. S. PARKES: Hypophysectomy of birds. III. Effect on gonads, accessory organs and head furnishings. Proc. roy. Soc. B **116**, 221—236 (1934).
HINGERTY, D.: The role of magnesium in adrenal insufficiency. Biochem. J. **66**, 429—431 (1957).
HIRSCHOWITZ, B. I.: Pepsinogen: its origins, secretion and excretion. Physiol. Rev. **37**, 475—511 (1957).
HOAGLAND, H.: Studies of brain metabolism and electrical activity in relation to adrenocortical physiology. Recent Progr. Hormone Res. **10**, 29—58 (1954).
HOFMANN, F. G.: Observations on in vitro adrenal steroid synthesis in the albino mouse. Endocrinology **59**, 712—715 (1956).
HOGNESS, J. R., N. D. LEE and R. H. WILLIAMS: Effects of amphenone B on thyroid and adrenal functions. Endocrinology **52**, 378—389 (1953).
HOLLANDER, W., JR., R. W. WINTERS, T. F. WILLIAMS, J. BRADLEY, J. OLIVER and L. G. WELT: Defect in the renal tubular reabsorption of water associated with potassium depletion in rats. Amer. J. Physiol. **189**, 557—563 (1957).

HOLMES, W. N.: Studies on the hormonal control of sodium metabolism in the rainbow trout (Salmo gairdneri). Acta endocr. (Kbh.) **31**, 587—602 (1959).
HOLZBAUER, M.: The effect of oestrogens on adrenocortical function in rats. J. Physiol. (Lond.) **138**, 45—46 P (1957).
— and M. VOGT: Restoration of normal secretion in the hexoestrol-inhibited adrenal cortex of the rat. J. Physiol. (Lond.) **138**, 460—472 (1957).
HOUSSAY, B. A., and J. T. LEWIS: The relative importance to life of cortex and medulla of the adrenal glands. Amer. J. Physiol. **64**, 512—521 (1923).
HSIEH, K.-M.: Effect of desoxycorticosterone and pitressin on the water balance of amphibia. Fed. Proc. **9**, 63—64 (1950).
HYMAN, C., and R. CHAMBERS: Effect of adrenal cortical compounds on edema formation of frogs' hind limbs. Endocrinology **32**, 310—318 (1943).
IDLER, D. R., A. P. RONALD and P. J. SCHMIDT: Biochemical studies on sockeye salmon during spawning migration. VII. Steroid hormones in plasma. Canad. J. Biochem. **37**, 1227—1238 (1959).
INGLE, D. J., and E. C. KENDALL: Influence of amorphous fraction from adrenal cortex on efficiency of muscle. Proc. Soc. exp. Biol. (N. Y.) **45**, 602—606 (1940).
— H. W. NILSON and E. C. KENDALL: The effect of cortin on the concentrations of some constituents of the blood of adrenalectomized rats. Amer. J. Physiol. **118**, 302—308 (1937).
— M. C. PRESTRUD and K. L. RICE: The effect of cortisone acetate upon the growth of the Walker rat carcinoma and upon urinary non-protein nitrogen, sodium, chloride and potassium. Endocrinology **46**, 510—513 (1950).
— J. E. NEZAMIS and E. H. MORLEY: Comparative value of corticosterone, hydrocortisone and adrenal cortex extract given by continuous intravenous injection in sustaining the ability of adrenalectomized rat to work. Amer. J. Physiol. **170**, 498—501 (1952).
JENKINS, J. S., J. W. MEAKIN, D. H. NELSON and G. W. THORN: Inhibition of adrenal steroid 11-oxygenation in the dog. Science **128**, 478—480 (1958).
— L. POTHIER, W. J. REDDY, D. H. NELSON and G. W. THORN: Clinical experience with selective inhibition of adrenal function. Brit. med. J. **1959 i**, 398—401.
JOHNSON, V. K., and H. H. USSING: The influence of corticotropic hormone from ox on the active salt uptake in the axolotl. Acta physiol. scand. **17**, 38—43 (1949).
JONES, E. S.: Cellular electrolytes and adrenal steroids. Nature (Lond.) **176**, 269 (1955).
JØRGENSEN, C. B.: Influence of adenohypophysectomy on the transfer of salt across the frog skin. Nature (Lond.) **160**, 872 (1947).
KAGAWA, C. M., and C. G. VAN ARMAN: Sodium retaining activity of 19-norsteroids in adrenalectomized rats. Proc. Soc. exp. Biol. (N. Y.) **94**, 444—447 (1957).
— J. A. CELLA and C. G. VAN ARMAN: Actions of new steroids in blocking effects of aldosterone and desoxycorticosterone on salt. Science **126**, 1015—1016 (1957).
KAHNT, F. W., R. NEHER u. A. WETTSTEIN: Umwandlung von radioaktivem Cortexon in Aldosteron durch Nebennieren-Enzyme. Experientia (Basel) **11**, 446—447 (1955).
KELLEWAY, C. H., and H. S. COWELL: Concentration of blood and effects of histamine in adrenal insufficiency. J. Physiol. (Lond.) **57**, 82—99 (1923).
KENDALL, E. C.: The function of the adrenal cortex. Proc. Mayo Clin. **15**, 297—304 (1940).
— H. L. MASON, W. J. HOELN and B. F. MCKENZIE: Further investigation of the suprarenal cortex. J. biol. Chem. **119**, 106—107 (1937).
KLEIN, R., P. TAYLOR, C. PAPADOTOS, Z. LARON, D. KEELE, J. FORTUNATO and C. BYERS: Sodium-losing material in the urine of newborn infants. A. M. A. J. Dis. Child **94**, 498—499 (1957).
— — — — — — — and C. BILLINGS: Sodium losing material in human urine. Proc. Soc. exp. Biol. (N. Y.) **98**, 863—866 (1958).
KLEINHOLZ, L. H.: Studies in reptilian colour changes. II. The pituitary and adrenal glands in the regulation of melanophores of *Anolis carolinensis*. J. exp. Biol. **15**, 474—491 (1938).
KNIGHT, R. P., JR., D. S. KORNFELD, G. H. GLASER and P. K. BONDY: Effects of intravenous hydrocortisone on electrolytes of serum and urine in man. J. clin. Endocr. **15**, 176—181 (1955).
KNOWLTON, A. I., and E. N. LOEB: Desoxycorticosterone acetate; the potentiation of its activity by sodium chloride. J. Ark. med. Soc. **43**, 187—198 (1947).
— — H. C. STOERK and B. C. SEEGAL: Desoxycorticosterone acetate. The potentiation of its activity by sodium chloride. J. exp. Med. **85**, 187—198 (1947).
— — — J. P. WHITE and J. F. HEFFERMAN: Induction of arterial hypertension in normal and adrenalectomized rats given cortisone acetate. J. exp. Med. **96**, 187—205 (1952).
KUHLMAN, D., C. RAGAN, J. W. FERREBEE, D. W. ATCHLEY and R. F. LOEB: Toxic effects of desoxycorticosterone esters in dogs. Science **90**, 496—497 (1939).
KUIZENGA, M. H.: The isolation and chemistry of the adrenal hormones. In: F. R. MOULTON, Ed., The chemistry and physiology of hormones, 57—68. Washington: Amer. Ass. Advanc. Sci. 1944.

Laidlaw, J. C., J. F. Dingman, W. L. Arons, J. T. Finkenstaedt and G. W. Thorn: Comparison of the metabolic effects of cortisone and hydrocortisone in man. Ann. N. Y. Acad. Sci. **61**, 315—323 (1955).

Leroy, P., et J. Benoit: Surrénalectomie bilatérale chez le canard mâle adulte; action sur le testicule. J. Physiol. Path. gén. **46**, 422—428 (1954).

Levitt, M. F., and M. E. Bader: Effect of cortisone and ACTH on fluid and electrolyte distribution in man. Amer. J. Med. **11**, 715—723 (1951).

Lichtlen, P.: Die Wirkung von Cortisol, Cortison und Prednison am normalen und am nebennierenlosen Hund unter Kochsalz- und Wasserbelastung. Acta endocr. (Kbh.) **27**, 1—35 (1958).

Liddle, G. W.: Sodium diuresis induced by steroidal antagonists of aldosterone. Science **126**, 1016—1018 (1957).

— J. E. Richard and G. M. Tomkins: Studies of structure function relationships of steroids: the 2-methyl-corticosteroids. Metabolism **5**, 384—294 (1956).

— D. Island, E. M. Lance and A. P. Harris: Alterations of adrenal steroid patterns in man resulting from treatment with a chemical inhibitor of 11β-hydroxylation. J. clin. Endocr. **18**, 906—911 (1958a).

— — H. Estep and G. M. Tomkins: Modification of adrenal steroid patterns in man by means of an inhibitor of 11β-hydroxylase. J. clin. Invest. **37**, 912 (1958b).

Liling, M., and R. Gaunt: Acquired resistance to water intoxication. Amer. J. Physiol. **144**, 571—577 (1945).

Llaurado, J. G., R. Neher and A. Wettstein: Chemical identification of aldosterone in postoperative urine. Clin. chim. Acta **1**, 236—241 (1956).

Lloyd, C. W.: Some clinical aspects of adrenal cortical and fluid metabolism. Recent Progr. Hormone Res. **7**, 469—500 (1952).

Locke, W., N. B. Talbot, H. S. Jones and J. Worcester: Studies on combined use of measurement of sweat electrolyte composition and rate of sweating as index of adrenal cortical activity. J. clin. Invest. **30**, 325—337 (1951).

Loeb, R. F.: Chemical changes in blood in Addison's disease. Science **76**, 420—421 (1932).

— Effect of sodium chloride in treatment of a patient with Addison's disease. Proc. Soc. exp. Biol. (N. Y.) **30**, 808—812 (1933).

— D. W. Atchley, E. M. Benedict and J. Leland: Electrolyte balance studies in adrenalectomized dogs with particular reference to the excretion of sodium. J. exp. Med. **57**, 775—792 (1933).

Lowenstein, B. E., and R. L. Zwemer: Resistance of rats to potassium poisoning after administration of thyroid or of desoxycorticosterone acetate. Endocrinology **33**, 361—365 (1943).

Lucas, G. H. W.: Blood and urine findings in desuprarenalized dogs. Amer. J. Physiol. **77**, 114—125 (1926).

Luetscher, J. A., jr.: Studies of aldosterone in relation to water and electrolyte balance in man. Recent Progr. Hormone Res. **12**, 175—184 (1956).

— Aldosterone. In: W. Dock and I. Snapper, Eds., Advances in Internal Medicine, Vol. 8, 155—203. Chicago: Year Book Publishers 1956.

— A. Dowdy, J. Harvey, R. Neher and A. Wettstein: Isolation of crystalline aldosterone from the urine of a child with the nephrotic syndrome. J. biol. Chem. **217**, 505—512 (1955a).

— C. F. Piel and R. H. Curtis: The nephrotic syndrome. J. chron. Dis. **1**, 442—445 (1955b).

Luft, R., and B. Sjögren: Metabolic studies with desoxycorticosterone acetate (DCA). Acta endocr. (Kbh.) **10**, 49—65 (1952).

— — D. Ikkos, H. Ljunggren and H. Tarukoski: Clinical studies on electrolyte and fluid metabolism. Recent Progr. Hormone Res. **10**, 425—457 (1954).

Lyster, S. C., L. E. Barnes, G. H. Lund, M. M. Meinzinger and W. W. Byrnes: Adrenal corticoid activities of 6-methyl-Δ^1-hydrocortisone. Proc. Soc. exp. Biol. (N. Y.) **94**, 159 to 162 (1957).

MacFarland, W. E.: The vital necessity of adrenal cortical tissue in a mammal and the effects of proliferation of cortical cells from dormant coelomic mesothelium. Anat. Rec. **93**, 233—249 (1945).

Mach, R. S., et A. F. Muller: L'inhibition de la sécrétion de l'aldostérone par l'amphenone. Schweiz. med. Wschr. Supp. **14**, 406—408 (1957).

— J. Fabre, A. F. Muller et R. Neher: Oedèmes par rétention de chlorure de sodium avec hyperaldostéronurie. Schweiz. med. Wschr. **85**, 1229—1234 (1955).

Mader, I. J., and L. T. Iseri: Spontaneous hypopotassemia, hypomagnesemia, alkalosis and tetany due to hypersecretion of corticosterone-like mineralocorticoid. Amer. J. Med. **19**, 976—988 (1955).

MAES, J.: Etude du syndrome surrénoprivé de la grenouille. Arch. int. Physiol. **45**, 135—188 (1937).
MARENZI, A. D., and O. FUSTINONI: Potassium of the blood and tissues of the adrenalectomized toad (in Spanish). Rev. Soc. argent. Biol. **14**, 118—122 (1938).
MARINE, D., and E. J. BAUMANN: Duration of life after suprarenalectomy in cats and attempts to prolong it by injections of solutions containing sodium salts, glucose and glycerol. Amer. J. Physiol. **81**, 86—100 (1927).
MARSHALL, E. K., JR., and D. M. DAVIS: The influence of the adrenals on the kidneys. J. Pharmacol. **8**, 525—550 (1916).
MARTIN, H. F., and H. L. WHITE: Endocrine influences on rate of excretion of a standard salt load. Amer. J. Physiol. **170**, 532—535 (1952).
MASON, H. L.: Chemistry of the adrenal cortical hormone. Endocrinology **25**, 405—412 (1939).
— C. S. MYERS and E. C. KENDALL: Chemical studies of suprarenal cortex; identification of substance which possesses qualitative action of cortin; its conversion into diketone closely related to androstenedione. J. biol. Chem. **116**, 267—276 (1936).
MASON, J. W., and J. V. BRADY: Plasma 17-hydroxycorticosteroid changes related to reserpine effects on emotional behavior. Science **124**, 983—984 (1956).
MASSON, G. M. C., A. C. CORCORAN and I. H. PAGE: Experimental production of a syndrome resembling toxemia of pregnancy. J. Lab. clin. Med. **38**, 213—226 (1951).
MATTOX, V. R.: Personal communication (1959).
— H. L. MASON and A. ALBERT: Isolation of a sodium-retaining substance from beef adrenal extract. Proc. Mayo Clin. **28**, 569—576 (1953).
MCALLISTER, F. F., and G. W. THORN: Effect of adrenal cortical hormone on reduction of plasma volume resulting from etherization. Proc. Soc. exp. Biol. (N. Y.) **36**, 736—737 (1937).
MCCANN, S. M., A. FRUIT and B. D. FULFORD: Studies on the loci of action of cortical hormones in inhibiting the release of adrenocorticotrophin. Endocrinology **63**, 29—42 (1958).
MCQUARRIE, I., J. A. ANDERSON and M. R. ZIEGLER: Observations on the antagonistic effects of posterior pituitary and cortico-adrenal hormones on the epileptic subject. J. clin. Endocr. **2**, 406—410 (1942).
MILLER, H. C., and D. C. DARROW: Relation of muscle electrolyte to alterations in serum potassium and to the toxic effects of injected potassium chloride. Amer. J. Physiol. **130**, 747—758 (1940).
MILLER, M. R.: The normal histology and experimental alteration of the adrenal of the viviparous lizard, *Xantusia vigilis*. Anat. Rec. **113**, 309—323 (1952).
— Experimental alteration of the adrenal histology of the urodele amphibian, *Triturus torosus*. Anat. Rec. **116**, 205—221 (1953).
MILLER, R. A., and O. RIDDLE: The cytology of the adrenal cortex of normal pigeons and in experimentally induced atrophy and hypertrophy. Amer. J. Anat. **71**, 311—341 (1942).
— — Effects of prolactin and cortical hormones on body weight and food intake of adrenalectomized pigeons. Proc. Soc. exp. Biol. (N. Y.) **52**, 231—233 (1943).
MITCHELL, J. B., JR.: Experimental studies on bird hypophysis. I. Effects of hypophysectomy in the brown Leghorn fowl. Physiol. Zoöl. **2**, 411—437 (1929).
MONTASTRUC, P.: Contribution expérimentale à l'étude de la régulation hormonale du métabolisme de l'eau (étude des effets de la suppression simultanée des hormones antidiurétique et cortico-surrénales). Toulouse: Laboratoire de Physiologie de la Faculté de Médecine 1954.
MOREL, F.: Action synergique de la cortisone et de la désoxycorticostérone sur la diurèse provoquée du rat surrénalectomisé. C. R. Soc. Biol. (Paris) **146**, 202—207 (1952).
— Quelques aspects de la régulation endocrinienne de l'équilibre hydrominéral enregistrés chez le rat à l'aide du radio-sodium Na^{24}. Bull. biol. Suppl. **39**, 1—110 (1955).
MOYER, J., ED.: Hypertension. Philadelphia: Saunders 1959.
MULLER, A. F., and C. M. O'CONNOR: Eds.: An International Symposium on Aldosterone. Boston: Little Brown; London: J. & A. Churchill 1958.
MUNSON, P. L., and F. N. BRIGGS: The mechanism of stimulation of ACTH secretion. Recent Progr. Hormone Res. **11**, 83—117 (1955).
MUNTWYLER, E., R. C. MELLORS and F. R. MAUTZ: Electrolyte and water equilibria in the dog. I. Equilibria in the blood in adrenal insufficiency. J. biol. Chem. **134**, 345—365 (1940a).
— — — and G. H. MANGUN: II. Electrolyte and water exchange between skeletal muscle and blood in adrenal insufficiency. J. biol. Chem. **134**, 367—387 (1940b).
— — — — III. Electrolyte and water exchange between tendon and blood. J. biol. Chem. **134**, 389—395 (1940c).
MUSHETT, C. W., C. C. PORTER and R. H. SILBER: Effects of cortisone administered subcutaneously and orally to dogs. Fed. Proc. **10**, 366 (1951).

NABARRO, J. D. N.: The adrenal cortex and renal function. In: F. R. WINTON: Ed., Modern Views on the Secretion of Urine, 148—185. Boston: Little, Brown 1956.
NALBANDOV, A., and L. E. CARD: Effect of hypophysectomy of growing chicks upon their basal metabolism. Proc. Soc. exp. Biol. (N. Y.) **51**, 294—296 (1942).
NANDI, J., and H. A. BERN: Corticosteroid production by interrenal tissue of teleost fishes. Endocrinology **66**, 295—303 (1960).
NEHER, R., P. DESAULLES, E. VISCHER, P. WIELAND u. A. WETTSTEIN: Isolierung, Konstitution und Synthese eines neuen Steroides aus Nebennieren. Helv. chim. Acta **41**, 1667—1692 (1958).
NICHOLS, J., and G. HENNIGAR: Studies on the adrenal during simultaneous administration of DDD and amphenone. Endocrinology **61**, 226—234 (1957a).
— — Studies on DDD, 2,2-bis(parachlorophenyl)-1,1-dichloroethane. Exp. Med. Surg. **15**, 310—316 (1957b).
— and H. L. SHEEHAN: Effect of partial adrenal cortical atrophy on the course of alloxan diabetes. Endocrinology **51**, 362—377 (1952).
— C. DAVIS and H. D. GREEN: Effect of hypophysectomy, DDD treatment, and surgical trauma on the oxygen consumption of the various zones of the adrenal cortex. Endocrinology **53**, 541—548 (1953).
OLEESKY, A., and M. B. MANC: A specific water diuresis test for adrenocortical insufficiency. Lancet **1953**, 769—770.
OLIVEREAU, M., et H. FROMENTIN: Influence de l'hypophysectomie sur l'histologie de l'interrénal antérieur de l'anguille (*Anguilla anguilla* L.). Ann. Endocr. (Paris) **15**, 805—826 (1954).
OVERMAN, R. R., A. K. DAVIS and A. C. BASS: Effects of cortisone and DCA on radiosodium transport in normal and adrenalectomized dogs. Amer. J. Physiol. **167**, 333—340 (1951).
PARKES, A. S., and H. SELYE: Adrenalectomy of birds. J. Physiol. (Lond.) **86**, 35—37 (1936).
PARKINS, W. M.: An experimental study of bilateral adrenalectomy in the fowl. Anat. Rec. **51**, 39 (1931).
PERERA, G. A.: Effect of continued desoxycorticosterone administration in hypertensive subjects. Proc. Soc. exp. Biol. (N. Y.) **68**, 48—50 (1948).
PERLMAN, P. L., J. W. CASSIDY and A. TALI: Some metabolic effects of $\Delta^{1,4}$-pregnadiene-17α,21-diol-3,11,20-trione (Meticorten) administered orally to dogs. Proc. Soc. exp. Biol. (N. Y.) **91**, 551—554 (1956).
PERRINE, J. W., P. H. BELL, L. BORTLE, E. HEYDER, E. K. ROSS and I. RINGLER: The biological activity of triamcinolone, 16α-hydroxy-Δ^1,9α-fluorohydrocortisone (Aristocort) and related steroids. J. Pharmacol. **122**, 60 A (1958).
PETERSON, R. E., R. HERTZ and H. A. LUBS: Suppression of biosynthesis of adrenal cortical steroids in man by amphenone. Proc. Soc. exp. Biol. (N. Y.) **94**, 421—425 (1957).
PFIFFNER, J. J., O. WINTERSTEINER and H. M. VARS: Chemical studies on the adrenal cortex. I. Fractionation studies on hormone concentrates. J. biol. Chem. **111**, 585—597 (1953).
PHILLIPS, J. G., and I. CHESTER JONES: The identity of adrenocortical secretions in lower vertebrates. J. Endocr. **16**, 3 (1957).
— and P. J. MULROW: Corticosteroid production in vitro by the interrenal tissue of killifish, *Fundulus heteroclitus* (Linn). Proc. Soc. exp. Biol. (N. Y.) **101**, 262—264 (1959).
— W. N. HOLMES and P. K. BONDY: Adrenocorticosteroids in salmon plasma (*Oncorhynchus nerka*). Endocrinology **65**, 811—818 (1959).
PICKFORD, G. E.: Disturbance of mineral metabolism and osmoregulation in hypophysectomized *Fundulus*. Anat. Rec. **115**, 409 (1953a).
— A study of the hypophysectomized male killifish, *Fundulus heteroclitus* Linn. Bull. Bingham oceanogr. Coll. **14**, 5—41 (1953b).
— and J. W. ATZ: The physiology of the pituitary gland of fishes. New York: New York Zoological Society 1957.
PORAK, R., et H. CHABANIER: Altération de la sécrétion rénale après l'ablation des glandes surrénales. C. R. Soc. Biol. (Paris) **77**, 440—442 (1914).
POUSIAKA, J. W., B. G. H. THOMAS and C. R. LINEGAR: Effect of fluorohydrocortisone on gastrointestinal and renal excretion of cations by the dog. Proc. Soc. exp. Biol. (N. Y.) **96**, 712—714 (1957).
PRADER, A., A. SPAHR u. R. NEHER: Erhöhte Aldosteronausscheidung beim kongenitalen adrenogenitalen Syndrom. Schweiz. med. Wschr. **85**, 1085—1088 (1955).
PULLMAN, T. N., and W. W. MCCLURE: The relative roles of sodium and chloride in the salt-retaining action of desoxycorticosterone in man. Metabolism **3**, 240—246 (1954).
RAGAN, C., J. W. FERREBEE and G. W. FISH: Effect of desoxycorticosterone acetate upon plasma volume in patients during ether anesthesia and surgical operation. Proc. Soc. exp. Biol. (N. Y.) **42**, 712—716 (1939).
— — P. PHYFE, D. W. ATCHLEY and R. F. LOEB: A syndrome of polydipsia and polyuria induced in normal animals by desoxycorticosterone acetate. Amer. J. Physiol. **131**, 73—78 (1940).

RAUSCHKOLB, E. W., and G. L. FARRELL: Evidence for diencephalic regulation of aldosterone secretion. Endocrinology **59**, 526—531 (1956).
REICHSTEIN, T., u. J. V. v. EUW: Über Bestandteile der Nebennierenrinde. 20. Isolierung der Substanzen *Q* (Desoxycorticosteron) und *R* sowie weiterer Stoffe. Helv. chim. Acta **21**, 1197—1210 (1938).
RELMAN, A. S., and W. B. SCHWARTZ: The effect of DOCA on electrolyte balance in normal man and its relation to sodium chloride intake. Yale J. Biol. Med. **24**, 540—558 (1952).
RENOLD, A. E., N. A. HAYDAR, W. J. REDDY, A. GOLDFIEN, J. R. ST. MARC and J. C. LAIDLAW: Biological effects of fluorinated derivatives of hydrocortisone and progesterone in man. Ann. N. Y. Acad. Sci. **61**, 582—590 (1955).
— J. CRABBÉ, L. HERNANDO-AVENDANO, D. H. NELSON, E. J. ROSS, K. EMERSON, JR., and G. W. THORN: Inhibition of aldosterone secretion by amphenone in man. New Engl. J. Med. **256**, 16—21 (1957).
RENZI, A. A., and R. GAUNT: Apresoline (1-hydrazinophthalazine) in the experimental "eclampsia-like" syndrome and related aspects of water metabolism. Amer. J. Physiol. **175**, 313—318 (1953).
— M. GILMAN and R. GAUNT: ACTH-suppressing action of aldosterone. Proc. Soc. exp. Biol. (N. Y.) **87**, 144—145 (1954).
— M. RENZI, J. J. CHART and R. GAUNT: The effects of aldosterone and other steroids on water intoxication and renal function. Acta endocr. (Kbh.) **21**, 47—56 (1956).
RICHTER, C. P., and J. F. ECKERT: Mineral metabolism of adrenalectomized rats studied by the appetite method. Endocrinology **22**, 214—224 (1938).
RICHTER, H. S.: Action of antidiuretic hormone in potassium-depleted rats; relation to aldosteronism. Proc. Soc. exp. Biol. (N. Y.) **97**, 141—144 (1958).
RIDDLE, O., G. C. SMITH and R. A. MILLER: The effect of adrenalectomy on heat production in young pigeons. Amer. J. Physiol. **141**, 151—157 (1944).
RIEMER, A. D.: The effect of prednisone in the treatment of refractory cardiac edema. Johns Hopk. Hosp. Bull. **98**, 445—453 (1956).
RIGLER, R.: "Water intoxication" und Wasserdiurese bei der Nebenniereninsuffizienz; Die Bedeutung der Nebenniere für die Osmoregulation. Klin. Wschr. **14**, 1—4 (1935).
ROBERTS, K. E., and H. T. RANDALL: The effect of adrenal steroids on renal mechanisms of electrolyte excretion. Ann. N. Y. Acad. Sci. **61**, 306—314 (1955).
ROBINSON, F. J., M. H. POWER and E. J. KEPLER: Two new procedures to assist in the recognition and exclusion of Addison's disease: a preliminary report. Proc. Mayo Clin. **16**, 577—583 (1941).
ROGOFF, J. M., and G. N. STEWART: Studies on adrenal insufficiency in dogs; blood studies in control of animals not subjected to treatment. Amer. J. Physiol. **78**, 711—729 (1926).
— — Studies on adrenal insufficiency: The influence of intravenous injections of Ringer's solution upon the survival period in adrenalectomized dogs. Amer. J. Physiol. **84**, 649—659 (1928a).
— — Studies on adrenal insufficiency: Influence of "heat" on survival period of dogs after adrenalectomy. Amer. J. Physiol. **86**, 20—24 (1928b).
ROSENFELD, G., and W. D. BASCOM: The inhibition of steroidogenesis by amphenone B: Studies in vitro with the perfused calf adrenal. J. biol. Chem. **222**, 565—580 (1956).
ROWNTREE, L. G.: Eight patients with adrenal cortical hypofunction surviving more than fifteen years. J. Amer. med. Ass. **159**, 1427—1529 (1955).
— and A. M. SNELL: A Clinical Study of Addison's Disease. Philadelphia: Saunders 1931.
RUBIN, M. I., and E. T. KRICK: The salt and water metabolism of adrenal insufficiency and partial starvation in rats. J. clin. Invest. **15**, 685—695 (1936).
SAWYER, W. H.: The hormonal control of water and salt electrolyte with special reference to the amphibia. Mem. Soc. Endocr. **5**, 44—59 (1956).
— D. F. TRAVIS and N. G. LEVINSKY: Identity and specificity of frog water balance principle of posterior pituitary extract. Amer. J. Physiol. **163**, 364—369 (1950).
SCHAEFFER, W. H.: Hypophysectomy and thyroidectomy of snakes. Proc. Soc. exp. Biol. (N. Y.) 30, 1363—1365 (1933).
SCHATZMANN, H. J.: Die Wirkung von Desoxycorticosteron auf den aktiven Kationenaustausch an Rattenblutzellen. Experientia (Basel) **10**, 189—190 (1954).
SCHEMM, F. R., and A. A. CAMARA: The correction of electrolyte deficits in cardiovascular-renal disease. Circulation **10**, 430—441 (1954).
SCHMIDLIN, J., G. ANNER, J. R. BILLETER u. A. WETTSTEIN: Über Synthesen in der Aldosteronreihe. 1. Totalsynthese des racemischen Aldosterons. Experientia (Basel) **11**, 365—368 (1955).
SCHOOLEY, J. P.: Technic for hypophysectomy of pigeons. Endocrinology **25**, 372—378 (1939).

SCHOTTÉ, O. E., and W. B. CHRISTIANSEN: Influence of amphenone B — a cortical inhibitor — upon regeneration in normal and in hypophysectomized newts. Anat. Rec. **128**, 619—620 (1957).

SCHULER, W., P. DESAULLES u. R. MEIER: Elektrocortinwirkung im Glykogentest. Experientia (Basel) **10**, 142 (1954).

SELDIN, D. W., L. G. WELT and J. H. CORT: The role of sodium salts and adrenal steroids in the production of hypokalemic alkalosis. Yale J. Biol. Med. **29**, 229—247 (1956).

SELYE, H., and H. STONE: Influence of sodium chloride upon the actions of desoxycorticosterone acetate. Amer. Heart J. **37**, 1009—1016 (1949).

SEXTON, A. W.: Factors influencing the uptake of sodium against a diffusion in the goldfish gill. Ph. D. Dissertation, Univ. Missouri. Diss. Abstr. **15**, 2270—2271 (1955).

SHEPPARD, H., T. F. MOWLES, W. L. BENCZE, M. J. ALLEN, J. J. CHART and R. GAUNT: An inhibitor of adrenocorticosteroid synthesis. Proc. 133rd Meet. Amer. chem. Soc., 22c (1958).

SILVETTE, H., and S. W. BRITTON: Effects of adrenalectomy and corticoadrenal extract on renal excretion and tissue fluids. Amer. J. Physiol. **104**, 399—411 (1933).

— — Carbohydrate and electrolyte changes in opossum and marmot following adrenalectomy. Amer. J. Physiol. **115**, 618—626 (1936).

SIMPSON, S. A., and J. F. TAIT: Dose response studies of the effect of desoxycorticosterone acetate (DOCA) on the sodium excretion of adrenalectomized rats. Endocrinology **47**, 308—310 (1950).

— — A quantitative method for the bioassay of the effect of adrenal cortical steroids on mineral metabolism. Endocrinology **50**, 150—161 (1952).

— — Recent progress in methods of isolation, chemistry, and physiology of aldosterone. Recent Progr. Hormone Res. **11**, 183—210 (1955).

— — A. WETTSTEIN, R. NEHER, J. v. EUW u. T. REICHSTEIN: Isolierung eines neuen kristallisierten Hormons aus Nebennieren mit besonders hoher Wirksamkeit auf den Mineralstoffwechsel. Experientia (Basel) **9**, 333—335 (1953).

— — — — — O. SCHINDLER u. T. REICHSTEIN: Konstitution des Aldosterons, des neuen Mineralcorticoid. Experientia (Basel) **10**, 132—133 (1954).

SINGER, B.: Aldosterone in the adrenal vein blood of nephrotic rats. Endocrinology **60**, 420—422 (1957).

— and E. H. VENNING: Methods of assay of a sodium-retaining factor in human urine. Endocrinology **52**, 623—633 (1953).

— and J. WENER: Excretion of sodium-retaining substances in patients with congestive heart failure. Amer. Heart J. **45**, 795—801 (1953).

SIREK, O. V., and C. H. BEST: Intramuscular cortisone administration to dogs. Proc. Soc. exp. Biol. (N. Y.) **80**, 594—598 (1952).

SJÖGREN, B.: Changes in total body sodium and potassium during prolonged administration of cortisone and desoxycorticosterone acetate. Acta endocr. (Kbh.) **10**, 40—48 (1952).

SKELTON, F. R.: Adrenal regeneration and adrenal-regeneration hypertension. Physiol. Rev. **39**, 162—182 (1959).

SLESSOR, A.: Studies concerning the mechanism of water retention in Addison's disease and in hypopituitarism. J. clin. Endocr. **2**, 700—723 (1951).

SMITH, D. C. W.: The role of the endocrine organs in the salinity tolerance of trout. Mem. Soc. Endocr. **5**, 83—101 (1956).

SMITH, H.: Salt and water volume receptors. Amer. J. Med. **23**, 623—652 (1957).

SNYDER, J. G., and L. C. WYMAN: Sodium and potassium of blood and urine in adrenalectomized golden hamsters. Amer. J. Physiol. **167**, 328—332 (1951).

SOFFER, L. J., G. LESNICK, S. Z. SORKIN, H. H. SOBOTKA and M. JACOBS: Utilization of intravenously injected salt in normals and in patients with Cushing's syndrome before and after the administration of desoxycorticosterone acetate. J. clin. Invest. **23**, 51—54 (1944).

— A. GUTMAN, J. GELLER and J. L. GABRILOVE: The role of adrenal steroids on renal function and electrolyte metabolism. Bull. N. Y. Acad. Med. **33**, 665—680 (1957).

STACHENKO, J., and C. J. P. GIROUD: Functional zonation of the adrenal cortex: Pathways of corticosteroid biogenesis. Endocrinology **64**, 730—742 (1959a).

— — Site of ACTH action. Endocrinology **64**, 743—752 (1959b).

STAFFORD, R. O., L. E. BARNES, B. J. BOWMAN and M. M. MEINZINGER: Glucocorticoid and mineralocorticoid activites of Δ^1-fluorohydrocortisone. Proc. Soc. exp. Biol. (N. Y.) **89**, 371—374 (1955).

STAHL, J., D. W. ATCHLEY and R. F. LOEB: Observations on adrenal insufficiency. J. clin. Invest. **15**, 41—46 (1936).

STEIGER, M., u. T. REICHSTEIN: Desoxy-cortico-steron (21-Oxy-progesteron) aus Δ^5-3-Oxy-ätio-cholensäure. Helv. chim. Acta **20**, 1164 (1937).

STEIN, L., and E. WERTHEIMER: Effect of adrenalectomy on intestinal absorption involving osmotic work in rats. Proc. Soc. exp. Biol. (N. Y.) **46**, 172—174 (1941).

STEWART, G. N.: Adrenalectomy and the relation of the adrenal bodies to metabolism. Phys. id. Rev. **4**, 163—190 (1924).

— and J. M. ROGOFF: Studies on adrenal insufficiency. Proc. Soc. exp. Biol. (N. Y.) **22**, 394—397 (1925).

STREETEN, D. H. P., and A. K. SOLOMON: The effect of ACTH and adrenal steroids on K transport in human erythrocytes. J. gen. Physiol. **37**, 643—661 (1954).

SWEET, A. Y., M. F. LEVITT and L. B. TURNER: Effect of DCA on electrolyte distribution. Fed. Proc. **13**, 149 (1954).

SWINGLE, W. W.: Studies on the functional significance of the suprarenal cortex. I. Blood changes following bilateral epinephrectomy in cats. Amer. J. Physiol. **79**, 666—678 (1927).

— and A. J. EISENMAN: II. The acid-base equilibrium of epinephrectomized cats. Amer. J. Physiol. **79**, 679—688 (1927).

— and J. J. PFIFFNER: An aqueous extract of the suprarenal cortex which maintains the life of bilaterally adrenalectomized cats. Science **71**, 321—322 (1930).

— and J. W. REMINGTON: The role of the adrenal cortex in physiological processes. Physiol. Rev. **24**, 89—127 (1944).

— J. J. PFIFFNER, N. M. VARS, P. A. BOTT and W. M. PARKINS: The function of the adrenal cortical hormone and the cause of death from adrenal insufficiency. Science **77**, 58—64 (1933).

— W. M. PARKINS, A. R. TAYLOR and H. W. HAYS: A study of water intoxication in the intact and adrenalectomized dog and the influence of adrenal cortical hormone upon fluid and electrolyte distribution. Amer. J. Physiol. **119**, 557—566 (1937a).

— — — — The influence of adrenal cortical hormone upon electrolyte and fluid distribution in adrenalectomized dogs maintained on a sodium and chloride free diet. Amer. J. Physiol. **119**, 684—691 (1937b).

— — — — and J. A. MORRELL: Effect of oestrus (pseudopregnancy) and certain pituitary hormones on life-span of adrenalectomized animals. Amer. J. Physiol. **119**, 675—683 (1937c).

— J. W. REMINGTON, H. W. HAYS and W. D. COLLINGS: The effectiveness of priming doses of desoxycorticosterone acetate in protecting the adrenalectomized dog against water intoxication. Endocrinology **28**, 531—634 (1941).

— E. J. FEDOR, M. BEN, R. MAXWELL, C. BAKER and G. BARLOW: Induction of diabetes insipidus in adrenalectomized dogs with cortisone. Proc. Soc. exp. Biol. (N. Y.) **82**, 571 to 573 (1953).

— L. J. BRANNICK, A. F. PARLOW, C. BAKER and S. LEBRIE: Life maintenance and mineral-corticoid activity of some new derivatives of cortisone, hydrocortisone and DOC. Endocrinology **59**, 226—232 (1956).

— — — High potency of 2-methyl-9α-fluorohydrocortisone for life maintenance and revival from insufficiency of adrenalectomized dogs. Endocrinology **60**, 658—663 (1957).

— J. DAVANZO, E. FEDOR, H. C. CROSSFIELD, D. GLENISTER, M. OSBORNE and G. WAGLE: Plasma volume and electrolyte changes induced by 2-methyl-9-alpha-fluorohydrocortisone in fasted adrenalectomized dogs. Proc. Soc. exp. Biol. (N. Y.) **97**, 416—419 (1958a).

— — H. C. CROSSFIELD, D. GLENISTER, M. OSBORNE and G. WAGLE: Effect of various adrenal steroids on internal fluid and electrolyte shifts of fasted adrenalectomized dogs. Proc. Soc. exp. Biol. (N. Y.) **99**, 75—78 (1958b).

TAIT, J. F., S. A. SIMPSON and H. GRUNDY: The effect of adrenal extract on mineral metabolism. Lancet **1952 I**, 122—124.

TAUBENHAUS, M., I. B. FRITZ and J. V. MORTON: *In vitro* effects of steroids upon electrolyte transfer through frog skin. Endocrinology **59**, 458—462 (1956).

THORN, G. W., and H. EISENBERG: Studies on desoxy-corticosterone. Endocrinology **25**, 39—46 (1939).

— D. JENKINS, J. C. LAIDLAW, F. C. GOETZ, J. F. DINGMAN, W. L. ARONS, D. H. P. STREETEN and B. H. MCCRACKEN: Pharmacologic aspects of adrenocortical steroids and ACTH in man. New Engl. J. Med. **248**, 232—245, 284—294, 323—337, 369—378, 414—423, 588 to 601, 632—646 (1953).

TRUSZKOWSKI, R., and J. DUSZYNSKA: Protection of mice against potassium poisoning by corticoadrenal hormones. Endocrinology **27**, 117—124 (1940).

VENNING, E. G., B. SINGER and G. A. SIMPSON: Adrenocortical function in toxemia of pregnancy. Amer. J. Obstet. Gynec. **67**, 542—548 (1954).

VESIN, P., P. FRUMUSAN et R. CATTAN: Action spectaculaire de la delta-cortisone dans les oedèmes de carence post-gastrectomie. Sem. Hôp. Paris **33**, 1—4 (1957a).

— C. JACQUIER, E.-E. BAULIEU et R. CATTAN: Hyperaldosteronisme et oedèmes cirrhotiques. Action de la Δ-cortisone. Bull. Soc. méd. Hôp. Paris **14—15**, 383—398 (1957b).

VOGT, M.: The effect of "amphenone" (1,2 bis(p-aminophenyl)-2 methyl propanone-1) on adrenocortical secretion in the rat. J. Endocr. **14**, 25—26 (1956).
— Morphological changes in the adrenal cortex in relation to concentration of steroids in adrenal vein blood. Ciba Found. Coll. Endocrin. **11**, 193—199 (1957).
WARMING-LARSEN, A., C. HAMBURGER and M. SPRECHLER: The influence of ACTH on the sodium and potassium concentrations of human saliva. Acta endocr. (Kbh.) **11**, 400—412 (1952).
WELLS, B. B., and E. C. KENDALL: A qualitative difference in the effects of compounds separated from the adrenal cortex on distribution of electrolytes and on atrophy of the adrenal and thymus glands of rats. Proc. Mayo Clin. **15**, 133—139 (1940a).
— — Influence of the adrenal cortex in phlorizin diabetes. Proc. Mayo Clin. **15**, 565—573 (1940b).
WELLS, H., F. N. BRIGGS and P. L. MUNSON: The inhibitory effect of reserpine on ACTH secretion in response to stressful stimuli. Endocrinology **59**, 571—579 (1956).
WELT, L. G., D. W. SELDIN and J. H. CORT: The effects of the pituitary and adrenal hormones on the metabolism and excretion of sodium and water. J. clin. Invest. **30**, 682 (1951).
WHEELER, T. D., and S. VINCENT: The question as to the relative importance to life of the cortex and medulla of the adrenal bodies. Trans. roy. Soc. Can. **11**, 125—127 (1917).
WHITE, H. L., and D. ROLF: Tissue electrolytes in normal and adrenalectomized ratst. Fed. Proc. **13**, 164 (1954).
WINTERSTEINER, O., H. M. VARS and J. J. PFIFFNER: Chemical investigations on the cortical hormone of the adrenal gland. J. biol. Chem. **105**, 100—101 (1934).
WOLFF, H. P., K. R. KOCZOREK, W. JESCH u. E. BUCHBORN: Untersuchungen über die Aldosteronausscheidung bei Leberkranken. Klin. Wschr. **34**, 366—371 (1956).
— — and E. BUCHBORN: Hyperaldosteronism in heart-disease. Lancet **1957 ii**, 63—66.
WOODBURY, D. M.: Extrarenal actions of desoxycorticosterone (DCA) and adrenocorticotrophic hormone (ACTH) on tissue electrolytes. Fed. Proc. **10**, 149 (1951).
— Effects of chronic administration of anticonvulsant drugs, alone and on combination with desoxycorticosterone, on electroshock seizure threshold and tissue electrolytes. J. Pharmacol. **105**, 46—57 (1952a).
— Effect of adrenocortical steroids and adrenocorticotrophic hormone on electroshock seizure threshold. J. Pharmacol. **105**, 27—36 (1952b).
— Effect of hormones on brain excitability and electrolytes. Recent Progr. Hormone Res. **10**, 65—104 (1954).
— and A. KOCH: Effects of aldosterone and desoxycorticosterone on tissue electrolytes. Proc. Soc. exp. Biol. (N. Y.) **94**, 720—723 (1957).
— C. P. CHENG, G. SAYERS and L. S. GOODMAN: Antagonism of adrenocorticotrophic hormone and adrenal cortical extract to desoxycorticosterone: electrolytes and electroshock threshold. Amer. J. Physiol. **160**, 217—227 (1950).
WRIGHT, A., and I. CHESTER JONES: The adrenal gland in lizards and snakes. J. Endocr. **15**, 83—99 (1957).
WYMAN, L. C., and C. TUM SUDEN: Studies on suprarenal insufficiency. VIII. The blood volume of the rat in adrenal insufficiency, anaphylactic shock and histamine shock. Amer. J. Physiol. **94**, 579—585 (1930).
ZAFFARONI, A., and R. B. BURTON: Identification of corticosteroids of beef adrenal extract by paper chromatography. J. biol. Chem. **193**, 749 (1951).
ZIERLER, K. L., and J. L. LILIENTHAL: Sodium loss in man induced by desoxycorticosterone acetate; study in a subject with myotonic dystrophy. Amer. J. Med. **4**, 186—192 (1948).
ZIFF, M., J. SIMSON and J. J. BUNIM: Effect of ACTH on water distribution in man as measured by antipyrine, T 1824 and bromide. J. clin. Invest. **31**, 829—839 (1952).
ZWEMER, R. L., and R. TRUSZOWSKI: The importance of corticoadrenal regulation of potassium metabolism. Endocrinology **21**, 40—49 (1937).

Chapter 5

Sex hormone secretion by the adrenal cortex*

By

Evelyn Howard and Claude J. Migeon

With 13 figures

Introduction

The production of sex hormones by congenital hyperplasias or malignancies of the adrenal cortex has long been recognized as suggestive of a biochemical relationship between the adrenal cortex and the gonads. The embryological origin of these tissues from adjacent portions of the dorsal mesentery provided a plausible background for similarities (Poll 1906, Politzer 1937, Gruenwald 1942). Present developments of enzyme chemistry point toward a qualitative similarity of the biosynthetic systems in all tissues forming steroid hormones (Samuels 1958). Thus, the concept is emerging that these tissues have, to a great degree, the same set of biosynthetic enzymes, but differ in the relative activities or amounts of different individual enzymes. In this view, the biochemical behavior of malignancies need not involve the presence of "embryonic cell rests" of another tissue, or the *de novo* formation of new enzymes, but simply an abnormally high concentration of certain enzymes normally present in small amounts. In this chapter, we shall summarize the evidence regarding the nature and amount of sex hormones secreted by the adrenal cortex, with primary emphasis on physiological rather than malignant states.

The hormones to be considered will include the steroids that either have been isolated from adrenal tissue, or are thought to be intermediaries in the biosynthesis of other steroids by the gland, or are the result of extra-adrenal metabolism of corticoids, and that are called sex hormones because they have some degree of biological activity on reproductive tract structures or on other organs that have a pronounced sex dimorphism. This classification, however, should not be interpreted to imply that effects of these hormones on other target systems may not turn out to be more significant than their effects on sexually dimorphic structures.

Adrenalectomy is known to produce a metabolic imbalance in the spheres of electrolytes and carbohydrates which have been clearly related to the absence of two groups of hormones: the electrolyte-regulating factor (aldosterone) and the glucocorticoids. On the other hand, the importance of the sex hormones from the normal adrenal cortex has not been well defined. It is clear, however, that the adrenal is not able to effect a substitution for the endocrine role of the gonads that approaches completeness.

The difficulties encountered in evaluating the role of the sex hormones of the adrenal gland are numerous. The direct method of determining glandular activity

* Supported in part by research grants (numbers H-2414, A-2609 and A-180) from the National Heart Institute and the Arthritis and Metabolic Diseases Division of the United States Public Health Service. The chapter was completed late in 1959, but a few additions were made on the proof in the Spring of 1961.

by the measurement of the hormonal content of the efferent vein blood of the gland is often difficult, and even when possible, the surgical trauma resulting from such procedure may alter the glandular activity. The isolation of hormones from the gland may be somewhat misleading because some of the products identified might be only intermediaries in the synthesis of the final active compounds, or even products of the extraction procedures; on the other hand, frequently the glands store little or no actual hormones. The determination of hormonal levels in the peripheral circulation is of the greatest physiological importance, but steroids are rapidly metabolized so that intact free hormones are often present in blood in very low concentrations, with the consequence that in many cases their measurement is difficult. Furthermore, there is a possibility that the active compound at the cellular level may be different from that occurring in blood. Urinary determinations have the advantage of dealing with larger amounts of material and give some idea of the hormonal production per unit of time, but the compounds found in urine are in large part catabolites of the products originally secreted, and it is not always easy to relate the catabolites to their precursors.

In the discussion to follow we will successively consider the androgens, the progestins and the estrogens. In each case, we will evaluate the evidence presented for the production of these hormones by the adrenal cortex, and will discuss their physiological role in man and experimental animals.

In spite of the immense technical progress of recent years, it will be seen that our knowledge is still rather limited as to the amount and nature of the sex hormones of adrenal origin and that, as a consequence, our understanding of their physiological importance is extremely meager, leaving open for the future a vast field of research.

Androgens

A. Nature of adrenal androgens

The pathway of biosynthesis of the C_{19} adrenal steroids has not been definitively established. Theoretically they could arise either directly from acetate or via a 21-carbon atom compound (17-hydroxypregnenolone, 17-hydroxyprogesterone or cortisol). If adrenal androgens are synthesized directly from acetate or via 17-hydroxypregnenolone, dehydroepiandrosterone would occupy a unique position as precursor of the other androgens (Fig. 1), and the variations of its level might be indicative of important aspects of adrenocortical function.

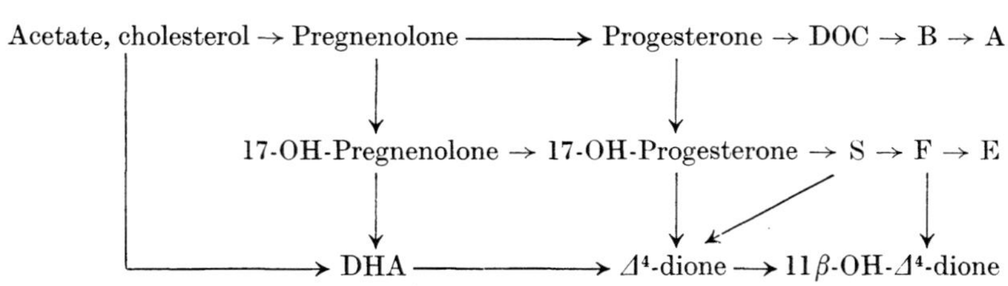

Abbreviations: A = 11-Dehydrocorticosterone; B = Corticosterone; DHA = Dehydroepiandrosterone; Δ^4-dione = Δ^4-Androstene-3,17-dione; DOC = 11-Deoxycorticosterone; E = Cortisone; F = Cortisol; 11β-OH-Δ^4-dione = 11β-Hydroxy-Δ^4-androstene-3,17-dione; S = 11-Deoxycortisol.

Fig. 1. Outline of possible pathways of biosynthesis of adrenal androgens

Dehydroepiandrosterone is considered to be primarily of adrenal origin. Although it has never been isolated from normal adult glands, it has been identified in human fetal adrenal tissue (Bloch et al. 1956) and in human adrenal vein blood from a psychiatric patient (LOMBARDO et al. 1959), and from a patient with hirsutism (BUSH and MAHESH 1959). Indirect evidence has been given by the work of CALLOW and CALLOW (1940) and of HIRSCHMANN (1940), who isolated this steroid from the urine of gonadectomized males and females. Ovariectomy was followed by a very minor decrease in dehydroepiandrosterone excretion, whereas it disappeared from the urine following adrenalectomy (DAVIS and PLOTZ 1956b). In addition, dehydroepiandrosterone has been repeatedly isolated or detected in large amounts in the urine of subjects with virilizing adrenal tumors (R. CALLOW 1936, TALBOT et al. 1940, WOLFE et al. 1941, FRAME 1944, PATTERSON 1947, L. GARDNER and MIGEON 1952) along with more reduced derivatives, such as Δ^5-androstene-3β,17β-diol (MASON and KEPLER 1945b, HIRSCHMANN and HIRSCHMANN 1945, SCHILLER et al. 1945) and Δ^5-androstene-3β,16α,17β-triol (HIRSCHMANN 1943). Dehydroepiandrosterone has also been isolated from the virilizing adrenocortical tumor of a 62-year-old female patient (PLANTIN et al. 1957), from the adrenal vein blood of a subject with post-puberal virilism (BUSH et al. 1956) and following incubation of the adrenal of a patient with the adrenogenital syndrome (BLOCH et al. 1957). In normal humans, it has been isolated from peripheral plasma (MIGEON and PLAGER 1954), where it is present in a conjugated form — probably as a sulfate, since the best yields are obtained by a 48-hour continuous ether extraction at p_H 1 (MIGEON and PLAGER 1955). Another piece of evidence for the adrenal origin of this C_{19} steroid is that its plasma concentration increases following ACTH administration, although not in all cases (MIGEON 1955, BAULIEU 1960). Quantitatively, dehydroepiandrosterone is the chief component of the 17-ketosteroids in normal human plasma. Its physiological significance has not yet been determined.

VANDE WIELE and LIEBERMAN (1960) injected tracer doses of dehydroepiandrosterone labeled with tritium at C-7 into human subjects, and on the basis of equal specific activities of urinary dehydroepiandrosterone, androsterone and etiocholanolone, concluded that practically all the urinary androsterone and etiocholanolone in normal subjects must be derived from dehydroepiandrosterone. From the dilution of the labeled dehydroepiandrosterone and its metabolites, it was calculated that this steroid is normally secreted at a rate of 25 mg daily.

The excretion of 17-ketosteroids by man is known to follow a diurnal rhythm (PINCUS 1943), and a similar cycle was demonstrated for the urinary excretion of corticosteroids (PINCUS et al. 1948, LAIDLAW et al. 1954) and for the plasma levels of 17-hydroxycorticosteroids (E. BLISS et al. 1953, MIGEON et al. 1956). It is interesting to note that the plasma levels of dehydroepiandrosterone show a similar variation (MIGEON et al. 1957).

Compounds related to dehydroepiandrosterone — namely, Δ^5-androstene-3β, 16α, 17-triol and Δ^5-androstene-3β, 17α-diol — have been isolated following incubation of rabbit liver slices with dehydroepiandrosterone (SCHNEIDER and MASON 1948). It has been suggested that pregnenolone is the precursor of the various Δ^5-steroids isolated in urine (LIEBERMAN and TEICH 1953).

More recently, OKADA et al. (1959) have reported the isolation of several other 3β-hydroxy-Δ^5 steroids from the urine of a patient with adrenal carcinoma, namely, 3β, 7α, 16α-trihydroxy-Δ^5-androstene-17-one, 3β, 16α-dihydroxy-Δ^5-androstene-7,17-dione, 3β-7α-dihydroxy-Δ^5-androstene-17-one and 3β, 16α-dihydroxy-Δ^5-androstene-17-one. The latter two compounds are also present in normal men's urine. OKADA et al. suggest that this evidence of *in vivo* oxygenation of dehydroepiandrosterone may indicate a more significant biological function for this steroid than has hitherto been supposed. It is known that oxygenation of progesterone, for example, changes the molecule from a progestin to a glucocorticoid.

A potential precursor of dehydroepiandrosterone, 17α-hydroxy-Δ^5-pregnenolone, has been found in the adrenal venous blood of dogs given ACTH (Carstensen et al. 1959). These authors consider that at least part of the circulating dehydroepiandrosterone arises from the peripheral metabolism of this precursor, a possibility that requires further study.

In the adrenal vein blood of cat, rat and sheep (Bush 1953), Δ^4-androstene-11β-ol-3,17-dione has been detected. Its concentration, however, is variable. This steroid might have been the unidentified androgenic compound detected by the chick comb response in adrenal blood of cows receiving 600 mg ACTH intravenously (Gassner et al. 1951). It has been isolated from human adrenal vein blood, along with Δ^4-androstene-3,17-dione (Romanoff et al. 1953, Lombardo et al. 1959). On the other hand, Liu et al. (1955) have failed to detect appreciable amounts of androgen in adrenal vein blood of dogs by means of the chick comb assay. The chick comb assay is described on page 575.

The 11-oxygenated androgens are not only secreted by the adrenals but are also formed during the metabolism of the corticosteroids possessing a 17-hydroxyl group. This subject has been reviewed by Roberts and Szego (1955) and Samuels (1955). From 5 to 10% of a given dose of cortisol is recovered in urine as 17-ketosteroid, and Gallagher (1957) has even reported a transformation of up to 20%.

Dorfman (1954), in a review on neutral steroid metabolism, has pointed out certain apparent directive effects of substituents at carbons 11 and 17 on the isomerism of the products formed by reduction of the Δ^4-3 ketone grouping. (See also Chapter 3, p. 481 ff.) Most of the 17-ketosteroids present in human urine are in a conjugated form, the ring A of the molecule being saturated. Such compounds are the end products of the metabolism of the active hormones. The reduction of the Δ^4-3 ketone compounds can give rise to products having an α or β hydrogen on carbon 5. From his own data and that of others, Dorfman concluded that (1) the C_{19} steroids with a Δ^4-3 ketone and an oxygen atom at carbon atom 17 are reduced at carbons 4 and 5, yielding only slightly more 5β- than 5α-isomers (ratio 5β/5α: 1.3—1.9), while the compounds with an additional oxygen atom at carbon 11 give rise to significantly more 5α- than 5β-isomers (ratio 5β/5α: 0.2); (2) the C_{21} steroids possessing a Δ^4-3 ketone and an oxygen atom at 17 and 21 are converted in part to C_{19} saturated 17-ketosteroids, the 5β isomers being present in greater amounts (ratio 5β/5α: 4.1—11.8). These conclusions must be regarded as somewhat tentative. Furthermore, certain conditions, such as the thyroid status of the subject, can modify the ratio 5β/5α (Bradlow et al. 1956).

The saturated C_{19} 17-ketosteroids with a 5β-hydrogen being biologically much less androgenic than the 5α-isomers, the considerations reported above suggest that the greater part of the 17-ketosteroid metabolites of cortisol are not significantly androgenic. However, Harrison et al. (1955) found that treatment of adrenalectomized castrated men with 300 mg/day of cortisone or cortisol was followed by a sharp rise in urinary androgens determined by chick comb assay; 25—50 mg/day of these corticoids did not increase the biologically determined urinary androgen over the basal level.

In addition to dehydroepiandrosterone and the 11-oxygenated 17-ketosteroids, there is good evidence that part of the urinary 11-deoxy 17-ketosteroids other than dehydroepiandrosterone are also synthesized by the adrenals (Fig. 1). Dehydroepiandrosterone has been shown to be metabolized by the *in vitro* bovine adrenal into Δ^4-androstene-3,17-dione and its 11β-hydroxy- and 11-keto-derivatives (Jeanloz et al. 1953, Meyer et al. 1955). Androstenedione has been isolated from adrenal vein blood of man (ROMANOFF et al. 1953), and from fetal adrenal tissue after incubation (SOLOMON et al. 1958). The Δ^4-androstene compounds are known to be reduced *in vivo* to the corresponding androstane and etiocholane derivatives (UNGAR et al. 1954). Consequently, Δ^4-androstene-3,17-dione gives rise to androsterone and etiocholanolone, these two latter steroids being also the main products of the catabolism of testosterone, the testicular hormone. Furthermore, exogenous dehydroepiandrosterone is excreted partly as androsterone and etiocholanolone (MASON and KEPLER 1945a, MILLER et al. 1950). It is, therefore, difficult to determine the respective contribution of the testis and the adrenal to the urinary androsterone and etiocholanolone in the intact male. However, see VANDE WIELE and LIEBERMAN (1960).

In the female, it has frequently been assumed that the total urinary 17-ketosteroids have an adrenal origin (MASON and ENGSTROM 1950), but recent work suggests that the ovaries are probably able to synthesize androgens. SOLOMON et al. (1956) found that progesterone incubated with bovine ovarian tissue was converted to 17α-hydroxyprogesterone and Δ^4-androstene-3,17-dione. In a case of arrhenoblastoma, incubation of the tumor with C^{14}-progesterone has demonstrated the conversion of this C_{21}-steroid into small amounts of Δ^4-androstene-3,17-dione, testosterone and possibly 11β-hydroxy-Δ^4-androstene-3,17-dione (SAVARD et al., unpublished). The ovarian vein blood from an adrenal rest tumor of the ovary has also been shown to contain significant amounts of Δ^4-androstene-3,17-dione (MIGEON et al., unpublished). However, the contribution of the ovarian secretion of the normal female to the total urinary excretion of 17-ketosteroids is not yet known.

B. Biological properties of the C_{19} steroids of adrenal origin

An androgen is considered to be a substance that promotes the growth or the structural and functional maintenance of the male reproductive tract or of other structures having a pronounced sex dimorphism, such as the cock's comb or the mouse submaxillary gland. In mammals, the ventral lobe of the prostate gland or the seminal vesicles have been frequently used as indicator tissues because their size and histological condition show a prompt response to the addition or withdrawal of testicular hormones, within certain limits of hormone concentration. In addition to the effects on adult structures, the gonadal hormones have an important role in sex differentiation: this subject has recently been reviewed (BURNS 1961). Certain of the conspicuous morphological effects of the gonadal hormones that have been used in bioassays have become the operational definitions of the respective sex hormones. These morphological manifestations are presumably endpoints of complex series of metabolic events in which the hormones participate. However, in addition to their effects on the specific sex structures, all three classes of sex hormones are known to have metabolic effects which are not limited to the reproductive system. The androgens promote nitrogen retention, increase the size of the kidneys and certain muscles, stimulate overall body growth under some circumstances, and accelerate ossification (KOCHAKIAN 1944, TALBOT and SOBEL 1947, GASSNER et al. 1958). The sex hormones presumably influence sex behavior patterns by direct influences on the central nervous system (HARRIS et al. 1958, PHOENIX et al. 1959). Accordingly any single method of bioassay reflects only a small part of the spectrum of physiological situations in which the sex hormones play a role. To what extent the different androgenic steroids have specific and distinct physiological roles is at present unknown.

I. Androgen bioassay methods

The literature on androgen bioassay has been reviewed recently by DORFMAN and SHIPLEY (1956) and EMMENS (1950), who may be consulted for the background of this material. Several test methods are available, which have varying degrees of sensitivity and accuracy; however the attainment of high sensitivity combined with high accuracy has yet to be reached. Furthermore, in view of the many factors which may influence a bioassay, it appears that no deep significance can be attached to minor differences in relative potencies, even for pure substances. Statistical methods for evaluating the significance of assay results have been discussed by EMMENS (1950) and by GADDUM (1953).

Chicken's comb

The capon's comb has been used extensively as a test object for the assay of androgenic activity. The comb is normally under the control of the testicular hormone, and after castration of the cock the comb involutes and maintains a constant size for some time. The administration of androgens can restore the comb to its normal size. Newly hatched chicks may be used in place of capons, as the comb in either sex is highly responsive to administered androgens. The use of chicks increases the sensitivity of the method and avoids the necessity for castration (DORFMAN and SHIPLEY 1956). If both sexes are used it is necessary to apply a correction for the sex difference in response: some workers prefer to use male chicks only (RAKOFF et al. 1944). White Leghorn chicks are much more sensitive than some other breeds. In the dosage range of 20 to 50 μg androsterone, using 16 to 20 chicks per assay, the mean error of the assays is 13% (DORFMAN and SHIPLEY 1956).

Chick comb test. Chicks are used at 2 to 3 days of age and are kept in a brooder, with thermostatic control, at 31—35°C. The material to be tested is applied to the lateral surface of the comb in an oil or ether solution, by means of a fine needle and a syringe fitted with a micrometer screw, or a microsyringe. The solution is administered daily for 7 days. On the 8th day the chicks are killed by means of chloroform, and the comb is removed with a scalpel, blotted lightly to remove blood, and weighed immediately on a torsion balance. Comb weights per 100 g body weight are obtained, using 10 chicks for each dilution. Each assay includes a control group of chicks receiving the solvent only and 3 groups receiving standard solutions of androsterone at 3 dose levels.

Male accessory sex organs

The growth of the accessory sex organs of the mammal is a useful criterion of androgenic activity (DORFMAN and SHIPLEY 1956). Although the mammalian endpoints in current use are less sensitive to androgens than are those of the fowl, they have practical as well as theoretical advantages. Anatomical considerations indicate that the seminal vesicles (vesicular glands) can be isolated more easily than the prostate in rodents. Seminal vesicle weights in immature intact rats were found by GREENE and BURRILL (1941) to be less variable than prostate weights and to show more uniform changes with testosterone dose. However, HERSHBERGER et al. (1953) furnished very satisfactory evidence that, in immature castrated rats, the ventral lobe of the prostate and the seminal vesicle weights increase in parallel and linearly with respect to the log of the testosterone dose. There is evidence that the ventral prostate is more sensitive than the seminal vesicles to androsterone (DEANESLY and PARKES 1936). The anatomy of the male accessory sex glands in the rat has been illustrated by R. CALLOW and DEANESLY (1935).

Seminal vesicle growth. Procedural details recommended by MATHIESON and HAYS (1945), as revised slightly by WILLS et al. (1949) and DORFMAN and SHIPLEY (1956) for the seminal vesicle assay, are given herewith. Rats are castrated at 26—28 days of age, and at least 14 days are allowed to elapse before use. Time intervals of several months between castration and use are permissable if there is a balanced distribution between groups with respect to the period since castration. The material to be tested, dissolved in 0.1 ml of oil, is injected intramuscularly into each hind leg, using a 26 gauge 15 mm needle and a microsyringe. After 72 hours, the animals are killed quickly, the ventral surface is opened, and the seminal vesicles are isolated under a dissecting microscope. The coagulating glands are teased from the seminal vesicles and the latter are incised at a point nearest the ejaculatory ducts. If secretion is present, the vesicles are cut open with a scalpel, the fluid is expressed and blotted on a damp paper towel. The vesicles are briefly immersed in saline, dried on blotting paper for a few seconds and weighed on a torsion balance. Weight loss from drying and weight gain from prolonged exposure even to isotonic saline must be minimized by adopting a dissection and weighing procedure that is as uniform and rapid as possible.

For quantitative work the simultaneous test of a standard and an unknown at least at 2 doses each are recommended. However, 2 points on a dose-response curve do not prove linearity. The dosages must be selected so that they are within a range in which the seminal vesicle weights increase linearly with respect to the logarithm of the dose, and quantitative comparisons are facilitated if the standard and the unknown produce parallel lines in the weight:log dose plot. If these conditions are fulfilled, the results may be calculated as follows: let S_1 and S_2 and U_1 and U_2 represent the vesicle weights obtained after treatment with the two doses of the standard and unknown respectively. The differences between the mean weights of the groups are then obtained as:

$$D_1 = U_2 - S_2$$
$$D_2 = U_1 - S_1$$
$$D_3 = U_2 - S_1$$
$$D_4 = S_2 - S_1$$

Then $Y_1 = D_1 + D_2$ and $Y_2 = D_3 + D_4$.
Let T_1 = sum of Y_1 for the respective groups and T_2 the sum of the Y_2.

Then $M' = \frac{I\,T_1}{T_2}$, where $I = \log \frac{\text{larger dose}}{\text{smaller dose}}$ and M' is the logarithm of the relative potency.

The log of the estimated potency, M, is then

$$M = \log\ (\text{potency of standard}) + M'.$$

Other details pertaining to the calculation are given by C. Bliss (1944). A valuable general discussion of mathematical methods useful in bioassays is given by Gaddum (1953).

The lowest dose of testosterone which gives a utilizable seminal vesicle response is about 0.2 mg per rat; hence an adequate quantitative assay using about 8 rats per group requires several mg of hormone.

Prostate growth. The test is carried out in the same manner as the seminal vesicle test. At autopsy the ventral lobe of the prostate is isolated and weighed (Deanesly and Parkes 1936, Hershberger et al. 1953). A particularly delicate index of androgenic hormone activity is the cytology of the prostate secretory epithelium (Moore et al. 1930).

Myotrophic activity

The weight of the muscles of the perineal complex has been found to be more responsive to testosterone than is the overall body weight. The *levator ani* has been selected for quantitative measurements as a representative of the perineal musculature, since it is easily separated from other tissues. Comparisons of steroids shows a lack of parallelism between activity in the seminal vesicle or prostate weight test and the *levator ani* test. However, those steroids which are most potent in producing nitrogen retention in rats are most potent in the *levator ani* test. The *levator ani* weight is also increased by the growth hormone of the anterior pituitary (Eisenberg and Gordon 1950, Shemano et al. 1951).

Levator ani test (Hershberger et al. 1953). Male rats are castrated at 21 days of age, and given daily subcutaneous injections of oil solutions of the substances to be tested. Injections are continued for 7 days, beginning on the day of castration. On the eighth day the animals are sacrificed. Dissection of the *levator ani* muscle is effected after removing the skin in the scrotal area. The posterior aspect of the perineal complex is cleared of fat and connective tissue with forceps, disclosing the constrictions at either end of the *levator ani* where it joins the bulbocavernosus muscle. The rectum is transected just caudally to the point where the *levator ani* loops around it dorsally. The body of the *levator ani* is then freed of the rectum and removed by incisions at the points of attachment to the bulbo-cavernosus muscle. The *levator ani* muscle is then cleaned of connective tissue and weighed to the nearest 0.1 mg on a torsion balance.

It has been found that while 19-nortestosterone shows a minimum of androgenic activity on the prostate, its myotrophic activity is approximately equal to that of testosterone (Hershberger et al. 1953).

II. Biological studies with adrenal androgens

Under normal circumstances, the adrenal steroids with 19 carbon atoms appear to be of little specific significance for the reproductive tract as substitutes for gonadal hormones. However, it is well known that in congenital adrenal hyperplasia, in cases of virilizing adrenal tumors and in cases of Cushing's syndrome there are marked signs of hyperproduction of androgens.

Many steroids are known to undergo metabolic conversion into other compounds with appreciably different physiological properties. Hence, the net physiological resultant of treatment with a given steroid would be expected to depend on relative rates of a number of different processes: e. g., absorption, utilization or binding at the active site, conversion, conjugation and excretion. Since we know that many steroids are excreted rapidly, it is possible that excretion rates may significantly exceed rates of conversion at times. If utilization rates exceed conversion rates, it would be expected that some interconversion would be compatible with a considerable difference in biological reactivity. Furthermore, when the end result depends on the relative rates of a number of processes, the route of administration and the size of the dose, as well as a number of other physiological variables, might be expected to produce considerable differences in the biological responses. Consequently, comparisons between different sets of data with respect to relative effectiveness of different steroids must be made with due reservations.

The biological activities of the C_{19} steroids derived from the adrenal are considerably less than those of testosterone in the usual assays for androgenic activity. However, these assays were developed in the course of studies of the biological properties of testicular hormones, so that the performance of the adrenal C_{19} products in these tests may not give us much insight into their physiological significance. The relative activities on the capon's comb, in terms of μg required to produce a standard response, are: testosterone, 15–30; androsterone and Δ^4-androstene-3,17-dione, 60–120; dehydroepiandrosterone, 120–250; etiocholan-3α-ol-17-one, over 1000 (DORFMAN and SHIPLEY 1956). The activities of these compounds vary in the same order when tested on immature rat seminal vesicles. It is of interest that Δ^4-androstene-3,17-dione and androsterone show relatively lower activity when given systemically than when given locally, suggesting that they are metabolized and/or excreted relatively more rapidly than testosterone.

On the basis of their effect on the weight of the muscles of the neck and shoulder of castrated guinea pigs, steroids can be classified in the following order of decreasing activity: testosterone, Δ^4-androstene-3,17-dione and dehydroepiandrosterone (KOCHAKIAN and TILLOTSON 1957). The relative renotrophic activity of these three steroids in castrated mice is similar, although dehydroepiandrosterone has an almost negligible activity. It has also little effect on the seminal vesicles and prostates of castrated mice, its activity per mole being only 2% of that of testosterone (KOCHAKIAN 1944). On the other hand, dehydroepiandrosterone does stimulate the rate of formation of ossification centers in infantile mice with an activity of the same order of magnitude as that of testosterone; hence a given amount of bone maturation can be produced by dehydroepiandrosterone with much less virilization than when testosterone is used (HOWARD, unpublished).

The biological effects of the 11-oxy-C_{19} steroids have been much less studied. Δ^4-Androstene-3,11,17-trione has an activity on the chick comb of a low order of magnitude comparable to that of androstanedione or of dehydroepiandrosterone (DORFMAN and SHIPLEY 1956). 11β-Hydroxyandrosterone was reported to be 0.25 as active as androsterone on the chick comb (MILLER et al. 1946). Tests of the anabolic effectiveness of a number of C_{19} steroids in nitrogen balance studies in humans indicated that 11β-hydroxylation results in loss of activity (PECHET and HENNEMAN 1957).

However, in another experimental situation, some biological activity of 11β-hydroxyandrostenedione is observed. In Table 1 data are presented indicating that, in doses having relatively minor effects on the seminal vesicles in infantile castrated mice, both 11β-hydroxy-Δ^4-androstenedione and dehydroepiandrosterone have considerable stimulating effects on the growth of the phallus (HOWARD 1959). The stimulus is roughly proportional to the dose, and consists of grossly normal maturation of the glans and the associated structures, including the phallic ossicle, illustrated in Fig. 2. The apparent decline in reactivity with increasing age from 4 to 22 days is complicated by the fact that the dose was not adjusted to body size. With a 10-fold increase in the dose, some reactivity at 92 days of age is manifest. The action of dehydroepiandrosterone was also manifest in young females when treatment was begun shortly after birth. These observations indicate that, although dehydroepiandrosterone and 11β-hydroxyandrostenedione have very low activity in promoting the growth of the seminal vesicles in castrated animals, nevertheless they have considerable effect on phallic growth. Hence abnormal increase in the rate of production of these or similar weak androgens in embryonic life might account for the genital abnormalities that occur with congenital adrenal hyperplasia which have been described by WILKINS (1957). These two androgens have been shown to be present in the human fetal adrenal together with androstenedione (BLOCH et al. 1956), in maximum amounts in embryos of 55 to 95 mm.

Table 1. *The preferential effect of certain adrenal androgens on phallic growth in male mice* (HOWARD 1959)

Status[1]	Dose, mg/day	Vesicle wt. (mg) × 10 / Body wt. (gm)	P[1]	Vesicle increment, % of intact[2]	Phallic wt. (mg) × 10 / Body wt. (gm)	P	Phallic increment, % of intact[2]
C-4	—	0.519		0.0	2.14		0.0
C-4: DHA-4	0.14	1.618	.01	3.0	7.92	.001	47.4
C-4: DHA-12	0.14	1.356	.01	2.3	6.36	.001	34.6
C-4: DHA-22	0.14	0.832	.04	0.9	3.07	.01	7.6
Intact	—	36.8		100.0	14.34		100.0
C-12	—	0.915		0.0	4.43		0.0
C-12: 11-OH-Δ^4-12	0.34	2.767	.01	5.2	10.99	.001	66.2
C-4	—	—		—	1.85		0.0
C-4: DHA-92	1.5	—		—	4.95	.01	25.4

[1] *Abbreviations:* C = castration at age (in days) given by suffix. DHA = subcutaneous implantation of a pellet of dehydroepiandrosterone at age (in days) given by suffix. 11-OH-Δ^4 = subcutaneous implantation of a pellet of 11β-hydroxy-Δ^4-androstene-3,17-dione. P = probability of chance distribution, values equal to or less than indicated, in comparison with untreated controls.

[2] Calculated as observed wt. — castrate wt./intact wt. — castrate wt.

Mice were autopsied at 6 weeks of age in the first and second experiments and at 17 weeks of age in last one.

It seems possible that adrenal androgens contribute to the development of the phallus in normal embryos. By the time the human embryo is 16 to 17 mm in length, the genital tubercle has elongated into a narrow conical organ, which continues to enlarge for some time and constitutes a very prominent structure of roughly the same size in embryos of both sexes. Even though sex differentiation becomes more definite with progressing age, there is no sex difference in phallic size even when the fetus reaches 100 mm in length (SPAULDING 1921).

Another aspect of the possible role of dehydroepiandrosterone derives from the following: dehydroepiandrosterone does not interfere with the vaginal effects of estrogens in ovariectomized mice in doses in which testosterone prevents the response (ROBSON 1938). Dehydroepiandrosterone promotes the growth of the uterine cervix and the vagina in intact prepuberal mice in contrast to testosterone, which interferes with the growth of these structures (HOWARD, unpublished).

Dehydroepiandrosterone does not potentiate the action of estrogens in the absence of the ovaries (EVANS et al. 1941), hence it could be postulated that dehydroepiandrosterone may be metabolized to estrogen by the ovaries more readily than is testosterone.

Since progesterone may be a component of the adrenal secretion, it is of interest to consider its androgenic potency. A marked androgenic effect on the ventral prostate of castrated immature rats has been observed with a daily dose of 3 mg progesterone (GREENE et al. 1939). WITSCHI and CHANG (1950) have described masculinizing effects in frog larvae, even to the point of complete sex reversal, by treatment with progesterone or deoxycorticosterone at dosages that approached lethal levels. Pregnenolone and cortisone were also masculinizing at similar dose levels. Metabolic conversion of these C_{21} compounds to C_{19} androgens could account for these effects (SAVARD et al. 1956). PRICE et al. (1955) have reported that the androgenic activity of 25 mg daily of progesterone, when administered to adult castrated male rats, is approximately equivalent to 5 μg daily of testosterone.

Dehydroepiandrosterone shares with certain other steroids an ability to suppress the psychomotor effects of lysergic acid diethylamide in rats. One mg doses of steroids were given each day for 3 days, and on the test day 40 μg of the lysergic acid diethylamide was given. In a vertical-rope-climbing test to obtain food, the climbing time was temporarily increased by the lysergic acid. Steroids partly prevented this lysergic acid-induced retardation (BERGEN et al. 1960).

A psychological effect of adrenal androgens in human females has been suggested (WAXENBERG et al. 1959). However, verification of this effect would require additional controls. (Compare p. 581.)

Since androgens have protein-anabolic effects, it has been postulated by various workers, notably ALBRIGHT (1942/43), that the adrenal C_{19} steroids might have some general systemic importance in protein anabolism. Children afflicted with the adrenogenital syndrome, in whom there is excess production of adrenal androgens, exhibit rapid growth, increased muscular development and advanced bone age with, eventually, precocious closing of the epiphyses. However, deficiency of the adrenal androgens does not result in subnormal growth; mice adrenalectomized at 12 days of age achieve grossly normal somatic growth on deoxycorticosterone without other steroids (HOWARD, unpublished); and in Addison's disease, growth and ossification may not be impaired if electrolyte balance is controlled and if there are no secondary complications (ALBRIGHT 1947). The growth of axillary and pubic hair

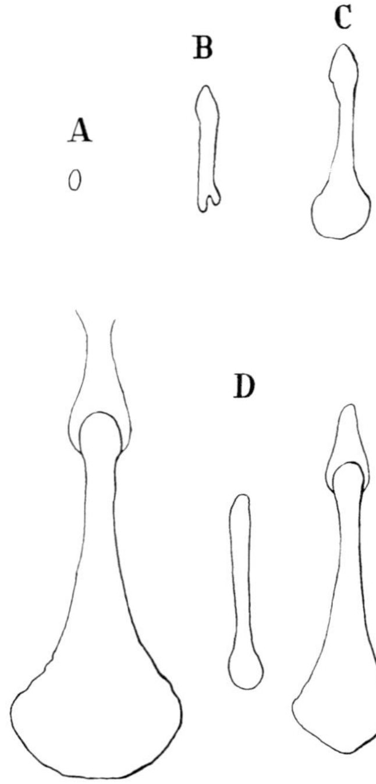

Fig. 2. The effect of dehydroepiandrosterone on the phallic bone in mice. All × 10. A. Oval type of phallic bone in female. From a mouse ovariectomized at 12 days, autopsied at 41 days of age. This is typical of its form in adult female mice whether intact or ovariectomized. Clitoris weight same as specimen shown in B. B. Elongated type of phallic bone in female, presumably a genetic variant on the small oval type shown in A. From a mouse ovariectomized at 4 days of age and autopsied at 32 days. Clitoris weight 2.5 mg. C. From a mouse ovariectomized at 4 days of age, implanted with 2 pellets of dehydroepiandrosterone, and autopsied at 32 days. Total steroid absorbed 7.9 mg in 28 days. Clitoris weight 7.8 mg. Litter mate to specimen shown in B. Note the considerable transformation toward the male shape, as shown in D, left, as well as the increase in size. D. Phallic bones from litter-mate male mice autopsied at 41 days of age. Left, from an intact male, showing normal adult form; center, from a male castrated at 4 days of age; right, from a male castrated at 4 days and implanted with a single pellet of dehydroepiandrosterone at 12 days. Steroid absorbed, 4 mg in 29 days. Phallic weights were 46.6, 6.3 and 18.1 mg, respectively. The distal ends (upper) retain the characteristic cartilagenous extensions. [Reproduced by permission of the publisher, Charles C. Thomas, from HOWARD: Endocrinology **65**, 785 (1959)]

in females seems to be a somewhat specific sign of the presence of normal amounts of adrenal androgens, since this hair may be absent in female Addisonians even with overall body growth within the normal limits (ALBRIGHT 1947, KINSELL et al. 1954).

On the basis of our present knowledge, mineralo- and glucocorticoids seem to substitute for the adrenal fairly well. Evaluation of the physiological significance of the adrenal androgens in the normal economy is largely a chapter for the future.

C. Physiological and pathological considerations in man

In an early survey, the lack of relation between urinary androgens and the sexual condition of the subject led R. CALLOW (1938) to the assumption that the androgens in urine are largely derived from sources other than the gonads, probably the adrenals. CALLOW noted that the androgens in the urine of castrated men and women did not differ significantly in amount from those in the urine of intact individuals and that the values for these substances were extremely high in cases of adrenal tumors. It has generally been believed that the urinary 17-ketosteroids derive from the gonads as well as the adrenals. They comprise a variety of compounds, some of which are much less active in the capon comb test than are others. Consequently, it has been found important to fractionate urinary 17-ketosteroids in order to study their pattern in detail. Various modifications (LAKSHMANAN and LIEBERMAN 1953) of the original method of DINGEMANSE et al. (1952) are used for this purpose.

Fractionation of urinary 17-ketosteroids. Since urinary steroids are present in urine in a conjugated form, the first step of the method is a hydrolysis using β-glucuronidase, followed by continuous extraction with ether at pH 1.0 for 48 to 72 hours and by strong acid hydrolysis. After proper purification of the extract, the ketonic steroids are usually separated from the non-ketonic fraction, using Girard's reagent T. The ketonic fraction is then applied on a column of alumina and the 17-ketosteroids are eluted by the technique of gradient elution. A total of 80 to 100 fractions are collected and their content in Zimmermann chromogens is determined. This method permits the separation of dehydroepiandrosterone, androsterone, etiocholanolone and of the 11-oxygenated-17-ketosteroids. It is important to identify each steroid by its infrared pattern (see Chapter 2).

These techniques permit one to differentiate dehydroepiandrosterone and 11-oxygenated-17-ketosteroids of exclusively adrenal origin from the 11-deoxy-17-ketosteroids that arise from both adrenal and gonadal precursors.

I. Effects of adrenocorticotrophin and adrenalectomy-gonadectomy

It is well established that ACTH administration causes an increase in the urinary excretion of 17-ketosteroids. When ACTH is given intramuscularly in doses ranging from 60 to 210 mg per 24 hours, the total urinary neutral 17-ketosteroid level increases from 44 to 105% above the control values (VENNING et al. 1950). When ACTH is given intravenously over an 8 hour period, the response is related to the amount administered, with a maximum reached at 20 mg (RENOLD et al. 1952).

When ACTH gel is administered on successive days (80 mg intramuscularly daily for 7–9 days), the effects on the various 17-ketosteroid fractions have been observed to be comparable in male and female (DOBRINER 1953, KAPPAS and GALLAGHER 1955). There is in normal adults a progressive increase of all the fractions. The augmentation of the 11-oxysteroids is not much greater than that of androsterone plus etiocholanolone; in certain subjects dehydroepiandrosterone shows the greatest rise. JAYLE et al. (1959) have presented a recent study of the latter point. The response obtained in young and old patients appears to be similar (DOBRINER 1954).

There is some information as to the action of ACTH on the urinary androgens in castrated patients. The effects of castration with the concomitant increases in gonadotrophins have received some attention.

SCOTT and VERMEULEN (1942), in studies of patients with prostatic cancer, reported that castration resulted in clinical changes indicative of gonadal insufficiency, in spite of the fact that there was only a moderate drop in 17-ketosteroids after operation, with an apparent tendency for the values to rise to or above the preoperative values during the 3-month period following surgery. Similar results have been obtained by FRAME and JEWETT (1944), BIRKE et al. (1954) and other investigators. It is not clear whether this rise in 17-ketosteroids is due to a specific compensatory activation of the adrenal or whether it is a non-specific result of the general post-operative relief from pain, improved appetite, etc. However, fractionation of the steroids shows that the androsterone fraction is greatly decreased while the other fractions are not. There may be a slight increase in 11-oxy-17-ketosteroids (BIRKE et al. 1954, JOHNSEN 1956).

MUNSON et al. (1954) and HARRISON et al. (1955) have reported androgen assays of urinary fractions by a chick-comb technique in patients with prostatic carcinoma who were subjected to castration followed by adrenalectomy. They reported preoperative values of 8–13 I. U./24 hrs., a low normal for elderly men, and post-castration decreases of 34 to 65%. The intravenous administration of 25 mg of ACTH to these castrated patients was followed by considerable increases in the urinary androgens, from 110 to 500% increases in one or two days. Subsequent total adrenalectomy resulted in further reductions of the biologically determined androgens to less than 2 I. U./24 hrs. These values persisted during maintenance on 50 mg of cortisone per day, while withdrawal of cortisone for 5 days did not further lower either androgen or 17-ketosteroids. Evidence of acute adrenal insufficiency developed when cortisone was withdrawn. In 2 patients who came to autopsy, no accessory adrenal or testicular tissue was found. The source of the residual androgen, whether real or artifactitious, remains uncertain. However, ACTH given to the patients after adrenalectomy did not cause any increase in 17-ketosteroids.

An apparent reduction in libido associated with adrenalectomy-ovariectomy in women with metastatic breast cancer was reported by WAXENBERG et al. (1959). The trauma of previous mastectomy and ovariectomy did not produce the same reduction in libido in these patients (see p. 603).

It is well established that the secretion of endogenous ACTH tends to be suppressed by the administration of cortisol or cortisone, as discussed in Chapters 1 and 10. Cortisone treatment also results in the suppression of the adrenal production of 17-ketosteroids. This has been amply documented for the adrenogenital syndrome (Chapter 9, also p. 585), where the output of endogenous ACTH is presumably excessive. The same effect is manifest in other conditions, for example, in rheumatoid arthritis, where cortisone therapy is accompanied by disappearance or decrease in urinary dehydroepiandrosterone, androsterone and etiocholanolone, but by an increase in 11-oxy-17-ketosteroids (BIRKE 1954).

II. Effects of menstruation and pregnancy

The menstrual cycle does not seem to change the pattern of the various urinary 17-ketosteroids (KAPPAS and GALLAGHER 1955, DAVIS and PLOTZ 1956a). Similar conclusions have been reported from studies of plasma androgens (MIGEON et al. 1957).

During late pregnancy, the 11-oxygenated-17-ketosteroids are excreted in smaller amounts than in early pregnancy. This is also true for dehydroepiandrosterone and androsterone, while the reduction is not so marked for etiocholanolone (DOBRINER et al. 1948, DAVIS and PLOTZ 1956a). Plasma 17-ketosteroid concentrations have also been found to be decreased in pregnancy at term (L. GARDNER et al. 1954, MIGEON et al. 1957). It has been noted that high doses of progesterone (100–200 mg daily, parenterally) administered during early pregnancy produce a decrease of the various urinary 17-ketosteroids similar to that observed during late pregnancy (PLOTZ and DAVIS 1956). It has been postulated that the decrease of certain 17-ketosteroids during normal pregnancy is related to the increased progesterone secretion by the ovary and placenta.

III. Effects of age

Although the secretion of the various adrenal androgens does respond to ACTH, as indicated above, the regulation of androgen production by the adrenal appears to be mediated by physiological factors that operate to some degree independently of the factors controlling glucocorticoid production. The circumstances involved are not well understood, but certain observations are of considerable relevance. Thus, TALBOT et al. (1950) have compared the urinary output of 17-ketosteroids and corticosteroids by normal individuals of various ages. With the possible exception of early infancy, the output of corticosteroids is within the same range for persons of all ages, when expressed as mg per square meter of body surface per day. By contrast, except for the neonatal period, the 17-ketosteroid output of children is very low up to about 8 years of age, and thereafter shows an approximately 10-fold increase up to adult levels at the beginning of the third decade of life. LELONG et al. (1957a) have also reported that urinary 17-ketosteroids change very little during the first 3 years of life, increase slowly between 3 and 9 years and then increase rapidly toward adult levels, while urinary 17-hydroxycorticosteroid levels show a linear progression from 0.5 mg at 6 months to 4.5 mg at 16 years. Although some of the urinary 17-ketosteroids probably come from gonadal sources, a considerable part presumably comes from the adrenals. Children are capable of secreting increased amounts of 17-ketosteroids if their adrenals are stimulated by ACTH (TALBOT et al. 1950, LELONG et al. 1957b). Hence, TALBOT et al. concluded that the appropriate stimulus is not present during early childhood, and postulated that there are two types of pituitary trophic hormones influencing the adrenal, one controlling 11-17-oxy-

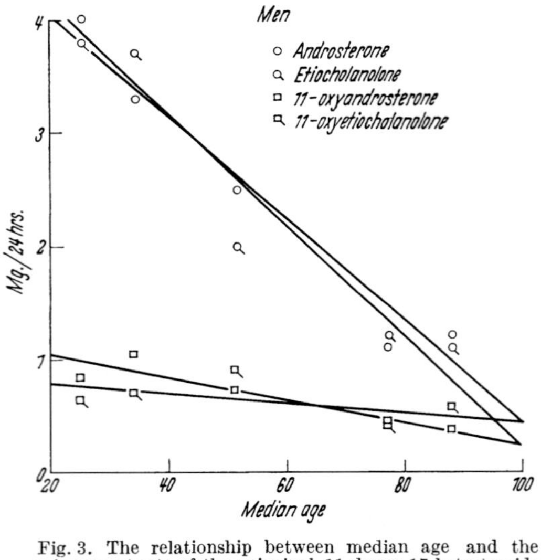

Fig. 3. The relationship between median age and the urinary outputs of the principal 11-deoxy-17-ketosteroids (androsterone and etiocholanolone) and the 11-oxygenated-17-ketosteroids in man. [Reproduced by permission from PINCUS et al.: Recent Progr. Hormone Res. 11, 307 (1955) (Academic Press).] The data illustrated are on male subjects: females show the same relationship, but the output of the 11-deoxy-17-ketosteroids is about 30% lower at age 20

corticoid production and one controlling 17-ketosteroid production. If this is so, both factors are apparently present in the usual ACTH extracts in current use.

PINCUS et al. (1954, 1955) have reported comparisons between the C_{19}-11-oxysteroids and the C_{19}-11-deoxyketosteroids in a number of subjects between 20 and 80 years of age. In both men and women, androsterone and etiocholanolone show a marked linear decrement throughout the four age groups presented, whereas the two corresponding C_{19}-11-oxysteroids show a relatively insignificant decline (see Fig. 3). Similarly the output of neutral reducing lipid, taken as an index of glucocorticoid activity, shows only a very slight decline between the ages of 20 and 80 years. PINCUS et al. (1955) concluded that the capacity for

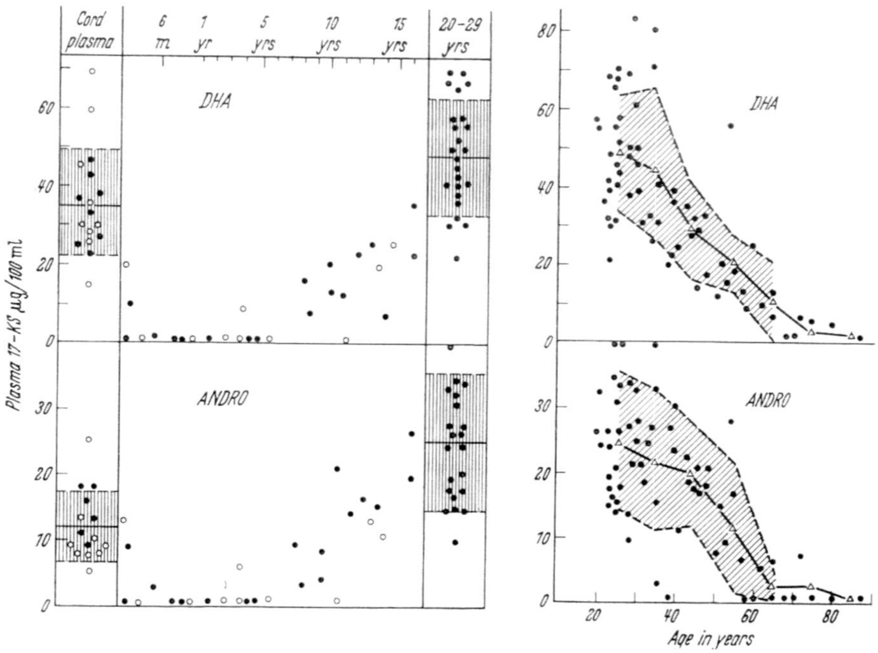

Fig. 4. Changes with age of the plasma levels of dehydroepiandrosterone (DHA) and androsterone (andro) in the human. [Reproduced by permission from MIGEON et al.: J. clin. Endocr. 17, 1051 (1957) (Charles C. Thomas).] On the left, changes during childhood and adolescence. The open circles are the values obtained in females and the black circles are those obtained in males. These values are compared with those found in umbilical cord blood and the blood of normal adult males, the full line being the average level and the dotted lines showing the value of one standard deviation. On the right, changes in adult males. The black circles represent individual values. The open triangles are the mean values at the various decades of age, and the shaded area shows the value of one standard deviation from the mean

synthesis of corticosteroids is sustained with advancing years, whereas the capacity for androgen synthesis declines: "Since side-chain scission at carbon 17 is the initial step in androgen synthesis, then aging in steroidogenic terms must involve decreased activity of those processes which effect the removal of the side-chain. Yet since both corticosteroidogenesis and androgenesis occur in the adrenal cortex, we must conclude that the adrenal cortex continues to be relatively youthful as a corticosteroid producer but ages notably as an androgen producer." Results in 8 female subjects over an age range of 22 to 62 years reported by KAPPAS and GALLAGHER (1955) support the conclusions of PINCUS et al. (1955). Thus, in advancing years there appears to be a reversal of the trend of steroid patterns toward that characterizing the first decade of life.

The two chief C_{19} steroids in human plasma, dehydroepiandrosterone and androsterone, show pronounced quantitative changes at various ages, with the same pattern as shown by the excretion rates of the C_{19}-11-deoxy-compounds. As illustrated in Fig. 4 the levels are high at birth, decrease rapidly during the first few weeks of life, and do not rise again until the onset of puberty; they attain their maximum height between the ages of 20 and 30 years in both men and women, after which they tend to decrease progressively toward zero values by the age of 60 to 70 years (MIGEON et al. 1957). These findings extend and confirm those of L. GARDNER and WALTON (1954) on total plasma 17-ketosteroids. Since dehydroepiandrosterone is presumably of adrenal origin, this cycle seems to indicate considerable qualitative changes in adrenal function at various stages of life.

The changes in adrenal androgen production throughout life appear to be roughly correlated with histological observations as to the size of the zona reticularis at various decades. The histological observations will be discussed in section D, p. 586.

IV. Hyperadrenocorticism

It is pertinent to discuss briefly the androgen secretion in cases of hyperadrenocorticism, even though this question will be reviewed in greater detail in Chapter 9. As shown in Table 2, the secretion of androgens is often increased in hyperadrenocorticism.

In Cushing's syndrome the findings reported by A. FORBES and ALBRIGHT (1951) are in accord with the results obtained by other investigators. These findings can be summarized as follows: (1) in cases due to bilateral adrenal hyperplasia, the

Table 2. *Syndromes of hyperadrenocorticism with their respective etiology and hormonal dysfunction* (from MIGEON and STEMPFEL 1957)

Syndrome	Etiology	Hormonal dysfunction: Androgens	Corticosteroids	Electrolyte regulating factor	Estrogens
Cushing's syndrome	1. Bilateral adrenal hyperplasia (related or not to hypothalamic lesion or "basophilic" adenoma of pituitary)	N or +	+++	+ ?	N or +
	2. Unilateral adrenal adenoma	— or N	+++		
	3. Unilateral adrenal carcinoma	+ to +++	+++		
Adrenogenital syndrome	1. Bilat. adrenal hyperplasia, a) Congenital, Non-salt loser	+++	— or N	N to +	+
	Salt loser	+++		—	
	Hypertensive	+++		+ (DOC)	
	b) Acquired	+++		N	
	2. Virilizing adrenal tumor	+++	N	N	+
Feminizing tumor	Adrenal tumor	N to +	N	N	+ to +++
Primary hyperaldosteronism	Adrenal tumor	N	N	+ to +++	N

Abbreviations: N = normal levels; — = lower than normal levels; + to +++ = higher than normal levels.

excretion of urinary 17-ketosteroids is usually moderately elevated with a relatively marked increase in the 11-oxygenated fraction, (2) in cases of unilateral adrenal adenoma, the urinary 17-ketosteroids may be somewhat decreased, (3) while in patients with adrenocortical carcinoma the excretion may be exceedingly high.

In patients with either primary hyperaldosteronism or a feminizing tumor, the urinary 17-ketosteroids levels are frequently normal. However, in some cases of feminizing adrenal tumor the levels may be high (Wallach et al. 1957).

In the adrenogenital syndrome there is, by definition, a greatly increased production of androgens by the adrenals.

In cases of adrenocortical carcinomas, most of the urinary steroid metabolites are compounds with a 3β-hydroxyl group and a double bond between carbons 5 and 6, such as dehydroepiandrosterone and related compounds, as previously mentioned (p. 572).

In bilateral adrenal hyperplasia of congenital origin, there is reason to believe that the defective synthesis of corticosteroids is associated with a deficiency in one or more of the enzymes that transform pregnenolone into cortisol (Bongiovanni et al. 1954). Under these circumstances the pituitary gland responds by liberating excessive amounts of ACTH (Sydnor et al. 1953), which presumably stimulates the adrenal metabolism to such extent that the gland secretes increased amounts of cortisol precursors, such as progesterone, 17-hydroxyprogesterone, 21-deoxycortisol, or 17-hydroxy-11-deoxycorticosterone (Bartter et al. 1951, Wilkins et al. 1952, Eberlein and Bongiovanni 1956). Under these circumstances, greatly increased amounts of androgens are produced. When this metabolic deviation occurs during embryonic life, it produces, in female embryos, various degrees of masculinization of the external genitalia, including female pseudohermaphroditism.

It is usually considered that there are three main forms of congenital adrenal hyperplasia: the non-salt-losing form, the salt-losing form and the hypertensive form (Wilkins 1957). Cases with hypoglycemic episodes have been reported, but are extremely rare.

The pathogenesis of virilizing adrenal hyperplasia has been the subject of a great deal of research. A large number of studies have been published on this problem and several theories proposed to explain the findings reported (Jailer et al. 1955, Dorfman 1955, Masuda 1957, Fukushima and Gallagher 1957). It is generally admitted that in the hypertensive form the main enzymatic defect of the adrenal is at the level of the 11β-hydroxylation (Eberlein and Bongiovanni 1956), while in the normotensive form the main defect is thought to be at the level of the 21-hydroxylation. The pathogenesis of the salt-losing form remains to be elucidated.

Treatment of these patients with moderate amounts of cortisone sufficient to supply the organism with its ordinary requirement results in a prompt reduction in the excessive urinary output of 17-ketosteroids. This is presumably a consequence of the maintenance of physiological blood levels of corticosteroids which reduces the activity of the overactive pituitary; hence, the adrenal output of excessive amounts of androgenic substances is decreased. This change results in a gradual regression of the syndrome of virilism and permits normal gonadal maturation.

V. Regulation of the production of adrenal androgens

Androgen production by the adrenal appears to be controlled at least in part, by factors independent from those controlling corticosteroid synthesis. This is suggested by the following observations.

(1) There is a lack of correlation between the trend followed by urinary 17-ketosteroids and corticosteroids through the life cycle (see p. 582.)

(2) As noted above, in congenital adrenal hyperplasia, hyperproduction of androgens is concomitant with a low-normal production of 17-hydroxycorticosteroids.

(3) In patients with hirsutism, an excessive urinary excretion of androsterone and etiocholanolone, with a more moderate increase in the C_{19}-11-oxysteroids, has been reported by GALLAGHER (1958). In one of these cases, the administration of ACTH produced only a slight increase of the C_{19}-11-deoxysteroids above their already high control levels, while there was a pronounced step-wise increase in the C_{19}-11-oxygenated steroids to a level about 4-fold that of the control periods. In this case, then, there was, for reasons which are not obvious, a dissociation between the rates of production of adrenal androgens and of glucocorticoids.

(4) Amphenone therapy for an adrenal cancer produces a marked decrease in urinary excretion of C_{19}-11-oxysteroids, while androsterone and etiocholanolone excretion is only slightly altered (GALLAGHER et al. 1956). PETERSON et al. (1957) have also reported that amphenone inhibits corticosteroid secretion by the adrenal, but not androgen secretion.

Such evidence has suggested the existence of two adrenocorticotrophic hormones, one acting primarily on androgen secretion, the other one on glucocorticoid synthesis (TALBOT et al. 1950). However, this hypothesis might not be necessary to explain the data at hand. The hormonal inbalance of congenital adrenal hyperplasia has been shown to be due to specific enzymatic defects affecting the biosynthesis of corticosteroids. It is quite possible that other enzymatic defects might be at the origin of the syndrome of hirsutism. It is also easy to imagine that amphenone would act on one system of enzymes more than on another and that enzymatic systems might go through a definite cycle through the life span of an individual. BUSH and MAHESH (1959) have described a case with a sudden onset of hirsutism associated with psychic stress. There was an abnormally increased ratio of 11β-hydroxyandrosterone to 11β-hydroxy- and 11-ketoetiocholanolone. The patient's monozygotic unstressed twin, without hirsutism, had the same abnormal ratio of $5\alpha : 5\beta$ C_{19}-11-oxyketosteroids, but the excretion of all 17-ketosteroids was only one-half that of the patient. BUSH and MAHESH propose that hirsutism in some cases may be due to an acceleration of the secretion rate of an adrenal with an abnormality in the cortisol : androgen secretion ratio.

D. Structural variations in the inner cortex related to hormonal status

The hypothesis that the fetal reticularis[1] of the primate cortex, the X zone of the mouse and to some extent the reticularis of the adult adrenal may be especially, though not exclusively, concerned with the secretion of androgens has received a measure of support from recent developments. This view does not exclude 17-ketosteroid production by other portions of the cortex, nor the production of 17-ketosteroids through somatic metabolism of glucocorticoids. Actual analysis of adrenal tissue separated by dissection into its three zones has been reported for a whale *(Physeter catadon)*. While the anatomical separation was imperfect, and interchange and alteration of steroids may have occurred post-mortem, characteristic differences appeared. Recovery of 17-hydroxycorticoids was as follows, in μg from 6 glands: subcapsular tissue, 234; mid-cortical, 261; inner cortical, 118. The 17-ketosteroids (Zimmermann chromogens) were: sub-

[1] In Chapter 1, the fetal reticularis is termed the "fetal" zone (Sub-ed.).

capsular, 1810; mid-cortical, 2160; inner cortical, 4910. No aldosterone was found in the inner zone (RACE and WU 1961).

Visible histological differentiations between the zona fasciculata and the inner zone in various species under various conditions may not have the same functional basis, so that interpretation of histological appearances must be made with caution.

Present information indicates that adrenocortical regression at birth, so dramatic in certain primates, can be detected to a lesser degree in other species. Similarly, the histological changes in the mouse adrenal that are temporally related to sexual maturation may have various counterparts in other species: however, there are so many variations that the different species must be considered separately.

I. Primates

1. Fetal reticularis

The remarkable alterations which the human adrenal undergoes during fetal and neonatal life have been the occasion of a great deal of comment. The comprehensive reviews by TÄHKÄ (1951), LANMAN (1953), BACHMANN (1954) and L. GARDNER (1956b) render it unnecessary here to cover the subject as a whole, and we will restrict ourselves for the most part to comment on recent work pertinent to the problem of possible production of C_{19} steroids by the transitory fetal reticularis.

In man, the adrenal of the fetus, relative to body size, is especially large: it reaches a maximum of 0.52% of body weight during the 4th and 5th prenatal months (EKHOLM and NIEMINEVA 1950), declining to 0.27 at birth (LANMAN 1957), when it is still about 10 times larger than the adult adrenal, relative to body weight. (See Chapter 1, Table 3, p. 20.) LANMAN (1957) points out that the adrenal of the chimpanzee and that of the Congo potto have a similar ratio to body weight at birth, with a similar decline in adults, and that the histological structure of these glands is quite comparable to that of the human. Hence, it now appears that the human and some other primates share these peculiarities of adrenal development. The adrenal is also relatively large in the fetal rhesus monkey but, apparently, the reticularis is not so well differentiated as that of the human and starts to regress earlier. Furthermore, the fetal cells are gradually transformed into adult type cells without the striking necrosis characteristic of man. The comparative development of the fetal reticularis in the human, the chimpanzee and the rhesus monkey is paralleled by the relative amounts of urinary chorionic gonadotrophin (LANMAN 1957).

It is generally recognized that the embryonic human adrenal becomes differentiated into a central area of relatively large acidophilic cells, which eventually become the fetal reticularis, and an outer rim of small basophilic cells, which give rise to the definitive fasciculata and glomerulosa. This differentiation is apparent in 17 to 20 mm human embryos. The adrenal of a human fetus is illustrated in Figure 5A. Two views as to the histogenesis of this differentiation have been expressed.

HETT (1925) describes the adrenocortical anlage in embryos of 12 to 16 mm as consisting of essentially a uniform type of cell, with a rather small amount of web-like cytoplasm and indistinct boundaries, except for some "giant cells." Mitoses are observed throughout the anlage. Subsequent differentiation of the adrenal then results from two processes that take place concurrently: first, the migration of columns and groups of sympathetic cells into the central part of the anlage; and second, the differentiation of the cortical cells. The cells of the outer zone remain small with lightly staining cytoplasm, and active mitosis continues. The cells occupying the inner area acquire an abundant fine-grained cytoplasm that stains intensely with acid dyes, and mitoses are not observed in the inner cells. The two cell types exist in contiguous groups without any sharp boundary, and the presence of transitional cells is suggested. According to this description, the histogenesis of the cells of the inner zone in the human is comparable to the histogenesis of the mouse X zone (see p. 593).

The second view, formulated tentatively by KEENE and HEWER (1927) and by UOTILA (1940), is that the cells that first arise from the coelomic mesothelium

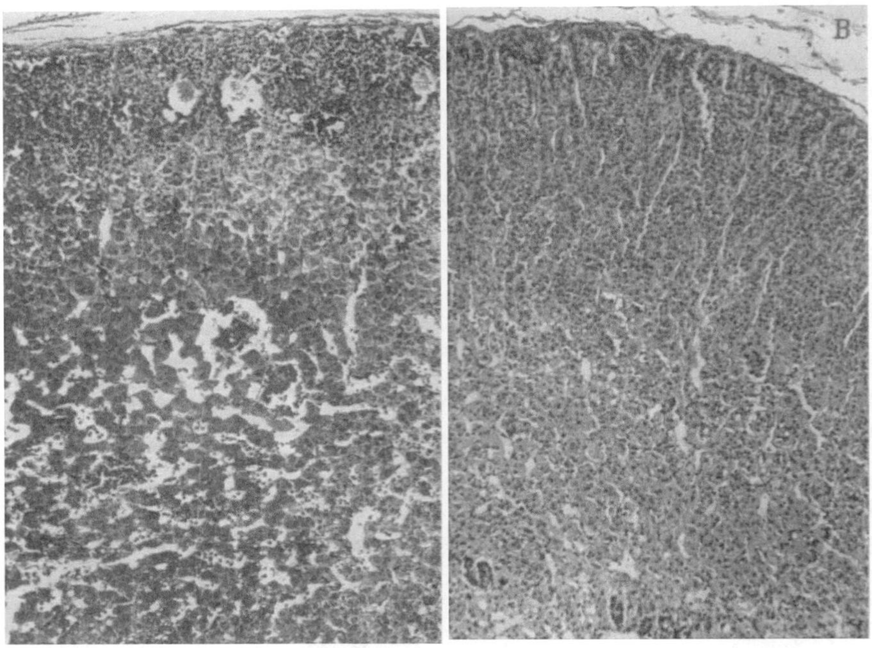

Fig. 5 A—C. Comparison of fetal, infant and adult human adrenal glands, all × 75

A. Adrenal of a normal male fetus of 18 cm CR length (19 weeks gestational age). Definitive cortex above. Note the large eosinophilic cells of the fetal reticularis in the lower two-thirds of the photograph. Hematoxylin and eosin. [Reproduced by permission from BENIRSCHKE et al.: Endocrinology **58**, 598 (1956) (Charles C. Thomas).] B. Adrenal of a male infant 6 days of age; death due to meningeal trauma. Degenerative vacuolation in the fetal reticularis is present. Note the absence of a sharp zonal transition between the outer and inner areas. A few clumps of darkly stained medullary cells are present at the bottom of the photograph. Hematoxylin and eosin. [Reproduced by permission from BENUA and HOWARD: Johns Hopk. Hosp. Bull. **86**, 200 (1950) (The Johns Hopkins Press).]

to form the cortical anlage become the inner large cells, and that the cells that arise secondarily remain on the periphery to constitute the small cells. However, GRUENWALD (1946) was not able to confirm the "alleged separate origin of the 'permanent' cortex by a second proliferation from the coelomic lining." He concluded: "the adrenal cortex develops by rapid differentiation of mesenchyme of the coelomic wall during the early phase and a slow continuation during a subsequent phase, rather than a succession of two separate proliferations."

BENNER (1940), referring to neonatal adrenals, states that there is considerable variation in the distinctness of the boundary between the "true" and "fetal" cortices; a few glands show a clear line of demarcation but the majority show an indistinct transition.

The adrenal of a six-day-old infant is illustrated in Fig. 5B and shows the absence of any sharp zonal demarcation; there is an intermingling of cell types. The fetal reticularis has been said to have the same cellular components as the adult cortex on the basis of a study of nuclear characteristics (CROWDER 1957).

2. Postnatal reticularis

BLACKMAN (1946) concluded that, although by the end of the first month of postnatal life the majority of the cells of the fetal reticular zone have disappeared, a few scattered large cells, typical of this zone, were found in some children as

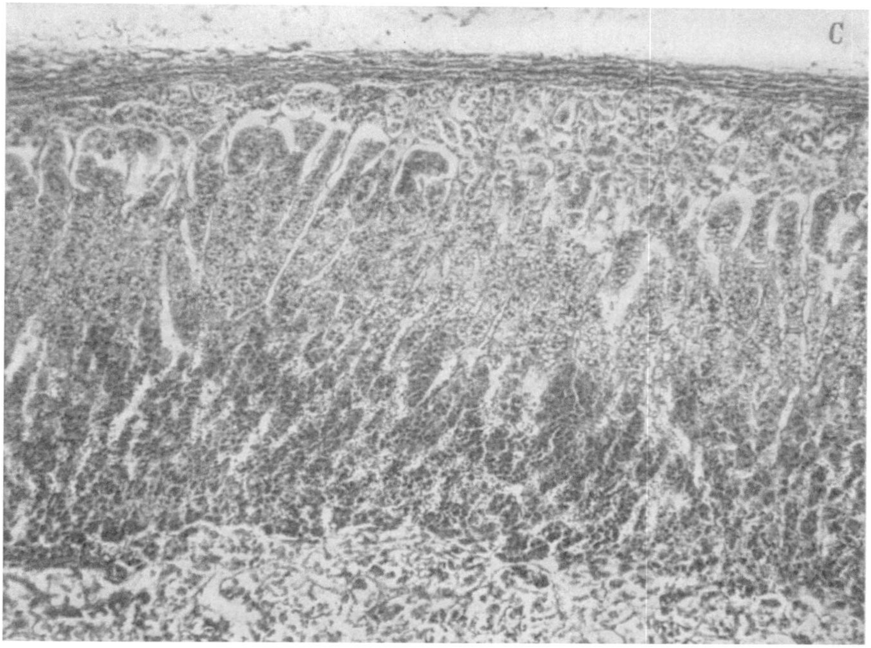

Fig. 5C. Normal adrenal from a woman aged 21. Autopsy specimen: sudden death. The three classical zones are present, with a strip of medulla at the bottom of the photograph. Note that the middle zona fasciculata, with its characteristic foamy cytoplasm, is relatively narrow, and that the strongly eosinophilic cells of the inner reticularis constitute a zone at least as wide as the fasciculata. Hematoxylin and eosin. [Reproduced by permission from JONES and JONES: Amer. J. Obst. Gynec. 68, 1330 (1954) (C. V. Mosby).]

old as 3 years. Furthermore, by the end of the first week of life, a very narrow irregular zone of fetal reticularis cells, lying next to the fascicular zone, had become different in appearance from the larger, involuting, central cells. "From the age of about 6 weeks to 3 years the reticular zone was a very thin layer ... After the age of 3 years the zone became gradually thicker" (BLACKMAN 1946). ROTTER (1949a, b) also reported that the reticularis was relatively small during childhood. However, quantitative measurements with an estimate of variability have not been published.

Age changes in the adrenal structure of adults have been studied by STIEVE (1947), who found that the text-book picture of the 3 adrenal layers could be recognized plainly in the 3rd to the 5th decades (see Fig. 5C). However, the reticularis was maximally developed during the 3rd decade and underwent a relative decrease in size during the 4th and 5th decades. During the 6th decade, sharp changes became evident in both men and women, the reticularis becoming very much reduced in width relative to the fasciculata. In women of the 6th decade who had not menstruated for at least a year, the cortex consisted almost entirely of fasciculata.

3. Metabolic considerations

Biochemical evidence regarding the normal human fetal adrenal is beginning to emerge. BLOCH et al. (1956) studied the steroid content of presumably normal human fetal adrenals of 9 to 21 gestational weeks, obtained immediately following therapeutic interruption of pregnancy. Steroids were characterized by paper chromatography. The adrenals from both male and female fetuses 16.5 to 21 weeks old contained about 0.8 μg of cortisol and corticosterone per g of tissue. Adrenals from younger fetuses contained less than 0.2 μg of each of these steroids. However, considerable amounts of 17-ketosteroids were found, including Δ^4-androstenedione, dehydroepiandrosterone and 11β-hydroxy-Δ^4-androstene-3,17-dione. The ratio of the sum of the C_{19} steroids to corticosteroids decreased with advancing age, between 9 and 21 weeks, from about 100 to 1 in the youngest stages to 5 or 10 to 1 in the older fetuses. The findings thus suggest that the human fetal adrenal synthesizes weak androgens predominantly. A sodium-retaining factor was also detected as early as 9–11 weeks of gestation.

Δ^4-Androstenedione has also been isolated from human fetal adrenal tissue after incubation with progesterone-4-C^{14} (SOLOMON et al. 1958). In further studies, BLOCH and BENIRSCHKE (1959) incubated adrenal slices from human fetuses, 12 to 22 weeks gestation age, with 1-C^{14} acetate. Both the definitive cortex and the fetal reticularis synthesized steroids from acetate. Dehydroepiandrosterone, Δ^4-androstenedione, 11β-hydroxy-Δ^4-androstenedione and 3β-hydroxy-Δ^5-pregnen-20-one and cortisol were isolated by carrier dilution methods from the incubations.

Studies on steroid hydroxylating enzymes in homogenates of human fetal adrenals revealed no difference with age in the hydroxylation of progesterone, between the ages of 11 and 21 weeks, per mg of adrenal tissue (VILLEE et al. 1959). Progesterone-C^{14} was converted to labeled cortisol by a fetal adrenal extract. However, VILLEE et al. (1959) were unable to demonstrate the conversion of pregnenolone to cortisol by fetal adrenal extracts, which might suggest either a lack or an extreme lability of the enzyme 3β-ol dehydrogenase.

Studies on the steroid content of umbilical cord blood at birth tend to support the evidence from analyses of fetal glands. The level of dehydroepiandrosterone in cord blood has been reported to be 39 ± 9 μg per 100 ml of plasma. In every case it has been found to be higher than the level in the blood of the respective mother, as shown in Fig. 6. In most instances the 17-hydroxycorticosteroid levels in cord plasma were lower than those of the mother. L. GARDNER (1956a) has reported that the plasma level of 17-ketosteroids is lower in pregnant women than in non-pregnant adults, and there is evidence that certain groups of urinary 17-ketosteroids are decreased during pregnancy. It is impossible at present to decide whether the fetal adrenal cortex is producing large amounts of dehydroepiandrosterone, or whether the high concentration in the fetal blood is due to accumulation in the fetal circulation, with contributions from the placenta. However, certain facts favor the first hypothesis. One such piece of evidence is that dehydroepiandrosterone is often detectable in the urine of premature infants (EAGLE 1952). Furthermore, the urinary output of 17-ketosteroids is relatively high during the first weeks of life, being followed by a downward trend (ZEISEL and PRESSLER 1953). The values fall from a mean of 1.5 mg/24 hrs during the first week to 0.5 during the third week. When excretion rates are referred to body surface, a sharp decline is also manifest from an initial 5 mg/24 hrs/m² to a value of 2 mg during the second month, with a further decline to 1 mg/24 hrs/m² during the second half of the first year; the latter level is then maintained during childhood. A downward trend in the urinary 17-ketosteroid excretion during the first 9 days of life was also observed by READ et al. (1950), who added the observation that ACTH administration was followed by a rise in urinary 17-ketosteroids during the neonatal period. The reduction in urinary 17-ketosteroid excretion and the downward trend of the plasma concentrations of total 17-ketosteroids (L. GARDNER and WALTON 1954a, b) and dehydroepiandrosterone (MIGEON et al. 1957) parallel the involution of the inner portion of the newborn adrenal (SCAMMON 1925/26, TÄHKÄ 1951). Evidence that dehydroepiandrosterone is not produced by the placenta has been recorded: i. e., in a pregnant patient with metastatic mammary carcinoma, ovariectomy produced no change in urinary output of dehydroepiandrosterone and a slight decrease in androsterone and etiocholanone; a subsequent adrenalectomy brought a marked decrease in all these steroids (DAVIS and PLOTZ 1956b). Furthermore, dehydroepiandrosterone has been isolated from meconium excreted by full term male infants during the first 24 hours after birth, in a yield of 50 μg/100 g, whereas etiocholanolone and androsterone were not detected (FRANCIS et al. 1960).

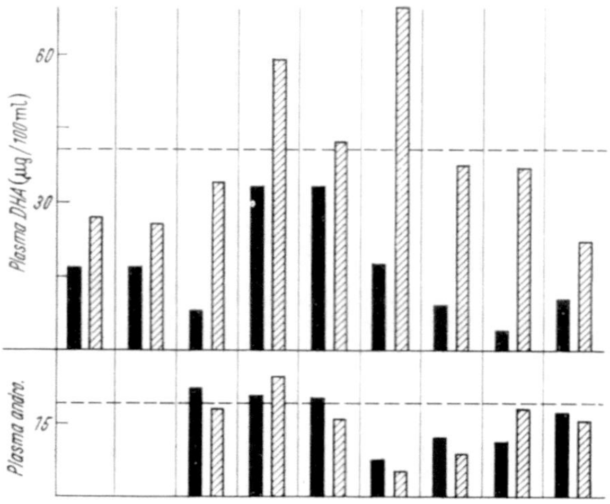

Fig. 6. Comparison of plasma levels of dehydroepiandrosterone (DHA) and androsterone (andro) in maternal and umbilical cord plasma. The horizontal dotted lines represent the mean values for non-pregnant women, 20 to 30 years of age. Maternal values are represented by the solid columns and cord values by the shaded columns, for individual cases. [Reproduced by permission from MIGEON et al.: Johns Hopk. Hosp. Bull. 97, 415 (1955) (The Johns Hopkins Press)]

By contrast, corticoid excretion, when referred to surface area, shows during the first 4 weeks of life a much less marked drop; thereafter, the excretion is fairly stable until 5 years of age (ZEISEL et al. 1953). VENNING et al. (1949) reported that the absolute amounts of urinary glucocorticoids, determined by bioassay, actually rise during the first few weeks of life. Similarly, ULSTROM et al. (1960) found that the urinary 17-hydroxycorticoids tend to rise during the first week after birth.

In anencephaly, observations recently reviewed by BENIRSCHKE (1956) show that adrenal atrophy characteristic of this condition is related to the premature involution of the "fetal" zone (Figure 7); in some cases of brain abnormalities associated with defects in the hypothalamus, similar findings have been reported. However, before the 20th week of gestation the "fetal" zone is normally developed. These findings are compatible with the suggestion of CHESTER JONES (1955) that in the early stages of gestation the "fetal" zone is stimulated by chorionic gonadotrophin and also with the comparative observations of LANMAN (1957). After the decline in chorionic gonadotrophin secretion, the atrophy of the "fetal" zone inanencephaly is thought to be the result of a failure of the fetal pituitary to secrete some trophic hormone as a result of a defect in the hypothalamus.

Since the fetal reticular zone normally regresses after birth, it is clear that the infant's pituitary is a necessary but not a sufficient condition for the maintenance of the fetal reticularis and that some stimulating factor must be withdrawn at birth. It has been suggested that this factor is LH from the fetal pituitary, the production of which is stimulated via the fetal hypothalamus by the action of placental estrogen (BENIRSCHKE et al. 1956).

There is no reason to suspect that withdrawal of maternal ACTH at birth is responsible for the postnatal adrenal regression. For example, an infant treated

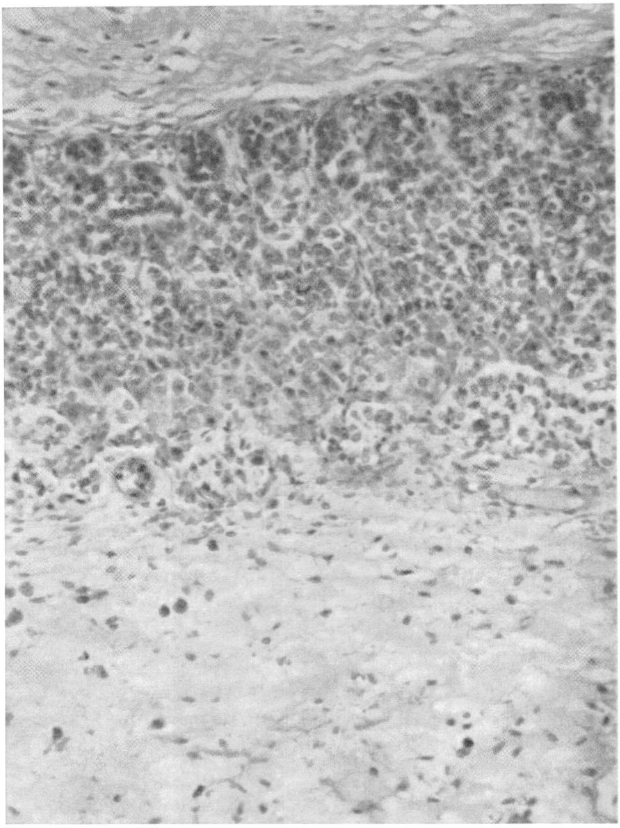

Fig. 7. Adrenal from a case of anencephaly at term. The entire cortex consists of deeply-staining cells similar to those of the outer portion of the normal fetal adrenal. The inner space is devoid of fetal reticularis and is formed of edematous loose mesenchyme. Hematoxylin and eosin, × 130. [Reproduced by permission from NICHOLS et al.: J. clin. Endocr. **18**, 444 (1958) (Charles C. Thomas)]

with large amounts of ACTH for retrolental fibroplasia and dying at 100 days of age, exhibited the usual regression of the fetal adrenal reticularis (LANMAN 1953).

At term, there is a rather constant relationship between maternal and cord levels of plasma 17-hydroxycorticosteroids, the cord values being 2 to 5 times lower than maternal values (MIGEON et al. 1955). Radioactive cortisol can cross the placental barrier, but the cord levels remain lower than the maternal values (MIGEON et al. 1957). From these and other experiments, it has been suggested that the fetal adrenal is relatively inactive in producing 17-hydroxycorticosteroids. NICHOLS et al. (1958) have reported that, in anencephaly, the maternal: cord ratio for plasma 17-hydroxycorticosteroid levels is similar to that observed in normal pregnancy; this finding precludes the possibility that the fetal reticularis is a necessary source for these hormones in fetal blood. On the other hand, the concentrations of 17-ketosteroids in the cord blood of two cases of anencephaly were found to be low or zero (NICHOLS et al. 1958), in conformity with other evidence suggesting that the fetal reticularis is the site of active production of adrenal androgens. However, DI GEORGE et al. (1956) have reported that an anencephalic fetus at 9 months gestation had a high titer of 17-ketosteroids in the plasma. It is difficult to reconcile these findings, but it is evident that no definite conclusions can presently be drawn from a study of only 3 cases.

In two female pseudohermaphrodites with congenital adrenal hyperplasia, each with an enlarged phallus and a single external vagino-urethral opening, BLACKMAN (1946) noted that there was, at 1 and 6 weeks of age, "fetal reticular tissue in stages of involution which were normal in degree for the age of each infant." However, these adrenals "contained postnatal reticular zones which were thicker than normal." JONES and JONES (1954), in a study of 5 such infants and 10 older cases, concluded that there was hyperplasia of the reticular zone of the adult type, but that the fetal reticularis did not participate in the hyperplastic process. Hence, it seems that adrenal hyperplasia need not necessarily be associated with failure of the fetal reticularis to undergo post-natal involution. On the other hand, some types of adrenal tumors have been thought to bear a close morphological resemblance to the cells of the fetal reticularis. An example of such a case was described by SOBEL et al. (1953), in which there was a considerable urinary output of dehydroepiandrosterone.

In conclusion, the evidence that the human fetal reticularis produces androgens appears to be fairly good, although some aspects of the situation merit further study. In children and adults, the degree of development of the reticularis seems to parallel the output of androgens.

II. Rodents

1. The X zone of the mouse adrenal

The term X zone was introduced as a provisional designation for a morphologically distinctive area in the mouse adrenal. This zone undergoes characteristic changes related to age and hormonal status (HOWARD 1927).

The X zone is recognized as a juxtamedullary area of irregularly arranged cells, somewhat smaller than the cells of the zona fasciculata. These cells have a granular cytoplasm staining more intensely with various acid stains than the cells of the fasciculata, but they lack lipide droplets (HOWARD 1927, DEANESLY 1928, CHESTER JONES 1949a, BENUA and HOWARD 1950). However, the X zone apparently contains ascorbic acid (LEBLOND and GARDNER 1938). The zona reticularis proper is not well developed in adult mice. The X zone appears as a rather narrow zone in both sexes at 21 days of age. It regresses in males before sexual maturity. In females it progressively enlarges, and may reach a volume of about half that of the definitive cortex at the beginning of sexual maturity. Thereafter it undergoes either gradual regression in unmated females or rapid regression during the first pregnancy. It hypertrophies following castration of prepuberal males, but it is not influenced by ovariectomy (MASUI and TAMURA 1924, HOWARD 1927, DEANESLY 1928). Ultimately, the X zone degenerates in gonadectomized mice of either sex. There are considerable variations in different strains of mice as to the maximum size attained by the X zone and as to its rate of degeneration. The histological

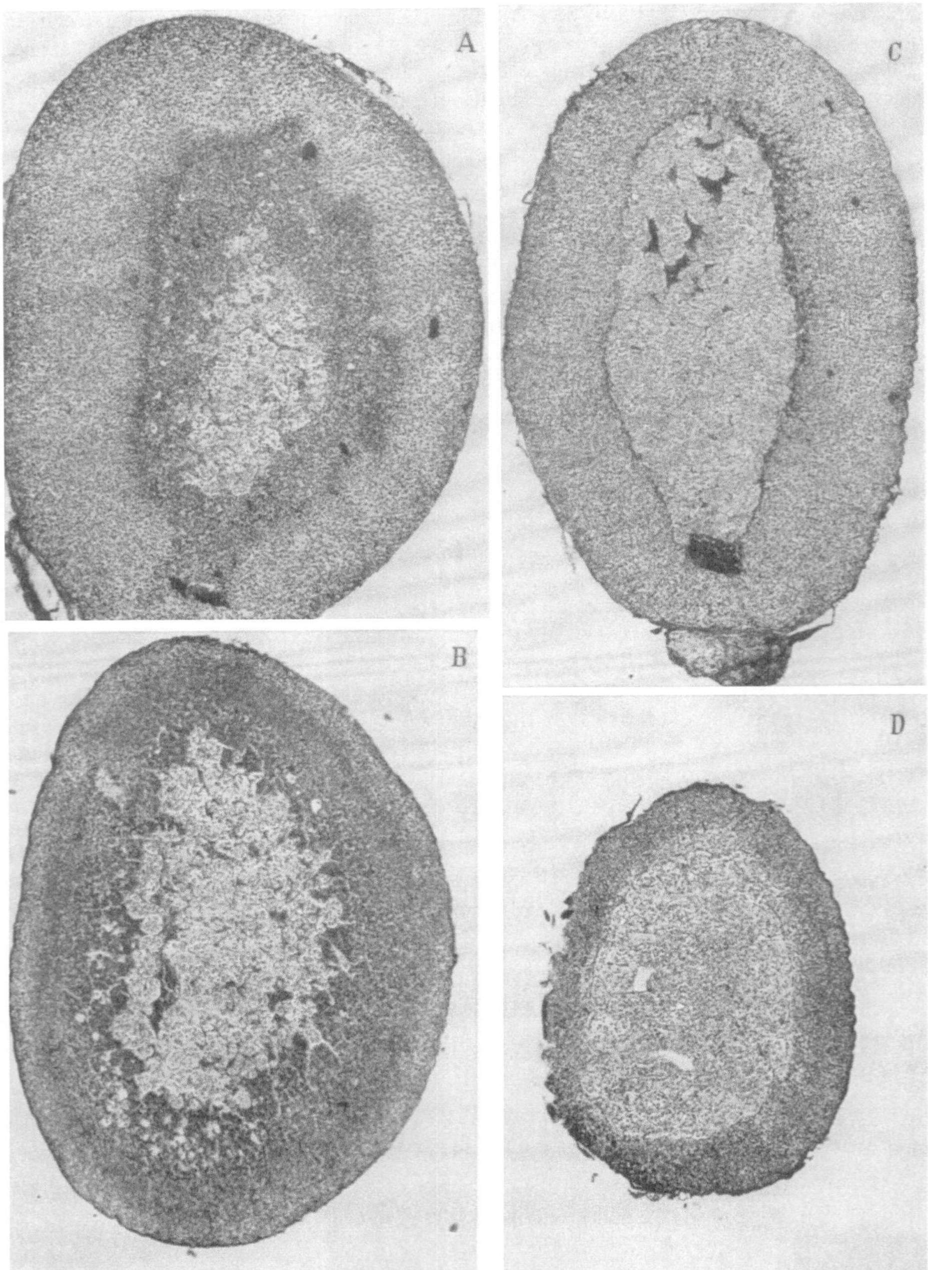

Fig. 8 A—D. Mouse adrenal glands, from females 6 weeks of age, showing effects of corticosterone and dehydroepiandrosterone, singly and in combination, for a 13 day period. Azan stain, × 50. [Reproduced by permission from HOWARD: Endocrinology **65**, 785 (1959) (Charles C. Thomas)]. A. Untreated control. Note the deeply stained X zone of moderate size which contrasts with the more lightly stained definitive cortex, and the light central medulla. B. Corticosterone pellet; total steroid absorbed 5.4 mg. There has been a reduction in the weight of the adrenal but the X zone has not been suppressed. C. Dehydroepiandrosterone pellet; 2.1 mg absorbed. There has been a complete suppression of the X zone, and the gland has assumed its typical mature structure. D. Combined treatment with corticosterone (5.4 mg) and dehydroepiandrosterone (2.1 mg) has resulted in a reduction of the cortex to a narrow band of small crowded cells. The entire central light area is medulla. D is from a litter mate of specimen B

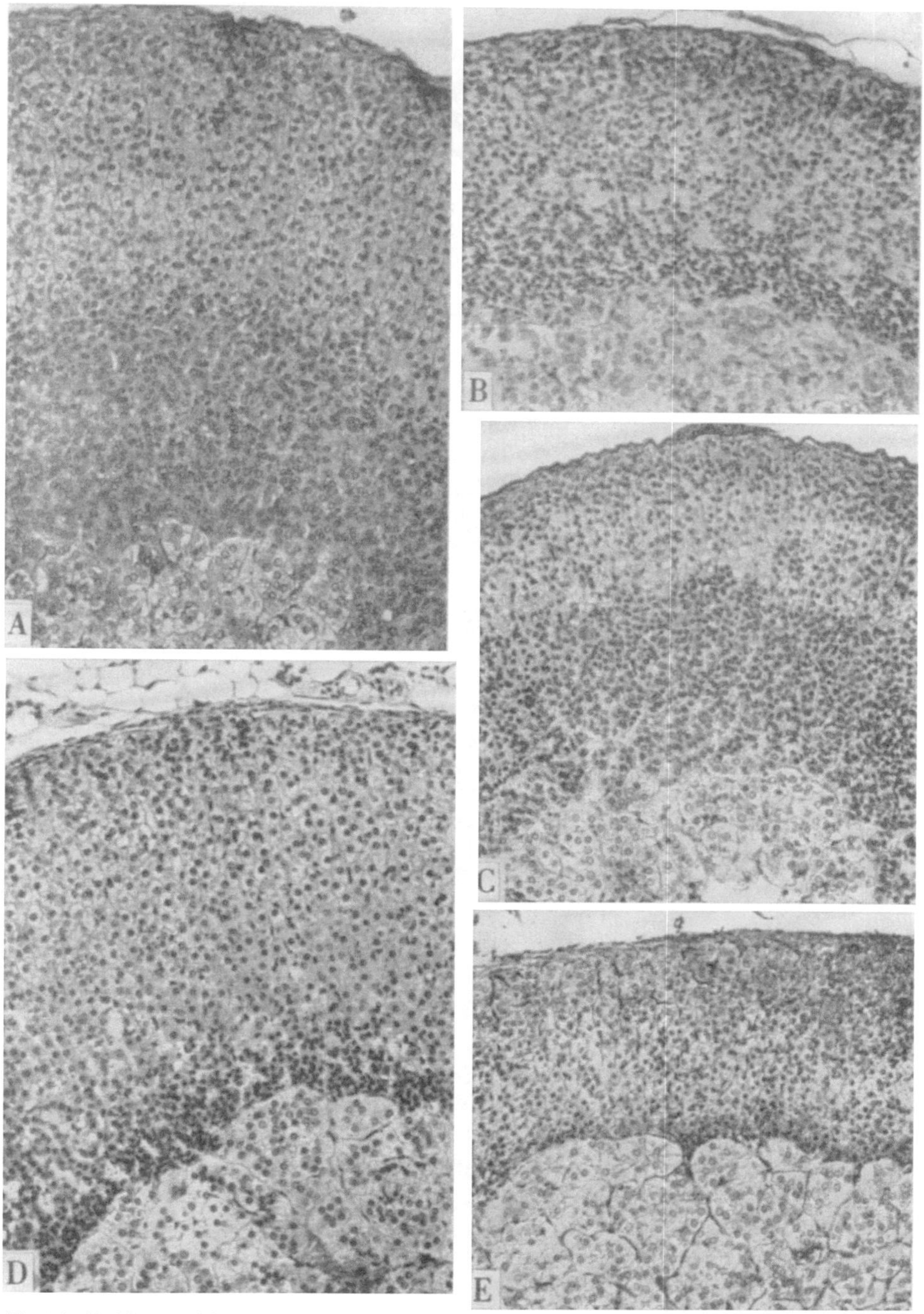

Fig. 9 A—E. Adrenals of female mice, illustrating the effects of hypophysectomy and pituitary hormones. Hematoxylin and eosin, × 180. [Reproduced by permission from CHESTER JONES: Endocrinology **45**, 514 (1949) (Charles C. Thomas).] A. Untreated control, aged 55 days. The darkly-staining X zone contrasts with the lighter cells of the definitive cortex. At the bottom of the photograph is a small amount of medulla. B. Mouse hypophysectomized 14 days prior to autopsy, at 54 days of age. The zona glomerulosa is not markedly different from that of the control. The degeneration of the zona fasciculata is well advanced. Areas of degeneration are particularly apparent in the region of the inner fasciculata. The X zone is represented by a layer of pyknotic nuclei above the medulla. C. Mouse ovarietomized at 43 days of age, hypophysectomized at 45 days. Injected with a gonadotrophic pituitary extract for 12 days prior to autopsy at 59 days of age. The X zone is maintained with nuclei normal in appearance and the cell cytoplasm staining with eosin. The zona glomerulosa is normal in appearance. The zona fasciculata is degenerating and narrow. D. Mouse hypophysectomized 14 days earlier and injected with ACTH for 12 days. The zona fasciculata is wide and well maintained. The X zone is collapsed, its nuclei pyknotic, the cell cytoplasm shrunken and non-eosinophilic. E. Mouse hypophysectomized 12 days earlier and injected with human chorionic gonadotrophin for 10 days, the ovaries being present. The X zone has disappeared

details of the degenerative process are also variable (HOWARD 1930, HOWARD 1938, HOWARD and GENGRADOM 1940, DAUGHADAY 1941).

X-zone suppression can be observed following administration of androgens (MARTIN 1930, POLL 1933, DEANESLY and PARKES 1937, STARKEY and SCHMIDT 1938, HOWARD 1940, LEATHEM 1949, CHESTER JONES 1949a, NISHIDA and MOCHIZUKI 1955). The suppression of the X zone by testosterone appears to be compatible with a stimulation of the zona fasciculata, this becoming apparent as the dosage is increased (ROY et al. 1959). Certain C_{19} compounds suppress the X zone at dose levels that produce little or no androgenic effects on male accessory glands (DEANESLY and PARKES 1937, HOWARD 1956). Conversely, the administration of deoxycorticosterone (HOWARD 1940, LEATHEM 1949) or cortisone (SAKIZ 1953, ALLEN 1954) fails to suppress the X zone. Corticosterone, the chief glucocorticoid produced by the mouse adrenal (HOFMANN 1956), does not suppress the X zone at a dose of 0.4 mg/day (HOWARD 1956, 1959). This dose produces a 29% reduction in adrenal weight, reduces the width of the definitive cortex and restricts somatic growth. On the other hand, dehydroepiandrosterone at a dose of 0.14 mg/day completely suppresses the X zone and causes an adrenal weight reduction of 29%. Corticosterone and dehydroepiandrosterone administered together produce a roughly additive effect with an adrenal weight reduction of 69% (see Fig. 8). These observations are compatible with the concept that the definitive cortex and the X zone are controlled independently. From studies on mitotic counts, WHITEHEAD (1933) concluded earlier that the X zone retains the ability to grow independently of the definitive cortex.

A strain of pituitary-deficient dwarf mice does not develop an X zone DEANESLY 1938), suggesting that this zone is dependent on a pituitary factor. This was shown to be the case by CHESTER JONES (1949b, 1950), who observed that both definitive cortex and X zone regressed following hypophysectomy. Furthermore, ACTH stimulated the growth of the definitive cortex, but it did not induce X-zone growth, whereas the X zone was stimulated by a gonadotrophic extract that was thought to contain chiefly LH (see Fig. 9). The X zone was suppressed by a low protein diet that did not result in a comparable atrophy of the definitive cortex (HOWARD and BENUA 1950). This was interpreted as related to a depression in the pituitary output of gonadotrophins, which has been observed in inanition states by others (ERSHOFF 1952). As pointed out by CHESTER JONES (1957) in a recent review, the mouse presents the only unequivocal case where gonadotrophins have been shown to have a direct immediate effect on the adrenal cortex.

ACTH not only fails to induce X-zone growth in hypophysectomized mice, but actually stimulates X-zone regression in gonadectomized male and female mice. This observation has been interpreted as evidence of an effect of androgens produced by the definitive cortex in response to the administered ACTH, since androgens are the only hormones known to cause X-zone regression (DEANESLY 1958).

MOOG et al. (1954) found a decreased rate of growth of the mouse adrenal immediately after birth and a marked decline in the ratio of the adrenal weight to body weight during the first postnatal week, without the dramatic zonal regression characteristic of the human gland in the neonatal period. In this connection it must be noted that the mouse is born at a much earlier stage of development than the human. Histological studies showed that the entire cortex stains with Sudan black at 2 days of age, but that after 16 days of age the X zone is negative to Sudan black (MOOG et al. 1954). These histochemical studies support the views of HOWARD (1939) that between the 5th and 15th days of age differentiation of the X zone occurs by means of a gradual accentuation of the properties which distinguish it from the permanent cortex.

The fact that the X zone of the mouse hypertrophies following castration (MASUI and TAMURA 1924, HOWARD 1927, 1930) made it reasonable to suppose that this might be in the nature of a compensatory reaction associated with the secretion of androgens by the adrenal. Hence the term "androgenic zone" was introduced by GROLLMAN (1936). Although no consistent androgenic effect of the adrenals on the seminal vesicles could be detected (HOWARD 1946), the fact that the X zone is suppressed by dehydroepiandrosterone at doses that do not exert any appreciable stimulation of the seminal vesicles leads to the hypothesis that the X zone is producing a substance of similar properties and that it undergoes diffuse atrophy when such a compound is provided exogenously (HOWARD 1956). Evidence suggesting that the mouse adrenal produces systemically appreciable amounts of androgens is now available (see p. 602).

2. A presumptive X zone in voles

Several subfamilies of voles appear to have a large X zone very similar to that of mice. The European field mouse or vole *(Microtus arvalis)* in its wild state breeds during the spring and summer, but in winter is in a state of sexual inactivity, when there is a marked reduction in the size of the male genital tract and absence of spermatogenesis. DELOST (1951) has observed that during winter the adrenals of male voles have a juxtamedullary zone somewhat similar to the mouse X zone, whereas this zone is absent in the adrenals of male voles trapped during the summer months. After surgical castration, a large X zone develops in the male vole (DELOST 1956). Furthermore, in *Clethrionomys glareolus* and in *Pitymys subterraneus* (DELOST and VINCENT 1955), adult females have large, clearly-differentiated X zones. However, in these forms the zone is not suppressed by pregnancy as it is in mice, but may even share in the general cortical hypertrophy associated with pregnancy and lactation. DELOST (1956) reported that cortisone suppresses this zone in voles *(Microtus)*, which is also contrary to the situation in mice. It is possible that the cortisone used was metabolized in the voles to androgens sufficient to suppress the X zone. Whether the zone in voles is analogous to the mouse X zone, or whether the morphological similarity is deceptive, remains for future study to answer.

3. A presumptive X zone in the golden hamster

In young golden hamsters, the female normally develops a special juxtamedullary zone, histologically similar to the mouse X zone, although not as large or as sharply differentiated from the overlying cortex. The parous female and the intact male do not exhibit such a zone. In males castrated the day after birth, an X zone is present at 7 weeks of age (LA VELLE 1951), but castration at 17 days of age is not followed by the appearance of an X zone (HOLMES 1955). Testosterone injection causes the zone to regress in ovariectomized females, with the development of a connective tissue capsule around the medulla. Thus, although the zone in question is rather small in the hamster, it appears to react to hormonal status in a manner similar to that of the mouse X zone.

4. The rat

The laboratory rat does not have a morphologically defined X zone. There is evidence, however, that the immature rat adrenal produces more biologically active androgens than does the mature adrenal (see p. 599) and that it has a histological appearance different from that of the mature adrenal (HOWARD 1938). The wild Norway rat has a much larger adrenal than the laboratory rat, and groups of cells in the juxtamedullary region have been thought to bear some resemblance to X zone cells (MOSIER 1957). A 25% postnatal shrinkage of the rat cortex has been described (DEANE 1956) (see Chapter 1, Fig. 62, p. 123).

III. Other mammals

A study of the fetal cat adrenal has indicated the presence of a zone morphologically comparable to the human fetal reticularis (DAVIES 1937). This zone is well developed in subadult females. The observations of DAVIES suggest that the cat may be somewhat intermediate between the mouse and the rhesus monkey and point to the desirability of further study of this species. In male cats the reticularis is small both absolutely and relatively to the fasciculata; after castration the reticularis enlarges; it can be suppressed by testosterone injections (LOBBAN 1952). The reticularis is small in immature or anestrous females, but enlarges at estrus and during pregnancy and pseudopregnancy. LANMAN (1953) cites other pertinent observations.

The immature male shrew *(Sorex araneus)* has been reported to have a large X zone similar to that of the mouse, while the sexually mature males lack this zone (DELOST 1957).

IV. Summary

The hypertrophy of a morphologically distinctive juxtamedullary zone after castration in males has been described in the mouse, voles, cat and golden hamster. In the rat, the cortex enlarges after castration, but without distinctive morphological differentiation. Pregnancy appears to have different effects: in the mouse and hamster, the zone is present before breeding, but disappears with the first pregnancy, while in the vole and cat, it has a tendency to enlarge.

The cat appears to have a fetal zone similar to that of primates, but less developed. Quantitative work has shown that an acute shrinkage of the cortex immediately after birth is not restricted to the primates, but occurs to a lesser extent in rodents.

In the human, the fetal reticularis largely degenerates in the neonatal period. The reticular zone enlarges during the second decade of life, remains well developed during the period of sexual activity, and practically disappears after the end of the 5th decade. The situation as a whole suggests that the morphological changes of the adrenal cortex are related to qualitative variations in hormone secretion.

E. Experimental studies in animals

I. Effects of adrenalectomy in castrated animals

One approach to an evaluation of the systemic significance of the C_{19} adrenal steroids is an examination of the accessory reproductive glands after castration followed by adrenalectomy. The animal must be in good general health so that secondary nonspecific effects will not interfere with the detection of possible specific androgenic effects. In the studies to be considered, the animals have usually been maintained on NaCl or on small amounts of deoxycorticosterone that have been shown not to be androgenic under the conditions employed. Theoretically, animals should be maintained on a constant intake of mineralocorticoid and glucocorticoid while being studied. However, the glucocorticoids are in part metabolized to androgens, so that the distinction between the physiological effects of the glucocorticoids and the androgens may not always be clear-cut.

In an earlier review, PONSE (1950) commented that «le cortex révèle sa bipotentialité sexuelle avec d'autant plus facilité et d'intensité que l'organisme est plus jeune ou castré plus précocement». PONSE referred to several examples of paradoxial masculinization in lower vertebrates following castration as larvae or

1. The rat

It is unquestionable that the adult prostate gland depends on the presence of testicular hormone to maintain its normal size and secretory state. It was demonstrated by MOORE et al. (1930) that 4 days after castration, cytological changes can be seen which become progressively greater. It was thus of considerable interest when PRICE (1936) reported that, in rats castrated at 2 to 6 days of age, the ventral prostate continues to grow and differentiate until 30 days of age. During this time the epithelial cells become columnar and develop the characteristic Golgi zone which is indicative of the presence of androgens. However, this development of the prostate in infant castrates ceases fairly abruptly; regression sets in after about 30 days of age, and by 52 days of age the prostate of the rat castrated at 2 days is comparable to that of the adult 21 days after castration. These observations were confirmed by HOWARD (1937, 1938), who suggested that a temporary phase of androgen secretion by the adrenal of the immature rat might be contributing to the differentiation and maintenance of the prostate in these prepuberal rats.

Experimental support for this concept was furnished by BURRILL and GREENE (1939, 1940). When male rats were castrated at 16 days of age, and subsequently adrenalectomized at 21 days, the ventral prostates were about half the weight of those in simple castrates by 26 days and showed evidence of histological regression to a state approximately equivalent to that of castrated adults. However, the body weights of the adrenalectomized rats, given salt therapy while being nursed by normal mothers, were inferior to those of the controls. Animals that were only adrenalectomized did not show significant prostatic regression, indicating that the testicular hormones were capable of maintaining the prostate at this age. The apparent androgenic activity of the adrenal declined markedly with advancing age of the rats: thus, the per cent differences between the prostate weight: body weight ratios of the castrates and the adrenalectomized castrates were 54% at 26 days, 34% at 31 days, 12% at 36 days and 11% at 41 days.

Further support for these conclusions was furnished by HOWARD (1941), who studied rats castrated at 17 days of age, adrenalectomized at 20 days of age and maintained on pellets of deoxycorticosterone which liberated 0.3 mg of steroid per day. These rats grew at an essentially normal rate, but on autopsy at 26 days the ventral prostates of the adrenalectomized castrates were half the size of those of the litter-mate castrates and showed histological regression. Adrenalectomized castrates supplied with NaCl without deoxycorticosterone grew at an appreciably slower rate, but the prostate/body weight ratio was not different from that of the group on deoxycorticosterone and the glands showed a similar degree of histological regression (see Fig. 10). MCPHAIL (1944) confirmed these findings.

PRICE (1939) observed the presence of a prostate gland in about 2% of the females in a particular rat colony. During the period between 21 and 40 days of age, the prostates of these females exhibited a well-differentiated columnar epithelium, with the development of large Golgi zones characteristic of androgenic

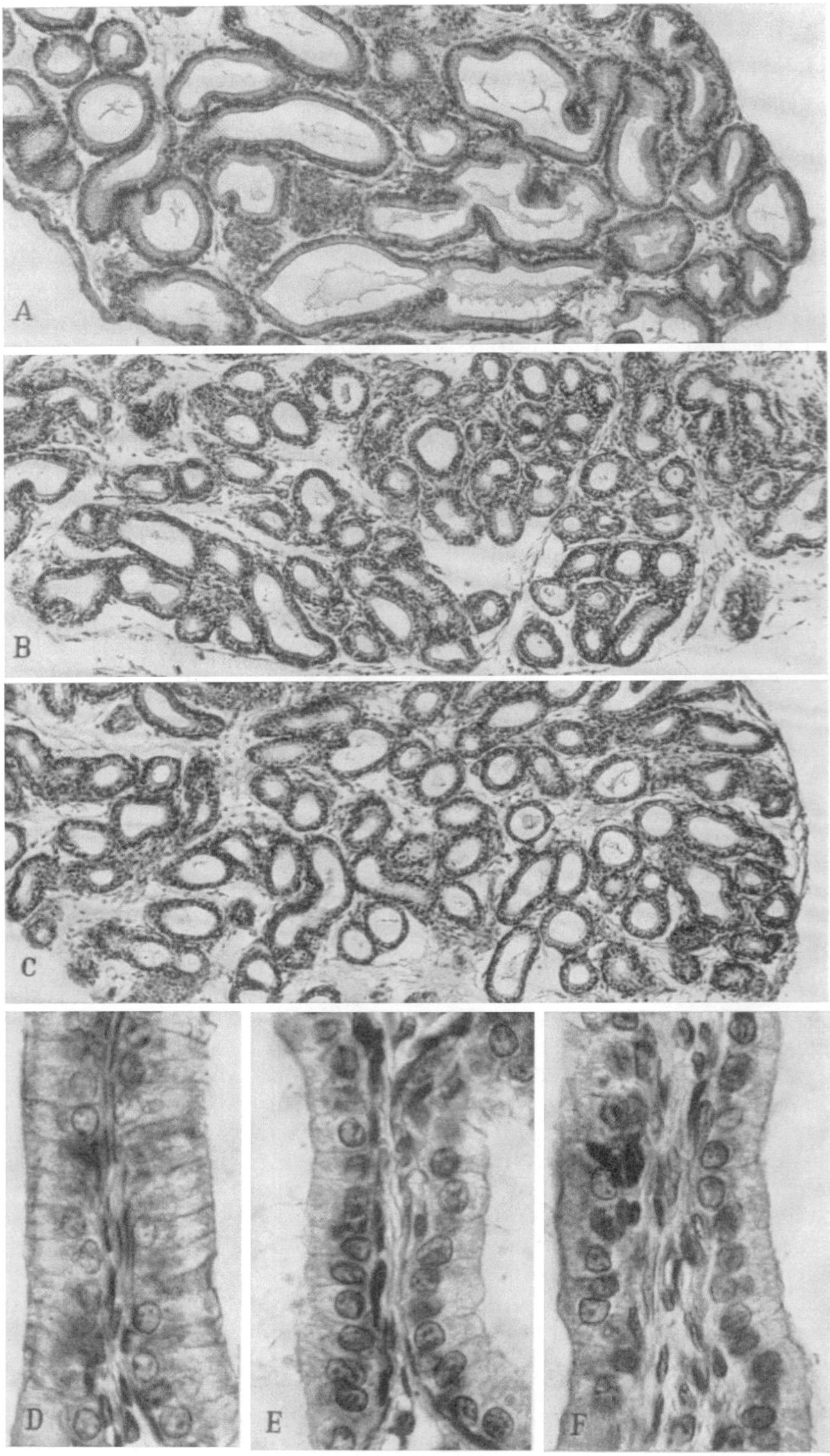

Fig. 10 A—F. The effects of adrenalectomy on the ventral prostate gland of infant rats. [Reproduced by permission from HOWARD: Endocrinology **29**, 746 (1941) (Charles C. Thomas).] The animals were all castrated at 17 days of age and autopsied at 26 days; adrenalectomies were performed at 20 days; the animals were not

stimulation. After this period the gland regressed as it did in male castrates. PRICE (1939, 1947) suggested that the prostate in the young female rat was exhibiting evidence of circulating adrenal androgens.

BURRILL and GREENE (1941 a) obtained a strain of rats having incidence of prostates in about 75% of the females, and furnished further evidence of adrenal androgen secretion in young female rats. There was a highly significant reduction in the size of the prostates in adrenalectomized-ovariectomized rats in comparison with ovariectomized rats. When an adrenalectomized group was compared with the normal intact group, no reduction in prostatic size was observed, thus confirming the previous findings in the males that adrenal hormones are not necessary for the maintenance of the prostate in the presence of the gonads, at least during short-term experiments. In further work, BURRILL and GREENE (1941 b) reported observations on the androgenic activity of adrenals grafted to the posterior surface of the stomach, suggesting that the adrenal androgens are not completely inactivated by passage through the liver in immature rats. Since androgens from testicular transplants were inactivated, it was suggested that testicular and adrenal androgens are metabolized differently under these conditions.

Failure to detect adrenal androgen secretion in rats has been reported by GERSH and GROLLMAN (1939), but BURRILL and GREENE (1940) and HOWARD (1941) have pointed out certain technical defects in this work. The histological description of GERSH and GROLLMAN (1939) suggested the presence of stratified epithelium in the prostates of the experimental animals. Since such an appearance is a typical effect of estrogen administration, it appears probable that the adrenal extracts used for maintenance contained estrogens, which interfered with the manifestation of endogenous adrenal androgens. Estrogens have been isolated from beef adrenals (BEALL 1939).

MOORE (1953) failed to observe any effects of adrenalectomy on the accessory organs of rats castrated at 50 days of age and autopsied at 65 days. MOORE also reported observations on 11 rats which had been castrated and adrenalectomized at 20–21 days of age, and autopsied at 29–34 days. The prostate: body weight ratios averaged 18% less in the adrenalectomized castrates than in litter mate castrates, and histologically there was evidence of prostatic regression in the adrenalectomized rats. The findings did not impress MOORE, but the weight differences are in accordance with the much more extensive data of BURRILL and GREENE (1940), who showed that the adrenal effect declined rapidly in rats older than 26 days.

An essential point which has sometimes been overlooked is the very transitory aspect of the adrenal secretion of androgens in amounts which can be detected by prostate reactivity in rats (BURRILL and GREENE 1939, 1940, 1941 a, b, and HOWARD 1941). It is noteworthy that the seminal vesicles remain uninfluenced by the adrenal androgen at this age period.

The androgenic activity of extracts of rat adrenals at different ages has been assayed by the chick-comb growth test (HÉBERT et al. 1954). The activity of glands from rats 1 day of age was considerably higher than from adults. This

weaned. A—C, × 100. All from litter mates. A. Prostate from a castrated control, showing moderately large alveoli and columnar epithelium. B. Prostate from an untreated adrenalectomized castrate. The size of the alveoli has been reduced and the epithelium is cuboidal. C. Prostate from an adrenalectomized castrate treated with deoxycorticosterone. The appearance is similar to that of the gland in B. D—F, × 750. All from litter mates. D. Columnar epithelium from prostate of a castrate. The light Golgi zones midway between nucleus and lumen border of cells may be seen. E. An example of the best-developed epithelium from the prostate of an untreated adrenalectomized castrate. The cell height has been reduced in comparison with D, and Golgi zones are absent. Most of the epithelium in the glands from these animals was cuboidal. F. An example of the best-developed epithelium from the prostate of an adrenalectomized castrate treated with deoxycorticosterone. The appearance is similar to that of the gland shown in E.

activity dropped markedly by 8 days of age, and by 11 days was only little higher than in adult adrenals. Differences in fuschsinophilia of the adrenocortical cells appeared to be correlated with the results of the assays.

The complexity of the relations between the adrenal and the physiological actions of androgens is illustrated in recent studies on the role of the adrenals in seasonal variations in the sensitivity of the dorsolateral prostate of the rat to testosterone (GUNN and GOULD 1959). Using the rate of uptake of Zn^{65} as an index of prostate activity, they observed seasonal variations in the response of the prostate to testosterone, in castrated rats, which can be correlated with the known seasonal breeding periods of wild rodents: i. e., there is a period of maximal prostate response in June—July, a period of waning response in August—September, followed by a period of relatively low response in October and later months. During the period of maximal response of the prostate, adrenalectomy reduced the response, but during the period of waning response, adrenalectomy enhanced the response, as if an inhibitory effect of the adrenal secretions had been removed. During the period of low response, adrenalectomy had no effect. The authors did not commit themselves as to what components of the adrenal secretions might be involved in this diphasic effect. The observations invite comparison with the evidence suggesting cyclic variations in adrenal activity in females. (See p. 622.)

2. The mouse

In spite of the fact that prepuberally-castrated mice exhibit relatively good maintenance of the seminal vesicle epithelium (HOWARD 1937, 1939), it has not been possible to demonstrate a consistent effect of adrenalectomy on the seminal vesicles of castrated mice. In different experiments the animals were maintained for short periods in good general health by adrenal transplants (MCPHAIL 1944), moderate amounts of deoxycorticosterone or a high Na^{+}/K^{+} ratio in the diet (HOWARD 1946). However, in over half of the adrenalectomized castrates not given adrenal substitution therapy, a clear cut histological regression of the seminal vesicles was observed (HOWARD 1946). Hence, it may be concluded that various sequelae of adrenal insufficiency can influence the state of the seminal vesicle epithelium.

The ventral prostate in castrated immature mice is rather small for satisfactory weighing, and exhibits considerable histological variation. GERSH and GROLLMAN (1939a) and MCPHAIL (1944) failed to detect an influence of the adrenal on the mouse prostate. However, NISHIDA and MOCHIZUKI (1955) reported a histological regression of the ventral prostate in castrated adrenalectomized mice, greater than that following castration alone. This effect of adrenalectomy was more distinct in adults than in immature mice. Following castration, the well-developed Golgi zones disappeared from most of the epithelial cells of the prostate, but when adrenalectomy was performed on mice previously castrated, the height of the epithelial cells was further reduced and the Golgi zones became totally indistinguishable. Adrenalectomy without castration had no apparent effect on the prostate gland.

RAYNAUD (1954) has studied the effect of adrenalectomy on the structure of the submaxillary gland of castrated mice. This gland is known to be considerably influenced by sex hormones. One month after gonadectomy, adult mice were adrenalectomized and maintained with implants of deoxycorticosterone. After further intervals of 25 to 41 days, it was found that the adrenalectomized castrates showed marked histological regression of the submaxillary gland in comparison with the mice which had been castrated only. It was pointed out by the author that the effect might be due to lack of either glucocorticoids or androgenic hormones.

In the congenital form of the adrenogenital syndrome in human females, masculinization is frequently confined to the external genitalia, with the internal duct derivatives being relatively normal (WILKINS 1957). This observation, together with the influence of dehydroepiandrosterone and 11β-hydroxy-Δ^4-androstenedione on phallic growth (see Table 1, p. 578) led to an examination of the effects of adrenalectomy on the phallic bone in young female mice. As illustrated in Fig. 2 (p. 579), the phallic bone in females appears either as a small fairly regular ovoid or as an elongated bone of more irregular form, the differences being presumably genetic. For quantitative measurements mice having the ovoid form were selected. Under magnification, the anterior-posterior and sagittal diameters of the phallic bone were measured. Mice were adrenalectomized and ovariectomized at 12 days of age, maintained on minimal amounts of deoxycorticosterone and autopsied at 40 days. Under these circumstances, body weight increased at a normal rate. The phallic bones of the adrenalectomized-ovariectomized mice were 44% smaller than those of the ovariectomized mice, a difference that was statistically highly significant (HOWARD, unpublished). Hence, it may be concluded that systemically appreciable amounts of adrenal androgens are produced in the mouse. The presence of $C_{19}O_3$ steroids in mouse blood has been reported, in amounts roughly equal to that of corticosterone (WILSON et al. 1958).

3. Voles

The ventral prostate of voles has been reported to regress for a period of 3 weeks following castration. After this initial regression it undergoes marked stimulation, so that from 1 to 3 months later the epithelium appears in a good functional condition (DELOST 1956). When adrenalectomy with deoxycorticosterone or salt maintenance followed castration, the animals exhibited a marked regression of the ventral prostate within 15 days. This regression could be prevented by administration of testosterone. In addition, cortisone was reported to improve the condition of the epithelium of the ventral prostate. However, the dose used, 25 mg of cortisone per day for 15 days, is in excess of the physiological range, and would be expected to give rise to considerable amounts of biologically active androgenic steroids (MUNSON et al. 1954). Changes were also described in the vas deferens and in other portions of the reproductive tract of the vole following castration and adrenalectomy (DELOST 1956) which are in many respects difficult to interpret.

4. The golden hamster

The persistence of many components of sexual behavior for a considerable time after castration is commonly observed in male individuals of many vertebrate species. An explanation frequently advanced is that, in castrated males, androgen from the adrenal cortex is responsible for the continuing sexual activity (literature cited by WARREN and ARONSON 1956). This hypothesis was tested by WARREN and ARONSON (1956), who selected hamsters as experimental animals because they are known to exhibit a considerable amount of sexual activity after gonadectomy. Furthermore, the seminal vesicle epithelium in hamsters has been observed to present a fairly normal appearance for about 2 months after castration, which suggested to ORTIZ (1953) a possible androgenic influence of the adrenal. WARREN and ARONSON conducted tests of sexual behavior before and after operation, their studies being limited to mature males. They found that castrated-adrenalectomized males maintained on deoxycortisterone behaved similarly to the castrated controls, and concluded that the sexual behavior that persists after castration is not dependent on androgens from the adrenal cortex. Deoxycorticosterone was not found to be androgenic in doses 3 times as great as those necessary for maintenance. The adrenalectomized animals were shown to be devoid of adrenal tissue by withdrawal of hormone therapy, whereupon they succumbed within 2 weeks. The seminal vesicles in the adrenalectomized castrates were similar to those in the castrate controls.

II. Effects of adrenals grafted to target organs

Androgenic stimulation of seminal vesicle by adrenal tissue grafted onto the seminal-vesicle wall was reported by KATSH et al. (1948). However, MOORE (1953)

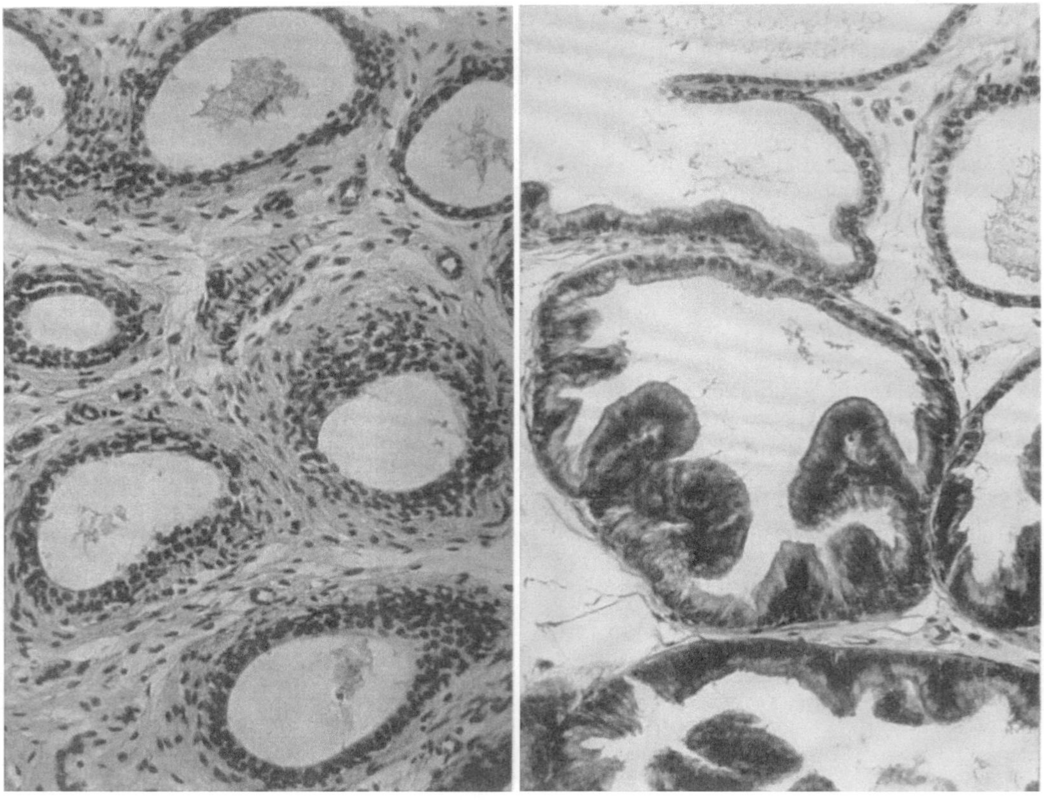

A B

Fig. 11 A u. B. The effects of an adrenal graft on the ventral prostate gland of the rat. × 250 (PRICE and INGLE 1957; courtesy Dr. DOROTHY PRICE). A. Ventral prostate gland of a rat that had been castrated for 3 months showing the involuted epithelium. B. Ventral prostate of a rat castrated 3 months previously and implanted with two adrenal glands, which were attached close to the acini here shown. The epithelium is columnar in type, and the cells contain large clear "Golgi zones"

failed to detect evidence of androgenic activity of adrenals grafted onto seminal vesicles of rats that had been castrated 1 to 3 months before the grafting. JOST (1948) and JOST and GELOSO (1954) used host rats which had been castrated for

3 or 8 months and reported that «Enfin, à côté de l'un des greffons laissés 30 jours, l'épithélium est nettement plus élevé qu'ailleurs». Studies on graft effects in voles (DELOST 1956) did not yield consistently satisfactory evidence of adrenal andromimetic activity.

PRICE and INGLE (1957) overcame certain technical difficulties, and evidence of considerable local androgenic activity of adrenal grafts was observed, as illustrated in Fig. 11. The procedure of PRICE and INGLE was as follows: adult rats were castrated and both adrenals were removed and sutured to the tip of the seminal vesicles or to the surface of the ventral prostate. No significant increase in the weight of the accessory glands occurred, but the adrenal grafts caused local androgenic stimulation of seminal vesicles, coagulating glands and prostates. The effects varied from the production of a small amount of secretion in the acini near the graft to a marked secretory activity and high epithelium, with development of conspicuous Golgi zones. The magnitude of the effect was related to the closeness of the graft to the glandular epithelium and to the nature and thickness of the intervening tissues. The stimulation was evident after the graft had been implanted for 1 month, but more marked after 3 months. The grafting technique is thus capable of detecting quantities of androgens that would be sub-threshold for the reproductive tract with the adrenals *in situ*.

Evidence that the adrenals of young chicks contain appreciable amounts of androgens is provided by a study in which adrenals from donors of various ages were implanted into an incision in the base of the comb of 3-day-old host chicks (ARRINGTON et al. 1952). Muscle was used as a nonspecific control implant. It was found that the adrenals from donors of 7 to 15 days of age produced an effect only slightly inferior to that of testicular tissue; adrenals from adults produced no appreciable effect under these circumstances. The finding that the adrenal in young chicks produces more androgen than it does in adults is in agreement with observations reported for other species, i. e., the rat and the human.

III. Adrenal transplants in chick embryos

In contrast to the marked activity of gonad grafts in chick embryos, adrenal grafts were not found to have any striking effect on the differentiation of the reproductive tract. WOLFF et al. (1951) transplanted adrenal tissue from chick embryo donors of 9–13 days into the coelomic cavity of hosts of 50 to 60 hours incubation age. These were autopsied at 14 to 16 days of incubation. No abnormalities were observed in male embryos. In some of the females, there was a persistence of the right Müllerian duct to a greater degree than normal. In the chick, the left Müllerian duct becomes the oviduct, while the right degenerates under the influence of ovarian hormones. Hence, the retardation in the regression of the right oviduct suggests the possibility of some hormonal effect of the graft.

IV. Androgen production by adrenocortical tumors in castrated animals

The development of adrenal cortical tumors in male guinea pigs 2 to 4 years after castration was described by SPIEGEL (1939/40). The animals were castrated 1 to 4 days after birth, and after intervals of 2 to 4 years they began to show evidence of growth of the external genitalia. At this time they effected deposition of vaginal plugs in estrous females. At autopsy the vas deferens and prostate glands were found to be as well developed as in intact males, and adrenal tumors were present. In one case the tumor appeared to be confined to the reticular zone; in other cases the alterations were so far advanced that the zone of origin of the tumors was uncertain.

The development of adrenal tumors following early gonadectomy in mice has been described in a series of papers by WOOLLEY and associates. This work has been reviewed recently (WOOLLEY 1950, 1958). The nodular hyperplasia of the adrenal occurs in gonadectomized mice of either sex, although the changes tend to be more marked in the females. Adrenocortical carcinomas were also produced. On the basis of the systemic effects, the tumors appeared to secrete both androgens and estrogens. Any one tumor usually produced a predominance of one or the other hormone, judging by the biological effects observed. An

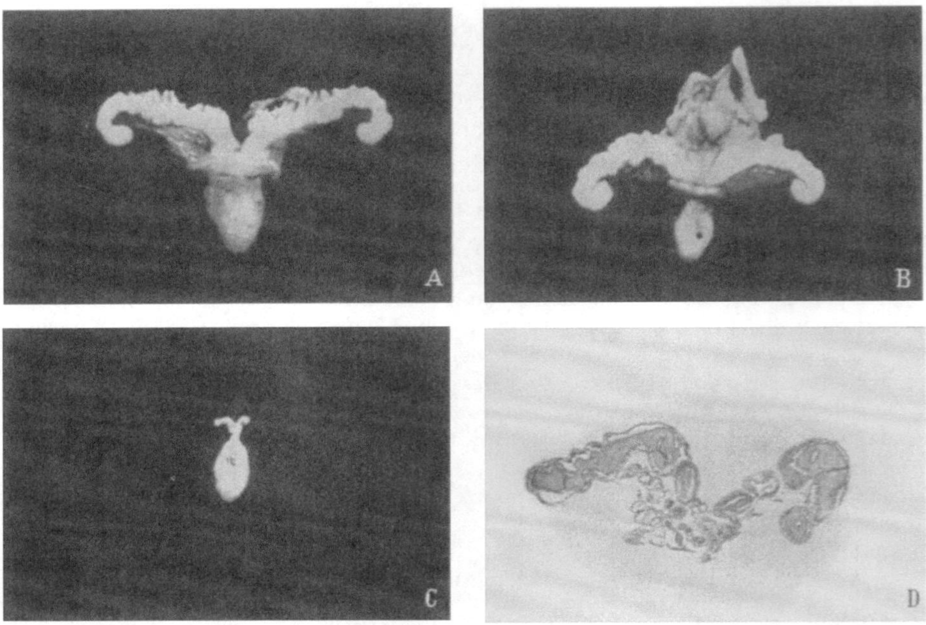

Fig. 12 A—D. The effect of a transplanted adrenal tumor on the seminal vesicles of the castrated mouse. A—C, × 3 approximately. [Reproduced by permission from WOOLLEY and LITTLE: Cancer Research **6**, 712 (1946). (The University of Chicago Press).] A. Accessory reproductive organs from an intact male mouse 8 months of age. The seminal vesicles are the prominent curved structures. The large round organ is the urinary bladder. B. Accessory reproductive organs from a castrated male bearing a transplanted adrenal tumor. The seminal vesicles are almost normal in size. C. Accessory reproductive organs of a castrated CE male before a spontaneous adrenal tumor had developed. The seminal vesicles are visible as the two small paired structures above the urinary bladder D. A section of the seminal vesicles from a castrated male bearing a transplanted adrenal tumor, showing the presence of secretion in the lumen. × 5

example of the effects of transplanted adrenal cortical carcinoma into a castrated host is presented in Fig. 12, which illustrates the resulting enlargement of the seminal vesicles (WOOLLEY and LITTLE 1946b). The estrogenic effects of these tumors will be further discussed in the section on estrogens (p. 618ff.). The adrenal changes were first apparent several months after gonadectomy, and consisted of groups of small, densely arranged polyhedral cells immediately below the capsule. As the subcapsular cells increased in number they interrupted the normal arrangement of the cells of the fasciculata, and later invaded and replaced the cells of all three zones of the cortex (WOOLLEY 1950). However, incubation of adrenal tumors from ovariectomized mice with progesterone-4-C^{14} failed to reveal the production of appreciable amounts of Δ^4-androstenedione (HOFMANN et al. 1960), although the submaxillary glands and the renal tubular epithelium of the tumor-bearing animals exhibited histological changes characteristic of androgenic effects.

Frantz and Kirschbaum (1949) have also observed the development of adrenal tumors in castrated mice of various strains, some of which produced androgenic hormones and others estrogens or mixtures of the two. These authors also found that 2-year old intact male mice of the Bagg Albino strain developed adrenal tumors producing predominantly androgens.

On the other hand, not all rodent tumors produce sex hormones in amounts sufficient to be systemically apparent (Cohen et al. 1957, Bahn et al. 1957, Wilson et al. 1958). While the output of corticosterone was doubled by ACTH-producing pituitary tumors, some of the C_{19}-steroids were also increased (Wilson et al. 1958). Although the C_{19}-steroids produced no detectable stimulation of seminal vesicle epithelium in castrated hosts, they may have contributed to vaginal mucification (Bahn et al. 1957). In these mice the adrenal X zone was not present, hence it is clear that the X zone is not a necessary site of production of urinary C_{19}-steroids in mice.

In conclusion, it has been shown that castration increases the incidence of androgen-producing tumors of the adrenal in several species of rodents. Hence, the withdrawal of gonadal hormones appears to have some influence on the adrenal, either directly or by virtue of the secondary high levels of pituitary gonadotrophins. Direct evidence of a stimulation of the adrenal X zone by gonadotrophins has been presented for the mouse (Chester Jones 1949b). The administration of gonadotrophins appears to modify the adrenal response to ACTH in guinea pigs, in that the 17-ketosteroid response is maintained, but the 17-hydroxycorticosteroid response is lessened (Clayton and Hammant 1955). Analogously, the corticosterone in adrenal venous blood of castrated rats was only one-fifth that of intact rats (Dean et al. 1959). More information as to the possible effects of gonadotrophins on adrenal metabolism would be of interest.

V. Effects of ACTH in castrated animals

Since ACTH has frequently been observed to increase the level of urinary 17-ketosteroids in man (Mason et al. 1948, Forsham et al. 1948, Harrison et al. 1955, Gallagher 1958), experimenters have attempted to obtain evidence of androgen secretion by castrated animals treated with ACTH.

Guyenot et al. (1932) observed a masculinization of ovariectomized guinea pigs following treatment with an alkaline extract of the anterior pituitary which produced a marked loss of body weight. In an often quoted paper, Davidson and Moon (1936) subjected 21-day-old rats to castration and injected them for 10 days with an ACTH extract which contained some lactogenic hormone. At the end of this treatment there was a 3-fold increase in adrenal size, and a 2-fold increase in the size of the accessory sex glands. Davidson (1937) reported that this stimulation of the accessory glands could be obtained in hypophysectomized castrates, but not in adrenalectomized castrates. Moon (1937a) observed that treatment of ovariectomized rats with ACTH for 23 days resulted in marked weight loss and in a 2-fold increase in the size of the preputial glands. These glands are known to be responsive to androgens, including dehydroepiandrosterone, while their size is markedly reduced by hypophysectomy (Huggins et al. 1955). W. Nelson (1941) reported that ACTH produced an enlargement of the accessory sex glands in immature castrated male rats, but not in adrenalectomized castrates.

Moore (1953) treated a number of castrated rats of various ages with an ACTH preparation. The only animals with any significant adrenal enlargement were the rats treated during the critical ages of 21–23 days (see p. 599); and these showed a prostatic enlargement averaging 56%. BAKER et al. (1950) injected ACTH, 1 mg/day, for 21 days into adult rats castrated some time previously. They observed no evidence whatever of any androgenic effect. However, they pointed out that since the glucocorticoids produce a negative nitrogen balance, the increased secretion of these steroids might have obscured the biological manifestation of any possible secretion of adrenal androgens.

Inhibition of the effects of testosterone, particularly its myotrophic effect, by simultaneous administration of cortisol and corticosterone has been observed by PINCUS and DORFMAN (1955), when the corticosteroids were administered at a dose 20 times greater than that of testosterone. An observation suggesting that the net systemic effect of the steroids secreted by the rat adrenal may be inhibitory to androgenic action has been made by PONSE (1954). Male ventral prostate tissue was transplanted into the ovarian bursa of female rats, gonadotrophic hormones were administered and a study was made of the effects on the grafts and the clitoris. Pronounced evidence of ovarian androgen secretion was observed, and the androgenic effects developed more rapidly and were more striking if the hosts were hypophysectomized or adrenalectomized, or both. Also, ARAI (1960) noted a reduction in the response of prostate and seminal vesicles to testosterone, 20 μg/day, when cortisone acetate, 2.5 mg/day, or deoxycorticosterone acetate, 5 mg/day, was given to adult castrated rats. Furthermore, rats were castrated and after 2 weeks were adrenalectomized and a 0.5 mg pellet of cortisone or deoxycorticosterone was implanted in one seminal vesicle. After 5 days the rats were given 20 μg of testosterone propionate daily for 10 days. It was found that the seminal vesicle carrying the hormone pellet was significantly smaller than the contralateral vesicle, while a control pellet of cholesterol caused no change in vesicle weight. Hence the interference was produced at the target-organ level.

Recently, LI et al. (1957) and LOSTROH and LI (1957) have examined the effect of a number of highly purified pituitary preparations on the accessory sex glands of male rats castrated at 43 days of age, hypophysectomized at 44 days and injected daily for 10 days. Alpha-corticotrophin, a polypeptide consisting of 39 amino-acid residues, when given in a dose sufficient to increase the weight of the adrenals 93% above that of intact rats, produced no statistically significant increase in the weight of the ventral prostate. LI et al. (1957) stated that some evidence of prostatic stimulation could be seen histologically, but the photographs published do not show a clear-cut indication of the specific effects of androgens in increasing the cell height, in comparison to the effects illustrated in Fig. 10 and 11. A number of C_{21}-steroids were shown to be without effect on prostate weights, at a dose of 0.5 mg/day. However, 11-deoxycortisol produced a significant increase in seminal vesicle weight, and 11-dehydrocorticosterone a slightly smaller increase.

It must be noted that the adrenal enlargement obtained by LI et al. (1957) was only about one-sixth as great as that studied by DAVIDSON and MOON (1936). Furthermore, it is unfortunate that the rats studied by LI et al. (1957) and by LOSTROH and LI (1957) were beyond the age period during which evidence of adrenal androgen secretion was obtained by DAVIDSON and MOON (1936), DAVIDSON (1937), BURRILL and GREENE (1939, 1940) and HOWARD (1941) (see p. 599). The evidence as a whole suggests that there may be considerable differences in the ratio of androgens to corticosteroids which the adrenal produces at different age periods. The same considerations apply to evaluation of the work of BORELL et al. (1951), who reported negative findings in experiments in which ACTH was administered to rats castrated at 27 days.

Administration to castrated infantile golden hamsters of a dose of ACTH, 50 μg/day, producing an adrenal enlargement of 33%, had no effect on the reproductive tract; larger doses of ACTH (100–200 μg/day) proved toxic (LA VELLE 1951).

More recently, DEANESLY (1960) has reported that exogenous ACTH stimulates the production of biologically detectable amounts of androgens in rats. She gave 10 U of Armour's Acthar gel twice daily for 12 days to rats that had been castrated when immature. The adrenals were increased in size by 286%, compared to untreated controls, and the ventral prostates and seminal vesicles showed evidence of stimulation intermediate between that produced by androsterone at doses of 0.05 and 0.1 mg/day in oil.

Significant adrenal-dependent stimulation of the immature rat prostate by ACTH (20 U for 2 weeks) has been reported also by TULLNER (1961). The response to ACTH was potentiated by prolactin (20 I. U. daily). Furthermore, in hypophysectomized-castrated male rats ACTH alone was ineffective, but D, L thyroxine (20 μg) in daily combination with ACTH (20 U) and prolactin (40 I.U.) produced a substantial increase in prostate weight after 8 days. These observations offer further reasons for the failure of earlier workers to obtain significant prostatic response to ACTH in hypophysectomized rats, in addition to the age factor mentioned above.

F. Conclusion

Androgens are derived from the adrenal in part as metabolites of the glucocorticoids and in part independently. The adrenal androgens are much weaker than the testicular androgens in the usual biological tests, but exhibit some effects on reproductive tract structures and on skeletal maturation in mammals. The output of androgens by the adrenal is partially independent of the corticoid production, as a result of factors which are incompletely defined. Immature adrenals in some species seem to produce more androgens than do glands from mature individuals. In man, during embryonic life, the adrenal appears to produce relatively large amounts of androgens; the blood concentrations are low during childhood, rise at puberty and decline progressively after the third decade. The physiological significance of the adrenal androgens is not well understood.

Progestins

A. Metabolic considerations

Progesterone occurs in adrenal tissue in sufficient quantity to permit its isolation (BEALL and REICHSTEIN 1938). There is good evidence to indicate that progesterone is a key intermediate in the reaction sequence which converts acetate and/or cholesterol into cortisol and corticosterone (HECHTER et al. 1951), the latter two compounds appearing to be the main corticosteroids produced by normal adrenals (REICH et al. 1950, 1951, BUSH and SANDBERG 1953, ROMANOFF et al. 1953 – see Chapters 1 and 3). Although attempts to isolate progesterone from human adrenal vein blood have been unsuccessful (PINCUS and ROMANOFF 1955), LOMBARDO et al. (1959) have reported the isolation of 17α-hydroxyprogesterone from human adrenal vein blood. And in an abstract without quantitative details, SHORT (1959) stated that "progesterone has been detected in the adrenal venous blood of the human, cow, sheep and pig," and "17α-hydroxyprogesterone is probably present in the adrenal venous blood of the human and the cow." Δ^4-Pregnen-20α-ol-3-one was detected in the adrenal venous blood of a young calf, but not in the adult cow. Progesterone has been found in the peripheral blood of the males of a number of vertebrates by means of a biological assay (LAZO-WASEM et al. 1954 — see p. 611 for assay methods). There is also indirect evidence that the adrenal may release a progesterone like substance into the circulation.

One of the main metabolites of progesterone is 5β-pregnane-3α,20α-diol (MARRIAN 1949), and variations in the urinary excretion of this metabolite under various experimental conditions might reflect in part changes in progesterone secretion by the adrenal. Progesterone is produced also by the ovary, the placenta and perhaps the testis; consequently, these sources must obviously be excluded before one can attribute urinary pregnanediol to an adrenal precursor.

A further consideration in the interpretation of such observations is that progesterone is not the only steroid to give rise to 5 β-pregnane-3α,20α-diol during its catabolism, since deoxycorticosterone can also produce this compound (SCHNEIDER and HORSTMANN 1951, DORFMAN 1955). Although deoxycorticosterone has been isolated from beef adrenals in very small amounts (REICHSTEIN and VON EUW 1938) it was not detected in human adrenal vein blood by PINCUS and ROMANOFF (1955). However the urinary excretion of pregnanediol and pregnane-3α, 21-diol-20-one in the hypertensive form of congenital adrenal hyperplasia suggested that deoxycorticosterone may be secreted by the human adrenal under some conditions (EBERLEIN and BONGIOVANNI 1955). Furthermore, small amounts of deoxycorticosterone have been reported in the adrenal vein blood of dogs (FARRELL et al. 1954).

Deoxycorticosterone has a progestational activity equal to one-third to one-tenth that of progesterone when administered systemically to rabbits, but it has no activity when applied directly on the uterine mucosa in mice (HOOKER and FORBES 1947). This variation in activity seems to be related to the fact that deoxycorticosterone can be converted into progesterone *in vivo* (HORWITT et al. 1944, ZARROW et al. 1950, DORFMAN 1955). To date, no other instance of 21-dehydroxylation *in vivo* has been observed (FUKUSHIMA and GALLAGHER 1958).

The indirect evidence that the adrenal produces progesterone is as follows:

(1) Small amounts of authentic 5β-pregnane-3α,20α-diol have been isolated from a large pool of normal male urine (ENGEL et al. 1941); similar findings were reported by HEARD and HOFFMAN (1941) and by WESTPHAL (1944), the latter estimating the pregnanediol content of male urine to be 0.7 mg/l. BERGSTRAND and GEMZELL (1957) reported a similar excretion rate for children under 10 years of age, and KLOPPER et al. (1957) found an average value of 1.1 mg per 24 hours for adult men. Since pregnanediol is one of the metabolites of progesterone, such observations indicate that the ovary and placenta are not the only sites of progesterone production. *In vitro* studies have demonstrated that rat testicular tissue (SLAUNWHITE and SAMUELS 1956) may transform progesterone into 17α-hydroxyprogesterone, Δ^4-androstene-3,17-dione and testosterone. If the *in vivo* synthesis of testicular androgens follows a similar pathway, it is conceivable that the male gonad secretes some progesterone. Indeed, the bird testis has been thought to be a site of progesterone production (FRAPS et al. 1949). However, LAZO-WASEM et al. (1954) found no change in the blood levels of progesterone in male rats following castration but observed a marked reduction following adrenalectomy. Therefore, the testis does not seem to be a major site of progesterone secretion.

(2) It has been observed that urinary pregnanediol declines to very low levels within a week following bilateral adrenalectomy in man (KLOPPER et al. 1957). In castrated, adrenalectomized rats, no progesterone-like activity could be

detected in the blood by means of the Hooker-Forbes bioassay (see p. 612) (ZARROW and LAZO-WASEM 1955).

(3) The administration of ACTH to menopausal women was found to produce an increase in their urinary pregnanediol (DAVIS et al. 1953). Similar results were observed in normal adult males (KLOPPER et al. 1957). The failure of DE WATTEVILLE et al. (1951) to detect any urinary pregnanediol after the administration of 125 mg ACTH intramuscularly over a 3 day period might be due to their use of a relatively insensitive assay, as pointed out by KLOPPER et al.

The studies of ZARROW and LAZO-WASEM (1955) have indicated the release of considerable amounts of progestin by the ACTH-stimulated adrenal. Castrated rats were given a single intramuscular injection of 40 U of ACTH and blood samples were obtained at intervals. Results were expressed as μg equivalents of progesterone. The values rose from a mean control of 0.3 μg per ml of serum to 4.7 μg 2 hours after injection and to 8.0 μg at 5 hours; 10 hours after injection the values were below pre-injection levels. No progestin was found in the serum of castrated, adrenalectomized rats before or after treatment with 40 U of ACTH. Chorionic gonadotrophin failed to increase the serum progestin levels significantly in the castrated rat, when injected in amounts of 500 and 1000 I. U. Similar results were obtained with rabbits.

(4) Surgical procedures are known to increase the functional activity of the adrenal gland. Following surgery, an increase in urinary pregnanediol in man has been observed (ZIMMERMANN and WOJACK 1952, KLOPPER et al. 1957).

In the rat, exposure to cold (2°) produced an increase in blood progestin, but the increment was much less than that noted following ACTH administration to castrated rats (ZARROW and LAZO-WASEM 1955).

(5) The urinary output of pregnanediol has been reported to be elevated in some cases of congenital adrenal hyperplasia and virilizing adrenal tumor (VENNING et al. 1939, FINKLER 1941, MALLEY and BRADSHAW 1941, SALMON et al. 1941, GENITIS and BRONSTEIN 1942, TALBOT et al. 1942, ANDERSON et al. 1943, LLOYD et al. 1951) as well as in cases of feminizing adrenal tumor (DICZFALUZY and LUFT 1952, WALLACH et al. 1957). Progesterone was isolated (SALHANICK and BERLINER 1957) from the extract of the tumor of the patient of WALLACH et al. These findings suggest a very large production of progesterone by the adrenals, but, as pointed out by BONGIOVANNI and CLAYTON (1954), much of the urinary material previously measured was probably pregnanetriol instead of pregnanediol. Using column chromatography as a means of separation of the two compounds, the latter authors have found only a moderate increase of pregnanediol with large amounts of 3α, 17α, 20α-pregnanetriol in the urine of cases of congenital adrenal hyperplasia. Pregnanetriol is a metabolite of 17-hydroxyprogesterone. (See Chapter 3.)

B. Progesterone bioassay methods

The stimulation of the endometrium of the uterus normally caused by the hormones of the corpus luteum is the basis for the usual bioassays for progesterone. Species differences in uterine reactions necessitate modifications of procedure according to the choice of experimental animal. Evidence that progesterone has some systemic effect as well as its uterine effect is found in the observation that its administration causes a rise in basal body temperature in women surgically deprived of ovaries and uterus (DAVIS and FUGO 1948) as well as in ovariectomized cows (WRENN et al. 1959). (Compare section on androgen bioassay, p. 575, regarding general procedures.)

Rabbit endometrial reaction

Recently the traditional Clauberg method of evaluating the progesterone effect on the uterine endometrium of the rabbit has been developed and put on a more quantitative basis by PINCUS et al. (1957). Immature albino rabbits, 1.0 to 1.5 kg body weight, were given 5 μg of estradiol once daily for 6 days, subcutaneously. After this period of estrogen priming, the rabbits were injected once daily for 5 days with the various doses of progesterone or unknowns dissolved in sesame oil. The animals were sacrificed 24 hours after the last injection. Small pieces of the midsections of both uterine horns were fixed in Bouin's fluid for histological examination. In cross section, the areas of the total mucosa (M) and the glandular portion of the mucosa (G) were measured with a planimeter and the G/M ratio was calculated. This ratio is an indicator of progestational stimulation, and it increased linearly when plotted against the log of the total progesterone dose between 0.5 and 2.0 mg per rabbit. Simultaneous determinations of the carbonic anhydrase of the uterine endometrium by the method of PHILPOT and PHILPOT (1936) showed that the enzyme activity per g of tissue also increased linearly with the log of the progesterone dose, but through a somewhat larger range, so that it was concluded that the enzyme measurements had useful advantages for quantitative bioassay. These methods have since been used in testing a series of compounds (MIYAKE and PINCUS 1958).

Rat decidual reaction

In the rat, uterine changes during pseudopregnancy and after progesterone injections are by no means as striking as in the rabbit. However, the properly sensitized uterine mucosa in the presence of progesterone responds to a slight traumatic stimulus with a marked proliferation analogous to that which initiates the formation of the maternal portion of the placenta. This decidual reaction is a specific indicator of corpus luteum activity that is useful in the assay of progesterone. If ovariectomized rats are first primed with estrogen and then injected with progesterone it is difficult to obtain a decidual reaction. However, progesterone is fully active in promoting the growth of deciduomata after the removal of the ovaries once the endometrium has been rendered sensitive to trauma by the normally functioning ovaries of the pseudopregnant animal (ASTWOOD 1939).

Procedure. Adult female rats, weighing 150—200 grams, are given electrical stimulation of the cervix on the day when they show fully cornified vaginal smears. The cervix is exposed with an aural speculum and electrodes, spaced at 2 mm, are applied at any place on the body of the cervix. The stimulus, a faradic current from an induction coil, is sufficiently intense to produce convulsive contractions of the hind quarters. The stimulus is applied continuously for 5 to 10 seconds. Four days after the stimulation, the rats are ovariectomized by the lumbar route. At the time of operation, the endometrium of one uterine horn is traumatized by inserting a needle through the cut tubal end of the uterus as far as the cervix on that side. By withdrawing the needle at a slant, the antimesometrial aspect of the endometrium is scratched throughout its entire length by the needle point. The injections of progesterone are given for 3 days beginning immediately after operation, and the animals are sacrificed 72 hours after the first injection. Uterine enlargement was originally estimated from diameter measurements, but comparisons of the weights of the traumatized with the untraumatized uterine horn permit a more quantitative treatment of the findings (HISAW and VELARDO 1955). Histological examination of the uteri does not increase the sensitivity of the test, which is of the same order of magnitude as the Clauberg rabbit test.

Mouse endometrial nuclear reaction

The HOOKER and FORBES (1947) bioassay for progestins, a highly sensitive technique, consists of the injection of a standard volume (0.6 μl) of the solution to be assayed into the lumen of a ligated uterine segment of an ovariectomized mouse. After 48 hours, the uterus is removed and examined histologically in 6 μ sections, the criteria of activity being evidence of a stimulation of the nuclei of the endometrial stromal cells. The nuclei become enlarged and assume a smooth, slightly elongated, oval outline. The chromatin particles become fine and evenly distributed, and the nucleolus becomes conspicuous. All of these conditions must be present before the response can be considered positive. It is necessary to distinguish between the stromal nuclei and the nuclei of the glandular epithelium. The latter nuclei may appear to have no cytoplasm, but they can be recognized as belonging to the uterine glands by examining serial sections. The nuclei of the glandular epithelium do not appear to be modified by progesterone. The response to progesterone in the mouse under these conditions is much simpler morphologically than the complex glandular changes which occur in the rabbit.

The use of serial dilutions in a series of mice and determination of the minimal effective concentration permit the results to be expressed quantitatively. The minimum amount of progesterone which can be detected is 0.0002 μg. This test is not modified by estrogens, which stimulate the uterine epithelium but do not influence the reaction of the stromal nuclei under the conditions of this test. Furthermore, deoxycorticosterone, which has a progesterone-like activity when given systemically, does not produce the characteristic change in the stromal nuclei when injected into the uterine lumen, even in an amount 3000 times that of the minimal effective dose of progesterone. Similarly, testosterone, which has a slight luteoid activity when given systemically, does not react positively by the HOOKER-FORBES criteria.

In recent years chemical assays have been devised for small amounts of progesterone. BUTT et al. (1951) use a polarigraphic technique. EDGAR (1953) and SHORT (1957) separate progesterone by paper chromatography and measure the steroid by ultraviolet absorption methods. ZANDER and SIMMER (1954) have developed a technique with a sensitivity of 0.05 μg progesterone/ml of blood, using infrared spectrophotometry for the quantitative determination.

When results obtained by the Hooker-Forbes test are compared with those obtained by chemical techniques, the values found by the former method are much higher than those with the latter. ZANDER (1957) has shown that two naturally occuring metabolites of progesterone, the 20-hydroxy-derivatives, are also present in blood extracts. Both compounds give positive responses in the Hooker-Forbes test and are about twice as active as progesterone (T. FORBES 1957). The discrepancy between the results obtained by biological and chemical assays appears, therefore, to be related to the presence of the progesterone derivatives.

C. Systemic significance of adrenal progesterone

Progesterone has a key position in the synthesis of steroids, but whether it has any hormonal role in addition to its progestational one is unknown.

Since the adrenal appears to produce progesterone, a question is relevant: does the adrenal secrete amounts of progesterone sufficient to play a role in reproductive tract physiology? Briefly, it seems that the adrenal does secrete appreciable amounts of progesterone following ACTH stimulation, as will be discussed below. However, the corticoids tend to inhibit the characteristic uterine effects of progesterone, so that the net effect of adrenal activity under ordinary circumstances is probably to diminish the uterine responses to endogenous progesterone.

The adrenal may be stimulated to secrete considerable amounts of progesterone-like products, since deciduomata can be produced in hypophysectomized-ovariectomized rats treated with ACTH (LYONS et al. 1953, LI et al. 1957). This effect was manifest following the daily administration for 7 days of α-corticotrophin. At a dose of 15 μg per day, adrenal weights were comparable to those in intact rats. With 22.5 μg there were no gross deciduomata, but sections showed "beginning deciduomal reactions in the form of enlarged stromal cells in some of the animals," while with 75 μg small deciduomata were evident in the gross. These results were obtained in spite of the fact that corticoids, which are known to be inhibitory to progestin, were also secreted in large amounts by the abnormally enlarged and over-stimulated adrenals.

POMEAU-DELILLE (1949) had earlier presented evidence indicating that the net effect of the adrenal is inhibitory to the formation of deciduomata when the adrenal is activated by endogenous corticotrophins. This author studied the formation of deciduomata in ovariectomized rats given 2.5 μg estradiol benzoate daily for 3 days and 0.6 mg progesterone per day for 10 days; the uteri were threaded on the 6th day. Under these conditions, the ovariectomized rats did not show a decidual reaction or an appreciable increase in uterine size. However, ovariectomized-adrenalectomized rats maintained on 1% NaCl (but demonstrating some mild adrenal insufficiency) showed in all cases the formation of deciduomata. In adrenalectomized rats without signs of insufficiency the results were variable depending on the degree of hypertrophy of accessory adrenal tissue. Under these conditions, the adrenal appeared to be an antagonist of administered progesterone.

This general conclusion has been substantiated and extended by HISAW and VELARDO (1951, 1955) and by VELARDO et al. (1953), using somewhat different experimental conditions. HISAW and VELARDO (1955) utilized rats rendered pseudopregnant by electrical stimulation of the cervix at estrus. On the 5th day of pseudopregnancy, the rats were ovariectomized, and one uterine horn was traumatized by scratching its entire length; at the same time, some of the animals were adrenalectomized. They were then given graded doses of progesterone for 3 days and killed after 72 hours. The decidual reaction was evaluated by the increase in weight of the traumatized horn over that of the control horn. With doses of progesterone of 0.25, 0.5 and 0.75 mg, significantly greater decidual reactions were produced in the adrenalectomized rats, the differences being greatest with the smallest dose. At doses of 1 to 2 mg of progesterone, there was no clear difference between the two groups. Hence, the inhibiting effect of the adrenal was neutralized by increased amounts of progesterone. When ACTH was administered, a very marked diminution of the decidual response was obtained.

The decidual reaction can be completely inhibited in ovariectomized pseudopregnant rats receiving 1.5 mg progesterone daily, by concomitant injections of 3 mg of deoxycorticosterone acetate or 4.5 mg of cortisone (HISAW and VELARDO 1951). Deoxycorticosterone is completely inhibitory to progesterone in rats, at doses which alone produce almost no uterine enlargement. In mice, cortisol and cortisone are inhibitory to progesterone, while deoxycorticosterone is not (VELARDO 1957).

These experiments of HISAW and VELARDO are in apparent contradiction to the observations of LI et al. However, there were differences in the experimental conditions. The inhibition by ACTH of the progesterone-induced decidual response was associated with an adrenal enlargement of 35% in the experiments of HISAW and VELARDO (1955), whereas the grossly evident decidual response to adrenal progestin reported by LI et al. (1957) was associated with the production of an adrenal enlargement of 160 to 270%. A further difference in the experimental conditions was that HISAW and VELARDO used rats at 100–120 days of age, whereas LI et al. used rats at 30–38 days of age. In the experiments of LI et al. (1957) large amounts of corticoids and presumably large amounts of progesterone were produced. It is possible that in these circumstances the inhibiting effect of high levels of corticoids was neutralized by the high concentrations of progesterone.

Progestin has been demonstrated in the follicular fluid of preovulatory follicles (HOOKER and FORBES 1947) and, therefore, the adrenal is not necessarily responsible for the production of progestins appearing in the absence of corpora lutea in the ovaries. Consequently, the origin of the small amounts of progestins occuring in peripheral blood of monkeys during the follicular phase of the cycle (BRYANS 1951) and of immature mice treated with gonadotrophins (GREEN 1952) need not be attributed to the adrenal.

On the other hand, under the experimental conditions used by VELARDO et al. (1953), the development of the deciduomata of pseudopregnancy artificially induced after adrenalectomy was not appreciably altered in comparison with intact stimulated controls; i. e., the adrenal did not inhibit the growth of deciduomata resulting from endogenous (ovarian) progestins. However, the weight of the adrenals decreased during the growth of deciduomata, suggesting that adrenal activity tended to be suppressed by the circumstances which initiate and maintain pseudopregnancy.

The possibility that the adrenal contributes sufficient progestins to maintain pregnancy in the absence of the ovary in some species, such as the sheep, has been suggested, since plasma progesterone levels did not drop following ovariectomy in pregnant ewes, and since progesterone could not be extracted from the placenta in this species (SHORT 1957). If this proves true, one would have to postulate some mechanism in the ewe whereby progesterone secretion by the adrenal can be stimulated during pregnancy, perhaps independently of the corticoid output.

In the pregnant rabbit, in contrast to the ewe, the serum progestin drops following ovariectomy (NEHER and ZARROW 1953, 1954), as might be expected since neither the adrenal nor the placenta is able to substitute for the ovary in this species (REYNOLDS 1949).

A curious observation on the effect of ovariectomy in the armadillo suggests the possibility of an adaptive production of progestins by the adrenal. In the armadillo, there is a 4 month delay between mating and implantation. If females are ovariectomized toward the end of this delay period, implantation occurs at the time when it would normally, but resorption or abortion follows shortly. However, if ovariectomy is performed in the middle of the 4-month delay period, normal implantation occurs 1 month later (BUCHANAN et al. 1956).

D. Conclusion

In conclusion, it appears that progesterone, an important intermediary in the biosynthesis of the corticosteroids, is probably secreted by the adrenal. The physiological significance of this secretion is difficult to evaluate, because of the antagonistic action of the glucocorticoids on the progestational effects of progesterone. With moderate stimulation by ACTH, the net effect of adrenal secretory activity is to diminish the uterine response to progesterone.

Estrogens

Estrone has been isolated from extracts of beef adrenals (BEALL 1939). This poses the question of the degree to which the adrenal may supplement the supply of estrogens secreted by the gonads. As in the case of progesterone, the adrenal is by no means an *alter ego* of the ovary. It is well-known that the uterus and other reproductive tract structures undergo marked regression after ovariectomy in adults, or fail to develop in size and functional capacity after prepuberal ovariectomy. Furthermore, uterine weights in ovariectomized mice and in ovariectomized-adrenalectomized animals are similar (HOWARD unpublished). Hence, the stimulus of ovariectomy is not associated with any immediate, grossly-detectable substitution by the adrenals, with respect to uterine maintenance. However, neoplastic alterations of the adrenal may be associated with considerable evidence of estrogen liberation.

A. Estrogen production in cases of adrenal tumor or hyperplasia in man

Feminizing effects of adrenal tumors have been described in various species. In the human, a case of gynecomastia associated with an adrenal tumor in a male was described in 1919 by BITTORF. Twelve cases have been reviewed by WILKINS (1948) and 22 additional cases by WALLACH et al. (1957). Most of these tumors were highly malignant; but some of them were relatively benign, since in one case the symptoms had been present for 16 years prior to surgery (LANDAU et al. 1954).

Urinary estrogens, measured fluorometrically and biologically, were elevated in 14 cases of congenital adrenal hyperplasia and in 3 cases of adrenocortical tumors of virilizing or hirsutizing type (MIGEON and GARDNER 1952). In cases of Cushing's syndrome they were increased in 1 patient and within the normal limits in 3 other patients. Cortisone therapy to patients with virilizing adrenal hyperplasia caused a prompt decrease of both androgen and estrogen excretion, while it did not modify the levels in patients with virilizing adrenal tumors. GASTINEAU et al. (1953) reported a similar reduction in estrogen excretion following cortisone therapy in one case of macrogenitosomia praecox. When urinary estrogens were fractionated by countercurrent distribution, there was a rather constant ratio between fractions resembling estrone, estradiol-17β and estriol in 4 normal subjects and in 4 patients with adrenal hyperplasia (MIGEON 1953). It is of interest that whatever the secreting organ may be, and whatever the hormone secreted, the ratio of the various urinary estrogen fractions appears to be the same.

In females with congenital adrenal hyperplasia, the signs of virilism are marked and, at puberty, these subjects do not show breast development and menstruation. As noted above, the adrenal glands of such patients secrete abnormally large amounts of estrogen in addition to androgens. The estrogens, however, appear unable to exert an effect on the sex organs, probably because of the antagonism by the much greater amounts of androgens. The presence in blood of large concentrations of estrogens and androgens probably inhibit the formation of gonadotrophins by the pituitary gland and, therefore, the ovary remains immature. The administration of cortisone inhibits the excess ACTH produced in this syndrome (BARTTER et al. 1951), putting the hyperplastic adrenal at rest. Thereafter, a normal production of pituitary gonadotrophins permits the ovary to mature. This interpretation explains why the untreated cases of female pseudohermaphroditism do not show signs of feminization in spite of the presence of large urinary excretion of estrogens, and why they feminize when cortisone therapy is instituted (WILKINS et al. 1952).

The increased excretion of estrogens in congenital adrenal hyperplasia has been related to a hypersecretion of these steroids by the adrenal. However, some experiments in both animals and man have suggested that androgens may be metabolized to estrogens. STEINACH et al. (1936, 1937) reported an increased urinary excretion of biologically measurable estrogens after administration of androgens to normal and castrated rats and to normal men. These observations have been confirmed by other workers (N. CALLOW et al. 1939, NATHANSON and TOWNE 1939, HOSKINS et al. 1939). DORFMAN and HAMILTON (1939) showed that the material measured was phenolic in nature, and NATHANSON et al. (1952) reported an increase of estrone, estradiol-17β and estriol in urine of breast-cancer patients treated with androgen. More recently, it has been demonstrated that following administration of 4-C^{14}-testosterone to the pregnant mare, radioactive-estrone could be isolated from the urine (HEARD et al. 1955). WEST et al. (1956) have also observed the conversion of testosterone propionate to estrogens in castrated, adrenalectomized human females. It would appear from these studies that the increased estrogen excretion noted in cases of virilizing hyperadrenocorticism might well arise from the peripheral metabolism of androgens.

The observation of BULBROOK and GREENWOOD (1957) should also be mentioned. In cancer patients, ovariectomy reduced estrogen excretion in premenopausal women; adrenalectomy carried out after ovariectomy further reduced but did not always abolish estrogen excretion. However, the presence of accessory adrenal tissue was not excluded in these patients. Ovarian tissue removed before the menopause (BAGGETT et al. 1955) or after (WOTIZ et al. 1956) and incubated with 4-C^{14}-testosterone metabolizes this to estrogens. In cases of feminizing adrenal tumors in human males, in which the estrogens are present in larger amounts than the androgens, the known rate of conversion of androgen to estrogen would not account for the estrogen : androgen ratio. Hence, there is no reason to exclude production of estrogens by an adrenal tumor.

In a case of Cushing's syndrome due to adrenal carcinoma, the urinary excretion of 6 μg/day of estrone and 34 μg of estriol was observed (WEST et al. 1958). The administration of ACTH resulted in a 12-fold increase of estrone excretion and a 5-fold increase of estriol. However, some adrenal carcinomas may produce no appreciable amounts of estrogens, even when stimulated by ACTH.

B. Estrogen bioassay methods

The occurrence of vaginal cornification in rodents shortly after ovulation is a conspicuous phenomenon which is the basis of the Allen-Doisy test for estrogenic activity. Either mice or rats may be used: mice are usually preferred since they require smaller amounts of active material. (Compare section on androgen bioassay, p. 575, regarding general technique.)

Allen-Doisy test procedure (EMMENS 1950). Weanling mice are ovariectomized via bilateral lumbar incisions under any convenient anesthetic and rested for 2 weeks following the surgery. They are then given a priming dose of 0.5 μg of estrone, which increases the sensitivity of the subsequent test. Thereafter it is not necessary to prime them unless a particular group has not responded positively during the past 6 weeks. It is wise to use the whole colony for a determination of the dose:response line, as a general check on technique and uniformity and sensitivity of the animals.

In an assay, mice are injected subcutaneously at 10 A. M. on Monday and Tuesday. Vaginal smears are taken at 4 P. M. on Wednesday and discarded, then at 11 P. M. Wednesday and at 10 A. M. and 4 P. M. on Thursday. Smears are taken as gently as possible from the dorsal vaginal wall with a fine metal spatula, transferred to a glass slide and stained for 10 minutes with methylene blue, washed and evaluated when dry. A positive smear contains nucleated or cornified epithelial cells but no leukocytes; a full proestrous smear is thus scored as a positive. The response is treated quantally, and is positive if any of the 3 smears from an animal is positive. Assays using such a method should have at least 2 groups of mice on both the standard and unknown, each group containing not fewer than 20 animals. If the dose ratio is about 2:1 on each substance, a valid assay will usually be performed, with fiducial limits of error between about 70 and 140% — or even less, if the dose:response lines from previous assays are available for the estimation of the slope, which does not normally vary. The most frequent source of trouble is an unexplainable change in the sensitivity of the colony of animals. Total doses from 0.1 to 0.25 μg of estriol per mouse give a satisfactory range for the standard response.

A variation of the Allen-Doisy test utilizes intravaginal application of the test material. This procedure increases the sensitivity greatly, the minimum effective dose of estriol being only 0.0008 μg. However, the technique of application is more difficult than is subcutaneous injection, so that apparently more animals per group would be needed to obtain a precision comparable with a normal Allen-Doisy test. Nonspecific responses must be carefully controlled.

Uterine weight

Assays based on uterine weight have various advantages. The response is intrinsically capable of gradation, so that the whole quantitation does not depend on scatter within a population, as it does in the vaginal cornification test. Furthermore, uterine growth is a summation effect, whereas vaginal cornification is a self-limited process.

Four-day uterine-weight test. This may be used with either immature intact or ovariectomized mice or rats. As used by RUBIN et al. (1951), mice aged 23—25 days were injected subcutaneously once daily for 3 days with a solution of the hormone dissolved in 0.1 ml of oil. Twenty-four hours after the last injection the animals were killed and the body and uterine weights obtained. The uteri were severed at the cervix, the utero-tubal junction was severed, and the mesometrial adnexa were trimmed away. The uteri were opened and any fluid in the lumen was blotted away. The uteri were dipped briefly in saline, blotted on gauze or filter paper, and weighed quickly on a torsion balance. The experimental design of C. BLISS (1944) was followed, as described briefly in the androgen section (p. 575). Using 20 animals on the standard and 20 on the unknown, estrone was measured with a standard error of 13% of the potency ratio. It could be detected in a dose of 0.04 μg. RUBIN et al. (1951) furnish a critical review of the work of various authors on estrogen bioassay.

Six-hour uterine-weight test. ASTWOOD (1938) showed that, following a single subcutaneous injection of 0.1 μg estradiol in oil, the uterine weights of immature (21—23 days of age) intact rats reach an initial maximum at 6 hours, and a second somewhat higher maximum at 30 hours. The first maximum is due in large part to an increase in the water content of the uteri. The 6-hour response shows a linear relationship to the dosage between the limits of 0.006 and 0.1 μg of estradiol.

C. Estrogen production by experimental tumors in animals

The first experimental test of the adrenal response to ovariectomy appears to have been performed by FEODOSSIEW at the University of Kazan in 1906. Dogs were gonadectomized and autopsied after various intervals. Local foci of adreno-cortical proliferation were described; after several months the proliferation reached and penetrated the medulla and also progressed through the capsule, eventually forming a mushroom-like growth, which remained united to the cells of the peripheral layer.

Observations suggestive of estrogen production by a similar adrenal nodular hyperplasia in mice were described in 1939 by WOOLLEY et al.; and W. GARDNER (1941) noted estrogenic effects from adrenal tumors in mice. Tumor-bearing ovariectomized mice excreted about 4 times as much estrogenic material as did intact normal females (DORFMAN and GARDNER 1944). In a series of papers, WOOLLEY et al. have studied this problem in detail. Neoplastic changes in the adrenals of intact mice are extremely rare, but following gonadectomy they occur with high frequency, especially in some strains. In strain JAX-CE, ovariectomy soon after birth produces adrenal cortical carcinomas in 100% of the animals at 6 months of age (WOOLLEY and LITTLE 1946a). However, the development of the neoplastic changes in the adrenals is a prolonged process, and for some months following ovariectomy there is no evidence of estrogen production. In the DBA strain, the signs of estrogen effects on the uterus, vagina and mammary glands can be detected only 8 to 10 months following ovariectomy and are maximal after 11 to 16 months (FEKETE et al. 1941). These authors do not give the uterine weights, but some of the histological effects are illustrated in Fig. 13. One may recall that mice are sexually mature at 6 to 7 weeks of age, and are past their reproductive prime at about 1 year of age. Estrous vaginal smears and estrous mating behavior have been reported in 10—12 month old mice ovariectomized at birth and exhibiting primary adrenal neoplasia (CHRISTY et al. 1950). Adrenal tumors arise also in gonadectomized males, and they are frequently associated with evidence of predominantly androgenic hormone production, as discussed earlier (p. 605).

The tumors may become very large (over 1.5 cm in diameter) and they have been transplanted successfully. It is notable that the neoplastic change appears to start as a nodular hyperplasia in the region of the glomerulosa. Further genetic study of these tumors has been reported (WOOLLEY et al. 1953, DICKIE and LANE 1956). An attempt to obtain evidence of the production of estrogens by these tumors on incubation with progesterone-4-C^{14} was unsuccessful (HOFMANN et al. 1960).

A similar adrenal hyperplasia occurs in castrated rats. B. HOUSSAY et al. (1954) and A. B. HOUSSAY et al. (1954), summarizing a series of papers, reported that following ovariectomy, either at birth or later, alterations of the glomerulosa leading to nodular hyperplasia are manifest after 1 or 2 months. Adenomas appear 12 to 14 months after ovariectomy and they develop progressive estrogenic activity. The uteri enlarge and exhibit muscular and glandular hypertrophy and in some cases metaplastic cornification of the epithelium. Vaginal cornification generally occurs; estrus is prolonged and often continuous. Exceptionally, the vaginal epithelium is mucified. One year after ovariectomy adrenal adenomas are observed in 24 to 45% of the rats. Tumors are not found to arise from adrenal accessories. Unilateral adrenalectomy impedes the appearance of the tumors. Either removal of the adenomatous adrenals or hypophysectomy suppresses the estrogenic effects. The picture of estrogenization reappears on treatment of hypophysectomized rats with serial doses of 50 I. U. chorionic gonadotrophin. It is notable that the administration of ACTH did not provoke estrogenization.

In gonadectomized golden hamsters, 17 to 24 months of age, adrenal cortical tumors are also associated with pronounced stimulation of the accessory reproductive organs (WOOLLEY 1953). However, a nodular hyperplasia of the glomerulosa 1 year after ovariectomy was associated with an infantile, unstimulated uterus (AGATE 1952). Examination of the serum sodium levels in these animals revealed a 13 meq. excess over the intact females, suggesting that the nodules of the glomerulosa produced excessive amounts of Na-retaining steroids rather than estrogens in this case.

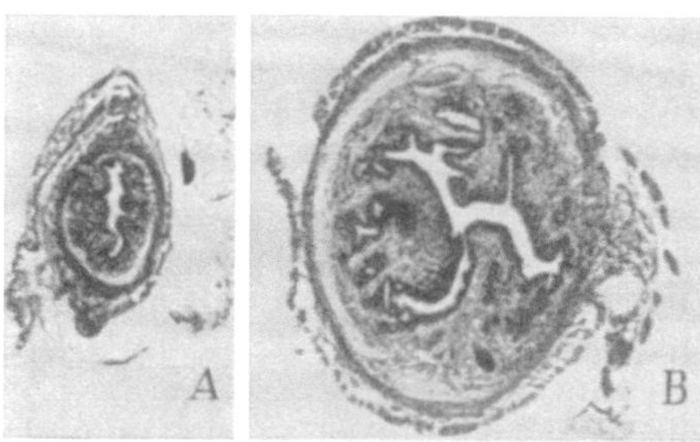

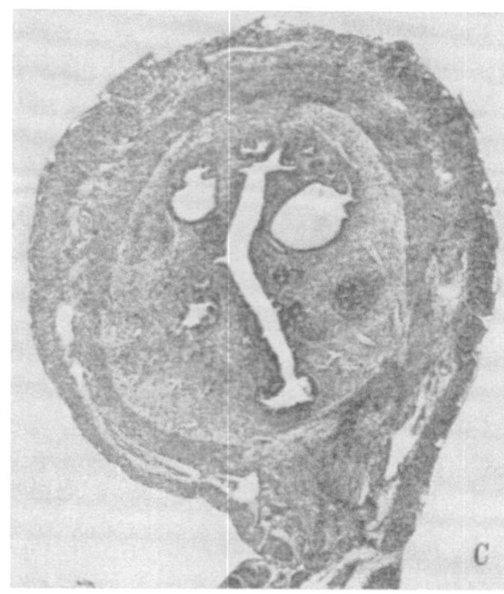

Fig. 13 A—C. Uteri from ovariectomized mice of the DBA strain showing the influence of the development of nodular hyperplasia in the adrenals. Magnification × 20. [Reproduced by permission from FEKETE et al.; J. exp. Med. 74, 1 (1941). (The Rockefeller Institute).] A. Cross section of the uterus of an ovariectomized mouse 4 months of age, showing typical undeveloped state prior to the development of nodular hyperplasia in the adrenals. B. Cross section of the uterus of an ovariectomized mouse $10^{1}/_{2}$ months of age, with nodular hyperplasia of the adrenals. Note the increased diameter of the uterus, with thickening of the endometrium and the muscular layers. C. Uterus of an ovariectomized mouse 15 months of age, showing more advanced hypertrophy than the specimen in B. Note dilated endometrial glands. The epithelial cells showed considerable mitotic activity, and there was edema of the lamina propria.

These observations of the long-term effects of gonadectomy suggest that certain metabolic qualities of the adrenal are modified by prolonged exposure to the increased amounts of pituitary gonadotrophins that result from the gonadectomy. Other evidence of adrenal alteration following gonadectomy is the reduced capacity to secrete corticosterone (DEAN et al. 1959). (See p. 623.) However, exogenous chorionic gonadotrophin administered to mice during the third week following ovariectomy produces no detectable effect on the involuted uterus or vagina (SAKIZ 1955).

Estrogen production by tumors of the adrenal thus has been demonstrated biochemically or biologically in the human, the mouse, the rat and the hamster. We will consider what evidence is available that physiologically significant amounts of estrogens are secreted by the normal adrenal.

D. Does the normal adrenal produce systemically significant amounts of estrogens?

I. Estrogen:glucocorticoid antagonism

As in the case of progesterone, we are confronted with the fact that under some circumstances the corticosteroids interfere with the manifestation of estrogen effects. This subject has recently been reviewed by ROBERTS and SZEGO (1953). These authors studied rats ovariectomized for 2 weeks and given intravenous injections of cortisol and estradiol in doses of 2.5 mg and 0.5 μg per 100 g of body weight, respectively. Four hours after injection, cortisol markedly diminished the increase in uterine water content caused by estradiol alone. Twenty hours after injection, the uterine growth was less inhibited by cortisol, but the estradiol-stimulated incorporation of C^{14}-alanine into uterine proteins was still reduced. Cortisol antagonism to estrogen was similarly demonstrated for uterine oxygen consumption, aerobic glucose utilization, and anaerobic glycolysis. The stimulating effect of estradiol on the β-glucuronidase activity in the preputial gland was blocked by cortisol.

FAWCETT and DEANE (1951) also studied the problem of cortisone : estradiol antagonism. Rats, ovariectomized for 40 days, were given a 4- or 7-day treatment of subcutaneous injections of cortisone and estradiol in doses of 2.5 mg and 0.15 μg per 100 g of body weight, respectively. A moderate antagonism on uterine growth was observed in one group of rats but not in a second. The conditions under which glucocorticoid:estrogen antagonism is manifested have been further explored by VELARDO and STURGIS (1956), who studied the 72-hour growth of the uterus of ovariectomized rats treated with estradiol-17 β. The uterine growth induced by the injection of 0.1 μg estradiol daily for 3 days was reduced appreciably by the simultaneous administration of as little as 0.08 mg cortisol; 40 to 50% growth reduction was produced by 0.32 mg of cortisol, but larger doses (up to 5.0 mg) produced no further reduction. Purified ACTH in doses 0.1 to 0.5 mg/day also caused a 35 to 50% reduction in estradiol-induced uterine growth in ovariectomized rats, but was without effect in estradiol-treated, ovariectomized, adrenalectomized animals. SZEGO and ROBERTS (1948) found that 2.5 to 4.0 mg ACTH in a single intravenous or intraperitoneal dose reduced significantly the 4-hour uterine response to estradiol-17 β (1.0 μg per 100 mg body weight, intravenously). Consequently, it appears that the uterine growth induced by estradiol is impeded but not completely blocked by cortisol within a very large dose range.

ROBERTS and SZEGO (1953) have cited additional observations suggesting an adrenal-inhibitory influence on estrogenic action. Immediately following a sham operation the 4-hour uterine response to estradiol is less than it is 48 hours later, an observation consistent with the interpretation that after the stress of surgery the adrenals secrete amounts of corticosteroids sufficient to interfere with the response to 1 μg of estradiol/100 mg body weight (SZEGO and ROBERTS 1948). The 48-hour response of an adrenalectomized group was slightly greater than the response of a sham-operated group.

In contrast, a situation in which the corticoids are probably necessary for the manifestation of the full effect of high doses of estrogens has been described by HOWARD and ALLEN (1957). Adrenalectomy reduced the connective-tissue hyperplasia in the seminal vesicles induced by estrogen in castrated mice. These adrenalectomized animals were maintained on small amounts of deoxycorticosterone and exhibited normal body growth. Treatment with corticosterone, 0.02 mg/day, permitted the manifestation of the connective-tissue hyperplasia.

II. Apparent estrogenization following ACTH administration

Although the short-term administration of purified ACTH preparations reduces the uterine-growth response to estradiol, some older observations of an opposite implication should be mentioned. W. NELSON (1941), using large doses of crude ACTH, 10 to 25 mg daily for 10 to 30 days, reported that in immature rats of either sex, it is possible to induce enlargement of sex accessories. This was observed in the absence of the gonads and the hypophysis, but not in the absence of the gonads and adrenals. These results were presented in abstract only and no experimental details are available. MOON (1937b), using periods of treatment of 16 to 25 days, injected crude ACTH, 1.5 to 1.9 units per day, into immature spayed female rats and produced greatly hypertrophied adrenals, associated with vaginal opening when the vaginas of the uninjected controls were not open. Unfortunately, there were no adrenalectomized controls, and no vehicle-injected controls. The question arises as to whether the early ACTH preparations contained a factor which is lacking in more highly purified preparations. Under what circumstances this factor might be a component of the pituitary secretion is unknown. Certain aspects of this question have been discussed by ZUCKERMAN (1953).

III. Functional relationships between adrenal and ovary

A number of authors have recognized that complete adrenalectomy in the untreated female rat results in suppression of the estrous cycle and eventual death of the animals. However, if the adrenalectomized rats were maintained in good health on saline, estrus appeared at normal intervals (KUTZ et al. 1934). Earlier observations are cited by SWINGLE et al. (1951). Although some writers have attempted to interpret the defects in reproductive function as being due to loss of adrenal sex hormones, it is clear that validation of such an interpretation would require evidence that the defect could not be repaired by salt and glucocorticoid therapy. RING (1945) studied the induction of mating response in ovariectomized rats compared to adrenalectomized-ovariectomized rats. The rats were adult at the time of surgery; they were maintained only with saline, and survived 13 to 58 days after adrenalectomy. A mating response could be induced by estradiol alone in the ovariectomized rats, but was induced more easily if progesterone was administered concomitantly. The adrenalectomized-ovariectomized rats required more progesterone as well as more estradiol to produce a mating response. It was suggested that the animals were lacking the progesterone normally produced by the adrenal gland. Unfortunately, RING was unable, in 1945, to test the effects of corticosteroids on the dosage of progesterone required to precipitate the mating response.

Good evidence exists that ovarian sensitivity to gonadotrophins is reduced after adrenalectomy, and that this defect can be repaired by the administration of corticosteroids. This subject will be discussed elsewhere in this Handbuch (Chapter 8), but in order to round out the present discussion, certain observations will be mentioned here. PAYNE (1951) studied the influence of the adrenal on the ovarian hyperemic reaction to chorionic gonadotrophin. Rats weighing 35—45 g were adrenalectomized 16 to 20 hours prior to use and were killed 4 hours after the test injections, so that secondary complications of adrenal insufficiency were minimal. It was found that the adrenalectomized rats did not show the ovarian hyperemic response to human chorionic gonadotrophin, at the standard dose of 0.02 mg. As little as 10 μg of cortisone or cortisol was effective in restoring the response. This effect of adrenalectomy has been confirmed by SMITH (1955).

In an extended study of the effects of human chorionic gonadotrophin and pregnant mare serum after adrenalectomy, MANDL (1954) utilized both immature and adult rats, maintained on saline for 4 to 10 days after adrenalectomy. The adrenalectomized rats had smaller ovaries, fewer large follicles and a smaller number of recently formed corpora lutea than their sham-operated litter mates. The ovarian-weight response to injected pregnant mare serum was less in the adrenalectomized animals. A normal response was restored by cortisone administration but not by deoxycorticosterone. Small amounts of accessory adrenal tissue appreciably influenced the response of the ovaries to pregnant mare serum. MANDL concluded that the ovarian reaction to this gonadotrophin is not normal unless the level of circulating cortisone is above a certain minimal value. The ovarian-weight response to chorionic gonadotrophin was less in immature adrenalectomized animals than in their sham-operated litter mates, but the ovaries of mature adrenalectomized rats responded to this gonadotrophin to an even greater extent than did their controls. [See also MANDL (1957).]

After X-ray treatment of the ovaries, although the animals may be infertile, certain manifestations of estrogenic effects continue. The possibility that such effects were produced by adrenal estrogen has been disproved (MANDL and ZUCKERMAN 1956), who found that these estrogenic effects continued after adrenalectomy, in rats on saline, but ceased after surgical removal of the X-irradiated ovaries.

Maturation effects which appear to be retarded by adrenalectomy are not necessarily due to the absence of sex hormones. For example, it has been noted that the glucocorticoids facilitate the rate of accumulation of phosphatase in the intestinal mucosa of the developing mouse (MOOG 1953, MOOG and NEHARI 1954) and hasten the eruption of rat incisors (PARMER et al. 1951). This type of acceleration of maturation processes should be kept in mind in evaluating such work as that of WADE and HASELWOOD (1941), who reported that adrenalectomy retarded the age of vaginal opening in ovariectomized rats. These rats were maintained on a low K diet, and did not exhibit normal body growth. It would be preferable to supply adequate amounts of a salt-retaining factor and a glucocorticoid to achieve normal body growth before concluding that a developmental retardation is caused by the absence of estrogens secreted by the adrenal.

The possibility that the adrenal exerts an inhibitory effect on endogenous estrogens has been considered by some authors. PARKES (1945) cited evidence of somewhat irregular but persistent, moderate changes in vaginal cell types, in the absence of the ovaries, which might possibly be due to cyclic changes in adrenal activity. The occurence of cyclical changes in the vagina of the spayed rat was subsequently confirmed (MANDL 1951). ZUCKERMAN (1937, 1941) and BOURNE and ZUCKERMAN (1941 a) studied the presence of cyclic changes in ovariectomized rats and monkeys, which were injected daily with a constant small dose of estrone. These cyclic vaginal changes in rats almost disappeared following adrenalectomy and maintenance with deoxycorticosterone. But DEL CASTILLO and DI PAOLA (1940) have reported that the vaginal effects of estrogens are reduced by deoxycorticosterone. Ovariectomized monkeys, injected with 100 I. U. estrone per day for a year, exhibited periods of uterine bleeding at intervals of 5 to 7 weeks (ZUCKERMAN 1937, 1941). Considered as estrogen withdrawal bleeding, these could be occasioned either by (1) a rhythmic rise in estrogen threshold, due to the cyclic liberation of an antagonist, or (2) a cyclic liberation of supplementary endogenous estrogen and/or progesterone, possibly from the adrenals, or (3) an inherent cyclic rhythm in the uterus. Further study, particularly on estrogen and corticosteroid excretion rates and, where possible, on hormonal levels in blood, will be necessary before deciding between the various possibilities mentioned above.

In the rat the adrenal weight is greater at estrus than at other times of the cycle, being minimal at diestrus (ANDERSEN and KENNEDY 1932); in the pigeon, adrenal size is greatest at ovulation (RIDDLE 1923). (See Chapter I, p. 106.) It was reported by ZUCKERMAN et al. (1938) and BOURNE and ZUCKERMAN (1941 b) that in ovariectomized rats injected daily with a uniform small dose of estrone, the adrenal fluctuates in size in an approximately 5 day cycle in conformity with vaginal cell changes.

A considerable estrous variation in the corticosterone content of adrenal venous blood has been observed in rats under ether anesthesia (DEAN et al. 1959). The amounts of corticosterone, in μg/100 ml adrenal venous blood, were as follows:

Intact females in	(i) proestrus	210 ± 17
	(ii) estrus	126 ± 8
	(iii) metestrus	72 ± 6
	(iv) diestrus	68 ± 8
Ovariectomized		28 ± 4

The effect of the ovarian hormones on the maximal secretory capacity of the adrenal thus appears to be even greater than might have been expected on the basis of the cyclic variations in size described by ANDERSEN and KENNEDY (1932), which were at best 35% between estrus and diestrus.

IV. Adrenal grafts to the uterus

Adrenal autografts to the uterine wall in spayed rats would appear to provide a relatively direct test of the liberation of small amounts of estrogens by non-tumorous adrenal tissue (WEINSTEIN et al. 1950). Curiously, an equivocal histological response was observed in the uteri, but the vaginas showed prominent and prolific stratification of the epithelium in the area of the vaginal orifice, although the grafts were situated on the wall of the uterus above the cervix. This response was less marked in the upper vagina, as if there was a gradient of sensitivity. The apparent activity of the graft at such a distance from its attachment is not in agreement with the observations of PRICE and INGLE (1957), who found the androgenic effects of adrenal grafts to be restricted to the area immediately around the graft, and failed to observe any significant increases in the weight of the whole target organ. Consequently, further study of the possible estrogenic effects of adrenal grafts would be welcome.

E. Conclusion

The normal adrenal cortex may secrete estrogens, but in such small amounts that their physiological significance is equivocal. Adrenal tumors and hyperplastic glands may produce large amounts of estrogens, their influence may be conspicuous or it may be offset by the concomitant secretion of androgens, of glucocorticoids, or of both.

Epilogue

In view of the close biochemical relationships between the hormones of the gonads and the adrenal cortex, it is noteworthy that there is so little evidence of functional overlap between the normal secretions of these glands. However, as knowledge develops regarding the reaction mechanisms that are influenced by steroids, more subtle physiological relationships between the various hormones may become manifest. Furthermore, questions of phylogeny remain: which steroid-producing organ was evolved first? Is one a derivative of the other, or are both the derivatives of a single protovertebrate tissue? Invertebrates possess specialized gonads but are not known to have a tissue comparable to the adrenal cortex. There is reason to believe that the protovertebrate ancestry was hermaphroditic at some stage (BERRILL 1955). Could the protovertebrate hermaphroditic gonad have been the precursor from which were derived not only the sexual gonads, but also the adrenal cortex specializing in steroids with metabolic functions useful to either sex? It is possible that comparative studies of steroid biosynthetic mechanisms will provide a plausible approach to these questions.

Literature

AGATE, F. J.: The possible role of the adrenal cortex in fluid shifts in the hamster following renal damage. Ann. N. Y. Acad. Sci. **55**, 404—411 (1952).

ALBRIGHT, F.: Cushing's Syndrome. Harvey Lect. **38**, 122—186 (1942/43).

— The effect of hormones on osteogenesis in man. Recent Progr. Hormone Res. **1**, 293—353 (1947).

ALLEN, J. C.: Effects of cortisone on the morphology of the mouse adrenal with special reference to changes in the X zone. Johns Hopk. Hosp. Bull. **94**, 349—350 (1954).

ANDERSEN, D. H., and H. S. KENNEDY: Studies on the physiology of reproduction. IV. Changes in the adrenal gland of the female rat associated with the oestrous cycle. J. Physiol. (Lond.) **76**, 247—261 (1932).

ANDERSON, A. F., A. M. HAINE and J. PATTERSON: A case of adrenal carcinoma and its hormone diagnosis. J. Path. Bact. **55**, 341—349 (1943).

ARAI, Y.: Effects of adrenocorticoid on response of reproductive accessories to androgen in castrated male rats. J. Fac. Sci. Tokyo, Sec. IV, **9**, 111—115 (1960).

ARRINGTON, L. R., M. H. FOX and H. A. BERN: Androgen content of testis and adrenal of white leghorn cockerels. Endocrinology **51**, 226—236 (1952).

ASTWOOD, E. B.: A six hour assay for the quantitative determination of estrogen. Endocrinology **23**, 25—31 (1938).

— An assay method for progesterone based upon the decidual reaction in the rat. J. Endocr. **1**, 49—55 (1939).

BACHMANN, R.: Die Nebenniere. Handb. mikr. Anat. Mensch. VI (5), 1—952 (1954).

BAGGETT, B., L. L. ENGEL, K. SAVARD and R. I. DORFMAN: Formation of estradiol-17β-C^{14} from testosterone-3-C^{14} by surviving human ovarian slices. Fed. Proc. **14**, 175—176 (1955).

BAHN, R., H. WILSON, L. KEDDA, J. FURTH and E. ANDERSON: Metabolic changes associated with transplantable ACTH producing pituitary, tumors of the mouse. Canad. J. Biochem. **25**, 1057 to 1065 (1957).

BAKER, B. L., M. A. SCHAIRER, D. J. INGLE and C. H. LI: The induction of involution in the male reproductive system by treatment with adrenocorticotropin. Anat. Rec. **106**, 345 to 360 (1950).

BARTTER, F. C., F. ALBRIGHT, A. P. FORBES, A. LEAF, E. DEMPSEY and E. CARROLL: The effects of adrenocorticotrophic hormone and cortisone in the adrenogenital syndrome associated with congenital adrenal hyperplasia: an attempt to explain and correct its disordered hormonal pattern. J. clin. Invest. **30**, 237—251 (1951).

BAULIEU, E. J.: Three sulphate esters of 17-ketosteroids in the plasma of normal subjects and after administration of ACTH. J. clin. Endocr. **20**, 900—904 (1960).

BEALL, D.: Isolation of estrone from the adrenal gland. Nature (Lond.) **144**, 76 (1939).

— and T. REICHSTEIN: Isolation of progesterone and allopregnanolone from the adrenal. Nature (Lond.) **142**, 479 (1938).

BENIRSCHKE, K.: Adrenals in anencephaly and hydrocephaly. Obstet. and Gynec. **8**, 412—425 (1956).

— E. BLOCH and A. T. HERTIG: Concerning the function of the fetal zone of the human adrenal gland. Endocrinology **58**, 598—625 (1956).

BENNER, M. C.: Studies on the involution of the fetal cortex of the adrenal glands. Amer. J. Path. **16**, 787—797 (1940).

BENUA, R. S., and E. HOWARD: A carbonyl reaction differentiating the fetal zona reticularis of the human adrenal cortex from the mouse X zone. Johns Hopk. Hosp. Bull. **86**, 200 to 213 (1950).

BERGEN, J. R., D. KRUS and G. PINCUS: Suppression of LSD-25 effects in rats by steroids. Proc. Soc. exp. Biol. (N. Y.) **105**, 254—256 (1960).

BERGSTRAND, C. G., and C. A. GEMZELL: Pregnanediol excretion in normal children and in children with various endocrine disorders, including congenital adrenal hyperplasia. J. clin. Endocr. **17**, 870—877 (1957).

BERRILL, N. J.: The Origin of Vertebrates. Oxford: The Clarendon Press 1955.

BIRKE, G.: The 17-ketosteroid excretion in rheumatoid arthritis. Acta med. scand. **148**, Suppl. **291**, 83—95 (1954).

BIRKE, G.: C. FRANKSSON and L. O. PLANTIN: On the excretion of androgens in carcinoma of the prostate. I. The effect of orchiectomy and adrenalectomy. Acta endocr. (Kbh.) Suppl. **17**, 3—18 (1954).
BITTORF, A.: Nebennierentumor und Geschlechtsdrüsenausfall beim Manne. Berl. klin. Wschr. **56**, 776 (1919).
BLACKMAN, S. S.: Concerning the function and origin of the reticular zone of the adrenal cortex. Johns Hopk. Hosp. Bull. **78**, 180—208 (1946).
BLISS, C. I.: Relative potency as applied to the assay of penicillin. Science **100**, 577—578 (1944).
BLISS, E. L., A. A. SANDBERG, D. H. NELSON and K. EIK-NES: The normal levels of 17-hydroxycorticosteroids in the peripheral blood of man. J. clin. Invest. **32**, 818—823 (1953).
BLOCH, E., and K. BENIRSCHKE: Synthesis in vitro of steroids by human fetal adrenal gland slices. J. biol. Chem. **234**, 1085—1089 (1959).
— — and E. ROSEMBERG: C_{19}-steroids, 17α-OH-corticosterone and a sodium retaining factor in human fetal adrenal glands. Endocrinology **58**, 626—633 (1956).
— R. I. DORFMAN and G. PINCUS: The conversion of acetate to C_{19} steroids by human adrenal gland slices. J. biol. Chem. **224**, 737—750 (1957).
BONGIOVANNI, A. M., and G. W. CLAYTON, JR.: A simplified method for the routine determination of pregnanediol and pregnanetriol in urine. Johns Hopk. Hosp. Bull. **94**, 180—186 (1954).
— W. R. EBERLEIN and J. CARA: Studies on the metabolism of adrenal steroids in the adrenogenital syndrome. J. clin. Endocr. **14**, 409—422 (1954).
BORELL, U., E. DICZFALUSY and A. WESTMAN: On the andromimetic action of the adrenals of immature hypophysectomized-gonadectomized rats. Acta endocr. (Kbh.) **8**, 116—124 (1951).
BOURNE, G., and S. ZUCKERMAN: The influence of the adrenals on cyclic changes in accessory reproductive organs of female rats. J. Endocr. **2**, 268—282 (1941a).
— — Changes in the adrenals in relation to the normal and artificial threshold oestrous cycles in the rat. J. Endocr. **2**, 283—310 (1941b).
BRADLOW, H. L., L. HELLMAN, B. ZUMOFF and T. F. GALLAGHER: Interaction of hormonal effects: influence of triiodothyronine on androgen metabolism. Science **124**, 1206—1207 (1956).
BRYANS, F. E.: Progesterone of the blood in the menstrual cycle of the monkey. Endocrinology **48**, 733—740 (1951).
BUCHANAN, G. P., A. C. ENDERS and R. V. TALMADGE: Implantation in armadillos ovariectomized during the period of delayed implantation. J. Endocr. **14**, 121—128 (1956).
BULBROOK, R. D., and F. C. GREENWOOD: Persistence of urinary oestrogen excretion after oophorectomy and adrenalectomy. Brit. med. J. **1957 i**, 662—666.
BURNS, R. K.: The role of hormones in the differentiation of sex. In: W. C. YOUNG, Ed., Sex and Internal Secretions, Ed. 3. Baltimore: Williams & Wilkins 1961.
BURRILL, M. W., and R. R. GREENE: Androgenic function of the adrenals in the immature male castrate rat. Proc. Soc. exp. Biol. (N. Y.) **40**, 327—330 (1939).
— — Further studies on the andromimetic function of the immature male rat adrenal. Endocrinology **26**, 645—650 (1940).
— — Androgen production in the female rat: the ovary and the adrenal in the immature rat. Endocrinology **28**, 871—873 (1941a).
— — The liver and the adrenal androgen of the rat. Endocrinology **28**, 874—876 (1941b).
BUSH, I. E.: Species differences in adrenocortical secretion. J. Endocr. **9**, 95—100 (1953).
— and V. B. MAHESH: Adrenocortical hyperfunction with sudden onset of hirsutism. J. Endocr. **18**, 1—25 (1959).
— and A. A. SANDBERG: Adrenocortical hormones in human plasma. J. biol. Chem. **205**, 783 to 793 (1953).
— J. SWALE and J. PATTERSON: Analysis of steroids in adrenal venous blood and in urine in a case of virilism. Biochem. J. **62**, 16P (1956).
BUTT, W. R., P. MORRIS, C. J. O. R. MORRIS and D. C. WILLIAMS: The polarographic estimation of steroid hormones. Five determinations of progesterone in blood. Biochem. J. **49**, 434—438 (1951).
CALLOW, N. H., and R. K. CALLOW: Excretion of androgens by eunuchs: the isolation of 17-ketosteroids from the urine. Biochem. J. **34**, 276—279 (1940).
— — and C. W. EMMENS: The effect of the administration of testosterone propionate on the urinary excretion of compounds allied to the steroid hormones. J. Endocr. **1**, 99—107 (1939).
CALLOW, R. K.: Isolation of the male hormone present in the urine of a patient with an adrenal tumor. Chem. and Ind. **55**, 1030 (1936).
— The significance of the excretion of sex hormones in urine. Proc. roy. Soc. Med. **31**, 841 to 855 (1938).
— and R. DEANESLY: Effect of androsterone and of male hormone concentrate on the accessory reproductive organs of castrated rats, mice and guinea pigs. Biochem. J. **29**, 1424 to 1445 (1935).

CARSTENSEN, H., G. W. OERTEL and K. B. EIK-NES: Secretion of 17α-hydroxy-Δ^5-pregnenolone by the canine adrenal gland during stimulation with ACTH. J. biol. Chem. **234**, 2570—2577 (1959).
CHESTER JONES, I.: The action of testosterone on the adrenal cortex of the hypophysectomized, prepuberally castrated male mouse. Endocrinology **44**, 427—438 (1949a).
— The relationship of the mouse adrenal cortex to the pituitary. Endocrinology **45**, 514 to 536 (1949b).
— The effect of hypophysectomy on the adrenal cortex of the immature mouse. Amer. J. Anat. **86**, 371—403 (1950).
— The adrenal cortex in reproduction. Brit. med. Bull. **11**, 156—160 (1955).
— The Adrenal Cortex. Cambridge: University Press. 1957.
CHRISTY, N. P., M. M. DICKIE and G. W. WOOLLEY: Estrus and mating in gonadectomized female mice with adrenal cortical abnormalities. Endocrinology **47**, 129—130 (1950).
CLAYTON, B. E., and J. E. HAMMANT: Modification of response to ACTH by simultaneous administration of gonadotrophic hormones. Nature (Lond.) **176**, 401 (1955).
COHEN, A. I., J. FURTH and R. F. BUFFETT: Histologic and physiologic characteristics of hormone secreting transplantable adrenal tumors in mice and rats. Amer. J. Path. **33**, 631—651 (1957).
CROWDER, R. E.: The development of the adrenal gland in man, with special reference to origin and ultimate location of cell types and evidence in favor of the "cell migration" theory. Contr. Embryol. Carneg. Instn. **36**, 195—210 (1957).
DAUGHADAY, W.: A comparison of the X zone of the adrenal cortex in two inbred strains of mice. Cancer Res. **1**, 883—885 (1941).
DAVIDSON, C. S.: Effects of adrenotrophic extract upon the accessory reproductive organs of castrated rats. Proc. Soc. exp. Biol. (N. Y.) **36**, 703—705 (1937).
— and H. D. MOON: Effect of adrenocorticotrophic extracts on accesssory reproductive organs of the castrate rat. Proc. Soc. exp. Biol. (N. Y.) **35**, 281—282 (1936).
DAVIES, S.: The development of the adrenal gland of the cat. Quart. J. micr. Sci. **80**, 81—98 (1937).
DAVIS, M. E., and N. W. FUGO: The cause of the physiologic basal temperature changes in women. J. clin. Endocr. 8, 550—563 (1948).
— and E. J. PLOTZ: The excretion of neutral steroids in the urine of normal non-pregnant and pregnant women. Acta endocr. (Kbh.) **21**, 245—258 (1956a).
— — Hormonal interrelationships between maternal adrenal, placental and fetal functions. Obstet. Surv. (Baltimore) **11**, 1—43 (1956b).
— C. E. TEST, C. A. NAVORI, B. HRYSE, R. E. POTTINGER and F. DUNKLE: Some effects of ACTH in various hypo-ovarian states. II. Primary amenorrhea, surgical castration and the menopause. J. clin. Endocr. **13**, 181—188 (1953).
DEAN, F. D., P. M. COLE and I. CHESTER JONES: Relative rates of corticosterone secretion in intact and gonadectomized male and female rats. J. Endocr. **18**, iii—iv (1959).
DEANE, H. W.: Histophysiology of the adrenal cortex in fetal and prepuberal rats, with comments on the carbonyl reactivity in the cortical lipide droplets. In: L. I. GARDNER, Ed., Adrenal Function in Infants and Children, 4—20. New York: Grune and Stratton 1956.
DEANESLY, R.: A study of the adrenal cortex in the mouse and its relation to the gonads. Proc. roy. Soc. B. **103**, 523—546 (1928).
— Adrenal cortex differences in male and female mice. Nature (Lond.) **141**, 79 (1938).
— Secretion of androgens by the adrenal cortex of the mouse. Nature (Lond.) **182**, 262—263 (1958).
— A test of androgen production by the rat adrenal. J. Endocr. **20**, vii-vii (1960).
— and A. S. PARKES: Comparative activities of compounds of the androsterone-testosterone series. Biochem. J. **30**, 291—303 (1936).
— — Multiple activities of androgenic compounds. Quart. J. exp. Physiol. **26**, 393—402 (1937).
DEL CASTILLO, E. B., et G. DI PAOLA: Action de la desoxycorticostérone sur la réponse vaginale à l'oestradiol. C. R. Soc. Biol. (Paris) **133**, 734—736 (1940).
DELOST, P.: Structure histologique et étude de l'activité phosphatasique alcaline du canal déférent du campagnol des champs (*Microtus arvalis* P.): état du cortex surrénal. C. R. Soc. Biol. (Paris) **145**, 373—377 (1951).
— Les corrélations génito-surrénaliennes chez le campagnol des champs (Microtus arvalis P.) Ann. Sci. Nat., Zool. **18**, 391—567 (1956).
— Recherches sur le cortex surrénale et la zone X des Musaraignes (*Sorex araneus* L. et *Crocidura russula* H.) C. R. Soc. Biol. (Paris) **151**, 484—487 (1957).
— et A. M. VINCENT: Structure du cortex surrénal du campagnol souterrain: présence d'une zone X. J. Physiol. (Paris) **47**, 167—169 (1955).
DICKIE, M. M., and P. W. LANE: Adrenal tumors, pituitary tumors and other pathological changes in F_1 hybrids of strain DE × strain DBA. Cancer Res. **16**, 48—52 (1956).

DICZFALUSY, E., and R. LUFT: Quantitative study on oestrogens excreted by male patient with adrenal cortical tumor. Acta endocr. (Kbh.) **9**, 327—332 (1952).
DI GEORGE, A. M., J. B. AREY and A. M. BONGIOVANNI: Plasma 17-ketosteroids in an anencephalic infant. J. clin. Endocr. **16**, 1281—1282 (1956).
DINGEMANSE, E., L. G. HUIS IN'T VELD and S. L. HARTOGH-KATZ: Clinical method for the chromatographic-colorimetric determination of urinary 17-ketosteroids. II. Normal adults. J. clin. Endocr. **12**, 66 (1952).
DOBRINER, K.: Studies in steroid metabolism XIX. The α-ketosteroid excretion pattern in normal males J. clin. Invest. **32**, 940 (1953).
— XXII. The adrenal response to ACTH in young and old men. J. clin. Invest. **33**, 222—229 (1954).
— S. LIEBERMAN, C. P. RHOADS and H. C. TAYLOR JR.: Urinary excretion of ketosteroids in pregnancy. Obstet. Surv. (Baltimore) **3**, 677—679 (1948).
DORFMAN, R. I.: Neutral steroid hormone metabolites. Recent Progr. Hormone Res. **9**, 5—26 (1954).
— Steroid hormone metabolism. In: G. PINCUS and K. V. THIMANN, Eds., The Hormones, Vol. III, 589—664. New York: Academic Press 1955.
— and J. B. HAMILTON: The urinary excretion of estrogenic substances after the administration of testosterone propionate. Endocrinology **25**, 33—38 (1939).
— and W. U. GARDNER: Metabolism of steroid hormones. Excretion of estrogenic material by ovariectomized mice bearing adrenal tumors. Endocrinology **34**, 421—423 (1944).
— and R. A. SHIPLEY: Androgens: Biochemistry, Physiology and Clinical Significance. New York: John Wiley 1956.
EAGLE, J. F.: Dehydroepiandrosterone and similar substances in urine of premature infants. Proc. Soc. exp. Biol. (N. Y.) **81**, 571—573 (1952).
EBERLEIN, W. R., and A. M. BONGIOVANNI: Congenital adrenal hyperplasia with hypertension: Unusual steroid pattern in blood and urine. J. clin. Endocr. **15**, 1531—1533 (1955).
— — Plasma and urinary corticosteroids in the hypertensive form of congenital adrenal hyperplasia. J. biol. Chem. **223**, 85—94 (1956).
EDGAR, D. G.: The chemical assay of progesterone. Biochem. J. **54**, 50—55 (1953).
EISENBERG, E., and G. S. GORDAN: The levator ani muscle of the rat as an index of myotrophic activity of steroidal hormones. J. Pharmacol. **99**, 38—44 (1950).
EKHOLM, E., and K. NIEMINEVA: On the prenatal changes in the relative weights of the human adrenals, the thymus and the thyroid gland. Acta paediat. (Stockh.) **39**, 67—86 (1950).
EMMENS, C. W.: Hormone Assay. New York: Academic Press 1950.
ENGEL, L. L., G. W. THORN and R. A. LEWIS: The urinary excretion of steroid compounds. I. Normal male subjects. J. biol. Chem. **137**, 205—215 (1941).
ERSHOFF, B. H.: Nutrition and the anterior pituitary, with special reference to the general adaptation syndrome. Vitam. and Horm. **10**, 79—140 (1952).
EVANS, J. S., R. F. VARNEY and F. C. KOCH: The mouse uterine weight method for the assay of estrogens. Endocrinology **28**, 747—752 (1941).
FARRELL, G. L., E. W. RAUSCHKOLB, P. C. ROYCE and H. HIRSCHMANN: Isolation of desoxycorticosterone from adrenal venous blood of the dog; effect of hypophysectomy and ACTH. Proc. Soc. exp. Biol. (N. Y.) **87**, 587—590 (1954).
FAWCETT, D. W., and H. W. DEANE: The effect of cortisone on uterine growth in ovariectomized rats receiving estradiol. Quart. J. micr. Sci. **92**, 385—392 (1951).
FEKETE, E., G. W. WOOLLEY and C. C. LITTLE: Histological changes following ovariectomy in mice. I. *dba* high tumor strain. J. exp. Med. **74**, 1—8 (1941).
FEODOSSIEW, N. E.: Hypertrophy of the adrenals after ovariectomy (in Russian). Russk. Vrach. **5**, 135 (1906).
FINKLER, R. S.: Pseudohermaphroditism; pregnanediol glucuronidate excretion. J. clin. Endocr. **1**, 151—153 (1941).
FORBES, A. P., and F. ALBRIGHT: A comparison of the 17-ketosteroid excretion in Cushing's Syndrome associated with adrenal tumor and with adrenal hyperplasia. J. clin. Endocr. **11**, 926—935 (1951).
FORBES, T. R.: Discussion of paper by DAVIS, M. E., and E. J. PLOTZ: Rec. Progr. Hormone Res. **13**, 380—381 (1957).
FORSHAM, P. H., G. W. THORN, F. T. G. PRUNTY and A. G. HILLS: Clinical studies with pituitary adrenocorticotrophin. J. clin. Endocr. **8**, 15—66 (1948).
FRAME, E. G.: A micro-method for separation of 17-ketosteroids into alpha and beta fractions. Endocrinology **34**, 175—180 (1944).
— and H. J. JEWETT: The excretion of 17-ketosteroids in carcinoma of the prostate. J. Urol. (Baltimore) **52**, 330—333 (1944).

FRANCIS, F. E., N. C. SHEN and R. A. KINSELLA: Enteric excretion of metabolites of steroid hormones in the human. II. Isolation of Δ^5-androsten-3β-ol-17-one from meconium. J. biol. Chem. **235**, 1957—1959 (1960).
FRANTZ, M. J., and A. KIRSCHBAUM: Sex hormone secretion by tumors of the adrenal cortex of mice. Cancer Res. **9**, 257—266 (1949).
FRAPS, R. M., C. W. HOOKER and T. R. FORBES: Progesterone in blood plasma of cocks and non-ovulating hens. Science **109**, 493 (1949).
FUKUSHIMA, D. K., and T. F. GALLAGHER: Steroid isolation studies in congenital adrenal hyperplasia. J. biol. Chem. **229**, 85—92 (1957).
— — The absence of 21-dehydroxylation in congenital adrenal hyperplasia. J. clin. Endocr. **18**, 694—698 (1958).
GADDUM, J. H.: Bioassays and mathematics. Pharmacol. Rev. **5**, 87—134 (1953).
GALLAGHER, T. F.: On alterations in adrenal function, especially with adrenocortical carcinoma. Cancer Res. **17**, 520—529 (1957).
— Steroid hormone metabolism and the control of adrenal secretion. Harvey Lect. **52**, 1—22 (1958).
— A. KAPPAS, H. SPENCER and D. LASZLO: Influences of invasiveness, hormones and amphenone on steroids in adrenal carcinoma. Science **124**, 487—489 (1956).
GARDNER, L. I.: Plasma neutral 17-ketosteroids in pregnancy. In: L. I. GARDNER, Ed., Adrenal Function in Infants and Children, 205—211. New York: Grune and Stratton 1956 (a).
— Adrenocortical metabolism of the fetus, infant and child. Pediatrics **17**, 897—924 (1956b).
— and C. J. MIGEON: Urinary dehydroisoandrosterone in hyperadrenocorticism; influence of cortisone, hydrocortisone and ACTH. J. clin. Endocr. **12**, 1117—1139 (1952).
— and R. L. WALTON: Plasma 17-ketosteroids of the human fetus; demonstration of concentration gradient between cord and maternal circulations. Helv. paediat. Acta **9**, 311 to 315 (1954a).
— — Plasma 17-ketosteroids of full term and premature infants. J. clin. Invest. **33**, 1642 to 1645 (1954b).
— — W. ELLIS and E. C. HUGHES: Diminished plasma 17-ketosteroid concentration in pregnancy. Proc. Soc. exp. Biol. (N. Y.) **86**, 804—805 (1954).
GARDNER, W. U.: Estrogenic effects of adrenal tumors of ovariectomized mice. Cancer Res. **1**, 632—637 (1941).
GASSNER, F. X., D. H. NELSON, H. REICH, R. T. RAPALA and L. T. SAMUELS: Isolation of an androgenic compound from the adrenal venous blood of cows. Proc. Soc. exp. Biol. (N. Y.) **77**, 829—831 (1951).
— E. C. REIFENSTEIN, JR., J. W. ALGEO and W. E. MATTOX: Effects of hormones on growth, fattening and meat production potential of livestock. Recent Progr. Hormone Res. **14**, 183—209 (1958).
GASTINEAU, C. F., G. B. LOGAN, A. ALBERT and H. L. MASON: Use of cortisone for congenital adrenal hyperplasia in the male; report of case. J. clin. Endocr. **13**, 724—729 (1953).
GENITIS, V. E., and I. P. BRONSTEIN: Pregnanediol excretion in female pseudohermaphroditism. J. Amer. med. Ass. **119**, 704—706 (1942).
GERSH, I., and A. GROLLMAN: The relation of the adrenal cortex to the male reproductive system. Amer. J. Physiol. **126**, 368—374 (1939).
GREEN, J. A.: The secretion of progesterone in the absence of luteal tissue in the immature mouse. Anat. Rec. **112**, 335 (1952).
GREENE, R. R., and M. W. BURRILL: Forty-eight hour response of the immature male rat to androgens. Endocrinology **29**, 402—408 (1941).
— — and A. C. IVY: Progesterone is androgenic. Endocrinology **24**, 351—357 (1939).
GROLLMAN, A.: The Adrenals. Baltimore: Williams and Wilkins 1936.
GRUENWALD, P.: Common traits in the development and structure of the organs originating from the coelomic wall. J. Morph. **70**, 353—379 (1942).
— Embryonic and postnatal development of the adrenal cortex, particularly the zona glomerulosa and accessory nodules. Anat. Rec. **95**, 391—422 (1946).
GUNN, S. A., and T. C. GOULD: Role of the adrenals on seasonal variations in sensitivity response of the rat dorsolateral prostate to sex hormones. Acta endocr. (Kbh.) **32**, 54—58 (1959).
GUYÉNOT, E., K. PONSE and J. WIETRZYKOWSKA: Lutéinisation de l'ovaire et masculinisation chez le cobaye. C. R. Acad. Sci. (Paris) **194**, 1051—1053 (1932).
HARRIS, G. W., R. P. MICHAEL and P. P. SCOTT: Neurological site of action of stilbestrol in eliciting sexual behaviour. Ciba Found. Symp. on the Neurological Basis of Behaviour, 236—251 (1958).

HARRISON, J. H., C. B. LEMAN, P. L. MUNSON and J. C. LAIDLAW: Androgen, gonadotrophin and steroid excretion in man before and after castration and adrenalectomy. J. Urol. (Baltimore) **73**, 580—584 (1955).
HEARD, R. D. H., and M. M. HOFFMAN: Steroids. IV. The fate in man of injected α-estradiol. J. biol. Chem. **141**, 329—342 (1941).
— P. H. JELLINCK and V. J. O'DONNELL: Biogenesis of the estrogens; the conversion of testosterone-4-C^{14} to estrone in the pregnant mare. Endocrinology **57**, 200—204 (1955).
HÉBERT, S., M. TRAN-VY et J. VERNE: L'activité androgène du cortex surrénal chez le rat nouveau-né. Ann Endocr. (Paris) **15**, 493—498 (1954).
HECHTER, O., A. ZAFFARONI, R. P. JACOBSEN, H. LEVY, R. W. JEANLOZ, V. SCHENKER and G. PINCUS: The nature and the biogenesis of the adrenal secretory product. Recent Progr. Hormone Res. **6**, 215—246 (1951).
HERSHBERGER, L. G., E. G. SHIPLEY and R. K. MEYER: Myotrophic activity of 19-nortestosterone and other steroids determined by modified levator ani muscle method. Proc. Soc. exp. Biol. (N. Y.) **83**, 175—180 (1953).
HETT, J.: Ein Beitrag zur Histogenese der menschlichen Nebenniere. Z. mikr.-anat. Forsch. **3**, 179—282 (1925).
HIRSCHMANN, H.: Steroids of urine of ovariectomized women. J. biol. Chem. **136**, 483—501 (1940).
— Steroid excretion in a case of adrenocortical carcinoma: I. The isolation of a Δ^5-androstenetriol-3(β),16,17. J. biol. Chem. **150**, 363—379 (1943).
— and F. B. HIRSCHMANN: Steroid excretion in a case of adrenocortical carcinoma. II. An examination of the non-ketonic fraction precipitable with digitonin. J. biol. Chem. **157**, 601—612 (1945).
HISAW, F. L., and J. T. VELARDO: Inhibition of progesterone in decidual development by steroid compounds. Endocrinology **49**, 732—741 (1951).
— — Influence of adrenal on decidual development. Proc. Soc. exp. Biol. (N. Y.) **89**, 253 to 254 (1955).
HOFMANN, F. G.: Observations on in vitro adrenal steroid synthesis in the albino mouse. Endocrinology **59**, 712—715 (1956).
— M. M. DICKIE and N. P. CHRISTY: Studies on gonadectomized mice bearing adrenocortical tumors. Acta endocr. (Kbh.) **34**, 84—96 (1960).
HOLMES, W. N.: Histological variations in the adrenal cortex of the golden hamster with special reference to the X zone. Anat. Rec. **122**, 271—293 (1955).
HOOKER, C. W., and T. R. FORBES: A bioassay for minute amounts of progesterone. Endocrinology **41**, 158—169 (1947).
HORWITT, B. N., R. I. DORFMAN, R. A. SHIPLEY and W. R. FISH: Metabolism of the steroid hormones. IV. Conversion of desoxycorticosterone to pregnanediol-3α-20α in man and the chimpanzee. J. biol. Chem. **155**, 213—218 (1944).
HOSKINS, W. H., J. R. COFFMAN, F. C. KOCH and A. T. KENYON: The effect of testosterone propionate on the urinary excretion of androgens and estrogens in eunuchoidism. Endocrinology **24**, 702—710 (1939).
HOUSSAY, A. B., F. A. CARDEZA, B. A. HOUSSAY et R. M. PINTO: Facteurs susceptibles de modifier l'apparition et la fonction des tumeurs surrénales chez les rats castrés. C. R. Soc. Biol. (Paris) **148**, 2123—2125 (1954).
HOUSSAY, B. A., F. A. CARDEZA, V. G. FOGLIA, A. B. HOUSSAY et R. M. PINTO: Evolution anatomique et fonctionelle des tumeurs surrénales chez les rats castrés. C. R. Soc. Biol. (Paris) **148**, 918—919 (1954).
HOWARD, E. (HOWARD-MILLER): A transitory zone in the adrenal cortex which shows age and sex relationships. Amer. J. Anat. **40**, 251—293 (1927).
— The X zone of the suprarenal cortex in relation to gonadal maturation in monkeys and mice and to epiphyseal unions in monkeys. Anat. Rec. **46**, 93—104 (1930).
— Is the adrenal X zone andromimetic? Amer. J. Physiol. **119**, 339—340 (1937).
— The representation of the adrenal X zone in rats, in the light of observations on X zone variability in mice. Amer. J. Anat. **62**, 351—373 (1938).
— Effects of castration on the seminal vesicles as influenced by age, considered in relation to the degree of development of the adrenal X zone. Amer. J. Anat. **65**, 105—149 (1939).
— Regarding the effects of desoxycorticosterone and testosterone on the adrenal X zone. Anat. Rec. **77**, 181—188 (1940).
— Effects of adrenalectomy and desoxycorticosterone substitution therapy on the castrated rat prostate. Evidence for an andromimetic function of the immature rat adrenal. Endocrinology **29**, 746—754 (1941).
— The effect of adrenalectomy on the accessory reproductive glands of mice castrated for short periods. Endocrinology **38**, 156—164 (1946).
— Biological effects of dehydroepiandrosterone. Fed. Proc. **15**, 97 (1956).

HOWARD, E.: A complementary action of corticosterone and dehydroepiandrosterone on the mouse adrenal, with observations on the reactivity of reproductive tract structures to dehydroepiandrosterone and 11-hydroxy-androstenedione. Endocrinology **65**, 785—802 (1959).
— and J. C. ALLEN: The retarding influences of adrenalectomy and age on estradiol-induced connective tissue hyperplasia in seminal vesicles. Endocrinology **60**, 785—796 (1957).
— and R. S. BENUA: The effect of protein deficiency and other dietary factors on the X zone of the mouse adrenal. J. Nutr. **42**, 157—174 (1950).
— and S. GENGRADOM: The effects of ovariectomy and administration of progesterone on the adrenal X zone and uterus. Endocrinology **26**, 1048—1052 (1940).
HUGGINS, C., F. M. PARSONS and E. V. JENSEN: Promotion of the growth of preputial glands by steroids and the pituitary growth hormone. Endocrinology **57**, 25—32 (1955).
JAILER, J. W., J. J. GOLD, R. VANDE WIELE and S. LIEBERMAN: 17-α-Hydroxyprogesterone and 21-deoxyhydrocortisone: their metabolism and possible role in congenital adrenal virilism. J. clin. Invest. **34**, 1639—1646 (1955).
JAYLE, M. F., D. MALASSIS et H. PINAUD: Excrétion de la DHA avant et après administration d'ACTH en pathologie surrénalienne. Acta endocr. (Kbh.) **31**, 1—32 (1959).
JEANLOZ, R. W., H. LEVY, R. P. JACOBSEN, O. HECHTER, V. SCHENKER and G. PINCUS: Chemical transformations of steroids by adrenal perfusion. III. Δ^4-Androstene-3,17-dione. J. biol. Chem. **203**, 453—461 (1953).
JOHNSEN, S. G.: Fractionation of urinary 17-ketosteroids. III. The clinical significance of the simplified fractionated 17-ketosteroid determination. A preliminary survey. Acta endocr. (Kbh.) **21**, 157—176 (1956).
JONES, H. W., and G. E. S. JONES: The gynecological aspects of adrenal hyperplasia and allied disorders. Amer. J. Obstet. Gynec. **68**, 1330—1365 (1954).
JOST, A.: Activité androgène du testicule foetal de rat greffé sur l'adulte castré. C. R. Soc. Biol. (Paris) **142**, 196—198 (1948).
— et J. P. GELOSO: Recherche de l'activité androgène de la surrénale par la greffe sur la vésicule séminale de rats castrés. C. R. Soc. Biol. (Paris) **148**, 474—477 (1954).
KAPPAS, A., and T. F. GALLAGHER: Studies on steroid metabolism. XXVIII. The α-ketosteroid excretion pattern in normal females and the response to ACTH. J. clin. Invest. **34**, 1566—1572 (1955).
KATSH, S., A. S. GORDON and H. A. CHARIPPER: The andromimetic action of adrenal cortical transplants to the seminal vesicle of the adult rat. Anat. Rec. **101**, 47—57 (1948).
KEENE, M. F. L., and E. E. HEWER: Observations on the development of the human suprarenal gland. J. Anat. (Lond.) **61**, 302—324 (1927).
KINSELL, L., D. BRYANT and F. ALBRIGHT: Rate of growth of axillary hair as a diagnostic index. J. clin. Endocr. **14**, 897—909 (1954).
KLOPPER, A., J. A. STRONG and L. R. COOK: The excretion of pregnanediol and adrenocortical activity. J. Endocr. **15**, 180—189 (1957).
KOCHAKIAN, C. D.: A comparison of the renotrophic with the androgenic activity of various steroids. Amer. J. Physiol. **142**, 315—325 (1944).
— and C. TILLOTSON: Influence of several C-19 steroids on the growth of individual muscles of the guinea pig. Endocrinology **60**, 607—618 (1957).
KUTZ, R. L., T. MCKEOWN and H. SELYE: Effect of salt treatment on certain changes following adrenalectomy. Proc. Soc. exp. Biol. (N. Y.) **32**, 331—332 (1934).
LAIDLAW, J. C., D. JENKINS, W. J. REDDY and T. JAKOBSON: The diurnal variation in adrenocortical secretion. J. clin. Invest. **33**, 950 (1954).
LAKSHMANAN, T. K., and S. LIEBERMAN: Application of principle of gradient elution to analysis of urinary ketosteroids. Fed. Proc. **12**, 235 (1953).
LANDAU, R. L., B. F. STIMMEL, E. HUMPHRYS and D. E. CLARK: Gynecomastia and retarded sexual development resulting from a long-standing estrogen-secreting adrenal tumor. J. clin. Endocr. **14**, 1097—1112 (1954).
LANMAN, J. T.: The fetal zone of the adrenal gland; its developmental course, comparative anatomy and possible physiological functions. Medicine **32**, 389—430 (1953).
— The adrenal fetal zone: its occurrence in primates and a possible relationship to chorionic gonadotrophin. Endocrinology **61**, 684—691 (1957).
LAVELLE, F. W.: A study of hormonal factors in the early sex development of the golden hamster. Contr. Embryol. Carneg. Instn. **34**, 19—53 (1951).
LAZO-WASEM, E. A., G. M. NEHER, R. L. SHOGER and M. X. ZARROW: Gelbkörperhormon im Blute männlicher Wirbeltiere. Endokrinologie **31**, 166—171 (1954).
LEATHEM, J. H.: The response of the mouse adrenal gland to steroid hormones. Trans. N. Y. Acad. Sci., Ser. II, **2**, 239—243 (1949).
LEBLOND, C. P., et W. U. GARDNER: Comparison de la zone X et des autres parties de la surrénale de la souris, à l'aide du nitrate d'argent. C. R. Soc. Biol. (Paris) **127**, 775—776 (1938).

LELONG, M., M. F. JAYLE, R. SCHOLLER, P. CANLORBE, P. BORNICHE et SZPER: Etude de l'élimination des stéroïdes urinaires chez l'enfant. Sem. Hôp. Paris (Ann. Péd.) Nr. **33**, 2738/P. 446—2745/P. 453 (1957a).

— — P. CANLORBE, R. SCHOLLER and P. BORNICHE: L'exploration par l'ATCH de la cortico-surrénale de l'enfant normal. Sem. Hôp. Paris (Ann. Péd.) Nr. **33**, 2745/P. 453—2756/P. 464 (1957b).

LI, C. H., P. FONSS-BECH, I. I. GESCHWIND, T. HAYASHIDA, G. F. HUNGERFORD, A. J. LOSTROH, W. R. LYONS, H. D. MOON, W. O. REINHARDT and M. SIDEMAN: Corticotrophins (ACTH). X. Biological investigations on α-corticotrophins. J. exp. Med. **105**, 335—359 (1957).

LIEBERMAN, S., and S. TEICH: Metabolic precursors of urinary dehydroisoandrosterone. J. clin. Endocr. **13**, 1141—1148 (1953).

LIU, T. Y., K. A. BROWNELL and F. A. HARTMAN: Production of androgenic hormone by the adrenal. Amer. J. Physiol. **180**, 50—52 (1955).

LLOYD, C. W., J. LOBOTSKY, J. JONES, J. FREDERICKS and T. C. WYATT: Hormone studies in a case of adrenogenitalism due to neoplasm of the adrenal cortex. J. clin. Endocr. **11**, 857—865 (1951).

LOBBAN, M. C.: Structural variations in the adrenal cortex of the adult cat. J. Physiol. (Lond.) **118**, 565—574 (1952).

LOMBARDO, M. E., C. MCMORRIS and P. B. HUDSON: The isolation of steroidal substances from human adrenal vein blood. Endocrinology **65**, 426—432 (1959).

LOSTROH, A. J., and C. H. LI: Stimulation of the sex accessories of hypophysectomized male rats by non-gonadotrophic hormones of the pituitary gland. Acta endocr. (Kbh.) **25**, 1—16 (1957).

LYONS, W. R., C. H. LI, R. E. JOHNSON and R. D. COLE: Evidence for progestogen secretion by ACTH stimulated adrenals. Proc. Soc. exp. Biol. (N. Y.) **84**, 356—358 (1953).

MALLEY, K., and T. G. BRADSHAW: Case of adrenocortical syndrome. Brit. med. J. **1941 I**, 454.

MANDL, A. M.: Cyclical changes in the vaginal smear of adult ovariectomized rats. J. exp. Biol. **28**, 585—592 (1951).

— The sensitivity of adrenalectomized rats to gonadotrophins. J. Endocr. **11**, 359—376 (1954).

— Factors influencing ovarian sensitivity to gonadotrophins. J. Endocr. **15**, 448—457 (1957).

— and S. ZUCKERMAN: The reactivity of the X-irradiated ovary of the rat. J. Endocr. **13**, 243—261 (1956).

MARRIAN, G. F.: Some aspects of progesterone metabolism. Recent Progr. Hormone Res. **4**, 3—23 (1949).

MARTIN, S. J.: Effect of certain endocrine secretions on the X zone of the adrenal cortex of the mouse. Proc. Soc. exp. Biol. (N. Y.) **28**, 41—42 (1930).

MASON, H. L., and W. W. ENGSTROM: The 17-ketosteroids: their origin, determination and significance. Physiol. Rev. **30**, 321—374 (1950).

— and E. J. KEPLER: Isolation of androsterone, etiocholan-3α-ol-17-one, and Δ^5-androstene-3α,17α-diol from the urine after administration of dehydroisoandrosterone to a man. J. biol. Chem. **160**, 255—264 (1945a).

— — Isolation of steroids from the urine of patients with adrenal cortical tumors and adrenal cortical hyperplasia: a new 17-ketosteroid, androstane-3(α),11-diol-17-one. J. biol. Chem. **161**, 235—257 (1945b).

— M. H. POWER, E. H. RYNEARSON, L. C. CIARAMELLI, C. H. LI and H. M. EVANS: Results of administration of pituitary adrenocorticotropic hormone to a normal human subject. J. clin. Endocr. **8**, 1—14 (1948).

MASUDA, M.: Urinary ketosteroid excretion patterns in congenital adrenal hyperplasia. J. clin. Endocr. **17**, 1181—1190 (1957).

MASUI, K., and Y. TAMURA: The effect of gonadectomy on the structure of the suprarenal glands of mice, with especial reference to the functional relation between this gland and the sex gland of the female (in Japanese). Jap. J. zootech. Sci. **1**, 55—79 (1924).

MATHIESON, D. R., and H. W. HAYS: The use of castrate male rats for the assay of testosterone propionate. Endocrinology **37**, 275—279 (1945).

MCPHAIL, M. K.: The androgenic activity of the adrenal gland with special reference to the X zone. Rev. canad. Biol. **3**, 312—327 (1944).

MEYER, A. S., M. HAYANO, M. C. LINDBERG, M. GUT and O. G. RODGERS: The conversion of Δ^4-androstene-3,17-dione-4-C^{14} and dehydroepiandrosterone by bovine adrenal homogenate preparations. Acta endocr. (Kbh.) **18**, 148—168 (1955).

MIGEON, C. J.: Fractionation by countercurrent distribution of urinary estrogens in normal individuals and in patients with hyperadrenocorticism. J. clin. Endocr. **13**, 674—687 (1953).

— Adrenocortical function and plasma 17-ketosteroids in man. Ciba Found. Coll. Endocr. **8**, 141—155 (1955).

MIGEON, C. J. and L. I. GARDNER: Urinary estrogens (measured fluorometrically and biologically) in hyperadreconsrticiom: influence of cortisone, compound F, compo und Band ACTH. J. clin. Endocr. **12**, 1513—1531 (1952).
— and J. E. PLAGER: Identification and isolation of dehydroisoandrosterone from peripheral human plasma. J. biol. Chem. **209**, 767—772 (1954).
— — A method for the fractionation and measurement of 17-ketosteroids in human plasma. J. clin. Endocr. **15**, 702—714 (1955).
— and R. S. STEMPFEL: Laboratory diagnosis in pediatric endocrinology. Pediat. Clin. N. Amer. **1957**, 959—980 (1957).
— A. R. KELLER and E. G. HOLMSTROM: Dehydroepiandrosterone, androsterone and 17-hydroxycorticosteroid levels in maternal and cord plasma in cases of vaginal delivery. Johns Hopk. Hosp. Bull. **97**, 415—421 (1955).
— F. H. TYLER, J. P. MAHONEY, A. A. FLORENTIN, H. CASTLE, E. L. BLISS and L. T. SAMUELS: The diurnal variation of plasma levels and urinary excretion of 17-hydroxycorticosteroids in normal subjects, night workers and blind subjects. J. clin. Endocr. **16**, 622—633 (1956).
— A. R. KELLER, B. LAWRENCE and T. H. SHEPARD: Dehydroepiandrosterone and androsterone levels in human plasma. Effect of age and sex; day to day and diurnal variations. J. clin. Endocr. **17**, 1051—1062 (1957).
MILLER, A. M., R. I. DORFMAN and E. L. SEVERINGHAUS: Metabolism of the steroid hormones: the isolation of an androgen from human urine containing an 11-oxygen substitution in the steroid ring. Endocrinology **38**, 19—25 (1946).
— — and M. MILLER: Metabolism of the steroid hormones; metabolism of dehydroisoandrosterone. Endocrinology **46**, 105—110 (1950).
MIYAKE, T., and G. PINCUS: Progestational activity of certain 19-nor steroids and progesterone derivatives. Endocrinology **63**, 816—824 (1958).
MOOG, F.: The influence of the pituitary-adrenal system on the differentiation of phosphatase in the duodenum of the suckling mouse. J. exp. Zool. **124**, 329—346 (1953).
— and V. NEHARI: The influence of hydrocortisone on epithelial phosphatase of embryonic intestine in vitro. Science **119**, 809—810 (1954).
— C. J. BENNETT and C. M. DEAN, JR.: Growth and cytochemistry of the adrenal gland of the mouse from birth to maturity. Anat. Rec. **120**, 873—891 (1954).
MOON, H. D.: Inhibition of somatic growth in castrate rats with pituitary extracts. Proc. Soc. exp. Biol. (N.Y.) **37**, 34—36 (1937a).
— Effect of ACTH on the sexual development of spayed rats. Proc. Soc. exp. Biol. (N. Y.) **37**, 36—37 (1937b).
MOORE, C. R.: Adrenal cortical secretions in relation to the reproductive system of rats. J. clin. Endocr. **13**, 330—368 (1953).
— D. PRICE and T. F. GALLAGHER: Rat-prostate cytology as a testis-hormone indicator and the prevention of castration changes by testis extract injections. Amer. J. Anat. **45**, 71 to 107 (1930).
MOSIER, H. D.: Comparative histological study of the adrenal cortex of the wild and domesticated Norway rat. Endocrinology **60**, 460—469 (1957).
MUNSON, P. L., F. G. GOETZ, J. C. LAIDLAW, J. H. HARRISON and G. W. THORN: Effect of adrenocortical steroids on androgen excretion by adrenalectomized-orchidectomized men. J. clin. Endocr. **14**, 495—508 (1954).
NATHANSON, I. T., and L. E. TOWNE: The urinary excretion of estrogens, androgens and FSH following the adminstration of testosterone to human female castrates. Endocrinology **25**, 754—758 (1939).
— L. L. ENGEL and R. M. KELLY: The effect of androgens on the urinary excretion of ketosteroids, nonketonic alcohols and estrogens. J. clin. Endocr. **12**, 1172—1186 (1952).
NEHER, G. M., and M. X. ZARROW: Serum progesterone levels in the pregnant rabbit. Fed. Proc. **12**, 102—103 (1953).
— — Concentration of progestin in the serum of the non-pregnant, pregnant and postpartum ewe. J. Endocr. **11**, 323—330 (1954).
NELSON, W. O.: Production of sex hormones in the adrenals. Anat. Rec. **81**, 97 (1941).
NICHOLS, J., O. L. LESCURE and C. J. MIGEON: The levels of 17-hydroxycorticosteroids and 17-ketosteroids in maternal and cord plasma in term anencephaly. J. clin. Endocr. **18**, 444—452 (1958).
NISHIDA, S., and K. MOCHIZUKI: Effects of adrenalectomy on the ventral prostate glands in mice. Endocr. jap. **2**, 289—296 (1955).
OKADA, M., D. K. FUKUSHIMA and T. F. GALLAGHER: Isolation and characterization of 3β-hydroxy-Δ^5-steroids in adrenal carcinoma. J. biol. Chem. **234**, 1688—1692 (1959).
ORTIZ, E.: The effects of castration on the reproductive system of the golden hamster. Anat. Rec. **117**, 65—87 (1953).

PARKES, A. S.: The adrenal-gonad relationship. Physiol. Rev. **25**, 203—254 (1945).
PARMER, L. G., F. KATONAH and A. A. ANGRIST: Comparative effects of ACTH, cortisone, deoxycorticosterone, pregnenolone on growth and development of infant rats. Proc. Soc. exp. Biol. (N. Y.) **77**, 215—218 (1951).
PATTERSON, J.: Diagnosis of adrenal tumors, a new chemical test. Lancet **1947 II**, 580—581.
PAYNE, R. W.: Relationship of pituitary and adrenal cortex to ovarian hyperemia reaction in the rat. Proc. Soc. exp. Biol. (N. Y.) **77**, 242—244 (1951).
PECHET, M. M., and P. H. HENNEMAN: The effects of 11-beta hydroxylation on the anabolic activities of steroids in man. Proc. 39th Meeting of The Endocrine Soc., 65 (1957).
PETERSON, R. E., R. HERTZ and H. A. LUBS: Suppression of biosynthesis of adrenal cortical steroids in man by amphenone. Proc. Soc. exp. Biol. (N. Y.) **94**, 421—425 (1957).
PHILPOT, F. J., and J. ST. L. PHILPOT: A modified colorimetric estimation of carbonic anhydrase. Biochem. J. **30**, 2191—2193 (1936).
PHOENIX, C. H., R. W. GOY, A. A. GERALL and W. C. YOUNG: Organizing action of prenatally administered testosterone propionate on the tissues mediating mating behavior in the female guinea pig. Endocrinology **65**, 369—382 (1959).
PINCUS, G.: A diurnal rhythm in the excretion of urinary ketosteroids by young men. J. clin. Endocr. **3**, 195—199 (1943).
— and R. I. DORFMAN: Effects of corticoids, progestins and certain of their derivatives on the androgenic and myotrophic properties of testosterone. Fed. Proc. **14**, 115 (1955).
— and E. B. ROMANOFF: The synthesis of corticosteroids by the human adrenal cortex. Ciba Found. Coll. Endocr. 8, 97—106 (1955).
— L. P. ROMANOFF and J. CARLO: A diurnal rhythm in the excretion of neutral reducing lipids by man and its relation to the 17-ketosteroid rhythm. J. clin. Endocr. 8, 221—226 (1948).
— — — The excretion of urinary steroids by men and women of various ages. J. Geront. **9**, 113—132 (1954).
— R. I. DORFMAN, L. P. ROMANOFF, B. L. RUBIN, E. BLOCH, J. CARLO and H. FREEMAN: Steroid metabolism in aging men and women. Recent Progr. Hormone Res. **11**, 307—341 (1955).
— T. MIYAKE, A. P. MERRILL and P. LONGO: The bioassay of progesterone. Endocrinology **61**, 528—533 (1957).
PLANTIN, L. O., E. DICZFALUSY and G. BIRKE: Isolation of dehydroepiandrosterone from an adrenal cortical tumor. Nature (Lond.) **179**, 421 (1957).
PLOTZ, E. J., and M. E. DAVIS: The excretion of neutral steroids in the urine of pregnant women following the administration of large doses of progesterone. Acta endocr. (Kbh.) **21**, 259—267 (1956).
POLITZER, G.: Über die Frühentwicklung der Nebennierenrinde beim Menschen. Z. Anat. Entwickl.-Gesch. **106**, 40—48 (1937).
POLL, H.: Die vergleichende Entwicklungsgeschichte der Nebennierensysteme der Wirbeltiere. In: O. HERTWIG, Herausgeb. Handbuch der vergleichenden und experimentellen Entwickelungslehre der Wirbeltiere **3**, 443—616. Jena: Fischer 1906.
— Über die Einwirkung von Proviron auf die männliche Nebennierenrinde. Anat. Anz. **77**, 113—123 (1933).
POMEAU-DELILLE, G.: Déciduome provoqué chez la rate sans surrénales et sans surrénales accessoires. C. R. Soc. Biol. (Paris) **143**, 1251—1252 (1949).
PONSE, K.: La fonction sexuelle de la cortico-surrénale. Schweiz. med. Wchnschr. **80**, 170—173 (1950).
— Masculinisation paradoxale de rats par les extraits gonadotropes gravidiques en fonction de l'hypophyse et de la surrénale. Bull. schweiz. Akad. med. Wiss. **10**, 1—11 (1954).
PRICE, D.: Normal development of the prostate and seminal vesicles of the rat, with a study of experimental post-natal modifications. Amer. J. Anat. **60**, 79—125 (1936).
— Normal development and regression of the prostate gland of the female rat. Proc. Soc. exp. Biol. (N. Y.) **41**, 580—583 (1939).
— An analysis of the factors influencing growth and development of the mammalian reproductive tract. Physiol. Zoöl. **20**, 213—247 (1947).
— and D. J. INGLE: Androgenic effects of autotransplants of adrenals in the accessory reproductive glands of adult castrated rats. Rev. suisse Zool. **64**, 743—755 (1957).
— T. MANN and C. LUTWAK-MANN: The stimulating effect of female hormones on the metabolic activity and histological structure of male rat accessory glands. Anat. Rec. **122**, 363 to 375 (1955).
RACE, G. J., and H. M. WU: Adrenal cortical functional zonation in the whale *(Physeter catadon)*. Endocrinology **68**, 156—158 (1961).
RAKOFF, A. E., K. E. PASCHKIS and A. CANTAROW: Androgen assay on three-day-old male chicks. Proc. Soc. exp. Biol. (N. Y.) **55**, 124—127 (1944).

RAYNAUD, J.: Effets de la surrénalectomie, associée à la castration, sur la structure de la glande sous-maxillaire de la souris mâle. C. R. Soc. Biol. (Paris) **148**, 1939—1942 (1954).

READ, C. H., E. H. VENNING and M. P. RIPSTEIN: Adrenal cortical function in newly born infants. J. clin. Endocr. **10**, 845—857 (1950).

REICH, H., D. H. NELSON and A. ZAFFARONI: Isolation of 17-hydroxycorticosterone from blood obtained from adrenal veins of dogs. J. biol. Chem. **187**, 411—417 (1950).

— — and F. X. GASSNER: Isolation of 17-hydroxycorticosterone and other steroid ketones from adrenal venous blood of cows. Fed. Proc. **10**, 331 (1951).

REICHSTEIN, T., u. J. VON EUW: Über Bestandteile der Nebennierenrinde. 20. Isolierung der Substanzen Q (Desoxy-corticosteron) und R sowie weiterer Stoffe. Helv. chim. Acta **21**, 1197—1210 (1938).

RENOLD, A. E., D. JENKINS, P. H. FORSHAM and G. W. THORN: The use of intravenous ACTH: a study in quantitative adrenocortical stimulation. J. clin. Endocr. **12**, 763—797 (1952).

REYNOLDS, S. R. M.: The Physiology of the Uterus. New York: Hoeber 1949.

RIDDLE, O.: Studies on the physiology of reproduction in birds. XIV. Suprarenal hypertrophy coincident with ovulation. Amer. J. Physiol. **66**, 322—339 (1923).

RING, J. R.: The hormonal induction of mating responses in the spayed-adrenalectomized female rat. Endocrinology **37**, 237—244 (1945).

ROBERTS, S., and C. M. SZEGO: Steroid interaction in the metabolism of reproductive target organs. Physiol. Rev. **33**, 593—629 (1953).

— — Biochemistry of the steroid hormones. Ann. Rev. Biochem. **24**, 543—596 (1955).

ROBSON, J. M.: Quantitative data on the inhibition of oestrus by testosterone, progesterone and certain other compounds. J. Physiol. (Lond.) **92**, 371—382 (1938).

ROMANOFF, E. B., P. HUDSON and G. PINCUS: Isolation of hydrocortisone and corticosterone from human adrenal vein blood. J. clin. Endocr. **13**, 1546—1547 (1953).

ROTTER, W.: Die Entwicklung der fötalen und kindlichen Nebennierenrinde. Virchows Arch. Path. Anat. **316**, 590—618 (1949a).

— Das Wachstum der foetalen und kindlichen Nebennierenrinde. Z. Zellforsch. **34**, 547—561 (1949b).

ROY, S. N., J. N. KARKUN and B. MUKERJI: Effect of graded doses of testosterone propionate on the adrenal cortex of prepuberally castrated male mice. Ind. J. med. Res. **47**, 30—35 (1959).

RUBIN, B. L., A. S. DORFMAN, L. BLACK and R. I. DORFMAN: Bioassay of estrogens using the mouse uterine response. Endocrinology **49**, 429—439 (1951).

SAKIZ, E.: Action de la cortisone sur la zone X surrénalienne de la souris. C. R. Soc. Biol. (Paris) **147**, 1222—1223 (1953).

— La gonadotrophine choriale ne provoque pas de sécrétion oestrogène surrénalienne chez la souris femelle castrée. C. R. Soc. Biol. (Paris) **149**, 1744—1746 (1955).

SALHANICK, H. A., and D. L. BERLINER: Isolation of steroids from a feminizing adrenal carcinoma. J. biol. Chem. **227**, 583—590 (1957).

SALMON, U. J., S. H. GEIST and A. A. SALMON: Excretion of pregnandiol in women with virilism. Proc. Soc. exp. Biol. (N. Y.) **47**, 279—280 (1941).

SAMUELS, L. T.: Metabolism of the steroid hormones. In: HOLMAN, R. T., W. O. LUNDBERG and T. MALKIN, Eds., Progress in the Chemistry of Fats and other Lipids. Vol. III, 395 to 447. London: Pergamon Press 1955.

— Presidential Address before the Endocrine Society. Endocrinology **63**, 260—268 (1958).

SAVARD, K., R. I. DORFMAN, B. BAGGETT and L. L. ENGEL: Biosynthesis of androgens from progesterone by human testicular tissue *in vitro*. J. clin. Endocr. **16**, 1629—1630 (1956).

SCAMMON, R. E.: The prenatal growth and natal involution of the human suprarenal gland. Proc. Soc. exp. Biol. (N. Y.) **23**, 809—811 (1925/26).

SCHILLER, S., A. M. MILLER, R. I. DORFMAN, E. L. SEVERINGHAUS and E. P. MCCULLAGH: Metabolism of steroid hormones; isolation of Δ^5-androstene-3β,17α and Δ^5-pregnenediol-3β, 20α. Endocrinology **37**, 322—328 (1945).

SCHNEIDER, J. J., and P. M. HORSTMANN: Effects of incubating desoxycorticosterone with various rat tissues. J. biol. Chem. **191**, 327—338 (1951).

— and H. L. MASON: Studies on intermediary steroid metabolism. I. Isolation of Δ^5-androstene-3β,17α-diol and Δ^5-androstene-3β,16β,17α-triol following the incubation of dehydroisoandrosterone with surviving rabbit liver slices. J. biol. Chem. **172**, 771—782 (1948).

SCOTT, W. W., and C. VERMEULEN: Studies on prostatic cancer. V. Excretion of 17-ketosteroids, estrogens and gonadotropins, before and after castration. J. clin. Endocr. **2**, 450 to 456 (1942).

SHEMANO, I., G. S. GORDON and E. EISENBERG: The singularly brief anabolic activity of testololactone, a compound lacking androgenicity. Proc. Soc. exp. Biol. (N. Y.) **78**, 612 to 613 (1951).

SHORT, R. V.: Progesterone and related steroids in the blood of domestic animals. Ciba Found. Coll. Endocr. **11**, 362—375 (1957).
— The secretion of sex hormones by the adrenal gland. Biochem. J. **71**, 35 P (1959).
SLAUNWHITE, W. R., JR., and L. T. SAMUELS: Progesterone as a precursor of testicular androgens. J. biol. Chem. **220**, 341—352 (1956).
SMITH, E. K.: Interrelationships of anterior pituitary and adrenal cortex in the rat ovarian hyperemia reaction. Endocrinology **56**, 567—574 (1955).
SOBEL, E. H., C. M. LEE, JR., V. M. ESSELBORN and L. L. CLARK, JR.: Functioning adrenal tumors in childhood. Amer. J. Dis. Child. **86**, 733—751 (1953).
SOLOMON, S., R. VANDE WIELE and S. LIEBERMAN: The *in vitro* synthesis of 17α-hydroxyprogesterone and Δ^4-androstene-3,17-dione from progesterone by bovine ovarian tissue. J. Amer. chem. Soc. **78**, 5453—5454 (1956).
— J. T. LANMAN, J. LIND and S. LIEBERMAN: The biosynthesis of Δ^4-androstenedione and 17α-hydroxyprogesterone from progesterone by surviving human fetal adrenals. J. biol. Chem. **233**, 1084—1088 (1958).
SPAULDING, M. H.: The development of the external genitalia in the human embryo. Contr. Embryol. Carneg. Instn. **13**, 69—88 (1921).
SPIEGEL, A.: Über das Auftreten von Geschwulsten der Nebennierenrinde mit vermännlichender Wirkung bei frühkastrierten Meerschweinchenmännchen. Virchows Arch. Path. Anat. **305**, 367—393 (1939/40).
STARKEY, W. F., and E. C. N. SCHMIDT, JR.: The effect of testosterone propionate on the X zone of the mouse adrenal. Endocrinology **23**, 339—344 (1938).
STEINACH, E., and H. KUN: Transformation of male sex hormones into a substance with the action of a female hormone. Lancet **1937 II**, 845.
— — u. O. PECZENIK: Beiträge zur Analyse der Sexualhormonwirkungen. Tierexperimentelle und klinische Untersuchungen. Wien. klin. Wschr. **49**, 899—903 (1936).
STIEVE, H.: Über physiologische und pathologische Veränderungen der Nebennierenrinde des Menschen und ihre Abhängigkeit von der Tätigkeit der Keimdrüsen. Z. Geburtsh. Gynäk. **127**, 209—231 (1947).
SWINGLE, W. W., E. J. FEDOR, G. BARLOW, JR., E. J. COLLINS and J. PERLMUTT: Induction of pseudopregnancy in rats following adrenal removal. Amer. J. Physiol. **167**, 593—598 (1951).
SYDNOR, K. L., V. C. KELLEY, R. B. RAILE, R. S. ELY and G. SAYERS: Blood ACTH in children with congenital adrenal hyperplasia. Proc. Soc. exp. Biol. (N. Y.) **82**, 695—697 (1953).
SZEGO, C. M., and S. ROBERTS: Pituitary-adrenal cortical antagonism to estrogenic stimulation of the uterus of the ovariectomized rat. Amer. J. Physiol. **152**, 131—140 (1948).
TÄHKÄ, H.: On the weight and structure of the adrenal glands and the factors affecting them, in children of 0—2 years. Acta paediat. (Stockh.) **40**, Supp. **81**, 1—95 (1951).
TALBOT, N. B., and E. H. SOBEL: Certain factors which influence the rate of growth and the duration of growth of children. Recent Progr. Hormone Res. **1**, 355—363 (1947).
— A. M. BUTLER and E. A. MACLACHLAN: Alpha and beta neutral 17-ketosteroids (androgens). Preliminary observations on their normal urinary excretion and the clinical usefulness of their assay in differential diagnosis. New Engl. J. Med. **223**, 369—373 (1940).
— — and R. A. BERMAN: Adrenal cortical hyperplasia with virilism; diagnosis, course and treatment. J. clin. Invest. **21**, 559—570 (1942).
— A. N. ZYGMUNTOWICZ, M. WOOD and E. CHRISTO: Observations on adrenal cortical "sugar-fat-nitrogen" hormone (11-17-OCS) and "17-keto-steroid precursor" production by normal and abnormal individuals of various ages with comments on the fact that (a) there may be two ACTH's and (b) the normal adrenal cortex may not produce true androgens. Proc. First Clinical ACTH Conference, pp. 32—43. Philadelphia: Blakiston Co. 1950.
TULLNER, W. F.: Hormonal factors in the adrenal-dependent growth of the rat ventral prostate. Proc. 43rd Meet. of the Endocrine Soc., 26—27 (1961).
ULSTROM, R. A., E. COLLE, J. BURLEY and R. GUNVILLE: Adrenocortical steroid metabolism in newborn infants. I. Urinary excretion of free and conjugated 17-hydroxycorticosteroids in normal fullterm infants. J. clin. Endocr. **20**, 1066—1079 (1960).
UNGAR, F., A. M. MILLER and R. I. DORFMAN: Metabolism of dehydroepiandrosterone and Δ^5-androstene-3β,17β-diol. J. biol. Chem. **206**, 597—605 (1954).
UOTILA, U. U.: The early embryological development of the fetal and permanent adrenal cortex in man. Anat. Rec. **76**, 183—203 (1940).
VANDE WIELE, R., and S. LIEBERMAN: The metabolism of dehydroepiandrosterone. In: G. PINCUS and E. P. VOLLMER, Eds., Biological activities of steroids in relation to cancer, 93 to 105. New York: Academic Press, 1960.
VELARDO, J. T.: Influence of purified adrenal hormones and adrenocorticotropin on progesterone in decidual tissue formation. Amer. J. Physiol. **190**, 408—412 (1957).

VELARDO, J. T., and S. H. STURGIS: Suppression of uterine growth by purified hydrocortisone acetate, 9α-fluorohydrocortisone acetate and corticotrophin. J. clin. Endocr. **16**, 496—506 (1956).
— F. L. HISAW and A. B. DAWSON: Growth of deciduomata in adrenalectomized rats. Proc. Soc. exp. Biol. (N. Y.) **83**, 509—511 (1953).
VENNING, E. H., P. G. WEIL and J. S. L. BROWN: Excretion of sodium pregnanediol glucuronidate in the adrenogenital syndrome. J. biol. Chem. **128**, CVII (1939).
— J. P. RANDALL and P. GYORGY: Excretion of glucocorticoids in the newborn. Endocrinology **45**, 430—434 (1949).
— V. E. KAZMIN, M. RIPSTEIN, H. T. MCALPINE and M. M. HOFFMAN: The effect of adrenocorticotropin on the excretion of adrenal metabolites in normal human subjects. J. clin. Endocr. **10**, 583—593 (1950).
VILLEE, D. B., L. L. ENGEL and C. A. VILLEE: Steroid hydroxylation in human fetal adrenals. Endocrinology **65**, 465—474 (1959).
WADE, N. J., and L. A. HASELWOOD: Effect of removal of ovaries and adrenals on opening of the vagina in the albino rat. Endocrinology **28**, 624—628 (1941).
WALLACH, S., H. BROWN, E. ENGLERT and K. EIK-NES: Adrenocortical carcinoma with gynecomastia: a case report and review of the literature. J. clin. Endocr. **17**, 945—958 (1957).
WARREN, R. P., and L. R. ARONSON: Sexual behaviour in castrated-adrenalectomized hamsters maintained on DCA. Endocrinology **58**, 293—304 (1956).
DE WATTEVILLE, H., R. BORTH, R. S. MACH and E. MUSSO: Failure to detect urinary pregnanediol following adrenal stimulation by ACTH. Acta endocr. (Kbh.) **8**, 319—326 (1951).
WAXENBERG, S. E., M. G. DRELLICH and A. M. SUTHERLAND: The role of hormones in human behaviour. I. Changes in female sexuality after adrenalectomy. J. clin. Endocr. **19**, 193 to 202 (1959).
WEINSTEIN, M. J., J. SCHILLER and H. A. CHARIPPER: Estrogenic activity of adrenal transplants to the uterus of ovariectomized rats. Anat. Rec. **108**, 441—455 (1950).
WEST, C. D., B. L. DAMAST, S. D. SARRO and O. H. PEARSON: Conversion of testosterone to estrogens in castrated, adrenalectomized human females. J. biol. Chem. **218**, 409—418 (1956).
— — and O. H. PEARSON: Urinary estrogen excretion in Cushing's syndrome. J. clin. Endocr. **18**, 15—27 (1958).
WESTPHAL, U.: Über den Nachweis von Pregnandiolglukuronid im Harn normaler Frauen und Männer. Hoppe-Seylers Z. Physiol. Chem. **281**, 14—24 (1944).
WHITEHEAD, R.: Growth and mitosis in the mouse suprarenal. J. Anat. (Lond.) **67**, 399—408 (1933).
WILKINS, L.: A feminizing adrenal tumor causing gynecomastia in a boy of five years contrasted with a virilizing tumor in a five-year-old girl. Classification of seventy cases of adrenal tumor in children according to their hormonal manifestations and a review of eleven cases of feminizing adrenal tumor in adults. J. clin. Endocr. **8**, 111—132 (1948).
— Diagnosis and Treatment of Endocrine Disorders in Childhood and Adolescence, Ed. 2. Springfield, Ill.: Charles C. Thomas 1957.
— J. F. CRIGLER, JR., S. H. SILVERMAN, L. I. GARDNER and C. J. MIGEON: Further studies on the treatment of congenital adrenal hyperplasia with cortisone. II. The effects of cortisone on sexual and somatic development, with an hypothesis concerning the mechanism of feminization. J. clin. Endocr. **12**, 277—295 (1952).
WILLS, C. G., S. E. RAMPTON and L. I. PUGSLEY: Variables affecting the assay of testosterone propionate using the seminal vesicle response of juvenile castrated male rat. Endocrinology **44**, 251—258 (1949).
WILSON, H., J. J. BORRIS and R. C. BAHN: Steroids in the blood and urine of female mice bearing an ACTH-producing pituitary tumor. Endocrinology **62**, 135—149 (1958).
WITSCHI, E., and G. Y. CHANG: Cortisone induced transformation of ovaries into testes in larval frogs. Proc. Soc. exp. Biol. (N. Y.) **75**, 715—718 (1950).
WOLFE, J. K., L. F. FIESER and F. B. FRIEDGOOD: Nature of the androgens in female adrenal tumor urine. J. Amer. chem. Soc. **63**, 582—593 (1941).
WOLFF, E., K. HAFFEN, Y. LUTZ-OSTERTAG et B. SALZGEBER: Les facteurs qui conditionnent le maintien des canaux de Muller chez l'embryon de poulet. Arch. Anat. micr. Morph. exp. **40**, 168—182 (1951).
WOOLLEY, G. W.: Experimental endocrine tumors with special reference to the adrenal cortex. Recent Progr. Hormone Res. **5**, 383—405 (1950).
— Evidence that adrenal cortical tumors in aged, gonadectomized hamsters produce sex-like hormones. Anat. Rec. **115**, 381—382 (1953).
— Tumors of the adrenal cortex. Ciba Found. Colloq. Endocr. **12**, 122—138 (1958).

WOOLLEY, G. W., and C. C. LITTLE: Prevention of adrenal, cortical carcinoma by diethylstilbestrol. Proc. nat. Acad. Sci. (Wash.) **32**, 239—240 (1946a).
— — Transplantation of an adrenal cortical carcinoma. Cancer Res. **6**, 712—717 (1946b).
— E. FEKETE and C. C. LITTLE: Mammary tumor development in mice ovariectomized at birth. Proc. nat. Acad. Sci. (Wash.) **25**, 277—279 (1939).
— M. M. DICKIE and C. C. LITTLE: Adrenal tumors and other pathological changes in reciprocal crosses in mice. II. An introduction to results of four reciprocal crosses. Cancer Res. **13**, 231—245 (1953).
WOTIZ, H. H., J. W. DAVIS, H. M. LEMON and M. GUT: Studies in steroid metabolism: V. The conversion of testosterone-4-C^{14} to estrogens by human ovarian tissue. J. biol. Chem. **222**, 487—495 (1956).
WRENN, T. R., J. BITMAN and J. F. SYKES: The thermogenic influence of progesterone in ovariectomized cows. Endocrinology **65**, 317—321 (1959).
ZANDER, J.: Discussion of paper by M. E. DAVIS and E. J. PLOTZ: Rec. Progr. Hormone Res. **13**, 379 to 380 (1957).
— u. H. SIMMER: Die chemische Bestimmung von Progesteron in organischen Substraten. Klin. Wschr. **32**, 529—540 (1954).
ZARROW, M. X., and E. A. LAZO-WASEM: The release of a progesterone-like substance from the adrenal gland. Acta endocr. (Kbh.) **18**, 273—280 (1955).
— F. L. HISAW and F. BRYANS: Conversion of desoxycorticosterone acetate to progesterone in vivo. Endocrinology **46**, 403 (1950).
ZEISEL, H., u. M. PRESSLER: Die corticoide und neutralen C-17-ketosteroids im Harn des Kindes. 1. Die Steroidausscheidung im Säuglingsalter. Z. Kinderheilk. **72**, 675—681 (1953).
— W. ENGEL u. M. PRESSLER: Die Corticoide und neutralen C-17 Ketosteroide im Harn des Kindes. 2. Die Steroidausscheidung im Kindesalter. Z. Kinderheilk. **72**, 682—688 (1953).
ZIMMERMANN, B., and A. J. WOJACK: Effect of surgery and adrenal stimulation on the excretion of pregnane-diol-like material in men. J. clin. Endocr. **12**, 972 (1952).
ZUCKERMAN, S.: Cyclic fluctuations in oestrin threshold. Nature (Lond.) **139**, 628 (1937).
— Periodic uterine bleeding in spayed rhesus monkeys injected daily with a constant threshold dose of oestrone. J. Endocr. **2**, 263 (1941).
— The adreno-genital relationship. In: J. M. YOFFEY, Ed., The Suprarenal Cortex, 69—83. London: Butterworths 1953.
— G. BOURNE and D. LEWES: Cyclical changes in the adrenal glands of spayed rats. Nature (Lond.) **142**, 754 (1938).

Chapter 6

Psychological manifestations of hyper- and hypoactivity of the adrenal cortex*

By

M. BLEULER and W. A. STOLL

This brief chapter aims to provide a basis for understanding the psychological aberrations observed in patients with either hyper- or hypoadrenocorticism. Our knowledge of endocrinological psychiatry is drawn from a very large number of small contributions. Observations in regard to the psychology of patients with endocrine disturbances are scattered throughout the medical literature. There are, on the other hand, few papers that deal systematically with the topic and are of themselves of outstanding significance. Only a few of the possible references are cited here — some report original cases, some survey the preceding literature. For reviews in which the extensive literature is cited more extensively, see our earlier works (M. BLEULER 1954, 1956, STOLL 1953) and the listed sections on psychiatry in the Encyclopédie Médico-chirurgicale.

Psychopathology of hyperadrenocorticism

By M. BLEULER

A. Excess of glucocorticoids

I. Cushing's syndrome

We can discuss the psychopathology of Cushing's syndrome without considering the question of its origin. No differences are known in the psychopathology of patients with Cushing's syndrome due to pituitary or to adrenal disease. It seems reasonable to assume that psychopathology in Cushing's syndrome does not depend on the genesis of the disease but depends on the hypersecretion of glucocorticoids. On the other hand, the psychic sequelae of hypersecretion of sex hormones by the adrenal cortex are very different from those of hypersecretion of glucocorticoids (see p. 650).

Most patients with long-standing or severe Cushing's syndrome are psychologically affected (SCHWÖBEL 1954). In the majority of cases, the psychological affection is a mild one and does not reach the grade of true psychosis. A considerable portion of the patients, however, develop true psychoses. Most of these psychoses are acute and episodic and have a good prognosis. Chronic psychoses are much rarer. It is impossible to give exact information concerning the frequency of severe mental derangement in Cushing's syndrome: large series of cases have never been examined carefully from the psychiatric point of view. Furthermore, there exist borderline

* This chapter was completed in 1959; some additions were inserted in 1960.

cases between real Cushing's syndrome and constitutional peculiarities of a similar kind. Statistics on the frequency of mental derangement would vary in relation to the inclusion or exclusion of these borderline cases.

The milder psychological affections, which arise in the great majority of patients with Cushing's syndrome, concern *mood, mental activity,* and *instincts.*

Mood can be altered in quite opposite directions — depressions on the one hand and euphoria on the other are frequent. Even more common, however, are other types of mood change, such as anxiety, irritability, aggressiveness, restlessness, or indifference and apathy. In some individuals, alterations of mood are of long-standing duration, but sudden changes are even more characteristic. The duration of mood swing is different — short mood swings, of some hours' or some days' duration, are common. As a rule, no psychological reason for the mood shifts can be elicited. They arise suddenly, to the perplexity of both the patient and the people around him.

Mental activity is frequently altered. It varies, of course, with mood. In some cases, however, the mental affection is better described as a change in activity than as a change in mood. Both morbid excess of mental activity and morbid inactivity are encountered. In many cases, there are abrupt changes from one condition to the other. Overactivity is generally not combined with flights of ideas and euphoria, as it is in manic states, but rather with an uneasy restlessness. On the other hand, euphoria is often combined with an apathetic state of mind.

The combination of euphoria and apathy was first described by FRANKL v. HOCHWART (1912) under the term „Hypophysärstimmung". Since then it has been shown that the name is erroneous, since the condition arises not only in pituitary diseases but also in many other endocrine diseases and also in cerebral disorders.

Among *instincts,* hunger is frequently increased. Sexuality is in the majority of cases weakened or seems even to disappear. Less commonly there are periods of sexual irritation. Maternal feelings and maternal attitudes are often very marked. In one patient, we saw a morbid exaggeration of maternal instinct combined with a morbid lactation, as has been observed in acromegaly and other endocrine derangements with increased secretion of lactogenic hormone (RICHET et al. 1948, BLICKENSTORFER 1949, 1952, BLICKENSTORFER et al. 1952, M. BLEULER 1954, SCHWÖBEL 1954). Other instincts and drives show different alterations from patient to patient — for instance, thirst or its absence, desire to move or rest, aggressiveness or submissiveness, desire to wander or to stay at home, etc.

In these mild forms of mental derangement, the intellectual functions are not, or not primarily, disturbed. If there are intellectual disturbances, they arise in connection with the state of mood. Such patients are generally very egocentric in attitude. In apathetic states, they lose intellectual activity and liveliness. Their conversation is often without wit and seems somewhat annoying and uninteresting.

Such patients are by no means psychotic. To people who know them, they seem to be moody and odd individuals. In some of them, social irresponsibility is predominant — they neglect their work in moody spells, they live without hesitation at the cost of other people, they are egotistic in following their impulses. In a few of them, dipsomania or drug addiction are a prevalent manifestation. Periods occur with desire to wander around or with even characteristic poriomania.

In many of these patients, an *amnestic syndrome* develops. It is particularly frequent in older patients, who develop early senility. But even younger patients with long-standing Cushing's syndrome suffer from mild difficulty in retention and memory, as in the beginning of senile dementia.

These mild forms of mental derangement show, in a considerable minority of cases, exacerbations to the grade of *true psychoses. Acute psychoses* of short duration (some days or weeks) are more frequent than chronic psychoses. Acute psychoses may be of a mere affective nature. Manic and depressive states occur. More frequent are irritable and apathetic conditions of a psychotic degree. These are often complicated by paranoid ideas and by hallucinations. As a rule, delusions and hallucinations conform with the prevalent mood — in manic states, ideas of grandeur occur; in depressive states, vague ideas of persecution. Delusions and hallucinations can dominate the clinical picture, with the patients presenting delirious and dreamy states. In these conditions, the patients become confused.

Chronic psychoses are rarer. Many of these result from cerebral arteriosclerosis. In other cases, the delusional and hallucinatory contents of the acute episodes are not fully corrected after the agitation has calmed down. Residual delusions („Restwahn") persist between the acute episodes.

Epileptic fits in Cushing's syndrome have been observed. They are not common, and their relationship to the syndrome has not yet been determined (Glaser 1953, Hess 1954, Toratani and Oku 1956). Abnormal alterations in the electroencephalogram have been described in some cases (Hess 1954).

II. Treatment with ACTH or glucocorticoids

Treatment with ACTH *or with cortisone has, in the large majority of cases, some influence on psychic function* (v. Zerssen 1957). The changes are, as a rule, mild and without great clinical significance. However, severe psychoses occur in a minority of cases.

The usual psychic effects of these hormones are changes of *mood*. It has been maintained that they regularly cause *euphoria*. As a matter of fact, euphoria is frequently observed during treatment and is, perhaps, the usual phenomenon at the beginning. Changes of mood in other directions, however, are by no means rare. All kinds of mood states may be observed — anxiety, irritability, aggressiveness, restlessness, apathy. As a rule, one mood does not last long, but the mood changes both during treatment and immediately after its cessation. Quality, degree, and temporal sequence of the mood states vary considerably from case to case. In one patient, for instance, depression is the immediate reaction to the beginning of treatment; in another, depression starts only after prolonged treatment; while in a third, euphoria dominates the whole period of treatment, and depression develops after its cessation.

There is but one fairly regular influence on *instinct:* hunger is increased[1]. This has been observed even in babies. There may also arise odd desires for particular foods, as in pregnancy. The sexual functions are often inhibited. Other changes in instinct are unknown.

Many publications have dealt with *psychoses* during treatment with these hormones (see M. Bleuler 1954). No exact data concerning frequency of psychosis are available. Some earlier statistics indicated psychotic complications in as many as 5%, or even more, of the patients treated. More recent statistics and the general experience of clinicians show that this was too pessimistic a percentage. We judge that real psychoses occur only in about 1% of these patients.

The symptomatology of these psychoses is well known — moody states, excitement or apathy are often the principal or even the only signs. As in Cushing's syndrome, the moodiness can be very different in kind. In the severe cases, the

[1] The weight loss due to the catabolic influence of glucocorticoids probably causes this increase in appetite. See Chapter 7 Sub-ed.

affective psychosis changes into a dreamy state, with delusional thinking, hallucinations, and confusion.

Most frequently, the duration of psychosis during hormone treatment is short. It generally lasts but a few hours, days, or weeks. As a rule, recovery follows rapidly on cessation of treatment or even on reduction of the dosage of the hormone. There have been cases, however, in which the recovery from psychosis did not follow cessation of the hormone treatment that seemed to cause them. These patients were, as a rule, finally treated by electroshock. It is very important to state that *every psychosis following treatment with* ACTH *or cortisone in patients who had previously been mentally healty has been cured*, even when it lasted many weeks or months and had to be treated by more active methods than mere cessation of hormone treatment.

The psychopathological sequelae of treatment with ACTH and cortisone are not in simple and direct correlation with the dosage of the hormone or with the duration of treatment. Some severe psychoses have been observed at the beginning of treatment with moderate doses; and, on the other hand, long-standing treatment with high doses has frequently been supported without significant psychic damage. If there is any correlation between dosage and duration of treatment and psychiatric complication, the relation is very loose. It is also of interest that the same psychiatric complications that, in one patient, arise during treatment and disappear with its cessation may, in another, arise just after cessation of treatment and present themselves as a withdrawal symptom.

On what do occurrence, grade, and type of psychiatric complication depend, if not on dosage and duration of treatment? We do not have sufficient knowledge to answer the question fully. One factor, however, is certain — combination of treatment with ACTH or cortisone with other conditions that might lead to psychological damage in themselves is particularly dangerous. Such a combination is frequent — high fever or disseminated lupus erythematosus or any general severe metabolic disturbance can cause delirious states, excitement, apathy, or other acute psychic damage. They can also complicate conditions that are treated by ACTH or cortisone (TRETHOWAN 1954, BRODY 1956, CLARK and BAILEY 1956). We also know that modern drugs used in the therapy of tuberculosis, particularly Iproniazid, can cause psychoses of the same kind as cortisone. Simultaneous treatment with such drugs and cortisone or ACTH is likely to produce psychiatric complications. (It is interesting to add that such drugs can also elicit the physical symptoms of Cushing's disease, as can ACTH and cortisone.)

In patients who have a history of earlier psychiatric derangement, the incidence of psychotic complications on treatment with ACTH or cortisone is higher than average. Some cases of relapse of a former psychosis that took a chronic and unfavorable course have been described (DENCKER et al. 1954). However, even after recovery from a psychosis, the majority of patients support treatment without developing a new psychosis (COPE 1953, DELAY et al. 1954a, b, LEWIS and FLEMINGER 1954, MCGEHEE and MACLEAN 1954). The administration of these hormones to psychotic patients has, as a rule, no, or only an insignificant, psychological effect.

One important factor as a reason for psychiatric damage in association with hormone treatment was long neglected — patients under treatment are very much influenced by their contact with nurses, physicians, and relatives. In most, treatment causes only a predisposition to react with moody states to difficulties in interhuman relations, not with moody states independent of the environmental situation. Psychotherapy influences such patients very readily.

Epileptic fits have been observed by several authors as a further complication of cortisone and ACTH treatment, but only in a small minority of cases. They occur mostly in children and young people (ARNDT et al. 1954, HAZARD 1954, WAYNE 1954, SCHROEDER and SHERMAN 1956). Both hormones frequently have a marked influence on the electroencephalogram. Its alterations are, however, variable.

Although some individuals have exhibited different reactions to the two types of hormone, *in general no differences in psychiatric complications with* ACTH *and with cortisone treatment have been observed.*

Psychiatric experience with treatment with *cortisol* and with *prednisone* is still rather restricted. As far as we know, their psychiatric complications are of about the same nature as those seen with treatment with cortisone or ACTH.

III. Strong cortical activity within normal limits

The activation of adrenal function is increased (and, finally, exhausted) during the adaptation reactions to stress. The work of SELYE on the general adaptation syndrome (1946) is generally known and it is unnecessary for me to summarize it here.

Hypercorticoidism under stress became evident at first from experiments on animals. Within the last few years it has been carefully studied in the human. The most varied kinds of stress have an influence on adrenocortical function: battle stress (HOWARD et al. 1955), fear of the psychiatrist of suicidal impulse of his patients (HETZEL et al. 1955), tension before jumping of parachutists (BASOWITZ et al. 1955), excitement on account of accident of a relative (BLISS et al. 1956), examination of students (BLISS et al. 1956), fear of an operation (FRANKSSON et al. 1954, 1955). Anxiety is best known as a stressor and therefore as an activator of adrenocortical function.

Adrenocortical activation under stress was formerly demonstrated particularly by the decrease of the eosinophilic leucocytes. Within the last few years it has also been shown that plasma hydrocortisone and the urine hydrocorticoids increase during anxiety. Proof of ACTH increase is much more difficult to obtain. A correlation between the grade of anxiety and the amount of ACTH secretion seems only probable in very strong anxiety. It has also been claimed that only one factor of the adrenocorticotrophic hormone (the blood ascorbic depletion factor) is increased in anxiety (PERSKY et al. 1959).

It is certain that emotional stress activates adrenocortical function; but it is not known if adrenocortical function within normal limits has an influence on the emotional life of healthy human beings.

IV. Discussion

The psychopathology of Cushing's syndrome and the psychopathology complicating treatment with ACTH or cortisone are identical. We must, therefore, conclude that *the main factor in the genesis of such psychopathology lies in the excess of hormones with glucocorticoid activity.*

How can the psychiatric action of an excess of glucocorticoids be explained?

The hypothesis that these hormones act directly only on the body and that all the sequelae on psychic function are to be explained as secondary psychological reactions to somatic influences cannot be accepted. Some workers have tried, for instance, to explain the depression in Cushing's disease as a simple psychic reaction to the bodily deformation, or the euphoria during treatment with cortisone as a very simple psychic reaction to the improvement of the somatic disease. The arguments against such hypotheses are overwhelming. I shall mention only the most important. In Cushing's syndrome, changes in mental attitude are often observed long before the patient is aware of somatic symptoms. The grade and type of psychic alteration are not correlated with the grade and type of somatic symptoms. In addition, the course of somatic phenomena with psychological significance is little related to the course of psychopathological phenomena. During treatment of organic disease by cortisone or ACTH, it can frequently be observed that the psychological reaction has nothing to do with somatic influence of the treatment on the disease. Euphoria may arise without any physical improvement. On the other hand, good physical results can be accompanied by depression, irritability, or even severe psychosis. *The types of psychopathology arising during an excess of glucocorticoids are quite similar in nature to those in other severe endocrine disorders*, and they conform much less with the tremendous variability of psychological reactions to somatic pain and somatic changes of cosmetic importance. As will be discussed later on, psychological reactions play an important role in the genesis of the psychopathology in question, but they play by no means the only role.

Some *have tried to correlate the psychopathology of adrenocortical hyperfunction with one definite aspect of the changes in metabolism*. For instance, the question has been raised as to whether this psychopathology is dependent on a deficit of potassium ions (see HOAGLAND 1950, GOLDBERG 1954). No regular and direct correlation of psychic damage and hypokalemia was found. The same is true for any other individual metabolic change due to hyperadrenocorticism — we cannot sufficiently explain the psychopathologic sequelae by changes in the metabolism of water, NaCl, sugar, or albumins, for instance. The metabolism of any of these substances can be greatly disturbed without psychopathological sequelae; on the other hand, very severe psychopathological phenomena may be present in mild Cushing's disease or after treatment with cortisone in which metabolic changes are slight (GEPPERT et al. 1952, DENCKER et al. 1954, QUARTON et al. 1955, etc.). It is also impossible to correlate the usual psychopathology of Cushing's syndrome with hypertension. Hypertension and cerebral arteriosclerosis are very important factors in the genesis of the amnestic syndrome and the severe organic psychoses in advanced cases of Cushing's syndrome, but they do not play an essential role in the more frequent milder personality changes and in the majority of the acute psychoses. Neither is it demonstrable that alterations in cerebral blood flow or in oxygen uptake are responsible for the psychiatric complications (ALMAN and FAZEKAS 1951, SENSENBACH et al. 1953). As far as we know, the psychopathological effect of an excess of glucocorticoids must act directly on the brain and psychic function, rather than through some single definite alteration in metabolism.

To approach an understanding of the psychopathological effects of an excess of glucocorticoids, we must emphasize general laws in endocrinological psychiatry. Any endocrine trouble can influence psychic function in two different ways:

(1) The endocrine alteration may have, quite independently of the gland and hormones in question, a nonspecific influence.

(2) Each hormone may also have a somewhat more specific psychological and psychopathological action, depending on its specific physiological role.

As a rule, former authors considered only the latter, more specific aspect of endocrine imbalance and derangement. Until recently most clinicians were not willing to accept the fact that many of the psychological repercussions of endocrine changes are quite nonspecific and highly independent of the nature of the endocrine change or disease.

(1) *The nonspecific repercussions are quite evident in the psychopathological aspects of Cushing's disease and of treatment with cortisone or* ACTH. This becomes

clear, for instance, when we compare them with the psychopathology of Addison's disease. The older, naive, and merely speculative expectation was based on the following argument — physiologically, adrenocortical hyperfunction in Cushing's syndrome and adrenocortical hypofunction in Addison's syndrome are essentially opposite conditions. It was usual to conclude that the psychopathology of Cushing's syndrome and of Addison's disease, too, had to be of opposite nature. Clinical observation shows that this is not the case. The psychopathology of Cushing's syndrome and of Addison's disease is quite similar (M. BLEULER 1953, 1954, STOLL 1953, PERKOFF et al. 1955). In both conditions we observe the same variety of changes in mood and instincts; in both conditions we observe amnestic symptoms; in both conditions acute psychoses show the same clinical picture. It would not be true to assert that the psychopathology of both conditions is absolutely identical. Inactivity, tiredness, and depression may be more frequent in Addison's disease; the course of psychic complications in Cushing's disease is, on the average, more variable and more eventful than in Addison's disease. However, there cannot be any doubt that the similarities are remarkable and that the dissimilarities are of far less importance.

The general thesis that endocrine changes can have a psychological or psychopathological effect that is not specific for the gland or hormone in question can be demonstrated best in a survey of psychopathology in all endocrine diseases, which review would be beyond the scope of this chapter. In our experience, similar psychopathologic changes during over- and under-production of one hormone are quite common. Pathology of the adrenal cortex by no means provides the sole example of this phenomenon.

Many features of the psychopathology of conditions with an excess of glucocorticoids are not specific. We observe acute psychoses with exactly the same psychopathology as in Cushing's syndrome in many severe and acute endocrinopathological episodes, for instance, during Graves' disease, after severe lesions of the parathyroids, after thyroid surgery, in acute myxedema, etc. We even observe the same psychoses in non-endocrine diseases. They are characteristic psychic sequelae of acute and severe metabolic and cerebral damage of any kind. They occur, for instance, in many exogenous intoxications, in uremia, in many general and cerebral infections. They belong to the *acute exogenous reaction type*, as BONHOEFFER (1909) termed psychotic reactions to any severe, acute physical damage. This is also demonstrated by the epileptic attacks that occur in certain patients.

The amnestic syndrome in many cases of Cushing's disease, too, is not due to specific damage. It must be explained by diffuse cerebral damage. It is certainly often caused by cerebral arteriosclerosis. In a few patients, it has been caused by therapeutic X-ray treatment that was directed to the hypophysis but damaged the brain. However, it is probable that a chronic excess of glucocorticoids might also produce directly (without cerebrovascular damage) degenerative brain processes, as do most severe, chronic endocrine derangements.

Also the milder psychopathologic phenomena that are observed in Cushing's disease and during treatment with cortisone or ACTH, i.e., changes in mood, activity and some instincts, are to a large degree nonspecific responses to the endocrine derangement. We have already mentioned that they are very similar to the psychopathological phenomena in Addison's disease. They are also similar to those in acromegaly, in mild hyperthyroidism, in hypoparathyroidism, etc.

(2) Despite our earlier assertion that the psychopathology of Cushing's syndrome and of patients under the influence of cortisone therapy is largely nonspecific, there are *some features* that *seem specifically related to a high level of glucocorticoids*. We

have mentioned that stimulation of activity is more frequent than in many other endocrine derangements, for instance, than in Addison's disease, or after castration, or in hypothyroidism. (Hyperadrenocorticism has this frequent stimulation of activity in common with hyperthyroidism; in hyperthyroidism it is even more marked and more frequent.) Hypothetically it seems probable that the "adaptation syndrome," which can be elicited both by physical and by emotional stress, influences adaptation on both the physical and emotional plane. The psychologic effect of adrenocortical hyperfunction can perhaps be regarded as such an adaptation to stress — angry moods prepare for aggression, anxious moods for flight, apathetic moods could be correlated with the tendency to hide and wait in a dangerous and stressful situation; the diminution of the sexual desire could signify the greater urgency for securing individual life under stress as compared with reproduction.

It is by no means astonishing that emotional adaptation to stress is much more variable from individual to individual and from situation to situation than metabolic adaptation. While metabolic adaptation is essentially the same in all individuals (and even similar in men and animals), emotional response to stress differs very greatly and ranges from anger to fear, from overactivity to apathy, from alertness to somnolence. The reason is clear — each individual reacts to a given situation in a personal way. The same stressor might convey the same metabolic signal to all individuals but greatly varies in psychological meaning from individual to individual. It seems reasonable to assume that emotional reactions to glucocorticoids can be partly understood through the assumption that they are originally a specific aspect of a "general adaptation syndrome" but individually worked out by psychic functions.

Like any endocrine disease, Cushing's syndrome affects not one gland alone (the adrenal cortex and the secretion of glucocorticoids). It has repercussions on the endocrine system as a whole. Some psychopathological aspects of Cushing's syndrome may be connected with these other endocrine alterations, for instance with the atrophy of the gonads. It is also known that in some endocrine diseases a primitive and morbid maternal instinct can be elicited by a hypersecretion of lactogenic hormone. As far as we know, such a hypersecretion has never been proved in Cushing's syndrome. The fact that in some cases of Cushing's syndrome galactorrhea has been observed suggests, however, that a relative over-secretion of lactogenic hormone might occur. It is possible that the stimulation of maternal feelings observed in some cases of Cushing's syndrome is related to such a hypersecretion.

It is probable that many endocrine influences on psychic function act through close relations of the hormones to the brain. There is both clinical and anatomical evidence for such a theory: Clinical observations demonstrate that the changes in mood, activity, and instincts due to endocrine derangements are similar to those caused by localized brain lesions, especially in the diencephalon. And in many cases of endocrine disease, localized cerebral degenerations have been discovered by pathologists (see M. Bleuler 1954). The assumption seems reasonable that endocrine alterations lead at first to functional alterations of certain portions of the central nervous system. In more severe of chronic endocrine diseases, prolonged functional alterations of parts of the brain might lead to their degeneration. The explanation of the symptomatological identity of the endocrine psychosyndrome and of the psychosyndrome of localized brain disease might therefore be quite simple — the endocrine disease causes localized alterations in the brain, at first of functional, finally of anatomical nature.

Some preliminary studies on animals support such an assumption. Neurosecretory activity of the diencephalon is influenced by adrenalectomy and by the application of cortisone. This has been detected microscopically (EICHNER 1953, MALANDRA and CORBETTA 1953). It has been shown in other experiments that a hormone [e.g., thyroid hormone (STURM and WERNITZ 1956), 5-hydroxytryptamine, intermedin] may have an affinity for certain parts of the central nervous system. Such studies make it possible to assume that a functional relation exists between some endocrine glands and some parts of the brain and that this functional relation leads to a morphological interdependence. It is a fertile working hypothesis to consider the genesis of the psychopathology of conditions with an excess of glucocorticoids from this point of view.

Many other recent experiences sustain the assumption that there are close interrelationships between endocrine activities and the function of certain cerebral systems: Observations after the intracerebral instillation of hormones prove that some parts of the brain are sensitive to certain hormones (ENGELHARDT 1957). The same has been demonstrated with the well known technique of OLDS (1960) with electrical self-stimulation of rats.

The emotional symptoms both of patients with Cushing's syndrome and of patients treated with cortisone or ACTH are not simple psychological reactions to somatic damage but they *are influenced to a high degree by psychological reactions*. Their great variability from individual to individual and from situation to situation in the same individual can only be understood by the personal evaluation of somatic influences on psychic function. We are not dealing simply with an expression of an endocrine stimulus on psychic function, but with highly personal reactions to this stimulus. These reactions depend not on the endocrine stimulus alone, but also on many other factors — the personal constitution, the personality development during the whole life, the present situation of the patient, and his intellectual and emotional evaluation of this situation. In other words, glucocorticoids (like other hormones) can influence the psychic life, but never is the psychic life or any aspect of it dominated by this hormonal influence alone.

Even for the genesis of psychotic states, psychogenic influences are of great importance. They also influence the symptomatology and often the degree of the psychosis. Some of the psychoses in question are not psychoses of the acute exogenous reaction type of BONHOEFFER. They can resemble hysterical twilight states, the reactive psychoses of prisoners, the stuporous states occurring after a catastrophe, or any type of psychogenic psychosis. The emotional lability of persons with Cushing's syndrome or under cortisone treatment is a predisposition for the development of such reactive psychoses. As a rule, we are faced with acute psychoses that develop under psychological *and* somatic influences.

The direct emotional sequelae of hyperadrenocorticism may also serve as defenses against the strain of the illness. Apathy or euphoria may prevent despair over the ugliness, pains and impotence in Cushing's patients. Apathy or euphoria may be partly due to the direct influence of the hormone on brain and psychic function, partly also to the psychological defensive reaction against the torturing experience of the disease.

The question has been asked, *could Cushing's syndrome itself* (and not only its psychiatric complications) *be in some cases a symptom of a neurotic personality development*? As a matter of fact, in several cases that have been thorougly studied from the psychological point of view, the patients had suffered earlier from a good deal of misery and mental stress. The examiners were impressed by the temporal interdependence of the development of Cushing's syndrome and of the neurosis. They were inclined to consider Cushing's syndrome in these cases as a "psychosomatic" disease in the simple sense that psychological stress might have caused it (WEBER 1926, BAUER 1930, FOGGIE and MONTGOMERY 1934, KEHRER 1937,

Benedek 1951, Trethowan and Cobb 1952, Foucas 1954, Schwöbel 1954). Similar theories are accepted by many clinicians for Grave's disease. If psychogenesis of Grave's disease is seriously discussed, a possible psychogenesis of certain cases of Cushing's syndrome ought also to be studied. This has not been done in an adequate number of cases.

The older literature often mentioned that the psychopathology of Cushing's syndrome resembles *schizophrenic and manic-depressive* psychoses. But we must not overestimate such conclusions. We know that the mere psychopathological diagnosis of acute psychotic states is very difficult and often impossible. For example, general paresis can start with an acute psychotic episode that cannot be distinguished from confused acute schizophrenia or from mania or depression as an aspect of manic-depressive psychosis without neurological and serological examination. Symptomatological similarity of acute psychoses in conditions with excess of glucocorticoids and schizophrenia or manic-depressive psychosis alone proves neither the schizophrenic nor manic-depressive nature of these psychoses nor an endocrine origin of schizophrenia and manic-depressive psychoses.

There are important arguments for the assumption that the psychoses observed in Cushing's syndrome and during cortisone and ACTH treatment have nothing to do with a schizophrenic reaction or with a manic-depressive psychosis. Firstly, the course and outcome are quite different. The acute psychoses in question never lead to a chronic schizophrenic deterioration. Secondly, much work with the most modern methods has been done within the last 10 years in the hope of discovering the etiology of schizophrenia by studying the adrenocortical function — it has failed to demonstrate irregularity in adrenocortical function as an important specific etiologic factor of schizophrenia and manic-depressive psychosis (see M. Bleuler 1954).

Even if one does not believe in the schizophrenic or manic-depressive nature of the psychiatric complications of Cushing's syndrome and of cortisone and ACTH treatment, one can not deny that occasional borderline cases exist. These may represent simple casual coincidences of two quite independent morbid conditions. The question, however, needs to be studied more carefully.

V. Therapeutic conclusions

In Cushing's syndrome with psychiatric complications the somatic therapy is of the greatest importance not only for the somatic, but also for the psychic condition. An improvement of the psychiatric complications can be expected when surgery or irradiation succeeds in improving the morbid adrenocortical hyperfunction. In some cases, a complete somatic and psychic recovery is accomplished by somatic treatment alone. We have observed this, for instance, in a 5-year-old child, who was operated on for adrenal carcinoma (Schwöbel, 1954). In many cases, however, somatic treatment is not fully successful and recovery does not occur — only some improvement or a checking of further impairment. As a rule, the psychiatric symptoms are more resistant to somatic therapy than many somatic symptoms. After partially successful somatic treatment, many patients suffer much more from their psychiatric complications than from somatic derangements. These patients need psychiatric treatment and advice. There exists, of course, no specific psychiatric therapy for Cushing's patients. All psychiatric treatments have to be considered according to individual needs. In many cases, constant social advice and assistance are urgently needed. Many patients are helped very much by psychotherapy in which mere contact and friendship with the doctor are the outstanding features. Little experience exists in regard to real psychoanalysis with Cushing's patients. Some patients need treatments for complications such as alcoholism or drug addiction or suicidal attempts. In cases in which cerebral arteriosclerosis develops, the treatment is the same as in other cases of cerebral arteriosclerosis. The therapy of the acute psychotic episodes has not been studied sufficiently. It is very difficult to evaluate therapeutic measures, because many of these episodes are of short duration even without therapy. It seems important not to irritate the patients during their psychoses, but to apply the modern psychiatric hospital approach, which strictly avoids restraint and which stresses the over-all importance of the personal relationships of the patient with the physician, with nurses, and with other patients. Electroshock treatment ought not to be applied except in extreme emergency — the amnestic disturbances of Cushing's syndrome might be exacerbated by electroshock, and alteration of bone structure, high blood pressure, and heart complications are in many cases strict contraindications.

Therapeutic adrenalectomy in Cushing's disease has most frequently a good effect on the psychological symptoms. The personality of adrenalectomized patients, however, seems somewhat affected. Their alteration is but mild and has frequently a happy aspect. The patients are less moody than before, they seem to be more resistant to the worries and troubles of everyday life, they can develop a stable and wise attitude to life (FURGER 1961). It would almost seem as though the constancy of therapeutic hormone application (and the absence of ups and downs in adrenocortical function due to emotions) leads to stability and to flattening of emotional responses. Such a statement, however, is merely speculative. The dependency on drugs of adrenalectomized patients is a severe psychological problem for most of them; as a rule, they are able to deal successfully with it. For a few of them, however, it represents a long-standing worry.

Loss of sexual drive in adrenalectomized women has been treated with success with androgens.

An important question concerns psychiatric contraindications for cortisone or ACTH *treatment for organic diseases.* It is clear that the psychoses developing with these treatments are a serious argument against their application. We have to bear in mind, however, that serious psychiatric complications are rather rare and that even severe psychoses due to such treatment have a good prognosis. For this reason, the psychiatric danger of these treatments is not an absolute contraindication. If relief from otherwise incurable, severe somatic diseases can only be expected from treatment with cortisone or ACTH, this treatment should certainly be applied despite the danger of psychiatric complications. On the other hand, these psychiatric complications are serious enough to preclude cortisone and ACTH therapy without urgent indication.

Does emotional lability or a history of former psychiatric derangement prohibit cortisone or ACTH treatment, when somatically indicated? We believe not. Even emotionally labile patients and patients with psychotic episodes in the past do not as a rule develop psychiatric complications with treatment. The benefit of treatment in some severe somatic conditions should, therefore, not be denied these patients.

In evaluating the reasons for and against cortisone and ACTH therapy, it should be remembered that psychiatric complications are more frequent in severe febrile or cachectic conditions and also when cortisone is combined with some anti-tuberculosis drugs.

Severe psychiatric complications cannot be prevented with cautious dosage alone. They also occur with small doses.

The importance of close contact of the doctor and patient during cortisone or ACTH treatment must be emphasized. It serves a double purpose. (1) Firstly, it enables the physician to recognize early the danger of a psychiatric complication. If he concerns himself only with the patient's somatic aspects, the risk is great that he will notice the psychiatric affection only when the whole tragedy of a severe psychosis has developed. Psychosis is regularly preceded by milder emotional changes, which increase from day to day. If the cortisone or ACTH administration is stopped in this phase, psychosis may be prevented. The question of stopping treatment depends, of course, on the urgency of the somatic indications. (2) Close contact with the patient has not only diagnostic value, it can also be useful for prophylaxis and for therapy. There is no doubt that the emotional lability during cortisone or ACTH treatment is very much influenced by the personal relationship between doctor and patient. The patient must have confidence in his doctor, he must not be irritated by measures he neither understands nor approves, his worries must be discussed with him, he must feel constantly supported and helped by the doctor's sympathy. If the relationship with the doctor is good, irritable moods, suspicion, ideas of persecution, as well as depression and fear can be prevented or moderated, and euphoria with its best aspects is free to develop. It has not been established, however, that psychoses can be prevented by psychotherapy. But when one has watched how many psychotic states develop during treatment in connection with misunderstandings of the therapeutic situation and irritations during contact with doctors and nurses, one is inclined to hope that good psychological understanding of the patient during hormonal treatment might also have its usefulness in preventing or moderating cortisone or ACTH psychoses.

Psychoses from treatment with cortisone or ACTH require very careful nursing, just as any acute severe psychosis. Suicidal attempts or self-mutilation must be prevented. Difficult problems of feeding and of general hygiene may arise which have to be handled as in other psychoses. The importance of personal contact has already been mentioned. Skillful occupational therapy is also valuable in acute organic psychoses. Most of the psychoses in question do not last long, even without special therapy. Psychoses of longer duration than a few days have frequently been treated by electroshock. As a rule, this treatment had excellent success. In spite of such favorable experience, we should not like to recommend electroshock without hesitation: it always seems somewhat dangerous to use electroshock in a condition that arises from organic damage. When the brain has already been damaged, one hesitates on theoretical grounds to add new damage. A better treatment, however, is still unknown. Treatment with growth hormone has been recommended (Nordström and Paulsen 1954), but not enough is known as yet to make an evaluation possible. Treatment with Chlorpromazine or Reserpine has not had decisive effects in our experience. The appropriate therapy has still to be discovered.

It has been proposed that psychotherapy be used not only for the psychiatric complications of Cushing's syndrome, but for the Cushing's syndrome itself, just as Graves' disease can be treated by psychotherapy. This problem must be studied in the future.

Some have tried to use *cortisone or* ACTH *for the therapy of psychiatric derangements.* As long as it was believed that these hormones regularly produce euphoria, it was logical to try to make use of this action. Since then we have learned that euphoria during cortisone or ACTH therapy does not occur regularly, and that, furthermore, many unpleasant mood swings often occur during such treatment. It is, therefore, not surprising that these hormones have proved of no value in the treatment of depressive conditions. It is true that at times depressive conditions of most varied nature clear up during treatment with these hormones. But this is only observed in some rather exceptional cases; in the majority of cases these hormones do not improve the depressive state essentially and in some cases they even impair it. Even in cases that have improved under treatment, improvement did not last. As such hormonal treatment is not without danger, it cannot be recommended even for trial in the treatment of depressions.

Cortisone or ACTH have also been tried in the treatment of schizophrenic psychoses, in some instances with success (Fang et al. 1952, Chance et al. 1954, Constantinides and Loukatos 1955, Kabanov 1957, and our own experience). These successes, however, are rare exceptions. The experience of most clinicians who have tried such treatment is discouraging; the large majority of schizophrenics show no improvement.

The only indication for cortisone in psychiatric therapy lies in conditions involving refusal of food, particularly in some cases of anorexia nervosa, which are not accessible to a more causal psychotherapy. Appetite is increased by cortisone in many of these patients; in certain phases of the disease this can be of great help, although cortisone application is mere symptomatic therapy.

B. Excess of adrenocortical androgens

I. Virilizing congenital adrenocortical hyperplasia in females (with pseudohermaphroditism and pseudopubertas precox)

Up to a few years ago, the psychopathology of patients with a virilizing adrenogenital syndrome was mainly judged on the basis of hypothetical expectations and a few observations of exceptional cases corresponding to these expectations. It was assumed (without sufficient empirical foundation) that women with masculine signs were homosexual and that physically premature girls were also premature in their intellectual, personal and psychosexual development. It was also assumed that pseudohermaphroditism was frequently combined with feeblemindedness or with psychosis. All these assumptions have proved to be wrong.

Today our experience regarding the personality of females with congenital adrenogenital syndrome is extensive. Good psychological descriptions of unselected cases are given by SECKEL et al. (1949, SECKEL 1950), by ZÜBLIN (1953), by MARZI and TEODORI (1953), by STUTTE (1954), and by MONEY and co-workers (MONEY 1955, MONEY and HAMPSON 1955, MONEY et al. 1955a, b, 1956, HAMPSON and MONEY 1955, HAMPSON et al. 1956). Furthermore, many descriptions of single cases have been published, and many papers dealing with the physical aspects of the condition contain brief psychological observations (summarized by ELLIS 1945). The following description is drawn from personal observations of patients from our Clinic and of the Pediatric Clinic in Zürich and has been checked and found to accord with an extensive literature.

The large majority of girls with a virilizing adrenogenital syndrome do not show accelerated personality development; on the contrary, their emotional development is rather retarded. They are frequently shy, seeking protection, and more dependent on parents or other older people than other children of their age. As a rule they have feelings of inferiority resulting from their physical abnormality. They try to hide it, and to behave and to be like normal girls (when they have been educated as girls).

Their intellectual development corresponds to their age. Feeble-mindedness is hardly more common than among the average population. Partial acceleration of intellectual development may occur, but this is not the rule.

Their psychosexual development is frequently retarded. Psychosexuality in many cases never reaches a mature level, but remains feeble and immature. Homosexuality in girls with congenital adrenocortical hyperplasia brought up as girls is a rare occurrence. However, when the patients are educated as boys, which is sometimes the case when the external genitals are mainly of male form, sex feelings tend rather to be of male nature and to be directed toward girls.

Their general interests are, as a rule, in accordance with the socially adopted sex and with the age. Exceptions do occur, but relatively seldom.

In some cases a tendency to mood swings and to abrupt changes in appetite and in the need for motion or rest are apparent.

Severe psychic abnormalities, particularly psychoses, are rare. (The adrenogenital syndrome is rarely met with in psychiatric hospitals, while hirsutism without adrenal hyperplasia is more frequent in psychotic than in healthy women.)

The personality of girls with adrenogenital syndrome arising from congenital adrenocortical hyperplasia as just outlined is not different from the personality of girls with premature puberty or early masculinization with other etiology.

There do exist exceptions to the usual personality type in females with adrenogenital syndrome as just described. An acceleration of intellectual development has been observed in rare cases. The literature also describes a few cases of homosexuality.

During successful treatment with cortisone, the psychological development can be accelerated and the emotional peculiarities improved (personal communication from PRADER, Zürich). However, the outbreak of a psychosis during cortisone treatment of adrenogenital syndrome has also been described (GOLDBERG 1954); this is not astonishing, since cortisone therapy for the most varied indications can be complicated by psychoses in rare cases (p. 640 ff.).

The *therapy* in patients brought up as girls consists in continued cortisone administration. (This is firstly a substitution therapy, since cortisone is not secreted in a normal way in adrenocortical hyperplasia; secondly, cortisone prevents the increased secretion of ACTH and therefore also of adrenocortical androgens. See Chapters 5 and 9). Frequently the extirpation of the enlarged clitoris is indicated for psychological reasons. [Clitoridectomy does not render orgasm impossible (HAMPSON 1955).] Therapy must be started as early as possible.

In the rare cases, however, in which the patients have been educated as boys and want to keep the socially adopted sex, forced changes in the social sex role and attempts to feminize the body are not always indicated. The forced change in the socially adopted sex role after early infancy has led to very serious psychological damage.

The most impressive conclusion drawn from our present knowledge of the personality of females with adrenogenital syndrome is as follows: this severe endocrine disturbance has no marked significance on psychic development.

Even the mild psychic peculiarities that are the rule in females with the adrenogenital syndrome can be explained to a large extend as mere psychological sequelae of their abnormal physical aspects: it is easy to understand that girls who appear ugly and perverse are shy and retiring with strangers. It is also a common experience that emotional stress of any kind retards psychosexual development.

It would be wrong, however, to claim that the experience obtained with patients with the adrenogenital syndrome proves the complete independence of the psychological (and particularly the psychosexual) development from endocrine function. As a matter of fact, this experience alone cannot settle the question as to whether the tendency to immaturity in the adrenogenital syndrome should be explained only as a psychological reaction to the ugly and abnormal bodily condition, or if it is partly due to more direct action of the abnormal endocrine secretion on psychic development. Other experience in endocrine psychiatry conclusively indicates direct endocrine influences on psychic function. (Even endocrine abnormalities not affecting conspicuously the appearance and the social acceptance of a patient are often combined with a retarded development of emotionality and of psychosexuality.)

II. Virilizing congenital adrenocortical hyperplasia in males (with pseudopubertas precox)

Since a virilizing adrenogenital syndrome in males is much less frequently diagnosed than in females, our experience with the personality of boys with this disease is very meagre. Only a few descriptions in the literature mention psychological observations. Our own experience (3 cases described by ZÜBLIN 1953) is also very limited.

The few boys with adrenogenital syndrome whose personalities have been studied show traits similar to the girls with the same condition. In spite of the physical pseudopubertas precox, intellectual development corresponds to their age, and emotional and psychosexual development is rather retarded. The physical strength and the premature physical masculinity form a strange contrast to the shy and rather childish mental attitude. Shyness and a tendency to retire have even been observed in adult patients.

Severe psychic aberrations have not been observed.

III. Adrenogenital syndrome from adrenocortical tumor or acquired adrenocortical hyperplasia

Systematic psychological studies of unselected cases of acquired adrenogenital syndrome in women with adrenocortical tumors have not been published as yet. The majority of patients described in the literature showed a marked personality change. It is by no means uniform in kind or degree; on the contrary, the most varied changes have been observed. Shyness, loss of self-confidence, and depression are frequently mentioned. Sexuality can disappear or be exaggerated. When it is exaggerated, this may be in the heterosexual or homosexual direction. Affective and paranoid psychoses have been attributed to the tumor leading to an adrenocortical syndrome (ALLEN and BROSTER 1945).

These observations of severe personality changes in the course of the masculinization due to an adrenocortical tumor should not be generalized. Selected cases have been chosen for publication; since mental changes seemed to be interesting they are frequently mentioned in the literature, while it was not found necessary to mention the psychology of patients who were not altered mentally by an active tumor. It is probable that severe personality changes are exceptions and that somatic masculinization due to an adrenocortical tumor is not followed by dramatic psychological changes in a great number of cases. However, the question needs to be studied carefully in a large number of unselected cases.

In very rare cases, adrenocortical tumors cause spontaneous hypoglycemia with the same severe psychopathological consequences as other forms of hypoglycemia (lethargy, agitation, delirious and incoherent episodes, coma) (SCHAMAUN et al. 1957).

IV. Discussion

As a whole, the psychic alterations of patients with the adrenogenital syndrome are less frequent and less severe than those of patients with Cushing's syndrome. The influence of hypersecretion of adrenocortical androgens on psychic function is less marked than the influence of hypersecretion of glucocorticoids. This fact is in accordance with the experience in hormonal therapy: therapy with glucocorticoids frequently leads to changes of mood and instincts and in some cases even to psychoses; therapy with male sex hormones, on the contrary, is never complicated by psychoses and leads less regularly to mood changes.

The observation that psychosexuality is not increased in the majority of patients with the adrenogenital syndrome (particularly young girls) is in opposition to the old speculative assumption that a simple and regular correlation exists between psychosexuality and sex hormones. It is one of the principal results of recent studies in the whole field of endocrinological psychiatry that we know today of no absolute and unconditional influence of sex hormones on personality as a whole and psychosexuality in particular. Any hormonal influence on personality and especially the influences of sex hormones on psychosexuality are shaped, and exaggerated, diminished or extinguished by the personality. The inundation of the otherwise normal child with sex hormones either of adrenocortical, of gonadal or of pharmaceutic origin only occasionally accelerates the development of the personality and of psychosexuality, and even more frequently may retard it. (On the other hand, in many cases of infantilism, sex hormones accelerate psychic maturation together with somatic maturation.)

There has been one reason to suspect that the adrenocortical androgens play a role in normal female psychosexuality: after hypophysectomy of women with cancer of the breast, in the majority of patients sexual desire and sexual satisfaction disappear or are diminished. This is not due to hypothyroidism, nor to lack of cortisone, since substitution therapy with thyroid gland or cortisone does not restore psychosexuality. The action of hypophysectomy on psychosexuality of women seems to be more regular than the effect of castration. It might, therefore, rather be attributable to the lack of ACTH and the resultant lack of adrenocortical androgens than to the lack of gonadotrophins (SCHON and SUTHERLAND 1960). Female sexuality, however, is *not* in a simple and regular correlation with the production of androgens by the adrenal cortex. There is, as has been mentioned, conclusive evidence against such an assumption.

C. Excess of estrogens

Only few cases of feminizing adrenal tumors are known. Their psychopathology has not been sufficiently studied. In adult males with feminizing adrenal carcinoma, loss or diminution of psychosexuality seems to be the rule (WILKINS 1948, MYHRE 1952). A boy, 5 years old, with marked gynecomastia and an adenoma of the adrenal cortex, behaved "like a normal boy and showed no feminine traits of the personality" (WILKINS 1948).

D. Excess of mineralocorticoids

The psychopathology of patients with CONN's *syndrome* (see Chapter 9) *is unknown as yet.* No systematic psychiatric studies on the few cases diagnosed up to now have been published. Severe psychopathology has not been mentioned. Milder and uncharacteristic psychic troubles, like irritability and nervousness, are briefly mentioned in some cases.

Deoxycorticosterone (cortexone) has no such marked influence on psychic function as cortisone. No psychoses following therapy with deoxycorticosterone (DOC) are known. On the other hand, mood changes in patients treated with DOC have frequently been observed. Patients with adrenocortical hypofunction react with euphoria and mild agitation to DOC when their somatic conditions improves. In physically normal people, the influence of DOC on mood and activity is mild and irregular, or entirely lacking.

Large doses of DOC, like several other steroids, have in animals an anesthetic and hypnotic action (SELYE 1941). This is probably due to their depressing effect on cerebral metabolism (GORDAN and ELLIOTT 1947). The anesthetic and hypnotic action of such steroids seems to be independent of their physiological action as hormones, since there have been steroids synthesized which have no hormonal, but marked anesthetic action. One of these (known as Viadril) is now frequently used for general anesthesia in the human. Another one (Cetadiol) has been recommended as a tranquilizing agent (LEMERE 1957).

Deoxycortisol (Reichstein's Compound S) has, as far as we know, little or only minor psychiatric significance.

Psychopathology of hypoadrenocorticism

By

W. A. STOLL

A. Addison's disease

These remarks are based in the main on the examination of 45 Addisonian patients at the Psychiatric Clinic of the University of Zürich (CHRIST 1952, A. MEYER 1952, A. E. MEYER 1954, STOLL 1953, 1958). A previous review (STOLL 1953) gives an exhaustive bibliography, so that only more recent studies are cited here.

Psychic alterations are extremely frequent in cases of Addison's disease. It may be said that it is an exception for the patient's psyche to remain unchanged as compared with the period preceding disease. This held true for cases that were treated before cortisone and other new hormones had been introduced. Even when the somatic condition seemed quite satisfactory, the patients might be abnormal psychologically. Cortisone now often seems to normalize the psychic condition. For the moment this is to be assumed from isolated observations, since exhaustive tests have as yet scarcely been reported (CLEGHORN 1953, 1954).

In Addison's disease, 3 main types of mental illness (see E. BLEULER 1960) are discernable:

(1) *the acute exogenous reaction type* (acute organic type) of BONHOEFFER (1912),

(2) *the endocrine psychosyndrome* (localized brain psychosyndrome) of M. BLEULER (1954)

(3) *the amnestic psychosyndrome* (diffuse brain psychosyndrome) of E. BLEULER (1960).

(1) The *acute exogenous reaction type* of psychopathology consists of more or less severe acute psychoses with varyingly pronounced alterations of consciousness, delirious and dreamy states, often with delusions and hallucinations in great variety. These disturbances last for hours or days. They are observed in numerous cases of severe physical damage, as well as in the most varied endocrine diseases. In the preceding section, these disturbances were described for Cushing's syndrome (p. 639). They are also observed in Addison's disease in acute and terminal failure of the adrenal cortex, that is to say, in the Addisonian crisis. ADDISON himself (1855) reported such observations.

(2) The *endocrine psychosyndrome* also comprises a nonspecific group of psychological alterations, to be seen in most endocrine diseases. (It is also observed in nonendocrine disturbances, particularly when there is localized brain damage.) The changes in mood, mental activity, and instincts described earlier for Cushing's syndrome (p. 639) are also characteristic for Addison's disease, particularly in its *chronic* phases. Despite the differences of the physiological alterations in Cushing's syndrome and Addison's disease, the psychosyndrome is on the whole very similar in both (pp. 643—644). Common to both is the fact that mood, mental activity, and instincts are regularly altered as compared with the premorbid condition. Within these limits there are numerous variations, which, however, are determined not by the disease, but in particular by the personality of the patient and its development.

The *mood* of the chronic Addisonian was described in former works as always depressive and apathetic. More recent reports contain similar observations (CLEGHORN and PATTEE 1954, DRAKE 1957). Our own experience has shown that in about a quarter of the cases the mood is depressive, and in another quarter apathetic, but that the remaining 50% exhibit more or less marked euphoric trends. Characteristic is the alteration in mood in itself, not its direction. Depression, apathy, or euphoria are usually complicated by anxiety, irritability, tension, indifference, etc., in many varieties. The fundamental mood is often replaced by mood swings lasting for hours or days. These occur suddenly, and their direction and type cannot be predicted.

The *mental activity* of a patient with Addison's disease is generally reduced. Interest in daily duties and occurrences is low, although at the same time the patient may exhibit euphoric swings. However, such more or less marked inactivity regarding the outer world does not necessarily signify that the patient is also inwardly inactive or without interests. Many patients read more, develop their religious life, or enrich their inner life in some other way.

Instincts show alterations in about three-quarters of the cases observed, and again in the most varied manner. Patients either sleep more than before, or they sleep less. They have either no appetite or a craving for food; they often have particular longings for salty or acid or sweet foods. Thirst may be increased. Sexuality is usually, though not always, reduced. The need for warmth is generally increased. Maternal feelings are often reduced. Beside the need for rest, we often observe restlessness or even a desire to wander.

Like Cushing's patients with mild psychopathological changes (p. 639), Addisonians are not psychotic individuals because of their endocrine psychosyndrome. They are simply odd people. There are no regular connections between the degree and development of the endocrine psychosyndrome with the severity of the somatic disease.

(3) On the other hand, a parallel does exist between mental and somatic disturbances in the case of the third psychosyndrome that occurs in Addison's disease, namely the *amnestic psychosyndrome*. Nearly all patients suffering from severe hypoadrenocorticism manifest disturbances in memory. Organic forgetfulness may even be a particularly prominent symptom and influence the patient's efficiency, long before the somatic diagnosis is made. A banker's clerk was an impressive example in this respect. The amnestic psychosyndrome is subjectively spoken of as soon as the patient is questioned about it. It may also be observed in the usual tests. It would seem that marked insufficiency of the adrenal glands causes diffuse brain damage, such as is evident in senile dementia. Like other patients suffering from severe endocrine disease, Addisonians sometimes exhibit premature senility. As soon as the physical condition has somewhat improved, the amnestic disturbances usually regress.

The endocrine psychosyndrome, the amnestic psychosyndrome and the acute exogenous reactions are often combined. In rare cases they may even become aggravated to form *severe chronic psychoses* with hallucinations and delusions (QUANDT and ROSENTHAL 1953, DE GENNES et al. 1954, MEDVECKY 1957). In general such psychoses may easily be distinguished from schizophrenias, for instance when amnestic disturbances are demonstrable. It has never been observed as yet that an Addisonian psychosis terminated in schizophrenic deterioration. Psychoses (mostly of short duration) are also observed at times during *therapy with cortisone* (BRACELAND 1954, CLEGHORN and PATTEE 1954, LAROCHE et al. 1956). So far as we know, peculiar differences have not been observed as compared with other cortisone psychoses (p. 640).

As in other severe endocrine diseases, *convulsive seizures* are sometimes observed in Addison's disease (STORRIE 1955). Even without seizures, the electroencephalogram may be altered in an epileptic direction (BRICAIRE et al. 1953). For the rest, it shows principally a slowing up of the basal rhythm (CLOCHE et al. 1952).

Reports on Addisonians in whom a psychogenesis of the disease must be considered are of especial interest (KLEINSORGE 1954, WALLERSTEIN et al. 1954). Particular importance is given to chronic emotional stress. An early case history by MARAÑON (1929) is especially striking. An extremely neurasthenic young man witnessed how a friend was taken ill with Addison's disease, whereupon he believed himself to be suffering from the same disease, although no organic findings were to be observed, at any rate with the methods possible at the time. After living for 12 years, he died after acute illness as the result of a proved Addisonian crisis.

When we consider such psychosomatic connections, we see clearly that a *psychotherapeutic* attitude also applies to the treatment of Addison's disease. Everything mentioned above (p. 647) regarding psychotherapy in Cushing's syndrome also applies here.

B. Mild hypoadrenocorticism

As might be expected, in slight hypoadrenocorticism (Addisonism), certain psychic disturbances may be found that characterize the endocrine psychosyndrome, accompanied by slight amnestic disturbances. Patients exhibit irritability, apathy, depression, mood swings, some forgetfulness. It is to be assumed that there are also patients with subclinical Addisonism who exhibit a euphoric basal mood. They do not of course consult a physician. They have no psychic reasons for doing so, and they do not suffer from severe somatic disturbances. This is probably why predominantly dysphoric and depressive pictures are reported; in other words we encounter the same type of complaint usually heard from neurasthenics.

May neurasthenia be regarded without further question as hypoadrenocorticism? May the asthenic syndrome represent Selye's "adaptation disease" in the stage of exhaustion? The 17-ketosteroid excretion of neurasthenics has been examined from this point of view (Birket-Smith 1955, Albeaux-Fernet et al. 1957); Høyrup (1956) performed certain other laboratory tests as well. It may be concluded that a large number of neurasthenics do indeed exhibit a somewhat lowered adrenocortical function. However, in half of his cases Høyrup found no indication of hypoadrenocorticism.

If we consider the psychogenesis of Addison's disease as possible, we are all the more likely to assume this genesis in *slight* deficiencies of the adrenal cortex. The demonstrability of such disturbances thus in no way proves the primarily organic nature of such neurasthenia. But it should also never be excluded. It is just in the case of moderate organic findings that the close and alternating connection between the psyche and the whole endocrine system should be borne in mind, and psychic and hormonal treatment combined.

Literature

Addison, T.: On the Constitutional and local effects of disease of the suprarenal capsules. London: Highley 1855.

Albeaux-Fernet, M., P. Bugard and J. D. Romani: Excretion of urinary corticoids in conditions of chronic asthenia. J. clin. Endocr. **17**, 519—533 (1957).

Allen, C., and L. R. Broster: A further case of paranoid psychosis successfully treated by adrenalectomy. Brit. med. J. **1945**i, 696.

Alman, R. W., and J. F. Fazekas: Effects of ACTH on cerebral blood flow and oxygen consumption. Arch. Neurol. Psychiat. (Chicago) **65**, 680—682 (1951).

Arndt, T., E. F. Pfeiffer u. K. Schöffling: Über die Beeinflußbarkeit des menschlichen Hirnstrombildes durch das adrenocorticotrope Hormon (ACTH) des Hypophysenvorderlappens. Arch. Psychiat. Nervenkr. **191**, 493—507 (1954).

Basowitz, H., H. Persky, S. J. Korchin and R. R. Grinker: Anxiety and Stress. An interdisciplinary study of a life situation. New York-Toronto-London: McGraw-Hill, blakison Div. 1955.

Bauer, J.: Überfunktion des gesamten Nebennierensystems ohne anatomischen Befund. Wien klin. Wschr. **43**, 582—586 (1930).

Benedek, T.: Die Funktionen des Sexualapparates und ihre Störungen. In: F. Alexander, Herausgeb., Psychosomatische Medizin. Berlin: de Gruyter 1951.

Birket-Smith, E.: Excretion of neutral 17-ketosteroids in normal and neurasthenic subjects with special reference to its diurnal variations. Acta psychiat. (Kbh.) **29**, 423—440 (1955).

Bleuler, E.: Lehrbuch der Psychiatrie, 10. Aufl. umgearbeitet von M. Bleuler, 177—182. Heidelberg: Springer 1960.

Bleuler, M.: Die Ähnlichkeit zwischen den endokrinen und hirnlokalen Psychosyndromen und ihre theoretische Bedeutung. Folia psychiat. neerl. **56**, 593—601 (1953).

— Endokrinologische Psychiatrie. Stuttgart: Thieme 1954.

— Endocrinological psychiatry (in Spanish). Buenos Aires: Finchelman 1956.

Blickenstorfer, E.: Mutterinstinkte bei einem Manne mit krankhafter Bildung von laktotropem Hypophysenhormon. Arch. Psychiat. Nervenkr. **182**, 536—542 (1949).

— Neugestaltung kranken Daseins unter dem Einfluß der Mutterschaftshormone. Psyche (Stuttgart) **6**, 321—335 (1952).

BLICKENSTORFER, E., P. ISLER, M. MARTI u. C. HEDINGER: Vermütterlichung eines erwachsenen Mannes mit Chorionepitheliom. Dtsch. Arch. klin. Med. **199**, 462—480 (1952).
BLISS, E. L., C. J. MIGEON, C. H. HARDIN BRANCH and L. T. SAMUELS: Reaction of the adrenal cortex to emotional stress. Psychosomat. Med. (N. Y.) **18**, 56—76 (1956).
BONHOEFFER, K.: Die Psychosen im Gefolge von akuten Infectionen, Allgemeinerkrankungen und inneren Erkrankungen. In: G. ASCHAFFENBURG, Herausgeb., Handbuch der Psychiatrie, Spez. Teil, 3. Abtg., 2—334. Leipzig: Deuticke 1912.
BRACELAND, F. J., Hormones and their influence in the emotions. In: R. L. CRAIG, Ed., Hormones in health and disease, 331—343. New York: Macmillan 1954.
BRICAIRE, H., R. MISES, C. DREYFUS-BRISAC et H. FISCHGOLD: Initiation à l'électro-encéphalographie. E. E. G. d'un cas de maladie d'Addison. Presse méd. **61**, 750—751 (1953) [abstract: Excerpta med. (Amst.), Sect. VIII, **7**, 363 (1954).]
BRODY, S.: Psychological factors associated with disseminated lupus erythematosus and effects of cortisone and ACTH. Psychiat. Quart. **30**, 44—60 (1956).
CHANCE, J., E. LOTSOF, I. PINE, R. M. PATTERSON and J. CRAIG: Effects of cortisone on psychiatric patients. Psychosom. Med. **16**, 516—525 (1954).
CHRIST, J.: Zur Psychopathologie des Morbus Addison. Medical Dissertation, Zürich 1952.
CLARK, E. D., and A. A. BAILEY: Neurological and psychiatric signs associated with systematic lupus erythematosus. J. Amer. med. Ass. **160**, 455—457 (1956).
CLEGHORN, R. A.: Psychologic changes in Addison's disease. J. clin. Endocr. **13**, 1291—1293 (1953).
— The interplay between endocrine and psychological dysfunction. In: E. D. WITTKOWER and R. A. CLEGHORN, Eds., Recent developments in psychosomatic medicine, 157—200. London: Pitman 1954.
— and C. J. PATTEE: Psychological changes in 3 cases of Addison's disease during treatment with cortisone. J. clin. Endocr. **14**, 344—352 (1954).
CLOCHE, R., E. AZERAD et STUHL: Modifications des traces électro-encéphalographiques altérées chez les Addisoniens en cours de traitement. Bull. Soc. Méd. Paris **68**, 486—491 (1952).
CONSTANTINIDES, et G. LOUKATOS: Résultat de la surrénalectomie partielle et de l'application de cortisone sur cas de virilisme avec schizophrénie. Rev. Hosp. N. S. Aparecida (S. Paulo) 8, 219—222 (1955).
COPE, C. L.: Cortisone and A.C.T.H. Brit. med. J. **1953 ii**, 332—334.
DELAY, J., L. BERTAGNA et A. LAURAS: ACTH, cortisone et psychisme. Presse méd. **63**, 1037—1039 (1954a).
— — — ACTH et cortisone en psychiatrie. Ann. méd.-psychol. **112**, 536—540 (1954b).
DENCKER, S. J., R. SCHLAUG and W. SILFVERSKIÖLD: Psychosen als Komplikationen bei ACTH- und Cortisonbehandlung. Ein Bericht von 4 Fällen. Nervenarzt **25**, 273—276 (1954).
DRAKE, F. R.: Neuropsychiatric-like symptomatology of Addison's disease: A review. Amer. J. med. Sci. **234**, 106—113 (1957).
EICHNER, D.: Über den morphologischen Ausdruck funktioneller Beziehungen zwischen Nebennierenrinde und neurosekretorischem Zwischenhirnsystem der Ratte. Z. Zellforsch. **38**, 488—508 (1953).
ELLIS, A.: The sexual psychology of human hermaphrodites. Psychosom. Med. **7**, 108—125 (1945).
Encyclopédie Médico-chirugicale: Psychiatrie. Vol. I, 37240 A 10, 8; 37240 C 10; 37280 A 10, 7; 37295 E 10, 5; Vol. II, 37640 A 10; 37640 G 10, 3; Vol. III, 37860 I 20.
ENGELHARDT, FR.: Über die Wirkung von Gonadotropinen nach gezielter intracerebraler Instillation bei der Ratte. In: H. NOWAKOWSKI: Die partielle Hypophysenvorderlappen-Insuffizienz. Berlin: Springer. 1957.
FANG, A. D., M. E. MARTIN and S. KATZENELBOGEN: Research in cortisone therapy. Quart. Rev. Psychiat. **7**, 169—173 (1952).
FOGGIE, W. E., and G. L. MONTGOMERY: The adreno-genital syndrome. Edinb. med. J. **41**, 29—42 (1934).
FOUCAS, M.: Sur un cas de syndrome de Cushing d'origine psychogène. Gynaecologia (Basel) **138**, 604—610 (1954).
FRANKL VON HOCHWART, L.: Über den Einfluß der inneren Sekretion auf die Psyche. Med. Klin. 8, 1953—1958 (1912).
FRANKSSON, C., and C. A. GEMZELL: Adrenocortical activity in the preoperative period. J. clin. Endocr. **15**, 1069—1072 (1955).
— — and U. S. VON EULER: Cortical and medullary activity in surgical and allied conditions. J. clin. Endocr. **14**, 608 (1954).
FURGER R.: Psychiatrische Untersuchungen beim Cushing-Syndrom. Schweiz. Arch. Neurol. Psychiat. **88**, 9—39 (1961).
GENNES, L., DE, P. BENSOUSSAN et F. KLEIN: Maladie d'Addison: Troubles mentaux aigus et délire chronique. Ann. méd.-psychol. **112**, 522—527 (1954).

Geppert, L. J., A. D. Dietrich, E. H. Johnson and B. Lind: Fatal convulsive seizures associated with cortisone therapy. Amer. J. Dis. Child. **84**, 416—420 (1952).

Glaser, G. H.: The pituitary gland in relation to cerebral metabolism and metabolic disorders of the nervous system. Res. Publ. Ass. nerv. ment. Dis. **32**, 21—39 (1953).

Goldberg, M. B.: Experience with long-term cortisone therapy in congenital adrenocortical hyperplasia. J. clin. Endocr. **14**, 389—408 (1954).

Gordan, G. S., and H. W. Elliott: The action of diethylstilbestrol and some steroids on the respiration of rat brain homogenates. Endocrinology **41**, 517—518 (1947).

Hampson, J. G.: Hermaphroditic genital appearance, rearing and eroticism in hyperadrenocorticism. Johns Hopk. Hosp. Bull. **96**, 265—273 (1955).

— and J. Money: Idiopathic sexual precocity in the female. Psychosom. Med. **17**, 16—35 (1955).

— — and J. L. Hampson: Hermaphroditism: Recommendations concerning case management. J. clin. Endocr. **16**, 547—556 (1956).

Hazard, J.: Accidents de la corticothérapie surrénale. Bull. méd. (Paris) (1954) [abstract: Ann. méd.-psychol. **112**, 824 (1954)].

Hess, R.: Das Elektroencephalogramm bei endokrinen Störungen. In: M. Bleuler Endokrinologische Psychiatrie, 362—371. Stuttgart: Thieme 1954.

Hetzel, B. S., W. W. Schottstaedt, W. J. Grace and H. G. Wolff: Changes in urinary 17-hydroxycorticosteroid excretion during stressful life experiences in man. J. clin. Endocr. **15**, 1057—1068 (1955).

Hoagland, H.: Stress and the adrenal cortex with special reference to K metabolism. Psychosom. Med. **12**, 142—148 (1950).

Høyrup, E.: Impaired adrenal function as a contributory factor in constitutional asthenia. Acta psychiat (Kbh.) Suppl. **108**, 185—196 (1956).

Howard, J. M., J. M. Olney, J. P. Frawley, R. E. Peterson, L. H. Smith, J. H. Davis, S. Guerra and W. H. Dibrell: Studies of adrenal function in combat and wounded soldiers (a study in the Korean theatre). Ann. Surg. **141**, 3 (1955); An. Cir. (B. Aires) **32**, 3, 314—320 (1955).

Kabanov, M. M.: L'expérience de l'application de l'adréno-corticotrophine (ACTH) dans certaines affections mentales. Z. Neuropat. Psikiat. **57**, 1009—1014 (1957).

Kehrer, E.: Endocrinologie für den Frauenarzt. Stuttgart: Enke 1937.

Kleinsorge, H.: Funktioneller Morbus Addison nach psychischem Trauma. Z. Psychother. med. Psychol. **4**, 18—21 (1954).

Laroche, C., A. Buge, J. Lerat et Y. le Quintrec: Troubles psychiques chez une addisonienne traitée par la cortisone. Ann. Endocr. (Paris) **17**, 720—724 (1956).

Lemere, F.: New steroid hormone tranquilizing agent (Cetadiol). Amer. J. Psychiat. **113**, 930 (1957).

Lewis, A., and J. J. Fleminger: The psychiatric risk from corticotrophin and cortisone. Lancet **1954 i**, 383—386.

Malandra, B., and S. Corbetta: The Gomori-positive material in the neurohypophysis of the rat after adrenalectomy and treatment with adrenocortical hormones and salt. Z. Zellforsch. **39**, 318—327 (1953).

Marañon, G.: On the etiology of Addison's disease (in Spanish). Siglo med. 8, 30 (1929) [quoted by Wallerstein et al. (1954)].

Marzi, A., and U. Teodori: Preliminary studies on the personality in disturbances of puberal development (in Italian). Folia endocr. (Pisa) **6**, 279—314 (1953).

McGehee, E. H., and K. S. MacLean: Experience with A.C.T.H. and cortisone. A note on long-term therapy. Brit. med. J. **1954 i**, 1171—1175.

Medvecky, J.: Psychotic state in Addison's disease (in Czech). Bratislavské lékárs. Listy **37**, 557—559 (1957) [abstract: Excerpta med. (Amst.), Sect. VIII, **11**, 500 (1958)].

Meyer, A.: Zur Psychiatrie des Morbus Addison. Schweiz. Arch. Neurol. Psychiat. **70**, 58—72 (1952).

Meyer, A. E.: Psychische Störungen bei Addison'scher Krankheit. Medical dissertation, Zürich 1954.

Money, J.: Hermaphroditism, gender and precocity in hyperadrenocorticism: Psychologic findings. Johns Hopk. Hosp. Bull. **96**, 253—264 (1955).

— and J. G. Hampson: Idiopathic sexual precocity in the male. Psychosom. Med. **17**, 1—15 (1955).

— — and J. L. Hampson: Hermaphroditism: Recommendations concerning assignment of sex, change of sex, and psychologic management. Johns Hopk. Hosp. Bull. **97**, 284—300 (1955a).

— — — An examination of some basic sexual concepts: The evidence of human hermaphroditism. Johns Hopk. Hosp. Bull. **97**, 301—319 (1955b).

— — — Sexual incongruities and psychopathology: The evidence of human hermaphroditism. Johns Hopk. Hosp. Bull. **98**, 43—57 (1956).

Myhre, J.: Feminizing adreno-cortical carcinoma and carcinoma of the prostata. Acta endocr. (Kbh.) **10**, 233—242 (1952).

NORDSTRÖM, S., and F. PAULSEN: Treatment of ACTH-psychosis with growth hormone (in Swedish). Svenska Läk. Tidn. **51**, 1468—1470 (1954).
OLDS, J.: Positive emotional systems studied by techniques of self-stimulation. In: L. J. WEST and M. GREENBLATT, Eds., Explorations in the physiology of emotions, Psychiatric Res. Rep. Amer. Psychiat. Ass., **12**, 1960.
PERKOFF, G. T., B. V. JAGER and F. H. TYLER: Complications in the management of Cushing's syndrome, including anaphylactic reactions to intravenous adrenocorticotropin after subtotal adrenalectomy. J. clin. Endocr. **15**, 362—372 (1955).
PERSKY, H., J. MAROC, E. CONRAD, A. DEN BREEIJEN: Blood corticotropin and adrenal weight-maintenance factor levels of anxious patients and normal subjects. Psychosom Med. **21**, 379 (1959).
QUANDT, V. V., and S. ROSENTHAL: Addison's disease with protracted unusual mental changes. Wis. med. J. **52**, 225—228 (1953) [abstract: Excerpta med. (Amst.) Sect. VIII, **7**, 297 (1954)].
QUARTON, G. C., L. D. CLARK, S. COBB and W. BAUER: Mental disturbances associated with ACTH and cortisone: A review of explanatory hypotheses. Medicine (Baltimore) **34**, 13—50 (1955).
RICHET, C., G. MARAÑON et M. RYMER: Pathologie de l'hypophyse (données récentes). Paris: Baillière 1948.
SCHAMAUN, M., F. DEUCHER u. S. GABLINGER: Spontane Hypoglykämie bei großem Nebennierenrindentumor. Schweiz. med. Wschr. **87**, 1348—1352 (1957).
SCHON, M., and A. M. SUTHERLAND: The role of hormones in human behavior. III. Changes in female sexuality after hypophysectomy. J. clin. Endocr. **20**, 833 (1960).
SCHROEDER, F. F., and J. L. SHERMAN: Reaction to prednisone masquerading as cer ebra vascular accident. J. Amer. med. Ass. **160**, 1354—1355 (1956).
SCHWÖBEL, G.: Die Psychopathologie des Cushing-Syndroms. Schweiz. Arch. Neurol.Psychiat. **74**, 382—403 (1954).
SECKEL, H. P. G.: Six examples of precocious sexual development. II. Studies in growth and maturation. Amer. J. Dis. Child. **79**, 278—309 (1950).
— W. W. SCOTT and E. P. BENDITT: I. Studies in diagnosis and pathogenesis. Amer. J. Dis. Child. **78**, 484—515 (1949).
SELYE, H.: Anaesthetic effect of steroid hormones. Proc. Soc. exp. Biol. (N. Y.) **46**, 116—121 (1941).
— The general adaptation syndrome and the disease of adaptation. J. clin. Endocr. **6**, 117 (1946).
SENSENBACH, W., L. MADISON and L. OCHS: The effect of ACTH and cortisone on cerebral blood flow and metabolism. J. clin. Invest. **32**, 372—380 (1953).
STOLL, W. A.: Die Psychiatrie des Morbus Addison, insbesondere seiner chronischen Formen. Stuttgart: Thieme 1953.
— Das endokrine und das amnestische Psychosyndrom bei Morbus Addison. In: H. NOWAKOWSKI, Herausgeb., Hormone and Psyche. Die Endokrinologie des alternden Menschen, 13—17. Heidelberg: Springer 1958.
STORRIE, V. M.: Convulsions in Addison's disease. Bull. New Engl. med. Cent. **1**, 229—237 (1955) [abstract: Excerpta med. (Amst.) Sect. VIII. **10**, 38 (1957)].
STURM, A., u. W. WERNITZ: Hormonjagd im Gehirn. Acta neuroveg. (Wien) **13**, 50—62 (1956).
STUTTE, H.: Pubertas praecox. In: H. GIESE, Herausgeb., Die Sexualität des Menschen. Handbuch der medizinischen Sexualforschung, 474—505. Stuttgart: Enke 1954.
TORATINI, Y., and T. OKU: A case of Cushing's syndrome accompanied with nervous disorders. Wakayama med. Rep. **3**, 1—8 (1956).
TRETHOWAN, W. H.: Psychosis in relation to treatment with corticotrophin and cortisone. Acta psychiat (Kbh.) **29**, 243—259 (1954).
— and S. COBB: Neuropsychiatric aspects of Cushing's syndrome. Arch. Neurol. Psychiat. (Chicago) **67**, 283—309 (1954).
WALLERSTEIN, R. S., R. L. SUTHERLAND and J. LYONS: Some psychosomatic considerations in Addison's disease. Psychosom. Med. **16**, 67—76 (1954).
WAYNE, H. L.: Convulsive seizures complicating cortisone and ACTH therapy: Clinical and electroencephalographic observations. J. clin. Endocr. **14**, 1038—1045 (1954).
WEBER, P.: Cutaneous striae, purpura, high blood pressure, amenorrhoea and obesity, of the type sometimes connected with cortical tumours of the adrenal glands, occurring in the absence of any such tumour — with some remarks on the morphogenetic and hormonic effects of true hypernephromata of the adrenal cortex. Brit. J. Derm. **38**, 1—19 (1926).
WILKINS, L.: A feminizing adrenal tumor causing gynecomastia in a boy of five years contrasted with a virilizing tumor in a five-year-old girl. J. clin. Endocr. **8**, 111—132 (1948).
ZERSSEN, D. VON: Die psychischen Nebenwirkungen der Pharmakotherapie mit Hormonen des Hypophysen-Nebennierenrinden-Systems. Z. psychosom. Med. **3**, 172—248; **4**, 1—10 (1957).
ZÜBLIN, W.: Zur Psychiatrie des adrenogenitalen Syndroms bei kongenitaler Nebennierenrindenhyperplasie. Helv. paediat. Acta. **8**, 117—135 (1953).

Author Index

Page numbers in *italics* refer to the bibliography

Abarbanel, A. R. see Engel, L. L. 427, *503*
Abelin, I. see Minder, W. H. 55, *171*
Abelous, J. E. *555*
— and P. Langlois 551, *555*
Abelson, D., and P. K. Bondy 367, *381*
— F. Ulrich and C. N. H. Long 460, *499*
— see Bondy, P. 367, *384*
Abelson, H. see Bondy, P. 367, *384*
Aboim, A. N. 6, 53, 56, *145*
Adams, A. E., and M. Harland 552, *556*
— see Kingsbury, J. W. 60, 145, *165*
Adams, E., and M. Baxter 49, 51, 54, 92, 94, *145*
Adams, W. E. 30, 40, *145*
Adams, W. J., D. N. Kirk, D. K. Patel, V. Petrow and I. A. Stuart-Webb 218, *381*
Addis, T., and H. Gray 18, 20, *145*
Addison, T. 654, *656*
Adesman, J. see Hellman, L. 471, *505*
Adolf, E. F. *145*
Agate, F. J. 94, *145*, 619, *624*
Agnello, E. J., B. L. Bloom and G. D. Laubach 277, 279, *381*
— and G. D. Laubach 267, 279, *381*
Albeaux-Fernet, M., P. Bugard and J. D. Romani 656, *656*
Albert, A. see Gastineau, C. F. 616, *628*
— see Mattox, V. R. 198, 313, *398*, 520, *564*
Albertin, R. H. see Hartman, F. A. 13, 14, *161*
Albright, F. 579, 580, *624*
— see Bartter, F. C. 531, *556*, 585, 616, *624*
— see Forbes, A. P. 584, *627*
— see Fraser, R. W. 363, *390*
— see Kinsell, L. 580, *630*
Aldrich, R. A. see Sweat, M. L. 443, 485, 489, *511*
Alexander, G. J., and E. Schwenk 438, *499*
— see Schwenk, E. 436, 438, *511*
Alexander, J. see Engel, L. L. 304, 306, *389*
Alfert, E. see Alfert, M. 30, *145*
Alfert, M., H. A. Bern and E. Alfert 30, *145*
Alfsen, A. see Weinmann, S. 434, *512*
Algeo, J. W. see Gassner, F. X. 574, *628*
Allen C., and L. R. Broster 652, *656*
Allen, J. C. 596, *624*
— see Howard, E. 620, *630*
Allen, J. M. 62, 63, 64, 65, 66, 81, 98, 101, 109, 110, 114, 115, 128, *145*
— and J. J. Slater 64, 65, 96, *145*
Allen, M. J., and A. H. Corwin 522, *556*
— see Bencze, W. L. 136, *147*
— see Chart, J. J. 457, *502*, 522, 523, *557*
— see Hertz, R. 136, *161*, 456, *505*, 523, *561*
— see Sheppard, H. 523, *567*
Allen, W. M. 364, 366, *381*
— S. J. Hayward and A. Pinto 364, *381*
Allen, W. S., and S. Bernstein 243, 267, 274, *381*
— — and R. Littell 244, *381*
— see Bernstein, S. 236, 267, 272, 284, *383*
Allers, W. D., and E. C. Kendall 530, *556*
Alman, R. W., and J. F. Fazekas 643, *656*
Alpert, M. 57, 94, 107, *145*
— see Glick, D. 58, 59, *159*
— see Marks, B. H. 137, *170*, 457, *507*
Altrock, J. R. see Bondy, P. K. 475, *500*
Altszuler, N. see Bacchus, H. 88, 114, 116, *146*
Amdur, B. H., H. Rilling and K. Bloch 438, *499*
Amelung, D., H. J. Hübener and L. Roka 463, 497, *500*
— — — and G. Meyerheim 463, 497, *500*
Amendolla, C., G. Rosenkranz and F. Sondheimer 236, *381*
— see Sondheimer, F. 279, *405*
Ames, A. see Herrmann, L. G. 293, *392*
Anason, A. see Brown, J. H. U. 58, 108, 136, *149*
Andersen, D. H., and H. S. Kennedy 622, 623, *624*
Anderson, A. F., A. M. Haine and J. Patterson 611, *624*
Anderson, A. J., and F. L. Warren 293, *381*
Anderson, C. H., M. McCallay and G. L. Farrell 115, *146*
— see Farrell, G. L. 103, 111, *156*
Anderson, D. H., and N. B. Woodall 350, *381*
Anderson, E., and M. Joseph 529, *556*
— see Bahn, R. 607, *624*
— see Bahn, R. C. 102, 108, 129, 133, *146*
— see Laqueur, G. L. 92, 104, 105, *167*
Anderson, H. see Mann, K. M. 493, *507*
Anderson, H. V., E. R. Garrett, F. H. Lincoln, A. H. Nathan and J. A. Hogg 227, *381*
— see Mann, K. M. 261, *397*
Anderson, J. A. see McQuarrie, I. 534, *564*
Anderson, R. G. see Clatworthy, H. W. jr. 39, 42, 43, *151*
Ando, T. 30, 88, 109, 114, 115, *146*
Andreč, K., and M. Kreč 334, *381*
— and L. Strauch 306, *381*
— see Savard, K. 431, 451, *510*

Angelico, R., and M. Quintilani 139, *146*
Angerer, C. A., and H. Angerer 551, *556*
— see Hartman, F. A. 548, *561*
Angerer, H. see Angerer, C. A. 551, *556*
Angrist, A. A. see Parmer, L. G. 622, *633*
Anker, H. S., and K. Bloch 436, *500*
Anliker, R., O. Rohr and M. Marti 204, *381*, 414, 415, 420, *500*
— — and L. Ruzicka 478, *500*
Anner, G., and K. Miescher 261, *382*
— see Schmidlin, J. 238, 262, 264, *404*, 520, *566*
— see Wettstein, A. 415, *513*
Anson, B. J., E. W. Cauldwell, J. W. Pick and L. E. Beaton 23, *146*
Anton, H. V. see Zimmermann, W. 363, *410*
Antonchak, N. see Gaunt, R. 80, 96, 100, 107, 110, *158*
Antoniades, H. N. see Sandberg, A. A. 292, *403*, *510*
Antonucci, R., S. Bernstein, M. Heller, R. Lenhard, R. Littel and J. Williams 225, 361, *382*
Appleby, J. I., G. Gibson, J. K. Norymberski and R. D. Stubbs 341, *382*
— and J. K. Norymberski 341, 366, *382*
Aprile, M., E. G. Bligh, J. L. Webb and R. D. H. Heard 421, 442, *500*
Arai, Y. 608, *624*
Arey, J. B. see Di George, A. M. 593, *627*
Arigoni, D. see Buchschacher, P. 237, *384*
— see Cainelli, G. 236, *385*
— see Riniker, B. 191, *402*
Armitage, P. see Clayton, R. E. 454, *502*
Arndt, T., E. F. Pfeiffer and K. Schöffling 642, *656*
Arnesen, K. 35, *146*
Aron, M. see Kayser, C. 21, 94, 128, *165*
Arons, W. L. see Laidlaw, J. C. 518, 533, 535, *563*
— see Thorn, G. W. 518, 540, *568*
Aronson, L. R. see Warren, R. P. 603, *636*
Arreguin, B. see Bonner, J. *501*
Arrington, L. R., M. H. Fox and H. A. Bern 605, *624*
Arroyave, G., and L. R. Axelrod 328, *382*
— see Axelrod, L. R. 486, *500*
Arth, G. E., J. Fried, D. B. R. Johnston, D. R. Hoff, L. H. Sarett, R. H. Silber, H. C. Stoerk and C. A. Winter 285, *382*
— D. B. R. Johnston, J. Fried, W. W. Spooncer, D. R. Hoff and L. H. Sarett 285, *382*
— see Poos, G. I. 242, *401*
— see Sarett, L. H. 262, *404*
Arvy, L. and A. Bonichon 65, 66, *146*
Arx, E. von, and R. Neher 313, *382*
Ascoli, G., and T. Legnani 78, 99, *146*
Ascoli, I. see Folch, J. 49, *156*
Ashburn, L. L. see McQueeney, A. J. 138, *170*
Ashmore, J., W. H. Elliott, E. A. Doisy, jr. and E. A. Doisy 480, *500*
Asling, C. W. see Harrison, R. G. 23, *161*
Assling, C. W. see Walker, D. G. 78, *183*
Astwood, E. B. 611, 618, *624*
Asworth, E. see Fukushima, D. K. 431, 432, *504*
Atchley, D. W. see Ferrebee, J. W. 531, 532, 534, *558*
— see Kuhlman, D. 534, *562*
— see Loeb, R. F. 528, 529, *563*
— see Ragan, C. 533, 534, *565*
— see Stahl, J. 535, *567*
Aterman, K. see Greenberg, S. M. 82, *159*
Atherden, S. M. see Fotherby, K. 428, 429, 446, 478, *504*
— see Marrian, G. F. 296, *397*
Atkinson, W. B., and M. M. Dickie 134, *146*
Attenburrow, J., A. F. B. Cameron, J. H. Chapman, R. M. Evans, B. A. Hems, A. B. A. Jansen and T. Walker 242, *382*
Atwater, N. W. 287, *382*
Atz, J. W. see Pickford, G. E. 548, *565*
August, J. T., D. H. Nelson and G. W. Thorn 138, *146*
Avigan, J. see Steinberg, D. 140, *180*, 455, *511*
Avivi, P., S. A. Simpson, J. F. Tait and J. K. Whitehead 371, *382*
Axelrad, B. J. see Luetscher, J. A., jr. *507*
Axelrod, J. see Isselbacher, K. J. 452, 461, 463, *506*
Axelrod, L. R. 320, 328, 341, 342, 346, *382*
— and G. Arroyave 486, *500*
— and L. L. Miller 461, 478, 486, 488, 495, *500*
— — and F. Herling 486, *500*
— and A. Zaffaroni 300, 301, *382*
— see Arroyave, G. 328, *382*
— see Ganis, F. M. *157*, 461, *504*
— see Miller, L. L. 462, 472, *508*
Ayres, P. J., J. Barlow, O. Garrod, S. A. S. Tait, J. F. Tait and G. Walker 76, *146*
— O. Garrod, S. A. S. Tait, J. F. Tait, G. Walker and W. H. Pearlman 423, 447, *500*
— — S. A. Simpson and J. F. Tait 313, 318, 372, *382*
— R. P. Gould, S. A. Simpson and J. F. Tait 109, *146*
— — S. A. S. Simpson and J. F. Tait 447, 451, *500*
— O. Hechter, N. Saba, S. A. Simpson and J. F. Tait 423, 447, *500*
— S. A. Simpson and J. F. Tait 367, *382*
Azerad, E. see Cloche, R. 655, *657*

Babcock, J. C. see Hogg, J. A. 272, *393*
— see Johnson, J. L. 245, *394*
— see Pederson, R. L. 281, 283, *401*, 487, *509*
Baber, A. see Mandel, L. 80, 91, *170*
Bacchus, H. 60, 71, 96, 104, 112, 114, 115, *146*
— M. H. Heiffer and N. Altszuler 88, 114, 116, *146*
Bachman, W. E., W. Cole and A. R. Wilds 261, *382*
Bachmann, R. 4, 23, 27, 28, 32, 35, 45, 53, 57, 90, 103, 130, *146*, 587, *624*
Baclesse, M. see Courrier, R. 106, 107, 124, *152*
Bacq, Z. M., and P. Fischer 125, *146*
Bader, M. E. see Levitt, M. F. 537, 538, *563*
Badinez, O., and H. Croxatto 112, 115, *146*
Baecker, R. 7, *146*

Baggett, B., L. L. Engel, L. Balderas, G. Lansan, K. Savard and R. I. Dorfman 412, 451, *500*
— — and L. L. Fielding 338, *382*
— — K. Savard and R. I. Dorfman 616, *624*
— see Engel, L. L. 484, *503*
— see Savard, K. 413, 488, *510*, 579, *634*
Bagnoli, E. see Savard, K. 434, *510*
Bahn, R., H. Wilson, L. Kedda, J. Furth and E. Anderson 607, *624*
Bahn, R. C., J. Furth, E. Anderson and E. Gadsden 102, 108, 129, 133, *146*
— and D. Glick 59, 80, 85, 86, *146*
— H. E. Storino and R. W. Schmit 3, 29, 79, 88, 109, 110, *146*
— see Glick, D. 3, *159*
— see Wilson, H. 75, 108, *184*, 603, 607, *636*
Bahner, F. see Tonutti, E. 31, 88, *182*
Bailey, A. A. see Clark, E. D. 641, *657*
Bailey, G. see Hirschmann, R. F. 269, 271, *393*
Bailey, R. E. see Ganong, W. F. 110, *157*
Baïsset, A., L. Douste-Blazy and P. Montastruc 116, *146*
— — H. Planel and P. Montastruc 115, 116, *146*
Baker, B. L. 29, 45, 118, *147*
— M. A. Schairer, D. J. Ingle and C. H. Li 608, *624*
— see Coutinho, H. B. 66, 98, 109, *152*
Baker, C. see Swingle, W. W. 99, *181*, 521, 524, 527, 537, *568*
Baker, C. P., F. Dobson and S. W. Stroud 313, *382*
Baker, D. D. 18, 106, *147*
Baker, G. D. see Beher, W. T. 336, *383*
— see Parsons, J. 336, *400*
Baker, H. S. see Longwell, B. B. 423, *507*
Baker, J. R. see Fishman, W. H. 66, *156*
Balderas, L. see Baggett, B. 412, 451, *500*
Baldini, J. T. see Zarrow, M. X. 58, 85, 144, *185*
Balfour, W. W., R. S. Comline and R. V. Short 75, *147*
Ball, W. C. jr. see Davis, J. O. 114, 115, *152*, 531, *558*
Ballin, H. M. see Gellhorn, E. 534, *560*
Bänder, A. 50, 93, 96, *147*
Bandy, H. E. see Southcott, C. M. 423, 425, *511*
Banerjee, S. see Mukerjee, A. K. 104, *171*
Banes, D. 314, *382*
Banide, P. see Planel, H. 133, *174*
Banks, R. C. see Farrell, G. L. 74, 81, 101, 110, *156*, *503*, 517, 521, *558*
Baran, J. S. 234, *382*
Barbarin, Y. see Hébert, S. 66, *161*
Barber, G., and M. Ehrenstein 204, 238, 239, 282, 290, 361, *382*
Barber, J. K. see Bartter, F. C. 114, *147*
— see Liddle, G. W. *506*
Barger, A. C., F. P. Muldowney and M. R. Liebowitz 531, *556*
— see Deane, H. W. 34, 114, *153*
Bargmann, W. 23, *147*
Bariety, M., A. Gajdos, C. Coury and M. Kovacevic 546, *556*
Barker, A. D. see Southcott, C. M. 423, 425, *511*
Barker, W. F. see Zabin, I. 436, *513*
Barlow, G. jr. see Swingle, W. 535, *568*, 621, *635*
Barlow, J. see Ayres, P. J. 76, *146*
Barnawell, E. B. see Westberg, J. A. 106, *184*
Barnes, L. E. see Lyster, S. C. 521, *563*
— see Stafford, R. O. 521, *567*
Barnes, L. F. see Byrnes, W. W. 521, *557*
Barnes, R. B. see Dobriner, K. 301, 303, *388*
Barnett, J., A. A. Henley and C. J. O. R. Morris 381, *382*
Barnett, S. A., and G. Bourne 59, 60, *147*
— and R. B. Fisher 60, *147*
Barrnett, R. J. 66, *147*
— see Deane, H. W. 63, 64, 66, *153*
Barry M. C., M. L. Eidinoff, K. Dobriner and T. F. Gallagher 480, *500*
— see Gallagher 478, *504*
Bartter, F. C. *500*
— F. Albright, A. P. Forbes, A. Leaf, E. Dempsey and E. Carroll 585, 616, *624*
— E. G. Biglieri, P. Pronove and C. S. Delea 114, *147*
— and P. Fourman 532, *556*
— — F. Albright, A. P. Forbes, W. McK. Jefferies, G. Griswold, E. Dempsey, D. Bryant and E. Carross 531, *556*
— G. W. Liddle, L. E. Duncan jr., J. K. Barber and C. Delea 114, *147*
— see Chen, P. S. jr. 447, *502*
— see Gann, D. S. 115, *157*
— see Liddle, G. W. 111, 113, 114, 115, *168*, *506*
— see Mills, I. H. 115, *171*
Barton, D. H. R. 193, 243, 246, 308, *382*
— and R. C. Cookson 193, *382*
— and W. C. Taylor 278, *383*
— R. M. Evans, J. C. Hamlet, P. G. Jones and T. Walker 228, *383*
— D. A. Ives, and B. R. Thomas 246, *383*
Barucci, M., and F. Mori 104, *147*
Bascom, W. D. see Rosenfeld, G. 76, 137, *176*, 457, *510*, 522, *566*
Basowitz, H., H. Persky, S. J. Korchin and R. R. Grinker 642, *656*
Bass, A. C. see Davis, A. K. 539, *558*
— see Overman, R. R. 537, 538, *565*
Bass, L. W. see Sherman, F. E. 129, *179*
Bassil, G. T., and A. M. Hain 365, *383*
Bata, G. see Endröczi, E. 75, 76, 103, *155*
Bates, R. W., O. Riddle and R. A. Miller 553, *556*
Batres, E. see Ringold, H. J. 272, *402*
— see Sondheimer, F. *405*
Bauer, J. 646, *656*
— and J. Karl 308, *383*
Bauer, V. see Staudinger, H. 296, *406*
Bauer, W. see Quarton, G. C. 643, *659*
Bauld, W. L., and R. D. H. Heard 431, *500*
Bauld, W. S. see Heard, R. D. H. 484, *505*
— see Marrian, G. F. 426, *507*
Baulieu, E. see Jayle, M. F. 434, *506*

Baulieu, E. E. see Chemama, R. 434, *502*
— see Jayle, M. F. 295, *394*
— see Mallassis, D. 434, *507*
— see Mialhe-Voloss, C. 111, *170*
— see Vesin, P. 521, 546, *568*
— see Weinman, S. 434, *512*
Baulieu, E. J. 572, *624*
Baum, G. J., and R. K. Meyer 553, *556*
Baumann, E. J., and S. Kurland 515, 529, *556*
— see Marine, D. 516, 528, *564*
Baumgartner, G. see Tamm, C. *407*
Baxter, M. see Adams, E. 49, 51, 92, *145*
Baxter, J. S. 32, 45, 90, 101, *147*
— see Yoffey, J. M. 53, 101, *185*
Bayan, A. P. see Perlman, D. 494, *509*
Bayliss, M. J. see Saffran, M. 77, *177*, 453, *510*
Bayliss, R. I. S. 296, *383*
Beal, P. F. see Hogg, J. A. 208, 256, 257, 277, *393*
Beall, D. 412, 420, *500*, 601, 615, *624*
— and T. Reichstein 420, *500*, 609, *624*
Bean, W. B., R. E. Hodges and K. Daum 139, *147*
— see Schedl, H. P. 363, *404*
Beary, D. F. see Ingle, D. J. 491, *506*
Beaton, L. E. see Anson, B. J. 23, *146*
Beatty, C. H., R. M. Bocek and R. D. Peterson 542, *556*
Bec, P. see Planel, H. 28, 57, 65, 96, 133, *174*
Beck, E. M. see Dumm, M. E. 458, *503*
Beck, J. C., E. E. McGarry, I. Dyrenfurth and E. H. Venning 116, *147*
— see Venning, E. H. 110, 111, *182*, 333, *407*
Beckers, C., and De Visscher, M. 140, *154*
Beckman, I. see Pontius, D. 337, *401*
Beckmann, I. see Schroeder, W. *404*, 418, *511*
— see Tamm, J. 293, *407*, 422, *423*, *511*
— see Voigt, K. D. 303, 304, *408*, 418, *512*
Beer, C. T., and T. F. Gallagher 484, *500*
Beher, W. T., J. Parsons and G. D. Baker 336, *383*
— see Parsons, J. 336, *400*
Beigelman, P. M., M. A. Slusher, G. G. Slater and S. Roberts 74, 75, *147*
Bejdl, W. 65, *147*
Belisle, J. see Schwenk, E. 233, *404*
Bell, J. C. see Venning, E. H. 475, *512*
Bell, J. T. jr., and A. F. Weber 94, *147*
— see Weber, A. F. 135, *184*
Bell, M. P. see Heigl, J. J. 357, *392*
Bell, P. H. see Perrine, J. W. 521, *565*
Bellamy, L. J. 348, *383*
Belt, W. D. 24, 32, 33, 54, *147*
— and E. R. Hayes 56, *147*
— and D. C. Pease 38, *147*
Ben, M. see Swingle, W. W. 99, *181*, 535, *568*
Beneze, W. L., and M. J. Allen 136, *147*
— see Chart, J. J. 457, *502*, 522, 523, *557*
— see Sheppard, H. 523, *567*
Bendas, H. see Djerassi, C. 289, *388*
Benditt, E. P. see Seckel, H. P. G. 650, *659*
Benedek, T. 647, *656*
Benedict, E. M. see Loeb, R. F. 528, 529, *563*
Benirschke, K. 41, 42, 43, 44, 123, 126, 127, *147*, 591, *624*
— E. Bloch and A. T. Hertig 121, 125, *147*, 414, 415, *500*, 588, 592, *624*
— see Bloch, E. 72, 122, 125, *148*, 415, *500*, 571, 578, 590, *625*
Benner, M. C. 27, *147*, 588, *624*
Bennett, C. J. see Moog, F. 20, 43, 125, *171*, 596, *632*
Bennett, H. S. 32, 37, 45, 55, *147*
— J. H. Luft and J. C. Hampton 23, *147*
— and L. Kilham 3, 21, *147*
Bennett, L. L. see Hazelwood, R. L. 139, *161*
Benoit, J. see Leroy, P. 13, 143, *168*, 554, *563*
Benoussan, P. see de Gennes, L. 655, *657*
Benua, R. S. and, E. Howard 48, *147*, 588, 593, *624*
— see Howard, E. 596, *630*
Berg, B. N. see Bloch, K. 436, *500*
Bergel'son, L. see Nazarow, I. 186, *399*
Bergen, J. R., D. Krus and G. Pincus 579, *624*
— see Caspi, E. 459, *501*
Bergenstal, D. M., R. Hertz, M. B. Lipsett and R. H. Moy 523, *556*
— see Hertz, R. 137, *162*
— see Werbin, H. 435, 441, *513*
Berger, E. Y., G. P. Quinn and M. A. Homer 540, *556*
Bergman, A. J. see Mixner, J. P. 18, *171*
Bergner, G. E., and H. W. Deane 93, 101, 110, *148*
Bergstrand, C. G. 81, 107, *148*
— and C. A. Gemzell 610, *624*
Bergström, S., and A. Norman *500*
— K. Pääbo and J. A. Rumpf *500*
— M. Rottenberg and J. Sjövall *500*
— and J. Sjövall *500*
— — and J. Voltz *500*
Berlin, J. see Djerassi, C. 281, *387*
— see Rosenkranz, G. 255, *403*
Berliner, D. see Gomez-Mont, F. 459, *504*
Berliner, D. L. *383*
— O. V. Dominguez and G. Westenkow 369, 372, *383*
— and T. F. Dougherty 461, 462, *500*
— J. E. Jones and H. A. Salhanick 422, *500*
— and H. A. Salhanick 372, *383*
— see Dougherty, T. F. 460, *503*
— see Nabors, C. J. jr. 61, *171*
— see Salhanick, H. A. 412, 420, 451, *510*, 611, 634
Berman, R. A. see Friedgood, H. B. 337, *390*
— see Talbot, N. B. 363, *406*, 611, *635*
Bern, H. A., S. Nandi, R. A. Campbell and L. E. Pissoti 57, 133, *148*
— see Alfert, M. 30, *145*
— see Arrington, L. R. 605, *624*
— see Campbell, R. A. 106, *149*
— see Nandi, J. 77, *172*, 550, *565*
— see Westberg, J. A. 106, *184*
Berne, R. M. see Driscol, T. E. 114, *154*
Bernhard, W. G. see Robinson, H. J. 106, *176*

Bernstein, D. E. 106, *148*
Bernstein, S., W. S. Allen, C. E. Linden and J. Clemente 236, *383*
— M. Heller, R. Littell, S. M. Stolar, R. H. Lemhard, W. S. Allen and J. Ringler 267, 284, *383*
— — and S. M. Stolar 272, *383*
— and R. H. Lenhard 275, 362, *383*
— — W. S. Allen, M. Heller, R. Littell, S. M. Stolar, L. I. Feldman and R. H. Blake 267, 272, *383*
— see Allen, W. S. 243, 267, 274, *381*
— see Antonucci, R. 225, 361, *382*
— see Littel, R. 228, 277, *397*
Berret, R. see Velluz, L. 221, *407*
Berrill, N. J. 623, *624*
Bertagna, L. see Delay, J. 641, *657*
Berthet, L. see Haynes, R. C. jr. 453, *505*
Berthrong, M. see Rich, A. R. 90, *175*
Bertrand, J. see Migeon, C. J. 471, *508*
Best, C. H. see Sirek. O. V. 535, *567*
Bethune, J. E. see Sobel, E. H. 135, 137, *180*
Beyer, K. F., and L. T. Samuels 61, *148*, 496, *500*
Beyler, R. E., and F. Hoffman 222, 272, 274, *383*
— R. Moriarty, F. Hoffman and L. H. Sarett 225, *383*
— see Hirschmann, R. F. 267, 277, *393*
— see Hoffman, F. 286, *393*
— see Poos, G. I. 242, *401*
— see Sarett, L. H. 262, *404*
Biggs, M. W., R. M. Lemmon and F. T. Pierce jr. 436, *500*
Biglieri, E. G. see Bartter, F. C. 114, *147*
Bijvoet, J. M. *383*
Billeter, J. R. see Schmidlin, J. 238, 262, 264, *404*, 520, *566*
Billings, C. see Klein, R. 521, *562*
Bimes, C., H. Planel and A. Guilhem 20, 35, 133, *148*
— — — and P. Girard 57, *148*
Bimmer, E. 11, 12, *148*
Birch, A. J. 280, *383*
Birke, G. 581, *624*
— C. Franksson and L. O. Plantin 581, 625
Birke, G., and L. O. Plantin 459, *500*
— see Diczfalusy, E. 364, *387*, 412, *502*
— see Plantin, L. O. 414, 415, 428, *509*, 572, *633*
Birket-Smith, E. 656, *656*
Birmingham, M. K., F. H. Elliot and P. H. L. Valere 454, *500*
— and E. Kurlents 77, 101, *148*, 452, *500*
— see Schönbaum, E. 114, *178*, 454, *511*
Birnbaum, J. P. see Tobin, C. E. 57, *182*
Birnie, J. H. 544, *556*
— W. J. Eversole, W. R. Boss, C. M. Osborn and R. Gaunt 545, *556*
— — and R. Gaunt 537, 546, *556*
— see Gaunt, R. 530, 531, 542, 543, *559*
Bischoff, F. see Gray, C. L. 484, *504*
Bishop, D. M., and J. H. Leathem 128, *148*
Biskind, G. R. see Glick, D. 59, 66, *158*
Bitman, J., and S. L. Cohen 292, 294, *383*
— J. P. Rosselet, A. M. Reddy and S. Lieberman 228, 250, 359, *383*
— see Wrenn, T. R. 611, *637*
Bittner, J. J. see Halberg, F. 32, *160*
Bittorf, A. 615, *625*
Black, E. G. see Cope, C. L. 475, *502*
Black, L. see Rubin, B. L. 617, *634*
Black, M. M. see Zweifach, B. W. 63, *185*
Blackman, S. S. 43, *148*, 589, 593, *625*
Bladon, P., H. B. Henbest, E. R. H. Jones, B. L. Lovell and G. F. Woods 275, *383*
— — and G. W. Wood 275, 360, *383*
Blake, R. H. see Bernstein, S. 267, 272, *383*
Blakemore, W. S. see Cooper, D. Y. 422, 423, *502*
Blanz, E. J. jr. see Dauben, W. G. 245, *387*
Blecher, M., and S. Gurin 436, *500*
Bleiberg, M. J., and P. S. Larson 135, 136, *148*, 458, *500*
Bleuler, E. 654, *656*
Bleuler, M. 638, 639, 640, 644, 645, 647, 654, *656*
Blickenstorfer, E. 639, *656*
— P. Isler, M. Marti and C. Hedinger 639, *657*
Bligh, E. G., R. D. H. Heard, V. J. O'Donnell, J. L. Webb, M. Saffran and E. Schonbaum 449, 451, *500*
— see Aprile, M. 421, 424, *500*
— see Heard, R. D. H. 420, 421, 446, 449, 451, 453, *505*
Bliss, C. I. 576, 617, *625*
Bliss, E. L., C. J. Migeon, C. H. Hardin Branch and L. T. Samuels 642, *657*
— A. A. Sandberg, D. H. Nelson and K. Eik-Nes 572, *625*
— see Migeon, C. J. 572, *632*
Blix, G., A. Tiselius and H. Svensson 292, *383*
Bloch, E. and K. Benirschke 590, *625*
— — and R. I. Dorfman 415, *500*
— — and E. Rosemberg 72, 122, 125, *148*, 571, 578, 590, *625*
— R. I. Dorfman and G. Pincus 415, 449, 451, *500*, 572, *625*
— G. Pincus and R. I. Dorfman 204, *383*
— see Benirschke, K. 121, 125, *147*, 415, *500*, 588, 592, *624*
— see Cohen, A. I. 101, 102, 108, 135, *151*
— see Pincus, G. 583, *633*
Bloch, H. S. see Zimmermann B. 135, *185*, 458, *513*
Bloch, K. 62, *148*, 434, 435, 436, 437, *500*
— B. N. Berg and D. Rittenberg 436, *500*
— E. Borek and D. Rittenberg 437, *500*
— C. Clark and I. Harary 436, *500*
— and D. Rittenberg 436, 437, *500*
— see Amdur, B. H. 438, *499*
— see Anker, H. S. 436, *500*
— see Clayton, R. B. 436, 438 *502*
— see Gautschi, F. 439, *504*
— see Johnston, J. D. 436, 438, *506*
— see Langdon, R. G. 436, 437, 438, *506*

Bloch, K. see Olson, J. A. 436, 438, *508*
— see Tchen, T. T. 438, *512*
— see Woodward, R. 436, 437, 438, *513*
— see Wuersch, J. 437, *513*
— see Zabin, I. 436, *513*
Block, R. J., E. L. Durrum and G. Zweig 304, 313, *383*
Bloom, B. M., M. Hayano, A. Saito, D. Stone and R. I. Dorfman 495, *500*
— and G. M. Shull *383*, 495, *500*
Bloom, B. L. see Agnello, E. J. 277, 279, *381*
Bloom, B. R. see Ungar, F. 496, *512*
Bloor, W. R. see Oleson, M. C. 49, 51, 84, 92, 94, *173*
Blumenthal, H. T. 32, *148*
— and L. Loeb 69, *148*
— see Joshi, R. A. 120, *164*
Boas, N. F. see Jailer, J. W. 58, 82, 85, 144, *164*
Bocek, R. M. see Beatty, C. H. 542, *556*
Bock, R. M., and N. S. Ling 310, *384*
Bodánszky, A., and J. Kollonitsch 346, *384*
Bodo, R. C. de see Earle, D. P. jr. 99, *155*
— see Lane, N. 110, *167*
Boerner, D. 2, 39, *148*
Boguth, W., H. Langendorf and E. Tonutti 88, *148*
Bohle, K. see Schroeder, W. 21, *178*
Bois, P., and H. Selye 68, 83, *148*
Bollet, A. J. see Bunim, J. J. 520, *557*
Bollinger, F. see Wendler, N. L. 279, *408*
Bompard, F. see Rubin, M. 214, *403*
Bonanni, R. see Lamedica, G. 476, *506*
Bonanno, S. see Thoma, R. W. 259, 274, *407*
Bonati, B., G. C. Manfredi and G. B. Pancati 81, 96, 108, 110, *148*
Bondy, P. K., D. Abelson, J. Scheuer, T. K. L. Tseu, V. Upton, H. Abelson and A. Clarke-Lewis 367, *384*
— and J. R. Altrock 475, *500*
— and G. V. Upton 367, *384*
— — and G. E. Pickford 550, *556*
— see Abelson, D. 367, *381*
— see Knight, R. P. jr. 535, *562*
Bondy, P. K. see Phillips, J. G. 75, *174*, 550, *565*
Bongiovanni, A. M., and G. W. Clayton 611, *625*
— and W. R. Eberlein 301, 364, 365, *384*, 425, 432, 460, *501*
— — and J. Cara 585, *625*
— — M. M. Grumbach, J. J. van Wyk and G. Clayton 296, 301, *384*, 424, *501*
— — P. Z. Thomas 338, *384*
— see Di George, A. M. 593, *627*
— see Eberlein, W. R. 328, 333, 334, *308*, 423, 427, 431, 432, *503*, 585, 610, *627*
Bonhoeffer, K. 644, 646, 654, *657*
Bonichon, A. see Arvy, L. 65 66, *146*
Bonomolo, A. see Fidanza, A. 51, *156*
Bonner, J., and B. Arreguin *501*
Bonnycastle, D. D. see Costa, P. J. 138, *152*
Borck, E., and D. Rittenberg 436, *501*
Borek, E. see Bloch, K. 437, *500*
Borell, U. 455, *501*
— E. Diczfalusy and A. Westman 608, *625*
Borgstrom, E., and T. F. Gallagher 251, *384*
Borman, A. see Fried, J. 258, 267, 268, 272, 274, 277, 278, *390*, 487, 491, 492, 493, 494, *504*, 520, *559*
Borniche, P. see Lelong, M. 582, *631*
Borris, J. J. see Wilson H. 75, 108, *184*, 603, 607, *636*
Borth, R. 364, *384*
— M. Gesell and H. de Watteville 455, *501*
— see de Watteville, H. 611, *636*
Bortle, L. see Perrine, J. W. 521, *565*
Boscott, R. J. 337, *384*
Boss, W. R. see Birnie, J. H. 545, *556*
Bosshard, H. see Dutler, H. 278, *388*
Boswell, T. see Woodruff, M. F. A. 120, *185*
Botella Llusia, J. 108, *148*
Bott, P. A. see Swingle, W. W. 529, 535, 545, *568*
Bourguet, J., and R. Debauche 531, *556*
Bourliere, F. see Verne, J. 17, *183*
Bourne, G. H. 4, 27, 30, 35, 48, 60, 65, *148*
— see Barnett, S. A. 59, 60, *147*
— and S. Zuckerman 622, *625*
— see Hoch-Ligeti C. 60, *162*
— see Malaty, H. A. 66, *169*
— see Zuckerman, S. 622, *637*
Bowers, A., and H. Ringold 272, *384*
— see Djerassi, C. 232, 288, *388*
— H. Ringold and R. Dorfman *384*
Bowman, B. J. see Byrnes, W. W. 521, *557*
— see Dulin, W. E. 267, 284, *388*
— see Stafford, R. O. 521, *567*
Boyd, G. S. 292, *384*
Boyer, F. see Chedid, L. 100, *150*
Boyle, A. J., see Epstein E. 337, *389*
Braceland, F. J. 655, *657*
Brachet, J. 91, *148*
Brackney, E. L. see Clark, W. G. 143, *151*, 552, *557*
Bradley, J. see Hollander, W. jr. 534, 543. *561*
Bradlow, H. L., and T. F. Gallagher 478, 480, 482, *501*
— L. Hellman, B. Zumoff and T. F. Gallagher 483, *501*, 573, *625*
— see Fukushima, D. K. 225, *390*, 433, 459, 460, 478, 482, 483, *504*
— see Hellman, L. 471, *505*
Bradshaw, T. G. see Malley, K. 611, *631*
Brady, J. V. see Mason, J. W. 105, *170*, 522, *564*
Brady, R. O., and S. Gurin 436, *501*
Bragdon, D. E. 12, *148*
Branin, M. L. 3, 8, 9, *148*
Brannick, L. J. see Swingle, W. W. 521, 524, 527, 535, *568*
Braude, E. A. 358, *384*
den Breeiyen, A. see Persky, H. 642, *659*
Breneman, W. R. 13, *149*
Brenneisen, K., C. Tamm and T. Reichstein 191, *384*
Brenner, R. M., D. I. Patt and L. C. Wyman 118, 119, *149*
Breuer, H. see Dirscherl, W. 337, *387*
Březina, M. see Zuman, P. 381, *410*

Bricaire, H., R. Mises, C. Dreyfus-Brisac and H. Fischgold 655, *657*
— see Gennes, L. de 546, *560*
Briggs, F. N., and W. Toepel 60, 85, *149*, 454, *501*
— see Knobil, E. 50, 51, 58, 59, 121, 124, *166*
— see Munson, P. L. 522, *564*
— see Wells, H. 522, *569*
Brink, N. G. see Ham, E. A. 238, *391*
— see Harman, R. E. 318, *391*, 421, 446, *504*
Britton, S. W. 515, 516, 525, 526, *556*
— and H. Silvette 528, *556*
— see Silvette H. 526, 542, *567*
Brizzee, K. R. see Eik-Nes, K. 79, 100, 103, 140, *155*
Broadbent, I. E., J. P. Newhouse and W. Klyne 339, *384*
Brobeck, J. R. see Gershberg, H. 59, *158*
Brodish, A., and C. N. H. Long 83, *149*
Brody, S. 641, *657*
Brogi, M. P. 120, *149*
— and C. Pellegrino 120, *149*
Brondel see Fraisse 546, *559*
Bronstein, I. P. see Genitis, V. E. 611, *628*
Brooks, C. J. W., and Norymberski 215, 222, 339, 340, *384*
Brooks, R. V., and W. Klyne 428, *501*
— — and E. Miller 214, 308, 310, *384*
— and F. T. G. Prunty 456, *501*
Brooksbank, B. W. L., and G. A. D. Haslewood 427, *501*
— see Savard, K. 431, 451, 510
Broster, L. R. see Allen, C. 652, *656*
Brown, B. M. see Knouff, R. A. 49, 51, 54, 92, 94, *166*
Brown, D. J. see Brown, K. I. 553, 554, 555, *556*
Brown, H., D. C. Willardson, L. T. Samuels and F. H. Tyler 471, *501*
— see Samuels, L. T. 475, *510*
— see Tyler, F. H. 475, *512*
— see Wallach, S. 134, *183*, 412, *512*, *585*, 611, 615, *636*
Brown, H. E. see Dougherty, T. F. 460, *503*
Brown, J. B., R. A. Knouff, M. M. Coulin and B. M. Schneider 49, 51, 54, *149*
Brown, J. B., and G. F. Marrian 484, *501*
— see Strong, J. A. 412, *511*
Brown, J. H. U. 523, *556*
— J. Griffin and R. B. Smith *501*
— — and R. B. Smith III 523, *556*
— J. B. Griffin, R. B. Smith III and A. Anason 58, 108, 136, *149*
— see Cueto, C. 135, 136, *152*, 523, *558*
Brown, J. S. L. see Venning, E. H. 611, *636*
Brown, K. I. 553, 554, *556*
— D. J. Brown and R. K. Meyer 553, 554, 555, *556*
Brown, R. see Lynn W. S. jr. *507*
Brown-Grant, K. 140, *149*
— see Forchielli, E. 498, 499, *504*
Browne, J. S. L. see Hoffman, M. M. 459, *505*
— see Venning, E. H. 431, 434, *512*
Brownell, K. A. see Hartman, F. A. 4, 5, 13, 14, 18, 21, 72, 73, 107, *161*, 516, 526, 548, *561*
— see Liu, T. Y. 573, *631*
Brownie, A. C., and J. K. Grant 62, *149*
— — and D. W. Davidson 443, 489, *501*
Browning, H. C. 134, *149*
Bruce, J. see Strong, J. A. 412, *511*
Brugin, H. see Robinson, H. J. 106, *176*
Brune, H., and W. Eger 103, *149*
Brunner, M. P. see Mann, K. M. 261, *397*, 493, *507*
Bryans, F. E. 614, *625*
— see Zarrow, M. X. 610, *637*
Bryant, D. see Bartter, F. C. 531, *556*
— see Kinsell, L. 580, *630*
Buchanan, G. P., A. C. Enders and R. V. Talmadge 615, *625*
Buchborn, E. see Wolff, H. P. 114, *184*, 532, *569*
Buchem, F. S. P. van 533, *556*
— H. Doorenbos and H. S. Ellings 135, *149*
Bucher, N. L. R. 436, 437, *501*
— and K. McGarrahan 436, *501*
— see Frantz, I. D. jr. 436, *504*
Buchschacher, P., J. Kalvoda, D. Arigoni and O. Jeger 237, *384*
Buck, R. R. see Mader, W. J. 340, 364, *397*
Bueding, E. see Lipschitz, W. L. 452, *507*
Buehler, H. J., P. A. Katzman and E. A. Doisy 293, 295, 297, *384*
Buffett, R. F. see Cohen, A. I. 101, 114, 115, 134, 135, *151*, 607, *626*
Bugard, P. see Albeaux-Fernet, M. 656, *656*
Buge, A. see Laroche, C. 655, *658*
Bulaschenko, H. see Richardson, E. M. 459, 460, *509*
— see Touchstone, J. C. 432, *512*
Bülbring, E. 143, *149*, 554, *557*
Bulbrook, R. D., and F. C. Greenwood 412, *501*, 616, *625*
— — and P. C. Williams 412, *501*
— see Greenwood, F. C. 412, *504*
Bunim, J. J., M. M. Pechet and A. J. Bollet 520, *557*
— see Johnson, D. F. 476, *506*
— see Ziff, M. 538, *569*
Burgers, A. C. J. see Carstensen, H. 555, *557*
Burgos, M. H. 9, 79, 143, 144, *149*
Burley, J. see Ulstrom, R. A. 591, *635*
Burn, D. see Weisenborn, F. 237, *408*
Burns, J. K., and A. J. Hale 36, 60, 65, 67, 90, 93, 96, 98, *149*
— — and W. C. Hutchison 88, *149*
— see Hutchison, W. C. 30, 88, *163*
Burns, R. K. 574, *625*
Burrill, M. W., and R. R. Greene 599, 601, 608, *625*
— see Greene, R. R. 575, 579, *628*
Burstein, S. 233, 234, 296, 338, *384*, 433, 445, 455, *501*
— and R. I. Dorfman 234, *384*, 428, 433, 460, *501*
— and E. M. Nadel 421, 424, 441, 454, *501*
— R. I. Dorfman and E. M. Nadel 454, 460, *501*
— K. Savard and R. I. Dorfman 459, 472, *501*

Burstein, S., F. Ungar, M. Gut and R.I. Dorfman 480, *501*
— see Nadel, E. M. 71, *172*, 432, 433, *508*
— see Savard, K. 311, *404*, 478, 480, 482, *510*
— see Zondek, B. 455, 456, *513*
Burton, R. B., E. H. Keutmann and C. Waterhouse 459, *501*
— A. Zaffaroni and E. H. Keutmann 319, 327, 339, 342, 346, *384*
— see Izzo, A. J. 340, *394*
— see Zaffaroni, A. 318, 342, *409*, 423, *513*, 517, *569*
Busch, R. D. see Silber, R. H. 341, 366, *405*
Busch, W. 23, *149*
Bush, I. E. 71, 74, 75, 77, *149*, 310, 313, 318, 319, 328, 342, 346, *384*, 424, 433, 482, *501*, 517, 518, *557*, 573, *625*
— and V. B. Mahesh 572, 586, *625*
— and A. A. Sandberg 311, *385*, 609, *625*
— and M. Willoughby 433, *501*
— J. Swale and J. Patterson 75, *149*, 572, *625*
— see Simpson, S. A. 423, *511*
Butenandt, A. 426, 427, 434, *501*
— and U. Westphal 427, *501*
Butler, A. M. see Talbot, N. B. 572, 611, *635*
Butler, G. C., and G. F. Marrian 431, *501*
Butler, L. C., and A. F. Morgan 139, *149*
Butt, W. R., A. A. Henley and C. J. O. R. Morris 303, *385*
— P. Morris, C. J. O. R. Morris and D. C. Williams 315, 318, 381, *385*, 613, *625*
Buzzetti, F., W. Wicki, J. Kalvoda and O. Jeger 237, *385*
Byers, C. see Klein, R. 296, *395*, 425, *506*, 521, *562*
Byrnes, W. W., L. F. Barnes, B. J. Bowman, W. E. Dulin, E. H. Morley and R. O. Stafford 521, *557*
— see Lyster, S. C. 521, *563*

Cafruny, E. J. see Tanabe, T. 100, 103, *181*
Cahen, R. L., and W. T. Salter 363, *385*
Cahn, J. see Georges, G. 82, *158*
Cain, A. J. 53, 55, 56, *149*
— and Harrison, R. G. 54, 92, *149*
— see Harrison, R. G. 54, *161*
Cainelli, G., M. L. Mihailovic, D. Arigoni and O. Jeger 236, *385*
Caldwell, J. R. see Therien, B. 135, *181*
Calhoun, M. L. see Smith, E. M. 104, *179*
Callamand, O. 549, *557*
Callow, N. H. 427, 478, *501*
— and R. K. Callow 427, *501*, 572, *625*
— R. K. Callow and C. W. Emmens 363, *385*, 616, *625*
Callow, R. K. 572, 580, *625*
— and R. Deanesly 575, *625*
— and V. H. T. James 228, *385*
— see Callow, N. H. 363, *385*, 427, *501*, 572, 616, *625*
Camain, R. see Quenum, A. 60, 62, 67, 92, *175*
Camara, A. A. see Schemm, F. R. 545, *566*
Camerino, B., B. Patelli and A. Vercellone 272, *385*
— and A. Vercellone 274, *385*
Cameron, A. F. B. see Attenburrow, J. 242, *382*
Campbell, J. A. see Hogg, J. A. 272, *393*
— see Pederson, R. L. 281, 283, *401*, 487, *509*
Campbell, R. A., H. A. Bern and K. B. DeOme 106, *149*
— see Bern, H. A. 57, 133, *148*
Campillo, C. C. see Zaffaroni, A. 494, *513*
Canlorbe, P. see Lelong, M. 582, *631*
Cann, M. C. see Heard, R. D. H. 420, 421, 446, 449, 451, 453, *505*
Cantarow, A. see Hansen, L. P. 338, *391*
— see Paschkis, K. E. 480, *508*
— see Pearlman, W. H. 484, *508*
— see Rakoff, A. E. 575, *633*
Cara, J. see Bongiovanni, A. M. 585, *625*
Caramazza, F. see Pellegrino, C. 120, *173*
Card, L. E. see Nalbandov, A. 553, *565*
— see Nalbandov, A. V. 143, *172*
Cardeza, A. F. 134, *149*
— see Houssay, A. B. 134, *162*, 618, *629*,
— see Houssay, B. A. 134, *163*, 618, *629*
— see Penhos, J. C. 79, 143, *174*
Cardoba, F. see Zaffaroni, A. 494, *513*
Cardwell, H. M., J. W. Cornforth, S. R. Duff, H. Holterman and R. Robinson 261, *385*
Carlo, J. see Pincus, G. 572, 583, *633*
Carlon, F. E. see Nussbaum, A. L. 225, *400*
Carnes, W. H. see Ferrebee, J. W. 532, 534, *558*
Carol, J., J. C. Molitor and E. O. Haenni 357, *385*
Carpenter, R. K., L. D. MacLeod, and M. Reiss 83, 92, *149*
Carr, I. A. 23, 38, *149*
Carroll, E. see Bartter, F. C. 585, 616, *624*
Carroll, K. K. 70, *150*
— and R. L. Noble 70, *150*
Carross, E. see Bartter, F. C. 531, *556*
Carstensen, H. *385*, 423, *501*
— A. C. J. Burgers and C. H. Li 555, *557*
— G. W. Oertel and K. B. Eik-Nes 573, *626*
Carter, A. C. see Wallace, E. Z. 455, *512*
— S. Weisenfeld and M. G. Goldner 107, *150*
Carter, P. see Engel, L. L. 304, 306, *389*, 426, 451, 460, 484, *503*
— see Wilson, H. 338, *409*
Cartland, G. F., and Kuizenga, M. H. 519, *557*
— see Kuizenga, M. H. 420, 421, *506*
Carvajal, F. see Herzog, H. L. 260, *392*
Carver, J. E. see Hartman, F. A. 14, *161*
Carvina, G. see Fidanza, A. 458, *503*
Casanova, R., C. W. Shoppee and G. Summers 280, *385*
Casas, C. B. see King, J. T. 133, *165*
Cascarano, J., and B. W. Zweifach 63, *150*

Case, J. F. 58, 143, *150*
Casper, A. see Mills, I. H. 115, *171*
Caspi, E. 461, 462, *501*
— and J. R. Bergen 459, *501*
— and O. Hechter 461, 462, *501*
— and M. M. Pechet 476, 498, *501*
— M. C. Lindberg, M. Hayano, J. L. Cohen, M. Matsuba, H. Rosenkrantz and R. I. Dorfman 441, 461, 462, *501*
— G. Rosenfeld and R. I. Dorfman 223, *385*, 421, 435, 441, *501*
— F. Ungar and R. I. Dorfman 435, 441, *501*
— see Vermeulen, A. 476, 497, *512*
Cassang, I. see Salvadori, B. 423, *510*
Casselman, W. G. B. see Schönbaum, E. 93, *178*
Cassidy, J. see Gould, D. 241, *391*
Cassidy, J. W. see Perlman, P. L. 535, *565*
Castells, J., and G. D. Meakins 249, *385*
Castle, H. see Migeon, C. J. 572, *632*
Castleman, B. see Cohen, R. B. 44, 51, 107, 129, 133, *151*
Cater, D. B., and J. D. Lever 9, 32, 36, 60, 87, 100, *150*
— and M. P. Stack-Dunne 32, 68, 79, 83, 90, 94, 100, 109, 110, *150*
— see Howard, A. N. 71, 104, *163*
Cattan, R., and P. Vesin 546, *557*
— see Vesin, P. 521, 546, 447, *568*
Cauldwell, E. W. see Anson, B. J. 23, *146*
Cauwenberge, H. van see Vliers, M. 424, *512*
Cavallero, C., E. Dova and L. Rossi 544, *557*
Cavina, G. 306, 334, *385*
— and L. Tentori 306, *385*
Celestino da Costa, A. 35, 36, 44, 57, *150*
Cella, J. A. see Kagawa, C. M. 524, *562*
Celozzi, E. see Cohen, A. I. 101, 102, 108, 135, *151*
Cerceo, E. see Pearlman, W. H. 293, 338, *401*
Cerviño, J. M., J. M. Manaro, E. Pollack and A. Proto 101, 135, *150*
Cessi, T. see Vannini, E. 38, *182*
Chabanier, H. see Porak, R. 515, *565*
Chaikoff, I. L. see Morris, M. D. 50, *171*
— see Siperstein, M. D. *511*
— see Tomkins, G. M. 455, *512*
— see Werbin, H. 435, *513*
Chamberlin, E. M., E. Tristram, T. Utne and J. M. Chemerda 254, *385*
Chambers, R. see Hyman, C. 552, *562*
Chan, T. K. 93, *150*
Chance, J., E. Lotsof, I. Pine, R. M. Patterson and J. Craig 649, *657*
Chang, D. W. see Melampy, R. M. 105, 139, *170*
Chang, E., and W. R. Slaunwhite 363, *385*
Chang, G. Y., and E. Witschi 142, 144, *150*
— see Witschi, E. 579, *636*
Channon, H. J. 437, *502*
Chapman, J. H., J. Elks, G. H. Phillips and L. J. Wyman 252, *385*
— see Attenburrow, J. 242, *382*
Chapman, W. B. see Cohen, R. B. 44, 51, 107, 129, 133, *151*
Charipper, H. A. see D'Angelo, S. A. 69, *152*
— see Hebard, W. B. 10, 11, 12, 53, *161*
— see Katsh, S. 604, *630*
— see Meyers, M. W. 18, 21, 27, 29, 57, 133, *170*
— see Stenger, A. H. 10, *180*
— see Weinstein, M. J. 623, *636*
Charney, W. see Gould, D. 267, 278, 279, *391*
— see Herzog, H. L. 260, 277, *392*, 494, *505*
— see Greenspan, G. 260, *391*
— see Nobile, A. 277, *399*
Chart, J. J. 521, *557*
— N. Hetzel and R. Gaunt 518, 521, 533, 543, *557*
— H. Sheppard, M. J. Allen, W. L. Bencze and R. Gaunt 457, *502*, 522, 523, *557*
— and E. S. Shipley 520, *557*
— — and E. S. Gordon 520, *557*
— see Gaunt, R. 518, 530, 543, 544, 545, 547, 548, *559*
— see Gordon, E. S. 520, *560*
— see Renzi, A. A. 531, 543, *566*
— see Sheppard, H. 523, *567*
Chavin, W. 4, 80, 88, 89, 90, 141, 144, *150*, 549, *557*
Chedid, L., and F. Boyer 100, *150*
Chemama, R., E. E. Baulieu, O. Crepy and M. F. Jayle 434, *502*
Chemerda, J. see Hirschmann, R. F. 271, *393*
Chemerda, J. M. see Chamberlin, E. M. 254, *385*
— see Steinberg, N. G. 287, *406*
Chen, C., S. M. Voegtli, and S. Freeman 312, *385*
— J. Wheeler and H. E. Tewell 340, *385*
— see Wheeler, J. 364, *409*
Chen, P. S., jr., H. P. Schedl, G. Rosenfeld and F. C. Bartter 447, *502*
Cheng, C. P. see Woodbury, D. M. 529, 531, 534, *569*
Cheng, C. S. see Nachlas, M. M. 63, *172*
Chester Jones, I. 4, 7, 9, 12, 18, 21, 27, 30, 31, 36, 47, 48, 57, 60, 78, 87, 100, 109, 110, 127, 128, 141, 143, *150*, 548, 549, 551, 552, 555, *557*, 591, 593, 595, 596, 607, *626*
— J. G. Phillips and W. N. Holmes 71, 75, 141, *151*, 416, *502*, *557*
— and C. C. Roby 78, 99, *150*
— and A. Wright 79, 116, 120, *150*
— see Christianson, M. 122, 123, 124, *151*
— see Dean, F. D. 607, 619, 623, *626*
— see Fowler, M. A. 551, *559*
— see Greep, R. O. 87, 104, 106, 107, *160*
— see Phillips, J. G. 555, *565*
— see Wright, A. 11, 12, 30, 53, 60, 78, 80, 83, 143, 144, *185*, 552, *569*
Chieffi, G. 6, 10, 53, 60, *151*
Chirvan-Nia, P. see Delost, P. 47, 48, 127, 128, *154*
Christ, J. 653, *657*
Christian, J. J. 104, *151*
Christiansen, R. G. see Clinton, R. O. 288, *386*
Christiansen, W. B. see Schotté, O. E. 523, *567*
Christianson, M., and I. Chester Jones 122, 123, 124, *151*
Christo, E. see Talbot, N. B. 582, 586, *635*
— see Zygmuntowicz, A. S. 308, *410*

Christy, N. P., M. M. Dickie and G. W. Woolley 618, *626*
— A. Donn and J. W. Jailer 105, *151*
— E. Z. Wallace, W. E. L. Gordon and J. W. Jailer 106, *151*
— — and J. W. Jailer 80, 100, 103, *151*
— see Hofmann, F. G. 606, 618, *629*
— see Jailer, J. W. 106, *164*
Ciaramelli, L. C. see Mason, H. L. 607, *631*
Cicardo, V. H. 552, *557*
Cinti, G., and E. Mignani 118, *151*
Clark, C. see Bloch, K. 436, *500*
Clark, D. E. see Landau, R.L. 615, *630*
Clark, E. D., and A. A. Bailey 641, *657*
Clark, L. C. jr., C. D. Kochakian and J. Lobotsky 478, *502*
— and H. Thompson 338, *386*
— see Sobel, E. H. 129, 135, *180*, 593, *635*
Clark, L. D. see Quarton, G. C. 643, *659*
Clark, W. G. 540, *557*
— E. L. Brackney and R. A. Miliner 143, *151*, 552, *557*
Clarke, J. R. 105, *151*
Clarke, R. L., K. Dobriner, A. Mooradian and C. M. Martini 233, 234, *386*
Clarke-Lewis, A. see Bondy, P. 367, *384*
Claton, G. W. see Johnson, J. E. jr. 457, *506*
Clatworthy, H. W. jr., and R. G. Anderson 39, 42, 43, *151*
Claus, J. L. see Llaurado, J. G. 107, *169*
Clayton, B. E., and J. E. Hammant 96, 106, *151*, 607, *626*
— see Prunty, F. T. G. 71, 105, *175*
Clayton, G. W. see Bongiovanni, A. M. 296, 301, *384*, 424, *501*, 611, *625*
— see Guillemin, R. 77, 85, 100, 103, *160*
Clayton, R. B., and K. Bloch 436, 438, *502*
Clayton, R. E., J. E. Hammant and P. Armitage 454, *502*
Cleghorn, R. A. 654, *657*
— and C. J. Pattee 654, 655, *657*
Clemente, J. see Bernstein, S. 236, *383*
Clinton, M. jr., G. W. Thorn, H. Eisenberg and K. E. Stein 537, *557*
Clinton, R. O., R. G. Christiansen, H. C. Neuman and S. C. Laskowski 288, *386*
— H. C. Neuman, A. J. Manson, S. C. Laskowski and R. G. Christiansen 288, *386*
Cloche, R., E. Azerad and Stuhl 655, *657*
Coates, V. J. see Williams, V. Z. 354, *409*
Cobb, S. see Quarton, G. C. 643, *659*
— see Trethowan, W. H. 647, *659*
Cobey, F. A., I. Taliaferro and H. H. Haag 136, *151*
— see Dituri, F. 436, 438, *502*
— see Rabinowitz, J. L. 438, *509*
— see Taliaferro, I. 455, *511*
Code, C. F. see Mattox, V. R. 198, 313, *398*
Coffman, J. R. 310, *386*
— see Hoskins, W. H. 616, *629*
Cohen, A. 44, 58, 59, 121, 123, *151*
— see Jost, A. 124, *164*
Cohen, A. I., E. Bloch and E. Celozzi 101, 102, 108, 135, *151*
— J. Furth and R. F. Buffett 101, 114, 115, 134, 135, *151*, 607, *626*
Cohen, H., H. H. Freedman, W. Kleinberg, N. R. Schechter and G. J. Martin 82, 83, *151*
Cohen, J. L. see Caspi, E. 441, 461, 462, *501*
Cohen, R. B. 63, *151*
— W. B. Chapman and B. Castleman 44, 51, 107, 129, 133, *151*
— see Rutenburg, A. M. 67, *177*
— see Seligman, A. M. 66, *178*
Cohen, S. L. 295, 296, 297, 299, *386*
— and A. Kaluszyner 338, *386*
— and I. B. Oneson 295, *386*
— see Bitman 292, 294, *383*
Colas, A. see Fotherby, K. 428, 429, 446, 478, *504*
Cole, A. R. H. 347, 350, *386*
— R. N. Jones and K. Dobriner *386*
— see Jones, R. N. 355, 356, *394*
Cole, D. F. 532, *557*
Cole, E. see Ulstrom, R. A. 591, *635*
Cole, P, M. see Dean, F. D. 607, 619, 623, *626*
Cole, R. D. see Lyons, W. R. 613, *631*
Cole, W., and P. Julian 273, *386*
— see Bachman, W. E. 261, *382*
— see Julian, P. L. 216, 225, 245, 274, *395*
Collings, W. D. see Swingle, W. W. 546, *568*
Collins, A. R. see Graff, M.M. 337, *391*
— see Holecheck, J. J. 337, *393*
Collins, E. J. 530, *557*
— see Swingle, W. W. 621, *635*
Collip, J. B. see Selye, H. 100, 106, *179*, 455, *511*
Cologne, A. see Courrier, R. 124, *152*
Colover, J., and L. E. Glynn 141, *151*
Colton, F. G., W. R. Nes, D. A. van Dorp, H. L. Mason, and E. C. Kendall 274, *386*
Combescot, C. 143, *151*
Comline, R. S. see Balfour, W. W. 75, 147
Compere, E. L. see Hastings, A. B. 529, *561*
Conbere, J. see Leanza, W. 225, *396*
Conlin, M. M. see Brown, J.B. 49, 51, 54, *149*
Conn, J. W. 135, *152*, 540, *557*
— and L. H. Louis 531, 535, 547, *557*
— S. S. Fajans and L. H. Louis 518, *557*
Conrad, E. see Persky, H. 642, *659*
Considine, W. J. see Rosenfeld, R. S. 436, 489, *510*
— see Soloway, A. H. 218, 225, *405*
Constable, B. J. see Howard, A. N. 51, 58, 84, 85, 144, *163*
Constantin, J. M., and L. H. Sarett 269, *386*
— see Sarett, L. H. 262, *404*

Constantinides, and G. Loukatos 649, *657*
Constantinides, P., C. Fortier and F. R. Skelton 69, 70, *152*
— see Fortier, C. 50, 58, 59, *157*
Conway, E. J. 532, *557*
Cook, E. R., B. Dell and D. J. Wareham 316, 317, *386*
— and M. E. Rooks 297, 299, 363, *386*
— S. R. Stitch, A. E. Hall and M. P. Feldman 314, *386*
Cook, J. W. see Dorfman, R. I. 478, *503*
Cook, L. R. see Klopper, A. 463, *506*, 610, 611, *630*
Cook, R. P., A. Kliman and L. F. Fieser 436, *502*
Cookson, R. C. see Barton, D. H. R. 193, *382*
Cooley, G., B. Ellis, F. Hartley and V. Petrow 273, *386*
Coons, A. H. see Hill, A. G. S. 118, *162*
— see Kaplan, M. H. 118, *164*
Cooper, D. Y., M. Kasparow, W. S. Blakemore and O. Rosenthal 422, 423, *502*
— J. C. Touchstone and J. M. Roberts 449, *502*
— see Touchstone, J. C. 75, *182*, 449, *512*
Cope, C. L. 641, *657*
— and E. G. Black 475, *502*
— and B. Hurlock 459, 460, 475, *502*
— and C. E. Sewell 422, *502*
Coppage, W. S., D. Island, M. Smith and G. W. Liddle 523, *558*
Corbetta, S. see Malandra, B. 646, *658*
Corcoran, A. C., H. P. Dustan and I. H. Page 365, *386*
— and I. H. Page 340, *386*
— see Masson, G. M. C. 532, *564*
— see Savard, K. 424, *510*
Corey, E. J., and W. E. Hertler 237, *386*
Cornforth, J. W. 261, *386*
— R. H. Cornforth, A. Pelter, M. G. Horning and G. Popjak 368, *386*
— — G. Popjak and I. Youhotsky-Gore 438, *502*
— I. Y. Gore and G. Popjak 437, 438, *502*
— G. D. Hunter and G. Popjak 437, *502*
— and G. Popjak 437, 438, *502*
— — and I. Y. Gore 437, *502*
Cornforth, J. W., I. Youhotsky and G. Popjak 191, *386*
— see Cardwell, H. M. 261, *385*
Cornforth, R. H. see Cornforth, J. W. 368, *386*, 438, *502*
Correale, L. see Lamedica, G. 476, *506*
Cort, J. H. see Seldin, D. W. 532, *567*
— see Welt, L. G. 530, 569
Corwin, A. H. see Allen, M. J. 522, *556*
Costa, P. J., D. D. Bonnycastle and I. J. Seligman 138, *152*
Coupland, R. E. 118, 120, *152*
— and R. L. Holmes 26, *152*
Courrier, R., M. Baclesse and M. Marois 106, 107, *152*
— A. Cologne and M. Baclesse 124, *152*
Coury, C. see Bariety, M. 546, 556
Coutinho, H. B., B. L. Baker, J. F. Kent, E. C. Pliske and J. G. van Dyke 66, 98, 109, *152*
Covian, M. R. 59, 86, *152*
Cowell, H. S. see Kelleway, C. H. 515, *562*
Cox, R. I. 340, 365, *386*
— and G. F. Marrian 318, *386*, 431, *502*
Crabbé, J. see Hernando, L. 457, *505*, 531, *561*
— see Renold, A. E. 137, *175*, 457, *509*, 522, *566*
Craig, D. see Craig, L. C. 296, 304, *386*
Craig, J. see Chance, J. 649, *657*
Craig, J. M., and B. H. Landing 129, *152*
Craig, L. C., D. Craig and E. G. Scheibel 296, 304, *386*
Crane, K. F. see Reich, H. 310, *402*
Crawshaw, A., H. B. Henbest and E. R. H. Jones 228, *386*
Crépy, O., M. F. Jayle and F. Meslin 310, *387*
— see Chemama, R. 434, *502*
— see Jayle, M. F. 293, *394*, 434, *506*
— see Mallassis, D. 434, *507*
Creutzfeldt, W., M. Husten and K. Haager 49, 51, 84, 88, 93, 94, 104, *152*
Criegee, R., B. Marchand and H. Wannowius 250, *387*
Crigler, J. F. jr., S. H. Silverman and L. Wilkins 521, *558*
— see Wilkins, L. 585, 616, *636*
Cristol, S. J., and H. J. Haller 135, *152*
Crooke, A. C., and J. R. Gilmour 99, 109, *152*
Crooks, H. M. see Marker, R. E. 431, *507*
Cross, J. M. see Tansey, R. P. 338, *407*
Crossfield, H. C. see Swingle, W. W. 529, 536, 547, *568*
Crowder, R. E. 589, *626*
Croxatto, H. see Badinez, O. 112, 115, *146*
Cueto, C., and J. H. U. Brown 135, 136, *152*, 523, *558*
Cumming, G. R. see Perry, W. F. 92, *174*
Curran, G. L. 436, *502*
— and D. Rittenberg 436, *502*
Curran, J. F. see Tomkins, G. M. 488, *512*
Curtis, F. R. see Vincent, S. 7, *183*
Curtis, J. M. see Doisy, E. A. 426, *502*
Curtis, R. H. see Luetscher, J. A. jr. 110, 113, 114, *169*, 546, *563*
Cushing, H. 107, 133, *152*
Cutuly, E. 78, 79, *152*
— and E. C. Cutuly 133, *152*
Cutuly, E. C. see Cutuly, E. 133, *152*

Dacanay, J. H. see Heming, A. E. 523, *561*
Daft, F. S. see Mc Queeney, A. J. 138, *170*
Daily, W. J. R. see Ganong, W. F. 110, *157*
Dallwig, R. 88, 116, *152*
Dalton, A. J., E. R. Mitchell, B. F. Jones and V. B. Peters 86, *152*
Damast, B. L. see West, C. D. 412, *513*, 616, 617, *636*
D'Angelo, S. A., A. S. Gordon and H. Charipper 69, *152*
Dann, S. see Gutner, L. B. 546, *560*
Dannenberg, H. 62, *152*
D'Arcy, P. F. 135, *152*
Darrach, M. see Southcott, C. M. 423, 425, *511*
Darrow, D. C., H. E. Harrison and M. Taffel 529, 538, *558*
— see Harrison 529, *560*
— see Miller, H. C. 532, *564*

Das Gupta, D., and C. J. P. Giroud 73, 114, *152*
Dauben, H. J., B. Löcken and H. J. Ringold 244, *387*
Dauben, W. G., E. J. Blanz jr., J. Jiv and R. A. Micheli 245, *387*
— D. Dickel, O. Jeger and V. Prelog 191, *387*
— and J. P. Eastham 233, 244, *387*
— G. J. Fonken and D. S. Noyce 245, *387*
— and T. W. Hutton 437, 502
— R. A. Micheli and J. F. Eastham *387*
— and K. H. Takemura 437, *502*
Daughaday, W. 596, *626*
Daughaday, W. H. 106, *152*
— H. Jaffe and R. H. Williams 296, *387*
— see Levin, M. E. 105, *168*
Daum, K. see Bean, W. B. 139, *147*
Daus, M. see Hirschmann, H. 273, *393*
Da Vanzo, J. see Swingle, W. W. 529, 538, 547, *568*
Davenport, H. A. see MacFarland, W. E. 4, 25, *169*
Dávid, M. A. see Kovács, K. 112, 113, *166*
Davidson, A. G., E. G. Dulit and I. D. Frantz jr. 436, *502*
Davidson, C. S. 607, 608, *626*
— and H. D. Moon 607, 608, *626*
Davidson, D. W. see Brownie, A. C. 443, 489, *501*
Davidson, J. N. see Symington, T. 36, 63, 65, 66, 90, 92, 104, *181*
Davies, G. see Levin, B. 293, *396*
Davies, J. N. P. see Stoner, H. B. 125, *181*
Davies, S. 48, *152*, 598, *626*
Davis, A. K., A. C. Bass and R. R. Overman 539, *558*
— see Flanagan, J. B. 536, 537, *558*
— see Overman, R. R. 537, 538, *565*
Davis, C. see Nichols, J. 457, *508*, 523, *565*
Davis, D. M. see Marshall, E. K. jr. 515, *564*
Davis, E. see Werbin, H. 435, 441, 449, *513*
Davis, J. H. see Howard, J. M. 642, *658*
Davis, J. O., and W. C. Ball jr. 115, *152*
Davis, J. O., M. J. Goodkind, M. M. Pechet and W. C. Ball jr. 114, *152*, 531, *558*
— and D. S. Howell 531, *558*
— B. Kliman, N. A. Yankopoulos and R. E. Peterson 114, *153*
— M. M. Pechet, W. C. Ball jr. and M. J. Goodkind 114, *153*
— see Yankopoulos, N. A. 117, *185*
Davis, J. W. see Ungar, F. 460, *512*
— see Wotiz, H. H. 616, *637*
Davis, M. E., and N. W. Fugo 611, *626*
— and E. J. Plotz 124, *153*, 572, 581, 582, 591, *626*
— — and E. Plotz 86, *153*
— C. E. Test, C. A. Navori, B. Hryse, R. E. Pottinger and F. Dunkle 611, *626*
— see Plotz, E. J. 582, *633*
Davis, R. B. see McMahon, J. M. 342, *398*
Davisson, J. W. see Shull, G. M. 493, *511*
Dawson, A. B. 17, 53, 60, *153*
— see Velardo, J. T. 614, *636*
Dawson, I. M. P. 45, *153*
De, N. N. see Kar, A. B. 107, *164*, *165*
Dean, F. D., P. M. Cole and I. Chester Jones 607, 619, 623, *626*
Dean C. M. jr., see Moog, F. 20, 43, 125, *171*, 596, *632*
Dean, J. A. see Willard, H. H. 358, 368, *409*
Deane, H. W. 20, 40, 43, 45, 49, 50, 53, 54, 56, 57, 58, 79, 84, 92, 116, 123, 125, *153*, 598, *626*
— and A. C. Barger 34, 114, *153*
— R. J. Barrnett and A. M. Seligman 63, 64, 66 *153*
— and D. W. Fawcett 57, *153*
— and R. O. Greep 28, 37, 38, 54, 78, 79, 88, 92, 99, 100, 104, 105, 109, 110, 132, 140, *153*
— F. G. Hofmann, A. K. Solomon and G. B. Wislocki 117, 118, *153*
— and J. M. McKibbin 105, 131, 138, 139, *153*
— and C. P. Lyman 19, 28, 35, 54, 106, 107, *153*
— and G. M. C. Masson 88, 112, 115, 116, 131, *153*, 547, *558*
Deane H. W., and A. Morse 59, 60, 87, 96, *153*
— R. H. Schneiweiss and L. I. Gidez 114, *153*
— and A. M. Seligman 55, 95, *153*
— and J. H. Shaw 93, *153*
— — and R. O. Greep 92, 94, 112, 113, 114, 115, *153*
— see Bergner, G. E. 93, 101, 110, *148*
— see Dempsey, E. W. 65, 96, 109, *154*
— see Dorp, A. W. V. van 31, 32, 40, 41, 45, 48, 125, *154*
— see Fawcett, D. W. 620, *627*
— see Greep, R. O. 1, 36, 45, 53, 95, 101, 110, 115, 116, 118, 119, 120, 139, *160*
— see Hill, A. G. S. 118, *162*
— see Josimovich, J. B. 44, 121, 125, *164*
— see Kaplan, M. H. 118, *164*
— see Karnovsky, M. L. 49, 50, 54, 55, 56, 78, 92, 94, 95, 109, *165*
— see Ladman, A. J. 125, *167*
— see Levy, H. 61, 63, 96, 97, 109, *168*
— see Olson, R. E. *173*
Deanesly, R. 21, 88, 92, 104, 128, *153*, 593, 596, 609, *626*
— and A. S. Parkes 575, 576, 596, *626*
— see Callow, R. K. 575, *625*
Debauche, R. see Bourguet, J. 531, *556*
De Bruin, C. H. see Sweat, M. L. 443, 485, 489, *511*
De Cicco, A. see Fidanza, A. 457, 458, *503*
De Courcy, C. 341, *387*
— and C. H. Gray 364, *387*
— and J. J. Schneider 461, 497, *502*
Defendi, V. see Pearson, B. 66, *173*
Dejardin, M. 26, *153*
De Jongh, E. S. see Dingemanse, E. 426, *502*
Delay, J., L. Bertagna and A. Lauras 641, *657*
Del Castillo, E. B., and G. Di Paola 622, *626*
— and R. Sammartino 106, *153*
Delea, C. S. see Bartter, F. C. 114, *147*
— see Liddle, G. W. *506*
Delgatte, C. see Tuchmann-Duplessis, H. 9, 65, 80, 88, 90, 96, 143, 144, *182*

Delgatte, F. see Tuchmann-Duplessis, H. 9, 65, 80, 88, 90, 96, 143, 144, *182*
Dell, B. see Cook, E. R. 316, 317, *386*
Delost, H. see Delost, P. 48, 106, 118, *154*
Delost, P. 48, 128, *153*, *154*, 597, 598, 603, 605, *626*
— and P. Chirvan-Nia 47, 48, 127, 128, *154*
— and H. Delost 48, 106, *154*
— and A. M. Vincent 597, *626*
— H. Delost and T. Terroine 118, *154*
Delsol, M. see Verne, J. 9, *183*
De Meio, R. H., and C. Lewycka 452, 478, *502*
— — and M. Wizerkaniuk 478, *502*
— see Pearlman, W. H. 484, *508*
Demetriou, D. A. see Eik-Nes, K. 422, *503*
Demey, L. see Vliers, M. 424, *512*
Demoisson, F. F. see Weinmann, S. 434, *512*
Dempsey, E. see Bartter, F. C. 531, *556*, 585, 616, *624*
Dempsey, E. W., R. O. Greep and H. W. Deane 65, 96, 109, *154*
Dempster, W. J. 118, 120, *154*
Denber, H. C. B. 21, *154*
Dencker, S. J., R. Schlaug and W. Silfverskiöld 641, 643, *657*
Dennis, C., and E. H. Wood 540, *558*
Dennison, M. see Korenchevsky, V. 455, *506*
Denton, D. A. see Goding, J. R. 541, *560*
DeOme, K. B. see Campbell, R. A. 106, *149*
Deramond, M. see Verne, J. 17, *183*
Desaulles, P. see Neher, R. 204, 206, *399*, *565*
— see Schuler, W. 524, *567*
— see Wettstein, A. 206, *408*
Desaulles, P. A. 521, *558*
— W. Schuler and R. Meier 518, 532, 533, 558
— J. Tripod and W. Schuler 532, *558*
Desclaux, P. see Soulairac, A. 98, *180*
Des Marais, A., L. P. Dugal and P. M. Gagnon 92, *154*
— and P. M. Gagnon 59, *154*
De Souza, E. see Nachlas, M. M. 63, *172*
Dettbarn, W., and F. Gross 531, *558*
Deucher, F. see Schamaun, M. 652, *659*
Deulofeu, V., and J. Ferrari 426, *502*
De Venuto, F., S. J. Mule and U. F. Westphal 300, *387*
Devis, R. 292, 293, *387*
— see Vanek, R. 295, *407*
De Visscher, M., and C. Beckers 140, *154*
De Young, J. see Harman, R. E. 421, 446, *504*
Dhyse, F. G. see Hertz, R. 137, *161*, 456, *505*, 522, *561*
Dibrell, W. H. see Howard, J. M. 642, *658*
Dickel, D. see Dauben, W. G. 191, *387*
Dickie, M. M., and P. W. Lane 134, *154*, 618, *626*
— see Atkinson, W. B. 134, *146*
— see Christy, N. P. 618, *626*
— see Hofmann, F. G. 606, 618, *629*
— see Woolley, G. W. 134, *185*, 618, *637*
Diczfalusy, E., G. Birke, C. Franksson, J. Hellstrom, S. Hultberg, L. O. Plantin and A. Westman 412, *502*
— G. Birke and L. O. Plantin 364, *387*
— and R. Luft 412, *502*, 611, *627*
— see Borell, U. 608, *625*
— see Plantin, L. O. 414, 415, *509*, 572, *633*
Dietrich, A. D. see Geppert, L. J. 643, *658*
Dieckert, J. W., and R. Reiser 310, 341, *387*
Di George, A. M., J. B. Arey and A. M. Bongiovanni 593, *627*
Dillon, R. N. see Majnarich, J. J. 433, *507*
Di Mauro, S. see Hollander, V. P. 365, *393*
Dingman, J. F. 545, *558*
— see Laidlaw, J. C. 518, 533, 535, *563*
— see Thorn, G. W. 518, 540, *568*
Dingemanse, E., S. E. De Jongh, S. Kober and E. Laqueur 426, *502*
— L. G. Huis in't Veld and S. L. Hartogh-Katz 308, 309, *387*, 580, *627*
— — and B. M. de Laat 308, *387*
Dingemanse, E., E. Laqueur and O. Muhlboch 426, *502*
— see Huis in't Veld, L. G. 412, *505*
Di Paola, G. see Del Castillo, E. B. 622, *626*
Dirscherl, W. 280, *387*
— and H. Breuer 337, *387*
— and E. Gerhards 319, *387*
— W. Korus and H. Schriefers 314, *387*
— see Shriefers, H. 476, *511*
Dittus, P. 6, 8, 79, 80, 90, 141, 144, 145, *154*
Dituri, F., F. Cobey, J. V. B. Warms and S. Gurin 436, 438, *502*
— see Rabinowitz, J. L. 438, *509*
Djerassi, C. 242, 251, 374, 375, *387*
— R. Engle and A. Bowers 232, 288, *388*
— E. W. Foltz and A. E. Lippman 374, *388*
— L. B. High, J. Fried and E. F. Sabo 487, 492, 493, *502*
— — T. T. Grossnickle, R. Ehrlich, J. A. Moore and R. B. Scott 336, *388*
— and W. Klyne 375, *387*
— and C. T. Lenk 220, *387*
— A. Lippman and J. Grossman 287, 288, 376, *388*
— A. Manson and H. Bendas 289, *388*
— L. Miramontes and G. Rosenkranz 280, *387*
— — — and F. Sondheimer 280, *387*
— R. Riniker and B. Riniker 281, 288, 290, *388*
— G. Rosenkranz, J. Iriarte, J. Berlin and J. Romo 281, *387*
— — J. Romo, J. Pataki and S. Kaufmann 288, *387*
— see Foltz, E. W. 205, 376, *389*
— see Kaufman, S. 279, *395*
— see Mancera, O. *397*, 492, *507*
— see Pataki, J. 204, *400*
— see Romo, J. 361, *403*
— see Rosenkranz, G. 255, 257, *403*
— see Sandoval, A. 281, 404
— see Sondheimer, F. 211, 227, 246, *405*
— see Stone, D. 491, *511*
— see Zaffaroni, A. 281, *409*
Dobriner, K. 431, *502*, 580, 581, *627*

Dobriner, K., E. R. Katzenellenbogen and R. N. Jones 205, 347, 350, *388*
— — and R. Schneider 420, 421, *502*
— and S. Lieberman 427, 432, 433, 459, *502*
— — and C. P. Rhoads 301, 303, 357, *388*
— — — R. N. Jones, V. Z. Williams and R. B. Barnes 301, 303, *388*
— — — and H. C. Taylor jr. 582, *627*
— C. P. Rhoads, S. Lieberman, B. R. Hill and L. F. Fieser 431, *502*
— see Barry, M. C. 480, *500*
— see Clarke, R. L. 233, 234, *386*
— see Cole, A. R. H. *386*
— see Fukushima, D. K. 480, 482, *504*
— see Gallagher, T. F. 478, *504*
— see Jones, R. N. 215, 347, 353, 354, 355, 357, *394*
— see Katzenellenbogen, E. R. 312, 314, *395*
— see Lieberman, S. 217, 307, *396*, 427, 428, 431, 432, *507*
— see Salamon, I. I. 429, *510*
Dobriner, S. see Fukushima, D. K. 217, *390*
— see Wendler, N. L. 217, *408*
Dobson, F., and S. S. Randall *300*, *388*
— see Baker, C. P. 313, *382*
Dobson, M. M. see Talalay, P. 496, *511*
Dodson, R. M., A. H. Goldkamp and R. D. Muir 233, 260, *388*, 487, *502*
— P. Sollman and B. Riegel 288, *388*
Doering, W. von E. see Hayano, M. 443, 485, 489, *505*
Doglin, A. E. see Wexler, B. C. 85, 100, 101, *184*
Dohan, F. C., J. C. Touchstone and E. M. Richardson 432, 433, *502*
— see Richardson, E. M. 432, 459, 460, *509*
— see Touchstone, J. C. 432, *512*
Doisy, E. A., S. A. Thayer, L. Levin and J. M. Curtis 426, *502*
— C. D. Veler and S. A. Thayer 426, *502*
— see Ashmore, J. 478, *500*
— see Buehler, H. J. 293, 295, 297, *384*
Doisy, E. A., see Smith, G. V. S. 426, *511*
Doisy, E. A. jr. see Ashmore, J. 480, *500*
Dominguez, O. V. see Berliner, D. L. 369, 372, *383*
— see Samuels, L. T. 475, *510*
Domm, L. V., and P. Leroy 123, *154*
Donaldson, H. H. 20, *154*
Donaldson, L. see Humphreys, E. M. 136, *163*
Donáth, T. 23, *154*
Done, A. K., R. S. Ely, L. R. Heiselt and V. C. Kelley 454, *502*
— — L. J. Olsen and V. C. Kelley 471, *502*
— — R. B. Raile and V. C. Kelley 75, *154*
— see Ely, R. S. 471, *503*
— see Good, T. A. 75, 82, 100, *159*
Donia, R. A. see Slomp, G. 357, *405*
Donin, M. N. see Fried, J. 491, 492, *504*
Donn, A. see Christy, N. P. 105, *151*
Doorenbos, H. see Buchem, F. S. P. van 135, *149*
Dörffler, P. see Schaumkell, K. W. 43, 44, *178*
Dorfman, A. S. see Rubin, B. L. 617, *634*
Dorfman, H. see Rosenkrantz, H. 478, *510*
Dorfman, L. 359, 360, *388*
Dorfman, R. I. 439, 451, 459, 460, 463, 472, 478, 481, 482, *502*, 573, 585, 610, *627*
— J. W. Cook and J. B. Hamilton 478, *503*
— and W. U. Gardner 618, *627*
— and J. B. Hamilton 478, *503*, 616, *627*
— M. Hayano, R. Haynes and K. Savard *503*
— and R. A. Shipley 575, 577, *627*
— and F. Ungar 463, *503*
— J. Wise and R. A. Shipley 478, *503*
— — and G. Van Wagenen 484, *503*
— see Baggett, B. 412, 451, *500*, 616, *624*
— see Bloch, E. 204, *383*, 415, 449, 451, *500*, 572, *625*
— see Bloom, B. M. 459, *500*
— see Bowers, H. *384*
— see Burstein, S. 234, *384*, 428, 433, 454, 459, 460, 472, 480, *501*
Dorfman, R. I. see Caspi, E. 223, *385*, 421, 435, 441, 461, 462, *501*
— see Engel L. L. 427, *503*
— see Fish, W. R. 459, 484, *503*
— see Forchielli, E. 462, 495, 497, 498, 499, *504*
— see Goldstein, M. 414, *504*
— see Hayano, M. 61, 62, *161*, 420, 435, 441, 443, 449, 454, 461, 485, 486, 489, 490, 491, 494, *504*
— see Haynes, R. C. jr. 77, *161*, 442, *505*
— see Horwitt, B. N. 459, *505*, 610, *629*
— see Koritz, S. B. 454, *506*
— see Miller, A. M. 219, *399*, 427, 428, 478, *508*, 574, 577, *632*
— see Nadel, E. M. 433, *508*
— see Peron, F. 414, 427, 428, 429, 433, 445, *509*
— see Pincus, G. 583, 608, *633*
— see Rosemberg, E. 446, 447, *509*
— see Rosenfeld, G. 76, 111, 113, *176*, 443, 449, 490, *510*
— see Rubin, B. L. 61, *177*, 250, 312, 328, *403*, 496, *510*, 617, *634*
— see Savard, K. 311, *404*, 413, 431, 434, 451, 460, 478, 480, 482, 488, *510*, 579, *634*
— see Schiller, S. 427, 431, *510*, 572, *634*
— see Stone, D. 491, *511*
— see Ungar, F. 427, 432, 435, 441, 442, 449, 460, 463, 477, 478, 486, 489, 495, 496, 497, *512*, 574, *635*
Dorp, A. W. V. van, and H. W. Deane 31, 32, 40, 41, 45, 48, 125, *154*
Dorp, D. A. van see Colton, F. G. 274, *386*
Dougherty, T. F., H. E. Brown and D. L. Berliner 460, *503*
— see Berliner, D. L. 461, 462, *500*
Douglas, J. F., and C. G. King 452, *503*
Douglas, M. see Strong, J. A. 412, *511*
Douste-Blazy, L. see Baïsset, A. 115, 116, *146*
Dova, E. see Cavallero, C. 544, *557*
Do Valle, J. R. 552, *558*
Dow, D., and S. Zuckerman 550, *558*

Dowben, R. M., and J. Rabinowitz 484, *503*
Dowdy, A. see Luetscher, J. A. jr. 520, *563*
Drake, B. 310, *388*
Drake, F. R. 654, *657*
Drake, N. A., W. J. Haines, R. E. Knauff and E. D. Nielson 341, *388*
Dreiding, A. C., W. J. Pummer and A. J. Tomasewski 287, *388*
Drekter, I. J., A. Heisler, R. G. Scism, S. Stern, S. Pearson and T. H. McGavack 302, *388*
Drellich, M. G. see Waxenberg, S. E. 579, 581, *636*
Drewes, A. A. see Repetto, O. M. 297, *402*
Dreyfus-Brisac, C. see Bricaire, H. 655, *657*
Dribben, I. S., and J. M. Wolfe 27, 29, *154*
Drilhon, A. 549, *558*
Driscol, T. E., M. M. Maultsby, G. L. Farrell and R. M. Berne 114, *154*
Druzhinia, K. V. see Udaev, N. A. 445, *512*
Dryenfurth, I. see Venning, E. H. 110, 111, *182*
Ducommun, P., and R. S. Mach 96, 101, 110, *154*
Duff, S. R. see Cardwell, H. M. 261, *385*
Dugal, L. P., and M. Therien 70, *154*
— see Des Marais, A. 92, *154*
Duguid, W. P. see Symington, T. 36, 63, 65, 66, 90, 92, 104, *181*
Dulaney, E. L., E. O. Stapley and C. Hlavac 492, *503*
Dulin, W. E. 144, *154*
— B. J. Bowman and R. O. Stafford 267, 284, *388*
— see Byrnes, W. W. 521, *557*
Dulit, E. G. see Davidson, A. G. 436, *502*
Dumm, M. E. and E. P., Ralli 139, *154*
— H. Gerschberg, E. M. Beck and E. P. Ralli 458, *503*
— B. Laken and E. P. Ralli 69, 71, *154*
— P. Orvando, P. Roth and E. P. Ralli 139, *154*
Duncan, I. W. 54, 105, *154*,
Duncan, L. E. jr. see Bartter, F. C. 114, *147*
— see Liddle, G. W. 111, 113, 114, 115, *168*, *506*
Dunkle, F. see Davis, M. E. 611, *626*
Dünnenberger, M., and M. Ehrenstein 240, *388*
Durey, J. M. 106, *154*
Durrum, E. L. see Block, R. J. 304, 313, *383*
— see Paul M. H. 304, *400*
Dury, A. see Dury, M. 71, *154*
Dury, M., and A. Dury 71, 154
Dustan, H. P. see Corcoran, A. C. 365, *386*
Duszynska, J. see Truszowski, R. 530, *568*
Dutler, H., H. Bosshard and O. Jeger 278, *388*
Dutton, G. J., and I. D. E. Storey 452, *503*
— see Storey, I. D. E. 452, *511*
Dyke, J. G. van see Continho, H. B. 66, 98, 109, *152*
Dyrenfurth, I., and E. H. Venning 296, *388*
— see Beck, J. C. 116, *147*
— see Venning, E. H. 304, 333, *407*, 475, *512*

Eades, H. C. jr., R. L. Pollack and J. S. King jr. 418, *503*
Eagle, J. F. 590, *627*
Earle, D. P. jr., R. C. de Bodo, E. L. Schwartz, S. J. Farber, M. Kurts and J. Greenberg 99, *155*
Eastheim, J. F. see Dauben, W. G. *387*
Eastham, J. P. see Dauben, W. G. 233, 244, *387*
Eberlein, W. R. 313, *388*
— and A. M. Bongiovanni 328, 333, 334, *388*, 423, 427, 431, 432, *503*, 585, 610, *627*
— see Bongiovanni, A. M. 296, 301, 338, 364, 365, *384*, 424, 425, 432, 460, *501*, 585, *625*
Eichhorn, J., and O. Hechter 441, *503*
Eckert, J. F. see Richter, C. P. 529, *566*
Eckstein, B. see Perek, M. 58, 85, *174*
Edelman, I. S., H. L. Young and J. B. Harris 550, *558*
Edelmann, A. see Eversole, W. J. 542, *558*
Edgar, D. G. 613, *627*
Edwards, R. W. H., and A. E. Kellie 310, 340, *388*
— — and A. P. Wade 293, *388*
Efron, D. H. 538, 540, *558*
Egdahl, R. H., and J. B. Richards 105, 106, *155*
Egdahl, R. H., J. B. Richards and D. M. Hume 522, *558*
— see Nelson, D. H. 73, 75, 76, 103, 106, *172*
Eger, W., W. Gottesleben and M. Tiefjen 64, 65, 96, *155*
— see Brune, H. 103, *149*
Ehrenbrand, F. 44, 104, *155*
Ehrenstein, M., and T. Stevens 234, 282, *388*
— see Barber, G. 204, 238, 239, 282, 290, 361, *382*
— see Dünnenberger, M. 240, *388*
— see Florey, K. 204, 214, 234, 360, 374, *389*
— see Herzig, P. T. 221, 234, 235, *392*
Ehrlich, R. see Djerassi, C. 336, *388*
Eichenberger, K. see Heusser, H. 212, *393*
Eichner, D. 88, 92, 94, 112, 115, *155*, 646, *657*
Eidinoff, M. L., J. E. Knoll, D. K. Fukushima and T. F. Gallagher 250, *388*
— R. S. Rosenfeld, J. E. Knoll, B. J. Marano and L. Hellman 436, 438, *503*
— see Barry, M. C. 480, *500*
— see Fukushima, D. K. 250, *390*
Eik-Nes, K. B. 304, *388*
— and K. R. Brizzee 79, 100, 103, 140, *155*
— D. A. Demetriou, Y. C. Mayne and R. S. Jones 422, *503*
— D. H. Nelson and L. T. Samuels 312, *388*
— A. A. Sandberg, F. H. Tyler and L. T. Samuels 82, *155*
— J. A. Schellmann, R. Lumry and L. T. Samuels *503*
— see Bliss, E. L. 572, *625*
— see Carstensen, H. 573, *626*
— see Jones, R. S. 85, *164*
— see Oertel, G. W. 75, *173*, 413, 425, 463, *508*
— see Samuels, L. T. 475, *510*
— see Sandberg, A. A. 475, *510*
— see Tyler, F. H. 475, *512*
— see Wallach, S. 134, *183*, 412, *512*, 585, 611, 615, *636*
Eilat, A. see Perek, M. 85, *174*
Eiseman, B., R. Hughes, W. Summers and F. Traylor 118, *155*
Eisenberg, E., and G. S. Gordan 576, *627*
— see Shemano, I. 576, *634*

Eisenberg, F., and S. Gurin 452, *503*
Eisenberg, H. see Clinton, M. jr. 537, *557*
— see Thorn, G. W. 537, *568*
Eisenman, A. J. see Swingle, W. W. 515, *568*
Eisenstein, A. B. 115, 139, *155*, 458, 463, *503*
— and P. M. Hartroft 113, 114, *155*
— and R. E. Shank 71, 105, *155*
— and A. S. Wenneker 458, *503*
— see Hartroft, P. M. 112, 114, 133, *161*
— see Wenneker, A. S. 71, 105, 139, *155*
Eisler, M. see Gould, D. 267, 278, 279, *391*
— see Oliveto, E. P. 285, *400*
— see Swingle, W. W. 99, *181*
Ekholm, E., and K. Niemineva 587, *627*
Elftman, H. 65, *155*
Elias, H. 26, 28, 45, *155*
— and J. E. Pauly 23, 27, 28, 45, 130, *155*
Elks, J. see Chapman, J. H. 252, *385*
Ellings, H. S. see Buchem, F. S. P. van 135, *149*
Elliott, F. H., and A. V. Schally 420, 421, *503*
— see Birmingham, M. K. 454, *500*
Elliott, H. A. see Gofman, J. W. 292, *391*
Elliott, H. W. see Gordan, G. S. 653, *658*
Elliott, J. see Engel, L. L. 304, 306, *389*
Elliott, W. H. see Ashmore, J. 480, *500*
Ellis, A. 650, *657*
Ellis, B., F. Hartley, J. Petrow and D. Wedlake 273, 274, *388*
— see Cooley, G. 273, *386*
Ellis, W. see Gardner, L. I. 582, *628*
Ellison, R. G. see Hamilton, W. F. 531, *560*
Ellsworth, R. see Harrop, G. A. 528, 529, 535, *560*
Elton, R. L., I. G. Zarrow and M. X. Zarrow 51, 58, 84, 85, 106, *155*
Ely, R. S., A. Done and V. C. Kelley 471, *503*
— E. R. Hughes and V. C. Kelley 311, *389*
— see Done, A. K. 75, *154*, 454, 471, *502*
Ely, R. S., see Good, T. A. 75, 82, 100, *159*
— see Sydnor, K. L. 585, *635*
Emerson, K. jr. see Renold, A. E. 137, *175*, 457, *509*, 522, *566*
Emery, J. L. see Stoner, H. B. 49, 104, 110, 125, *181*
Emery, S. L. see Kingbury, J. W. 60, 145, *165*
Emmens, C. W. 575, 617, *627*
— see Callow, N. H. 363, *385*, 616, *625*
Enders, A. C. see Buchanan, G. P. 615, *625*
Endröczi, E., G. Bata and J. Martin 75, 76, 103, *155*
— J. Szalay and K. Lissák 125, *155*
Engel, C. 286, *389*
— and G. Just 244, 286, *389*
— see Heusser, H. 286, *393*
Engel, F. L. see Savard, K. 413, 488, *510*
Engel, L. see Turner, R. B. 228, *407*
Engel, L. L. 293, 295, 339, 363, *389*
— J. Alexander, P. Carter, J. Elliott and M. Webster 304, 306, *389*
— B. Baggett and P. Carter 484, *503*
— P. Carter and L. L. Fielding 304, *389*, 460, *503*
— R. I. Dorfman and A. R. Abarbanel 427, *503*
— H. R. Patterson and H. Wilson 303, 338, *389*
— W. R. Slaunwhite jr., P. Carter and I. T. Nathanson 451, *503*
— G. W. Thorn and R. A. Lewis 427, 431, *503*, 610, *627*
— R. Wilson, W. R. Slaunwhite jr., P. Carter, P. C. Olmsted and I. T. Nathanson 426, *503*
— see Baggett, B. 338, *382*, 412, 451, *500*, 616, *624*
— see Nathanson, I. T. 616, *632*
— see Ryan, K. J. 62, *177*, 443, 484, 488, 490, *510*
— see Savard, K. 413, 488, *510*, 579, *634*
— see Villee, D. B. 62, 63, *183*, 590, *636*
Engel, W. see Zeisel, H. 591, *637*
Engelhardt, Fr. 646, *657*
Engle, R. see Djerassi, C. 232, 288, *388*
Englert, E. see Wallach, S. 134, *183*, 412, *512*, 585, 611, 615, *636*
Engström, H. 32, *155*
Engstrom, W. W., and H. L. Mason 338, *389*
— see Mason, H. L. 363, *398*, 574, *631*
Eposito, L. see Fidanza, A. 457, *503*
Eppstein, S. H., P. D. Meister, H. C. Murray and D. H. Peterson 258, 259, *389*, 487, 492, 493, 494, *503*
— — D. H. Peterson, H. C. Murray, H. M. Leigh Osborne, D. A. Lyttle, L. M. Reineke and A. Weintraub 491, 492, *503*
— — — H. C. Murray, L. M. Reineke and A. Weintraub 487, *503*
— D. H. Peterson, H. M. Leigh Osborne, H. C. Murray, A. Weintraub, L. M. Reineke, P. D. Meister 491, 492, *503*
— see Meister, P. D. 487, 491, 492, 493, 494, *508*
— see Pederson, R. L. 281, 283, *401*, 487, *509*
— see Peterson, D. H. 259, 260, *401*, 491, 492, 493, *509*
Epstein, E., W. O. Maddock and A. J. Boyle 337, *389*
Epstein, S. see Utewski, A. 67, *182*
Eränkö, O. 3, 59, 60, 92, 114, *155*
Ershoff, B. H. 139, *155*, 596, *627*
Erway, W. F. see McShan, W. H. 49, 68, *170*
Erwin, H. L. see Greer, M. A. 137, *160*
Eschenmoser, A. see Schreiber, J. 228, 242, *404*
Eshkol, Z. see Perek, M. 58, 85, *174*
Esselborn, V. M. see Sobel, E. H. 129, 135, *180*, 593, *635*
Estep, H. see Liddle, G. W. 523, *563*
von Euler, U. S. see Franksson, C. 475, *504*, 642, *657*
Euw, J. von, and T. Reichstein 227, *389*, 415, 420, *503*
— A. Lardon and T. Reichstein 242, *389*
— C. Meystre, R. Neher, T. Reichstein and A. Wettstein 198, 204, *389*
— R. Neher and T. Reichstein 196, 264, *389*

Euw, J. von, see Finkelstein, M. 428, 432, *503*
— see Reichstein, T. 221, *402*, 413, 415, 420, 421, *509*, 517, *566*, 610, *634*
— see Simpson, S. A. 198, 204, 238, 240, 317, 354, *405*, 421, 446, *511*, 520, *567*
Evans, D. D., and C. W. Shoppee *389*
Evans, H. M. see Mason, H. L. 607, *631*
— see Simpson, M. E. 87, 94, 100, 101, *179*
— see Walker, D. G. 78, *183*
— see Zizine, L. 107, *185*
Evans, J. S., R. F. Varney and F. C. Koch 579, *627*
Evans, R. M. see Attenburrow, J. 242, *382*
— see Barton, D. H. R. 228, *383*
Evelyn, K. A. see Gibson, J. G. *390*
Everett, N. B. 118, 120, *155*
Everett, J. W. 53, *155*
Eversole, W. J., A. Edelmann and R. Gaunt 542, *558*
— R. Gaunt and E. C. Kendall 519, *558*
— see Birnie, J. H. 537, 545, 546, *556*
— see Gaunt, R. 515, 519, 530, 531, 542, 543, *559*
Eyestone, W. H. see Mulay, A. S. 101, *171*

Fabre, J. see Mach, R. S. 531, *563*
Fairbanks, R. see Wilson, H. 340, 341, 365, *409*
Fajans, S. S. see Conn, J. W. 518, *557*
— see O'Donnell, W. M. 110, *173*
Fang, A. D., M. E. Martin and S. Katzenelbogen 649, *657*
Farah, A. see Green, D. M. *560*
Farber, S. J. see Earle, D. P. jr. 99, *155*
Farkas, T. G. see Rudney, H. *510*
Farquhar, M. G. 3, *155*
Farrell, G. L. 75, 114, 115, 116, *156*, 517, 522, 544, *558*
— R. C. Banks and S. Koletsky 74, 81, 101, 110, *156*, *503*, 517, 521, *558*
— R. B. Fleming, E. W. Rauschkolb, F. M. Yatsu, M. McCally and C. H. Anderson 103, 111, *156*
Farrell, G. L., S. Koletsky and L. W. Lapham 117, *156*
— E. W. Rauschkolb and P. C. Royce 100, 103, 110, 111, *156*, *503*
— — — and H. Hirschmann 423, *503*, 610, *627*
— R. S. Rosnagle and E. W. Rauschkolb 74, 76, *156*
— see Anderson, C. H. 115, *146*
— see Driscol, T. E. 114, *154*
— see Newman, A. E. *172*
— see Rauschkolb, E. W. 74, 100, 109, 110, *175*, 544, *566*
— see Rosnagle, R. S. 113, *176*
— see Sweat, M. A. 75, 100, *181*
— see Travis, R. H. 447, *512*
Fassbender, H. G. 128, *156*
Faulkner, R. see McQueeney, A. J. 138, *170*
Fawcett, D. W. 53, 67, *156*
— and H. W. Deane 620, *627*
— see Deane, H. W. 57, *153*
Fazekas, I. G. 422, *503*
Fazekas, J. F. see Alman, R. W. 643, *656*
Fedor, E. see Swingle, W. W. 529, 536, *568*
Fedor, E. J. see Swingle, W. W. 535, *568*, 621, *635*
Fekete, E., G. W. Woolley and C. C. Little 618, 619, *627*
— see Woolley, G. W. 618, *637*
Feldman, J. D. 54, 100, *156*
Feldman, L. J. see Bernstein, S. 267, 272, *383*
Feldman, M. P. see Cook, E. R. 314, *386*
Feldstein, M. see Klendshoj, N. C. 363, *395*
Fenn, W. O. see Gerschman, R. 59, *158*
Feodossiew, N. E. 618, *627*
Feriozi, D. see Landing, B. H. 80, *167*
Fernholz, H. see Lettré, H. R. 186, *396*
Ferrari, J. see Deulofeu, V. 426, *502*
Ferrebee, J. W., D. Parker, W. H. Carnes, M. K. Gerity, D. W. Atchley and R. F. Loeb 532, 534, *558*
— C. Ragan, D. W. Atchley and R. F. Loeb 531, *558*
— see Kuhlman, D. 534, *562*
— see Ragan, C. 533, 534, 537, *565*
Fetterman, G. H. see Sherman, F. E. 129, *179*
Fetzer, S. 55, *156*
Feuer, M. see Heusser, H. 212, *392*
Feulgen, R., and K. Voit 55, *156*
Fiala, A. see Fiala, S. 90, 91, *156*
Fiala, S., E. E. Sproul and A. Fiala 90, 91, *156*
Fidanza, A., and A. Bonomolo 51, *156*
— G. Carvina and A. DeCicco 458, *503*
— A. DeCicco and L. Eposito 457, *503*
Fielding, L. L. see Baggett, B. 338, *382*
— see Engel, L. L. 304, *389*, 460, *503*
Fields, M. see Fieser, L. F. 340, *389*
Fieser, L. F. 232, 234, 240, 242, 336, *389*, 434, *503*
— and M. Fieser 186, 194, 208, 217, 225, 232, 360, 373, 374, *389*
— and S. Rajagopalan 242, *389*
— and M. Romero 234, *389*
— M. Fields and S. Lieberman 340, *389*
— see Cook, R. P. 436, *502*
— see Dobriner, K. 431, *502*
— see Hershberg, E. B. 381, *392*
— see Klass, D. L. 247, *395*
— see Lieberman, S. 217, 307, *396*, 427, 428, 431, *507*
— see Wolfe, J. K. 381, *409*, 572, *636*
Fieser, M. see Fieser, L. F. 186, 194, 208, 217, 225, 232, 360, 373, 374, *389*
— see Klass, D. L. 247, *395*
Fillerup, D. L., and J. E. Mead 311, *389*
Finckenor, L. see Gould, D. 241, *391*
Finerty, J. C. see Hess, M. *162*
— see Rennels, E. G. 106, 107, *175*
Finkelstein, M., and S. Goldberg 341, *389*
— J. von Euw and T. Reichstein 428, 432, *503*
Finkenstaedt, J. T. see Laidlaw, J. C. 518, 533, 535, *563*
Finkler, R. S. 611, *627*
Finnegan, J. K. see Larson, P. S. 135, 136, *167*, 458, *506*

Fiori, C. 98, *156*
Firschein, H. E. see Pearce, E. M. 370, *401*
— see Westphal, U. 292, *408*
Fischer, P. see Bacq, Z. M. 125, *146*
Fischgold, H. see Bricaire, H. 655, *657*
Fish, C. A., M. Hayano and G. Pincus 463, 497, *503*
— see Levy, L. K. 484, *506*
— see Schwenk, E. 436, 438, *511*
Fish, G. W. see Ragan, C. 537, *565*
Fish, W. A. see Stokes, W. M. 310, 406, *511*
Fish, W. R., and R. I. Dorfman 484, *503*
— B. N. Horwitt and R. I. Dorfman 459, *503*
— see Horwitt, B. N. 459, *505*, 610, *629*
Fisher, G. R. see Schricker, J. A. 137, *178*
Fisher, R. B. see Barnett, S. A. 60, *147*
Fishman, W. H., and J. R. Baker 66, *156*
— see Talalay, P. 66, *181*
Fitch, W. M. see Pearce, E. M. 370, *401*
Flaks, J. 134, *156*
Flanagan, J. B., A. K. Davis and R. R. Overman 536, 537, *558*
Fleischer, G. A. see Mattox, V. R. 278, *398*
Fleming, R. B. see Farrell, G. L. 111, *156*
Fleminger, J. J. see Lewis, A. 641, *658*
Flexner, L. B., and A. Grollman 3, 56, 82, 92, *156*
Flickinger, D. D. 13, 106, 144, *156*
Flint, J. M. 23, 27, 28, *156*
Florentin, A. A. see Migeon, C. J. 572, *632*
Florey, K., and M. Ehrenstein 204, 214, 234, 360, 374, *389*
— see Fried, J. 267, 272, 277, 278, *390*
— see Tennent, D. 313, *407*
Flux, D. S. 81, *156*
Foggie, W. E., and G. L. Montgomery 646, *657*
Foglia, V. G., B. Günther and R. M. Pinto 20, *156*
— see Houssay, B. A. 134, *163*, 618, *629*
Folch, J., J. Ascoli, M. Lees, J. A. Meath and F. N. Le Baron 49, *156*
Foley, R. C. see Greenstein, J. S. 23, *159*
Foltz, E. W., A. Lippman and C. Djerassi 205, 376, *389*
— see Djerassi, C. 374, *388*
Fonken, A. E. see Johnson, J. L. 245, *394*
Fonken, G. J. see Dauben, W. G. 245, *387*
Fonken, G. S., J. L. Thompson and R. H. Levin 243, *390*
Fønns-Bech, P. see Li, C. H. 82, 83, 100, 101, 103, 117, *168*, 608, 613, 614, *631*
Fontaine, M. 549, *559*
— and J. Hatey 8, 141, *156*, 549, *559*
— and R. Motais 549, *559*
— J. Hatey and M. Olivereau 88, 90, *157*
Fontana, P. jr. see Pavoa, H. jr. 91, *173*
Forbes, A. P., and F. Albright 584, *627*
— see Bartter, F. C. 531, *556*, 585, 616, *624*
— see Fraser, R. W. 363, *390*
Forbes, T. R. 11, *157*, 613, *627*
— and J. Zander 423, *503*
— see Fraps, R. M. 610, *628*
— see Hooker, C. W. 610, 612, 613, 614, *629*
Forchielli, E. 462, 495, 497, *503*
— and R. I. Dorfman 462, 495, 497, 498, *504*
— K. Brown-Grant and R. I. Dorfman 498, 499, *504*
— H. Rosenkrantz and R. I. Dorfman 462, 497, *504*
Ford, P. 7, *157*
Forrest, A. P. M. see Grant, J. K. 75, *159*
Forsham, P. H., G. W. Thorn, F. T. G. Prunty and A. G. Hills 607, *627*
— see Renold, A. E. 580, *634*
Forsyth, B. T. 531, *559*
Fortier, C. 544, *559*
— and J. de Groot 120, *157*
— F. R. Skelton, P. Constantinides, P. O. Timiras, M. Herlant and H. Selye 50, 58, 59, *157*
— see Constantinides, P. 69, 70, *152*
— see de Groot, J. 29, 118, 119, *160*
— see Guillemin, R. 103, *160*
— see Skelton, F. R. 107, *179*
Fortunato, J. see Klein, R. 72, 125, *165*, 296, *395*, 423, 425, *506*, 521, *562*
Fotherby, K. 429, *504*
— A. Colas, S. M. Atherden and G. F. Marrian 428, 429, 446, 478, *504*
Foucas, M. 647, *657*
Fourman, P. see Bartter, F. C. 531, 532, *556*
Fowler, M. A., and I. Chester Jones 551, *559*
Fowlks, W. L. see Sweat, M. L. 443, 485, 489, *511*
Fox, M. H. see Arrington, L. R. 605, *624*
Fraisse, Plan and Brondel 546, *559*
Frame, E. G. 303, *390*, 572, *627*
— and H. J. Jewett 581, *627*
Francis, F. E., N. C. Shen and R. A. Kinsella 591, *628*
Frankl von Hochwart, L. 639, *657*
Franksson, C., and C. A. Gemzell 642, *657*
— — and U. S. von Euler 475, *504*, 642, *657*
— see Birke, G. 581, *625*
— see Diczfalusy, E. 412, *502*
Frantz, I. D. jr. see Davidson, A. G. 436, *502*
— and N. L. R. Bucher 436, *504*
Frantz, M. J., and A. Kirschbaum 134, *157*, 607, *628*
— see Halberg, F. 32, *160*
— see Kirschbaum, A. 134, *165*
Franze, F. 17, *157*
Franzl, R. E. see Rapport, M. M. 55, *175*
Fraps, R. M., C. W. Hooker and T. R. Forbes 610, *628*
Fraser, R. W., A. P. Forbes, F. Albright, H. Sulkowitsch and E. Reifenstein jr. 363, *390*
Frawley, J. P. see Howard, J. M. 642, *658*
Frawley, T. F., and G. W. Thorn 541, *559*
Frazão, J. V. 27, 28, 35, 36, 37, 45, 60, 121, *157*
Fredericks, J. see Lloyd, C. W. 611, *631*
Frederickson, D. S., R. E. Peterson and D. Steinberg 140, *157*, 455, *504*
— see Steinberg, D. 140, *180*, 455, *511*
Freedman, H. H. see Cohen, H. 82, 83, *151*
Freeman, H. see Pincus, G. 583, *633*
Freeman, S. see Chen, C. 312, *385*
— see Wheeler, J. 364, *409*

Fregly, M. J. 530, *559*
— see Knobil, E. 71, *166*
Freier, E. F. see Glick, D. 68, *159*
French, T. H. see Popjak, G. 436, *509*
Frenkel, J. K. 132, 141, *157*
Frey, H. see Meystre, C. 254, 277, *398*
Fried, John see Arth, G. E. 285, *382*
Fried, Josef 521, *559*
— and E. Sabo 221, 250, 265, 267, 268, 276, 361, *390*
— and E. F. Sabo 491, *504*
— A. Borman, W. B. Kessler, P. Grabowich and E. F. Sabo 274, *390*
— K. Florey, E. Sabo, J. Herz, A. Restivo, A. Borman and F. Singer 267, 272, 277, 278, *390*
— J. Herz, E. Sabo, A. Borman, F. Singer and P. Numerof 267, 268, *390*
— R. W. Thoma, J. R. Gerke, J. E. Herz, M. N. Donin and D. Perlman 491, 492, *504*
— — and A. Klingsberg 260, *390*
— — D. Perlman, J. E. Herz and A. Borman 258, *390*, 487, 491, 492, 493, 494, *504*, 520, *559*
— see Djerassi, C. 487, 492, 493, *502*
— see Herz, J. 267, 270, 271, *392*
— see Perlman, D. 494, *509*
— see Thoma, R. W. 259, 274, *407*, 487, *512*
Friedgood, F. B. see Wolfe, J. K. 572, *636*
Friedgood, H. B., and R. A. Berman 337, *390*
— E. H. Taylor and M. L. Wright 294, *390*
— and H. L. Whidden 337, *390*
Friedman, N. B. 132, 140, 141, *157*
Frilley, M. see Raynaud, A. 121, 122, *175*
Frisch, E., and L. J. Gardner 136, *157*
Fritz, I. B. see Taubenhaus, M. *568*
Fromentin, H. see Olivereau, M. 88, 141, *173*, 549, *565*
Frost, R. C., and R. V. Talmage 537, 542, *559*
Fruit, A. see McCann, S. M. 545, *564*
Frumusan, P. see Vesin, P. 546, 547, *568*
Fry, E. G. see Gershberg, H. 59, *158*
— see Sayers, G. 83, *177*
Fugo, N. W. see Davis, M. E. 611, *626*
Führer, J. see Prelog, V. 451, *509*
Fujimoto, G. I. 232, *390*
Fukushima, D. K., H. L. Bradlow, K. Dobriner and T. F. Gallagher 478, 482, *504*
— K. Dobriner and T. F. Gallagher 480, *504*
— S. Dobriner, M. S. Haffler, T. H. Kritchevsky, F. Herling and G. Roberts 217, *390*
— and T. F. Gallagher 217, 219, 250, 295, *390*, 427, 431, 432, 459, 465, *504*, 585, 610, *628*
— A. D. Kemp, R. Schneider, M. B. Stokem and T. F. Gallagher 431, 460, *504*
— T. H. Kritchevsky, M. L. Eidinoff and T. F. Gallagher 250, *390*
— N. S. Leeds, H. L. Bradlow, T. H. Kritchevsky, M. B. Stokem and T. F. Gallagher 225, *390*, 433, 459, 460, 483, *504*
— C. D. Meyer, E. Ashworth and T. F. Gallagher 431, 432, *504*
— see Eidinoff, M. L. 250, *388*
— see Gallagher, T. F. 478, *504*
— see Hellman, L. 471, *505*
— see Hübener, H. J. 461, 462, 463, 497, *505*
— see Leeds, N. S. 212, *396*
— see Lieberman, S. 217, 294, 297, 307, *396*, 428, *507*
— see Miller, W. R. 312, *399*
— see Okada, M. 572, *632*
— see Soloway, A. H. 218, 225, *405*
— see Wendler, N. L. 217, *408*
Fulford, B. D. see McCann, S. M. 545, *564*
Fuller, R. C. 369, *390*
Furger, R. 648, *657*
Furman, M. see Rosselet, J. P. 432, 459, 460, *510*
Fürst, A. see Heusser, H. 288, *393*
Furth, J. 29, 102, *157*
— see Bahn, R. 607, *624*
— see Bahn, R. C. 102, 108, 129, 133, *146*
Furth, J. see Cohen, A. I. 101, 114, 115, 134, 135, *151*, 607, *626*
— see Upton, A. C. 108, 133, *182*
Furtunato, J. see Klein, R. 423, 425, *506*
Fustinoni, O. 143, *157*, 552, *559*
— and J. Porto 9, *157*
— see Marenzi, A. D. 551, *564*

Gaarde, F. see Williams, V. Z. 354, *409*
Gabe, M. 9, *157*
— see Verne, J. 57, 133, *183*
Gablinger, S. see Schamaun, M. 652, *659*
Gabrilove, J. L. see Soffer, L. J. 531, 542, 543, *567*
Gaddum, J. H. 575, 576, *628*
Gadsden, E. see Bahn, R. C. 102, 108, 129, 133, *146*
Gagnon, P. M. see DesMarais, A. 59, 92, *154*
Gagnon, R. 23, *157*
Gaillard, L. 515, *559*
Gajdos, A. see Bariety, M. 546, *556*
Gallagher, B. S. see Roberts, G. 347, *402*
Gallagher, T. F. 72, 137, *157*, 251, *390*, 456, 460, 482, *504*, 573, 586, 607, *628*
— D. K. Fukushima, M. C. Barry and K. Dobriner 478, *504*
— A. Kappas, H. Spencer and D. Lazlo 456, *504*, 586, *628*
— and T. H. Kritchevsky 211, 212, *390*
— see Barry, M. C. 480, *500*
— see Beer, C. T. 484, *500*
— see Borgstrom, E. 251, *384*
— see Bradlow, H. L. 478, 480, 482, 483, *501*, 573, *625*
— see Eidinoff, M. L. 250, *388*
— see Fukushima, D. K. 217, 219, 225, 250, 295, *390*, 427, 428, 431, 432, 433, 459, 460, 465, 480, 482, 483, *504*, 585, 610, *628*
— see Hellman, L. 436, 471, *505*
— see Hübener, H. J. 461, 462, 463, 497, *505*
— see Kappas, A. 295, *395*, 580, 581, 583, *630*
— see Kemp, A. D. 228, *395*, 428, 460, *506*
— see Kraychy, S. 484, *506*
— see Kritchevsky, T. H. 215, 228, 242, 255, *396*

Gallagher, T. F., see Leeds, N. S. 212, *396*
— see Moore, C. R. 576, 599, *632*
— see Munson, P. L. 337, 364, *399*, 434, *508*
— see Okada, M. 572, *632*
— see Rosenfeld, R. S. 233, *403*, 436, 489, *510*
— see Soloway, A. H. 218, 225, *405*
Gandossi, S. K. see Southcott, C. M. 423, 425, *511*
Ganis, F. M., L. R. Axelrod and L. L. Miller 461, *504*
— L. L. Miller and L. R. Axelrod *157*
— and M. Radakovich 435, *504*
Gann, D. S., I. H. Mills and F. C. Bartter 115, *157*
Ganong, W. F. 82, *157*
— and D. M. Hume 74, 76, 79, 100, 105, *157*
— A. H. Lieberman, W. J. R. Daily, V. S. Yuen, P. J. Mulrow, J. A. Luetscher jr. and R. E. Bailey 110, *157*
Garberi, J. C. see Pasqualini, R. Q. 58, *173*
Garcia, M. see Zaffaroni, A. 238, *409*
Gardner, L. I. 40, 108, 125, 126, 133, *157*, 295, *390*, 587, 590, *628*
— and C. J. Migeon 572, *628*
— and A. A. Tice 421, *504*
— and R. L. Walton 126, *157*, 584, 591, *628*
— — W. Ellis and E. C. Hughes 582, *628*
— see Frisch, E. 136, *157*
— see Migeon, C. J. 615, *632*
— see Nelson, D. H. 74, 75, *172*
— see Nichols, J. 135, 136, *172*, 457, *508*
— see Wilkins, L. 585, 616, *636*
Gardner, W. U. 618, *628*
— see Dorfman, R. I. 618, *627*
— see Leblond, C. P. 60, *168*, 593, *630*
Garm, O. 19, 94, *157*
Garmaise, D. see Kritchevsky, T. H. 228, 242, *396*
Garrod, O., S. A. Simpson and J. F. Tait 115, 116, *157*
— see Ayres, P. J. 76, *146*, 313, 318, 372, *382*, 423, 447, *500*
Garrett, E. R. see Anderson, H. V. 227, *381*
Garrett, F. D. 7, *157*
Gassner, F. X., D. H. Nelson, H. Reich, R. T. Rapala and L. T. Samuels 573, *628*
— E. C. Reifenstein jr., J. W. Algeo and W. E. Mattox 547, *628*
— see Reich, H. 609, *634*
Gastineau, C. F., G. B. Logan, A. Albert and H. L. Mason 616, *628*
Gatica, J. see Rosenkranz, G. 255, 257, *403*
Gatzi, K. see Reichstein, T. 413, 420, *509*
Gaudino, M., and M. F. Levitt 537, *559*
Gaunt, J. H. see Gaunt, R. 21, *158*, 526, *559*
Gaunt, R. 21, *158*, 526, 527, 534, 540, 546, *559*
— J. H. Birnie and W. J. Eversole 530, 531, 542, 543, *559*
— and W. J. Eversole 515, *559*
— — and E. C. Kendall 519, *559*
— J. H. Gaunt and C. E. Tobin 21, *158*, 526, *559*
— F. Gross, A. A. Renzi and J. J. Chart 547, 548, *559*
— and H. W. Hays 526, 527, *559*
— J. H. Leathem, C. Howell and N. Antonchak 80, *158*
— M. Liling and C. W. Mushett 139, *158*
— C. W. Lloyd and J. Chart 518, 543, 544, 545, *559*
— H. E. Potts and E. Loomis 542, *559*
— A. A. Renzi and J. J. Chart 530, *559*
— C. H. Tuthill, N. Antonchak and J. H. Leathem 96, 100, 107, 110, *158*
— see Birnie, J. H. 537, 545, 546, *556*
— see Chart, J. J. 457, *502*, 518, 521, 522, 523, 533, 543, *557*
— see Eversole, W. J. 519, 542, *558*
— see Liling, M. 546, *563*
— see Renzi, A. A. 521, 531, 532, 543, *566*
— see Sheppard, H. 523, *567*
Gautschi, F., and K. Bloch 439, *504*
Gaylord, N. G. 245, *390*
Geary, J. R. jr. see Grunbaum, B. W. 49, *160*
Gebert, W. see Oliveto, E. P. 285, *400*
Geist, S. H. see Salmon, U. J. 611, *634*
Geloso, J. P. see Jost, A. 604, *630*
Geller, J. see Soffer, L. J. 531, 542, 543, *567*
Gellhorn, E., and H. M. Ballin 534, *560*
Gemzell, C. A. 100, 105, 106, *158*, 364, *390*, 455, *504*
— see Bergstrand, C. G. 610, *624*
— see Franksson, C. 475, *504*, 642, *657*
Genesi, M., and G. F. Melloni 96, *158*
Genest, J. see Nowaczynski, W. 236, 300, 333, 362, 364, *399*
Gengradom, S. see Howard, E. 596, *630*
Genitis, V. E., and I. P. Bronstein 611, *628*
Gennes, L. de, P. Bensoussan and F. Klein 655, *657*
— B. M. de Moreau, H. Bricaire, R. Tourneur and Janson 546, *560*
Gentles, M. J. see Gould, D. 267, 278, 279, *391*
— see Herzog, H. L. 219, 260, *392*
Georges, G., and J. Cahn 82, *158*
Geppert, L. J., A. D. Dietrich, E. H. Johnson and B. Lind 643, *658*
Gerall, A. A. see Phoenix, C. H. 574, *633*
Gérard, P. 6, 53, *158*
Gerhards, E. see Dirscherl, W. 319, *387*
Gerity, M. K. see Ferrebee, J. W. 532, 534, *558*
Gerke, J. R. see Fried, J. 491, 492, *504*
Gerold, C. see Hershberg, E. B. 242, *392*
— see Oliveto, E. P. 204, 228, *400*
Gerschberg, H. see Dumm, M. E. 458, *503*
Gerschman, R., and W. O. Fenn 59, *158*
Gersh, I., and A. Grollman 23, *158*, 601, 602, *628*
Gershberg, H., E. G. Fry, J. R. Brobeck and C. N. H. Long 59, *158*
Geschwind, I. 85, 116, 117, *158*
— see Li, C. H. 82, 83, 100, 101, 103, 117, *168*, 608, 613, 614, *631*

Gesell, M. see Borth, R. 455, *501*
Geyer, G. 475, *504*
Geyer, R. P. see Lyon, I. 436, *507*
Giacomini, E. 5, 6, 13, *158*, 548, *560*
Gibbs, M. H. see Tavormina, P. A. 436, 437, 438, *511*
Gibson, G. see Appleby, J. I. 341, *382*
Gibson, J. G., and K. A. Evelyn *390*
Gibson, R. B. see Robbie, W. A. 295, *402*
Gidez, L. I. see Deane, H. W. 114, *153*
Gillissen, G., and I. Gillissen 107, *158*
Gillissen, I. see Gillissen, G. 107, *158*
Gilman, M. see Renzi, A. A. 521, *566*
Gilmore, M. see Gould, D. 267, 278, 279, *391*
Gilmour, J. R. see Crooke, A. C. 99, 109, *152*
Ginsburg, M. 544, 545, *560*
Giraldi, P. see Logemann, W. 265, *397*
Girard, P. see Bimes, C. 57, *148*
Giroud, A. *158*
— and C. P. Leblond 58, *158*
— and M. Martinet 59, *158*
— and A. R. Ratsimamanga 59, *158*
Giroud, C. J. P., M. Saffran, A. V. Schally, J. Stachenko and E. H. Venning 77, 111, 113, *158*, 447, *504*
— J. Stachenko and P. Piletta 109, 111, *158*, 447, 451, *504*
— — and E. H. Venning 109, *158*
— see Das Gupta, D. 73, 114, *152*
— see Stachenko, J, 109, 111, *180*, 517, *567*
— see Venning, E. H. 333, *407*, 475, 512
Glascock, R. F. 369, *391*
Glaser, G. H. 640, *658*
— see Knight, R. P. jr. 535, *562*
Glaviano, V. V. see Hilton, J. G. 76, *162*
Glazer, L. see Touchstone, J. C. 75, *182*, 449, *512*
Glenister, D. see Swingle, W. W. 529, 536, 547, *568*
Glenn, E. M., and E. Heftmann 335, *391*
— and D. H. Nelson, 295, 296, *391*
Glenn, E. M. see Sandberg, A. A. 475, *510*
Glick, D., M. Alpert and H. R. Stecklein 58, 59, *159*
— and G. R. Biskind 59, 66 *158*
— E. F. Freier and M. J. Ochs 68, *159*
— and L. J. Greenberg 63, 92, *158*
— and S. N. Nayyar 63, *159*
— and M. J. Ochs 51, 80, 84, 94, *159*
— and H. R. Stecklein 67, 98, *159*
— H. R. Stecklein and B. G. Malmstrom 67, *159*
— R. H. Swigart, R. C. Bahn and D. W. Hastings 3, *159*
— — S. N. Nayyar and H. R. Stecklein 50, 68, *159*
— see Bahn, R. C. 59, 80, 85, 86, *146*
— see Greenberg, L. J. 62, 96, *159*
— see Grunbaum, B. W. 49, *160*
— see Jardetzky, C. D. 63, *164*
— see Malmström, B. G. 62, *170*
— see Nayyar, S. N. 66, 98, *172*
— see Wattenberg, L. W. 66, *183*
Glick, J. H. jr. 422, *504*
— see Kinsella, R. A. jr. 296, *395*
Glock, G. E., and P. McLean 62, *159*
Glynn, L. E. see Colover, J. 141, *151*
Goding, J. R., and D. A. Denton 541, *560*
Goetz, F. G. see Munson, P.L. 581, 603, *632*
— see Thorn, G. W. 518, 540, *568*
Gofman, J. W., F. T. Lindgren and H. A. Elliott 292, *391*
Gofstein, R. see Rutenburg 340, *403*
Gold, J. J. 75, 103, *159*, 364, *391*
— see Jailer, J. W. 107, 135, *164*, 465, *506*, 585, *630*
Gold, N. I., E. Singleton, D. A. Macfarlane and F. D. Moore 475, *504*
Goldberg, M. B. 643, 651, *658*
Goldberg, S. see Finkelstein, M. 341, *389*
Golden, A., and R. R. Overman 130, 132, *159*
Goldfaden, S. H. see Savard, K. 460, *510*
Goldfien, A. see Renold, A. E. 521, *566*
— see Thorn, G. W. 137, *181*, 457, *512*
Goldkamp, A. H. see Dodson, R. M. 233, 260, *388*, 487, *502*
Goldman, H. see Sapirstein, L. A. 73, *177*
Goldman, M. L., E. Ronzoni and H. A. Schroeder 112, 113, *159*
Goldner, M. see Nowaczynski, W. 364, *400*
Goldner, M. G. see Carter, A. C. 107, *150*
Goldstein, M., M. Gut and R. I. Dorfman 414, *504*
Gomez-Maestro, D. V. 455, *504*
Gomez-Mont, F., and D. Berliner 459, *504*
Gomori, G. 50, 56, *159*
Gompertz, D. 106, *159*
Gonken, G. S. see Levin, R. H. 227, *396*
Gonzalez, I. E. 27, *159*
Good, T. A., R. S. Ely, L. R. Heiselt, A. K. Done and V. C. Kelley 75, 82, 100, *159*
Goodkind, M. J. see Davis, J. O. 114, *152*, 531, *558*
Goodman, A. D. 139, *159*
Goodman, L. S. see Woodbury, D. M. 529, 531, 534, *569*
Goodsell, J. E. 82, *159*
Goormaghtigh, N. 43, 129, *159*
Gorbman, A. 548, *560*
Gordan, G. S. 534, *560*
— and H. W. Elliott 653, *658*
— see Eisenberg, E. 576, *627*
Gordon, A. S. see D'Angelo, S. A. 69, *152*
— see Katsh, S. 604, *630*
Gordon, E. S., J. J. Chart, D. Hagedorn and E. G. Shipley 520, *560*
— see Chart, J. J. 520, *557*
Gordon, G. S. see Shemano, I. 576, *634*
Gordon, J. J. 62, *159*
Gordon, M. F. see Rapela, C. E. 9, 79, 145, *175*
Gordon, M. L. 119, *159*
Gordon, W. E. L. see Christy, N. P. 106, *151*
— see Jailer, J. W. 106, *164*

Gordy, W. 377, *391*
Gore, I. Y. see Cornforth, J. W. 437, 438, *502*
Gornall, A. G., and C. Gwilliam 333, *391*
— and P. M. Mac Donald 341, *391*
Gottesleben, W. see Eger, W. 64, 65, 96, *155*
Gottschau, M. 45, *159*
Gould, D., L. Finckenor, E. B. Hershberg, J. Cassidy and P. I. Perlman 241, *391*
— E. L. Shapiro, H. L. Herzog, M. J. Gentles, E. B. Hersherg, W. Charney, M. Gilmore, S. Tolksdorf, M. Eisler, P. L. Perlman and M. M. Pechet 267, 278, 279, *391*
— see Herzog, H. L. 267, 277, *392*
— see Oliveto, E. P. 243, *400*
— see Shapiro, E. L. 214, *405*
Gould, R. G. 437, *504*
— and G. Popjak 436, *504*
— see Werbin, H. 435, 441, *513*
Gould, R. P. see Ayres, P. J. 109, *146*, 447, 451, *500*
Gould, T. C. see Gunn, S. A. 602, *628*
Gourfein, D. 554, *560*
Goy, R. W. see Phoenix, C. H. 574, *633*
Graber, R. P. see Wendler, N. L. 223, 279, *408*
Grabowich, P. see Fried, J. 274, *390*
— see Herz, J. 267, 270, 271, *392*
— see Thoma, R. W. 259, 274, *407*
Grace, W. J. see Hetzel, B. S. 642, *658*
Grad, B. 541, *560*
— see Saffran, M. 77, *177*
Gradinesco, A., and N. Santa 515, *560*
Graff, M. M., J. J. Holecheck and A. R. Collins 337, *391*
— J. T. McElroy and A. L. Mooney 337, *391*
— see Hertz, R. 456, *505*
— see Tullner, W. W. 137, *182*, 456, *512*
Grandou, P. see Jost, A. 107, *164*
Grant, J. K., A. P. M. Forrest and T. Symington 75, *159*
— see Brownie, A. C. 62, *149*, 443, 489, *501*
Grauber, R. C. see Saier, E. 338, *403*
Grauer, R. C. see Wilson, D. 207, *409*
Graumann, W. 67, *159*
Gray, C. H. see De Courcy, C. 364, *387*
— see Holness, N. J. 432, 433, *505*
Gray, C. L., and F. Bischoff 484, *504*
— M. A. S. Green, J. J. Holness and J. B. Lunnon 476, *504*
Gray, H. see Addis, T. 18, 20, *145*
Gray, I. see Kritchevsky, D. 436, *506*
Gray, S. J., and C. G. Ramsey 540, *560*
Gregory, R. see Johnson, J. E. 457, *506*
Green, A. A. see Savard, K. 443, 489, *510*
Green, D. M., A. Farah and W. W. Klemperer *560*
— see Picha, G. M. 461, *509*
Green, E. L. see Hartman, F. A. 5, *161*
Green, H. D. see Nichols, J. 136, *172*, 457, *508*, 523, *565*
Green, J. A. 614, *628*
Green, M. A. S. see Gray, C. L. 476, *504*
Green, R. F. see Race, G. J. 32, 88, 111, 112, *175*
Greenberg, J. see Earle, D. P. jr. 99, *155*
— see Taylor, D. J. 51, 84, 144, *181*
Greenberg, L. J., and D. Glick 62, 96, 139, *159*
— see Glick, D. 92, *158*
Greenberg, S. M., and K. Aterman 82, *159*
Greene, R. R., and M. W. Burrill 575, *628*
— M. W. Burrill and A. C. Ivy 579, *628*
— see Burrill, M. W. 599, 601, 608, *625*
Greenfield, R. B. see Perlman, D. 494, *509*
Greenspan, G., C. P. Schaffner, W. Charney, H. L. Herzog and E. B. Hershberg 260, *391*
Greenstein, J. S., and R. C. Foley 23, *159*
Greenwood, F. C., and R. D. Bulbrook 412, *504*
— see Bulbrook, R. D. 412, *501*, 616, *625*
Greep, R. O., and I. Chester Jones 87, 104, 106, 107, *160*
Greep, R. O. and H. W. Deane 36, 45, 46, 47, 53, 95, 101, 110, 115, 116, 118, 119, 120, 139, *160*
— see Deane, H. W. 28, 37, 38, 54, 78, 79, 87, 92, 94, 99, 100, 104, 105, 109, 110, 111, 113, 114, 115, 132, 140, *153*
— see Dempsey, E. W. 65, 96, 109, *154*
— see Knobil, E. 78, 80, 87, 99, 101, 107, 109, 110, 115, 116, *166*
Greer, M. A., and H. L. Erwin 137, *160*
Greuter, F., J. Kalvoda and O. Jeger 237, *391*
Griboff, S. I. 135, *160*
Griffin, A. C. see Wexler, B. C. 110, *184*
Griffin, J. see Brown, J. H. U. *501*, 523, *556*
Griffin, J. B. see Brown, J. H. U. 58, 108, 136, *149*
Grinker, R. R. see Basowitz, H. 642, *656*
Griswold, G. see Bartter, F. C. 531, *556*
Groat, R. A. 21, *160*, 526, *560*
Grollman, A. 118, *160*, 597, *628*
— see Flexner, L. B. 3, 56, 82, 92, *156*
— see Gersh, I. 23, *158*, 601, 602, *628*
Groot, C. A. de see Schaberg, A. 121, *178*
Groot, J. de, and C. Fortier 29, 118, 119, *160*
— see Fortier, C. 120, *157*
Gross, F. 520, 524, 530, 542, 547, *560*
— and H. Gysel 524, *560*
— and P. Lichtlen 547, *560*
— and R. Meier 524, 527, 560
— and H. Schmidt 533, *560*
— see Dettbarn, W. 531, *558*
— see Gaunt, R. 547, 548, *559*
Grossman, J. see Djerassi, C. 287, 288, 376, *388*
Grossnickle, T. T. see Djerassi C. 336, *388*
Grostic, M. F. see Johnson, J. L. 357, *394*
Gruenwald, P. 2, 3, 39, *160*, 570, 588, *628*
— and W. M. Konikov 27, 28, 45, *160*
Grumbach, J. J. see Bongiovanni, A. M. 296, 301, *384*
Grumbach, M. M. see Bongiovanni, A. M. 424, *501*
Grunbaum, B. W., J. R. Geary jr. and D. Glick 49, *160*

Grundy, H. see Tait, J. F. 524, *568*
Gruner, P. see Heni, F. 88, *161*
Grynfeltt, E., J. Turchini and S. Turchini 53, 60, *160*
Guerra, S. see Howard, J. M. 642, *658*
Guerra, S. L. see Peterson, R. E. 364, *401*
Guggenheim, K. 139, *160*
Guilhem, A. see Bimes, C. 20, 35, 57, 133, *148*
— see Planel, H. 20, 28, 57, 65, 67, 96, 133, *174*
Gunn, S. A., and T. C. Gould 602, *628*
Gunville, R. see Ulstrom, R. A. 591, *635*
Günther, B. see Foglia, V. G. 20, *156*
Gurd, F. R. N. see Oncley, J. L. 292, *400*
Gurin, S. see Blecher, M. 436, *500*
— see Brady, R. O. 436, *501*
— see Dituri, F. 436, 438, *502*
— see Eisenberg, F. 452, *503*
— see Lynn, W. S. jr. 439, 499, *507*
Guillemin, R. 59, *160*
— G. W. Clayton, J. D. Smith and H. S. Lipscomb 77, 85, 100, 103, *160*
— C. Fortier and H. S. Lipscomb 103, *160*
— see Rabinowitz, J. L. 436, 438, *509*
— see Staple, E. 420, 435, 439, 441, *511*
Gurpide, E. H. G. de 100, *160*
Gut, M. *391*
— see Burstein, S. 480, *501*
— see Goldstein, M. 414, *504*
— see Hayano, M. 485, *505*
— see Meyer, A. S. 415, 449, 486, *508*, 573, *631*
— see Schwenk, E. 233, *404*
— see Ungar, F. 496, *512*
— see Wotiz, H. H. 616, *637*
Gutman, J. see Soffer, L. J. 531, 542, 543, *567*
Gutner, L. B., J. B. Moses, S. Damm and H. S. Kupperman 546, *560*
Gutowsky, H. S. 377, *391*
— see Meyer, L. H. 377, *398*
Gutsell, E. S. Levin, R. H. 227, *396*
Guyénot, E., K. Ponse and J. Wietrzykowska 607, *628*
Gwilliam, C. see Gornall, A. G. 333, *391*
Gyorgy, P. see Venning, E. H. 591, 636
Gysel, H. see Gross. F. 524, *560*
Haag, H. B. see Larson, P. S. 135, 136, *167*, 458, *506*
Haag, H. H. see Cobey, F. A. 136, *151*
Haager, K. see Creutzfeldt, W. 49, 51, 84, 88, 93, 94, 104, *152*
Haase, J. 32, 59, 60, 96, 104, *160*
de Hachen, D. P. C. 423, *504*
Hadd, H. E., and W. H. Perloff 297, *391*
Hadházy, C., and L. Karmazsin 118, *160*
Hadler, H. I. 248, *391*
Hadlow, W. J. 141, *160*
Haenni, E. O. see Carol, J. 357, *385*
Haffen, K. see Wolff, E. 605, *636*
Haffler, M. S. see Fukushima, D. K. 217, *390*
Hagedorn, D. see Gordon, E. S. 520, *560*
Hagemann, U. 92, 104, *160*
Hagiwara, H. see Nishikawa, M. 493, 494, *508*
Hagney, M. G. see Knobil, E. 50, *166*
Hagopian, M. 484, 487, *504*
— and L. K. Levy 487, *504*
— see Levy, L. K. 484, *506*
Hague, E. E. see Hamilton, W. F. 531, *560*
Hain, A. M. see Bassil, G. T. 365, *383*
— see Anderson, A. F. 611, *624*
Haines, W. J. 301, 314, 315, *391*, 442, 443, 488, 489, *504*
— and J. N. Karnemaat 313, 314, 318, *391*
— see Drake, N. A. 341, *388*
— see Johnson, R. H. 421, *506*
— see Knauff, R. E. 198, *395*
Halász, B. 80, 88, *160*
Halberg, F., M. J. Frantz and J. J. Bittner 32, *160*
— R. E. Peterson and R. H. Silber 32, 93, 96, *160*
Hale, A. J. see Burns, J. K. 36, 60, 65, 67, 88, 90, 93, 96, *149*
— see Hutchison, W. C. 30, 88, *163*
Halkerston, I. D. K., J. Hillman, D. Palmer and A. Rundle 455, *504*
— see Stitch, S. R. 295, *406*, 434, *511*
Hall, A. E. see Cook, E. R. 314, *386*
Haller, H. J. see Cristol, S. 135, *152*
Hallman, L. F. see Hertz, R. 137, *161*, 456, *505*, 522, *561*
Ham, E. A., R. E. Harman, N. G. Brink and L. H. Sarett 238, *391*
— see Harman, R. E. 318, *391*, 421, 446, *504*
Hamburger, C. 522, *560*
— see Warming-Larsen, A. 541, *569*
Hamilton, J. B. see Dorfman, R. I. 478, *503*, 616, *627*
Hamilton, J. G. jr., and R. T. Holman 312, *391*
Hamilton, W. F., R. G. Ellison, R. W. Pickering, E. E. Hague and J. T. Tucker 531, *560*
Hamlet, J. C. see Barton, D. H. R. 228, *383*
Hammant, J. E. see Clayton, B. E. 96, 106, *151*, 607, *626*
— see Clayton, R. E. 454, *502*
Hampson, J. G. 651, *658*
— and J. Money 650, *658*
— — and J. L. Hampson 650, *658*
— see Money, J. 650, *658*
Hampson, J. L. see Hampson, J. G. 650, *658*
— see Money, J. 650, *658*
Hampton, J. C. see Bennett, H. S. 23, *147*
Hancock, J. E. H. see Hayano, M. 443, 485, 489, *505*
Hane, S. see Rinfret, A. P. 59, 125, *176*
Hansbury, E. see Longwell, B. B. 139, *169*, 422, 423, *507*
Hansen, L. P., A. Cantarow, A. E. Rakoff and K. E. Paschkis 338, *391*
— see Paschkis, K. E. 480, *508*
Hanson, F. R. see Mann, K. M. 261, *397*, 493, *507*
— see O'Connel, P. W. 259, *400*
— see O'Donnell, P. W. 493, *508*
Hanze, A. R., see Hogg, J. A. 208, 256, 257, 277, *393*
— see Levin, R. H. 227, *396*
— see Spero, G. B. 267, 283, *406*
Harary, I. see Bloch, K. 436, *500*
Hardin Branch, C. H. see Bliss, E. L. 642, *657*

Hardy, J. D., and M. D. Turner 475, *504*
— V. B. Ward, M. D. Turner and L. P. Sampson 75, *160*
Hariton, L. B. see Lieberman, S. 217, 294, 297, *396*
Harkness, D. M., E. Muntwyler, F. R. Mautz and R. C. Mellors 537, *560*
Harland, M. see Adams, A. E. 552, *556*
Harman, P. J. 25
Harman, R. E., E. A. Ham, J. De Young, N. G. Brink and L. H. Sarett 421, 446, *504*
— — J. J. de Young, N. G. Brink and L. H. Sarett 318, *391*
— see Ham, E. A. 238, *391*
Harris, A. P. see Liddle, G. W. 138, *168*, 523, *563*
Harris, G. W. 544, *560*
— R. P. Michael and P. P. Scott 574, *628*
Harris, J. B. see Edelman, J. S. 550, *558*
Harris, R. S. see Sebrell, W. H. 139, *178*
Harrison, H. E., and D. C. Darrow 529, *560*
— see Darrow, D. C. 529, 538, *558*
Harrison, J. H., C. B. Leman, P. L. Munson and J. C. Laidlaw 573, 581, 607, *629*
— see Munson, P. L. 581, 603, *623*
Harrison, R. G. 4, 23, 86, 140, *161*
— and C. W. Asling 23, *161*
— and A. J. Cain 54, *161*
— see Cain, A. J. 54, 92, *149*
Harrop, G. 529, 535, *560*
— L. J. Soffer, R. Ellsworth and J. H. Trescher 528, 529, 535, *560*
— — W. M. Nicholson and M. Strauss 525, 530, *561*
— A. Weinstein, L. J. Soffer and J. H. Trescher 528, *561*
Hartley, F. see Cooley, G. 273, *386*
— see Ellis, B. 273, 274, *388*
Hartman, F. A. 18, *161*
— and R. H. Albertin 13, 14, *161*
— and K. A. Brownell 4, 5, 12, 13, 14, 18, 21, 107, *161*, 526, *561*
Hartman, F. A., K. A. Brownell and W. E. Hartman 516, *561*
— — and T. Y. Liu 72, 73, *161*
— R. A. Knouff and G. A. Howard 80. 144, 145, *161*
— — A. W. McNutt and J. E. Carver 14, *161*
— L. A. Lewis, K. A. Brownell, C. A. Angerer and F. F. Shelden 548, *561*
— F. F. Shelden and E. L. Green 5, *161*
— D. E. Smith and L. A. Lewis 526, *561*
— and H. J. Spoor 519, *561*
— see Knouff, R. A. 13, 14, 15, 16, 17, 53, *166*
— see Liu, T. Y. 573, *631*
Hartman, W. E. see Hartman, F. A. 516, *561*
Hartmann, M., and F. Locher 431, *504*
Hartogh-Katz, S. L. see Dingemanse, E. 308, 309, *387*, 580, *627*
Hartroft, P. M., and A. B. Eisenstein 112, 114, 133, *161*
— and W. S. Hartroft 109, 112, 115, *161*
— see Eisenstein, A. B. 113, 114, *155*
— see Pitcock, J. A. 112, *174*
Hartroft, W. S. see Hartroft, P. M. 109, 112, 115, *161*
Harvey, J. see Luetscher, J. A. jr. 520, *563*
Harwood, C. T., and J. W. Mason 364, *391*
Haselwood, L. A. see Wade, N. J. 622, *636*
Haslam, R., and W. Klyne *391*
Haslewood, G. A. D. see Brooksbank, B. W. L. 427, *501*
Hastings, A. B., and E. L. Compere 529, *561*
Hastings, D. W. see Glick, D. 3, *159*
Hatey, J. 58, 78, *161*
— see Fontaine, M. 88, 141, *156*, *157*, 549, *559*
Haussler, E. P. 426, *504*
Hayano, M., and R. I. Dorfman 61, 62, *161*, 443, 449, 485, 486, 489, 490, *504*
— M. Gut, R. I. Dorfman, O. K. Sebek and D. H. Peterson 485, *505*
— M. C. Lindberg, R. I. Dorfman, J. E. H. Hancock and W. von E. Doering 443, 485, 489, *505*
— — M. Wiener, H. Rosenkrantz and R. I. Dorfman 461, *505*
Hayano, M., N. Saba, R. I. Dorfman and O. Hechter 420, 435, 441, 454, 485, *505*
— A. Saito, D. Stone and R. I. Dorfman 420, 435, 441, 485, 491, 494, *505*
— see Bloom, B. M. 495, *500*
— see Caspi, E. 441, 461, 462, *501*
— see Dorfman, R. I. *503*
— see Fish, C. A. 463, 497, *503*
— see Meyer, A. S. 415, 449, 486, *508*, 573, *631*
— see Stone, D. 491, *511*
Hayashida, T. see Li. C. H. 82, 83, 100, 101, 103, 117, *168*, 608, 613, 614, *631*
Haydar, N. A., J. C. Laidlaw, W. J. Reddy and G. W. Thorn 425, *505*
— see Reddy, W. J. 293, *402*, 423, *509*
— see Renold, A. E. 521, *566*
Hayden, A. L. 351, 355, *391*
— see Johnson, D. F. 316, *394*
Hayes, E. R. 55, 67, *161*
— see Belt, W. D. 56, *147*
Haynes, R. C. jr. *161*, 453, *505*
— and L. Berthet 453, *505*
— S. Koritz and F. Peron 453, *505*
— K. Savard and R. I. Dorfman 77, *161*, 442, *505*
— see Dorfman, R. I. *503*
Hays, H. W. 537, *561*
— and D. R. Mathieson 546, *561*
— see Gaunt, R. 526, 527, *559*
— see Mathieson, D. R. 575, *631*
— see Swingle, W. W. 527, 536, 542, 545, *568*
Hayward, S. J. see Allen, W. M. 364, *381*
Hazard, J. 642, *658*
Hazelwood, R. L., L. L. Bennett and M. M. Nelson 139, *161*
Heagy, F. C. see Nicholls, D. 91, *172*
— see Logan, J. E. 91, *169*
Heard, R. D. H. 197, 207, *391*
— W. S. Bauld and M. M. Hoffman 484, *505*
— E. G. Bligh, M. C. Cann, P. H. Jellinck, V. J. O'Donnell, B. G. Rao and J. L. Webb 420, 421, 446, 449, 451, 453, *505*
— and M. M. Hoffman 431, 451, 484, *505*, 610, *629*
— — and G. E. Mack 434, *505*

Heard, R. D. H., P. H. Jellinck and V. J. O'Donnell 451, *505*, 616, *629*
— and A. F. McKay 428, *505*
— and H. Sobel 339, 340, 364, *391*
— — and E. H. Venning 296, *391*
— see Aprile, M. 421, 442, *500*
— see Bauld, W. L. 431, *500*
— see Bligh, E. G. 449, 451, *500*
— see Rao, B. G. 443, 447, 449, 465, *509*
Hebard, W. B., and H. A. Charipper 10, 11, 12, 53, *161*
Hébert, S., and Y. Barbarin 66, *161*
— M. Tran-Vy and J. Verne 43, 108, *161*, 601, *629*
— see Verne, J. 35, 65, 96, *183*
Hechter, O. 445, 446, 447, *505*
— R. P. Jacobsen, R. W. Jeanloz, H. Levy, C. W. Marshall, G. Pincus and V. J. Schenker 442, 489, *505*
— V. Schenker and G. Pincus 442, 489, *505*
— A. Zaffaroni, R. P. Jacobsen, H. Levy, R. W. Jeanloz, V. Schenker and G. Pincus 311, *392*, 441, 442, 486, 489, 490, *505*, 517, *561*, 609, *629*
— see Ayres, P. J. 423, 447, *500*
— see Caspi, E. 461, 462, *501*
— see Eichhorn, J. 441, *503*
— see Hayano, M. 420, 435, 441, 454, 485, *505*
— see Jeanloz, R. W. 449, 486, 489, *506*, 573, *630*
— see Kass, E. H. 75, *165*
— see Levy, H. 441, 442, 489, 490, *506*
— see Luft, J. 24, 38, *169*
— see Macchi, I. A. 21, 76, *169*
— see Saba, N. 420, 421, 435, 439, 441, 443, 489, 499, *510*
— see Stone, D. 446, 491, *511*
— see Zaffaroni, A. 435, 441, *513*
Hecker, E. 304, *392*
Hedinger, C. see Blickenstorfer, E. 639, *657*
Heer, J., and A. Wettstein 254, *392*
Hefferman, I. F. see Knowlton, A. I. 547, *562*
Heffron, R. 118, *161*
Heftmann, E. 313, *392*
Heftmann, E., and D. F. Johnson 299, 300, 316, 318, *392*
— and E. Mosettig 186, *392*
— see Glenn, E. M. 335, *391*
— see Johnson, D. F. 316, *394*, 476, *506*
Hegedüs, H., C. Tamm and T. Reichstein 317, *392*
Hegnauer, A. H., and E. J. Robinson 535, *561*
Hegner, P., and T. Reichstein 270, *392*
Heidorn, G. H., and F. R. Schemm 546, *561*
— — and J. A. Layne 546, *561*
Heiffner, M. H. see Bacchus, H. 88, 114, 116, *146*
Heigl, J. J., M. P. Bell and J. V. White 357, *392*
Heiselt, L. R. see Done, A. K. 454, *502*
— see Good, T. A. 75, 82, 100, *159*
— see Sweat, M. L. 443, 485, 489, *511*
Heisler, A. see Drekter, I. J. 302, *388*
Heller, H. 543, *561*
— T. F. Hewer, J. M. Hughes and H. Schnieden 547, *561*
Heller, M. see Antonucci, R. 225, 361, *382*
— see Bernstein, S. 267, 272, 284, *383*
Hellman, L., H. L. Bradlow, J. Adesman, D. K. Fukushima, J. L. Kulp and T. F. Gallagher 471, *505*
— R. S. Rosenfeld and T. F. Gallagher 436, 471, *505*
— see Bradlow, H. L. 483, *501*, 573, *625*
— see Eidinoff, M. L. 436, 438, *503*
— see Rosenfeld, R. S. 436, 489, *510*
Helmreich, M. L. see Samuels, L. T. 96, *177*, 496, *510*
Hellstrom, J. see Diczfalusy, E. 412, *502*
Heming, A. E., D. E. Holtkamp, J. F. Kerwin, L. F. Mansor and J. H. Dacanay 523, *561*
Hems, B. A. see Attenburrow, J. 242, *382*
Henbest, H. B., E. R. H. Jones, A. A. Wagland and T. I. Wrigley 228, *392*
— G. D. Meakins and G. W. Wood 355, *392*
— see Bladon, P. 275, 360, *383*
— see Crawshaw, A. 228, *386*
Heneage, P. see Sourkes, T. L. *180*
Heni, F., P. Gruner and H. Mast 88, 116, *161*
Henly, A. A. 340, *392*
— and M. Potter 365, *392*
— see Barnett, J. 381, *382*
— see Butt, W. R. 303, *385*
Henneman, P. H. see Pechet, M. M. 577, *633*
Hennigar, G. R. see Larson, P. S. 135, 136, *167*, 458, *506*
— see Nichols, J. 135, 136, 137, *172*, 458, *508*, 523, *565*
Henry, R. 363, 364, *392*
— M. Thevenet and P. Jarrige 295, *392*, 434, *505*
Holness, J. J. see Gray, C. L. 476, *504*
Herbst, A. L. see Yates, F. E. 498, 499, *513*
Herlant, M. see Fortier, C. 50, 58, 59, *157*
Herling, F. see Axelrod, L. R. 486, *500*
— see Fukushima, D. K. 217, *390*
— see Jones, R. N. 215, 351, 353, 354, 355, *394*
— see Kemp, A. D. 228, *395*, 428, 460, *506*
Hernando-Avendano, L. see Renold, A. E. 137, *175*, 457, *509*, 522, *566*
Hernando, L., J. Crabbé, E. J. Ross, W. J. Reddy, A. E. Renold, D. H. Nelson and G. W. Thorn 457, *505*, 531, *561*
Herr, M. E., and F. W. Heyl 211, 254, *392*
— see Johnson, J. L. 245, *394*
Herrick, E. H., and O. Torstveit 143, *161*, 554, *561*
Herrmann, L. G., A. Ames and R. J. Tapke 293, *392*
Hersch, L. S. see Wallace, E. Z. 455, *512*
Hershberg, E. B., C. Gerold and E. P. Oliveto 242, *392*
— M. Rubin and E. Schwenk 281, *392*
— and J. K. Wolfe 297, 299, *392*
— J. K. Wolfe and L. F. Fieser 381, *392*
— see Gould, D. 241, 267, 278, 279, *391*
— see Greenspan, G. 260, *391*

Hershberg, E. B. see Herzog, H. L. 219, 227, 246, 260, 267, 277, *392*, 494, *509*
— see Nobile, A. 277, *399*
— see Oliveto, E. P. 204, 216, 227, 228, 243, 272, 285, *400*
— see Shapiro, E. L. 214, *405*
— see Wolfe, J. K. *381*, 409
Hershberger, L. G., E. G. Shipley and R. K. Meyer 575, 576, *629*
Hertig, A. T. see Benirschke, K. 121, 125, *147*, 415, *500*, 588, 592, *624*
Hertler, W. E. see Corey, E. J. 237, *386*
Hertz, R., M. J. Allen and W. W. Tullner 136, *161*
— D. M. Bergenstal, H. A. Lubs and S. J. Jackson 137, *162*
— J. A. Pittman and M. M. Graff 456, *505*
— W. W. Tullner and M. J. Allen 136, *161*, 456, *505*, 523, *561*
— W. W. Tullner, J. A. Schricker, F. G. Dhyse and L. F. Hallman 137, *161*, 456, *505*, 522, *561*
— see Bergenstal, D. M. 523, *556*
— see Peterson, R. E. 457, *509*, 522, *565*, 586, *633*
— see Schricker, J. A. 137, *178*
— see Thorn, G. W. 137, *181*, 457, *512*
— see Tullner, W. W. 136, 137, *182*, 456, *512*
Herz, J. E., J. Fried, P. Grabowich and E. Sabo 267, 270, 271, *392*
— — and E. Sabo 267, 270, *392*
— see Fried, J. 258, 267, 268, 272, 277, 278, *390*, 487, 491, 492, 493, 494, *504*, 520, *559*
Herzig, P. T., and M. Ehrenstein 221, 234, 235, *393*
— see Heusser, H. 286, 288, *392*, *393*
— see Ruzicka, L. 288, *403*
Herzog, H. L., M. J. Gentles, E. B. Hershberg, F. Carvajal, D. Sutter, W. Charney and C. P. Schaffner 260, *392*
Herzog, H. L., M. Jevnik and E. B. Hershberg 227, 246, *392*
— C. C. Joyner, M. J. Gentles, M. T. Hughes, E. P. Oliveto, E. B. Hershberg 219, *392*
— A. Nobile, S. Tolksdorf, W. Charney, E. B. Hershberg, P. L. Perlman and M. M. Pechet 277, *392*, 494, *505*
— C. C. Payne, M. A. Jevnik, D. Gould, F. L. Shapiro, E. P. Oliveto and E. B. Hershberg 267, 277, *392*
— see Gould, D. 267, 278, 279, *391*
— see Greenspan, G. 260, *391*
— see Nobile, A. 277, *399*
— see Oliveto, E. P. 285, *400*
Hess, M., E. G. Rennels and J. C. Finerty *162*
— see Rennels, E. G. 106, 107, *175*
Hess, R. 640, *658*
Hess, W. C. see Hilmer, P. E. 310, *393*
Hett, J. 23, 39, *162*, 587, *629*
Hetzel, B. S., W. W. Schottstaedt, W. J. Grace and H. G. Wolff 642, *658*
Hetzel, N. see Chart, J. J. 518, 521, 533, 543, *557*
Heusghem, C. see Vliers, M. 424, *512*
Heusler, K., and A. Wettstein 273, 274, *393*
— see Schmidlin, J. 238, 264, *404*
— see Wettstein, A. 206, *408*
— see Woodward, R. B. 261 *409*
Heusser, H., C. Engel, P. Herzig and P. Plattner 286, *393*
— M. Feuer, K. Eichenberger and V. Prelog 212, *393*
— P. T. Herzig, A. Fürst and P. Plattner 288, *393*
— N. Wahba and F. Winternitz 288, *393*
— see Julia, S. A. 275, *394*
— see Ruzicka, L. 288, *403*
Hewer, E. E. see Keene, M. F. L. 39, 125, 126, *165*, 588, *630*
Hewer, T. F. see Heller, H. 547
Heyder, E. see Perrine, J. W. 521, *565*
Heyl, F. W. see Herr, M. E. 211, 254, *392*
— see Johnson, J. L. 245, *394*
Hickey, F. C. see Stokes, W. M. 310, *406*, *511*
Higgins, G. M. see Ingle, D. J. 46, 82, *163*
High, L. B. see Djerassi, C. 336, *388*, 487, 492, 493, *502*
Highhouse, F. see Nadel, E. M. 297, *399*
Hill, A. G. S., H. W. Deane and A. H. Coons 118, *162*
Hill, B. R. 297, *393*
— see Dobriner, K. 431, *502*
— see Lieberman, S. 217, 307, *396*, 427, 428, 431, *507*
Hill, R. T., and A. S. Parkes 553, *561*
Hill, W. C. O. 42, 43, *162*
Hilgar, A. C. see Nadel, E. M. 71, *172*
Hillman, J. see Halkerston, J. D. K. 455, *504*
— see Stitch, S. R. 295, *406*
Hilmer, P. E., and W. C. Hess 310, *393*
Hills, A. G. see Forsham, P. H. 607, *627*
Hilton, J. G., L. F. Scian, C. D. Westermann and O. R. Kruesi 76, *162*
— D. C. Weaver, C. Muelheims, V. V. Glaviano and R. Wégria 76, *162*
Hingerty, D. 529, *561*
Hirschmann, F. B. see Hirschmann, H. 218, 241, 273, 374, *393*, 427, 431, 432, 445, *505*, 572, *629*
Hirschmann, H. 210, 236, 357, *393*, 427, 428, 431, *505*, 572, *629*
— and F. B. Hirschmann 218, 241, 374, *393*, 427, 431, 432, 445, *505*, 572, 629
— — and M. Daus 273, *393*
— and O. Wintersteiner 426, *505*
— see Farrell, G. L. 423, *503*, 610, *627*
Hirschmann, R. see Steinberg, N. G. 287, *406*
Hirschmann, R. F., G. Bailey, G. Poos, R. Walker and J. Chemerda 271, *393*
— — J. Wood and R. E. Jones 269, *393*
— R. Miller, R. E. Beyler, L. H. Sarett and M. Tishler 267, 277, *393*
Hirschowitz, B. 540, *561*
Hisaw, F. L., and J. T. Velardo 612, 614, *629*
— see Velardo, J. T. 614, *636*
— see Zarrow, M. X. 610, *637*

Hitchcock, C. R. see Zimmerman, N. B. 458, *513*
Hitchcock, S. R. see Zimmermann, B. 135, *185*
Hlavac, C. see Dulaney, E. L. 492, *503*
Hoagland, H. 534, *561*, 643, *658*
Hoch-Ligeti, C., and G. H. Bourne 60, *162*
Hodge, H. C. see MacLachlan, P. L. 49, 51, 54, 92, *169*
Hodges, R. E. see Bean, W. B. 139, *147*
Hoede, K. 81, 88, 101, 129, *162*
Hoehn, W. M. see Mason, H. L. 420, 421, *507*
Hoeln, W. J. see Kendall, E. C. 519, *562*
Hoelscher, B. see Zimmermann, B. 135, *185*, 458, *513*
Hoerr, N. L. 32, 37, 45, 54, 56, 90, 92, *162*
Hoet, J. J. see Sobel, E. H. 135, 137, *180*
Hoff, D. R. see Arth, G. E. 285, *382*
Hoffman, F., R. E. Beyler and M. Tishler 286, *393*
— see Beyler, R. E. 222, 225, 272, 274, *383*
Hoffman, M. M., V. E. Kazmin and J. S. L. Browne 459, *505*
— see Heard, R. D. H. 431, 434, 451, 484, *505*, 610, *629*
— see Venning, E. H. 434, *512*, 580, *636*
Hoffman, M. N., and M. H. Lott 428, *505*
Hoffmann, R. A. see Schmidt, R. A. 101, 122, 123, 124, *178*
Hofmann, F. G. 77, *162*, 518, *561*, 596, *629*
— M. M. Dickie and N. P. Christy 606, 618, *629*
— see Deane, H. W. 117, 118, *153*
— see Knobil, E. 78, 80, 87, 99, 101, 107, 109, 110, 115, *166*
Hofmann, H., and H. Staudinger 342, 346, *393*
Hofsommer, R. D. see Taub, D. 222, 267, 276, 285, *407*
Hogg, J. A., P. F. Beal, A. H. Nathan, F. H. Lincoln, W. F. Schneider, B. J. Magerlein, A. R. Hanze and R. W. Jackson 208, 256, 257, *393*
Hogg, J. A., F. H. Lincoln, R. W. Jackson and W. P. Schneider 256, 267, 284, 361, *393*
— — A. H. Nathan, A. R. Hanze, W. P. Schneider, P. F. Beal and J. Korman 256, 277, *393*
— G. B. Spero, J. L. Thompson, B. J. Magerlein, W. P. Schneider, D. H. Peterson, O. K. Sebek, H. C. Murray, J. C. Babcock, R. L. Pederson and J. A. Campbell 272, *393*
— see Anderson, H. V. 227, *381*
— see Magerlein, B. J. 281, 283, *397*
— see Spero, G. B. 267, 271, 283, *406*
Hogness, J. R., N. D. Lee and R. H. Williams 523, *561*
Hohlweg, W., and U. Laschet 86, 96, *162*
Hokin, L. E. see Hokin, M. R. 92, *162*
Hokin, M. R., L. E. Hokin, M. Saffran, A. V. Schally and B. U. Zimmermann 92, *162*
Holechek, J. J., and A. R. Collins 337, *393*
— see Graff, M. M. 337, *391*
Holland, R. C. 122, 123, 124, *162*
Hollander, V. P., S. DiMauro and O. H. Pearson 365, *393*
Hollander, W. jr., R. W. Winters, T. F. Williams, J. Bradley, J. Oliver and L. G. Welt 534, 543, *561*
Hollings, H. L. see Winter, C. A. *184*
Holman, R. T. see Hamilton, J. G. jr. 312, *391*
Holmdahl, S. see Lundlin, P. M. 123, *169*
Holmes, R. L. see Coupland, R. E. 26, *152*
Holmes, W., and D. E. Moorhouse 5, 53, *162*
Holmes, W. N. 48, 58, 85, 128, *162*, 550, *562*, 597, *629*
— see Chester Jones, I. 71, 75, 141, *151*, 416, *502*, *557*
— see Phillips, J. G. 75, *174*, 550, *565*
Holmstrom, E. G. see Migeon, C. J. 591, 592, *632*
Holness, N. J., J. B. Lunnon and C. H. Gray 432, 433, *505*
Holterman, H. see Cardwell, H. M. 261, *385*
Holtkamp, D. E. see Heming, A. E. 522, *561*
Holtorff, A. G., and F. C. Koch 363, *393*
Holtz, A. H. 427, 429, *505*
Holub, D. A. see Kitay, J. I. 85, *165*
Holysz, R. P. 257, *393*
— see Slomp, G. 357, *405*
Holzbauer, M. 54, 71, 75, 105, 106, *162*, 454, *505*, 523, *562*
— and M. Vogt 76, 107, *162*, 456, *502*, 523, *562*
Homer, M. A. see Berger, E. Y. 540, *556*
Hooker, C. W., and T. R. Forbes 610, 612, 613, 614, *629*
— see Fraps, R. M. 610, *628*
Horeau, A., J. Jaques, J. P. Mathieu and A. Petit 196, *393*
Horger, L. M. see Zarrow, M. X. 140, *185*
Horn, E. H. 69, *162*
Horning, M. G. see Cornforth, J. W. 368, *386*
Horstmann, P. M. see Schneider, J. J. 461, *510*, 610, *634*
Horwitt, B. N. 296, *394*
— R. I. Dorfman, R. A. Shipley and W. R. Fish 459, *505*, 610, *629*
— see Fish, W. R. 459, *503*
Hoskins, W. H., J. R. Coffman, F. C. Koch and A. T. Kenyon 616, *629*
Houssay, A. B., A. F. Cardeza, B. A. Houssay and R. M. Pinto 134, *162*, 618, *629*
— see Houssay, B. A. 134, *163*, 618, *629*
Houssay, B. A. 142, *162*
— F. A. Cardezy, V. G. Foglia, A. B. Houssay and R. M. Pinto 618, *629*
— A. B. Houssay, A. F. Cardeza, V. G. Foglia and R. M. Pinto 134, *163*
— — — and R. M. Pinto 134, *163*
— and J. T. Lewis 516, *562*
— and R. Sammartino 46, 78, 79, 99, 109, *163*
— see Houssay, A. B. 134, *162*, 618, *629*
Houssay, H. E. J. 79, *163*
Houssay, R. H. 120, *163*
Howard, A. N., and D. B. Cater 71, 104, *163*
— and B. J. Constable 51, 58, 84, 85, *163*

Howard, E. (Howard-Miller) 47, 48, *163*, 577, 578, 579, 593, 594, 596, 597, 598, 599, 600, 601, 602, 603, 608, 615, *629*, *630*
— and J. C. Allen 620, *630*
— and R. S. Benua 596, *630*
— and S. Gengradom 596, *630*
— see Benua, R. S. 48, *147*, 588, 593, *624*
Howard, G. A. see Hartman, F. A. 80, 144, 145, *161*
Howard, J. M., J. M. Olney, J. P. Frawley, R. E. Peterson, L. H. Smith, J. H. Davis, S. Guerra and W. H. Dibrell 642, *658*
Howard, P. J. see Therien, B. 135, *181*
Howell, C. see Gaunt, R. 80, *158*
Howell, D. S. see Davis, J. O. 531, *558*
Høyrup, E. 656, *658*
Hryse, B. see Davis, M. E. 611, *626*
Hsieh, K. M. 550, *562*
Hsu, C. T. see Touchstone J. C. 346, *407*
Huang, R. L. see Wuersch, J. 437, *513*
Huang-Minlon 246, *394*
— E. Wilson, N. L. Wendler and M. Tishler 216, *394*
Hübener, H. J. 461, 462, 463, *505*
— D. K. Fukushima and T. F. Gallagher 461, 462, 463, 497, *505*
— see Amelung, D. 463, 497, 500
Hubotter, F. see Sandritter, W. 90, 94, *177*
Hudson, P. see Romanoff, E. B. 75, *176*, 422, *509*, 573, 574, 609, *634*
Hudson, P. B., and M. E. Lombardo 418, *505*
— see Lombardo, M. E. 75, 77, *169*, 300, *397*, 443, 459, 490, *507*, 572, 573, 609, *631*
Huff, J. W. see Tavormina, P. A. 436, 437, *511*
Huffman, M. N. see Smith, G. V. S. 426, *511*
Hügin, F. see Verzár, F. 72, *183*
Huggins, C., F. M. Parsons and E. V. Jensen 607, *630*
— see Talalay, P. 66, *181*
Hughes, E. C. see Gardner, L. I. 582, *628*
Hughes, E. R. see Ely, R. S. 311, *389*
Hughes, J. M. see Heller, H. 547, *561*
Hughes, M. T. see Herzog, H. L. 219, *393*
Hughes, R. see Eiseman, B. 118, *155*
Huhn, A., and P. Schneppenheim 104, 130, 133, *163*
— see Schneppenheim, P. 82, 88, 133, *178*
Huis in't Veld, L. G. 308, *394*
— and E. Dingemanse 412, *505*
— see Dingemanse, E. 308, 309, *387*, 580, *627*
— see Jayle, M. F. 434, *506*
Hultberg, S. see Diczfalusy, E. 412, *502*
Hume, D. M., and D. H. Nelson 456, *505*
— see Egdahl, R. H. 522, *558*
— see Ganong, W. F. 74, 76, 79, 100, 105, *157*
— see Nelson, D. H. 73, 75, 76, 103, 106, *172*
Hummel, J. P. see Ruliffson, W. S. 371, *403*
Hummel, K. P. 21, *163*
Humphreys, E. M., and L. Donaldson 136, *163*
Humphries, P. see Jones, R. N. 215, 353, 355, *394*
— see Lieberman, S. 427, 428, 431, 432, *507*
Humphrys, E. see Landau, R. L. 615, *630*
Hunger, A., and T. Reichstein 239, *394*
Hungerford, G. F. 112, 114, *163*
— see Li, C. H. 82, 83, 100, 101, 103, 117, *168*, 608, 613, 614, *631*
Hunt, E. A. see Hunt, T. E. 67, *163*
Hunt, T. E., and E. A. Hunt 67, *163*
Hunter, G. D. see Cornforth, J. W. 437, *502*
— see Popjak, G. 436, *509*
Hunter, J. H. see Moffet, R. B. 227, *399*
Hurley, L. S., and A. F. Morgan 139, *163*
Hurlock, B., and P. Talalay 338, *394*
— see Cope, C. L. 459, 460, 475, *502*
Husten, M. see Creutzfeldt, W. 49, 51, 84, 88, 93, 94, 104, *152*
Hutchison, W. C., J. K. Burns and A. J. Hale 30, 88, *163*
Hutchison, W. C. see Burns, J. K. 88, *149*
Hutton, T. W. see Dauben, W. G. 437, *502*
Hyman, C., and R. Chambers 552, *562*

Ibayashi, H. see Okinaka, S. 456, *508*
Ide, T. see Ishizuka, N. 461 *506*
Idler, D. R., A. P. Ronald and P. J. Schmidt 550, *562*
Iglesias, R., and E. Mardones 135, *163*
Ikeda, M. see Okinaka, S. 456, *508*
Ikkos, D. see Luft, R. *529*, 531, 538, *563*
Inhoffen, H. see Woodward, R. B. 287, *409*
Ingle, D. J. 82, 118, 120, *163*
— D. F. Beary and A. Purmalis 491, 506
— and G. M. Higgins 46, *163*
— — and E. C. Kendall 82, *163*
— and E. C. Kendall 519, *562*
— J. E. Nezamis and E. H. Morley 519, 528, *562*
— H. W. Nilson and E. C. Kendall 535, *562*
— M. C. Prestrud and C. H. Li 103, *163*
— — and K. L. Rice 531, *562*
— see Baker, B. L. 608, *624*
— see Price, D. 108, *175*, 604, 605, 623, *633*
Iriarte, J. see Djerassi, C. 281 *387*
Iseri, L. T. see Mader, J. J. 529, *563*
Ishizuka, N., M. Maeyama, T. Ide and M. Kawata 461, *506*
Isler, P. see Blickenstorfer, E. 639, *657*
Island, D. see Coppage, W. S. 529, *558*
— see Liddle, G. W. 138, *168*, 523, *563*
Isselbacher, K. J. 452, 461, *506*
— and J. Axelrod 452, 461, 463, *506*
— see Tomkins, G. M. 495, 498, *512*
Ito, T. 27, 32, 37, 48, 128, *163*
— see Togari, C. 23, *182*
Ives, D. A. see Barton, D. H. R. 246, *383*
Ivy, A. C. see Greene, R. R. 579, *628*
Izzo, A. J., E. H. Keutmann and R. B. Burton 340, *394*

Jackson, C. M. 30, 31, 32, 41, 69, *163*
Jackson, R. W. see Hogg, J. A. 208, 256, 257, 267, 284, 361, *393*
Jackson, S. J. see Hertz, R. 137, *162*
Jacob, M. see Mandel, P. 90, *170*
Jacobs, M. see Soffer, L. J. 531, *567*
Jacobs, W. A. see Sato, Y. 238, *404*
Jacobsen, R. P. see Hechter, O. 311, *392*, 441, 442, 486, 489, 490, *505*, 517, *561*, 609, *629*
— see Jeanloz, R. W. 449, 486, 489, *506*, 573, *630*
— see Levy, H. 441, 442, 489, 490, *506*
Jacquier, C. see Vesin, P. 521, 546, *568*
Jacquot, R. 121, *163*
— see Jost, A. 124, *164*
Jaffe, H. see Daughaday, W. H. 296, *387*
— see Rogers, W. F. jr. 337, *403*
Jager, B. V. see Perkoff, G. T. 644, *659*
Jailer, J. W. 59, 125, *163*
— and N. F. Boas 58, 82, 85, 144, *164*
— N. P. Christy, D. Longson, E. Z. Wallace and W. E. L. Gordon 106, *164*
— J. J. Gold, R. Vande Wiele and S. Lieberman 465, *506*, 585, *630*
— — and E. Z. Wallace 107, 135, *164*
— see Christy, N. P. 80, 100, 103, 105, 106, *151*
— see Kitay, J. I. 85, *165*
— see Rosselet, J. P. 219, *403*, 432, 459, 465, *510*
Jakobovits, A. see K. Kovács 120, *166*
Jakobson, T. see Laidlaw, J. C. 572, *630*
James, V. H. T. see Callow, R. K. 228, *385*
Jansen, A. B. A. see Attenburrow, J. 242, *382*
Janson see Gennes, L. de 546, *560*
Jansson, J. I., and J. Kyrki 363, *394*
Jaques, J. see Horeau, A. 196, *393*
Jardetzky, C. D., and D. Glick 63, *164*
Jarrige, P. see Henry, R. 295, *392*, 434, *505*
Järvinen, P. A. see Uuspää, V. J. 106, *182*
Jayle, M. F., and E. E. Baulieu 295, *394*
— and O. Crépy 293, *394*
— L. G. Huis in't Veld, E. Baulieu and O. Crepy 434, *506*
— D. Malassis and H. Pinaud 580, *630*
— see Chemama, R. 434, *502*
— see Crépy, O. 310, *387*
— see Lelong, M. 582, *631*
— see Mallassis, D. 434, *507*
— see Weinmann, S. 434, *512*
Jayne, E. P. 27, 57, 133, *164*
Jeanloz, R. W., H. Levy, R. P. Jacobsen, O. Hechter, V. Schenker and G. Pincus 449, 486, 489, *506*, 573, *630*
— see Hechter, O. 311, *392*, 441, 442, 486, 489, 490, *505*, 517, *561*, 609, *629*
— see Levy, H. 441, 442, 489, 490, *506*
— see Meyer, A. S. 486, 489, *508*
Jefferies, W. McK. see Bartter, F. C. 531, *556*
Jeger, O. see Buchschacher, P. 237, *384*
— see Buzzetti, F. 237, *385*
— see Cainelli, G. 236, *385*
— see Dauben, W. G. 191, *387*
— see Dutler, H. 278, *388*
— see Greuter, F. 237, *391*
— see Riniker, B. 191, *402*
Jellinck, P. H. see Heard, P. H. 420, 421, 446, 449, 451, 453, *505*
— see Heard, R. D. H. 616, *629*
Jenkins, D. see Laidlaw, J. C. 572, *630*
— see Reddy, W. J. 292, *402*
— see Renold, A. E. 580, *634*
— see Thorn, G. W. 518, 540, *568*
Jenkins, J. S., J. W. Meakin and D. H. Nelson 137, 138, *164*
— — — and G. W. Thorn 138, *164*, 523, *562*
— L. Pothier, W. J. Reddy, D. H. Nelson and G. W. Thorn 523, *562*
Jensen, A. O. see Johnson, J. L. 357, *394*
— see Slomp, G. 357, *405*
Jensen, C. C., and L. E. Tötterman 294, *394*
Jensen, E. V. see Huggins, C. 607, *630*
— see Tannhauser, P. 271, *407*
Jesch, W. see Wolff, H. P. 532, *569*
Jevnik, M. A. see Herzog, H. L. 227, 246, 267, 277, *392*
— see Nobile, A. 277, *399*
Jewett, H. J. see Frame, E. G. 581, *627*
Jiv, J. see Dauben, W. G. 245, *387*
Johannes, A., H. Staudinger and H. Wilz 58, 85, *164*
Johns, W. F. see Sarett, L. H. 262, *404*
Johnsen, S. G. 581, *630*
Johnson, B. A. see Slomp, G. 357, *405*
Johnson, B. B., A. H. Liebermann and P. J. Mulrow *506*
Johnson, D. F., E. Heftmann and J. J. Bunim 476, *506*
— — and A. L. Hayden 316, *394*
— see Heftmann, E. 299, 300, 316, 318, *392*
Johnson, E. H. see Geppert, L. J. 643, *658*
Johnson, J. A. see Wilds, A. *409*
Johnson, J. E. jr. A. R. Remmers, G. W. Clayton and R. Gregory 457, *506*
Johnson, J. L., M. F. Grostic and A. O. Jensen 357, *394*
— M. E. Herr, J. C. Babcock, A. E. Fonken, J. E. Stafford and F. W. Heyl 245, *394*
— see Slomp, G. 357, *405*
Johnson, R. B. see Kelly, T. L. 63, *165*
Johnson, R. E. see Lyons, W. R. 613, *631*
Johnson, R. H., and W. J. Haines 421, 457, *506*
Johnson, V. K., and H. H. Ussing 550, *562*
Johnston, C. D. 310, *394*
Johnston, D. B. R. see Arth, G. E. 285, *382*
Johnston, J. D., and K. Bloch 436, 438, *506*
Johnstone, F. R. C. 23, *164*
de Jongh, S. E., S. Kober and E. Laqueur 426, *506*
Jones, B. F. see Dalton, A. J. 87, *152*
Jones, E. see Werbin, H. 435, *513*

Jones, E. R. H. see Bladon, P. 275, *383*
— see Crawshaw, A. 228, *386*
— see Henbest, H. B. 228, *392*
Jones, E. S. 539, *562*
Jones, G. E. S. see Jones, H. W. 589, 593, *630*
Jones, H. S. see Locke, W. 540 541, *563*
Jones, H. W., and G. E. S. Jones 589, 593, *630*
Jones, H. W. jr. see Ledogar, J. A. 484, *506*
Jones, J. E. see Berliner, D. L. 422, *500*
Jones, J. K. N., and S. R. Stitch 314, *394*
Jones, J. M., C. W. Lloyd and T. C. Wyalt 51, 59, *164*
— see Lloyd, C. W. 611, *631*
Jones, M. E. see Munson, P. L. 337, 364, *399*
Jones, P. G. see Barton, D. H. R. 228, *383*
Jones, R. E., and S. A. Robinson 225, *394*
— see Hirschmann, R. F. 269, *393*
Jones, R. N. 349, 351, *394*
— and A. R. H. Cole 355, 356, *394*
— — and B. Nolin *394*
— and K. Dobriner 347, 353 357, *394*
— and F. Herling 351, 354, *394*
— — and E. Katzenellenbogen 353, *394*
— P. Humphries, F. Herling and K. Dobriner 215, 353, 355, *394*
— D. A. Ramsay, D. S. Keir and K. Dobriner 354, *394*
— and C. Sandorfy 348, *394*
— see Cole, A. R. H. *386*
— see Dobriner, K. 205, 301, 303, 347, *388*
— see Roberts, G. 347, *402*
Jones, R. S., Y. C. Mayne and K. Eik-Nes 85, *164*
— see Eik-Nes, K. 422, *503*
Jordan, E. J. see Race, G. J. 83, 101, 108, 111, 112, *175*
Jørgensen, C. B. 551, *562*
Jorgensen, H. see Oliveto, E. P. 204, 228, *400*
Joseph, M. see Anderson, E. 529, 556
Josephson, E. S. see Taylor, D. J. 51, 84, 144, *181*
Joshi, R. A., and H. T. Blumenthal 120, *164*
Josimovich, J. B., A. J. Ladman and H. W. Deane 44, 45, 121, 125, *164*
Jost, A. 121, 123, *164*, 604, *630*
— and J. P. Geloso 604, *630*
— and P. Grandou 107, *164*
— and R. Jacquot 121, *164*
— — and A. Cohen 124, *164*
Jouan, P. see Roche, J. *176*
Jouanneteau, J. see Zwingelstein, G. 310, *410*
Joyner, C. C. see Herzog, H. L. 219, *393*
Julia, S. A., P. Plattner and H. Heusser 275, *394*
Julian, P. L., W. Cole, E. W. Meyer and B. M. Regan 274, *395*
— E. Meyer, W. Karpel and W. Cole 216, 225, 245, *395*
— — and I. Ryden 255, *395*
— — — and I. R. Waller 255, *395*
— see Cole, W. 273, *386*
Junkmann, K. 68, *164*, 221, *395*
Just, G. see Engel, C. 244, 286, *389*

Kabanov, M. M. 649, *658*
Kagawa, C. M., J. A. Cella and C. G. Van Arman 524, *562*
— and C. G. Van Arman 521, *562*
Kägi, H. see Miescher, K. 212, *398*
Kahn, F. see Klotz, H. P. *165*
Kahnt, F. W., C. Meystre, R. Neher, E. Vischer and A. Wettstein 491, 492, *506*
— R. Neher and A. Wettstein 234, 236, 238, *395*, 413, 443, 446, 447, 490, *506*, 517, *562*
Kalant, O. J. 72, 92, 107, *164*
— and E. A. Sellers 62, 107, *164*
Kalliomaki, L. see Vasama, R. 87, *182*
Kalnitsky, G. see McMahon, J. M. 342, *398*
Kaluszyner, A. see Cohen, S. L. 338, *386*
Kalvoda, J. see Buchschacher P. 237, *384*
— see Buzzetti, F. 237, *385*
— see Greuter, F. 237, *391*
Kamen, M. D. 367, *395*
Kamm, O. see Marker, R. E. 209, *397*, 431, *507*
Kampmeier, O. F. 129, *164*
Kaplan, M. H., A. H. Coons and H. W. Deane 118, *164*
Kappas, A. and T. F. Gallagher 295, *395*, 580, 581, 583, *630*
— see Gallagher, T. F. 456, *504*, 586, *628*
— see Kemp, A. 228, *395*, 418, 460, *506*
Kar, A. B. 17, 53, 60, 65, 92, *164*, *165*
— J. N. Karkun and N. N. De 107, *165*
Karkun, J. N. see Kar, A. B. 107, *164*, *165*
— see Roy, S. N. 107, *177*, 596, *634*
Karl, J. see Bauer, J. 308, *383*
Karmazin, L. see Hadházy, C. 118, *160*
Karnemaat, J. M. see Mann, K. M. 261, *397*
Karnemaat, J. N. see Haines, W. J. 313, 314, 318, *391*
— see Mann, K. M. 493, *507*
Karnovsky, M. L., and H. W. Deane 49, 50, 54, 55, 56, 78, 84, 92, 94, 95, 109, *165*
Karpel, W. see Julian, P. L. 216, 225, 245, 255, *395*
Karrer, A. see Peterson, R. E. 364, *401*
Karrer, P. 434, *506*
Kasparow, M. see Cooper, D. Y. 422, 423, *502*
Kass, E. H., V. Hechter, I. A. Macchi and T. W. Mou 75, *165*
Katonah, F. see Parmer, L. G. 622, *633*
Katsh, S., A. S. Gordon and H. A. Charipper 604, *630*
Katsura, H. 62, *165*
Katzenelbogen, S. see Fang, A. D. 649, *657*
Katzenellenbogen, E. R., K. Dobriner and T. H. Kritchevsky 312, 314, *395*
— see Dobriner, K. 205, 347, 350, *388*, 420, 421, *502*
— see Jones, R. N. 355, *394*
Katzman, P. A., see Buehler, H. J. 293, 295, 297, *384*
Kaufman, E. H. see Sokoloff, L. 20, 51, 80, *180*
Kaufmann, E. see Fassbender H. G. 128, *156*
Kaufmann, S. see Djerassi, C. 288, *387*
— see Löken, B. 273, *397*
— J. Pataki, G. Rosenkranz, J. Romo and C. Djerassi 279, *395*
— see Rosenkranz, G. 255, *403*
— see Sondheimer, F. 233, 361, *405*

Kawata, M. see Ishizuka, N. 461, *506*
Kay, W. W. and R. Whitehead 51, 70, *165*
Kayser, C., and M. Aron 21, 94, 128, *165*
Kazmin, V. E., see Hoffman M. M. 459, *505*
— see Venning, E. H. 304, *407*, 580, *636*
Keating, V. J. see Stirling, G. A. 19, 69, *181*
Kedda, L. see Bahn, R. 607, *624*
Keele, D. see Klein, R. 521, *562*
Keene, M. F. L., and E. E. Hewer 39, 125, 126, *165*, 588, *630*
Kehrer, E. 646, *658*
Keir, D. S. see Jones, R. N. 354, *394*
Keller, A. R. see Migeon, C. J. 572, 581, 582, 583, 584, 591, 592, *632*
Keller, M. 421, 427, *506*
Kelleway, C. H., and H. S. Cowell 515, *562*
Kelley, V. C. see Done, A. K. 75, *154*, 454, 471, *502*
— see Ely, R. S. 311, *389*, 471, *503*
— see Good, T. A. 75, 82, *159*
— see Sydnor, K. L. 585, *635*
Kellie, A. E. 363, *395*
— and E. R. Smith 318, *395*
— — and A. P. Wade 338, *395*
— and A. P. Wade 293, 310, *395*
— see Edwards, R. W. H. 293, 310, 340, *388*
Kelsey, W. M. see Little, J. M. 136, *169*
Kelly, R. M. see Nathanson, I. T. 616, *632*
Kelly, T. L., E. D. Nielson, R. B. Johnson and C. S. Vestling 63, *165*
Kemp, A. D., A. Kappas, I. Salamon, F. Herling and T. Gallagher 228, *395*, 428, 460, *506*
— see Fukushima, D. K. 431, 460, *504*
Kendall, E. C. 519, *562*
— H. L. Mason, W. J. Hoeln and B. F. McKenzie 519, *562*
— see Allers, W. D. 530, *556*
— see Colton, F. G. 274, *386*
— see Eversole 519, *558*
Kendall, E. C. see Gaunt, R. 519, *559*
— see Ingle, D. J. 82, *163*, 519, 535, *562*
— see Mason, H. L. 223, *398*, 420, 421, *507*, 519, *564*
— see Mattox, V. R. 221, 278, *398*
— see McGuckin, W. F. 229, 257, *398*
— see Turner, R. B. 228, *407*
— see Wells, B. B. 519, *569*
Kennedy, H. S. see Andersen, D. H. 622, 623, *624*
Kenny, A. D. see Munson, P. L. 337, 363, *399*
Kent, J. F. see Coutinho, H. B. 66, 98, 109, *152*
Kenyon, A. T. 134, *165*
— see Hoskins, W. H. 616, *629*
Kepler, E. J. see Mason, H. L. 412, 427, 431, 434, 478, *507*, 572, 574, *631*
— see Robinson, F. J. 542, *566*
Kerwin, J. F. see Heming, A. E. 523, *561*
Kessler, W. B. see Fried, J. 274, *390*
Keston, A. S., D. Rittenberg and R. Schoenheimer 368, *395*
— see Sonenberg, M. 82, *180*
Keutmann, E. H. see Burton, R. B. 319, 327, 339, 342, *384*, 459, *501*
— see Izzo, A. J. 340, *394*
— see Zaffaroni, A. 318, 342, *409*
Kilman, L. see Bennett, H. S. 3, *147*
King, C. G. see Douglas, J. F. 452, *503*
King, D. C. see Weichselbaum T. E. 366, *408*
King, J. S. jr. see Eades, H. C. jr. 418, *503*
King, J. T., C. B. Casas and M. B. Visscher 133, *165*
Kingsbury, J. W., S. L. Emery and A. E. Adams 60, 145, *165*
Kinsell, L., D. Bryant and F. Albright 580, *630*
Kinsella, R. A. jr., and J. H. Glick jr. 296, *395*
Kinsella, R. A. see Francis, F. E. 591, *628*
Kirk, D., D. Patel and V. Petrow 214, 288, *395*
Kirk, D. N. see Adams, W. J. 218, *381*
Kirk, M. A. see Kritchevsky, D. 310, 342, 345, *396*
Kirschbaum, A., M. Frantz and W. L. Williams 134, *165*
— see Frantz, M. J. 134, *157*, 607, *628*
Kiss, T. 26, *165*
Kissman, H. M., A. M. Small and M. J. Weiss 272, *395*
Kita, D. A. see Shull, G. M. 493, *511*
Kitay, J. I., D. A. Holub and J. W. Jailer 85, *165*
Kitchell, R. L. 124, *165*
— and L. J. Wells 122, 123, 124, *165*
Klärner, P. 88, 90, 93, 101, 111, 133, *165*
Klass, D. L., M. Fieser and L. F. Fieser 247, *395*
Klein, F. see de Gennes, L. 655, *657*
Klein, R., J. Fortunato, Z. Laron and C. Papadatos 423, *506*
— — and C. Papadatos 72, 125, *165*
— C. Papadatos, J. Fortunato and C. Byers 296, *395*, 425, *506*
— and A. Rovnanek 125, *165*
— P. Taylor, C. Papadotos, Z. Laron, D. Keele, J. Fortunato and C. Byers 521, *562*
— — — Z. Laron, D. Keele, J. Fortunato, C. Byers and C. Billings 521, *562*
Kleinberg, W. see Cohen, H. 82, 83, *151*
Kleinholz, L. H. 552, *562*
Kleinsorge, H. 655, *658*
Klemperer, W. W. see Green, D. M. *560*
Klendshoj, N. C., M. Feldstein and A. S. Sprague 363, *395*
Kliman, A. see Cook, R. P. 436, *502*
Kliman, B. see Davis, J. O. 114, *153*
— see Peterson, R. E. 432, 460, *509*
— see Yankopoulos, N. A. 117, *185*
Kline, I. T. 70, *165*
Kling, A. see Schwartz, N. B. 18, 58, 85, *178*
Klingsberg, A. see Fried, J. 260, *290*
Kloos, K., and H. J. Staemmler 54, 122, 123, *165*
Klopper, A., J. A. Strong and L. R. Cook 463, *506*, 610, 611, *630*

Klopper, A. J. see Strong, J. A. 412, *511*
Klose, H. G. 9, *165*
Klotz, H. P., and F. Kahn 117, *165*
Klyne, W. 186, 193, 214, 215, 218, 219, 372, 373, *395*
— see Broadbent, I. E. 339, *384*
— see Brooks, R. V. 214, 308, 310, *384*, 428, *501*
— see Djerassi, C. 375, *387*
— see Haslam, R. *391*
Knab, J. 13, 15, 60, *165*
Knauff, R. E. see Drake, N. A. 341, *388*
— E. D. Neilson and W. J. Haines 198, *395*
Knigge, K. M. 18, 37, 54, 78, 79, 88, 91, 92, 101, 104, 105, 106, 109, 115, *165*, *166*
— see Schindler, W. J. 73, 74, 75, 76, *178*
Knight, R. P. jr., D. S. Kornfeld, G. H. Glaser and P. K. Bondy 535, *562*
Knobil, E. 34, 66, 98, *166*
— and F. N. Briggs 50, 51, 58, 59, 121, 124, *166*
— and M. J. Fregly 71, *166*
— and R. O. Greep 99, *166*
— M. G. Hagney, E. J. Wilder and F. N. Briggs 50, *166*
— F. G. Hofmann and R. O. Greep 107, *166*
— A. Morse, F. G. Hofmann and R. O. Greep 78, 80, 87, 99, 101, 109, 110, 115, *166*
— — R. C. Wolf and R. O. Greep 116, *166*
Knoll, J. E. see Eidinoff, M. L. 250, *388*, 436, 438, *503*
Knouff, R. A., J. B. Brown and B. M. Schneider 49, 51, 54, 92, 94, *166*
— and F. A. Hartman 13, 14, 15, 16, 17, 53, *166*
— see Brown, J. B. 49, 51, 54, *149*
— see Hartman, F. A. 14, 80, 144, 145, *161*
Knowlton, A. I. 475, *506*
— and E. N. Loeb *562*
— — B. C. Seegal and H. C. Stoerk 115, *166*
— — H. C. Stoerk and B. C. Seegal 115, *166*, 531, *562*
— — — J. P. White and J. F. Hefferman 547, *562*
Kober, S. 337, *395*
— see Dingemanse, E. 426, *502*
— see de Jongh, S. E. 426, *506*
Koch, A. see Woodbury, D. M. 531, 538, *569*
Koch, F. C. 434, *506*
— see Evans, J. S. 579, *627*
— see Holtorff, A. G. 363, *393*
— see Hoskins, W. H. 616, *629*
— see Munson, P. L. 434, *508*
Koch, G. see Neumann, K. 63, *172*
Kochakian, C. D. 574, 577, *630*
— and C. Tillotson 577, *630*
— see Clark, L. C. jr. 478, *502*
— and G. Stidworthy 327, 342, *396*
Koczorek, K. R. see Wolff, H. P. 114, *184*, 532, *569*
Koenig, V. L., F. Melzer, C. M. Szego and L. T. Samuels 337, *396*
Kohno, S. 17, 23, 35, *166*
Koiw, E. see Nowaczynski, W. 236, 333, 334, *400*
Koletsky, S. see Farrell, G. L. 74, 81, 101, 110, 117, *156*, *503*, 517, 521, *558*
— see Rauschkolb, E. W. 74, 100, 109, 110, *175*
Kolff, W. see Savard, K. 424, *510*
Kollonitsch, J. see Bodánszky, A. 346, *384*
Kolmer, W. 27, 35, *166*
Kolthoff, I. M., and J. J. Lingane 380, *396*
Konrad, E. L., and L. C. Wyman 120, *166*
Konikov, W. M. see Gruenwald, P. 27, 28, 45, *160*
Korchin, S. J. see Basowitz, H. 642, *656*
Korenchevsky, V. 20, *166*
— and M. Dennison 455, *506*
Kornfeld, D. S. see Knight, R. P. jr. 535, *562*
Kornfeld, W. 13, *166*
Koritz, S. B., and F. G. Peron 453, *506*
— — and R. I. Dorfman 454, *506*
— see Haynes, R. C. jr. 453, *505*
— see Masson, G. M. C. 120, *170*
— see Peron, F. 77, *174*, 454, *509*
Korman, J. see Hogg, J. A. 256, 277, *393*
Korpássy, B., J. Török and K. Kovács 83, *166*
Korus, W. see Dirscherl, W. 314, *387*
— see Shriefers, H. 476, *511*
Kosto, B. see Pickford, G. E. 144, *174*
Koupreianow, A. see Talbot, N. B. *407*
Kovacevic, M. see Bariety, M. 546, *556*
Kovács, K., and M. A. Dávid 112, 113, *166*
— and A. Jakobovits 120, *166*
— see Korpássy, B. 83, *166*
Kowalewski, K. 140, *166*
Kozlowski, M. A. see McAleer, W. J. 341, 342, *398*
Kracht, J. 88, 100, *166*
— and M. Spaethe 3, 88, 107, 140, *166*, *167*
Kraemer, A. see Osorio, J. A. 18, 135, *173*
Krahl, W. A. see Schultz, R. B. 138, 139, *178*
Krauter, D. 7, 8, *167*
Kraychy, S., and T. Gallagher 484, *506*
Kreč, M. see Andreč, K. 234, *381*
Krehl, W. A. see Winters, R. W. 105, 139, *184*, 458, *513*
Krick, E. T. see Rubin, M. I. 529, *566*
Krisch, K., and H. Staudinger 63, 85, *167*
Kritchevsky, D., and I. Gray 436, *506*
— and M. R. Kirk 310, 342, 345, *396*
Kritchevsky, T. H., and T. F. Gallagher 215, 255, *396*
— D. Garmaise and T. H. Gallagher 228, 242, *396*
— and A. Tiselius 328, *396*
— see Fukushima, D. K. 217, 225, 250, 390, 443, 459, 460, 483, *504*
— see Gallagher T. F. 211, 212, *390*
— see Katzenellenbogen, E. R. 312, 314, *395*
Krohn, P. L. 18, 80, 82, 83, *167*
Kruesi, O. R. see Hilton, J. G. 76, *162*
Kruger, F. A. see Marks, B. H. 137, *170*, 457, *507*
Krus, D. see Bergen, J. R. 579 *624*
Kučera, J., Ž. Próchozka and K. Veres 328, *396*
Küchmeister, H., G. Pliess and W. Wilhelm 135, 136, *167*
Kuether, C. A. see Roe, J. H. 58, *176*
Kuhlman, D., C. Ragan, J. W. Ferrebee, D. W. Atchley and R. F. Loeb 534, *562*

Kuizenga, M. H. 518, 519, 527, 562
— and G. F. Cartland 420, 421, *506*
— see Cartland, G. F. 519, *557*
Kulp, J. L. see Hellman, L. 471, *505*
Kumita 24, *167*
Kun, H. see Steinach, E. 616, *635*
Kupperman, H. S. see Gutner, L. B. 546, *560*
Kurland, S. see Baumann, E. J. 515, 529, *556*
Kurlents, E. see Birmingham, M. K. 77, 101, *148*, 452, *500*
Kurts, M. see Earle, D. P. jr. 99, *155*
Kutz, R. L., T. McKeown and H. Selye 621, *630*
Kushinsky, S. 442, 449, 488, *506*
— see Levy, H. 238, 300, *396*, 442, 490, *506*
Kyrki, J. see Jansson, J. I. 363, *394*

Laat, B. M. de see Dingemanse, E. 308, *387*
Labrosse, E. H. 350, *396*
Ladman, A. J., and H. W. Deane 125, *167*
— see Josimovich, J. B. 44, 121, 125, *164*
Laeschke, R. 30, *167*
La Flamme, A. see Nowaczynski, W. 236, 333, *400*
Laforce, R. C. see Weaver, H. E. 368, *408*
Laidlaw, J. C., J. F. Dingman, W. L. Arons, J. T. Finkenstaedt and G. W. Thorn 518, 533, 535, *563*
— D. Jenkins, W. J. Reddy and T. Jakobson 572, *630*
— see Harrison, J. H. 573, 581, 607, *629*
see Haydar, N. A. 425, 505
— see Munson, P. L. 581, 603, *632*
— see Reddy, W. J. 293, *402*, 423, *509*
— see Renold, A. E. 521, *566*
— see Thorn, G. W. 518, 540 *568*
Lake, K. H. see Marrian, G. F. 426, *507*
Laken, B. see Dumm, M. E. 69, 71, *154*
Lakshmanan, T. K., and S. Lieberman 309, 310, *396*, 580, *630*
Lamedica, G., L. Correale and R. Bonanni, 476, *506*
— and G. Missale 446, *506*
Lance, E. M. see Liddle, G. W. 138, *168*, 523, *563*
Landau, R. L. 363, *396*
— B. F. Stimmel, E. Humphrys and D. E. Clark 615, *630*
Landing, B. H., and D. Feriozi 80, 111, *167*
— see Craig, J. M. 129, *152*
— see Shepard, T. H. 141, *179*
Lane, N., and R. C. de Bodo 110, *167*
Lane, P. W. see Dickie, M. M. 134, *154*, 618, *626*
Lang, H. M. see Ruliffson, W. S. 371, *403*
Langbein, G. see Schubert, A. 260, *404*
Langdon, R. G., and K. Bloch 436, 437, 438, *506*
— see Nicolaides, N. 436, *508*
Langecker, H., and R. Lurie 81, *167*
Langendorf, H. see Boguth, W. 88, *148*
Langlois, P. see Abelous, J. E. 551, *555*
Langstroth, G. O. see Talbot, N. B. 297, *406*
Lanman, J. T. 19, 27, 44, 125, *167*, 587, 591, 592, 598, *630*
— and L. M. Silverman 42, 72 *167*
— see Solomon, S. 72, 126, *180*, 574, 590, *635*
Lansan, G. see Baggett, B. 412, 451, *500*
Lapham, L. W. see Farrell, G. L. 117, *156*
Laqueur, E. see Dingemanse, E. 426, *502*
— see de Jongh, S. E. 426, *506*
Laqueur, G. L., S. M. McCann, L. H. Schreiner, E. Rosenberg, D. M. Rioch and E. Anderson 92, 104, 105, *167*
Laragh, J. H., and H. C. Stoerk 114, *167*
Lardon, A., and T. Reichstein 191, 200, 220, 228, *396*
— O. Schindler, and T. Reichstein 196, *396*
— see Euw, J. v. 242, *389*
— see Reber, F. 241, *402*
Laron, Z. see Klein, R. 423, *506*, 521, *562*
Laroche, C., A. Buge, J. Lerat and Y. le Quintrec 655, *658*
Larroche, J. see Tuchmann-Duplessis, H. 123, 126, *182*
Larson, H. see Woodward, R. B. 287, *409*
Larson, P. S., G. R. Henningar, J. K. Finnegan, R. B. Smith jr. and H. B. Haag 135, 136, *167*
— — J. K. Finnegan, R. B. J. Smith and H. B. Haag 458, *506*
— see Bleiberg, M. J. 135, 136, *148*, 458, *500*
Lasater, M. B. see Samuels, L. T. 61, *177*, 496, *510*
Laschet, U. see Hohlweg, W. 86, 96, *162*
Laskowski, S. C. see Clinton, R. O. 288, *386*
Laszlo, D. see Gallagher, T. F. 456, *504*, 586, *628*
Latimer, H. B. 13, 18, 20, *167*
— and J. A. Rosenbaum 13, *167*
— and P. B. Sawin 18, *167*
Laubach, G. D. see Agnello, E. J. 267, 277, 279, *381*
Laucius, J. F. see Marker, R. E. 209, *397*
Lauras, A. see Delay, J. 641, *657*
La Velle, F. W. 49, *167*, 597 609, *630*
Lawrence, B. see Migeon, C. J. 572, 581, 582, 583, 584, 591, 592, *632*
Lawson, E. J. see Marker, R. E. 209, *397*, 431, 433, *507*
Layne, J. A. see Heidorn, G. H. 546, *561*
Lazo-Wasem, E. A. 140, *167*
— G. M. Neher, R. L. Shoger and M. X. Zarrow 610, *630*
— see Zarrow, M X. 611, *637*
Leaf, A. see Bartter, F. C. 585, 616, *624*
Leanza, W., J. Conbere, E. F. Rogers and K. Pfister 225, *396*
Leathem, J. H. 69, 107, 128, *168*, 596, *630*
— and L. A. Stauber 94, 96, *168*
— see Bishop, D. M. 128, *148*
— see Gaunt, R. 96, 80, 100, 107, 110, *158*
Le Baron, F. N. see Folch, J. 49, *156*
Leblond, C. P. and W. U. Gardner 60, *168*, 593, *630*
— see Giroud, A. 58, *158*
Le Brie, S. J. see Swingle, W. W. 99, *181*, 521, 524, 527, 535, *568*
Lederer, E., and M. Lederer 306, 307, 313, *396*

Lederer, M. 304, *396*
— see Lederer, E. 306, 307, 313, *396*
Ledogar, J. A., and H. W. Jones jr. 484, *506*
Lee, C. M. jr. see Sobel, E. H. 129, 135, *180*, 593, *635*
Lee, N. D. see Hogness, J. R. 523, *561*
Leeds, N. S., D. K. Fukushima and T. F. Gallagher 212, *396*
— see Fukushima, D. K. 225, *390*, 433, 459, 460, 483, *504*
Lees, F. see Wood, K. F. 129, *185*
Lees, M. see Folch, J. 49, *156*
Legnani, T. see Ascoli, G. 78, 99, *146*
Lehninger, A. L. see Reich, E. 420, 421, 435, 441, *509*
Leigh, H. M. see Meister, P. D. 487, 492, 493, *508*
— see Peterson, D. H. 259, 260, *401*, 491, 492, 493, *509*
Leigh Osborne, H. M. see Eppstein, S. H. 491, 492, *503*
Leland, J. see Loeb, R. F. 528, 529, *563*
Lelong, M., M. F. Jayle, P. Canlorbe, R. Scholler and P. Borniche 582, *631*
— — R. Scholler, P. Canlorbe, P. Borniche and Szper 582, *631*
Leloup-Hatey, J. 51, 54, 75, 141, *168*
Leman, C. B. see Harrison, J. H. 573, 581, 607, *629*
Lemere, F. 653, *658*
Lemmon, R. M. see Biggs, M. W. 436, *500*
Lemon, H. M. see Wotiz, H. H. 616, *637*
Lenhard, R. H. see Antonucci, R. 225, 361, *382*
— see Bernstein, S. 267, 272, 275, 284, 362, *383*
Lenk, C. T. see Djerassi, C. 220, *387*
Lenniham, R. see Steenburg, R. W. 475, *511*
Lenz, A. L. see Solomon, S. 435, 439, 441, *511*
Leone, L. see Taliaferro, I. 136, *181*, 455, *511*
Leone, U. see Salvadori, B. 423, *510*
Lerat, J. see Laroche, C. 655, *658*
LeRoy, G. V. see Werbin, H. 435, 441, 449, *512*
Leroy, P., and J. Benoit 13, 143, *168*, 554, *563*
— see Domm, L. V. 123, *154*
Lescure, O. L. see Nichols, J. 127, *173*, 592, 593, *632*
Lesnick, G. see Soffer, L. J. 531, *567*
Leszynsky, H. E. see Lowenthal, M. 129, *169*
Lettré, H. R., Tschesche and H. Fernholz 186, *396*
Leulier, A., and L. Revol 51, *168*
Lever, J. D. 23, 24, 25, 38, 39, 60, 91, 110, 111, *168*
— see Cater, D. B. 9, 32, 36, 87, 100, *150*
Levin, B., and G. Davies 293, *396*
Levin, L. 426, *506*
— see Doisy, E. A. 426, *502*
Levin, M. E., and W. H. Daughaday 105, *168*
Levin, R. H., B. J. Magerlein, A. V. Mc Intosh, A. R. Hanze, G. S. Gonken, J. L. Thompson, A. M. Searcy, M. A. Scheri and E. S. Gutsell 227, *396*
— see Fonken, G. S. 243, *390*
— see Magerlein, B. M. 227, *397*
— see Spero, G. B. 247, *406*
Levinsky, N. G., and W. H. Sawyer 143, *168*
— see Sawyer, W. H. *566*
Levitan, P. see Schneider, J. J. 460, *511*
— see Solomon, S. 421, 435, 439, *511*
Levitt, M. F., and M. E. Bader 537, 538, *563*
— see Gaudino, M. 537, *559*
— see Sweet, A. Y. 537, *568*
Levitz, M., J. R. Spitzer and G. H. Twombly 484, *506*
Levy, Harold, R. W. Jeanloz, R. P. Jacobsen, O. Hechter, V. Schenker and G. Pincus 441, 442, 489, 490, *506*
— — C. W. Marshall, R. P. Jacobsen, O. Hechter, V. Schenker and G. Pincus 442, 489, *506*
— and S. Kushinsky 238, 300, *396*, 442, 490, *506*
— see Hechter, O. 311, *392*, 441, 442, 486, 489, 490, *505*, 517, *561*, 609, *629*
— see Jeanloz, R. W. 449, 486, 489, *506*, 573, *630*
Levy, Helena, H. W. Deane and B. L. Rubin 61, 63, 96, 97, 109, *168*
Levy, L. K., M. Hagopian and C. A. Fish 484, *506*
— see Hagopian, M. 487, *504*
Lewbart, M. L., and J. J. Schneider 334, *396*
— see Schneider, J. J. 452, 460, 478, *510*
Lewes, D. see Zuckerman, S. 622, *637*
Lewis, A., and J. J. Fleminger 641, *658*
Lewis, B. 346, *396*
Lewis, J. T. see Houssay, B. A. 516, *562*
Lewis, L. A. see Hartman, F. A. 526, 548, *561*
— see Savard, K. 443, 489, *510*
Lewis, R. A., E. Rosemberg and L. Wilkins 80, *168*
— see Engel, L. L. 427, 431, *503*, 610, *627*
Lewycka, C. see De Meio, R. H. 452, 478, *502*
Li, C. H., P. Fønss-Bech, I. I. Geschwind, T. Hayashida, G. F. Hungerford, A. J. Lostroh, W. R. Lyons, H. D. Moon, W. O. Reinhardt, and M. Sideman 82, 83, 100, 101, 103, 117, *168*, 608, 613, 614, *631*
— see Baker, B. L. 608, *624*
— see Carstensen, H. 555, *557*
— see Ingle, D. J. 103, *163*
— see Lostroh, A. J. 68, *169*, 608, *631*
— see Lyons, W. R. 613, *631*
— see Mason, H. L. 607, *631*
— see Simpson, M. E. 87, 94, 100, 101, *179*
Li, M. C. see Mac Lean, J. P. 110, 113, *169*
Liang, T. Y. see Sayers, G. 51, 58, 84, 85, *177*, 454, *510*
Liebegott, G. 3, 32, 84, 90, 94, 104, 129, 133, *168*
Lieberman, A. H. see Ganong, W. F. 110, *157*
Lieberman, S., and K. Dobriner 428, *507*
— — B. R. Hill, L. F. Fieser and C. P. Rhoads 217, 307, *396*
— — — L. L. Fieser and C. P. Rhoads 427, 428, 431, *507*
— D. K. Fukushima and K. Dobriner 307, 397, 428, *507*
— L. B. Hariton and D. K. Fukushima 217, 294, 297, *396*
— B. Mond and E. Smyles 293, 297, *397*
— B. Praetz, P. Humphries and K. Dobriner 437, 428, 431, 432, *507*

Lieberman, S., and S. Teich 572, *631*
— see Bitman, J. 228, 250, 359, *383*
— see Dobriner, K. 301, 303, 357, *388*, 427, 431, 432, 433, 459, *502*, 582, *627*
— see Fieser, L. F. 340, *389*
— see Jailer, J. W. 465, *506*, 585, *630*
— see Lakshmanan, T. K. 309, 310, *396*, 580, *630*
— see Rosselet, J. P. 219, *403*, 432, 459, 465, *510*
— see Schneider, J. J. 460, *511*
— see Solomon, S. 72, 126, *180*, 421, 435, 439, 441, *511*, 574, 590, *635*
— see Ulick, S. 236, *407*, 433, 476, *512*
— see Vande Wiele, R. 572, 574, *635*
Liebermann, A. H. see Johnson, B. B. *506*
Liebowitz, M. R. see Barger, A. C. 531, *556*
Lichtlen, P. 528, *563*
— see Gross, F. 547, *560*
Liddle, G. W. 527, *563*
— F. C. Bartter, L. E. Duncan jr. J. K. Barber and C. Delea *506*
— L. E. Duncan jr. and F. C. Bartter 111, 113, 114, 115, *168*
— D. Island, H. Estep and G. M. Tomkins 523, *563*
— — E. M. Lance and A. P. Harris 138, *168*, 523, *563*
— J. E. Richard and G. M. Tomkins 521, 533, *563*
— see Bartter, F. C. 114, *147*
— see Coppage, W. S. 523, *558*
Lilienthal, J. L. see Zierler, K. L. 531, *569*
Liling, M., and R. Gaunt 546, *563*
— see Gaunt, R. 139, *158*
Lillie, R. D. 57, *168*
Lincoln, F. H. see Anderson, H. V. 227, *381*
— see Hogg, J. A. 208, 256, 257, 267, 277, 284, 361, *393*
— see Spero, G. B. 267, 271, 283, *406*
Lind, B. see Geppert, L. J. 643, *658*
Lind, J. see Solomon, S. 72, 126, *180*, 574, 590, *635*
Lindberg, M. C. see Caspi, E. 441, 461, 462, *501*
— see Hayano, M. 443, 461, 485, 489, *505*
— see Meyer, A. S. 340, *398*, 415, 449, 486, *508*, 573, *631*
Linden, C. E. see Bernstein, S. 236, *383*
Lindgren, F. T. see Gofman, J. W. 292, *391*
Linegar, C. R. see Pousiaka, J. W. 535, 540, *565*
Linford, J. H. 312, 337, 362, *397*
Ling, N. S. see Bock, R. M. 310, *384*
Lingane, J. J. see Kolthoff, I. M. 380, *396*
Lippman, A. see Djerassi, C. 287, 288, 374, 376, *388*
— see Foltz, E. W. 205, 376, *389*
Lipscomb, H. S., and D. H. Nelson 76, 85, 103, *168*
— see Guillemin, R. 77, 85, 100, 103, *160*
Lipschitz, W. L., and E. Bueding 452, *507*
Lipsett, M. B., C. D. West, J. P. MacLean and O. H. Pearson 100, *168*
— see Bergenstal D. M. 523, *556*
— see MacLean, J. P. 110, 113, *169*
Lisboa, B. see Weinmann, S. 434, *512*
Lison, L. 50, 51, 56, *169*
Lissák, K. see Endröczi, E. 125, *155*
Lister, L. M. see Savard, K. 413, 488, *510*
Littel, R., and S. Bernstein 228, 277, *397*
— see Antonucci, R. 225, 361, *382*
— see Allen, W. S. 244, *381*
— see Bernstein, S. 267, 272, 284, *383*
Litteria, M. see Recknagel, R. O. 340, *402*
Little, B., and E. Rossi 422, *507*
Little, C. C. see Fekete, E. 618, 619, *627*
— see Woolley, G. W. 134, *185*, 606, 618, *637*
Little, J. M., W. M. Kelsey and E. H. Yount jr. 136, *169*
Liu, H. C., and R. B. Maneely 10, *169*
Liu, T. Y., K. A. Brownell and F. A. Hartman 573, *631*
— see Hartman, F. A. 72, 73, *161*
Li Voti, P. 140, *169*
Ljunggren, H. see Luft, R. 529, 531, 538, *563*
Llaurado, J. G. 110, 116, *169*
— R. Neher and A. Wettstein *507*, 520, *563*
— J. B. Trunnell and J. L. Claus 107, *169*
— see Wettstein, A. 433, *513*
— and M. F. A. Woodruff 116, *169*
Lloyd, C. W. 542, 544, *563*
— and J. Lobotsky 304, *397*
— — J. Jones, J. Fredericks and T. C. Wyatt 611, *631*
— see Gaunt, R. 518, 543, 544, 545, *559*
— see Jones, J. M. 51, 59, *164*
Lobban, M. C. 43, *169*, 598, *631*
Lobotsky, J. see Clark, L. C. jr. 478, *502*
— see Lloyd, C. W. 304, *397*, 611, *631*
Locher, F. see Hartmann, M. 431, *504*
Locke, W., N. B. Talbot, H. S. Jones and J. Worcester 540, 541, 563
Loeb, E. N. see Knowlton, A. I. 115, *166*, 531, 547, *562*
Loeb, L. see Blumenthal, H. T. 69, *148*
Loeb, R. F. 528, *563*
— D. W. Atchley, E. M. Benedict and J. Leland 528, 529, *563*
— see Ferrebee, J. W. 531, 532, 534, *558*
— see Kuhlman, D. 534, *562*
see Ragan, C. 533, 534, *565*
— see Stahl, J. 535, *567*
Loewus, F. A. see Talalay, P. 312, *406*
Logan, G. B. see Gastineau, C. F. 616, *628*
Logan, J. E., F. C. Heagy and R. J. Rossiter 91, *169*
— see Riedel, B. E. *176*
Logeman, W. 291, *397*
— and P. Giraldi 265, *397*
Löhr, G. see Schultz, A. 53, *178*
Loke, K. H., and G. F. Marrian 426, 451, *507*
Löken, B., S. Kaufman, G. Rosenkranz and F. Sondheimer 273, *397*
— see Dauben, H. J. 244, *387*
Lombardo, M. E., and P. B. Hudson 75, 77, *169*, 459, *507*
— P. H. Mann, T. A. Viscelli and P. Hudson 300, *397*
— C. McMorris and P. B. Hudson 75, *169*, 572, 573, 609, *631*
— E. Roitman and P. B. Hudson 443, 490, *507*

Lombardo, M. E., T. A. Viscelli, A. Mittelman and P. B. Hudson 300, *397*
— see Hudson, P. B. 418, *505*
Long, C. N. H. 454, *507*
— see Abelson, D. 460, *499*
— see Brodish, A. 83, *149*
— see Gershberg, H. 59, *158*
— see Sayers, G. 51, 58, 83, 84, 85, *177*, 454, *510*
Long, M. E. see Schweizer, M. 78, 79, 100, 109, 110, *178*
Longo, P. see Pincus, G. 612, *633*
Longson, D. see Jailer, J. W. 106, *164*
Longwell, B. B., A. E. Reif and E. Hansbury 139, *169* 422, 423, *507*
— — — and H. S. Baker 423, *507*
— see Reif, A. E. 74, 75, *175*
— see Sterling, R. E. 51, 69, 70, *180*
Loomis, E. see Gaunt, R. 542, *559*
Loraine, J. A. see Stone, J. A. *511*
Lostroh, A. J. 83, 100, 111, *169*
— and C. H. Li 68, *169*, 608, *631*
— and P. Woodward 68, 83, 100, 101, 111, *169*
— see Li, C. H. 82, 83, 100, 101, 103, 117, *168*, 608, 613, 614, *631*
Lotsof, E. see Chance, J. 649, *657*
Lott, H. M. see Hoffman, M. N. 428, *505*
Louis, L. H. see Conn, J. W. 518, 531, 535, 547, *557*
Loukatos, G. see Constantinides 649, *657*
Loustalot, P. 133, *169*
Lovell, B. J. see Bladon, P. 275, *383*
Lowenstein, B. E., and R. L. Zwemer 530, *563*
Lowenthal, M., H. E. Leszynsky, M. Marcus and H. Zondek 129, *169*
Lubs, H. A. see Hertz, R. 137, *162*
— see Perterson, R. E. 457, *509*, 522, *565*, 586, *633*
Lucas, G. H. W. 515, *563*
Lucis, O. J. see Venning, E. H. 111, *182*
Lüders, C. J. 57, 129, *169*
Luetscher, J. A. jr. *507*, 519, 520, 521, *563*
Luetscher, J. A. jr., and B. J. Axelrad *507*
— and R. H. Curtis 110, 113, 114, *169*
— A. Dowdy, J. Harvey, R. Neher and A. Wettstein 520, *563*
— R. Neher and A. Wettstein 433, *507*
— C. F. Piel and R. H. Curtis 546, *563*
— see Ganong, W. F. 110, *157*
Luft, J., and O. Hechter 24, 36, 38, *169*
— see Bennett, H. S. 23, *147*
Luft, R., and B. Sjögren 531, 537, *563*
— — D. Ikkos, H. Ljunggren and H. Tarukoski 529, 531, 538, *563*
— see Diczfalusy, E. 412, *502*, 611, *627*
Lukes, R. M. see Sarett, L. H. 262, *404*
Lumry, R. see Eik-Nes, K. *503*
Lund, G. H. see Lyster, S. C. 521, *563*
Lundlin, P. M., and S. Holmdahl 123, *169*
Lunnon, J. B. see Gray, C. L. 467, *504*
— see Holness, N. J. 432, 433, *505*
Lur'i, F. A. see Madaeva, O. S. 212, *397*
Lurie, R. see Langecker, H. 81, *167*
Lutwak-Mann, C. see Price, D. 579, *633*
Lutz-Ostertag, Y. see Wolff, E. 605, *636*
Lyman, C. P. see Deane, H. W. 28, 35, 54, 94, 106, 107, *153*
Lynn, W. S. jr. *507*
— and R. Brown *507*
— E. Staple and S. Gurin 439, 499, *507*
— see Staple, E. 420, 435, 439, 441, *511*
Lyon, I., R. P. Geyer and L. D. Marshall 436, *507*
Lyons, J. see Wallerstein, R. S. 655, *659*
Lyons, W. R., C. H. Li, R. E. Johnson and R. D. Cole 613, *631*
— see Li, C. H. 82, 83, 100, 101, 103, 117, *168*, 608, 613, 614, *631*
Lyster, S. C., L. E. Barnes, G. H. Lund, M. M. Meinzinger and W. W. Byrnes 521, *563*
Lyttle, D. A. see Eppstein, S. H. 491, 492, *503*

Macchi, I. A., and O. Hechter 21, 76, *169*
— see Kass, E. H. 75, *165*
MacCorquodale, D. W. see Smith, G. V. S. 426, *511*
MacDonald, P. M. see Gornall, A. G. 341, *391*
MacFarland, W. E. 21, *169*, 526, *563*
— and H. A. Davenport 4, 25, *169*
Macfarlane, D. A. see Gold, N. I. 475, *504*
Mach, R. S., J. Fabre, A. F. Muller and R. Neher 531, *563*
— and A. F. Muller 457, *507*, 522, *563*
— see Ducommun, P. 96, 101, *154*
— see de Watteville, H. 611, *636*
Macheboeuf, M. A. 292, *397*
Mack, G. E. see Heard, R. D. H. 434, *505*
MacLachlan, E. A. see Talbot, N. B. 363, *406*, 573, *635*
MacLachlan, P. L., H. C. Hodge and R. Whitehead 49, 51, 54, 92, *169*
MacLean, J. P., M. B. Lipsett, M. C. Li, C. D. West and O. H. Pearson 110, 113, *169*
— see Lipsett, M. B. 100, *168*
MacLean, K. S. see McGehee, E. H. 641, *658*
MacLeod, L. D. see Carpenter, R. K. 83, 92, *149*
Madaeva, O. S., and F. A. Lur'i 212, *397*
Maddock, W. O. see Epstein, E. 337, *389*
Mader, I. J., and L. T. Iseri 529, *563*
Mader, W. J., and R. R. Buck 340, 364, *397*
— see Szalkowski, C. R. 337, *406*
Madison, L. see Sensenbach, W. 643, *659*
Maes, J. 143, *169*, 551, *564*
Maeyama, M. see Ishizuka, N. 461, *506*
Magerlein, B. J., and J. A. Hogg 281, 283, *397*
— and Levin, R. H., 227, *397*
— see Hogg, J. A. 208, 256, 257, 272, *393*
— see Levin, R. H. 227, *396*
— see Peterson, D. H. 492, 493, *509*

Magerlein, B. J. see Spero, G. B. 267, 283, *406*
Mahesh, V. B. see Bush, I. E. 586, *625*
Mahoney, J. P. see Migeon, C. J. 572, *632*
Majnarich, J. J., and R. N. Dillon 433, *507*
Malandra, B., and S. Corbetta 646, *658*
Malaty, H. A., and G. H. Bourne 66, *169*
Mallassis, D., E. E. Baulieu, S. Weinmann, O. Crepy and M. F. Jayle 434, *507*
— see Jayle, M. F. 580, *630*
Malley, K., and T. G. Bradshaw 611, *631*
Malmström, B. G., and D. Glick 62, *170*
— see Glick, D. 67, *159*
Mamoli, L., and A. Vercellone 259, *397*
Manaro, J. M., and A. Zygmuntowicz 346, *397*
— see Cerviño, J. M. 101, 135, *150*
Manc, M. B. see Oleesky, A. 542, *565*
Mancera, O., H. J. Ringold, C. Djerassi, G. Rosenkranz and F. Sondheimer *397*
— A. Zaffaroni, B. A. Rubin, F. Sondheimer, G. Rosenkranz and C. Djerassi 492, *507*
— see Ringold, H. J. 272, *402*
— see Rosenkranz, G. 233, 234, 255, 257, *403*
— see Sondheimer, F. 211, 213, 227, 234, 236, 241, 246, 361, *405*
Mandel, L., A. Baber and P. Mandel 80, 91, *170*
— see Mandel, P. 90, *170*
Mandel, P., M. Jacob and L. Mandel 90, *170*
— see Mandel, L. 80, 91, *170*
Mandl, A. M. 621, 622, *631*
— and S. Zuckerman 622 *631*,
Maneely, R. B. see Liu, H. C. 10, *169*
Manfredi, G. C. see Bonati, B. 81, 96, 108, 110, *148*
Mangun, G. H. see Muntwyler, E. 538, *564*
Manheimer, L. M. see Seligman, A. M. 65, *178*
Mann, K. M., F. R. Hanson and P. W. O'Connell 493, *507*
— — — H. V. Anderson, M. P. Brunner and J. M. Karnemaat 261, *397*
— — — H. Anderson, M. P. Brunner and J. N. Karnemaat 493, *507*
— see O'Connell, P. W. 259, *400*, 493, *508*
Mann, P. H. see Lombardo, M. E. 300, *397*
Mann, T. see Price, D. 579, *633*
Manson, A. see Djerassi, C. 289, *388*
Manson, A. J. see Clinton, R. O. 288, *386*
Mansor, L. F. see Heming, A. E. 523, *561*
Maqsood, M. 140, *170*
Marano, B. J. see Eidinoff, M. L. 436, 438, *503*
Marañon, G. 655, *658*
— see Richet, C. 639, *659*
Maraud, R., and R. Stoll *170*
Marc, J. R. St. see Renold, A. E. 521, *566*
Marchand, B. see Criegee, R. 250, *387*
Marchi, P. see Stigliani, R. 87, 131, *181*
Marcus, M. see Lowenthal, M. 129, *169*
Marcus, P. I., and P. Talalay 496, *507*
— see Talalay, P. 496, *511*
Mardones, E. see Iglesias, R. 135, *163*
Marenzi, A. D., and O. Fustinoni 551, *564*
Margraf, H. W. see Weichselbaum, T. E. 366, *408*, 425, *512*
Mariani, A., and C. Vicari 306, *397*
Marine, D., and E. J. Baumann 516, 528, *564*
Marker, R. E. 254, *397*, 426, 427, 431, *507*
— and O. Kamm 431, *507*
— — and R. V. McGrew 431, *507*
— — E. L. Wittle, T. S. Oakwood, E. J. Lawson and J. F. Laucius 209, *397*
— and E. J. Lawson 433, *507*
— — E. L. Wittle and H. M. Crooks 431, *507*
— and E. Rohrmann 431, *507*
Marks, B. H., M. Alpert and F. A. Kruger 137, *170*, 457, *507*
Marks, V. 107, 133, *170*
Marlow, H. W. 337, *397*
Maroc, J. see Persky, H. 642, *659*
Marois, M. see Courrier, R. 106, 107, *152*
Marrian, G. F. 296, *397*, 426, 428, 431, *507*, 610, *631*
— and S. M. Atherden 296, *397*
— and W. S. Bauld 426, *507*
— K. H. Lake, E. J. D. Watson and M. Panattoni 426, *507*
— J. Y. F. Paterson and S. M. Atherden 296, *397*
— see Brown, J. B. 484, *501*
— see Butler, G. C. 431, *501*
— see Cox, R. I. 318, *386*, 431, *502*
— see Fotherby, K. 428, 429, 446, 478, *504*
— see Loke, K. H. 426, 451, *507*
— see Paterson, J. Y. F. 365, *400*
— see Schacter, B. 434, *510*
— see Sutherland, E. S. 434, *511*
— see Watson, E. J. D. 426, *512*
Marshall, C. W. see Hechter, O. 442, 489, *505*
— see Levy, H. 442, 489, *506*
Marshall, E. K. jr., and D. M. Davis 515, *564*
Marshall, L. D. see Lyon, I. 436, *507*
Marthaler, T. M. see Mühlemann, H. R. 32, *171*
Marti, M. 308, *397*
— see Anliker, R. 204, *381*, 414, 415, 420, *500*
— see Blickenstorfer, E. 639, *657*
Martin, G. J. see Cohen, H. 82, 83, *151*
Martin, H. F., and H. L. White 531, *564*
Martin, J. see Endröczi, E. 75, 76, 103, *155*
Martin, M. E. see Fang, A. D. 649, *657*
Martin, R. P. 328, *397*
Martin, S. J. 596, *631*
Martindale, W. E. see Olsen, N. S. 139, *173*
Martinet, M. see Giroud, A. 59, *158*
Martinez, H. see Sondheimer, F. 233, 361, *405*
Martini, C. M. see Clarke, R. L. 233, 234, *386*
Martinovitch, P. N. 46, 118, 120, 121, *170*

Marzi, A., and U. Teodori 650, *658*
Mason, D. G. see Shepard, T. H. 141, *179*
Mason, J. W., and J. V. Brady 522, *564*
— — and M. Sidman 105, *170*
— see Harwood, C. T. 364, *391*
Mason, H. L. 295, 304, 337, *398*, 427, 428, 459, 478, 507, *564*
— and W. W. Engstrom 363, *398*, 574, *631*
— W. M. Hoehn and E. C. Kendall 421, *507*
— — B. F. McKenzie and E. C. Kendall 420, *507*
— and E. C. Kendall 223, *398*
— and E. J. Kepler 412, 427, 431, 434, 478, *507*, 572, 574, *631*
— C. S. Meyers and E. C. Kendall 421, *507*
— C. S. Myers and E. C. Kendall 519, *564*
— M. H. Power, E. H. Rynearson, L. C. Ciaramelli, C. H. Li and H. M. Evans 607, *631*
— and J. J. Schneider 427, *507*
— and H. S. Strickler 431, *507*
— see Colton, F. G. 274, *386*
— see Engstrom, W. W. 338, *389*
— see Gastineau, C. F. 616, *628*
— see Kendall, E. C. 519, *562*
— see Schneider, J. J. 478, 486, *511*, 572, *634*
— see Mattox, V. R. 198, 313, 318, *398*, 420, 421, *507*, 520, *564*
— see McGuckin, W. F. 274, 280, *398*
Mason, H. S. see Sweat, M. L. 443, 485, 489, *511*
Masoni, A. 73, *170*
Masson, G. M. C., A. C. Corcoran and I. H. Page 532, *564*
— S. B. Koritz and F. G. Peron 120, *170*
— see Deane, H. W. 88, 112, 115, 116, 131, *153*, 547, *558*
Mast, H. see Heni, F. 88, *161*
Masuda, M. 585, *631*
— and H. C. Thuline 363, *398*
Masui, K., and Y. Tamura 593, 597, *631*
Materna, A. 18, 20, *170*
Mathieson, D. R., and H. W. Hays 575, *631*
Mathieson, D. R. see Hays, H. W. 546, *561*
Mathieu, J. P., and A. Petit 373, *398*
— see Horeau, A. 196, *393*
Matsuba, M. see Caspi, E. 441, 461, 462, *501*
Mattox, V. R. 204, 223, *398*, 420, *507*, 517, *564*
— and E. C. Kendall 221, *398*
— and H. L. Mason 318, *398*, 420, 421, *507*
— — and A. Albert 520, *564*
— — — and C. F. Code 198, 313, *398*
— E. L. Woroch, G. A. Fleischer and E. C. Kendall 278, *398*
— see Turner, R. B. 228, *407*
Mattox, W. E. see Gassner, F. X. 574, *628*
Mauer, S. I. see Ritterson, A. L. 93, 104, *176*
Maultsby, M. M. see Driscol, T. E. 114, *154*
Mautz, F. R. see Harkness, D. M. 537, *560*
— see Muntwyler, E. 535, 538, *564*
Maxwell, R. see Swingle, W. W. 99, *181*, 535, *568*
Mayer, C. 484, *507*
Mayne, Y. C. see Eik-Nes, K. 422, *503*
— see Jones, R. S. 85, *164*
Mazur, Y. see Sondheimer, F. 287, *406*
McAleer, W. J., and M. A. Kozlowski 341, 342, *398*
McAllister, F. F., and G. W. Thorn 537, *564*
McAlpine see Venning, E. H. 580, *636*
McCall, P. J. see Munson, P. L. 337, 364, *399*
McCalley, M. see Anderson, C. H. 115, *146*
— see Farrell, G. L. 103, 111, *156*
McCann, S. M., A. Fruit and B. D. Fulford 545, *564*
— see Laqueur, G. L. 92, 104, 105, *167*
McCarthy, J. L. see Zarrow, M. X. 140, *185*
McClure, W. W. see Pullman, T. N. 529, *565*
McCracken, B. H. see Thorn, G. W. 518, 540, *568*
McCullagh, E. P., and H. A. Tretbar 457, *507*
— see Schiller, S. 427, 431, *510*, 572, *634*
McDonald, H. J. 304, *398*
McDonald, I. R., and M. Reich 73, 86, 103, 113, *170*
McElroy, J. T. see Graff, M. M. 337, *391*
McGarrahan, K. see Bucher, N. L. R. 436, *501*
McGarry, E. E. see Beck, J. C. 116, *147*
McGavack, T. H. see Drekter, I. J. 302, *388*
McGehee, E. H., and K. S. MacLean 641, *658*
McGinty, D. A., G. N. Smith, M. L. Wilson and C. S. Worrel 443, 489, *507*
McGrew, R. V. see Marker, R. E. 431, *507*
McGuckin, W. F., and E. C. Kendall 229, 257, *398*
— and H. L. Mason 274, 280, *398*
McIntosh, A. see Spero, G. B. 247, *406*
McIntosh, A. V. see Levin, R. H. 227, *396*
McIntosh, H. see Southcott, C. M. 423, 425, *511*
McKay, A. F. see Heard, R. D. H. 428, *505*
McKeever, S. 18, 21, *170*
McKenzie, B. F. see Hoehn, W. M. 420, *507*
— see Kendall, E. C. 519, *562*
— see Turner, R. B. 228, *407*
McKeown, T. see Kutz, R. L. 621, *630*
McKerns, K. W. 106, 107, *170*
— and E. Nordstrand 77, *170*
McKibbin, J. M. see Deane, H. W. 105, 131, 138, 139, *153*
McKinley, W. P. 304, *398*
McLamore, W. see Woodward, R. B. 261, *409*
McLean, P. see Glock, G. E. 62, *159*
McMahon, J. M., R. B. Davis and G. Kalnitsky 342, *398*
McMorris, C. see Lombardo, M. E. 75, *169*, 572, 573, 609, *631*
McNutt, A. W. see Hartman, F. A. 14, *161*
McNutt, S. H. see Weber, A. F. 23, 28, *183*
McPhail, M. K. 599, 602, *631*
— and H. C. Read 128, *170*
McQuarrie, I., J. A. Anderson and M. R. Ziegler 534, *564*
McQueeney, A. J., L. L. Ashburn, F. S. Daft and R. Faulkner 138, *170*

McShan, W. H., W. F. Erway and R. K. Meyer 49, 68, *170*
McSwiney, R. R. see Prunty, F. T. G. 71, 105, *175*
Mead, J. F. see Fillerup, D. L. 311, *389*
Meakin, J. W. see Jenkins, J. 137, 138, *164*, 523, *562*
Meakins, G. D. see Castellis, J. 249, *385*
— see Henbest, H. B. 355, *392*
Meath, J. A. see Folch, J. 49, *156*
Medawar, P. B., and P. S. Russell 118, 120, *170*
Medvecky, J. 655, *658*
Meeks, R. C. see Meister, P. D. 491, 492, 494, *508*
— see Pederson, R. L. 281, 283, *401*, 487, *509*
Meier, R. see Desaulles, P. A. 532, 533, *558*
— see Gross, F. 523, *560*
— see Schuler, W. 524, *567*
Meinzinger, M. M. see Lyster, S. C. 521, *563*
— see Stafford, R. O. 521, *567*
Meister, P. D., S. H. Eppstein, D. H. Peterson, H. C. Murray, H. M. Leigh, A. Weintraub and L. M. Reineke 487, 493, *508*
— D. H. Peterson, H. C. Murray, S. H. Eppstein, L. M. Reineke, A. Weintraub and H. M. Leigh 492, *508*
— — — G. B. Spero, S. H. Eppstein, A. Weintraub, L. M. Reineke and H. M. Leigh 492, *508*
— L. M. Reineke, R. C. Meeks, H. C. Murray, S. H. Eppstein, H. M. L. Osborn, A. Weintraub and D. H. Peterson 491, 492, 494, *508*
— see Eppstein, S. H. 258, 259, *389*, 487, 491, 492, 493, 494, *503*
— see Pederson, R. L. 281, 283, *401*
— see Peterson, D. H. 259, 260, *401*, 491, 492, 493, *509*
Meister, L. M. see Pederson, R. L. 487, *509*
Melampy, R. M., D. W. Chang and L. C. Northrup 105, 139, *170*
Melin, M. see Oncley, J. L. 292, *400*
Mellinger, R. C. see Therien, B. 135, *181*
Melloni, G. F. see Genesi, M. 96, *158*
Mellors, R. C. see Harkness, D. M. 537, *560*
— see Muntwyler, E. 535, 538, *564*
Melzer, F. see Koenig, V. L. 337, *396*
Menten, M. L. see Weber, E. J. 135, *184*
Menzel, K. see Woodhard, R. B. 287, *409*
Merriam, R. W. 3, *170*
Merrill, A. P. see Pincus, G. 612, *633*
Merrill, E. J. see Sakal, E. H. 312, *403*
Merritt, L. L. jr. see Willard, H. H. 358, 368, *409*
Meslin, F. see Crepy, O. 310, *387*
Meyer, A. 653, *658*
Meyer, A. E. 653, *658*
Meyer, A. S. 233, 238, 239, 300, 362, *398*, 412, 413, 415, 449, 451, 486, *508*
— M. Hayano, M. C. Lindberg, M. Gut and O. G. Rodgers 415, 449, 486, *508*, 573, *631*
— R. W. Jeanloz and G. Pincus 486, 489, *508*
— and M. C. Lindberg 340, *398*
— O. G. Rodgers and G. Pincus 442, 490, *508*
— H. Rosenkrantz and G. Pincus 360, *398*
Meyer, C. D. see Fukushima, D. K. 431, 432, *504*
Meyer, E. see Julian, P. L. 216, 225, 245, 255, 274, *395*
Meyer, L. H., A. Saika and H. S. Gutowsky 377, *398*
Meyer, R. 126, *170*
Meyer, R. K. see Baum, G. J. 553, *556*
— see Brown, K. I. 553, 554, 555, *556*
— see Hershberger, L. G. 575, 576, *629*
— see McShan, W. H. 49, 68, *170*
Meyerheim, G. see Amelung, D. 463, 497, *500*
Meyers, C. S. see Mason, H. L. 421, *507*
Meyers, M. W., and H. A. Charipper 18, 21, 27, 29, 37, 57, 133, *170*
Meystre 280, *398*
Meystre, C., H. Frey, W. Voser and A. Wettstein 277, *398*
— — A. Wettstein and K. Miescher 254, *398*
— E. Tschopp and A. Wettstein *398*
— E. Vischer and A. Wettstein 491, 493, 494, *508*
— and A. Wettstein 280, *398*
— see Euw, J. v. 198, 204, *389*
— see Kahnt, F. W. 491, 492, *506*
— see Neher, C. 198, 204, *399*
— see Vischer, E. 277, *408*
Mialhe-Voloss, C., and E. E. Baulieu 111, *170*
Michael, P. J. see Tomkins, G. M. 488, *512*
Michael, R. P. see Harris, G. W. 574, *628*
Michel, R. see Roche, J. *176*
Micheli, R. A. see Dauben, W. G. 245, *387*
Miescher, K., and H. Kägi 212, *398*
— and J. Schmidlin 256, *399*
— see Anner, G. 261, *382*
— see Meystre, C. 254, *398*
— see Wieland, P. 209, *409*
Migeon, C. J. 422, *508*, 572, 616, *631*
— J. Bertrand, P. E. Wall, R. S. Stempfel and H. Prystowsky 471, *508*
— and L. I. Gardner 615, *632*
— A. R. Keller and E. G. Holmstrom 591, 592, *632*
— — B. Lawrence and T. H. Shepard 572, 581, 582, 583, 584, 591, 592, *632*
— and J. E. Plager 422, *508*, 572, *632*
— and R. S. Stempfel 584, *632*
— F. H. Tyler, J. P. Mahoney, A. A. Florentin, H. Castle, E. L. Bliss and L. T. Samuels 572, *632*
— see Bliss, E. L. 642, *657*
— see Gardner, L. I. 572, *628*
— see Nichols, J. 127, *173*, 592, 593, *632*
— see Wilkins, L. 585, 616, *636*
Mighorst, J. C. A. see Sluiter, J. W. 79, 143, *179*
Miglioretto, P. see Viscontini, M. 191, *408*
Mignani, E. see Cinti, G. 118, *151*
Mihailovic, M. L. see Cainelli, G. 236, *385*

Miliner, R. A. see Clark, W. G. 143, *151*, 552, *557*
Milković, K., and S. Milković 123, 125, *170*
Milković, S. see Milković, K. 123, 125, *170*
Miller, A. M., and R. I. Dorfman 219, *399*, 427, *508*
— — and M. Miller 478, *508*, 574, *632*
— — and E. L. Sevringhaus 428, *508*, 577, *632*
— H. Rosenkrantz and R. I. Dorfman 427, *508*
— see Schiller, S. 427, 431, *510*, 572, *634*
— see Ungar, F. 478, 486, *512*, 574, *635*
Miller, E. see Brooks, R. V. 214, 308, 310, *384*
Miller, H. C., and D. C. Darrow 532, *564*
Miller, L. L., and L. R. Axelrod 462, 472, *508*
— see Axelrod, L. R. 461, 478, 486, 488, 495, *500*
— see Ganis, F. M. *157*, 461, *504*
Miller, M. see Miller, A. M. 478, *508*, 574, *632*
— see Rubin, B. L. *510*
Miller, M. R. 8, 9, 11, 12, 79, 80, 94, 141, 143, 144, 145, *170*, 551, 552, *564*
— and M. E. Robbins 9, *170*
Miller, R. see Hirschmann, R. F. 267, 277, *393*
Miller, R. A. 30, 37, 90, 91, 92, 94, 100, 101, 104, 109, 111, 115, 118, *171*
— and O. Riddle 14, 17, 53, 57, 78, 143, 144, 145, *171*, 553, 554, *564*
— see Bates, R. W. 553, *556*
— see Riddle, O. 554, *566*
Miller, W. R., C. W. Turner, D. K. Fukushima and I. I. Salamon 312, *399*
Mills, G. T., R. Ondarza and E. E. B. Smith 452, *508*
— see Smith, E. E. B. 452, *511*
Mills, I. H., A. Caspar and F. C. Bartter 115, *171*
— see Gann, D. S. 115, *157*
— see Prunty, F. T. G. 71, 105, *175*
Minder, W. H., and I. Abelin 55, *171*
Miramontes, L. see Djerassi, C. 280, *387*
Mises, R. see Bricaire, H. 655, *657*
Missale, G. see Lamedica, G. 446, *506*
Mitamura, T. see Suzuki, T. 74, 105, *181*
Mitchell, E. R. see Dalton, A. J. 86, *152*
Mitchell, J. B. jr. 553, *564*
Mitchell, R. M. 32, *171*
Mittelman, A. see Lombardo, M. E. 300, *397*
Mixner, J. P., A. J. Bergman and C. W. Turner 18, *171*
Miyake, T., and G. Pincus 612, *632*
— see Pincus, G. 612, *633*
Mochizuki, K. see Nishida, S. 596, 602, *632*
Moeri, E. 130, *171*
Moffet, R. B., and J. H. Hunter 227, *399*
Molitor, J. C. see Carol, J. 357, *385*
Mond, B. see Lieberman, S. 293, 297, *397*
Money, J. 650, *658*
— and J. G. Hampson 650, *658*
— — and J. L. Hampson 650, *658*
— see Hampson, J. G. 650, *658*
Money, W. L. 100, 140, *171*
— see Sonenberg, M. 82, *180*
Montanari, L. see Rossi, C. A. 107, *176*
Montandon, A. 88, *171*
Montastruc, P. 543, *564*
— see Baïsset, A. 116, *146*
Montella, A. see Pellegrino, C. 118, 119, 120, *174*
Montgomery, G. L. see Foggie, W. E. 646, *657*
Moog, F. 20, 43, 45, 121, 125, *171*, 622, *632*
— C. J. Bennett and C. M. Dean jr. 20, 43, 125, *171*, 596, *632*
— and V. Nehari 622, *632*
Moolenaar, A. J. 341, *399*
Moon, H. D. 125, *171*, 607, 621, *632*
— see Davidson, C. S. 607, 608, *626*
— see Li, C. H. 82, 83, 100, 101, 103, 117, *168*, 608, 613, 614, *631*
— see Reese, J. D. 37, 92, *175*
Mooney, A. L. see Graff, M. M. 337, *391*
Mooradian, A. see Clarke, R. L. 233, 234, *386*
Moore, C. R. 601, 604, 607, *632*
— D. Price and T. F. Gallagher 576, 599, *632*
Moore, F. D. see Gold, N. I. 475, *504*
— see Steenburg, R. W. 475, *511*
Moore, J. A. see Djerassi, C. 336, *388*
— see Wagner, R. B. 225, 256, *408*
Moorhouse, D. E. 94, 141, 144, *171*
— see Holmes, W. 5, 53, *162*
de Moreau, B. M. see Gennes, L. de 546, *560*
Morel, F. 543, *564*
Morgan, A. F. see Butler, L. C. 139, *149*
— see Hurley, L. S. 139, *163*
Morgan, B. B. see Weber, A. F. 23, 28, *183*
Mori, F. see Barucci, M. 104, *147*
Morin, I. see Nowaczynski, W. 236, 333, *400*
Morley, E. H. see Byrnes, W. W. 521, *557*
— see Ingle, D. J. 519, 528, *562*
Morrell, J. A. see Swingle, W. W. 527, *568*
Morris, C. J. O. R. 315, 381, *399*
— and D. C. Williams 315, *399*
— see Barnett, I. 381, *382*
— see Butt, W. R. 303, 315, 318, 381, *385*, 613, *625*
Morris, M. D., and I. L. Chaikoff 50, *171*
Morris, P. see Butt, W. R. 315, 318, 381, *385*, 613, *625*
Morse, A. 34, *171*
— see Deane, H. W. 59, 60, 87, 96, *153*
— see Knobil, E. 78, 80, 87, 99, 101, 109, 110, 115, 116, *166*
Morton, J. V. see Taubenhaus, M. *568*
Moriarty, R. see Beyler, R. E. 225, *383*
Moses, J. B. see Gutner, L. B. 546, *560*
Mosettig, E. see Heftmann, E. 186, *392*
Mosier, H. D. 3, 141, *171*, 598, *632*
— and C. P. Richter 115, *171*
Mosimann, W. 30, *171*
Motais, R. see Fontaine, M. 549, *559*
Mou, T. W. see Kass, E. H. 75, *165*
Mowles, T. F. see Sheppard, H. 523, *567*

Moxham, A., and J. D. N. Nabarro 475, *508*
— see Nabarro, J. D. N. 103, 107, 108, 133, 135, *171*
Moy, R. H. see Bergenstal, D. M. 523, *556*
Moyer, J. 547, *564*
Muelheims, C. see Hilton, J. G. 76, *162*
Mueller, G. C., and G. Rumney 484, 486, *508*
Muhlboch, O. see Dingmanse, E. 426, *502*
Mühlemann, H. R., T. M. Marthaler and K. H. Rateischak 32, *171*
Muir, R. D. see Dobson, R. M. 487, *502*
— see Dodson, R. M. 233, 260, *388*
Mukerjee, A. K., and S. Banerjee 104, *171*
Mukerji, B. see Roy, S. N. 107, *177*, 596, *634*
Mulay, A. S., and W. H. Eyestone 101, *171*
Muldowney, F. P. see Barger, A. C. 531, *556*
Mule, S. J. see De Venuto, F. 300, *387*
Mulinos, M. G., and L. Pomerantz 69, *171*
Muller, A. F., and C. M. O'Connor 520, *564*
— see Mach, R. S. 457, *507*, 522, 531, *563*
— see Riondel, A. M. 366, *402*
— see Wolff, H. P. 114, *184*
Mulrow, P. J. see Ganong, W. F. 110, *157*
— see Johnson, B. B. *506*
— see Phillips, J. G. 77, *174*, 550, 555, *565*
Munch-Petersen, A. see Smith, E. E. B. 452, *511*
Munson, P. L., and F. N. Briggs 522, *564*
— T. F. Gallagher and F. C. Koch 434, *508*
— F. G. Goetz, J. C. Laidlaw, J. H. Harrison and G. W. Thorn 581, 603, *632*
— M. E. Jones, P. J. McCall and T. F. Gallagher 337, 364, *399*
— and A. D. Kenny 337, 363, *399*
— see Harrison, J. H. 573, 581, 607, *629*
— see Wells, H. 522, *569*
Muntwyler, E., R. C. Mellors and F. R. Mautz 535, 538, *564*
— — — and G. H. Mangun 538, *564*
Muntwyler, E. see Harkness, D. M. 537, *560*
Murray, H. C., and D. H. Peterson 487, 491, 492, 493, *508*
— see Eppstein, S. H. 258, 259, *389*, 487, 491, 492, 493, 494, *503*
— see Hogg, J. A. 272, *393*
— see Meister, P. D. 487, 491, 492, 493, 494, *508*
— see Pederson, R. L. 281, 283, *401*, 487, *509*
— see Peterson, D. H. 259, 260, *401*, 491, 492, 493, *509*
— see Spero, G. B. 267, 283, *406*
Murray, M. F. see Ott, A. C. 241, *400*
Muschke, E. see Tonutti, E. 31, 88, *182*
Mushett, C. W., C. C. Porter and R. H. Silber 535, *564*
— see Gaunt, R. 139, *158*
Musso, E. see de Watteville, H. 611, *636*
Myers, C. S. see Mason, H. L. 519, *564*
Myhre, J. 653, *658*

Näätanen, E. see Vasama, R. 87, *182*
Nabarro, J. D. N. 543, *565*
— A. Moxham and G. Walker 103, 107, 108, 133, 135, *171*
— see Moxham, A. 475, *508*
Nabors, C. J. jr., and D. L. Berliner 61, *171*
Nace H. R. *399*
Nachlas, M. M., and A. M. Seligman 66, *171*
— K. C. Tsou, E. De Souza, C. S. Cheng and A. M. Seligman 63, *172*
Nadel, E. M., and S. Burstein 432, 433, *508*
— — and R. I. Dorfman 433, *508*
— and F. Highhouse 297, *399*
— and J. J. Schneider 71, 105, *172*, 454, *508*
— B. G. Young, A. C. Hilgar and S. Burstein 71, *172*
— see Burstein, S. 421, 424, 441, 454, 460, *501*
— see Taylor, D. J. 51, 84, 144, *181*
Nakao, K. see Okinaka, S. 456, *508*
Nalbandov, A. V., and L. E. Card 143, *172*, 553, *565*
Nandi, J., and H. A. Bern 77, *172*, 550, *565*
Nandi, S. see Bern, H. A. 57, 133, *148*
Nathan, A. H. see Anderson, H. V. 227, *381*
— see Hogg, J. A. 208, 256, 257, 277, *393*
— see Peterson, D. H. 492, *509*
Nathanson, I. T., L. L. Engel and R. M. Kelly 616, *632*
— and L. E. Towne 616, *632*
— and H. Wilson 363, *399*
— see Engel, L. L. 426, 451, *503*
— see Wilson, H. 338, *409*
Navori, C. A. see Davis, M. E. 611, *626*
Nayyar, S. N., and D. Glick 66, 98, *172*
— see Glick, D. 50, 63, 68, *159*
Nazarow, I., and Bergel'son 186, *399*
Neely, L. see Slaunwhite, W. R. jr. 350, *405*
Nehari, V. see Moog, F. 622, *632*
Neher, G. M., and M. X. Zarrow 615, *632*
— see Lazo-Wasem, E. A. 610, *630*
Neher, H. see Wettstein, A. 433, *513*
— P. Desaulles, E. Vischer, P. Wieland and A. Wettstein 204, 206, *399*, *565*
— C. Meystre and A. Wettstein 198, 204, *399*
— and A. Wettstein 198, 204, 238, 320, 327, 328, 333, 341, 342, 345, *399*, 420, 421, 442, 443, 499, *508*
— see Arx, E. von 313, *382*
— see Euw, J. von 196, 198, 204, 264, *389*
— see Kahnt, F. W. 234, 236, 238, *395*, 413, 443, 446, 447, 490, 491, 492, *506*, 517, *562*
— see Llaurado, J. G. *507*, 520, *563*
— see Luetscher, J. A. jr. 433, *507*, 520, *563*
— see Mach, R. S. *563*
— see Prader, A. 521, *565*
— see Simpson, S. A. 198, 204, 238, 240, 317, 354, *405*, 421, 446, *511*, 520, *567*
— see Wettstein, A. 204, 206, *408*
Neilson, E. D. see Knauff, R. E. 198, *395*
Nelson, A. A., and G. Woodward 135, *172*, 457, *508*
Nelson, D. H. 74, 75, *172*

Nelson, D. H., R. H. Egdahl and D. M. Hume 73, 75, 76, 103, 106, *172*
— and D. M. Hume 76, 103, *172*
— and L. T. Samuels 296, 300, 312, 366, *399*
— see August, J. T. 138, *146*
— see Bliss, E. L. 572, *625*
— see Eik-Nes, K. 312, *388*
— see Gassner, F. X. 573, *628*
— see Glenn, E. M. 295, 296, *391*
— see Hernando, L. 457, *505*, 531, *561*
— see Hume, D. M. 456, *505*
— see Jenkins, J. J. 137, 138, *164*
— see Jenkins, J. S. 523, *562*
— see Lipscomb, H. S. 76, 85, 103, *168*
— see Reich, H. 341, *402*, 609, *634*
— see Renold, A. E. 137, *175*, 457, *509*, 522, *566*
— see Sandberg, A. A. 475, *510*
— see Thorn, G. W. 137, *181*, 457, *512*
Nelson, M. M. see Hazelwood, R. L. 139, *161*
Nelson, N. A. see Wilds, A. L. 280, 281, *409*
Nelson, W. O. 607, 621, *632*
Nes, W. R. see Colton, F. G. 274, *386*
Neterval, C. R. 297, *399*
Neuman, H. C. see Clinton, R. O. 288, *386*
Neumann, K., and G. Koch 63, *172*
Neuss, J. D. see Schulz, E. P. 338, *404*
Newhouse, J. P. 297, *399*
— see Broadbent, I. E. 339, *384*
Newman, A. E., E. S. Redgate and G. Farrell *172*
Newson, S. E. see Southcott, C. M. 423, 425, *511*
Nezamis, J. E. see Ingle, D. J. 519, 528, *562*
Nicander, L. 27, 35, 53, 55, 64, 65, 67, *172*
Nicholls, D. 71, *172*
— F. C. Heagy and R. J. Rossiter 91, *172*
— and R. J. Rossiter 91, *172*
— see Rossiter, R. J. 91, *176*
Nichols, J. 3, 50, 51, 84, 94, 112, 115, 126, *172*
— C. Davis and H. D. Green 457, *508*, 523, *565*
Nichols, J., and L. I. Gardner 135, 136, *172*, 457, *508*
— and H. D. Green 136, *172*
— and G. Hennigar 135, 136, 137, *172*, 458, *508*, 523, *565*
— O. L. Lescure and C. J. Migeon 127, *173*, 592, 593, *632*
— and H. L. Sheehan 135, 136, *173*, 457, *508*, 523, *565*
— see Sheehan, H. L. 458, *511*
Nicholson, W. M. see Harrop, G. H. 525, 530, *561*
Nickey, W. M. see Race, G. J. 83, 101, 108, 111, 112, *175*
Nicolaides, N., O. K. Reess and R. G. Langdon 436, *508*
Nielson, E. D. see Drake, N. A. 341, *388*
— see Kelly, T. L. 63, *165*
— see O'Connell, P. W. 259, *400*
— see O'Donnell, P. W. 493, *508*
Niemineva, K. see Ekholm, E. 587, *627*
Nigeon-Dureuil, M., and A. R. Ratsimamanga 71, *173*
Nilson, H. W. see Ingle, D. J. 535, *562*
Nims, L. F. see Tepperman, J. 50, 51, *181*
Nishida, S., and K. Mochizuki 596, 602, *632*
Nishikawa, M., and H. Hagiwara 493, 494, *508*
— see Okinaka, S. 456, *508*
Nobile, A., W. Charney, P. L. Perlman, H. L. Herzog, C. C. Payne, M. E. Tully, M. A. Jevnik and E. B. Hershberg 277, *399*
— see Herzog, H. L. 277, *392*, 494, *505*
Noble, N. L., and E. Papageorge 67, *173*
Noble, R. L. see Carroll, K. K. 70, *150*
Nolin, B. see Jones, R. N. *394*
Nordstrand, E. see McKerns, K. W. 77, *170*
Nordström, S., and F. Paulsen 649, *659*
Norman, A. see Bergström, S. *500*
North, H. B. see Pfiffner, J. J. 420, *509*
Northrup, L. C. see Melampy, R. M. 105, 139, *170*
Norton, J. M. see Robinson, A. M. 296, 297, 299, *403*
Norymberski, J. K. 225, 340, *399*
— and A. T. Shermin 366, *399*
— and R. D. Stubbs 216, *399*
— — and H. F. West 365, *399*
— and G. F. Woods 246, *399*
— see Appleby, J. I. 341, 366, *382*
— see Brooks, C. J. W. 215, 222, 339, 340, *384*
Noumura, T. 44, 45, 121, 122, 123, 124, 125, *173*
Novelli, G. 62, *173*
Novikoff, A. B. 62, 64, *173*
Nowaczynski, W., and J. Genest 300, 362, *399*
— M. Goldner and J. Genest 364, *400*
— and E. Koiw 334, *400*
— — J. Genest, R. Tellier, I. Morin, A. La Flamme and P. Robinson 236, 333, *400*
— T. Sandor, E. Koiw and J. Genest *400*
— and P. R. Steyermark *400*
Nowell, N. W. 79, *173*
Noyce, D. S. see Dauben, W. G. 245, *387*
Nubel, R. C. see Shull, G. M. 310, 342, *405*
Numerof, P. see Fried, J. 267, 268, *390*
Nussbaum, A. L., and F. E. Carlon 225, *400*
— see Oliveto, E. P. 285, *400*

Oakwood, T. S. see Marker, R. E. 209, *397*
O'Brien, E. see Perlman, D. 494, *509*
O'Brien, M. G. see Szalkowski, C. R. 337, *406*
Ochs, L. see Sensenbach, W. 643, *659*
Ochs, M. J. see Glick, D. 51, 68, 80, 84, 94, *159*
O'Connel, P. W., K. M. Mann, E. D. Nielson and F. R. Hanson 259, *400*
— see Mann, K. M. 261, *397*, 493, *507*
O'Connor, C. M. see Muller, A. F. 520, *564*
— see Wolff, H. P. 114, *184*
O'Donnell, M. see Stimpson, M. M. 350, *406*
O'Donnell, P. W., K. M. Mann, E. D. Nielson and F. R. Hanson 493, *508*
O'Donnell, V. J. see Bligh, E. G. 449, 451, *500*
— see Heard, R. D. H. 420, 421, 446, 449, 451, 453, *505*, 616, *629*

O'Donnell, W. M., S. S. Fajans and J. G. Weinbaum 110, *173*
Oertel, G. W. 313, 340, 346, *400*
— and K. Eik-Nes 75, 108, *173*, 413, 425, 463, *508*
— J. Schmidt-Thomé and R. Reuber 291, *400*
— see Carstensen, H. 573, *626*
— see Reuber, R. 186, *402*
Offe, H. A. see Reuber, R. 186, *402*
Okada, M., D. K. Fukushima and T. F. Gallagher 572, *632*
Okinaka, S., K. Nakao, M. Nishikawa, M. Ikeda and H. Ibayashi 456, *508*
Oku, T. see Torantini, Y. 640, *659*
Olds, J. *659*
Oleesky, A., and M. B. Manc 542, *565*
Oleson, M. C., and W. R. Bloor 49, 51, 84, 92, *173*
Oliver, J. see Hollander, W. jr. 534, 543, *561*
Olivereau, M., and H. Fromentin 88, 141, *173*, 549, *565*
— see Fontaine, M. 88, 90, *157*
Oliveto, E. P., C. Gerold, R. Rausser and E. B. Hershberg 228, *400*
— — L. Weber, H. Jorgensen R. Rausser and E. B. Hershberg 204, 228, *400*
— and E. B. Hershberg 216, 227, *400*
— R. Rausser, H. L. Herzog, E. B. Hershberg, S. Tolksdorf, M. Eisler, P. L. Perlman and M. M. Pechet 285, *400*
— — A. L. Nussbaum, W. Gebert, E. B. Hershberg, S. Tolksdorf, M. Eisler, P. L. Perlman and M. M. Pechet 285, *400*
— — L. Weber, E. Shapiro, D. Gould and E. B. Hershberg 243, *400*
— H. Q. Smith, C. Gerold, R. Rausser and E. B. Hershberg 228, *400*
— — and E. B. Hershberg 272, *400*
— — R. Rausser and E. B. Hershberg 216, 272, *400*
— see Hershberg, E. B. 242, *392*
Oliveto, E. P. see Herzog, H. L. 219, 267, 277, *392*
Olmsted, P. C. see Engel, L. L. 426, *503*
Olney, J. M. see Howard, J. M. 642, *658*
Olsen, L. J. see Done, A. K. 471, *502*
Olsen, N. S., and W. E. Martindale 139, *173*
— see Rudolph, G. G. 96, *177*
Olson, J. A., and K. Bloch 436, 438, *508*
Olson, R. E., and H. W. Deane 116, *173*
Oncley, J. L., F. R. N. Gurd and M. Melin 292, *400*
Ondarza, R. see Mills, G. T. 452, *508*
Oneson, I. B. see Cohen, S. L. 295, *386*
Ong, S. see Werbin, H. 363, *408*
van Oordt, G. J. see Sluiter, J. W. 79, 143, *179*
Opdyke, M. see Silberberg, R. 134, *179*
Oppenauer, R. 427, 428, *508*
Ortiz, E. 603, *632*
Orvando, P. see Dumm, M. E. 139, *154*
Osborn, C. M. see Birnie, J. H. 545, *556*
Osborn, H. M. L. see Meister, P. D. 491, 492, 494, *508*
Osborn, M. L. see Peterson, D. H. 491, *509*
Osborne, M. see Swingle, W. W. 529, 536, 547, *568*
Osorio, J. A., and A. Kraemer 18, 135, *173*
Ott, A. C., M. F. Murray and R. L. Pederson 241, *400*
— see Slomp, G. 357, *405*
Overbeeke, A. P. van 145, *173*
Overland, L. see Rosselet, J. P. 432, 465, *510*
Overman, R. R., A. K. Davis and A. C. Bass 537, 538, *565*
— see Davis, A. K. 539, *558*
— see Flanagan, J. B. 536, 537, *558*
— see Golden, A. 130, 132, *159*

Pääbo, K. see Bergström, S. *500*
Pacheco, H. see Zwingelstein, G. 310, *410*
Padykula, H. A. 63, *173*
Page, I. H. see Corcoran, A. C. 340, 365, *386*
— see Masson, G. M. C. 532, *564*
Palmer, D. see Halkerston, I. D. K. 455, *504*
Panattoni, M. see Marrian, G. F. 426, *507*
Pancati, G. B. see Bonati, B. 81, 96, 108, 110, *148*
Papadatos, C. see Klein, R. 72, 125, *165*, 296, *395*, 423, 425, *506*, 521, *562*
Papageorge, E see Noble, N. L. 67, *173*
Pappo, R. 237, *400*
Parker, D. see Ferrebee, J. W. 532, 534, *558*
Parkes, A. S. 120, *173*, 622, *633*
— and H. Selye 554, *565*
— see Deanesly, R. 575, 576, 596, *626*
— see Hill, R. T. 553, *561*
Parkins, W. M. 554, *565*
— see Swingle, W. W. 527, 529, 535, 536, 542, 545, *568*
Parlow, A. F. see Swingle, W. W. 521, 524, 527, 535, *568*
Parmer, L. G., F. Katonah and A. A. Angrist 622, *633*
Parsons, F. M. see Huggins, C. 607, *630*
Parsons, J. W., T. Beher and G. D. Baker 336, *400*
Parsons, J. see Beher, W. T. 336, *383*
Paschkis, K. E., A. Cantarow, A. E. Rakoff, L. Hansen and A. A. Walkling 480, *508*
— see Hansen, L. P. 338, *391*
— see Pearlman, W. H. 484, *508*
— see Rakoff, A. E. 575, *633*
Pasqualini, C. D. de see Pasqualini, R. Q. 58, *173*
Pasqualini, R. Q., C. D. de Pasqualini and J. C. Garberi 58, *173*
Paszkowski, M. see Rivoire, R. 295, *402*
Pataki, J.. G. Rosenkranz and C. Djerassi 204, *400*
— see Djerassi, C. 288, *387*
— see Kaufman, S. 279, *395*
— see Rosenkranz, G. 255, *403*
Patel, D. K. see Adams, W. J. 218, *381*
— see Kirk, D. 214, 288, *395*
Patelli, B. see Camerino, B. 272, *385*
Paterson, J. Y. F., and G. F. Marrian 365, *400*
— see Marrian, G. F. 296, *397*
Patt, D. I. see Brenner, R. M. 118, 119, *149*

Pattee, C. J. see Cleghorn, R. A. 654, 655, *657*
Patterson, J. 572, *633*
— see Anderson, A. F. 611, *624*
— see Bush, I. E. 75, *149*, 572, *625*
Patterson, H. R. see Engel, L. L. 303, 338, *389*
Patterson, R. M. see Chance, J. 649, *657*
Patton, B. W. see Tepperman, J. 50, 51, *181*
Patzelt, V. 9, *173*
Paul, M. H., and E. L. Durrum 304, *400*
Paulsen, F. see Nordström, S. 649, *659*
Pauly, J. E. 22, 23, 27, *173*
— see Elias, H. 23, 27, 45, *155*
Pankratz, D. S. 39, *173*
Pavoa, H. jr., and P. Fontana jr. 91, *173*
Payne, C. C. see Herzog, H. L. 267, 277, *392*
— see Nobile, A. 277, *399*
Payne, F. 17, 57, *173*
Payne, R. W. 621, *633*
Peace, R. F. 23, *173*
Pearce, E. M., F. de Venuto, W. M. Fitch, H. E. Firschein and U. Westphal 370, *401*
— see Westphal, U. 292, *408*
Pearlman, W. H. 304, *401*, 427, *508*
— and E. Cerceo 293, 338, *401*
— and R. H. de Meio 484, *508*
— and G. Pincus 484, *508*
— — and N. T. Werthessen 431, *508*
— A. E. Rakoff, K. E. Paschkis, A. Cantarow and A. A. Walkling 484, *508*
— see Ayres, P. J. 423, 447, *500*
— see Pincus, G. 302, *401*
Pearse, A. G. E. 57, *173*
Pearson, B., and V. Defendi 66, *173*
Pearson, O. H. see Hollander, V. P. 365, *393*
— see Lipsett, M. B. 100, *168*
— see MacLean, J. P. 110, 113, *169*
— see West, C. D. 412, *513*, 616, 617, *636*
Pearson, S. see Drekter, I. J. 302, *388*
Pease, D. C. see Belt, W. D. 38, *147*
Pechet, M. M. 320, 333, *401*
— and P. H. Henneman 577, *633*
— see Bunim, J. J. 520, *557*
Pechet, M. M. see Caspi, E. 476, 498, *501*
— see Davis, J. O. 114, *152*, 531, *558*
— see Gould, D. 267, 278, 279, *391*
— see Herzog, H. L. 277, *392*, 494, *505*
— see Oliveto, E. P. 285, *400*
Peczenik, O. see Steinach, E. 616, *635*
Pederson, R. L., J. A. Campbell, J. C. Babcock, S. H. Eppstein, H. C. Murray, A. Weintraub, R. C. Meeks, P. D. Meister, L. M. Reineke and D. H. Peterson 281, 283, *401*
— — — S. H. Eppstein, H. C. Murray, A. Weintraub, R. C. Meeks, L. M. Meister, L. M. Reineke and D. H. Peterson 487, *509*
— see Hogg, J. A. 272, *393*
— see Ott, A. C. 241, *400*
— see Slomp, G. 357, *405*
Pekkarinen, A. see Uotila, U. U. 18, 59, 85, *182*
Pellegrino, C. 46, 119, 120, *173*
— and F. Caramazza 120, *173*
— and A. Montella 118, 119, 120, *174*
— and A. Torcigliani 119, 120, *174*
— see Brogi, M. P. 120, *149*
Pelter, A. see Cornforth, J. W. 368, *386*
Penhos, J. C., and A. F. Cardeza 79, 143, *174*
Pennacchietti, M. 40, *174*
Perek, M., B. Eckstein and Z. Eshkol 58, 85, *174*
— and A. Eilat 85, *174*
Perera, G. A. 531, *565*
Perkoff, G. T., B. V. Jager and F. H. Tyler 644, *659*
Perlman, D., E. O'Brien, A. P. Bayan and R. B. Greenfield 494, *509*
— E. Titus and J. Fried 494, *509*
— see Fried, J. 258, *390*, 487, 491, 492, 493, 494, *504*, 520, *559*
Perlman, P. L., J. W. Cassidy and A. Tali 535, *565*
— see Gould, D. 241, 267, 278, 279, *391*
— see Herzog, H. L. 277, *392*, 494, *505*
— see Nobile, A. 277, *399*
— see Oliveto, E. P. 285, *400*
Perlmutt, J. see Swingle, W. W. 621, *635*
Perloff, W. H. see Hadd, H. E. 297, *391*
Peron, F. G., and R. I. Dorfman 414, 427, 428, 429, 433, 445, *509*
— and S. B. Koritz 77, *174* 454, *509*
— see Haynes, R. C. jr. 453, *505*
— see Koritz, S. B. 452, *506*
— see Masson, G. M. C. 120, *170*
Perrine, J. W., P. H. Bell, L. Bortle, E. Heyder, E. K. Ross and I. Ringler 521, *565*
Perry, W. F., and G. R. Cumming 92, *174*
Persky, H., J. Maroc, E. Conrad, A. den Breeijen 642, *659*
— see Basowitz, H. 642, *656*
Pescetto, G. see Rossi, F. 63, 65, 66, *176*
Pesez, M. see Velluz, L. 221, *407*
Peschel, E., and G. J. Race 112, *174*
Peters, V. B. see Dalton, A. J. 87, *152*
Peterson, D. H. 494, *509*
— S. H. Eppstein, P. D. Meister, B. J. Magerlein, H. C. Murray, H. M. Leigh, A. Weintraub and L. M. Reineke 492, 493, *509*
— — — H. C. Murray, H. M. Leigh, A. Weintraub and L. M. Reineke 260, *401*
— P. D. Meister, A. Weintraub, L. M. Reineke, S. H. Eppstein, H. C. Murray and H. M. L. Osborn *491*, *509*
— and H. C. Murray 259, *401*, 492, *509*
— — S. H. Eppstein, L. M. Reineke, A. Weintraub, P, D. Meister and H. M. Leigh 259, *401*, 491, 492, *509*
— A. H. Nathan, P. D. Meister, S. H. Eppstein, H. C, Murray, A. Weintraub, L. M. Reineke and H. M. Leigh 492, *509*
— see Eppstein, S. H. 258, 259, *389*, 487, 491, 492, 493, 494, *503*
— see Hayano, M. 485, *505*
— see Hogg, J. A. 272, *393*
— see Meister, P. D. 487, 491, 492, 493, 494, *508*

Peterson, D. H., see Murray, H. C. 487, 491, 492, 493, *508*
— see Pederson, R. L. 281, 283, *401*, 487, *509*
Peterson, R. D. see Beatty, C. H. 542, *556*
Peterson, R. E. 372, *401*
— R. Hertz and H. A. Lubs 457, *509*, 522, *565*, 586, *633*
— A. Karrer and S. L. Guerra 364, *401*
— C. E. Pierce and B. Kliman 432, 460, *509*
— and J. B. Wyngaarden 471, 475, *509*
— see Davis, J. O. 114, *153*
— see Fredrickson, D. S. 140, *157*, 455, *504*
— see Halberg, F. 32, 93, 96, *160*
— see Howard, J. M. 642, *658*
— see Pierce, C. E. 297, *401*
— see Yankopoulos, N. A. 117, *185*
Petit, A. see Horeau, A. 196, *393*
— see Mathieu, J. P. 373, *398*
— see Velluz, L. 221, *407*
Petrow, J. see Ellis, B. 273, 274, *388*
Petrow, V. see Adams, W. J. 218, *381*
— see Cooley, G. 273, *386*
— see Kirk, D. 214, 288, *395*
Pettebone, R. H. see Taub, D. 223, *407*
Pfeffer, K. H., W. Ruppel, Hj. Staudinger and L. Weissbecker *401*
Pfeiffer, E. F. see Arndt, T. 642, *656*
Pfiffner, J. J. 197, *401*
— and H. B. North 420, *509*
— O. Wintersteiner and H. M. Vars 519, *565*
— see Swingle, W. W. 21, *181*, 516, 529, 535, 545, *568*
— see Wintersteiner, O. 337, *409*, 420, 421, *513*, 519, *569*
Pfister, H. 338, *401*
Pfister, K. see Leanza, W. 225, *396*
Phillips, G. H. see Chapman, J. H. 252, *385*
Phillips, J. G., and I. Chester Jones 555, *565*
— W. N. Holmes and P. K. Bondy 75, *174*, 550, *565*
— and P. J. Mulrow 77, *174*, 550, 555, *565*
— see Chester Jones, I. 71, 75, 141, *151*, 416, *502*, *557*
— see Wright, A. 30, 31, 53, *185*
Philpot, F. J., and J. St. L. Philpot 612, *633*
Philpot, J. St. L. see Philpot, F. J. 612, *633*
Phoenix, C. H., R. W. Goy, A. A. Gerall and W. C. Young 574, *633*
Phyfe, P. see Ragan, C. 533, 534, *565*
Picha, G. M., F. J. Saunders and D. M. Green 461, *509*
Pick, J. W. see Anson, B. J. 23, *146*
Pickering, R. W. see Hamilton, W. F. 531, *560*
Pickford, G. E. 5, 7, 141, 144, *174*, 549, *565*
— and J. W. Atz 548, *565*
— and B. Kosto 144, *174*
— see Bondy, P. K. 550, *556*
Pickford, M., and M. Vogt 100, *174*
Picon, L. O. 124, *174*
Piel, C. F. see Luetscher, J. A. jr. 546, *563*
Pierce, C. E., and R. E. Peterson 297, *401*
— see Peterson, R. E. 432, 460, *509*
Pierce, F. T. jr. see Biggs, M. W. 436, *500*
Piletta, P. see Giroud, C. J. P. 109, 111, *158*, 447, 451, *504*
Pinaud, H. see Jayle, M. F. 580, *630*
Pincus, G. 72, *174*, 338, 364, *401*, 441, 490, *509*, 572, *633*
— and R. I. Dorfman 608, *633*
— — L. P. Romanoff, B. L. Rubin, E. Bloch, J. Carlo and H. Freeman 583, *633*
— T. Miyake, A. P. Merrill and P. Longo 612, *633*
— and W. H. Pearlman 302, *401*
— and E. B. Romanoff 422, 423, *509*, 609, 610, *633*
— L. P. Romanoff and J. Carlo 572, 583, *633*
— see Bergen, J. R. 579, *624*
— see Bloch, E. 204, *383*, 415, 449, 451, *500*, 572, *625*
— see Fish, C. A. 463, 497, *503*
— see Hechter, O. 311, *392*, 441, 442, 486, 489, 490, *505*, 517, *561*, 609, *629*
— see Jeanloz, R. W. 449, 486, 489, *506*, 573, *630*
— see Levy, H. 441, 442, 489, 490, *506*
— see Meyer, A. S. 360, *398*, 442, 486, 489, 490, *508*
— see Miyake, T. 612, *632*
— see Pearlman, W. H. 431, 484, *508*
Pincus, see Romanoff, E. B. 75, *176*, 422, *509*, 573, 574, 609, *634*
— see Romanoff, L. P. 482, *509*
— see Rosenfeld, G. 443, 449, 490, *510*
— see Rubin, B. L. 250, 312, 328, *403*
— see Ungar, F. 442, 489, *512*
— see Zaffaroni, A. 435, 441, *513*
Pine, I. see Chance, J. 649, *657*
Pinto, A. see Allen, W. M. 364, *381*
Pinto, R. M. 106, *174*
— see Foglia, V. G. 20, *156*
— see Houssay, A. B. 134, *162*, 618, *629*
— see Houssay, B. A. 134, *163*, 618, *629*
Pischinger, A. 55, *174*
Pissoti, L. E. see Bern, H. A. 57, 133, *148*
Pitcock, J. A., and P. M. Hartroft 112, *174*
Pittman, J. A. see Hertz, R. 456, *505*
Plager, J. E., and L. T. Samuels 443, 490, *509*
— see Migeon, C. J. 422, *508*, 572, *632*
Plan see Fraisse 546, *559*
Planel, H., and A. Guilhem 20, 67, 133, *174*
— — and P. Banide 133, *174*
— — and P. Bec 28, 57, 65, 96, 133, *174*
— see Baïsset, A. 115, 116, *146*
— see Bimes, C. 20, 35, 57, 133, *148*
Plantin, L. O. 348, *401*
— and G. Birke 428, *509*
— E. Diczfalusy and G. Birke 414, 415, *509*, 572, *633*
— see Birke, G. 459, *500*, 581, *625*
—see Diczfalusy, E. 364, *387*, 412, *502*
Plate, W. P. 455, *509*
Plattner, P. see Heusser, H. 286, 288, *393*
see Julia, S. A. 275, *394*
Pliess, G. see Küchmeister, H. 135, 136, *167*
Pliske, E. C. see Coutinho, H. B. 66, 98, 109, *152*
Plotz, E. see Davis, M. E. 86, *153*
Plotz, E. J., and M. E. Davis 582, *633*
— see Davis, M. E. 86, 124, *153*, 572, 581, 582, 591, *626*

Plotz, J. see Werbin, H. 435, 441, 449, *513*
Politzer, G. 39, *174*, 570, *633*
Poll, H. 570, 596, *633*
Pollack, E. see Cerviño, J. M. 101, 135, *150*
Pollack, R. L. see Eades, H. C. jr. 418, *503*
Pomeau-Delille, G. 613, *633*
Pomerantz, L. see Mulinos, M. G. 69, *171*
Pond, M. H. 294, 308, *401*
Ponse, K. 599, 608, *633*
— see Guyénot, E. 607, *628*
Pontius, D., I. Beckman and K. D. Voigt 337, *401*
— see Zimmermann, W. 363, *410*
Poos, G. I. 216, 225, *401*
— G. E. Arth, R. E. Beyler and L. H. Sarett 242, *401*
— and L. H. Sarett 265, *401*
— see Hirschmann, R. F. 271, *393*
— see Sarett, L. H. 262, *404*
Popjak, G. 438, *509*
— G. D. Hunter and T. H. French 436, *509*
— see Cornforth, J. W. 191, 368, *386*, 437, 438, *502*
— see Gould, R. G. 436, *504*
Porak, R., and H. Chabanier 515, *565*
Porter, C. C., and R. H. Silber 340, *401*
— see Mushett, C. W. 535, *564*
— see Silber, R. H. 341, 366, *405*
Porto, J. 9, 79, 91, 92, 94, 142, 144, *174*
— see Fustinoni, O. 9, *157*
Pothier, L. see Jenkins, J. S. 523, *562*
Potter, M. see Henly, A. A. 365, *392*
Pottinger, R. E. see Davis, M. E. 611, *626*
Potts, H. E. see Gaunt, R. 542, *559*
Poulton, B. R., and R. P. Reese 50, 51, 106, *174*
Pousiaka, J. W., B. G. H. Thomas and C. R. Linegar 535, 540, *565*
Power, M. H. see Mason, H. L. 607, *631*
— see Robinson, F. J. 542, *566*
Prader, A. 651
— A. Spahr and R. Neher 521, *565*
— and R. E. Siebenmann 138, *175*
Praetz, B. see Lieberman, S. 427, 428, 431, 432, *507*
Pratt, R. J. see Tannhauser, P. 271, *407*
Prelog, V., and J. Führer 451, *509*
— and G. Tsatsas 209, *401*
— see Dauben, W. G. *387*
— see Heusser, H. 212, *393*
Pressler, M. see Zeisel, H. 590, 591, *637*
Prestrud, M. C. see Ingle, D. J. 103, *163*, 531, *562*
Price, D. 599, 601, 604, *633*
— and D. J. Ingle 108, *175*, 604, 605, 623, *633*
— T. Mann and C. Lutwak-Mann 579, *633*
— see Moore, C. R. 576, 599, *632*
Price, T. D., and D. Rittenberg 436, *509*
Price, W. J. 369, *401*
Prins, D. A., and T. Reichstein 210, *401*
Pritchard, W. R. see Weber, A. F. 115, *184*
Procházka, Ž. 341, 342, *401*
— see Kučera, J. 328, *396*
Pronove, P. see Bartter, F. C. 114, *147*
Proskauer, E. S. see Weissberger, A. 296, *408*
Proto, A. see Cerviño, J. M. 101, 135, *150*
Pruitt, R. L. see Richard, J. B. 73, 81, *175*
Prunty, F. T. G., B. E. Clayton, R. R. McSwiney and I. H. Mills 71, 105, *175*
— see Brooks, R. V. 456, *501*
— see Forsham, P. H. 607, *627*
Prystowsky, H. see Migeon, C. J. 471, *508*
Pugsley, L. I. see Wills, C. G. 575, *636*
Pullman, T. N., and W. W. McClure 529, *565*
Pummer, A. J. see Dreiding, A. C. 287, *388*
Purmalis, A. see Ingle, D. J. 491, *506*

Quandt, V. V., and S. Rosenthal 655, *659*
Quarton, G. C., L. D. Clark, S. Cobb and W. Bauer 643, *659*
Quenum, A., and R. Camain 60, 65, 67, 92, *175*
Quinn, G. P. see Berger, E. Y. 540, *556*
Quintilani, M. see Angelico, R. 139, *146*
le Qintrec, Y. see Laroche, C. 655, *658*
Rabinovici, N. 130, 141, *175*
Rabinowitz, J. see Dowben, R. M. 484, *503*
Rabinowitz, J. L. 438, *509*
— F. Dituri, F. Cobey and S. Gurin 438, *509*
— and S. Gurin 436, 438, *509*
Race, G. J., and R. F. Green 32, 88, 111, 112, *175*
— W. M. Nickey, P. S. Wolf and E. J. Jordan 83, 101, 108, 111, 112, *175*
— and H. M. Wu 587, *633*
— see Peschel, E. 112, *174*
Radakovich, M. see Ganis, F. M. 435, *504*
Ragan, C., J. W. Ferrebee and G. W. Fish 537, *565*
— — P. Phyfe, D. W. Atchley and R. F. Loeb 533, 534, *565*
— see Ferrebee, J. W. 531, *558*
— see Kuhlman, D. 534, *562*
Raile, R. B. see Done, A. K. 75, *154*
— see Sydnor, K. L. 585, *635*
Rajagopalan, S. see Fieser, L. F. 242, *389*
Rakoff, A. E., K. E. Paschkis and A. Cantarow 575, *633*
— see Hansen, L. P. 338, *391*
— see Paschkis, K. E. 480, *508*
— see Pearlman, W. H. 484, *508*
Ralli, E. P. see Dumm, M. E. 69, 71, 139, *154*, 458, *503*
Rampton, S. E. see Wills, C. G. 575, *636*
Ramsay, D. A. 354, *401*
— see Jones, R. N. 354, *394*
Ramsey, C. G. see Gray, S. J. 540, *560*
Randall, H. T. see Roberts, K. E. 533, *566*
Randall, J. P. see Venning, E. H. 591, *636*
Randall, S. S. see Dobson, F. 300, *388*
Rao, B. G., and R. D. H. Heard 443, 447, 449, 465, *509*
— see Heard, R. D. H. 420, 421, 446, 449, 451, 453, *505*
Rapala, R. T. see Gassner, F. X. 573, *628*
Rapela, C. E., and M. F. Gordon 9, 79, 145, *175*
Rapport, M. M., and R. E. Franzl 55, *175*
Rasquin, P. 7, *175*
Rateischak, K. H. see Mühlemann, H. R. 32, *171*
Rather, L. J. 21, 133, *175*

Ratsimamanga, A. R. see Giroud, A. 59, *158*
— see Nigeon-Dureuil, M. 71, *173*
Rauschkolb, E. W., and G. L. Farrell 117, *175*, 544, *566*
— — and S. Koletsky 74, 100, 109, 110, *175*
— see Farrell, G. L. 74, 76, 100, 103, 110, 111, *156*, 423, 503, 610, *627*
Rausser, R. see Oliveto, E. P. 204, 216, 228, 243, 272, 285, *400*
Raynaud, A., and M. Frilley 121, 122, *175*
Raynaud, J. 602, *634*
Read, C. H., E. H. Venning and M. P. Ripstein 591, *634*
Read, H. C. see McPhail, M. K. 128, *170*
Reale, E. see Rossi, F. 63, 65, 66, *176*
Reber, F., A. Lardon and T. Reichstein 241, *402*
Recknagel, R. O., and M. Litteria 340, *402*
von Reddy, A. M. see Bitman, J. 228, 250, 359, *383*
Reddy, E. J. see Hernando, L. 457, *505*
Reddy, W. J. 292, *402*
— N. A. Haydar and J. C. Laidlaw 423, *509*
— — — A. Renold and G. W. Thorn 293, *402*
— D. Jenkins and G. W. Thorn 292, *402*
— see Haydar, N. A. 425, *505*
— see Hernando, L. 531, *561*
— see Jenkins, J. S. 523, *562*
— see Laidlaw, J. C. 572, *630*
— see Renold, A. E. 521, *566*
— see Sobel, E. H. 135, 137, *180*
— see Thorn, G. W. 137, *181*, 457, *512*
Regate, E. S. see Newmann, A. E. *172*
Reece, R. P. see Poulton, B. R. 50, *174*
Reese, J. D., and H. D. Moon 37, 92, *175*
Reess, O. K. see Nicolaides, N. 436, *508*
Regan, B. M. see Julian, P. L. 274, *395*
Reich, D. H. see Gassner, F. X. 573, *628*
Reich, E., and A. L. Lehninger 420, 421, 435, 441, *509*
Reich, H., D. H. Nelson and F. X. Gassner 609, *634*
— — and A. Zaffaroni 341, *402*, 609, *634*
Reich H. and T. Reichstein 242, *402*
— and B. K. Samuels 221, 225, *402*
— S. J. Sanfilippo and K. F. Crane 310, *402*
— see Samuels, L. T. 61, *177*, 496, *510*
Reich, M. see McDonald, I. R. 73, 86, 103, 113, *170*
Reichstein, T. 99, *175*, 197, 198, 204, 227, *402*, 411, 413, 415, 420, 421, *509*, *520*
— — and J. v. Euw 221, *402*, 413, 415, 420, 421, *509*, 517, *566*, 610, *634*
— and K. Gatzi 413, 420, *509*
— and C. W. Shoppee 197, 198, 204, 307, 308, 336, 337, *402*
— see Beall, D. 420, *500*, 609, *624*
— see Brenneisen, K. 191, *384*
— see Euw, J. von 196, 198, 204, 227, 242, 264, *389*, 415, 420, *503*
— see Finkelstein, M. 428, 432, *503*
— see Hegedüs, H. 317, *392*
— see Hegner, P. 270, *392*
— see Hunger, A. 239, *394*
— see Lardon, A. 191, 196, 200, 220, 228, *396*
— see Prins, D. A. 210, *401*
— see Reber, F. 241, *402*
— see Reich, H. 242, *402*
— see Simpson, S. A. 198, 204, 238, 240, 317, 354, *405*, 421, 446, *511*, 520, *567*
— see Sorkin, M. 206, *406*
— see Steiger, M. 219, 220, *406*, 420, *511*, 517, *567*
Reif, A. E., and B. B. Longwell 74, 75, *175*
— see Longwell, B. B. 139, *169*, 422, 423, *507*
Reifenstein, E. C. jr. see Fraser, R. W. 363, *390*
— see Gassner, F. X. 574, *628*
Reineke, E. P. see Smith, E. M. 104, *179*
Reineke, L. M. 320, 327, 333, *402*
— see Eppstein, S. H. 487, 491, 492, *503*
— see Meister, P. D. 487, 491, 492, 493, 494, *508*
— see Pederson, R. L. 281, 283, *401*, 487, *509*
— see Peterson, D. H. 259, 260, *401*, 491, 492, 493, *509*
Reiner, C. B. 53, 59, 60, *175*
Reinhardt, W. O. see Li, C. H. 82, 83, 100, 101, 103, 117, *168*, 608, 613, 614, *631*
Reiser, R. see Dieckert, J. W. 310, 341, *387*
Reiss, M. see Carpenter, R. K. 83, 92, *149*
Relman, A. S., and W. B. Schwartz 531, *566*
Remington, J. W. see Swingle, W. W. 99, *181*, 530, 537, 538, 545, *568*
Remmers, A. R. see Johnson, J. E. jr. 457, *506*
Rennels, E. G. 107, *175*
— M. Hess and J. C. Finerty 106, 107, *175*
— see Hess, M. *162*
Renold, A. E., J. Crabbé, L. Hermando-Avendano, D. H. Nelson, E. J. Ross, K. Emerson jr. and G. W. Thorn 137, *175*, 457, *509*, 522, *566*
— N. A. Haydar, W. J. Reddy, A. Goldfien, J. R. St. Marc and J. C. Laidlaw 521, *566*
— D. Jenkins, P. H. Forsham and G. W. Thorn 580, *634*
— see Hernando, L. 457, *505*, 531, *561*
— see Reddy, W. J. 293, *402*
— see Sobel, E. H. 135, 137, *180*
— see Thorn, G. W. 137, *181*, 457, *512*
Renzi, A. A., and R. Gaunt 521, 532, *566*
— M. Gilman and R. Gaunt *521*, *566*
— M. Renzi, J. J. Chart and R. Gaunt 531, 543, *566*
— see Gaunt, R. 530, 547, 548, *559*
Renzi, M. see Renzi, A. A. 531, 543, *566*
Repetto, O. M., and A. A. Drewes 297, *402*
Restivo, A. see Fried, J. 267, 272, 277, 278, *390*
Restle, H. see Schiedt, U. 350, *404*
Retzlaff, E. W. 11, *175*
Reuber, R., R. Tschesche, J. Schmidt-Thome, G. Oertel and H. A. Offe 186, *402*
— see Oertel, G. 291, *400*
Revol, L. see Leulier, A. 51, *168*
Reyneri, C. see Savard, K. 431, 451, *510*
Reynolds, S. R. M. 615, *634*

Rhoads, C. P. see Dobriner, K. 301, 303, 357, *388*, 431, *502*, 582, *627*
— see Lieberman, S. 217, 307, *396*, 427, 428, 431, *507*
Ribaudo, C. A. 141, *175*
Rice, K. L. see Ingle, D. J. 531, *562*
Rich, A. R. 130, *175*
— and M. Berthrong 90, *175*
— see Wilbur, O. M. jr. 130, 133, *184*
Richard, J. E. see Liddle, G. W. 521, 533, *563*
Richards, J. B., and R. L. Pruitt 73, 81, *175*
— see Egdahl, R. H. 105, 106, *155*, 522, *558*
Richardson, E. M., H. Bulaschenko and H. C. Dohan 459, 460, *509*
— J. C. Touchstone and F. C. Dohan 432, 459, 460, *509*
— see Dohan, F. C. 432, 433, *502*
— see Touchstone, J. C. 432, *512*
Richardson, H. L. see Wexler, B. C. 110, *184*
Richet, C., G. Marañon and M. Rymer 639, *659*
Richter, C. P., and J. F. Eckert 529, *566*
— see Mosier, H. D. 115, *171*
Richter, H. S. 534, 543, *566*
Richter, W. R. 45, 129, *175*
Riddick, J. A. see Weissberger, A. 296, *408*
Riddle, O. 622, *634*
— see Bates, R. W. 553, *556*
— see Miller, R. A. 14, 17, 53, 57, 78, 143, *171*, 553, 554, *564*
— G. C. Smith and R. A. Miller 554, *566*
Riedel, B. E., J. E. Logan, R. J. Rossiter *176*
Riegel, B. see Dodson, R. M. 288, *388*
Riemer, A. D. 546, *566*
Rigler, R. 542, 545, *566*
Rilling, H. see Amdur, B. H. 438, *499*
Rinaldi, L. M. 69, *176*
Rinfret, A. P., and S. Hane 59, 125, *176*
— see Wexler, B. C. 110, *184*
Ring, J. R. 621, *634*
Ringler, I. see Bernstein, S. 267, 284, *383*
— see Perrine, J. W. 521, *565*
Ringold, H. J., E. Batres, O. Mancera and G. Rosenkranz 272, *402*
Ringold, H. J., G. Rosenkranz and F. Sondheimer 221, 232, 241, 287, *402*
— and G. Stork 258, *402*
— see Bowers, A. 272, *384*
— see Dauben, H. J. 244, *387*
— see Mancera, O. *397*
— see Zaffaroni, A. 281, *409*
Riniker, B., D. Arigoni and O. Jeger 191, *402*
— see Djerassi, C. 281, 288, 290, *388*
Riniker, R. see Djerassi, C. 281, 288, 290, *388*
Rioch, D. M. see Laqueur, G. L. 92, 104, 105, *167*
Riondel, A. M., M. C. Sanz and A. F. Muller 366, *402*
Ripstein, M. P. see Read, C. H. 591, *634*
— see Venning, E. H. 580, *636*
Rittenberg, D., and R. Schoenheimer 436, *509*
— see Bloch, K. 436, 437, *500*
— see Borck, E. 436, *501*
— see Curran, G. L. 436, *502*
— see Keston, A. S. 368, *395*
— see Price, T. D. 436, *509*
Ritterson, A. L., and S. I. Mauer 93, 104, *176*
Rivoire, R., and M. Paszkowski 295, *402*
Roaf, R. 48, *176*
Robbie, W. A., and R. B. Gibson 295, *402*
Robbins, M. E. see Miller, M. R. 9, *170*
Roberts, G. 350, *402*
— B. S. Gallagher and R. N. Jones 347, *402*
— see Fukushima, D. K. 217, *390*
Roberts, J. M. see Cooper, D. Y. 449, *502*
— see Touchstone, J. C. 75, *182*, 449, *512*
Roberts, K. E., and H. T. Randall 533, *566*
Roberts, S. 77, *176*
— and C. M. Szego 573, 620, *634*
— see Beigelman, P. M. 74, 75, *147*
— see Slusher, M. A. 60, 85, *179*, 454, *511*
— see Szego, C. M. 620, *635*
Robertson, A. J. B. 368, *402*
Robertson, J. M. 338, *402*
Robertson, O. H., and B. C. Wexler 144, 145, *176*
Robinson, A. M. 308, *403*
— and J. M. Norton 296, 297, 299, *403*
Robinson, C. R. see Stone, D. 491, *511*
Robinson, E. J. see Hegnauer, A. H. 535, *561*
Robinson, F. J., M. H. Power and E. J. Kepler 542, *566*
Robinson, F. B., and J. M. Yoffey 104, *176*
Robinson, H. J., W. G. Bernhard, H. Brugin, H. Wanner, G. W. Sewekow and R. H. Silber 106, *176*
Robinson, P. see Nowaczynski, W. 236, 333, *400*
Robinson, R. see Cardwell, H. M. 261, *385*
Robinson, S. A. see Jones, R. E. 225, *394*
Robson, J. M. 578, *634*
Roby, C. C. see Chester Jones, I. 78, 99, *150*
Roche, J., R. Michel and P. Jouan *176*
Rodgers, O. G. see Meyer, A. S. 415, 442, 449, 486, 490, *508*, 573, *631*
Rodriguez, R. see Romanoff, L. P. 482, *509*
Roe, J. H., and C. A. Kuether 58, *176*
Roepke, M. L. 60, 85, 96, *176*
Rogers, E. F. see Leanza, W. 225, *396*
Rogers, M. T. see Shoolery, J. N. 377, 378, 379, *405*
Rogers, W. F. jr., and H. Jaffe 337, *403*
— and R. H. Williams 44, 51, 57, 84, *176*
Rogoff, J. M. 21, 38, 72, 83, 140, *176*
— and G. N. Stewart 515, 527, *566*
— see Stewart, G. N. 515, 525, *568*
Rohr, O. see Anliker, R. 204, *381*, 414, 415, 420, 478, *500*
Rohrmann, E. see Marker, R. E. 431, *507*
Roitman, E. see Lombardo, M. E. 443, 490, *507*
Roka, L. see Amelung, D. 463, 497, *500*
Rolf, D. see White, H. L. 537, *569*
Romani, J. D. see Albeaux-Fernet, M. 656, *656*
Romanoff, E. B., P. Hudson and G. Pincus 75, *176*, 422, *509*, 573, 574, 609, *634*
— see Pincus, G. 422, 423, *509*, 609, 610, *633*
— see Ungar, F. 442, 489, *512*

Romanoff, L. P., J. Seelye, R. Rodriguez and G. Pincus 482, *509*
— and R. S. Wolf 311, *403*
— see Pincus, G. 572, 583, *633*
Romer, A. S. 4, *176*
Romero, M. see Fieser, L. F. 234, *389*
Romo, J., and A. R. de Vivar 273, *403*
— G. Rosenkranz, C. Djerassi and F. Sondheimer 361, *403*
— G. Rosenkranz, F. Sondheimer 235, *403*
— see Djerassi, C. 281, 288, *387*
— see Kaufman, S. 279, *395*
— see Sondheimer, F. 233, 361, *405*
Ronald, A. P. see Idler, D. R. 550, *562*
Ronzoni, E. see Goldman, M. L. 112, 113, *159*
Rooks, M. E. see Cook, E. R. 297, 299, 363, *386*
Rosemberg, E., G. Rosenfeld, F. Ungar and R. I. Dorfman 446, 447, *509*
— see Bloch, E. 72, 122, 125, *148*, 571, 578, 590, *625*
— see Laqueur, G. L. 92, 104, 105, *167*
— see Lewis, R. A. 80, *168*
— see Rosenfeld, G. 76, 111, 113, *176*, *510*
Rosenbaum, J. A. see Latimer, H. B. 13, *167*
Rosenblum, C. 370, *403*
Rosenfeld, G. 19, 51, 58, 61, 62, 68, 103, 137, *176*, 456, 458, *510*
— and W. D. Bascom 76, 137, *176*, 457, *510*, 522, *566*
— E. Rosemberg, F. Ungar and R. I. Dorfman 76, 111, 113, *176*, *510*
— F. Ungar, R. I. Dorfman and G. Pincus 443, 449, 490, *510*
— see Caspi, E. 223, *385*, 421, 435, 441, *501*
— see Chen, P. S. jr. 447, *502*
— see Rosemberg, E. 446, 447, *509*
Rosenfeld, R. S. 216, *403*
— and T. H. Gallagher 233, *403*
— L. Hellman, W. J. Considine and T. F. Gallagher 436, 489, *510*
— see Eidinoff, M. L. 436, 438, *503*
— see Hellman, L. 436, 471, *505*
Rosenheim 187
Rosenkrantz 100
Rosenkrantz, H. 327, 341, 345, 347, *403*
— and R. I. Dorfman 478, *510*
— and L. Zablow 349, *403*
— see Caspi, E. 441, 461, 462, *501*
— see Forchielli, E. 462, 497, *504*
— see Hayano, M. 461, *505*
— see Meyer, A. S. 360, *398*
— see Miller, A. M. 427, *508*
— see Rubin, B. L. 312, *403*
— see Savard, K. 311, *404*, 478, 480, 482, *510*
— see Ungar, F. 460, *512*
Rosenkranz, G., O. Mancera, J. Gatica and C. Djerassi 255, 257, *403*
— — and F. Sondheimer 233, 234, *403*
— J. Pataki, S. Kaufmann, J. Berlin and C. Djerassi 255, *403*
— and F. Sondheimer 251, *403*
— see Amendolla, C. 236, *381*
— see Djerassi, C. 280, 281, 288, *387*
— see Kaufman, S. 279, *395*
— see Löken, B. 273, *397*
— see Mancera, O. *397*, 492, *507*
— see Pataki, J. 204, *400*
— see Ringold, H. J. 221, 232, 241, 272, 287, *402*
— see Romo, J. 234, 361, *403*
— see Sandoval, A. 281, 404
— see Sondheimer, F. 211, 213, 227, 233, 234, 236, 241, 246, 279, 361, *405*
— see Zaffaroni, A. 281, *409*, 494, *513*
Rosenthal, F. D. see Wood, K. F. 129, *185*
Rosenthal, O. see Cooper, D. Y. 422, 423, *502*
Rosenthal, S. see Quandt, V. V. 655, *659*
Rosnagle, R. S., and G. L. Farrell 113, *176*
— see Farrell, G. L. 74, 76, *156*
Ross, E. J. see Hernando, L. 457, *505*, 531, *561*
— see Renold, A. E. 137, *175*, 457, *509*, 522, *566*
Ross, E. K. see Perrine, J. W. 521, *565*
Rosselet, J. P., M. Furman, S. Lieberman and J. W. Jailer 432, 459, 465, *510*
Rosselett, J. P., J. W. Jailer and S. Lieberman 219, *403*, 459, 465, *510*
— L. Overland, J. W. Jailer and S. Lieberman 432, 465, *510*
— see Bitman, J. 228, 250, 359, *383*
Rossi, C. A., and L. Montanari 107, *176*
Rossi, E. see Little, B. 422, *507*
Rossi, F., G. Pescetto and E. Reale 63, 65, 66, *176*
Rossi, L. see Cavallero, C. 544, *557*
Rossi, M. 363, *403*
Rossiter, R. J., and D. Nicholls 91, *176*
— see Logan, J. E. 91, *169*
— see Nicholls, D. 91, *172*
— see Riedel, B. E. *176*
Rottenburg, M. see Bergström, M. *500*
Rotter, W. 43, 45, 121, 126, *176*, *177*, 589, *634*
Roth, P. see Dumm, M. E. 139, *154*
Routien, J. B. see Shull, G. M. 492, *511*
Rovnanek, A. see Klein, R. 125, *165*
Rowntree, L. G. 519, 528, *566*
— and A. M. Snell 542, *566*
Roy, A. B. 434, *510*
Roy, S. N., J. N. Karkun and B. Mukerji 107, *177*, 596, *634*
Royce, P. C., and G. Sayers 75, 85, *177*
— see Farrell, G. L. 100, 103, 110, 111, *156*, 423, *503*, 610, *627*
Rubin, B. A. see Mancera, O. 492, *507*
Rubin, B. L. 1, *177*, 411, 478, 495, 496, *510*
— and R. I. Dorfman 61, *177*, 496, *510*
— A. S. Dorfman, L. Black and R. I. Dorfman 617, *634*
— R. I. Dorfman and M. Miller *510*
— — and G. Pincus 250, 328, *403*
— H. Rosenkrantz, R. I. Dorfman and G. Pincus 312, *403*
— see Levy, H. 61, 63, 96, 97, 109, *168*
— see Pincus, G. 583, *633*
Rubin, M., H. Wishinsky and F. Bompard 214, *403*
— see Hershberg, E. B. 281, *392*

Rubin, M. I., and E. T. Krick 529, *566*
Rudy, H. 68, *177*
Rudney, H. 438, *510*
— and T. G. Farkas *510*
Rudolph, G. G., and N. S. Olsen 96, *177*
Rudolph, H. 374, *403*
Ruliffson, W. S., H. M. Lang and J. P. Hummel 371, *403*
Rumney, G. 484, *510*
— see Mueller, G. C. 484, 486, *508*
Rumpf, J. A. see Bergström, S. *500*
Rumpf, K. see Bergström, S. *500*
Rundle, A. see Halkerston, I. D. K. 455, *504*
Ruppel, W. see Pfeffer, K. K. *401*
Ruschig, H. *403*
Russell, P. S. see Medawar, P. B. 118, 120, *170*
Rutenburg, A. M., R. B. Cohen and A. M. Seligman 67, *177*
— R. Gofstein and A. M. Seligman 340, *403*
— and A. M. Seligman 65, *177*
Rutenburg, S. H. see Seligman, A. M. 66, *178*
Ruzicka, L., N. Wahba, P. Herzig and H. Heusser 288, *403*
— and A. Wettstein 204, *403*
— see Anliker, R. 478, *500*
Ryan, K. J. 443, 488, 490, *510*
— and L. L. Engel 62, *177*, 443, 484, 488, 490, *510*
Ryden, I. see Julian, P. L. 255, *395*
Rymer, M. see Richet, C. 639, *659*
Rynearson, E. H. see Mason, H. L. 607, *631*

Saba, N., and O. Hechter 420, 421, 439, 443, 489, *510*
— — and D. Stone 435, 439, 441, 499, *510*
— see Ayres, P. J. 423, 447, *500*
— see Hayano, M. 420, 435, 441, 454, 485, *505*
Sabbadin, A. see Vannini, E. 2, 9, *182*
Sabo, E. F. see Fried, J. 221, 250, 265, 267, 268, 272, 274, 276, 277, 278, 361, *390*, 491, *504*
Sabo, E. F. see Herz, J. 267, 270, 271, *392*
— see Djerassi, C. 487, 492, 493, *502*
Sadownikow, W. 133, *177*
Saffran, M., and M. J. Bayliss 453, *510*
— B. Grad and M. J. Bayliss 77, *177*
— and A. V. Schally 77, *177*
— see Bligh, E. G. 449, 451, *500*
— see Giroud, C. J. P. 77, 111, 113, *158*, 447, *504*
— see Hokin, M. R. 92, *162*
— see Schönbaum, E. 114, *178*, 454, *511*
Saier, E., M. Warga and R. C. Grauber 338, 403
— see Wilson, D. 297, *409*
Saika, A. see Meyer, L. H. 377, *398*
Saito, A. see Bloom, B. M. 495, *500*
— see Hayano, M. 420, 435, 441, 485, 491, 494, *505*
Sakal, E. H., and E. J. Merrill 312, *403*
Sakiz, E. 47, 81, 127, 128, 177, 596, 619, *634*
Salamon, I. I., and K. Dobriner 429, *510*
— see Kemp, A. D. 228, *395*, 428, 460, *506*
— see Miller, W. R. 312, *399*
Salgado, E. see Selye, H. 133, *179*
Salhanick, H. A., and D. L. Berliner 412, 420, 451, *510*, 611, *634*
— see Berliner, D. L. 372, *383*, 422, *500*
Salmon, A. A. see Salmon, U. J. 611, *634*
Salmon, U. J., S. H. Geist and A. A. Salmon 611, *634*
Salter, W. T. see Cahen, R. L. 363, *385*
Saltzman, A. H. see Talbot, N. B. 296, 302, 304, 339, *407*
Salvadori, B., U. Leone and I. Cassang 423, *510*
Salzgeber, B. see Wolff, E. 605, *636*
Sammartino, R. see Del Castillo, E. B. 106, *153*
— see Houssay, B. A. 46, 78, 79, 99, 109, *163*
Sampson, L. P. see Hardy, J. D. 75, *160*
Samuels, B. K. see Reich, H. 221, 225, *402*
Samuels, L. T. 439, *510*, 570, 573, *634*
— H. Brown, K. Eik-Nes, F. H. Tyler and O. V. Dominguez 475, *510*
— and M. L. Helmreich 96, *177*, 496, *510*
— — M. B. Lasater and H. Reich 61, *177*, 496, *510*
— see Beyer, K. F. 61, *148*, 496, *500*
— see Bliss, E. L. 642, *657*
— see Brown, H. 471, *501*
— see Eik-Nes, K. 82, *155*, 312, *388*, *503*
— see Gassner, F. X. 573, *628*
— see Koenig, V. L. 337, *396*
— see Migeon, C. J. 572, *632*
— see Nelson, D. H. 296, 300, 312, 366, *399*
— see Plager, J. E. 443, *509*
— see Sandberg, A. A. 475, *510*
— see Slaunwhite, W. R. jr. 610, *635*
— see Tyler, F. H. 475, *512*
Sandberg, A. A., K. Eik-Nes, L. T. Samuels and F. H. Tyler 475, *510*
— D. H. Nelson, E. M. Glenn, F. H. Tyler and L. T. Samuels 475, *510*
— and W. R. Slaunwhite 106, *177*, 480, *510*
— — and H. N. Antoniades 292, 403, *510*
— see Bliss, E. L. 572, *625*
— see Bush, I. E. 311, *385*, 609, *625*
— see Eik-nes, K. 82, *155*
— see Slaunwhite, W. R. jr. 484, *511*
Sandermann, W., and H. Stockmann 510
Sandor, T. see Nowaczynski, W. *400*
Sandorfy, C. see Jones, R. N. 348, *394*
Sandoval, A., G. H. Thomas, C. Djerassi, G. Rosenkranz and F. Sondheimer 281, 404
Sandritter, W. 92, 94, *177*
— and F. Hübotter 90, 94, *177*
Sanfilippo, S. J. see Reich, H. 310, *402*
Santa, N. see Gradinesco, A. 515, *560*
Santisteban, G. A., and C. A. Swinyard 105, *177*
Sanz, M. C. see Riondel, A. M. 366, *402*

Sapirstein, L. A., and H. Goldman 73, *177*
Sarason, E. L. 46, 69, 100, 101, 109, 115, *177*
Sardinas, J. L. see Shull, G. M. 310, 342, *405*, 492, 493, *511*
Sarter, J. 26, *177*
Sarett, L. H. 204, 210, 214, 222, 225, 242, 256, 263, 374, *404*
— G. E. Arth, R. M. Lukes, R. E. Beyler, G. I. Poos, W. F. Johns and J. M. Constantin 262, *404*
— see Arth, G. E. 285, *382*
— see Beyler, R. E. 225, *383*
— see Constantin, J. M. 269, *386*
— see Ham, E. A. *391*
— see Harman, R. E. 318, *391*, 421, 446, *504*
— see Hirschmann, R. F. 267, 277, *393*
— see Poos, G. I. 242, 265, *401*
Sarro, S. D. see West, C. D. 610, *636*
Sato, Y., and W. A. Jacobs 238, *404*
Saunders, F. J. see Picha, G. M. 461, *509*
Savard, K. 319, 320, 327, 328, *404*, 422, 423, *510*
— K. Andreč, B. W. L. Brooksbank, C. Reyneri and R. I. Dorfman 431, 451, *510*
— E. Bagnoli and R. I. Dorfman 434, *510*
— S. Burstein, H. Rosenkrantz and R. I. Dorfman 311, *404*, 478, 480, 482, *510*
— and R. I. Dorfman 460, 478, *510*
— — B. Baggett and L. L. Engel 413, 488, *510*, 579, *634*
— — — L. L. Engel, L. M. Lister and F. L. Engel 413, 488, *510*
— and S. H. Goldfaden 460, *510*
— A. A. Green and L. A. Lewis 443, 489, *510*
— W. Kolff and A. C. Corcoran 424, *510*
— see Baggett, B. 412, 451, *500*, 616, *624*
— see Burstein, S. 459, 472, *501*
— see Dorfman, R. I. *503*
— see Haynes, R. C. jr. 77, *161*, 442, *505*
Sawin, P. B. see Latimer, H. B. 18, *167*
Sawyer, W. H. 550, 551, *566*
— D. F. Travis and N. G. Levinsky *566*
— see Levinsky, N. G. 143, *168*
Sayers, G. 59, 82, 83, 84, 85, 93, 103, *177*
— see Royce, P. C. 75, 85, *177*
— and M. A. Sayers 82, 83, 84, 85, 93, 99, 101, 110, *177*
— — E. G. Fry, A. White and C. N. H. Long 83, *177*
— — T. Y. Liang and C. N. H. Long 51, 58, 84, 85, *177*, 454, *510*
— see Sydnor, K. L. 585, *635*
— see Woodbury, D. M. 529, 531, 534, *569*
Sayers, M. A. see Sayers, G. 51, 58, 82, 83, 84, 85, 93, 99, 101, 110, *177*, 454, *510*
Scammon, R. E. 591, *634*
Schaberg, A. 46, 47, 118, 121, 130, *177*, *178*
— and C. A. de Groot 121, *178*
— and M. Straub 69, *178*
Schacter, B., and G. F. Marrian 434, *510*
Schaeffer, W. H. 79, 143, *178*, 552, *566*
Schaffner, C. P. see Greenspan, G. 260, *391*
— see Herzog, H. L. 260, *392*
Schairer, M. A. see Baker, B. L. 608, *624*
Schally, A. V. see Elliott, F. H. 420, 421, *503*
— see Giroud, C. J. P. 77, 111, 113, *158*, 447, *504*
— see Hokin, M. R. 92, *162*
— see Saffran, M. 77, *177*
Schamaun, M., F. Deucher and S. Gablinger 652, *659*
Schapiro, S., and L. Stjarne 73, *178*
Schatzmann, H. J. 539, *566*
Schaumkell, K. W. 45, *178*
— H. H. Stange and P. Dörffler 43, 44, *178*
Schechter, N. R. see Cohen, H. 82, 83, *151*
Schedl, H. P., W. B. Bean, B. M. Stevenson and E. R. Schumacher 363, *404*
— see Chen, P. S. jr. 447, *502*
Scheibel, E. G. see Craig, L. C. 296, 304, *386*
Schellmann, J. A. see Eik-Nes, K. *503*
Schemm, F. R., and A. A. Camara 545, *566*
— see Heidorn, G. H. 546, *561*
Schenker, V. see Hechter, O. 311, *392*, 441, 442, 486, 489, 490, *505*, 517, *561*, 609, *629*
— see Jeanloz, R. W. 449, 486, 489, *506*, 573, *630*
— see Levy, H. 441, 442, 489, 490, *506*
Scheri, M. A. see Levin, R. H. 227, *396*
Scheuer, J. see Bondy, P. 367, *384*
Schiedt, U., and H. Restle 350, *404*
Schild, W. 366, *404*
Schiller, J. see Weinstein, M. J. 623, *636*
Schiller, S., A. M. Miller, R. I. Dorfman, E. L. Sevringhaus and E. P. McCullagh 427, 431, *510*, 572, *634*
Schindler, O. see Lardon, A. 196, *396*
— see Simpson, S. A. 198, 204, 238, 240, 317, 354, *405*, 421, 446, 511, 520, *567*
Schindler, W. J. 83, 100, *178*
— and K. M. Knigge 73, 74, 75, 77, *178*
Schlaug, R. see Dencker, S. J. 641, 643, *657*
Schmeiser, K. 367, *404*
Schmeisser, M. see Staudinger, H. 340, 364, *406*
Schmid, O. J., and K. D. Voigt 304, *404*
Schmidlin, J., G. Anner, J. R. Billeter, K. Heusler, H. Ueberwasser, P. Wieland and A. Wettstein 238, 264, *404*
— — — and A. Wettstein 262, *404*, 520, *566*
— see Miescher, K. 256, *399*
— Schmidlin, J. see Vischer, E. 264, *408*, 494, *512*
— see Wettstein, A. 206, *408*
Schmidt, C. D. see Tyler, F. H. 475, *512*
Schmidt, E. C. N. jr. see Starkey, W. F. 596, *635*
Schmidt, H., and H. Staudinger 333, *404*
— see Gross, F. 533, *560*
Schmidt, I. G. 21, *178*
— and R. A. Hoffmann 101, 122, 123, 124, *178*

Schmidt, I. G., and L. H. Schmidt 22, 104, *178*
Schmidt, L. H. see Schmidt, I. G. 32, 104, *178*
Schmidt, P. J. see Idler, D. R. 550, *562*
Schmidt-Thomé, J. 215, *404*
— see Oertel, G. 291, *400*
— see Reuber, R. 186, *402*
Schmit, R. W. see Bahn, R. C. 3, 29, 79, 88, 109, 110, *146*
Schmulovitz, M. J. see Wylie, H. B. 297, *409*
Schneider, B. M. see Brown, J. B. 49, 51, 54, *149*
— see Knouff, R. A. 49, 51, 54, 92, 94, *166*
Schneider, J. J. 432, 433, 499, *510*
— and M. L. Lewbart 452, 478, *510*
— and P. M. Horstmann 461, 510, 610, *634*
— M. Lewbart, P. Levitan and S. Lieberman 460, *511*
— and H. L. Mason 478, 486, *511*, 572, *634*
— see De Courcy, D. 461, 497, *502*
— see Lewbart, M. L. 334, *396*
— see Mason, H. L. 427, *507*
— see Nadel, E. M. 71, 105, *172*, 454, *508*
Schneider, R. see Dobriner, K. 420, 421, *502*
— see Fukushima, D. K. 431, 460, *504*
Schneider, W. P. see Hogg, J. A. 208, 256, 257, 267, 272, 277, 284, 361, *393*
— see Spero, G. B. 267, 271, 283, *406*
Schneiweiss, R. H. see Deane, H. W. 114, *153*
Schneppenheim, P., and A. Huhn 82, 88, 104, 133, *178*
— see Huhn, A. 104, 130, 133, *163*
Schnieden, H. see Heller, H. 547, *561*
Schoenheimer, R. see Keston, A. S. 368, *395*
— see Rittenberg, D. 436, *509*
Schöffling, K. see Arndt, T. 642, *656*
Schon, M., and A. M. Sutherland 653, *659*
Schönbaum, E., M. K. Birmingham and M. Saffran 114, *178*, 454, *511*
— and W. G. B. Casselman 93, *178*
Schönbaum, E. M. see Bligh, E. G. 449, 451, *500*
Scholler, R. see Lelong, M. 582, *631*
Schooley, J. P. 549, *566*
Schotté, O. E., and W. B. Christiansen 523, *567*
Schottstaedt, W. W. see Hetzel, B. S. 642, *658*
Schramm, B. see Verne, J. 57, 133, *183*
Schreiber, J., and A. Eschenmoser 228, 242, *404*
Schreiner, L. H. see Laqueur, G. L. 92, 104, 105, *167*
Schricker, F. G. see Hertz, R. 137, *161*
Schricker, J. A., G. R. Fisher, W. W. Tullner and R. Hertz 137, *178*
— see Hertz, R. 522, *561*
Schriefers, H. see Dirscherl, W.. 314, *387*
Schroeder, F. F., and J. L. Sherman 642, *659*
Schroeder, H. A. see Goldman, M. L. 112, 113, *159*
Schroeder, W., and K. Bohle 21, *178*
— K. D. Voigt and I. Beckmann *404*
— and K. D. Voigt 203, *404*
— — H. van der Werth and I. Beckmann 418, *511*
— see Voigt, K. D. 303, *408*, 418, *512*
Schubert, A., G. Langbein and R. Siebert 260, *404*
Schuler, W., P. Desaulles and R. Meier 524, *567*
— see Desaulles, P. A. 532, 533, *558*
Schultz, A., and G. Löhr 53, *178*
Schultz, R. B., R. W. Winters and W. A. Krahl 138, 139, *178*
— see Winters, R. 105, 139, *184*, 458, *513*
Schulz, E. P., and J. D. Neuss 338, *404*
Schumacher, E. R. see Schedl, H. P. 363 404
Schwartz, E. L. see Earle, D. P. jr. 99, *155*
Schwartz, N. B., and A. Kling 18, 58, 85, *178*
Schwartz, W. B. see Relman, A. S. 531, *566*
Schwarz, V. 313, 342, *404*
Schweizer, M., and M. E. Long 78, 79, 100, 109, 110, *178*
Schwenk, E., G. J. Alexander, C. A. Fish and T. H. Stoudt 438, *511*
— — T. H. Stoudt and C. A. Fish 430, 438, *511*
— M. Gut and J. Belisle 233, *404*
— and N. T. Werthessen 250, *404*
— see Alexander, G. J. 438, *499*
— see Hershberg, E. B. 281, *392*
— see Wintersteiner, O. 426, *513*
Schwöbel, G. 638, 639, 647, *659*
Scian, L. F. see Hilton, J. G. 76, *162*
Scism, R. G. see Drekter, I. J. 302, *388*
Scott, P. P. see Harris, G. W. 574, *628*
Scott, R. B. see Djerassi, C. 336, *388*
Scott, W. W., and C. Vermeulen 581, *634*
— see Seckel, H. P. G. 650, *659*
Searcy, A. M. see Levin, R. H. 227, *396*
Sebek, O. K. see Hayano, M. 485, *505*
— see Hogg, J. A. 272, *393*
— see Spero, G. B. 267, 283, *406*
Sebrell, W. H. jr., and R. S. Harris 139, *178*
Seckel, H. P. G. 650, *659*
— W. W. Scott and E. P. Benditt 650, *659*
Seegal, B. C. see Knowlton, A. I. 115, *166*, 531, *562*
Seelye, J. see Romanoff, L. P. 482, *509*
Segal, V. see Simonnet, H. 50, 107, *179*
Selby, C. C. 3, *178*
Seldin, D. W., L. G. Welt and J. H. Cort 532, *567*
— see Welt, L. G. 530, *569*
Seligman, A. M., and L. M. Manheimer 65, *178*
— K. C. Tsou, S. H. Rutenburg and R. B. Cohen 66, *178*
— see Deane, H. W. 55, 63, 64, 66, 95, *153*
— see Nachlas, M. M. 63, 66, *171*, *172*
— see Rutenburg, A. M. 65, 67, *177*, 340, *403*
Seligman, I. J. see Costa, P. J. 138, *152*

Sellers, A. F. see Weber, A. F. 115, 135, *184*
Sellers, E. A. see Kalant, O. J. 107, *164*
Selye, H. 69, 100, 103, 114, *178*, 642, 653, *659*
— and J. B. Collip 100, 106, *179*, 455, *511*
— — and D. L. Thomson 455, *511*
— and E. Salgado 133, *179*
— and H. Stone 129, 130, 133, *179*, 531, *567*
— see Bois, P. 68, 83, *148*
— see Fortier, C. 50, 58, 59, *157*
— see Kutz, R. L. 621, *630*
— see Parkes, A. S. 554, *565*
— see Skelton, F. R. 107, *179*
Sensenbach, W., L. Madison and L. Ochs 643, *659*
Serchi, G. 426, *511*
Sevringhaus, E. L. see Miller, A. M. 428, *508*, 577, *632*
— see Schiller, S. 427, 431, *510*, 572, *634*
Sewekow, G. W. see Robinson, H. J. 106, *176*
Sewell, C. E. see Cope, C. L. 422, *502*
Sexton, A. W. 549, *567*
Shank, R. E. see Eisenstein, A. B. 71, 105, *155*
Shapiro, E. L., D. Gould and E. B. Hershberg 214, *405*
— see Gould, D. 267, 278, 279, *391*
— see Herzog, H. L. 267, 277, *392*
— see Oliveto, E. P. 243, *400*
Sharp, J. T. see Sokoloff, L. 20, 51, 80, *180*
Shaw, J. H. see Deane, H. W. 92, 93, 94, 112, 113, 114, 115, *153*
Shealy, Y. F. see Slomp, G. 357, *405*
Shelden, F. F. see Hartman, F. A. 5, *161*, 548, *561*
Sheehan, H. L., V. K. Summers and J. Nichols 458, *511*
— see Nichols, J. 135, 136, *173*, 457, 508, 523, *565*
Shemano, I., G. S. Gordon and E. Eisenberg 576, *634*
Shen, N. C. see Francis, F. E. 591, *628*
Shepard, T. H., B. H. Landing and D. G. Mason 141, *179*
— see Migeon, C. J. 572, 581, 582, 583, 584, 591, 592, *632*
Sheppard, H., T. F. Mowles, W. L. Bencze, M. J. Allen, J. J. Chart and R. Gaunt 523, *567*
— see Chart, J. J. 457, *502*, 522, 523, *557*
Sheppard, H. H. see Tomkins, G. M. 455, *512*
Sherman, F. E., L. W. Bass and G. H. Fetterman 129, *179*
Sherman, J. L. see Schroeder, F. F. 642, *659*
Shermin, A. T. see Norymberski, J. K. 366, *399*
Shipley, E. G. see Hershberger, L. G. 575, 576, *629*
Shipley, E. S. see Chart, J. J. 520, *557*
— see Gordon, E. S. 520, *560*
Shipley, R. A. see Dorfman, R. I. 478, *503*, 575, 577, *627*
— see Horwitt, B. N. 459, *505*, 610, *629*
Shoger, R. L. see Lazo-Wasem, E. A. 610, *630*
Shoolery, J. N. 377, *405*
— and M. T. Rogers 377, 378, 379, *405*
Shoppee, C. W. 218, 251, 374, *405*
— and G. R. Summers 232, *405*
— see Casanova, R. *385*
— see Evans, D. D. *389*
— see Reichstein, T. 197, 198, 307, 308, 336, 337, *402*
Shorr, E. see Zweifach, B. W. 63, *185*
Short, R. V. 423, *511*, 609, 613, 615, *635*
— see Balfour, W. W. 75, *147*
Shricker, J. A. see Hertz, R. 456, *505*
Shriefers, H., W. Korus and W. Dirscherl 476, *511*
Shull, G. M. 258, 328, *405*
— and D. A. Kita 493, *511*
— J. L. Sardinas and J. W. Davisson 493, *511*
— — and R. C. Nubel 310, 342, *405*
— — and J. B. Routien 492, *511*
— see Bloom, B. M. *383*, 495, *500*
Siebenmann, R. E. 54, 101, 135, 138, *179*
— see Prader, A. 138, *175*
Siebert, R. see Schubert, A. 260, *404*
Sideman, M. see Li, C. H. 82, 83, 100, 101, 103, 117, *168*, 608, 613, 614, *631*
— see Mason, J. W. 105, *170*
Silber, R. H. 475, *511*
— and R. D. Busch 341, 366, *405*
— and C. C. Porter 341, 366, *405*
— see Arth, G. E. 285, *382*
— see Halberg, F. 32, 93, 96, *160*
— see Mushett, C. W. 535, *564*
— see Porter, C. C. 340, *401*
— see Robinson, H. J. 106, *176*
— see Winter, C. A. 80, 84, 86, *184*
Silberberg, M. see Silberberg, R. 134, *179*
Silberberg, R., M. Silberberg and M. Opdyke 134, *179*
Silfverskiöld, W. see Dencker, S. J. 641, 643, *657*
Silverman, L. M. see Lanman, J. T. 42, 72, *167*
Silverman, S. H. see Crigler jr. J. F. 521, *558*
— see Wilkins, L. 585, 616, *636*
Silvette, H., and S. W. Britton 526, 542, *567*
— see Britton, S. W. 528, *556*
Simmer, H. see Zander, J. 613, *637*
Simmonds, M. 78, *179*
Simonnet, H., L. Thieblot and V. Segal *50*, 107, *179*
Simpson, G. A. see Venning, E. H. *520*, 568
Simpson, G. G. 4, *179*
Simpson, M. E., H. M. Evans and C. H. Li 87, 94, 100, 101, *179*
— see Walker, D. G. 78, *183*
— see Zizine, L. 107, *185*
Simpson, S. A., and J. F. Tait 272, 333, 367, 372, *405*, 518, 520, 531, 541, *567*
— — and I. E. Bush 423, *511*
— — A. Wettstein, R. Neher, J. v. Euw and T. Reichstein 198, *405*, 520, *567*
— — — R. Neher, J. v. Euw, O. Schindler and T. Reichstein 198, 204, 238, 240, 317, 354, *405*, 421, 446, *511*, 520, *567*
— see Avivi, P. 371, *382*
— see Ayres, P. J. 109, *146*, 313, 318, 367, 372, *382*, 423, 447, 451, *500*
— see Garrod, O. 115, 116, *157*
— see Tait, J. F. 524, *568*

Simson, J. see Ziff, M. 538, *569*
Singer, B. 113, 114, *179*, 532, *567*
— and M. P. Stack-Dunne 73, 74, 75, 77, 100, 103, 110, 111, 113, 115, 116, *179*
— and E. H. Venning 520, *567*
— and J. Wener 520, *567*
— see Venning, E. H. 520, *568*
Singer, F. see Fried, J. 267, 268, 272, 277, 278, *390*
Singleton, E. see Gold, N. I. 475, *504*
Siperstein, M. D., and I. L. Chaikoff *511*
Sirek, O. V., and C. H. Best 535, *567*
Sjögren, B. 531, *567*
— see Luft, R. 529, 531, 537, 538, *563*
Sjöstrand, T. 86, *179*
Sjövall, J. see Bergström, S. *500*
Skelton, F. R. 69, 70, 120, *179*, 548, *567*
— C. Fortier and H. Selye 107, *179*
— see Constantinides, P. 69, 70, *152*
— see Fortier, C. 50, 58, 59, *157*
Skowron, S., and S. Zajaczek 21, *179*
Slater, G. G. 69, *179*
— see Beigelman, P. M. 74, 75, *147*
Slater, J. J. see Allen, J. M. 64, 65, 96, *145*
Slates, H. L. see Taub. D. 285, *407*
— see Wendler, N. L. 288, *408*
Slaunwhite, W. R. jr., and L. Neely 350, *405*
— and L. T. Samuels 610, *635*
— and A. A. Sandberg 484, *511*
— see Chang, E. 363, *385*
— see Engel, L. L. 426, 451, *503*
— see Sandberg, A. A. 106, *177*, 292, *403*, 480, *510*
Slessor, A. 542, *567*
Slomp, G., Y. F. Shealy, J. L. Johnson, R. A. Donia, B. A. Johnson, R. P. Hohysz, R. L. Pederson, A. O. Jensen and A. C. Ott 357, *405*
Sluiter, J. W., J. C. A. Mighorst and G. J. van Oordt 79, 143, *179*
Slusher, M. A. 86, *179*
— and S. Roberts 60, 85, *179*, 454, *511*
— see Beigelman, P. M. 74, 75, *147*
Small, A. M. see Kissman, H. M. 272, *395*
Smith, D. C. see Tompsett, S. L. 295, 296, *407*
Smith, D. C. W. 549, *567*
Smith, D. E. see Hartman, F. A. 526, *561*
Smith, E. E. B., and G. T. Mills 452, *511*
— A. Munch-Petersen and G. T. Mills 452, *511*
— see Mills, G. T. 452, *508*
Smith, E. K. 621, *635*
Smith, E. M., M. L. Calhoun and E. P. Reineke 104, *179*
Smith, E. R. see Kellie, A. E. 318, 338, *395*
Smith, G. C. see Riddle, O. 554, *566*
Smith, G. N. see McGinty, D. A. 443, 489, *507*
Smith, G. V. S., O. W. Smith, M. N. Huffman, D. W. MacCorquodale, S. A. Thayer and E. A. Doisy 426, *511*
Smith, H. 514, *567*
— and H. Staudinger 454, *511*
Smith, H. Q. see Oliveto, E. P. 216, 228, 272, *400*
Smith, I. P. see Smith, P. F. 142, 144, *180*
Smith, J. D. see Guillemin, R. 77, 85, 100, 103, *160*
Smith, L. H. see Howard, J. M. 642, *658*
Smith, M. see Coppage, W. S. 529, *558*
Smith, O. W. see Smith, G. V. S. 426, *511*
Smith, P. E. 18, 78, 79, 100, 106, *179*, *180*
— and I. P. Smith 142, 144, *180*
Smith, R. B. jr. see Larson, P. S. 135, 136, *167*, 458, *506*
Smith, R. B. III see Brown, J. H. U. 58, 108, 136, *149*, 501, 523, *556*
Smollich, A. 17, *180*
Smyles, E. see Lieberman, S. 293, 297, *397*
Snaith, A. H. 134, *180*
Sneeden, R. P. A., and R. B. Turner 282, *405*
Snell, A. M. see Rowntree, L. G. 542, *566*
Snell, K. C., and H. L. Stewart 101, 129, 135, *180*
Snoddy C. S. jr., see Wendler, N. L. 279, *408*
Snyder, J. G., and L. C. Wyman 526, *567*
Sobel, E. H., C. M. Lee jr., V. M. Esselborn and L. C. Clark jr. 129, 135, *180*, 593, *635*
— A. E. Renold, J. E. Bethune, J. J. Hoet, W. J. Reddy and G. W. Thorn 135, 137, *180*
— see Talbot, N. B. 574, *635*
Sobel, H. see Heard, R. D. H. 296, 339, 340, 364, *391*
Sobotka, H. H. see Soffer, L. J. 531, *567*
Soffer, L. J. 412, *511*
— A. Gutman, J. Geller and J. L. Gabrilove 531, 542, 543, *567*
— G. Lesnick, S. Z. Sorkin, H. H. Sobotka and M. Jacobs 531, *567*
— see Harrop, G. 525, 528, 529, 530, 535, *560*, *561*
Sokoloff, L., J. T. Sharp and E. H. Kaufman 20, 51, 80, *180*
Sollman, P. see Dodson, R. M. 288, *388*
Solomon, A. K. see Deane, H. W. 117, 118, *153*
— see Streeten, D. H. P. 539, *568*
Solomon, S., J. T. Lanman, J. Lind and S. Lieberman 72, 126, *180*, 574, 590, *635*
— A. L. Lenz, R. Vande Wiele and S. Lieberman 435, 439, 441, *511*
— P. Levitan and S. Lieberman 421, 435, 439, *511*
— R. Vande Wiele and S. Lieberman 574, *635*
Soloway, A. H., W. J. Considine, D. K. Fukushima and T. F. Gallagher 218, 225, *405*
Sonderhoff, R., and H. Thomas 436, *511*
Sondheimer, F., C. Amendolla and G. Rosenkranz 279, *405*
— S. Kaufman, J. Romo, H. Martinez and G. Rosenkranz 233, 361, *405*

Sondheimer, F., O. Mancera, G. Rosenkranz 234, 236, 241, 361, *405*
— — — and C. Djerasssi 227, 246, *405*
— — M. Urquiza and G. Rosenkranz 213, *406*
— and Y. Mazur 287, *406*
— G. Rosenkranz, O. Mancera and C. Djerassi 211, *406*
— M. Velasco, E. Batres and G. Rosenkranz *405*
— see Amendolla, C. 236, *381*
— see Djerassi, C. 280, *387*
— see Löken, B. 273, *397*
— see Mancera, O. *397*, 492, *507*
— see Ringold, H. J. 221, 232, 241, 287, *402*
— see Romo, J. 235, 361, *403*
— see Rosenkranz, G. 233, 234, 251, *403*
— see Sandoval, A. 281, *404*
— see Woodward, R. B. 261, *409*
— see Zaffaroni, A. 281, *409*
Sonenberg, M., A. S. Keston and W. L. Money 82, *180*
Sorkin, M., and T. Reichstein 206, *406*
Sorkin, S. Z. see Soffer, L. J. 531, *567*
Sosa, J. M. 59, *180*
Soulairac, A. P. Desclaux and J. Teysseyre 98, *180*
Sourkes, T. L., and P. Heneage 62, *180*
Southcott, C. M., H. E. Bandy, S. E. Newsom and M. Darrach 423, 425, *511*
— S. K. Gandossi, A. D. Barker, H. E. Bandy, H. McIntosh and M. Darrach 423, 425, *511*
Spaethe, M. see Kracht, J. 3, 88, 107, 140, *166*, *167*
Spahr, A. see Prader, A. 521, *565*
Spalding 549
Spaulding, M. H. 578, *635*
Spector, W. S. 9, 12, 18, 19, 20, *180*
Spencer, H. see Gallagher, T. F. 456, *504*, 586, *628*
Spero, G. B., A. McIntosh jr., and R. H. Levin 247, *406*
— J. L. Thompson, F. H. Lincoln, W. P. Schneider and J. A. Hogg 267, 271, 283, *406*
— — B. J. Magerlein, A. R. Hanze, H. C. Murray, O. K. Sebek and J. A. Hogg 267, 283, *406*
Spero, G. B., see Hogg, J. A. 272, *393*
— see Meister, P. D. 492, *508*
Sperry, W. M. 293, *406*
— and M. Webb 50, *180*
Spiegel, A. 134, *180*, 605, *635*
Spiers, R. S. 21, *180*
Spitzer, J. R. see Levitz, M. 484, *506*
Spooncer, W. W. see Arth, G. E. 285, *382*
Spoor, H. J. see Hartman, F. A. 519, *561*
Sprague, A. see Klendshoj, N. C. 363, *395*
Sprechler, M. 364, *406*
— see Warming-Larsen, A. 541, *569*
Sprenkel, H. B. van der 4, *180*
Sproul, E. E. see Fiala, S. 90, 91, *156*
Stachenko, J., and C. J. P. Giroud 109, 111, *180*, 517, *567*
— see Giroud, C. J. P. 77, 109, 111, 113, *158*, 447, 451, *504*
Stack-Dunne, M. P. see Cater, D. B. 32, 68, 79, 83, 90, 94, 100, 109, *150*
— see Singer, B. 73, 74, 75, 76, 100, 103, 110, 111, 113, 115, 116, *179*
Staemmler, H. J. see Kloos, K. 54, 122, 123, *165*
Stafford, J. E. see Johnson, J. L. 245, *394*
Stafford, R. O., L. E. Barnes, B. J. Bowman and M. M. Meinzinger 521, *567*
— see Byrnes, W. W. 521, *557*
— see Dulin, W. E. 267, 284, *388*
Stahl, J., D. W. Atchley and R. F. Loeb 535, *567*
Stange, H. H. see Schaumkell, K. W. 43/44, *178*
Staple, E., W. S. Lynn jr. and S. Gurin 420, 435, 439, 441, *511*
— see Lynn, W. S. jr. 439, 499, *507*
Stapley, E. O. see Dulaney, E. L. 492, *503*
Starkey, W. F., and E. C. N. Schmidt jr. 596, *635*
Stary, Z. 68, *180*
Stauber, L. A. see Leathem, J. H. 96, *168*
Staudinger, H., and V. Bauer 296, *406*
— and M. Schmeisser 340, 364, *406*
Staudinger, H. see Hofmann, H. 342, 346, *393*
— see Johannes, A. 58, 85, *164*
— see Krisch, K. 63, 85, *167*
— see Schmidt, H. 333, *404*, 454, *511*
— see Weissbecker, L. 304, *408*
Staudinger, Hj. see Pfeffer, K. H. *401*
Stebbins, R. B. 82, 139, *180*
— see Winter, C. A. *184*
Stecklein, H. R. see Glick, D. 50, 58, 59, 67, 68, 98, *159*
Steenburg, R. W., R. Lenniham and F. D. Moore 475, *511*
Steiger, M., and T. Reichstein 219, 220, *406*, 420, *511*, 517, *567*
Stein, K. E. see Clinton, M. jr. 537, *557*
Stein, L., and E. Wertheimer, 540, *568*
Steinach, E., and H. Kun 616, *635*
— — and O. Peczenik 616, *635*
Steinberg, D., and D. S. Fredrickson 455, *511*
— — and J. Avigan 140, *180*, 455, *511*
— and S. Udenfriend 370, *406*
— see Fredrickson, D. S. 140, *157*, 455, *504*
Steinberg, N. G., R. Hirschmann and J. M. Chemerda 287, *406*
Stempfel, R. S. see Migeon, C. J. 471, *508*, 584, *632*
Stenger, A. H., and H. A. Charipper 10, *180*
Sterba, G. 5, *180*
Sterling, R. E., and B. B. Longwell 51, 69, 70, *180*
Stern, S. see Drekter, I. J. 302, *388*
Stewart, G. N. 515, 516, 526, *568*
— and J. M. Rogoff 515, 525, *568*
— see Rogoff, J. M. 515, 527, *566*
Stewart, H. L. see Snell, K. C. 101, 129, 135, *180*
Stevens, T. see Ehrenstein, M. 234, 282, *388*
Stevenson, B. M. see Schedl, H. P. 363, *404*
Steyermark, P. see Nowaczynski, W. *400*
Stidworthy, G. see Kochakian, D. C. 327, 342, *396*

Stieve, H. 589, *635*
Stigliani, R., and P. Marchi 87, 131, *181*
Stilling, H. 9, *181*
Stimmel, B. F. 295, 365, *406*, 484, *511*
— see Landau, R. L. 615, *630*
Stimpson, M. M., and M. J. O'Donnell 350, *406*
Stirling, G. A. 19, 69, *181*
— and V. J. Keating 19, 69, *181*
Stitch, S. R., and I. D. Halkerston 295, *406*, 434, *511*
— — and J. Hillman 295, *406*
— see Cook, E. R. 314, *386*
— see Jones, J. K. N. 314, *394*
Stjarne, L. see Schapiro, S. 73, *178*
Stockmann, H. see Sandermann, W. *510*
Stoerk, H. C. see Arth, G. E. 285, *382*
— see Knowlton, A. I. 115, *166*, 531, 547, *562*
— see Laragh, J. H. 114, *167*
— see Winter, C. A. 80, 84, 86, *184*
Stöhr, P. jr. 26, *181*
Stokem, M. B. see Fukushima, D. K. 225, *390*, 431, 433, 459, 460, 483, *504*
Stokes, W. M., W. A. Fish and F. C. Hickey 310, *406*, *511*
Stolar, S. M. see Bernstein, S. 267, 272, 284, *383*
Stoll, R. see Maraud, R. *170*
Stoll, W. A. 638, 644, 653, *659*
Stone, D., M. Hayano, R. I. Dorfman, O. Hechter, C. R. Robinson and C. Djerassi 491, *511*
— and O. Hechter 446, *511*
— see Bloom, B. M. 495, *500*
— see Hayano, M. 420, 435, 441, 485, 491, 494, *505*
— see Saba, N. 435, 439, 441, 499, *510*
Stone, H. see Selye, H. 129, 130, 133, *179*, 531, *567*
Stone, S. see Witter, R. F. 346, *409*
Stoner, H. B., J. N. P. Davies, H. J. Whiteley, and J. L. Emery 125, *181*
— and H. J. Whiteley 80, *181*
— — and J. L. Emery 49, 104, 110, 125, *181*
— see Whiteley, H. J. 106, *184*
Storey, I. D. E., and G. J. Dutton 452, *511*
— see Dutton, G. J. 452, *503*
Storrie, V. M. 655, *659*
Storino, H. E. see Bahn, R. C. 3, 29, 79, 88, 109, 110, *146*
Stork, G. see Ringold, H. J. 258, *402*
Stoudt, T. H. see Schwenk, E. 436, 438, *511*
Straub, M. see Schaberg, A. 69, *178*
Strauch, L. see Andreč, K. 306, *381*
Strauss, M. see Harrop, G. H. 525, 530, *561*
Streeten, D. H. P., and A. K. Solomon 539, *568*
— see Thorn, G. W. 518, 540, *568*
Strickler, H. S. see Mason, H. L. 431, *507*
Strong, J. A., J. B. Brown, J. Bruce, M. Douglas, A. J. Klopper and J. A. Loraine 412, *511*
— see Klopper, A. 463, *506*, 610, 611, *630*
Stroud, S. W. 484, *511*
— see Backer, C. P. 313, *382*
Stuart-Webb, I. A. see Adams W. J. 218, *381*
Stubbs, R. D. see Appleby, J. I. 341, *382*
— see Norymberski, J. K. 216, 365, *399*
Stuhl see Cloche, R. 655, *657*
Sturgis, S. H. see Velardo, J. T. 620, *636*
Sturm, A., and W. Wernitz 646, *659*
Sturtevant, F. M. 100, 107, *181*
Stutinsky, F. 133, *181*
Stutte, H. 650, *659*
Sulkowitch, H. see Fraser, R. W. 363, *390*
Summer, H. see Zander, J. 423, *513*
Summers, G. R. see Casanova, R. 280, *385*
— see Shoppee, C. W. 232, *405*
Summers, V. K. see Sheehan, H. L. 458, *511*
Summers, W. see Eiseman, B. 118, *155*
Sutherland, A. M. see Schon, M. 653, *659*
— see Waxenberg, S. E. 579, 581, *636*
Sutherland, E. S., and G. F. Marrian 434, *511*
Sutherland, R. L. see Wallerstein, R. S. 655, *659*
Sutter, D. see Herzog, H. L. 260, *392*
Sutton, R. E. see Wilds, A. *409*
Suzuki, T., K. Yamashita, and T. Mitamura 74, 105, *181*
Svensson, H, see Blix, G. 292, *383*
Swale, J. see Bush, I. E. 75, *149*, 572, *625*
Swann, H. G. 46, 73, 99, *181*
Sweet, A. Y., M. F. Levitt and L. B. Turner 537, *568*
Sweat, M. A., and G. L. Farrell 75, 100, *181*
Sweat, M. L. 311, 312, 358, 366, *406*
— R. A. Aldrich, C. H. De Bruin, W. L. Fowlks, L.R. Heiselt and H. S. Mason 443, 485, 489, *511*
Swern, D. 249, *406*
Swigart, R. H. see Glick, D. 3, 50, 68, *159*
Swingle, W. W. 515, *568*
— L. J. Brannick and A. F. Parlow 527, *568*
— — — C. Baker and S. Le Brie 521, 524, 527, 535, *568*
— J. Da Vanzo, H. C. Crossfield, D. Glenister, M. Osborne and G. Wagle 529, 536, 547, *568*
— — E. Fedor, H. C. Crossfield, D. Glenister, M. Osborne and G. Wagle 529, 536, *568*
— and A. J. Eisenman 515, *568*
— E. J. Fedor, G. Barlow jr., E. J. Collins and J. Perlmutt 621, *635*
— — M. Ben, R. Maxwell, C. Baker and G. Barlow 535, *568*
— R. Maxwell, M. Ben, C. Baker, S. J. Le Brie and M. Eisler 99, *181*
— W. M. Parkins, A. R. Taylor and H. W. Hays 536, 542, *568*
— — — H. W. Hays and J. A. Morrell 527, *568*
— and J. J. Pfiffner 21, *181*, 516, *568*
— — N. M. Vars, P. A. Bott and W. M. Parkins 529, 535, 545, *568*
— and J. W. Remington 73, 99, *181*, 530, 537, 538, *568*
— — H. W. Hays and W. D. Collings 546, *568*

Swinyard, C. A. 3, 25, 28, 42, 43, *181*
— see Santisteban, G. A. 105, *177*
Sydnor, K. L., V. C. Kelley, R. B. Raile, R. S. Ely and G. Sayers 585, *635*
Sykes, J. F. see Wrenn, T. R. 611, *637*
Symington, T., and J. N. Davidson 36, 90, 92, *181*
— W. P. Duguid and J. N. Davidson 36, 63, 65, 66, 90, 92, 104, *181*
— see Grant, J. K. 75, *159*
Szalay, J. see Endröczi, E. 125, *155*
Szalkowski, C. R., M. G. O'Brien and W. J. Mader 337, *406*
Szego, C. M., and S. Roberts 620, *635*
— see Koenig, V. L. 337, *396*
— see Roberts, S. 573, 620, *634*
Szent-Györgyi 58
Szper, see Lelong, M. 582, *631*

Taffel, M. see Darrow, D. C. 529, 538, *558*
Tähkä, H. 43, 51, *181*, 587, 591, *635*
Tait, J. F., S. A. Simpson and H. Grundy 524, *568*
— see Avivi, P. 371, *382*
— see Ayres, P. J. 76, 109, *146*, 313, 318, 367, 372, *382*, 423, 447, 451, *500*
— see Garrod, O. 115, 116, *157*
— see Simpson, S. A. 198, 204, 238, 240, 272, 317, 333, 354, 367, 372, *405*, 421, 423, 446, *511*, 518, 520, 531, 541, *567*
Tait, S. A. S. see Ayres, P. J. 76, *146*, 423, 447, *500*
Takemura, K. H. see Dauben, W. G. 437, *502*
Talalay, P. 497, *511*
— M. M. Dobson and D. F. Tapley 496, *511*
— W. H. Fishman, and C. Huggins 66, *181*
— F. A. Loewus and B. Vennesland 312, *406*
— and P. I. Marcus 496, *511*
— see Hurlock, B. 338, *394*
— see Marcus, P. I. 496, *507*
Talbot, N. B., S. Ulick, A. Koupreianow and A. Zygmuntowicz *407*
— R. A. Berman and E. A. Mac Lachlan 363, *406*
Talbot, N. B., A. M. Butler and R. A. Berman 611, *635*
— — and E. A. Maclachlan 572, *635*
— and G. O. Langstroth 297, *406*
— A. H. Saltzman, R. L. Wixon and J. K. Wolfe 296, 302, 304, 339, 407
— and E. H. Sobel 574, *635*
— A. N. Zygmuntowicz, M. Wood and E. Christo 582, 586, *635*
— see Locke, W. 540, 541, *563*
— see Zygmuntowicz, A. S. 308, *410*
Tali, A. see Perlman, P. L. 535, *565*
Taliaferro, I., F. Cobey and L. Leone 455, *511*
— and L. Leone 136, *181*, *511*
— see Cobey, F. A. 136, *151*
Talmadge, R. V. see Buchanan, G. P. 615, *625*
— see Frost, R. C. 537, 542, 559
Tamm, C., G. Volpp and G. Baumgartner 283, *407*
— see Brenneisen, K. 191, *384*
— see Hegedüs, H. 317, *392*
Tamm, J., I. Beckmann and K. D. Voigt 293, *407*, 422, 423, *511*
Tamura, Y. 47, *181*
— see Masui, K. 593, *631*
Tanabe, T., and E. J. Cafruny 100, 103, *181*
Tannhauser, P., R. J. Pratt and E. V. Jensen 271, *407*
Tansey, R. P., and J. M. Cross 338, *407*
Tapke, R. J. see Herrmann, L. G. 293, *392*
Tapley, D. F. see Talalay, P. 496, *511*
Tarpley, W., and M. Yudis 336, *407*
Tarukoski, H. see Luft, R. 529, 531, 538, *563*
Taub, D., R. D. Hofsommer, H. L. Slates and N. L. Wendler 285, *407*
— — and N. L. Wendler 222, 267, 276, *407*
— R. H. Pettebone, N. L. Wendler and M. Tishler 223, *407*
— see Wendler, N. L. 217, 273, 288, *408*
— see Woodward, R. B. 261, *409*
Taubenhaus, M., I. B. Fritz and J. V. Morton *568*
Tauber, H. 337, *407*
Tavormina, P. A., and M. H. Gibbs 438, *511*
— — and J. W. Huff 436, 437, *511*
Taylor, A. R. see Swingle, W. W. 527, 536, 542, *568*
Taylor, D. J., J. Greenberg. E. S. Josephson and E. M. Nadel 51, 84, 144, *181*
Taylor, E. H. see Friedgood, H. B. 294, *390*
Taylor, H. C. jr. see Dobriner, K. 582, *627*
Taylor, P. see Klein, R. 521, *562*
Taylor, W. 461, *512*
Taylor, W. C. see Barton, D. H. R. 278, *383*
Tchen, T. T., and K. Bloch 438, *512*
Teich, S. see Lieberman, S. 572, *631*
Tellier, R. see Nowaczynski, W. 236, 333, *400*
Tennent, D., J. B. Whitla and K. Florey 313, *407*
Tentori, L. see Cavina, G. 306, *385*
Tenygl, J. see Zieman, P. 381, *410*
Teodori, U. see Marzi, A. 650, *658*
Tepperman, H. M. see Tepperman, J. 50, 51, *181*
Tepperman, J., H. M. Tepperman, B. W. Patton and L. F. Nims 50, 51, *181*
Terroine, T. see Delost, P. 118, *154*
Test, C. E. see Davis, M. E. 611, *626*
Tewell, H. E. see Chen, C. 340, *385*
Teysseyre, J. see Soulairac, A. 98, *180*
Thayer, S. A. see Doisy, E. A. 426, *502*
— see Smith, G. V. S. 426, *511*
Therien, B., R. C. Mellinger, J. R. Caldwell and P. J. Howard 135, *181*
Therien, M. see Dugal, L. P. 70, *154*
Thevenet, M. see Henry, R. 295, *392*, 434, *505*
Thieblot, L. see Simonnet, H. 50, 107, *179*
Thoma, R. W., and J. Fried 487, *512*

Thoma, R. W., J. Fried, S. Bonanno and P. Grabowich 259, 274, *407*
— see Fried, J. 258, 260, *390*, 487, 491, 492, 493, 494, *504*, 520, *559*
Thomas, B. G. H. see Pousiaka, J. W. 535, 540, *565*
Thomas, B. R. see Barton, D. H. R. 246, *383*
Thomas, G. H. see Sandoval, A. 281, 404
— see Zaffaroni, A. 281, *409*
Thomas, H. see Sonderhoff, R. 436, *511*
Thomas, P. Z. see Bongiovanni, A. M. 338, *384*
Thompson, H. see Clark, L. C. jr. 338, *386*
Thompson, J. L. see Fonken, G. S. 243, *390*
— see Hogg, J. A. 272, *393*
— see Levin, R. H. 227, *396*
— see Spero, G. B. 267, 271, 283, *406*
Thomson, D. L. see Selye H. 455, *511*
Thorn, G. W., and H. Eisenberg 537, *568*
— D. Jenkins, J. C. Laidlaw, F. C. Goetz, J. F. Dingman, W. L. Arons, D. H. P. Streeten and B. H. Mc Cracken 518, 540, *568*
— A. E. Renold, A. Goldfein, D. H. Nelson, W. J. Reddy and R. Hertz 137, *181*
— — Goldfien, D. H., Nelson, W. J. Reddy and R. Hertz 457, *512*
— see August, J. T. 138, *146*
— see Clinton, M. jr. 537, *557*
— see Engel, L. L. 427, 431, *503*, 610, *627*
— see Forsham, P. H. 607, *627*
— see Frawley, T. F. 541, *559*
— see Haydar, N. A. 425, *505*
— see Hernando, L. 457, *505*, 531, *561*
— see Jenkins, J. J. 138, *164*, 523, *562*
— see Laidlaw, J. C. 518, 533, 535, *563*
— see McAllister, F. F. 537, *564*
— see Munson, P. L. 581, 603, *632*
— see Reddy, W. J. 292, 293, *402*
— see Renold, A. E. 137, *175*, 457, *509*, 522, *566*, 457, *509*, 522, *566*, 580, *634*
Thorn, G. W. see Sobel, E. H. 135, 137, *180*
Thuline, H. C. see Masuda, M. 363, *398*
Tice, A. A. see Gardner, L. I. 412, *504*
Tiefjen, M. see Eger, W. 64, 65, 96, *155*
Tillotson, C. see Kochakian, C. D. 577, *630*
Timiras, P. O. see Fortier, C. 50, 58, 59, *157*
Timiras, P. S., and D. M. Woodbury 105, *181*
Tiselius, A. see Blix, G. 292, *383*
— see Kritchevsky, T. H. 328, *396*
Tishler, M. 265, *407*
— see Hirschmann, R. F. 267, 277, *393*
— see Hoffman, F. 286, *393*
— see Huang-Minlon 216, *394*
— see Taub, D. 223, *407*
Titus, E. see Perlman, D. 494, *509*
Tobin, C. E., and J. P. Birnbaum 57, *182*
— see Gaunt, J. H. 526, *559*
— see Gaunt, R. 21, *158*
Toepel, W. see Briggs, F. N. 60, 85, *149*, 454, *501*
Togari, C., and T. Ito 23, *182*
Tolbert, B. M. see Weaver, H. E. 368, *408*
Tolksdorf, S. see Gould, D. 267, 278, 279, *391*
— see Herzog, H. L. 277, *392*, 494, *505*
— see Oliveto, E. P. 285, *400*
Tomasewski, A. J. see Dreiding, A. C. 287, *388*
Tomkins, G. M. 461, 463, 476, 488, 489, 495, 497, 498, *512*
— J. F. Curran and P. J. Michael 488, *512*
— and K. J. Isselbacher 495, 498, *512*
— P. J. Michael and J. F. Curran 488, *512*
— H. H. Sheppard and I. L. Chaikoff 455, *512*
— see Liddle, G. W. 521, 523, 533, *563*
Tompsett, S. L. 293, *407*
— and D. C. Smith 295, 296, *407*
Tonutti, E. 47, 129, *182*
— F. Bahner and E. Muschke 31, 88, *182*
— see Boguth, W. 88, *148*
Toops, E. E. jr. see Weissbecker, A. 296, *408*
Toratini, Y., and T. Oku 640, *659*
Torcigliani, A. see Pellegrino, C. 119, 120, *174*
Török, J. see Korpássy, B. 83, *166*
Torstveit, O. see Herrick, E. H. 143, *161*, 554, *561*
Tötterman, L. E. see Jensen, C. C. 294, *394*
Touchstone, J. C. 423, *512*
— H. Bulaschenko, E. M. Richardson and F. C. Dohan 432, *512*
— L. Glazer, D. Y. Cooper and J. M. Roberts 75, *182*, 449, *512*
— and C. T. Hsu 346, *407*
— see Cooper, D. Y. 449, *502*
— see Dohan, F. C. 432, 433, *502*
— see Richardson, E. M. 432, 459, 460, *509*
Tourneur, R. see Gennes, L. de 546, *560*
Towne, L. E. see Nathanson, I. T. 616, *632*
Tran-Vy, M. see Hébert, S. 43, 108, *161*, 601, *629*
Travis, D. F. see Sawyer, W. H. *566*
Travis, R. H., and G. L. Farrell 447, *512*
Traylor, F. see Eisemann, B. 118, *155*
Trescher, J. H. see Harrop, G. A. 528, 529, 535, *560*, *561*
Tretbar, H. A. see Mc Cullagh, E. P. 457, *507*
Trethowan, W. H. 641, *659*
— and S. Cobb 647, *659*
Tripod, J. see Desaulles, P. A. 532, *558*
Tristram, E. see Chamberlin, E. M. 254, *385*
Troop, R. C. 497, *512*
Troncoso, V. see Zaffaroni, A. 238, *409*
Trunnell, J. B. see Llaurado, J. G. 107, *169*
Truszkowski, R., and J. Duszynska 530, *568*
— see Zwemer, R. L. 529, *569*
Tryczynski, E. see Wexler, B. C. 85, 100, 101, *184*
Tsatsas, G. see Prelog, V. 209, *401*
Tschesche, see Lettré, H. R. 186, *396*
Tschesche, R. see Reuber, R. 186, *402*
Tschopp, E. see Meystre, C. *398*

Tseu, T. K. L. see Bondy, P. 367, *384*
Tsou, K. C. see Nachlas, M. M. 63, *172*
— see Seligman, A. M. 66, *178*
Tuchmann-Duplessis, H. 80, 100, 101, *182*
— C. Delgatte, and F. Delgatte 9, 65, 80, 88, 90, 96, 143, 144, *182*
— and J. Larroche 123, 126, *182*
Tucker, J. T. see Hamilton, W. F. 531, *560*
Tullner, W. 609, *635*
— M. M. Graff, and R. Hertz 137, *182*, 456, *512*
— and R. Hertz 136, *182*
— see Hertz, R. 136, 137, *161*, 456, *505*, 522, 523, *561*
— see Schricker, J. A. 137, *187*
— see Vilar, O. 136, *183*
Tully, M. E. see Nobile, A. 277, *399*
Tum Suden, C. see Wyman, L. C. 515, 529, *569*
Turchini, J. see Grynfeltt, E. 53, 60, *160*
Turchini, S. see Grynfeltt, E. 53, 60, *160*
Turner, C. W. see Milller, W. R. 312, *399*
— see Mixner, J. P. 18, *171*
Turner, M. D. see Hardy, J. D. 75, *160*, 475, *504*
Turner, L. B. see Sweet, A. Y. 537, *568*
Turner, R. B. 189, 212, 215 216, 217, 228, 232, 283, *407*
— V. Mattox, L. Engel, B. Mc Kenzie and E. Kendall 228, *407*
— see Sneeden, R. P. A. 282, *405*
Tuthill, C. H. see Gaunt, R. 96, 100, 107, 110, *158*
Twombly, G. H. see Levitz, M. 484, *506*
Tyler, F. H., C. D. Schmidt, K. Eik-Nes, H. Brown and L. T. Samuels 475, *512*
— see Brown, H. 471, *501*
— see Eik-Nes, K. 82, *155*
— see Migeon, C. J. 572, *632*
— see Perkoff, G. T. 644, *659*
— see Samuels, L. T. 475, *510*
— see Sandberg, A. A. 475, *510*

Überwasser, H. see Schmidlin, J. 238, 264, *404*
— see Wettstein, A. 206, *408*

Udaev, N. A., and K. V. Druzhinia 445, *512*
Udenfriend, S. see Steinberg, D. 370, *406*
Ulick, S., and S. Lieberman 236, *407*, 433, 476, *512*
— see Talbot, N. B. *407*
Ulrich, F. see Abelson, D. 460, *499*
Ulstrom, R. A., E. Cole, J. Burley and R. Gunville 591, *635*
Umberger, E. J. 338, *407*
Ungar, F., and B. R. Bloom 496, *512*
— J. W. Davis, H. Rosenkrantz and R. I. Dorfman 460, *512*
— and R. I. Dorfman 427, 432, 435, 441, 449, 463, 477, 495, 496, 497, *512*
— M. Gut and R. I. Dorfman 496, *512*
— A. M. Miller and R. I. Dorfman 478, 486, *512*, 574, *635*
— E. B. Romanoff, R. I. Dorfman and G. Pincus 442, 489, *512*
— see Burstein, S. 480, *501*
— see Caspi, E. 435, 441, *501*
— see Dorfman, R. I. 463, *503*
— see Rosemberg, E. 446, 447, *509*
— see Rosenfeld, G. 76, 111, 113, *176*, 443, 449, 490, *510*
Uotila, U. U. 17, 39, 42, *182*, 588, *635*
— and A. Pekkarinen 18, 59, 85, *182*
Upton, A. C., and J. Furth 108, 133, *182*
Upton, G. V. see Bondy, P. K. 367, *384*, 550, *556*
Urech, H. J. see Wettstein, A. 204, *408*
Urquhart, J. see Yates, F. E. 498, 499, *513*
Urquiza, M. see Sondheimer, F. 213, *406*
Ussing, H. H. see Johnson, V. K. 550, *562*
Utewski, A., and S. Epstein 67, *182*
Utne, T. see Chamberlin, E. M. 254, *385*
Uuspää, V. J., and P. A. Järvinen 106, *182*

Vaccarezza, A. J. 36, 46, 57, 118, *182*
Valere, P. H. L. see Birmingham, M. K. 454, *500*
Van Arman, C. G. see Kagawa, C. M. 521, 524, *562*
Vande Wiele, R., and S. Lieberman 572, 574, *635*
— see Jailer, J. W. 465, *506* 585, *630*
— see Solomon, S. 435, 439, 441, *511*, 574, *635*
Vanek, R., and R. Devis 295, *407*
Vannini, E., and T. Cessi 38, *182*
— and A. Sabbadin 2, 9, *182*
Van Wagenen, G. see Dorfman R. I. 484, *503*
Van Wyk, J. J. see Bongiovanni, A. M. 424, *501*
Varney, R. F. see Evans, J. S. 579, *627*
Vars, H. M. see Pfiffner, J. J. 519, *565*
— see Wintersteiner, O. 519, *569*
Vars, N. M. see Swingle, W. W. 529, 535, 545, *568*
Vasama, R., L. Kalliomaki, R. Vasama and E. Näätanen 87, *182*
Velardo, J. T. 614, *635*
— F. L. Hisaw and A. B. Dawson 614, *636*
— and S. H. Sturgis 620, *636*
— see Hisaw, F. L. 612, 614, *629*
Velasco, M. see Sondheimer, F. *405*
Veler, C. D. see Doisy, E. A. 426, *502*
Velican, C. 23, 38, 39, 42, *182*
Velluz, L., A. Petit, M. Pesez and R. Berret 221, *407*
Vennesland, B. see Talalay, P. 312, *406*
Venning, E. H. 125, *182*, *512*
— and J. S. L. Browne 431, 434, *512*
— I. Dyrenfurth and J. C. Beck 110, *182*
— — and V. E. Kazmin 304, *407*
— C. J. P. Giroud, I. Dyrenfurth and J. C. Beck 333, *407*
— — — and J. C. Bell 475, *512*
— M. M. Hoffman and J. S. L. Browne 434, *512*
— V. E. Kazmin, M. Ripstein, H. T. McAlpine and M. M. Hoffman 580, *636*
— O. J. Lucis, I. Dyrenfurth and J. C. Beck 111, *182*
— J. P. Randall and P. Gyorgy 591, *636*
— B. Singer and G. A. Simpson 520, *568*

Venning, E. H., P. G. Weil and J. S. L. Brown 611, *636*
— see Beck, J. C. 116, *147*
— see Dyrenfurth, I. 296, *388*
— see Giroud, C. J. P. 77, 109, 111, 113, *158*, 447, *504*
— see Heard, R. D. H. 296, *391*
— see Read, C. H. 591, *634*
— see Singer, B. 520, *567*
Venske, W. G. 13, *183*
de Venuto, F. see Pearce, E. M. 370, *401*
Vercellone, A. see Camerino, B. 272, 274, *385*
— see Mamoli, L. 259, *397*
Veres, K. see Kučera, J. 328, *396*
Vermeulen, A. 423, 425, 460, 476, *512*
— and E. Caspi 476, 497, *512*
Vermeulen, C. see Scott, W. W. 581, *634*
Verne, J. 55, 66, 98, *183*
— F. Bourliere and M. Deramond 17, 35, *183*
— and M. Delsol 9, *183*
— M. Gabe and B. Schramm 57, 133, *183*
— and S. Hébert 35, 65, 96, 120, *183*
— see Hébert, S. 43, 108, *161*, 601, *629*
Verzár, F., and F. Hügin 72, *183*
Vescelli, T. A. see Lombardo, M. E. 300, *397*
Vesin, P. P., Frumusan and R. Cattan 546, 547, *568*
— C. Jacquier, E. E. Baulieu and R. Cattan 521, 546, *568*
— see Cattan, R. 546, *557*
Vestling, C. S. see Kelly, T. L. 63, *165*
Vicari, C. see Mariani, A. 306, *397*
Vicari, E. M. 35, *183*
Vies, J. van der 77, *183*
Vilar, O., and W. W. Tullner 136, *183*
Villee, C. A. see Villee, D. B. 62, 63, *183*, 590, *636*
Villee, D. B., L. L. Engel and C. A. Villee 62, 63, *183*, 590, *636*
Vincent, A. M. see Delost, P. 597, *626*
Vincent, S., and F. R. Curtis 7, *183*
— see Wheeler, T. D. 516, *569*
Vines, H. W. C. 43, 108, *183*
Vischer, E., C. Meystre and A. Wettstein 277, *408*
— J. Schmidlin and A. Wettstein 264, *408*, 494, *512*
— and A. Wettstein 260, *408*
— see Kahnt, F. W. 491, 492, *506*
— see Meystre, C. 491, 493, 494, *508*
— see Neher, R. 204, 206, *399*, *565*
Viscontini, M., and P. Miglioretto 191, *408*
Visscher, M. B. see King, J. T. 133, *165*
de Vivar, A. R. see Romo, J. 273, *403*
Vliers, M., L. Demey, C. Heusghem and H. van Cauwenberge 424, *512*
Voegtli, S. M. see Chen, C. 312, *385*
Vogt, M. 70, 71, 74, 75, 76, 107, 137, *183*, 423, 455, 456, *512*, 523, 524, *569*
— see Holzbauer, M. 76, 107, *162*, 456, *505*, 523, *562*
— see Pickford, M. 100, *174*
Voigt, K. D., and I. Beckmann 304, *408*
— — and H. van der Werth 418, *512*
— and W. Schroeder 418, *512*
— — and I. Beckmann 303, *408*
— see Pontius, D. 337, *401*
— see Schmid, O. J. 304, *404*,
— see Schroeder, W. 303, *404*, 418, *511*
— see Tamm, J. 293, *407*, 422, 423, *511*
Voit, K. see Feulgen, R. 55, *156*
Volpp, G. see Tamm, C. *407*
Voltz, J. see Bergström, S. *500*
Voser, W. see Meystre, C. 277, *398*
Voss, H. 55, *183*

Wade, A. P. see Edwards, R. W. H. 293, *388*
— see Kellie, A. E. 293, 310, 338, *395*
Wade, N. J., and L. A. Haselwood 622, *636*
Wagland, A. A. see Henbest, H. B. 228, *392*
Wagle, G. see Swingle, W. W. 529, 536, 547, *568*
Wagner, R. B., and J. A. Moore 225, 256, *408*
Wahba, N. see Ruzicka, L. 288, *403*
Walker, D. G., M. E. Simpson, C. W. Asling and H. M. Evans 78, *183*
Walker, G. see Ayres, P. J. 76, *146*, 423, 447, *500*
— see Nabarro, J. D. N. 103, 107, 108, 133, 135, *171*
Walker, R. see Hirschmann, R. F. 271, *393*
Walker, T. see Attenburrow, J. 242, *382*
— see Barton, D. H. R. 228, *383*
Walkling, A. A. see Paschkis, K. E. 480, *508*
— see Pearlman, W. H. 484, *508*
Wall, P. E. see Migeon, C. J. 471, *508*
Wallace, E. Z., H. I. Wilverberg, A. C. Carter and L. S. Hersch 455, *512*
— see Christy, N. P. 80, 100, 103, 106, *151*
— see Jailer, J. W. 106, 107, 135, *164*
Wallace, G. J. 4, *183*
Wallach, E. 412
Wallach, S., H. Brown, E. Englert and K. Eik-Nes 412, *512*, 585, 611, 615, *636*
— — E. Englert jr. and K. Eik-Nes 134, *183*
Waller, I. R. see Julian, P. L. 255, *395*
Wallerstein, R. S., R. L. Sutherland and J. Lyons 655, *659*
Wallraff, J. 60, *183*
Walter, K. 363, *408*
Walthard, B. 117, 118, *183*
Walton, R. L. see Gardner, L. I. 126, *157*, 582, 584, 591, *628*
Wanner, H. see Robinson, H. J. 106, *176*
Wannowius, H. see Criegee, R. 250, *387*
Ward, V. B. see Hardy, J. D. 75, *160*
Wareham, D. J. see Cook, E. R. 316, 317, *386*
Warren, F. L. see Anderson, A. J. 293, *381*
Warren, R. P., and L. R. Aronson 603, *636*
Warga, M. see Saier, E. 338, *403*
Waring, H. 39, 47, 128, *183*
Warming-Larsen, A., C. Hamburger and M. Sprechler 541, *569*
Warms, J. V. B. see Dituri, F. 436, 438, *502*

Wattenberg, L. W. 61, 63, *183*
— and D. Glick 66, *183*
Waterhouse, C. see Burton, R. B. 459, *501*
Watteville, H. de, R. Borth, R. S. Mach and E. Musso 611, *636*
— see Borth, R. 455, *501*
Watson, E. J. D., and G. F. Marrian 426, *512*
— see Marrian, G. F. 426, *507*
Waxenberg, S. E., M. G. Drellich and A. M. Sutherland 579, 581, *636*
Wayne, H. L. 642, *659*
Weaver, D. C. see Hilton, J. G. 76, *162*
Weaver, H. E., B. M. Tolbert and R. C. Laforce 368, *408*
Weaver, J. A. 93, 101, *183*
Webb, J. L. see Aprile, M. 421, 442, *500*
— see Bligh, E. G. 449, 451, *500*
— see Heard, R. D. H. 420, 421, 446, 449, 451, 453, *505*
Webb, M. see Sperry, W. M. 50, *180*
Weber, A. F., J. T. Bell jr. and A. F. Sellers 135, *184*
— S. H. McNutt and B. B. Morgan 23, 28, *183*
— W. R. Pritchard and A. F. Sellers 115, *184*
— see Bell J. T. jr. 94, *147*
Weber, E. J., and M. L. Menten 135, *184*
Weber, L. see Oliveto, E. P. 204, 228, 243, *400*
Weber, P. 646, *659*
Webster, M. see Engel, L. L. 304, 306, *389*
Wedlake, D. see Ellis, B. 273, 274, *388*
Wegmann, R. 54, *184*
Wégria, R. see Hilton, J. G. 76, *162*
Weichselbaum, T. E., and H. W. Margraf 425, *512*
— — and D. C. King 366, *408*
Weil, P. G. see Venning, E. H. 611, *636*
Weinbaum, J. G. see O'Donnell, W. M. 110, *173*
Weinmann, S., E. E. Baulieu, A. Alfsen, B. Lisboa and M. F. Jayle 434, *512*
— F. F. Demoisson, E. E. Baulieu and M. F. Jayle 434, *512*
— see Mallassis, D. 434, *507*
Weinstein, A. see Harrop, G. H. 528, *561*
Weinstein, M. J., J. Schiller and H. A. Charipper 623, *636*
Weintraub, A. see Eppstein, S. H. 487, 491, 492, *503*
— see Meister, P. D. 487, 491, 492, 493, 494, *508*
— see Pederson, R. L. 281, 283, *401*, 487, *509*
— see Peterson, D. H. 259, 206, *401*, 491, 492, 493, *509*
Weisenborn, F., and D. Burn 237, *408*
Weisenfeld, S. see Carter, A. C. 107, *150*
Weiss, M. J. see Kissman, H. M. 272, *395*
Weissbecker, L., and H. Staudinger 304, *408*
— see Pfeffer, K. H. *401*
Weissberger, A., E. S. Proskauer, J. A. Riddick and E. E. Toops jr. 296, *408*
Wells, B. B., and E. C. Kendall 519, *569*
Wells, H., F. N. Briggs and P. L. Munson 522, *569*
Wells, L. J. 121, 123, *184*
— see Kitchell, R. L. 122, 123, 124, *165*
— see Yakaitis, A. A. 124, *185*
Welt, L. G., D. W. Seldin and J. H. Cort 530, *569*
— see Hollander, W. jr. 534, 543, *561*
— see Seldin, D. W. 532, *567*
Wendler, N. L., and R. P. Graber 223, *408*
— — C. S. Snoddy jr. and F. Bollinger 279, *408*
— and D. Taub 217, 273, *408*
— — S. Dobriner and D. Fukushima 217, *408*
— — and H. L. Slates 288, *408*
— see Huang-Minlon 216, *394*
— see Taub, D. 222, 223, 267, 276, 285, *407*
Wenneker, A. S. see Eisenstein, A. B. 71, 105, *155*, 458, *503*
Wener, J. see Singer, B. 520, 567
Werbin, H., D. M. Bergenstal, R. G. Gould and G. V. Le Roy 435, 441, *513*
— I. L. Chaikoff and E. Jones 435, *513*
— and G. V. Le Roy 435, 441, 449, *512*
— and S. Ong 363, *408*
— J. Plotz, G. V. Le Roy and E. Davis 435, 441, 449, *513*
Wernitz, W. see Sturm, A. 646, *659*
Werth, H. van der see Schroeder, W. 418, *511*
— see Voigt, K. D. 418, *512*
Werthessen, N. T. see Pearlman, W. H. 431, *508*
— see Schwenk, E. 250, *404*
Wertheimer, E. see Stein, L. 540, *568*
West, C. D., B. Damast and O. H. Pearson 412, *513*
— B. L. Damast and O. H. Pearson 617, *636*
— — S. D. Sarro and O. H. Pearson 616, *636*
— see Lipsett, M. B. 100, *168*
— see MacLean, J. P. 110, 113, *169*
West, H. F. see Norymberski, J. K. 365, *399*
West, W. 358, 359, *408*
Westberg, J. A., H. A. Bern and E. B. Barnawell 106, *184*
Westenskow, G. see Berliner, D. L. 369, 372, *383*
Westermann, C. D. see Hilton, J. G. 76, *162*
Westman, A. see Borell, U. 601, *625*
— see Diczfalusy, E. 412, *502*
Westphal, U. 335, *408*, 459, *513*, 610, *636*
— H. E. Firschein and E. M. Pearce 292, *408*
— see Butenandt, A. 427, 501
— see Pearce, E. M. 370, *401*
Westphal, U. F. see De Venuto, F. 300, *387*
Wettstein, A. 258, 279, *408*, 411, 491, 494, *513*
— and G. Anner 415, *513*
— P. Desaulles, K. Heusler, R. Neher, J. Schmidlin, H. Überwasser and P. Wieland 206, *408*
— H. Neher, and J. G. Llaurado 433, *513*
— R. Neher and H. J. Urech 204, *408*
— see Euw, J. von 198, 204, *389*
— see Heer, J. 254, *392*
— see Heusler, K. 273, 274, *393*
— see Kahnt, F. W. 234, 236, 238, *395*, 413, 443, 446, 447, 490, 491, 492, *506*, 517, *562*
— see Llaurado, J. G. *507*, *520*, *563*
— see Luetscher, J. A. jr. 433, *507*, 520, *563*
— see Meystre, C. 254, 277, 280, *398*, 491, 493, 494, *508*

Wettstein, A. see Neher, R. 198, 204, 206, 238, 320, 327, 328, 341, 342, 345, *399*, 420, 421, 442, 443, 499, *508*, *565*
— see Ruzicka, L. 204, *403*
— see Schmidlin, J. 238, 262, 264, *404*, 520, *566*
— see Simpson, S. A. 198, 204, 238, 240, 317, 354, *405*, 421, 446, *511*, 520, *567*
— see Vischer, E. 260, 264, 277, *408*, 494, *512*
Wexler, B. C. 115, *184*
— A. E. Doglin and E. W. Tryczynski 85, 100, *184*
— — J. F. Zaroslinski and E. S. Tryczynski 101, *184*
— and A. P. Rinfret 110, *184*
— — A. C. Griffin and H. L. Richardson 110, *184*
— see Robertson, O. H. 144, 145, *176*
Whaba, N. see Heusser, H. 288, *393*
Wheeler, J., S. Freeman and C. Chen 364, *409*
— see Chen, C. 340, *385*
Wheeler, T. D., and S. Vincent 516, *569*
Whidden, H. L. see Friedgood, H. B. 337, *390*
White, A. see Sayers, G. 83, *177*
White, H. L., and D. Rolf 537, *569*
— see Martin, H. F. 531, *564*
White, J. P. see Knowlton, A. I. 547, *562*
White, J. V. see Heigl, J. J. 357, *392*
Whitehead, J. K. see Avivi, P. 371, *382*
Whitehead, R. 32, 35, 47, 48, 53, 133, *184*, 596, *636*
— see Kay, W. W. 51, 70, *165*
— see MacLachlan, P. L. 49, 51, 54, 92, *169*
Whiteley, H. J., and H. B. Stoner 106, *184*
— see Stoner, H. B. 49, 80, 104, 110, 125, *181*
Whitla, J. B. see Tennent, D. 313, *407*
Whitman, B. see Wintersteiner, O. 426, *513*
Wicki, W. see Buzzetti, F. 237, *385*
Wieland, P. 186, *409*
— and K. Miescher 209, *409*
— see Neher, R. 204, 206, *399*, *565*
Wieland, P. see Schmidlin, J. 238, 264, *404*
— see Wettstein, A. 206, *408*
Wiener, M. see Hayano, M. 461, *505*
Wiest, W. G. 423, *513*
Wietrzykowska, J. see Guyénot, E. 607, *628*
Wilbur, O. M. jr. 130, *184*
— and A. R. Rich 130, 133, *184*
Wilder, E. J. see Knobil, E. 50, *166*
Wilds, A., J. A. Johnson and R. E. Sutton *409*
Wilds, A. L., and N. A. Nelson 280, 281, *409*
Wilds, A. R. see Bachman, W. E. 261, *382*
Wilhelm, W. see Küchmeister H. 135, 136, *167*
Wilkins, L. 578, 585, 603, 615, *636*, 653, *659*
— J. F. Crigler jr., S. H. Silverman, L. I. Gardner and C. J. Migeon 585, 616, *636*
— see Crigler, J. F. jr. 521, *558*
— see Lewis, R. A. 80, *168*
Willard, H. H., L. L. Meritt, jr. and J. A. Dean 358, 368, *409*
Willardson, D. C. see Brown, H. 471, *501*
— see Butt, W. R. 315, 318, 381, *385*, 613, *625*
— see Morris, C. J. O. R. 315, *399*
Williams, J. see Antonucci, R. 225, 361, *382*
Williams, P. C. see Bulbrook, R. D. 412, *501*
Williams, R. G. 46, 47, 118, *184*
Williams, R. H. see Daughaday, W. H. 296, *387*
— see Hogness, J. R. 523, *561*
— see Rogers, W. F. jr. 44, 51, 84, *176*
Williams, T. F. see Hollander, W. jr. 534, 543, *561*
Williams, V. Z., V. J. Coates and F. Gaarde 354, *409*
— see Dobriner, K. 301, 303, *388*
Williams, W. L. see Kirschbaum, A. 134, *165*
— see Zimmermann, B. 135, *185*, 485, *513*
Willoughby, M. see Bush, I. E. 433, *501*
Wills, C. G., S. E. Rampton and L. I. Pugsley 575, *636*
Wilson, D., R. C. Grauer and E. Saier 297, *409*
Wilson, E. see Huang-Minlon 216, *394*
Wilson, H. 338, 363, 365, *409*
— J. J. Borris and R. C. Bahn 75, 108, *184*, 603, 607, *636*
— and P. Carter 338, *409*
— and R. Fairbanks 340, 341, 365, *409*
— and I. T. Nathanson 338, *409*
— see Bahn, R. 607, *624*
— see Engel, L. L. 303, 338, *389*
— see Nathanson, I. T. 363, *399*
Wilson, M. L. see McGinty, D. A. 443, 489, *507*
Wilson, R. see Engel, L. L. 426, *503*
Wilverberg, H. I. see Wallace, E. Z. 455, *512*
Wilz, H. see Johannes, A. 58, 85, *164*
Windaus 187, *409*
Winter, C. A., H. L. Hollings and R. B. Stebbins *184*
— R. A. Silber and H. C. Stoerk 80, 84, 86, *184*
— see Arth, G. E. 285, *382*
Winternitz, F. see Heusser, H. 288, *393*
Winters, R. W., R. B. Schultz and W. A. Krehl 105, 139, *184*, 458, *512*
— see Hollander, W. jr. 534, 543, *561*
— see Schultz, R. B. 138, 139, *178*
Wintersteiner, O. 197, *409*, 411, *513*
— and J. J. Pfiffner 337, *409*, 420, 421, *513*
— E. Schwenk and B. Whitman 426, *513*
— H. M. Vars and J. J. Pfiffner 519, *569*
— see Hirschmann, H. 426, *505*
— see Pfiffner, J. J. 519, *565*
Wise, J. see Dorfman, R. I. 478, 484, *503*
Wishinsky, H. see Rubin, M. 214, *403*
Wislocki, G. B. see Deane, H. W. 117, 118, *153*
Witschi, E. 9, 17, 38, 39, 121, *184*
— and G. Y. Chang 579, *636*
— see Chang, G. Y. 142, 144, *150*
Witter, R. F., and S. Stone 346, *409*
Wittle, E. L. see Marker, R. E. 209, *397*, 431, *507*

Wixom, R. L. see Talbot, N. B. 296, 302, 304, 339, *407*
Wizerkaniuk, M. see De Meio, R. H. 478, *502*
Wojack, A. J. see Zimmermann, B. 611, *637*
Wolf, P. S. see Race, G. J. 83, 101, 108, 111, 112, *175*
Wolf, R. S. see Romanoff, L. P. 311, *403*
— see Knobil, E. 116, *166*
Wolfe, J. M. see Dribben, I. S. 27, 29, *154*
Wolfe, J. K., L. F. Fieser and F. B. Friedgood 572, *636*
— E. B. Hershberg and L. F. Fieser 381, *409*
— see Hershberg, E. B. 297, 299, 381, *392*
— see Talbot, N. B. 296, 302, 304, 339, *407*
Wolff, E., K. Haffen, Y. Lutz-Ostertag and B. Salzgeber 605, *636*
Wolff, H. G. see Hetzel, B. S. 642, *658*
Wolff, H. P., K. R. Koczorek and E. Buchborn 114, *184*, *569*
— — W. Jesch and E. Buchborn 532, *569*
Wood, E. H. see Dennies, C. 540, *558*
Wood, G. W. see Bladon, P. 275, 360, *383*
— see Henbest, H. B. 355, *392*
Wood, J. see Hirschmann, R. F. 269, *393*
Wood, K. F., F. Lees and F. D. Rosenthal 129, *185*
Wood, M. see Talbot, N. B. 582, 586, *635*
— see Zygmuntowics, A. S. 308, *410*
Woodall, N. B. see Anderson, D. H. 350, *381*
Woodbury, D. M. 534, 537, 538, *569*
— C. P. Cheng, G. Sayers and L. S. Goodman 529, 531, 534, *569*
— and A. Koch 531, 538, *569*
— see Timiras, P. S. 105, *181*
Woodruff, M. F. A., and T. Boswell 118, 120, *185*
— see Llaurado, J. G. 116, *169*
Woods, G. F. see Bladon, P. 275, *383*
— see Norymberski, J. K. 246, *399*
Woods, J. W. 18, 58, 85, 106, *185*
Woodward, G. see Nelson, A. A. 135, *172*, 457, *508*
Woodward, P. see Lostroh, A. J. 68, 83, 100, 101, 111, *169*
Woodward, R. B. 262, 360, *409*
— and K. Bloch 436, 437, 438, *513*
— H. Inhoffen, H. Larson and K. Menzel 287, *409*
— Sondheimer, F. D. Taub, K. Heusler and W. McLamore 261, *409*
Woolley, G. W. 134, *185*, 606, 619, *636*
— M. M. Dickie and C. C. Little 134, *185*, 618, *637*
— E. Fekete and C. C. Little 618, *637*
— and C. C. Little 606, 618, *637*
— see Christy, N. P. 618, *626*
— see Fekete, E. 618, 619, *627*
Worcester, J. see Locke, W. 540 541, *563*
Woroch, E. L. see Mattox, V. R. 278, *398*
Worrel, C. S. see McGinty, D. A. 443, 489, *507*
Wotiz, H. H., J. W. Davis, H. M. Lemon and M. Gut 616, *637*
Wrenn, T. R., J. Bitman and J. F. Sykes 611, *637*
Wright, A., and I. Chester Jones 11, 12, 53, 60,78, 80, 83, 143, 144, *185*, 552, *569*
— I. Chester Jones and J. G. Phillips 30, 31, 53, *185*
— see Chester Jones, I. 79, 116, 120, *150*
Wright, M. L. see Friedgood, H. B. 294, *390*
Wright, N. 357, *409*
Wrigley, T. I. see Henbest, H. B. 28, *392*
Wu, H. M. see Race, G. J. 587, *633*
Wuersch, J., R. L. Huang and K. Bloch 437, *513*
Wyatt, T. C. see Jones, J. M. 51, 59, *164*
— see Lloyd, C. W. 611, *631*
van Wyk, J. J. see Bongiovanni, A. M. 296, 301, *384*
Wylie, H. B., and M. J. Schmulovitz 297, *409*
Wyman, L. C., and C. Tum Suden 515, 521, *569*
— see Brenner, R. M. 118, 119, *149*
— see Konrad, E. L. 120, *166*
— see Snyder, J. G. 526, *567*
Wyman, L. J. see Chapman, J. H. 252, *385*
Wyngaarden, J. B. see Peterson, R. E. 471, 475, *509*

Yakaitis, A. A., and L. J. Wells 124, *185*
Yamashita, K. see Suzuki, T. 74, 105, *181*
Yankopoulos, N. A., J. O. Davis, B. Kliman and R. E. Peterson 117, *185*
— see Davis, J. O. 114, *153*
Yates, F. E., A. L. Herbst, and J. Urquhart 498, 499, *513*
Yatsu, F. M. see Farrell, G. L. 103, 111, *156*
Yoffey, J. M. 47, 65, 108, *185*
— and J. S. Baxter 53, 101, *185*
— see Robinson, F. B. 104, *176*
Yoshimura, F. 37, 80, 91, 100, *185*
Youhotsky, I. see Cornforth, J. W. 191, *386*
Youhotsky-Gore, I. see Cornforth, J. W. 438, *502*
Young, B. G. see Nadel, E. M. 71, *172*
Young, H. L. see Edelman, I. S. 550, *558*
de Young, J. J. see Harman, R. E. 318, *391*
Young, J. Z. 6, *185*
Young, W. C. see Phoenix, C. H. 574, *633*
Yount, E. H. jr. see Little, J. M. 136, *169*
Yudis, M. see Tarpley, W. 336, *407*
Yuen, V. S. see Ganong, W. F. 110, *157*

Zabin, I., and W. F. Barker 436, *513*
— and K. Bloch, 436, *513*
Zablow, L. see Rosenkrantz, H. 349, *403*
Zaffaroni, A. 300, 313, 318, 319, 320, 362, *409*
— and R. B. Burton 423, *513*, 517, *569*
— — and E. H. Keutmann 318, 342, *409*
— C. C. Campillo, F. Cordoba and G. Rosenkranz 494, *513*
— O. Hechter and G. Pincus 435, 441, *513*
— H. J. Ringold, G. Rosenkranz, F. Sondheimer, G. H. Thomas and C. Djerassi 281, *409*
— V. Troncoso and M. Garcia 238, *409*

Zaffaroni, A. see Axelrod, L. R. 300, 301, *382*
— see Burton, R. B. 319, 327, 339, 342, *384*
— see Hechter, O. 311, *392*, 441, 442, 486, 487, 490, *505*, 517, *561*, 609, *629*
— see Mancera, O. 492, *507*
— see Reich, H. 341, *402*, 609, *634*
Zajaczek, S. see Skowron, S. 21, *179*
Zalesky, M. 19, 21, *185*
Zamcheck, N. 130, *185*
Zander, J. 613, *637*
— and H. Simmer 613, *637*
— and H. Summer 423, *513*
— see Forbes, T. R. 423, *503*
Zaroslinski, J. F. see Wexler, B. C. 101, *184*
Zarrow, I. G. see Elton, R. L. 51, 58, 84, 85, 106, *155*
— see Zarrow, M. X. 140, *185*
Zarrow, M. X., and J. T. Baldini 58, 85, 144, *185*
— F. L. Hisaw and F. Bryans 610, *637*
— L. M. Horger and J. L. McCarthy 140, *185*
— and E. A. Lazo-Wasem 611, *637*
— and I. G. Zarrow 140, *185*
— see Elton, R. L. 51, 58, 84, 85, 106, *155*
Zarrow, M. X. see Lazo-Wasem, E. A. 610, *630*
— see Neher, G. M. 615, *632*
Zeisel, H., W. Engel and M. Pressler 591, *637*
— and M. Pressler 590, *637*
Zelander, T. 3, 23, 24, 27, 37, 38, 58, 91, 133, *185*
Zerssen, D. von 640, *659*
Ziegler, M. R. see McQuarrie, I. 534, *564*
Zierler, K. L., and J. L. Lilienthal 531, *569*
Ziff, M., J. Simson and J. J. Bunim 538, *569*
Zimmermann, B., H. S. Bloch, W. L. Williams, S. R. Hitchcock and B. Hoelscher 135, *185*, 458, *513*
— and A. J. Wojack 611, *637*
Zimmermann, B. U. see Hokin, M. R. 92, *162*
Zimmermann, W. 291, 337, 363, *409*
— H. V. Anton and D. Pontius 363, *410*
— and D. Pontius 363, *410*
Zinsser, A. D., and H. H. Zinsser 43, *185*
Zinsser, H. H. see Zinsser, A. D. 43, *185*
Zizine, L. 59, 107, *185*
— M. E. Simpson and H. M. Evans 107, *185*
Zondek, B., and S. Burstein 455, 456, *513*
Zondek, H. see Lowenthal, M. 129, *169*
Zuckerman, S. 621, 622, *637*
— G. Bourne and D. Lewes 622, *637*
— see Bourne, G. 622, *625*
— see Dow, D. 550, *558*
— see Mandl, A. M. 622, *631*
Züblin, W. 650, 651, *659*
Zuman, P., J. Tenygl and M. Březina 381, *410*
Zumoff, B. see Bradlow, H. L. 483, *501*, 573, *625*
Zweifach, B. W., M. M. Black and E. Shorr 63, *185*
— see Cascarano, J. 63, *150*
Zweig, G. see Block, R. J. 304, 313, *383*
Zwemer, R. L., and R. Truszowski 529, *569*
— see Lowenstein, B. E. 530, *563*
Zwingelstein, G., H. Pacheco and J. Jouanneteau 310, *410*
Zygmuntowicz, A. N., M. Wood, E. Christo and N. B. Talbot 308, *410*
— see Manaro, J. M. 346, *397*
— see Talbot, N. B. *407*, 582, 586, *635*

Subject Index

Accessory adrenocortical tissue 21, 29, 526, 619, 622
Acetate, conversion of to steroids 436—442, 446, 449—451, 571, 609
Acetylcholine administration 456
Acid phosphatase in adrenal glands 65, 98
Acidophilia, see Fuchsinophilia
ACTH, effects of on adrenal gland VII, 47, 68—71, 73, 82—85, 88—92, 96—99, 101—103, 110, 111, 118, 121—125, 127, 129, 130, 133, 137, 140, 141, 144, 145, 453, 552, 553, 592, 595, 596, 609
—, increased secretion of VII, 107, 108, 116, 133, 137, 139, 424, 638; see also Cushing's disease, Stress
— -like actions 77, 85, 100, 107
—, secretion of VII, 80—82, 110, 111, 121, 123—125, 140,517, 521, 522, 544, 581, 585, 616
—, stimulation of steroid secretion by 75—77, 103, 108—111, 141, 411, 424, 425, 450, 452—456, 458, 471, 475, 498, 517, 534, 538, 540, 541, 546, 549, 551, 555, 580—582, 607—609, 611, 613—615, 617, 619—621, 640—644, 646—649
Addison's disease V, VI, 132, 140, 141, 424, 455, 475, 517, 531, 540, 542, 579, 644, 645, 653—656
Adenomas, adrenocortical, see Tumors
Adrenal cortex, steroid content of VI, VII, 54, 197—204, 411—421, 516, 517, 519, 571, 572; see also Adrenocortical steroids, biosynthesis, chemical nature
— medulla V, VI, 1—3, 5—14, 16, 17, 21, 27, 29, 30, 33—35, 39—41, 43, 44, 48, 49, 52, 61, 62, 68, 82, 87, 516, 548, 553
— transplants 46, 120, 121, 604, 605, 623
— -venous blood, hormones in 73, 74—76, 103, 105, 107, 137, 139, 140, 422—424, 516, 518, 555, 572, 609, 610; see also Adrenocortical steroids in blood
— virilism (Adrenogenital syndrome, Congenital adrenal hyperplasia) VIII, 108, 129, 133, 138, 413—415, 423, 424, 449, 451, 465, 475, 482, 521, 540, 572, 576, 579, 584, 585, 593, 603, 610, 611, 615, 616, 650—652
Adrenalectomy VI, 12, 21, 98, 122, 141, 143, 144, 412, 482, 515, 519, 525—531, 533, 536, 537, 545, 546, 551, 552, 554,. 555, 579—581, 591, 601, 602, 610, 614, 615, 620, 621, 645, 648
Adrenaline (Epinephrine) administration 86, 104, 105, 139
Adrenocortical cells, see also Adrenocortical tissue, anatomy, microscopic, electronmicroscopic
— —, atrophy of 80—82, 87, 88, 99—102, 110, 143
— —, Golgi apparatus of 2, 3, 9, 12, 24, 33, 37, 54, 60, 92, 145
— —, hypertrophy of 88, 101, 102, 104
— —, lipide droplets of VI, 2, 3, 5—12, 15—17, 24, 32—36, 44, 45, 48, 50, 52—58, 63, 67, 80, 81, 90, 92—96, 101, 104, 107, 110—115, 117—119, 121, 125, 137, 138, 140, 141, 143, 145
— —, mitochondria of 2—7, 9, 12, 15—17, 24, 33, 37, 38, 45, 54, 55, 60, 62, 91, 92, 104, 110, 145; see also Adrenocortical tissue, mitochondrial fraction
— —, mitosis of 31, 32, 45, 47, 88, 90, 104, 110, 111, 118, 129, 130, 133—135, 138, 587
— —, nuclei of 2, 3, 5, 10—12, 15, 24, 30—32, 45, 47, 54, 80, 81, 88, 99, 101, 104, 109, 111, 112, 114, 115, 130, 589
— —, nucleoli of 90, 99, 101, 109
— extracts, biological activity of VI 80, 82, 516, 517, 519, 525, 536—538, 540, 550, 552
— insufficiency, see Insufficiency, adrenocortical
—: neurohypophyseal interrelationship VII, 543—545
— sex hormones VII, 570—624; see also Adrenocortical steroids, sex hormone activity, Androgens, Estrogens, Progestins
— steroids, see also individual compounds
— — and metabolites in urine 125, 136, 197, 210, 291—346, 411, 412, 414, 419, 421, 422, 425—434, 441, 445, 457, 459, 460, 463, 572, 574, 582, 583, 585, 586, 590, 591, 611, 615, 616, 656; see also Adrenocortical steroids, catabolism
— —, biosynthesis of VII, 108, 137—139, 434—458, 571—574; see also Incubations with adrenocortical tissue, Perfusion
— —, catabolism of 458—499, 576, 577; *in vitro*, 461—463, 478, 484; *in vivo* 459, 460, 478, 482, 484
— —, chemical nature and reactions of 186—291, 411—434
— —, common names of 202—204, 412, 414, 417—421
— —, exogenous, see individual hormones, biological activity

Adrenocortical steroids, extraction of, see isolation,
— —, glucocorticoid activity of VI, 69, 80—82, 100, 101, 110, 124, 132, 206, 514, 517, 527, 536, 538, 546, 547, 583, 584, 596, 599, 608, 620—622, 638, 640, 645; see also Corticosterone, Cortisol, Cortisone, Dehydrocorticosterone, Glucocorticoids, Prednisolone, Prednisone
— —, identification 334—381
— —, — by color reactions 337, 346, 362—367
— —, — by crystalline properties 335—337
— —, — by instrumental methods 346—367; infrared spectrophotometry 202—204, 347—357; ultraviolet and visible-light spectrophotometry and fluorescence 358—367
— —, — by mobility 334, 335
— —, — by molecular rotation differences 373, 374
— —, — by nuclear magnetic resonance 376—380
— —, — by optical rotatory dispersion 374—376
— —, — by polarography 380, 381
— —, — with isotopic compounds 367—372
— — in bile 422, 480
— — in blood 74, 125, 291—346, 422—425, 455, 456, 471, 555, 572—574, 583, 584, 590—593, 609, 610; see also Adrenal-venous blood
— — in feces 471, 480
— —, isolation of 197, 292—334, 411—434
— —, melting points of 202—204
— —, mineralocorticoid activity of VII, 514—521, 524—541; see also Aldosterone, Deoxycorticosterone, Halogen-substituted corticoids, Mineralocorticoids
— —, partition coefficients of 305, 306
— —, secretion of, see biosynthesis, and in blood
— —, separation of 301—334
— —, — by adsorption chromatography 306—312
— —, — by electrochromatography 304, 306
— —, — by partition chromatography 54, 74, 76, 313, 314; on columns 313—318, on paper 318—334
— —, — by solvent distribution 304—306, 315, 317, 320
— —, sex hormone activity of VI, VII, 134, 135, 571—623; see also Androgens, Estrogens, Progestins
— —, side chains of 187, 206—210
— —, structures of 197—239, 411—421
— —, synthesis of 251—264; partial, chemical 251—258; partial, microbial 258—261; total, chemical 261—264
— —, systematic names of 202—204
Adrenocortical tissue, anatomy of V, VI, 1—48
— —, — of, developmental 2, 3, 5—10, 12—14, 17, 20, 27, 30—32, 38—45, 47, 48, 121—127, 570, 587—589, 595—598; see also "Fetal" zone, Postnatal development of adrenal gland, Prenatal development of adrenal gland, X zone
— —, — of, electronmicroscopic 3, 9, 23, 24, 32, 33, 36—38, 58, 110, 111
— —, — of, gross 1, 3, 5—13, 17—21, 77—83, 141
— —, — of, methods of study 3, 4, 50, 52, 54—57, 59, 61, 63—67
— —, — of, microscopic 1—3, 5—17, 22—48, 50, 52—67, 80, 81, 86—104, 109—121, 137, 138, 143, 586—598
— —, — of, pathological alteration of 128—141, 522, 523
— —, atrophy of VII, 77—82, 99, 100, 135, 136, 140, 145, 523; see also Addison's disease, Hypophysectomy, Simmonds' disease
— —, blood flow through 72, 73, 86
— —, blood supply of 1—3, 5—14, 19, 21—24, 28, 30, 37, 39, 44, 118, 140, 141, 143
— —, chemistry of 1, 48—68; see also physiology
— —, — of, by classical methods 48, 49, 54, 55, 58, 59, 61, 62, 66
— —, — of, by quantitative histochemistry 49—51, 59, 62, 63, 66, 67
— —, — of, by visualization techniques 49, 50, 52—61, 63—67
— —, color of 2, 6, 9, 17
— —, hypertrophy and hyperplasia of 127, 129, 133, 137—139, 145, 455, 523, 524, 549, 552—555, 585, 605, 615, 618; see also Stress
— —, lipides in 49—58, 68, 92—96, 103
— —, microsomal fraction of 61—63, 443, 461, 463, 484; see also Ribonucleoproteins
— —, mitochondrial fraction of 435, 441—443, 449, 489, 490; see also Adrenocortical cells, mitochondria
— —, nerve supply of 2, 4, 6, 11, 13, 24—26, 39, 456
— —, physiology of VI, 68—145
— —, — of, and associated anatomical changes VI, 69, 70, 77—83, 86—98, 104, 586—598
— —, — of, and associated chemical changes 70, 83—86, 91—96, 98, 103
— —, secretory activity of 70, 71, 73—77, 85, 93, 97, 100, 103, 105—109, 125; see also Adrenocortical steroids, biosynthesis
Adrenocorticotrophic hormone, see ACTH
Adrenogenital syndrome, see Adrenal virilism, Androgens

Adrenosterone (Δ^4-Androstene-3,11,17-trione) 196, 202, 414, 415, 480, 481, 573, 577
Agama, see Lizards, Reptiles
Age differences in anatomy and physiology of adrenal glands 20, 47, 48, 51, 58, 66, 67, 72, 98, 108, 114, 128, 133, 463, 582—584, 587—591, 608; see also Adrenocortical tissue, anatomy, developmental, gross
Agnatha (Cyclostomes; Lamprey eel, *Petromyzon*), glands of, anatomy of 1, 5
—, — of, chemistry of 50
3-Alcohols, see 3-Hydroxy compounds
11-Alcohols, see 11-Hydroxy compounds
17-Alcohols, see 17-Hydroxy compounds
20-Alcohols, see 20-Hydroxy compounds
Aldosterone (11 β,21-Dihydroxy-3,20-diketo-Δ^4-pregnen-18-al), antagonists of 524
—, biosynthesis of 446—449
—, catabolism of 475, 476
—, chemistry of 196, 203, 205, 206, 238, 239, 264, 305, 372, 419
—, excretion of 111, 115, 116, 433, 475
—, physiological activities of 99, 101, 519—521, 523—525, 527, 530—533, 535, 537, 541, 543, 544, 546, 547, 550, 555
—, secretion of 71, 74—76, 101, 103, 108 to 118, 120, 135, 137, 140, 141, 423
Aldosteronism, primary (Conn's syndrome) VII, 135, 531, 532, 535, 547, 584, 585, 653
—, secondary VII, 114, 532, 541; see also Edema
Alkaline phosphatase in adrenal gland 64, 65, 96, 98, 120
Allopregnane, see 5α-Pregnane
Amnestic syndrome 639, 643, 644, 648, 654, 655
"Amorphous" fraction of adrenocortical extracts VI, 519
Amphenone [1,1-bis(p-aminophenyl)-1-methyl propanone], 107, 136, 137, 140, 145, 456, 457, 522—524, 586
Amphibians, see also Anurans, Urodeles
—, glands of, anatomy of VI, 1, 8—10
—, — of, chemistry of 50
—, — of, physiology of 75, 80, 142—144
—, mineralocorticoid action in 550—552, 555
Amphiuma, see Urodeles
Anas, see Duck
Androgen: estrogen interaction 578, 616
— : glucocorticoid interaction 608
"Androgenic" zone of adrenal cortex 44, 108, 586—598
Androgens, adrenocortical 71, 126—128, 133, 134, 196, 414, 424, 426, 571—609, 616, 650, 651; see also Adrenal virilism, Adrenosterone, Δ^4-Androstenedione, C_{19} steroids, Dehydroepiandrosterone, 11 β-Hydroxy-Δ^4-androstenedione
—, —, biological activity of 574—580, 650, 651
—, —, biosynthesis of 108, 449—451, 571
—, —, catabolism of 478—481, 572—574
Androgens, adrenocortical, conversion of to estrogens 451, 616
—, effects of on adrenal gland 14, 68, 107, 128
—, relative activities of 577
Androstane (5α-) 194
— (5β-) see Etiocholane
Androstane-3,17-dione 306, 343, 427, 479, 577
Δ^4-Androstene-3,17-dione 202, 343, 414, 415, 422, 427, 479, 571, 573, 574, 576, 578, 590, 606, 610
Δ^5-Androstene-3 β,17 β-diol 572
Δ^5-Androstene-3 β,16 α,17 β-triol 572
Δ^5-Androsten-3 β-ol-17-one, see Dehydroepiandrosterone
Androsterone (3 α-Hydroxyandrostan-17-one), biological activity of 577
—, biosynthesis of 441, 449, 462, 466, 469, 478, 479, 574, 590
—, catabolism of 486, 489, 496
—, chemistry of 195, 196, 294, 306, 321, 343
— in blood 75, 422, 583, 584, 591
— in urine 294, 427, 452, 459, 460, 572, 582, 583, 586
Anencephaly 123, 126, 127, 591—593
Anguilla see Eel, silver
Anorexia nervosa 69
Anterior interrenals 7; see also Fishes, glands of, anatomy
Antidiuretic hormone (Vasopressin) 543—545
Anti-inflammatory activity of glucocorticoids 132, 517, 520, 533
Anurans (Frogs, Toads), glands of, anatomy of 1, 2, 6, 9, 10
—, — of, chemistry of 53
—, — of, physiology of 142
Apes (*Pan*, etc.), adrenocortical steroids of, catabolism of 459
—, glands of, anatomy of 17, 18, 42
—, neurosecretory material in hypothalamus of 588
Apoda, see Gymnophiona
Armadillo, see Edentata
Artiodactyla, see also Goat, Ox, Pig, Ruminants, Sheep
—, glands of, anatomy of 2, 19, 25, 35
Aryl sulfatase (Phenolsulfatase) in adrenal gland 66, 67, 98
Ascites 114; see also Aldosteronism, secondary
Ascorbic acid as index of adrenocortical activity 80, 84—86, 93, 96, 103, 104, 120, 123, 125, 137, 140
— — in adrenal gland 48, 58—60
— —, influence of on adrenal gland 70, 454
Aves, see Birds
Axolotl, see Amphibians

Bacterial toxins 103, 104, 129, 458; see also Stress
Bats (Chiroptera), glands of, anatomy of 18
Beef, see Ox
Bioassay of androgens 465, 573—576, 601
— of estrogens 617, 618
— of glucocorticoids 73
— of mineralocorticoids 73, 519, 520, 524—527, 533

Bioassay of progestins 611—613
Biosynthesis of steroids, see Adrenocortical steroids, biosynthesis
Birds (Aves), adrenocortical sex steroids in 610
—, — steroids of 417
—, glands of, anatomy of V, 2, 4, 13—17
—, — of, chemistry of 50, 53, 58, 60
—, — of, physiology of 75, 80, 143, 144
—, mineralocorticoid action in 553—555
Birefringence of adrenal lipides 52, 53, 83, 92, 93, 117, 121, 125
Bisnor steroid 413
Blood bicarbonate level 515, 528, 531
— electrolytes, see Electrolytes, plasma
— flow to brain, influence of adrenocortical hormones on 534, 643, 644
— pressure VI, 116, 514, 525, 536, 547; see also Hypertension
— steroids, see Adrenocortical steroids in blood
— volume (Hemoconcentration), influence of adrenocortical steroids on 114, 115, 515, 529, 535—537
Bos, see Ox
Bovine, see Ox
Brain function, glucocorticoids and 644—646
Breeding season, influence of on adrenal glands 9, 18, 21
Bufo, see Toads, Anurans

C_{17} steroids 413
C_{18} steroids 202, 210—213, 411—413; see also Estrogens
C_{19} steroids 202, 210—213, 413—415; see also Androgens
— —, secretion of 71, 72, 75, 108
C_{21} steroids 202—204, 416—421, 458—477; see also Glucocorticoids, Mineralocorticoids, Progestins
C_{27} steroids 204, 416—421; see also Cholesterol
Calf, see Ox
Canine, see Dog
Canis, see Dog
Capon, see Chicken
Capra, see Goat
Capsular nests of cortical tissue 29, 45, 46
Carassius, see Goldfish
Carbohydrate level in diet, influence of on adrenal gland 69, 70
Carbonyl groups of steroids, reactions of 243—247
Cardiac insufficiency, experimental 114, 531; see also Aldosteronism, secondary
Carnivores, see also Cat, Dog, Ferret, *Panthera*, Seals; glands of, anatomy of 17, 18, 25, 28, 598
Carp, see Teleosts
Cat *(Felis domesticus)*, adrenocortical sex steroids in 398, 573
—, — steroids of 424
—, gland of, anatomy of 18, 20, 21, 23, 25, 26, 32, 35, 37, 45
—, — of, chemistry of 51, 55, 58, 65
—, — of, physiology of 68, 74—76, 85, 104, 105, 140
Cat, mineralocorticoid action in 525, 542
Cattle, see Ox
Caucasians, gland of, anatomy of 18, 19; see also Negroes
Cavia, see Guinea pig
Cell-migration hypothesis 45, 46
Cetaceans, see Whales
Chelonia (Turtles, Terrapins), glands of, anatomy of 10
—, — of, physiology of 143
Chicken *(Gallus domesticus)*, adrenocortical sex steroids in 605
—, — steroids of, catabolism of 496
—, gland of, anatomy of 13, 14, 16, 17
—, — of, chemistry of 51, 52, 57, 58
—, — of, physiology of 71, 75, 83—85, 143, 145
—, mineralocorticoid action in, 553—555
Chlorosteroids, see Halogen-substituted steroids
Δ^4-Cholestenone 139, 455
Cholesterol as adrenocortical hormone precursor 435, 439, 441, 447, 449—451, 571, 609
— as index of adrenocortical activity 80, 83, 84, 92—94, 103, 104, 107, 120, 121, 137—140
—, chemistry of 48, 50—54, 187, 204, 343, 416
— in diet, influence of on adrenal 70
—, synthesis of 436—438, 523
Cholinesterase, histochemical demonstration of in adrenal gland 26
Chorionic gonadotrophin, influence of on adrenal gland 126, 454, 455, 587, 591, 611, 619, 621, 622
Chromaffin tissue, see Adrenal medulla
Chromatography, see Adrenocortical steroids, isolation, identification
Citellus, see Squirrel, ground
Clitoris, effect of steroids on, 579; see also Phallus
Coenzyme A, 62, 96, 139, 437, 458
Colchicine, influence of on adrenal gland 90
Colinus, see Quail, Virginia
Color reactions of steroids, see Adrenocortical steroids, identification
Columba, see Pigeon
Compound A, see 11-Dehydrocorticosterone
— B, see Corticosterone
— E, see Cortisone
— F, see Cortisol
— S (of Reichstein), see 11-Deoxycortisol
Configuration of steroids 189—191, 289—291
— of 5α compounds 481, 482
— of 5β compounds 481, 482
Conformation of steroids 191—193; axial and equatorial substituents 193
Congenital adrenal hyperplasia, see Adrenal virilism
Conjugated steroids 292—296, 424, 425, 434, 452, 460—463
— —, extraction of 292, 293
— —, glucosiduronates (glucuronates) 292—296, 424, 425, 434, 452, 460—463, 475, 480

Conjugated steroids, hydrolysis of 293—296, 434
— —, steroidoproteins 292, 293
— —, sulfates 292—296, 434, 452, 572
Conn's syndrome, see Aldosteronism, primary
Cortexone, see Deoxycorticosterone
Corticoids, see Adrenocortical steroids
Corticosterone (11 β,21-Dihydroxy-Δ⁴-pregnene-3,20-dione, Compound B), biological activity of 99, 518, 524, 527, 534, 546, 594, 596
—, biosynthesis of 444, 523, 571, 609
—, catabolism of 460, 470, 494
—, chemistry of, 196, 203, 205, 305, 345, 358, 418
— in urine 432
—, secretion of 71, 74—77, 86, 96, 108, 109, 120, 137, 140, 141, 423, 467, 550, 555, 590, 603, 619, 623
Cortisol (11 β,17 α,21-Trihydroxy-Δ⁴-pregnene-3,20-dione, Compound F), biological activity of 81, 101, 517, 518, 521, 524, 527, 528, 533—535, 538, 544—546, 581, 592, 620, 642
—, biosynthesis of, 440—445, 489, 490, 493, 494, 571, 609
—, catabolism of 459—462, 471—475, 495, 497, 573
—, chemistry of 196, 197, 203, 205, 305, 345, 351, 358, 419, 422, 423
— in urine 432, 433, 459
—, secretion of 71, 74—76, 100, 136, 138, 141, 422, 423, 472, 475, 522, 523, 550, 555, 590
Cortisone (17 α,21-Dihydroxy-Δ⁴-pregnene-3,11,20-trione, Compound E), biological activity of 80, 81, 86, 91, 94, 96, 122, 125, 127, 128, 135, 137, 143, 518, 519, 521, 524, 527, 528, 531—535, 538, 539, 544, 546, 549, 550, 552, 554, 579, 581, 596, 597, 603, 607, 608, 614—616, 620, 622, 641—644, 646—649, 651, 654, 655
—, biosynthesis of 442, 443, 490, 493, 571
—, catabolism of 459, 462, 463, 472, 474, 497
—, chemistry of 196, 197, 203, 205, 262, 263, 305, 342, 345, 419, 422, 423
— in urine 432, 459
—, secretion of 71, 75, 423, 471, 550, 555
Cortol (Pregnane-3 α,11 β,17 α,20 α,21-pentol) 433, 434, 460, 473
β-Cortol (Pregnane-3 α,11 β,17 α,20 β,21-pentol) 433, 434, 460, 461, 473
Cortolone (3 α,17 α,20 α,21-Tetrahydroxy-pregnan-11-one) 433, 434, 459, 460, 461, 474, 497
β-Cortolone (3 α,17 α,20 β,21-Tetrahydroxy-pregnan-11-one) 433, 434, 459, 460, 461, 474, 497
Corticotrophin, see ACTH
Cow, see Ox
Cricetomys (Giant rat), gland of, chemistry of 60, 65, 67
Cricetulus, see Hamster, Chinese
Cricetus, see Hamster, European
Crocodilia, glands of, anatomy of 10
—, — of, chemistry of 65, 66
Cushing's disease VII, 101, 107, 130, 133, 413—415, 424, 435, 457, 458, 531, 540, 576, 584, 615, 617, 638—640, 642—649, 652, 655
Cyclic secretion of adrenal sex steroids 622, 623
Cyclostomes, see Agnatha
Cyprinus (Carp), see Teleosts
Cytoplasmic basophilia, see Ribonucleoproteins
Cytotoxic agents for adrenocortical tissue 129, 135, 136, 141, 457, 523

Daily production of adrenocortical steroids 76, 109, 412, 472—475, 482, 483, 582, 583
DDD (Bis-chlorophenyl-dichloroethane) 135, 136, 141, 457, 458, 523
"Definitive" cortex, see Adrenocortical tissue, anatomy of, developmental, "Fetal" zone
Δ¹-Dehydro derivatives 266, 277, 278; see also Prednisone, Prednisolone
Dehydro derivatives, other 266, 277—280
11-Dehydrocorticosterone (21-Hydroxy-Δ⁴-pregnene-3,11,20-trione, Compound A) 75, 196, 203, 205, 305, 345, 418, 423, 467, 471, 518, 524, 527, 534, 571, 608
Dehydroepiandrosterone (3 β-Hydroxy-Δ⁵-androsten-17-one), biological activity of 577—579, 594, 596, 597, 603, 607
—, biosynthesis of 441, 449—451, 571
—, catabolism of 61, 478, 479, 486, 487, 571, 572, 574
—, chemistry of 196, 202, 294, 343, 364, 414, 415
— in urine 294, 427, 572, 574, 580, 585
—, secretion of 75, 422, 583, 584, 590, 591
Dehydrogenases in adrenocortical tissue:
DPNH 62, 63
Glucose-6-phosphate (TPNH-generating) 62, 63, 96, 453
Succinic 63, 91, 92, 107
— and reductases, steroid, in hormoneproducing glands:
Δ⁵-3 β-Hydroxy-(Steroid-3 β-ol dehydrogenase), 61, 96, 97, 128, 137, 413, 439, 496, 497, 590
20 β-Hydroxy-, 61, 62
— — —, —, in microorganisms:
Δ¹-, 260, 277, 497
3 α- and 3 β-Hydroxy-, 260, 496
Δ⁴-5 α- and 5 β-Hydroxy-, 260
7 α-Hydroxy-, 260
12 α-Hydroxy-, 260
17 α- and 17 β-Hydroxy-, 260
20-keto-, 260
— — —, —, in peripheral tissues, mammalian:
Δ¹-, 497, 498
3 α- and 3 β-Hydroxy-, 495, 496
Δ⁴-5 α- and 5 β-Hydroxy-, 498, 499
11 β-Hydroxy-, 497
20-keto-, 497

21-Dehydroxylation 610
11-Deoxycorticosterone (21-Hydroxy-Δ^4-pregnene-3,20-dione, DOC, Cortexone), biological activity of 82, 90, 101, 115, 143, 145, 517, 524, 527, 530—547, 596, 598, 599, 601—604, 608, 613, 614, 620, 622, 653
—, biosynthesis of 443, 444, 448, 571
—, catabolism of 432, 440, 442, 444, 448, 466, 468, 571
—, chemistry of 196, 202, 205, 305, 344, 418
—, secretion of 75, 76, 138, 423, 523, 610
11-Deoxycortisol (17α,21-Dihydroxy-Δ^4-pregnene-3,20-dione, Reichstein's Compound S), biological activity of 518, 524, 527, 534, 608, 653
—, biosynthesis of 441—444, 571
—, catabolism of 432, 444, 467, 469
—, chemistry of 197, 203, 205, 305, 345, 418
—, secretion of 76, 82, 138, 423, 523, 585
21-Deoxycortisol 432, 433, 442, 444, 585
17,20-Desmolase 416, 451, 465, 499
20,22-Desmolase 416, 439, 499
Diabetes insipidus-like syndrome 517, 534, 535
Didelphis, see Opossum
Diethylstilbestrol, see Estrogens, synthetic
3β,16α-Dihydroxy-Δ^5-androsten-17-one 572
Dihydroxyketones 222—225, 339—341; see also Cortisol, Cortisone, 11-Deoxycortisol, Glucocorticoids
Diols 218, 219, 221, 222, 321, 322, 329, 330, 344, 420, 427—429, 431—433, 435, 449, 459, 464, 478, 479, 483, 484
Diones 321, 322, 329, 343, 344, 420, 422, 427—430, 432, 442, 443, 449, 450, 464, 476, 478, 479
Dipnoi (Lungfishes), glands of, anatomy of VI, 1, 2, 6
—, — of, chemistry of 50, 53
—, — of, physiology of 141
—, mineralocorticoid action in 555
Diurnal variation in adrenocortical tissue 50, 62, 93, 96
— — in steroid production 475, 541, 572
DOC, see Deoxycorticosterone
Dog *(Canis familiaris)*, adrenocortical sex steroids in 573, 610, 618
—, — steroids of 420, 421, 423, 424
—, — — of, biosynthesis of 435, 439, 441, 457, 458
—, — — of, catabolism of 461, 463, 480, 484, 486
—, gland of, anatomy of 18, 20, 21, 23, 25 to 29, 33, 34, 45
—, — of, chemistry of 51, 58, 63, 65—68
—, — of, physiology of 71, 72, 74—76, 78, 79, 81—83, 85, 88, 100, 104—106, 108—112, 114, 116—118, 135—138, 140, 141
—, mineralocorticoid action in 515, 519, 521, 523—525, 527—531, 533—537, 539, 540, 542, 645
Dogfish, see Elasmobranchs
Double bonds, reactions of 247—250

Drake, see Duck
Duck *(Anas platyrhynchus)*, gland of, anatomy of 13
—, — of, physiology of 143
—, mineralocorticoid action in 553, 554

Edema 109, 114, 517, 520, 532, 533, 538, 546, 552; see also Aldosteronism, secondary
Edentata (*Armadillo*, Sloths), adrenocortical sex steroids in 615
—, glands of, anatomy of 18
—, mineralocorticoid action in 526
Eel, lamprey, see Agnatha
—, silver *(Anguilla anguilla)*, gland of, anatomy of 8
—, —, — of, chemistry of 51, 58
—, —, — of, physiology of 78
—, —, mineralocorticoid action in 549
Elasmobranchs (Dogfish, Sharks, Rays, *Torpedo*), glands of, anatomy of VI, 1, 5—7
—, — of, chemistry of 50, 53, 60
—, — of, physiology of 75, 79, 80, 141
—, mineralocorticoid action in 548, 555
Electrocortin, see Aldosterone
Electroencephalogram, 534, 640, 642, 655
Electrolytes, see also Sodium, Potassium
—, in adrenocortical tissue 68
— in plasma 73, (76), 98, 99, 115, 118, 136, 141, 515, 521, 526, 528—532, 534—537, 548, 551, 554, 643
— in saliva 535, 540
— in sweat 73, 535, 540
— in tissues 143, 532, 538—540, 549, 552
— in urine, VI, 73, 520, 524, 534, 535, 543, 553, 554
—, metabolism of 514—521, 529—533
Electroshock 534, 648, 649
Enantiomorphs, see Stereoisomerism
Endoplasmic reticulum 37, 54, 92; see also Adrenocortical tissue, microsomal fraction, Ribonucleoproteins
Enzyme deficiencies in adrenal virilism 585, 586
Enzymes in adrenocortical tissue, see Dehydrogenases, Hydroxylases, Hydrolytic enzymes
Eosinopenia, see Eosinophils
Eosinophils (leukocytes) 71, 73, 102, 103, 105, 136, 139, 517
Epiandrosterone (3β-Hydroxyandrostan-17-one), biosynthesis of 478
—, catabolism of 489, 496
—, chemistry of 197, 262, 306, 321, 343, 429
— in urine 427, 452
Epilepsy 534, 640
Epimers, see Stereoisomerism
Epinephrine, see Adrenaline
Equilenin and Equilin 412, 451
Equus, see Horse
Erinaceus, see Insectivores
Esterase, nonspecific, in adrenocortical tissue, 65, 81, 98, 115
Estradiol-17β [$\Delta^{1,3,5(10)}$-Estratriene-3,17β-diol], biological activity of 96, 106, 613, 618

Estradiol-17-β, biosynthesis of 483
—, catabolism of 425, 483, 484
—, chemistry of 196, 321, 329, 343
— in urine, 334, 412, 425, 426, 484, 616
Estrane 194
Estriol [$\Delta^{1,3,5(10)}$-Estratriene-3,16α,17β-triol], biosynthesis of 451
—, catabolism of 425, 484
—, chemistry of 196, 321
— in urine 334, 412, 425, 426, 484, 616, 617
Estrogen: glucocorticoid interaction 620—622
Estrogens, see also C_{18} steroids, Estradiol, Estriol, Estrone; biological activity of 68, 103, 106, 126, 128, 412, 455, 497, 499, 585, 592, 601, 606, 612, 613, 617—619, 623, 652
—, biosynthesis of 109, 412, 451, 483, 579, 616
—, catabolism of 425, 426, 483, 484, 496
—, chemistry of 321, 343, 425
— in urine 411, 412, 425, 426, 484, 615—617
—, secretion of 75, 134, 411, 584, 606, 607, 616, 618
—, synthetic, biological activity of 107, 455, 456, 523, 524
Estrone [3-Hydroxy-$\Delta^{1,3,5(10)}$-estratrien-17-one], biological activity of 106, 455, 617
—, biosynthesis of 451, 479
—, catabolism of 425, 483—485
—, chemistry of 196, 202, 301, 321, 329, 343, 412
— in urine 334, 412, 425, 426, 484, 616, 617
—, secretion of 412, 615
Etianic acid 194
Etiocholane 194
Etiocholanolone (3α-Hydroxyetiocholan-17-one) 197, 306, 422, 466, 469, 479, 572, 574, 576, 580—583, 586, 691; see also 5β/5α Reduced metabolites
Extracellular: intracellular fluid compartments 529, 535—539, 544
Extraction of steroids, see Adrenocortical steroids, isolation

Fasciculata, see Zona fasciculata
Felidae, see Cat, *Panthera*
Ferret *(Mustela furo)*, adrenocortical steroids of 424
—, mineralocorticoid action in 526, 527
Fetal glands, see also Adrenocortical tissue, anatomy of, developmental
— —, chemistry of 50, 51, 54, 58, 59, 65, 66
— —, secretory activity of 54, 109, 125, 415, 421, 471, 574, 578, 584, 587—593, 605
"Fetal" zone 20, 27, 41, 45, 59, 72, 109, 123, 125—127, 129, 587—589, 593, 598; see also Fuchsinophilia
Fishes, see also Agnatha, Dipnoi, Elasmobranchs, Holostei, Teleosts; adrenocortical steroids in 417
—, glands of, anatomy of VI, 1, 5—8
—, — of physiology of 71, 75, 82, 89, 141, 143, 145
—, mineralocorticoid activity in 548—550, 555
Fluorosteroids, see Halogen-substituted steroids
Formaldehydogenic steroids 77, 340
Fowl, see Chicken
Frogs (*Rana*, *Hyla*), glands of, anatomy of 9, 10
—, — of, chemistry of 51, 58, 60, 65
—, — of, physiology of 76, 79, 80, 85, 90, 96, 143
—, mineralocorticoid action in 550, 551, 555
Fuchsinophilia of adrenocortical tissue 2, 7, 9, 17, 41—44, 108, 109, 602
Fundulus, see Teleosts

Gallus, see Chicken
Ganoid fishes, see Holostei
Gastrointestinal function, influence of corticoids on 540, 622
Giant cells in adrenal cortex 32, 129, 587; see also Adrenocortical tissue, hypertrophy and hyperplasia
— rat, see *Cricetomys*
β-Globulin, ACTH-like action of 77
Glomerular filtration rate, 543
Glomerulosa, see Zona glomerulosa
"Glomerulotropin" VII, 115, 117, 522, 544
Glucocorticoids, deficiency of 136, 457, 523; see also Addison's disease, Adrenocortical tissue, atrophy, Hypophysectomy
—, secretion of 71—77, 80—82, 99—103, 105, 108, 110, 111, 114, 115, 119, 122, 133, 135, 140, 457, 523; see also Adrenocortical steroids, glucocorticoid activity, Adrenocortical tissue, physiology, C_{21} steroids, Cushing's disease, Inner zones
Gluconeogenesis VI, 73, 99, 102, 106, 122, 139, 524; see also Adrenocortical steroids, glucocorticoid activity
Glucosiduronates, see Conjugated steroids
β-Glucuronidase in adrenocortical tissue 66, 67, 98
Glutathione in adrenocortical tissue 68
Glycogen in adrenocortical tissue 67
Glycols, see Diols
Goat *(Capra hircus)*, adrenocortical steroids of 423
—, gland of, anatomy of 17, 30, 35
—, — of, chemistry of 65, 67
—, — of, physiology of 129
Goldfish *(Carassius auratus)*, gland of, anatomy of 4, 7, 8
—, — of, physiology of 89, 90
—, mineralocorticoid action in 549
Gonadectomy 127—129, 134, 411, 412, 572, 573, 579, 580, 593, 597, 599, 601—603, 605—608, 610, 615, 616, 618—622
Gonadotrophins, control of secretion of 616
—, influence of on adrenal gland 127, 128, 134, 498, 592, 596, 607, 619, 621; see also Chorionic gonadotrophin, Luteinizing hormone, Luteotrophic hormone
Grafts, adrenal, see Adrenal transplants
Graves' disease 644, 645, 647, 649; see also Hyperthyroidism

Growth hormone 68, 90, 100, 110, 111, 116, 498, 576
Guinea pig *(Cavia cobaya)*, adrenocortical sex steroids in 577, 604, 607
— —, — steroids of, biosynthesis of 435, 439, 441, 445, 452, 454, 456, 458
— —, — steroids of, catabolism of 460, 461, 480, 484, 496
— —, — steroids of, chemistry of 414, 421, 422, 424, 427—430, 432, 433
— —, gland of, anatomy of 17, 18, 20, 21, 28, 32, 35, 37, 39, 40, 45, 47
— —, — of, chemistry of 49, 50, 51, 54, 57—60, 62, 63, 65—68
— —, — of, physiology of 69, 71, 75, 78, 81, 82, 84, 85, 88, 90, 92, 94, 96, 100, 104—106, 109, 129
— —, mineralocorticoid action in 523, 526
Gymnophiona (Apoda), glands of, anatomy of 8

Hair growth 579, 580
Halogen-substituted steroids 265—272, 424, 520, 521, 527, 533, 536, 540, 550
Hamster, Chinese *(Cricetulus griseus)*, gland of, physiology of 104
—, European *(Cricetus cricetus)*, gland of, anatomy of 21, 35
—, —, — of, physiology of 94, 128
—, golden *(Mesocricetus auratus)*, adrenocortical sex steroids in 597, 603, 604, 609, 619
—, —, — steroids in, biosynthesis of 457
—, —, — steroids in, catabolism of 496
—, —, gland of, anatomy of 18, 19, 21, 27, 32, 34, 35, 37, 48
—, —, — of, chemistry of 54, 57, 58, 63, 65, 67
—, —, — of, physiology of 73, 74, 75, 78, 83, 85, 92, 94, 96, 104, 105, 109, 115, 128, 132, 137
"Head" kidney of fishes 4, 7
Heart failure, see Cardiac insufficiency
Hematopoiesis in adrenocortical gland 23
Hemoconcentration, see Blood volume
Hemorrhage in adrenal gland 129—132
Hepatic, see Liver
Hexestrol, see Estrogens, synthetic
Hibernating mammals, see also Hamster, European and golden, Insectivores, Marmot, Squirrel, ground
— —, glands of, anatomy of 2, 19, 35
— —, — —, physiology of 72, 106, 107
Hirsutism 586, 650
Histochemistry, see Adrenocortical tissue, chemistry of
Hog, see Pig
Holostei (Ganoid fishes), glands of, anatomy of 7
—, — of, chemistry of 50
Homo, see Man
D-Homosteroids 195, 288, 289
Horse *(Equus caballus)*, adrenocortical sex steroids in 616
Horse, adrenocortical steroids of 426—428, 431
—, — — of, biosynthesis of 451
—, gland of, anatomy of 19, 33, 37
—, — of, chemistry of 51, 59, 65, 67
Human being, see Man
Hydrocortisone, see Cortisol
Hydrolysis and extraction of steroids, see Adrenocortical steroids, isolation, Conjugated steroids
— artifacts 218, 294, 465, 466
Hydrolytic enzymes in adrenocortical tissue 64—67, 96—98
Hydroxy compounds:
  2-, 233, 234, 433, 445
  3-, see Androgens, Androsterone, Cholesterol, Cortol, Cortolone, Dehydroepiandrosterone, Epiandrosterone, Estradiol, Estriol, Estrone, Etiocholanolone, 11 β-Hydroxyandrosterone, 11 β-Hydroxyetiocholanolone, 17 α-Hydroxypregnanolone, Pregnane-3 α,20 α-diol, Pregnanetriol, Pregnanolone, Pregnenolone, Urocortisol, Urocortisone
  3-, saturated 231
  Δ^5-3 β-, 231, 232, 337
  6-, 234—236, 414, 415, 430, 433, 445, 459—463
  7-, 491, 495, 572
  11-, 228; see also Aldosterone, Cortisol, Corticosterone, 11 β-Hydroxyandrostenedione, 11 β-Hydroxyandrosterone, 11 β-Hydroxyetiocholanolone, 11 β-Hydroxyprogesterone
  16-, 236, 266, 426, 429, 430, 433, 445, 446; see also 16 α-Hydroxyprogesterone
  17-, 212, 213; see also Cortisol, Cortisone, 17-Hydroxycorticoids, 17 α-Hydroxypregnenolone, 17 α-Hydroxyprogesterone, Testosterone
  18-, 415, 433, 445, 451
  19-, 414, 415, 433, 445
  20-, 214, 423, 613
  21-, 445; see also Aldosterone, Corticosterone, Cortisol, Cortisone, Cortol, Cortolone, Deoxycorticosterone, Prednisolone, Prednisone, Urocortisol, Urocortisone
— — and derivatives, other 275—277, 459—463
11 β-Hydroxyandrostenedione 75, 108, 414, 422, 480, 481, 571, 573, 574, 577, 578, 590, 603
11 β-Hydroxyandrosterone 422, 481, 577, 586
17-Hydroxycorticoids 572, 582, 586, 590, 592, 593, 607; see also Glucocorticoids
11-Hydroxyestrone, see Estrogens
18-Hydroxyestrone, see Estrogens
11 β-Hydroxyetiocholanolone 294, 422, 481
Hydroxyketones 339—341, 445
17 α-Hydroxy-20-ketones 215—218
21-Hydroxy-20-ketones 219—221
Hydroxyl groups, reactions of 240—243

Hydroxylases, steroid, in hormone-producing glands
6α- and 6β-, 486, 488
11β-, 61, 137, 413, 440, 453, 485, 486, 488, 489, 523, 585
16α-, 486, 487
17α-, 137, 440, 451, 453, 489, 490
18-, 447, 451, 486, 487, 490
19-, 486, 488, 490
21-, 62, 137, 440, 453, 490, 585
—, —, in liver
2β-, 486, 488
6β-, 486, 488, 495
16α-, 486, 488
—, —, in microorganisms 258, 259
1-, 260, 487, 488
2β-, 260, 487, 488
6β-, 260, 485, 487, 488, 491
7α- and 7β-, 260, 487, 488, 491
8-, 260, 491
9-, 260, 491
10-, 260, 487, 488
11α- and 11β-, 259—261, 485, 487, 488, 491—493
12-, 260
14α-, 260, 277, 487, 488, 493
15α- and 15β-, 260, 487, 488, 493
16α-, 260, 487, 494
17α-, 260, 485, 494
19-, 494
21-, 260, 485, 494
11β-Hydroxylation, mechanism of 485, 488
17α-Hydroxypregnenolone 432, 450, 571, 573
11β-Hydroxyprogesterone 431, 432, 441, 444, 448, 466, 467, 533
16α-Hydroxyprogesterone 443, 445, 465, 466
17α-Hydroxyprogesterone 431, 441, 443, 444, 450, 465, 550, 571, 574, 585, 609—611
Δ^5-3β-Hydroxysteroid dehydrogenase, see Dehydrogenases and reductases, steroid
Hyla, see Frogs
Hyperactivity, adrenocortical VII, 129, 411, 414, 584, 585, 638—647, 650—653; see also Adrenocortical tissue, hypertrophy, Cushing's disease, Stress
Hypertension 120, 423, 517, 547, 548, 643; see also Blood pressure
Hyperthyroidism, endogenous 103
—, exogenous 90, 103—105, 110, 133, 609
Hypertrophy and hyperplasia of adrenocortical tissue, see Adrenocortical tissue, hypertrophy and hyperplasia, Tumors
Hypoglycemia 98, 100
Hyponatremia 112; see also Electrolytes in plasma
Hypothalamic lesions 86, 544, 588
Hypophysectomy 68, 70, 78, 79, 84, 86, 88—92, 94—99, 101, 109, 110, 113, 118, 122, 124, 125, 127, 132, 133, 135, 141—144, 456, 517, 549, 551—554, 619
Hypothyroidism 68, 110, 140, 145, 483, (522)

Immunological factor in regeneration of adrenocortical tissue 120
— — in adrenocortical insufficiency 141
Incubations with adrenocortical tissue 77, 109, 111, 113, 435, 439, 441—443, 446, 447, 449, 453, 486, 489, 490, 572, 590; see also Adrenocortical steroids, biosynthesis, catabolism
Infection of adrenocortical tissue 132, 141
Infrared spectra of adrenocortical steroids 202—204, 347—357
— spectroscopy of adrenal tissue 54
Inner zones of adrenal cortex, VII, 47, 65, 80, 81, 94, 96, 99—110, 124, 132, 133, 135, 136, 139—141, 143, 145, 447, 586—593, 595, 596; see also "Fetal" zone, X zone, Zona fasciculata, Zona reticularis
Insectivores (*Erinaceus*, Hedgehog, Shrews), adrenocortical sex steroids in 598
—, glands of, anatomy of 18, 21, 30, 35, 40, 48
—, — of, chemistry of 129
Insufficiency, adrenocortical, VI, 129, 132, 135, 136, 138—140, 515—520, 525, 535—537, 540, 542, 548, 653—656; see also Addison's disease, Adrenocortical tissue, atrophy, Cytotoxic agents, Hypophysectomy Infection, Ischemia
Insufficiency, pituitary, see Pituitary, anterior, insufficiency
Insulin, effect of on adrenal gland 145
Interrenal tissue, see Adrenocortical tissue, Lower vertebrates
Ischemia of adrenocortical tissue 140
Isoandrosterone, see Epiandrosterone
Isomerase, $\Delta^5 \rightarrow \Delta^4$ 497
Isotopic analysis of steroids 367—372
i-Steroids 294

Kepler's test, see water loading
Keto compounds
3-, 230, 231, 445
Δ^4-3-, 299, 230; see also individual adrenocortical steroids
7-, 572
11-, 227, 228, 426, 445, 467; see also Adrenosterone, Cortisone, 11-Dehydrocorticosterone
17-, 210—212, 426, 445, 455; see 17-Ketosteroids
20-, 213, 445
17-Ketogenic steroids 340, 341, 365, 366, 472
α-Ketols, see Hydroxyketones
16,17-α-Ketols 446
Ketonic and non-ketonic steroids, separation of 302, 303
11-Ketopregnanediol 467
11-Ketoprogesterone 467
Ketosteroids, reactions for 337—339, 342
—, question of histochemical tests for 55, 56
17-Ketosteroids, see also Androgens, Androsterone, Epiandrosterone, Etiocholanolone, 11-Oxy-17-ketosteroids, Testosterone, catabolism
—, chemistry of 293—295, 572, 573
— in urine, 81, 136, 137, 139, 457, 572, 573, 580—586, 591, 607, 656
—, secretion of 125, 127, 574, 590, 591, 593

Kidney function, influence of corticoids on 514, 515, 532, 535, 537, 541, 543
— lesions, effect of on adrenal gland 112, 116, 532
—, steroid transformations in 452, 453, 461—463, 465

Lacertilia, see Lizards
Lactogenic hormone, see Luteotrophic hormone
Lagomorpha, glands of, anatomy of 18; see also Rabbit
Lamprey eel, see Agnatha
Larvae 5, 8
LH, see Luteinizing hormone
Life maintenance 516, 519, 524—528, 552, 553
Lion, see *Panthera*
Lipides in diet, influence of on adrenal gland 70
Lipofuscin 17, 45, 56—58, 67, 133
Liver cirrhosis, see Aldosteronism, secondary
—, steroid transformations in 412, 424, 425, 461—463, 472, 476, 478, 484, 486, 495—497, 499
Lizards (Lacertilia), see also Reptiles
—, glands of, anatomy of 11
—, — —, physiology of 80, 143
—, mineralocorticoid action in 552
Lower vertebrates, see also Fishes, Amphibians, Reptiles, Birds
— —, adrenocortical sex steroids in 599
— —, — steroids of 417
— —, glands of, anatomy of, V, VI, 1—17
— —, — of, chemistry of 56, 68
— —, — of, physiology of 141—145
— —, mineralocorticoid action in, 548—555
Lungfishes, see Dipnoi
Luteinizing hormone (LH), effect of on adrenal gland 47, 126—128, 592, 596
Luteotrophic hormone (LTH) 607, 609, 639
Lymphatic drainage of adrenal 24
Lymphopenia 73, 103, 139
Lysergic acid diethylamide (LSD) 579

Macaca, see Monkeys
Malnutrition 69; see also Stress
Mammals, see also individual species
—, glands of, anatomy of V 1—4, 17—48
—, — —, chemistry of 50, 53, 56, 58, 60, 68
—, — —, physiology of 72, 77, 78, 98, 99, 103, 109, 110, 121, 133, 141, 145
Man *(Homo sapiens)*, see also Caucasians, Negroes
—, adrenocortical sex steroids of 571—574, 580—593, 598, 603, 607, 609—611, 615—617, 638—656
—, — steroids of 412—416, 420—424, 426—429, 431—433
—, — — of, biosynthesis of 435, 439, 441—443, 449, 451, 455—458
—, — — of, catabolism of 459, 460, 463, 467, 471, 472, 475, 476, 478, 480, 482, 484, 497
—, gland of, anatomy of 17, 21, 23—28, 33, 39, 41—45, 48, 587—593
Man, gland of, chemistry of 49, 51, 54, 57, 59, 60, 63, 65—68
—, — of, physiology of 69, 71, 72, 75, 76, 78, 80, 84, 85, 89, 90, 92, 94, 100, 101, 104, 105, 107, 108, 110—113, 115—118, 121—123, 125—127, 129, 130, 132, 133, 135—140, 587—593
—, mineralocorticoid action in 517, 518, 523, 528—531, 533, 535, 537, 538, 540—543, 546
—, psychopathology in with adrenal disease 638—656
Marine mammals 17, 19; see also Seals, Sirenia, Whales
Marmot (Woodchuck, *Marmota monax*), adrenocortical lipides in 35
—, mineralocorticoid action in 526
Marsupials, see also Opossum; glands of, anatomy of 18, 30
Meadow mice, see Voles
Medulla, Medullary tissue, see Adrenal medulla
Menopause 463, 611, 616
Menstruation 581
Mesocricetus, see Hamster, golden
Methyl compounds
2α-, 267, 284, 285
4-, 287
6α-, 266, 283, 284
9α-, 286
16α-, 285
17-, 286
1-Methyl,19-nor compounds 287, 288
Microbial transformations of steroids, see Dehydrogenases and reductases, Hydroxylases
"Microbodies" in adrenocortical cells 33
Microsomes, see Adrenocortical tissue, microsomal fraction, Ribonucleoproteins
Microtus, see Voles
Microvilli on adrenocortical cells 38
Mineralocorticoids, see also Adrenocortical steroids, mineralocorticoid activity, Aldosterone, Aldosteronism, C_{21} steroids, 11-Deoxycorticosterone, Zona glomerulosa
—, chemistry of 206
—, definition of 514—521
—, secretion of 73—77, 99, 109—118, 120, 135, 144
Mobilities of steroids, see Relative mobilities
Modified hormones 264—291, 520, 521, 527, 533, 535, 536; see also Dehydro derivatives, Halogen-substituted steroids, Hydroxy compounds, Methyl compounds
Monkeys (*Macaca* spp.), adrenocortical steroids of 423, 424
—, — — of, catabolism of 484
—, glands of, anatomy of 3, 17—19, 21, 23, 25, 34, 35, 42, 45
—, — of, chemistry of 50, 51, 59, 66—68
—, — of, physiology of 75, 78, 80, 84—87, 104, 105, 109, 110, 115, 122
—, mineralocorticoid action in 522

Monkeys, neurosecretory material in hypothalamus of 588, 598, 622
Monotremata (Platypus, Echidna), glands of, anatomy of 29—31
Mouse, laboratory *(Mus musculus)*, adrenocortical sex steroids in 577—579, 586, 587, 593—598, 602, 603, 605, 607, 610, 612, 613, 617—620
—, —, — steroids of 423
—, —, — of, biosynthesis of 437
—, —, — of, catabolism of 460, 480, 484, 496
—, —, gland of, anatomy of 18, 20, 21, 23, 31, 32, 35, 37, 39, 43, 45, 47
—, —, gland of, chemistry of 50, 57, 58, 60, 62—67
—, —, — of, physiology of 71, 74, 78, 81, 83, 85, 86, 88, 93, 96, 98, 100—102, 104—106, 108—111, 115, 118, 120, 122, 125, 127, 128, 134, 135
—, —, mineralocorticoid action in 517, 518, 524, 526, 542, 545
Muscular dystrophy 531
Myotrophic activity of androgens 574—576

Natremia, see Sodium, Electrolytes, plasma
Natriuresis, see Sodium excretion
Natrix, see Reptiles, Snakes
Necrosis of adrenal gland 129, 138, 140, 145
Negroes, see also Caucasians; gland of, anatomy of 19
—, — —, physiology of 69
Nephrotic syndrome 114, 520, 532, 546; see also Aldosteronism, secondary
Neurasthenia 656
Neurohypophysis, see Pituitary, posterior
Neurosecretory material in hypothalamus 544, 545
Newts, see Urodeles
Nitrogen-containing steroids 418
Nocturnal animals 32, 50, 93, 96; see also Diurnal variation
Nomenclature of steroids, see Steroid nomenclature
Non-cholesterol pathway to corticoids 446
18-Norsteroids 413
19-Norsteroids, 239, 280—283, 287, 288, 355, 413
19-Nortestosterone 576

Opossum *(Didelphis virginianus)*, gland of, anatomy of 18, 23
—, — of, chemistry of 51, 58
—, — of, physiology of 85
—, mineralocorticoid activity in 526
Optical rotation of steroids 202—204
— rotatory dispersion spectra 202—204
Ornithorhynchus, see Montotremata
Oryctolagus, see Rabbit
Osmiophilia of adrenocortical lipides, see Adrenocortical cells, lipide droplets
Osmoregulation 549, 550
Ossification and adrenocortical androgens 574, 577—579, 603
Ovarian androgens 574
Ovis, see Sheep
Ox *(Bos taurus)*, adrenocortical sex steroids in 573, 574, 609, 610, 615
—, — steroids of 412, 415, 416, 420—423, 426, 427, 429, 431
—, — — of, biosynthesis of, 435, 439—443, 447, 449, 456, 457
—, — — of, catabolism of 461, 463, 466, 478, 484, 486, 497
—, gland of, anatomy of 17, 19, 21, 23, 25—27, 35, 37, 38, 45
—, — of, chemistry of 49, 51, 54, 58, 59, 62, 63, 65—67
—, — of, physiology of 76, 85, 94, 109, 111, 113, 115, 135, 137
Oxidative enzymes, see Dehydrogenases and reductases
18-Oxygenated compounds 236—238
19-Oxygenated compounds 238—240
11-Oxy-17-ketosteroids 580—583, 585, 586

Pan, see Apes
Panthera spp. (Leopard, Lion), glands of, anatomy of, 18, 42, 43
Pantothenic acid 62, 71, 96, 105, 138, 139, 458; see also Coenzyme A
Paraganglia VI, 2; see also Adrenal medulla
Parasitization of adrenocortical cells 132
Parent hydrocarbons 194
Partition coefficients of steroids 305, 306
Passerine birds, see Birds
Pathological changes in adrenal gland, 128—141; see also Adrenal virilism, Adrenocortical tissue, anatomy, atrophy, Cytotoxic agents, Infection, Ischemia, Tumors
Pelican, brown *(Pelecanus occidentalis)*, gland of, anatomy of 13, 17
—, —, — of, physiology of 80, 144
—, — mineralocorticoid action in 553
Pentols 326, 332, 419, 420, 433, 460, 461, 473
Perfusion of adrenals 76, 111, 113, 137, 435, 441—443, 447, 449, 486, 488—490, 516; see also Adrenocortical steroids, biosynthesis
Perissodactyla, see Horse
Petromyzon, see Agnatha
Phagocytes in adrenal gland 23, 58
Phallus 578, 579, 603; see also Clitoris
Phenolsulfatase, see Aryl sulfatase
Phosphatides in adrenocortical tissue 54, 55, 91, 92
Phosphorus turnover in adrenal gland 103, 107
Phosphorylase in adrenal gland 67, 453
Physical properties of steroids 202—204, 266, 267
Pig *(Sus scrofa)*, adrenocortical sex steroids of 609
—, — steroids of 415, 416, 420—423, 431
—, — — of, biosynthesis of 438—440, 442, 443, 447, 449, 454
—, — — of, catabolism of 461—463, 497
—, gland of, anatomy of 19, 35
—, — of, chemistry of 58, 65, 67
—, — of, physiology of 85

Pigeon *(Columba livia)*, gland of, anatomy of 13—15, 17
—, — of, chemistry of 57, 65
—, — of, physiology of 78, 143, 145
—, mineralocorticoid action in 553, 554
Pinacolones, see Su 4885
Pineal gland, see "Glomerulotropin"
Pituitary, anterior, hormones of, see ACTH, Gonadotrophins, Growth hormone
—, —, hyperactivity or tumor of 101, 102, 107, 108, 129, 133, 607; see also Cushing's disease, Stress
—, —, insufficiency VII; see also ACTH, secretion of, Anencephaly, Anorexia nervosa, Hypophysectomy Simmonds' disease
—, posterior 79, 116, 543—545; see also ACTH secretion, control, Antidiuretic hormone, Vasopressin
Placental secretion of steroids 414, 421, 422, 435, 484, 592
Plasma levels of corticoids, see Adrenocortical steroids in blood
Plasmalogens in adrenocortical tissue 55
Porcine, see Pig
Posterior interrenals (Stannius corpuscles) 7
Postnatal development of adrenal gland 62, 125, 128, 587, 589; see also Adrenocortical tissue, anatomy, developmental, X zone, Zona reticularis
— involution of adrenal gland 20, 27, 42, 43, 48, 121, 123, 125, 126, 587—590, 598; see also "Fetal" zone
Potassium deficiency 115
— excess 114
— excretion 514—520, 524, 532, 533, 555
— retention 515, 525, 528, 529, 555; see also Electrolytes, Sodium:potassium ratio
Prednisolone (Δ^1-Dehydrocortisol) 81, 266, 277, 476, 477, 520, 521, 527, 543, 546
Prednisone (Δ^1-Dehydrocortisone) 81, 266, 277, 476, 477, 520, 527, 528, 535, 543, 642
Pregnancy 47, 48, 86, 106, 128, 435, 471, 518, 582, 597, 598, 615
5α-Pregnane (Allopregnane) 194
5β-Pregnane 194
Pregnane-3α,20α-diol 431, 435, 464, 610, 611
Pregnane-3α,17α,20α-triol 344, 431, 466, 611
Pregnanolone (3α-Hydroxypregnan-20-one) 344, 431
Pregnenolone (3β-Hydroxy-Δ^5-pregnen-20-one) 344, 417, 431—435, 440, 441, 443, 444, 448, 450, 463, 464, 571, 579, 585, 590
Prenatal development of adrenal gland, see Adrenocortical tissue, anatomy, developmental
Primates, see also Apes, Man, Monkeys; glands of, anatomy of 3, 17, 18, 25, 26, 28, 38, 42
—, — of, physiology of 125, 126
Progesterone (Δ^4-Pregnene-3,20-dione) 75, 109, 196, 202, 205, 344, 417, 423, 431, 435, 441—444, 448, 450, 456, 463, 464, 468, 550, 571, 574, 579, 585, 590, 606, 609—615, 620—622
Progestins 441—443, 609—615; see also C_{21} steroids, Progesterone
—, antagonism of by corticoids 613—615
Prolactin, see Luteotrophic hormone
Proportion of cortical tissue in adrenal gland 14, 21
Prostate gland, response of to androgens 576, 579, 599—605, 608, 609
Protein level in diet, influence of on adrenal gland 69
— synthesis in adrenal gland 91, 454
Pseudohermaphroditism 593, 616, 650, 651
Pseudoplasmals in adrenocortical tissue 54—56, 80, 94—96
Pseudotubular formation in adrenocortical tissue 129, 130, 141, 145
Psychopathology with adrenocortical dysfunction VIII, 638—656
Psychoses, acute 639, 640, 641, 644, 646, 647, 649, 651, 654—656
—, chronic 640, 654
Psychosomatic adrenocortical dysfunction 646, 647, 655
Pyridoxine deficiency, effect of on adrenal gland 71, 139

Quail, California *(Lophortyx californica)*, gland of, anatomy of 13
—, Virginia (*Colinus virginianus*, Bobwhite), gland of, anatomy of 14
—, —, — of, chemistry of 58
—, —, — of, physiology of, 85
Quantitative histochemical technique, see Adrenocortical tissue, chemistry

R_F, see Relative mobilites of steroids
R_T, see Relative mobilities of steroids
Rabbit *(Oryctolagus cuniculus)*, adrenocortical sex steroids in 610—612, 615
—, — steroids of, biosynthesis of 440, 452, 455
—, — — of, catabolism of 459, 461, 463, 478, 484, 486, 495, 496
—, gland of, anatomy of 18, 21, 23, 25—27, 35, 45
—, — of, chemistry of 49, 51, 54, 61—63, 65—67
—, — of, physiology of 70, 74, 75, 81, 85, 92, 107, 121, 123, 133, 135, 136, 140
—, mineralocorticoid action in 517, 518, 526
Radioactive steroids, synthesis of 250, 251
Rana, see Frogs, Anurans
Rat, laboratory strain *(Rattus norwegicus)*, adrenocortical sex steroids in 573, 579, 597—602, 604—614, 616—623
—, — —, — steroids of 420—422, 424
—, — —, — steroids of, biosynthesis of 437—440, 442, 447, 452—456, 458
—, — —, — steroids of, catabolism of 459—463, 467, 476, 478, 480, 486, 495—499
—, — —, gland of, anatomy of 3, 18—23, 26—33, 36—40, 43—45

Rat, laboratory strain, gland of, chemistry of 49—51, 53—60, 62—68
—, — —, gland of, physiology of 69—74, 76—86, 88—101, 104—125, 129, 131, 132, 134—139
—, — —, mineralocorticoid action in 517—520, 523, 524, 526, 527, 530, 532—535, 540, 547, 548
—, wild strain *(Rattus norwegicus)*, gland of, anatomy of 3, 19
—, — —, — of, chemistry of 51, 58
—, — —, — of, physiology of 84, 85, 106
Rats, other, see *Cricetomys* (Giant rat)
5 β/5 α Reduced metabolites 481, 482, 498, 573
"Reducing" steroids 123; see also Adrenocortical steroids, identification
Reductases, steroid, see Dehydrogenases and reductases
Regeneration of adrenocortical tissue 21, 46, 47, 96, 118—121, 547
Relative mobilities of steroids 320—332, 334; see also Adrenocortical steroids, identification
— volume of cortical tissue in adrenal gland 14, 21, 79, 119
Renal, see Kidney
Renin administration 116, 532, 547
Renotropic activity of androgens 574, 577, 606
Reptiles, see also Chelonia, Lizards, Snakes; glands of, anatomy of VI, 2, 10—12
—, — —, chemistry of 50, 53, 60
—, —, physiology of 75, 78, 80, 83, 143, 144
—, mineralocorticoid action in 552, 553, 555
Reticularis, see Zona reticularis
Riboflavin in adrenal gland 68
Ribonucleoproteins in adrenocortical tissuo 3, 36, 37, 68, 80, 90, 91, 104; see alse Adrenocortical tissue, microsomal fraction
Rodents, see also Guinea pig, Hamsters, Marmot, Mouse, Rats, Squirrels Voles
—, glands of, anatomy of 2, 3, 18, 20, 21, 25, 27—29, 38, 42
—, — —, physiology of 82, 129, 134
Ruminants, see also Artiodactyla, Goat, Horse, Ox, Sheep, Ungulates; glands of, anatomy of 2, 25, 35

Salamanders, see Urodeles
Salinity, effect of on fishes 549
Salivary glands, 535, 574, 602, 606
Salmo spp. (Salmon, Trout), glands of, chemistry of 51, 58
—, — —, physiology of 75, 90
—, mineralocorticoid action in 549, 550
Salt-losing syndrome 521, 585
Sayers' test 85
Scurvy 71, 103, 105, 454; see also Ascorbic acid
Scyliorhinus, see Elasmobranchs
Seals 17
Seasonal differences in adrenocortical function 9, 13, 17, 21, 78, 79, 143, 551, 602
Selachians, see Elasmobranchs
Seminal vesicles, response of to androgens 575, 578, 602—609
Serum, see Plasma
Sex differences in metabolism of steroids 495—499
— — in the adrenal gland 5, 13, 18—21, 35, 50, 62, 65, 66, 72, 80, 81, 583
— reversal, frogs 579
— steroids, see Adrenocortical sex steroids, Adrenocortical steroids, sex hormone activity, Androgens, Androsterone, Estrogens, Progestins, Testosterone
Sharks, see Elasmobranchs
Sheep *(Ovis aries)*, adrenocortical sex steroids in 573, 609, 614, 615
—, — steroids of 423, 424
—, gland of, anatomy of 17, 35
—, — of, chemistry of 51, 65, 67
—, — of, physiology of 75, 86
—, mineralocorticoid action in 541
Shock (Hemostasis) 516, 525, 537, 546
Shrews, see Insectivores
Simmonds' disease VII
Sirenia (Manatees), gland of, anatomy of 19
Skates, see Elasmobranchs
Skeletal modifications of steroids 268—272, 277—291
Sloths, see Edentata
Snakes (Squamata), glands of, anatomy of 12
—, — of, physiology of 71, 75, 80, 83, 143
—, mineralocorticoid action in 552
Sodium, see also Electrolytes
— deficiency 86, 112—114, 133 515, 536, 537
— excess or supplement 99, 115, 120, 515, 516, 525, 528, 530, 537, 547, 554
— excretion VII, 115, 206, 516, 518, 521, 523, 528, 529, 531—533, 536, 542, 546, 555
Sodium: potassium ratio in diet 109, 516, 525, 602
— — — in plasma 113—115, 136, 516, 525, 529, 530, 535, 549, 551
— — — in saliva 541
— — — in urine VI, 520, 524, 532, 555
— retention 101, 109, 206, 446, 449, 516—520, 522—524, 528, 530—533, 535, 542, 554, 555
Species differences, 65, 67, 71, 75, 85, 106; see also Adrenocortical tissue, anatomy, chemistry, physiology; and individual species
— specificity of ACTHs 85
Spirolactones 524
Squalene 70, 436, 437—439
Squamata, see Snakes
Squirrel, ground *(Citellus tridecemlineatus)*, gland of, anatomy of 21, 35
—, —, mineralocorticoid action in 526
Stannius corpuscles (Posterior interrenals) 7
Steer, see Ox
Stereoisomerism of hydroxyl and halogen substituents on steroids 195—197, 495—497

Stereoisomerism of steroid nucleus 186—196, 481, 482, 498, 499, 573
Steroid hormones and metabolites, see Adrenocortical sex steroids, Adrenocortical steroids, and individual compounds
— nomenclature 188—197
Steroid-3β-ol dehydrogenase, see Dehydrogenases and reductases, steroid
Steroidoproteins, see Conjugated steroids
Steroids and sterols, side chains of 187, 206—210, 253—256, 260—264, 289, 290
Stilbestrol, see Estrogens, synthetic
"Stress", VI, 3, 36, 50, 69—71, 73, 75, 80, 81, 83, 84, 86, 88, 90, 92, 94, 96, 98, 100, 101, 103—105, 110, 116, 122, 123, 125, 130, 133, 136, 137, 139, 144, 145, 424, 475, 519, 522, 530, 547, 611, 642, 645, 656; see also ACTH, increased secretion, Adrenocortical tissue, hypertrophy and hyperplasia
Stroma of adrenal gland 2, 3, 4, 9, 10, 12, 14, 17, 26, 27, 46, 47
Su 4885 [2-Methyl-1,2-bis(3-pyridyl)-1-propanone] 136, 138, 456, 457, 523
Subglomerulosal zone of adrenal cortex 27, 31, 32, 35—37, 47, 63, 87, 100, 101, 103, 104, 107, 124, 131
Sudanophilia, see Adrenocortical tissue, lipides
Sudanophobic zone, see Subglomerulosal zone
Sulfates, see Conjugated steroids
"Summer" cells 6, 9
Suramin (Germanin*, Bayer) 136
Sus, see Pig
Swine, see Pig
Synthetic corticoids, see Modified hormones

Taricha (Newt), see Urodeles
Teleosts (Bony fishes), glands of, anatomy of VI, 1, 2, 4, 7, 8
—, — of, chemistry of 50, 54, 60
—, — of, physiology of 75, 80, 89, 141
—, mineralocorticoid action in 549, 550, 555
Terrapins, see Chelonia
Testane, see Etiocholane
Testosterone (17β-Hydroxy-Δ^4-androsten-3-one), biological activity of 576—579, 596—598, 603, 608, 613, 616; see also Androgens
—, biosynthesis of 450, 478, 610
—, catabolism of 479, 486, 487, 574, 616
—, chemistry of 196, 202, 205, 287, 321, 334, 414, 415
Tetrahydrocortisol, see Urocortisol
Tetrahydrocortisone, see Urocortisone
Tetrahydro-E, see Urocortisone
Tetrahydro-F, see Urocortisol
Tetrols 324, 331, 418, 420, 432, 445, 462, 469, 482
Thiamine deficiency, effect of on adrenal gland 118
Thymic weight as index of adrenocortical activity 70, 101—103, 105, 106, 108, 137, 140
Thyroid hormone administration, see Hyperthyroidism
Toads (*Bufo*, *Xenopus*), glands of, anatomy of 10
—, — of, chemistry of 51, 58
—, — of, physiology of 68, 75, 79, 85, 143, 145
—, mineralocorticoid action in 551, 552
Torpedo, see Elasmobranchs
Toxins, effects of on adrenal gland, see Bacterial toxins, Cytotoxic agents, DDD
TPNH generation, see Dehydrogenases (Glucose-6-phosphate)
Transformation-field hypothesis 47
Transitional zone, see Subglomerulosal zone
Transplants of adrenocortical tissue, see Adrenal transplants
Triglycerides in adrenocortical tissue 53, 92
Triols 222—226, 322, 330, 341, 344, 417, 418, 420, 425, 426, 428, 430—433, 460, 465, 466, 468, 470, 476, 483, 484
Triones 321, 322, 330, 428, 430, 449, 461, 467, 476, 477, 480, 481
Triterpenoid analogues 291
Triturus (Newt), see Urodeles
Trout, see *Salmo*, Teleosts
Tumor, ovarian 574
—, testicular 413
Tumors, adrenocortical, structure and function of 101, 102, 129, 134, 135, 412, 414, 415, 421, 424, 441, 449, 451, 458, 572, 574, 584—586, 605—607, 611, 615—619, 652
—, pituitary, see Pituitary, anterior, hyperactivity
Turtles, see Chelonia

Ultraviolet absorption, see Adrenocortical steroids, identification
Ungulates, glands of, anatomy of 17, 28, 45; see also Goat, Horse, Ox, Pig, Sheep
Unilateral adrenalectomy, effect of on remaining gland 82, 103, 105, 118, 553
Unsaturated fatty acids in adrenal lipides 56, 94—96, 133
Urinary steroids, see Adrenocortical steroids and metabolites in urine
Urocortisol (3α,11β,17α,21-Tetrahydroxypregnan-20-one, Tetrahydrocortisol) 197, 332, 419, 422, 423, 435, 441, 459, 460, 461, 473
—: urocortisone ratio 475
Urocortisone (3α,17α,21-Trihydroxypregnane-11,20-dione, Tetrahydrocortisone) 197, 331, 421—423, 425, 435, 441, 452, 459—461, 474
Urodeles (Newts, Salamanders), glands of, anatomy of 8, 9
—, — of, physiology of 142

Vascularization of adrenal gland, see Adrenocortical tissue, blood supply
Vasopressin 531, 534, 543, 544; see also Antidiuretic hormone, Pituitary, posterior
Virilism, adrenal, see Adrenal virilism
Vitamin A in adrenal gland 68

Vitamin D excess, effect of on adrenal gland 103
Vitamins, effects of on adrenal gland 71, 103—105; see also Ascorbic acid, Pantothenic acid, Pyridoxine, Thiamine, Vitamin D
Voles (Meadow mice, *Microtus*), adrenocortical sex steroids in 597, 598, 603
—, glands of, anatomy of 18, 21, 35, 48
—, — of, physiology of 105, 128

Water content of adrenocortical tissue 68, 83
— deprivation 116
— diuresis 139, 517, 529, 530, 531, 533, 542, 543, 546, 554
— loading 531, 534, 542, 545, 546
— metabolism 514—521, 529, 534, 535, 537, 542—547, 550, 553, 554
— retention 515—520, 531, 550
Waterhouse-Friderichsen syndrome 130
Weights of adrenal glands 3, 17, 20, 68, 77—83, 123—125, 135, 137; see also Adrenocortical tissue, anatomy; and individual species
Whales (Cetaceans), adrenocortical steroids in 586, 587
—, glands of, anatomy of 17

X zone 27, 47, 48, 60, 68, 127, 128, 586, 589, 593—598
Xantusia vigilis (Yucca night lizard), see Reptiles, Lizards
Xenopus laevis (South African toad), see Toads

Zona fasciculata 15, 27, 28, 30—37, 43, 44, 47, 50, 51, 63, 65, 67, 87, 99—108, 111, 130, 133, 135, 136, 587, 588, 595; see also Inner zones
Zona glomerulosa VII, 15, 27, 28, 30—37, 44, 45, 47, 63, 65—67, 99, 101, 108—118, 124, 130, 136, 138, 141, 143—145, 447, 449, 451, 517, 587, 618, 619
— intermedia, see Subglomerulosal zone
— reticularis 15, 28, 30—36, 43, 44, 47, 58, 63, 65, 67, 99—108, 133, 135, 586—593; see also "Fetal" zone, Inner zones, X zone
Zonal hypothesis of adrenocortical function VII, 46, 47, 98—118, 145, 447
Zonation of adrenal cortex 9, 15, 16, 27—30, 143—145, 586—598; see also Adrenocortical tissue, chemistry, physiology
Zones of adrenal cortex, volumes 28, 29, 88, 109, 116
— — — —, widths 28, 87, 99—104, 109, 111—113, 115—118

FSC
www.fsc.org
MIX
Papier | Fördert gute Waldnutzung
FSC® C083411

Kolibri 360 GmbH
Ferdinand-Jühlke-Straße 7
99095 Erfurt, Deutschland
produktsicherheit@kolibri360.de